NANOBIOSENSORS

NANOBIOSENSORS

Nanotechnology in the Agri-Food Industry, Volume 8

Edited by

ALEXANDRU MIHAI GRUMEZESCU

Department of Science and Engineering of Oxide Materials and Nanomaterials, Faculty of Applied Chemistry and Materials Science, University Politehnica of Bucharest, Bucharest, Romania

ELSEVIER

AMSTERDAM • BOSTON • HEIDELBERG • LONDON • NEW YORK • OXFORD
PARIS • SAN DIEGO • SAN FRANCISCO • SINGAPORE • SYDNEY • TOKYO
Academic Press is an imprint of Elsevier

Academic Press is an imprint of Elsevier
125 London Wall, London EC2Y 5AS, United Kingdom
525 B Street, Suite 1800, San Diego, CA 92101-4495, United States
50 Hampshire Street, 5th Floor, Cambridge, MA 02139, United States
The Boulevard, Langford Lane, Kidlington, Oxford OX5 1GB, United Kingdom

Library of Congress Cataloging-in-Publication Data
A catalog record for this book is available from the Library of Congress

British Library Cataloguing-in-Publication Data
A catalogue record for this book is available from the British Library

ISBN: 978-0-12-804301-1

Publisher: Nikki Levy
Acquisition Editor: Patricia Osborn
Editorial Project Manager: Jaclyn Truesdell
Production Project Manager: Caroline Johnson
Designer: Mark Rogers

Typeset by Thomson Digital

CONTENTS

LIST OF CONTRIBUTORS

Nóra Adányi
National Agricultural Research and Innovation Center, Food Science Research Institute, Budapest, Hungary

Sruthi Ann Alex
VIT University, Centre for Nanobiotechnology, Vellore, India

Gustavo A. Alonso
Autonomous University of Guerrero, Chilpancingo, Guerrero, Mexico

Elif Burcu Bahadır
Namık Kemal University, Scientific and Technological Research Center, Tekirdağ, Turkey

Rinti Banerjee
Nanomedicine Lab, Department of Biosciences and Bioengineering, Indian Institute of Technology, Mumbai, India

Santanu Basu
Panjab University, Dr. S. S. Bhatnagar University Institute of Chemical Engineering & Technology, Chandigarh, India

Carl A. Batt
Cornell University, Department of Food Science, Ithaca, NY, United States

Bhaswati Bhattacharya
National Institute of Food Technology Entrepreneurship and Management, Kundli, Haryana, India

Prakash Singh Bisen
Jiwaji University, School of Studies in Biotechnology; Tropilite Foods Pvt. Ltd., R&D Division, Gwalior, Madhya Pradesh, India

Hikmet Budak
Sabanci University, Faculty of Engineering and Natural Sciences; Sabanci University, Nanotechnology Research and Application Center, Istanbul, Turkey

Nathaniel C. Cady
SUNY Polytechnic Institute, Colleges of Nanoscale Science & Engineering, Albany, NY, United States

Thiago Caon
Department of Pharmaceutical Sciences, Federal University of Santa Catarina, Florianópolis, Santa Catarina, Brazil

N. Chandrasekaran
VIT University, Centre for Nanobiotechnology, Vellore, India

Daniel S. Correa
National Nanotechnology Laboratory for Agribusiness (LNNA), Embrapa Instrumentation (CNPDIA), São Carlos, São Paulo, Brazil

Anisha A. D'Souza
Nanomedicine Lab, Department of Biosciences and Bioengineering, Indian Institute of Technology, Mumbai, India

Rachatida Det-udom
Chulalongkorn University, Faculty of Science, Food Research and Testing Laboratory (FRTL), Bangkok, Thailand

Rocio B. Dominguez
Advanced Materials Research Center, S.C., Chihuahua, Chihuahua, Mexico

Fatma Dridi
University of Lyon, Lyon 1, Institute of Analytical Sciences, Villeurbanne, France; University of Carthage, National Institute of Applied Sciences and Technology, Laboratory of Ecology and Microbial Technology, Tunis, Tunisia

Gennady A. Evtugyn
Kazan Federal University, A.M. Butlerov' Chemistry Institute, Kazan, Russian Federation

Farayde Matta Fakhouri
Faculty of Engineering, Federal University of Grande Dourados, Campo Grande, Mato Grosso do Sul, Brazil

Vincenzina Fusco
National Research Council of Italy, Institute of Sciences of Food Production, CNR-ISPA, Bari, Italy

Mohamed Gargouri
University of Carthage, National Institute of Applied Sciences and Technology, Laboratory of Ecology and Microbial Technology, Tunis, Tunisia

Rükan Genç
Mersin University, Functional Nanomaterials Lab, Chemical Engineering Department, Yenişehir/Mersin, Turkey

Juan M. Gutiérrez
CINVESTAV-IPN, Department of Electrical Engineering, Bioelectronics Section, Mexico DF, Mexico

Akhtar Hayat
COMSATS Institute of Information Technology (CIIT), Interdisciplinary Research Centre in Biomedical Materials (IRCBM), Lahore, Pakistan

Tibor Hianik
Comenius University, Department of Nuclear Physics and Biophysics, Bratislava, Slovakia

Babar Hussain
Sabanci University, Faculty of Engineering and Natural Sciences, Istanbul, Turkey

Nicole Jaffrezic-Renault
University of Lyon, Lyon 1, Institute of Analytical Sciences, Villeurbanne, France

Suresh Kumar Kailasa
S. V. National Institute of Technology, Department of Applied Chemistry, Surat, Gujarat, India

D. Karthiga
VIT University, Centre for Nanobiotechnology, Vellore, India

Niha Mohan Kulshreshtha
Jaipur National University, School of Life Sciences, Jaipur, Rajasthan, India

D. Nanda Kumar
VIT University, Centre for Nanobiotechnology, Vellore, India

Durga Kumari
Nanomedicine Lab, Department of Biosciences and Bioengineering, Indian Institute of Technology, Mumbai, India

Florence Lagarde
University of Lyon, Lyon 1, Institute of Analytical Sciences, Villeurbanne, France

Rashmi Madhuri
Indian School of Mines, Department of Applied Chemistry, Dhanbad, Jharkhand, India

Krisztina Majer-Baranyi
National Agricultural Research and Innovation Center, Food Science Research Institute, Budapest, Hungary

Mouna Marrakchi
University of Carthage, National Institute of Applied Sciences and Technology, Laboratory of Ecology and Microbial Technology; Tunis El Manar University, Higher Institute of Applied Biological Science (ISSBAT), Tunis, Tunisia

Ramón Martínez-Máñez
Polytechnic University of Valencia, Inter-University Research Institute of Molecular Recognition and Technological Development (IDM), Mixed Unit-Polytechnic University of Valencia University of Valencia, Valencia; CIBER of Bioengineering, Biomaterials and Nanomedicine (CIBER - BBN), Zaragoza, Spain

Silvia Maria Martelli
Faculty of Engineering, Federal University of Grande Dourados, Campo Grande, Mato Grosso do Sul, Brazil

Jean-Louis Marty
Université de Perpignan via Domitia, IMAGES, Perpignan Cedex, France

Giuseppe Maruccio
University of Salento, Department of Mathematics and Physics "Ennio De Giorgi," CNR NANOTEC—Institute of Nanotechnology, Lecce, Italy

Luiz H.C. Mattoso
National Nanotechnology Laboratory for Agribusiness (LNNA), Embrapa Instrumentation (CNPDIA), São Carlos, São Paulo, Brazil

Luiza A. Mercante
National Nanotechnology Laboratory for Agribusiness (LNNA), Embrapa Instrumentation (CNPDIA), São Carlos, São Paulo, Brazil

Krystian Mistewicz
Silesian University of Technology, Institute of Physics, Center for Science and Education, Katowice, Poland

Roberto Muñoz
CINVESTAV-IPN, Department of Electrical Engineering, Bioelectronics Section, Mexico DF, Mexico

Amitava Mukherjee
VIT University, Centre for Nanobiotechnology, Vellore, India

Marian Nowak
Silesian University of Technology, Institute of Physics, Center for Science and Education, Katowice, Poland

Juliano E. Oliveira
Federal University of Lavras, Engineering Department, Lavras, Minas Gerais, Brazil

Santanu Patra
Indian School of Mines, Department of Applied Chemistry, Dhanbad, Jharkhand, India

Ida Evangeline Paul
VIT University, Centre for Nanobiotechnology, Vellore, India

Adriana Pavinatto
National Nanotechnology Laboratory for Agribusiness (LNNA), Embrapa Instrumentation (CNPDIA), São Carlos, São Paulo, Brazil

Cheunjit Prakitchaiwattana
Chulalongkorn University, Faculty of Science, Department of Food Technology, Bangkok, Thailand

Elisabetta Primiceri
University of Salento, Department of Mathematics and Physics "Ennio De Giorgi," CNR NANOTEC—Institute of Nanotechnology, Lecce, Italy

Ashok M. Raichur
Indian Institute of Science, Department of Materials Engineering, Bangalore, India; Nanotechnology and Water Sustainability Unit, University of South Africa, Florida Park, Johannesburg, South Africa

A. Rajeshwari
VIT University, Centre for Nanobiotechnology, Vellore, India

Antonio Riul, Jr
University of Campinas, Institute of Physics Gleb Wataghin, Campinas, São Paulo, Brazil

Jigneshkumar V. Rohit
S. V. National Institute of Technology, Department of Applied Chemistry, Surat, Gujarat, India

Jose Vicente Ros-Lis
Polytechnic University of Valencia, Institute of Food Engineering for Development, Valencia, Spain

Kobun Rovina
University Malaysia Sabah, Biotechnology Research Institute, Sabah, Malaysia

Ekta Roy
Indian School of Mines, Department of Applied Chemistry, Dhanbad, Jharkhand, India

Joelle Saulnier
University of Lyon, Lyon 1, Institute of Analytical Sciences, Villeurbanne, France

Mustafa Kemal Sezgintürk
Namık Kemal University, Faculty of Science, Chemistry Department, Biochemistry Division, Tekirdağ, Turkey

Rezeda V. Shamagsumova
Kazan Federal University, A.M. Butlerov' Chemistry Institute, Kazan, Russian Federation

Prashant K. Sharma
Indian School of Mines, Department of Applied Physics, Functional Nanomaterials Research Laboratory, Dhanbad, Jharkhand, India

Divya Shrivastava
Jaipur National University, School of Life Sciences, Jaipur, Rajasthan, India

Shafiquzzaman Siddiquee
University Malaysia Sabah, Biotechnology Research Institute, Sabah, Malaysia

Siddhartha Singha
National Institute of Food Technology Entrepreneurship and Management, Kundli, Haryana, India

Rakesh Kumar Singhal
Bhabha Atomic Research Center, Analytical Chemistry Division, Mumbai, Maharashtra, India

András Székács
National Agricultural Research and Innovation Center, Agro-Environmental Research Institute, Budapest, Hungary

Naimat Ullah
Sabanci University, Faculty of Engineering and Natural Sciences, Istanbul, Turkey

Jose-Luis Vivancos
Polytechnic University of Valencia, Inter-University Research Institute of Molecular Recognition and Technological Development (IDM), Mixed Unit-Polytechnic University of Valencia University of Valencia, Valencia; CIBER of Bioengineering, Biomaterials and Nanomedicine (CIBER - BBN), Zaragoza, Spain

Meral Yüce
Sabanci University, Nanotechnology Research and Application Center, Istanbul, Turkey

SERIES FOREWORD

The emergence of nanotechnology has reached impressive heights in recent years and the development of special nanodevices and nanomaterials has found intriguing applications in agriculture and food sector. Most of the investigated nanotechnological approaches initially aimed to solve evolving problems in the agri-food industry in order to impact on the economic potential. Soon after the implementation of new technologies and approaches that were using nanostructured materials, the worldwide concern was rapidly extended to numerous applications that could be developed by using the science of nanosized materials. Smart materials, biosensors, packaging materials, nutraceuticals, and nanodevices have been designed to address numerous agri-food related issues with direct impact in health, economy, ecology, and industry. As the engineering of nanostructures has constantly progressed and extended its applications, there is virtually unlimited potential in this sector. However, the widely differing opinions on the applicability and usefulness of nanotechnology between both specialists and the general public has hampered progress. The main concern manifested by people is related to the potential risk for health and the environmental impact of the recently developed nanoengineered materials and devices. Therefore, current approaches are strictly considering these concerns when designing nanotechnological solutions for agriculture and food sectors.

This multivolume series was developed by the constant need to discover current inquiries and approaches on the field of agri-food science and also to learn about the most recent progress, approaches, and applications that have emerged through nanotechnology.

As agriculture is the backbone of most developing countries, nanotechnology has the potential to revolutionize the agriculture and food sector by promoting productivity through genetic improvement of plant and animal foods. It can also ensure the delivery of drugs, genes, and pesticides to specific sites at cellular levels in targeted plants and animals, by limiting side effects. Nanotechnology can be used to evaluate gene expression under different stress condition for both plant and animal foods through the development of nanoarray-based gene-technologies. Additionally, this technology can detect fertilizers, pesticides with high precision by smart nanosenors for an adequate management of the natural resources. Moreover, numerous industrial-related applications with direct impact on economy have emerged. For example, nano- and micro-structured arrays can detect the early presence of pathogens, contaminants, and food spoilage factors. Other applications for this technology are smart integration systems for food processing and packaging, as well as nanoemulsion-based decontaminants for food equipment and storage compartments, and nanoparticles that facilitate the bioavailability and delivery of nutrients directly to cells.

The potential benefits of nanotechnology for agriculture, food, fisheries, and aquaculture were identified and supported by many countries, which invested a significant amount of money in the development of applications. Also, numerous campaigns are currently trying to increase awareness on the developing process and recent technologies in order to influence the acceptance of customers. Although nanoagri-food industrialized concept could help to find a sustainable solution for the current global food crisis, the offered advantages should balance the concerns regarding soil, water, environment, and health related issues that such approach could bring.

The series entitled *Nanotechnology in the Agri-Food Industry* brings comprehensive and recent knowledge regarding the impact of the science of nanometer-sized materials on the field of agriculture and food industry, but also discuss the current inquiries regarding risks of these applications in all relevant fields such as environment and health, aiming to increase awareness to a wider amount of readers.

Dr Alina Maria Holban
Department of Microbiology and Immunology, Faculty of Biology, University of Bucharest, Romania

Dr Monica Cartelle Gestal
Harvill Lab, University Of Georgia, Athens, GA, United States

SERIES PREFACE

About the Series (Volumes I–X)

In a permanently changing society, health and well being remain the key drivers for the food industry. Despite the technological progress made in the agri-food industry, a true food crisis emerges in several areas of the globe. This can be explained by insufficient food but mostly by inadequate food for a very distinct range of consumers. In this context, innovative technologies represent the core throughout the whole food chain from raw materials/ingredient sourcing, food processing, quality control of finished products, and packaging. Nanotechnology, coupled with novel interdisciplinary approaches and processing methods, has enabled some important advances recently flourishing in many of these areas. The science of nanosized materials can improve and even resolve the huge challenges faced by the food and bioprocessing industries for developing and implementing systems that can produce qualitative and quantitative foods that are safe, sustainable, environment friendly, and efficient. This emerging tool finds its applications in various fields and represents an endless approach for the development of innovative strategies in food development, processing, and packaging.

This multivolume set aims to bring together the most recent and innovative applications of nanotechnology in the agri-food industry, but also to present the future perspectives in the design of new or alternative foods.

The series contains 200 chapters organized in 10 volumes, prepared by outstanding research groups that made significant impacts on the field of nanotechnology and food-related research sectors. This comprehensive set represents an updated and highly structured material for undergraduate and postgraduate students in food science, biotechnological, engineering fields, but also a valuable resource of recent scientific progress, along with most known applications of nanomaterials on the food industry to be used by researchers, engineers, and academia. Moreover, novel opportunities and ideas for developing or improving technologies in the agri-food industry by innovative companies, biotechnological industries, and other economical structures are highlighted and their potential is widely dissected. This series may be also valuable for the wide audience interested in recent nanotechnological progress in the agri-food field worldwide.

These 10 volumes cover almost all aspects related to the applications of *Nanotechnology in the Agri-Food Industry* and are named as:

Volume I Novel Approaches
Volume II Encapsulations
Volume III Emulsions
Volume IV Nutraceuticals
Volume V Nutrient Delivery
Volume VI Food Preservation
Volume VII Food Packaging
Volume VIII NanoBioSensors
Volume IX Water Purification
Volume X New Pesticides and Soil Sensors

Each volume contains 20 chapters, which were carefully composed and illustrated to highlight the most innovative and intensively investigated applications of nanotechnology on particular wide interest domains of the agri-food industry field.

Alexandru Mihai Grumezescu
University Politehnica of Bucharest,
Department of Science and Engineering
of Oxide Materials and Nanomaterials,
Faculty of Applied Chemistry and
Materials Science, Bucharest, Romania
http://grumezescu.com/

VOLUME PREFACE

There is an increasing demand for robust, rapid, inexpensive, and safe alternative technologies for monitoring, testing and detection of harmful or potentially dangerous foods. Due to their high sensitivity and selectivity, nanobiosensors have attracted attention for monitoring biological contaminants in food but also various chemical and physical hazards. This book offers a broad overview regarding the current progress made on the field of nanosensors, the cutting-edge technologies progress, and the impact of these devices in food industry. A special attention is given to the detection of microbial contaminants and harmful metabolites, such as toxins and hormones, which have a great impact both on humans and animal health and feed.

Volume VIII contains 20 chapters, prepared by outstanding international researchers from the USA, Brazil, Mexico, Spain, France, Italy, Tunisia, Russia, Poland, Slovakia, Hungary, Turkey, India, Malaysia, and Thailand.

In Chapter 1, entitled *Nanomaterials-Based Optoelectronic Noses for Food Monitoring and Classification*, Jose Vicente Ros-Lis et al., present a compilation of the results obtained using arrays of chromogenic indicators with different chemical recognition properties and applications, varying from how they can be used to follow the evolution of chicken ageing, detection of fresh pork sausages spoilage, boiled marinated turkey freshness, and identification of blue cheese varieties.

Gennady A. Evtugyn et al., in Chapter 2, *Biosensors for Detection of Mycotoxins and Pathogenic Bacteria in Food*, focus on the critical analysis of existing biosensors for the detection of food toxicants, such as mycotoxins and pathogenic bacteria by means of antibodies and DNA/RNA aptamers. Future trends are also discussed.

Chapter 3, prepared by Babar Hussain et al., *Bioconjugated Nanomaterials for Monitoring Food Contamination*, reviews the application of aptamer-conjugated nanomaterials in food safety surveillance as well as the conventional techniques used for food safety monitoring in order to provide a comprehensive and comparative approach.

In Chapter 4, *Detection of Food Contaminants by Gold and Silver Nanoparticles*, Ida Evangeline Paul et al., discuss the mechanisms of detecting chemical hazardous factors and microbial contaminants in food systems using gold and silver plasmonic nanoparticles.

In Chapter 5, *Nanomaterial-Based Electrochemical Biosensors for Food Safety and Quality Assessment*, Fatma Dridi et al. review new trends in nanomaterial-based electrochemical biosensors for food analysis, focusing on the determination of major contaminants, that is, foodborne pathogens, toxins, veterinary drug residues and pesticides.

Chapter 6, *Chemical Sensors-Based on Hybrid Nanomaterials for Food Analysis*, prepared by Daniel S. Correa et al., presents an overview on distinct nanostructured materials and their molecular architectures employed as active layers for chemical sensors, including thin films of conjugated polymers, nanofibers, metallic nanoparticles, carbon nanotubes, and enzymes. Such materials can drastically improve both sensitivity and limit the detection of sensors, including conventional potentiometric sensors, biosensors, electronic tongue-, and electronic nose-type sensors.

Krystian Mistewicz et al., in Chapter 7, *Prevention of Food Spoilage Using Nanoscale Sensors*, discuss the monitoring of humidity and excess gasses in production, packaging, and storage processes. Also, the recently fabricated conductive and photoconductive humidity nanosensors are deeply reviewed within this book. These high sensitive, small sized, and low power devices can address the requirements of food industry.

Elif Burcu Bahadır et al., in the Chapter 8, *Biosensor Technologies for Analyses of Food Contaminants*, give a general overview about the possible application of biosensors in the food analysis field. In this chapter, the transduction principles, immobilization techniques, and analytical performances of these biosensors are summarized.

Kobun Rovina et al., in Chapter 9, *Analytical and Advances Methods-Based Determination of Melamine in Food Products*, discuss the current progress in the development of advanced detection techniques of melamine and its compounds. Other section of this chapter deals with the current and past melamine contamination incidents together with modern instrument analytical methods for the present determination of melamine and its analogs.

Rükan Genç, in Chapter 10, *Nanomaterial-Based Sensor Platforms for Facile Detection of Food Contaminants*, reveals recent advances obtained in nanotechnology employed for the detection of foodborne pathogens and associated toxins in the last 5 years. Also, updated outbreak statistics and traditional methods are briefly discussed in this chapter. In addition to covering the technical aspects of the use of nanotechnology in food industry focused applications, sample pretreatment, signal transduction, and enhancement have been discussed separately in terms of their role on the sensor platform.

Nóra Adányi et al., in Chapter 11, *Evanescent field effect based nanobiosensors for agro-environmental and food safety*, focus on the impact in food industry of some nanobiosensors, their preparation and characterization through a series of techniques, such as reflectometric interference spectroscopy, interferometry, optical waveguide lightmode spectroscopy, resonant mirrors, fiber optics, total internal reflection ellipsometry, and total internal reflection fluorescence spectroscopy. Novel nanobiosensors are presented through their applications in food safety.

Chapter 12, *Micro and Nanotechnology-Based Approaches to Detect Pathogenic Agents in Food*, prepared by Nathaniel C. Cady et al., summarizes the latest micro- and nanotechnology-based approaches to detect pathogenic agents in food, as well highlighting the most promising technologies with regard to their future exploitation. The goal of this chapter is to provide the reader with new insights into how and when these technologies can be exploited to ensure the enumeration and accurate detection of pathogenic bacteria in foods.

Chapter 13, *Contaminant sensors: Nanosensor, an Efficient Alarm for Food Pathogen Detection*, prepared by Cheunjit Prakitchaiwattana et al., provides a comprehensive review of current developments in application of nanotechnology on food contaminants focusing specifically on pathogen detection and monitoring. The review includes: (1) a brief overview of significant foodborne pathogens and their key properties in relation to detection and (2) nanosensors/assays developed for detection or monitoring of pathogens which include detail on types, method efficiency, and applications as well as current and future status of the methods.

Chapter 14, *Nanotechnology-Based Contaminant Sensors*, prepared by Niha Mohan Kulshreshtha et al., presents recent advancements in the field of food nanosensing including novel nanosensory approaches called E-nose and E-tongue, Lab-on-chip nanotechnology-based pathogen sensors, surface enhanced Raman scattering-based sensors, and aptamer-based sensors along with their current and potential applications.

Chapter 15, prepared by Anisha A. D'Souza, *Nanocomposite Biosensors for Point of Care—Evaluation of Food Quality and Safety*, gives an overview of the contribution of nanotechnology-based biosensors in the food industry, ongoing research, technology advancements, regulatory guidelines, future challenges, and industrial outlook.

Suresh Kumar Kailasa et al., in Chapter 16, entitled *Plasmonic Nanoparticles and Quantum Dots in the Identification of Inorganic and Organic Contaminants in Food Samples,* provide an overview of nanomaterials-based UV-visible and fluorescence spectrometric strategies that have been applied in the identification of inorganic (metal ions) and organic contaminants in various food samples, discussing their advantages and drawbacks. This proposed book chapter offers a rational vision of the main achievements yielded in the relevant literature according to this exciting and creative nanomaterials-based analytical field in foodomics.

Santanu Patra et al., in Chapter 17, *A Technique Comes to Life for Security of Life: The Food Contaminant Sensors,* highlight the various advancements made so far in the development of nanosensors for the trace level determination of various food contaminants. Along with this, the pros and cons of nanomaterials on environment as well as human health are also discussed. The chapter also revealed the novel and more beneficial prospects of nanosensors which must be biocompatible and/or ecofriendly.

Chapter 18, *New Trends in Food Industry: Application of Nanosensors in Food Packaging,* prepared by Thiago Caon et al., focuses on design, application, and constituents of three main types of nanosensors—radio frequency identification, for sensing gas and nanobiosensors. A variety of nanosensor designs for a same analyte are revealed by this work, and also their different levels of sensitivity and specificity. The main challenges (eg, complexity of food matrices) that make the commercialization of these devices difficult are presented and discussed. In addition to the development of novel materials to be included in nanosensors, different materials presenting well-known chemical and physical properties may be combined to optimize the operation of these devices and overcome some of their limitations.

Rocio B. Dominguez et al., in Chapter 19, *Nanomaterial-Based Biosensors for Food Contaminant Assessment,* describe the basis of biosensors and the impact of nanotechnology for developing novel nanomaterial based biosensors (NBB). Later, concrete applications of NBB for food toxin detection (ie, pesticides and biotoxin) are presented. Finally, the role of nanomaterials for sensor arrays in food quality and risk assessment is described and some new perspectives are highlighted.

Chapter 20 prepared by Bhaswati Bhattacharya et al., *Fluorescent Nanosensors: Rapid Tool for Detection of Food Contaminants*, discusses the potential of fluorescent-based quantum dots in food contaminant detection and is broadly divided into three thematic areas. First, theoretical explanation of the fluorescence phenomena including the sensing mechanism is provided followed by the use of optical nanosensors like quantum dots and finally the utilization of these quantum dots for sensing food pathogens and toxins are covered.

Alexandru Mihai Grumezescu
University Politehnica of Bucharest,
Department of Science and Engineering
of Oxide Materials and Nanomaterials,
Faculty of Applied Chemistry and
Materials Science, Bucharest, Romania
http://grumezescu.com/

1

NANOMATERIALS-BASED OPTOELECTRONIC NOSES FOR FOOD MONITORING AND CLASSIFICATION

Jose Vicente Ros-Lis*, Jose-Luis Vivancos,†, Ramón Martínez-Máñez**,†**

**Polytechnic University of Valencia, Institute of Food Engineering for Development, Valencia, Spain; **Polytechnic University of Valencia, Inter-University Research Institute of Molecular Recognition and Technological Development (IDM), Mixed Unit-Polytechnic University of Valencia University of Valencia, Valencia, Spain; †CIBER of Bioengineering, Biomaterials and Nanomedicine (CIBER - BBN), Zaragoza, Spain*

1 Introduction

The food industry is one of the most important industrial sectors and its proper functioning plays a key role in health and prosperity in the society. According to the European Commission, in 2009, it was the second largest sector in the manufacturing industry in value added terms and it accounted for 13.7% of employment (over 4.5 million workers). The relative relevance of the sector has increased since then due to the economic crisis and its well-known economic buffer effect, being nowadays the largest manufacturing sector in terms of turnover and employment (European Technology Platform Food for Life, 2014). Simultaneously, the food industry is claimed to be a key actor in one of the most challenging trends of the 21st century, that is, population growth. World population is expected to increase to about 9.6 billion people by 2050 and, accordingly, the global food production will need to be 70% higher to fulfill future food demand. Hence, the food industry now faces the huge challenge of finding sustainable options for feeding a growing global population. However, despite its significance, the food industry is not recognized as being particularly innovative compared to other industries. One of the best strategies to improve innovation in a sector such as food

Nanobiosensors. http://dx.doi.org/10.1016/B978-0-12-804301-1.00001-1

is collaboration and cooperation with other sectors or technologies. The knowledge enabling different technologies and the solutions developed for other fields can help to solve problems in the food sector and to provide solutions to meet the needs and expectations of society.

Nanotechnology is emerging as a critical technology and potential applications of nanoscale sciences to food and agriculture are limited only by the imagination (Chen et al., 2014). Some areas in which nanotechnology has been applied to food industry is by means of the development of nanoscale biomaterials, encapsulation of active compounds, controlled release of active ingredients, packaging to enhance the shelf life of foods, and sensors to detect undesirable compounds present in foods. Public food safety crises create a high degree of concern among consumers, and cause huge economic losses. In this particular area, nanoscale sensors and intelligent packaging are potential important tools for food quality and safety. The use of sensors gives better traceability for the food manufacturer, reducing waste consumption caused by and assuring trust within the delivery chain.

Among food processing subsectors, meat is the most relevant since it accounts for 22% of the turnover. Also dairy products are highly significant with 15% of the food sector turnover (Commission of the European Communities, 2009). Among the various kinds of meat, pork and poultry are the most consumed in the world (United States Department of Agriculture, 2011). Besides, it is well-established that a country's meat consumption increases in accordance with its degree of development, therefore meat consumption is expected to rise in the following years. However, meat also has disadvantages as a source of calories since it is comparatively expensive and it has a short shelf life.

Fresh meat aging is a complex issue because many biological processes occur depending on storage time, packaging conditions, microbial loading, humidity, and so forth. Spoilage is observed when meat develops off-flavors and/or off-odors and other organoleptic modifications such as variations in appearance (slime, discoloration) making a product undesirable for human consumption. Modified atmosphere packaging (MAP) is one of the more readily accepted systems to delay meat spoilage, but it does not provide knowledge of freshness by mere visual inspection to alert consumers of contamination during the packaging process or eventual alterations of the cooling chain. Although use-by date is incorporated into the package, this is an approach that does not inform about the state of every package, and it is not useful to report improper meat treatment in distribution or by end users' handling. Furthermore, the use-by date tends to be conservative in order to ensure a margin of security that avoids any risk to the consumer. As a consequence, meat suitable for consumption is disposed, increasing food waste and the environmental impact of the industry (ie, 1 kg of pork or chicken requires 5.988 or 4.325 litres of water to produce, respectively) (Institution of Mechanical Engineers, 2013) and threatening food security in a framework of increased demand for food.

Current analytical methodologies to determine meat aging include the use of sensorial panels, detection of metabolite concentrations (ie, glucose, ATP trimethylamine, or other biogenic amines), ion mobility spectrometry (IMS), microbiological techniques, diffuse reflectance spectroscopy, near-infrared spectroscopy (NIR), and fluorescence spectroscopy (Bota and Harrington, 2006; Grau et al., 2011; Sahar et al., 2011; Lin et al., 2004). Regardless of the precision of some of these procedures, they are usually time consuming, destructive, employ relatively expensive instrumentation, usually cannot be used for in situ determinations, or require the participation of qualified personnel. These techniques are therefore appropriate for food safety agencies, but not for use in supermarkets or at consumers' homes. In this sense, Ellis and Goodacre (2001) advanced, in their compilation about rapid and quantitative methods for the detection of the microbial spoilage of muscle foods, that despite the fact that processes involved in the microbial spoilage of meats and the metabolites have been well established for decades, there is a lack of systems that could be really applied to the industries. It requires ideally being accurate, nondestructive, and giving answers in real time.

On the other hand, in recent years it has been observed that increasing consumer demand for reliable information about quality of food and both ingredients. Furthermore, production methods influence quality, sensorial properties, and market prices. Any alteration can imply fraud and legal implications and must be controlled. As a strategy to increase consumers' information and confidence and to protect agricultural activity, the European Union developed diverse regulations. Some of those include protected geographical indication (PGI), protected designation of origin (PDO), and traditional specialities guaranteed (TSG). PDO is especially common in cheeses and is usually related to microbiological species, the kind of milk, maturation conditions, and heat treatments which are factors of final quality and organoleptic properties, including texture and appearance. However, in order to ensure its effectiveness, this regulatory mark must be complemented with analytical methodologies that allow the monitoring of the behavior of producers and retailers. Panels of skilled experts is a well-suited strategy for such purpose, but limitations in the number of samples that an expert can analyze and the high costs of trained specialists with strong sensory capabilities demand alternative instrumental approaches. Instrumental methods (Wetzel and Charalambous, 1998) offer accuracy but they are not suitable for supermarkets or at consumers' homes. Food origin and composition are integrated into the package label. However, this is a generic approach that is not appropriate to report food fraud during distribution. Therefore, improved easy and disposable approximations are necessary.

Cheeses that use the mold Penicillium roquefortii as flavor-producing agent are grouped under the category of blue cheeses (Roquefort, Blue Stilton, etc.). Differences in the origin of milk (cow's, ewe's), Penicillium roquefortii strains and manufacturing practices results in differences in gross composition, volatile composition and differences in texture and flavor among cheese varieties (Gallois and Langlois, 1990; Karlshøj and Larsen, 2005). Nevertheless, blue cheeses are quite similar from a sensorial point of view, and consumers cannot be sure that the products they are consuming are the same as those they have paid for. Taking into account this, the development of simple-to-use systems able to offer individual information to end users at any point of the distribution channel may be of relevance.

Cheeses develop their characteristic flavor, texture and color during the ripening process and it has been reported that volatile derivatives offer characteristic fingerprints that can be employed to identify and classify them (Lawlor et al., 2003; Varming et al., 2013; Trihaas et al., 2005). Both lipolysis and proteolysis are more pronounced in blue cheeses than in other cheese kinds (Larsen and Jensen, 1999), and their physicochemical environment differs from other cheese varieties. Blue cheeses develop a superior number of volatile compounds than bacterially ripened cheeses (Cakmakci et al., 2013) which makes blue cheeses particularly suitable for classification by headspace analyses based instrumental or sensorial procedures (Gkatzionis et al., 2013).

Among techniques to develop easy handling disposable systems that could be printed on food packaging and inform about meat aging or food classification, the use of chromogenic chemosensors (Martínez-Máñez and Sancenón, 2003; Amendola et al., 2010; Mohr, 2006) is, perhaps, one of the most interesting as they are cheap, can be printed, and color modulations can be easily measured using image-capturing systems or by the naked eye. In fact, few technologies are as advanced or inexpensive as visual imaging. In the case of food freshness, time temperature indicators (TTI) (Kerry et al., 2006) have been studied; in these systems a increase of the temperature above the threshold induces a permanent change of color that indicates a break in the cold chain. However, some relationships can be established between temperature and freshness, TTI do not really offer information on the biochemical processes occurring in food.

From another point of view, one potent and versatile approach to analyze complex systems is the use of arrays of sensors displaying diverse responses and sensibilities. Widely known as electronic noses or tongues (Deisingh et al., 2004), they can work online without sample treatment and are usually low cost. Gardner and Bartlett (1994) defined the electronic nose as an instrument that comprises an array of electronic chemical sensors with partial specificity and an appropriate pattern recognition (PR) system, capable of recognizing simple or complex odors. Electronic noses have been used to classify and predict via monitorization of flavor of several dairy products. In particular, studies using an electronic nose applied to meat and cheese have been used to ensure meat freshness and cheese classification (Cevolia et al., 2011; Musatov et al., 2010). Although several types of sensors and chromatographic methods have been studied in electronic noses, they are generally based on electrochemical properties and/or need expensive instrumentation, offering low selectivity or high complexity respectively. On the contrary, chromogenic arrays (optoelectronic noses), built by an array of dyes capable of offering information through suitable color modulations, have been much less studied in spite of the advantages noted above for optical systems (ie, low cost, versatility, and printability). Moreover, their disposable character makes sensor drift and irreversible poisoning irrelevant.

2 Principles of Chromogenic Arrays

Chemists have a natural tendency to assume that for any complex mixture (ie, food), the analytical goal is a complete component-by-component analysis. Thus, a full analysis with several techniques is necessary to obtain the composition of a sample. In contrast the mammalian olfactory system gives a merged response to complex mixtures and, for many applications, this kind of response may be not only sufficient, but actually preferred. Besides, it is important to note that selective recognition tools (chemosensors) cannot be easily developed for a previously unknown analyte (ie, unexpected pollutants). In these cases, one could consider a complementary sensor-array approach (Anzenbacher et al., 2010) that mimics a mammalian olfactory system. This approach treats the mixture as a single analyte and collects a combined response simultaneously on an array of sensors. Compared with traditional analytical technologies, this approach can be less expensive and easier to operate, making it potentially attractive for many industrial applications.

Chromogenic arrays are the basis of optoelectronic noses. They are the recognition element that interacts with the gaseous sample generating a color change that is collected by a camera and further treated by multivariate analysis. Simply by taking the difference before and after exposure of scanned images of the array, it is possible to obtain unique color change signatures that can be related with both qualitative recognition and quantitative analysis.

Individual dyes involved in the molecular recognition processes are key elements of the chromogenic array. Preferably they must offer accessible positions of interaction with analytes, show large spectral shifts upon contact with the sample, and present intense coloration. The kind of interaction with the analyte is usually well established and may involve bond formation and coordination, acid–base interactions, hydrogen bonding, molecular complexation, dipolar and multipolar interactions, van der Waals forces, and physical adsorption (Janzen et al., 2006). Typically, electronic noses (not based in color) are limited to the weakest and least specific intermolecular interaction forces, typically van der Waals and physical adsorption interactions between the analyte and the probe. Therefore, they usually offer limited sensitivity for compounds at low concentrations and low selectivity to discriminate between compounds in complex matrixes. In contrast, Lewis acid/base dyes (ie, metal–ion containing dyes), Bronsted acidic or basic dyes (ie, pH indicators), and dyes with large permanent dipoles (ie, zwitterionic solvatochromic dyes) usually display stronger interaction with analytes (Zhang and Suslick, 2005; Anzenbacher et al., 2010).

A complex task for the design of chromogenic arrays is the selection of individual chemosensors and the array size. This can be solved by different approaches, which depend largely on the knowledge of the target analyte (if known) and whether a qualitative or quantitative analysis is sought. Whether or not the analyte composition of the matrix is known, it is advisable to perform preliminary experiments to select those dyes that exhibit a suitable response (color change). Sensing in multianalyte environments may be achieved by sensors with a high affinity for one specific target, by nonselective indicators, or with a combination of both classes. However, note that single chemosensors are, in general, not selective but often display certain responses to several analytes (cross-sensitivity). In fact, the design of sensor arrays utilizing cross-reactive sensor elements offer advantages. The key factor of such devices is centered on the perturbation that the sample induces on a large number of nonspecific sensors resulting in the formation of a specific pattern. The advantage of the cross-reactive arrays is the possibility of avoiding the costly preparation of selective sensors (if available). The inconvenience is that the information content that each individual sensor generates is often low and only the simultaneous analysis of a number of such sensor elements generates a differential response pattern with enough information.

3 Multivariate Analysis Methodologies

Once the array has been in contact with the sample (ie, with the vapors generated by the sample in a certain headspace) the information must be collected and analyzed. Data collection benefits nowadays from popularization, reduced size, and the low cost of digital cameras and the recent advances in software for image analysis. This allows an automatic measurement of the color even with portable devices. In general, optoelectronic noses are based on color differences. The output signal from each sensor element after analyte application is directly compared to the output of the same element before analyte application and the difference is then used in data processing using PR and multivariate analysis tools.

The origin of multivariate analysis started in the late 19th century with work on the development of the correlation coefficient carried out by Galton and Pearson. Moreover, in the last decades, the wide availability of personal computers and statistical software has allowed multivariate analysis to be applied routinely in a number of applications (Everitt and Hothorn, 2011). Multivariate analysis consists of a collection of methods that can be used when several measurements (variables) are made on each individual or object (observations) in one or more samples. One possible approach is to isolate each variable and to study separately, however, the variables are not only correlated with each other, but each variable is usually influenced by other variables. Using multivariate analysis, the variables can be examined simultaneously, enabling the exploration of the joint performance of the variables and the determination of the effect of each variable in the presence of the others (Rencher and Christensen, 2002). The goal of many multivariate approaches is simplification, in terms of a reduced set of dimensions. Several multivariate techniques have been designed to reduce the dimensionality of the data. One of the most used data reduction techniques is Principal Component Analysis (PCA). Another approach is to categorize objects into clusters or cluster objects into categories (cluster analysis) (Timm, 2004).

While the testing of hypotheses always requires a prior assumption (hypothesis) about the data, exploratory data analysis (ie, PCA, HCA vide infra) is not based on any prior assumptions. In a typical exploratory data analysis, all variables are taken into account using both graphical and formal methods to search for systematic patterns. These analyses seek to explore the data to identify the trends and clustering behaviors that could reflect a large number of variables (Anzenbacher et al., 2010). Descriptive or exploratory methods are usually characterized by the importance of graphical displays and visualization (Everitt and Hothorn, 2011), which are used to represent the data relationships as dependencies both within a set of variables and dependencies with other variables.

PCA permits the reinterpretation of the data into a lower dimensional space (dimensionality reduction) and the comparison of the vector responses direction and magnitude for the set of measurements. When singular value decomposition (SVD) is applied directly to the covariance matrix, it results in score and loading matrices, which are orthonormal and formed by eigenvectors termed principal components (PC), representing the projection of the covariance matrix into the new PC eigenspace. The PC are ordered so that PC1 displays the largest amount of variance, followed by the next largest, PC2, and so forth. The columns of the score matrix can be used to project the covariance matrix into a lower dimensional space for exploration of patterns in the data (clustering). This representation is called score plot. The main feature of PCA is the correlation of the data in the new base (scores plot) and the contribution of each component of the sensors (loads plot). PCA can be used to identify the features (variables) that are most important for the description of the chemical behavior of the sensor. Such important features could be leveraged for other analytical tasks, for example, improving signal to noise ratio and increasing the reliability of classification, reduction of the data set for simpler analysis, removing sensor elements that have a poor contribution, and so on (Anzenbacher et al., 2010).

In a clustering task, the goal is to learn the structural relationships between different samples. Hierarchical clustering analysis (HCA) is an unsupervized method of multivariate analysis, which pursues sample classification by measuring the interpoint distances among all samples in the n-dimensional space resulting from n-numbers of studied features (Musto et al., 2009; Lim et al., 2009). In contrast to classic classification methods such as PCA, HCA considers the complete dimensionality of the data and provides graphic output, usually in the form of a dendrogram.

Regression techniques intend to establish predictive models that correlate the set of independent variables (eg, array response) with a second set of variables that are the properties of the sample analyzed (eg, concentration, quality). Partial Least Square (PLS) regression is a multivariate projection procedure that models the relation between an array of dependent variables with another array of independent variables (Y and X, respectively). PLS technique finds the components of the matrix of input (X) that describe relevant variations in input variables while achieving the highest correlation with the objectives (Y) and providing the lowest weight to changes that are irrelevant related to noise at the same time (Berrueta et al., 2007). In general, both the measured and predicted values are plotted together to evaluate the accuracy and precision of the created prediction models. By using a simple linear model a more rigorous analysis is achieved by a linear fitting. Slope of the fitting line and intercept with the y axis represent accuracy in prediction; meanwhile, the regression coefficient can relate to the PLS model's precision.

Classification tasks address the problem of identifying an unknown sample and to assign it to a certain set of previously learned categorized samples. Typical classification models used are partial least square discriminant analysis (PLS-DA) and minimal distance classifiers, such as K-nearest neighbor (KNN).

PLS-DA is particularly adequate when it is necessary to model the relationships between a multivariate X data matrix and a univariate or a multivariate Y data matrix (Wold et al., 2001; Matthew and William, 2003). It allows data sets with many correlated X variables to be handled, provides excellent results, even when there are fewer observations than the number of predictor (X) variables, and produces parsimonious models by extracting only the portion of the X variance that is related to the Y variance (Parente et al., 2012).

Minimal distance classifiers, such as, KNN and learning vector quantization (Ciosek and Wróblewski, 2006; O'Farrell et al., 2005) are based on the rule that similar patterns belong to the same class. KNN classifies an unknown sample based in the distance to the "k" closest samples (equal to the number of closest neighbors) in a reference set and the majority indicates the class affinity of the test object. If no majority is found, the unknown is assigned to the class of the closest sample. KNN models include the probabilities that each sample belongs to each class in the model classification probability field. The probability that a sample belongs to class "A" is calculated as the fraction of nearest neighbors that have that particular class.

4 Examples of Application of Chromogenic Optoelectronic Noses

The principles shown above in relation to the preparation and use of optoelectronic noses will be applied in the following section. In particular selection of sensing materials and how they can be used for meat freshness (chicken, fresh sausages, and boiled marinated turkey) monitorization and for cheese classification will be detailed as examples of applications of optoelectronic noses. Also diverse techniques of data analysis will be employed depending on the data and the kind of response (qualitative or quantitative expected).

4.1 Sensing Materials

Ideally, sensing materials should be responsive to complex VOC mixtures generated by food matrixes and be able to monitor headspace composition changes. In general, meat spoilage at refrigeration temperatures is attributed mainly to microbial by-products and not to autolytic products originating from the tissue. Decay is accompanied by the generation of several microbiologically induced biochemical processes that result in the formation of a wide range of volatile compounds, including hydrogen sulfide, dimethyl disulfide, indole, lactic acid, acetic acid, other fatty acids (such as propionic, isobutyric, isovaleric, *n*-butyric), organic volatiles (such as strains or branched primary and secondary C2–C5 alcohols, C6–C8 hydrocarbons, C3–C4 ketones), diacetyl/acetoin, putrescine, cadaverine, tyramine, and other biogenic amines and ammonia (Smolander et al., 2002; Garcia de Fernando et al., 1995). The spoilage and the materials response is temperature dependent, thus the sensing materials must be tested at the working temperature.

The aroma of blue cheeses is also very complex, and includes more than 100 substances (Curioni and Bosset, 2002). Volatile compounds in cheese derive mainly from three major metabolic pathways: lipid catabolism, lactose catabolism, and protein catabolism (Cakmakci et al., 2013; Qian et al., 2002). The aroma of blue cheeses is dominated largely by lipid metabolism, and larger proportions of free fatty acids accumulate in blue cheese than in most other cheese types during ripening (McSweeney and Sousa, 2000). Ketones, together with alcohols, esters, and organic acids, are the dominant compounds in mold-ripened chesses (Gallois and Langlois, 1990; Varming et al., 2013; Day and Anderson, 1965). Other products such as aldehydes, dicarbonyls, acetaldehyde, diacetyl, amino acids, hydroxy acids, indole, ammonia, α-keto acids, amines, amides, phenolic and sulfur compounds, ethers, benzene, toluene, furfural, lactones, acetoine, terpenes, and anisol, are also present.

Inspired by the metabolites present in the headspace, and according to our own experience in designing colorimetric probes (Ros-Lis et al., 2004; Ros-Lis et al., 2005; Ros-Lis et al., 2011), a series of cross-sensitive indicators that can be potentially used to develop the array and that could offer selectivity towards some of the volatile compounds were selected. The indicators are based on small dyes supported on different materials. Silica gel, UVM-7, MCM-41, and alumina were used as supports given their versatility, optical transparency in the visible range, and increased dye stability. Furthermore the materials offer opposite acid-base properties: alumina is basic while silica is slightly acid and this can favor modulation of dye response. Similar dye-containing materials have been reported to be suitable solids in vapor detection (Descalzo et al., 2007). UVM-7 typically displays a specific surface of 1200 $m^2 g^{-1}$ and is formed by ordered small mesoporous particles (of c. 12–17 nm) that originate in a bimodal pore system of 3.0–3.2 nm mesopores and 20–70 nm textural pores, which improves diffusion. The mesoporous material MCM-41 was selected due to its nanoscopic structure and, although it does not offers textural pores, it contains mesopores with a diameter of round 2 nm and a large surface area. Besides, basic alumina (particle size 63–200 μm, specific surface 120 $m^2 g^{-1}$) was tested as a support for pH indicators in order to improve the array response to volatile acids. Silica gel was also employed as a support for certain complexes.

Dyes can be incorporated to the inorganic solids by simply stirring the corresponding dye solution in the presence of the solid for c. 24 h in order to guarantee maximum dye adsorption, followed by solvent removal.

In order to select the sensing materials more adequate for each application, a screening of over more than 100 materials containing pH indicators, natural extracted dyes, chemical reagents for metabolites, transition metal salts and complexes, solvatochromic dyes, and food-allowed colorants were tested. From them and in agreement with the intensity of the color changes and the response time observed under the experimental conditions, certain sensing materials were selected for each food (Table 1.1). As it can be seen in the table the final arrays used comprise a wide set of materials including four supports and eighteen dyes. A full description of the materials composition can be found in the authors' previous work (Salinas et al.,2012, 2014a,b; Zaragozá et al., 2015).

The dyes selected cover a wide range of possible dye-analyte interaction and among the dyes used we can find pH indicators

Table 1.1 General Description of the Chromogenic Sensing Materials and List in Which Have Been Used

Sensing Material	Indicator	Support	Chicken Meat	Fresh Pork Sausages	Boiled Marinated Turkey	Blue Cheese
1	Phenol red	UVM-7	X			
2	2,6-diphenyl-4-[2-(4-*N*,*N*-dimethylaminophenyl)vinyl]-pyryliumtetrafluoroborate	UVM-7	X	X		
3	Dimethyl yellow	UVM-7	X		X	
4	4-(4-*N*,*N*-dimethylphenyl)-2,6-diphenylpyryliumperchlorate	UVM-7	X		X	
5	Dinuclear rhodium complex	Silica gel	X		X	
6	Malachite green	UVM-7	X	X	X	
7	Malachite green	Alumina	X		X	
8	1-butyl-3-(4-nitro-phenylazo)-phenylthiourea	UVM-7	x			
9	Carminic acid	UVM-7	X		X	
10	m-cresol purple	UVM-7	X	X	X	
11	m-cresol purple	alumina	X	X		
12	curcumin	UVM-7	X			
13	Br-cresol purple	UVM-7	X		X	
14	Br-cresol purple	alumina	X	X		X
15	phenolphtalein	alumina	X		X	
16	litmus	UVM-7	X			
17	Bis (4-dibutylaminophenyl) Squaraine	UVM-7		X	X	
18	Methanil yellow	UVM-7			X	
19	Brilliant yellow	UVM-7			X	
20	Thymol blue	UVM-7			X	
21	m-cresolsulfonephtalein	Alumina				X
22	m-cresolsulfonephtalein	MCM-41				X
23	Thymol blue	Alumina				X
24	Thymol blue	MCM-41				X

(phenolphthalein, malachite green, m-cresol purple, phenol red, br-cresol purple, dimethyl yellow, methanil yellow, brilliant yellow, thymol blue, and m-cresol sulfonephtalein), Lewis acids (dinuclear rhodium complex), compounds able to give hydrogen bonding interactions (1-butyl-3-(4-nitro-phenylazo)-phenylthiourea) and reactands reactive to the presence of thiols and amine functional groups (2,6-diphenyl-4-(2-(4-*N*,*N*-dimethylaminophenyl) vinyl)-pyrilium tetrafluoroborate, 4-(4-*N*,*N*-dimethylphenyl)-2,6-diphenylpyrilium perchlorate and bis-4-dibutylaminophenyl-squaraine). Moreover, some natural dyes were tested (curcumin, carminic acid, litmus). Most of them were used only in one support, in particular reactants specific for certain metabolites, but some of them, mainly pH indicators (ie, m-cresol-purple, Br-cresol purple, or litmus) were tested in diverse materials in order to benefit from the potential acid-base modulation properties of the support. In some cases also the load of dye was a variable. It was expected that response of the sensing material would depend on the concentration of indicator because a higher metabolite concentration is necessary to induce a full change of color if the amount of dye increases. In particular, two materials of bis-4-dibutylaminophenyl-squaraine adsorbed on UVM-7, one with a concentration 3.5 higher of squaraine than the other, were used in the chromogenic arrays for fresh pork sausages and boiled marinated turkey. A similar approach was followed in the case of 2,6-diphenyl-4-(2-(4-*N*,*N*-dimethylaminophenyl)vinyl)-pyrilium tetrafluoroborate adsorbed on UVM-7 in the array used for boiled marinated turkey with good results in all the cases (vide infra).

4.2 Chromogenic Array Response

The arrays were prepared by placing each material into an ELISA microplate. The samples were packaged together with the array in order to favor the contact of the array with the metabolites generated by the food matrix. In the case of the meat the packaging consisted in polystyrene boxes under 30% CO_2, 70% N_2 MAP conditions, and kept refrigerated at 4°C throughout the experiment in order to simulate the ordinary packaging conditions. On the contrary, cheeses were packaged with the array at room temperature because the static headspace direct exposure to the volatile compounds at this temperature simulates real consumption conditions as closely as possible.

For each sampling time a photograph of the array was taken inside a light box under controlled illumination conditions and Lab and RGB coordinates were extracted with Photoshop image processing software. Unlike RGB, the Lab system offers a uniform

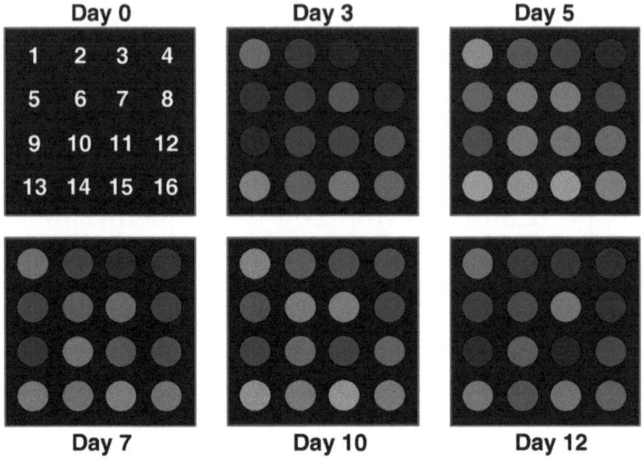

Figure 1.1. Color change profiles after exposure to the atmosphere generated by the chicken aging. Numbers indicate the sensing material. Days indicate the days after packaging. Reproduced from Salinas et al. (2012), with permission of The Royal Society of Chemistry.

nonlinear color system in which equal color differences correspond to equal distances in Euclidean space. Moreover, Lab data are usually more appropriate to study color differences on surfaces than RGB coordinates (Gilabert, 2007).

For visualization purposes, difference maps obtained by the difference of the red, green, and blue (RGB) coordinates of each compound from day n and the initial values measured on day 0 were obtained. To facilitate visualization, the color palette of the difference map was enhanced by expanding the color range from the maximum values of variation to 0–255. Fig. 1.1 shows as an example the difference vector for chicken aging visualized as a map of color changes; each spot corresponds to a sensing material (the numbers indicate the sensing material in relation to Table 1.1). Day 0 shows a black color corresponding to 0,0,0 RGB coordinates. Most interesting is that Fig. 1.1 clearly shows characteristic color fingerprints for each day, confirming the possibility of using this optoelectronic nose to monitor the chicken meat aging process. Two tendencies in color differences can be highlighted: (1) almost all the dyes change color and (2) color changes strength during aging. The former suggested that all dyes are affected by atmospheric changes during the aging process, and all of them contribute to outlining the patterns. The latter is in agreement with a preferential interaction of the dyes with the metabolites generated during chicken decaying. A similar behavior can be observed in the case of blue cheese classification at 30 min (Fig. 1.2), Roquefort in particular but also Blue Stilton and blue cheese with leaves (respectively marked as B, C, and D

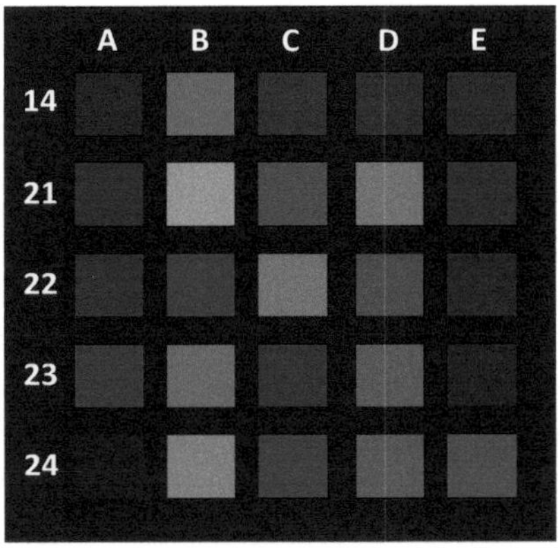

Figure 1.2. Color change profiles after exposure to the atmosphere generated by the blue cheeses. Numbers indicate the sensing materials. Letters denote the kind of cheese: Cheddar (A), Roquefort (B), Blue Stilton (C), blue cheese with leaves (D) and blue cheese spread (E). Reproduced from Zaragozá et al. (2015), with permission of Elsevier.

in Fig. 1.2) showed a brighter color, indicating a stronger response of the sensing materials to the packaging headspace. Also a similar behavior was observed in the case of the other two kinds of meats (fresh pork sausages and boiled marinated turkey), indicating that in all the cases the array generates a fingerprint that can be useful for the generation of an optoelectronic nose.

5 Freshness Monitoring

As noted in the previous section, the prepared chromogenic arrays seem well suited for monitoring of chicken, fresh pork sausages, and boiled marinated turkey freshness. In a further step, color differences obtained from the arrays were analyzed with PCA, HCA, and KNN techniques.

Spoilage is a safety and organoleptic issue related with human consumption accompanied by microbiological growth and sensorial effects. CO_2 containing MAP in absence of oxygen increases meat shelf life through the suppression of aerobial bacterial spoilage and, in general, a significant reduction of the growth of other typical bacteria such as lactic acid bacteria. Metabolites generated during aging depend on flora. Moreover, whereas lactic acid bacteria produces only minor changes in flavor, aerobics induce the degradation sensation mostly associated with the formation of sulfur-containing metabolites (Garcia de Fernando et al., 1995). Based on this, a suitable strategy to follow spoilage process and obtain data that can be correlated with the response of the chromogenic arrays is to perform simultaneously sensorial analysis and microbiological counts of aerobic bacteria. This methodology has been applied to boiled marinated turkey monitoring and fresh pork sausages. However, in the case of chicken meat packaged under MAP and stored under refrigeration, freshness was directly correlated with storage time.

5.1 Chicken Meat

In the case of chicken, an array of 16 sensing materials was selected (Table 1.1) and PCA was applied to the color data as a simple method to project data to a three-dimensional plane. Six sampling dates were selected (0, 3, 5, 7, 10, and 12 days after packaging). In this case, the *a* and *b* color coordinates were used as the original variables and the mean centering preprocessing technique was applied to the data set. PCA study showed a high degree of dispersion among the independent dimensions and nine PCs are necessary to account for 95% of variance. The first principal component contained 50.8% of variance of the data and the first three components represented 78.8% of total variance. This large number of independent dimensions compared to other systems is typical of optoelectronic noses and helps discriminate among highly related samples (eg, different days) of the complex matrixes (meat aging). This is also in agreement with the use of a wide range of chemical responsive probes in the array that are based in several forms of intermolecular interactions between dyes and the volatile compounds generated during chicken decay.

Typically, chicken refrigerated under 30% CO_2–70% N_2 MAP conditions reaches its limit of acceptability at around days 10–12; although sensorial decaying, biogenic amines, and other off-odor generating substances are noted around days 5–7 (Jiménez et al., 1997; Balamatsia et al., 2006, 2007; Smolander et al., 2002). Thus, three groups, days 0–3 (fresh chicken), 5–7 (beginning of the decay), and 10–12 (nonfresh chicken), can be established. PC scores can be used in the discrimination analysis to assign each sample to a particular group. The first three PCs, representing the largest fraction of overall variability in the samples (78.8% of the variance), were plotted on the *x*, *y*, and *z* axes. Fig. 1.3 shows the resulting PCA for the six sampling days (three replicates). As observed, each sampling day can be clearly clustered in a region diverse from the rest, confirming that it was possible to discriminate among these days. When control data were introduced into the PCA analysis, all dyes were placed around day 0, which agrees with the systems' ability to differentiate the control of the chicken-exposed arrays.

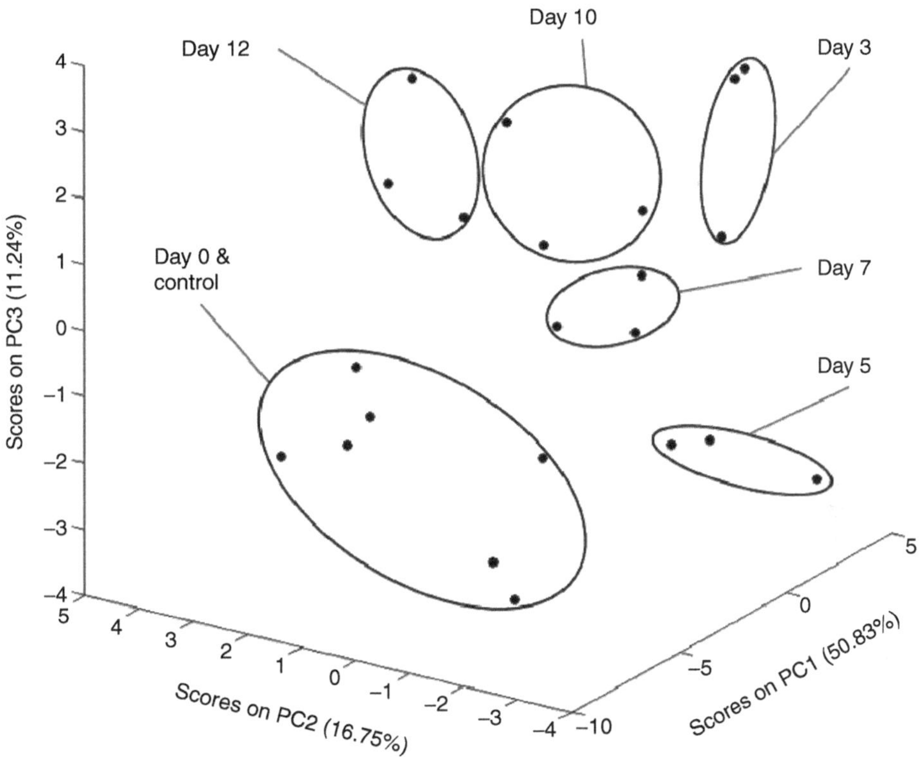

Figure 1.3. The principal component analysis (PCA) score plot for the diverse chicken aging days. Data are shown for three different trials. Reproduced from Salinas et al. (2012), with permission of The Royal Society of Chemistry.

Further analysis of the contribution of each PC to monitoring the aging process and each indicator to the three PC through loadings analysis was performed. The differences in the mean scores on each day (Table 1.2) show, from packaging (day 0) until day 3, a strong variation in PC1. Decaying was followed primarily by PC2 and PC3, where PC3 showed stronger changes on days 3–5, and PC2 from days 10 to 12, with a similar displacement shown in both components from days 5 to 10. Once more, it can be noted that PC2 accumulated more variance than PC3. This is in agreement with the preferential monitorization of spoilage over the final days, though the array was able also to monitor intermediate decaying.

Some more information can be obtained if we take into account the contribution of each indicator to PCs. While PC1 has active participation of pH indicators, PC2 and PC3, which offered the main changes of the aging process, were mainly influenced by compounds based on compounds able to interact with metabolites such as amines, carboxylic acids, or other possible

Table 1.2 Chicken Meat Aging Array: Differences of Mean Scores on Each Day by Component

Days	PC1[a]	PC2[a]	PC3[a]
3–0	8.826	–3.432	3.334
5–3	1.879	0.765	–5.422
7–5	–0.789	1.923	1.931
10–7	0.812	1.186	1.412
12–10	–0.590	2.405	0.135

[a] Absolute values.

nucleophilic molecules originated during chicken aging (materials 2, 4, 5, and 8) and by supports containing natural products (compounds 9, 12, and 16) (Salinas et al., 2012).

5.2 Fresh Pork Sausages

Fresh pork sausages are mainly composed by variable proportions of muscle pork meat and fat, which are more or less coarsely minced, mixed with diverse nonmeat ingredients, such as aromas, species, salt, conservative additives, and coloring agents depending on local preparations (Tarté, 2009; Pearson and Gillet, 1996). The sausage mixture is stuffed into natural casings and ready-made products have to be kept under cold storage conditions until consumed because, due to the high water contents, the lack of thermal processing, the comminuted nature of the raw materials, and the high fat content, such a product is prone to spoilage by both lipid oxidation and microbial contamination (Georgantelis et al., 2007).

In this case, an array composed of seven sensing materials was tested (Table 1.1 and Salinas et al., 2014a). Also, the sampling period was longer than in the case of chicken and 9 sampling points were taken during 52 days (0, 3, 10, 17, 24, 31, 38, 45, and 52 days). Since we had no reference values of the shelf life of these fresh pork sausages in particular, analyses of the content of aerobic bacteria and sensorial analysis were performed in order to follow the aging process and validate the results offered by the optoelectronic nose.

Microbiological analysis of aerobic content revealed that our samples, stored under MAP (30% CO_2–70% N_2) at 4°C, maintained a level of aerobics of round 10^5CFU g^{-1} until day 17. Although there are no specific regulations on fresh pork sausages that fix a maximum level of aerobics, in general, values of c. 10^6–10^7CFU $g^{-1}$are admitted as the limit for the shelf life of fresh sausages (Georgantelis et al., 2007; Martínez et al., 2005, 2006a,b; Insausti et al., 2001). In our case, this limit was reached at around days 31–38. Correspondingly, the sensory scores followed the microbiology evolution. Sausages maintained a perfect odor and appearance up to day 24, exceeding the sensory score threshold after day 38 when a concentration of 10^7CFU g^{-1} of aerobic counts was present.

In this case, also PCA was applied to color data from the array as an initial approach to see the capability of the optoelectronic nose to differentiate between sampling days through a visual tridimensional representation. The original variables, in this case, were the responses of the seven sensing materials used in the array. Only chroma coordinates (*a* and *b*) were employed in the statistical analysis and the mean centering preprocessing technique was applied to the dataset. Again PCA revealed a high degree of dispersion among the independent dimensions. The first principal component contained only 40.48% of data variance. The first three components represented 80.6% of total variance and six PC were required to account for 95% of variance. Fig. 1.4 shows the PCA plot for the nine sampling days. As can be seen, it was possible to discriminate among most sampling days.

Using the differences in the mean scores on each day (Table 1.3) some conclusions were reached. First of all it must be noted that the accumulated value of PC1, the principal component with the highest variance (40.48%), was able to monitor the total aging process, with a variation of up to 8 points in the score and a continuous increase up to day 45 (Table 1.3). However, in certain moments of the aging process (ie, during spoilage on days 31–45) in which the variation of PC1 is relatively low, it is of importance to also take into account PC2 and PC3. In particular, PC2 changes mainly until day 31 when growth of bacterial population was observed, but it was not accompanied by sausage spoilage. The beginning of the spoilage phase, from day 31 to day 38 with aerobic counts reaching 10^7CFU g^{-1} and the sensory score rising to 4, was followed primarily by PC3. Minor contribution of PC3 is observed also in the next phase (up to day 45, in which spoilage was completed). Finally, at the postspoilage stage (days 45–52), there was almost no variations in all three PCs, confirming the selectivity of the chromogenic array to monitor the aging process.

The reduced number of probes, the fact that the sensing materials did not vary in only one part of the aging process, and the diversity of metabolites generated during spoilage does not allow in

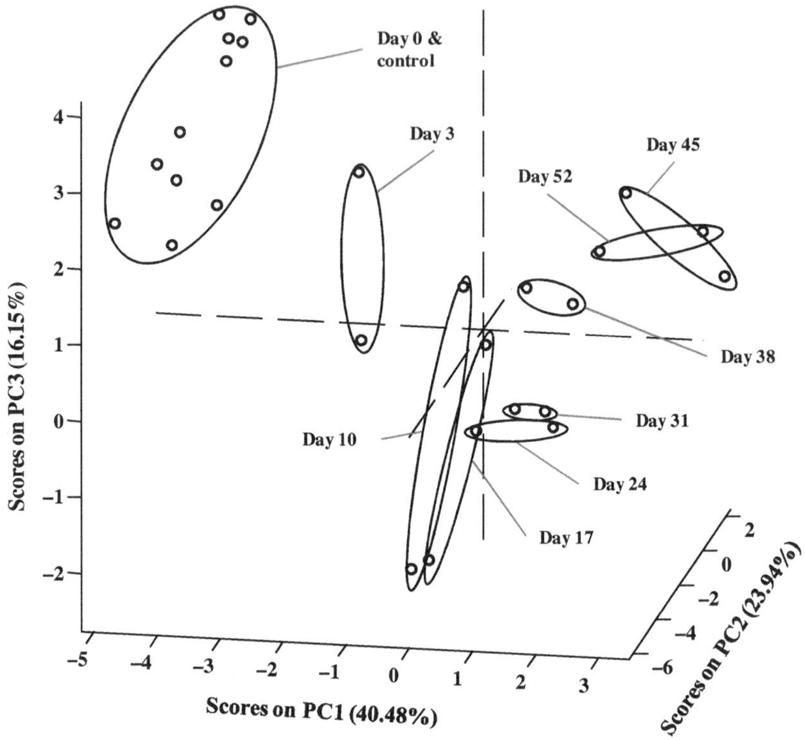

Figure 1.4. The principal component analysis (PCA) score plot for the diverse fresh pork sausage aging days. Data are shown for two different trials. Reproduced from Salinas et al. (2014a), with permission of Elsevier.

this case establishing an exact correlation between the principal components and the indicators. Moreover, some conclusions can be extracted by taking into account the sensing properties of the dyes and their contribution to each PC. Material 2 (Table 1.1), containing the dye (2,6-diphenyl-4-(2-(4-*N*,*N*-dimethylaminophenyl) vinyl)-pyrilium), which is known to respond to biogenic amines and other aminated compounds (García-Acosta et al., 2006), contributed to PC2 and PC3, which is in agreement with the generation of these compounds at the beginning of the spoilage process. Conversely, material 17 of Table 1.1, based on the dye bis(4-dibutylamino-phenyl) squaraine known to react with thiols, contributed more to PC1 and PC3, which are the components that mainly varied after day 31, during the decaying process and during the generation of main off-odors compounds such as sulfur-containing substances (Ros-Lis et al., 2004).

In order to check the viability of the array not only to generate a 3D representation of the data such as that obtained with

Table 1.3 Fresh Pork Sausages Aging Array: Differences of Mean Scores on Each Day by Component

Days	PC1[a]	PC2[a]	PC3[a]	Accumulated PC1 Difference[b]
3–0	2.73	1.22	0.73	2.73
10–3	1.47	1.37	1.89	4.20
17–10	0.56	1.30	0.01	4.76
24–17	0.59	1.74	0.05	5.35
31–24	0.04	1.07	0.01	5.39
38–31	0.42	0.59	1.65	5.81
45–38	2.16	0.85	1.09	7.97
52–45	0.31	0.43	0.00	7.66

[a] Absolute values
[b] Mean score on day *n*—mean score on day 0

PCA, but also classify the samples into groups, a minimal distance classifier, such as KNN (O'Farrell et al., 2005; Ciosek and Wróblewski, 2006) was applied to the data set. KNN is based on the rule that similar patterns belong to the same class. Using the data obtained with the dyes, the samples were grouped in three categories depending of the storage time; days 0 and 3 (including control), days 10–38, and days 45 and 52. Table 1.4 summarizes the percentage of each of the classifications for the test set. In general a good classification was obtained. It is important to note that none of the controls or samples up to day 38 was misclassified as aged sausages (days 45–52).

5.3 Boiled Marinated Turkey

Boiled marinated meat is a traditional Spanish product that combines two common traditional approaches to extend meat consuming period: boiling and marination. During marination, spices, salt and herbs are added to prolong shelf life and improve sensory properties such as meat tenderness, flavor, and juiciness (Mielnik et al., 2008). Besides, thermal treatment further increases product shelf-life and defines organoleptic properties of the final product.

Table 1.4 Percentage of Each Class, as Classified by K-Nearest Neighbor Classifier, for Each of the Stages of the Days After Packing (the Test set is Made of Control, Days 0 and 3; Days 10, 17, 24, 31, and 38; and Days 45 and 52)

	Actual Sample		
Predicted as	Control, Days 0–3	Days 10–38	Days 45–52
Control, days 0–3	85	13	0
Days 10–38	15	88	17
Days 45–52	0	0	83

An array of 16 sensing materials (Table 1.1 and Salinas et al., 2014b) was selected. Eight sampling days (days 0, 3, 10, 17, 24, 31, 38, and 45 postpackaging) were defined during a sampling period of 45 days. Also in this case the analysis of aerobic bacteria counts and sensorial analysis were performed at each sampling time in order to follow aging process and validate the results offered by the chromogenic array.

According to the European Commission for such kind of meat, it is recommended not to exceed the value of c. 5 log CFU g^{-1} of aerobic mesophilic bacteria. In our case, under MAP atmosphere (30% CO_2–70% N_2) and refrigerated at 4°C, such values were reached on day 38. In concordance with bacteria growth, in general a good odor and appearance was maintained up to day 24. Sensory analysis performed to the turkey meat showed a moderate increase of the off-edible properties of the samples on days 31–38 postpackage. The presence of lixiviated and off-odor was more significant after day 38, and additional color alterations and increased viscosity clearly indicating off-edible properties were observed on day 45 postpackage (6.9 log CFU g^{-1}). Based on these sensory evaluation and on the microbiological analysis, the samples can be classified in three groups: (1) "fresh" on days 3–24, with high sensory scores and a microbiological flora characterized by a lactic acid bacteria growing phase, (2) "not fresh" for days 31 and 38 (beginning of the spoilage process), with sensory values close to the limit and the amount of aerobic mesophilic bacteria around the limit of the EC regulation, and (3) "not edible" on day 45, with

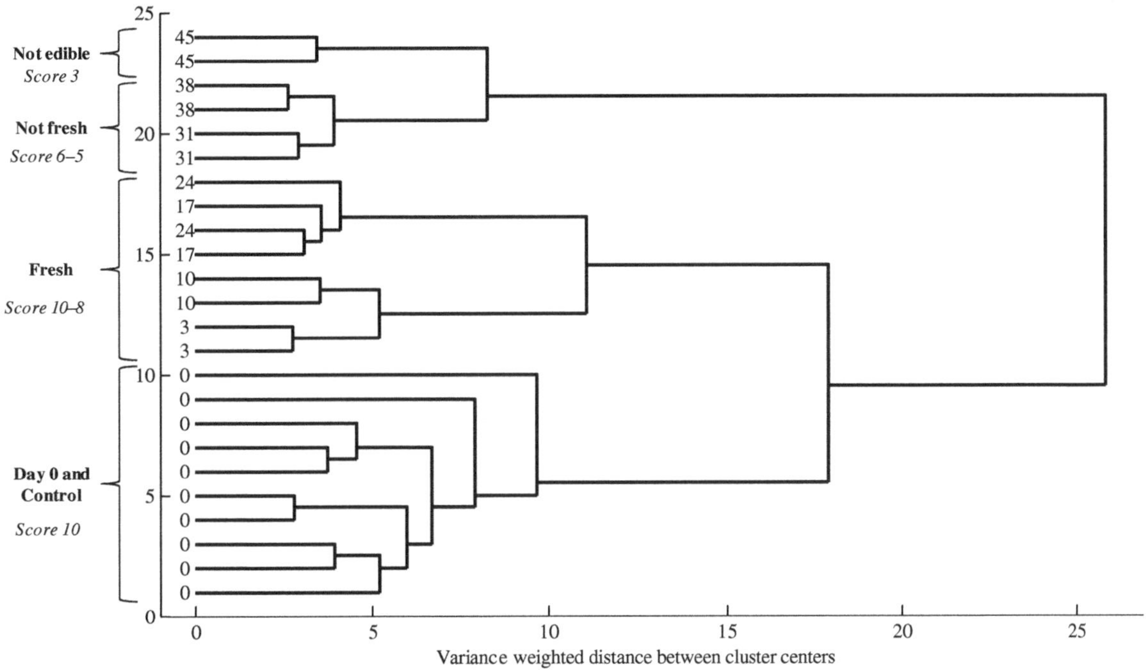

Figure 1.5. Hierarchical cluster analysis for the boiled marinated turkey samples using Ward's method (two samples). Numbers indicate the day after packaging. Reproduced from Salinas et al. (2014b), with permission of Elsevier.

high presence of aerobic mesophilic bacteria and a sensory score under the threshold with clearly recognizable disgusting color and off-odor.

In the case of turkey, the colorimetric response of the array was studied by HCA. In our case the Ward's minimum variance method and an external parameter orthogonalization (EPO) approach (Roger et al., 2003) were applied to the *a* and *b* chroma coordinates of the 16 sensing materials to generate the HCA dendrogram (Fig. 1.5). As seen in Fig. 1.5 the dendrogram groups the samples into three main clusters: (1) days 0 and control, (2) days 3–24, and (3) days 31–45. The latter, however, can be divided into two subclusters (days 31–38 and day 45). The dendrogram correlates well with the above described sensory evaluation and microbiological analysis. In fact there is a quite good relationship between the clustering observed in Fig. 1.5 and the classification "fresh" (days 0–24), "notfresh" (days 31–38), and "not edible" (day 45). As expected, most of duplicate trays are clustered together. Also, day 0 and control samples (stored in absence of meat) are grouped in the same cluster, clearly indicating that no color changes were observed in the array in the absence of the boiled marinated turkey.

6 Quantification of Microbiological Loading, Storage Days, and Sensory Score

As we have seen above, PCA, HCA, or KNN showed that the data of the three optoelectronic noses clustered in accordance with the time elapsed after the packaging and allowed us to obtain good classification models. In this section, we will analyze whether the chroma coordinates from the chromogenic arrays can be employed to predict the concentration of aerobics, sensory scores, and postpackaging days. As a possibility to achieve this goal, the partial least square (PLS) regression technique was considered. In the case of chicken meat, only storage time data was available, therefore only a PLS model for such parameters was developed.

Typically, in order to evaluate the accuracy and precision of the created PLS prediction model, measured versus predicted values are plotted on a graph. Fig. 1.6 shows as an example the PLS graph of aging time for chicken meat. The measured values represent experimental values, while the predicted values are the values calculated according to the PLS algorithm. The results may be evaluated by a visual inspection of the difference between measured and predicted values. In the case of Fig. 1.6 all points are

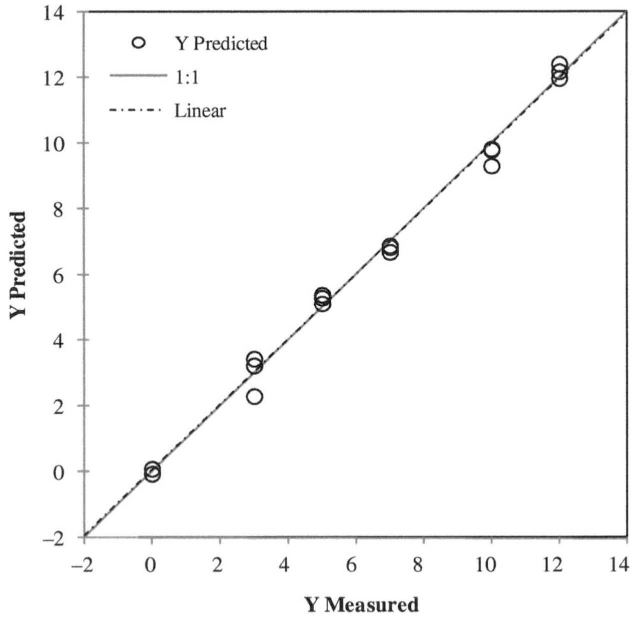

Figure 1.6. The plot score of the prediction model of chicken aging for the calibration set and linear fitting. Reproduced from Salinas et al. (2012), with permission of The Royal Society of Chemistry.

Table 1.5 Adjusting Parameters (*p*1, *p*2) From the PLS Prediction Models for Levels of Mesophilic Aerobic Counts, Sensory Score, and Storage Time for Diverse Kinds of Meat

	Turkey		Sausages		Chicken	
Parameter	*p1*	*p2*	*p1*	*p2*	*p1*	*p2*
Storage time	0.9424	1.3832	0.8839	2.8381	0.9937	0.0389
Mesophilic aerobic counts	0.9185	0.3418	0.8972	0.6491	—	—
Sensory score	0.9038	0.6910	0.8799	0.3001	—	—

close to the line, indicating that the model fits the experimental data. Moreover, a further analysis can be carried out by linearly fitting the experimental points. Here by using a simple linear model, namely $y = p1^* x + p2$, a fitting line and the fitting parameters ($p1$, $p2$) were obtained. $p1$ (slope of the fitting line) and $p2$ (intercept with the y-axis) are related with accuracy in prediction. In an ideal situation, the predicted values should lie along the diagonal line ($p1 = 1$ and $p2 = 0$), which would indicate that predicted and actual values are the same. In particular the PLS prediction model for storage time on chicken meat gave values of 0.9937 and 0.0389 for the slope ($p1$) and the intercept ($p2$) respectively, which is in agreement with a good fit between predicted and measured values. The same approach was applied to data collected from the other optoelectronic arrays and aerobic, sensory score, and storage time parameters. Table 1.5 summarizes the obtained values for slope ($p1$) and intercept ($p2$) for the diverse parameters. In all the cases a good fit is observed between predicted and measured values. In the case of boiled marinated turkey values observed for the slope are in all the cases over 0.9 (0.94 for the storage time, 0.92 for aerobic counts and 0.90 for the sensory score). This is a remarkable behavior and indicted the potential use of chromogenic arrays. Values of $p1$ for the sensory score are slightly lower than for storage time and aerobic counts, but nevertheless $p1$ values are reasonable taking into account that sensory analysis is subjective and strongly dependent on the panel list. In the case of fresh pork sausages, values are slightly worse than for the other two kinds of meat, but they are very close to 0.9 in all cases. These results reinforce the possibility to use this approach to develop models to predict these three parameters (ie, storage time, aerobic counts, and sensory score) in different kinds of meat during aging.

7 Determination of Blue Cheese Origin

Five cheeses (four blue cheeses and Cheddar) were used in this study. Three of the selected blue cheeses were from diverse origin (Blue Stilton, Rochefort, and blue cheese with leaves) and the fourth was blue cheese spread. Cheddar was tested as control (a nonblue cheese). It has been reported that times of up to 16 h or temperatures above room temperature are necessary to reach the equilibrium in the headspace (Cakmakci et al., 2013; Frank et al., 2004). So although the array may be suitable as a fast screening method to classify or identify blue cheeses, the array response may vary with time, and this factor should be controlled to avoid misclassification. Thus in this study time was controlled, and color data from the array was taken at 0.5, 1.5, 3.5, 5.5, and 8.5 h after packaging. The optoelectronic array for the analysis of the origin of blue cheeses was composed by five sensing materials (Table 1.1). They consist of three pH indicators (br-cresol purple, m-cresolsulfonphthalein, thymol blue), which were combined with two support materials: aluminium oxide (basic) and MCM-41 (slightly acid). The objective was to design an optoelectronic nose that was as simple as possible that could be used by consumers and retailers. For such purposes pH indicators offer advantages over other indicators since they are cheap and commercially available, and can be easily incorporated to support ensuring array scalability.

Color differences were analyzed by PCA. Lab-based color data generated by the responses of the five sensing materials used in the chromogenic array was used to perform the analysis and an autoscaled technique was applied to the data set. PCA study at 30 min revealed a high degree of dispersion among the independent dimensions. PC1 contained only 37.03% of data variance and the first eight PCs were required to account for 95% of variance. The first three components, representing the main part of total variance (70.60%), were plotted on the x, y, and z axes as a method to project data to a three-dimensional hyperplane. Fig. 1.7 shows the resulting PCA for the five cheeses (three replicates) 30 min after packaging. As observed, all cheeses clustered in a completely different zone, confirming that the chromogenic array generates a characteristic fingerprint for each blue cheese.

Moreover the data set was also analyzed by PLS-DA. The output of PLS-DA consists in assigning each sample to one of the

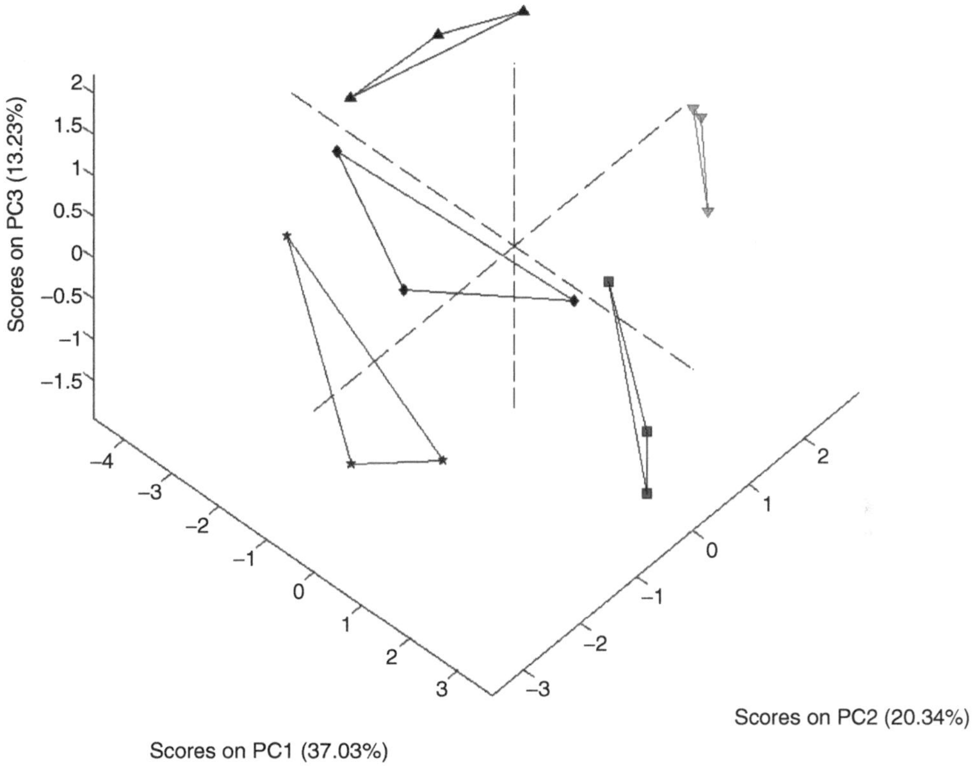

Figure 1.7. The principal components analysis (PCA) score plot of the diverse cheeses using the data obtained 0.5 h after packaging. Data are shown for three different trials: (♦) Blue Stilton; (■) blue cheese with leaves; (▲) Cheddar; (▼) Roquefort; (★) blue cheese spread. Reproduced from Zaragozá et al. (2015), with permission of Elsevier.

predefined categories that, in line with our general purpose of obtaining an array capable of making classifications between blue cheeses, was each one of the cheeses. Five classes (one for each blue cheese and one for the Cheddar control) were fixed and the model was applied to the data collected after only 30 min (see first entry on Table 1.6). An average of 67% of correct prediction was obtained at 30 min. Although 67% of proper assignment is in line with the results obtained with other technologies for the same purpose (Kulmyrzaev et al., 2008), it is probably too low for practical use. Classifications close to 100% are required to ensure users' confidence in this technology.

We must take into account that we are using only five sensing materials to classify five cheeses. Cross-sensitive sensors tend to generate a limited amount of information, which is the reason why, in general, the arrays tend to require a large number of indicators. Since we were interested in keeping the array as simple

Table 1.6 Percentage of Samples Correctly Classified by the PLS-DA Model Using Five Classes (Cheddar, Roquefort, Blue Stilton, Blue Cheese With Leaves, and Blue Cheese Spread): Time Indicates That Color Data up to the Time Spent in the PLS-DA Model

Time	LV	Blue Stilton	Blue Cheese With Leaves	Roquefort	Blue Cheese Spread	Cheddar
0.5 h	4	0%	67%	67%	100%	100%
1.5 h	4	33%	33%	100%	100%	100%
3.5 h	4	33%	67%	100%	67%	100%
5.5 h	4	100%	100%	100%	100%	100%
8.5 h	4	100%	100%	100%	100%	100%

as possible, and as a strategy to increase the information and the number of independent variables, the PLS-DA model of our chromogenic array was fed with not only the data collected at a certain time, but also with the data measured up to the labelled time (ie, for 1.5 h, the data measured at 0.5 and 1.5 were incorporated into the model). As can be seen on Table 1.6, as more information was incorporated, the prediction ability improved from 67% at 30 min to 100% at 5.5 h and 8.5 h. This perfect prediction at 5.5 h confirms the capacity of the chromogenic array to classify blue cheeses and opens up the use of optoelectronic noses as a real, easy-to-use, and rapid classification tool for cheeses.

Given our interest in developing a tool as quickly as possible and the above-reported results at 30 min, also a PLS-DA model with a lower number of classes was run. If three classes are used: (1) Roquefort, (2) Cheddar, and (3) the remaining blue cheeses. All three classes were correctly classified by 30 min (100% of correct assignments), which suggests that the array can be a useful rapid Roquefort classification test.

8 Conclusions

Optoelectronic noses are a suitable approach for monitoring meat aging (chicken meat, fresh pork sausages, and boiled marinated turkey), and food classification (blue cheeses). All chromogenic arrays described here have in common the use of chromogenic sensing materials developed by the combination of dyes into inorganic supports with diverse acidities and topologies (UVM-7, MCM-41, silica gel, and alumina). Dyes included pH indicators, Lewis acids, donor hydrogen bonding derivatives, reactants selective to metabolites, and natural dyes. Color modulations were statistically tested using diverse techniques. PCA and HCA were used with the full set of data as initial tools for simplification purposes and to determine the potential of the array to differentiate between the samples and clustering them. PCA was used in the case of chicken meat, fresh pork sausages, and blue cheese classification, while HCA was used in the case of boiled marinated turkey. In all the cases good results were obtained and categories (aging time or cheese origin) where clearly differentiated between them. Moreover PLS was used to correlate, for a given sample, the amount of mesophilic aerobic counts, sensory score, and storage days after packaging. In all cases good PLS prediction models were obtained. Finally, considering that in the case of the blue cheeses the composition of the headspace is time dependent, a PLS-DA model that incorporates data at different times was developed. The model, incorporating data up to 5.5 h is able to classify samples in one of the five classes of cheese with 100% correct classification. All these results suggest that optoelectronic noses are suitable to develop simple, cheap systems that can be applied for the improvement of safety on the food chain and for the detection of food fraud. Moreover a number of other applications are envisioned.

References

Amendola, V., Fabbrizzi, L., Mosca, L., 2010. Anion recognition by hydrogen bonding: urea-based receptors. Chem. Soc. Rev. 39, 3889–3915.

Anzenbacher, P., Lubal, P., Bucˇek, P., Palacios, M.A., Kozelkova, M.E., 2010. A practical approach to optical cross-reactive sensor arrays. Chem. Soc. Rev. 39, 3954–3979.

Balamatsia, C.C., Paleologos, E.K., Kontominas, M.G., Savvaidis, I.N., 2006. Correlation between microbial flora, sensory changes, and biogenic amines formation in fresh chicken meat stored aerobically or under modified atmosphere packaging at 4°C: possible role of biogenic amines as spoilage indicators. Antonie van Leeuwenhoek 89, 9–17.

Balamatsia, C.C., Patsias, A., Kontominas, M.G., Savvaidis, I.N., 2007. Possible role of volatile amines as quality-indicating metabolites in modified

atmosphere-packaged chicken fillets: correlation with microbiological and sensory attributes. Food Chem. 104, 1622–1628.
Berrueta, L.A., Alonso-Salces, R.M., Héberger, K., 2007. Supervized pattern recognition in food analysis. J. Chromatogr. A. 1158 (1–2), 196–214.
Bota, G.M., Harrington, P.B., 2006. Direct detection of trimethylamine in meat food products using ion mobility spectrometry. Talanta 68, 629–635.
Cakmakci, S., Dagdemir, E., Hayaloglu, A.A., Gurses, M., Cetin, B., Tahmas-Kahyaoglu, D., 2013. Effect of *Penicillium roqueforti* and incorporation of whey cheese on volatile profiles and sensory characteristics of mold-ripened Civil cheese. Int. J. Dairy Tech. 66, 512–526.
Cevolia, C., Cerretania, L., Gori, A., Caboni, M.F., Gallina Toschi, T., Fabbri, A., 2011. Classification of Pecorino cheeses using electronic nose combined with artificial neural network and comparison with GC–MS analysis of volatile compounds. Food Chem. 129, 1315–1319.
Chen, H., Seiber, J.N., Hotze, M., 2014. ACS select on nanotechnology in food and agriculture: a perspective on implications and applications. J. Agric. Food Chem. 62, 1209–1212, 2014.
Ciosek, P., Wróblewski, W., 2006. The analysis of sensor array data with various pattern recognition techniques. Sensor. Actuat. B-Chem. 114, 85–93.
Commission of the European Communities, 2009. Brussels, 30.7.2009 SEC (2009) 1111 final Commission staff working document. European Industry in a Changing World. Updated Sectoral Overview 2009.
Curioni, P.M.G., Bosset, J.O., 2002. Key odorants in various cheese types as determined by gas chromatography-olfactometry. Int. Dairy J. 12, 959–984.
Day, E.A., Anderson, D.F., 1965. Cheese flavor, gas chromatographic, and mass spectral identification of neutral components of aroma faction and blue cheese. J. Agric. Food Chem. 13, 2–4.
Deisingh, A.K., Stone, D.C., Thompson, M., 2004. Applications of electronic noses and tongues in food analysis. Int. J. Food Sci. Tech. 39, 587–604.
Descalzo, A.B., Marcos, M.D., Monte, C., Martínez-Máñez, R., Rurack, K., 2007. Mesoporous silica materials with covalently anchored phenoxazinone dyes as fluorescent hybrid materials for vapour sensing. J. Mat. Chem. 17, 4716–4723.
Ellis, D.I., Goodacre, R., 2001. Rapid and quantitative detection of the microbial spoilage of muscle foods: current status and future trends. Trends Food Sci. Technol. 12, 414–424.
European Technology Platform Food for Life., 2014. Strategic Research and Innovation Agenda 2015–2020 and beyond.
Everitt, B., Hothorn, T., 2011. An introduction to applied multivariate analysis with R. Springer Science & Business Media, New York.
Frank, D.C., Owen, C.M., Patterson, J., 2004. Solid phase microextraction (SPME) combined with gas-chromatography and olfactometry-mass spectrometry for characterization of cheese aroma compounds. LWT Food Sci. Tech. 37, 139–154.
Gallois, A., Langlois, D., 1990. New results in the volatile odorous compounds of French cheeses. Lait 70, 89–106.
Garcia de Fernando, G.D., Nychas, G.J.E., Peck, M.W., Ordóñez, J.A., 1995. Growth/survival of psychrotrophic pathogens on meat packaged under modified atmospheres. Int. J. Food Microbiol. 28, 221–231.
García-Acosta, B., Comes, M., Bricks, J.L., Kudinova, M.A., Kurdyukov, V.V., Tolmachev, A.I., Descalzo, A.B., Marcos, M.D., Martínez-Máñez, R., Moreno, A., Sancenón, F., Soto, J., Villaescusa, L.A., Rurack, K., Barat, J.M., Escriche, I., Amorós, P., 2006. Sensory hybrid host materials for the selective chromo-fluorogenic detection of biogenic amines. Chem. Commun., 2239–2241.
Gardner, J.W., Bartlett, P.N., 1994. A brief history of electronic noses. Sensor. Actuat. B-Chem. 18 (1), 210–211.

Georgantelis, D., Ambrosiadis, I., Katikou, P., Blekas, G., Georgakis, S.A., 2007. Effect of rosemary extract, chitosan, and a-tocopherol on microbiological parameters and lipid oxidation of fresh pork sausages stored at 4°C. Meat Sci. 76, 172–181.
Gilabert, E.J., 2007. Medida del color. Universidad Politécnica de Valencia, Spain.
Gkatzionis, K., Hewson, L., Hollowood, T., Hort, J., Dodd, C.E.R., Linforth, R.S.T., 2013. Effect of *Yarrowia lipolytica* on blue cheese odour development: flash profile sensory evaluation of microbiological models and cheeses. Int. Dairy J. 30, 8–13.
Grau, R., Sanchez, A.J., Giron, J., Iborra, E., Fuentes, A., Barat, J.M., 2011. Nondestructive assessment of freshness in packaged sliced chicken breasts using SW-NIR spectroscopy. Food Res. Int. 44, 331–337.
Insausti, K., Beriain, M.J., Purroy, A., Alberti, P., Gorraiz, C., Alzueta, M.J., 2001. Shelf life of beef from local Spanish cattle breeds stored under modified atmosphere. Meat Sci. 57, 273–281.
Institution of Mechanical Engineers. 2013. Global food: waste not, want not.
Janzen, M.C., Ponder, J.B., Bailey, D.P., Ingison, C.K., Suslick, K.S., 2006. Colorimetric sensor arrays for volatile organic compounds. Anal. Chem. 78, 3591–3600.
Jiménez, S.M., Salsi, M.S., Tiburzi, M.C., Rafaghelli, R.C., Tessi, M.A., Coutaz, V.R., 1997. Spoilage microflora in fresh chicken breast stored at 4°C: influence of packaging methods. J. Appl. Microbiol. 83, 613–618.
Karlshøj, K., Larsen, T.O., 2005. Differentiation of species from the *Penicillium roqueforti* group by volatile metabolite profiling. J. Agric. Food Chem. 53, 708–715.
Kerry, J.P., ÓGrady, M.N., Hogan, S.A., 2006. Past, current, and potential utilisation of active and intelligent packaging systems for meat and muscle-based products: a review. Meat Sci. 74, 113–130.
Kulmyrzaev, A., Bertrand, D., Dufour, É., 2008. Characterization of different blue cheeses using a custom-design multispectral imager. Dairy Sci. Tech. 88, 537–548.
Larsen, M.D., Jensen, K., 1999. The effects of environmental conditions on the lipolytic activity of strains of *Penicillium roqueforti*. Int. J. Food Microbiol. 46, 159–166.
Lawlor, J.B., Delahunty, C.M., Sheehan, J., Wilkinson, M.G., 2003. Relationships between sensory attributes and the volatile compounds, non-volatile and gross compositional constituents of six blue-type cheeses. Int. Dairy J. 13, 481–494.
Lim, S.H., Kemling, J.W., Feng, L., Suslick, K.S., 2009. A colorimetric sensor array of porous pigments. Analyst 134, 2453–2457.
Lin, M., Al-Holy, M., Mousavi-Hesary, M., Al-Qadiri, H., Cavinato, A.G., Rasco, B.A., 2004. Rapid and quantitative detection of the microbial spoilage in chicken meat by diffuse reflectance spectroscopy (600–1100 nm). Lett. Appl. Microbiol. 39, 148–155.
Martínez, L., Dejenane, D., Cilla, I., Beltrán, J.A., Roncales, P., 2005. Effect of different concentrations of carbon dioxide and low concentration of carbon monoxide on the shelf-life of fresh pork sausages packaged in modified atmosphere. Meat Sci. 71, 563–570.
Martínez, L., Cilla, I., Beltrán, J.A., Roncalés, P., 2006a. Combined effect of modified atmosphere packaging and addition of rosemary (*Rosmarinus officinalis*), ascorbic acid, red beet root (*Beta vulgaris*), and sodium lactate and their mixtures on the stability of fresh pork sausages. J. Agric. Food Chem. 54, 4674–4680.
Martínez, L., Djenane, D., Cilla, I., Beltrán, J.A., Roncales, P., 2006b. Effect of varying oxygen concentrations on the shelf-life of fresh pork sausages packaged in modified atmosphere. Food Chem. 94, 219–225.

Martínez-Máñez, R., Sancenón, F., 2003. Fluorogenic and chromogenic chemosensors and reagents for anions. Chem. Rev. 103, 4419–4476.
Matthew, B., William, R., 2003. Partial least squares for discrimination. J. Chemom. 17, 166–173.
McSweeney, P.L.H., Sousa, M.J., 2000. Biochemical pathways for the production of flavour compounds in cheeses during ripening: a review. Lait 80, 293–324.
Mielnik, M.B., Sem, S., Egelandsdal, B., Skrede, G., 2008. By-products from herbs: essential oil production as ingredient in marinade for turkey thighs. LWT Food Sci. Tech. 41, 93.
Mohr, G.J., 2006. New chromogenic and fluorogenic reagents and sensors for neutral and ionic analytes based on covalent bond formation: a review of recent developments. Anal. Bioanal. Chem. 386, 1201–1214.
Musatov, V.Y., Sysoeva, V.V., Sommerb, M., Kiselevb, I., 2010. Assessment of meat freshness with metal oxide sensor microarray electronic nose: a practical approach. Sensor. Actuat. B-Chem. 144 (1), 99–103.
Musto, C.J., Lim, S.H., Suslick, K.S., 2009. An optoelectronic nose for the detection of toxic gases. Anal. Chem. 81, 6526–6533.
O'Farrell, M., Lewis, E., Flanagan, C., Lyons, W., Jackman, N., 2005. Comparison of k-NN and neural network methods in the classification of spectral data from an optical fibre-based sensor system used for quality control in the food industry. Sensor. Actuat. B-Chem. 111-112, 354–362.
Parente, E., Patel, H., Caldeo, V., Piraino, P., McSweeney, P.H.L., 2012. RP-HPLC peptide profiling of cheese extracts: a study of sources of variation, repeatability, and reproducibility. Food Chem. 131, 1552–1560.
Pearson, A.M., Gillet, T.A., 1996. Processed Meats. Springer, USA.
Qian, M., Nelson, C., Bloomer, S., 2002. Evaluation of fat-derived aroma compounds in blue cheese by dynamic headspace GC/Olfactometry-MS. J. Am. Oil Chemists' Soc. 79, 663–667.
Rencher, A.C., Christensen, W.F., 2002. Methods of Multivariate Analysis, third ed. Wiley Series in Probability and Statistics. John Wiley & Sons, Hoboken, New Jersey, pp. 759–768.
Roger, J., Chauchard, F., Bellon-Maurel, V., 2003. EPO–PLS external parameter orthogonalization of PLS application to temperature-independent measurement of sugar content of intact fruits. Chemometr. Intell. Lab. 66, 191–204.
Ros-Lis, J.V., García, B., Jiménez, D., Martínez-Máñez, R., Sancenón, F., Soto, J., Gonzalvo, F., Valldecabres, M.C., 2004. Squaraines as fluoro-chromogenic probes for thiol-containing compounds and their application to the detection of biorelevant thiols. J. Am. Chem. Soc. 126, 4064–4065.
Ros-Lis, J.V., Martínez-Máñez, R., Soto, J., 2005. Colorimetric signaling of large aromatic hydrocarbons via the enhancement of aggregation processes. Org. Lett. 7, 2337–2339.
Ros-Lis, J.V., Martínez-Máñez, R., Soto, J., Villaescusa, L.A., Rurack, K., 2011. Squaraine "ships" in the Y zeolite "bottle": a chromogenic sensing material for the detection of volatile amines and thiols. J. Mater. Chem. 21, 5004–5010.
Sahar, A., Boubellouta, T., Dufour, E., 2011. Synchronous front-face fluorescence spectroscopy as a promising tool for the rapid determination of spoilage bacteria on chicken breast fillet. Food Res. Int. 44, 471–480.
Salinas, Y., Ros-Lis, J.V., Vivancos, J.L., Martínez-Máñez, R., Marcos, M.D., Aucejo, S., Herranz, N., Lorente, I., 2012. Monitoring of chicken meat freshness by means of a colorimetric sensor array. Analyst 137, 3635–3643.
Salinas, Y., Ros-Lis, J.V., Vivancos, J.L., Martínez-Máñez, R., Marcos, M.D., Aucejo, S., Herranz, N., Lorente, I., Garcia, E., 2014a. A novel colorimetric sensor array for monitoring fresh pork sausages spoilage. Food Control. 35, 166–176.

Salinas, Y., Ros-Lis, J.V., Vivancos, J.L., Martínez-Máñez, R., Aucejo, S., Herranz, N., Herranz, N., Lorente, I., Garcia, E., 2014b. A chromogenic sensor array for boiled marinated turkey freshness monitoring. Sensor. Actuat. B-Chem. 190, 326–333.
Smolander, M., Hurme, E., Latva-Kala, K., Luoma, T., Alakomi, H.-L., Ahvenainen, R., 2002. Myoglobin-based indicators for the evaluation of freshness of unmarinated broiler cuts. Innov. Food Sci. Emerg. 3, 279–288.
Tarté, R., 2009. Ingredients in Meat Products. Springer-Verlag, New York.
Timm, N.H., 2004. Applied Multivariate Analysis. Springer, NY.
Trihaas, J., van den Tempel, P.V., Nielsen, T., 2005. Electronic nose technology in quality assessment: predicting volatile composition of danish blue cheese during ripening. J. Food Sci. 70, e392–e400.
United States Department of Agriculture, 2011. Foreing Agricultural Service. Livestock and Poultry: World Markets and Trade, April.
Varming, C., Andersen, L.T., Petersen, M.A., Ardö, Y., 2013. Flavor compounds and sensory characteristics of cheese powders made from matured cheeses. Int. Dairy J. 30, 19–28.
Wetzel, D.L.B., Charalambous, G., 1998. Instrumental methods in food and beverage analysis. Elsevier Science B.V., Armsterdam, The Netherlands.
Wold, S., Sjostrom, M., Eriksson, L., 2001. PLS-regression: a basic tool of chemometrics. Chemometr. Intel. Lab. 58, 109–130.
Zaragozá, P., Ros-Lis, J., Vivancos, J.-L., Martínez-Máñez, R., 2015. Proof of concept of using chromogenic arrays as a tool to identify blue cheese varieties. Food Chem. 172, 823–830.
Zhang, C., Suslick, S., 2005. A colorimetric sensor array for organics in water. J. Am. Chem. Soc. 127, 11548–11549.

2

BIOSENSORS FOR DETECTION MYCOTOXINS AND PATHOGENIC BACTERIA IN FOOD

Gennady A. Evtugyn*, Rezeda V. Shamagsumova*, Tibor Hianik**

**Kazan Federal University, A.M. Butlerov' Chemistry Institute, Kazan, Russian Federation; **Comenius University, Department of Nuclear Physics and Biophysics, Bratislava, Slovakia*

1 Introduction

There is an urgent need for the development of low-cost and reliable devices for the detection of toxic substances and microbial contamination, which are needed in various areas of human activities, especially agriculture and food safety control. Besides chemical contamination mainly caused by industry, foodborne pathogens are of great concern. This is a group of microorganisms capable of infecting humans via contaminated food or water (Dwivedi and Jaykus, 2011). Together with mycotoxins, secondary metabolites of some fungi cause an important public health problem in the world (Zhao et al., 2014). In accordance with the risk assessment provided by the European Food Safety Authority (EFSA) within the period from 2002 and 2012, 15% of risk is related to mycotoxins versus 20% of persistent organic pollutants (Alexander et al., 2002).

The urgency of overall control of these factors is caused by the large scale of the agriculture production and complications related to a big number of agricultural manufacturers, many of whom cannot afford adequate measures preventing biological and chemical contamination due to low production culture and lack of financial resources. The variety of conditions of cultivation and harvesting as well as long storage and transportation of raw materials and foodstuffs do not promote food safety, either. The globalization of world economics resulted in huge traffic flows, which creates favorable conditions for fast dissemination of contaminated products around the world, so that the potential source

Nanobiosensors. http://dx.doi.org/10.1016/B978-0-12-804301-1.00002-3

of mycotoxins and foodborne pathogens can be very distant from incidents of foodborne poisoning (Sekhon, 2014). This increases the necessity of portable measurement techniques devoted to the detection of potential hazards out of well-equipped laboratories, in field, barn, mill, or brewery.

Depending on assembly, price, and function, such warning devices can solve many analytical tasks, beginning with signal on/off systems for early detection of toxins and leading to quantitative analysis of individual contaminants and their main metabolites in complex matrices. Their development and further progress in types of analytical equipment are mainly directed to increasing sensitivity of detection and improved selectivity of the response. Besides, usability and extended field of application should be taken into account.

In this review, the application of biosensors, especially those including various nanomaterials, has been described for the detection of mycotoxins and pathogenic bacteria as food contaminants.

2 Biosensors and Nanomaterials: General Consideration

Biosensors have been known since 1962 and many efforts have been undertaken to develop reliable reproducible devices with high selectivity and sensitivity toward various analytes. Before considering mycotoxins and bacteria detection, it seems useful to give a short description of main principles of their signal generation with particular emphasis to immunosensors and DNA sensors mainly used for food quality testing. The number of original articles in this area is growing swiftly, so that only some recent reviews are cited to simplify the following consideration of appropriate biosensors for particular targets.

Biosensors utilizing target biochemical interactions for detection and quantification of various hazards should meet requirements mostly required for food contamination detection, especially selectivity of response and direct application in complex media (McGrath et al., 2012). In biosensors, the characteristics required for effective contamination control are achieved by high specificity of analyte interactions involving natural receptor molecules like antibodies (Prieto-Simón and Campàs, 2009), enzymes (Mehrvar and Abdi, 2004), or nucleic acids and their fragments (Cruz-Aguado and Penner, 2008). Starting from the first commercialization success related to glucose sensing with glucose oxidase (Clark and Lyons, 1962), biosensors offer great opportunities for the detection of metabolites, disease biomarkers,

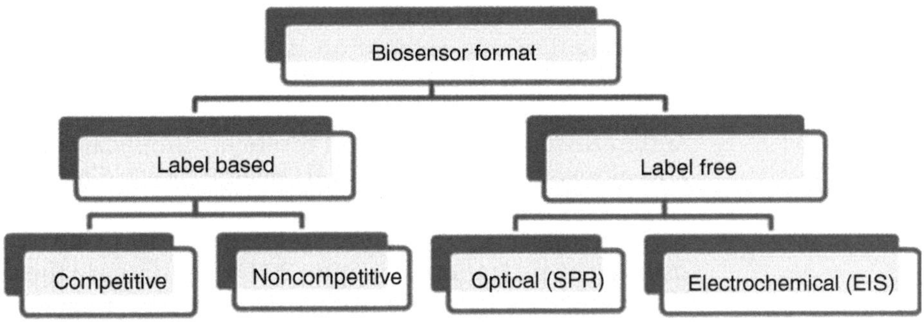

Figure 2.1. Classification of biosensors based on the measurement mode.

nutrients, and contaminants including those of food and feed. However, contrary to many metabolites like glucose, the biosensor for detection of mycotoxins is complicated by extremely strict requirements to the detectable concentrations of food contaminants, which are commonly on nanomolar or subnanomolar level (Scognamiglio et al., 2014). For this reason, the world biosensor market, estimated to be €57 billion in 2011, involved mainly medical devices, and only 1% of the market could be referred to other purposes like environmental monitoring and food safety control.

High sensitivity of the biosensor response is mainly achieved by the progress in the instrumental approaches to the detection of target biomolecular interactions taken place on the transducer surface. An outline of biosensor formats mostly used for the detection of biological hazards like mycotoxins and pathogenic microorganisms is presented in Fig. 2.1.

Many of the approaches currently applied were derived from the conventional immunochemistry techniques. They are based on labeling biochemical components and further accumulation of easily detectable labels on the transducer interface. ELISA employing enzyme as specific label is one of the most popular examples of such an approach (Voller et al., 1978). *Label-based* measurement protocols are subdivided into two groups, that is, *competitive* and *noncompetitive* assay.

In the first approach, target molecules are labeled and compete with nonlabeled analyte present in the sample for binding site on the transducer interface. Increasing concentration of an analyte decreases the amount of labeled analog bonded to the transducer. In sandwich assay, two types of bioreceptors are used. The capture receptor is immobilized on the transducer surface and binds target molecule. Signaling receptor-bearing label interacts with the complex formed by capture receptor and analyte. Increasing concentration of an analyte results in a more intensive signal of label.

Comparing with a competitive assay, sandwich protocol provides accumulation and separation of the target molecules from the sample so that sensitivity of their detection can be sufficiently higher. On the other hand, the necessity in two different receptors and multistage measurement protocol complicate the procedure and elongate the analysis up to several hours. This is hardly collocated with the biosensor format.

More recently, *label-free techniques* have been introduced in biosensor investigations (Cooper, 2009). In this case the changes in the characteristics of the surface layer caused by an analyte binding are monitored. Some measurement techniques like surface plasmon resonance (SPR) analysis (Guo, 2012a) or electrochemical impedance spectroscopy (EIS) (Daniels and Pourmand, 2007) make it possible to record specific shifts of light scattering or electroconductivity depending on interactions between immobilized receptor and biological target. Label-free techniques assume separation of the reactants in a thin surface layer, which result in significant changes in its properties. They are faster and simpler than label-based modes, though often they are not as accurate in sensitivity. Label-free approaches assuming a one-step measurement protocol are denoted as *direct assay* subsequently.

It should be mentioned that the approaches described previously employ both antibodies (Ab) and oligonucleotides able to have specific interactions with target species. In the latter case, *aptamers* are mainly assumed (Mascini, 2008). DNA or RNA aptamers are synthetic oligonucleotides obtained by combinatorial chemistry coupled with affine chromatography to specify nucleotide sequences able to specifically bind target molecules. They can have primary and spatial structure with loops or guanine quadruplexes different from natural DNA molecules. Aptamers are more stable toward hydrolysis and oxidative denaturation than Ab and often exert similar affinity toward targets. For this reason, they were also called synthetic antibodies.

The recognition followed by binding of target analytes with Ab or aptamers does not change chemical structure of an analyte molecule and formally belongs to reversible reaction. Hence, the detection of such interactions is performed in conditions corresponding to equilibrium state with maximal degree of an analyte binding. For this reason, the overall duration of biosensor-based detection of many biological hazards requires rather long time periods comparable with those of conventional ELISA. The use of flow-through setups and multiplex sensors with many recognition elements positioned on the same support improve the productivity of biosensors and makes them more competitive with commercially available test systems.

Table 2.1 Nanomaterials Applied in Biosensor Assembly and Examples of Their Application

Function in Biosensor Assembly	Nanomaterials	Examples
Enhancement of working surface/number of binding sites	Carbonaceous materials, mesoporous films, metal nanoparticles	Carbon nanotubes (CNTs), carbon black, Au nanoparticles, electroconductive polymers, mesoporous alumina films
Improved electron transduction	Carbonaceous materials, metal nanoparticles, metal complexes with transient metals	CNTs, carbon black, graphene, fullerene, Au and Ag nanoparticles, Co phenanthroline, Ru bipyridine complex
Support for immobilization of bioreceptors	Carbonaceous materials, metal oxides, dendrimers	CNTs, carbon black, graphene, ZnO nanoparticles, PAMAM dendrimer
Labels/signal generation system	Metal nanoparticles, quantum dots, metal complexes with transient metals	CdTe nanodots, Au, Ag nanoparticles, Ru bipyridine complexes

Many recent achievements in biosensor developed are related to the application of *nanomaterials* able to significantly improve the performance of existed assemblies and offer new opportunities in miniaturization and automation of newly proposed biosensors. Table 2.1 summarizes the functions of nanomaterials applied in the assembly of modern biosensors. It should be noted that most of the examples are related to electrochemical biosensors mainly presented in the literature. The specific functions—receptor immobilization and stabilization—can be easily extended to other types of signal detection. However, this does not exclude unique properties of nanomaterials involved in optical detection systems, that is, plasmonic effect of Au and Ag nanoparticles used in surface-enhanced Raman spectroscopy (SERS) (Wu et al., 2012c) and SPR-related techniques (Li et al., 2012a). Application of nanoparticles to signal amplification in mass-sensitive detectors—for example, quartz crystal microbalances (Spinella et al., 2014), and microcantilevers (Ricciardi et al., 2013a)—is described mainly for the detection of biopolymers. Thus, such systems are rarely used in the detection of relatively small molecules of mycotoxins.

Metal nanoparticles are mainly applied in the assembly of electrochemical and to some extent optical biosensors (Wang and Hu, 2009). They are obtained by chemical or electrochemical reduction of appropriate precursors in conditions preventing growth of primary formed metal nuclei. This is reached by addition of stabilizers that decrease rate of the amalgamation of nanocrystals. Stability of some colloidal dispersion is sufficient for their commercial distribution. The Ag and Au nanoparticles and their dendrites are also synthesized directly on the electrode surface by electrolysis. The variation of electrolyte content and polarization conditions makes it possible to synthesize alloys nanoparticles of several metals or coat primary metal with a thin shell of a second one. Au nanoparticles play a significant role in SPR- and SERS-based detection systems (Bedford et al., 2012). They also exert electrocatalytic effect in redox conversion of various substrates and are frequently used for immobilization of biomolecules and auxiliary reagents via spontaneously formed Au—S (Ag—S) bonds. The inclusion of such nanoparticles on the transducer surface increases the number of binding sites and decreases steric limitations of receptor attachment (Wilson, 2008). Silver shows similar functions as gold but is less stable toward chemical oxidation and can initiate undesirable photochemical reactions in the surface layer. Nanoparticles of Ru, Rh, Pd, and Pt and their alloys are applied mainly as electrocatalysts and mediators of electron transfer (Guo et al., 2011). They reduce the overvoltage of working potential necessary for electrochemical reaction and promote chemical reactions of free radicals formed in primary electron transfer.

Metal complexes with organic ligands are utilized as mediators of electron transfer (Welch and Compton, 2006). Some of them can specifically interact with analytes or signal generating reagents. Thus, flat aromatic ligands make complexes able to intercalate double-stranded oligonucleotides and native DNA. This reaction changes the electrochemical properties of the complexes and simplifies oxidation of selected nucleotides with reactive oxygen species generated in solution during electrochemical and chemical reactions. This can be used for the detection of specific DNA interactions. Bipyridine complex of Ru(II) is mostly applied in electrochemiluminescence due to its ability to generate irradiation within the electron transfer stage (Hu and Xu, 2010).

Metal oxides play an auxiliary role in the biosensor assembly. Some of them, such as ZrO_2 or ZnO, promote immobilization of biomolecules due to their hydrophilic structure and protecting effect (Arya et al., 2012). The oxides are obtained directly on the transducer surface by chemical conversion of metal/metal complex nanoparticles (NiO) or by deposition of insoluble products of appropriate ion exchange reactions (ZnO, Fe_3O_4). Ferrite particles are applied for magnetic separation of biomolecules adsorbed on their surface or involved in specific interactions with immobilized bioreceptors (Rocha-Santos, 2014). The oxides of transient metals

(CoO_x, Ni[O]OH) can be used as redox active modifiers to increase the signal of oxidation of specific substrates.

Carbonaceous nanomaterials offer a big variety of physical-chemical properties, which depend on their synthesis protocol, purity, and implementation methods (Vamvakaki and Chaniotakis, 2007). Single *CNTs*, especially single-walled, are applied in microelectronic devices, for example, field effect transistors developed for the detection of biomolecular interaction with the receptors placed on their sidewalls (Allen et al., 2007). Target interactions change charge distribution and conducting properties of the CNT molecules. Multiwalled CNTs in the assembly of voltammetric sensors and biosensors show electroconductivity sufficient for electron transduction and current amplification. Defects of the sidewalls and terminal carboxylic groups formed on the stage of oxidative treatment of CNTs can be used for immobilization and electric wiring of various biomolecules (Jacobs et al., 2010). CNTs as well as *carbon black* particles increase the roughness of the electrode surface and hence the currents recorded in the presence of analytes.

Graphene, a rather new nanomaterial that became popular in the past decade, and relative materials of similar structure (reduced graphene oxide) can be obtained in different forms able to dispense in aqueous solutions and used as labels of bioreceptors (Kuila et al., 2011). The amplification of the signal is achieved not only due to enhanced square of the surface but also because of the increased number of redox-centers showing electrocatalytic properties toward target species. All carbonaceous materials are easily compatible with metal nanoparticles, which often exert synergic effect on the currents recorded.

Dendrimers and hyperbranched polymers are applied as supports for both bioreceptors and labels including other nanomaterials like metal nanoparticles. They protect bioreceptor from chemical oxidation and denaturing agents on the stages of immobilization and further biosensor operation (Satija et al., 2011). Dendrimer molecules are big enough to serve as molecular nanocontainers that keep the biomolecules in their internal cavities. Besides, the electrostatic interaction with the surface charged groups and covalent attachment to them were carried out for biosensor assembling. Polyamidoamine (PAMAM) dendrimers variously functionalized by terminal groups are mostly used in bioanalysis (Jang and Kataoka, 2005).

Similar functions are described for thin *polymer films with nano-sized pores* or 3D-structure, which can be obtained by deposition from organic solvent or electropolymerization (Kocer et al., 2012; Melde and Johnson, 2010). Some of them, for example, polyaniline and polypyrrole, exert their own redox activity and optical properties varied depending on the microenvironment, specifically, with pH or redox potential changes (Rajesh et al., 2009). Affine biomolecular interactions affect such changes due to introduction of bulky charged molecules of proteins and oligonucleotides in the polymer layer and hence change the signal of appropriate biosensor.

Many of the nanomaterials shortly characterized previously can be combined. Thus, dendrimers modified with organic ligands are successfully applied for the formation of metal complexes or stabilization of metal nanoparticles. Electropolymerization is widely used for physical entrapment of nanoparticles of noble metals, their oxides, and CNTs in the surface layer of a biosensor (Budnikov et al., 2012). Moreover, nanoparticles predetermine the structure of the electrolysis products, the shape and size distribution of pores and clusters.

It should also be mentioned that many biomolecules used in biosensors have their own size quite comparable with that of common nanoparticles. Their immobilization within a thin surface layer often results in the changes of mass transfer and accessibility, similar to those achieved by nanostructuring or implementing nanoparticles. This refers to the advantages received from enhanced interface roughness, elongation of the reaction zone from the surface to the solution, favorable charge distribution, and so forth. For this reason, the review also includes some examples of biosensors based on protein conjugates and DNA oligonucleotides, if their functioning assumes heterogeneous conditions of target interactions and use of nanoscale advantages presented previously.

3 Mycotoxins

3.1 General Characteristics of Common Mycotoxins

Mycotoxins are a group of secondary metabolites of fungi, which belong to sesquiterpenoids and exert toxic effect on vertebrates and other living beings. Among many other toxic compounds, most important mycotoxins belong to *Aspergillus*, *Fusarium*, and *Penicillium* spp. (Peraica et al., 1999). Chemical structures of selected mycotoxins are depicted in Fig. 2.2.

From toxicological and legislative point of view, aflatoxins, ochratoxins, some trichothecene toxins (fumonisins, deoxynivalenol, T-2, HT-2), zearalenone, patulin, and citrinin are most important. Mycotoxicoses arise if environmental, social, and economic conditions combine with humidity and increased temperature, which promote the growth of molds.

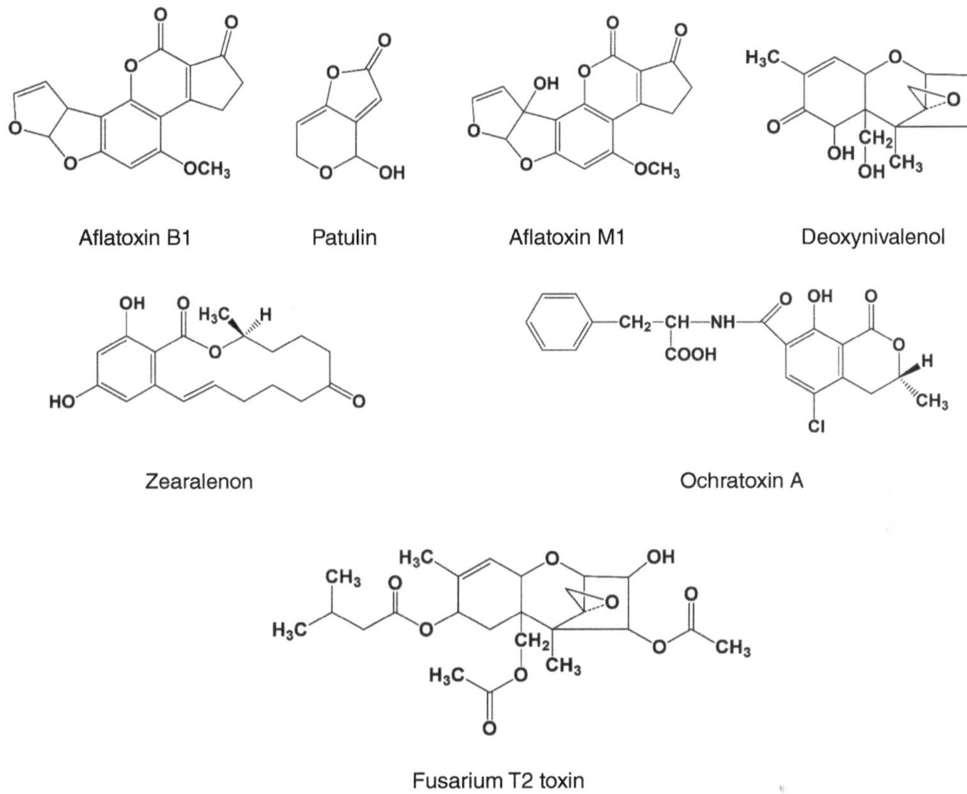

Figure 2.2. Chemical structures of most important mycotoxins.

Mycotoxins affect a broad range of agricultural products including cereals, cereal-based foods, dried fruits, wine, milk, coffee beans, cocoa bakery, and meat products (Shephard et al., 2012). High-performance liquid chromatography (HPLC) with fluorescence or mass detectors are mainly recommended for use in accredited laboratories. Analytical methods of mycotoxins determination were the subject of several recent reviews (Köppen et al., 2010; Cigić and Prosen, 2009; Zheng et al., 2006).

Aflatoxins occur in nuts, cereals, and rice. Two major *Aspergillus* sp. producing aflatoxins are *A. flavus* (only B aflatoxins) and *A. parasiticus* (B and G aflatoxins). Aflatoxins B1 and B2 are converted in metabolic oxidative activation to aflatoxins M1 and M2, respectively. They are less toxic than primary aflatoxins but quite stable in conditions of milk pasteurization, storage, and preparation of various dairy products. Aflatoxicol is a reductive metabolite of aflatoxin B1. Aflatoxins are acutely toxic, immunosuppressive, mutagenic, teratogenic, and carcinogenic compounds. The main target organ for toxicity and carcinogenicity is the liver (first hazard class in accordance with the classification of the International Agency for Research on Cancer) (Do and Choi, 2007). Indeed, aflatoxin B1 is metabolically activated in living beings by cytochrome P450 monooxygenase to epoxide, a carcinogenic and mutagenic product. The maximum permitted amount of aflatoxins established by European Community legislation is 4 μg/kg (2 μg/kg of aflatoxin B1) in groundnuts, nuts, dried fruits, and processed products for direct human consumption and up to 15 μg/kg (8 μg/kg of aflatoxin B1) for those undergoing physical processing prior to consumption. US FDA action levels are equal to 20 μg/kg of aflatoxins in all foods. The appropriate permissible levels for aflatoxin residues in feed are much milder. Thus, EU limits aflatoxin B1 content in complementary feeding stuffs for cattle, sheep, and goats to 50 μg/kg and the US FDA limits that in maize and peanut products intended for finishing beef cattle to 300 μg/kg (Zheng et al., 2006).

Ochratoxin A (OTA) is a secondary metabolite produced mainly by *Aspergillus ochraceus* and *Penicillium verrucosum* sp. OTA formation occurs mainly after harvesting on insufficiently dried cereal and cereal products. It can be found in maize, barley, wheat, oats, rye, hay, and mixed feed. Exposure to OTA contaminated feed reduces animal growth rates and affects animal production, especially that of pigs and poultry. OTA occurrence in human food commodities of vegetal and animal origin can result in nephropathy, renal diseases, and increased frequency of tumor incidents (Denli and Perez, 2010). World Health Organization proposed maximal permissible OTA content of 5 μg/kg in cereals and 3 μg/kg in cereal products (World Health Organization, 2002). Maximal admissible levels established in EU for feed commodities vary from 50 to 250 μg/kg for cereals and feedstuffs for pigs and poultry.

Trichothecene mycotoxins form a chemically diverse group of compounds mainly produced by *Fusarium* spp. and affected animals and human through contaminated grains (wheat, oats, barley, maize, and rice) (Bennett and Klich, 2003). Trichothecenes are categorized as nonmacrocyclic and macrocyclic compounds. In accordance with the structure and producing organisms, four groups of trichothecene mycotoxins are selected. The type A includes T-2 toxin, HT-2 toxin, neosolaniol, and diacetoxyscirpenol. Deoxynivalenol, nivalenol, 3-acetyldeoxynivalenol, and fusarenon belong to type B. Type C representatives contain second epoxide group (crotocin and baccharin) and those of type D macrocyclic ring system with two ester linkages (satratoxin G, H, roridin A, and verrucarin A) (Meneely et al., 2011).

Trichothecenes are commonly found on cereals grown in the temperate regions of Europe, America, and Asia (wheat, rye, barley, oats, and corn) (Yazar and Omurtag, 2008). Clinical signs of trichothecene infection in animals include feed refusal and vomiting, growth retardation, reproductive disorders, blood disorders, dermatitis, and depression of the immune response. As in previous case, pigs and poultry are most sensitive toward trichothecene mycotoxins while ruminants are able to metabolize trichothecenes into less toxic metabolites (Rocha et al., 2005). Contamination with trichothecenes causes a number of illnesses in human such as anorexia, depression, inhibition on immune system function, and hematoxicity.

Zearalenone is a nonsteroidal estrogenic mycotoxin produced by *Fusarium* spp., which occurs in the form of four hydroxyl derivatives. Contrary to nonestrogenic trichothecenes, zaeralenone and some of its metabolites competitively bind to estrogen receptors (Wood, 1992). Zearalenone is a stable compound during storage, milling, and cooking of food. It is associated with reproductive problems in specific animals and possibly in humans. In 1993, it was evaluated by the International Agency or Research on Cancer and allocated together with other *Fusarium* toxins in group 3 (not classifiable as to their carcinogenicity to humans). The scientific committee on food of the European Commission on Health and Consumer Protection arrived at a temporary tolerable daily intake for zearalenone of 0.2 μg/kg of body weight, while the joint FAO/WHO Expert Committee on Food Additives recommended a provisional maximum tolerable daily intake of 0.5 ug/kg (Mitterbauer et al., 2003).

3.2 Aptasensors for Mycotoxins

As was mentioned before, aptamers are synthetic oligonucleotides randomly synthesized from DNA/RNA library and selected against target analyte by affine chromatography (SELEX technology) (Famulok and Mayer, 1999). Regarding mycotoxins assay, the selection of aptamers is complicated by low molecular mass of analytes (<1 kDa). First aptamer against OTA was synthesized in 2008 (Yang et al., 2013). So far, aptamers against aflatoxin B1, aflatoxin B2, OTA, fumonisin B1, zearalenone, and deoxynivalenol have been selected.

Fig. 2.3 outlines typical schemes for the label-based detection of aptamer–mycotoxin interactions. Displacement mode (Fig. 2.3a) is similar to sandwich assay commonly used in immunoanalysis. In this protocol, two DNA probes are used, both of which are partially complementary to aptamer structure. Capture DNA probe is

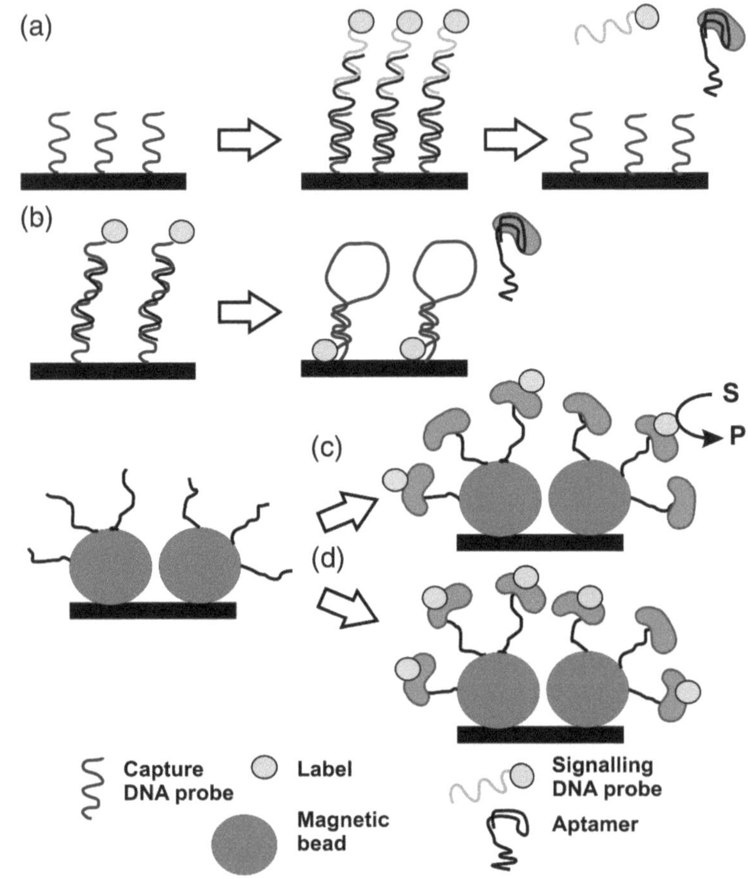

Figure 2.3. Schematic outlines of mycotoxin measurement protocols based on label application. (a) Displacement mode; (b) signal-on biosensors; (c) direct competitive assay; (d) sandwich assay.

immobilized on the surface of nanoparticles or transducer. Then, hybridization with aptamer and second signaling DNA probe are performed in separate steps. Signaling aptamer contains label, which signal can be recorded either electrochemically or by emission level in fluorescent analysis. Then biosensor is incubated in the analyte solution. The affinity of an aptamer toward target molecules is higher than that to DNA probes so that double-stranded fragments of the surface product are dissociated and a new complex of an aptamer with mycotoxin is formed. As a result, labeled DNA probe is moved from the electrode surface and can be removed by washing. Resulting signal decreases in accordance with the mycotoxin concentration. For this reason, appropriate sensor is often called a signal-off biosensor. Changes in the conformation of aptamers are used in signal-on biosensors (Fig. 2.3b). In this case, capture DNA probe takes terminal label. Prior to contact with target species, aptamers are hybridized with capture aptamer so that rather rigid double-stranded fragment prevents approaching label to the transducer surface. Dissociation of hybridized fragment followed by formation of aptamer–mycotoxin complex increases flexibility of labeled DNA probe, which can undergo partial self-hybridization with formation of a loop tightly fixed label to the transducer. In a simpler manner, similar effect is attained without self-hybridization due to a higher flexibility of single-stranded DNA probe bearing label. In both cases, signal increases with the concentration of the analyte molecules. More sophisticated displacement schemes include exonucleases that shorten capture DNA probes and simplify following hybridization of labeled fragments to achieve higher density of the surface layer and label concentration recorded (Tong et al., 2011). Increased sensitivity of fluorescence biosensor can be achieved by the use of quenching properties of reactants.

In direct competitive assay (Fig. 2.3c), labeled mycotoxins are added to the sample tested. The reaction with aptamers attached to carrier resulted in binding both labeled and nonlabeled molecules so that label concentration is inverse to the analyte concentration. In sandwich assay (Fig. 2.3d) labels are attached to aptamer–analyte complexes on the transducer surface. The use of magnetic beads increases accumulation of the species on the biosensor interface and enhances performance of appropriate biosensors. The examples of application of various measurement protocols to mycotoxin aptasensors are summarized in Table 2.2.

As could be seen from the Table 2.2, most attention has been paid to OTA. To some extent, this is due to success in selection of appropriate aptamers. It changes in conformation from linear to G_4-quadruplex in OTA binding (Hayat et al., 2013). This simplifies the label-free detection of target interactions because conformation changes affect permeability of the surface layer for the redox probe (ferricyanide ions commonly applied in EIS measurements). Electrostatic repulsion of negatively charged phosphate residues of the aptamers and ferricyanide ions also promotes high sensitivity of signal detection. The applications of magnetic beads is stimulated by their commercial availability including preparations modified with biotin groups compatible with various biomolecules via avidin (streptavidin)–biotin binding. Indeed, the size of commercial products, for example, Dynabeads (http://www.lifetechnologies.com/ru/ru/home/brands/product-brand/dynal/dynabeads-technology.html), SureBeads (http://www.bio-rad.com/en-us/product/surebeads-protein-g-magnetic-beads) and so forth, exceeds nanoscale (commonly about 1 μm) they are

Table 2.2 Analytical Characteristics of Aptasensors for Mycotoxin Determination

Nanomaterial	Surface Layer Assembly	Signal Measurement Protocol, Samples Tested	LOD, Linearity Range	References
Aflatoxin B1				
PAMAM G4 dendrimer	Cystamine–aptamer layer on Au electrode with aptamer assembled by glutaraldehyde cross-linking	Changes in DC voltammetry and EIS in the presence of $[Fe(CN)_6]^{3-/4-}$ redox probe, peanut-corn flacks	LOD 0.04 nM, conc. range 0.1–10 nM	Castillo et al. (2015)
Hemin/DNAzyme	Hemin/DNAzyme attached to coating protein	Competitive assay, chemiluminescence induced by luminol oxidation	LOD 0.11 ng/mL, conc. range 0.1–10 ng/mL	Shim et al. (2014)
Calix[4]arene macrocycle bearing phenazine dye	Self-assembling of polymeric Neutral red, carboxylated macrocycle followed by covalent binding of phenazine dye and aminated aptamer	Changes in DC voltammetry and EIS in the presence of $[Fe(CN)_6]^{3-/4-}$ redox probe, peanuts, cashew nuts, white wine and soy sauce	LOD 0.1 nM (DC), 0.05 nM (EIS), conc. range 0.1–100 nM	Evtugyn et al. (2014)
Magnetic nanoparticles, silica encapsulated Au hollow nanoparticles on Co core	Magnetic beads conjugated with capture Ab, Silica covered Au nanoparticles with malachite green isothiocyanate as label and signaling Ab	Sandwich assay with detection of label Raman spectra after magnetic separation	LOD 0.1 ng/mL, conc.range (logistic curve) 1 ng/mL–100 μg/mL	Ko et al. (2015)
Aflatoxin M1				
Au nanoparticles	Cystamine–Au nanoparticles monolayer on Au electrode with self-assembled thiolated aptamer	EIS measurements in the presence of $[Fe(CN)_6]^{3-/4-}$ redox probe	Conc.range 1–14 ng/mL	Dinçkaya et al. (2011)
Magnetic nanoparticles	Fe_3O_4-polyaniline composite followed by aptamers immobilization via cross-linking with glutaraldehyde	EIS measurements in the presence of $[Fe(CN)_6]^{3-/4-}$ redox probe	LOD 1.98 mg/L, conc.range 6–60 ng/L	Nguyen et al. (2013)

Table 2.2 Analytical Characteristics of Aptasensors for Mycotoxin Determination (*cont.*)

Nanomaterial	Surface Layer Assembly	Signal Measurement Protocol, Samples Tested	LOD, Linearity Range	References
Ochratoxin A				
Polyaniline–stearic acid composite	Aptamer covalently attached to hybrid film by glutaraldehyde cross-linking	EIS measurements in the presence of $[Fe(CN)_6]^{3-/4-}$ redox probe	LOD 0.1 ng/mL, conc.range 0.1–10 ng/mL and 1–25 μg/mL	Prabhakar et al. (2011)
Nonmodified Au nanoparticles	Salt induced aggregation of Au nanoparticles	Changes in the color of suspension caused by conformational changes of aptamer molecules interacting with analyte and followed by aggregation of Au nanoparticles	LOD 20 nM, conc. range 20–625 nM (UV–vis detection)	Yang et al. (2011)
Au nanoparticles with immobilized DNA probe	Self-assembling of thiolated DNA probe, aptamer and capture probe on glassy carbon electrode	Displacement assay with Au nanoparticles labeled DNA probe and detection of methylene blue signal	LOD 30 pg/mL, conc. range 0.1–20 ng/mL	Kuang et al. (2010)
Au nanoparticles in hyperbranched polymers	Self-assembling of thiolated aptamers on Au nanoparticles in dendrimeric polymer assembled onto electropolymerized neutral red	EIS measurements in the presence of $[Fe(CN)_6]^{3-/4-}$ redox probe, beer	LOD 0.02 nM, conc. range 0.1–100 nM	Evtugyn et al. (2013a)
Magnetic particles	Streptavidin-biotin binding of aptamer to magnetic nanoparticles	Competitive assay with alkaline phosphatase or horseradish peroxidase (HRP) as labels with visual and electrochemical detection of the products of enzymatic reactions, wine	LOD 1.1 and 0.11 ng/mL, conc. range 1.1–50 and 0.11–15 ng/mL for indirect and direct competitive assay, respectively	Barthelmebs et al. (2011)

(*Continued*)

Table 2.2 Analytical Characteristics of Aptasensors for Mycotoxin Determination (*cont.*)

Nanomaterial	Surface Layer Assembly	Signal Measurement Protocol, Samples Tested	LOD, Linearity Range	References
Magnetic particles	Streptavidin-biotin binding of aptamer to magnetic nanoparticles	Competitive assay with OTA–HRP conjugate, hydroquinone signal, wheat	LOD 0.07 ng/mL, conc. range 0.78–8.74 ng/mL	Bonel et al. (2011)
Magnetic particles	Carboxylated magnetic beads modified with aminated aptamers by Invitrogen protocol	Competitive assay in flow regime with alkaline phosphatase modified OTA, electrochemical detection of naphtol-1, beer samples	LOD 0.05 μg/L	Rhouati et al. (2013)
Magnetic nanoparticles	Fe_3O_4 modified with aptamers via avidin-streptavidin binding	Displacement assay with Tb^{3+} enhanced fluorescence recorded at 540 nm, wheat samples	LOD 20 pg/mL, conc. range 0.1–1.0 ng/mL	Zhang et al. (2013)
Magnetic nanoparticles	Fe_3O_4 modified with aptamers via avidin-streptavidin binding	Displacement assay with complementary DNA probe labeled with fluorescein derivative, evanescent optic fiber sensor and magnetic separation, oat-spiked samples	LOD 3 nM, conc. range 6–500 nM	Wang et al. (2015b)
Ag nanoparticles stabilized with macrocyclic ligands	Glassy carbon electrode modified with poly(neutral red) and Ag nanoparticles modified with macrocyclic ligands bearing catechol fragments and	EIS measurements in the presence of $[Fe(CN)_6]^{3-/4-}$ redox probe, beer	LOD 0.05 nM, conc. range 0.3–30 nM	Evtugyn et al. (2013b)
Magnetic particles, Ag dendrites	Fe_3O_4 modified with aptamers via avidin-streptavidin binding	Displacement assay with DNA probe partially hybridized with aptamer. Changes in the rate of Ag dendrites formation promoted by DNA probe are recorded by light emission at 632 nm, wheat	LOD 2 pg/mL, conc. range 0.01–0.30 ng/mL	Chen et al. (2014)

Table 2.2 Analytical Characteristics of Aptasensors for Mycotoxin Determination (*cont.*)

Nanomaterial	Surface Layer Assembly	Signal Measurement Protocol, Samples Tested	LOD, Linearity Range	References
Au nanoparticles	Self-assembling of thiolated aptamers on Au nanoparticles and complementary probe on bare Au	Displacement assay with Au labeled DNA aptamers, electroluminescence of luminol derivative oxidation, naturally contaminated wheat	LOD 0.007 ng/mL, conc. range 0.02–3.0 ng/mL	Wang et al. (2010)
Single-walled CNTs	Aptamers modified with carboxyfluorescein	Displacement assay, CNTs as quencher of label fluorescence, real samples of beer	LOD 24.1 nM, conc. range 25–200 nM	Guo and Wang (2011)
Graphene oxide	Aptamers modified with carboxyfluorescein	Displacement assay, graphene oxide as quencher of label fluorescence, red wine	LOD 1.9 μM, conc. range 2–35 μM	Sheng et al. (2011)
Reduced graphene oxide–Au nanocomposite	Capture DNA probe covalently attached to Au electrode, reduced graphene oxide and Au nanoparticles as labels of signaling DNA probe	Displacement assay, EIS measurements of charge transfer resistance caused by dissociation of triple complex (capture DNA probe–aptamer–signaling DNA probe) in OTA presence, real red wine samples	LOD 0.3 pg/mL, conc. range 1 pg/mL–50 ng/mL	Jiang et al. (2014)
Reduced graphene oxide–Au nanocomposite	Au electrode modified with cysteamine and reduced graphene oxide with Au nanoparticles covalently attached via 2-aqminothiophenol	EIS measurements in the presence of $[Fe(CN)_6]^{3-/4-}$ redox probe	LOD 0.03 ng/mL, conc. range 1.0–200 ng/mL	Qian et al. (2014)

considered together with nanomodifiers of biosensor assembly. Low molecular thiolated compounds, for example, cysteamine, mostly suppress unfavorable sorption on the transducer surface.

Few examples of aptasensors with dendrimers and supramolecular carriers that both stabilize recognition elements and participate in electron transduction in rather thick surface layer are reported for aflatoxins. Though the changes in conformation and hence their influence on the surface reaction for antiaflatoxin aptamer are lower than for anti-OTA aptamers, the sensitivity of appropriate aptasensors is sufficient for detection of amounts comparable with those with immunosensors.

3.3 Immunosensors for Mycotoxins

The immunosensors developed for direct detection of mycotoxins are presented in Table 2.3. As in the case of aptasensors, main attention is paid to nanomaterials employed in biosensor assembly and measurement protocol.

Recent investigations on immunosensors are mainly focused on the use of monoclonal Ab, which exert higher affinity and selectivity toward target molecules than polyclonal Ab. They are implemented in conjugates with various labels and supports via carbodiimide binding or glutaraldehyde cross-linking. In case of Au nanoparticles as immobilization substrate, the use of rather long linkers between the support surface and conjugate improves the sensitivity of the signal due to lower steric hindrance of an analyte access. The selectivity of the immune response is established on the stage of Ab selection and production, which is not described in the articles devoted to the immunosensors themselves. In some publications, the influence of other mycotoxins on target signal is considered, for example, the effect of OTA on the signal of aflatoxin B1 in their mixtures (Ko et al., 2015; Singh et al., 2013), or comparison of individual signals of the previously mentioned mycotoxins in separate solutions (Liu et al., 2006), or the possible interferences of matrix components are described. Most attention is paid to cross-selectivity of aflatoxin B1 detection. Being most toxic, aflatoxin B1 is commonly present with its metabolites and relative toxins. The specificity of the response was proved by appropriate measurement of relative binding constants or direct experiments with standard solutions (Singh et al., 2013). Regarding the assay of spiked and real samples, the use of extraction technique previously optimized for HPLC analysis provides both high efficiency and biosensor recovery.

Measurement schemes described for electrochemical immunosensors are similar to those previously considered for

Table 2.3 Analytical Characteristics of Immunosensors for Mycotoxin Determination

Nanomaterial	Surface Layer Assembly	Signal Measurement Protocol, Samples Tested	LOD, Linearity Range	References
Aflatoxin B1				
Carboxylated multiwalled CNTs	Deposition of a monolayer of CNTs, carbodiimide binding of monoclonal antiaflatoxin Ab	Direct assay. DC measurements of $[Fe(CN)_6]^{3-/4-}$ redox probe	LOD 0.08 ng/mL, conc. range 0.25–1.375 ng/mL	Singh et al. (2013)
Multiwalled CNTs	Preoxidized CNTs were mixed together with ionic liquid and antiaflatoxin Ab and placed on glassy carbon electrode	Direct assay. DC measurements of $[Fe(CN)_6]^{3-/4-}$ redox probe, olive oil	LOD 0.03 ng/mL, conc. range 0.1–10 ng/mL	Yu et al. (2015)
Au nanoparticles	Gold electrode modified with cysteamine and BSA–aflatoxin conjugate, Au nanoparticles coated with secondary Ab	Sandwich assay. Quartz crystal microbalance (mass increase recording), spiked milk samples	LOD 0.01 ng/mL, conc. range 0.1–100 ng/mL	Jin et al. (2009)
Au nanoparticles	Interdigitated microelectrode covered with 2-aminoethane thiol monolayer, Au nanoparticles, HRP and antiaflatoxin antibodies	Direct assay. Conductometric detection of HRP activity with iodide as substrate, measurement of inhibition degree caused by aflatoxin binding	LOD 0.1 ng/mL, conc. range 0.5–10 ng/mL	Liu et al. (2006)
Au nanoparticles	Au nanoparticles covered with Ab and BSA conjugate	Competitive flow-lateral assay with polyclonal Ab, binding analyte on nitrocellulose membrane, visual and smartphone based absorbance detection, maize	LOD 5 μg/kg, conc.range up to 1000 5 μg/kg (smartphone based detection)	Lee et al. (2013)
Au nanoparticles, Fe_3O_4 nanoparticles with gold shell	Glassy carbon electrode covered with suspension of Au nanoparticles in chitosan with attached Ab against aflatoxin B1, Fe_3O_4 nanoparticles covered with Au and modified with covalently attached catechol derivative	Indirect competitive immunoassay with DC measurement of catechol redox activity as label signal	LOD 0.2 ng/mL, conc. range 0.6–110 ng/mL	Masoomi et al. (2013)

(*Continued*)

Table 2.3 Analytical Characteristics of Immunosensors for Mycotoxin Determination (*cont.*)

Nanomaterial	Surface Layer Assembly	Signal Measurement Protocol, Samples Tested	LOD, Linearity Range	References
Au-Ag nanoparticles	Gate area of FET covered with BSA–aflatoxin conjugate by 3-aminopropyltriethoxysilane; Au nanoparticles are modified with monoclonal antiaflatoxin Ab	Indirect competitive assay. Photocurrent produced under photodiode illumination depends on the Au-Ag nanoparticles concentration in the gate area	Conc. range 0–15 μg/L	Park et al. (2011)
Gold binding protein–protein G (GBP-PG) fusion protein	Au film of SPR chip covered with self-assembled GBP-PG monolayer and antiaflatoxin Ab	Direct assay. Changes in reflectance/dissipation properties by SPR analyzer, corn kernels	LOD 1 μg/mL, conc. range 1–15 μg/mL	Park et al. (2014)
Reduced graphene oxide	Electrophoretically deposited reduced graphene oxide on transparent conductive glass followed by Ab immobilization in the layer of polylysine	Field effect capacitor measurements of the changes in surface layer capacitance	LOD 0.1 fg/mL, conc. range 0.1 fg/mL–1 pg/mL	Basu et al. (2015)
Au-reduced graphene nanocomposite	Graphene oxide and $AuCl_4^-$ reduced with citrate to form Au/reduced graphene layer applied then for carbodiimide binding of monoclonal Ab	Direct assay, measurement of electrocatalytic current in voltammetrc mode	LOD 0.1 ng/mL, conc. range 0.1–12 ng/mL	Srivastava et al. (2015)
Reduced graphene and Au nanoparticles	Reduced graphene and Au nanoparticles consecutively deposited on glassy carbon electrode followed by immobilization of Ab in chitosan containing ionic liquid	Direct assay, measurement of $[Fe(CN)_6]^{3-}$ current in voltammetrc mode, rice, milk, flower, soybean	LOD 1 fM, conc. range 3.2 fM–0.32 pM	Linting et al. (2012)

Table 2.3 Analytical Characteristics of Immunosensors for Mycotoxin Determination (*cont.*)

Nanomaterial	Surface Layer Assembly	Signal Measurement Protocol, Samples Tested	LOD, Linearity Range	References
Reduced graphene oxide–polypyrrole	Glassy carbon electrode covered with graphene oxide reduced with ascorbic acid and polypyrrole–polypyrrolpropionic acid copolymer by electropolymerization, Ab immobilized to side carboxylic groups via carbodiimide binding	Direct assay. DC measurements of $[Fe(CN)_6]^{3-/4-}$ redox probe, corn	LOD 10 fg/mL, conc. range 10 fg/mL–10 pg/mL	Wang et al. (2015a)
Aflatoxin M1				
Fe_3O_4–graphene oxide, CNT–quantum dots	Magnetic separation of Fe_3O_4-graphene–capture Ab composite on screen-printed electrode, signaling Ab attached to CNT-CdTe nanoparticles	Sandwich assay with electrochemilumenscence detection related to quantum dots concentration, milk	LOD 0.3 pg/mL, conc. range 1.0 pg/mL–100 ng/mL	Gan et al. (2013)
Ochratoxin A				
Au nanoparticles	Au electrode modified with monolayer of 1,6-hexandithiol, Au nanoparticles and OTA–ovalbumin conjugate	Indirect competitive assay with OTA–alkaline phosphatase conjugate, detection of current related to oxidation of naphtol formed in enzymatic reaction of napthylphosphate, corn samples	LOD 8.2 pg/mL, conc. range 10 pg/mL–100 ng/mL	Liu et al. (2009)
Magnetic nanoparticles	Screen-printed carbon electrode covered with magnetically accumulated magnetic beads modified with protein G and monoclonal anti-OTA Ab	Direct assay with OTA–HRP conjugates added to the sample. Amperometric measurement of pyrocatechol formed in HRP reaction, red wine	LOD 0.008 μg/L, conc. range 0.01–20 μg/L	Perrotta et al. (2012)

(*Continued*)

Table 2.3 Analytical Characteristics of Immunosensors for Mycotoxin Determination (*cont.*)

Nanomaterial	Surface Layer Assembly	Signal Measurement Protocol, Samples Tested	LOD, Linearity Range	References
Magnetic nanoparticles	Screen-printed carbon electrode covered with magnetically accumulated magnetic beads modified with streptavidin and monoclonal anti-OTA Ab	Direct assay with OTA–HRP conjugates added to the sample. Amperometric measurement of *p*-benzoquinone formed in HRP reaction, red wine	LOD 0.11 ng/L (0.12 ng/L in wine), conc. range 0.26–8.87 ng/L (0.24–8.33 in wine)	Vidal et al. (2012)
Magnetic nanoparticles	Au electrode covered with aminothiol–BSA conjugate to prevent nonspecific adsorption, magnetic nanoparticles modified with Ab by carbodiimide binding attached onto BSA layer	Direct assay with DC, SPR, and EIS signal measurement in the presence of $[Fe(CN)_6]^{3-/4-}$ redox probe	Conc. range 0.01–10 ng/mL	Zamfir et al. (2011)
ZnO nanoparticles	ZnO deposited on ITO electrode followed by coimmobilization of BSA and anti-OTA Ab	Direct assay. EIS measurements in the presence of $[Fe(CN)_6]^{3-/4-}$ redox probe, beer	LOD 0.006 nM, conc. range 0.006–0.01 nM	Ansari et al. (2010)
Deoxynivalenol				
F_0F_1–ATPase rotator	Conjugation of ATPase rotator with monoclonal antideoxynivalenol Ab via avidin-biotin binding	Direct assay. Changes in the rate of ATP synthesized recorded by chemiluminescence in luminol–H_2O_2 reaction	LOD 10^{-7} mg/L, conc. range 10^{-7}–10^{-1} mg/L	Zhao et al. (2012)
Fullerene/ferrocene	Fullerene/ferrocene/ionic liquid composite obtained on glassy carbon electrode	Direct assay. EIS measurements in the presence of $[Fe(CN)_6]^{3-/4-}$ redox probe, wheat, barley, maize, flour, soybean, rice, peanut	LOD 0.3 pg/mL, conc. range 1 pg/mL–0.3 ng/mL	Zhilei et al. (2011)

Table 2.3 Analytical Characteristics of Immunosensors for Mycotoxin Determination (*cont.*)

Nanomaterial	Surface Layer Assembly	Signal Measurement Protocol, Samples Tested	LOD, Linearity Range	References
Au nanoparticles	Au nanoparticles covered with BSA-Ab mixture	Direct assay. Refractometric measurements with glass fiber coated with mycotoxin–BSA conjugate followed by binding Ab labeled with Au nanoparticles, wheat four	LOD 0.09 mg/kg, conc. range 0.35–10 mg/kg	Maragos (2012)
Zearalenone				
Mesoporous carbon, Au at AgPt nanorattles	Glassy carbon electrode covered with dispersion of mesoporous carbon and Au at AgPt nanorattles and dried. Antizearalenone Ab were physically adsorbed on modified electrode	Direct assay. Square-wave voltammetry of $[Fe(CN)_6]^{3-}$ ions, milk	LOD 1.7 pg/mL, conc. range 0.005–15 ng/mL	Liu et al. (2014a)
Nitrogen doped graphene sheets, PtCo nanoparticles	Capture Ab on nitrogen doped graphene nanosheets, Pt-Co alloys nanoparticles as labels of secondary Ab	Sandwich assay. Amperometric H_2O_2 signal affected by Pt-Co alloy measured in DC mode, pig feed samples	LOD 2.1 pg/mL, conc. range 0.005–25 ng/mL	Feng et al. (2013)
Multiwalled CNTs	Glassy carbon electrode covered with CNTs, anti-zearalenon Ab immobilized on rotating disk.	Competitive flow-through assay with Zearalenon–HRP conjugate and detection of 4-tret-butylcatechol, a product of enzymatic reaction, on CNTs modified electrode, corn silage	LOD 0.77 μg/L, conc. range up to 500 μg/L	Panini et al. (2010)
Au nanoparticles	Au nanoparticles modified with adsorbed bipyridine as SERS label and antizearalenone monoclonal Ab	Competitive assay with detection of the label Raman signal on glass support, natural and spiked feed samples	LOD 1 pg/mL, conc. range 1 pg/mL–50 μg/mL	Liu et al. (2014b)

aptasensors. The electrochemical ELISA employs alkaline phosphatase and horseradish peroxidase (HRP), which offer a variety of substrates and detection conditions. In most cases, electrochemical reaction takes place on the electrode that works as support for immobilization of Ab conjugates. In flow-through assay, the immune reaction and signal generation can be separated as shown in (Panini et al., 2010) for HRP-based ELISA detection of zearalenone.

The examples of optical immunosensors developed for mycotoxin determination are presented in Fig. 2.4. In SPR biosensors (Fig. 2.4a) as in reflectometers, changes in dielectric properties are recorded in a thin (about 100 nm) layer approaching the interface with thin Au layer. The important peculiarity of this sensor is that the reactions on the metal surface and the interaction with the light take place on the opposite sides of the metal film. Thus, the detection is not affected by the optical characteristics of the solution. Maximal dissipation of the energy observed at the angle of incident light, θ, changes with biomolecular interactions. The shift of the angle is expressed in appropriate values of refractive index measured with sensitivity of about 1×10^{-6} arbitrary units. It corresponds to the detection of about 1 pg/mm^2. Thus, the sensitivity of SPR biosensor increases with the molar mass of guest part of the immunocomplex. This gives advantage to sandwich assay and application of rather heavy labels, for example, Au nanoparticles.

Fig. 2.4b depicts SERS based immunoassay with two nanoparticles (Masoomi et al., 2013). Magnetic particles are used for immobilization of capture Ab with following separation of the immunocomplex formed. Au nanoparticles modified with silica layer bear secondary Ab to be attached to capture Ab in the presence of an analyte, that is, aflatoxin B1. Resulting signal is related to Raman spectrum recorded in the presence of Au nanoparticles participating in interaction with laser light.

Fig. 2.4c shows a typical flow-lateral test strip consisting of porous nitrocellulose strip with immobilized Ab. This is one-spot sample of the device, which can be developed in many forms that are different in number of test lines and immunoassay format. In this particular case, a sandwich assay is used with detection of color changes corresponded to formation of triple complex between capture Ab, aflatoxin B1, and Au labeled signaling Ab with Samsung smartphone (Lee et al., 2013). Although the detection limits attained were sufficiently higher than those of other immunosensor described, simple use and intuitively understandable readout format make such test strip devices very attractive for unqualified users. Other immunochromatographic test strips operated on

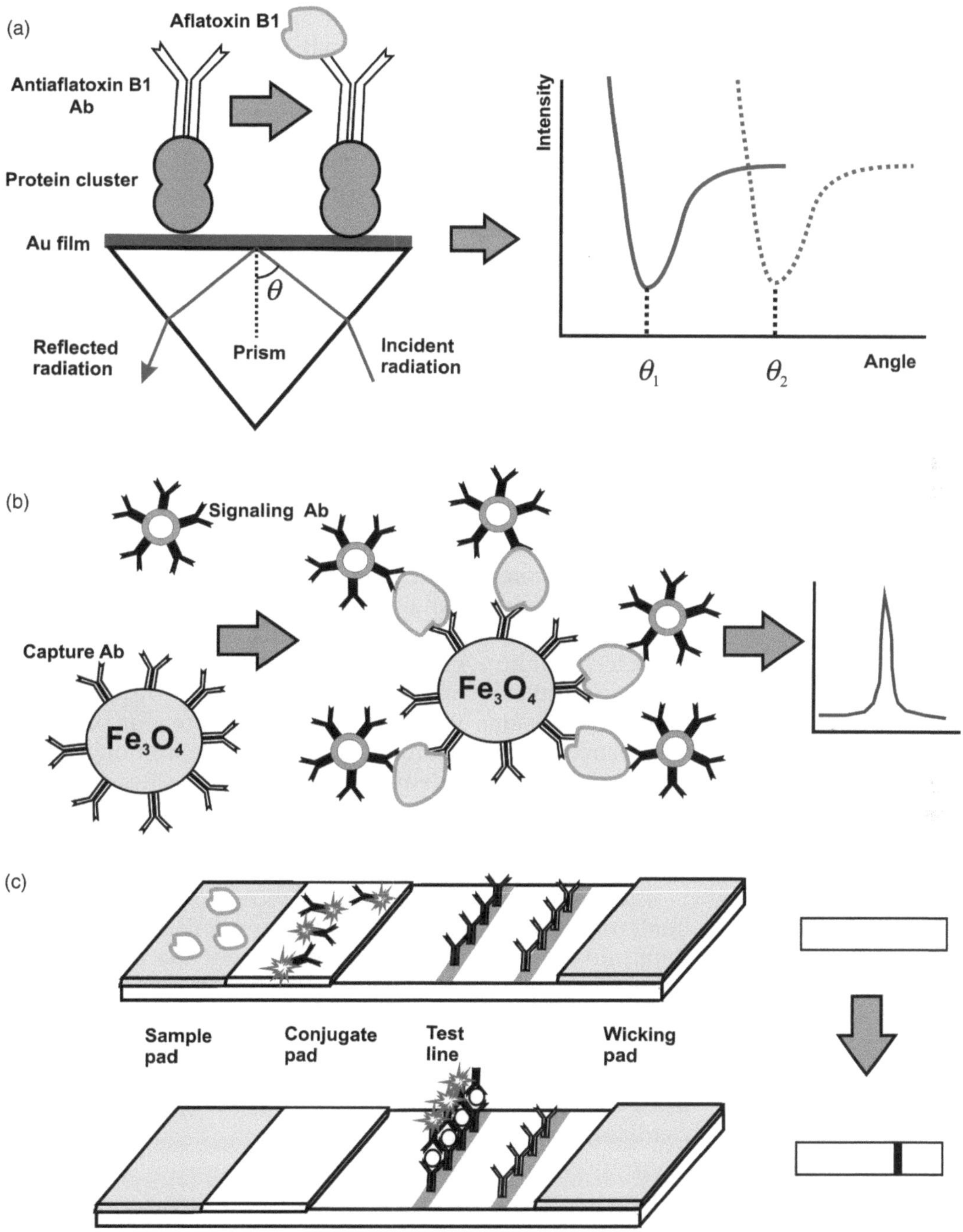

Figure 2.4. Optical biosensors used for mycotoxin determination. (a) SPR biosensor; (b) SERS based immunoassay; (c) flow-lateral test strip. Partially adapted from Lee et al. (2013).

similar principles were reported for some mycotoxins. Thus, aflatoxin B1 can be detected in milk with visual detection limit of 0.3 ng/mL (Zhang et al., 2012a). T-2 toxin was determined in extracts from wheat and oat with the cutoff level of 100 μg/kg within 4 min with test strip based on monoclonal Ab labeled with Au nanoparticles (Molinelli et al., 2008). Three monoclonal Ab labeled with Au nanoparticles were used in dipstick tests for simultaneous detection of aflatoxin B1, zearalenone, and deoxynivalenol (Song et al., 2014). The visual limits of detection (LODs) were 0.03, 1.6, and 10 μg/kg and cutoff levels 1, 50, and 60 μg/kg, respectively. Some other examples of multiplex test strips based on labeled Ab for the detection of mycotoxins are described elsewhere (Lattanzio et al., 2012).

Most immunosensors for mycotoxins detection involve rather universal stages of immunoseparation and following signal generation. This is particularly true for label-free assay, which can be rather easily adapted to new targets by replacing the recognition element. This might be useful for the detection of multiplex biosensors intended for the detection of several analytes from a single measurement. Such an approach can reduce overall duration of the analysis, which is often critical for application of the immunosensors for real sample testing. Surprisingly, the number of such sensor arrays is rather modest.

Microcantilever resonator systems with two Ab against OTA and total aflatoxins were described for the detection of 6 and 3 ng/mL, respectively (Ricciardi et al., 2013b). The immobilization was performed by glutaraldehyde cross-linking on silica layer. The formation of Ab-mycotoxin complex decreased resonator frequency registered with laser beam. All the measurements are performed after drying the sensor in vacuum.

Commercially available SPR device (flow-through SPREETA analyzer, Texas Instruments, USA) was applied for simultaneous determination of four mycotoxins, namely, aflatoxin B1, zearalenone, fumonisin B1, and deoxynivalenol by indirect competitive assay. In this method, the sensor surface was modified with mycotoxin derivatives, antimycotoxin Ab were mixed with the sample and excess of Ab was bonded to the sensor surface. The LODs achieved were equal to 0.2, 0.01, 50, and 0.5 ng/g, respectively (van der Gaag et al., 2003).

3.4 Other Biosensors

Although affinity biosensors offer excellent selectivity of mycotoxin binding, some other principles of their determination have been also proposed. They involve application of enzymes, whole cells and artificial receptors (molecularly imprinted polymers, MIPs).

Enzyme biosensors. Aflatoxins exert inhibiting effect on acetylcholinesterase (AChE), a key enzyme providing transduction of nerve impulses in vertebrates. The enzyme catalysis hydrolysis of acetylcholine (2.1) resulting in formation of acetic acid and choline.

$$\underset{\text{Acetylcholine}}{(CH_3)_3N^+CH_2CH_2OC(=O)CH_3} + H_2O \xrightarrow{\text{AChE}} \underset{\text{Choline}}{(CH_3)_3N^+CH_2CH_2OH} + CH_3COOH \tag{2.1}$$

In the presence of aflatoxins the rate of reaction (2.1) and hence the concentration of the products decreases to a degree depending on the toxin concentration and reaction conditions. For signal quantification, all the approaches previously developed for the detection of anticholinesterase agents, for example, chemical weapons or organophosphate pesticides, are acceptable. Thus, the amounts of acetic acid can be recorded using conductometric device with planar electrodes (Stepurska et al., 2015), or with pH sensitive field-effect transistor (Starodub et al., 2012), or spectrophotometrically (Pohanka, 2013). The pH shift used in these devices as a measure of enzyme is rather sensitive to the content of the sample and buffer capacity of the working solution. The use of thiocholine derivatives instead of natural substrates increases selectivity of the response. In such biosensors, thiocholine is detected by formation of yellow products with Ellman reagent (2.2) (Arduini et al., 2007) or directly on electrode by the current of its oxidation.

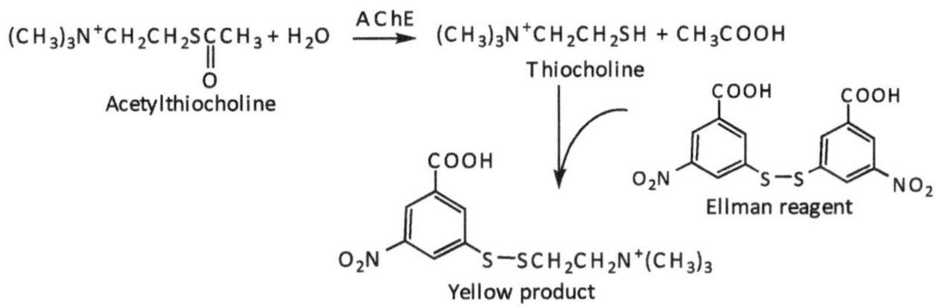

(2.2)

Similar response was applied in dipstick test with acetylcholinesterase included in the ink printed on paper covered with polyvinyl alcohol protective layer (Hossain et al., 2009). The LOD was

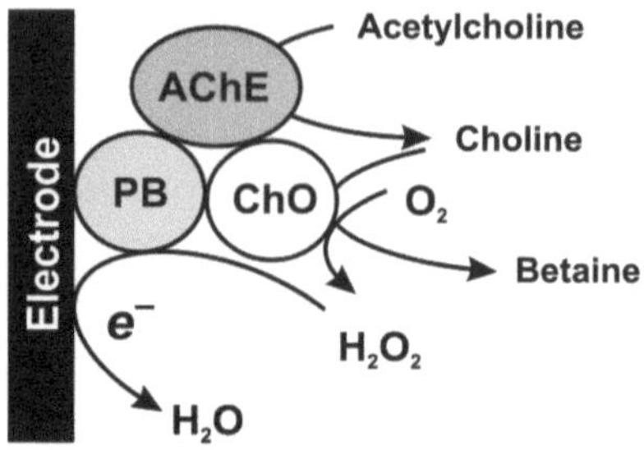

Figure 2.5. Determination of enzyme activity with acetylcholinesterase (AChE)–choline oxidase (ChO) biosensor based on glassy carbon electrode modified with Prussian blue (PB).

found to be 100 nM against 50 nM in previously described assay performed in solution.

Choline produced in the reaction (2.1) was detected with choline oxidase accelerating its oxidation to choline and hydrogen peroxide (2.3) (Rejeb et al., 2009). The reduction of hydrogen peroxide on electrode is mediated with nanosized particles of Prussian Blue that increase the current and decrease working potential to –0.05 V. General reaction scheme of the biosensor response is presented in Fig. 2.5.

$$\underset{\text{Choline}}{(CH_3)_3N^+CH_2CH_2OH} + O_2 + H_2O \xrightarrow{\text{Choline oxidase}} \underset{\text{Betaine}}{(CH_3)_3N^+CH_2C(=O)OH} + H_2O_2 \qquad (2.3)$$

The biosensor described makes it possible to detect 10–60 μg/kg of aflatoxin B1 and was tested with spiked olive oil samples. The detection of the toxin on permissible level requires preconcentration of the sample.

It should be mentioned that acetylcholinesterase shows a variety of inhibitors able to affect its activity after incubation. Besides mycotoxins, organophosphate and carbamate pesticides, detergents, metal cations and biogenic amines can affect the signal. This complicates possible application of acetylcholinesterase biosensors for mycotoxin determination. Nevertheless, they can be used as warning devices for fast preliminary control of dangerous pollution level with no respect of the origin of appropriate hazard.

Another example of inhibition measurements includes quantification of inhibition effect of aflatoxins B1 and G1 on neutrophil porcine elastase monitored with PCR (Cuccioloni et al., 2008). Succinyl-Ala-Ala-Ala-4-nitroanilide was used as enzyme substrate and relative decay of enzymatic activity as inhibition degree. The LOD obtained for aflatoxin B1 was 0.97 μg/kg and linear piece of calibration curve from 1.67 to 17.8 μg/kg. The SPR sensor was tested on maize extract.

Aflatoxin oxidase covalently attached to multiwalled CNTs was successfully applied for the detection of aflatoxin B1, which is converted to oxidized form with oxygen consumption and H_2O_2 generation (Li et al., 2011). The biosensor based on voltammetric or chronoamperometric recording of appropriate current exhibited a linear range from 3.2 to 721 nM (1–225 ng/mL).

In a similar manner, citrinin was determined with HRP-based biosensor using ferrocene mediation of H_2O_2 reduction (Zachetti et al., 2013). Glassy carbon electrode was covered with composite obtained by mechanical mixing multiwalled CNTs, ferrocene, and enzyme and covered with Teflon membrane. The calibration curve was linear in the concentration range from 1.0 to 11.6 nM (LOD 0.25 nM). The biosensor was validated against fluorometric measurements in citrinin standard solutions and spiked rice samples.

HRP immobilized on polypyrrole matrix was involved in OTA oxidation followed by formation of H_2O_2 detected amperometrically (Alonso-Lomillo et al., 2010). The biosensor makes it possible to detect down to 0.1 ng/mL of OTA. The calibration range covered 0.24–2.06 nM. OTA biosensor was applied for the detection of mycotoxin in spiked beer and roasted coffee.

In *cell-based biosensors*, the response of microorganisms on toxicant is assessed by their reactions, including respiratory activity, integrity, adhesion properties, and so forth (Su et al., 2011). The main advantage of such an approach over conventional chemical analysis is that microorganisms respond to toxic species similarly to physiologic responses of the vulnerable subjects. The results obtained from toxin-cell interactions reveal functional information on mode of action, toxic potency, bioavailability, target tissue or organ, and so forth, about the toxin. Appropriate biosensors can involve both prokaryotic (bacteria) and eukaryotic (yeast, invertebrate, and vertebrate) cells that are integrated in the appropriate supports, mechanically keeping them at the transducer surface.

Most mycotoxins induce cytotoxicity in mammalian or higher eukaryotic cells. The cytotoxic effects can include release of intracellular chemicals, for example, enzymes, cell necrosis, or specific changes in cellular morphology. Some of them are detected in biosensor format. Thus, bioelectric recognition assay (BERA) has been proposed for aflatoxin M1 detection (Larou et al., 2013). In this method, electric signals from membrane-engineered whole cells suspended in a gel matrix are recorded in the presence of target analytes. Membrane engineered cells were created by electroinserting monoclonal Ab into the membrane of Vero cells. The BERA cellular sensor can detect down to 5 pg/mL of aflatoxin M1 within 3 min.

Estrogenic mycotoxins (zearalenone and its metabolites, such as, zearalanone, α-zearalanol, β-zearalanol, α-zearalenol, and β-zearalenol) were detected in milk by bioluminescent whole-cell biosensor with genetically modified *Saccharomyces cerevisiae* strain (Välimaa et al., 2010). In the presence of an estrogenic compound, cells produced firefly luciferase able to light production in the presence of D-luciferin as substrate. The resulting luminescence (yellowish light) can be measured very sensitively after an induction period of a few hours. The LOD values achieved varied from 1 (α-zearalanol and α-zearalenol) to 71 nM (β-zearalenol). Luminescent biosensor was tested on spiked samples of milk and coffee cream as well as 10% ethanol.

Individual and combined toxicity of zearalenone and deoxynivalenol was studied by EIS measurements with biosensor utilizing BEL-7402 cells in self-assembled layers (Gu et al., 2015). The biosensor was obtained by consecutive deposition of Au nanoparticles, *p*-aminothiophenol, and folic acid. Then BEL-7402 cell suspension was incubated to attach cells via folic receptors. Finally, collagen was placed onto the biofilm to maintain cell adhesion and viability. Incubation of the biosensor in mycotoxin solutions resulted in decrease of the charge transfer resistance, which was referred to partial leaching cells from the biosensor interface. This can be due to the inhibiting effect of zearalenone that prevents cell proliferation as well as oxidative stress exerted by both toxins on whole cells. Cytotoxicity effect increased with zearalenone concentration ranged from 0.1 to 20 μg/mL and that of deoxynivalenol from 0.1 to 50 μg/mL. Synergic effect was found in binary mixtures of analytes.

MIPs are obtained by polymerization of monomers in the presence of target molecules called templates. After polymerization, templates are removed from the matrix (Suryanarayanan et al., 2010). As a result, wholes that mimic target molecules including their shape, size, and charge distribution remain in the polymer. MIP materials can be obtained on the transducer surface by electropolymerization or by chemical synthesis. In the latter case, the polymer is milled and then implemented in polymeric carriers. MIPs are used for solid-phase extraction of traces (He et al., 2007) and for preconcentration of analytes on the transducers interface (Holthoff and Bright, 2007). Ideally, high selectivity of MIP–analyte interactions is established by a proper choice of monomers that should meet functional groups of the template molecule and steric factors affecting their spatial distribution in the product (Kotova et al., 2013). However, swelling in aqueous electrolytes and insufficient rigidity decrease selectivity of analyte binding. Best results of MIP selection were obtained for multifunctional compounds of rather high molecular mass. Mycotoxins were also applied as templates for their detection with appropriate sensors.

Thus, OTA was detected with electrochemical sensor consisting of glassy carbon electrode modified with multiwalled CNTs and polypyrrole electrosynthesized in the presence of analyte molecules (Pacheco et al., 2015). The accumulated OTA was then oxidized and the peak current recorded with DC and differential pulse voltammetry as analytical signal. The sensor made it possible to determine from 0.05 to 1.0 μM OTA (LOD 14 nM). The recovery in wine and beer testing was between 84% and 104%. Similar material was used for the solid phase extraction of OTA from red wine (Yu and Lai, 2006). Additional inclusion of quantum dots allows combining extraction and fluorescent detection of the analyte. In comparison with CNTs/polypyrrole matrix, the addition of quantum dots with CdSe core and ZnS shell increased the recovery of OTA from 1 ng/mL solution from 10 to about 60% (Wei et al., 2007). Other directions of MIP optimization for OTA extraction and analysis are summarized in review (Yu and Lai, 2010).

SPR sensor was developed for citrinin assay (Atar et al., 2015). Au film was consecutively coated with allylmercaptane and citrinin imprinted poly(2-hydroxyethyl methacrylate–methacryloylamidoglutamic acid). The SPR sensor detected from 0.005 to 1.0 ng/mL of citrinin (LOD 0.0017 ng/mL). Extracts from red yeast rice and cross-selectivity toward zaeralenone and OTA quantified. Similar SPR sensors have been proposed for the detection of zearalenone (Choi et al., 2009) and deoxynivalenol (Choi et al., 2011), with imprinted polypyrrole. The LOD values of 0.3 ng/g and 0.1 ng/mL, respectively, were achieved.

4 Pathogenic Microorganisms

Foodborne diseases have become a significant health problem over the last years due to the sharp increase in the number of incidences (Law et al., 2015). Generally, foodborne diseases are caused by the consumption of food or water contaminated with pathogens (bacteria, viruses, fungi and parasites) or their toxins. Although viruses are the primary causes of illnesses, bacteria are mostly responsible for patient hospitalizations and deaths (Scallan et al., 2011). The list of common foodborne pathogens includes *Listeria monocytogenes, Escherichia coli, Staphylococcus aureus, Salmonella enterica, Bacillus cereus, Vibrio* spp., *Campylobacter jejuni,* and *Clostridium perfringens.*

In conventional bacteriological analysis, pathogenic strains are transferred from contaminated water or food sample to the growth medium. The amplification of microbial cell numbers permits their quantitative determination (Mandal et al., 2011). The application of antibiotics to suppress the growth of other strains and the use of different media generate the colonies distinguished by their morphologies by ocular inspection. The identification of strains is then confirmed by biochemical or DNA-based assays. The analysis of pathogen-specific DNA by polymerase chain reaction (PCR) is the most reliable identification method that can be performed in real-time scale with minimal biomass requirements and multiple pathogen-assay formats (Maurer, 2011). For some bacteria, commercial PCR kits are available (*Salmonella* BAX PCR, ADIAFOOD, AnDiaTec and others, see Maurer, 2011 for more information). A detailed consideration of the DNA analysis of PCR products including hybridization based methods and application of the DNA arrays with fluorescence signal are beyond the scope of this review. Modern progress in this area is presented in some recent reviews (Law et al.,2015; Gehring and Tu, 2011; Severgnini et al., 2010). The aptamers against whole cells and immunoassay technique in biosensor format are more convenient for field application and point-of-care diagnostics devices.

4.1 Aptasensors for Detection of Pathogenic Microorganisms

Aptamers have been selected for all the pathogenic microorganisms mentioned previously. The summary of their structures as well as details of aptamer selection are considered in Wang et al. (2012). The characteristics of appropriate aptasensors including nanomaterials used in their assembly are summarized in Table 2.4. As in the case of mycotoxins, the greater part of the articles is devoted to the use of nanosized supports for aptamers, which provide higher density of the binding sites and increased sensitivity of detection methods based on estimation of permeability of the surface layer (EIS techniques). Ag and Au nanoparticles are easily compatible with thiolated aptamers and conventional transducers (Wu et al., 2012b; Ravindranath et al., 2011; Labib et al., 2012a,b; Yuan et al., 2014). The rather large size of whole cells simplifies the detection of target interactions, which do not require specific amplification, such as attachment of heavy or bulky labels. Many of the aptasensors developed are based on the determination of charge distribution in the surface layer. This is mainly caused by a combination of two factors, that is, own negative charge of aptamers due

Table 2.4 Analytical Characteristics of Aptasensors with Nanomaterials for Pathogenic Bacteria Detection

Nanomaterial	Surface Layer Assembly	Signal Measurement Protocol, Samples Tested	LOD, Linearity Range	References
Escherichia coli				
Single-walled CNTs	Deposition of suspension of single-walled CNTs followed by carbodiimide binding of aminated RNA aptamer against *E. coli* CECT 67	Potentiometric analysis, milk, apple juice	LOD 6 (milk) and 26 (juice) CFU/mL	Zelada-Guillen et al. (2012)
Single-walled CNTs	Single-walled CNTs in FET assembly were modified with aptamer by avidin-biotin binding	FET measurements of drain-source current	–	So et al. (2008)
Polydiacetylene vesicles	Aptamer against *E. coli* O157:H7 attached to the vesicles	Visual detection of red-blue transition of polydiacetylene	Conc. range 10^4–10^8 CFU/mL	Wu et al. (2012a)
Salmonella typhimurium				
Single-walled CNTs	Deposition of suspension of single-walled CNTs followed by carbodiimide binding of aminated RNA aptamer	Potentiometric analysis	LOD 0.2 CFU/mL, conc. range 0.2–10^3 CFU/mL	Zelada-Guillén et al. (2009)
Au nanoparticles	Target bacteria are adsorbed on Au nanoparticles	Interaction with aptamer results in self-aggregation and color change	LOD 10^5 CFU/mL	Wu et al. (2012b)
Au and Ag nanoparticles	Au nanoparticles modified with 4-mercapto-benzoic acid, Ag nanoparticles with 4,6-dimethyl-2-pyrimidinethiol and Ag–Au core-shell nanoparticles with 2-thiouracil as Raman reporters and with thiolated aptamers	SERS analysis after separation on commercial polyvinylidene fluoride membrane	LOD 10–100 CFU/mL	Ravindranath et al. (2011)

(*Continued*)

Table 2.4 Analytical Characteristics of Aptasensors with Nanomaterials for Pathogenic Bacteria Detection (*cont.*)

Nanomaterial	Surface Layer Assembly	Signal Measurement Protocol, Samples Tested	LOD, Linearity Range	References
Au nanoparticles	Screen-printed electrodes modified with Au nanoparticles and thhiolated aptamers	EIS measurements in the presence of $[Fe(CN)_6]^{3-/4-}$ redox probe	LOD 600 CFU/mL	Labib et al. (2012a)
Salmonella enteritidis				
Au nanoparticles	Screen-printed electrodes modified with Au nanoparticles and thhiolated aptamers	EIS measurements in the presence of $[Fe(CN)_6]^{3-/4-}$ redox probe	LOD 600 CFU/mL	Labib et al. (2012b)
Listeria monocytogenes				
Protamine–polycation sensitive membrane	Competitive assay with aptamer specific to bacteria and protamine	Potentiometric measurements with polycation sensitive membrane by suppression of protamine signal	LOD 10 CFU/mL	Ding et al. (2014)
Staphylococcus aureus				
Au nanoparticles	Aptamer–HRP conjugate immobilized on wells of microtiter plate	Sandwich assay, color changes caused by formation of Au nanoparticles in tyramine enzymatic reaction	LOD 9 CFU/mL, conc. range 10–10^6 CFU/mL	Yuan et al. (2014)
Graphene oxide and reduced graphene oxide	Graphene oxide covalently modified with aminate aptamer by carbodiimide binding, reduced graphene oxide with noncovalently attached pyrene modified aptamer	Potentiometric measurements of *S. aureus* binding	LOD 1 CFU/mL	Hernández et al. (2014)

Table 2.4 Analytical Characteristics of Aptasensors with Nanomaterials for Pathogenic Bacteria Detection (*cont.*)

Nanomaterial	Surface Layer Assembly	Signal Measurement Protocol, Samples Tested	LOD, Linearity Range	References
Reduced graphene oxide	Reduced graphene oxide attached to single-stranded DNA on glassy carbon and deposited with Au nanoparticles with thiolated aptamer	EIS measurements in the presence of $[Fe(CN)_6]^{3-/4-}$ redox probe	LOD 10 CFU/mL, conc.range 10–10^6 CFU/mL	Jia et al. (2014)
Ag nanoparticles, magnetic nanoparticles	Streptavidin coated magnetic beads modified with capture aptamers, An nanoparticles modified with thiolated signaling aptamer	Sandwich assay, electrochemical signal of Ag detected by differential pulse anodic stripping voltammetry	LOD 1 CFU/mL, conc. range 10–10^6 CFU/mL	Abbaspour et al. (2015)

to phosphate residues of main chain and negative charge of outer surface of cell membrane of the majority of pathogenic microorganisms. As a result, the electrostatic repulsion provides quite sensitive EIS detection of attached molecules. Moreover, changes in the biosensor interface charge can be rather easily monitored by standard potentiometric techniques (Zelada-Guillén et al., 2012; Zelada-Guillén et al., 2009) or microelectronic devices such as FETs (So et al., 2008). To distinguish processes responsible for aptamers immobilization and potentiometric response, it was suggested to use hydrophobic noncovalent immobilization of pyrene modified aptamers onto graphene plane (Hernández et al., 2014). The use of potentiometric methods is complicated by a rather high charge of reactants that decreases the sensitivity of appropriate calibration curves. To avoid this limitation related to the Nernst equation, an interesting approach relative to substitution titration has been proposed (Ding et al., 2014). In this method, ion-selective electrode with anion-selective membrane was used as a biosensor transducer. Polycationic protamine molecules simultaneously form with aptamers uncharged complexes, which are dissociated in the presence of bacteria. The release of protamine is detected as a measure of bacteria cells.

4.2 Immunosensors Including Nanomaterials for the Detection of Pathogenic Bacteria

The immunoassay of pathogenic microflora is a traditional area of immunochemical analysis with thousands of publications and many commercial test kits providing reliable and rather fast response to bacterial contamination of food, soil, water, and so forth. In this section, the immunosensors including various nanomaterials sensitive to whole cells are considered, which were developed within the past 10 years (Table 2.5).

The majority of such immunosensors utilize polyclonal and monoclonal Abs selected against target analyte, isolated, purified, and immobilized on appropriate support or signal transducer. Although it is not always mentioned in appropriate reports, membrane bonded proteins and receptor structures are real antigens, which specifically bind Ab molecules in a biosensor assembly. Conventional measurement format includes thermal killing of the bacteria prior to immunosensor incubation. In some cases, viable cells are tested. However, appropriate samples are first incubated

Table 2.5 Analytical Characteristics of Immunosensors with Nanomaterials for Pathogenic Bacteria Detection

Nanomaterial	Surface Layer Assembly	Signal Measurement Protocol, Samples Tested	LOD, Linearity Range	References
Escherichia coli				
Au nanoparticles	Au nanoparticles modified with capture Ab against *E. coli* O157:H7 immobilized on quartz crystal sensor	Sandwich assay with Au nanoparticles covered with signaling Ab, QCM measurements, wild blueberry	LOD 10 CFU/mL (with 8-hincubation)	Guo et al. (2012b)
Au nanoparticles	Screen-printed carbon electrode modified with ferrocene dicarboxylic acid and monoclonal capture Ab against *E. coli* O157:H7	Sandwich assay with signaling Ab conjugated with HRP, voltammetric measurement of H_2O_2 oxidation signal, milk	OD 6 (buffer) and 50 CFU/strip, conc. range 10^2–10^7 CFU/milk	Lin et al. (2008)

Table 2.5 Analytical Characteristics of Immunosensors with Nanomaterials for Pathogenic Bacteria Detection (*cont.*)

Nanomaterial	Surface Layer Assembly	Signal Measurement Protocol, Samples Tested	LOD, Linearity Range	References
Au nanoparticles beacon magnetic nanoparticles	Magnetic nanoparticles were prepared from $Fe^{2+/3+}$ solution in the presence of dextran and modified with capture Ab, suspension of Au nanoparticles was conjugated with streptavidin	E.coli O157:H7 were magnetically separated with magnetic nanoparticles and Au nanoparticles were bonded to the complex by avidin-biotin conjugation, catalytic growth of Au was attained by $HAuCl_4$–ascorbic acid mixture and resulted mass shift was recorded by QCM, milk	LOD 23 CFU/mL in buffer solution and 53 CFU/mL in milk	Shen et al. (2011)
Micro- and nanobeads (30–970 nm) made of various materials (magnetic, silica, polymer)	QCM sensor with immobilized capture Ab; streptavidin conjugated beads modified with signaling Ab against *E. coli* O157:H7	Sandwich assay, QCM measurement of bacteria attachment to Ab followed by nano- or microparticles binding	LOD 10^3 CFU/mL vs 10^6 CPU/mL without nanoparticles amplification	Jiang et al. (2011)
Fullerene C_{60}, Pt nanochains, Au/silica nanoparticles	Nano SiO_2 particles modified with 3-aminothrethoxysilane, ferrocene, and Au nanoparticles. Glassy carbon electrode modified with C_{60}, thiochitosan and Au/SiO_2 nanoparticles and capture Ab against *E. coli* O157:H7	Sandwich assay with signaling Ab immobilized on Pt nanochain–Au composite, signal of ferrocene after addition of glucose oxidase and glucose. Measurements with DC voltammetry, synthetic stool samples	LOD 15 CFU/mL, conc. range 32–3.2 × 10^6 CFU/mL	Li et al. (2013)

(*Continued*)

Table 2.5 Analytical Characteristics of Immunosensors with Nanomaterials for Pathogenic Bacteria Detection (*cont.*)

Nanomaterial	Surface Layer Assembly	Signal Measurement Protocol, Samples Tested	LOD, Linearity Range	References
Chitosan-multiwalled CNTs–SiO_2/thionine hybrid	Layer-by layer deposition of chitosan-multiwalled CNTs–SiO_2/thionine hybrid and AU nanoparticles on Au electrode modified with 11-amino-1-undecanethiol hydrochloride. Anti-*E. coli* O157:H7Ab immobilized on Au nanoparticles	Sandwich assay in ELISA format with HRP conjugate with signaling Ab, thionine signal measurement as enzyme substrate by direct current voltammetry	LOD 250 CFU/mL, conc. range 4.12×10^2 – 4.12×10^5	Li et al. (2012a,b)
Single-walled CNTs	Deposition of suspension of single-walled CNTs modified with 1-pyrenebutanoic acid succinimidyl ester as linker and physically adsorbed Ab against *E. coli* O157:H7	Source-drain current detection with field-effect transistor	LOD 820 CFU/mL, conc. range up to 8.2×10^8 CFU/mL	Zhang et al. (2012b)
CNT-Au nanoparticles hybrids	Au deposition on glassy carbon electrode modified with multiwalled CNTs, followed by adsorption of capture Ab against *E. coli* O157:H7	Sandwich assay with conjugate of polyclonal signaling Ab and HRP, voltammetric measurement of hydroquinone as enzyme signal	Conc. range 2×10^2–2×10^6 CFU/mL	Zhang et al. (2014)
Au nanoparticles on graphene paper	Au nanoparticles were electrodeposited on graphene paper and modified with Ab against *E. coli* O157:H7 by avidin-streptavidin binding	EIS measurements in the presence of $[Fe(CN)_6]^{3-/4-}$ redox probe	LOD 1.5×10^2, conc. range 1.5×10^2–1×10^7 CFU/mL	Wang et al. (2013)

Table 2.5 Analytical Characteristics of Immunosensors with Nanomaterials for Pathogenic Bacteria Detection (*cont.*)

Nanomaterial	Surface Layer Assembly	Signal Measurement Protocol, Samples Tested	LOD, Linearity Range	References
Au nanoparticles and magnetic beads	Screen-printed carbon electrode modified with Au nanoparticles and signaling Ab, magnetic beads modified with capture Ab	Sandwich assay, magnetic beads are incubated with bacteria and magnetically separated on electrode surface. Signaling Ab conjugated with Au nanoparticles are bonded on the surface and hydrogen evolution current is recorded in the presence of Au as electrocatalyst, minced beef and water samples	LOD 10^2 CFU/mL, conc, rane 5×10^2–10^5 CFU/mL	Hussein et al. (2015)
Cysteine 3D flowers	Au coated glass coated with cysteine nanostructures obtain by sonication of cysteine solution in basic media monoclonal Ab against *E. coli* O157:H7 immobilized by carbodiimide binding	EIS measurements in the presence of $[Fe(CN)_6]^{3-/4-}$ redox probe	LOD 964 (palm like structures) 4.70 (flower like structures). Conc. range 10^3–3×10^9 CFU/mL	Pandey et al. (2014)
Salmonella spp.				
Au nanoparticles	Au nanoparticles in a monolayer on glassy carbon electrode grafted with polyethylene imine, monoclonal Ab immobilized by carbodiimide binding	EIS measurements in the presence of $[Fe(CN)_6]^{3-/4-}$ redox probe, commercial pork samples	LOD 100 CFU/mL, conc. range 10^2–10^5 CFU/mL	Yang et al. (2009)

(*Continued*)

Table 2.5 Analytical Characteristics of Immunosensors with Nanomaterials for Pathogenic Bacteria Detection (*cont.*)

Nanomaterial	Surface Layer Assembly	Signal Measurement Protocol, Samples Tested	LOD, Linearity Range	References
Magnetic beads	Nanoporous alumina membrane modified with 3-aminopropyltriethoxysilane and capture Ab by carbodiimide binding	Polystyrene magnetic beads modified by signaling Ab. Bacteria are separated on nanoporous membrane, then magnetic beads are interacted to shift resistance measured by EIS	LOD 10 CFU/mL, conc. range 10–10^4 CFU/mL	Chan et al. (2013)
Salmonella pullorum				
Au nanoparticles	Au nanoparticles electrdeposited on carbon screen-printed electrode followed by deposition of HRP labeled Ab and ionic liquid	Voltammetric detection of thionine as enzyme substrate, wastewater plant treatment samples	LOD 548 CFU/mL, conc. range 10^2–10^9 CFU/mL	Wang et al. (2014)
Salmonella typhimurium				
As grown double-wall CNTs	Monoclonal Ab immobilized on electrode with CNTs forest by carbodiimide binding	Chronoamperometric measurement of the current of electrode with impregnated CNTs and bacteria	LOD 8.9 CFU/mL, conc. range 10^2–10^7 CFU/mL	Punbusayakul et al. (2013)
As grown CNTs	CNTs grown in gate area of FET on SiO_2 substrate and modified with polyclonal Ab	Source-drain current measurements with FET	LOD 10^3 CFU/mL, conc. range up to 10^8 CFU/mL	Lerner et al. (2011)
Au/PAMAM/multiwalled CNTs/chitosan composite	Glassy carbon electrode covered with PAMAM–CNTs dispersion and then with chemically synthesized Au nanoparticles followed by carbodiimide binding of Ab	EIS measurements in the presence of $[Fe(CN)_6]^{3-/4-}$ redox probe, milk	LOD 5 × 10^2 CFU/mL, conc. range 10^3–10^7 CFU/mL	Dong et al. (2013)

Table 2.5 Analytical Characteristics of Immunosensors with Nanomaterials for Pathogenic Bacteria Detection (*cont.*)

Nanomaterial	Surface Layer Assembly	Signal Measurement Protocol, Samples Tested	LOD, Linearity Range	References
Au nanoparticles	QCM sensor with monolayer of Mercaptohexadecanoicacid and carbodiimide bonded monoclonal capture Ab	Sandwich assay with signaling Ab attached to Au nanoparticles, precooked chicken meat	LOD 10–20 CFU/mL, conc. range 25–10^7 CFU/mL	Salam et al. (2013)
Au nanoparticles, Magnetic beads	Screen-printed electrode modified magnetic beads with capture Ab,	Sandwich assay with Au nanoparticles modified with signaling Ab, skimmed milk	LOD 143 CFU/mL, conc. range 10^3–10^6 CFU/mL	Afonso et al. (2013)
Fe_3O_4/Au core/shell, CdS nanoparticles	Fe at Au nanoparticles modified with cystamine, mercaptopropionic acid or mercaptoethanol/mecraptoundecanoic acid and capture Ab by carbodiimide binding	Fe at Au particles with Ab–bacteria conjugates magnetically separated on screen-printed electrode and treated with CdS-signaling Ab hybrid particles, Cd(II) signal (square-wave anodic stripping voltammetry)	LOD 13 cells/mL, conc. range 100–10^6 cells/mL	Freita et al. (2014)
Magnetic nanoparticles	Magnetic particles modified with capture Ab via biotin-streptavidin binding	Sandwich assay, signaling Ab conjugated with HRP, voltammetric signal of hydroquinone as enzyme substrate, milk	LOD 1.4 CFU/mL (8-h enrichment)	Liébana et al. (2009)
Superparamagnetic nanoparticles	Carboxylated magnetic particles modified with Ab by carbodiimide binding	Measurement of spin–spin relaxation by NMR resonator affected by aggregation of particles after bacteria binding	LOD 10^3 CFU/mL Conc. range 5×10^3–1×10^6 CFU/mL	Wang et al. (2015c)

(*Continued*)

Table 2.5 Analytical Characteristics of Immunosensors with Nanomaterials for Pathogenic Bacteria Detection (*cont.*)

Nanomaterial	Surface Layer Assembly	Signal Measurement Protocol, Samples Tested	LOD, Linearity Range	References
Magnetic silica nanotubes	SiO_2-Fe_3O_4 hybrids were modified with 3-aminopropyltriethoxysilane and electrostatically conjugated with bacteria	Interdigitated electrodes as EIS sensors were modified with mercaptoundecanoic acid and specific Ab, impedance measurements	Conc. range 1×10^3–1×10^7 CFU/mL	Nguyen et al. (2014)
Magnetic particles	Carboxylated micro- and nanosized magnetic particles modified with capture monoclonal Ab by carbodiimide binding	Sandwich assay with magnetic separation and signaling Ab–HRP conjugate, voltammetric measurement of hydroquinone as enzyme signal, milk	LOD 538 (micro-), 291 (nanocarrier) CFU/mL, 1-h incubation, milk: 0.04 CFU/mL, 8-h incubation	Brandão et al. (2015)
Listeria spp.				
Magnetic nanoparticles	Magnetic particles precoated with streptavidin modified with biotinylated Ab specific for *Listeria* spp.	Fiber optic sensor with absorbance detection at 450 nm, hot dogs	LOD 3×10^2 CFU/mL (*L. monocytogenes)*	Mendonça et al. (2012)
Listeria monocytogenes				
TiO_2 nanowire bundle	Ab adsorbed on the surface of TiO_2 nanowire of microelectrode resistor	Impedance measurements,	LOD 470 CFU/mL	Wang et al. (2008)
Streptococcus spp.				
Magnetic nanoparticles	Carboxylated, aminated or bearing aldehyde groups magnetic particles modified with Ab by chelate immobilization	Western blotting assay, distinguishing *S. aureus* and group G *Streptococcus* due to Fc region binding accessible in immobilization schemes tested	—	Xiao et al. (2007)

Table 2.5 Analytical Characteristics of Immunosensors with Nanomaterials for Pathogenic Bacteria Detection (*cont.*)

Nanomaterial	Surface Layer Assembly	Signal Measurement Protocol, Samples Tested	LOD, Linearity Range	References
Vibrio parahaemolyticus				
Au nanoparticles	Screen-printed electrode covered with agarose–Au nanoparticle membrane and HRP–Ab conjugate	Partial blocking the enzymatic oxidation of thionine caused by bacteria binding, voltammetry measurement of thionine signal as enzyme substrate	LOD 7374 CFU/mL, conc. range 10^5–10^9 CFU/mL	Zhao et al. (2007)

in favorable conditions in growing media to multiply the number of cells assessed. This increases the sensitivity of appropriate measurement protocol showing the LOD values down to several CFU/mL. Meanwhile the whole measurement period can exceed 24 h, though 1–8 h incubation is used more frequently. Such a long sample pretreatment as well as necessity of thermostating at 37–40°C levels the advantages of biosensors often announced, that is, fast response and portability of devices. Many efforts are paid to the immobilization of specific Ab, which should be rather tightly attached to the transducer surface but offer extended accessibility for bulky cells interacted with Ab. In some cases, the choice of immobilization protocol varies the selectivity of detection. Thus, separate detection of *S. aureus* and group G *Sreptococcus* has been achieved by implementation of Fe region of Ab molecules on the immobilization stage (Zhao et al., 2007). In other cases, the more common hydrophobic linkers are used especially in combination with Au nanoparticles as Ab supports. Variation in the linker length changes selectivity of the response and detectable quantities of bacterial cells (Freita et al., 2014).

Some of the common detection schemes realized in bacteria immunosensors are outlined in Fig. 2.6. Single-walled CNT FET was described for both Ab immobilization and *E. coli* O157:H7 detection. The shift of the current between source and drain is related to the charge of cells specifically bonded to the sidewalls of the CNT. The morphology of gate area and the formation of Ab complex with whole cells were confirmed by electron microscopy and AFM. As in the case of aptamers described previously, EIS measurements showed high efficiency in pathogenic bacteria detection (Fig. 2.6b). They can utilize interdigitated electrodes, which react on the bacteria binding by specific changes in the capacitance measured in noncurrent regime. Besides, more common Faradaic EIS with ferrocyanide redox probe was applied for the same purposes (Wang et al., 2013; Pandey et al., 2014).

Fig. 2.6c illustrates most common sandwich assay with Ab covalently attached to nanoparticles placed on the sensor surface. Au nanoparticles or commercial magnetic beads already modified with carboxylic groups or biotin can be applied. The detection of bacteria is attained by secondary (signaling) Abs bearing labels. In enzyme sensors, the appropriate reaction can be recorded by the

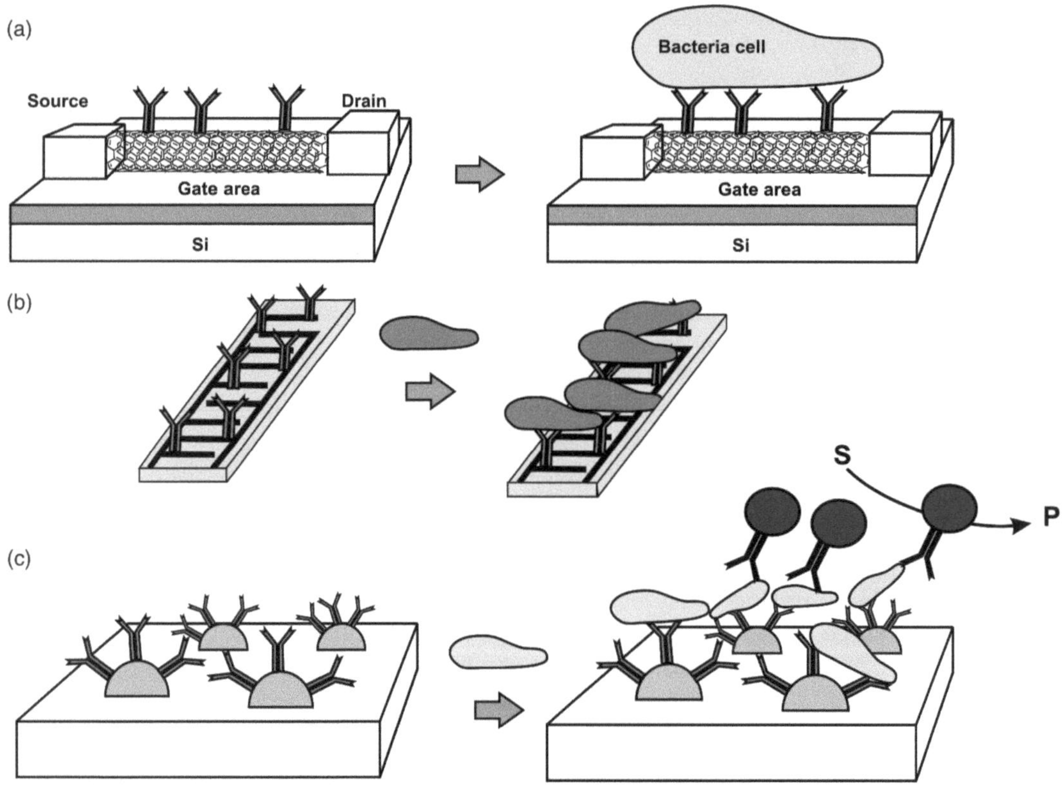

Figure 2.6. Schematic outlines of bacteria detection with immunosensors. (a) Single-walled CNT FET; (b) interdigital electrodes as resistor; (c) sandwich assay.

amounts of products formed near the electrode. Thus, in the case of HRP, hydroquinone (Zhang et al., 2014) or thionine (Li et al., 2012b) yielding well-defined remarkable current changes can be used. In a similar manner, sandwich assay with glucose oxidase as label and fullerene as mediator has been reported (Zhilei et al., 2011).

Sandwich assay is easily combined with magnetic separation and mass sensitive detection. In the latter case, a decreased number of magnetic particles decreases the LOD values by three orders of magnitude [*E. coli* O157:H7 detection (Jiang et al., 2011)]. From less trivial detection system, the use of catalytic processes can be mentioned. Thus, the sensitivity of QCM detection in sandwich assay was increased by additional deposition of Au on nanoparticles separated from analyte solution by magnetic nanoparticles. For this purpose, chemical reduction of $AuCl_4^-$ ions with ascorbic acid was performed on the last stage of the biosensor treatment (Shen et al., 2011). In other example, the current of electrocatalytic hydrogen evolution was recorded for specific detection of *E. coli* with screen-printed electrode after separation of the bacteria cells with signaling Ab labeled with Au nanoparticles (Hussein et al., 2015).

Selectivity of immunosensors for pathogenic bacteria detection is commonly rather high due to selection of Ab and minimal contribution of nonspecific sorption of reactants on the transducer surface. The latter advantage is directly related to the size of bacterial cells and blocking surface of transducers with inert proteins (BSA) or charged thiocarbonic acids (monolayers on gold). In many works, the selectivity is additionally proved by appropriate experiments performed with standard cultures of other pathogens introduced in the spiked samples prior to cultivation or prior to biosensor contact with enriched culture. The selection inside serovar types and between killed and viable cells are satisfactory for common microbiological testing of food- and feedstuffs.

Few numbers of multiplex assay were published. Thus, the Ab molecules against *E. coli* O157:H7, *Campylobacter* and *Salmonella* were coimmobilized on screen-printed carbon electrode modified with multiwalled CNTs (Viswanathan et al., 2012). Signaling Abs used in sandwich assay format were labeled with different nanocrystals (CdS, PbS, and CuS). After immunoseparation, the labels were dissolved in acidic media and the Cd(II), Pb(II), and Cu(II) ions determined by anodic stripping voltammetry. The selectivity of the response was achieved due to different potentials required for the detection of metal ions on the last stage of the assay. The principal scheme of the analysis is presented in Fig. 2.7. Colorimetric test system based on vesicles synthesized from 10,12-pentacosadyinoic acid and *N*-[(2-tetradecanamide)-ethyl]

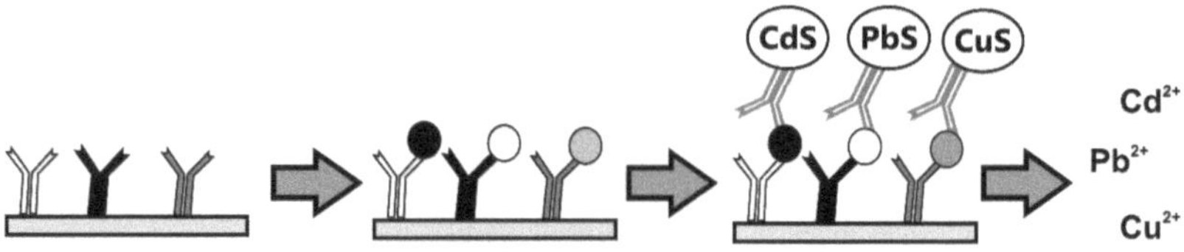

Figure 2.7. Multiplex determination of three bacteria using different labels.

ribonamide has been proposed for the detection of *S. aureus* and *E. coli* (Sapsford et al., 2006). The interaction of bacterial components with vesicles on cellulose strip resulted in the color change from blue to red. The test strip was applied for preliminary screening of apple juices.

5 Conclusions

Various biosensors based on aptamers, enzymes, and immunochemistry reagents have been developed for reliable and sensitive detection of mycotoxins and foodborne pathogens. Although all of them are directed for the analysis of similar samples covering food, agriculture products, soil, water, the requirements to their performance are rather difficult. Mycotoxin biosensors should provide a high sensitivity on the subnanomolar level resulted from modern legislation and tendencies to stricter limitations of permissible residuals of the previously mentioned hazards. Regarding whole pathogenic bacteria detection, the sensitivity of the response can be compensated by application of cultivation stages together with rather mild requirements to the duration of assay which covers commonly several hours. Nevertheless, reliability of the signal and portable design remain most important for such warning devices. This review shows many examples of approaches to the improved performance of appropriate biosensors which can be subdivided into several parts in accordance with the biosensor assembly and classification.

Bioreceptor. The selection of aptamers for mycotoxin detection continues to extend the number of analytes recognized. The choice of appropriate Ab does not seem a serious problem. Monoclonal Ab are available for capturing target cells and conjugates of polyclonal signaling Ab with labels for sandwich assay. Meanwhile some new variations of Ab can find application, for example, nanobodies consisting of heavy chains or fragments of conventional Ab (Fab, single-chain variable fragments, etc.). The bioreceptors for measurements in aqueous organic solvents used for analyte extraction is a problem that remains to be solved in the near future. To some extent immobilization increases the stability of bioreceptors in preferably nonaqueous media, but decreases the accessibility of analytes and can result in cell lysis.

Detection system. Main trends of the future progress in the detection system of biosensors for mycotoxin and pathogens detection involve a broader application of microelectronic devices, for example, FETs and microfluidics systems that combine sample sampling, its extraction, purification, and analysis. The use of carbonaceous nanomaterials, quantum dots, and semiconductor nanoparticles offers new opportunities for miniaturization of sensing part of biosensors and increased reproducibility of their properties. Although electrochemical devices dominate at the moment, the focus of future investigations will be shifted to optical devices based on SERS or FRET signals. The usability of mass sensitive sensors—for example, QCM or microcanilevers—is limited by their lower sensitivity and more complicated design.

Analytes. As mentioned previously, the number of potential targets for biosensors in food safety and human health is constantly increasing. The nature of specific interactions in affine biosensors that cover both aptasensors and immunosensors leaves room for optimism. Universal approaches to the assembling of biorecognition layers and signal transduction will result in the creation of several types of universal platforms easily adapted for new analytes on demand. To some extent we can see this kind of innovation in the example of fluorescent biosensor arrays or screen-printed electrodes with changeable content of the surface layer. In this case, the number of new biosensors in this area will be mainly dictated by market necessities than limited with some complications related to lack of sensitivity or insufficient stability. From the same reasons of economics, multiplex devices will receive priority in development and commercialization.

Acknowledgments

The work was performed according to the Russian Government Program of Competitive Growth of Kazan Federal University and supported by the Slovak Research and Development Agency under the contract No. APVV-14-0267 and by Science Grant Agency VEGA (project No. 1/0152/15).

References

Abbaspour, A., Norouz-Sarvestani, F., Noori, A., Soltani, N., 2015. Aptamer-conjugated silver nanoparticles for electrochemical dual-aptamer-based sandwich detection of *Staphylococcus aureus*. Biosens. Bioelectron. 68, 149–155.

Afonso, A.S., Pérez-López, B., Faria, R.C., Mattoso, L.H.C., Hernández-Herrero, M., Roig-Sagués, A.X., Maltez-da Costa, M., Merkoçi, A., 2013. Electrochemical detection of *Salmonella* using gold nanoparticles. Biosens. Bioelectron. 40, 121–126.

Alexander, J., Benford, D., Boobis, A., Eskola, M., Fink-Gremmels, J., Fürst, P., Heppner, C., Schlatter, J., van Leeuwen, R., 2002. Risk assessment of contaminants in food and feed. EFSA J. 10, s1004.

Allen, B.L., Kichambare, P.D., Star, A., 2007. Carbon nanotube field-effect-transistor-based biosensors. Adv. Mater. 19, 1439–1451.

Alonso-Lomillo, M.A., Domínguez-Renedo, O., Ferreira-Gonçalves, L., Arcos-Martíneza, M.J., 2010. Sensitive enzyme-biosensor based on screen-printed electrodes for Ochratoxin A. Biosens. Bioelectron. 25, 1333–1337.

Ansari, A.A., Kaushik, A., Solanki, P.R., Malhotra, B.D., 2010. Nanostructured zinc oxide platform for mycotoxin detection. Bioelectrochemistry 77, 75–81.

Arduini, F., Errico, I., Amine, A., Micheli, A., Palleschi, G., Moscone, D., 2007. Enzymatic spectrophotometric method for aflatoxin B detection based on acetylcholinesterase inhibition. Anal. Chem. 79, 3409–3415.

Arya, S.K., Saha, S., Ramirez-Vick, J.E., Gupta, V., Bhansali, S., Singh, S.P., 2012. Recent advances in ZnO nanostructures and thin films for biosensor applications: review. Anal. Chim. Acta. 737, 1–21.

Atar, N., Eren, T., Yola, M.L., 2015. A molecular imprinted SPR biosensor for sensitive determination of citrinin in red yeast rice. Food Chem. 184, 7–11.

Barthelmebs, L., Hayat, A., Limiadi, A.W., Marty, J.-L., Noguer, T., 2011. Electrochemical DNA aptamer-based biosensor for OTA detection, using superparamagnetic nanoparticles. Sens. Actuat. B-Chem. 156, 932–937.

Basu, J., Datta, S., Chaudhuri, S.R., 2015. A graphene field effect capacitive immunosensor for sub-femtomolar food toxin detection. Biosens. Bioelectron. 68, 544–549.

Bedford, E.E., Spadavecchia, J., Pradier, C.-M., Gu, F.X., 2012. Surface plasmon resonance biosensors incorporating gold nanoparticles. Macromol. Biosci. 12, 724–739.

Bennett, J.W., Klich, M., 2003. Mycotoxins. Clin. Microbiol. Rev. 16, 497–516.

Bonel, L., Vidal, J.C., Duato, P., Castillo, J.R., 2011. An electrochemical competitive biosensor for ochratoxin A based on a DNA biotinylated aptamers. Biosens. Bioelectron. 26, 3254–3259.

Brandão, D., Liébana, S., Campoy, S., Alegret, S., Pividori, M.I., 2015. Immunomagnetic separation of *Salmonella* with tailored magnetic micro and nanocarriers: a comparative study. Talanta 143, 198–204.

Budnikov, H.C., Evtugyn, G.A., Porfireva, A.V., 2012. Electrochemical DNA sensors based on electropolymerized materials. Talanta 102, 137–155.

Castillo, G., Spinella, K., Poturnayová, A., Snejdárková, M., Mosiello, L., Hianik, T., 2015. Detection of aflatoxin B1 by aptamer-based biosensor using PAMAM dendrimers as immobilization platform. Food Control 52, 9–18.

Chan, K.Y., Ye, W.W., Zhang, Y., Xiao, L.D., Leung, P.H.M., Li, Y., Yang, M., 2013. Ultrasensitive detection of *E. coli* O157:H7 with biofunctional magnetic bead concentration via nanoporous membrane based electrochemical immunosensor. Biosens. Bioelectron. 41, 532–537.

Chen, J., Zhang, X., Cai, S., Wu, D., Chen, M., Wang, S., Zhang, J., 2014. A fluorescent aptasensor based on DNA-scaffolded silver-nanocluster for ochratoxin A detection. Biosens. Bioelectron. 57, 226–231.

Choi, S.-W., Chang, H.-J., Lee, N., Kim, J.-H., Chun, H.S., 2009. Detection of mycoestrogen zearalenone by a molecularly imprinted polypyrrole-based surface plasmon resonance (SPR) sensor. J. Agric. Food Chem. 57, 1113–1118.

Choi, S.-W., Chang, H.-J., Lee, N., Chun, H.S., 2011. A surface plasmon resonance sensor for the detection of deoxynivalenol using a molecularly imprinted polymer. Sensors 11, 8654–8664.

Cigić, I.K., Prosen, H., 2009. An overview of conventional and emerging analytical methods for the determination of mycotoxins. Int. J. Mol. Sci. 10, 62–115.

Clark, L.C., Lyons, C., 1962. Electrode systems for continuous monitoring in cardiovascular surgery. Ann. N.Y. Acad. Sci. 148, 133–153.

Cooper, M.A. (Ed.), 2009. Label-Free Biosensors: Techniques and Applications. Cambridge University Press, New York, NY.

Cruz-Aguado, J.A., Penner, G., 2008. Determination of ochratoxin A with a DNA aptamers. J. Agric. Food Chem. 56, 10456–10461.

Cuccioloni, M., Mozzicafreddo, M., Barocci, S., Ciuti, F., Pecorelli, I., Eleuteri, A.M., Spina, M., Fioretti, E., Angeletti, M., 2008. Biosensor-based screening method for the detection of aflatoxins B1-G1. Anal. Chem. 80, 9250–9256.

Daniels, J.D., Pourmand, N., 2007. Label-free impedance biosensors: opportunities and challenges. Electroanalysis 19, 1239–1257.

Denli, M., Perez, J.F., 2010. Ochratoxins in feed, a risk for animal and human health: control strategies. Toxins 2, 1065–1077.

Dinçkaya, E., Kınık, Ö., Sezgintürk, M.K., Altuğ, Ç., Akkoca, A., 2011. Development of an impedimetric aflatoxin M1 biosensor based on a DNA probe and gold nanoparticles. Biosens. Bioelectron. 26, 3806–3811.

Ding, J., Lei, J., Ma, X., Gong, J., Qin, W., 2014. Potentiometric aptasensing of *Listeria monocytogenes* using protamine as an indicator. Anal. Chem. 86, 9412–9416.

Do, J.H., Choi, D.-K., 2007. Aflatoxins: detection, toxicity, and biosynthesis. Biotechnol. Bioprocess Eng. 12, 585–593.

Dong, J., Zhao, H., Xu, M., Ma, Q., Ai, S., 2013. A label-free electrochemical impedance immunosensor based on AuNPs/PAMAM-MWCNT-Chi nanocomposite modified glassy carbon electrode for detection of *Salmonella typhimurium* in milk. Food Chem. 141, 1980–1986.

Dwivedi, H.P., Jaykus, L.A., 2011. Detection of pathogens in foods: the current state-of-the-art and future directions. Crit. Rev. Microbiol. 37, 40–63.

Evtugyn, G., Porfireva, A., Stepanova, V., Kutyreva, M., Gataulina, A., Ulakhovich, N., Evtugyn, V., Hianik, T., 2013a. Impedimetric aptasensor for ochratoxin A determination based on Au nanoparticles stabilized with hyper-branched polymer. Sensors 13, 16129–16145.

Evtugyn, G., Porfireva, A., Sitdikov, R., Evtugyn, V., Stoikov, I., Antipin, I., Hianik, T., 2013b. Electrochemical aptasensor for the determination of ochratoxin A at the Au electrode modified with Ag nanoparticles decorated with macrocyclic ligand. Electroanalysis 25, 1847–1854.

Evtugyn, G., Porfireva, A., Stepanova, V., Sitdikov, R., Stoikov, I., Nikolelis, D., Hianik, T., 2014. Electrochemical aptasensor based on polycarboxylic macrocycle modified with neutral red for aflatoxin B1 detection. Electroanalysis 26, 2100–2109.

Famulok, M., Mayer, G., 1999. Aptamers as tools in molecular biology and immunology. Curr. Top. Microb. Immunol. 243, 123–136.

Feng, R., Zhang, Y., Ma, H., Wu, D., Fan, H., Wang, H., Li, H., Du, B., Wei, Q., 2013. Ultrasensitive nonenzymatic and nonmediator electrochemical biosensor using nitrogen-doped graphene sheets for signal amplification and nanoporous alloy as carrier. Electrochim. Acta 97, 105–111.

Freita, M., Viswanathan, S., Nouws, H.P.A., Oliveira, M.B.P.P., Delerue-Matos, C., 2014. Iron oxide/gold core/shell nanomagnetic probes and CdS biolabels for amplified electrochemical immunosensing of *Salmonella typhimurium*. Biosens. Bioelectron. 51, 195–200.

Gan, N., Zhou, J., Xiong, P., Hu, F., Cao, Y., Li, T., Jiang, Q., 2013. An ultrasensitive electrochemiluminescent immunoassay for aflatoxin M1 in milk, based on extraction by magnetic graphene and detection by antibody-labeled CdTe quantum dots-carbon nanotubes nanocomposite. Toxins 5, 865–883.

Gehring, A.G., Tu, S.-I., 2011. High-throughput biosensors for multiplexed foodborne pathogen detection. Annu. Rev. Anal. Chem. 4, 151–172.

Gu, W., Zhu, P., Jiang, D., He, X., Li, Y., Ji, J., Zhang, L., Sun, Y., Sun, X., 2015. A novel and simple cell-based electrochemical impedance biosensor for evaluating the combined toxicity of DON and ZEN. Biosens. Bioelectron. 70, 447–454.

Guo, X., 2012a. Surface plasmon resonance based biosensor technique: a review. J. Biophotonics 5, 483–501.

Guo, S., Wang, E., 2011. Noble metal nanomaterials: controllable synthesis and application in fuel cells and analytical sensors. Nano Today 6, 240–264.

Guo, Z., Ren, J., Wang, J., Wang, E., 2011. Single-walled carbon nanotubes based quenching of free FAM-aptamer for selective determination of ochratoxin A. Talanta 85, 2517–2521.

Guo, X., Lin, C.-S., Chen, S.-H., Ye, R., Wua, V.C.H., 2012b. A piezoelectric immunosensor for specific capture and enrichment of viable pathogens by quartz crystal microbalance sensor, followed by detection with antibody-functionalized gold nanoparticles. Biosens. Bioelectron. 38, 177–183.

Hayat, A., Yang, C., Rhouati, A., Marty, J.-L., 2013. Recent advances and achievements in nanomaterial-based, and structure switchable aptasensing platforms for ochratoxin A detection. Sensors 13, 15187–15208.

He, C., Long, Y., Pan, J., Li, K., Liu, F., 2007. Application of molecularly imprinted polymers to solid-phase extraction of analytes from real samples. J. Biochem. Biophys. Methods 70, 133–150.

Hernández, R., Vallés, C., Benito, A.M., Maser, W.K., Rius, F.X., Riu, J., 2014. Graphene-based potentiometric biosensor for the immediate detection of living bacteria. Biosens. Bioelectron. 54, 553–557.

Holthoff, E.L., Bright, F.V., 2007. Molecularly templated materials in chemical sensing. Anal. Chim. Acta. 594, 147–161.

Hossain, S.M.Z., Luckham, R.E., Smith, A.M., Lebert, J.M., Davies, L.M., Pelton, R.H., Filipe, C.D.M., Brennan, J.D., 2009. Development of a bioactive paper sensor for detection of neurotoxins using piezoelectric inkjet printing of sol-gel-derived bioinks. Anal. Chem. 81, 5474–5483.

Hu, L., Xu, G., 2010. Applications and trends in electrochemiluminescence. Chem. Soc. Rev. 39, 3275–3304.

Hussein, A.-R., Hassan, A.-A., de la Escosura-Muñiz, A., Merkoçi, A., 2015. Highly sensitive and rapid determination of *Escherichia coli* O157:H7 in minced beef and water using electrocatalytic gold nanoparticle tags. Biosens. Bioelectron. 67, 511–515.

Jacobs, C.B., Peairs, M.J., Venton, B.J., 2010. Review: carbon nanotube-based electrochemical sensors for biomolecules. Anal. Chim. Acta 662, 105–127.

Jang, W.-D., Kataoka, K., 2005. Bioinspired applications of functional dendrimers. J. Drug Deliv. Sci. Technol. 15, 19–30.

Jia, F., Duan, N., Wu, S., Ma, X., Xia, Y., Wang, Z., Wei, X., 2014. Impedimetric aptasensor for *Staphylococcus aureus* based on nanocomposite prepared from reduced graphene oxide and gold nanoparticles. Microchim. Acta. 181, 967–974.

Jiang, X., Wang, R., Wang, Y., Su, X., Ying, Y., Wang, J., Li, Y., 2011. Evaluation of different micro/nanobeads used as amplifiers in QCM immunosensor for more sensitive detection of *E. coli* O157:H7. Biosens. Bioelectron. 29, 23–28.

Jiang, L., Qian, J., Yang, X., Yan, Y., Liu, Q., Wang, K., Wang, K., 2014. Amplified impedimetric aptasensor based on gold nanoparticles covalently bound graphene sheet for the picomolar detection of ochratoxin A. Anal. Chim. Acta. 806, 128–135.
Jin, X., Jin, X., Chen, L., Jiang, J., Shen, G., Yu, R., 2009. Piezoelectric immunosensor with gold nanoparticles enhanced competitive immunoreaction technique for quantification of aflatoxin B1. Biosens. Bioelectron. 24, 2580–2585.
Ko, J., Lee, C., Choo, J., 2015. Highly sensitive SERS-based immunoassay of aflatoxin B1 using silica-encapsulated hollow gold nanoparticles. J. Hazard Mater. 285, 11–17.
Kocer, A., Tauk, L., Déjardin, P., 2012. Nanopore sensors: from hybrid to abiotic systems. Biosens. Bioelectron. 38, 1–10.
Köppen, R., Koch, M., Siegel, D., Merkel, S., Maul, R., Nehls, I., 2010. Determination of mycotoxins in foods: current state of analytical methods and limitations. Appl. Microbiol. Biotechnol. 86, 1595–1612.
Kotova, K., Hussain, M., Mustafa, G., Lieberzeit, P.A., 2013. MIP sensors on the way to biotech applications: targeting selectivity. Sens. Actuat. B-Chem. 189, 199–202.
Kuang, H., Chen, W., Xu, D., Xu, L., Zhu, Y., Liu, L., Chu, H., Peng, C., Xu, C., Zhu, Z., 2010. Fabricated aptamer-based electrochemical "signal-off" sensor of ochratoxin A. Biosens. Bioelectron. 26, 710–716.
Kuila, T., Bose, S., Khanra, P., Mishra, A.K., Kim, N.H., Lee, J.H., 2011. Recent advances in graphene-based biosensors. Biosens. Bioelectron. 26, 4637–4648.
Labib, M., Zamay, A.S., Kolovskay, O.S., Reshetneva, I.T., Zamay, G.S., Kibbee, R.J., Sattar, S.A., Zamay, T.N., Berezovski, M.V., 2012a. Aptamer-based viability impedimetric sensor for bacteria. Anal. Chem. 84, 8966–8969.
Labib, M., Zamay, A.S., Kolovskaya, O.S., Reshetneva, I.T., Zamay, G.S., Kibbee, R.J., Sattar, S.A., Zamay, T.N., Berezovski, M.V., 2012b. Aptamer-based impedimetric sensor for bacterial typing. Anal. Chem. 84, 8114–8117.
Larou, E., Yiakoumettis, I., Kaltsas, G., Petropoulos, A., Skandamis, P., Kintzios, S., 2013. High throughput cellular biosensor for the ultra-sensitive, ultra-rapid detection of aflatoxin M1. Food Control 29, 208–212.
Lattanzio, V.M.T., Nivarlet, N., Lippolis, V., Gatta, S.D., Huet, A.-C., Delahaut, P., Granier, B., Visconti, A., 2012. Multiplex dipstick immunoassay for semi-quantitative determination of Fusarium mycotoxins in cereals. Anal. Chim. Acta. 718, 99–108.
Law, J.W.-F., Ab Mutalib, N.-S., Chan, K.-G., Lee, L.-H., 2015. Rapid methods for the detection of foodborne bacterial pathogens: principles, applications, advantages and limitations. Frontiers Microbiol. 5, 770.
Lee, S., Kim, G., Moon, J., 2013. Performance improvement of the one-dot lateral flow immunoassay for aflatoxin B1 by using a smartphone-based reading system. Sensors 13, 5109–5116.
Lerner, M.B., Goldsmith, B.R., McMillon, R., Dailey, J., Pillai, S., Singh, S.R., Johnson, A.T.C., 2011. A carbon nanotube immunosensor for *Salmonella*. AIP Advances 1, 042127.
Li, S., Chen, J., Cao, H., Yao, D., Liu, D., 2011. Amperometric biosensor for aflatoxin B1 based on aflatoxin-oxidase immobilized on multiwalled carbon nanotubes. Food Control 22, 43–49.
Li, Y., Liu, X., Lin, Z., 2012a. Recent developments and applications of surface plasmon resonance biosensors for the detection of mycotoxins in foodstuffs. Food Chem. 132, 1549–1554.

Li, Y., Cheng, P., Gong, J., Fang, L., Deng, J., Liang, W., Zheng, J., 2012b. Amperometric immunosensor for the detection of *Escherichia coli* O157:H7 in food specimens. Anal. Biochem. 421, 227–233.

Li, Y., Fang, L., Cheng, P., Deng, J., Jiang, L., Huang, H., Zheng, J., 2013. An electrochemical immunosensor for sensitive detection of *Escherichia coli* O157:H7 using C_{60} based biocompatible platform and enzyme functionalized Pt nanochains tracing tag. Biosens. Bioelectron. 49, 485–491.

Liébana, S., Lermo, A., Campoy, S., Cortés, M.P., Alegret, S., Pividori, M.I., 2009. Rapid detection of *Salmonella* in milk by electrochemical magneto-immunosensing. Biosens. Bioelectron. 25, 510–513.

Lin, Y.-H., Chen, S.-H., Chuang, Y.-C., Lu, Y.-C., Shen, T.Y., Chang, C.A., Lin, C.-S., 2008. Disposable amperometric immunosensing strips fabricated by Au nanoparticles-modified screen-printed carbon electrodes for the detection of foodborne pathogen *Escherichia coli* O157:H7. Biosens. Bioelectron. 23 (2008), 1832–1837.

Linting, Z., Ruiyi, L., Zaijun, L., Qianfang, X., Yinjun, F., Junkang, L., 2012. An immunosensor for ultrasensitive detection of aflatoxin B1 with an enhanced electrochemical performance based on graphene/conducting polymer/gold nanoparticles/the ionic liquid composite film on modified gold electrode with electrodeposition. Sens. Actuat. B-Chem. 174, 359–365.

Liu, Y., Qin, Z., Wu, X., Jiang, H., 2006. Immune-biosensor for aflatoxin B1 based bio-electrocatalytic reaction on micro-comb electrode. Biochem. Eng. J. 32, 211–217.

Liu, X.-P., Deng, Y.-J., Jin, X.-Y., Chen, L.-G., Jiang, J.-H., Shen, G.-L., Yu, R.-Q., 2009. Ultrasensitive electrochemical immunosensor for ochratoxin A using gold colloid-mediated hapten immobilization. Anal. Biochem. 389, 63–68.

Liu, L., Chao, Y., Cao, W., Wang, Y., Luo, C., Pang, X., Fan, D., Wei, Q., 2014a. A label-free amperometric immunosensor for detection of zearalenone based on trimetallic Au-core/AgPt-shell nanorattles and mesoporous carbon. Anal. Chim. Acta. 847, 29–36.

Liu, J., Hu, Y., Zhang, T., 2014b. Highly sensitive detection of zearalenone in feed samples using competitive surface-enhanced Raman scattering immunoassay. J. Agric. Food Chem. 62, 8325–8332.

Mandal, P.K., Biswas, A.K., Choi, K., Pal, U.K., 2011. Methods for rapid detection of foodborne pathogens: an overview. Am. J. Food. Technol. 6, 87–102.

Maragos, C.M., 2012. Signal amplification using colloidal gold in a biolayer interferometry-based immunosensor for the mycotoxin deoxynivalenol. Food Addit. Contam. A 29, 1108–1117.

Mascini, M., 2008. Aptamers and their applications. Anal. Bioanal. Chem. 390, 987–988.

Masoomi, L., Sadeghi, O., Banitaba, M.H., Shahrjerdi, A., Davarani, S.S.H., 2013. A nonenzymatic nanomagnetic electro-immunosensor for determination of Aflatoxin B1 as a model antigen. Sens. Actuat. B-Chem. 177, 1122–1127.

Maurer, J.J., 2011. Rapid detection and limitations of molecular techniques. Annu. Rev. Food Sci. Technol. 2, 259–279.

McGrath, T.F., Elliott, C.T., Fodey, T.L., 2012. Biosensors for the analysis of microbiological and chemical contaminants in food. Anal. Bioanal. Chem. 403, 75–92.

Mehrvar, M., Abdi, M., 2004. Recent developments, characteristics, and potential applications of electrochemical biosensors. Anal. Sci. 20, 1113–1126.

Melde, B.J., Johnson, B.J., 2010. Mesoporous materials in sensing: morphology and functionality at the meso-interface. Anal. Bioanal. Chem. 398, 1565–1573.

Mendonça, M., Conrad, N.L., Conceição, F.R., Moreira, Â.N., da Silva, W.P., Aleixo, J.A.G., Bhunia, A.K., 2012. Highly specific fiber optic immunosensor coupled with immunomagnetic separation for detection of low levels of *Listeria monocytogenes* and *L. ivanovii*. BMC Microbiology 12, 275.

Meneely, J.P., Ricci, F., van Egmond, H.P., Elliott, C.T., 2011. Current methods of analysis for the determination of trichothecene mycotoxins in food. Trends in Anal. Chem. 30, 192–203.

Mitterbauer, R., Weindrofer, H., Safaie, N., Krska, R., Lemmens, M., Ruckenbauer, P., Kuchler, K., Adam, G., 2003. A sensitive and inexpensive yeast bioassay for the ycotoxin earalenone and other compounds with estrogenic activity. Appl. Environ. Microbiol. 69, 805–811.

Molinelli, A., Grossalber, K., Führer, M., Baumgartner, S., Sulyok, M., Krska, R., 2008. Development of qualitative and semiquantitative immunoassay-based rapid strip tests for the detection of T-2 toxin in wheat and oat. J. Agric. Food Chem. 56, 2589–2594.

Nguyen, B.H., Tran, L.D., Do, Q.P., Nguyen, H.L., Tran, N.H., Nguyen, P.X., 2013. Label-free detection of aflatoxin M1 with electrochemical Fe_3O_4/polyaniline-based aptasensor. Mater. Sci. Eng. C Mater. Biol. Appl. 33, 2229–2234.

Nguyen, P.-D., Tran, T.B., Nguyen, D.T.X., Min, J., 2014. Magnetic silica nanotube-assisted impedimetric immunosensor for the separation and label-free detection of *Salmonella typhimurium*. Sens. Actuat. B-Chem. 197, 314–320.

Pacheco, J.G., Castro, M., Machado, S., Barroso, M.F., Nouws, H.P.A., Delerue-Matos, C., 2015. Molecularly imprinted electrochemical sensor for ochratoxin A detection in food samples. Sens. Actuat. B-Chem. 215 (2015), 107–112.

Pandey, C.M., Sumana, G., Tiwari, I., 2014. Nanostructuring of hierarchical 3D cystine flowers for high-performance electrochemical immunosensor. Biosens. Bioelectron. 61, 328–335.

Panini, N.V., Bertolino, F.A., Salinas, E., Messina, G.A., Raba, J., 2010. Zearalenone determination in corn silage samples using an immunosensor in a continuous-flow/stopped-flow systems. Biochem. Eng. J. 51, 7–13.

Park, C.W., Yang, J.-H., Ah, C.S., Ahn, C.-G., Choi, Y.H., Chung, K.H., Kim, W.-J., Sung, G.J., 2011. Toxin detection by Si photosensitive biosensors with a new measurement scheme. Biosens. Bioelectron. 29, 219–223.

Park, J.H., Kim, Y.-P., Kim, I.-H., Ko, S., 2014. Rapid detection of aflatoxin B1 by a bifunctional protein cross-linker based surface plasmon resonance biosensor. Food Control 36, 183–190.

Peraica, M., Radić, B., Lucić, A., Pavlović, M., 1999. Toxic effects of mycotoxins in humans. WHO Bull. 77, 754–766.

Perrotta, P.R., Arevalo, F.J., Vettorazzi, N.R., Zon, M.A., Fernandez, H., 2012. Development of a very sensitive electrochemical magneto immunosensor for the direct determination of ochratoxin A in red wine. Sens. Actuat. B-Chem. 162, 327–333.

Pohanka, M., 2013. Spectrophotomeric assay of aflatoxin B1 using acetylcholinesterase immobilized on standard microplates. Anal. Lett. 46, 1306–1315.

Prabhakar, N., Matharu, Z., Malhotra, B.D., 2011. Polyaniline Langmuir-Blodgett film-based aptasensor for ochratoxin A detection. Biosens. Bioelectron. 26, 4006–4011.

Prieto-Simón, B., Campàs, M., 2009. Immunochemical tools for mycotoxin detection in food. Monatsh. Chem. 140, 915–920.

Punbusayakul, N., Talapatra, S., Ajayan, P.M., Surareungchai, W., 2013. Label-free as-grown double wall carbon nanotubes bundles for *Salmonella typhimurium* immunoassay. Chem. Central J. 7, 102.

Qian, J., Jiang, L., Yang, X., Yan, Y., Mao, H., Wang, K., 2014. Highly sensitive impedimetric aptasensor based on covalent binding of gold nanoparticles on reduced graphene oxide with good dispersity and high density. Analyst 139, 5587–5593.

Rajesh, Ahuja, T., Kumar, D., 2009. Recent progress in the development of nano-structured conducting polymers/nanocomposites for sensor applications. Sens. Actuat. B-Chem. 136, 275–286.

Ravindranath, S.P., Wang, Y., Irudayaraj, J., 2011. SERS driven cross-platform based multiplex pathogen detection. Sens. Actuat. B-Chem. 152, 183–190.

Rejeb, I.B., Arduini, F., Arvinte, A., Amine, A., Gargouri, M., Micheli, L., Bala, C., Moscone, D., Palleschi, G., 2009. Development of a bio-electrochemical assay for AFB1 detection in olive oil. Biosens. Bioelectron. 24, 1962–1968.

Rhouati, A., Hayat, A., Hernandez, D.B., Meraihi, Z., Munoz, R., Marty, J.-L., 2013. Development of an automated flow-based electrochemical aptasensor for on-line detection of Ochratoxin A. Sens. Actuat. B-Chem. 176, 1160–1166.

Ricciardi, C., Castagna, R., Ferrante, I., Frascella, F., Marasso, S.L., Ricci, G., Canavese, G., Lor, A., Prelle, A., Gullino, M.L., Spadaro, D., 2013a. Development of a microcantilever-based immunosensing method for mycotoxin detection. Biosens. Bioelectron. 40, 233–239.

Ricciardi, C., Castagna, R., Ferrante, I., Frascella, F., Marasso, S.L., Ricci, A., Canavese, G., Loré, A., Prelle, A., Gullino, M.L., Spadaro, D., 2013b. Development of a microcantilever-based immunosensing method for mycotoxin detection. Biosens. Bioelectron. 40, 233–239.

Rocha, O., Ansari, K., Doohan, F.M., 2005. Effects of trichothecene mycotoxins on eukaryotic cells: a review. Food Addit. Contam. 22, 369–378.

Rocha-Santos, T.A.P., 2014. Sensors and biosensors based on magnetic nanoparticles. Trends in Anal. Chem. 62, 28–36.

Salam, F., Uludag, Y., Tothill, I.E., 2013. Real-time and sensitive detection of *Salmonella typhimurium* using an automated quartz crystal microbalance (QCM) instrument with nanoparticles amplification. Talanta 115, 761–767.

Sapsford, K.E., Ngundi, M.M., Moore, M.H., Lassman, M.E., Shriver-Lake, L.C., Taitt, C.R., Ligler, F.S., 2006. Rapid detection of foodborne contaminants using an array biosensor. Sens. Actuat. Chem. 113, 599–607.

Satija, J., Sai, V.V.R., Mukherji, S., 2011. Dendrimers in biosensors: concept and applications. J. Mater. Chem. 21, 14367–14386.

Scallan, E., Hoekstra, R.M., Angulo, F.J., Tauxe, R.V., Widdowson, M.A., Roy, S.L., Jones, J.L., Griffin, P.M., 2011. Foodborne illness acquired in the United States: major pathogens. Emerg. Infect. Dis. 17, 7–15.

Scognamiglio, V., Arduini, F., Palleschi, G., Rea, G., 2014. Biosensing technology for sustainable food safety. Trends Anal. Chem. 62, 1–10.

Sekhon, B.S., 2014. Nanotechnology in agri-food production: an overview. Nanotechnol. Sci. Appl. 7, 31–53.

Severgnini, M., Cremonesi, P., Consolandi, C., De Bellis, G., Castiglioni, B., 2010. Advances in DNA microarray technology for the detection of foodborne pathogens. Food Bioprocess Technol. 4, 936–953.

Shen, Z.-Q., Wang, J.-F., Qiu, Z.-G., Jin, M., Wang, X.-W., Chen, Z.-L., Li, J.-W., Cao, F.-H., 2011. QCM immunosensor detection of *Escherichia coli* O157:H7 based on beacon immunomagnetic nanoparticles and catalytic growth of colloidal gold. Biosens. Bioelectron. 26, 3376–3381.

Sheng, L., Ren, J., Miao, Y., Wang, J., Wang, E., 2011. PVP-coated graphene oxide for selective determination of ochratoxin A via quenching fluorescence of free aptamer. Biosens. Bioelectron. 26, 3494–3499.

Shephard, G.S., Berthiller, F., Burdaspal, P.A., Crews, C., Jonker, M.A., Krska, P., McDonald, S., Malone, R.J., Maragos, C., Sabino, M., Solfrizzo, M., Van Egmond, H.P., Whitaker, T.B., 2012. Developments in mycotoxin analysis: an update for 2010–2011. World Mycotoxin J. 5, 3–30.

Shim, W.-B., Mun, H., Joung, H.-A., Ofori, J.A., Chung, D.-H., Kim, M.-G., 2014. Chemiluminescence competitive aptamer assay for the detection of aflatoxin B1 in corn samples. Food Control 36, 30–35.

Singh, C., Srivastavaa, S., Ali, Md.A., Gupta, T.K., Sumana, G., Srivastava, A., Mathur, R.B., Malhotra, B.D., 2013. Carboxylated multiwalled carbon nanotubes–based biosensor for aflatoxin detection. Sens. Actuat. B-Chem. 185, 258–264.

So, H.-M., Park, D.-W., Jeon, E.-K., Kim, Y.-H., Kim, B.S., Lee, C.-K., Choi, S.Y., Kim, S.C., Chang, H., Lee, J.-O, 2008. Detection and titer estimation of *Escherichia coli* using aptamer-functionalized single-walled carbon-nanotube field-effect transistors. Small 4, 197–201.

Song, S., Liu, N., Zhao, Z., Ediage, E.N., Wu, S., Sun, C., De Saeger, S., Wu, A., 2014. Multiplex lateral flow immunoassay for mycotoxin determination. Anal. Chem. 86, 4995–5001.

Spinella, K., Mosiello, L., Palleschi, G., Vitali, F., 2014. Development of a QCM (quartz crystal microbalance) biosensor for detection of mycotoxins. Lect. Notes Elect. Eng. 268, 195–198.

Srivastava, S., Abraham, S., Singh, C., Ali, Md. A., Srivastava, A., Sumana, A., Malhotra, B.D., 2015. Protein conjugated carboxylated gold@reduced graphene oxide for aflatoxin B1 detection. RSC Adv. 5, 5406–5414.

Starodub, N.F., Ogorodniichuk, Y.A., Sitnik, Y.A., Slishik, N.F., 2012. Biosensors for the control of some toxins, viral and microbial infections to prevent actions of bioterrorists. In: Nikolelis, D.P. (Ed.), Portable Chemical Sensors: Weapons Against Bioterrorism, NATO Science for Peace and Security Series A: Chemistry and Biology. Springer, Amsterdam, pp. 95–117.

Stepurska, K.V., Soldatkin, O.O., Kucherenko, I.S., Arkhypova, V.M., Dzyadevych, S.V., Soldatkin, A.P., 2015. Feasibility of application of conductometric biosensor based on acetylcholinesterase for the inhibitory analysis of toxic compounds of different nature. Anal. Chim. Acta. 854, 161–168.

Su, L., Jia, W., Hou, C., Lei, Y., 2011. Microbial biosensors: a review. Biosens. Bioelectron. 26, 1788–1799.

Suryanarayanan, V., Wu, C.-T., Ho, K.-C., 2010. Molecularly imprinted electrochemical sensors. Electroanalysis 22, 1795–1811.

Tong, P., Zhang, L., Xu, J.-J., Chen, H.Y., 2011. Simply amplified electrochemical aptasensor of Ochratoxin A based on exonuclease-catalyzed target recycling. Biosens. Bioelectron. 29, 97–101.

Välimaa, A.L., Kivistö, A.T., Leskinen, P.I., Karp, M.T., 2010. A novel biosensor for the detection of zearalenone family mycotoxins in milk. J. Microbiol. Methods 80, 44–48.

Vamvakaki, V., Chaniotakis, N.A., 2007. Carbon nanostructures as transducers in biosensors. Sens. Actuat. B-Chem. 126, 193–197.

van der Gaag, B., Spath, S., Dietrich, H., Stigter, E., Boonzaaijer, G., van Osenbruggen, T., Koopal, K., 2003. Biosensors and multiple mycotoxin analysis. Food Control 14, 251–254.

Vidal, J.C., Bonel, L., Ezquerra, A., Duato, P., Castillo, J.R., 2012. An electrochemical immunosensor for ochratoxin A determination in wines based on a monoclonal antibody and paramagnetic microbeads. Anal. Bioanal. Chem. 403, 1585–1593.

Viswanathan, S., Rani, C., Ho, J.A., 2012. Electrochemical immunosensor for multiplexed detection of food-borne pathogens using nanocrystal bioconjugates and MWCNT screen-printed electrode. Talanta 94, 315–319.

Voller, A., Bartlett, A., Bidwell, D.E., 1978. Enzyme immunoassays with special reference to ELISA techniques. J. Clin. Pathol. 31, 507–520.

Wang, F., Hu, S., 2009. Electrochemical sensors based on metal and semiconductor nanoparticles. Microchim. Acta. 165, 1–22.

Wang, R., Ruan, C., Kanayeva, D., Lassiter, K., Li, Y., 2008. TiO_2 nanowire bundle microelectrode based impedance immunosensor for rapid and sensitive detection of *Listeria monocytogenes*. Nano Lett. 8, 2625–2631.

Wang, Z., Duan, N., Hun, X., Wu, S., 2010. Electrochemiluminescent aptamer biosensor for the determination of ochratoxin A at a gold-nanoparticles-modified gold electrode using *N*-(aminobutyl)-*N*-ethylisoluminol as a luminescent label. Anal. Bioanal. Chem. 398, 2125–2132.

Wang, Y.-X., Ye, Z.-Z., Si, C.-Y., Ying, Y.-B., 2012. Application of aptamer based biosensors for detection of pathogenic microorganisms. Chin. J. Anal. Chem. 40, 634–642.

Wang, Y., Ping, J., Ye, Z., Wu, J., Ying, Y., 2013. Impedimetric immunosensor based on gold nanoparticles modified graphene paper for label-free detection of *Escherichia coli* O157:H7. Biosens. Bioelectron. 49, 492–498.

Wang, D., Dou, W., Zhao, D., Chen, Y., 2014. Immunosensor based on electrodeposition of gold-nanoparticles and ionic liquid composite for detection of *Salmonella pullorum*. J. Microbiol. Methods 106, 110–118.

Wang, D., Hu, W., Xiong, Y., Xu, Y., Li, C.M., 2015a. Multifunctionalized reduced graphene oxide-doped polypyrrole/pyrrolepropylic acid nanocomposite impedimetric immunosensor to ultra-sensitively detect small molecular aflatoxin B1. Biosens. Bioelectron. 63, 185–189.

Wang, R., Xiang, Y., Zhou, X., Liu, L., Shi, H., 2015b. A reusable aptamer-based evanescent wave all-fiber biosensor for highly sensitive detection of Ochratoxin A. Biosens. Bioelectron. 66, 11–18.

Wang, S., Zhang, Y., An, W., Wei, Y., Liu, N., Chen, Y., Shuang, S., 2015c. Magnetic relaxation switch immunosensor for the rapid detection of the foodborne pathogen *Salmonella enterica* in milk samples. Food Control 55, 43–48.

Wei, Y., Qiu, L., Owen, C., Lai, E.P.C., 2007. Encapsulation of quantum dots and carbon nanotubes with polypyrrole in a syringe needle for automated molecularly imprinted solid phase pre-concentration of ochratoxin A in red wine analysis. Sens. Instrum. Food Qual. 1, 133–141.

Welch, C.M., Compton, R.G., 2006. The use of nanoparticles in electroanalysis: a review. Anal. Bioanal. Chem. 384, 601–619.

Wilson, R., 2008. The use of gold nanoparticles in diagnostics and detection. Chem. Soc. Rev. 37, 2028–2045.

Wood, G.E., 1992. Mycotoxins in foods and feeds in the United States. J. Anim. Sci. 70, 3941–3949.

World Health Organization, 2002. Evaluation of certain mycotoxins in food. 56th Report of the Joint FAO/WHO Expert Committee on Food Additives, WHO Technical Report Series 906, Geneva, Switzerland, p. 70.

Wu, W., Zhang, J., Zheng, M., Zhong, Y., Yang, J., Zhao, Y., Wu, W., Ye, W., Wen, J., Wang, Q., Lu, J., 2012a. An aptamer-based biosensor for colorimetric detection of *Escherichia coli* O157:H7. PLOS One 7, e48999.

Wu, W., Li, M., Wang, Y., Ouyang, H., Wang, L., Li, C., Cao, Y., Meng, Q., Lu, J., 2012b. Aptasensors for rapid detection of *Escherichia coli* O157:H7 and *Salmonella typhimurium*. Nanoscale Res. Lett. 7, 658.

Wu, X., Gao, S., Wang, J.-S., Wang, H., Huang, Y.-W., Zhao, Y., 2012c. The surface-enhanced Raman spectra of aflatoxins: spectral analysis, density functional theory calculation, detection and differentiation. Analyst 137, 4226–4234.

Xiao, X., Yang, X., Liu, T., Chen, Z., Chen, L., Li, H., Deng, L., 2007. Preparing a highly specific inert immunomolecular-magnetic beads for rapid detection and separation of *S. aureus* and group G *Streptococcus*. Appl. Microbiol. Biotechnol. 75, 1209–1216.

Yang, G.-J., Huang, J.-L., Meng, W.-J., Shen, M., Jiao, X.-A., 2009. A reusable capacitive immunosensor for detection of *Salmonella* spp. based on grafted ethylene diamine and self-assembled gold nanoparticle monolayers. Anal. Chim. Acta. 647, 159–166.

Yang, C., Wang, Y., Marty, J.-L., Yang, X., 2011. Aptamer-based colorimetric biosensing of ochratoxin A using unmodified gold nanoparticles indicator. Biosens. Bioelectron. 26, 2724–2727.

Yang, X.-H., Kong, W.-J., Yang, M.-H., Zhao, M., Ouyang, Z., 2013. Application of aptamer identification technology in rapid analysis of mycotoxins. Chin. J. Anal. Chem. 41, 297–306.

Yazar, S., Omurtag, G.Z., 2008. Fumonisins, trichothecenes and zearalenone in cereals. Int. J. Mol. Sci. 9, 2062–2090.

Yu, J.C.C., Lai, E.P.C., 2006. Molecularly imprinted polypyrrole modified carbon nanotubes on stainless steel frit for selective micro solid phase pre-concentration of ochratoxin A. React. Funct. Polym. 66, 702–711.

Yu, J.C.C., Lai, E.P.C., 2010. Molecularly imprinted polymers for ochratoxin A extraction and analysis. Toxins 2, 1536–1553.

Yu, L., Zhang, Y., Hu, C., Wu, H., Yang, Y., Huang, C., Jia, N., 2015. Highly sensitive electrochemical impedance spectroscopy immunosensor for the detection of AFB1 in olive oil. Food Chem. 176, 22–26.

Yuan, J., Wu, S., Duan, N., Ma, X., Xia, Y., Chen, J., Ding, Z., Wang, Z., 2014. A sensitive gold nanoparticle-based colorimetric aptasensor for *Staphylococcus aureus*. Talanta 127, 163–168.

Zachetti, V.G.L., Granero, A.M., Robledo, S.N., Zon, M.A., Fernández, H., 2013. Development of an amperometric biosensor based on peroxidases to quantify citrinin in rice samples. Bioelectrochem. 91, 37–43.

Zamfir, L.-G., Geana, I., Bourigua, S., Rotariu, L., Bala, C., Errachid, F., Jaffrezic-Renault, N., 2011. Highly sensitive label-free immunosensor for ochratoxin A based on functionalized magnetic nanoparticles and EIS/SPR detection. Sens. Actuat. B-chem. 159, 178–184.

Zelada-Guillén, G.A., Riu, J., Düzgün, A., Rius, F.X., 2009. Immediate detection of living bacteria at ultralow concentrations using a carbon nanotube based potentiometric aptasensor. Angew. Chem. Int. Ed. 48, 7334–7337.

Zelada-Guillen, G.A., Bhosale, S.V., Riu, J., Rius, F.X., 2012. Real-time potentiometric detection of bacteria in complex samples. Anal. Chem. 82, 9254–9260.

Zhang, D.H., Li, P.W., Zhang, Q., Yang, Y., Zhang, W., Guan, B., Ding, X.X., 2012a. Extract-free immunochromatographic assay for on-site tests of aflatoxin M1 in milk. Anal. Methods 4, 3307–3313.

Zhang, X., Wang, D., Yang, D., Li, S., Shen, Z., 2012b. Electronic detection of *Escherichia coli* O157:H7 using single-walled carbon nanotubes field-effect transistor biosensor. Engineering 5, 94–98.

Zhang, J., Zhang, X., Yang, G., Chen, J., Wang, S., 2013. A signal-on fluorescent aptasensor based on Tb^{3+} and structure-switching aptamer for label-free detection of Ochratoxin A in wheat. Biosens. Bioelectron. 41, 704–709.

Zhang, X., Lu, W., Han, E., Wang, S., Shen, J., 2014. Hybrid nanostructure-based immunosensing for electrochemicall assay of *Escherichia coli* as indicator bacteria relevant to the recycling of urban sludge. Electrochim. Acta. 141, 384–390.

Zhao, G., Xing, F., Deng, S., 2007. A disposable amperometric enzyme immunosensor for rapid detection of *Vibrio parahaemolyticus* in food based on agarose/nano-Au membrane and screen-printed electrode. Electrochem. Commun. 9, 1263–1268.

Zhao, Y., Wang, P., Wang, F., Zhou, H., Li, W., Yue, J., Ha, Y., 2012. A novel biosensor regulated by the rotator of F0F1-ATPase to detect deoxynivalenol rapidly. Biochem. Biophys. Res. Commun. 423, 195–199.

Zhao, X., Lin, C.-W., Oh, D.H., 2014. Advances in rapid detection methods for foodborne pathogens. J. Microbiol. Biotechnol. 24, 297–312.

Zheng, M.Z., Richard, J.L., Binder, J., 2006. A review of rapid methods for the analysis of mycotoxins. Mycopathologia 161, 261–273.

Zhilei, W., Xiulan, S., Zaijun, L., Yinjun, F., Guoxiao, R., Yaru, H., Junkang, L., 2011. Highly sensitive deoxynivalenol immunosensor based on a glassy carbon electrode modified with a fullerene/ferrocene/ionic liquid composite. Microchim. Acta. 172, 365–371.

3

BIOCONJUGATED NANOMATERIALS FOR MONITORING FOOD CONTAMINATION

Babar Hussain*, Meral Yüce, Naimat Ullah*, Hikmet Budak*,****

**Sabanci University, Faculty of Engineering and Natural Sciences, Istanbul, Turkey; **Sabanci University, Nanotechnology Research and Application Center, Istanbul, Turkey*

1 Introduction

Food contaminants, especially pathogens, often go unchecked due to the presence of small quantities initially, and cause diseases in the consumers in later stages due to rapid replication during food transportation or the food supply chain. For example, *Staphylococcus aureus* is a foodborne bacterium that causes more than 0.5 million infections in the USA. Symptoms range from skin burn to pus formation in infected cells (Chang et al., 2013). Pathogens can transmit into the food through contaminated water, nearby soil, and even air. Another problem associated with food safety is the presence of toxins, heavy metals, and allergens in food, which also cause health problems in consumers. Several food toxins have been identified that are incorporated in foodstuff during growth, storage, or processing (Pollet et al., 2011). For example, selenium enters the food chain when microbes convert selenium from inorganic to organic form (Sors et al., 2005). Excessive uptake of selenium causes selenium toxicity or selenosis. This problem has been reported on a large scale in Chinese seleniferous region, and deformity, nail and hair loss, skin lesions, fatigue, garlic-like breath odor, diarrhea, and irritability are the most common symptoms of this disease.

Incorporation of contaminants from the environment is not the only way through which toxins are added to food. Several food

Nanobiosensors. http://dx.doi.org/10.1016/B978-0-12-804301-1.00003-5

toxins are found in crop plants that are harmful to human health, for example, prussic or hydrocyanic acid is constituted in peach, apple, and cherry plant tissues when cyanogenic glycosides are damaged, and come in contact with emulsion enzymes or beta-glycosidase. Consumption of food carrying this acid can result in respiratory arrest, trembling, cardiac arrest, and muscle noncoordination depending on its dose (Dolan et al., 2010). The presence of goiter genesis chemical in peaches, canola, cauliflower spinach, sweet potatoes, cassava, peas, and peanuts can cause swelling of throat or goiter if eaten uncooked. The erucic acid in rape, the most commonly used edible oil crop, can cause cholesterol accumulation in muscles leading to a heart problem and lectins in legumes can even cause agglutination of blood cells. Similarly, higher doses of oxalic acid in many leafy vegetables such as celery, broccoli, cabbage, turnip, beans, coffee, beans, peas, tomatoes, and potatoes can lead to vomiting, kidney stones, diarrhea, and blood clotting (Dolan et al., 2010).

Food adulterants are an issue that arises due to the professional dishonesty of some traders or farmers who mix inferior quality, some wild-type or closely related species' products with food in order to get more profit. On the other hand, this is a potential threat to food safety as some of the wild relatives of crop plants whose seeds or product is mixed with food containing potential harmful chemicals for human health. For example, in turmeric, moong bean, and chickpea pulse, methanol is added to add a yellow color to these foods, but is carcinogenic and causes stomach disorders. Similarly, malachite green (MG) is applied to green peas and green chilies to make their skin glow and is also carcinogenic (Lakshmi and Pradesh, 2012). Adding to the problem is the fact that nobody can identify these toxins, allergens, and adulterants with the naked eye, thus, the symptoms appear only after usage of the contaminated food.

From the earlier discussion, it is evident that food pathogens, allergens, and toxins are a serious threat to human health and cause many diseases and complications in consumers if they are part of the food. Therefore, rapid and accurate detection of the food contaminants is a crucial step toward the food safety and public health. Flow cytometry, polymerase chain reaction (PCR) and antibody-based immunoassay platforms (ELISA, LFI, IMS) have been conventionally used for the detection of food contamination for years (Zhao et al., 2014a).

Recently, aptamers have emerged as potential detection elements that can identify different target molecules ranging from small molecules to whole cells with an affinity and specificity similar to monoclonal antibodies. Aptamers are selected short oligonucleotides or peptide molecules that act as ligands due to their high affinity and specificity toward their targets. Aptamers can be used as probes by labeling, and their low cost of production makes them an excellent choice for sensing applications over the conventional probes, antibodies. A diverse collection of aptamers for hundreds of various targets has been developed from the random pools using either traditional or the modified selection techniques. As the applications vary from medical science to food security, a broad range of analytical approaches has been developed for the purpose (Song et al., 2012). Aptamers have found applications in food contaminant detection, food packaging, and food processing industries, and food monitoring during trading at ports by quarantine departments.

Employment of aptamers in sensor devices is usually achieved through incorporation of nanomaterials, such as a gold nanoparticles, quantum dots, carbon nanotubes (CNTs), graphene, and upconversion nanoparticles into the sensor platform as the signal transducers. Nanotechnology has gained considerable attention from numerous disciplines owing to the exceptional physical and chemical properties of nanomaterials, such as high thermal conductivity and high surface-to-volume ratios (Nel et al., 2009; Geim and Novoselov, 2007). Coupling of nanomaterials with aptamers has laid the foundation for a new vision in sensing, in which aptamer is used as the sensing reagent or capture molecules while the nanomaterial is used as the signal converter or amplifier. In this chapter, we review the food safety monitoring techniques that are mainly based on the aptamers and aptamer-conjugated nanomaterials. The conventional techniques are also mentioned in the chapter in order to provide a comprehensive as well as comparative approach to the topic.

2 Aptamers

Aptamers are short oligonucleotide sequences (ssDNA, RNA, or peptide), which can acquire three-dimensional structures in solutions to bind with a variety of target molecules such as ions, proteins, drugs, cells, and even nanoparticles. Aptamers are similar to monoclonal antibodies regarding their binding affinities, but they offer some advantages over antibodies, as listed in Table 3.1, which make them suitable for analytical applications.

Aptamers are selected from randomized oligonucleotide libraries by repetitive cycles of binding, partition, recovery, and amplification steps through a well-defined in vitro selection method known as systematic evolution of ligands by exponential enrichment (SELEX). The artificially synthesized random oligonucleotide

Table 3.1 General Differences Between Aptamers and Antibodies

Characters	Aptamers	Antibodies
Selection	In vitro under a variety of conditions	In vivo limited to physiologic conditions
Nature	Short RNA, DNA, or peptide sequences	Protein in nature
Stability	Tolerate various environmental conditions	Have tendency to degrade upon heating
Production process	In a test tube	In well-equipped research and development facilities
Expenses	Priced based on the number of nucleotides and modifications	Laborious and expensive
Batch activity	Negligible batch to batch variation	Varies
Affinity	Varies	Strong
Size	Small	Large
Development time	Several weeks	Several months
Shelf life	Prolonged	Limited
Targets applicable	Applicable for a wide variety of targets	Limited to immunogenic targets
Modification	Convenient to modification on exact site	Difficult to adjust modification
Target sites	Aptatopes can be determined by the investigator	Epitopes determine the target sites
Environmental monitoring	Can be used for environmental monitoring	Cannot work for environment

pool is considered as the starting point for the selection of an aptamer through SELEX, which usually consists of 10^{14-15} unique sequences that have 30–80 nt central random region flanked by primer binding sites for the amplification step. Random oligonucleotide library of complex structures can be synthesized according to combinatorial chemistry principle to perform a variety of functions such as the production of high-affinity aptamers. Unnatural bases can be also incorporated into the random libraries to increase the final diversity and the affinity of the sequences. The affinity of the selected aptamers depends on the method, formation of the three-dimensional structure upon binding to the target molecule, as well as the target molecule itself. Unlike antibody production, aptamer screening process is completed in weeks, and no animal sacrifice is required for the production. SELEX technique has become the basic process for screening of aptamers directly in vitro against the targets including pathogen cells, small biomolecules, and proteins.

2.1 Aptamer Screening

The concept of joining nucleic acids to protein originated from when it was found that some RNAs bound with viral proteins with high affinity and specificity in adenovirus. In adenovirus, virus-associated short RNA–ligand binding led to translational regulation (O'Malley et al., 1986). This in vivo discovery was a first step toward its evolution to an in vitro selection procedure, which was developed in 1990 by two research groups independently. The aptamer screening procedure, SELEX, was described by Tuerk and Gold (1990), when they were studying the interactions between T4 DNA polymerase and natural ribosome binding site of the mRNA that encodes the enzyme. In the same year when SELEX was introduced by Tuerk and Gold (1990), Ellington and Szostak (1990) independently applied the same strategy to isolate target specific RNA sequences from a random RNA library, and then named them as "aptamers".

The basic aim of the selection process is to obtain the tightest binding sequence from the randomized pool through the exponential enrichment, which includes repetitive cycles of target binding, partition of the target oligonucleotide complex, elution of the target bound oligonucleotides, amplification of the target specific oligonucleotides, and conditioning for the next SELEX cycle. By these repetitive cycles of selection and amplification, the affinity and specificity of an initial random oligonucleotide pool for the target is enriched, meaning that the total complexity of the randomized pool is reduced to relatively few sequence motifs. RNA aptamer selection process begins after the ssDNA library is converted into a dsDNA library with the help of sense and antisense primers consisting of T7 promoter at 5′ end. The dsDNA is then in vitro transcribed with the assistance of T7 RNA polymerase to form a randomized RNA library to start SELEX process. The selected RNA of each round is reverse transcribed into DNA and amplified for the following round of selection (Thiel et al., 2011). An illustration of the standard SELEX procedure is presented in Fig. 3.1 (Yüce et al., 2015).

Additional steps can be inserted into each round of SELEX procedure to enhance further the stringency of the process. For example, negative-SELEX steps are performed to prevent the enrichment of nonspecific oligonucleotides during the selection, in which the pool is incubated with the naive immobilization surface

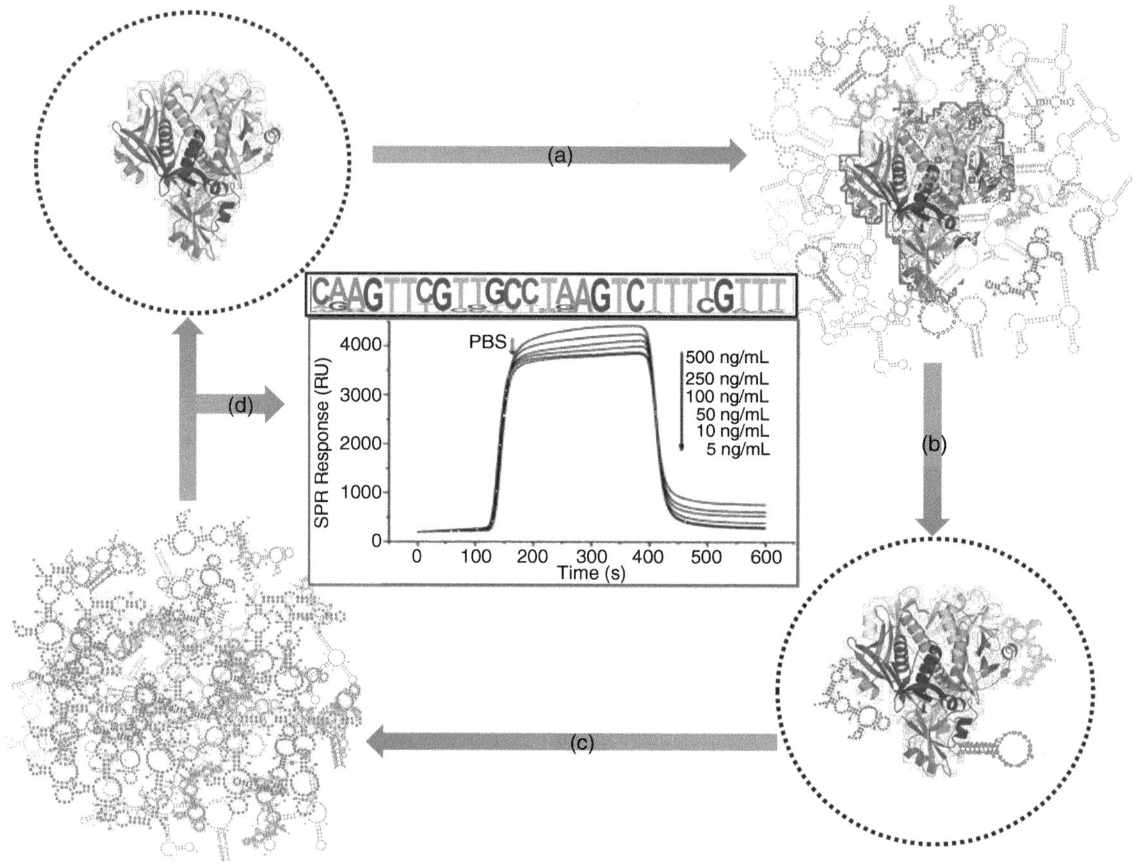

Figure 3.1. Illustration of a standard SELEX round. (a) Incubation of the target molecule with the random oligonucleotide library, (b) elution of the target bound oligonucleotides and elimination of the nonspecific binders, (c) amplification of the target bound oligonucleotides in which the tightly binding oligonucleotides dominate the pool, known as enrichment, (d) following X-cycles of the SELEX (a–c), the enriched pool is sequenced and analyzed for the consensus sequences whose binding abilities are tested through the well-known analytic methods such as surface plasmon resonance (Yüce et al., 2015). Copyright© 2015, Royal Society of Chemistry.

lacking the target of interest. The nonspecific oligonucleotides are those that bind to the immobilization surface (magnetic beads, columns, or chips) or blocking agents (BSA or ethanolamine) instead of the specified target. Counter-SELEX, on the other hand, can be applied to remove oligonucleotides that bind with the substrates that are structurally similar to the target of interest (ie, glycated hemoglobin, nonglycated hemoglobin).

Enrichment of the start pool is monitored over the course of selection to decide where to stop SELEX, which is followed by sequencing of the enriched final pool. Although Sanger sequencing method has been immensely used in SELEX, next generation sequencing (NGS) techniques have recently paved the way for a new era in aptamer sequencing because of the greatness of the received data allowing a thorough evaluation of the final pool so the SELEX process. In the current state, NGS data is processed using the tools such as AliBee (Sarell et al., 2014), Clustal W (Jing and Bowser, 2013), Clustal Omega (Wilson et al., 2014), MEME ChIP (Ma et al., 2014), mfold (Sung et al., 2014), RNAfold (Belter et al., 2014), or some local algorithms (Gupta et al., 2014). Processing of the NGS data from SELEX has been a general problem in the field due to lack of knowledge, expertise, and bioinformatics tools, however, there is a considerable progress being made to develop software, online prediction tools or scripts capable of sorting huge aptamer data sets, including, AptaMotif (Hoinka et al., 2012), AptaCluster (Hoinka et al., 2014), and Aptaligner (Lu et al., 2014).

Sequence analyzing is followed by the chemical synthesis of the selected sequences and the evaluation of the binding kinetics, which is typically identified by surface plasmon resonance, pull-down assays, or fluorescent-based techniques. Recently, microscale thermophoresis and backscattering interferometry methods have been also shown as valuable as the previous techniques to assess the ligand and analyte interactions (Olmsted et al., 2011; Seidel et al., 2014).

Since the inception, SELEX procedure has been substantially modified to increase the final specificity, binding affinity, and to reduce the labor, material consumption as well as the time spent for the process. Within the concept, magnetic bead-based selection, bead or chip-integrated microfluidic selection platforms, NGS-based screening procedures, surface plasmon resonance-based real-time selection, affinity column-based immobilization free selection and many other combined techniques have been introduced in the recent literature (Yüce et al., 2015). Among them, the magnetic bead-based selection has received the greatest attention due to ease of handling, ease of coupling with the target molecules (covalent or noncovalent), and the ability to be employed in automated and microfluidic based-selection platforms. For example, Tok and Fischer (2008) used only single magnetic bead functionalized with *Clostridium botulinum* neurotoxin-related targets Hc-peptide and the toxoid to be used in SELEX, in which they traced the single bead under the fluorescence stereomicroscope using fluorescently labelled random pool, and obtained target-specific aptamers with affinities in nanomolar range.

Postmodification of the selected aptamers can be performed to increase their stability, affinity, or shelve-live/serum half-life. Also, labeling with classical fluorophore molecules or fluorescent nanoparticles such as quantum dots and upconverting nanoparticles, allow their further utilization in biosensing and imaging applications as fluorescent probes. As the binding ability of the antibodies compromises after labeling or modification, their application in sensing is limited by their surviving capacity. Aptamers can be also modified with biotin or thiol tags via in vitro transcription or PCR on demand. Additionally, aptamers could tolerate the change in thermal conditions, enabling the recycle of the sensing element whereas antibodies go irreversible denaturation and aggregation upon heating that prevent the multiple uses of the biomaterial in sensing applications.

3 Conventional Methods for Food Contamination Detection

The conventional methods used for the identification of food contaminants include antibody-based immunoassays, different types of PCRs, and flow cytometry.

3.1 Immunoassay-Based Methods

The antibody-based contaminant detection is one of the most commonly used methods, and the fundamental principle of this method is the binding of a specific antibody to the target, followed by the antigen–antibody complex detection. Antibodies are produced in a living system in response to the specific pathogen, and they are being produced in large-scale by using different pathogens and toxins in laboratory animals. Antibodies are very specific and can identify a specific pathogen attached to the food in the presence of other pathogens, but they can only work if the pathogen gives stable expression of the antigen (Wacoo et al., 2014).

In an immunoassay, antibodies are usually attached to a solid base, and samples are added to see the interaction, followed by probe-based detection (Song et al., 2012). The major immunochemical assays are described in the subsequent sections.

3.1.1 Enzyme-Linked Immunosorbent Assay

Sandwich type enzyme-linked immunosorbent assay (ELISA) has the highest efficiency as it utilizes two antibodies to detect its target. In ELISA, the primary antibody is attached to ELISA plate, and antigen such as a pathogen or toxin is added and adhered to the antibody while unbound contaminant is removed. Now, a secondary antibody conjugated with the enzyme is added, which finds the antigen. In this way, the antigen is sandwiched between two antibodies and the antigen is detected by the change of substrate color in the presence of an enzyme (Zhao et al., 2014a). The most commonly used enzymes in ELISA include beta-galactosidase, alkaline phosphatase, and horseradish peroxidase (Yeni et al., 2014). Some foodborne pathogen and toxins that have been identified by ELISA include *Salmonella* spp. (Lee et al., 2015), staphylococcal enterotoxin (Cremonesi et al., 2007), *Escherichia coli* toxins, that is, enterotoxins and botulinum toxins (Aschfalk and Müller, 2011), *Vibrio parahaemolyticus* (Kumar et al., 2011), α, β, and ε toxin of *Clostridium perfringens* (Aschfalk and Müller, 2011), *E. coli* (Carvalho et al., 2014), *Campylobacter* spp. and *Listeria monocytogenes* (Gómez-govea et al., 2012). Several food types in which ELISA has provided successful detection include fruits, vegetable, fish, beef, pork, and cheese (Law et al., 2014).

3.1.2 Lateral Flow Immunoassay

Although ELISA has been the most popular immunoassay technique, it requires a skilled workforce and various equipments. Therefore, some simple, cheap, rapid, and yet reliable method of contaminant detection is needed. In recent years, LFI-based techniques such as immunofiltration, immune-chromatography, and dipsticks are becoming popular for food pathogen and mycotoxin detection (Wacoo et al., 2014). In LFI, the test sample streams along the immunoassay device comprising of four sections via capillary action. A colored reagent (antibody) loaded on gold particles is mixed with food sample solution containing contaminant while flowing through the tube. Most of LFI assays give visual results in 2–10 min, which is faster than ELISA (Gómez et al., 2010). Usually, immune-chromatographic strips are developed for quick and visual detection, and various food contaminants have been identified by using these strips such as avian influenza virus (Peng et al., 2008), *Streptococcus suis* serotype 2 (Ju et al., 2010), Porcine circovirus-2 (Jin et al., 2012), *Campylobacter jejuni* (Xu et al., 2013), *E. coli* (Sajid et al., 2015), *Salmonella* (Fang et al., 2014), *Riemerella anatipestifer* (Hou et al., 2015), and many other food contaminants. Nowadays, several commercial strips are available for rapid detection of food contaminants.

3.1.3 Immunomagnetic Separation

Immunomagnetic separation (IMS) method uses immunomagnetic beads (IMBs) for microbe identification. This assay is quite selective in nature that allows the growth of target pathogen while the growth of nontarget pathogens is suppressed. In this method, target cells are incubated with IMBs for 1 h, separated by a magnetic separator, followed by several cycles of washing to eliminate the nonspecific interactions. IMBs do not interact with the sample in assay and conjugation of various ranges of molecule to the magnetic beads provides an additional advantage (Zhao et al., 2014a). Unlike ELISA, which can take a number of days for identifying the pathogens, a highly efficient IMBs based fluorescence assay was developed which identified *E. coli* O157:H7 within 8 h (Decory et al., 2005). A unique way of this approach is coupling immune-affinity with mass spectrometry based on matrix-assisted laser desorption that helped in the detection of staphylococcal enterotoxin B (SEB) (Schlosser et al., 2007). Another unique method is to combine IMS with PCR for the detection of *Alicyclobacillus acidoterrestris* in apple juice. The immune-capture took 60 min with IMPs quantity at 2.5 mg mL^{-1} followed by pathogen detection with IMS-PCR that showed highly specific detection of *A. acidoterrestris* in 3–4 h only (Wang et al., 2013). However, IMS method does not produce a pure culture of microbes and requires to be coupled with other methods for reliable detection. Another disadvantage of this method is a requirement of large sample even up to 500 times more than the other assays during capture by sample recirculation to reduce detection time and enhance the sensitivity (Fedio et al., 2011).

3.2 PCR-Based Methods

Standard PCR, multiplex PCR, and quantitative PCR are the most commonly used nucleic acid amplification-based contaminant detection methods in food safety monitoring.

3.2.1 *Standard PCR*

In PCR, DNA strands are denatured at around 95°C followed by annealing of specific primers at 55–65°C, and finally extension step makes several copies of targeted DNA at 72°C by DNA polymerase enzyme, after several cycles of these three stages. The amplified DNA is separated by gel electrophoresis under voltage difference and action of negative charge, followed by DNA documentation as intensity bands through the gel staining. Reliability of this detection method can be enhanced by sequencing, which is quite an expensive method to date. However, this method is quite specific, sensitive, and rapid as compared to many assay-based detection methods (Khan et al., 2014).

The PCR-based assay detects the food pathogens by amplification of the genes coding for the toxins that are produced by a particular organism. For instance, *Salmonella* was identified in 150 samples of fish ($n = 6$), chicken ($n = 80$), and beef ($n = 64$) by 12-h length PCR. A total of 35 samples were positive by PCR against 32 by a culturing method with an evident advantage of less time consumption and along with 100 and 97.5% sensitivity and specificity, respectively. The amplification was done by Salm4 and Salm3 primers that amplified a 389-bp DNA fragment in *invA*, a conserved gene of *Salmonella* spp. (Hegazy AI, 2014). In another study, *Salmonella typhimurium* bacteria was identified in wheat, rice, snacks, juices, fish, and meal by PCR-based amplification of same *invA* gene (389 bp) by specific primers after 10 h preenrichment (Khan et al., 2014).

3.2.2 Multiplex PCR

Multiplex PCR is a particular kind of reaction that can amplify more than one locus linked to multiple microbes, hence, helps in simultaneous detection of various microorganisms by a single PCR reaction. This method is just similar to the previous method except that several specific primers are combined in a single PCR assay. The problems associated with multiplex PCR include interaction between different primers so primer designing and concentrations should be kept in mind while designing the primers in addition to a same annealing temperature (Zhou et al., 2013). Several multiplex PCR assays have been reported in which foodborne pathogens have been detected, for instance, a rapid multiplex PCR simultaneously detected five important pathogens including *L. monocytogenes, Shigella flexner, Salmonella enteritidis, E. coli* O157:H7, and *S. aureus* in artificially contaminated pork. This highly specific, efficient, and sensitive method also found that 80% meat samples were positive for these pathogens by using five primer sets specific to these microbes (Chen et al., 2012). A novel GeXP analyzer-based multiplex PCR was developed for detection of six pathogens including *L. monocytogenes, S. entries, E. coli* O157:H7, *C. jejuni, S. aureus,* and *Shigella* spp. simultaneously. They fused chimeric primers unique to both pMD19 plasmid and microbes, and PCR products were separated by capillary electrophoresis. DNA bands were visualized with fluorescence spectrophotometry while pMD19-T plasmid was used as a control in each reaction. This method did not need any enrichment and was specific, sensitive, rapid, and high throughput (Zhou et al., 2013). Recently, a triplex multiplex PCR identified three major raw poultry pathogens, that is, *Shigella, L. monocytogenes,* and *Salmonella* in 5.2, 6.2, and 26.9% of the samples studied respectively (Zhao et al., 2014b).

3.2.3 Quantitative PCR

Quantitative PCR (qPCR) or real-time PCR is an advanced reaction that can amplify and quantify or detect the target DNA simultaneously, thus, providing an advantage of observing the effect of environmental conditions on the microbe. As no postamplification treatments are required, it reduces the effort and time needed for food contamination detection (Law et al., 2014). The newly developed qPCR identified shiga toxin producing *E. coli* (STEC) in dairy products and minced beef with equal accuracy level to the methods mentioned earlier (Derzelle et al., 2011). A highly reliable and fast "most probable number (MPN) qPCR" was developed, which quantified *E. coli* O157:H7 and *Listeria monocytogenes* in fresh-cut vegetables. The method was validated by comparing the results with simple MPN method followed by biochemical assay and phenotyping. MPN-qPCR identified as low as 1 CFU g^{-1} of both pathogens within 48 h of artificial inoculation of minimally processed vegetables. Whereas, qPCR alone detected 101 CFU g^{-1} of both pathogens after 2-h enrichment as vegetables are perishable commodity, so such method requiring short time can help to monitor their food safety (Russo et al., 2014).

3.2.4 Flow Cytometry Method

In contrast to most of the immunoassays in which samples are fixed to the slide, flow cytometry is based on passing the samples through a liquid dye or sensing medium making it just like an automated fluorescence microscopy. In flow cytometry method (FCM), the sample is carried by laminar flow through a light whose wavelength is matched to the absorption spectrum of the stained dye in the sample cells. The fluorescence emitted by pathogen appears in the form of scattered light; that is, collected by detectors and converted to an electric signal. Analysis of scattered light from pathogen cells provides the insight about size, structure, and shape of microorganisms. FCM can analyze the thousands of cells simultaneously at a single cell level, showing its effectiveness (Mandal et al., 2011). However, real-time pathogen detection by this method has several limiting factors including the limited performance of instrument for particle size resolution, the complexity of sample matrix and ruggedness of the system (Buzatu et al., 2014). Moreover, FCM studies have been much limited to bacteria to date (Ambriz-aviña et al., 2014). A unique and improved FCM protocol eliminated the background noise and improved signal strength. This provided high sensitivity for detecting *E. coli* O157:H7 and other bacteria, and it required few minutes after enrichment (Buzatu et al., 2014).

4 Monitoring Food Contamination Through Bioconjugated Nanomaterials

Conventional methods for detection of contamination were based on various media, cultures, assays, and instruments that gave results in qualitative and quantitative forms. However, these techniques are comprised of series of steps to be followed to get the final results. In this way, these methods required several days to confirm the presence of a particular contaminant because the results were based on the pathogen ability to multiply and form detectable colonies (Biswas et al., 2008). Preparation of culture medium, inoculation, and counting of grown colonies to interpret the presence of a particular contaminant in a food sample makes the conventional techniques time consuming and labor intensive. Another complication related to the conventional methods is the separation of target pathogen or other contaminant from the food as food is a complex entity consisting of different macromolecules ranging from carbohydrates to fats (Doyle and Erickson, 2006). Even after the separation process, the contaminant/pathogen concentration may be too low for efficient and accurate detection that may lead to a wrong conclusion about the type of contaminant present in food (Bhunia, 2008).

The aforementioned limitations of the conventional contamination monitoring methods have resulted in the development of highly efficient and specific aptamer based biosensor platforms. These systems have several advantages over conventional ones; such as aptamers are cheap, stable, and can be easily modified. Other benefits include portability, low toxicity, purity, reproducibility, multipathogen or multiplexed detection, high level of specificity and efficacy, and real-time or fast detection (Song et al., 2012). In recent years, several nanoparticles have been conjugated with aptamers for more accurate, efficient, and real-time detection of food contaminants. The nanomaterials that have been conjugated with aptamers for the purpose include gold nanoparticles, CNTs, quantum dots, graphene, and upconversion nanoparticles. The unique properties of these particles such as high stability, physical and geometric properties, and larger surface area make them excellent material for improving the interaction of aptamers with biological molecules, for example, food contaminants. Further, nanomaterials are easy to be conjugated with aptamers and do not require any other reagent for signal transduction. Conjugation of NPs with nucleotides further prevents the nanoparticle aggregation due to electrostatic repulsion. A major advantage of this conjugation approach is that it prevents any direct modification of the aptamer sequence involved in the binding process. Some of the nanomaterials such as quantum dots, graphene oxide (GO), and upconversion nanoparticles, in addition to the aforementioned advantages, helps in accurate and rapid detection of food contaminants by enhancing or quenching the signal from contaminants (Javier et al., 2008). The fast detection of contaminants in foodstuff can help to screen large trade consignments for a number of targets simultaneously so that much needed early delivery of food to the consumer is achieved. In the subsequent sections, we describe nanoparticle-conjugated aptamers for detection of food contaminants.

4.1 Detection of Foodborne Pathogens

Keeping in view the advantages provided by aptamers conjugated with nanoparticles, several detection systems have been reported. For example, a non-PCR-based aptamer-conjugated gold nanoparticles (GNPs) system with low cost and ultrasensitivity for *S. aureus* bacterium identification has been developed. The pathogen signals were measured by utilizing the resonance light scattering, and the single cell detection was achieved within 1.5 h only. Two aptamers SA17 and SA61 were conjugated with GNPs showed higher binding ability and specificity with *S. aureus*. The GNP-based conjugation provided greater surface area for interaction with the sample including pathogen. These results were further confirmed by scanning electron microscopy, and fluorescence microscopy and the quantification was done through immunofluorescence assays. Both GNPs conjugated-aptamers were able to identify six different strains of *S. aureus* (ATCC: 12600, 25923, 6538DR, 6538P, 6538, 29213), but did not show any cross-reaction with 13 other bacteria studied, showing highly specific nature of the selected aptamers. However, SA61 showed a weaker interaction with *P. aeruginosa* and *S. epidermis* than *S. aureus* strains. It was also found that immobilization of both aptamers on GNPs significantly altered their functional structure. This advantageous change was documented in terms of K_d (binding constant) values as they changed from 129 and 35 nM in SA61 and SA17 to 9.9 and 3.03 nM for SA61-GNPs and SA17-GNPs, respectively. This showed a visible enhancement in pathogen signal due to conjugation of GNPs and increased affinity in comparison to free aptamers. The whole process is illustrated in Fig. 3.2 (Chang et al., 2013).

Another GNP-conjugated aptamer assay for rapid, simple, and cost-effective detection of *S. typhimurium* has been reported. SELEX identified 96 candidate sequences for pathogens, which were classified into six families (S1–S6) on the basis of high-sequence homology (75–85%). One aptamer per family was selected, and its binding affinity to *S. typhimurium* was studied by fluorescence spectroscopy. S2, S3, and S6 aptamers showed high binding affinity to the pathogen and S6 aptamer with the highest value. Aptamer S6 was conjugated with GNPs in colloidal gold solution, and the aptamer was denatured/renatured for the target specific 3D structure. Then aptamer-conjugated GNPs solution was treated with a

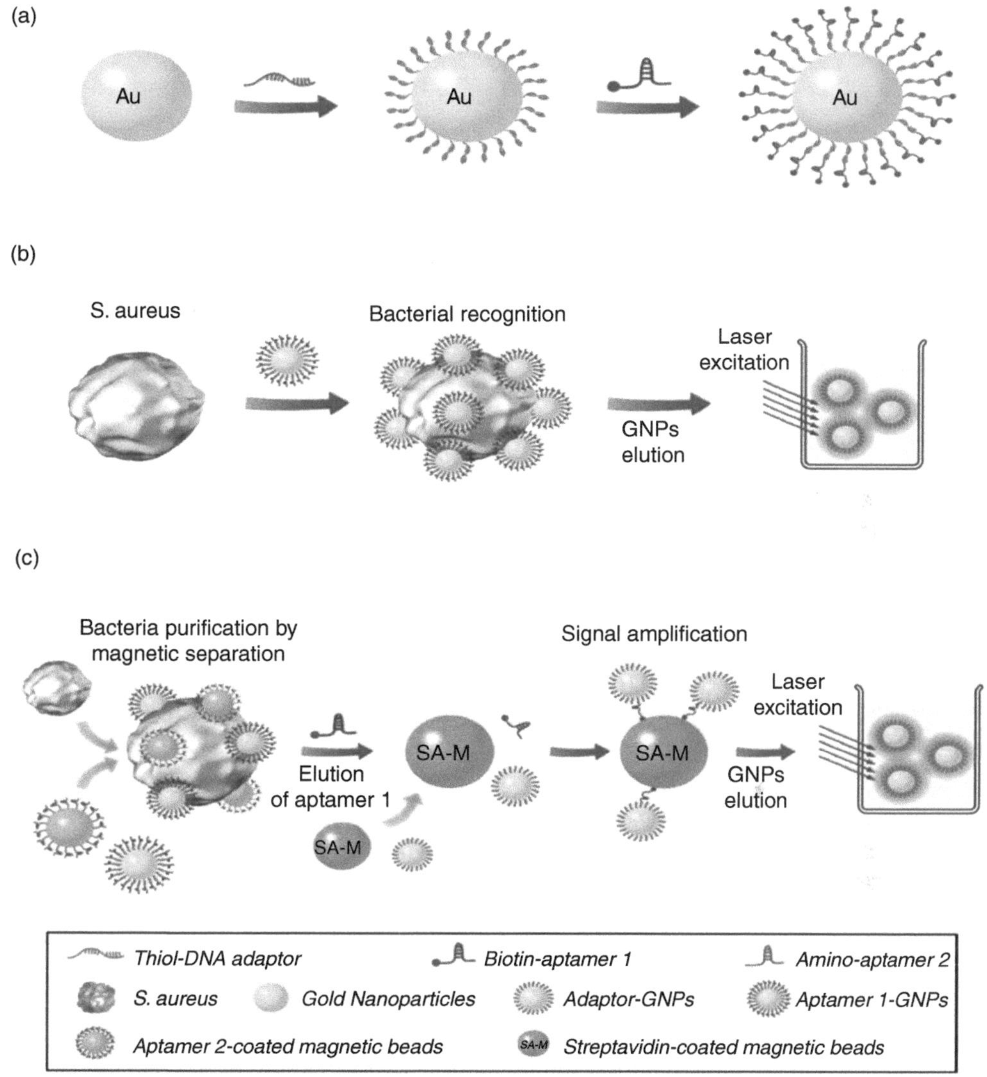

Figure 3.2. Flowchart of *S. aureus* detection using aptamer-conjugated GNPs. (a) Aptamers were conjugated onto 60-nm GNPs with thiol-DNA adaptors. (b) Aptamer-GNPs in the direct detection of *S. aureus*. 109 aptamer-GNPs were incubated with *S. aureus* cells. After removal of unbound aptamer-GNPs, bound aptamer-GNPs were eluted, and their light-scattering signals were analyzed. (c) Bead-based amplification for the detection of *S. aureus*. SA61-aptamers (biotin-aptamer 1) were conjugated onto 60-nm GNPs, and SA17-aptamers (aptamer 2) were conjugated to magnetic beads. Aptamer 1-GNPs and aptamer 2-magnetic beads interacted with *S. aureus* and the resulting complexes were isolated with a magnet. Bound biotin aptamer 1 was eluted by heating and further incubated with an excess of reporter-GNPs (conjugated with DNA adapter) and streptavidin (SA)-coated magnetic beads. The reporter-GNPs were then captured with SA-magnetic beads in the presence of biotin-aptamer 1. The bound reporter-GNPs were eluted with NaOH and their light-scattering signals were analyzed (Chang et al., 2013).

high concentration of salt (1.0 M) in the presence or absence of the target pathogen. GNPs helped in maintaining the balance between van der Waals attraction and electrostatic repulsion in solution. Thus, it provided stability to the detection system. Furthermore, high mobility of GNPs into a porous membrane, and less aggregation due to their binding with an aptamer, the aptamer–GNPs complex remained stable when *S. typhimurium* was added. When S6 aptamer was used in the GNP-based colorimetric procedure, even 106 CFU mL^{-1} of the pathogen was successfully detected; which was more efficient than the aptamer alone (Moon et al., 2014).

CNTs, especially single-walled carbon nanotubes (SWCNTs) have been conjugated with aptamers for simple detection of food pathogens. SWCNTs have exceptional thermal, electrical, physical, and mechanical properties. The unique properties of SWCNTs include outstanding mechanical properties, even better than commercially available carbon, steel, or aramid fibers; a low defect density or sound structural integrity; exceptional electrical conductivity of metallic SWCNTs, that is, 1000 times better than most commonly used copper. An SWCNT-conjugated aptamer system involving ssDNA-based Apt22 aptamer has been developed, which showed high specificity and binding affinity for *Salmonella paratyphi* with the K_d value of 47 ± 3 nM. For the conjugation, SWCNTs were combined with aptamer detection probe P0, which was labeled with DNAzyme. In the first step, SWCNTs-COOH complex was sonicated for 30 min in DMF until a homogeneous black solution was obtained and diluted into a working solution to get SWCNTs suspensions. In the second step, SWCNTs suspension was hatched with P0 (500 nM) for 4 h at 37°C. Then, centrifugation of the solution was done at 22,000 *g* for 15 min to obtain P0/SWNTs containing supernatant. Finally, the target pathogen and hemin were incubated with the supernatant for 15 h at 37°C to react. For detection, H_2O_2 and luminol were quickly mixed with the final mixture and fluorescence spectrophotometer was used to measure pathogen signal. This novel aptamer-based system conjugated with DNAzyme and SWCNTs by noncovalent self-assembly showed high specificity to the pathogen with 103 CFU mL^{-1} detection limit. A further advantage of this system is that it does not need the probe designing, synthesis, and gene extraction (Yang et al., 2013). Aptamer–SWCNTs complex has also been used for detection of one of the most common foodborne pathogens *E. coli* (So et al., 2008).

4.2 Detection of Food Toxins

Aptamers and aptamers conjugated with nanoparticles, have also been used to detect toxins in food. Danofloxacin is a synthetic fluoroquinolone, which is used in poultry and other animal husbandry systems as an antibacterial drug. It remains as a residue in the animal body and is transferred to humans as toxin via food. This can affect the nervous system, liver, muscles, and skin. SELEX was used to develop highly affinitive and specific RNA aptamers for danofloxacin having 2-fluoro-2-deoxyribonucleotide modified pyrimidine. The truncated aptamer (minimum sequence of the aptamer that is required to bind any other molecule), dano-I showed binding ability with danofloxacin, but did not show any binding affinity with tetracycline, thus, helped in discriminating danofloxacin from the rest of fluoroquinolones. Dano-I (36 and 24 mer) aptamers showed affinity to danofloxacin, but a 21-mer truncated dano-I, and dano-II aptamer did not show any binding (Han et al., 2014).

Ochratoxin A (OTA) is nephrotoxic while the International Agency for Research on Cancer has declared it carcinogenic to humans. More than 50% of humans are exposed to OTA by eating cereals and cereal products. SELEX procedure was used to screen aptamers, called A08 and B08, which had high binding affinity against OTA mycotoxin rather than the other toxins. The K_d values for both selected parameters were 290 ± 150 and 110 ± 50 nM, respectively. These aptamer sequences and the truncated versions were modified with SYBR Green I fluorescence schemes for OTA detection. This fluorescent-based platform was highly selective, sensitive, and rapid, which was capable of identifying 9 nM toxin (McKeague et al., 2014).

A chromatographic strip assay for rapid identification of OTA by conjugating aptamers with GNPs has been developed. GNPs were prepared, and thiolated aptamers were coated on GNP surface by a well-known self-assembly method. Two DNA probes were fixed on streptavidin by utilizing the 5′ biotin followed by the assembly of the strip. A plastic plate was used as a base on which nitrocellulose and glass fiber membranes, sample and absorbent pads, were placed as layers in said sequence. Now, both DNA probe conjugated with streptavidin were transferred to the strip and GNP-aptamer probe was attached to the glass fiber membrane as presented in Fig. 3.3. The working principle of the strip was based on the competitive reaction between the target OTA and test line or DNA probe 1. When OTA toxin was present in the sample to be tested, the aptamer–GNP conjugate was unable to hybridize with DNA probe 1; and thus the red color of the strip became weaker. But aptamer-GNP bound to DNA probe 2 even at lower OTA concentrations in the control line which validated the results. This strip had a limit of detection (LOD) value of 1 ng mL^{-1} in qualitative tests, but it also showed promise in semiquantitative analysis.

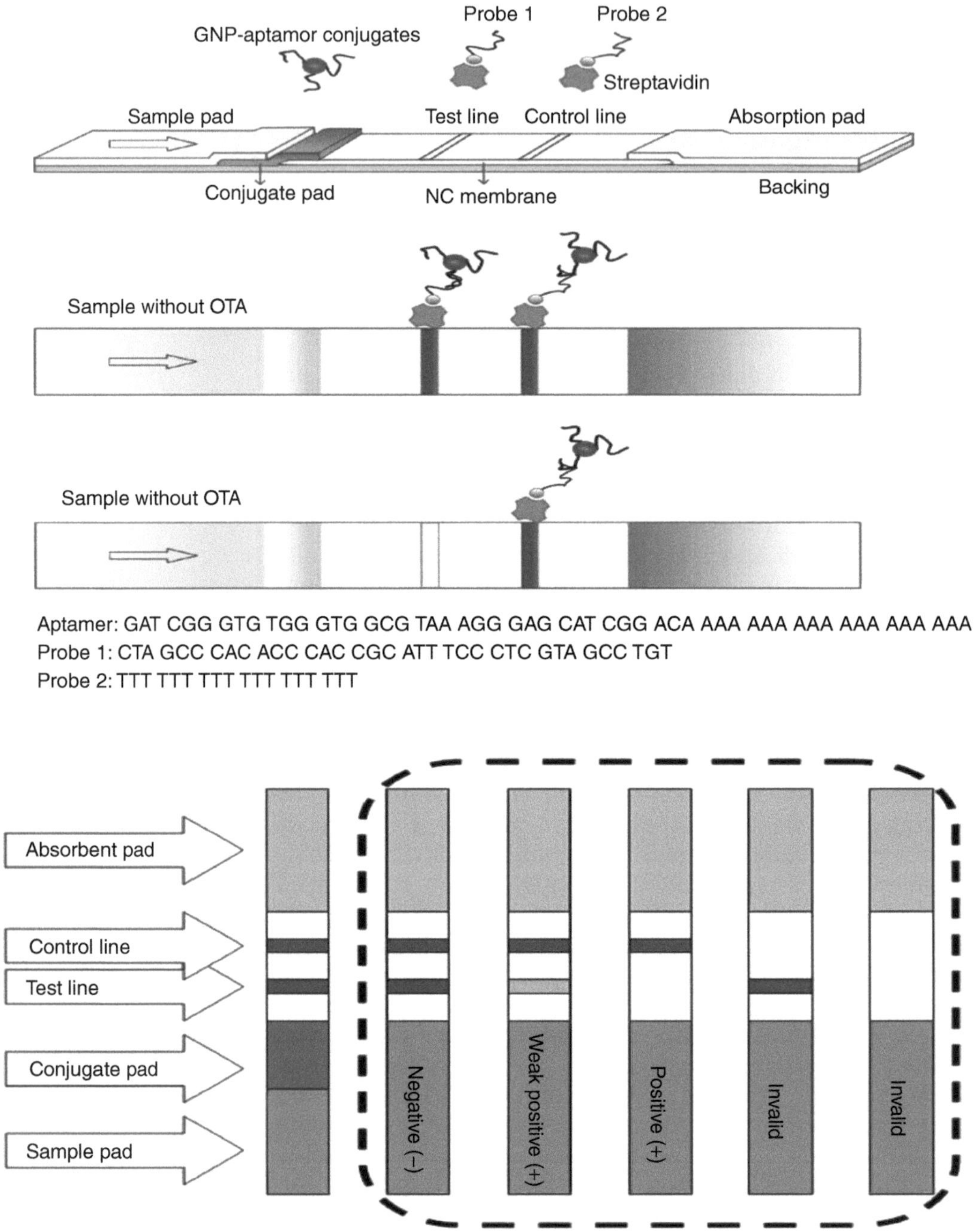

Figure 3.3. Working principle of the aptamer-based strip (Wang et al., 2011). Copyright© 2010, Elsevier B.V.

A quantitative calibration curve was constructed by using a scanning reader with an LOD of 0.18 ng mL^{-1}. This detection limit was better than ELISA, antibody-based strip, and aptamer without GNP conjugation, thus showing the greater efficacy of nanoparticle conjugates as compared to other detection systems. This strip did not show any binding to other food toxins including fumonisin B1, mycotoxins deoxynivalenol, microcystin-LR, and zearalenone showing specificity of the strip for OTA. In addition to higher efficiency and specificity of this nanoparticle-based detection system; other advantages of this strip were its stability after 30 days, 96–110% recoveries, fastness (less than 10 min for detection), reliability, and cost effectiveness (Wang et al., 2011).

In addition to earlier strip-based system, GNP-conjugated aptamers have also been reported for rapid, highly sensitive, specific, and reliable OTA detection by utilizing fluorescence resonance energy transfer aptasensor. This detection system was based on hybridization of a dye-tagged ssDNA aptamer conjugated with GNPs. The hybridization of the aptamer-conjugated GNPs with the dye-labeled ssDNA aptamer provided the advantage of fluorescence quenching of FAM. When OTA is present in a food sample, it gives fluorescence as a consequence of hybridization, and a quadruplex-OTA complex is formed; this complex now detaches from GNPs. The fluorescence intensity is directly proportional to the OTA concentration in food samples, which ranged from 5×10^{-12} to 5×10^{-9} g mL^{-1}, with an LOD value of 2×10^{-12} g mL^{-1}. The aforementioned system successfully measured OTA concentrations in naturally contaminated maize, and it was validated by using a commercially available ELISA method (Duan et al., 2012a).

Staphylococcal enterotoxins are secreted in food by *S. aureus*, a foodborne bacterium. Staphylococcal food poisoning occurred when SEs contaminated food is eaten, and it leads to severe abdominal pain, diarrhea, and emesis. A novel aptamer APTSEB1 showed high binding affinity and specificity against SEB, which isolated SEB from a complicated mixture of enterotoxins (DeGrasse, 2012). A GNP-based enhanced chemiluminescence biosensor has been developed for rapid and reliable detection of SEB in food. Use of GNPs gives rise to a stronger signal as compared to an alone antibody or aptamer. This enhanced chemiluminescence provides the additional benefit to detect minor quantities of SEB in samples; thus helping screening the foodstuff for this toxin with increased sensitivity. According to this system, the anti-SEB primary antibody was conjugated with GNPs through physical adsorption and this complex was fixed on a polycarbonate surface (Yanga et al., 2009). The earlier system has also been replicated by GNPs conjugated with ssDNA aptamer for SEB detection, with low cost and in less time. Several aptamer sequences linked to SEB were gotten by SELEX and five of them were selected for further analysis. Now, aptamers were conjugated with GNPs and a colorimetric procedure was followed to detect SEB. The main structure of the aptamers was found to be in stem-loop and hairpin form. One of the GNP-conjugated aptamer, that is, aptamer No. 15-1, detected the even lower concentration of SEB like 10 ng mL^{-1}. This aptamer also showed higher selectivity for both SEC1 and SEB toxin as compared to the nonconjugated aptamer (Liu et al., 2013).

In another study, aptamers selected through SELEX procedure were functionalized with GO nanoparticles. This GO-conjugated ssDNA aptamer system detected T-2 toxin in food with high binding affinity and sensitivity. The high binding affinity of the aptamer Seq.16 to the T-2 toxin was recorded by circular dichroism spectroscopy and dissociation constant for the interaction was 20.8 ± 3.1 nM. This aptamer was used to develop an aptamer-based fluorescent method for detection of T-2 toxin in beer with detection range varying between 0.5 to 37.5 μM (Chen et al., 2014). T-2 is a trichothecene mycotoxin that cause diarrhea, hemorrhaging, and feed refusal in livestock when ingested through contaminated grains such as cereal crops (McCormick et al., 2015).

4.3 Detection of Food Allergens

Allergy caused by peanuts is a notorious and life-threatening food contamination as it can lead to death by causing a lifelong disorder. *Arachis hypogaea* 1 (Ara h 1), a 63,000 Da molecule, is an important peanut allergen, and it is found as a homotrimeric complex in food. The particular 3D structure of this protein makes it stable, viable, and thus allergic even if it goes under heat and enzyme treatment. This protein has protease-resistant fragments, which make its denaturation and protease digestion difficult. Therefore, rapid and efficient detection system is required to isolate the infected food from a healthier one and to ensure food safety or consumer protection.

Although several immunoassays are available for Ara h 1 detection, such as ELISA, dipstick assay, lateral flow assay, and immunosensors, all these methods are based on antibodies whose generation is difficult and these assays are time-consuming as well (Pollet et al., 2011). Recently, a GNP-based capillary electrophoresis-based-SELEX method was used to develop Ara h1 specific ssDNA aptamers that additionally showed no affinity for Ara h 2 and other similar proteins. The dissociation constant of the aptamer having the highest binding affinity and specificity, that is, A1 aptamer, was 450 ± 60, 419 ± 63, and 353 ± 82 nM as measured by capillary electrophoresis, fluorescence anisotropy, and surface plasmon resonance, respectively. This selected aptamer was utilized to develop another biosensor called as FO-SPR or fiber optic-based surface plasmon resonance for detection of the same allergen in different food materials. FO-SPR device's robotic arm was used for binding in three steps as it was programmed for the purpose. Optimization was done with tris(hydroxyamino)methane-glycine-potassium (TGK) buffer and a baseline measurement was made, followed by Ara h 1 molecules' capture by the aptamer in 100 μL of food sample for 10 min. In the second step, FO-probe was dipped in TGK buffer having a rabbit polyclonal antibody (10 μg mL^{-1}), which had binding ability for Ara h 1. In the final step, the fiber was immersed in GNP solution for 10 min. Before the third step, GNPs were functionalized by adsorption of protein A on their surface. This was done by mixing 5 ml GNPs (1 OD, pH of 7.0) with 100 μM protein A in 100 μL of 0.2 M Na_2CO_3 buffer and the whole process is shown in Fig. 3.4. Protein A helped in the recognition of the antibody's Fc portion and along with GNPs amplified the signal by the FO-SPR sensor if the antibody was present. The FO-SPR assay proved to be a valuable procedure as it took less time and had larger linear dynamic range though the most signal amplification was gotten in SPR by the addition of GNPs. Gradual augmentation of the signal with increasing Ara h 1 concentration

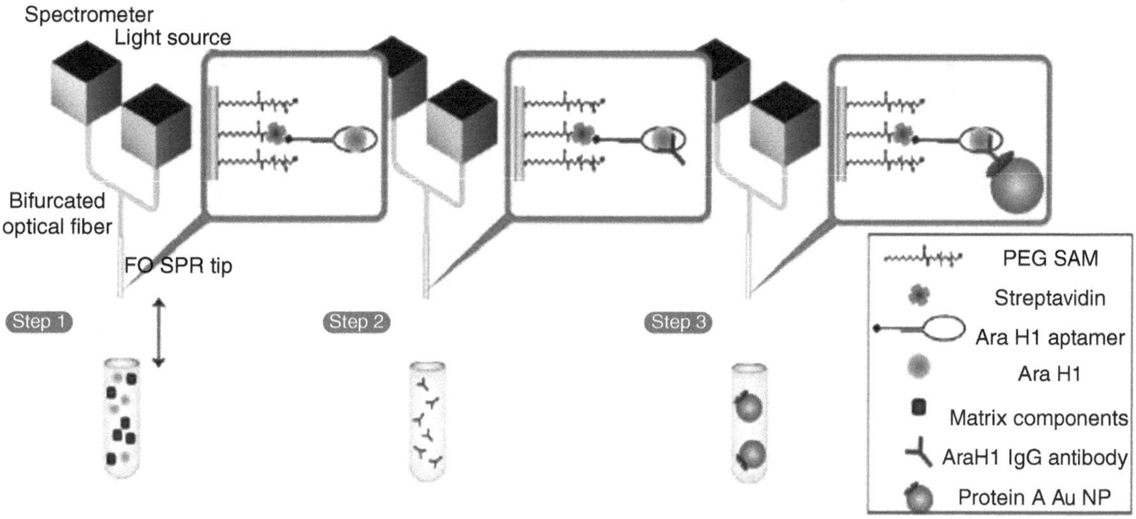

Figure 3.4. Schematic representation of multistep protocol for Arah1 protein detection in food matrix samples. Allergen from the food matrix is captured by aptamer on FO-SPR sensor tip followed by sandwich assay formation with polyclonal secondary antibodies which are used for further enhancement of the signal (Tran et al., 2013). Copyright© 2012, Elsevier B.V.

verified the multistep process. The dose response and signal amplification was also observed in the candy bar matrix, when GNPs were added. SPR was used to study the detection signal at each step, and when only antibody was used, the signal was unable to discriminate the Ara h 1 protein below the threshold level of 423 nM. But, GNP conjugation helped in the detection of Ara h 1 below this level and even in the food sample matrix. The lower sensitivity of the aptamer was a limitation, but this assay provided a strong advantage of the complete elimination of any manual handling (Tran et al., 2013).

Two DNA aptamers, named as aptamer 2 and 40, which showed high affinity and selectivity to flour allergen Lup an1 have also been developed by SELEX. The binding of the selected aptamers was monitored by ELISA and enzyme-linked oligonucleotide assay (ALISA). The selected aptamers were highly specific as they did not have any affinity for other conglutin proteins of lupin and other flour ingredients (Nadal et al., 2012). Lysozyme, also called as Gal d4, is an egg-white allergen that can cause allergies even in trace amounts. Capillary electrophoresis-based SELEX (CE-SELEX) method has been used to generate ssDNA aptamers, which showed high specificity and binding to lysozyme allergen. Aptamers, Apta1, Apta3, and Apta8, showed higher binding toward the lysozyme. The final aptamer showed K_d values of 52.9 ± 9.1, 6.1 ± 0.5, and 2.8 ± 0.3 nM as measured by capillary electrophoresis, SPR, and fluorescence anisotropy, respectively (Tran et al., 2010). But, no nanoparticle-conjugated aptamer-based detection system has been reported for both of aforementioned food allergens to date. However, GNP-conjugated aptamer for rapid and accurate detection of another life-threatening peanut allergen, that is, *A. hypogaea* 2 (Ara h 2) has been developed (Liu et al., 2010). This sensor comprised of an important IgE-binding epitope having binding ability to Ara h 2, which was attached to a GNP film on a pyrolytic graphite electrode. The LOD for the sensor was 5 pg mL^{-1} that was 60 and 100 times lower than HRP-Ab2-based amperometric and label-free impedimetric immunosensors, respectively.

4.4 Detection of Other Food Safety Hazards

Some genotoxic or carcinogenic dyes such as sulforhodamine B and melamine are suspected to be included in the processed foodstuff in order to expand the apparent protein content. An aptamer targeting sulforhodamine B was develop in 1998 by Wilson and Szostak research group (Wilson and Szostak, 1998) that showed 660 ± 60 nM affinity toward the dye in solution while the affinity was even lower around 190 ± 20 nM for immobilized sulforhodamine B. Later in the same year, the same research group published an RNA aptamer for sulforhodamine B target, but this time with a greater binding ability in solution around 310 nM (Holeman et al., 1998). Although the aptamers were selected for general applications in biotechnology, such as DNA staining, these aptamers might also be useful for food safety monitoring purposes. On the other hand, MG is used in aquaculture to prevent antifungal diseases in fish. MG and its metabolite named as leucomalachite green (LG) has been reported to induce toxicity extending to gene mutations in soil fungi and bacteria as well as in fish (Gopinathan et al., 2015). The first MG aptamer, called Clone MG-4, developed by Grate and Wilson (1999) is recently characterized by Stead et al. (2010) which showed significant affinity toward the MG rather than its metabolite LG. Recently, an electro-chemiluminescence sensor platform based on the aforementioned aptamers were designed using QDs and the luminol GNPs as the signal transducers, where the detection limit was found as low as 0.03 nM. Besides, the procedure was successfully applied to the real fish samples, indicating a high potential for the food analysis (Feng et al., 2015).

Human exposure to heavy metals, which is a grave threat to both environment and living organisms, could be through direct ingestion, such as direct water intake from the heavy metal contaminated soil or consumption of the crop products that are grown in heavy metal accumulated soil. A high-affinity DNA aptamer, named as Ars-3, was selected by Kim et al. (2009) for the detection and elimination of arsenic from Vietnamese groundwater. Ars-3 aptamer assays was successfully applied to the field samples where the arsenic concentrations between 28.1 and 739.2 $\mu g\ L^{-1}$ were eliminated after 5 min of exposure.

Bisphenol A (BPA) is a potential hazard to foetuses, infants, and young children, which has been used in common polymer plastic products since 1930s, as a monomer. An aptamer for BPA developed by Lee et al. (2011) detected BPA at very low concentrations Another BPA targeting DNA aptamer, named as called #3, R11, was also developed within the same year (Jo et al., 2011) that does not recognize the Bisphenol B, showing only one methyl group difference with BPA.

5 Multiplexed Biosensors

The simultaneous detection of several foodborne contaminants can save time and effort and also ensures fast and efficient monitoring of the food materials. A DNA-based electrochemical microarray platform identified four foodborne microbes (*E. coli* O157:H7, *S. aureus*, *Salmonella* spp., and *L. monocytogenes*) simultaneously (Farabullini et al., 2007). But the dependence of such detection systems on unstable enzymes has shifted the attention toward nanoparticle-conjugated aptamers for simultaneous detection of various food contaminants. Lab-on-a-chip and lateral flow devices are being used by aptamer-based food analysis procedures. A procedure that combines both lab-on-a-chip and multiplexed detection has been utilized by designing a polydimethylsiloxane or PDMS/paper/glass chip that detected *S. aureus* and *Salmonella enteric* simultaneously. Aptamers were labeled with fluorescent material and were attached to GO nanoparticles that quenched its fluorescence. When pathogen cells were present, aptamers separated from the GO surface, and the pathogen was detected by increased fluorescence. This method provides one-step and "turn on" method for detection of several pathogens together, and this microfluidic device takes around 10 min with an LOD of 11.0 CFU mL^{-1} for completion on biochip (Zuo et al., 2013).

Recently, a multiplexed aptamer system has been reported for simultaneous detection of *S. typhimurium* and *V. parahaemolyticus*. This detection system used unique amorphous carbon nanoparticles (CNPs) as acceptors and red and green light emitting quantum dots (rQDs and gQDs) as donors; and QDs provide dual fluorescence resonance energy transfer, as illustrated in Fig. 3.5. Aptamer Apt1 conjugated with gQDs was specific for *V. parahaemolyticus* and Apt2-conjugated with rQDs recognized *S. typhimurium*. Use of CNPs showed strong quenching effect on the fluorescence of both QDs. However, when a triplex complex is formed between aptamer, QDs and target, a decrease in quenching by CNPs was observed. The QDs-based fluorescence was directly proportional to the pathogens' concentration between 50 and 106 CFU mL^{-1}. This assay detected 35 and 25 CFU mL^{-1} of *S. typhimurium* and *V. parahaemolyticus* respectively in only 2 h 20 min in chicken and shrimp food samples (Duan et al., 2015).

Multicolor upconversion (or upconverting) nanoparticles (UCNPs) have also been conjugated with aptamers for simultaneous detection of food contaminants as they add more sensitivity to the luminescence process. An extremely sensitive aptamer conjugated magnetic nanoparticles (MNPs)-based system for simultaneous detection of *S. aureus* and *S. typhimurium* has been reported. This system utilized UCNPs as dual color fluorescence, which added a higher level of the sensitivity to the detection process. Apt1 and 2 specific for *S. typhimurium* and *S. aureus*, respectively, were conjugated to the surface of MNPs. In the next step, UCNPs modified with apt1 and 2 were attached to the captured bacteria to

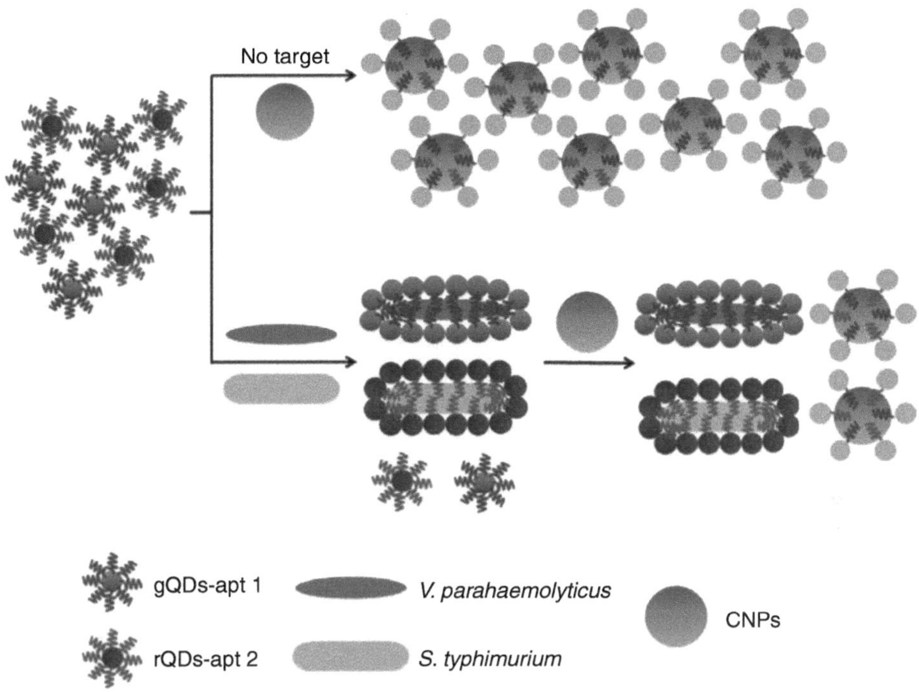

Figure 3.5. Schematic illustration of the dual fluorescence resonance energy transfer from QDs-apts to CNPs for the simultaneous detection of pathogenic bacteria (Duan et al., 2015). Copyright© 2014, Springer-Verlag Wien.

form a sandwich complex. The correlation between the signal and pathogen concentration was 0.9964 and 0.9936; and LOD was 5 and 8 CFU mL^{-1} for *S. typhimurium* and *S. aureus*, respectively. This system also detected the pathogens in real water samples mainly due to the improved sample concentration through magnetic separation and the sensitivity by UCNPs (Duan et al., 2012b).

In another study, three foodborne bacteria were detected simultaneously by aptamers-UCNPs conjugates. UNCPs played the role of luminescence labels and were developed by doping procedures using many rare earth ions. The aptamers were screened by SELEX targeting *S. typhimurium*, *V. parahemolyticus*, and *S. aureus* bacteria. Aptamers-UCNPs complex was used for the capture and quantification of aforementioned three target bacteria simultaneously and successfully on the basis of independent peaks for each bacterium. The quantity of bacterial detection ranged between 50 and 106 CFU mL^{-1} but it was later improved to 15, 10, and 25 CFU mL^{-1} for *S. typhimurium*, *V. parahemolyticus*, and *S. aureus*, respectively, by concentration effect of magnetic separation (Wu et al., 2014).

Pathogens are not the only food contaminants that have been detected by aptamer-conjugated nanomaterials, aptamer conjugated UCNPs, and GO nanoparticles have been developed that simultaneously detected two notorious food mycotoxins. This system was based on dual fluorescence resonance energy transfer between GO nanoparticles as acceptor and the multicolor UCNPs as donors. BaY0.78F5:Yb0.7, Tm0.02 and BaY0.78F5:Yb0.2, Er0.02 UCNPs were functionalized with aptamers specific to fumonisin B1 (FB1) and OTA toxins. The strong upconversion fluorescence of both UCNPs was completely quenched by the GO, due to a nice overlap between the absorption spectrum of GO and fluorescence emission of UCNPs. On the other hand, when both toxins were present, the aptamer bound to their related toxin leading to changes in aptamer formation that kept aptamer bound UCNPs away from the GO. Fluorescence intensity of both UCNPs was found to be directly proportional to FB1 and OTA concentrations in maize samples, which helped in direct detection and quantification of both toxins. The detection range was found to be between 0.1 and 500 ng mL^{-1} for FB1, and 0.05 and 100 ng mL^{-1} for LOD of 0.1 and 0.02 ng mL^{-1} for FB1 and OTA, respectively (Wu et al., 2012).

6 Summary

Foodborne pathogens cause severe infections across the world, while food toxins and allergens also cause health problems in humans. The resultant health problems range from minor health issues to diseases that could lead to fatalities. Therefore, an efficient and reliable food monitoring system is required for food safety and public health. Conventionally, flow cytometry, PCR, and particularly immunoassay-based detection of food contaminants have been carried out. But, time consumption, difficult handling, multiple steps, labor need, and nonstability of proteins used in these systems have limited the scope of the food safety monitoring. In recent years, aptamers have emerged as reliable and efficient probes for environmental food monitoring as they are inexpensive, stable, easily modifiable, portable, reproducible, and can allow real-time and multiplexed detection. Successful conjugations of aptamers with various nanoparticles have combined the additional advantages of high stability, physical and geometric properties, larger surface area for them. This also results in prevention of aptamer modification and provides enhanced detection signals that make them excellent biosensors. A list of aptamers that are used for food safety monitoring purposes is presented in Table 3.2, as a comprehensive summary of the literature reviewed.

Table 3.2 List of Aptamers Used for Food Safety Monitoring Purposes

Contaminant	Aptamer-NPs	5′-Aptamer Sequence-3′	K_d (nM)	References
S. typhimurium	S6-GNPs	GCGTGCGGAGCCAGGATG GGCGTGCGGAGCCAG	0.3	Moon et al. (2014)
S. aureus	SA17	TCCCTACGGCGCTAACCCC CCCAGTCCGTCCTCCCAGC CTCACACCGCCACCGTGCTACAAC	35	Chang et al. (2013)
S. aureus	SA17-GNPs	TCCCTACGGCGCTAACCCC CCCAGTCCGTCCTCCCAGC CTCACACCGCCACCGTGCTACAAC	3.03	Chang et al. (2013)
S. aureus	SA61	TCCCTACGGCGCTAACCTC CCAACCGCTCCACCCTGCC TCCGCCTCGCCACCGTGCTACAAC	129	Chang et al. (2013)
S. aureus	SA61-GNPs	TCCCTACGGCGCTAACCTC CCAACCGCTCCACCCTGCC TCCGCCTCGCCACCGTGCTACAAC	9.9	Chang et al. (2013)
S. paratyphi	Apt10	GATGATGGACGTATATCGTCTCC CATGAATTCAGTCGGACAGCG	73	Yang et al. (2013)
S. paratyphi	Apt45	ATGGACGAATATCGTCTCCCAGT GAATTCAGTCGGACAGC	68	Yang et al. (2013)
S. paratyphi	Apt60	CGCCCACCCATAATGGATCAGGG CGGGCACCACGATG	56	Yang et al. (2013)
S. typhimurium	Apt1-UCNPs	Biotin-C6-TATGGCGGCGTCACCCGA CGGGGACTTGACATTATGACAG	*5 cfu/mL*	Duan et al. (2012b)
S. aureus	Apt2-UCNPs	Biotin-C6-GCAATGGTACGGTACTT CCTCGGCACGTTCTCAGTAGCGCT CGCTGGTCAT CCCACAGCTACGTC AAAAGTGCACGCTACTTTGCTAA	*8 cfu/mL*	Duan et al. (2012b)
S. paratyphi	Apt22-SWCNT	ATGGACGAATATCGTCTCCCAGT GAATTCAGTCGGACAGCG	47	Yang et al. (2013)
Danofloxacin	dano-I	GGGUGACUUUUCCCUCAGGCU CCUGUGAAGCAACCGAAUG GACUGAAAUGAAACCGCCC	3–7	Han et al. (2014)

(*Continued*)

Table 3.2 List of Aptamers Selected for Food Safety Monitoring Purposes (*cont.*)

Contaminant	Aptamer-NPs	5′-Aptamer Sequence-3′	K_d (nM)	References
Ochratoxin A (OTA)	A08	GGCAGTGTGGGCGAATCTATG CGTACCGTTCGATATCGTG	290	McKeague et al. (2014)
Ochratoxin A (OTA)	B08	GGCGCATGATCATTCGGTGGGTA AGGTGGTGGTAACGTTG	110	McKeague et al. (2014)
Ochratoxin A (OTA)	Apt-GNPs	GATCGGGTGTGGGTGGCGTA AAGGGAGCATCGGACAAA AAAAAAAAAAAAAAAA-SH-	1 ng/mL	Wang et al. (2011)
Ochratoxin A (OTA)	Apt-GNPs	SH-GATCGGGTGTGGGTGGCGT AAAGGGAGCATCGGACA	2 pg/mL	Duan et al. (2012a)
Ochratoxin A (OTA)	Apt-UCNPs-GO	GATCGGGTGTGGGTGGCGTA AAGGGAGCATCGG-(CH_2)6-biotin	0.02 ng/mL	Wu et al. (2012)
Fumonisin B1	Apt-UCNPs-GO	Biotin-(CH_2)6-ATACCAGCTTAT TCAATTAATCGCATTACCTTATAC CAGCTTATTCAATTACGTCTGCA CATACCAGCTTATTCAATTAGATA GTAAGTGCAATCT	0.1 ng/mL	Wu et al. (2012)
Fumonisin mycotoxin	FB139	AATCGCATTACCTTATACCAGCT TATTCAATTACGTCTGCACATAC CAGCTTATTCAATT	100	McKeague et al. (2014)
Staphylococcal enterotoxin B	APTSEB1	GGTATTGAGGGTCGCATCCACT GGTCGTTGTTGTCTGTTGTCTGT TATGTTGTTTCGTGATGGCTCTA ACTCTCCTCT	—	DeGrasse (2012)
Staphylococcal enterotoxin B	15-1-GNPs	ACCGACCGTGCTGGACTCTAT TCATTACCCATCCGAGAACTGG TGGTCATATCTCTACAAGTATG AGCGAGCGTTGCG	10-50 ng/mL	Liu et al. (2013)
T-2 toxin	Seq.16-GO	GTATATCAAGCATCGCGTGTTTA CACATGCGAGAGGTGAA	21	Chen et al. (2014)
Gal d4 allergen	Apta1	GCAGCTAAGCAGGCGGCTCA CAAAACCATTCGCATGCGGC	53	Tran et al. (2010)
Gal d4 allergen	Apta3	GCGGGTCGGTTGCTCGCTTCG CCCGATCGGTCTAAGGGTG	47	Tran et al. (2010)

Table 3.2 List of Aptamers Selected for Food Safety Monitoring Purposes (*cont.*)

Contaminant	Aptamer-NPs	5′-Aptamer Sequence-3′	K_d (nM)	References
Gal d4 allergen	Apta8	GCACCTTGATGACATGATAGTCG TTGTGTATGCAGTTGGC	52	Tran et al. (2010)
Flour allergen Lup an 1	Apta2	AGCTGACACAGCAGGTTGGTGG GGGTGGCTCACATCATGGTAG AATGACTGAACAGCGTTGATT AAAAGGCACGAGTCGAGCAA TCTCGAAAT	515	Nadal et al. (2012)
Flour allergen Lup an 1	Apta40	AGCTGACACAGCAGGTTGGTG GGGGTGGCTTCCAGTTGGGTT GACAATAGTAGGGACACGAAG TCCAACCACGAGTCGAGCAA TCTCGAAAT	360	Nadal et al. (2012)
Ara h 1 allergen	A1-GNPs	GGGGGGGTCGAGCTGAGTGGA TGCGAATCTGTGGGTGGGC	353	Tran et al. (2013)

In this chapter, we reviewed the latest literature on aptamer-conjugated nanoparticles for rapid and accurate detection of major food pathogens (*S. aureus, P. aeruginosa, S. epidermis, S. typhimurium, S. paratyphi, E. coli*), toxins (danofloxacin, ochratoxin A, fumonisin B1, staphylococcal enterotoxin B, T-2 toxin), and allergens (*A. hypogaea* 1 or Ara h 1, Ara h 2, flour allergen Lup an1, and Gal d4). The nanoparticles conjugated with aptamers included gold nanoparticles, CNTs, quantum dots, GO, and UCNPs. Among these, GNP-based assays proved to be most frequent one due to its easily binding with aptamers through thiol interaction, color transform upon binding with the target as well as the signal enhancement ability. Multiplexed or simultaneous detection of the contaminants have been mostly achieved by aptamers that are conjugated with GO, quantum dots, and UCNPs, which has added more power to biosensor platforms as it reduces the time and another resource requirement. In future, these multiplexed systems based on fluorescent nanoparticles such as UCNPs or QDs could lead the field as they offer the fastest detection platforms with minimal sample consumption and labor.

Acknowledgments

The authors acknowledge the support from Scientific and Technological Research Council of Turkey (TUBITAK), Grant ID: 114Z379. The project 114379 was also carried out in the frame of European COST Action CM1403, "The European upconversion network from the design of photon-upconverting nanomaterials to biomedical applications." BH acknowledges TUBITAK 2215 programme.

References

Ambriz-aviña, V., Contreras-garduño, J.A., Pedraza-reyes, M., 2014. Applications of flow cytometry to characterize bacterial physiological responses. BioMed Res. Int. 2014, 461941.

Aschfalk, A., Müller, W., 2011. *Clostridium perfringens* toxin types from wild-caught Atlantic cod (*Gadus morhua* L.), determined by PCR and ELISA. Can. J. Microbiol. 48 (4), 365–368, http://www.nrcresearchpress.com/doi/abs/10.1139/w02-015#.VUE-BGiUeSp

Belter, A., et al., 2014. Mature MiRNAs form secondary structure, which suggests their function beyond RISC. PLoS One 9 (11), e113848, http://www.plosone.org/article/info:doi/10.1371/journal.pone.0113848#pone-0113848-g005

Bhunia, A.K., 2008. Biosensors and bio-based methods for the separation and detection of foodborne pathogens. Adv. Food Nutr. Res. 54, 1–44, http://www.ncbi.nlm.nih.gov/pubmed/18291303

Biswas, A.K., et al., 2008. Microbial profiles of frozen trimmings and silver sides prepared at Indian buffalo meat packing plants. Meat Sci. 80 (2), 418–422, http://www.ncbi.nlm.nih.gov/pubmed/22063348

Buzatu, D.A., et al., 2014. An integrated flow cytometry-based system for real-time, high sensitivity bacterial detection and identification. PLoS One 9 (4), e94254.

Carvalho, R.N., et al., 2014. PCR and ELISA (VIDAS ECO O157 ®) *Escherichia coli* O157: H7 identification in Minas Frescal cheese commercialized in Goiânia, GO. Brazil. J. Microbiol. 10, 7–10.

Chang, Y.-C., et al., 2013. Rapid single cell detection of *Staphylococcus aureus* by aptamer-conjugated gold nanoparticles. Sci. Rep. 3, 1863, http://www.pubmedcentral.nih.gov/articlerender.fcgi?artid=3659324&tool=pmcentrez&rendertype=abstract

Chen, J., et al., 2012. Development and evaluation of a multiplex PCR for simultaneous detection of five foodborne pathogens. J. Appl. Microbiol. 112 (4), 823–830.

Chen, X., et al., 2014. Screening and Identification of DNA aptamers against T-2 toxin assisted by graphene oxide. J. Agric. Food Chem. 62 (42), 10368–10374.

Cremonesi, P., et al., 2007. Detection of enterotoxigenic *Staphylococcus aureus* isolates in raw milk cheese. Lett. Appl. Microbiol. 45 (6), 586–591, http://www.ncbi.nlm.nih.gov/pubmed/17916131

Decory, T.R., et al., 2005. Development of an immunomagnetic bead-immunoliposome fluorescence assay for rapid detection of of *Escherichia coli* O157:H7 in aqueous samples and comparison of the assay with a standard microbiological method. Appl Environ Microbiol. 71 (4), 1856–1864.

DeGrasse, J.A., 2012. A single-stranded DNA aptamer that selectively binds to *Staphylococcus aureus* enterotoxin B. PLoS One 7 (3).

Derzelle, S., et al., 2011. A quantitative PCR assay for the detection and quantification of Shiga toxin-producing *Escherichia coli* (STEC) in minced beef and dairy products. Int. J. Food Microbiol. 151 (1), 44–51, http://dx.doi.org/10.1016/j.ijfoodmicro.2011.07.039

Dolan, L.C., Matulka, R.A., Burdock, G.A., 2010. Naturally occurring food toxins. Toxins 2 (9), 2289–2332, http://www.pubmedcentral.nih.gov/articlerender.fcgi?artid=3153292&tool=pmcentrez&rendertype=abstract

Doyle, M.P., Erickson, M.C., 2006. Reducing the carriage of foodborne pathogens in livestock and poultry. Poult. Sci. 85 (6), 960–973

Duan, N., Wu, S., Zhu, C., et al., 2012b. Dual-color upconversion fluorescence and aptamer-functionalized magnetic nanoparticles-based bioassay for the simultaneous detection of *Salmonella typhimurium* and *Staphylococcus aureus*. Anal. Chim. Acta 723, 1–6, http://www.sciencedirect.com/science/article/pii/S0003267012002620

Duan, N., Wu, S., Ma, X., et al., 2012a. Gold nanoparticle-based flourescence resonance energy transfer aptasensor for ochratoxin A detection. Anal. Lett. 45 (7), 714–723.

Duan, N., et al., 2015. Simultaneous detection of pathogenic bacteria using an aptamer based biosensor and dual fluorescence resonance energy transfer from quantum dots to carbon nanoparticles. Microchim. Acta 182 (5–6), 917–923, http://link.springer.com/10.1007/s00604-014-1406-3

Ellington, A.D., Szostak, J.W., 1990. In vitro selection of RNA molecules that bind specific ligands. Nature 346 (6287), 818–822.

Fang, Z., et al., 2014. Lateral flow biosensor for DNA extraction-free detection of *Salmonella* based on aptamer mediated strand displacement amplification. Biosens. Bioelectron. 56, 192–197, http://dx.doi.org/10.1016/j.bios.2014.01.015

Farabullini, F., et al., 2007. Disposable electrochemical genosensor for the simultaneous analysis of different bacterial food contaminants. Biosens. Bioelectron. 22 (7), 1544–1549.

Fedio, W.M., et al., 2011. Detection of *E. coli* O157:H7 in raw ground beef by Pathatrix (TM) immunomagnetic-separation, real-time PCR and cultural methods. Int. J. Food Microbiol. 148 (2), 87–92, http://dx.doi.org/10.1016/j.ijfoodmicro.2011.05.005

Feng, X., et al., 2015. A novel "dual-potential" electrochemiluminescence aptasensor array using CdS quantum dots and luminol-gold nanoparticles as labels for simultaneous detection of malachite green and chloramphenicol. Biosens. Bioelectron. 74, 587–593, http://www.sciencedirect.com/science/article/pii/S0956566315302153

Geim, A.K., Novoselov, K.S., 2007. The rise of graphene. Nat. Mater. 6 (3), 183–191, http://dx.doi.org/10.1038/nmat1849

Gómez, P., et al., 2010. A fast molecular nondestructive protocol for evaluating aerobic bacterial load on fresh-cut lettuce. Food Sci. Technol. Int. 16 (5), 409–415.

Gómez-govea, M., et al., 2012. Analysis of microbial contamination levels of fruits and vegetables at retail in Monterrey, Mexico. Food Agric. Environ. 10 (1), 152–156.

Gopinathan, R., Kanhere, J., Banerjee, J., 2015. Effect of malachite green toxicity on non target soil organisms. Chemosphere 120, 637–644, http://www.sciencedirect.com/science/article/pii/S0045653514011102

Grate, D., Wilson, C., 1999. Laser-mediated, site-specific inactivation of RNA transcripts. Proc. Natl. Acad. Sci. USA 96 (11), 6131–6136, http://www.pnas.org/content/96/11/6131.short

Gupta, S., et al., 2014. Chemically modified DNA aptamers bind interleukin-6 with high affinity and inhibit signaling by blocking its interaction with interleukin-6 receptor. J. Biol. Chem. 289 (12), 8706–8719.

Han, S.R., Yu, J., Lee, S.W., 2014. In vitro selection of RNA aptamers that selectively bind danofloxacin. Biochem. Biophys. Res. Commun. 448 (4), 397–402, http://dx.doi.org/10.1016/j.bbrc.2014.04.103

Hegazy AI, O.B.A., 2014. Detection of *Salmonella* in food samples by culture and polymerase chain reaction methods. J. Bacteriol. Parasitol. 05 (03), 5–7, http://www.omicsonline.org/open-access/detection-of-salmonella-in-food-samples-by-culture-and-polymerase-chain-reaction-methods-2155-9597.1000187.php?aid=28242

Hoinka, J., et al., 2012. Identification of sequence-structure RNA binding motifs for SELEX-derived aptamers. Bioinformatics 28 (12), 115–128.

Hoinka, J., et al., 2014. AptaCluster: a method to cluster HT-SELEX aptamer pools and lessons from its application. In Lecture Notes in Computer Science. Res. Comput. Mol. Biol. 8394, 115–128, http://link.springer.com/chapter/10.1007/978-3-319-05269-4_9

Holeman, L.A., et al., 1998. Isolation and characterization of fluorophore-binding RNA aptamers. Fold. Design 3 (6), 423–431, http://www.sciencedirect.com/science/article/pii/S1359027898000595

Hou, W., et al., 2015. Development of colloidal gold immunochromatographic strips for detection of *Riemerella anatipestifer*. PloS One 10 (3), e0122952, http://dx.plos.org/10.1371/journal.pone.0122952

Javier, D.J., et al., 2008. Aptamer-targeted gold nanoparticles as molecular-specific contrast agents for reflectance imaging. Bioconjug. Chem. 19 (6), 1309–1312.

Jin, Q., et al., 2012. Development of an immunochromatographic strip for the detection of antibodies against Porcine circovirus-2. J. Vet. Diag. Investig. 24 (6), 1151–1157.

Jing, M., Bowser, M.T., 2013. Tracking the emergence of high affinity aptamers for rhVEGF165 during capillary electrophoresis-systematic evolution of ligands by exponential enrichment using high throughput sequencing. Anal. Chem. 85 (22), 10761–10770, http://www.pubmedcentral.nih.gov/articlerender.fcgi?artid=3892959&tool=pmcentrez&rendertype=abstract

Jo, M., et al., 2011. Development of single-stranded DNA aptamers for specific bisphenol A detection. Oligonucleotides 21 (2), 85–91, http://online.liebertpub.com/doi/full/10.1089/oli.2010.0267

Ju, Y., et al., 2010. Development of colloidal gold-based immunochromatographic assay for rapid detection of *Streptococcus suis* serotype 2. Vet. Immunol. Immunopathol. 133 (2–4), 207–211.

Khan, A.B., et al., 2014. Rapid detection of *Salmonella* in food samples by polymerase chain reaction after a 10 h pre-enrichment. J. Food Saf. 34 (1), 79–86, http://doi.wiley.com/10.1111/jfs.12099

Kim, M., et al., 2009. Arsenic removal from Vietnamese groundwater using the arsenic-binding DNA aptamer. Environ. Sci. Technol. 43 (24), 9335–9340, http://dx.doi.org/10.1021/es902407g

Kumar, B.K., et al., 2011. Development of monoclonal antibody based sandwich ELISA for the rapid detection of pathogenic *Vibrio parahaemolyticus* in seafood. Int. J. Food Microbiol. 145 (1), 244–249, http://www.ncbi.nlm.nih.gov/pubmed/21276628

Lakshmi, V., Pradesh, A., 2012. Food adulteration. IJSIT 1 (2), 106–113.

Law, J.W.-F., et al., 2014. Rapid methods for the detection of foodborne bacterial pathogens: principles, applications, advantages and limitations. Front. Microbiol. 5, 770, http://www.pubmedcentral.nih.gov/articlerender.fcgi?artid=4290631&tool=pmcentrez&rendertype=abstract

Lee, J., et al., 2011. Aptamer sandwich-based carbon nanotube sensors for single-carbon-atomic-resolution detection of non-polar small molecular species. Lab Chip 11 (1), 52–56, http://www.ncbi.nlm.nih.gov/pubmed/20967396

Lee, K.-M., et al., 2015. Review of *Salmonella* detection and identification methods: aspects of rapid emergency response and food safety. Food Control 47, 264–276, http://linkinghub.elsevier.com/retrieve/pii/S0956713514003879

Liu, A., et al., 2013. Gold nanoparticle-based colorimetric detection of staphylococcal enterotoxin B using ssDNA aptamers. Eur. Food Res. Technol. 237 (3), 323–329.
Liu, H., et al., 2010. Electrochemical immunosensors for antibodies to peanut allergen Ara h2 using gold nanoparticle-peptide films. Anal. Chem. 82 (13), 5865–5871, http://pubs.acs.org/doi/abs/10.1021/ac101110q
Lu, E., et al., 2014. Aptaligner: automated software for aligning pseudorandom DNA X-aptamers from next-generation sequencing data. Biochemistry 53 (22), 3523–3525, http://dx.doi.org/10.1021/bi500443e
Ma, W., Noble, W.S., Bailey, T.L., 2014. Motif-based analysis of large nucleotide data sets using MEME-ChIP. Nat. Prot. 9 (6), 1428–1450, http://dx.doi.org/10.1038/nprot.2014.083
Mandal, P.K., et al., 2011. Methods for rapid detection of foodborne pathogens: an overview. Am. J. Food Technol. 6 (2), 87–102.
McCormick, S.P., et al., 2015. Anomericity of T-2 Toxin-glucoside: masked mycotoxin in cereal crops. J. Agric. Food Chem. 63 (2), 731–738, http://dx.doi.org/10.1021/jf504737f
McKeague, M., et al., 2014. Selection and characterization of a novel DNA aptamer for label-free fluorescence biosensing of ochratoxin A. Toxins 6 (8), 2435–2452, http://www.mdpi.com/2072-6651/6/8/2435
Moon, J., Kim, G., Park, S., 2014. Development of ssDNA aptamers for the capture and detection of *Salmonella typhimurium*. Anal. Meth. 6 (18), 7442–7448, http://pubs.rsc.org/en/Content/ArticleHTML/2014/AY/C4AY01035C
Nadal, P., et al., 2012. DNA aptamers against the lup an 1 food allergen. PLoS One 7 (4), 17–19.
Nel, A.E., et al., 2009. Understanding biophysicochemical interactions at the nano-bio interface. Nat. Mater. 8 (7), 543–557, http://www.ncbi.nlm.nih.gov/pubmed/19525947
O'Malley, R.P., et al., 1986. A mechanism for the control of protein synthesis by adenovirus VA RNAI. Cell 44 (3), 391–400, http://www.sciencedirect.com/science/article/pii/0092867486904605
Olmsted, I.R., et al., 2011. Measurement of aptamer–protein interactions with back-scattering interferometry. Anal. Chem. 83 (23), 8867–8870, http://dx.doi.org/10.1021/ac202823m
Peng, F., et al., 2008. Development of an immunochromatographic strip for rapid detection of H9 subtype avian influenza viruses. Clin. Vacc. Immunol. 15 (3), 569–574.
Pollet, J., et al., 2011. Fast and accurate peanut allergen detection with nanobead enhanced optical fiber SPR biosensor. Talanta 83 (5), 1436–1441, http://www.ncbi.nlm.nih.gov/pubmed/21238733
Russo, P., et al., 2014. A fast, reliable, and sensitive method for detection and quantification of *Listeria monocytogenes* and *Escherichia coli* O157 : H7 in ready-to-eat fresh-cut products by MPN-qPCR. BioMed. Res. Int. 2014 (608296), 9.
Sajid, M., Kawde, A.-N., Daud, M., 2015. Designs, formats and applications of lateral flow assay: a literature review. J. Saudi Chem. Soc. 19, 689–705, http://linkinghub.elsevier.com/retrieve/pii/S131961031400129X
Sarell, C.J., et al., 2014. Distinguishing closely related amyloid precursors using an an RNA aptamer. J. Biol. Chem. 289 (39), 26859–26871, http://www.jbc.org/content/289/39/26859.short
Schlosser, G., et al., 2007. Coupling immunomagnetic separation on magnetic beads with matrix-assisted laser desorption ionization-time of flight mass spectrometry for detection of staphylococcal enterotoxin B. Appl. Environ. Microbiol. 73 (21), 6945–6952.

Seidel, S.A.I., et al., 2014. Thermophoresis in nanoliter droplets to quantify aptamer binding. Angew. Chem. Int. Ed. 53 (30), 7948–7951, http://www.ncbi.nlm.nih.gov/pubmed/24895233

So, H.M., et al., 2008. Detection and titer estimation of *Escherichia coli* using aptamer-functionalized single-walled carbon-nanotube field-effect transistors. Small 4 (2), 197–201.

Stead, S.L., et al., 2010. An RNA-aptamer-based assay for the detection and analysis of malachite green and leucomalachite green residues in fish tissue. Anal. Chem. 82 (7), 2652–2660.

Song, K.-M., Lee, S., Ban, C., 2012. Aptamers and their biological applications. Sensors (Basel, Switzerland) 12 (1), 612–631, http://www.pubmedcentral.nih.gov/articlerender.fcgi?artid=3279232&tool=pmcentrez&rendertype=abstract

Sors, T.G., Ellis, D.R., Salt, D.E., 2005. Selenium uptake, translocation, assimilation and metabolic fate in plants. Photosynth. Res. 86 (3), 373–389, http://www.ncbi.nlm.nih.gov/pubmed/16307305

Sung, H.J., et al., 2014. Inhibition of human neutrophil activity by an RNA aptamer bound to interleukin-8. Biomaterials 35 (1), 578–589.

Thiel, W.H., et al., 2011. Nucleotide bias observed with a short SELEX RNA aptamer library. Nucl. Acid Therap. 21 (4), 253–263, http://www.pubmedcentral.nih.gov/articlerender.fcgi?artid=3198618&tool=pmcentrez&rendertype=abstract

Tok, J.B.-H., Fischer, N.O., 2008. Single microbead SELEX for efficient ssDNA aptamer generation against botulinum neurotoxin. Chem. Commun. (16), 1883–1885, http://pubs.rsc.org/en/content/articlehtml/2008/cc/b717936g

Tran, D.T., et al., 2010. Selection and characterization of DNA aptamers for egg white lysozyme. Molecules 15 (3), 1127–1140, http://www.ncbi.nlm.nih.gov/pubmed/20335968

Tran, D.T., et al., 2013. Selection of aptamers against Ara h 1 protein for FO-SPR biosensing of peanut allergens in food matrices. Biosens. Bioelectr. 43 (1), 245–251, http://dx.doi.org/10.1016/j.bios.2012.12.022

Tuerk, C., Gold, L., 1990. Systematic evolution of ligands by exponential enrichment: RNA ligands to bacteriophage T4 DNA polymerase. Science 249 (4968), 505–510.

Wacoo, A.P., et al., 2014. Methods for detection of aflatoxins in agricultural food crops. J. of Appl. Chem. 2014 (706291), 1–15, http://www.hindawi.com/journals/jac/2014/706291/

Wang, L., et al., 2011. An aptamer-based chromatographic strip assay for sensitive toxin semi-quantitative detection. Biosens. Bioelectron. 26 (6), 3059–3062, http://dx.doi.org/10.1016/j.bios.2010.11.040

Wang, Z., et al., 2013. Immunomagnetic separation combined with polymerase chain reaction for the detection of *Alicyclobacillus acidoterrestris* in apple juice. PLoS One 8 (12), 8–15.

Wilson, C., Szostak, J.W., 1998. Isolation of a fluorophore-specific DNA aptamer with weak redox activity. Chem. Biol. 5 (11), 609–617, http://www.ncbi.nlm.nih.gov/pubmed/9831529

Wilson, R., et al., 2014. Single-step selection of bivalent aptamers validated by comparison with SELEX using high-throughput sequencing. PLoS One 9 (6), e100572, http://www.pubmedcentral.nih.gov/articlerender.fcgi?artid=4070925&tool=pmcentrez&rendertype=abstract

Wu, S., et al., 2012. Multiplexed fluorescence resonance energy transfer aptasensor between upconversion nanoparticles and graphene oxide for the simultaneous determination of mycotoxins. Anal. Chem. 84 (14), 6263–6270.

Wu, S., et al., 2014. Simultaneous aptasensor for multiplex pathogenic bacteria detection based on multicolor upconversion nanoparticles labels. Anal. Chem. 86 (6), 3100–3107.

Xu, D., et al., 2013. Rapid detection of *Campylobacter jejuni* using fluorescent microspheres as label for immunochromatographic strip test. Food Sci. Biotechnol. 22 (2), 585–591.

Yang, M., et al., 2013. Highly specific and cost-efficient detection of *Salmonella paratyphi* A combining aptamers with single-walled carbon nanotubes. Sensors 13 (5), 6865–6881, http://www.pubmedcentral.nih.gov/articlerender.fcgi?artid=3690085&tool=pmcentrez&rendertype=abstract

Yanga, M., Kostova, Y., Hugh, A., Bruckb, A.R., 2009. Gold nanoparticle-based enhanced chemiluminescence immunosensor for detection of Staphylococcal Enterotoxin B (SEB) in food. Int. J. Food Microbiol. 133 (3), 265–271.

Yüce, M., Ullah, N., Budak, H., 2015. Trends in aptamer selection methods and applications. Analyst 140 (16), 5379–5399, http://pubs.rsc.org/en/content/articlehtml/2015/an/c5an00954e

Yeni, F., et al., 2014. Rapid and standardized methods for detection of foodborne pathogens and mycotoxins on fresh produce. Food Control 40 (1), 359–367, http://dx.doi.org/10.1016/j.foodcont.2013.12.020

Zhao, H.K., Wang, W.J., Zhao, P., Zhang, H.B., et al., 2014a. Simultaneous detection of *Salmonella, Listeria monocytogenes* and *Shigella* in poultry samples by Triplex PCR. J. Bacteriol. Mycol. 1 (1), 1–5.

Zhao, X., et al., 2014b. Advances in rapid detection methods for foodborne pathogens. J. Microbiol. Biotechnol. 24 (3), 297–312, http://www.ncbi.nlm.nih.gov/pubmed/24375418

Zhou, B., et al., 2013. Simultaneous detection of six food-borne pathogens by multiplex PCR with a GeXP analyzer. Food Control 32 (1), 198–204, http://dx.doi.org/10.1016/j.foodcont.2012.11.044

Zuo, P., et al., 2013. A PDMS/paper/glass hybrid microfluidic biochip integrated with aptamer-functionalized graphene oxide nano-biosensors for one-step multiplexed pathogen detection. Lab Chip 13 (19), 3921–3928, http://www.pubmedcentral.nih.gov/articlerender.fcgi?artid=3913183&tool=pmcentrez&rendertype=abstract

4

DETECTION OF FOOD CONTAMINANTS BY GOLD AND SILVER NANOPARTICLES

Ida Evangeline Paul*, D. Nanda Kumar*, A. Rajeshwari*, Sruthi Ann Alex*, D. Karthiga*, Ashok M. Raichur,†, N. Chandrasekaran*, Amitava Mukherjee***

**VIT University, Centre for Nanobiotechnology, Vellore, India; **Indian Institute of Science, Department of Materials Engineering, Bangalore, India; †Nanotechnology and Water Sustainability Unit, University of South Africa, Florida Park, Johannesburg, South Africa*

1 Introduction

Food contamination is the presence of harmful toxic chemicals and pathogens in food products that can cause diseases or illness in living beings. The food contaminants include compounds that come either from natural sources or agricultural or industrial waste. The food contaminants occurring in nature are mainly of microbiological origin that includes harmful bacteria and other bacterial and fungal toxins such as aflatoxin that is a contaminant in grains and peanuts. Contamination caused by the presence of toxic organic chemicals, metals, and their respective complexes and radionuclides come under the next category of food contaminants. The food products are contaminated as a result of various human activities like industry, mining, and agriculture. Food free of contamination is one of the most important criteria for public security as food is one of the fundamental parts of living beings. Therefore, steps have to be taken to ensure that the food products are free of physical, chemical, radioactive, and biological contamination that enter into the food system either through handling, processing, and distribution of the food.

The safety of the food products can be ensured by packing them with the nanomaterials that act as biosensors for alerting the consumers about contamination or food spoilage due to pathogens. The nanomaterials even help in monitoring the shelf life of the

Nanobiosensors. http://dx.doi.org/10.1016/B978-0-12-804301-1.00004-7

preserved food products. They can be designed to fluoresce when they come in contact with chemical contaminants or food pathogens, which serve as a signal for the detection of trace amount of contaminants. The duration required to obtain the sensor signal has to be curtailed to a few minutes or seconds, and this is one of the challenges that has been overcome with the design and development of nanomaterial-based probes for food contaminants. The chemicals produced as a result of food contamination can be sensed by the nanoprobes, which serve as an electronic nose/tongue and can be attached to the packaging material. Nanomaterials have unique chemical and physical properties as they have at least one dimension lesser than 100 nm. Due to their outstanding properties they are widely used in nanosensors, biosensors, and immune sensors. They are able to detect the food contaminants, both chemical and microbial adulterants, that are present in the food system (Sekhon, 2010).

In this chapter, we will discuss the mechanism of detecting chemical adulterants and microbial contaminants in food systems using plasmonic nanoparticles (NPs) (gold and silver).

1.1 Chemical Adulterants

Food products have been found to be adulterated with unwanted toxic chemicals or substances knowingly or unknowingly by unethical manufacturers. These toxic chemicals are commonly known as chemical adulterants that occur either naturally as a raw material or get added into the food items due to pollution or could be produced during the processing of food. Intake of these toxic chemicals for a prolonged period even in low quantities has led to many health problems in humans. Therefore food-based products need to be checked thoroughly and should be approved to be free from toxic chemicals before they are sent into the market and taken up by the consumers. The various toxic chemicals are melamine, acrylamide, ampicillin, 2,4-dinitrophenol, and pesticides like dichlorodiphenyl trichloroethane and carbofuran.

Food contamination due to pesticides can occur mainly in the process of keeping the insects and pests under control. Crop yield can be improved with the help of insecticides by decreasing the losses caused by insects. Herbicides, on the other hand, are used to control the growth of weeds. Both herbicides and insecticides are together called pesticides, and they enter into the food supply if they continue to remain in the food. All the pesticides used on the crops are monitored by the USDA (US Department of Agriculture) and must undergo several tests to check the effect of pesticides on humans. The test mainly determines whether the food products pose any health issues after being exposed to pesticides.

The second route of food contamination is through water. Water is primarily used for processing all food products. However, the most poisonous chemicals can also enter the food system through the water that is supplied to the crops. The various toxic chemicals that enter the food system through water are cadmium, mercury, benzene, lead, polychlorinatedbiphenyls (PCBs), and chloroform. The two important sources of contamination in the water supply are industrial waste and pesticides. Some of the pesticides are not readily biodegradable and, therefore the usage of these pesticides is strictly prohibited. Nature can easily recycle the food products after breaking them down into different chemical parts. However, some pesticides are washed away by the rain water into the streams and rivers and, further, to oceans and lakes, and they cannot be easily broken down. The exhausts released by the industries into the atmosphere reach the land by rain. Some of the waste is directly disposed into the water system. Therefore, the waste products from the industries and water sources should be frequently tested to ensure public safety (Ward and Ward, 2007).

In this section, more details regarding melamine (the most commonly used chemical adulterant) and malathion (a widely used pesticide) are discussed.

1.1.1 Melamine

Melamine is an organic chemical widely used in the manufacture of glues, paper, coatings, plastics, filters, flame retardants, and fertilizers (Yang et al., 2009b). Due to its high nitrogen content up to 66% by mass, the food products were adulterated with melamine to produce a false signal for protein content, which was measured using the standard Kjeldahl and Dumas tests. Consumption of melamine on a daily basis has slowly resulted in the formation of complex melamine cyanurate crystals. These complex crystals finally gave rise to various serious problems, including kidney failure, renal breakdown, and even fatality in some cases (Zhao et al., 2009). Two major outbreaks related to foodborne diseases in association with melamine have been reported recently. In Mar. 2007, there was a high rise in the death rate of pet dogs and cats after consuming food contaminated with melamine, which led to renal failure and finally death in animals. The international media on Sep. 2008 also reported kidney-related problems in thousands of infants after consuming milk and infant formulas that were adulterated with melamine (Dobson et al., 2008; Brown et al., 2007; Kennaway, 1921). Therefore, the tolerable daily intake proposed by the World Health Organization for melamine is 0.2 mg melamine per kilogram of body mass. Different maximum residue levels (MRLs) were also set by various countries for melamine. In the United States, the MRL was set to 1.0 mg/kg for infant formula and 2.5 mg/kg for milk and other milk-based products. For all food products in Europe containing more than 15% milk, the Food Safety Authority has set the safety level limit for melamine to 2.5 mg/kg (Zhao et al., 2009; Li et al., 2010b). Therefore, there is an increasing need to quality check the food products in order to ensure that they are free from melamine before they are released into the market.

1.1.2 Malathion

In the earlier decades, pesticides that contain organophosphorus (OP) groups and their derivatives are widely employed in agriculture (Upadhyay et al., 2009). Gerhard Schrader, a German chemist, first synthesized insecticides, and he accidentally discovered nerve agents, namely sarin and tabun and, hence, he has been called "the father of nerve agents" (Tucker, 2006). Nerve agents like soman, sarin, and tabun (widely used in warfare) and pesticides, such as carbofuran, dimethoate, and paraoxon possess OP groups. The excess use of nerve agents that contain OP groups inhibits the active sites of acetylcholinesterase (AChE) in living organisms (Munro et al., 1999).

The increase in the worldwide usage of OP pesticides have several effects on the living beings and environment and have also raised the threat of terrorist attacks by employing these OP compounds as chemical warfare agents (Burnworth et al., 2007). Therefore, there is an increasing concern in developing a simple, rapid, and sensitive detection technique for these AChE inhibitors without the use of sophisticated instruments (Verma and Bhardwaj, 2015). There are several analytical techniques that have been used for the detection of these compounds using spectroscopic methods and chromatographic methods coupled with different detectors. The disadvantages of these methods are the requirement for expensive instruments with well-skilled personnel, difficulty to carry out analysis in the field, and as it is time-consuming.

1.2 Bacterial Adulterants

Though the presence of chemical contaminants in the food system makes it toxic, they do not alter the food. While the presence of bacterial adulterants in food causes undesirable changes in the food system resulting in spoilage of food. Spoilage of food can be described with the help of the changes in its texture, appearance, odor, and/or flavor. Food spoilage can be defined as the deterioration of food to a level in which it is hard to consume by the humans. Deterioration of food is caused by the presence of enzymes and food contaminants that finally lead to food spoilage. The different signs of food spoilage are bad odor and flavor, mushy texture, cloudy appearance, discoloration, development of foams or gas bubbles in the food product, bulging of the can, soft breaks on the surface of the food (fruits and vegetables), development of slimy area, and fuzzy appearance on the food surface.

Consumption of food contaminated with toxic bacteria can lead to various diseases in human beings, as well as, in animals. The foodborne diseases through microbes occur majorly through two different ways. The first type of infection is caused due to the presence of microorganisms that multiply slowly inside the food system. The common pathogenic bacteria are *E. coli* O157:H7, *Bacillus cereus, Campylobacter, Listeria monocytogenes, Salmonella, Vibrio cholera,* brevetoxins, and *Staphylococcus aureus* (Gandhi and Chikindas, 2007; Park et al., 1999; Murphy et al., 2006). In the second type of infection, these microorganisms release a toxic substance during the process of multiplication, sporulation, or lyses. Bacterial strains like *Clostridium perfringens* and *E. coli* can release toxic substances that can cause diseases. Therefore, the detection and further prevention of harmful bacteria and the toxins released by them are very essential in food and food-based products. Herein, a detailed discussion about the bacterial adulterant, *E. coli* O157:H7, and the toxic substance, endotoxin, released by the pathogen (Ward and Ward, 2007) is provided.

1.2.2 E. coli *O157:H7*

The Gram-negative pathogenic bacteria, *E. coli*, inhabit in the intestinal walls of human beings. Food and food-based products are contaminated with *E. coli* during the time of growth and harvesting (horticulture food products) and collection of food like milk or slaughter of meat. Further, postharvest handling, processing, and transport of food, and unhygienic handling of food during the process of preparation can all lead to contamination of food by *E. coli*. The insufficient control of various processing factors like pH, cooking temperature, water activity, and storing in warm conditions for longer period can allow the growth of pathogen and contribute to the survival of *E. coli* in the food system. Few examples of food that can be contaminated by *E. coli* are raw meat or meat that are poorly processed (undercooked beef, fermented meat, etc.), unpasteurized milk, and dairy products like cheese, raw vegetables (mushrooms, spinach, lettuce, sprouted seeds, cantaloupe, etc.), and unpasteurized fruit juices.

E. coli is a foodborne pathogen that releases a toxic substance, commonly referred to as shiga-like toxins, with an infective dosage below 50–100 cells. Food poisoning caused by *E. coli* can lead to inflammation, peritonitis, hemorrhagic colitis, urinary tract infection, diarrhea, and even kidney breakdown in certain conditions (Nataro and Kaper, 1998; Tarr et al., 2005). This has drawn the interest of many researchers around the globe to detect and remove the pathogenic bacteria, *E. coli*, in order to make food and water free of bacterial contaminants (Li et al., 2010a). It was reported by the US Centers for Disease Control and Prevention (CDC) that about 76 million diseases, 325,000 hospitalizations, and 5,000 deaths occurred every year in United States.

1.2.3 Endotoxin

E. coli releases a toxic substance known as endotoxin that is present on the outer membrane walls of the bacteria (Kim et al., 2006). Endotoxin is also widely known as lipopolysaccharide (LPS). It is the most significant contaminant both in pharmaceutical and food industry. The endotoxin is released by bacteria due to damage of the bacterial cell wall or autocytolysis (Cho et al., 2012). The released endotoxin consists of three parts: first, a lipid structure (lipid A) that is unique in nature; second, a lipid attached to an oligosaccharide component; and third, an antigen O situated on the outer cell wall. The toxicity of the endotoxin is due to the presence of lipid A, and it is responsible for causing a large number of diseases in humans. The commonly encountered complaints in human beings are fever, vomiting, inflammatory reaction, diarrhea, low blood pressure, disseminated intravascular coagulation, and septic shock, which can occur as a result of continuous consumption of endotoxin. In the United States alone, 150,000 related disease cases have been reported every year due to septic shock. Therefore, it is essential to sense, control, and eliminate the endotoxin from food and drug, as intake of even small quantities will lead to various diseases in human beings (Kim et al., 2006; Limbut et al., 2007). Since endotoxin can withstand even high temperatures of up to 100°C, they are highly prevalent in food and water where proper sterilization and treatment facilities were not carried out (Uzarski and Mello, 2012). Therefore, it is essential to quality check the food products in order to make sure they are free of endotoxin before they are sent out to the market.

The currently available analytical methods used for analyzing chemical adulterants such as, melamine and malathion, and bacterial adulterants, *E. coli* O157:H7, are discussed further.

1.3 Conventional Methods

The analytical methods used for monitoring the presence of chemical adulterants in the system are liquid chromatography–mass spectroscopy (LC-MS/MS) (Ibáñez et al., 2009; Tran et al., 2009), high-performance liquid chromatography (HPLC) (Peng et al., 2011), gas chromatography-mass spectroscopy (GC-MS) (Li and Li, 2009), mass spectrum (Yang et al., 2009a,b; Tang et al., 2009; Klampfl et al., 2009; Cook et al., 2005), enzyme-linked immunosorbent assay (Garber, 2008), Raman spectrometry (Chen et al., 2010), absorption spectrophotometry (Rima et al., 2009), infrared spectroscopy (Mauer et al., 2009), chemiluminescence (CL) (Ding et al., 2010), and fluorescence sensing (Zhou et al., 2010). Separation methods like capillary zone electrophoresis were also used to detect melamine in food (Yan et al., 2009). Even though the currently available methods are highly sensitive, the cost of the instruments is very high and need highly trained technicians. Pretreatment of food samples is also a tedious process and a time-consuming step.

The available conventional methods for detecting pathogenic bacteria (*E. coli* O157:H7) include culturing techniques (March and Ratnam, 1986; Gracias and McKillip, 2004), polymerase chain reaction (Jofré et al., 2005; Liu et al., 2008; Cady et al., 2005; Savoye et al., 2011), enzyme-linked immunosorbent assays (Park et al., 2008; Tsai et al., 2010; Li et al., 2010a; Lazcka et al., 2007), surface plasmon resonance (SPR) (Eum et al., 2010; Jindou et al., 2004; Uchida et al., 2004; Homola, 2008), long-range surface plasmon-enhanced–fluorescence spectroscopy (LRSP-FS) (Huang et al., 2011), and inductively coupled plasma–mass spectrometry (ICP-MS) (Li et al., 2010a). The endotoxin released from the *E. coli* O157:H7 was also quantified using the limulus amebocyte lysate (LAL) method. This method is entirely based on the enzymatic blood coagulation of a marine arthropod (horse shoe crab) when endotoxin is present in the system. Though this method is very highly sensitive, it has few drawbacks as it is not sensitive to environmental conditions like variations in pH, temperature, and pyrogen present in water. This method also gives false signal in presence of other carbohydrate-like molecules (β-glucan) and is unable to sense the inactive endotoxin present in food system (Uzarski and Mello, 2012; Cheng et al., 2012). Fishing of horse shoe crabs was also prohibited by the authorities as they are on the verge of extinction, and this made the extraction of LAL highly impossible (Sun et al., 2012). Even though the available conventional methods are sensitive, they require high degrees of preconcentration, expensive instruments, and is also time-consuming. Therefore, efforts should be taken to develop a simple, cost-effective, and highly sensitive nanobased probe to detect food contaminants, both chemical and pathogenic bacteria, in the food system.

1.3.1 Plasmonic Nanoparticles

Plasmonic NPs are metal NPs such as gold (Au), silver (Ag), copper (Cu), platinum (Pt), palladium (Pd), ruthenium, and rhenium (Re), which have exceptional properties, both physical and chemical. These properties entirely differ from that of the bulk or atomic states as they exhibit quantum size effects and give rise to definite electronic structures. Among the metal NPs, gold NPs (AuNPs), and silver NPs (AgNPs) have drawn the attention of researchers due to their potential in diverse applications like biological and chemical sensor, biology, catalysis, optoelectronic materials, and storage of information (Homola, 2008; Ghosh et al., 2008; Eustis and El-Sayed, 2006; Song et al., 2010; Wang and Ma, 2009; Murphy et al., 2008; Daniel and Astruc, 2004). They also have outstanding biocompatibility in comparison to the other biomacromolecules (proteins and nucleic acids). AuNPs and AgNPs exhibit a characteristic optical property known as SPR due to the interaction between the incident light and free electrons on the metal NP surface, which causes the resonance of these surface plasmons (Vaccarello et al., 2012). SPR is known to be very highly sensitive to the environmental changes, and therefore, can be used as the basis for developing a sensor probe. Depending on the size, shape, and capping of AuNPs and AgNPs, their light scattering properties can be tuned. AgNPs and AuNPs are known to have high extinction coefficient range in the visible region of around 1×10^8–1×10^{10} $M^{-1}\ cm^{-1}$ (Wang et al., 2007; Eustis and El-Sayed, 2006; Prathna et al., 2011; Ravindran et al., 2010; Li and Bian, 2009).

2 Melamine Detection Using Gold Nanoparticles

2.1 Synthesis of Spherically Shaped Gold Nanoparticles

AuNP synthesis can be carried out using two different approaches, namely "top-down" and "bottom-up" approach. In top-down process, the AuNPs are obtained by systematically breaking down the bulk state of Au. In the case of bottom-up approach, AuNPs are synthesized from molecules individually either using a chemical or biological reducing agent (Nguyen et al., 2011). The bottom-up approach using a chemical as a reducing agent is widely used in order to obtain repeatability of AuNPs, for varying their shape and size, and to obtain desired functionalization (Parab et al., 2011). The synthesis of AuNPs involves two major steps: (1) reduction process with the help of chemicals like citric and oxalic acids, borohydrides, hydroxylamines, aminoboranes, formaldehydes, saturated and unsaturated alcohols, hydrazines, polyols, hydrogen peroxide (H_2O_2), carbon monoxide, sugars, acetylene, sulfites, hydrogen, and monoelectronic reducing agents; (2) stabilization process with the help of compounds like trisodium citrate dehydrate, dendrimers, surfactants (cetyltrimethylammonium bromide), polymers, phosphorus ligands, sulphur, oxygen, and nitrogen-based ligands, (Zhao et al., 2013).

$$AuHCl_4 + Na_3C_6H_5O_7 \xrightarrow{\Delta} Au$$

Herein, AuNPs are synthesized by Turkevich method, which is one of the most popular in situ synthesis processes for AuNPs (Frens, 1973). Trisodium citrate is known to reduce chloroauric salt to chloroaurous salt and then to Au^0. Further assembling of reduced aurous salt on Au^0 atoms leads to the formation of AuNPs. A visual color change from yellow to dark red can be observed using the naked eye, and this indicated the completion of the process. Trisodium citrate plays the role of both reducing and stabilizing agent.

The size of AuNPs was varied by Frens from 15 to 150 nm by simply varying the concentration ratio of $HAuCl_4$ and trisodium citrate (Frens, 1973). Kimling et al. (2006) also reported that as the concentration of trisodium citrate increased, the stabilization of AuNPs occurred faster for smaller sizes, whereas the aggregation of AuNPs was observed as the concentration of trisodium citrate was decreased gradually. Initial characterization of AuNPs was carried out using ultraviolet (UV)–visible spectroscopy by observing the strong SPR peak at 520 nm, which confirmed the formation of NPs. AuNPs are stable at a temperature of 40°C due to the electrostatic repulsion occurring between the negatively charged citrate ions found on the surface of AuNPs.

2.2 Size Characterization

The effect of different sizes of AuNPs on the colorimetric detection of melamine was studied to develop a simple AuNP probe with an optimized size that was able to detect low levels of melamine in real samples (Paul et al., 2015). The standard Turkevich method was employed to synthesize AuNPs (AuNPs-I, AuNPs-II, and AuNPs-III) of various sizes. The initial pH values of AuNPs after synthesizing were determined to be 4.0, 7.0, and 4.0 for AuNPs-I (15 nm), AuNPs-II (30 nm), and AuNPs-III (40 nm), respectively. The size distributions of AuNPs were determined using the scanning electron microscope (SEM) and dynamic light scattering (DLS). The AuNP size and morphological characteristics were studied using SEM, which showed that the AuNPs-I, AuNPs-II, and AuNPs-III were of spherical shape having a mean SEM size of 8 ± 1, 20 ± 1, and 22 ± 1 nm, respectively. The mean hydrodynamic diameters of AuNPs-I, AuNPs-II, and AuNPs-III measured using DLS were reported to be 15 ± 1, 30 ± 1, and 40 ± 1 nm. The mean hydrodynamic diameter of AuNPs was slightly higher than the theoretically calculated mean SEM size as the DLS method measured the ligand shell along with the hydrodynamic radius, while in the case of SEM, only the inner core is measured. The DLS technique is used widely in colloidal chemistry to study the aggregation of AuNPs; therefore, the mean hydrodynamic diameters of AuNPs (AuNPs-I: 15 nm, AuNPs-II: 30 nm, and AuNPs-III: 40 nm) were considered. Jans et al. (2009) also reported DLS technique to be the most effective method for studying the conjugation on AuNPs of different sizes and their aggregation state. In another report, Liu et al. (2009) also considered the DLS size for analyzing the aggregation of AuNPs after interaction with Hg (II).

The effect of various sizes of as-synthesized AuNPs capped with citrate has been studied for the colorimetric detection of melamine in aqueous solution.

2.3 Melamine Detection Using Different Sizes of AuNPs (15, 30, and 40 nm)

The developed colorimetric probe for melamine detection was entirely based on the aggregation of citrate-capped AuNPs. The presence of melamine in the aqueous solution induced the aggregation of AuNPs, and this was visualized by the change in solution color from wine red to blue color. The mechanism of AuNP aggregation was reported to be due to the ligand exchange between the amine groups ($—NH_2$) of melamine and citrate ions of AuNPs that are available on the outer surface (Li et al., 2010b; He et al., 2005; Fang et al., 2004; Thomas and Kamat, 2000) (Fig. 4.1). The effect of citrate-capped AuNPs of three different sizes, namely AuNPs-I (15 nm), AuNPs-II (30 nm), and AuNPs-III (40 nm) on melamine detection in aqueous solution was studied, and the sensitivity of detection was improved on using the optimized AuNP size. Previous reports are also available wherein the effect of AuNP size on sensing melamine was studied by Li et al. (2010a,b) and Zhou et al. (2011) using different synthesis approaches and with the help of a surface modifier. Zeng et al. also studied the aggregation of AuNPs of different sizes after interacting with protonated melamine, which can be induced by HCl, for which the AuNP probe was developed without using any surface functionalization or inducers (Zeng et al., 2012).

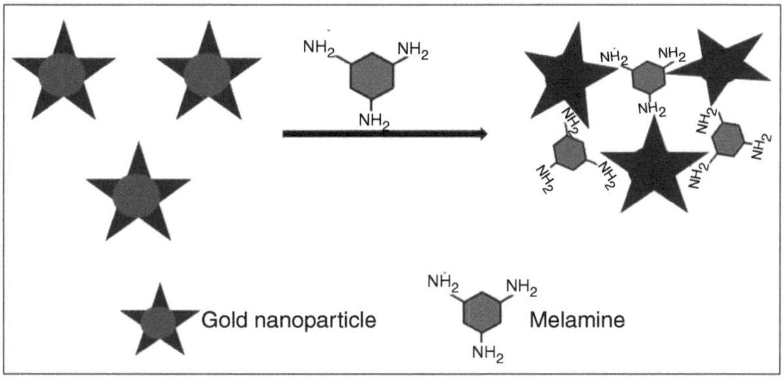

Figure 4.1. Schematic illustration of melamine detection using citrate-capped gold nanoparticles.

AuNPs-I, AuNPs-II, and AuNPs-III exhibited strong SPR bands at 527, 525, and 526 nm, respectively, as observed from the UV–visible spectra. A steady peak shift was noted to higher wavelengths of 638, 626, and 680 nm for the aggregated AuNPs-I, AuNPs-II, and AuNPs-III, respectively, after interaction with highest concentration (1 mM) of melamine. Fig. 4.2 shows the absorption spectra of citrate-capped AuNPs-II before and after interaction with melamine. The obtained λ_{max} of AuNPs-III (40 nm) was higher compared to the other sizes of AuNPs (AuNPs-I and AuNPs-II) as λ_{max} was known to increase as the size of AuNPs increases. However,

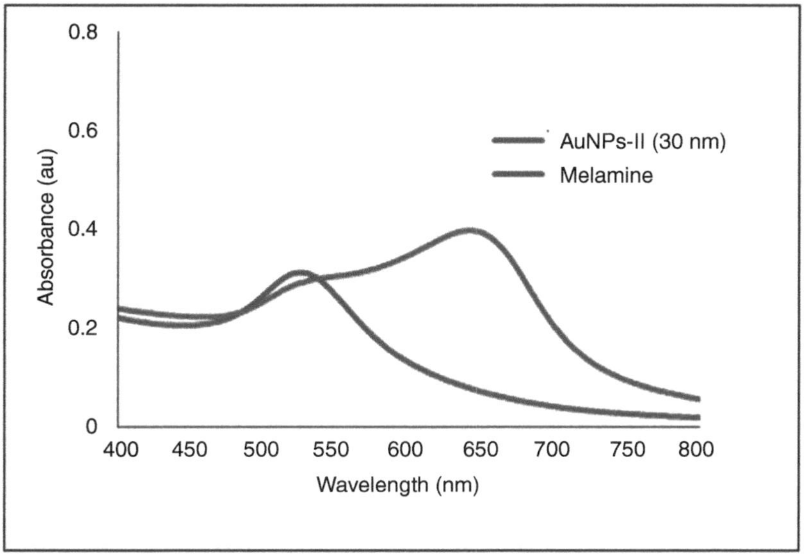

Figure 4.2. Absorption spectra of citrate-capped AuNPs before and after interaction of melamine.

in this case, the λ_{max} of AuNPs-I (15 nm) was higher than that of AuNPs-II (30 nm), regardless of having a lower size. This can be explained as the molar concentration of AuNPs-I was higher than that of AuNPs-II (Mie, 1976; Paul et al., 2015; Zeng et al., 2012).

The mean hydrodynamic diameter values of the aggregated AuNPs measured using DLS were noticed to be 1160 ± 2, 620 ± 2, and 116 ± 2 nm. The increase in the DLS size and the aggregation of NPs observed in the SEM image further confirmed the interaction of AuNPs with melamine. Under the optimized parameters (pH, volumetric ratio, and size of AuNPs), the sensitivity of melamine detection using the three different sizes of AuNPs (AuNPs-I, AuNPs-II, and AuNPs-III) was further studied.

2.4 Sensitivity of Melamine Detection

The effect of AuNP size on the sensitivity of melamine detection was studied. The sensitivity of the method was determined using AuNPs-I at pH 3 and under the optimized volumetric ratio, that is, 400 μL of AuNPs with 600 μL of melamine (pH 7) for a concentration range from 1.6 to 0.8 μM. The linear equation $y = 0.375x + 0.39$ ($R^2 = 0.9279$) was derived between the concentrations of melamine and their respective $A_{638/527}$ values. The limit of detection (LOD) was calculated theoretically, and it was noted to be 3.3×10^{-8} M using AuNPs-I. Melamine sensing using AuNPs-II was also studied under the same optimized volumetric ratio for concentrations ranging from 1.8 to 0.4 μM. From the linear equation $y = 0.526x + 0.1619$ ($R^2 = 0.9745$), the theoretically calculated LOD for detecting melamine was noted to be 2.37×10^{-8} M. The linearity was obtained between the different concentrations of melamine and their respective absorbance ratios $A_{626/525}$. In the case of AuNPs-III (pH 4), the sensitivity of melamine detection using the same volume ratio at pH 7 was determined for melamine concentrations ranging from 14 to 2 μM, and a linear relationship ($y = 0.0674x + 0.1112$; $R^2 = 0.9812$) between the melamine concentrations and $A_{680/526}$ ratios was obtained. The calculated LOD for melamine sensing using AuNPs-III was noted to be 8.9×10^{-8} M.

The sensitivity of detection was completely dependent on the size of AuNPs and their molar concentration. The sensitivity increased with increase in the AuNP size, but as the molar concentration of AuNPs-III (40 nm) was higher when compared to AuNPs-I and AuNPs-II, it led to lower melamine detection. As the LOD of AuNPs-II (30 nm) was lower, a better sensitivity was obtained in comparison to the other two sizes, namely AuNPs-I (15 nm) and AuNPs-III (40 nm) (Zeng et al., 2012). The particle size was also determined for AuNPs-II after interaction with different melamine concentrations ranging from 10^{-3} to 10^{-6} M using the DLS method, and it was noted to be 620 ± 2, 333 ± 2, 92 ± 2, and 72 ± 2 nm, respectively. The mean hydrodynamic size for AuNPs-II was found to have progressively increased after melamine addition when compared to the AuNP size in the absence of melamine (30 ± 1 nm). Wei et al. (2010) also reported that the mean hydrodynamic diameter of AuNPs after interaction with melamine was found to increase with increase in the concentration of melamine. The one-way ANOVA (one-way analysis of variance) test performed using Microsoft excel sheet, determined the sensitivity of the developed AuNPs-II probe for a wide range of melamine concentrations (10^{-3}–10^{-6} M), and it was observed to be statistically significant having a p-value <0.0001.

2.5 Interference Study

The interference studies showed that the aggregation of AuNPs-II was negligible in the presence of common ions like (Cl^-, Ca^{2+}, K^+, NH_4^+, Na^+, SO_4^{-2}, and PO_4^{3-}), amino acids [histidine (His), glycine (Gly), arginine (Arg) and tryptophan (Trp)], and other compounds like bovine serum albumin (BSA), ascorbic acid (AA), sucrose (SUC), and dextrose. The aggregation of AuNPs-II was only observed upon the addition of melamine, resulting in the change of color from wine red to blue, which can be clearly seen with the naked eye. The interaction studies showed a distinct red shift in the SPR band for AuNPs-II from 525 to 626 nm in the visible region only in the presence of melamine. However, the other interfering compounds did not cause any noticeable visual color change or shift in the SPR peak, which showed that the developed probe, AuNPs-II (30 nm), was specific for detecting melamine in aqueous solution. The AuNP probe that was developed was also used to detect the presence of melamine in milk matrices like milk powder, liquid milk, and raw milk after pretreatment.

2.6 Pretreatment of Milk Samples

Milk samples are complex in nature because of the presence of abundant proteins and carbohydrates that make the analysis of these samples difficult. Therefore, it is essential to carry out the effective isolation and extraction process to isolate melamine from the complex milk system before the samples are used for detection. The main advantage of carrying out the pretreatment of milk samples that involves the extraction, preconcentration, and derivatization of the samples is to obtain a lower limit of detection (LOD) by eliminating the unwanted matrix components that may interfere with the detection process

(Yuan et al., 2008). However, most of the pretreatment methods that include the extraction process followed by the clean-up steps consume time, and it may vary from 10 min to several hours.

2.6.1 Extraction of Melamine From Milk Sample

Melamine was found to hydrolyze in strong alkali and acid conditions as it is a weakly alkaline chemical compound. Therefore, the extraction of milk samples should be carried out in acidic, neutral, and alkaline pH conditions. The most common pH ranges used for the pretreatment of milk samples were ≤3.0 and 7.0. At pH 7.0 (neutral extraction), pretreatment can be done either using a mixture of acetonitrile-water or by methanol-water (Filigenzi et al., 2008). While in the case of pH ≤3.0 (acidic extraction), the pretreatment was done with the help of different mixtures or solvents like formic acid, trichloroacetic acid, acetic acid, trifluoroacetic acid, or hydrochloric acid (Sun et al., 2010). Among these solvents, trichloroacetic acid was widely used in the pretreatment. Since the milk system has abundant protein and fat, the sample defatting and precipitation of protein are also carried out along with the pretreatment process. The protein precipitation along with the extraction of melamine were carried out by the solvents, acetonitrile and trichloroacetic acid, for simplifying the purification steps (Sun et al., 2010).

2.6.2 Purification of Milk Samples

Purification of the milk samples after the extraction process is very important in order to remove the fats and proteins in milk that interfere with the detection of melamine in the milk system. Therefore, effective and suitable purification methods are adopted before they are analyzed. Purification of milk samples can be carried out by centrifugation, filtration, and derivatization. Liquid–liquid extraction and solid-phase extraction (Yokley et al., 2000; Tittlemier et al., 2009) are the most frequently and widely used purification processes for melamine extraction. In an acidic solution, melamine, a weakly alkaline chemical compound, is found to electrolyze to a cation (positive ion). The three different cation exchange solid-phase column extractors, namely mixed-mode cation exchange, strong cation exchange, and polystyrene cation exchange are employed to efficiently eliminate the acidic and neutral contaminants. Solid-phase extraction cartridges are used for the selective detection of melamine interferents and is one of the most widely used extraction and concentration process for many milk systems (Zhu et al., 2009; Tittlemier et al., 2009; Yokley et al., 2000). Several reports are also available where extraction was carried out without the help of solvents, for example single-drop microextraction and solid-phase microextraction processes. In brief, the milk samples were washed initially with an acidic solution followed with methanol as the acidic condition is ideal for combining the cation and melamine. Reports are also present where molecularly imprinted polymer is used for selectively removing the interferents from the milk system. The sensitivity of melamine detection is dependent on the extraction method and the order of washing and elution (Yang et al., 2009a; He et al., 2009; Sun et al., 2010).

2.7 Melamine Detection in Milk Samples Using AuNPs-II

Herein, the pretreatment of the milk samples were carried out at neutral pH 7.0 using trichloroacetic acid and chloroform. Casein, a phosphoprotein present in the milk system was removed from the system as it interferes with the detection of melamine. The calcium cations were separated by pretreating the milk samples, which leaves back the insoluble phosphoprotein in the milk system (Li et al., 2010b). The samples were then centrifuged and filtered using 0.22-μm filter to remove completely any protein residues.

The milk samples (liquid milk, milk powder, and raw milk) spiked with the melamine compound was used for sensing in a range of concentrations from 1.4 to 0.4 μM using the developed probe, and the recovery percentage was determined. The *t* values calculated using the Student *t*-test in the Microsoft excel sheet for liquid milk, raw milk, and milk powder were found to be 0.831, 0.745, and 0.965, which were much less than the tabulated *t*-value, that is, 2.22. The *t*-test showed that the plotted calibration curve for the different milk samples was insignificant with respect to the standard calibration curve of the aqueous solution. A clear change in color for the AuNPs-II was observed from wine red to blue when milk samples spiked with melamine was added to the AuNPs-II in the volumetric ratio 2:3. The melamine concentration was determined from the linear equation ($y = 525\,974x + 0.1619$; $R^2 = 0.9745$) by obtaining their respective $A_{626/525}$ ratio. The percentages of melamine recovered for liquid milk, raw milk, and milk powder were calculated to be in the range from 91% to 106%, 94%, and 95% to 104%, respectively.

3 Malathion Detection Using Silver Nanoparticles

Metal and semiconductor NPs are drawing more attention toward the probing of biorecognition process, and their applications have been extensively used in the field of biosensors for specific and efficient detection of analytes, as compared to the other nanomaterials. The NPs, particularly metallic NPs, possess unique optical, electrical, chemical, and magnetic properties (Lee and El-Sayed, 2006). The interesting properties of metal NPs have resulted in the extensive study of their optical spectroscopic properties, that is, optical biosensors that exploit light absorption, reflectance, fluorescence, luminescence, Raman scattering, and refractive index for potential technological applications (Long et al., 2013). The functional use of nanomaterial in biosensors is to sense biological materials, which are mainly proteins, enzymes, antibodies, and immunological molecules (Malik et al., 2013). The commonly employed nanomaterials for this purpose are NPs, nanowires, nanorods, and nanotubes (Jianrong et al., 2004). In recent years, the exploitation of NP-based biosensors promotes the determination of OPs also by exploiting their major advantages, which are rapid response, high sensitivity, real-time measurements, less time-consuming sample pretreatment steps, and sensitive response toward biological molecules (Simonian et al., 2005).

There are several enzyme-based biosensors that have been established based on amperometric technique and colorimetric and fluorescence spectroscopy for the detection of OP group-containing pesticides in the environment using NPs (Mulchandani et al., 2001). The emerging technology of nanosensors offers an attractive tool to detect a wide range of chemical or biological molecules and can be used for related applications. Biorecognition elements like enzymes, functional oligonucleotides, whole cells, and antibodies have also attracted more attention due to their high specificity and high affinity (Dorst et al., 2010). Enzymes are specific toward their substrate and catalyze the formation of specific products. Recently, enzyme-based biosensors that have been extensively studied for environmental monitoring is becoming very vital (Young et al., 2001). The enzyme (AChE) is used as the biosensing element, as it can act selectively on its substrate. The catalytic action of the enzyme can be explained by using the following equation:

$$\mathrm{S}+\mathrm{E} \underset{k_{-1}}{\overset{k_1}{\longleftrightarrow}} \mathrm{ES} \xrightarrow{k_2} \mathrm{E}+\mathrm{P}$$

where, S is substrate, E is enzyme, ES is enzyme–substrate complex, P is product, k_1 is the rate of enzyme–substrate complex formation, k_{-1} is the rate of enzyme–substrate complex dissociation, and k_2 is the rate of dissociation of enzyme–substrate complex to products. AChE inhibitors can decrease the enzymatic activity and affect the product formation, and thus, can be useful for the detection of these inhibitors (Amine et al., 2006).

The principle of enzyme-based biosensor has been established based on the measurement of an inhibitor by assessing the activity of the enzyme in the absence and presence of this inhibitor. Irreversible and reversible inhibition are the two types of inhibition, in which the irreversible inhibition results in the permanently inactivated enzyme due to the covalent binding of the inhibitor to the active site of enzyme. Reversible inhibitors are attached noncovalently to the enzyme, that is, the active site of the enzyme are weakly bound by the reversible inhibitor and can recover to its original activity by dilution or dialysis (Miao et al., 2010). The catalytic activity of these enzymes can be directly or indirectly affected by the analyte, which in turn affects the conversion of the enzyme substrate. The use of NPs makes probing of this reaction easier, as the altered levels of the product (of enzymatic catalysis) can cause variations in the NP stability (aggregation state), and thereby can cause visible color changes.

The potential use and development of biosensors is primarily targeted to help the agricultural industries, as they mainly focus on the determination of AChE inhibitors. This section mainly focuses on the colorimetric determination of OP pesticides in aqueous and food (vegetables, fruits) samples using NPs.

3.1 Biosensors for Pesticide Detection

A pesticide, as defined by the US EPA (2015) (US Environmental Protection Agency), is a chemical or biological substance that can be used to prevent, repel, or destroy pests. The term "pesticides" includes insecticides, fungicides, herbicides, and various other substances that are used to control pests. The presence of pesticidal residues in food, water, and soil can impact the human beings and environment. With respect to human health and environmental concern, the continuous monitoring of even low traces of pesticidal residue levels in food, soil, air, and water must be strictly followed.

Pesticides can pose severe problems and can be mainly classified into three important classes that are organophosphates, organochlorines, and carbamates. The commonly used OP members are methamidophos, fenitrothion, tetrachlorvinphos, malathion, parathion, methyl parathion, diazinon, chlorpyrifos, and dichlorvos. OP compounds include amides, esters, or thiol derivatives of phosphonic, phosphinic, or phosphoric acids (Čolović et al., 2013). There are various types of pesticides that are used in the form of fertilizers in order to enhance food and agricultural production; however, the excessive use of fertilizers can accumulate in human and animal bodies. Pesticides like OPs and carbamates are major AChE inhibitors. AChE can hydrolyze the neurotransmitter acetylcholine (ACh) into choline (Ch), and the hydrolysis can be inhibited by OP compounds (nerve agents and pesticides) and also several toxic chemicals such as, heavy metals and toxins. The action of OP compounds can be imputed to the inhibition of AChE, and the decreased level of Ch results in the termination of nerve impulses (Pundir and Chauhan, 2012).

The emergence of biosensors for the detection of pesticides using colorimetric method has become important, especially in field analysis due to their high efficiency, rapid detection, low cost, and detection of very trace amounts of target substances (pesticides). Biological molecules like AChE, butyrylcholinesterase, and organophosphorus-hydrolase are utilized in the design of colorimetric biosensors for pesticide detection. AChE are mostly used for the detection of OP pesticides because most of the OPs specifically bind to the active site of AChE. The active site of AChE possesses three amino acids, namely serine, His, and aspartic acid. For experimental purposes, acetylthiocholine is used instead of ACh, due to the presence of thiol group. The OP pesticides can covalently bind with the active site of serine group; the nucleophilic part of serine group can covalently bind to the phosphorus atom of the organophosphate and can inhibit the hydrolysis of ATCh into thiocholine (TCh) (Lin et al., 2006).

The characteristic interest for the development of biosensing methods based on the principle of enzyme inhibition has increased as these techniques can be used for the detection of ultralow concentrations of the analyte present in aqueous samples. George L. Ellman developed a photometric method to determine the activity of AChE. In this method, he used a chemical substance, that is, 5,5'-dithiobis-(2-nitrobenzoic acid) or DTNB (Ellman reagent), which can react with the thiol group of the thiocholine (the product of enzymatic hydrolysis) and result in the formation of TNB (2-nitro-5-thiobenzoate), which is a yellow-color product. The higher intensity of the yellow color produced is an indication of the AChE activity (Ellman et al., 1961). The reaction mechanism is as follows:

$$\text{Acetylthiocholine} \xrightarrow{\text{Acetylcholinesterase}} \text{thiocholine} + \text{acetate}$$
$$\text{thiocholine} + \text{dithiobisnitrobenzoate} \longrightarrow \text{yellow color}$$

Based on this principle, few colorimetric-based studies have been developed for the detection of OP pesticides using NPs. NPs (like AuNPs and AgNPs) have high extinction coefficient and unique size-dependent optical properties. In colorimetric assays, the detection of OPs are usually based on the changes in color, which occurs due to change in size and shape of the NPs and also due to the aggregation state of the NPs. The change in color can be further confirmed by the shift in absorption band in the visible region of the electromagnetic spectrum (Choi, 2004; Long et al., 2013). AuNPs and AgNPs are widely used and are the most favorable system for the colorimetric sensing of OP pesticides due to their unique physical and chemical properties and ease of synthesis using chemical reduction method. Based on the change in color of AuNPs, Liu et al. (2012) investigated different types of pesticides, including carbaryl, diazinon, malathion, and phorate using rhodamine B-covered gold nanoparticle (RB-AuNP) dual-readout assay. The mechanism of the system is mainly based on preventing the generation of thiocholine in the presence of inhibitors (pesticides) by decreasing the AChE activity. The addition of AChE and ATCh into RB-AuNP solution resulted in the decrease in absorption intensity, and in the presence of AChE inhibitors along with AChE and ATCh, the absorption band of RB-AuNPs increased significantly. The system achieved a good LOD of 0.1, 0.1, 0.3, and 1 μg/L for carbaryl, diazinon, malathion, and phorate, respectively. The developed system was also assessed for the detection of pesticides in agricultural products like residues of food samples and runoff water. Wang et al. (2009b) reported that the crosslinking/aggregation of AuNPs in the presence of AChE and ATCh resulted in the shift of absorption spectrum. The addition of ATCh alone into AuNP solution did not induce any change in absorption band at 520 nm, but in presence of AChE and ATCh, the absorption band at 520 nm decreased significantly with the appearance of a new broad absorption band at 600 nm with increase in the reaction time. However, the addition of inhibitor (tacrine) to the system resulted in lesser aggregation by suppressing the hydrolysis of ATCh into TCh. With addition of tacrine to the system, the color of AuNP solution remains unchanged (ie, aggregation of AuNPs was prevented).

Recently, Wang et al. (2014b) developed both colorimetric and fluorometric assays for the detection of AChE and its inhibitors, neostigmine and tacrine, that could induce the decrease in fluorescence signal by reducing the production of thiocholine. In this study, the fluorescence and absorption changes were studied by the 1,4-addition of the hydrolyzed product, thiocholine, to α, β-unsaturated ketone, which results in the formation of a thioether that results in the strong fluorescence enhancement and absorption change. The signal response is mainly based on the concentration of the analyte and the type of inhibited enzyme. The addition of neostigmine to the system results in the decrease of fluorescence intensity at 525 nm. The absorption intensity at

499 nm also increased when neostigmine was added to the system. Sun et al. (2011) developed an ultrasensitive sensor for detection of nerve agents and highly toxic OP pesticides by using AuNPs. Citrate-capped AuNPs were synthesized using a chemical reduction method and were found to improve the sensitivity of the system. The detection level of AuNPs can be further enhanced by displacing the citrate ions in order to provide lipoic acid functionalization, as the interaction between citrate and AuNPs was found to be weak. The enzymatic hydrolysis by AChE results in the formation of thiocholine, which causes the aggregation of LA-capped AuNPs with a change in color of the NP solution from red to steel-blue. As a result, an LOD of 4.52×10^4 pM could be achieved for paraoxon.

Virel et al. (2008) utilized AuNPs prepared using seed-mediated growth method for the detection of AChE inhibitors. The developed system was capable of detecting even trace level concentrations using thiol-modulated growth of Au–Ag NPs in the presence of AA. As the concentration of inhibitors [1,5-bis(4-allyldimethylammoniumphenyl)-pentane-3-one dibromide or paraoxon] increases, the catalytic growth of AuNPs was prevented. Moreover, they were able to develop a surface-immobilized assay using a functionalized glass substrate in order to confirm the effect of inhibitor on NP growth.

An increase in absorption spectra was obtained in the presence of paraoxon with AChE and ATCh using NP-immobilized glass plate. The sensing system renders a novel technique for the detection of nerve gases and OP pesticides. Recently, the same group reported the use of in situ-generated CdS quantum dots (QDs) with thiol-mediated stabilization for the detection of AChE inhibitors using a fluorimetric method. In this process, the hydrolysis of ATCh yields acetic acid and thiocholine (TCh). As the product is a thiol-bearing compound, it can bind to the surface of the growing CdS QDs via interaction between Cd^{2+} and S^{2-}. Upon increasing the concentration of inhibitor, an increase in quenching of the fluorescence intensity due to the thiol-mediated growth of CdS QDs was noted. They accomplished a detection limit of 0.06 and 115 nM for paraoxon and galantamine, respectively, due to the high sensitivity in the enzymatic activity (Garai-Ibabe et al., 2014).

Apart from AuNPs, there are several reports available on the use of AgNPs for the determination of OPs using enzyme-based method. AgNPs are capable of producing stronger plasmon resonance than AuNPs due to their shorter SPR wavelength as compared to AuNPs. This property allows the greater tunability of the resonance frequency for AgNPs when compared with AuNPs. AgNPs show higher extinction coefficients and higher ratio of scattering to extinction, and therefore, they have been utilized for the development of sensors based on surface-enhanced spectroscopic methods (Yong et al., 2006). There are only a limited number of reports for the detection of OP pesticides using unmodified AgNPs.

Vasimalai and John (2013) described a novel luminescent sensor for malathion using chitosan-capped AgNPs. To improve the sensitivity for malathion, the reaction was maintained at an acidic pH of 5.0, because in acidic pH, the amine group of chitosan will be protonated (—NH^{3+}). In acidic pH, the hydrolysis of malathion can cause its dissociation into two moieties, namely dimethyl phosphorothionic acid and 2-mercaptodiethylsuccinic acid. These two moieties have hydrogen bonding interaction with the —OH group of chitosan. These bonding interactions resulted in a decrease in both the absorbance and emission intensities of aggregated Chi-AgNPs. The proposed method was found to have an LOD of 94 fM for malathion, and the developed method was also applied for the detection of malathion residue in fruit and water samples. Recently, He et al. (2015) established a simple facile, and highly sensitive AgNP-based chemiluminescent (CL) sensor array for OP and carbamate pesticides. In this method, luminol is used to reduce the silver nitrate in order to form luminol-functionalized AgNPs (Lum-AgNPs), and the functionalization takes place via covalent interaction of the Ag—N bond. To produce CL emission, H_2O_2 is used along with Lum-AgNPs. When different pesticides are added to the Lum-AgNPs–H_2O_2 system, OP and carbamate pesticides will cause different CL responses. This developed sensor array for different pesticides could reach a detection limit of 24 μg/mL for dimethoate, dipterex, carbaryl, chlorpyrifos, and carbofuran. Additionally, more than 20 unknown pesticide samples were also validated by the CL sensor array to a level as low as 24 μg/mL with a precision of 95%.

All the aforementioned methods are for continuous colorimetric and fluorometric assays for the detection of OP pesticides using AuNPs and AgNPs. Though these methods can be used for the detection of OP pesticides, functionalization of NPs can help to enhance the sensitivity of the system. In a recent work, an indirect enzyme-based biosensor for the detection of malathion in the presence of AChE and ATCh using unmodified AgNPs as a probe has been tested (Fig. 4.3). When AChE and ATCh were added to AgNPs (without malathion addition), the aggregation of AgNPs and a sudden decrease in the peak intensity at 396 nm were observed with a new broad peak at 550 nm. The decrease in absorption band at 396 nm was due to the aggregation of AgNPs. In the presence of AChE and ATCh, citrate capping on the surface of AgNPs was

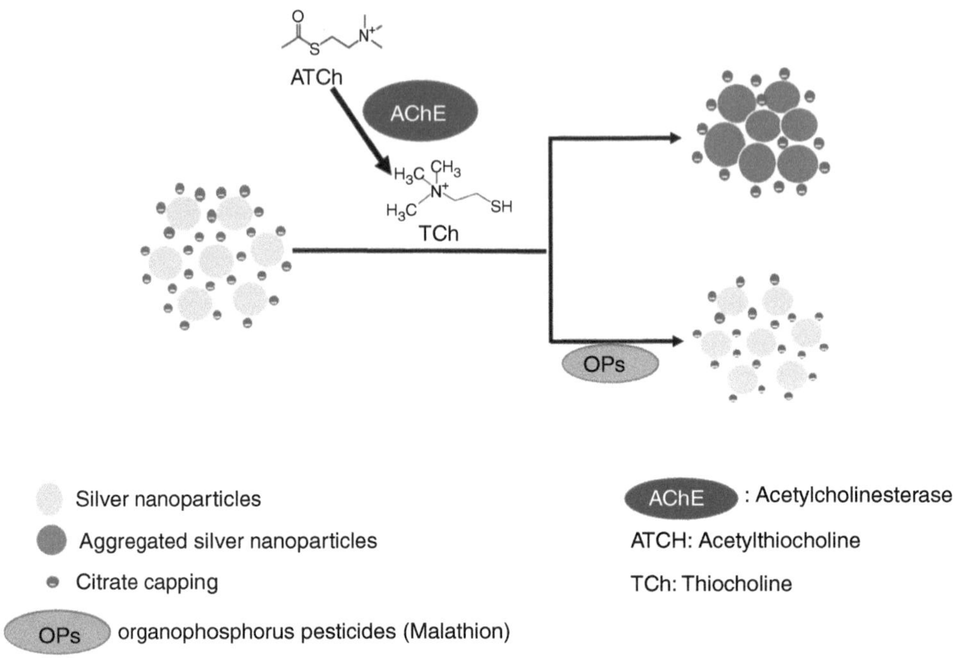

Figure 4.3. Detection of malathion using unmodified silver nanoparticles in the presence of acetylcholinesterase and acetylthiocholine.

replaced by the thiol group (—SH) of thiocholine. The addition of different concentrations ($1–10 \times 10^{-13}$ M) of malathion to the system inhibits the yield of thiol group and causes an increase in the absorption band at 396 nm with a decrease in the peak at 550 nm, respectively (Fig. 4.4). After a reaction time of 5 min, the color of the colloidal solution changed from yellow to pale pink and then to grey. The linear calibration curve was obtained by plotting the malathion concentrations versus their respective A_{396}/A_{550} ratio, and a good correlation constant $R^2 = 0.9962$ was obtained (Fig. 4.5). The LOD for malathion was found to be 0.1317×10^{-13} M with a signal-to-noise ratio of 3. Further, the system was demonstrated in real sample matrices like agricultural runoff water, lake water, cabbage, and apple extracts with high recovery rates.

3.2 Detection of Pesticides in Fruit and Vegetable Samples

Most of the pesticides are toxic, environmentally stable, and they have the capacity to accumulate in crops, that is, both vegetables and fruits. Pesticides can be carried by air, soil, water, land, and human beings. Therefore, the detection of pesticide residues

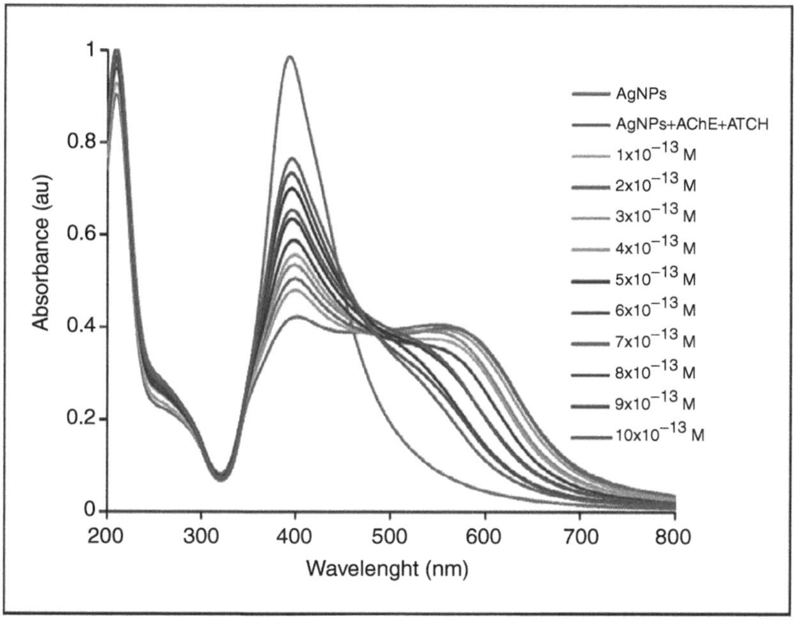

Figure 4.4. UV-visible absorption spectra of AgNPs in presence of different concentrations of malathion (2–100 × 10^{-13} M) with AChE (400 mU/mL) and ATCh (80 μM).

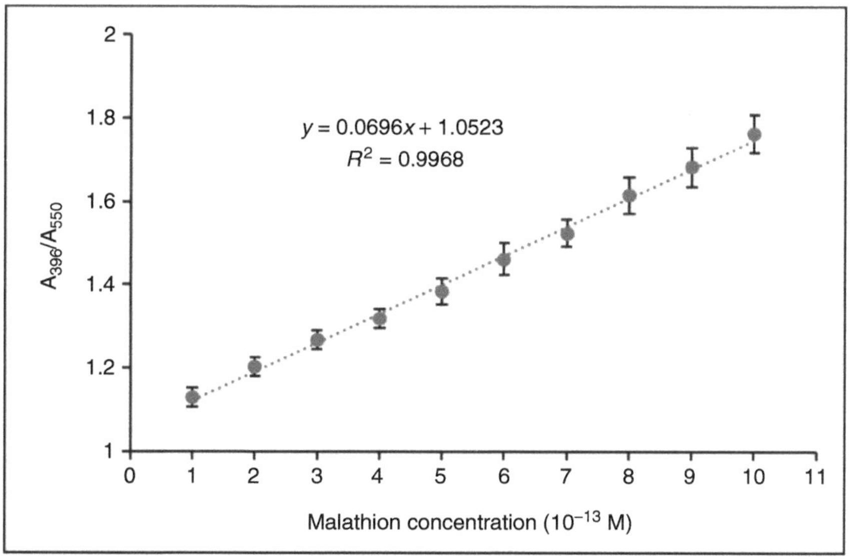

Figure 4.5. Linear correlation graph with absorption ratios of AgNPs (A_{396}/A_{550}) plotted against different concentrations of malathion (2–100 × 10^{-13} M).

has become important due to their toxicity. The sample preparation and extraction of pesticides from the samples are important steps. In the case of vegetables and fruits, some steps are required to remove any foreign contaminants by washing with distilled water and later dried at an appropriate temperature. Then, the edible parts of the samples were crushed with a pestle and mortar or ground in a mill. Further, the samples were mixed with a suitable quantity of distilled water (or suitable solvent) to extract the pesticides from the samples. The quantitative determination of pesticides in the samples using chromatographic methods like HPLC and GC is quite unsuitable. This is because the pesticide levels in vegetable and fruit samples are very low, and chromatographic techniques require highly skilled personnel and require minimum concentrations of the pesticide residue for the analysis to be carried out. The advantage of chromatographic method is the ability to determine individual analytes in the presence of other interferents. The separation of analytes can be attained with the help of chromatographic columns. These conventional methods require sophisticated instruments, preconcentration of samples, and are time-consuming due to the complexity of sample preparation. As compared with these conventional methods, an optical biosensor provides outstanding advantages for the detection of OP pesticides. Both colorimetric and fluorometric biosensors are highly sensitive and provide rapid response for single-molecule detection.

4 Colorimetric Detection of LPS Using Gold Nanorods (AuNRs)

4.1 Synthesis of Gold Nanorods

The seed-mediated silver-ion assisted method is the most widely used technique for the synthesis of gold nanorods (AuNRs) (Nikoobakht and El-Sayed, 2003). The size and shape of AuNRs can be controlled to a larger extent using this method in comparison to the in situ synthesis processes. AuNRs are usually synthesized in a two-step process. AuNP seeds of smaller size are initially synthesized and then further added to the growth solution. Gold seeds were synthesized by adding ice-cold sodium borohydride to chloroauric salt, and cetyltrimethylammonium bromide (CTAB) acts as a stabilizing agent. The hydrodynamic diameter of as-synthesized gold seeds was found to be less than 4 nm and was used for carrying out further growth. The formation of Au seeds was considered as the initial nucleation step.

The growth solution contains auric salt along with silver nitrate that acts a structure-directing agent, AA as a reducing agent, and

CTAB as a stabilizing agent. The reducing agent (AA) used in the growth process was very mild and, hence, was capable to reduce Au^{III} to Au^0 only after the addition of Au seeds, which catalyzes the reduction. The reduced Au^0 in the growth solution settles over the Au seeds and does not involve in the particle nucleation process. The growth of AuNRs was slower when compared to the formation of Au seeds, and hence, has to be incubated over night for the complete nanorod formation. The size, shape, and functionalization of AuNRs can be varied by simply varying the concentration and composition of the reducing and stabilizing agent and the Au seed ratio (Nikoobakht and El-Sayed, 2003).

AuNRs possess four {110} facets, which are completely absent in the AuNPs (Wang et al., 1999). The surfactant was known to bind to the {110} facet more strongly in comparison to the other available facets (Wang et al., 2000). In case of multiple-twined nanorods, affinity of the surfactant was noticed to be even much stronger due to its crystalline structure that was entirely different from nanorods having single crystal structure (Gai and Harmer, 2002; Johnson et al., 2002). The gold seeds were noticed to grow in the direction of *x*- and *y*-axis to attain the necessary length and width of AuNRs, when compared to the *z*-axis of AuNRs. The surfactant concentration and structure were known to significantly influence the width of AuNRs. The growth occurred parallel to the {001} plane along the longitudinal axis with the development of four unstable {110} facets and four {111} facets. The surfactant was known to bind to the facets once the gold seeds were developed fully to a significant size. The rate of growth of different facets played a vital role in determining the final structure of AuNRs after binding of the facets to the surfactant. The slow growth of the AuNR width was mainly due to the protection of the {110} facets by CTAB. Finally, the AuNR size and growth rate were completely dependent on the concentration of surfactant and the available gold concentration. Since the surfactant (CTAB) concentration is very much higher when compared to the gold, the growth of the AuNRs occurred until the supply of gold diminished (Nikoobakht and El-Sayed, 2003).

The primary characterization of AuNRs was carried out using UV–visible spectroscopy and a strong longitudinal plasmon band and a weaker transverse plasmon band were observed. These plasmon bands were formed due to the interaction of the incident light with electrons that are oscillating in two different directions. The scattering and absorption of light along the short and long axes of AuNRs correspond to the transverse and longitudinal plasmon band, respectively. The transverse band is very much similar to that of the spherical-shaped AuNPs (Yu and Irudayaraj, 2007;

Ho et al., 2012). The longitudinal band is highly sensitive to the variations in the dielectric properties of the environment (absorbates and solvents) and the interparticle distance of AuNRs (Ho et al., 2012; Chen et al., 2013). AuNRs also have a large absorption coefficient, which gradually diminishes or gives a red/blue shift after interaction with the analyte (Kitazaki et al., 2012). Based on the aggregation property of AuNRs, they can be used to detect pathogens (Wang and Irudayaraj, 2008), amino acids (Sudeep et al., 2005; Huang et al., 2013), disease markers (Huang et al., 2012; Truong et al., 2011), proteins (Yasun et al., 2012; Zhang and Shen, 2013; Zhu et al., 2011), metal ions (Wang et al., 2013; Liang et al., 2012; Anand et al., 2013), and few other chemicals. Therefore, AuNRs can be used in cancer photothermal therapy (Wang et al., 2010), medical diagnosis (Truong et al., 2011), biosensing (He et al., 2008; Yu and Irudayaraj, 2007), and cellular imaging (Wang et al., 2009a).

4.2 Characterization of Gold Nanorods (AuNRs)

Colorimetric detection of *E. coli* endotoxin was effectively carried out using rod-shaped AuNPs (AuNRs) (Paul et al., 2016). The synthesis procedure proposed by Nikoobakht and El-Sayed (2003) was employed to synthesize AuNRs having an aspect ratio of 2.0, that is, the ratio of the length to breadth of nanorods, and these measurements can be obtained from the transmission electron microscope (TEM) image. The synthesized AuNRs were blue in color (pH 3.0). The CTAB present on the outer surface of AuNRs made them highly stable and soluble in Millipore water as the presence of the NR^{4+} (tetramethylammonium ion) on AuNRs provided the positive charge (Nikoobakht and El-Sayed, 2001; Gole et al., 2004). The excess CTAB present in the colloidal solution were then removed by carrying out efficient centrifugation process, and then, redispersal of the pellet in Millipore water. The UV–visible graph of the synthesized AuNRs showed a distinct transverse surface plasmon peak at 511 nm and a longitudinal surface plasmon peak at 691 nm.

4.3 Extraction of LPS From *E. coli*

CTAB-capped AuNRs can be used for the detection of LPS that was initially extracted from the *E. coli* bacteria before carrying out the detection using AuNRs. The extraction of LPS from a reasonable and safe surrogate, *E. coli* K12 generic strain, was done as it was difficult to handle the pathogenic bacterial strain, *E. coli* O157:H7, in lab conditions (Radke and Alocilja, 2005; Pathanibul et al., 2009). The growth of the generic strain, *E. coli* K12, in nutrient broth was carried out at room temperature for 12 h. The bacterial cells were then harvested, and the LPS was extracted from these cells. The standard phenol micro extraction process was adopted for extracting LPS (Currie and Poxton, 1999). The extraction process involved the addition of phenol solution that resulted in the formation of two different phases. The LPS present in the clear upper phase was separated out and concentrated using acetone. The obtained LPS was then dispersed in phosphate buffer (10 mM; pH 7.0) and kept at 4°C in order to maintain the stability. The quantification of LPS was then carried out using phenol sulphuric acid method (Dubois et al., 1956), and the concentration of LPS was determined to be 209,000 ng/mL.

4.4 Sensitivity of the AuNR Probe

The developed CTAB-capped AuNR probe was used to detect the LPS in phosphate buffer (10 mM; pH 7.2). The synthesized AuNRs (pH 3.0) having an aspect ratio of 2.0 was interacted with LPS under the optimized volumetric ratio of 3:2, that is, 600 µL of AuNRs with 400 µL of LPS. The stability of AuNRs decreased in the presence of LPS, which could be visually observed by the human eye by a change in color from blue to black. The aggregation of AuNRs was the result of interaction between the tetraethylammonium ion (NR^{4+}) having a positive charge and the phosphate and carboxylic groups of LPS possessing a negative charge (Wang et al., 2014a) (Fig. 4.6). The aggregation of AuNRs in presence of LPS was studied using a scanning electron microscope, which clearly showed the clustering of AuNRs randomly.

The interaction of LPS with AuNRs was studied using the obtained UV–visible graph, which showed an apparent decline in the AuNR absorbance at the longitudinal band of 691 nm, which was dependent completely on the concentration of LPS. Fig. 4.7 shows the absorption spectra of CTAB-capped AuNRs before and after interaction with lipopolysaccharide. However, in the case of the transverse band at 511 nm, the absorbance remained unchanged and was found to be independent of the concentration of LPS. The A/A_0 values for the aggregated AuNRs after interacting with the respective LPS concentrations ranging from 200 to 5 ng/mL were studied, and a linear equation $y = -0.0027 + 0.9495$; $R^2 = 0.9725$ was obtained using the developed nanosensor probe. The absorbance intensities of AuNRs at 691 nm, before and after addition of LPS, were represented as A_0 and A. The theoretically calculated LOD obtained for detecting LPS was found to be 1.11 ng/mL.

The charges on the surface of AuNRs and LPS (200 and 5 ng/mL) were studied to be +66, −50.3, and −24 mV, respectively, using the zeta analyzer. The large difference in the surface charge between

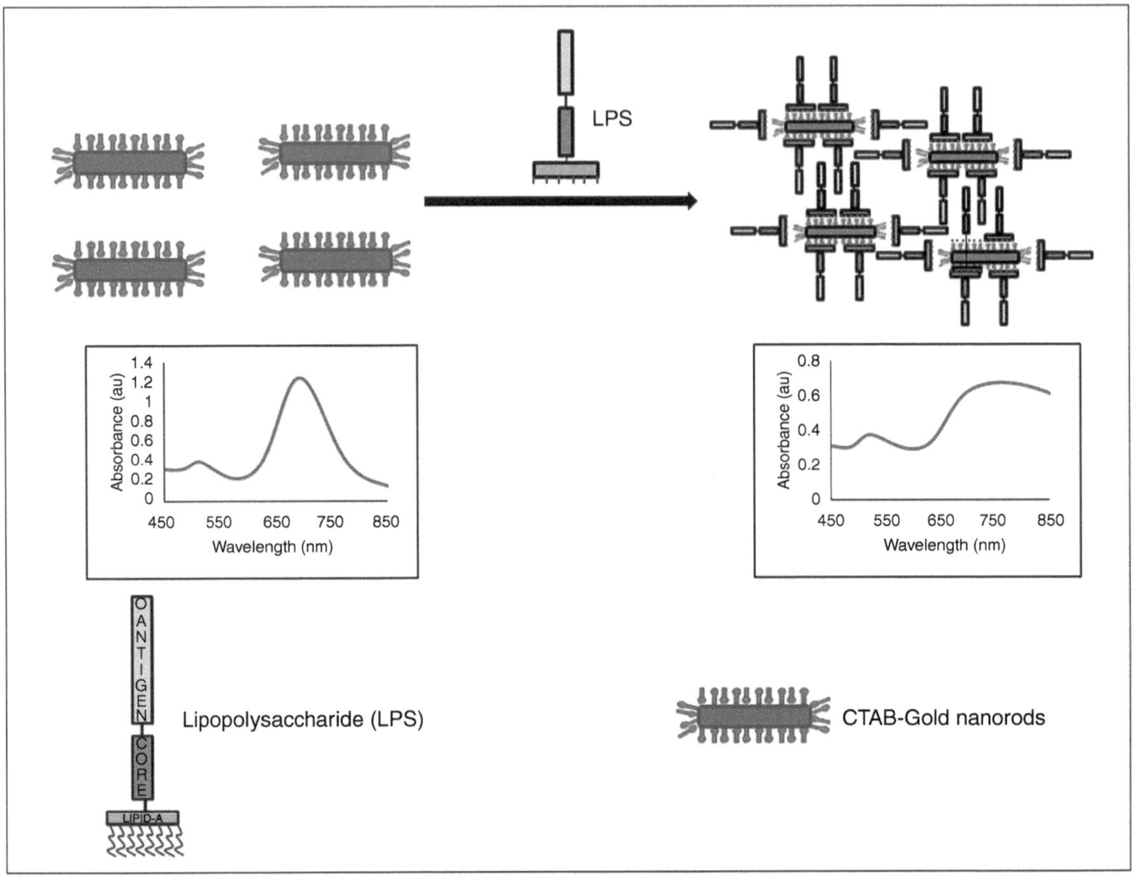

Figure 4.6. Schematic illustration of lipopolysaccharide detection using CTAB-capped gold nanorods based on the surface charge interaction.

the AuNRs and LPS (even for low concentrations) enabled the developed AuNR probe to detect LPS even at ultralow level. The surface charges of aggregated AuNRs were also studied using the zeta analyzer after the addition of different LPS concentrations (200 and 5 ng/mL) and the zeta potentials were determined to be +14.1 and +21.9 mV, respectively. The sensitivity of the developed probe was determined to be statistically significant (p value < 0.05) by running the one-way ANOVA test in the excel sheet for a wide range of LPS concentrations from 200 to 5 ng/mL.

4.5 Interference Study

The selectivity study was also performed to observe the interference of other molecules present along with LPS. The different

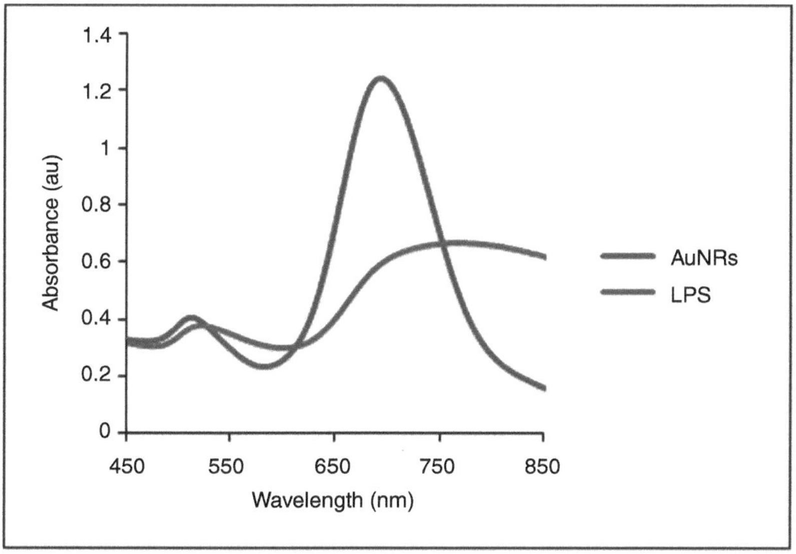

Figure 4.7. Absorption spectra of CTAB-capped AuNRs before and after interaction with lipopolysaccharide.

molecules studied, included common inorganic salts like NaH_2PO_4, NaCl, K_2HPO_4, and Na_2CO_3, and few food nutrients like AA, citric acid, SUC, and glucose that may interfere with the detection of LPS. Though the concentrations of LPS taken for the study were 100 times lower than that of the interferents, the aggregation of AuNRs was noticed only when LPS was added. The interferents dispersed in the phosphate buffer did not affect the stability of AuNRs. They did not even cause any visual color change or obvious decrease in the absorbance for AuNRs at the longitudinal peak (691 nm). Compounds like BSA and chitosan also could not give significant changes in the absorbance at 691 nm due to the lack of electrostatic interaction between the CTAB-capped AuNRs with polysaccharides and proteins. The stability of AuNRs was also studied in presence of compounds possessing a hydrophobic tail, namely polyvinyl alcohol (PVA, 160 kDa) and Tween-20. The absence of visual color change showed that their presence also did not affect the AuNR stability. The UV–visible graph for AuNRs interacted with these compounds showed very minute differences in the absorbance of the longitudinal band (691 nm). The calculated A/A_0 absorbance ratio was studied for AuNRs in presence of LPS and other interferents, which clearly showed that there was an evident decrease in the A/A_0 ratio of AuNRs when LPS was added in comparison to the other interfering compounds. Therefore, the developed probe had high selectivity, sensitivity, and reproducibility.

4.6 Application of CTAB-Capped AuNRs in Real Samples

The developed nanosensor probe was successfully used to detect LPS in food samples such as liquid milk, raw milk, and milk powder, and also in fruit juices like apple and mango juices. The milk samples were initially pretreated before analyzing the samples for LPS contamination. Proteins and fats present in the milk system interfere with the detection of LPS and, hence, pretreatment process was carried out using trichloroacetic acid and chloroform (Li et al., 2010b). Further purification process was done by repeated centrifugation and filtration of the milk samples using a 0.22-μm filter to remove any protein residues. In case of fruit samples, the fruits were first thoroughly washed with water, and then cut into very fine pieces. The pieces were ground using a mixer, and the juice was obtained after filtering it with the help of a nylon mesh. Proteins and fats were then removed from the fruit juices by carrying out the pretreatment process. The pretreatment process involved are very much similar to the pretreatment process carried out for the milk samples.

The pH of milk samples and fruit juices were maintained at pH 7.0 with the help of 1 M Na_2CO_3, and then, the LPS present in the real samples were detected using the developed probe. The CTAB-capped AuNRs produced a color change from blue to black when LPS was present in the real system, and this could be observed by the naked eye. The LPS concentration was then determined from the linear equation $y = -0.0027 + 0.9495$, $R^2 = 0.9725$ by calculating the A/A_0 ratio for the respective real samples. The recovery percentages for the milk samples and fruit juices were determined to be 102–103, 100–102, 103, 105–106, and 97–100% for apple juice, mango juice, raw milk, liquid milk, and milk powder, respectively.

5 Conclusions

The easy synthesis process of plasmonic nanoparticles (gold and silver) have led to the designing of simple, cheap, and rapid techniques for the colorimetric detection of chemical contaminants (melamine and pesticide malathion) and microbial contaminant (*E. coli* endotoxin) in aqueous solution. The size effect of spherically shaped AuNPs was studied for the detection of melamine. High sensitivity and selectivity were achieved with the optimized AuNP size (30 nm) to give an LOD of up to 2.37×10^{-8} M. AgNPs capped with citrate were used to detect the pesticide malathion, and the limit of detection of the designed probe was 0.1317×10^{-13} M.

Finally, rod-shaped AuNRs capped with CTAB were used to successfully detect the microbial contaminant *E. coli* endotoxin in aqueous solution having a sensitivity as low as 3 ng/mL.

Therefore, plasmonic NPs can be used for the detection of food contaminants in food matrices like milk (skimmed milk, liquid milk, and milk powder), fruit juices (apple and mango juice), and also in raw fruits and vegetables. In the coming years, an on-site assay can be further developed with these proposed techniques.

References

Amine, A., Mohammadi, H., Bourais, I., Palleschi, G., 2006. Enzyme inhibition-based biosensors for food safety and environmental monitoring. Biosens. Bioelectron. 21, 1405–1423.

Anand, G.S., et al., 2013. Development of a surface plasmon assisted label-free calorimetric method for sensitive detection of mercury based on functionalized gold nanorods. J. Anal. At. Spectrom. 28 (4), 488–498.

Brown, C.A., et al., 2007. Outbreaks of renal failure associated with melamine and cyanuric acid in dogs and cats in 2004 and 2007. J. Vet. Diag. Invest. 19 (5), 525–531.

Burnworth, M., Rowan, S.J., Weder, C., 2007. Fluorescent sensors for the detection of chemical warfare agents. Chem. Eur. J. 13, 7828–7836.

Cady, N.C., et al., 2005. Real-time PCR detection of *Listeria monocytogenes* using an integrated microfluidics platform. Sens. Actuat. B Chem. 107 (1), 332–341.

Chen, X., et al., 2010. β-Cyclodextrin-functionalized silver nanoparticles for the naked eye detection of aromatic isomers. ACS Nano 4 (11), 6387–6394.

Chen, H., et al., 2013. Gold nanorods and their plasmonic properties. Chem. Soc. Rev. 42 (7), 2679–2724.

Cheng, J.Y.W., Hui, E.L.C., Lau, A.P.S., 2012. Bioactive and total endotoxins in atmospheric aerosols in the Pearl River Delta region, China. Atmos. Environ. 47, 3–11.

Cho, M., et al., 2012. Sensitive electrochemical sensor for detection of lipopolysaccharide on metal complex immobilized gold electrode. Sens. Actuat. B Chem. 174, 490–494.

Choi, M.M., 2004. Progress in enzyme-based biosensors using optical transducers. Microchim. Acta 148, 107–132.

Čolović, M.B., Krstić, D.Z., Lazarević-Pašti, T.D., Bondžić, A.M., Vasić, V.M., 2013. Acetylcholinesterase inhibitors: pharmacology and toxicology. Curr. Neuropharmacol. 11, 315.

Cook, H.A., Klampfl, C.W., Buchberger, W., 2005. Analysis of melamine resins by capillary zone electrophoresis with electrospray ionization-mass spectrometric detection. Electrophoresis 26 (7–8), 1576–1583.

Currie, C.G., Poxton, I.R., 1999. The lipopolysaccharide core type of *Escherichia coli* O157: H7 and other non-O157 verotoxin-producing *E. coli*. FEMS Immunol. Med. Microbiol. 24 (1), 57–62.

Daniel, M.-C., Astruc, D., 2004. Gold nanoparticles: assembly, supramolecular chemistry, quantum-size-related properties, and applications toward biology, catalysis, and nanotechnology. Chem. Rev. 104 (1), 293–346.

Ding, N., et al., 2010. Colorimetric determination of melamine in dairy products by Fe_3O_4 magnetic nanoparticles—H_2O_2—ABTS detection system. Anal. Chem. 82 (13), 5897–5899.

Dobson, R.L., et al., 2008. Identification and characterization of toxicity of contaminants in pet food leading to an outbreak of renal toxicity in cats and dogs. Toxicol. Sci. 106 (1), 251–262.
Dorst, V., Bieke, Mehta, J., Bekaert, K., Rouah-Martin, E., Coen, W.D., Dubruel, P., Blust, R., Robbens, J., 2010. Recent advances in recognition elements of food and environmental biosensors: a review. Biosens. Bioelectron. 26, 1178–1194.
Dubois, M., et al., 1956. Colorimetric method for determination of sugars and related substances. Anal. Chem. 28 (3), 350–356.
Ellman, G.L., Courtney, K.D., Andres, V., Featherstone, R.M., 1961. A new and rapid colorimetric determination of acetylcholinesterase activity. Biochem. Pharmacol. 7, 88–95.
Eum, N.-S., et al., 2010. Enhancement of sensitivity using gold nanorods—antibody conjugator for detection of *E. coli* O157: H7. Sens. Actuat. B Chem. 143 (2), 784–788.
Eustis, S., El-Sayed, M.A., 2006. Why gold nanoparticles are more precious than pretty gold: noble metal surface plasmon resonance and its enhancement of the radiative and nonradiative properties of nanocrystals of different shapes. Chem. Soc. Rev. 35 (3), 209–217.
Fang, J., et al., 2004. Aggregation and surface-enhanced Raman activity study of dye-coated mixed silver–gold colloids. J. Raman Spectrosc. 35 (11), 914–920.
Filigenzi, M.S., et al., 2008. Diagnostic determination of melamine and related compounds in kidney tissue by liquid chromatography/tandem mass spectrometry. J. Agric. Food Chem. 56 (17), 7593–7599.
Frens, G., 1973. Controlled nucleation for the regulation of the particle size in monodisperse gold suspensions. Nature 241 (105), 20–22.
Gai, P.L., Harmer, M.A., 2002. Surface atomic defect structures and growth of gold nanorods. Nano Lett. 2 (7), 771–774.
Gandhi, M., Chikindas, M.L., 2007. Listeria: a foodborne pathogen that knows how to survive. Int. J. Food Microbiol. 113 (1), 1–15.
Garai-Ibabe, G., Saa, L., Pavlov, V., 2014. Thiocholine mediated stabilization of in situ produced CdS quantum dots: application for the detection of acetylcholinesterase activity and inhibitors. Analyst 139, 280–284.
Garber, E.A., 2008. Detection of melamine using commercial enzyme-linked immunosorbent assay technology. J. Food Protect. 71 (3), 590–594.
Ghosh, P., et al., 2008. Gold nanoparticles in delivery applications. Adv. Drug Deliv. Rev. 60 (11), 1307–1315.
Gole, A., Orendorff, C.J., Murphy, C.J., 2004. Immobilization of gold nanorods onto acid-terminated self-assembled monolayers via electrostatic interactions. Langmuir 20 (17), 7117–7122.
Gracias, K.S., McKillip, J.L., 2004. A review of conventional detection and enumeration methods for pathogenic bacteria in food. Can. J. Microbiol. 50 (11), 883–890.
He, X., et al., 2005. Gold nanoparticle-based fluorometric and colorimetric sensing of copper (II) ions. Adv. Mater. 17 (23), 2811–2815.
He, W., et al., 2008. One-step label-free optical genosensing system for sequence-specific dna related to the human immunodeficiency virus based on the measurements of light scattering signals of gold nanorods. Anal. Chem. 80 (22), 8424–8430.
He, L., et al., 2009. Novel cyromazine imprinted polymer applied to the solid-phase extraction of melamine from feed and milk samples. J. Chromatogr. A 1216 (34), 6196–6203.
He, Y., Xu, B., Li, W., Yu, H., 2015. Silver nanoparticle-based chemilumiscent sensor array for pesticide discrimination. J. Agri. Food Chem. 63, 2930–2934.
Ho, F.H., et al., 2012. A solution-based nanoplasmonic sensing technique by using gold nanorods. Analyst 137 (11), 2545–2548.

Homola, J., 2008. Surface plasmon resonance sensors for detection of chemical and biological species. Chem. Rev. 108 (2), 462–493.
Huang, C.-J., et al., 2011. Long-range surface plasmon-enhanced fluorescence spectroscopy biosensor for ultrasensitive detection of *E. coli* O157: H7. Anal. Chem. 83 (3), 674–677.
Huang, H., et al., 2012. Sensitive and simultaneous detection of different disease markers using multiplexed gold nanorods. Anal. Chim. Acta 755, 108–114.
Huang, H., et al., 2013. Multiplex plasmonic sensor for detection of different metal ions based on a single type of gold nanorod. Anal. Chem. 85 (4), 2312–2319.
Ibáñez, M., Sancho, J.V., Hernández, F., 2009. Determination of melamine in milk-based products and other food and beverage products by ion-pair liquid chromatography–tandem mass spectrometry. Anal. Chim. Acta 649 (1), 91–97.
Jans, H., et al., 2009. Dynamic light scattering as a powerful tool for gold nanoparticle bioconjugation and biomolecular binding studies. Anal. Chem. 81 (22), 9425–9432.
Jianrong, C., Yuqing, M., Nongyue, H., Xiaohua, W., Sijiao, L., 2004. Nanotechnology and biosensors. Biotechnol. Adv. 22, 505–518.
Jindou, S., et al., 2004. Cohesin-dockerin interactions within and between *Clostridium josui* and *Clostridium thermocellum* binding selectivity between cognate dockerin and cohesin domains and species specificity. J. Biol. Chem. 279 (11), 9867–9874.
Jofré, A., et al., 2005. Simultaneous detection of *Listeria monocytogenes* and *Salmonella* by multiplex PCR in cooked ham. Food Microbiol. 22 (1), 109–115.
Johnson, C.J., et al., 2002. Growth and form of gold nanorods prepared by seed-mediated, surfactant-directed synthesis. J. Mater. Chem. 12 (6), 1765–1770.
Kennaway, E.L., 1921. The estimation of non-protein nitrogen in blood by a micro-Kjeldahl method. Biochem. J. 15 (4), 510.
Kim, Y.-G., et al., 2006. Selection of peptides for lipopolysaccharide binding on to epoxy beads and selective detection of Gram-negative bacteria. Biotechnol. Lett. 28 (2), 79–84.
Kimling, J., et al., 2006. Turkevich method for gold nanoparticle synthesis revisited. J. Phys. Chem. B 110 (32), 15700–15707.
Kitazaki, H., et al., 2012. A colorimetric assay of protein kinase activity based on peptide-induced coagulation of gold nanorods. Coll. Surf. B 99, 7–11.
Klampfl, C.W., et al., 2009. Analysis of melamine in milk powder by CZE using UV detection and hyphenation with ESI quadrupole/TOF MS detection. Electrophoresis 30 (10), 1743–1746.
Lazcka, O., Campo, F.J.D., Muñoz, F.X., 2007. Pathogen detection: a perspective of traditional methods and biosensors. Biosens. Bioelectron. 22 (7), 1205–1217.
Lee, K.S., El-Sayed, M.A., 2006. Gold and silver nanoparticles in sensing and imaging: sensitivity of plasmon response to size, shape, and metal composition. J. Phys. Chem. B 110, 19220–19225.
Li, H., Bian, Y., 2009. Selective colorimetric sensing of histidine in aqueous solutions using cysteine modified silver nanoparticles in the presence of Hg^{2+}. Nanotechnology 20 (14), 145502.
Li, L., Li, B., 2009. Sensitive and selective detection of cysteine using gold nanoparticles as colorimetric probes. Analyst 134 (7), 1361–1365.
Li, F., et al., 2010a. Detection of *Escherichia coli* O157: H7 using gold nanoparticle labeling and inductively coupled plasma mass spectrometry. Anal. Chem. 82 (8), 3399–3403.
Li, L., et al., 2010b. Visual detection of melamine in raw milk using gold nanoparticles as colorimetric probe. Food Chem. 122 (3), 895–900.
Liang, G.-X., et al., 2012. A colorimetric probe for the rapid and selective determination of mercury(II) based on the disassembly of gold nanorods. Microchim. Acta 179 (3–4), 345–350.

Limbut, W., et al., 2007. Capacitive biosensor for detection of endotoxin. Anal. Bioanal. Chem. 389 (2), 517–525.

Lin, T.J., Huang, K.T., Liu, C.Y., 2006. Determination of organophosphorous pesticides by a novel biosensor based on localized surface plasmon resonance. Biosens. Bioelectron. 22, 513–518.

Liu, Y., et al., 2008. Detection of viable but nonculturable *Escherichia coli* O157:H7 bacteria in drinking water and river water. Appl. Environ. Microbiol. 74 (5), 1502–1507.

Liu, Z.D., et al., 2009. A localized surface plasmon resonance light-scattering assay of mercury (II) on the basis of Hg^{2+}–DNA complex induced aggregation of gold nanoparticles. Environ. Sci. Technol. 43 (13), 5022–5027.

Liu, D., Chen, W., Wei, J., Li, X., Wang, Z., Jiang, X., 2012. A highly sensitive, dual-readout assay based on gold nanoparticles for organophosphorus and carbamate pesticides. Anal. Chem. 84, 4185–4191.

Long, F., Zhu, A., Shi, H., 2013. Recent advances in optical biosensors for environmental monitoring and early warning. Sensors 13, 13928–13948.

Malik, P., Katyal, V., Malik, V., Asatkar, A., Inwati, G., Mukherjee, T.K., 2013. Nanobiosensors: concepts and variations. ISRN Nanomater.

March, S.B., Ratnam, S., 1986. Sorbitol-MacConkey medium for detection of *Escherichia coli* O157: H7 associated with hemorrhagic colitis. J. Clin. Microbiol. 23 (5), 869–872.

Mauer, L.J., et al., 2009. Melamine detection in infant formula powder using near- and mid-infrared spectroscopy. J. Agric. Food Chem. 57 (10), 3974–3980.

Miao, Y., He, N., Zhu, J.J., 2010. History and new developments of assays for cholinesterase activity and inhibition. Chem. Rev. 110, 5216–5234.

Mie, G., 1976. Contributions to the optics of turbid media, particularly of colloidal metal solutions. Contributions to the optics of turbid media, particularly of colloidal metal solutions. Transl. Ann. Phys. 25(3), 1908, 377–445.

Mulchandani, A., Chen, W., Mulchandani, P., Wang, J., Rogers, K.R., 2001. Biosensors for direct determination of organophosphate pesticides. Biosens. Bioelectron. 16, 225–230.

Munro, N.B., Talmage, S.S., Griffin, G.D., Waters, L.C., Watson, A.P., King, J.F., Hauschild, V., 1999. The sources, fate, and toxicity of chemical warfare agent degradation products. Environ. Health Perspect. 107, 933.

Murphy, C., Carroll, C., Jordan, K.N., 2006. Environmental survival mechanisms of the foodborne pathogen *Campylobacter jejuni*. J. Appl. Microbiol. 100 (4), 623–632.

Murphy, C.J., et al., 2008. Gold nanoparticles in biology: beyond toxicity to cellular imaging. Acc. Chem. Res. 41 (12), 1721–1730.

Nataro, J.P., Kaper, J.B., 1998. Diarrheagenic *Escherichia coli*. Clin. Microbiol. Rev. 11 (1), 142–201.

Nguyen, D.T., Kim, D.-J., Kim, K.-S., 2011. Controlled synthesis and biomolecular probe application of gold nanoparticles. Micron 42 (3), 207–227.

Nikoobakht, B., El-Sayed, M.A., 2001. Evidence for bilayer assembly of cationic surfactants on the surface of gold nanorods. Langmuir 17 (20), 6368–6374.

Nikoobakht, B., El-Sayed, M.A., 2003. Preparation and growth mechanism of gold nanorods (NRs) using seed-mediated growth method. Chem. Mater. 15 (10), 1957–1962.

Parab, H., et al., 2011. An anisotropic snowflake-like structural assembly of polymer-capped gold nanoparticles. J. Nanopart. Res. 13 (5), 2173–2180.

Park, S., Worobo, R.W., Durst, R.A., 1999. *Escherichia coli* O157: H7 as an emerging foodborne pathogen: a literature review. Crit. Rev. Food Sci. Nutr. 39 (6), 481–502.

Park, S., et al., 2008. Enzyme-linked immuno-strip biosensor to detect *Escherichia coli* O157: H7. Ultramicroscopy 108 (10), 1348–1351.

Pathanibul, P., et al., 2009. Inactivation of *Escherichia coli* and *Listeria innocua* in apple and carrot juices using high pressure homogenization and nisin. Int. J. Food Microbiol. 129 (3), 316–320.
Paul, I.E., Rajeshwari, A., et al., 2015a. Colorimetric detection of melamine based on the size effect of AuNPs. Anal. Methods 7 (4), 1453–1462.
Paul, I.E., Rajeshwari, A., et al., 2016. Label-free colorimetric detection of bacterial lipopolysaccharide in food samples using gold nanorods. Sens. Lett. 14 (1), 19–25.
Peng, X.-T., Shi, Z.-G., Feng, Y.-Q., 2011. Rapid and high-throughput determination of melamine in milk products and eggs by full automatic on-line polymer monolith microextraction coupled to high-performance liquid chromatography. Food Anal. Methods 4 (3), 381–388.
Prathna, T.C., et al., 2011. Biomimetic synthesis of silver nanoparticles by *Citrus limon* (lemon) aqueous extract and theoretical prediction of particle size. Coll. Surf. B 82 (1), 152–159.
Pundir, C.S., Chauhan, N., 2012. Acetylcholinesterase inhibition-based biosensors for pesticide determination: A review. Anal. Biochem. 429, 19–31.
Radke, S.M., Alocilja, E.C., 2005. A microfabricated biosensor for detecting foodborne bioterrorism agents. IEEE Sens. J. 5 (4), 744–750.
Ravindran, A., et al., 2010. Studies on interaction of colloidal Ag nanoparticles with bovine serum albumin (BSA). Coll. Surf. B 76 (1), 32–37.
Rima, J., et al., 2009. New spectrophotometric method for the quantitative determination of melamine using Mannich reaction. J. Food Comp. Anal. 22 (7), 689–693.
Savoye, F., et al., 2011. Comparative evaluation of a phage protein ligand assay with real-time PCR and a reference method for the detection of *Escherichia coli* O157: H7 in raw ground beef and trimmings. J. Food Protect. 74 (1), 6–12.
Sekhon, B.S., 2010. Food nanotechnology: an overview. Nanotechnol. Sci. Appl. 3, 1–15.
Simonian, A.L., Good, T.A., Wang, S.S., Wild, J.R., 2005. Nanoparticle–based optical biosensors for the direct detection of organophosphate chemical warfare agents and pesticides. Anal. Chim. Acta 534, 69–77.
Song, S., et al., 2010. Functional nanoprobes for ultrasensitive detection of biomolecules. Chem. Soc. Rev. 39 (11), 4234–4243.
Sudeep, P.K., Joseph, S.T.S., Thomas, K.G., 2005. Selective detection of cysteine and glutathione using gold nanorods. J. Am. Chem. Soc. 127 (18), 6516–6517.
Sun, F., et al., 2010. Analytical methods and recent developments in the detection of melamine. Trends Anal. Chem. 29 (11), 1239–1249.
Sun, J., Guo, L., Bao, Y., Xie, J., 2011. A simple, label-free AuNPs-based colorimetric ultrasensitive detection of nerve agents and highly toxic organophosphate pesticide. Biosens. Bioelectron. 28, 152–157.
Sun, J., et al., 2012. A facile assay for direct colorimetric visualization of lipopolysaccharides at low nanomolar level. Nano Res. 5 (7), 486–493.
Tang, H.-W., et al., 2009. Analysis of melamine cyanurate in urine using matrix-assisted laser desorption/ionization mass spectrometry. Anal. Chem. 81 (9), 3676–3682.
Tarr, P.I., Gordon, C.A., Chandler, W.L., 2005. Shiga-toxin-producing *Escherichia coli* and haemolytic uraemic syndrome. Lancet 365 (9464), 1073–1086.
Thomas, K.G., Kamat, P.V., 2000. Making gold nanoparticles glow: enhanced emission from a surface-bound fluoroprobe. J. Am. Chem. Soc. 122 (11), 2655–2656.
Tittlemier, S.A., et al., 2009. Melamine in infant formula sold in Canada: occurrence and risk assessment. J. Agric. Food Chem. 57 (12), 5340–5344.
Tran, B.N., et al., 2009. Use of methanol for the efficient extraction and analysis of melamine and cyanuric acid residues in dairy products and pet foods. J. Agric. Food Chem. 58 (1), 101–107.

Truong, P.L., et al., 2011. A new method for non-labeling attomolar detection of diseases based on an individual gold nanorod immunosensor. Lab Chip 11 (15), 2591–2597.
Tsai, Y.-T., Cheng, P.-C., Pan, T.-M., 2010. Immunomodulating activity of Lactobacillus paracasei subsp. paracasei NTU 101 in enterohemorrhagic *Escherichia coli* O157H7-infected mice. J. Agric. Food Chem. 58 (21), 11265–11272.
Tucker, J.B., 2006. War of Nerves: Chemical Warfare from World War I to Al-Qaeda. Pantheon Books, New York.
Uchida, H., et al., 2004. A new assay using surface plasmon resonance (SPR) to determine binding of the *Lactobacillus acidophilus* group to human colonic mucin. Biosci. Biotechnol. Biochem. 68 (5), 1004–1010.
Upadhyay, S., Rao, G.R., Sharma, M.K., Bhattacharya, B.K., Rao, V.K., Vijayaraghavan, R., 2009. Immobilization of acetylcholineesterase–choline oxidase on a gold–platinum bimetallic nanoparticles modified glassy carbon electrode for the sensitive detection of organophosphate pesticides, carbamates and nerve agents. Biosens. Bioelectron. 25, 832–838.
US EPA, O. of P.P., 2015. About pesticides | pesticides | US EPA. Available from: http://www.epa.gov/pesticides/about/#what_pesticide
Uzarski, J.R., Mello, C.M., 2012. Detection and classification of related lipopolysaccharides via a small array of immobilized antimicrobial peptides. Anal. Chem. 84 (17), 7359–7366.
Vaccarello, P., et al., 2012. Characterization of localized surface plasmon resonance transducers produced from Au 25 nanoparticle multilayers. Coll. Surf. A 402, 146–151.
Vasimalai, N., John, S.A., 2013. Biopolymer capped silver nanoparticles as fluorophore for ultrasensitive and selective determination of malathion. Talanta 115, 24–31.
Verma, N., Bhardwaj, A., 2015. Biosensor technology for pesticides: a review. Appl. Biochem. Biotechnol. 175, 3093–3119.
Virel, A., Saa, L., Pavlov, V., 2008. Modulated growth of nanoparticles: application for sensing nerve gases. Anal. Chem. 81 (1), 268–272.
Wang, C., Irudayaraj, J., 2008. Gold nanorod probes for the detection of multiple pathogens. Small 4 (12), 2204–2208.
Wang, Z., Ma, L., 2009. Gold nanoparticle probes. Coord. Chem. Rev. 253 (11), 1607–1618.
Wang, Z.L., et al., 1999. Crystallographic facets and shapes of gold nanorods of different aspect ratios. Surf. Sci. 440 (1), L809–L814.
Wang, Z.L., et al., 2000. Surface reconstruction of the unstable {110} surface in gold nanorods. J. Phys. Chem. B 104 (23), 5417–5420.
Wang, H., et al., 2007. Plasmonic nanostructures: artificial molecules. Acc. Chem. Res. 40 (1), 53–62.
Wang, C., et al., 2009a. Gold nanorod/Fe_3O_4 nanoparticle "nano-pearl-necklaces" for simultaneous targeting, dual-mode imaging, and photothermal ablation of cancer cells. Angew. Chem. Int. Ed. 48 (15), 2759–2763.
Wang, M., Gu, X., Zhang, G., Zhang, D., Zhu, D., 2009b. Continuous colorimetric assay for acetylcholinesterase and inhibitor screening with gold nanoparticles. Langmuir 25, 2504–2507.
Wang, C.-H., Chang, C.-W., Peng, C.-A., 2010. Gold nanorod stabilized by thiolated chitosan as photothermal absorber for cancer cell treatment. J. Nanopart. Res. 13 (7), 2749–2758.
Wang, S., et al., 2013. Label-free colorimetric sensing of copper(II) ions based on accelerating decomposition of H_2O_2 using gold nanorods as an indicator. Analyst 138 (7), 2080–2084.

Wang, Y., et al., 2014a. Facile colorimetric method for simple and rapid detection of endotoxin based on counterion-mediated gold nanorods aggregation. Biosens. Bioelectron. 55, 242–248.
Wang, B., Wang, H., Wang, F., Zhou, G., Wang, Y., Kambam, S., Chen, X., 2014b. Colorimetric and fluorometric assays for acetylcholinesterase and its inhibitors screening based on a fluorescein derivate. Bioorg. Med. Chem. Lett. 24, 552–555.
Ward, J.D., Ward, L.T., 2007. Principles of Food Science. Goodheart-Willcox Company, Tinley Park, IL.
Wei, F., et al., 2010. Rapid detection of melamine in whole milk mediated by unmodified gold nanoparticles. Appl. Phys. Lett. 96 (13), 133702.
Yan, N., et al., 2009. Determination of melamine in dairy products, fish feed, and fish by capillary zone electrophoresis with diode array detection. J. Agric. Food Chem. 57 (3), 807–811.
Yang, H.-H., et al., 2009a. Molecularly imprinted polymer as SPE sorbent for selective extraction of melamine in dairy products. Talanta 80 (2), 821–825.
Yang, S., et al., 2009b. Detection of melamine in milk products by surface desorption atmospheric pressure chemical ionization mass spectrometry. Anal. Chem. 81 (7), 2426–2436.
Yasun, E., et al., 2012. Enrichment and detection of rare proteins with aptamer-conjugated gold nanorods. Anal. Chem. 84 (14), 6008–6015.
Yokley, R.A., et al., 2000. Analytical method for the determination of cyromazine and melamine residues in soil using LC-UV and GC-MSD. J. Agric. Food Chem. 48 (8), 3352–3358.
Yong, K.T., Sahoo, Y., Swihart, M.T., Prasad, P.N., 2006. Synthesis and plasmonic properties of silver and gold nanoshells on polystyrene cores of different size and of gold–silver core–shell nanostructures. Coll. Surf. A 290, 89–105.
Young, S.J., Hart, J.P., Dowman, A.A., Cowell, D.C., 2001. The non-specific inhibition of enzymes by environmental pollutants: a study of a model system towards the development of electrochemical biosensor arrays. Biosens. Bioelectron. 16, 887–894.
Yu, C., Irudayaraj, J., 2007. Multiplex biosensor using gold nanorods. Anal. Chem. 79 (2), 572–579.
Yuan, Y., et al., 2008. Analytical methods for the detection of corticosteroids-residues in animal-derived foodstuffs. Crit. Rev. Anal. Chem. 38 (4), 227–241.
Zeng, S., et al., 2012. Size-dependent colorimetric visual detection of melamine in milk at 10 ppb level by citrate-stabilized Au nanoparticles. Anal. Methods 4 (8), 2499–2505.
Zhang, K., Shen, X., 2013. Cancer antigen 125 detection using the plasmon resonance scattering properties of gold nanorods. Analyst 138 (6), 1828–1834.
Zhao, B., et al., 2009. Silver microspheres for application as hydrogen peroxide sensor. Electrochem. Commun. 11 (8), 1707–1710.
Zhao, P., Li, N., Astruc, D., 2013. State of the art in gold nanoparticle synthesis. Coord. Chem. Rev. 257 (3), 638–665.
Zhou, Y.-Y., et al., 2010. A novel fluorometric determination of melamine using cucurbit [7] uril. JOL 130 (5), 817–820.
Zhou, Q., et al., 2011. Development of gold nanoparticle-based rapid detection kit for melamine in milk products. J. Agric. Food Chem. 59 (22), 12006–12011.
Zhu, X., et al., 2009. Determination of residues of cyromazine and its metabolite, melamine, in animal-derived food by gas chromatography-mass spectrometry with derivatization. J. Agric. Food Chem. 57 (23), 11075–11080.
Zhu, Y., et al., 2011. Simple, rapid and sensitive detection of antibiotics based on the side-by-side assembly of gold nanorod probes. Biosens. Bioelectron. 26 (11), 4387–4392.

5

NANOMATERIAL-BASED ELECTROCHEMICAL BIOSENSORS FOR FOOD SAFETY AND QUALITY ASSESSMENT

Fatma Dridi*,, Mouna Marrakchi**,†, Mohamed Gargouri**, Joelle Saulnier*, Nicole Jaffrezic-Renault*, Florence Lagarde***

**University of Lyon, Lyon 1, Institute of Analytical Sciences, Villeurbanne, France; **University of Carthage, National Institute of Applied Sciences and Technology, Laboratory of Ecology and Microbial Technology, Tunis, Tunisia; †Tunis El Manar University, Higher Institute of Applied Biological Science (ISSBAT), Tunis, Tunisia*

1 Introduction

Assuring food quality and safety has become a subject of major concern for authorities and professionals of the food supply chain. Food quality relates primarily to characteristics such as appearance, texture, taste, smell, nutritional value content, which have to be ensured to guarantee both nutritional quality and acceptability of the delivered foods by the consumers. Analysis of food composition helps controlling that all the desired constituents, including natural components (eg, sugars, amino acids, and alcohols) and additives (eg, vitamins and minerals), are present in a suitable range of concentration. Food analysis is also of uppermost importance to protect the consumers against any adulteration, spoilage, or contamination. Food may be contaminated at various points of the food chain, either chemically by small molecules (eg, toxins, pesticides, residues of veterinary drugs), or microbiologically by pathogenic bacteria. This may be the result of environmental pollution, animal medical care, or failures during the food production, handling, processing, storage, packing, and distribution. Introduced accidentally or deliberately in food commodities, chemical and biological hazardous contaminants represent a major threat for

Nanobiosensors. http://dx.doi.org/10.1016/B978-0-12-804301-1.00005-9

human health with serious consequences on the health-care system and economic productivity. One of the key challenges to both control foodstuff composition and detect chemical or biological contaminants is the availability of selective, sensitive, and reliable analytical methods. To date, most analyses are performed by skilled personnel in accredited laboratories using sophisticated and costly instrumentation such as liquid or gas chromatography coupled with mass spectrometry. Among the emerging and innovative technologies proposed for food analysis, biosensors are presented to date as very promising low-cost screening tools usable as complementary devices to conventional methods. Biosensors could help to select a rather limited number of suspect samples that would be further analyzed for confirmation by conventional techniques, reducing the cost and time of analysis (McGrath et al., 2012). Biosensors are self-contained integrated devices combining the selectivity of biological recognition elements, the so-called bioreceptor (eg, enzymes, antibodies, DNA, or microorganisms), with physical transducers (eg, electrochemical, optical, thermal, or piezoelectrical) able to convert the bioreceptor-analyte interaction process into a signal proportional to the analyte concentration (Perumal and Hashim, 2014). Among the existing modes of detection, electrochemical transduction offers high sensitivity and specificity and the possibility to miniaturize the required instrumentation providing compact and portable analysis devices (Ronkainen et al., 2010). Advances in nanotechnology, microelectronics, and microfluidics have permitted the miniaturization of electrochemical biosensors and the fabrication of high density arrays, particularly interesting for the real-time parallel monitoring of multiple chemicals or biological parameters, or the monitoring of one single parameter in several samples (Scognamiglio, 2013). The miniaturization of sensing platforms can provide a number of benefits including a reduction in both bioreceptor and sample amounts, an increase of sensitivity and high throughput analysis. Additionally, the small size is well-suited for the design of portable biosensors, especially for on-site monitoring applications. They can also be designed to be disposable for single use, reusable for several analyses or for continuous monitoring. However, the proposed biosensors do not always possess the required analytical performances, for example, sensitivity, selectivity, or stability. In this respect, the recent emergence of nanomaterials (NMs) and progress in nanotechnology have opened new horizons in designing integrated nanobioelectrochemical systems. The nanoscale modification of electrode surfaces is likely to provide novel and sometimes unique properties due to the ability of the intrinsic properties of the NMs employed, the ability to control the architecture of the electrode interface at the nanoscale,

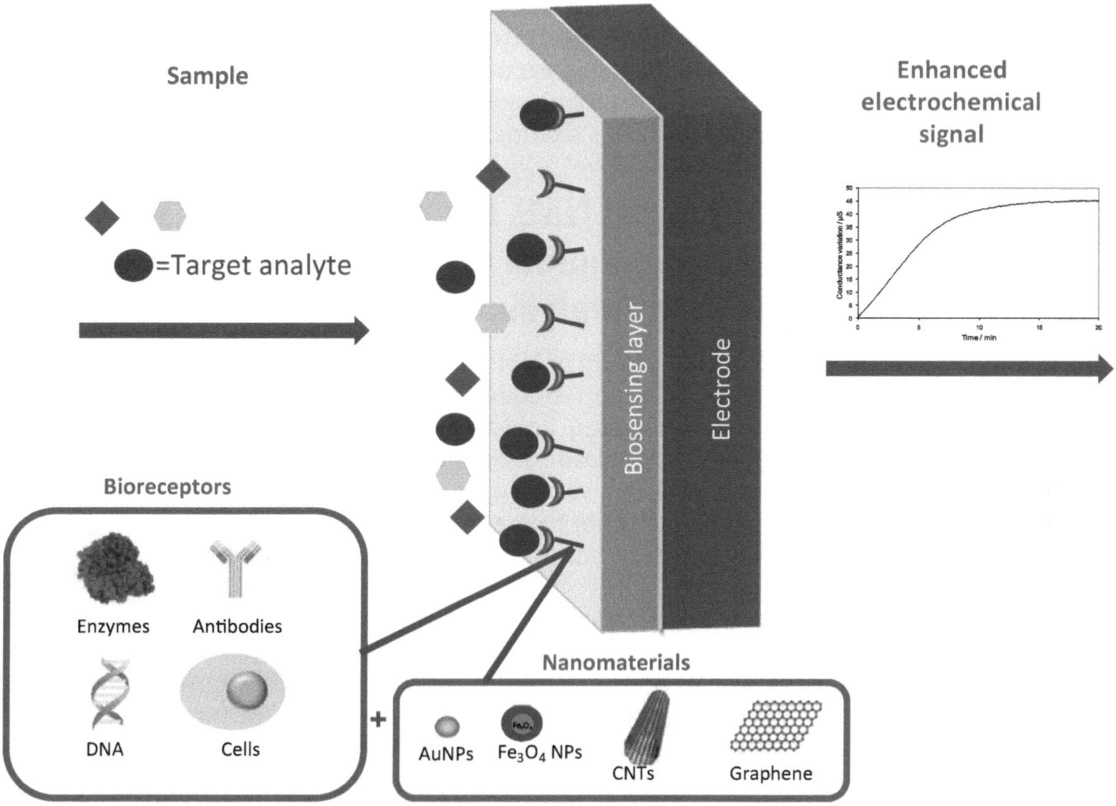

Figure 5.1. Main constituents of a NM-based electrochemical biosensor.

or both. Fig. 5.1 presents the main components and working principle of NM-based electrochemical biosensors. A large number of recent reviews have documented new trends in NMs applied to chemical sensing and biosensing (Justino et al., 2013; Holzinger et al., 2014; Sagadevan and Periasamy, 2014; Zhu et al., 2015a), some of them focusing more particularly on electrochemical transduction (Walcarius et al., 2013; Putzbach and Ronkainen, 2013; Hayat et al., 2014) and/or specific NMs such as metal NPs (Wang, 2012; Saie et al., 2013; Dimcheva and Horozova, 2015), carbon nanotubes (CNTs) (Vashist et al., 2011; Wang and Dai, 2015), graphene (Pumera et al., 2010; Gan and Zhao, 2015; Lawal, 2015; Vashist and Luong, 2015) or composites (Campuzano and Wang, 2011; Gan and Zhao, 2015; Vashist and Luong, 2015). Some other review papers dealing with NM-based biosensors are more specifically dedicated to an area of application such as environment (Wang et al., 2010), health (Kumar et al., 2015b; Lee et al., 2012; Vashist et al., 2011) or food (Perez-Lopez and Merkoci, 2011; Sharma et al., 2015).

In this chapter, the current state-of-the-art of NMs applied to the elaboration of electrochemical biosensors for food analysis are reviewed, focusing on new trends in the detection of some major contaminants, that is, pathogenic bacteria, toxins, pesticides, and veterinary drugs residues.

2 Typical NMs Used in New Electrochemical Biosensing Devices for Food Analysis

A wide variety of NMs (ie, matrices with at least one of their dimensions in the 1–100 nm range) have been reported in the literature for interesting analytical applications. Among these, gold NPs (AuNPs), iron oxide magnetic NPs (γ-Fe_2O_3/Fe_3O_4 NPs), carbon-based NMs (eg, graphene, CNTs, or nanowires), and various combinations of them have provided promising building blocks for the elaboration of nanoscale electrochemical biosensors due to their biocompatibility and the important combination of properties. These nanostructures offer high surface areas and can be used as platforms to enhance bioreceptor loading and stability. Conductive NMs can serve as efficient relays for electron transfer between the bioreceptors and the electrode surface, allowing electrochemical sensing to be performed with no need for electron transfer mediators and resulting in significant signal amplification. The ability to tailor the size and structure and hence the properties of NMs offers excellent prospects for designing novel sensing systems and enhancing the performance of the biosensors. Huge efforts have been devoted in the past 10 years to harness these attractive properties for the development of electrochemical biosensors in view of food contaminants analysis. Immobilization of the biological sensing element onto the transducer plays a key role in biosensor design as it affects the sensitivity, selectivity, and robustness of the analytical device by influencing bioreceptor orientation, stability, activity, and availability to the analyte (Putzbach and Ronkainen, 2013). The next paragraphs report some important features of the major NMs used in the elaboration of electrochemical biosensors for food analysis and the different strategies to elaborate operational, efficient, and robust NMs/bioreceptors sensing layers onto the electrode surface.

2.1 Nanoparticles

2.1.1 AuNPs

Among the various metal NPs proposed for the elaboration of electrochemical biosensors, AuNPs are the most popular and have been extensively used for signal amplification due to their simple production and easy modification by biomolecules (Wang, 2012; Walcarius et al., 2013; Putzbach and Ronkainen, 2013). Owing to their small size, typically some nm to some tenths nm, AuNPs possess electrical, physical, and chemical properties that are different from the bulk metal. These size- and shape-dependent properties may be easily modulated by choosing the appropriate method of synthesis, the strong tendency of NPs to agglomerate via van der Waals' forces being also easily prevented by covering them with an appropriate protecting shell. AuNPs combine high surface area and surface energy with high conductivity and offer a large number of functional groups at their surfaces, allowing easy functionalization of the particles by various kinds of biomolecules (Fig. 5.2) and/or covalent attachment to the electrode. Frequently, adsorption of proteins onto electrode surfaces results in a change in their conformation and a loss in biological activity. Owing to their high biocompatibility and surface free energy, NPs

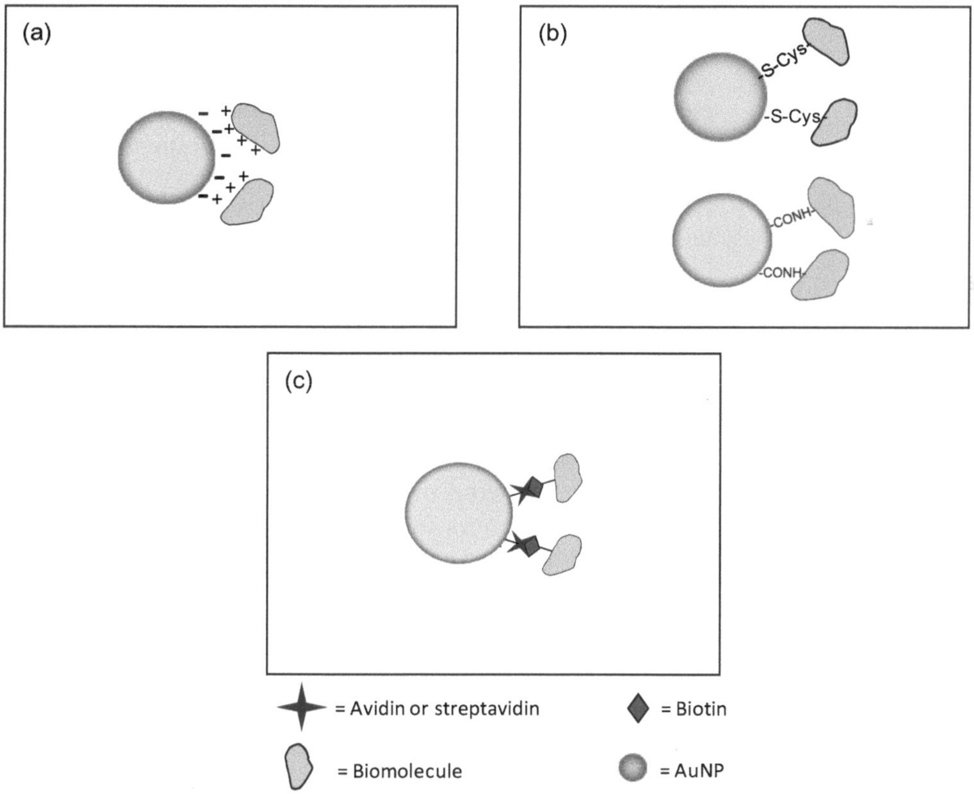

Figure 5.2. Schematic representation (not in scale) of the different approaches reported in the literature for AuNPs modification by biomolecules. (a) Ionic interactions, (b) covalent binding, and (c) affinity interactions.

help biomolecules retaining their bioactivity and bioreceptors loading can be improved. AuNPs not only serve as *platforms* for biosensing elements immobilization but also as efficient shuttles in enzyme biosensors, transferring electrons involved in the redox enzyme reaction from the active sites of the proteins to the electrode surface. AuNPs have also shown a high potential as labels in DNA sensors, immunosensors, and cell sensors (Walcarius et al., 2013). Different biosensor formats are possible, including direct (ie, sensing event is directly measured), competitive, and sandwich assays.

Various strategies, more or less sophisticated, have been proposed to modify electrodes with biosensing layers integrating AuNPs (Wang, 2012; Putzbach and Ronkainen, 2013). Bioconjugated NPs, prepared as described in Fig. 5.2, can be merely deposited onto the electrode surface and crosslinked in the presence of an inert protein, for example, bovine serum albumin (BSA), for the formation of a stable biomolecule/BSA/AuNPs network. This method is simple but can result in biomembrane leakage due to insufficient interactions with the electrode surface or a decrease in biomolecule activity due to crosslinking reaction. Another simple way consists in the direct electrodeposition of NPs onto the electrode surface followed by immobilization of the bioreceptor onto the NPs (Fig. 5.3a). Multilayer films of biological entities and AuNPs can be also prepared by layer-by-layer (LBL) technique, which has attracted much attention because of its simplicity and wide choice of methods (Fig. 5.3b). Another classical strategy relies on the formation of self-assembled monolayers (SAMs) of short-chain alkanethiols bearing free amino or thiol groups used as functional sites to immobilize AuNPs and enzymes (Fig. 5.3c). SAMs provide a simple and well-studied method of immobilization, allowing a high degree of control of the composition and thickness of the transducer surface. Entrapment of NMs and bioreceptors, such as cells or enzymes, into polymeric matrices has also attracted much attention in recent years (Fig. 5.3d). Conductive polymers (eg, polypyrrole or polyaniline), prepared by electropolymerization, or nonconductive polymers such as chitosan, polyvinylalcohol or sol-gel matrices, are often used for that purpose. The presence of NPs in the polymer matrix provides enhanced electronic transfer and prevents enzyme leakage, while confining the reaction products in the vicinity of the electrode.

2.1.2 Magnetic NPs (MNPs)

MNPs with adequate physicochemical and tailored surface properties have been synthesized under controlled conditions for a wide range of applications including sensors and biosensors (Xu and Wang, 2012; Rocha-Santos, 2014). MNPs exhibit their

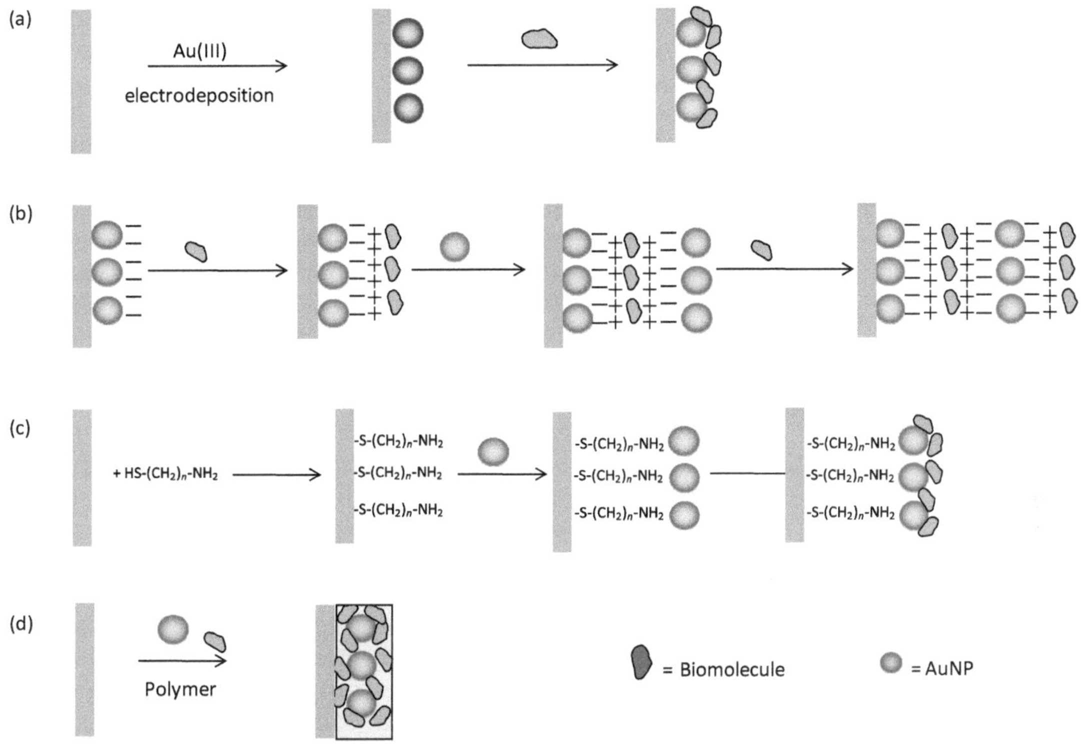

Figure 5.3. Schematic representation (not in scale) of different strategies to modify electrodes with AuNPs and biomolecules. (a) electrodeposition of AuNPs followed by biomolecules immobilization, (b) AuNPs deposition followed by layer-by-layer nanobiofilms elaboration, (c) immobilization of AuNPs and biomolecules onto SAMs, and (d) physical entrapment.

best performance at sizes of 10–20 nm due to supermagnetism, which makes them especially suitable when looking for a fast response due to applied magnetic fields. The particles can be easily functionalized with various bioreceptors using different synthetic pathways and fixed on the electrode surface by simply applying a magnetic field. Then MNPs can be redispersed in the solution after measurement by removing the magnet, enabling a facile purification and concentration of the sample before the detection event and an easy recycling of electrodes and MNPs afterward (Liebana et al., 2014). Moreover, as biological entities do not possess any magnetic properties, no noise or interference is expected during signal capturing. Batch and flow MNP-based biosensor configurations have been reported.

2.2 Carbon-Based NMs

This category of NMs includes one-dimensional nanostructures, such as CNTs, and two-dimensional graphene-based materials.

CNTs are fullerene-related molecules composed of graphene sheets rolled into a cylindrical tube. Single-walled nanotubes (SWCNTs) are cylindrical tubes of 0.5–1 nm diameter capped by hemispherical ends, while multiwalled CNTs (MWCNTs) comprise several concentric cylinders of typically 2–100 nm diameter with a layer spacing of 0.3–0.4 nm. CNTs can be produced by various methods, including arc discharge methods, laser ablation, or chemical vapor deposition (CVD), which has the advantage of allowing the control of the location and alignment of synthesized nanostructures (Yang et al., 2015). Owing to their unique electrical, electrocatalytic, and mechanical properties CNTs have been widely used for the elaboration of electrochemical biosensors, serving as a platform for biomolecule immobilization and improving electrochemical transduction (Wang and Dai, 2015; Yang et al., 2015). The presence of pentagonal domains in the hemispherical ends or in defects along the tubes produces regions with charge density higher than in the regular hexagonal network, thus increasing the electroactivity of CNTs.

Although electrical properties of CNTs are particularly attractive for electrochemical biosensing applications, these NMs suffer from spontaneous tendency to agglomerate and lack of solubility in aqueous media due to their hydrophobic nature. To overcome these issues and facilitate further conjugation with biomolecules, CNTs can be modified with water-soluble polymers or with ionic or polar functional groups, for example, amine or carboxyl moieties, using covalent and noncovalent approaches (Putzbach and Ronkainen, 2013).

CNTs may be further immobilized on the electrode surface using various methods. In the nonaligned approaches, bioconjugated CNTs can be physically adsorbed, chemically grafted, or entrapped into a polymer matrix at the electrode surface. In the aligned approach, density-controlled, aligned CNTs can be prepared, creating parallel nanoelectrodes at the transducer surface, effectively summing up the individual electric signals into an enhanced, detectable signal. Vertically aligned CNTs act as molecular wires (nanoconnectors), facilitating direct electron transfer between the underlying electrode and the redox centers of enzymes. A number of CNT-based biosensors have been elaborated using either the first or the second approach. The first one offers the advantage to be simple, while the second results in a significant enhancement of the electron transfer due to the high conductance of the edge plane of the nanotubes.

Graphene, a monolayer of carbon atoms perfectly arranged in a honeycomb lattice, and its main derivates (graphene oxide, GO, and reduced GO, rGO), are increasingly used for applications in various fields including biosensing. The large number of reviews published on graphene and derived materials in the past 5 years witnesses the huge attractiveness of these exciting new materials and their wide range of application fields, in particular biosensing (Artiles et al., 2011; Gan and Zhao, 2015; Lawal, 2015; Pumera et al., 2010; Vashist and Luong, 2015) and food safety and quality monitoring (Sundramoorthy and Gunasekaran, 2014). They possess extraordinary physicochemical properties such as large surface-area-to-volume ratio, strong mechanical strength, excellent thermal and electrical conductivity, and ease of production in comparison to other carbon based NMs. As CNTs, graphene-based NMs can be chemically modified with suitable functional groups such as hydroxyl, carboxylic, or epoxide, improving their compatibility with water solutions, enabling easy functionalization with biomolecules and facilitating electron transfers (Pavlidis et al., 2014).

2.3 Carbon NMs/NPs Hybrid Nanoarchitectures

In many cases, the combination of different types of NMs possess enhanced properties compared to those of individual components due to cooperative effect. A variety of assemblies and composites (eg, CNTs or graphene decorated with gold NPs) and their applications to electrochemical biosensors have thus been investigated, allowing improved detection performances (Campuzano and Wang, 2011). CNTs-AuNPs hybrid materials may be prepared using various strategies including deposition of AuNPs onto CNTs or direct formation at their surface. LBL technique is often used for the elaboration of graphene-AuNPs films.

3 NM-Based Electrochemical Biosensors for Food Contaminants Analysis

3.1 Pathogens

Foodborne diseases have become a major public health issue worldwide due to the significantly increased incidence of foodborne diseases over the past 20 years. Foodborne diseases generally result from the consumption of food or water contaminated with pathogens including bacteria, viruses, fungi, and parasites or their toxins. According to published reports, contaminated food is considered as the transmitting agent of about 250 known diseases. Thousands of foodborne pathogenic bacteria have been found to affect the health and safety of the world's populations of humans, animals, and plants. Among these, *Campylobacter*, *Salmonella* spp., *Listeria monocytogenes*, *Escherichia coli* O157:H7, *Staphylococcus aureus*, and *Bacillus cereus* are responsible for the majority of foodborne illness outbreaks. Among the methods explored for bacteria detection, the traditional culture and colony counting method, followed by confirmation by biochemical and serological tests, allows obtaining reliable results and is rather inexpensive, but it is labor intensive and time consuming, with completion ranging from 24 h for *E. coli* to 7 days for *L. monocytogenes*, hardly applicable in the quality control of semiperishable foods.

More rapid methods, including PCR and enzyme-linked immunosorbent assays (ELISA) have been proposed, taking 30 min or a few hours to achieve detection result (Law et al., 2015). However, there are still key issues that need to be considered in the development of rapid methods for the detection of foodborne pathogens, including differentiation of live and dead cells, automation, cost, simplicity, training, and accuracy.

The detection of bacterial pathogens using biosensor technology is rapidly expanding and has been reviewed by several authors in the past 5 years (Arora et al., 2011; Sharma and Mutharasan, 2013). Amperometry, potentiometry, and impedimetry are the most common transduction modes in electrochemical biosensors. Biorecognition process generally relies on the capture of the bacteria of interest by specific antibodies or nucleic acids immobilized on the electrode, making impedimetric detection particularly well suited (Wang et al., 2012). It has been shown that integrating NMs in biosensors can help improving the sensitivity and the rapidity of analysis (Koedrith et al., 2014; Liebana et al., 2014; Sharma et al., 2015; Perez-Lopez and Merkoci, 2011). As seen in Table 5.1, very low limits of detection (LODs), typically some CFU mL^{-1} to some tenths CFU mL^{-1}, can be achieved using a variety of NMs and hybrid assemblies.

For the detection of *Salmonella* spp., Ma et al. (2014) proposed an impedimetric biosensor based on specific recognition of the bacteria by a *Salmonella*-specific aptamer. A glassy carbon electrode (GCE) was used as transducer and modified with GO and AuNPs. For that, a GO solution was first spread onto the electrode and after a drying step, AuNPs were electrodeposited before incubation into an aptamer solution. Impedance signal increased between 2.4 and 2.4×10^3 CFU mL^{-1} and an LOD of 3 CFU mL^{-1} was achieved. Recoveries obtained for the detection of the bacteria in pork samples spiked with 10 to 1000 CFU mL^{-1} were close to 100%. The NM-based biosensor developed by Viswanathan et al. (2012) for the same microbial strain, was less performing, with an LOD of 400 CFU mL^{-1} but enabled the simultaneous detection of two other strains (multiplexing). The immunosensor was fabricated by

Table 5.1 Some Recent NM-Based Electrochemical Biosensors for Pathogens Detection

Bacteria	Bioreceptor	Transduction	NMs	Real Sample	LOD	References
Salmonella spp.	Antibodies	SWASV	MWCNT-COOH	Spiked bovine milk	400 CFU mL^{-1}	Viswanathan et al. (2012)
	Aptamers	EIS	GO/AuNPs	Spiked pork meat	3 CFU mL^{-1}	Ma et al. (2014)
Salmonella enteritidis	DNA	SWASV	AuNPs/MNPs	—	0.5 ng mL^{-1}	Zhang et al. (2010a,b)
E. coli O157:H7	Antibodies	CV	MNPs	—	6 CFU mL^{-1}	Setterington and Alocilja (2012)
		SWASV	MWCNT-COOH	Spiked bovine milk	800 CFU mL^{-1}	Viswanathan et al. (2012)
		EIS	MNPs/ nanoporous membrane	—	10 CFU mL^{-1}	Chan et al. (2013)
	DNA	DPV	MWCNT/Bi	Spiked beef meat	2×10^{-14} M	Abdalhai et al. (2015)
		EIS	GO/IONPs	—	10^{-14} M	Tiwari et al. (2015)
Staphylococcus aureus	DNA	DPV	MWCNT/Bi	Spiked beef meat	3.2×10^{-14} M	Abdalhai et al. (2014)
		DPV	AuNPs/ MWCNTs	Tap water	3.3×10^{-16} M	Sun et al. (2015)
	Aptamer	EIS	GO/AuNPs	Fish, water	10 CFU mL^{-1}	Jia et al. (2014)
Listeria monocytogenes	Antibodies	EIS	MNPs	lettuce, milk, and ground beef	10^3 CFU mL^{-1}	Kanayeva et al. (2012)
		CV	AuNPs	Blueberries	2 log CFU g^{-1}	Davis et al. (2013)
Campylobacter jejuni	Antibodies	SWASV	MWCNTs-COOH	Spiked bovine milk	400 CFU mL^{-1}	Viswanathan et al. (2012)
Bacillus cereus	Antibodies	CV	MNPs	—	40 CFU mL^{-1}	Setterington and Alocilja (2012)
		CA	AuNPs	Spiked milk	10 CFU mL^{-1}	Kang et al. (2013)
Brucella melitensis	Antibodies	EIS	AuNPs	Spiked milk	4×10^5 CFU mL^{-1}	Wu et al. (2013)

CA, Chronoamperometry; CS, chitosan; CV, cyclic voltammetry; EIS, electrochemical impedance spectroscopy; IONPs, iron oxide NPs; LOD, limit of detection; MWCNTs-COOH, carboxylated multiwalled carbon nanotubes; SWASV, square wave anodic stripping voltammetry.

immobilizing anti-*Salmonella*, anti-*E. coli* and anti-*Campylobacter* antibodies on the surface of an MWCNT-polyallylamine modified screen-printed electrode (MWCNT-PAH/SPE). A sandwich-type strategy was chosen, and was performed with antibodies conjugated with metal nanocrystals, which have releasable metal ions detectable by square wave anodic stripping voltammetry (SWASV). LODs were 400 CFU mL^{-1} for *Salmonella* and *Campylobacter*, and 800 CFU mL^{-1} for *E. coli*. A multiplexed determination of bacteria in milk samples could be achieved. *Campylobacter* spp. is an important cause of acute bacterial diseases in humans worldwide, causing diarrhea, cramping, abdominal pain, and fever. Campylobacteriosis is considered as one of the most common causes of diarrheal illness in the United States and affects an estimated 2.4 million people each year. Performances of other biosensors developed for *Campylobacter* spp. detection may be found in the recent review written by Yang et al. (2013).

Other electrochemical biosensing platforms have been reported for multiplexed determination of pathogenic bacteria. For example, a biobar-coded genosensor based on AuNPs and MNPs in a screen-printed carbon electrode chip was proposed by Zhang et al. (2010a). The biosensor was able to detect as low as 0.5 ng mL^{-1} of *Salmonella enteritidis* and 50 pg mL^{-1} of *Bacillus anthracis* in only 2.5 h. In a similar way, Dou et al. (2013) developed an immunosensor for *E. coli* O157:H7 and *Enterobacter sakazakii* detection using carbon screen-printed low-density arrays coated with MWCNTs/sodium alginate/carboxymethyl chitosan composite films and further functionalized with horseradish peroxidase (HRP)-labeled antibodies. LODs were 4.57×10^3 and 3.47×10^3 CFU mL^{-1} for *E. sakazakii* and *E. coli* O157:H7, respectively.

As Table 5.1 shows, other electrochemical biosensors integrating nanostructures recently reported for *E. coli* detection are principally immunosensors and DNA sensors. Setterington and Alocilja (2012) combined cyclic voltammetry (CV) with immunomagnetic separation in a rapid method requiring only 1 h. An immunofunctionalized magnetic/polyaniline core/shell NP was employed to extract target cells from the sample solution, magnetically position them on a screen-printed carbon electrode, and as a mediator for current flow. The NP consisted of an iron oxide core, imparting magnetic functionality, and a polyaniline shell, imparting electronic activity. The cell-binding event was accompanied by a decrease of current response in CV. The technique was evaluated for the detection of *E. coli* O157:H7 but also of *B. cereus* and low LODs were achieved in both cases (6 and 40 CFU mL^{-1}, respectively). The cost of the assay was estimated at $1 per disposable sensor and $0.20 per test for reagents.

Harnessing the separation capacities of polystyrene magnetic particles and combining them with adsorption properties of an alumina porous membrane, an ultrasensitive method for *E. coli* detection was reported by Chan et al. (2013). In this work, antibody-conjugated MNPs were used to capture the cells, which were further immobilized on a nanoporous alumina membrane functionalized with anti-*E. Coli* antibodies, forming a sandwich complex of the antibody-modified magnetic beads—*E. coli* O157:H7—antibody. The change in impedance across the nanoporous membrane resulting from the cells fixation enabled the detection of *E. coli* O157:H7 with low concentration levels. Compared with direct detection of *E. coli* O157:H7 using nanoporous alumina membrane, the LOD was lowered by one order from 10^2 to 10 CFU mL^{-1}.

A cost-effective, sensitive, label-free impedimetric aptasensor, based on AuNPs/rGO was also recently proposed by Jia et al. (2014) for the direct detection of *S. aureus*. Thiolated ssDNA was covalently linked to the AuNPs bound to rGO, and probe DNA was immobilized on the surface of an AuNP modified GCE to capture and concentrate bacterial cells. The probe DNA of the aptasensor selectively captured the target bacteria in its three-dimensional space, resulting in a tremendous increase in impedance and a 10 CFU mL^{-1} LOD. An excellent selectivity was observed toward *E. coli* and *Salmonella*. The recovery rates for fish and water samples spiked in the 20–320 CFU mL^{-1} varied from 92% to 114%.

In our group, we recently developed a highly performing conductometric biosensor for the determination of Gram-negative bacteria (El Ichi et al., 2014). These bacteria possess lipopolysaccharides (LPS) as conserved antigens on their membrane. Antibodies directed against LPS were used as bioreceptors and grafted to the surface of MNPs. A very sensitive detection of *E. coli* and *Serratia marcescens* strains (down to 1 CFU mL^{-1}), as well as of clinically relevant *Pseudomonas aeruginosa* and *Acinetobacter baumannii* strains (down to 10 CFU mL^{-1}), could be achieved. The biosensor response was not inhibited in 1% serum, paving the way for applications to the analysis of other types of samples such as food.

3.2 Toxins

Food toxins are natural substances covering a large variety of molecules, generated by fungi, algae, plants, or bacteria metabolism with harmful effects on humans or other vertebrates even at very low doses. Toxins are generally divided into three main categories: mycotoxins, algal toxins, and plant/bacteria toxins. Several recent reviews have been published on food toxin detection using biosensors (McGrath et al., 2012), NM-based biosensors (Malhotra et al., 2014) or more specifically NM-based electrochemical biosensors (Campas et al., 2012). For his part, Tothill (2011) discussed recent trends in NM-based biosensors for mycotoxins analysis.

3.2.1 Mycotoxins

Mycotoxins are low molecular weight toxicants exhibiting a large variety of chemical structures and produced by filamentous fungi belonging mainly to *Aspergillus, Penicillium,* or *Fusarium* genera. They may be found in a variety of agricultural products (eg, cereals, fruits, milk, coffee, meat) and in derived products. Some mycotoxins induce genetic disorders and processes of carcinogenesis, others have teratogenic, embryotoxic, dysbacteriosis, allergenic, dermonecrotic effects (Marin et al., 2013). Contamination of agricultural crops by mycotoxins can occur at various points (preharvest, harvest, drying, storage, or transport) in the food chain. From a toxicological and legislative point of view, the most relevant mycotoxins are aflatoxins, ochratoxins, *Fusarium* toxins, patulin, citrinin, and ergot alkaloids. The toxicity of these molecules has led many countries to set up strict regulations for their control in food and feed and the consequent establishment of legislation to control their possible contamination. In the European Union, the maximum allowed concentrations of ochratoxin A (OTA) are in the 0.5–80 $\mu g\ kg^{-1}$ range, depending on the kind of food, while aflatoxin B1 (AFB1) is regulated in the 0.1–12 $\mu g\ kg^{-1}$ range. Patulin contents should not exceed 10–50 $\mu g\ kg^{-1}$, while tolerated values are much higher for citrinin (2000 $\mu g\ kg^{-1}$) or *Fusarium* toxins (20–4000 $\mu g\ kg^{-1}$). The analysis of mycotoxins is challenging as these molecules are present in low concentrations in complex matrices, and they may occur in various combinations produced by a single or by several fungal species. Immunological techniques such as ELISAs and radioimmunoassay (RIAs) have been developed for screening analysis of mycotoxins and chromatographic techniques using mass spectrometry or fluorescence detection for the separation and quantification of the molecules (Pereira et al., 2014). A range of biosensors has been also reported in the literature for mycotoxins analysis, most of them targeting ochratoxins and aflatoxins (more particularly OTA and AFB1) and exploiting the affinity of the molecules for specific antibodies or aptamers, the transduction mode being mostly optical (Li et al., 2012), electrochemical (Vidal et al., 2013), or in a lesser extent mechanical (Chauhan et al., 2015). Biosensors based on the inhibition or activation of enzymes by mycotoxins (Alonso-Lomillo et al., 2010, 2011; Arduini et al., 2010; Li et al., 2011; Dridi et al. 2015; Stepurska et al., 2015), or using artificial receptors such as molecularly imprinted polymers as sensing element (Pacheco et al., 2015) have been also reported. The following paragraphs describe recent advances in NM-based electrochemical biosensors for ochratoxin, aflatoxins and *Fusarium* toxins. Tables 5.2–5.4 provide some features of recent electrochemical biosensors reported in the field.

3.2.1.1 Ochratoxins

OTA is the most prevalent and relevant fungal toxin of this group of mycotoxins and is produced by *Aspergillus* and *Penicillium* fungi (El Khoury and Atoui, 2012). OTA is highly toxic, usually considered as genotoxic, but also hepatotoxic, teratogenic, immunotoxic, and neurotoxic. Due to its high resistance to acidity and temperature, OTA is hard to remove from contaminated foodstuffs, and then seriously threatens human health.

Table 5.2 Some Recent NM-Based Electrochemical Biosensors for OTA Detection

Bioreceptor	Transduction	NMs	Real Sample	LOD (pg mL^{-1})	References
Enzymes	Conductometry	AuNPs	Spiked olive oil	400	Dridi et al. (2015)
Antibodies	EIS	MNPs	Spiked white wine	10	Zamfir et al. (2011)
	SWV	MNPs	Spiked red wine	8	Perrotta et al. (2012)
	DPV	AuNPs	Spiked corn	200	Liu et al. (2013)
			Spiked red wine	110	Vidal et al. (2012)
Aptamers	DPV	AuNPs	Spiked red wine	0.12	Xie et al. (2014)
		AuNPs	Spiked red wine	0.3	Yang et al. (2014)
		MNPs	Spiked red wine	110	Barthelmebs et al. (2011)
	CV	AuNPs	Spiked red wine	30	Kuang et al. (2010)
		GO/Ce NPs	Spiked corn	40	Bulbul et al. (2015)
	Amperometry	MNPs	Spiked beer	0.05	Rhouati et al. (2013)
	EIS	AuNPs	Spiked beer	8	Evtugyn et al. (2013)
		AuNPs/rGO	—	30	Qian et al. (2014)
		AuNPs/GO	Spiked red wine	0.3	Jiang et al. (2014)

CV, Cyclic voltammetry; DPV, differential pulse voltammetry; EIS, electrochemical impedance spectroscopy; LOD, limit of detection; SWV, square wave voltammetry.

Table 5.3 Some Recent NM-Based Electrochemical Biosensors for AFB1 and AFM1 Detection

Analyte	Bioreceptors	Detection	NMs	Real Sample	LOD ($pg\ mL^{-1}$)	References
AFB1	Antibodies	EIS	Au NWs	Spiked corn	19	Hu et al. (2015)
			rGO/AuNPs	Peanuts, rice, milk, flour, soybean	0.000312	Linting et al. (2012)
			rGO	spiked corn	0.01	Wang et al. (2015)
			GO	—	230	Srivastava et al. (2014)
		CV	MWCNTs-COOH	—	80	Singh et al. (2013)
			rGO/AuNPs	—	100	Srivastava et al. (2015)
AFM1	Aptamers	EIS	AuNPs	Spiked milk	1000	Dinckaya et al. (2012)
		SWV	MNPs	Spiked milk	1.98	Nguyen et al. (2013)
	Antibody	CA	MNPs	Spiked milk	10	Paniel et al. (2010)

AuNWs, Gold nanowires; CA, chronoamperometry; LOD, limit of detection; MWCNTs-COOH, carboxylated multiwalled carbon nanotubes; SWV, square wave voltammetry.

A number of electrochemical biosensors harnessing the selectivity of antibodies or combining the selectivity and stability of aptamers have been reported for OTA detection (Vidal et al., 2013), various strategies being proposed for signal enhancement using AuNPs, MNPs, and graphene derivatives. Table 5.2 gives some examples from the past 5 years. As shown in this table, the biosensors are, except in one case, applied to real samples, that is, alcoholic beverages (wines or beers), corn, or oil.

Different approaches to integrate AuNPs and bioreceptors in a polymeric matrix have been described. In our group, we recently developed an electrochemical biosensor based on thermolysin enzyme entrapped into a polyvinylalcohol/polyethylenimine/gold NPs (PVA/PEI/AuNPs) nanocomposite (Dridi et al., 2015). The enzyme/AuNPs/polymer blend was prepared by simply mixing solutions or suspensions containing each of the three components and the resulting suspension was casted onto gold interdigitated microelectrodes before crosslinking. A very low LOD of 1 nM (or 400 $pg\ mL^{-1}$) could be achieved for the proposed OTA biosensor, which could be attributed not only to the beneficial impact of NPs, but also to the very favorable aqueous environment created

Table 5.4 Some Recent NM-Based Electrochemical Biosensors for *Fusarium* Toxins Detection

Analyte	Bioreceptors	Detection	NMs	Real Sample	LOD ($pg\ mL^{-1}$)	References
ZEA	Antibodies	Amperometry	N-GrOS	Spiked pig feed	2.1	Feng et al. (2013)
			MWCNTs	Spiked corn silage	770	Panini et al. (2010)
		CV	MNPs	Spiked feedstuff	20	Panini et al. (2011)
		SWV	MWCNTs/AuNPs	Bovine serum	16	Regiart et al. (2015)
DON	Antibodies	EIS	Fullerene	Spiked food	0.3	Zhilei et al. (2011)
Fumonisins B1, B2, B3	Aptamers	EIS	AuNPs	Spiked maize	1.4	Chen et al. (2015)
		CV	AuNPs/GS	Spiked wheat	1	Shi et al. (2015)
	Antibodies	Amperometry	MNPs	Maize, spiked beer	330	Jodra et al. (2015)
		EIS	MWCNTs	Corn	0.46	Masikini et al. (2013)

LOD, Limit of detection; N-GrOS, nitrogen doped reduced graphene oxide sheet; SWV, square wave voltammetry.

around the enzyme through the use of PVA/PEI hydrogel. In comparison, LOD was 700 times higher when thermolysin was immobilized by crosslinking in presence of BSA at the surface of the electrode. The newly designed biosensor was successfully applied to the quantification of OTA in spiked olive oil samples without any sample pretreatment. Owing to the small size of the electrodes and the availability of portable instrumentation, this method may be easily adaptable for on-site analysis.

Another approach combining AuNPs and polymers for an efficient detection of OTA was reported by Evtugyn et al. (2013). In this work, an impedimetric aptasensor was prepared by drop casting a suspension of AuNPs stabilized by the dendrimeric polymer Botlorn H30 onto gold electrodes previously modified by electropolymerized neutral red. Thiolated aptamer specific to OTA was covalently attached to AuNPs via Au—S bonding. The interaction of the aptamer with OTA induced the conformational switch of the aptamer from linear to guanine quadruplex form followed by consolidation of the surface layer and an increase of the charge transfer resistance. The aptasensor enabled OTA detection in the 0.1–100 nM range in the presence of at least 50-fold excess of ochratoxin B and an LOD of 0.02 nM (or 8 pg mL^{-1}) was achieved. The applicability of the aptasensor for real sample assay was confirmed by testing spiked beer samples.

Wang group (Jiang et al., 2014; Qian et al., 2014; Yang et al., 2014) also recently elaborated several NM-based aptasensors using either single AuNPs (Yang et al., 2014) or AuNPs/rGO nanocomposites (Jiang et al., 2014; Qian et al., 2014) as signal amplifiers. In the first configuration, a two-level cascaded amplified electrochemical sandwich detection strategy using methylene blue (MB) as a redox indicator was proposed. By employing this novel strategy, an 8.5-fold amplification in signal intensity was achieved, with an extremely low LOD (0.3 pg mL^{-1}). In the second configuration, AuNPs/GrO composites were synthesized by covalently binding NPs to 2-aminothiophenol-functionalized rGO. Well-dispersed and controllable surface coverage of AuNPs could be achieved. The aptasensor sensitivity was significantly improved in presence of the composite.

Bulbul et al. (2015) recently reported an original nonenzymatic nanocatalyst based approach to construct an electrochemical aptasensor involving the synergistic contribution of a nanoceria (nCe) tag and GO for OTA detection. The aptamer was immobilized on the surface of a GO modified electrode. The target analyte was captured by the immobilized aptamer via a specific competitive mechanism between the free and the nCe labeled target. The electrochemical signal was generated by monitoring the electro-oxidation of a generic redox species upon reaction with the nCe tag. The signal was further amplified by the GO layer used as an electrode material to immobilize the aptamer and to increase the electron transfer at the electrode surface, further enhancing sensitivity of the assay. The aptasensor exhibited a linear response to OTA in the range 0.15–180 nM with a very low LOD (0.1 nM or 40 pg mL^{-1}).

A range of MNPs-based electrochemical biosensors has been also proposed for OTA detection using direct (Zamfir et al., 2011) or competitive formats (Rhouati et al., 2013; Barthelmebs et al., 2011) in static or flow systems. Recently, we developed a label-free immunosensor using gold electrodes modified with a BSA conjugate bearing a glutaraldehyde–thiolamine linker, creating a layer that prevents nonspecific binding of OTA on gold (Zamfir et al., 2011). The OTA antibodies were further attached to MNPs using the carbodiimide chemistry and immobilized on the modified gold electrode using a strong magnetic field. The impedance variation due to the specific antibody–OTA interaction was correlated with the OTA concentration in the samples. The increase in electron-transfer resistance values was proportional to the concentration of OTA in the 0.01 and 5 ng mL^{-1} range, with a LOD of 0.01 ng mL^{-1}.

Rhouati et al. (2013) developed a fully automated flow electrochemical aptasensor based on the use of a selective capture aptamer and modified magnetic beads. The proposed flow-based aptasensor was demonstrated under continuous/stopped flow conditions for the detection of OTA employing direct and indirect competitive strategies. In the direct format, labeled biotin and free OTA competed to bind with immobilized aptamer onto the surface of a screen-printed carbon electrode placed in a flow cell. In the indirect assay, immobilized OTA and free OTA in solution competed to bind with aptamer in solution. A lower LOD (0.05 μg L^{-1}) was obtained with the indirect flow-based aptasensor.

3.2.1.2 Aflatoxins

Aflatoxins are highly toxic compounds produced principally by three anamorphous species of the genus *Aspergillus*, that is, *A. flavus*, *A. parasiticus*, and *A. nomius* growing mostly on crops, such as grains and nuts. Among the 18 types of aflatoxins identified, AFB1 is predominant, is a potent human carcinogen and responsible for liver cancer in animals. Ingested by cows via the feed, AFB1 is transformed into aflatoxin M1 (AFM1), which is further secreted in milk. Biosensors developed for aflatoxins are therefore targeting mainly AFB1 and AFM1. A variety of biosensors and NM-based biosensors (Campas et al., 2012; Malhotra et al., 2014; Tothill, 2011) have been proposed. Some of recent and significant examples of NM-based electrochemical biosensors are in Table 5.3 As OTA biosensors, AFB1 biosensors are mainly immunosensors and aptasensors.

3.2.1.3 AFB1

As seen in Table 5.3, Malhotra team developed several immunosensors for AFB1 detection using either carboxylated MWCNT (MWCNTs-COOH), GO or AuNPs/rGO composite for signal amplification. In the first study (Singh et al., 2013), MWCNTs-COOH were produced by CVD followed by acidic treatment, while in the second the simultaneous reduction of GO and Au(III) salt allowed fabricating uniformly distributed AuNPs on rGO sheets through an easy and one-step synthesis (Srivastava et al., 2015). Both types of NMs were electrophoretically deposited onto indium tin oxide substrate and used for covalent attachment of monoclonal AFB1 antibodies (anti-AFB1), AFB1 being further detected by CV. In the third investigation, a GO modified Au electrode was produced and functionalized with anti-AFB1 for label-free electrochemical impedance spectroscopy (EIS) determination of AFB1 (Srivastava et al., 2014). LODs were respectively 80, 100, and 230 pg mL^{-1}. Neither selectivity nor applicability to the analysis of real samples was demonstrated. A remarkable LOD (0.3 fg mL^{-1}) could be achieved by Linting et al. (2012) who developed an impedimetric immunosensor for AFB1 detection by forming graphene/conducting polymer/AuNPs/ionic liquid(IL) composite films onto gold electrodes surface. Graphene oxide, 2,5-di-(2-thienyl)-1-pyrrole-1-(*p*-benzoic acid) (DPB) and chloroauric acid were electrodeposited on the surface of gold electrode in sequence. Anti-AFB1 antibody was covalently connected to the conducting polymer film with EDC/NHS as activator and recovered by IL and chitosan. It was demonstrated that IL plays a stabilizing role for anti-AFB1 antibodies. The biosensor was applied to AFB1 analysis in several real samples (peanuts, rice, milk, flour, and soybeans) and a good agreement was obtained between biosensor and HPLC results. Another strategy, adopted by Wang et al. (2015), consisted in combining EIS, an intrinsically very sensitive technique, with GO (conductivity and stability enhancer), electroactive polypyrrole (PPy) and pyrrole-propylic acid (PPa) for antibody attachment, to achieve a low LOD for AFB1 (0.01 pg mL^{-1}). This biosensor was applied to spiked corn analysis, with recoveries in the 80–112% range.

3.2.1.4 AFM1

For AFM1 determination, different approaches have been reported recently. Dinckaya et al. (2012) elaborated an aptamer biosensor by modifying gold electrodes by a cysteamine SAM and grafting AuNPs and then a thiolated aptamer binding specifically AFM1, LBL. EIS detection of AFM1 binding event provided a linear response in the 1–14 ng mL^{-1}, with a LOD of 1 ng mL^{-1}. The biosensor was subsequently applied to a real milk sample, with recoveries close to 100%. Other studies were performed using MNPs (Paniel et al., 2010) or MNPs/polyaniline (PANI) films (Nguyen et al., 2013) coated with specific antibodies or aptamers.

The first sensor was based on a competitive immunoassay using HRP as a tag. MNPs coated with antibody (anti-AFM1) were used to separate the bound and unbound fractions. The samples containing AFM1 were incubated with a fixed amount of antibody and tracer [AFM1 linked to HRP (conjugate)] until the system reached equilibrium. In the MNPs/PANI biosensor, aptamers were attached to PANI by crosslinking and AFM1 binding to the aptamers was detected in direct format by single wave voltammetry (SWV). The latter biosensor allowed achieving a better LOD (1.98 pg mL^{-1}) than the MNPs-based chronoamperometric biosensor (10 pg mL^{-1}). Both biosensors were also successfully applied to spiked milk analysis.

3.2.1.5 Fusarium Toxins

Another important class of mycotoxins is produced by *Fusarium* filamentous fungi and affect cereal crops. Among these, zearalenone (ZEA), some tricothecenes (ie, deoxynivalenol, DON, and the sum of T-2 and HT-2 toxins), as well as the sum of fumonisins B1 and B2 (FB1 and FB2) are regulated in Europe. Analytical methods for the determination of trichothecene mycotoxins in food tricothecenes have been recently reviewed (Meneely et al., 2011). These molecules are pathogenic to humans and animals and can induce growth retardation, wilting, chlorosis, necrosis, and inhibition of germination in plants. To date, there is still only few electrochemical biosensors elaborated for *Fusarium* toxins. Some oldest references may be found in Campas et al. (2012). New architectures integrating NMs and proposed in the past 5 years for DON, ZEA, and fumonisins (mainly FB1), are reported in Table 5.4. As seen in the table, only antibody-based and aptamer-based biosensors have been described.

3.2.1.6 ZEA

Regiart et al. (2015) described a sensitive immunosensor for ZEA determination in bovine serum. The biosensor was prepared by casting MWCNTs previously modified with poly(vinylpyridine) (PVP) onto screen-printed carbon electrodes. Further AuNPs electrodeposition on MWCNTs/PVP helped enlarging the surface available for subsequent immobilization of anti-ZEA antibodies. The biosensor signal, recorded by SWV, was linear in the 0.05–50 ng mL^{-1} range, with a LOD of 16 pg mL^{-1}. The time required for analysis was 12 min only, which compares quite favorably with the time (90 min) required by the conventional ELISA. The same team (Panini et al., 2010) proposed an immunosensor using GCEs modified with MWCNTs and integrated into a continuous-flow system for rapid and sensitive quantification of ZEA in corn silage samples. Mouse monoclonal anti-ZEA antibodies were immobilized on a rotating disk. ZEA in a corn sample was allowed to compete immunologically with ZEA bound to HRP for the immobilized antibodies. HRP in presence of H_2O_2 catalyzed the oxidation of 4-tert-butylcatechol (4-TBC), whose back electrochemical reduction was detected by amperometry at −0.15 V. The total assay time was 15 min. LOD was similar to that of the AuNPs/MWCNTs biosensor but lower than the standard ELISA method.

Feng et al. (2013) proposed a rapid and very sensitive immunosensor for ZEA detection using a sandwich-type format. GCE electrodes were first modified with nitrogen-rGO sheets (N-GrOS) functionalized with anti-ZEA primary antibodies (Ab1). ZEA, captured by Ab1, further interacted with secondary antibodies (Ab2) labeled with the nanoporous bimetallic PtCo alloy (NP-PtCo). NP-PtCo with three-dimensional bicontinuous spongy structures and various hollow channels was prepared by corrosion of PtCoAl alloy, and exhibited remarkably improved electrocatalytic activity toward H_2O_2. The chronoamperometric signal, recorded after H_2O_2 injection, was linear in the 0.005–25 ng mL^{-1} ZEA concentration range, and LOD was 2.1 pg mL^{-1}. Recoveries obtained on fortified pig feed samples were close to 100%, demonstrating the absence of matrix effect.

3.2.1.7 DON

Zhilei et al. (2011) reported a very sensitive impedimetric immunosensor for DON detection. The biosensor was based on anti-DON polyclonal antibodies covalently conjugated to the surface of a GCE previously modified with a fullerene (C60)/ferrocene/1.3-dibutylimidazolium bromide (IL) composite immobilized on crosslinked chitosan. C60 was used to improve electron transfer at the modified electrode surface, while IL provided a biocompatible favorable environment for the biomolecules. The immunosensor was stable over more than 180 days, the signal was linear in the 1 pg mL^{-1} to 0.3 ng mL^{-1} range, and LOD (0.3 pg mL^{-1}) was two-fold better than in absence of fullerene and IL.

3.2.1.8 Fumonisins

Chen et al. (2015) presented an aptasensor for the sensitive and label-free determination of fumonisin B1 (FB-1) by EIS. For that, a thiolated aptamer specific to FB-1 was anchored at the surface of AuNPs previously electrodeposited onto a GCE. A significant increase in the electron transfer resistance was observed following the interaction of FB-1 with the modified electrode in the 0.1 nM to 100 μM concentration range, and the detection limit was as low as 2 pM (or 1.4 pg mL^{-1}). The assay was applied to the analysis of maize samples spiked with FB-1, and recovery rates ranged from 91% to 105%. This LOD was very close to the value obtained by Shi et al. (2015) who also developed an aptasensor but chose a competitive mode and a dual amplification strategy using AuNPs and graphene/thionine nanocomposite. Thionine was loaded at the surface of the graphene sheet to serve as electrochemical probe and increase the biosensor electrochemical signal due to the excellent conductivity and the large surface area of the graphene sheet. The biosensor signal was proportional to FB-1 concentration in the 1–106 pg mL^{-1} range with a detection limit of 1 pg mL^{-1}. Besides, the developed aptasensor was successfully applied to the determination of FB1 in feed samples.

Both Jodra et al. (2015) and Masikini et al. (2013) reported immunosensors for fumonisins, the first biosensor being a magnetosensor and the second being based on poly(2.5-dimethoxya niline)-MWCNTs doped with palladium telluride quantum dots onto a glassy carbon surface. The composite was assembled by an LBL method to form a multilayer film of quantum dots and poly(2,5-dimethoxyaniline)-MWCNTs. The biosensor preparation further involved drop-coating of fumonisins antibody onto the composite modified GCE. The impedimetric signal was linear in the 7–49 ng L^{-1} range and LOD was 0.46 pg L^{-1}. LODs calculated for fumonisins in corn certified material was 0.014 and 0.011 ppm for FB-1, and FB-2 and FB-3, respectively. The results were lower than those obtained by ELISA but in good agreement with values certified for FB-1 + FB-2 + FB-3.

3.2.2 Algal Toxins

Algal toxins are secondary metabolites produced mostly by dinoflagellates and cyanobacteria, which are part of the phytoplankton. They reach shellfish and fish through the trophic chain and can easily contaminate human consumers. Toxic episodes occur as a result of the proliferation of some toxin-producing phytoplankton species or harmful algal blooms. Although marine toxins were initially classified according to the acute poisoning syndrome they induce in humans, nowadays a classification based on their chemical structure seems to be more widely accepted.

Until very recently, the regulations of most countries based the detection of marine toxins mainly on laboratory animal bioassays, except for the detection of domoic acid, which was performed by HPLC-UV technique. Although the desire to move away from these types of bioassays has been recognized in the legislation of many countries, the criteria for substitution by alternative techniques set by regulatory authorities are difficult to meet. Currently, the detection of saxitoxin and analogs by HPLC-fluorescence is officially accepted in some countries and an ELISA has also been published as an AOAC method for the detection of domoic acid. In addition to the officially accepted methods, there is a wide variety of techniques that have been developed for the detection of algal toxins, among them are HPLC-UV, HPLC-fluorescence, HPLC-MS or HPLC-MS/MS. Various biosensors, including electrochemical biosensors, have been also reported for some of these molecules, for example, okadaic acid, domoic acid, brevetoxins (Pb-Tx and PbTx-3), anatoxin, saxitoxin, microcystins, and tetrodotoxin (Campas et al., 2012; Campbell et al., 2011; Reverte et al., 2014; Vilarino et al., 2009, 2013). The following paragraphs provide some examples of the most recent electrochemical biosensors integrating NMs.

3.2.2.1 Okadaic Acid

Okadaic acid (OA) and its derivatives (dinophysis toxins) are responsible for the so-called diarrheic shellfish poisoning and are produced by acute intoxication after consumption of contaminated mollusks. These compounds are well-known inhibitors of protein phosphatases, mainly PP2A and PP1. They have a worldwide distribution with a higher occurrence in Europe and Japan. OA may be determined using different methods including biosensors (Vilarino et al., 2009; Sassolas et al., 2013). Biosensors reported for OA detection are either immunosensors or enzyme biosensors based on the inhibition of protein phosphatase PP2A. The inhibition of PP2A enzymatic activity is classically measured by electrochemical detection, by direct immobilization of the enzyme on a screen-printed electrode or by using a bienzyme amplification system with off-line enzymatic incubation with PP2A.

Eissa and Zourob (2012) recently developed a graphene-based voltammetric immunosensor for sensitive detection of OA. Graphene-modified screen-printed electrodes were functionalized with a carboxyphenyl film by electrochemical reduction of in situ generated 4-carboxyphenyl diazonium salt in acidic aqueous solution. Next, the anti-OA antibody was covalently immobilized on the carboxyphenyl modified graphene electrodes via carbodiimide chemistry. A competitive assay between OA and a fixed concentration of OA–ovalbumin conjugate (OA–OVA) for the immobilized antibodies was employed for OA detection by SWV. The biosensor signal was linear up to 5 mg L^{-1} and a LOD of 19 ng L^{-1} was achieved. Recoveries obtained by analyzing a spiked uncontaminated shellfish extract and a certified reference mussel sample were close to 100%, indicating no significant effect of the tissue matrix.

Two other electrochemical immunosensors, reported by Marty team, employed magnetic particles functionalized with anti-OA antibodies, either in an indirect competitive immunoassay format under automated flow conditions and amperometric detection (Dominguez et al., 2012), or in a direct affinity binding assay with differential pulse voltammetry (DPV) as transduction mode (Hayat et al., 2012). The incorporation of the modified magnetic beads into the flow device enabled to improve the LOD by a factor 2.5 as compared to the equivalent batch system (0.15 instead of 0.38 mg L^{-1}). The LOD was much better (0.5 μg L^{-1}) with the DPV biosensor. Both biosensors were validated through the analysis of a fortified mussel sample.

3.2.2.2 Brevetoxins

Brevetoxins (PbTx) are lipophilic polyether toxins responsible for neurotoxic shellfish poisoning. Apart from mammalian bioassays, several methods have been developed to detect brevetoxins, including other biological methods such as receptor binding assays, RIA, ELISA, and fluorimetric assays based on changes in membrane potential. Brevetoxins may be also detected by LC-MS and micellar electrokinetic capillary chromatography coupled to laser-induced fluorescence detection. In the biosensors field, only few electrochemical biosensors have been reported (Vilarino et al., 2009) and NM-based biosensors are more scarce.

Zhang et al. (2012) proposed to coimmobilize monoclonal mouse antibodies directed against brevetoxin B (BTX-2) and dinophysistoxin-1 (DTX-1) onto magnetic beads for the simultaneous determination of these two toxins by stripping voltammetry. A competitive-type immunoassay format was chosen, cadmium and copper nanoclusters being used as electrochemical tags for BSA–BTX-2 and BSA–DTX-1 conjugates, respectively. BTX-2 and DTX-1 could be determined in a single run with a wide working range of 0.005–5 ng mL^{-1} for the two marine toxins. LODs were 1.8 and 2.2 pg mL^{-1} for BTX-2 and DTX-1, respectively. Twelve spiked seafood samples with different levels of BTX-2 and DTX-1 were assessed using the multiplexed electrochemical immunoassay and commercial ELISA kit. No significant difference was observed between the two methods.

The same group (Tang et al., 2012) proposed a magneto-controlled immunosensing platform for sensitive electrochemical determination of BTX-2 in seafood by using guanine-assembled graphene nanoribbons (GGNRs) as molecular tags on a home-made magnetic carbon paste electrode. First, monoclonal mouse anti-BTX-2 antibodies were covalently immobilized on the surface of magnetic beads, which were used as the immunosensing probes for the capture of BTX-2. The recognition elements were prepared by chemical modification of BSA–BTX-2 conjugates (BSA–BTX-2) with the GGNRs used for signal amplification. Based on a competitive-type immunoassay format, the formed magnetic immunocomplex was integrated on the electrode with an external magnet, followed by BTX-2 determination by SWV. A LOD of 1 pg mL^{-1} could be achieved and a good agreement with ELISA was obtained on several samples fortified with BTX-2.

3.2.2.3 Microcystins

Microcystins (MCs) are a group of more than 90 hepatotoxins produced by cyanobacteria, of which microcystin-LR (MC-LR) is the most common. Several analytical and biochemical screening methods are currently available, including bioassays, immunoassays, chromatographic techniques, and biosensors.

Biosensors developed for MCs detection are mainly immunosensors and enzyme biosensors based on the inhibition of protein phosphatases (Singh et al., 2012). NM-based electrochemical biosensors reported in the past 5 years for the determination of MC-LR integrate various nanostructures including AuNPs, carbon NMs and composites.

Tong et al. (2011), for example, developed an immunosensor by immobilizing anti-MC-LR antibodies on a gold electrode coated with L-cysteine modified AuNPs. Hydroquinone was used as the electron mediator. The DPV immunosensor signal varied linearly in the 0.05 to 15 μg L^{-1} concentration range, with a LOD of 20 ng L^{-1}. The developed biosensor was used to determine MC-LR in spiked crude algae samples. Recoveries were in the 95.6–105% range.

An MWCNT-based impedimetric immunosensor was proposed by Han et al. (2013). The biosensor electrodes were fabricated using vertically well-aligned, dense, millimeter-long MWCNT arrays with a narrow size distribution, grown on patterned Si substrates by water-assisted CVD. MWCNT surface were further functionalized, thus providing the anchoring sites for linking monoclonal antibodies. A linear dependence of the electron-transfer resistance on the MC-LR concentration was observed in the range of 0.05–20 μg L^{-1}.

The exciting properties of various other carbon-based nanostructures such as carbon nanohorns (Zhang et al., 2010b), graphene (Eissa et al., 2014), or graphene combined with carbon nanospheres (Zhao et al., 2013), or with AuNPs (Ruiyi, 2013), have been also harnessed for the elaboration of performing electrochemical immunosensors or aptasensors for MC-LR detection. LODs ranged from 0.18 ng L^{-1} to 368 μg L^{-1}, the best LOD being achieved by Eissa et al. (2014) with screen-printed electrodes modified with aptamer-functionalized graphene. The aptasensor signal resulting from the interaction between the aptamers and the microcystin was recorded by SWV. The aptasensor was further applied to the analysis of spiked tap water and fish samples showing good recovery percentages.

3.2.3 Bacterial Toxins

Bacterial toxins may be divided into exotoxins (peptides and proteins) produced by both Gram-positive and Gram-negative bacterial pathogens and endotoxins (LPS) produced by Gram-negative bacteria. Various methods have been developed in the past for their determination, including chromatographic techniques, antibody-based assays and biosensors (Zhu et al., 2014). In the past few years, several NM-based aptasensors or immunosensors of high sensitivity have been produced to detect toxins A and B from *Clostridium difficile* (Luo et al., 2014; Zhu et al., 2015b) or epsilon toxin from *Clostridium perfringens* (Palaniappan et al., 2013). Signal could be efficiently amplified, either by AuNPs (Luo et al., 2014; Zhu et al., 2015b) or by aligned SWCNTs (Palaniappan et al., 2013).

3.3 Other Chemical Contaminants

As food contamination can result from a number of events occurring at all points of the food chain as well as from external environmental pollution, an almost unlimited list of chemicals can affect food quality and threaten human and animal health. Among these, pesticides and veterinary drug residues are of major concern and NM-based electrochemical biosensors have been developed for their detection.

3.3.1 Pesticides

Pesticides are widely used to improve the productivity in agricultural activities. This group of chemicals covers a variety of molecules that may be broadly classified into three categories (insecticides, herbicides, and fungicides) on the basis of a target organism (Kumar et al., 2015a). Among these, organophosphate and *N*-methyl carbamates play an important role in the environment and food contamination. The extensive use of pesticides has resulted in their appearance as residuals in crops, livestock, and poultry products. Their toxicity is based on the inhibition of acetylcholinesterase (AChE), an enzyme hydrolyzing acetylcholine neurotransmitter. Pesticides are classically measured using chromatographic techniques but a large number of biosensors, including electrochemical biosensors have been reported in the literature. These latter are mostly using enzymes and whole cells as biological recognition elements, and in a lesser extent antibodies or DNA (Lagarde and Jaffrezic-Renault, 2011; Sassolas et al., 2012).

Enzyme biosensors developed for pesticides analysis are based on either inhibition or activation of the enzyme by the chemicals. Cholinesterases (AChE or butylcholinesterase), often coupled with choline oxidase, have been extensively employed as bioreceptors in the elaboration of biosensors for organophoshate pesticides (OPs) and carbamates, although other compounds such as heavy metals or aflatoxins can also inhibit ChEs, resulting in a lack in selectivity. Recent works have extensively explored the possibilities of improving biosensors features by integrating into the assemblies NMs such as AuNPs, MNPs, graphene, GO and rGO, and composites (Kumar et al., 2015a; Liu et al., 2015; Zhang et al., 2014; Xia and Gao, 2015).

Yu et al. (2015) developed an efficient immobilization of AChE onto amino functionalized carbon nanotubes (CNT–NH2) for the elaboration of OPs biosensors. It was possible to achieve high enzyme loadings and a favorable orientation of the NTs onto the glass carbon electrode. The modified biosensor showed a high affinity to acetylthiocholine chloride (ATCh) and could catalyze the hydrolysis of ATCh. Using paraoxon as a model compound, a low LOD was obtained (0.08 nM) with satisfying reproducibility and stability and the biosensor was successfully employed for the determination of low concentrations of pesticides in real vegetable samples.

Using a boron-doped diamond electrode modified with a IL/AuNPs/porous carbon composite and AChE adsorbed at the surface, Wei and Wang (2015) were able to determine dichlorvos in the 10^{-4}–1 μg L^{-1} range with a LOD of 6.61×10^{-5} μg L^{-1}. DPV signal was 4.5 times higher in presence of the IL/AuNPs-porous carbon nanocomposite.

Zhao et al. (2015) also developed a very sensitive and selective sensing platform for OPs determination. The biosensor was elaborated by direct electrodeposition of rGO-AuNPs-β-cyclodextrin (β-CD) and Prussian blue-chitosan (PB-CS) on a GCE followed by AChE fixation. The PB-CS not only effectively catalyzed the oxidation of thiocholine (TCh), but also shifted its oxidation potential from 0.68 to 0.2 V, and accordingly the sensitivity of the amperometric biosensor was improved. The synergistic effect between rGO and AuNPs significantly promoted the electron transfer between PB and the electrode, and remarkably enhanced the electrochemical oxidation of TCh. Low LODs were obtained (4.14 and 1.15 pg mL^{-1} for malathion and carbaryl, respectively) and the biosensor was applied to spiked vegetable samples, with recoveries close to 100%.

Organophosphorus hydrolase (OPH) has a broad substrate specificity and is able to hydrolyze a number of OPs such as paraoxon, parathion, or methylparathion. Although OPH-based biosensors offer a lower sensitivity compared to the AChE-based biosensors and can detect only some OPs, they are more selective. Some examples of recent works published on NM-based biosensors using OPH as bioreceptor can be found in Sassolas et al. (2012), Kumar et al. (2015a), and Zhang et al. (2014).

Combining the advantages of OPH and AChE, a novel LBL assembled bi-enzyme biosensor has been recently proposed by Zhang et al. (2015) to discriminate organophosphorus and non-organophosphorus pesticides. This bi-enzyme biosensing system was composed of electrostatically interacted MWCNT–OPH and MWCNT–AChE along with a set of cushioning bilayers consisting of MWCNT–polyethyleneimine and MWCNT–DNA on GCE. The detection limit was found to be 0.5 μM for OP pesticide paraoxon and 1 μM for non-OP pesticide carbaryl, and the biosensor was linear in a wide range of concentration. The biosensor performance was also validated using apple samples. Remarkable discriminative and straightforward detection between OP and non-OP neurotoxins was successfully achieved with the proposed biosensor.

3.3.2 Veterinary Drug Residues

Veterinary drugs are regularly administrated to animals in the feed or in water for therapeutics or prophylactic purposes. It is known, for example, that antibiotics can improve animal health and growth and can enhance feed efficiency. However, the extensive use of antibiotics has triggered the development of bacterial resistance. In order to protect the consumer health, the European Union has set maximal residue values of veterinary drugs in various foodstuffs of animal origin, for example, liver, milk, egg, or meat, which enter the human food chain. The use of antibiotics as growth promoter has been also banned in Europe and, since 1996, monitoring of veterinary drug residues are governed by the Council Directive 96/23/EC (European Commission, 1996). Chafer-Pericas et al. (2010) recently gave an overview of the three most important screening approaches to detect antibiotics residues in food samples (ie, immunoassays, microbiological tests, and biosensors), while Huet et al. (2010) reviewed the developments performed in the field of biosensors for antibiotics determination, focusing more particularly on antibiotics belonging to seven groups, that is, aminoglycosides, quinolones, ß-lactams, macrolides, fenicols, tetracyclines, and sulfonamides and Babington et al. (2012) described bioanalytical methods for ß-lactams. However, new trends in NM-based biosensors were not addressed in this review, although this type of biosensors has been also developed for antibiotics, as for the other contaminants.

Most of the electrochemical biosensors proposed for antibiotics detection (integrating or not NMs) are immunosensors and signal amplification strategies using NMs are multiple, assembling antibodies with individual components or nanocomposites based on CNTs, graphene, GO, AuNPs, Au nanorods, CeO_2, MNPs (Cai et al., 2012; Gamella et al., 2013; Guo et al., 2015; Wu et al., 2012; Yang et al., 2011; Zang et al., 2013). Recently, an aptasensor was developed based on the MWCNTs-CS and CS-PB-graphene NPs assembled on a GCE for tetracycline detection by DPV. MWCNTs-CS combined with CS-PB-GR increased the sensitivity and stability of the electrode. LOD was 5.6 pM.

For their part, Wu et al. (2012) proposed an original biosensor consisting in a paper supported immunosensor, for the simple, inexpensive, fast and sensitive electrochemical analysis of antibiotic residues in milk samples. Well-dispersed SWNTs were impregnated with an antibody against neomycin to obtain a composite coating solution, followed by dipping the filtration paper in the solution to fabricate the sensitive biosensor which had high electrical conductivity. Chronoamperometric signal of the paper supported biosensor increased with neomycin concentrations in the 0.2–125 ng mL^{-1} range, and the LOD was 0.04 ng mL^{-1}, well below the EU regulations for neomycin in this matrix. The biosensor was successfully applied to the analysis of fortified milk samples.

4 Conclusions

Biosensors have been demonstrated as attractive analytical devices for rapid screening of food contaminants. Electrochemical detection offers several advantages, in particular easy miniaturization and possible integration into high density arrays for on-site multiplexed analysis using easy-to-use formats. Over the past decade, a variety of new assemblies integrating biomolecules and metal NPs or carbon-based nanostructures, acting as labels, signal enhancers, or nanostructured immobilization supports, have been proposed to produce electrochemical biosensors of enhanced analytical performances. The use of magnetic NPs and integration into microfluidic systems has also paved the way toward integrated devices where sample pretreatment and analysis could be done in simple, rapid, and easy-to-use formats.

However, in spite of the huge variety of biosensors reported in the literature for food analysis, and the major progress in their robustness and analytical performances observed during the past decade, transfer of the biosensor technology from the laboratory research to commercial products is still impeded by some major drawbacks. First, the long-standing trend for the development of biological sensing layers of ever increasing sensitivity has pushed the searchers to design architectures of higher complexity integrating costly components. The stability of the biological receptors immobilized within these structures is sometimes not maintained in real sample analysis conditions, which constitutes a stumbling block preventing biosensor commercialization. Research activities in the field of artificial receptors such as aptamers have increased in the past few years, but only a few of them have been produced for a limited number of target analytes. Efforts have to be done in this direction. Research of adequate sample treatments and of multiplexed compact portable platforms incorporating all required preparation and NM-based biosensing arrays of electrodes into automated and low-costs formats, has also to be strengthened.

References

Abdalhai, M.H., Maximiano Fernandes, A., Bashari, M., He, Q., Sun, X., Jing, J., 2014. A rapid and sensitive detection of foodborne pathogenic bacteria (*Staphylococcus aureus*) by using electrochemical DNA genomic biosensor and its application in beef products. J. Agric. Food Chem. 62, 12659–12667.

Abdalhai, M.H., Maximiano Fernandes, A., Xia, X., Musa, A., Ji, J., Sun, X., 2015. Electrochemical genosensor to detect pathogenic bacteria (*Escherichia coli* O157:H7) as applied in real food samples (fresh beef) to improve foods safety and quality control. J. Agric. Food Chem. 63, 5017–5025.

Alonso-Lomillo, M.A., Domínguez-Renedo, O., Ferreira-Gonçalves, L., Arcos-Martínez, M.J., 2010. Sensitive enzyme-biosensor based on screen-printed electrodes for ochratoxin A. Biosens. Bioelectron. 25, 1333–1337.

Alonso-Lomillo, M.A., Domínguez-Renedo, O., Román, L.D.T., Arcos-Martínez, M.J., 2011. Horseradish peroxidase-screen printed biosensors for determination of echratoxin A. Anal. Chim. Acta. 688, 49–53.

Arduini, F., Amine, A., Moscone, D., Palleschi, G., 2010. Biosensors based on cholinesterase inhibition for insecticides, nerve agents and aflatoxin B1 detection. Microchim. Acta. 170, 193–214.

Arora, P., Sindhu, A., Dilbaghi, N., Chaudhury, A., 2011. Biosensors as innovative tools for the detection of food borne pathogens. Biosens. Bioelectron. 28, 1–12.

Artiles, M.S., Rout, C.S., Fisher, T.S., 2011. Graphene-based hybrid materials and devices for biosensing. Adv. Drug Deliv. Rev. 63, 1352–1360.

Babington, R., Matas, S., Marco, M.-P., Galve, R., 2012. Current bioanalytical methods for detection of penicillins. Anal. Bioanal. Chem. 403, 1549–1566.

Barthelmebs, L., Hayat, A., Limiadi, A.W., Marty, J.-L., Noguer, T., 2011. Electrochemical DNA aptamer-based biosensor for OTA detection, using superparamagnetic nanoparticles. Sens. Actuat. B 156, 932–937.

Bulbul, G., Hayat, A., Andreescu, S., 2015. A generic amplification strategy for electrochemical aptasensors using a non-enzymatic nanoceria tag. Nanoscale 7, 13230–13238.

Cai, M., Zhu, L., Ding, Y., Wang, J., Li, J., Du, X., 2012. Determination of sulfamethoxazole in foods based on CeO_2/chitosan nanocomposite-modified electrodes. Mater. Sci. Eng. C 32, 2623–2627.

Campas, M., Garibo, D., Prieto-Simon, B., 2012. Novel nanobiotechnological concepts in electrochemical biosensors for the analysis of toxins. Analyst 137, 1055–1067.

Campbell, K., Vilarino, N., Botana, L.M., Elliott, C.T., 2011. A European perspective on progress in moving away from the mouse bioassay for marine-toxin analysis. Trends Anal. Chem. 30, 239–253.

Campuzano, S., Wang, J., 2011. Nanobioelectroanalysis based on carbon/inorganic hybrid nanoarchitectures. Electroanalysis 23, 1289–1300.

Chafer-Pericas, C., Maquieira, A., Puchades, R., 2010. Fast screening methods to detect antibiotic residues in food samples. Trends Anal. Chem. 29, 1038–1049.

Chan, K.Y., Ye, W.W., Zhang, Y., Xiao, L.D., Leung, P.H.M., Li, Y., Yang, M., 2013. Ultrasensitive detection of *E. coli* O157:H7 with biofunctional magnetic bead concentration via nanoporous membrane based electrochemical immunosensor. Biosens. Bioelectron. 41, 532–537.

Chauhan, R., Solanki, P.R., Singh, J., Mukherjee, I., Basu, T., Malhotra, B.D., 2015. A novel electrochemical piezoelectric label free immunosensor for aflatoxin B1 detection in groundnut. Food Control 52, 60–70.

Chen, X., Huang, Y., Ma, X., Jia, F., Guo, X., Wang, Z., 2015. Impedimetric aptamer-based determination of the mold toxin fumonisin B1. Microchim. Acta. 182, 1709–1714.

Davis, D., Guo, X., Musavi, L., Lin, C.-S., Chen, Sz-H., Wu, V.C.H., 2013. Gold nanoparticle-modified carbon electrode biosensor for the detection of *Listeria monocytogenes*. Ind. Biotechnol. 9, 31–36.

Dimcheva, N.D., Horozova, E.G., 2015. Electrochemical enzymatic biosensors based on metal micro-/nanoparticles-modified electrodes: a review. Chem. Paper 69, 17–26.

Dinckaya, E., Kınık, O., Sezgintürk, M.K., Altug, C., Akkoca, A., 2012. Development of an impedimetric aflatoxin M1 biosensor based on a DNA probe and gold nanoparticles. Biosens. Bioelectron. 26, 3806–3811.

Dominguez, R.B., Hayat, A., Sassolas, A., Alonso, G.A., Munoz, R., Marty, J.L., 2012. Automated flow-through amperometric immunosensor for highly sensitive and on-line detection of okadaic acid in mussel sample. Talanta 99, 232–237.

Dou, W., Tang, W., Zhao, G., 2013. A disposable electrochemical immunosensor arrays using 4-channel screen-printed carbon electrode for simultaneous detection of *Escherichia coli* O157:H7 and *Enterobacter sakazakii*. Electrochim. Acta. 97, 79–85.

Dridi, F., Marrakchi, M., Gargouri, M., Garcia-Cruz, A., Dzyadevych, S., Vocanson, F., Saulnier, J., Jaffrezic-Renault, N., Lagarde, F., 2015. Thermolysin entrapped in a gold nanoparticles/polymer composite for direct and sensitive conductometric biosensing of ochratoxin A in olive oil. Sens. Actuat. B 221, 480–490.

Eissa, S., Zourob, M., 2012. A graphene-based electrochemical competitive immunosensor for the sensitive detection of okadaic acid in shellfish. Nanoscale 4, 7593–7599.

Eissa, S., Ng, A., Siaj, M., Zourob, M., 2014. Label-free voltammetric aptasensor for the sensitive detection of microcystin-LR using graphene-modified electrodes. Anal. Chem. 86, 7551–7557.

El Ichi, S., Leon, F., Vossier, L., Marchandin, H., Errachid, A., Coste, J., Jaffrezic-Renault, N., Fournier-Wirth, C., 2014. Microconductometric immunosensor for label-free and sensitive detection of Gram-negative bacteria. Biosens. Bioelectron. 54, 378–384.

El Khoury, A., Atoui, A., 2012. Ochratoxin A: general overview and actual molecular status. Anal. Biochem. 429, 19–31.

European Commission, 1996. Council directive 96/23/EC of 29 April 1996. Off. J. Eur. Union L, 125, 10.

Evtugyn, G., Porfireva, A., Stepanova, V., Kutyreva, M., Gataulina, A., Ulakhovich, N., Evtugyn, V., Hianik, T., 2013. Impedimetric aptasensor for ochratoxin A determination based on Au nanoparticles stabilized with hyper-branched polymer. Sensors 13, 16129–16145.

Feng, R., Zhang, Y., Ma, H., Wu, D., Fan, H., Wang, H., Li, H., Du, B., Wie, Q., 2013. Ultrasensitive non-enzymatic and non-mediator electrochemical biosensor using nitrogen-doped graphene sheets for signal amplification and nanoporous alloy as carrier. Electrochim. Acta. 97, 105–111.

Gamella, M., Campuzano, S., Conzuelo, F., Esteban-Torres, M., de las Rivas, B., Reviejo, A.J., Munoz, R., Pingarron, J.M., 2013. An amperometric affinity penicillin-binding protein magnetosensor for the detection of b-lactam antibiotics in milk. Analyst 138, 2013–2022.

Gan, X.R., Zhao, H.M., 2015. A review: nanomaterials applied in graphene-based electrochemical biosensors. Sens. Mater. 27, 191–215.

Guo, Y., Shen, G., Sun, X., Wang, X., 2015. Electrochemical aptasensor based on multiwalled carbon nanotubes and graphene for tetracycline detection. IEEE Sens. J. 15, 1951–1958.

Han, C., Doepke, A., Cho, W., Likodimos, V., de la Cruz, A.A., Back, T., Heineman, W.R., Halsall, H.B., Shanov, V.N., Schulz, M.J., Falaras, P., Dionysiou, D.D., 2013. A multiwalled-carbon-nanotube-based biosensor for monitoring microcystin-LR in sources of drinking water supplies. Adv. Funct. Mater. 23, 1807–1816.

Hayat, A., Barthelmebs, L., Sassolas, A., Marty, J.-L., 2012. Development of a novel label-free amperometric immunosensor for the detection of okadaic acid. Anal. Chim. Acta. 724, 92–97.
Hayat, A., Catanante, G., Marty, J.L., 2014. Current trends in nanomaterial-based amperometric biosensors. Sensors 14, 23439–23461.
Holzinger, M., Le Goff, A., Cosnier, S., 2014. Nanomaterials for biosensing applications: a review. Frontiers Chem. 2, 1–10, Article 63.
Hu, H., Cao, L., Li, Q., Ma, K., Yan, P., Kirk, D.W., 2015. Fabrication and modeling of an ultrasensitive label free impedimetric immunosensor for aflatoxin B1 based on poly(*o*-phenylenediamine) modified gold 3D nano electrode ensembles. RSC Adv. 5, 55209–55217.
Huet, A.-C., Fodey, T., Haughey, S.A., Weigel, S., Elliott, C., Delahaut, P., 2010. Advances in biosensor-based analysis for antimicrobial residues in foods. Trends Anal. Chem. 29, 1281–1294.
Jia, F., Duan, N., Wu, S., Ma, X., Xia, Y., Wang, Z., Wei, X., 2014. Impedimetric aptasensor for *Staphylococcus aureus* based on nanocomposite prepared from reduced graphene oxide and gold nanoparticles. Microchim. Acta. 181, 967–974.
Jiang, L., Qian, J., Yang, X., Yan, Y., Liu, Q., Wang, K., Wang, K., 2014. Amplified impedimetric aptasensor based on gold nanoparticles covalently bound graphene sheet for the picomolar detection of ochratoxin A. Anal. Chim. Acta. 806, 128–135.
Jodra, A., López, M.A., Escarpa, A., 2015. Disposable and reliable electrochemical magnetoimmunosensor for fumonisins simplified determination in maize-based foodstuffs. Biosens. Bioelectron. 64, 633–638.
Justino, C.I.L., Rocha-Santos, T.A.P., Cardoso, S., Duarte, A.C., 2013. Strategies for enhancing the analytical performance of nanomaterial-based sensors. Trends Anal. Chem. 47, 27–36.
Kanayeva, D.A., Wang, R., Rhoads, D., Erf, G.F., Slavik, M.F., Tung, S., Li, Yanbin, 2012. Efficient separation and sensitive detection of *Listeria monocytogenes* using an impedance immunosensor based on magnetic nanoparticles, a microfluidic chip, and an interdigitated microelectrode. J. Food Prot. 75, 1951–1959.
Kang, X., Pang, G., Chen, Q., Liang, X., 2013. Fabrication of *Bacillus cereus* electrochemical immunosensor based on double-layer gold nanoparticles and chitosan. Sens. Actuat. B 177, 1010–1016.
Koedrith, P., Thasiphu, T., Tuitemwong, K., Boonprasert, R., Tuitemwong, P., 2014. Recent advances in potential nanoparticles and nanotechnology for sensing food-borne pathogens and their toxins in foods and crops: current technologies and limitations. Sens. Mater. 26, 711–736.
Kuang, H., Chen, W., Xu, D., Xu, L., Zhu, Y., Liu, L., Chu, H., Peng, C., Xu, C., Zhu, S., 2010. Fabricated aptamer-based electrochemical "signal-off" sensor of ochratoxin A. Biosens. Bioelectron. 26, 710–716.
Kumar, P., Kim, K.-H., Deep, A., 2015a. Recent advancements in sensing techniques based on functional materials for organophosphate pesticides. Biosens. Bioelectron. 70, 469–481.
Kumar, S., Ahlawat, W., Kumar, R., Dilbaghi, N., 2015b. Graphene, carbon nanotubes, zinc oxide and gold as elite nanomaterials for fabrication of biosensors for healthcare, 2015. Biosens. Bioelectron. 70, 498–503.
Lagarde, F., Jaffrezic-Renault, N., 2011. Cell-based electrochemical biosensors for water quality assessment. Anal. Bioanal. Chem. 400, 947–964.
Law, J.W.-F., Ab Mutalib, N.-S., Chan, K.-G., Lee, L.-H., 2015. Rapid methods for the detection of foodborne bacterial pathogens: principles, applications, advantages and limitations. Front. Microbiol. Food Microbiol. 5, 1–19, Article 770.

Lawal, A.T., 2015. Synthesis and utilization of grapheme for fabrication of electrochemical sensors. Talanta 131, 424–443.
Lee, S.H., Sung, J.H., Park, T.H., 2012. Nanomaterial-based biosensor as an emerging tool for biomedical applications. Ann. Biomed. Eng. 40, 1384–1397.
Li, S.C., Chen, J.H., Cao, H., Yao, D.S., Liu, D.L., 2011. Amperometric biosensor for aflatoxin B1 based on aflatoxin-oxidase immobilized on multiwalled carbon nanotubes. Food Cont. 22, 43–49.
Li, Y., Liu, X., Lin, Z., 2012. Recent developments and applications of surface plasmon resonance biosensors for the detection of mycotoxins in foodstuffs. Food Chem. 132, 1549–1554.
Liebana, S., Brandao, D., Alegret, S., Pividori, M.I., 2014. Electrochemical immunosensors, genosensors and phagosensors for *Salmonella* detection. Anal. Methods 6, 8858–8873.
Linting, Z., Ruiyi, L., Zaijun, L., Qianfang, X., Yinjund, F., Junkang, L., 2012. An immunosensor for ultrasensitive detection of aflatoxin B1 with an enhanced electrochemical performance based on graphene/conducting polymer/gold nanoparticles/the ionic liquid composite film on modified gold electrode with electrodeposition. Sens. Actuat. B 174, 359–365.
Liu, X., Yang, Z., Zhang, Y., Yu, R., 2013. A novel electrochemical immunosensor for ochratoxin A with hapten immobilization on thionine/gold nanoparticle modified glassy carbon electrode. Anal. Methods 5, 1481–1486.
Liu, Q., Fei, A., Huan, J., Mao, H., Wang, K., 2015. Effective amperometric biosensor for carbaryl detection based on covalent immobilization acetylcholinesterase on multiwall carbon nanotubes/graphene oxide nanoribbons nanostructure. J. Electroanal. Chem. 740, 8–13.
Luo, P., Liu, Y., Xia, Y., Xu, H., Xie, G., 2014. Aptamer biosensor for sensitive detection of toxin A of *Clostridium difficile* using gold nanoparticles synthesized by *Bacillus stearothermophilus*. Biosens. Bioelectron. 54, 217–221.
Ma, X., Jiang, Y., Jia, F., Yu, Y., Chen, J., Wang, Z., 2014. An aptamer-based electrochemical biosensor for the detection of *Salmonella*. J. Microbiol. Methods 98, 94–98.
Malhotra, B.D., Srivastava, S., Ali, M.A., Singh, C., 2014. Nanomaterial-based biosensors for food toxin detection. Appl. Biochem. Biotechnol. 174, 880–896.
Marin, S., Ramos, A.J., Cano-Sancho, G., Sanchis, V., 2013. Mycotoxins: occurrence, toxicology, and exposure assessment. Food Chem. Toxicol. 60, 218–237.
Masikini, M., Mailu, S.N., Tsegaye, A., Njomo, N., Molapo, K.M., Ikpo, C.O., Sunday, C.E., Rassie, C., Wilson, L., Baker, P.G.L., Iwuoha, E.I., 2013. A Fumonisins immunosensor based on polyanilino-carbon nanotubes doped with palladium telluride quantum dots. Electrochim. Acta. 90, 246–253.
McGrath, T.F., Elliott, C.T., Fodey, T.L., 2012. Biosensors for the analysis of microbiological and chemical contaminants in food. Anal. Bioanal. Chem. 403, 75–92.
Meneely, J.P., Ricci, F., van Egmond, H.P., Elliott, C.T., 2011. Current methods of analysis for the determination of trichothecene mycotoxins in food. Trends Anal. Chem. 30, 192–203.
Nguyen, B.H., Tran, L.D., Do, Q.P., Nguyen, H.L., Tran, N.H., Nguyen, P.X., 2013. Label-free detection of aflatoxin M1 with electrochemical Fe_3O_4/polyaniline-based aptasensor. Mater. Sci. Eng. C 33, 2229–2234.
Pacheco, J.G., Castro, M., Machado, S.M., Barroso, F., Nouws, H.P.A., Delerue-Matos, C., 2015. Molecularly imprinted electrochemical sensor for ochratoxin A detection in food samples. Sens. Actuat. B 215, 107–112.
Palaniappan, Al., Goh, W.H., Fam, D.W.H., Rajaseger, G., Chan, C.E.Z., Hanson, B.J., Moochhala, S.M., Mhaisalkar, S.G., Liedberg, B., 2013. Label-free electronic detection of bio-toxins using aligned carbon nanotubes. Biosens. Bioelectron. 43, 143–147.

Paniel, N., Radoi, A., Marty, J.-L., 2010. Development of an electrochemical biosensor for the detection of aflatoxin M1 in milk. Sensors 10, 9439–9448.
Panini, N.V., Bertolino, F.A., Salinas, E., Messina, G.A., Raba, J., 2010. Zearalenone determination in corn silage samples using an immunosensor in a continuous-flow/stopped-flow systems. Biochem. Eng. J. 51, 7–13.
Panini, N.V., Salinas, E., Messina, G.A., Raba, J., 2011. Modified paramagnetic beads in a microfluidic system for the determination of zearalenone in feedstuffs samples. Food Chem. 125, 791–796.
Pavlidis, I.V., Patila, M., Bornscheuer, U.T., Gournis, D., Stamatis, H., 2014. Graphene-based nanobiocatalytic systems: recent advances and future prospects. Trends Biotechnol. 32, 312–320.
Pereira, V.L., Fernandes, J.O., Cunha, S.C., 2014. Mycotoxins in cereals and related foodstuffs: a review on occurrence and recent methods of analysis. Trends Food Sci. Technol. 36, 96–136.
Perez-Lopez, B., Merkoci, A., 2011. Nanomaterials based biosensors for food analysis applications. Trends Food Sci. Technol. 22, 625–639.
Perrotta, P.R., Arévalo, F.J., Vettorazzi, N.R., Zón, M.A., Fernández, H., 2012. Development of a very sensitive electrochemical magneto immunosensor for the direct determination of ochratoxin A in red wine. Sens. Actuat. B-Chem. 162, 327–333.
Perumal, V., Hashim, U., 2014. Advances in biosensors: principle, architecture and applications. J. Appl. Biomed. 12, 1–15.
Pumera, M., Ambrosi, A., Bonanni, A., Chng, E.L.K., Poh, H.L., 2010. Graphene for electrochemical sensing and biosensing. Trends Anal. Chem. 29, 954–965.
Putzbach, W., Ronkainen, N.J., 2013. Immobilization techniques in the fabrication of nanomaterial-based electrochemical biosensors: a review. Sensors 13, 4811–4840.
Qian, J., Jiang, L., Yang, X., Yan, Y., Mao, H., Wang, K., 2014. Highly sensitive impedimetric aptasensor based on covalent binding of gold nanoparticles on reduced graphene oxide with good dispersity and high density. Analyst 139, 5587–5593.
Regiart, M., Seia, M.A., Messina, G.A., Bertolino, F.A., Raba, J., 2015. Electrochemical immunosensing using a nanostructured functional platform for determination of α-zearalanol. Microchim. Acta. 182, 531–538.
Reverte, L., Solino, L., Carnicer, O., Diogene, J., Campas, M., 2014. Alternative methods for the detection of emerging marine toxins: biosensors, biochemical assays and cell-based assays. Mar. Drugs 12, 5719–5763.
Rhouati, A., Hayat, A., Hernandez, D.B., Meraihi, Z., Munoz, R., Marty, J.-L., 2013. Development of an automated flow-based electrochemical aptasensor for on-line detection of Ochratoxin A. Sens. Actuat. B 176, 1160–1166.
Rocha-Santos, T.A.P., 2014. Sensors and biosensors based on magnetic nanoparticles. Trends Anal. Chem. 62, 28–36.
Ronkainen, N.J., Halsall, H.B., Heineman, W.R., 2010. Electrochemical biosensors. Chem. Soc. Rev. 39, 1747–1763.
Ruiyi, L., Qianfang, X., Zaijun, L., Xiulan, S., Junkang, L., 2013. Electrochemical immunosensor for ultrasensitive detection of microcystin-LR based on graphene–gold nanocomposite/functional conductingpolymer/goldnanoparticle/ionic liquid composite film with electrodeposition. Biosens. Bioelectron. 44, 235–240.
Sagadevan, S., Periasamy, M., 2014. Recent trends in nanobiosensors and their applications: a review. Rev. Adv. Mater. Sci. 36, 62–69.
Sassolas, A., Prieto-Simón, B., Marty, J.-L., 2012. Biosensors for pesticide detection: new trends. Am. J. Anal. Chem. 3, 210–232.
Sassolas, A., Hayat, A., Catanante, G., Marty, J.-L., 2013. Detection of the marine toxin okadaic acid: assessing seafood safety, 2013. Talanta 105, 306–316.

Scognamiglio, V., 2013. Nanotechnology in glucose monitoring: advances and challenges in the last 10 years. Biosens. Bioelectron. 47, 12–25.

Setterington, E.B., Alocilja, E.C., 2012. Electrochemical biosensor for rapid and sensitive detection of magnetically extracted bacterial pathogens. Biosensors 2, 15–31.

Sharma, H., Mutharasan, R., 2013. Review of biosensors for foodborne pathogens and toxins. Sens. Actuat. B 183, 535–549.

Sharma, T.K., Ramanathan, R., Rakwal, R., Agrawal, G.K., Bansa, I.V., 2015. Moving forward in plant food safety and security through nanobiosensors: adopt or adapt biomedical technologies? Proteomics 15, 1680–1692.

Shi, Z.-Y., Zheng, Y.-T., Zhang, H.-B., He, C.-H., Wu, W.-D., Zhang, H.-B., 2015. DNA electrochemical aptasensor for detecting fumonisins B1 based on graphene and thionine nanocomposite. Electroanalysis 27, 1097–1103.

Singh, S., Srivastava, A., Ohb, H.-M., Ahn, C.-Y., Choi, G.-G., Asthana, R.K., 2012. Recent trends in development of biosensors for detection of microcystin. Toxicon 60, 878–894.

Singh, C., Srivastava, S., Ali, Md.A., Gupta, T.K., Sumana, G., Srivastava, A., Mathur, R.B., Malhotra, B.D., 2013. Carboxylated multiwalled carbon nanotubes based biosensor for aflatoxin detection. Sens. Actuat. B 185, 258–264.

Srivastava, S., Ali, Md.A., Umrao, S., Parashkar, U.K., Srivastava, A., Sumana, G., Malhotra, B.D., Pandey, S.S., Hayase, S., 2014. Graphene oxide-based biosensor for food toxin detection. Appl. Biotech. Biotechnol. 174, 960–970.

Srivastava, S., Abraham, S., Singh, C., Ali, Md.A., Srivastava, A., Sumana, G., Malhotra, B.D., 2015. Protein conjugated carboxylated gold@reduced graphene oxide for aflatoxin B1 detection. RSC Adv. 5, 5406–5414.

Stepurska, K.V., Soldatkin, O.O., Arkhypova, V.M., Soldatkin, A.P., Lagarde, F., Jaffrezic-Renault, N., Dzyadevych, S.V., 2015. Development of novel enzyme potentiometric biosensor based on pH-sensitive field-effect transistors for aflatoxin B1 analysis in real samples. Talanta 144, 1079–1084.

Sun, Y., He, X., Ji, J., Jia, M., Wang, Z., Sun, X., 2015. A highly selective and sensitive electrochemical CS–MWCNTs/Au-NPs composite DNA biosensor for *Staphylococcus aureus* gene sequence detection. Talanta 141, 300–306.

Sundramoorthy, A.K., Gunasekaran, S., 2014. Applications of graphene in quality assurance and safety of food. Trends Anal. Chem. 60, 36–53.

Tang, J., Hou, L., Tang, D., Zhou, J., Wang, Z., Li, J., Chen, G., 2012. Magneto-controlled electrochemical immunoassay of brevetoxin B in seafood based on guanine-functionalized graphene nanoribbons. Biosens. Bioelectron. 38, 86–93.

Tiwari, I., Singh, M., Pandey, C.M., Sumana, G., 2015. Electrochemical genosensor based on graphene oxide modified iron oxide–chitosan hybrid nanocomposite for pathogen detection. Sens. Actuat. B 206, 276–283.

Tong, P., Tang, S., He, Y., Shao, Y., Zhang, L., Chen, G., 2011. Label-free immunosensing of microcystin-LR using a gold electrode modified with gold nanoparticles. Microchim. Acta. 173, 299–305.

Tothill, I.E., 2011. Biosensors and nanomaterials and their application for mycotoxin determination. World Mycotox. J. 4, 361–374.

Vashist, S.K., Luong, J.H.T., 2015. Recent advances in electrochemical biosensing schemes using graphene and graphene-based nanocomposites. Carbon 84, 519–550.

Vashist, S.K., Zheng, D., Al-Rubeaan, K., Luong, J.H.T., Sheu, F.-S., 2011. Advances in carbon nanotube based electrochemical sensors for bioanalytical applications. Biotechnol. Adv. 29, 169–188.

Vidal, J.C., Bonel, L., Ezquerra, A., Duato, P., Castillo, J.R., 2012. An electrochemical immunosensor for ochratoxin A determination in wines based on a monoclonal antibody and paramagnetic microbeads. Anal. Bioanal. Chem. 403, 1585–1593.

Vidal, J.C., Bonel, L., Ezquerra, A., Hernández, S., Bertolín, J.R., Cubel, C., Castillo, J.R., 2013. Electrochemical affinity biosensors for detection of mycotoxins: a review. Biosens. Bioelectron. 49, 146–158.
Vilarino, N., Fonfria, E.S., Louzao, M.C., Botana, L.M., 2009. Use of biosensors as alternatives to current regulatory methods for marine biotoxins. Sensors 9, 9414–9443.
Vilarino, N., Louzao, M.C., Fraga, M., Rodríguez, L.P., Botana, L.M., 2013. Innovative detection methods for aquatic algal toxins and their presence in the food chain. Anal. Bioanal. Chem. 405, 7719–7732.
Viswanathan, S., Rani, C., Hoc, J.A., 2012. Electrochemical immunosensor for multiplexed detection of food-borne pathogens using nanocrystal bioconjugates and MWCNT screen-printed electrode. Talanta 94, 315–319.
Walcarius, A., Minteer, S.D., Wang, J., Lin, Y., Merkoçi, A., 2013. Nanomaterials for bio-functionalized electrodes: recent trends. J. Mater. Chem. B 1, 4878–4908.
Wang, J., 2012. Electrochemical biosensing based on noble metal nanoparticles. Microchim. Acta. 177, 245–270.
Wang, Z., Dai, Z., 2015. Carbon nanomaterial-based electrochemical biosensors: an overview. Nanoscale 7, 6420–6431.
Wang, L., Ma, W., Xu, L., Chen, W., Zhu, Y., Xu, C., Kotov, N.A., 2010. Nanoparticle-based environmental sensors. Mater. Sci. Eng. R 70, 265–274.
Wang, Y., Ye, Z., Ying, Y., 2012. New trends in impedimetric biosensors for the detection of foodborne pathogenic bacteria. Sensors 12, 3449–3471.
Wang, D., Hu, W., Xiong, Y., Xu, Y., Li, C.M., 2015. Multifunctionalized reduced graphene oxide-doped polypyrrole/pyrrolepropylic acid nanocomposite impedimetric immunosensor to ultra-sensitively detect small molecular aflatoxin B1. Biosens. Bioelectron. 63, 185–189.
Wei, M., Wang, A.J., 2015. Novel acetylcholinesterase biosensor based on ionic liquids-AuNPs-porous carbon composite matrix for detection of organophosphate pesticides. Sens. Actuat. B 211, 290–296.
Wu, X., Kuang, H., Hao, C., Xing, C., Wang, L., Xu, C., 2012. Paper supported immunosensor for detection of antibiotics. Biosens. Bioelectron. 33, 309–312.
Wu, H., Zuo, Y., Cui, C., Yang, W., Ma, H., Wang, X., 2013. Rapid quantitative detection of *Brucella melitensis* by a label-free impedance immunosensor based on a gold nanoparticle-modified screen-printed carbon electrode. Sensors 13, 8551–8563.
Xia, N., Gao, Y., 2015. Carbon nanostructures for development of acetylcholinesterase electrochemical biosensors for determination of pesticides. Int. J. Electrochem. Sci. 10, 713–724.
Xie, S., Chai, Y., Yuan, Y., Bai, L., Yuan, R., 2014. Development of an electrochemical method for ochratoxin A detection based on aptamer and loop-mediated isothermal amplification. Biosens. Bioelectron. 55, 324–329.
Xu, Y., Wang, E., 2012. Electrochemical biosensors based on magnetic micro/nano particles. Electrochim. Acta. 84, 62–73.
Yang, G., Jin, W., Wu, L., Wang, Q., Shao, H., Qin, A., Yu, B., Li, D., Cai, B., 2011. Development of an impedimetric immunosensor for the determination of 3-amino-2-oxazolidone residue in food samples. Anal. Chim. Acta. 706, 120–127.
Yang, X., Kirsch, J., Simonian, A., 2013. *Campylobacter* spp. detection in the 21st century: a review of the recent achievements in biosensor development. J. Microbiol. Methods 95, 48–56.
Yang, X., Qian, J., Jiang, L., Yan, Y., Wang, K., Liu, Q., Wang, K., 2014. Ultrasensitive electrochemical aptasensor for ochratoxin A based on two-level cascaded signal amplification strategy. Bioelectrochemistry 96, 7–13.
Yang, N., Chen, X., Ren, T., Zhang, P., Yang, D., 2015. Carbon nanotube based biosensors. Sens. Actuat. B 207, 690–715.

Yu, G., Wu, W., Zhao, Q., Wei, X., Lu, Q., 2015. Efficient immobilization of acetylcholinesterase onto amino functionalized carbon nanotubes for the fabrication of high sensitive organophosphorus pesticides biosensors. Biosens. Bioelectron. 68, 288–294.
Zamfir, L.-G., Geana, I., Bourigua, S., Rotariu, L., Bala, C., Errachid, A., Jaffrezic-Renault, N., 2011. Highly sensitive label-free immunosensor for ochratoxin A based on functionalized magnetic nanoparticles and EIS/SPR detection. Sens. Actuat. B 159, 178–184.
Zang, S., Liu, Y., Lin, M., Kang, J., Sun, Y., Lei, H., 2013. A dual amplified electrochemical immunosensor for ofloxacin: polypyrrole film-Au nanocluster as the matrix and multi-enzyme-antibody functionalized gold nanorod as the label. Electrochim. Acta. 90, 246–253.
Zhang, D., Huarng, M.C., Alocilja, E.C., 2010a. A multiplex nanoparticle-based bio-barcoded DNA sensor for the simultaneous detection of multiple pathogens. Biosens. Bioelectron. 26, 1736–1742.
Zhang, J., Lei, J., Xu, C., Ding, L., Ju, H., 2010b. Carbon nanohorn sensitized electrochemical immunosensor for rapid detection of microcystin-LR. Anal. Chem. 82, 1117–1122.
Zhang, B., Hou, L., Tang, D., Liu, B., Li, J., Chen, G., 2012. Simultaneous multiplexed stripping voltammetric monitoring of marine toxins in seafood based on distinguishable metal nanocluster labeled molecular tags. J. Agric. Food Chem. 60, 8974–8982.
Zhang, W., Asiri, A.M., Liu, D., Du, D., Lin, Y., 2014. Nanomaterial-based biosensors for environmental and biological monitoring of organophosphorus pesticides and nerve agents. Trends Anal. Chem. 54, 1–10.
Zhang, Y., Arugula, M.A., Wales, M., Wild, J., Simonian, A.L., 2015. A novel layer-by-layer assembled multi-enzyme/CNT biosensor for discriminative detection between organophosphorus and non-organophosphorus pesticides. Biosens. Bioelectron. 67, 287–295.
Zhao, H., Tian, J., Quan, X., 2013. A graphene and multienzyme functionalized carbon nanosphere-based electrochemical immunosensor for microcystin-LR detection. Coll. Surf. B 103, 38–44.
Zhao, H., Ji, X., Wang, B., Wang, N., Li, X., Ni, R., Ren, J., 2015. An ultra-sensitive acetylcholinesterase biosensor based on reduced grapheme oxide-Au nanoparticles-β-cyclodextrin/Prussianblue-chitosan nanocomposites for organophosphorus pesticides detection. Biosens. Bioelectron. 65, 23–30.
Zhilei, W., Xiulan, S., Zaijun, L., Yinjun, F., Guoxiao, R., Yaru, H., Junkang, L., 2011. Highly sensitive deoxynivalenol immunosensor based on a glassy carbon electrode modified with a fullerene/ferrocene/ionic liquid composite. Microchim. Acta. 172, 365–371.
Zhu, K., Dietrich, R., Didier, A., Doyscher, D., Märtlbauer, E., 2014. Recent developments in antibody-based assays for the detection of bacterial toxins. Toxins 6, 1325–1348.
Zhu, C., Yang, G., Li, H., Du, D., Lin, Y., 2015a. Electrochemical sensors and biosensors based on nanomaterials and nanostructures. Anal. Chem. 87, 230–249.
Zhu, Z., Shi, L., Feng, H., Zhou, H.S., 2015b. Single domain antibody coated gold nanoparticles as enhancer for *Clostridium difficile* toxin detection by electrochemical impedance immunosensors. Bioelectrochemistry 101, 153–158.

6

CHEMICAL SENSORS BASED ON HYBRID NANOMATERIALS FOR FOOD ANALYSIS

Daniel S. Correa*, Adriana Pavinatto*, Luiza A. Mercante*, Luiz H.C. Mattoso*, Juliano E. Oliveira, Antonio Riul, Jr[†]**

**National Nanotechnology Laboratory for Agribusiness (LNNA), Embrapa Instrumentation (CNPDIA), São Carlos, São Paulo, Brazil; **Federal University of Lavras, Engineering Department, Lavras, Minas Gerais, Brazil; [†]University of Campinas, Institute of Physics Gleb Wataghin, Campinas, São Paulo, Brazil*

1 Introduction to Hybrid Nanomaterials

The synthesis, characterization, and application of nanostructured materials have expanded in recent years, once they presented performance improvement for several properties compared to bulk and microsized materials. The increase in surface area/volume ratio provides high chemical reactivity and size-dependent (optical, electrical) properties, which are highly suitable for sensors and devices (Bhadra et al., 2009; Huang et al., 2009; Lang et al., 2010; Nambiar and Yeow, 2011; Van Tassel, 2012; Gomez and Tigli, 2013; Prakash et al., 2013; Correa et al., 2014). In this context, designing chemical sensors based on nanomaterials that are sensitive, portable, and simple to fabricate is an important issue to be addressed for food industry and food safety. Such sensors can be employed for beverage and food quality control, allowing, for instance, monitoring of the chemical composition, pH, and taste (Riul et al. 2002; Bargon et al., 2003; Ishihara et al., 2005; Scampicchio et al., 2006; Steffens et al., 2010; Baldwin et al., 2011). Furthermore, such sensors can also be used to detect food contamination and/or degradation caused by bacteria and fungus.

One crucial step for designing reliable sensors lies on the choice of sensor active layer, which can be composed by traditional materials (ceramic, polymer, metal), biomolecules, nanocomposites, and hybrid materials. According to Kickelbick (2007), conventional materials can be combined or mixed up to generate hybrid

Nanobiosensors. http://dx.doi.org/10.1016/B978-0-12-804301-1.00006-0

materials that are multifunctional, presenting superior properties when compared to their counterparts. For chemical sensor applications, hybrid materials stand out once they combine synergistically electrical and optical properties of organic and inorganic materials, improving the sensor performance. For instance, conjugated polymers used to fabricate ultrathin films or nanofibers, can be surface- or bulky-modified by metallic nanoparticles, carbon nanotubes, or enzymes to improve sensor sensitivity and detection limit. This chapter gives an overview of several types of hybrid materials and the molecular architectures used for fabricating distinct chemical sensors, including potentiometric sensors and electronic tongue- and electronic nose-type sensors, discussed in detail in the next sections.

2 Chemical Sensors Aspects

2.1 Definition

Sensors are important in health care (Mignani and Baldini, 1997; Casalinuovo et al., 2006; Huang et al., 2012; Ponmozhi et al., 2012; Segev-Bar and Haick, 2013; Bandodkar and Wang, 2014; Matzeu et al., 2015), control of industrial process (Azad et al., 1992; Hierlemann et al., 2000; Kosterev and Tittel, 2002; Mor et al., 2003; Persaud, 2005; Mandenius and Gustavsson, 2015; Shao and Tian, 2015), and environmental analysis (Pina et al., 2000; Wilson et al., 2001; Waggoner and Craighead, 2007; von der Kammer et al., 2012; Luo et al., 2015; Wright et al., 2015). Chemical sensors can be employed, for example, for aroma and taste analysis (Berna et al., 2004; Ragazzo-Sanchez et al., 2006; Ghasemi-Varnamkhasti et al., 2009; Rudnitskaya et al., 2009a; Zakaria et al., 2011), for product quality control, or for detecting spoiled food (Panigrahi et al., 2006; Alimelli et al., 2007; Zhang et al., 2009; Sberveglieri et al., 2014; Zaragoza et al., 2015). Some decades ago, the term *chemical sensor* was used for the glassy pH electrode (Cheng, 1989; Vonau and Guth, 2006; Kaden, 2009). Nowadays, sensors are becoming crucial in our daily lives. The food industry is changing quickly and sensors have played an important role in this process (Bartlett et al., 1997; Ramstrom et al., 2001; James et al., 2005; Wang et al., 2006a; Granda Valdes et al., 2009; Neethirajan et al., 2009; Ruiz-Altisent et al., 2010; Perez-Lopez and Merkoci, 2011), with chemical sensors able to analyze our environment, that is, detect qualitatively and quantitatively substances present in a system.

The IUPAC definition is: "A chemical sensor is a device that transforms chemical information, ranging from the concentration

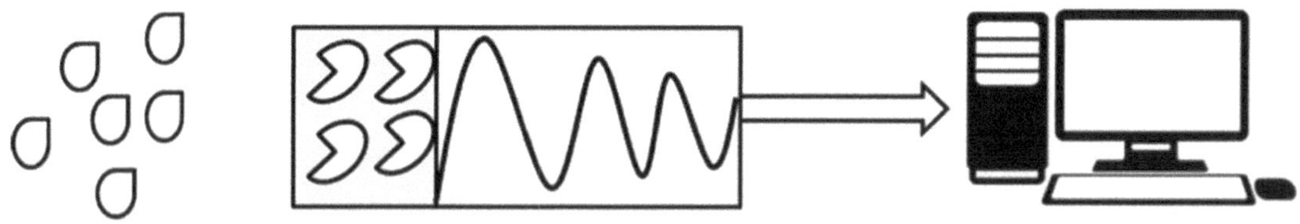

Figure 6.1. Schematic arrangement of a chemical sensor.

of a specific sample component to total composition analysis, into an analytically useful signal," as depicted in Fig. 6.1. The chemical information may originate from a chemical reaction of the analyte or from a physical property of the system investigated (Hulanicki et al., 1991; Buck et al., 2004; Vlasov et al., 2010; Loock and Wentzell, 2012).

The detection element has to be connected to a transducer generating an observable response. Detection elements are the most important component of any chemical sensor, once they ensure selectivity to a particular chemical, and therefore allows one to obtain reliable results with none or few interferences from other substances (Morrison, 1987; Bakker et al., 2000; Kwon et al., 2005).

The transducer can be defined as a device that converts an observed change (adsorption, reaction) into a measurable signal (Wohltjen and Snow, 1998; Lavrik et al., 2004; Bobacka et al., 2008; Grieshaber et al., 2008). In chemical sensors, the transduction of the concentration for a specific chemical is related to a signal whose magnitude is linearly proportional to the concentration. Transducers can be subdivided in the four main types: electrochemical (Trojanowicz and Hitchman, 1996; Badea et al., 2001; Palchetti and Mascini, 2008), optical (Kersey et al., 1997; Homola et al., 1999; Stewart et al., 2008; Mayer and Hafner, 2011), piezoelectric (O'Sullivan and Guilbault, 1999; Janshoff et al., 2000; Muralt, 2000; Wang et al., 2006b; Shrout and Zhang, 2007), and thermal (Ivnitski et al., 1999; Someya et al., 2005; Li et al., 2008) transducers.

2.2 Principles

As previously seen, the measurement or "response" of a sensor is generated by a specific change in a physical parameter, as a result of some chemical stimulation (Vlasov et al., 2010). The exact nature or type of physical change depends on both the transducer and the chemical stimulation analyzed and that is ordinarily

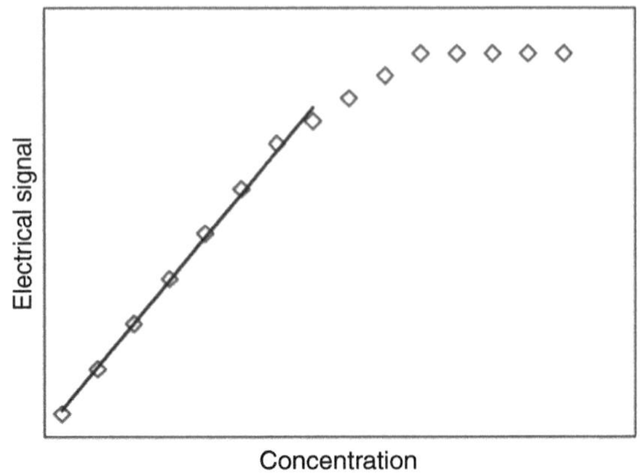

Figure 6.2. Schematic chemical sensor response.

called the sensor curve response or analytical curve (Fig. 6.2). In many cases, the detection element is a thin layer capable to interact with analyte molecules by an adsorption process, redox reactions, or other chemical interaction with the analyte (Ramstrom et al., 2001; Kosterev and Tittel, 2002; Persaud, 2005; Huang and Choi, 2007; Shrout and Zhang, 2007; Bobacka et al., 2008; Segev-Bar and Haick, 2013). Detection element layers can respond selectively to a particular chemical or group of substances. The molecular recognition term is used to describe this sensor behavior (Kubo et al., 1996; Valeur and Leray, 2000; Beer and Gale, 2001). Usually, sensor signals are processed mainly by electrical or optical instrumentation (Badea et al., 2001; Grieshaber et al., 2008; Bandodkar and Wang, 2014). Accordingly, most sensors include a transducing function where the actual measurement/quantification can be transformed into a property (voltage, current, resistance, capacitance, or impedance).

The performance of chemical sensors can be expressed in the form of parameters characterizing the sensor response (Tiemann, 2007; Wang et al., 2010; Huebert et al., 2011). The following list contains static and dynamic parameters used to characterize the performance of chemical sensors.

Sensitivity: represents the change in the measured signal per concentration unit of the analyte, that is, the slope of a calibration curve (Fig. 6.3).

Detection limit: represented by the lowest concentration value that can be detected by the sensor, above the system noise (Fig. 6.4).

$$\text{Detection limit} = \frac{3 \times (\text{standard deviation of response})}{\text{Sensitivity}}$$

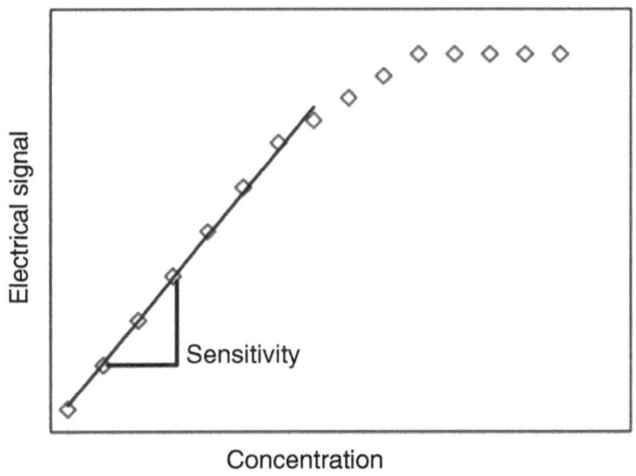

Figure 6.3. Schematic representation of sensor sensitivity.

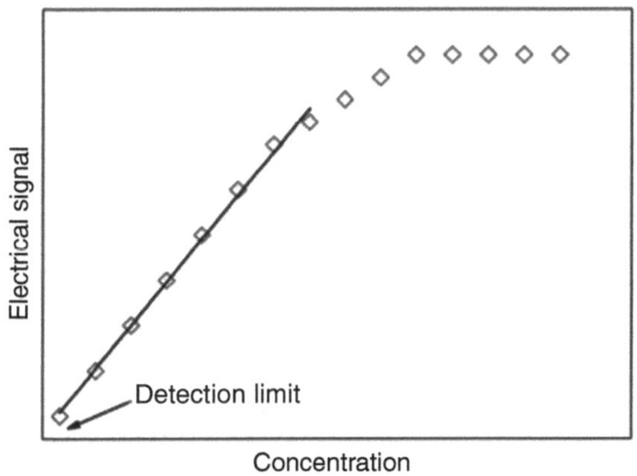

Figure 6.4. Schematic representation of detection limit.

Dynamic range: the concentration range between the detection limit and the upper limiting concentration that is within the linear range of response (Fig. 6.5).

Selectivity: one of the most important parameters for chemical sensing. It can be defined as the ability of a sensor to respond primarily to only one specie (analyte) in the presence of other species (interferents) (Fig. 6.6).

Linearity: the relative deviation of an experimentally determined analytical curve from an ideal straight line. Usually values for linearity are specified for a definite concentration range (Fig. 6.7).

Response time: is the response time from zero to a step change in concentration, until the definite average of the final value (Fig. 6.8).

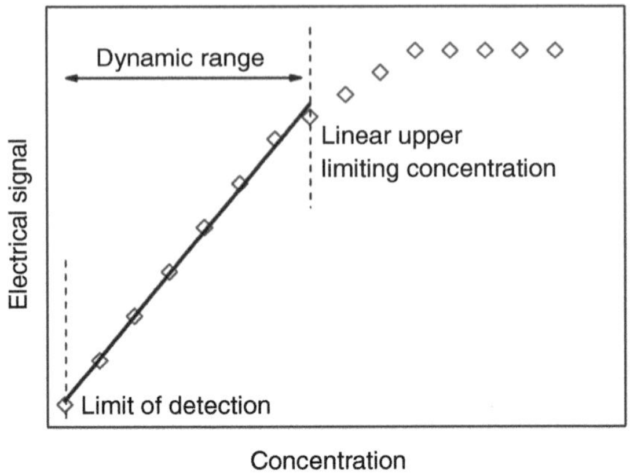

Figure 6.5. Schematic representation of dynamic range.

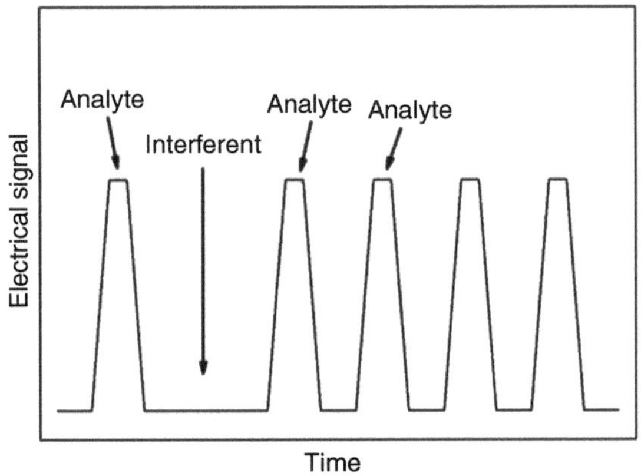

Figure 6.6. Schematic representation of selectivity.

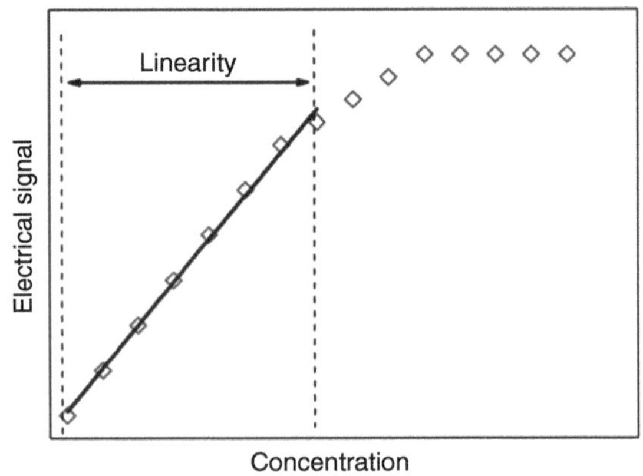

Figure 6.7. Schematic representation of linearity.

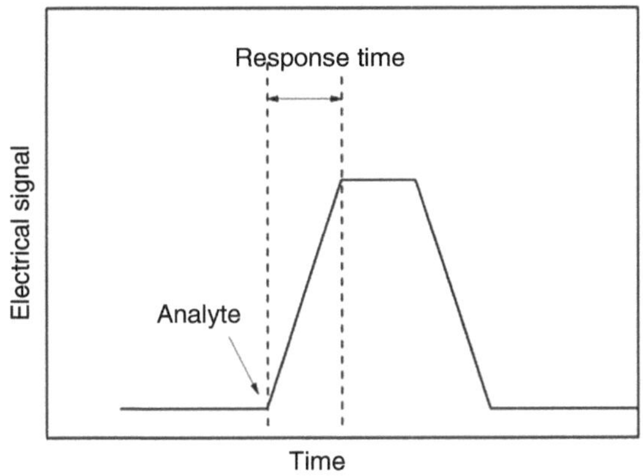

Figure 6.8. Schematic representation of response time.

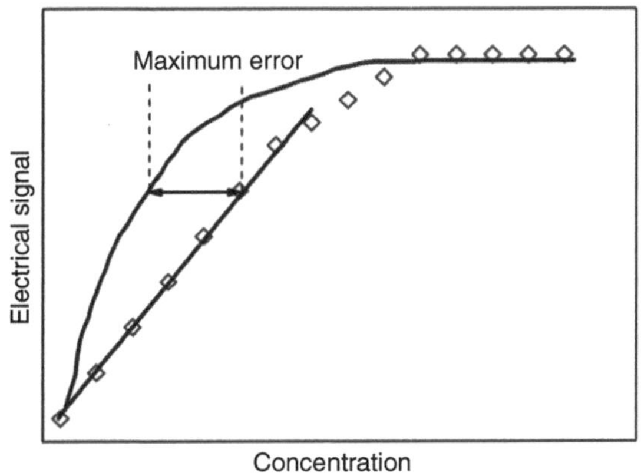

Figure 6.9. Schematic representation of hysteresis.

Hysteresis: The maximum changes of the input parameter regardless of the direction a change is made (Fig. 6.9).

Stability: deals with the degree to which sensor parameters remain constant over time. Changes in stability are also known as drift and can be due to sensor degradation, contamination, transducer failure, and also impregnation of surface sensor (Fig. 6.10).

Life cycle: the time length over which the sensor is able to operate. The maximum storage time (shelf life) must be distinguished from the maximum operating life. The latter can be specified either for continuous operation or for repeated on-off cycles (Fig. 6.11).

Accuracy: an expression of the agreement between the measured result (given as the average value of a series of experimental

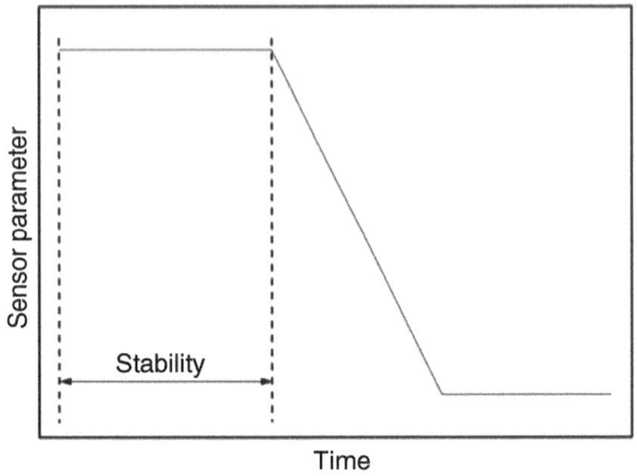

Figure 6.10. Schematic representation of stability.

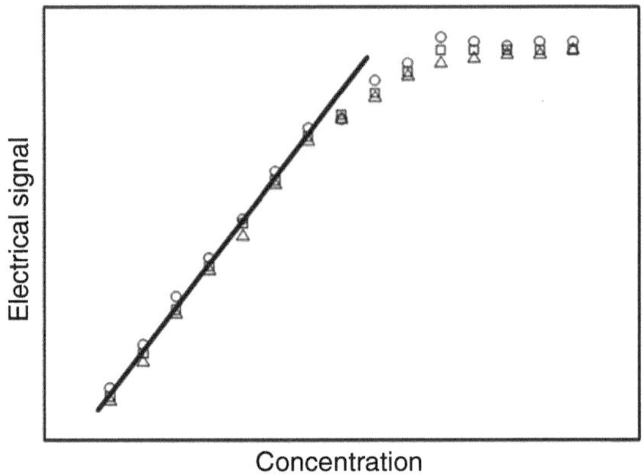

Figure 6.11. Schematic representation of life cycle.

measurements) and the true value. It is also a measurement of the systematic error, that is, deviation from the true value (normally given as a percentage).

Precision: an expression of the random error of a series of measurements, for example, the scattering of single values around the average value. The generally accepted way to express precision is with the standard deviation (STD).

2.3 Market Aspects and Food Safety

The number of sensors and the diversity of application in food industry can be expected to increase quickly over the next years. It is recognized that these interactions will depend on the industry,

government, and consumers and for this reason the following four key areas should be considered:
1. process monitoring and control;
2. food quality control;
3. food safety; and
4. environmental aspects.

Food safety is a vital key in public-health area connecting human health to farming and all areas of food production (Friedman, 1996; Shalaby, 1996; McLaughlin et al., 1999; Di Girolamo et al., 2015; Lago et al., 2015; Malhotra et al., 2015; Mizan et al., 2015). For instance, the extensive use of pesticides (Torres et al., 1996; van Lenteren, 2000; Hong and Moorman, 2005; Winter and Davis, 2006) and other chemical compounds (Haller et al., 2002; Malmauret et al., 2002; Knecht et al., 2004; Demirezen and Uruc, 2006; Alibabic et al., 2007) leads to an accumulation of chemical residues in food such as vegetables, meat, milk, and grains, decreasing food quality and exposing individuals to hazardous contaminants (Nasreddine and Parent-Massin, 2002; Alocilja and Radke, 2003; Piggott and Marsh, 2004; Schecter et al., 2004; Vitas et al., 2004). Some chemical residues have a relatively long half-life and may have direct toxic effects on consumers (Betarbet et al., 2000; Canli and Atli, 2003; Wang et al., 2005b; Yoon et al., 2006). Monitoring pesticides, heavy metals, and microbial pathogens in food helps to assess the potential risk of these products to human health (Alocilja and Radke, 2003; Wang et al., 2005b; Luong et al., 2008). Microbial contamination of food products is a common source of pathogenic microorganisms and can happen before and after processing (Chen et al., 2001; Franz et al., 2003; Kusumaningrum et al., 2003; Redmond and Griffith, 2003; Oliver et al., 2005; Heaton and Jones, 2008). Contaminations can result from a number of different factors including improper handling, insufficient processing, microbial infections, and contaminated soil and water. Researchers (Sivapalasingam et al., 2004; Adak et al., 2005; Cartwright et al., 2013; Painter et al., 2013) state that foodborne disease is a major cause of hospitalizations around the world that results from contaminated foods. Nowadays, 31 organisms are recognized as foodborne pathogens, and recent statistics indicate that millions of foodborne illnesses occur annually even in the USA, resulting in hospitalization or death (Olaimat and Holley, 2012; Mizan et al., 2015).

The presence of pathogenic microorganisms, and fecal contamination on a variety of food products are associated with many human health risks (Shephard et al., 1996; van den Bogaard and Stobberingh, 1999; Jarup, 2003; Redmond and

Griffith, 2003; Hites et al., 2004; Adak et al., 2005; Khan et al., 2008; Berger et al., 2010). Therefore, the development and implementation of new strategies for monitoring and controling food contaminants is crucial for public health.

3 Types of Nanomaterials Employed for Sensor Design

In the last decades, significant advances have been made in synthesis and application of nanomaterials with highly controllable size, shape, surface charge, and physicochemical characteristics (Chen and Chatterjee, 2013). Briefly, the ability to precisely engineer nanomaterials and to tailor their properties have culminated in highly robust and sensitive active layers. In this section we attempt to clarify how nanomaterials can be utilized in the design of high-performance sensors and biosensors, which currently assist in the detection of various food contaminants. In the following section, we provide perspectives on potential applications of nanomaterials based sensors and biosensors, drawing on some examples.

3.1 Polymeric Nanomaterials

Polymeric nanomaterials have been widely used as electron mediators and immobilization matrices in the design of chemical sensors. Biopolymers, functional and conducting polymers are the most employed types of polymers in the design of nanostructured devices (Moyo et al., 2012), which can be structured at the nanoscale allowing the miniaturization of devices. Thin films formation is the main successfully and most used approach to produce organized nanostructures at molecular level, where sensoactive layers can be achieved by different techniques and used as a sensing platform. In this context, the most commonly employed techniques for nanostructured films as sensor platform are the layer-by-layer (LbL), Langmuir–Bblodgett (LB), spin coating, and casting. These allow the formation of highly organized and morphologically controlled films providing suitable tools for sensors and biosensors. Furthermore, electrospinning has been shown to be a promising technique for producing polymeric nanofibers of varied sizes and geometries, which are interesting materials for sensor and biosensor platforms due to the increase of surface area/volume ratio, facilitating the immobilization of nanoparticles (Su et al., 2014; Andre et al., 2015).

Although biopolymers have a nonconductive nature, they are frequently used as sensor matrices due to the natural ability to form nanostructured films with tunable mechanical properties, low cost, availability and minimal environmental impacts associated. Polysaccharides extracted from natural gums, for instance, were used as building blocks for nanostructuration with phthalocyanines. The multilayer films formed from Chichá (*Sterculia striata*), Angico (*Anadenanthera colubrine*), and Caraia (*Sterculia urens*) displayed electroactivity, good chemical and electrochemical stability, being suitable as platform for sensing applications (Zampa et al., 2007; Eiras et al., 2010).

Functional polymers are those with chemical groups such as thiols, amine, carboxyl or carboxylic acid (–SH, $-NH_2$, –C = O, and –COOH) in their structure. These functional groups allow the ability to conduct electrons and act as mediators shuttling electrons between the electrode and the immobilized enzyme (Moyo et al., 2012). Chitosan (CS), a versatile functional biomaterial, has been extensively used as an enzyme immobilization platform for biosensing due to its interesting characteristics, such as biocompatibility, biodegradability, and polyelectrolyte nature in acidic medium, allowing heavy metal ions chelation and affinity to proteins (Krajewska, 2004). Within this context, chitosan-based sensors and biosensors have been developed for determination of food freshness and detection of heavy metals and food contaminants. Okuma et al. (1992) developed a biosensor with double enzyme reactors for fish freshness monitoring, by immobilizing the enzymes nucleoside phosphorylase, xanthine oxidase (XOD) and 5-nucleotidase (NT) simultaneously on CS porous beads. The sensor system was stable and tests with real sample (fish meat) presented good correlation with those obtained by liquid chromatography and ion-exchange column chromatography (Okuma et al., 1992). Ahmed and Fekry (2013) applied modified platinum electrodes in electrochemical detection of arsenic, lead, and nickel ions from standard and real aqueous solution. The proposed electrode was modified with a nanocomposite formed by super-paramagnetic iron oxide nanoparticles (α-Fe_3O_4) in CS film and used for sensing analysis employing cyclic voltammetry, linear scanning voltammetry, and impedance techniques. Authors stated that the nanocomposite film could be easily obtained, displaying ability for the determination and removal of heavy metals with remarkable current and fast response . Kandimalla and Ju (2006) produced an amperometric biosensor through the immobilization of acetylcholinesterase (AChE) in a multiwalled carbon nanotube-cross-linked chitosan

(CS-MWCNTs) composite for organophosphorous insecticide detection. Organophosphorous compounds are the most toxic substances used as highly effective insecticides, which can be lethal for human at minimal doses. Authors state that the platform formed by cross-linked CS-MWNTs favors covalently enzyme linkage and the as-prepared biosensor displayed good stability and reproducibility being successfully used for organophosphorous detection, with linearity ranging from 1.5 to 80 μM and a detection limit of 1.0 nM .

Conjugated polymers have attracted much attention for sensing and device applications once they present interesting electrical and optical properties (Correa et al., 2014). These materials are constituted by alternating sequence of σ and π bonds along its backbone (conjugated bonds), resulting in unpaired delocalized electrons along the polymer chain, given their characteristics, which are close to a semiconductor material. The existence of unpaired electrons (π-electrons) allows the formation of an electronic flow along the polymer backbone, and in such cases, high electrical conductivity (in some, cases becoming similar to a metal) can be achieved by chemical doping methods. In short, the chemical doping process consists on exposing such materials to reducing or oxidizing agents, for electrons addition or removal, by chemical reactions (Skotheim and Reynolds, 2006). In fact, depending on doping level and the structure, these organic materials can be classified as conductors and used in applications where conducting properties are required, for instance, in chemical sensors. The most common types of conjugated polymers used in nanostructured chemical sensors are polyaniline (PANI), polythiophenes (PTh), polypyrrole (PPy), poly(*para*-phenylene) (PPP), poly(*para*-phenylene-vinylene) (PPV), poly(*para*-phenilene-sulfide) (PPS), poly(3,4-ethylenedioxythiophene) (PEDOT), and polyfluorene (PFO) (McQuade et al., 2000).

A piezoelectric immunosensor was developed by Karaseva and Ermolaeva (2012) using a PPy electrochemically synthesized for the determination of antibiotic chloramphenicol in food samples, including meat, milk, egg, and honey. Li et al. (2015) produced a poly(*para*-aminobenzene sulphonic acid)-based electrochemical sensor for determination of Sudan I, a synthetic chemical colorant, in chili powder and ketchup. Authors stated that the excellent electrocatalytic activity of the conjugated polymer leads to a high sensitivity and good selectivity toward the detection, leading to a detection limit of 0.3 μg/L (Li et al., 2015).

3.2 Carbon-Based Materials

In the last decades, carbon-nanotubes (CNTs) and graphene (GN)-based nanomaterials have been widely employed for fabricating devices owing to their interesting field emission and electronic transport properties, mechanical strength and chemical properties (Rao et al., 2001; Ren and Cheng, 2014). CNTs (Iijima, 1991) research was boosted after Iijima's paper in 1991, and can be described as cylindrical graphite sheets, including both single-walled (SWCNTs), and multiwalled (MWCNTs) carbon nanotube structures. SWCNTs are formed by a single graphite sheet rolled up in a tube capped by hemispherical ends, while MWCNTs are formed by several graphite layers rolled up and arranged in multilayers with a layer space of 0.3–0.4 nm. Such materials can be prepared by various methods including electrochemical synthesis (arc-discharge method) and pyrolysis of precursor molecules (laser evaporation and chemical-vapor deposition) (Rao et al., 2001; Merkoci et al., 2005). Derived from fullerenes, CNTs have specific features such as high electrical conductivity, high surface area and aspect ratio, minimization of the surface fouling, high chemical stability, excellent adsorptive, high mechanical strength and biocompatibility, making them extremely attractive for chemical and electrochemical applications (Wang, 2005a; Wang and Lin, 2008). Besides its unique properties, the insertion of functional groups such as carboxyl groups in the surface of CNTs are frequent, improving its solubility in aqueous media and biomolecules immobilization, allowing CNTs-based biosensors and immunosensors (Wang and Lin, 2008). CNTs-based chemical sensors are widely reported including application in food samples analysis. A modified platinum-MWCNTs sensor was developed by Rasheed et al. (2015) for monitoring the superscale use of butylated hydroxyanisole (BHA), a phenolic antioxidant, in vegetable oil and mayonnaise samples. The MWCNTs-modified Pt electrode showed good performance in the electrochemical oxidation BHA detection, including in the presence of interferents, with a detection limit of 0.9 nM (Rasheed et al., 2015). Glassy carbon electrode (GCE) was modified with MWCNTs for determination of Sudan I in hot chili samples (powder and juice). Gan et al. (2008) stated that the excellent properties of MWCNTs remarkably enhances the oxidation peak of Sudan I, leading to a specific and sensitivity sensor. Liu et al. (2011) reported a high-sensitive carbon-nanotube-based amperometric immunosensor for determination of clenbuterol (CLB) detection. CLB is a bronchodilator drug sometimes administered to animals due to anabolic malfunctions; however, its use is not approved by the US Food and Drug Administration, and researches are focusing on its detection in meat or animal urine. MWCNTs were coated onto screen-printed electrodes as nanostructured layer and used as platform to CLB monoclonal antibody immobilization. CLB detection was performed in swine urine with results showing good consistency with ELISA and liquid chromatography–mass spectrometry (LC–MS) analysis, showing high stability, sensitivity, and accuracy in the detection.

Graphene (GN) is a single monolayer of planar-sp^2-bound carbon atoms arranged in condensed hexagonal rings. Discovered in 2004 (Novoselov et al., 2004), GN exhibits similar dimensions and elasticity as CNTs; however, they differ from each other in shape, colloidal stability, durability, and amount of impurities (Bussy et al., 2013). Both materials are interesting for technological applications due to good optical, electrical, mechanical, and thermal properties. GN can be produced from graphite through various methods, including mechanical and chemical cleavage and exfoliation or even from nongraphitic sources using chemical vapor deposition (Whitener and Sheehan, 2014) approach. GN, graphene oxide and reduced graphene oxide (rGO), have been extensively used in sensing and biosensing platforms, including the food safety area (Sundramoorthy and Gunasekaran, 2014). Gan et al. (2013) developed an electrochemical sensor based on graphene and titanium dioxide (TiO_2) assembled in a carbon paste electrode (CPE), for the trace determination of colorants (sunset yellow and tartrazine) in food. The results from this sensor platform are in good agreement with HPLC results, showing high sensitivity. Surface-enhanced Raman spectroscopy (SERS) was successfully applied by Xie et al. (2012) for detecting various prohibited colorants (amaranth, erythrosine, lemon yellow, and sunset yellow) in food samples using graphene and silver nanoparticles (AgNPs) films as platform. Authors stated that the Raman peaks were significantly enhanced on the graphene/Ag compared to Ag substrate alone and the limits of detection ranged from 10^{-5} to 10^{-7} M, depending of the colorant.

3.3 Metal and Metal Oxide Nanoparticles

Nanoparticles (NPs) have also played an important role in the development of new sensors/biosensors platforms. In fact, NPs are, in general, one of the most common nanotechnology-based approaches for the development of sensors and biosensors, due to simplicity, physicochemical features, and high surface areas (Doria et al., 2012). They can measure between 1 and 100 nm, have different shapes, and can be composed of one or more inorganic compounds. The majority of them exhibit size-related properties that differ significantly from those observed in microparticles or bulk materials. Depending on their size and composition, one can observe abnormal properties, such as quantum confinement in semiconductor nanocrystals, surface plasmon resonance in metal NPs, and superparamagnetism in magnetic materials.

Metal, oxide and semiconductor nanoparticles, and even composite nanoparticles have been widely used for sensors and biosensors applications (Luo et al., 2006). Metal NPs, for instance, have been explored for sensing applications through electrical transduction and design of a new generation of electronic devices (Wang, 2005b; Cao et al., 2011). The difference between electrical and electrochemical approach is that the former typically measures changes in the ohmic response of a circuit due to a target and subsequent nanoparticle binding (Wittenberg and Haynes, 2009). On the other hand, in electrochemical detection the current that indicates the presence of a target molecule is faradaic. More specifically, the current arises as a result of the oxidation or reduction process of either a redox probe in the detection medium or the redox activity of a conjugated electroactive nanoparticle. During electrochemical assays, the use of metal nanoparticles can decrease the overpotentials of many electroanalytical reactions and maintain the reversibility of redox reactions. Although the direct electrical and electrochemical detection of the analyte of interest is possible, many have benefited from the electroactive or catalytic properties of NPs for sensing/biosensing with unprecedented levels of sensitivity (Doria et al., 2012).

Glassy carbon electrodes (GCE) modified with gold nanoparticles, for example, were applied to detect nitrite. Nitrite ions in food samples can react in the acidic environment of human stomach with secondary and tertiary amines and amides to produce toxic and carcinogenic nitrosamine compounds. For this reason, it becomes important to develop specific, simple, and low-cost methods for nitrite determination (Santos et al., 2009). The prepared gold nanoparticles attached to glassy carbon electrode (Au/GCE) presented excellent catalytic ability toward the oxidation of nitrite. Compared with bare GCE and planar gold electrode, the Au/GCE decreased the overpotential of nitrite oxidation and improved the peak current. The Au/GCE was successfully applied to the electrochemical determination of nitrite in a real wastewater sample, showing excellent stability and antiinterference ability (Cui et al., 2007). Some impedimetric-based biosensors have also been reported. Yang et al. (2009) fabricated a capacitive immunosensor based on self-assembled gold NPs monolayer on GCE for the detection of bacteria using electrochemical impedance spectroscopy. The authors showed that the large surface area of gold nanoparticles increased the amount of antibody immobilized on the surface of glassy carbon electrodes, providing higher sensitivity and lower detection limit. This platform was successfully used for the detection of *Salmonella* spp. in lab-processed commercial pork samples and presented a detection limit of 100 CFU/mL (Yang et al., 2009). Metal nanoparticles based electrodes have also been used to detect different contaminants in water such as toxic ions (arsenic, mercury, antimony, and chromium) and pesticides (atrazine, methyl parathion, paraoxon ethyl, carbofuran, phoxim) (Saha et al., 2012).

Metal-oxide based nanoparticles have also been commonly used for sensing applications since they can offer functionality from electrically conducting to insulator and from highly catalytic to inert (Rassaei et al., 2011). Iron-oxide NPs have been used as the immobilizing matrix for applications in biosensing, due to their good biocompatibility, low toxicity, high electron-transfer capability, and high adsorption ability (Rassaei et al., 2011). Metal-oxide semiconductors (eg, WO_3, TiO_2, ZnO, and SnO_2) are widely used in gas sensors, presenting high detection ability and stability (Wang et al., 2010). Fang et al. (2014), developed SnO_2 and TiO_2 nanoparticle-modified screen-printed carbon electrodes that possess high sensitivity and low detection limit for the detection of *p*-ethylguaiacol, a compound present in the volatile signature of fruits and plants infected with the pathogenic fungus *Phytophthora cactorum*.

3.4 Hybrid Organic–Inorganic Nanomaterials

In order to improve the properties of the nanomaterials, considerable efforts have been directed toward constructing multifunctional hybrid nanostructures (Sanchez et al., 2011). The goal is to produce a new composite material that has distinct properties and cannot be observed in individual components. This may include either new or improved chemical properties that can be exploited for chemical or biological sensing (Nicole et al., 2014). Various strategies could be used in this direction. One-dimensional hybrid nanomaterials based platforms for electrochemical sensor and biosensor applications could be developed, for example, by combining electrospun nanofibers with nanostructured films (Oliveira et al., 2013), nanoparticles (Zhang and Yu, 2014), or carbon-based nanomaterials (Mercante et al., 2015a). Oliveira et al. (2012) reported the modification of nylon electrospun nanofibers with LbL films of conductive polypyrrole (PPy) and poly(*o*-ethoxyaniline) (POEA) that were assembled onto graphite interdigitated polyethylene terephthalate substrates, and used to detect pesticide in water. The electronic tongue apparatus, assembled to a flow analysis system was able to distinguish trace levels of hazardous pesticides contaminants at nanomolar concentrations (Oliveira et al., 2012).

The assembly of carbon nanostructures or nanoparticles into nanostructured films (LbL or LB films) is also an attractive approach for producing functionalized hybrid materials for (bio-)chemical sensing applications (Siqueira et al., 2009). Manzoli et al. (2014) reported the fabrication of LbL films based on AgCl–PANI nanocomposites. In contrast to conventional PANI-containing films, the AgCl–PANI/PSS LbL films deposited on interdigitated electrodes (IDEs) exhibited electrical resistance that was practically unaffected by changes in pH from 4 to 9. Therefore, these hybrid films can be used in electronic tongues (e-tongues) for analyzing both acidic and basic media, which could be very useful during the analysis of food samples (Manzoli et al., 2014).

Molecularly imprinted polymers (MIPs) are synthetic polymers possessing specific binding sites for a target compound, which are used as template during synthesis of the material (Song et al., 2014). Although the development of electrochemical sensing devices based on MIPs can have drawbacks due to the nonconductive nature of the polymers, their integration with electrochemical transducers, such as nanoparticles or carbon based nanomaterials, would produce robust devices capable of working in complicated matrices, overcoming many problems found in foodstuff (Azevedo et al., 2013). Deng et al. (2014) developed an electrochemical sensor based on an electrode modified with molecularly imprinted chitosan-graphene composite film for sensitive and selective detection of bisphenol A (BPA) in plastic bottled drinking water and canned beverages. The nanocomposite effectively increased the electrode active area and catalytic activity of the sensor, offering a fast response and sensitive BPA quantification with detection limit of 6.0 nM .

4 Types of Sensors and Methods of Detection

4.1 Electrochemical Sensors

Among the wide range of techniques used to detect different analytes, the electrochemistry ones, are certainly the most used, as verified by the vast number of studies published. By way of illustration, a search in the Web of Science with the keywords "electrochemical" and "sensors" retrieved over 27,000 papers and patents published from 2000 up to Jun. 2015. Electroanalytical techniques have as operating principle the monitoring of the electrical properties of analytes in solution, through transference of electrons that occurs at the interface electrode-solution.

The electroanalytical methods are versatile, once they make possible the control of reactions occurring in the electrode by modifying the electrode-solution interface, carefully selecting the cell potential applied (Lowinsohn and Bertotti, 2006). For instance, electrochemical biosensors combine the sensitivity of electroanalytical methods with the bioselectivity promoted by biological component immobilization, leading to high sensitive and selective devices (Ronkainen et al., 2010). Therefore, the electrode surface modification is a powerful tool in electroanalytics with a wide applicability, once it leads to improvement in recognition capabilities and/or in current signals amplification (improving sensibility), simultaneously with a more selective detection (Lowinsohn and Bertotti, 2006). Furthermore, some advantages are associated with electroanalytical techniques such as high sensitivity low cost, portability, ease of automation, and miniaturization (Li and Hu, 2011).

Electrochemical methods comprise a range of different specific techniques that can be mainly divided in two groups: the controlled-current techniques (galvanostatic), and controlled-potential techniques (potentiostatic). In the controlled-current techniques the current is adjusted (frequently maintained constant) and the potential is monitored as a function of time. Conversely, in controlled-potential techniques, the electrode potential is the independent variable (adjusted) and the current is monitored as a function of time or potential (Bard, 2001). The basic instrumentation necessary to perform electrochemical controlled-potential experiments includes a cell with three-electrode systems, namely working, reference, and auxiliary electrodes, a voltammetric analyzer formed by a potentiostatic circuitry, a voltage ramp generator and a plotter (Wang, 2000). Controlled-potential techniques are the commonly employed principle of detection in electrochemical sensors and biosensors, with specific techniques such as polarography, cyclic voltammetry, pulse voltammetry, chronoamperometry and electrochemical impedance spectroscopy. Briefly, such techniques differ from each other by the potential (excitation) method and consequently current response.

All modern voltammetric methods employed nowadays were originated from polarography. Developed in 1922, it had enormous impact on electroanalysis, awarding Heyrovsky with the Nobel Prize in Chemistry in 1959 (Wang, 2000). The use of dropping mercury electrode (DME) as the working electrode is the main characteristic of this technique, once it has a high hydrogen overvoltage and, consequently, a wide cathodic potential range window with a continuously renewable surface, eliminating passivation problems that may occur in solid electrodes. In conventional polarography (DC) the potential ramp is increased linearly, while the current is measured (Scholz, 2002).

Cyclic voltammetry (CV) is the most widely used electroanalytical technique in sensors and biosensors, being a powerful and rapid way to acquire qualitative information about electrochemical reactions, especially on the thermodynamics of redox process and on the kinetics of electron-transfer reactions. In CV measurements the potential applied in working electrode is continuously changed. The increase in potential is linear with time and the direction of the potential is reversed at the end of the scan (single or multiple scans can be used) resulting in a triangle waveform (Wang, 2000). Pulse voltammetry comprises a set of techniques where the potential stimulation is organized in sequences of increase and stabilization (pulse). Such techniques where developed aiming at lowering the detection limits by increasing the ratio between the faradaic and nonfaradaic currents, which promotes detection limits down to 10^{-8} M (Wang, 2000). Pulse voltammetry techniques differ each other by the waveform and current sampling regime, which main techniques named differential-pulse voltammetry (DPV), normal-pulse voltammetry, staircase voltammetry, and square-wave voltammetry (SWV).

In the Chronoamperometry method, the working electrode potential is stepped in a value in which no Faradaic current exists and only capacitive currents are detected and recorded as a function of time (Scholz, 2002). During measurements, a stationary working electrode and unstirred solution are used and the resulting current-time response is an exponential decreasing current.

Electrochemical impedance spectroscopy (EIS) is a powerful method for investigating electrical properties of conducting electrodes interfaces and materials. Relevant information for sensing field can be obtained with EIS data; for instance, kinetics of charges in bulk or interfacial regions, the efficiency in the charge (electron) transfer of ionic or mixed ionic-ionic conductors, semiconductors electrodes, and the corrosion inhibition of electrode processes. Therefore, EIS is an important technique employed in materials characterization and solid electrolyte such as solid state devices (Scholz, 2002). In EIS experiments, a potential should be defined or optimized in an open current potential (OCP) measurement and the amplitude value and range of frequencies should be defined. The impedance data response could be represented in a complex plane plot as *Z*"(imaginary part impedance) versus *Z*' (real part impedance), called Nyquist diagrams (Wang et al., 2005a).

As stated previously, the electroanalytical techniques are widely used for the development of sensors and biosensors, including analysis of food quality and safety (Scognarniglio et al., 2014; Sundramoorthy and Gunasekaran, 2014). Eksin et al. developed an electrochemical assay for gluten determination in flour samples using DPV method and pencil graphite electrode (PGE). They reported a detection limit of 7.11 μg/mL and successfully detected gluten in different flour samples beyond commercial vinegar and baker yeast (Eksin et al., 2015). Amperometric biosensors were developed by Goriushkina et al. (2009) for ethanol, glucose, and lactate determination in wine and must samples, with enzyme immobilization resulting in high selectivity detection. A screen-printed amperometric biosensor was successfully developed by Crew et al. (2011) for simultaneous determination of six organophosphate pesticides (Ops) in raw foods. Six different types of acetylcholinesterase (AChE) enzymes were immobilized in the screen-printed carbon electrode surfaces using glutaraldehyde. A schematic of the screen-printed electrode is shown in Fig. 6.12a. The biosensors exhibited qualitative and quantitative precision for in situ analysis of environmental samples and food extracts leading to a rapid detection system for early contamination warning in water and food (Crew et al., 2011). A typical range of chronoamperometric responses for the biosensors is shown in Fig. 6.12b.

DPV method was also used as electroanalytical technique for determination of caffeine in different beverages. Brunetti et al. (2007) developed a based Nafion-modified GCE to quantitative detection of caffeine in cola beverages and S'vorc et al. (2012) reported a sensitive and selective determination of caffeine in tea samples using bare boron-doped diamond electrode (BDDE) with relatively low detection limit of 0.15 μM. Fig. 6.13a shows that DPV technique was successfully used in the caffeine calibration curve construction, displaying good linearity between peak current and caffeine concentration, ranging from 4×10^{-7} to 2.5×10^{-5} M (inset). Also, the technique was used in the selective detection of caffeine in presence of different concentrations of theophelline (a structurally similar compound to caffeine), as shown in Fig. 6.13b.

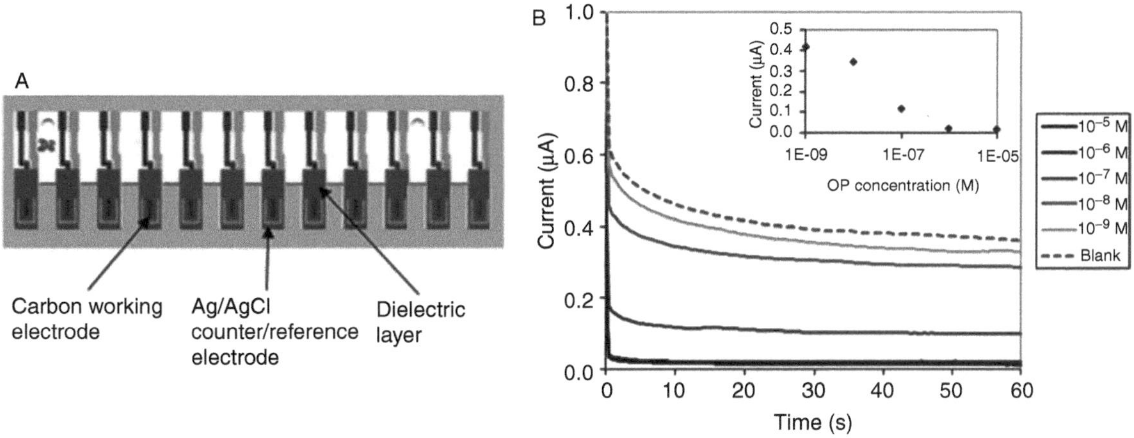

Figure 6.12. (a) Electrode array comprising 12 screen-printed carbon electrodes modified with CoPC and an Ag/AgCl counter/reference electrode printed on an alumina substrate. (b) Chronoamperograms produced from biosensors with the B04 enzyme inhibited with chlorfenvinphos (10^{-5} to 10^{-9} M). Reprinted with permission from: Crew, A., Lonsdale, D., Byrd, N., Pittson, R., Hart, J.P., 2011. Biosens. Bioelectron. 26, 2847–2851. Copyright (2011) Elsevier.

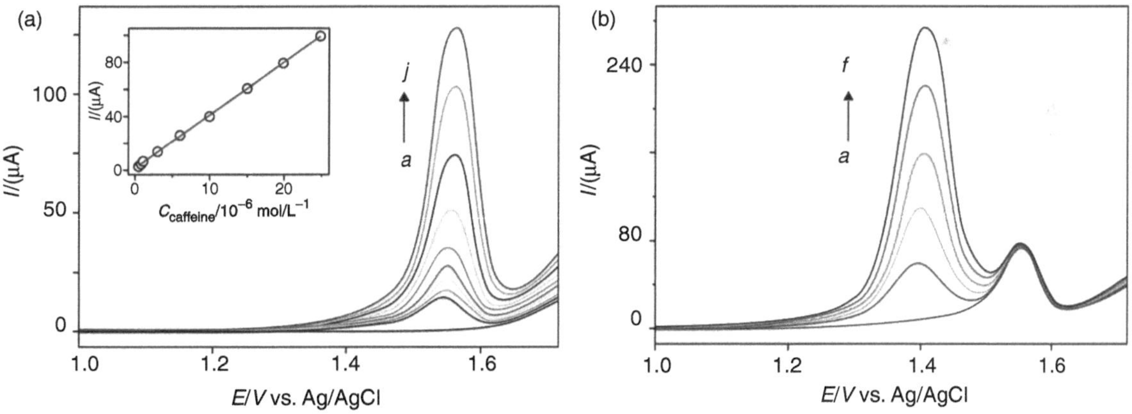

Figure 6.13. (a) Differential pulse voltammograms of caffeine solutions with various concentrations: (a) 0, (b) 4×10^{-7}, (c) 8×10^{-7}, (d) 1×10^{-6}, (e) 3×10^{-6}, (f) 6×10^{-6}, (g) 1×10^{-5}, (h) 1.5×10^{-5}, (i) 2×10^{-5}, and (j) 2.5×10^{-5} M (supporting electrolyte 0.4 M $HClO_4$) on bare BDD electrode at optimized DPV parameters: modulation amplitude of 0.05 V, modulation time 0.02 s and scan rate 0.05 V/s. The dependence between peak current (μA) and caffeine concentrations (M) appears in the inset. (b) Differential pulse voltammograms of 1×10^{-5} M caffeine (fixed concentration) and some different theophylline concentrations: (a) 0, (b) 2×10^{-5}, (c) 4×10^{-5}, (d) 6×10^{-5}, (e) 8×10^{-5}, and (f) 10×10^{-5} M (supporting electrolyte 0.4 M $HClO_4$) on bare BDD electrode at optimized DPV parameters: modulation amplitude of 0.05 V, modulation time 0.02 s, and scan rate 0.05 V/s. Reprinted with permission from Svorc, L., Tomcik, P., Svitkova, J., Rievaj, M., Bustin, D., 2012. Food Chem. 135, 1198–1204. Copyright (2012) Elsevier.

4.2 Electronic Nose and Electronic Tongue

Several sensors developed nowadays represent an effort to mimic the smell and taste functions of human beings, leading to the so-called electronic noses (e-noses) and tongues (e-tongues), respectively (Peris and Escuder-Gilabert, 2013). The general concepts of the e-nose and e-tongue used for analysis of gases and liquids, respectively, are similar. They involve application of an array of nonspecific or low-selective sensors with high stability and cross-sensitivity to different species, and an appropriate method for data processing in order to produce analytically useful signals during the multicomponent analysis (Song and Choi, 2015). The rationale for application of low-selective sensors is based on an analogy to biological organization of the olfactory and taste systems in mammals. There are millions of nonspecific receptors in the biological regions of the nose and tongue that respond to different substances present in the gas and liquid phases. However, only about 100 different types of olfactory receptors are known, while several dozens were identified in the taste buds on tongues of mammals. The taste and odor signals from the receptors are transmitted to the brain where they are processed by nets of neurons. As a result, the image of the sensed analyte is generated (Vlasov et al., 2005).

4.2.1 Electronic Tongues

Electronic tongues have emerged as a powerful tool for the rapid assessment of information of complex liquid systems. The term e-tongue was coined owing to the similarity with the human gustatory system, which is based on the concept of global selectivity. Global selectivity means the unique ability of the brain in grouping all the information received from the tongue in distinct patterns of response encoding the taste quality. An approach for mimicking the human tongue is the use of nonspecific sensor arrays able to recognize the five basic tastes: sweet, salty, sour, bitter, and umami (Sliwinska et al., 2014).

The first system for liquids analysis using a multisensory array was reported by Otto and Thomas (1985). After this seminal work several research groups have reported new technologies using different analytical methods, such as piezoelectric, electrochemical, and colorimetric sensors, being the electrochemical techniques (potentiometry, amperometry, and cyclic voltammetry) the most-used in e-tongues (Sliwinska et al., 2014). Potentiometric sensors, for example, have been used to monitor cheese fermentation (Esbensen et al., 2004), to evaluate the presence of phenol compounds in wine composition

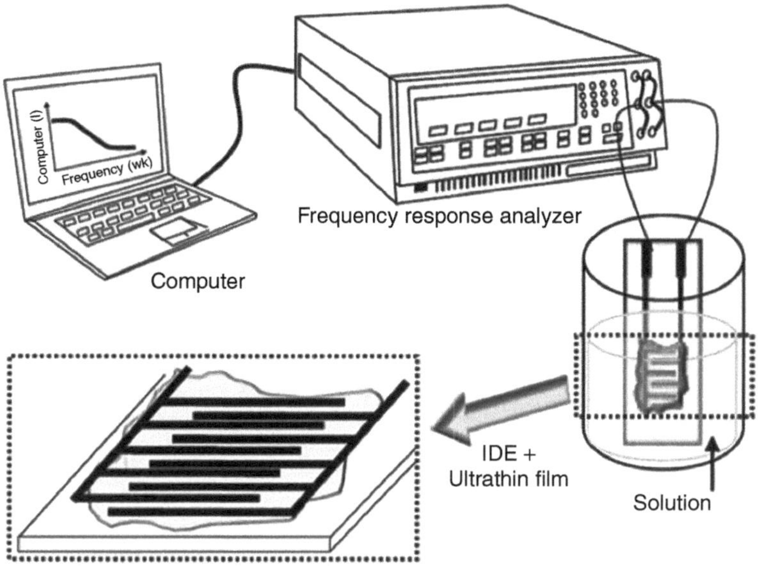

Figure 6.14. Experimental setup usually employed in impedance spectroscopy measurements. Reprinted with permission from Riul, Jr. A., Dantas, C. A. R., Miyazakic, C. M., Oliveira, Jr. O. N., 2010. Analyst 135, 2481–2495. Copyright (2010) The Royal Society of Chemistry.

(Rudnitskaya et al., 2009b), and to identify the botanical origin of honey (Dias et al., 2008).

Riul et al. (2002, 2010) reported the use of impedance spectroscopy for electronic tongues, which is advantageous because the materials composing the sensing units do not need to be electroactive and a reference electrode is not required, in contrast to the conventional electrochemical methods. In this technique, the impedance of the whole system is measured for varying frequencies of the signal applied on IDEs covered with ultra-thin films of different materials, as illustrated in Fig. 6.14. This system can be described by an equivalent electrical circuit, in which the double-layer formed at the electrode/electrolyte interface controls the response at low frequencies, the solution conductance and ultrathin films coating the electrodes rule the total impedance at intermediate frequencies, and the geometric capacitance is most relevant at higher frequencies (Taylor and Macdonald, 1987). Based on this technique, an electronic tongue composed by pure and composite LB films of conducting polymers was developed and showed good ability to detect trace amounts of different tastants and inorganic contaminants in water. The technique was also able to distinguish brands of several commercial beverages and three red wines with regard to vintage and chemicals added without complex laboratory analysis (Riul et al., 2003).

As mentioned before, most of sensing units reported for e-tongue systems are based on the concept of global selectivity, requiring no need for specific interactions between the analytes and the transducers. However, a step forward was taken in these systems by including sensing units with molecular recognition capability and specific interaction, characteristic of biological systems (Riul et al., 2010). Based on this concept, industrialized products containing tomatoes were assessed with an amperometric biosensor, in which an enzyme was immobilized in the sensing unit to recognize glutamic acid and monosodium glutamate, since these substances are responsible for the umami taste (Pauliukaite et al., 2006).

4.2.2 Electronic Noses

A general definition of an electronic nose is based on a device that comprises an array of heterogeneous electrochemical sensors with partial specificity and a pattern recognition system (Peris and Escuder-Gilabert, 2013). However, more recently, the term e-nose has also been used to describe several types of gas sensors that are able to change their sensor properties in a specific way as the ambient atmosphere is altered (Sliwinska et al., 2014). For the gas sensor development, materials that are commonly used include: metal oxides, intrinsically conducting polymers and (nano)composites. In addition to conductive sensors, gas detection can also be achieved by using quartz microbalance sensors, optical sensors, surface acoustic wave sensors and field effect transistors (Peris and Escuder-Gilabert, 2009). The applications of electronic noses have been numerous and have been focused on environmental monitoring, medical applications, and food industry. For instance, the application areas for food monitoring include detection of features such as freshness, adulteration, off-flavors, and bacteria detection (Loutfi et al., 2015), for distinct types of food such as fish, meat, milk, wine, coffee, and so forth. Manzoli et al. (2011) reported the fabrication of chemical sensors for monitoring banana maturity, while sensing units were composed by printed interdigitated graphite electrodes on tracing paper employing a graphite line–patterning technique. Such electrodes were coated with a thin layer of polyaniline by in situ polymerization to be used for the detection of ethylene, emitted during banana ripening. The array of three sensors was able to produce a distinct pattern of signals, taken as a signature (fingerprint) that can be used to characterize bananas ripeness (Manzoli et al., 2011).

4.3 Methods of Data Analysis

An important step during the use of artificial electronic sensors is the data analysis. Since the number of samples may be very large and many measurements are needed to distinguish the samples, the volume of data generated is tremendous. Therefore, the use of chemometrics or pattern recognition techniques is inevitable. Important methods used to process data obtained from sensor arrays are principal component analysis (PCA), multivariate linear regression (MLR), linear discriminant analysis (Goriushkina et al., 2009), principal component regression (PCR), partial least squares (PLS), cluster analysis, artificial neural networks (Manzoli et al., 2011), and fuzzy logic (Riul et al., 2010; Sliwinska et al., 2014). Among these, PCA is the most used methodology in the literature for e-tongues, which is based on a unsupervised method for determining similarity of the input data by comparing the relative location of its associated so-called principal components in a PCA-plot. The corresponding scores plots correlates the observations or experiments, grouping samples in score plots that can be used for further classification. For instance, Wei et al. (2013) reported the use of the PCA method to classify pasteurized milk with different storage time using a voltammetric e-tongue. While Mercante et al. (2015b) employed an impedimetric e-tongue to evaluate milk samples regarding fat content.

5 Novel Sensing Platforms Based on Microfluidics

As previously mentioned, food analysis is an essential concern to assure higher safety and quality of products, with an increasing global demand for analytical devices able to detect contaminants, residues and pathogens in a short time, simple and accurate way. Therefore, the complexity of the food matrix remains a major challenge for sample preparation, selectivity, and sensitivity in real test food evaluation. Gas chromatography, high performance liquid chromatography, and mass spectroscopy are the most commonly used sensitive screening techniques nowadays (Atalay et al., 2011; Eikel and Henion, 2011). However, such techniques are expensive, relatively slow, require trained personnel for sample preparation and operation, and in some cases can be quite time consuming for routine analysis. In this sense, emerging technologies have an enormous potential for improvements in the existing analytical methodologies, offering advantages as low cost, decreasing detection limits, high throughput and integration, paving the way for innovative applications in food analysis.

We report in this section some of these potential tools to target demands required by the food chain (production, packaging, and quality of products).

Microfluidics is one of those techniques and deals with the precise manipulation and control of liquids confined in channels at submillimeter dimensions, where gravity effects are small when compared to those caused by capillary and surface tension forces inside microchannels. It is a technological field lying at the interface of several traditional areas (physics, engineering, chemistry, biology, materials science, nanotechnology, etc.) that has moved forward acutely over the past two decades (Whitesides, 2006; Sackmann et al., 2014), occupying one of the frontiers of knowledge nowadays, offering some advantages such as:

1. small sample volumes (much lower than 10^{-9} L);
2. reduced costs and wastes;
3. easy integration and multiple functionalities;
4. miniaturized, portable systems;
5. high accuracy and reliability of analytical results; and
6. short-time analysis.

Microfluidic technologies started in 1990s (Verpoorte et al., 1994) and since then are rapidly advancing for the development of fast, simple, and low-cost devices in distinct applications (point-of-care, environmental, biological, and clinical uses) (Whitesides, 2006; Martinez et al., 2010; Sackmann et al., 2014). Briefly, microfluidic chips have been successfully used for detection of bacterial pathogens with no need of trained persons to process and interpret results (Safavieh et al., 2014), integrated with chemiluminescense methods for fast detection of mycotoxins in beer and wine analysis (Novo et al., 2013), and total sugar content in food and beverages (Alam et al., 2012). It has also been successfully applied for the detection of *Salmonella enterica serotype enteritides* at 10^5 cfu/mL within 40 min without sample preparation (Ricciardi et al., 2010) and for the development of fluidic batteries for sensing applications (Bae et al., 2013).

Microfluidic platforms fulfill all requirements for food research analysis, encouraging new analytical tools in food processing and diagnostics. Electrochemical detection and capillary electrophoresis have been intensively explored in this sense (Lee et al., 2008; Palchetti and Mascini, 2008; Atalay et al., 2011; Escarpa, 2014). Microfluidic capillary electrophoresis has been broadly applied in the detection of polyphenols, alkaloids and antioxidants using microchips without loss of analytical performance (Blasco et al., 2005), with an improved performance and remarkable sensitivity in the detection of a micotoxyn (*Zearalenona*) in baby foods (Escarpa, 2012). Also interesting in food analysis is the use of microfluidic devices applied for the recovery of small particles and bacteria from complex food samples (Clime et al., 2015), or the development of sensors to determine pesticide residues in soil and vegetables (Duford et al., 2013).

An interesting interface is the use of nanomaterials to improve performance analysis with potential increase in selectivity and sensitivity due to the use of biological elements (enzymes, DNA, antibodies, aptamers, etc.) coupled to the transduction system, or even the use of nanomaterials in polymer nanocomposites in food package developments. Biosensors have already demonstrated increased sensitivity in analytical detection; however, the simplicity and adaptability of nanomaterials in different systems/applications can bring some benefits to the agro-food sector such as producing low-cost, portable sensors integrated with multiple functionalities in a single device. To illustrate, an array of gold IDEs integrated with magnetic nanoparticle antibody conjugates detected easily *E. coli* O157:H7 in ground beef samples (Varshney et al., 2007) using impedance spectroscopy. More recently, a microfluidic taste sensor was fabricated using nanostructured thin films of different materials deposited inside polydimethylsiloxane (PDMS) microchannels sealed on the top of IDEs, discriminating basic tastes and also detecting suppression effects (Daikuzono et al., 2015).

Several developments have been made using PDMS, a low-viscosity elastomer largely employed in microfluidic devices because it is chemically and thermally stable, nontoxic and biocompatible, flexible, optically transparent, and permeable to gas and vapor (McDonald and Whitesides, 2002). A new class of 3D microfluidic cell culture chips are also impacting the literature simulating activities of organ systems (organ-on-a-chip), aiming at the replacement of animal testing by developing lung-on-a-chip, kidney-on-a-chip, heart-on-a-chip, and so on, devices. Therefore, despite fast curing and high fidelity in molding, PDMS permeability to vapor and gases together with its capability to absorb small molecules might be troublesome in real-time signaling in biological research (Sackmann et al., 2014). Alternative materials beyond the PDMS realm are highly keen in the microfluidic community and the surge of 3D-printing technologies (Hwang et al., 2015) can impact positively the field, both in innovative and cheaper developments, making more accessible methodologies able to attract higher attention from researches in other research areas.

Despite huge potential to miniaturize and automate analytical methods employing a coordinated and creative use of emergent technologies, researchers must always pursue reducing devices costs, using less time-consuming protocols. In addition, high-performance devices based on smart sensors coupled to a new generation of microfluidic platforms for food analysis must be sought for faster, easier measurements in routine analysis with minimal technical skills, "easy-to-buy" and "easy-to-use," even considering single comparative answers (yes/no responses). Nevertheless, due to the complexity of the food matrix, sample preparation remains a major challenge when thinking of integration and smart tools from emergent technologies to discover innovative solutions in the agro-food sector.

6 Final Remarks

Chemical sensors have been the subject of scientific research and technological development of numerous research groups and corporations all over the world, with potential for wide application in daily life needs. For instance, chemical sensors can be applied in medical diagnosis, food and beverage quality control, and environmental and homeland monitoring. Concerning food quality and safety, sensors have been used for the chemical monitoring of food industrial processes (eg, fermentation, acidity, taste, etc.) as well as mitigating problems associated with food contaminated by microorganisms. For reaching reliable and high performance miniaturized sensor devices, including microfluidics, newly and highly responsive materials have become of paramount importance. Hybrid nanostructured materials can accomplish such tasks, once they harmonically combine the properties of distinct materials (in physical–chemical nature), and therefore promote synergistic effects between them. In this chapter we demonstrated that the use of conjugated polymer-based sensors combined with metallic nanoparticles, nanofibers, carbon nanotubes, graphene, and other molecular elements have provided chemical sensors (including potentiometric, e-tongues, and e-noses, etc.) capable of performing host-guest interactions with extraordinary levels of sensitivity and selectivity, enlarging the range of potential applications for such devices. Other approaches are based on the use of molecular imprinted polymers, as well as the development of biomimetic sensors, which present sensibility to detect specific target molecules. Ultimately, the combined efforts of physicists, chemists, and engineers have succeeded in making novel hybrid nanostructured materials available for use in distinct sensors platforms, decreasing the fabrication costs, and facilitating its large-scale production.

Acknowledgments

The authors are indebted to FAPESP, CNPq, CAPES, EMBRAPA, and MCTI-SisNano from Brazil for financial support.

References

Adak, G.K., Meakins, S.M., Yip, H., Lopman, B.A., O'Brien, S.J., 2005. Disease risks from foods: England and Wales, 1996–2000. Emerg. Infect. Dis. 11 (3), 365–372.

Ahmed, R.A., Fekry, A.M., 2013. Preparation and characterization of a nanoparticles-modified chitosan sensor and its application for the determination of heavy metals from different aqueous media. Int. J. Electrochem. Sci. 8 (3), 6692–6708.

Alam, A.-M., Kamruzzaman, M., Trung-Dung, D., Lee, S.H., Kim, Y.H., Kim, G.-M., 2012. Enzymeless determination of total sugar by luminol-tetrachloroaurate chemiluminescence on chip to analyze food samples. Anal. Bioanal. Chem. 404 (3), 3165–3173.

Alibabic, V., Vahcic, N., Bajramovic, M., 2007. Bioaccumulation of metals in fish of salmonidae family and the impact on fish meat quality. Environ. Monitor. Assess. 131 (3), 349–364.

Alimelli, A., Pennazza, G., Santonico, M., Paolesse, R., Filippini, D., D'Amico, A., Lundstrom, I., Di Natale, C., 2007. Fish freshness detection by a computer screen photoassisted based gas sensor array. Anal. Chim. Acta 582 (3), 320–328.

Alocilja, E.C., Radke, S.M., 2003. Market analysis of biosensors for food safety. Biosens. Bioelect. 18 (3), 841–846.

Andre, R.S., Pavinatto, A., Mercante, L.A., Paris, E.C., Mattosoab, L.H.C., Correa, D.S., 2015. Improving the electrochemical properties of polyamide 6/polyaniline electrospun nanofibers by surface modification with ZnO nanoparticles. RSC Adv. 5, 73875–73881.

Atalay, Y.T., Vermeir, S., Witters, D., Vergauwe, N., Verbruggen, B., Verboven, P., Nicolai, B.M., Lammertyn, J., 2011. Microfluidic analytical systems for food analysis. Trends Food Sci. Technol. 22 (3), 386–404.

Azad, A.M., Akbar, S.A., Mhaisalkar, S.G., Birkefeld, L.D., Goto, K.S., 1992. Solid-state gas sensors: a review. J. Electrochem. Soc. 139 (3), 3690–3704.

Azevedo, S.D., Lakshmi, D., Chianella, I., Whitcombe, M.J., Karim, K., Ivanova-Mitseva, P.K., Subrahmanyam, S., Piletsky, S.A., 2013. Molecularly imprinted polymer-hybrid electrochemical sensor for the detection of beta-estradiol. Ind. Eng. Chem. Res. 52 (3), 13917–13923.

Badea, M., Amine, A., Palleschi, G., Moscone, D., Volpe, G., Curulli, A., 2001. New electrochemical sensors for detection of nitrites and nitrates. J. Electroanal. Chem. 509 (3), 66–72.

Bae, H., Selimovic, S., Dokmeci, M.R., Khademhosseini, A., 2013. Paper-based sickle cell disease test. Lab Chip 13 (3), 3305–3308.

Bakker, E., Pretsch, E., Buhlmann, P., 2000. Selectivity of potentiometric ion sensors. Anal. Chem. 72 (3), 1127–1133.

Baldwin, E.A., Bai, J.H., Plotto, A., Dea, S., 2011. Electronic noses and tongues: applications for the food and pharmaceutical industries. Sensors 11 (3), 4744–4766.

Bandodkar, A.J., Wang, J., 2014. Noninvasive wearable electrochemical sensors: a review. Trends Biotechnol. 32 (3), 363–371.

Bard, A.J.F.L.R., 2001. Electrochemical Methods: Fundamentals and Applications. John Wiley & Sons, New York.

Bargon, J., Braschoss, S., Florke, J., Herrmann, U., Klein, L., Loergen, J.W., Lopez, M., Maric, S., Parham, A.H., Piacenza, P., Schaefgen, H., Schalley, C.A., Silva,

G., Schlupp, M., Schwierz, H., Vogtle, F., Windscheif, G., 2003. Determination of the ripening state of Emmental cheese via quartz microbalances. Sensor. Actuat. B 95 (3), 6–19.
Bartlett, P.N., Elliott, J.M., Gardner, J.W., 1997. Electronic noses and their application in the food industry. Food Technol. 51 (3), 44–48.
Beer, P.D., Gale, P.A., 2001. Anion recognition and sensing: The state of the art and future perspectives. Angew. Chem.-Int. Ed. Engl. 40 (3), 486–516.
Berger, C.N., Sodha, S.V., Shaw, R.K., Griffin, P.M., Pink, D., Hand, P., Frankel, G., 2010. Fresh fruit and vegetables as vehicles for the transmission of human pathogens. Environ. Microbiol. 12 (3), 2385–2397.
Berna, A.Z., Lammertyn, J., Saevels, S., Di Natale, C., Nicolai, B.M., 2004. Electronic nose systems to study shelf life and cultivar effect on tomato aroma profile. Sensor. Actuat. B 97 (3), 324–333.
Betarbet, R., Sherer, T.B., MacKenzie, G., Garcia-Osuna, M., Panov, A.V., Greenamyre, J.T., 2000. Chronic systemic pesticide exposure reproduces features of Parkinson's disease. Nat. Neurosci. 3 (3), 1301–1306.
Bhadra, S., Khastgir, D., Singha, N.K., Lee, J.H., 2009. Progress in preparation, processing, and applications of polyaniline. Prog. Polym. Sci. 34 (8), 783–810.
Blasco, A.J., Barrigas, I., Gonzalez, M.C., Escarpa, A., 2005. Fast and simultaneous detection of prominent natural antioxidants using analytical microsystems for capillary electrophoresis with a glassy carbon electrode: a new gateway to food environments. Electrophoresis 26 (3), 4664–4673.
Bobacka, J., Ivaska, A., Lewenstam, A., 2008. Potentiometric ion sensors. Chem. Rev. 108 (3), 329–351.
Brunetti, B., Desimoni, E., Casati, P., 2007. Determination of caffeine at a Nafion-covered glassy carbon electrode. Electroanalysis 19 (3), 385–388.
Buck, R.P., Lindner, E., Kutner, W., Inzelt, G., 2004. Piezoelectric chemical sensors: IUPAC Technical Report. Pure Appl. Chem. 76 (3), 1139–1160.
Bussy, C., Ali-Boucetta, H., Kostarelos, K., 2013. Safety considerations for graphene: lessons learnt from carbon nanotubes. Account. Chem. Res. 46 (3), 692–701.
Canli, M., Atli, G., 2003. The relationships between heavy metal (Cd, Cr, Cu, Fe, Pb, Zn) levels and the size of six Mediterranean fish species. Environ. Pollut. 121 (3), 129–136.
Cao, X.D., Ye, Y.K., Liu, S.Q., 2011. Gold nanoparticle-based signal amplification for biosensing. Anal. Biochem. 417 (3), 1–16.
Cartwright, E.J., Jackson, K.A., Johnson, S.D., Graves, L.M., Silk, B.J., Mahon, B.E., 2013. Listeriosis outbreaks and associated food vehicles, United States, 1998–2008. Emerg. Infect. Dis. 19 (3), 1–9.
Casalinuovo, I.A., Di Pierro, D., Coletta, M., Di Francesco, P., 2006. Application of electronic noses for disease diagnosis and food spoilage detection. Sensors 6 (3), 1428–1439.
Chen, A.C., Chatterjee, S., 2013. Nanomaterials based electrochemical sensors for biomedical applications. Chem. Soc. Rev. 42 (3), 5425–5438.
Chen, Y.H., Jackson, K.M., Chea, F.P., Schaffner, D.W., 2001. Quantification and variability analysis of bacterial cross-contamination rates in common food service tasks. J. Food Protect. 64 (3), 72–80.
Cheng, K.L., 1989. PH glass-electrode and its mechanism. ACS Symposium Ser. 390, 286–302.
Clime, L., Hoa, X.D., Corneau, N., Morton, K.J., Luebbert, C., Mounier, M., Brassard, D., Geissler, M., Bidawid, S., Farber, J., Veres, T., 2015. Microfluidic filtration and extraction of pathogens from food samples by hydrodynamic focusing and inertial lateral migration. Biomed. Microdev. 17 (1), .
Correa, D.S., Medeiros, E.S., Oliveira, J.E., Paterno, L.G., Mattoso, L.H.C., 2014. Nanostructured conjugated polymers in chemical sensors: synthesis, properties, and applications. J. Nanosci. Nanotechnol. 14 (3), 6509–6527.

Crew, A., Lonsdale, D., Byrd, N., Pittson, R., Hart, J.P., 2011. A screen-printed, amperometric biosensor array incorporated into a novel automated system for the simultaneous determination of organophosphate pesticides. Biosens. Bioelect. 26 (3), 2847–2851.
Cui, Y.P., Yang, C.Z., Zeng, W., Oyama, M., Pu, W.H., Zhang, J.D., 2007. Electrochemical determination of nitrite using a gold nanoparticles-modifled glassy carbon electrode prepared by the seed-inediated growth technique. Anal. Sci. 23 (3), 1421–1425.
Daikuzono, C.M., Dantas, C.A.R., Volpati, D., Constantino, C.J.L., Piazzetta, M.H.O., Gobbi, A.L., Taylor, D.M., Oliveira, Jr., O.N., Riul, Jr., A., 2015. Microfluidic electronic tongue. Sensor. Actuat. B 207, 1129–1135.
Demirezen, D., Uruc, K., 2006. Comparative study of trace elements in certain fish, meat and meat products. Meat Sci. 74 (3), 255–260.
Deng, P.H., Xu, Z.F., Kuang, Y.F., 2014. Electrochemical determination of bisphenol A in plastic bottled drinking water and canned beverages using a molecularly imprinted chitosan-graphene composite film modified electrode. Food Chem. 157, 490–497.
Di Girolamo, F., Muraca, M., Mazzina, O., Lante, I., Dahdah, L., 2015. Proteomic applications in food allergy: food allergenomics. Curr. Opin. Allerg. Clin. Immunol. 15 (3), 259–266.
Dias, L.A., Peres, A.M., Vilas-Boas, M., Rocha, M.A., Estevinho, L., Machado, A.A.S.C., 2008. An electronic tongue for honey classification. Microchim. Acta 163 (3), 97–102.
Doria, G., Conde, J., Veigas, B., Giestas, L., Almeida, C., Assuncao, M., Rosa, J., Baptista, P.V., 2012. Noble metal nanoparticles for biosensing applications. Sensors 12 (3), 1657–1687.
Duford, D.A., Xi, Y., Salin, E.D., 2013. Enzyme inhibition-based determination of pesticide residues in vegetable and soil in centrifugal microfluidic devices. Anal. Chem. 85 (3), 7834–7841.
Eikel, D., Henion, J., 2011. Liquid extraction surface analysis (LESA) of food surfaces employing chip-based nano-electrospray mass spectrometry. Rapid Commun. Mass Spectrom. 25 (3), 2345–2354.
Eiras, C., Santos, A.C., Zampa, M.F., de Brito, A.C.F., Constantino, C.J.L., Zucolotto, V., dos Santos, J.R., 2010. Natural polysaccharides as active biomaterials in nanostructured films for sensing. J. Biomater. Sci. Polym. Ed. 21 (3), 1533–1543.
Eksin, E., Congur, G., Erdem, A., 2015. Electrochemical assay for determination of gluten in flour samples. Food Chem. 184, 183–187.
Esbensen, K., Kirsanov, D., Legin, A., Rudnitskaya, A., Mortensen, J., Pedersen, J., Vognsen, L., Makarychev-Mikhailov, S., Vlasov, Y., 2004. Fermentation monitoring using multisensor systems: feasibility study of the electronic tongue. Anal. Bioanal. Chem. 378 (3), 391–395.
Escarpa, A., 2012. Food electroanalysis: sense and simplicity. Chem. Record 12 (3), 72–91.
Escarpa, A., 2014. Lights and shadows on food microfluidics. Lab Chip 14 (3), 3213–3224.
Fang, Y., Umasankar, Y., Ramasamy, R.P., 2014. Electrochemical detection of *p*-ethylguaiacol, a fungi infected fruit volatile using metal oxide nanoparticles. Analyst 139 (3), 3804–3810.
Franz, C., Stiles, M.E., Schleifer, K.H., Holzapfel, W.H., 2003. *Enterococci* in foods: a conundrum for food safety. Int. J. Food Microbiol. 88 (3), 105–122.
Friedman, M., 1996. Food browning and its prevention: an overview. J. Agric. Food Chem. 44 (3), 631–653.
Gan, T., Li, K., Wu, K.B., 2008. Multiwall carbon nanotube-based electrochemical sensor for sensitive determination of Sudan I. Sensor. Actuat. B 132 (3), 134–139.

Gan, T., Sun, J.Y., Meng, W., Song, L., Zhang, Y.X., 2013. Electrochemical sensor based on graphene and mesoporous TiO_2 for the simultaneous determination of trace colorants in food. Food Chem. 141 (3), 3731–3737.
Ghasemi-Varnamkhasti, M., Mohtasebi, S.S., Siadat, M., Balasubramanian, S., 2009. Meat quality assessment by electronic nose (machine olfaction technology). Sensors 9 (3), 6058–6083.
Gomez, J.L., Tigli, O., 2013. Zinc oxide nanostructures: from growth to application. J. Mater. Sci. 48 (3), 612–624.
Goriushkina, T.B., Soldatkin, A.P., Dzyadevych, S.V., 2009. Application of amperometric biosensors for analysis of ethanol, glucose, and lactate in wine. J. Agric. Food Chem. 57 (3), 6528–6535.
Govindhan, M., Adhikari, B.R., Chen, A.C., 2014. Nanomaterials-based electrochemical detection of chemical contaminants. RSC Adv. 4 (3), 63741–63760.
Granda Valdes, M., Valdes Gonzalez, A.C., Garcia Calzon, J.A., Elena Diaz-Garcia, M., 2009. Analytical nanotechnology for food analysis. Microchim. Acta 166 (3), 1–19.
Grieshaber, D., MacKenzie, R., Voeroes, J., Reimhult, E., 2008. Electrochemical biosensors: sensor principles and architectures. Sensors 8 (3), 1400–1458.
Haller, M.Y., Muller, S.R., McArdell, C.S., Alder, A.C., Suter, M.J.F., 2002. Quantification of veterinary antibiotics (sulfonamides and trimethoprim) in animal manure by liquid chromatography-mass spectrometry. J. Chromatogr. A 952 (3), 111–120.
Heaton, J.C., Jones, K., 2008. Microbial contamination of fruit and vegetables and the behavior of enteropathogens in the phyllosphere: a review. J. Appl. Microbiol. 104 (3), 613–626.
Hierlemann, A., Lange, D., Hagleitner, C., Kerness, N., Koll, A., Brand, O., Baltes, H., 2000. Application-specific sensor systems based on CMOS chemical microsensors. Sensor. Actuat. B 70 (3), 2–11.
Hites, R.A., Foran, J.A., Carpenter, D.O., Hamilton, M.C., Knuth, B.A., Schwager, S.J., 2004. Global assessment of organic contaminants in farmed salmon. Science 303 (3), 226–229.
Homola, J., Yee, S.S., Gauglitz, G., 1999. Surface plasmon resonance sensors: review. Sensor. Actuat. B 54 (3), 3–15.
Hong, C.X., Moorman, G.W., 2005. Plant pathogens in irrigation water: challenges and opportunities. Crit. Rev. Plant Sci. 24 (3), 189–208.
Huang, X.-J., Choi, Y.-K., 2007. Chemical sensors based on nanostructured materials. Sensor. Actuat. B 122 (3), 659–671.
Huang, X.H., Neretina, S., El-Sayed, M.A., 2009. Gold nanorods: from synthesis and properties to biological and biomedical applications. Adv. Mater. 21 (3), 4880–4910.
Huang, X., Yeo, W.-H., Liu, Y., Rogers, J.A., 2012. Epidermal differential impedance sensor for conformal skin hydration monitoring. Biointerphases 7 (1–4).
Huebert, T., Boon-Brett, L., Black, G., Banach, U., 2011. Hydrogen sensors: a review. Sensor. Actuat. B 157 (3), 329–352.
Hulanicki, A., Glab, S., Ingman, F., 1991. Chemical sensors definitions and classification. Pure Appl. Chem. 63 (3), 1247–1250.
Hwang, Y., Paydar, O.H., Candler, R.N., 2015. 3D printed molds for nonplanar PDMS microfluidic channels. Sensor. Actuat. A 226, 137–142.
Iijima, S., 1991. Helical microtubules of graphitic carbon. Nature 354 (3), 56–58.
Ishihara, S., Ikeda, A., Citterio, D., Maruyama, K., Hagiwara, M., Suzuki, K., 2005. Smart chemical taste sensor for determination and prediction of taste qualities based on a two-phase optimized radial basis function network. Anal. Chem. 77 (3), 7908–7915.

Ivnitski, D., Abdel-Hamid, I., Atanasov, P., Wilkins, E., 1999. Biosensors for detection of pathogenic bacteria. Biosens. Bioelect. 14 (3), 599–624.
James, D., Scott, S.M., Ali, Z., O'Hare, W.T., 2005. Chemical sensors for electronic nose systems. Microchim. Acta 149 (3), 1–17.
Janshoff, A., Galla, H.J., Steinem, C., 2000. Piezoelectric mass-sensing devices as biosensors: an alternative to optical biosensors? Angew. Chem.Int. Ed. Engl. 39 (3), 4004–4032.
Jarup, L., 2003. Hazards of heavy metal contamination. Br. Med. Bull. 68, 167–182.
Kaden, H., 2009. The history of the glass electrode. Chem. Anal. 54 (3), 1089–1108.
Kandimalla, V.B., Ju, H.X., 2006. Binding of acetylcholinesterase to a multiwall carbon nanotube-cross-linked chitosan composite for flow-injection amperometric detection of an organophosphorous insecticide. Chem. Eur. J. 12 (3), 1074–1080.
Karaseva, N.A., Ermolaeva, T.N., 2012. A piezoelectric immunosensor for chloramphenicol detection in food. Talanta 93, 44–48.
Kersey, A.D., Davis, M.A., Patrick, H.J., LeBlanc, M., Koo, K.P., Askins, C.G., Putnam, M.A., Friebele, E.J., 1997. Fiber grating sensors. J. Lightw. Technol. 15 (3), 1442–1463.
Khan, S., Cao, Q., Zheng, Y.M., Huang, Y.Z., Zhu, Y.G., 2008. Health risks of heavy metals in contaminated soils and food crops irrigated with wastewater in Beijing, China. Environ. Pollut. 152 (3), 686–692.
Kickelbick, G., 2007. Hybrid Materials: synthesis, properties, and applications. Wiley-VCH Verlag GmbH & Co. KGaA, New York, NY.
Knecht, B.G., Strasser, A., Dietrich, R., Martlbauer, E., Niessner, R., Weller, M.G., 2004. Automated microarray system for the simultaneous detection of antibiotics in milk. Anal. Chem. 76 (3), 646–654.
Kosterev, A.A., Tittel, F.K., 2002. Chemical sensors based on quantum cascade lasers. IEEE J. Quantum Elect. 38 (3), 582–591.
Krajewska, B., 2004. Application of chitin- and chitosan-based materials for enzyme immobilizations: a review. Enzyme Microb. Tech. 35 (3), 126–139.
Kubo, Y., Maeda, S., Tokita, S., Kubo, M., 1996. Colorimetric chiral recognition by a molecular sensor. Nature 382 (3), 522–524.
Kusumaningrum, H.D., Riboldi, G., Hazeleger, W.C., Beumer, R.R., 2003. Survival of foodborne pathogens on stainless steel surfaces and cross-contamination to foods. Int. J. Food Microbiol. 85 (3), 227–236.
Kwon, J.Y., Jang, Y.J., Lee, Y.J., Kim, K.M., Seo, M.S., Nam, W., Yoon, J., 2005. A highly selective fluorescent chemosensor for Pb^{2+}. J. Am. Chem. Soc. 127 (3), 10107–10111.
Lago, M.A., Rodriguez-Bernaldo de Quiros, A., Sendon, R., Bustos, J., Nieto, M.T., Paseiro, P., 2015. Photoinitiators: a food safety review. Food Add. Contam. Part A 32 (3), 779–798.
Lang, H., Hegner, M., Gerber, C., 2010. Nanomechanical cantilever array sensors. In: Bhushan, B. (Ed.), Springer Handbook of Nanotechnology. Springer, Berlin, Heidelberg, pp. 427–452.
Lavrik, N.V., Sepaniak, M.J., Datskos, P.G., 2004. Cantilever transducers as a platform for chemical and biological sensors. Rev. Sci. Instrum. 75 (3), 2229–2253.
Lee, K.-S., Shiddiky, M.J.A., Park, S.-H., Park, D.-S., Shim, Y.-B., 2008. Electrophoretic analysis of food dyes using a miniaturized microfluidic system. Electrophoresis 29 (3), 1910–1917.
Li, T., Hu, W.P., 2011. Electrochemistry in nanoscopic volumes. Nanoscale 3 (3), 166–176.
Li, C., Thostenson, E.T., Chou, T.-W., 2008. Sensors and actuators based on carbon nanotubes and their composites: a review. Compos. Sci. Technol. 68 (3), 1227–1249.

Li, B.L., Luo, J.H., Luo, H.Q., Li, N.B., 2015. A novel conducting poly(*p*-aminobenzene sulphonic acid)-based electrochemical sensor for sensitive determination of Sudan I and its application for detection in food stuffs. Food Chem. 173, 594–599.
Liu, G., Chen, H.D., Peng, H.Z., Song, S.P., Gao, J.M., Lu, J.X., Ding, M., Li, L.Y., Ren, S.Z., Zou, Z.Y., Fan, C.H., 2011. A carbon nanotube-based high-sensitivity electrochemical immunosensor for rapid and portable detection of clenbuterol. Biosens. Bioelectron. 28 (3), 308–313.
Loock, H.-P., Wentzell, P.D., 2012. Detection limits of chemical sensors: applications and misapplications. Sensor. Actuat. B 173, 157–163.
Loutfi, A., Coradeschi, S., Mani, G.K., Shankar, P., Rayappan, J.B.B., 2015. Electronic noses for food quality: a review. J. Food Eng. 144, 103–111.
Lowinsohn, D., Bertotti, M., 2006. Electrochemical sensors: fundamentals and applications in microenvironments. Quim. Nova 29 (3), 1318–1325.
Luo, X.L., Morrin, A., Killard, A.J., Smyth, M.R., 2006. Application of nanoparticles in electrochemical sensors and biosensors. Electroanalysis 18 (3), 319–326.
Luo, X., Pan, J., Pan, K., Yu, Y., Zhong, A., Wei, S., Li, J., Shi, J., Li, X., 2015. An electrochemical sensor for hydrazine and nitrite based on graphene-cobalt hexacyanoferrate nanocomposite: toward environment and food detection. J. Electroanal. Chem. 745, 80–87.
Luong, J.H.T., Male, K.B., Glennon, J.D., 2008. Biosensor technology: technology push versus market pull. Biotechnol. Adv. 26 (3), 492–500.
Malhotra, B., Keshwani, A., Kharkwal, H., 2015. Antimicrobial food packaging: potential and pitfalls. Front. Microbiol. 6.
Malmauret, L., Parent-Massin, D., Hardy, J.L., Verger, P., 2002. Contaminants in organic and conventional foodstuffs in France. Food Addit. Contam. 19 (3), 524–532.
Mandenius, C.-F., Gustavsson, R., 2015. Mini-review: soft sensors as means for PAT in the manufacture of bio-therapeutics. J. Chem. Technol. Biotechnol. 90 (3), 215–227.
Manzoli, A., Steffens, C., Paschoalin, R.T., Correa, A.A., Alves, W.F., Leite, F.L., Herrmann, P.S.P., 2011. Low-cost gas sensors produced by the graphite line–patterning technique applied to monitoring banana ripeness. Sensors 11 (3), 6425–6434.
Manzoli, A., Shimizu, F.M., Mercante, L.A., Paris, E.C., Oliveira, O.N., Correa, D.S., Mattoso, L.H.C., 2014. Layer-by-layer fabrication of AgCl-PANI hybrid nanocomposite films for electronic tongues. Phys. Chem. Chem. Phys. 16 (3), 24275–24281.
Martinez, A.W., Phillips, S.T., Whitesides, G.M., Carrilho, E., 2010. Diagnostics for the developing world: microfluidic paper-based analytical devices. Anal. Chem. 82 (3), 3–10.
Matzeu, G., Florea, L., Diamond, D., 2015. Advances in wearable chemical sensor design for monitoring biological fluids. Sensor. Actuat. B 211, 403–418.
Mayer, K.M., Hafner, J.H., 2011. Localized surface plasmon resonance sensors. Chem. Rev. 111 (3), 3828–3857.
McDonald, J.C., Whitesides, G.M., 2002. Poly(dimethylsiloxane) as a material for fabricating microfluidic devices. Acc. Chem. Res. 35 (3), 491–499.
McLaughlin, M.J., Parker, D.R., Clarke, J.M., 1999. Metals and micronutrients: food safety issues. Field Crops Res. 60 (3), 143–163.
McQuade, D.T., Pullen, A.E., Swager, T.M., 2000. Conjugated polymer-based chemical sensors. Chem. Rev. 100 (3), 2537–2574.
Mercante, L.A., Pavinatto, A., Iwaki, L.E.O., Scagion, V.P., Zucolotto, V., Oliveira, O.N., Mattoso, L.H.C., Correa, D.S., 2015a. Electrospun polyamide 6/ poly(allylamine hydrochloride) nanofibers functionalized with carbon

nanotubes for electrochemical detection of dopamine. ACS Appl. Mater. Interf. 7 (3), 4784–4790.

Mercante, L.A., Scagion, V.P., Pavinatto, A., Sanfelice, R.C., Mattoso, L.H.C., Correa, D.S., 2015b. Electronic tongue based on nanostructured hybrid films of gold nanoparticles and phthalocyanines for milk analysis. J. Nanomater. 2015, 7, (Article ID 890637).

Merkoci, A., Pumera, M., Llopis, X., Perez, B., del Valle, M., Alegret, S., 2005. New materials for electrochemical sensing VI: carbon nanotubes. Trends Anal. Chem. 24 (3), 826–838.

Mignani, A.G., Baldini, F., 1997. Fibre-optic sensors in health care. Phys. Med. Biol. 42 (3), 967–979.

Mizan, M.F.R., Jahid, I.K., Ha, S.-D., 2015. Microbial biofilms in seafood: a food-hygiene challenge. Food Microbiol. 49, 41–55.

Mor, G.K., Varghese, O.K., Paulose, M., Grimes, C.A., 2003. A self-cleaning, room-temperature titania-nanotube hydrogen gas sensor. Sensor Lett. 1 (3), 42–46.

Morrison, S.R., 1987. Selectivity in semiconductor gas sensors. Sensor. Actuat. 12 (3), 425–440.

Moyo, M., Okonkwo, J.O., Agyei, N.M., 2012. Recent advances in polymeric materials used as electron mediators and immobilizing matrices in developing enzyme electrodes. Sensors. 12 (3), 923–953.

Muralt, P., 2000. Ferroelectric thin films for micro-sensors and actuators: a review. J. Micromech. Microeng. 10 (3), 136–146.

Nambiar, S., Yeow, J.T.W., 2011. Conductive polymer-based sensors for biomedical applications. Biosens. Bioelectron. 26 (3), 1825–1832.

Nasreddine, L., Parent-Massin, D., 2002. Food contamination by metals and pesticides in the European Union: should we worry? Toxicol. Lett. 127 (3), 29–41.

Neethirajan, S., Jayas, D.S., Sadistap, S., 2009. Carbon dioxide (CO_2) sensors for the agri-food industry: a review. Food Bioprocess Technol. 2 (3), 115–121.

Nicole, L., Laberty-Robert, C., Rozes, L., Sanchez, C., 2014. Hybrid materials science: a promised land for the integrative design of multifunctional materials. Nanoscale 6 (3), 6267–6292.

Novo, P., Moulas, G., Franca Prazeres, D.M., Chu, V., Conde, J.P., 2013. Detection of ochratoxin A in wine and beer by chemiluminescence-based ELISA in microfluidics with integrated photodiodes. Sensor. Actuat. B 176, 232–240.

Novoselov, K.S., Geim, A.K., Morozov, S.V., Jiang, D., Zhang, Y., Dubonos, S.V., Grigorieva, I.V., Firsov, A.A., 2004. Electric field effect in atomically thin carbon films. Science 306 (3), 666–669.

Okuma, H., Takahashi, H., Yazawa, S., Sekimukai, S., Watanabe, E., 1992. Development of a system with double enzyme reactors for the determination of fish freshness. Anal. Chim. Acta 260 (3), 93–98.

Olaimat, A.N., Holley, R.A., 2012. Factors influencing the microbial safety of fresh produce: a review. Food Microbiol. 32 (3), 1–19.

Oliveira, J.E., Scagion, V.P., Grassi, V., Correa, D.S., Mattoso, L.H.C., 2012. Modification of electrospun nylon nanofibers using layer-by-layer films for application in flow injection electronic tongue: detection of paraoxon pesticide in corn crop. Sensor. Actuat. B 171, 249–255.

Oliveira, J.E., Grassi, V., Scagion, V.P., Mattoso, L.H.C., Glenn, G.M., Medeiros, E.S., 2013. Sensor array for water analysis based on interdigitated electrodes modified with fiber films of poly(lactic acid)/multiwalled carbon nanotubes. IEEE Sensor. J. 13 (3), 759–766.

Oliver, S.P., Jayarao, B.M., Almeida, R.A., 2005. Foodborne pathogens in milk and the dairy farm environment: food safety and public health implications. Foodborne Pathog. Dis. 2 (3), 115–129.

O'Sullivan, C.K., Guilbault, G.G., 1999. Commercial quartz crystal microbalances: theory and applications. Biosens. Bioelectron. 14 (3), 663–670.
Otto, M., Thomas, J.D.R., 1985. Model studies on multiple channel analysis of free magnesium, calcium, sodium, and potassium at physiological concentration levels with ion-selective electrodes. Anal. Chem. 57 (3), 2647–2651.
Painter, J.A., Hoekstra, R.M., Ayers, T., Tauxe, R.V., Braden, C.R., Angulo, F.J., Griffin, P.M., 2013. Attribution of foodborne illnesses, hospitalizations, and deaths to food commodities by using outbreak data, United States, 1998–2008. Emerg. Infect. Dis. 19 (3), 407–415.
Palchetti, I., Mascini, M., 2008. Electroanalytical biosensors and their potential for food pathogen and toxin detection. Anal. Bioanal. Chem. 391 (3), 455–471.
Panigrahi, S., Balasubramanian, S., Gu, H., Logue, C., Marchello, M., 2006. Neural-network-integrated electronic nose system for identification of spoiled beef. LWT Food Sci. Technol. 39 (3), 135–145.
Pauliukaite, R., Zhylyak, G., Citterio, D., Spichiger-Keller, U.E., 2006. L-glutamate biosensor for estimation of the taste of tomato specimens. Anal. Bioanal. Chem. 386 (3), 220–227.
Perez-Lopez, B., Merkoci, A., 2011. Nanomaterials based biosensors for food analysis applications. Trends Food Sci. Technol. 22 (3), 625–639.
Peris, M., Escuder-Gilabert, L., 2009. A 21st-century technique for food control: electronic noses. Anal. Chim. Acta 638 (3), 1–15.
Peris, M., Escuder-Gilabert, L., 2013. On-line monitoring of food fermentation processes using electronic noses and electronic tongues: a review. Anal. Chim. Acta 804, 29–36.
Persaud, K.C., 2005. Polymers for chemical sensing. Mater. Today 8 (3), 38–44.
Piggott, N.E., Marsh, T.L., 2004. Does food safety information impact US meat demand? Am. J. Agric. Econ. 86 (3), 154–174.
Pina, F., Bernardo, M.A., Garcia-Espana, E., 2000. Fluorescent chemosensors containing polyamine receptors. Eur. J. Inorg. Chem. (10), 2143–2157.
Ponmozhi, J., Frias, C., Marques, T., Frazao, O., 2012. Smart sensors/actuators for biomedical applications: review. Measurement 45 (3), 1675–1688.
Prakash, S., Chakrabarty, T., Singh, A.K., Shahi, V.K., 2013. Polymer thin films embedded with metal nanoparticles for electrochemical biosensors applications. Biosens. Bioelectron. 41, 43–53.
Ragazzo-Sanchez, J.A., Chalier, P., Chevalier, D., Ghommidh, C., 2006. Electronic nose discrimination of aroma compounds in alcoholized solutions. Sensor. Actuat. B 114 (3), 665–673.
Ramstrom, O., Skudar, K., Haines, J., Patel, P., Bruggemann, O., 2001. Food analyses using molecularly imprinted polymers. J. Agric. Food Chem. 49 (3), 2105–2114.
Rao, C.N.R., Satishkumar, B.C., Govindaraj, A., Nath, M., 2001. Nanotubes. Chemphyschem 2 (3), 78–105.
Rasheed, Z., Vikraman, A.E., Thomas, D., Jagan, J.S., Kumar, K.G., 2015. Carbon-nanotube-based sensor for the determination of butylated hydroxyanisole in food samples. Food Anal. Method. 8 (3), 213–221.
Rassaei, L., Amiri, M., Cirtiu, C.M., Sillanpaa, M., Marken, F., Sillanpaa, M., 2011. Nanoparticles in electrochemical sensors for environmental monitoring. Trends Anal. Chem. 30 (3), 1704–1715.
Redmond, E.C., Griffith, C.J., 2003. Consumer food handling in the home: a review of food safety studies. J. Food Protect. 66 (3), 130–161.
Ren, W.C., Cheng, H.M., 2014. The global growth of graphene. Nat. Nanotechnol. 9 (3), 726–730.
Ricciardi, C., Canavese, G., Castagna, R., Digregorio, G., Ferrante, I., Marasso, S.L., Ricci, A., Alessandria, V., Rantsiou, K., Cocolin, L.S., 2010. Online portable

microcantilever biosensors for salmonella enterica serotype Enteritidis detection. Food Bioprocess Technol. 3 (3), 956–960.
Riul, A., Dos Santos, D.S., Wohnrath, K., Di Tommazo, R., Carvalho, A., Fonseca, F.J., Oliveira, O.N., Taylor, D.M., Mattoso, L.H.C., 2002. Artificial taste sensor: efficient combination of sensors made from Langmuir-Blodgett films of conducting polymers and a ruthenium complex and self-assembled films of an azobenzene-containing polymer. Langmuir 18 (3), 239–245.
Riul, A., Malmegrim, R.R., Fonseca, F.J., Mattoso, L.H.C., 2003. An artificial taste sensor based on conducting polymers. Biosens. Bioelectron. 18 (3), 1365–1369.
Riul, A., Dantas, C.A.R., Miyazaki, C.M., Oliveira, O.N., 2010. Recent advances in electronic tongues. Analyst 135 (3), 2481–2495.
Ronkainen, N.J., Halsall, H.B., Heineman, W.R., 2010. Electrochemical biosensors. Chem. Soc. Rev. 39 (3), 1747–1763.
Rudnitskaya, A., Polshin, E., Kirsanov, D., Lammertyn, J., Nicolai, B., Saison, D., Delvaux, F.R., Delvaux, F., Legin, A., 2009a. Instrumental measurement of beer taste attributes using an electronic tongue. Anal. Chim. Acta. 646 (3), 111–118.
Rudnitskaya, A., Schmidtke, L.M., Delgadillo, I., Legin, A., Scollary, G., 2009b. Study of the influence of micro-oxygenation and oak chip maceration on wine composition using an electronic tongue and chemical analysis. Anal. Chim. Acta. 642 (3), 235–245.
Ruiz-Altisent, M., Ruiz-Garcia, L., Moreda, G.P., Lu, R., Hernandez-Sanchez, N., Correa, E.C., Diezma, B., Nicolai, B., Garcia-Ramos, J., 2010. Sensors for product characterization and quality of specialty crops: a review. Comput. Electron. Agric. 74 (3), 176–194.
Sackmann, E.K., Fulton, A.L., Beebe, D.J., 2014. The present and future role of microfluidics in biomedical research. Nature 507 (3), 181–189.
Safavieh, M., Ahmed, M.U., Sokullu, E., Ng, A., Braescuac, L., Zourob, M., 2014. A simple cassette as point-of-care diagnostic device for naked-eye colorimetric bacteria detection. Analyst 139 (3), 482–487.
Saha, K., Agasti, S.S., Kim, C., Li, X.N., Rotello, V.M., 2012. Gold nanoparticles in chemical and biological sensing. Chem. Rev. 112 (3), 2739–2779.
Sanchez, C., Belleville, P., Popall, M., Nicole, L., 2011. Applications of advanced hybrid organic-inorganic nanomaterials: from laboratory to market. Chem. Soc. Rev. 40 (3), 696–753.
Santos, W.J.R., Lima, P.R., Tanaka, A.A., Tanaka, S.M.C.N., Kubota, L.T., 2009. Determination of nitrite in food samples by anodic voltammetry using a modified electrode. Food Chem. 113 (3), 1206–1211.
Sberveglieri, V., Carmona, E.N., Comini, E., Ponzoni, A., Zappa, D., Pirrotta, O., Pulvirenti, A., 2014. A novel electronic nose as adaptable device to judge microbiological quality and safety in foodstuff. Biomed. Res. Int. 2014, 529519.
Scampicchio, M., Benedetti, S., Brunetti, B., Mannino, S., 2006. Amperometric electronic tongue for the evaluation of the tea astringency. Electroanalysis 18 (3), 1643–1648.
Schecter, A., Papke, O., Tung, K.C., Staskal, D., Birnbaum, L., 2004. Polybrominated diphenyl ethers contamination of United States food. Environ. Sci. Technol. 38 (3), 5306–5311.
Scholz, F., 2002. Electroanalytical Methods: Guide to Experiments and Applications. Springer, Berlin, New York.
Scognarniglio, V., Arduini, F., Palleschi, G., Rea, G., 2014. Biosensing technology for sustainable food safety. Trends Anal. Chem. 62, 1–10.
Segev-Bar, M., Haick, H., 2013. Flexible Sensors Based on Nanoparticles. ACS Nano 7 (3), 8366–8378.
Shalaby, A.R., 1996. Significance of biogenic amines to food safety and human health. Food Res. Int. 29 (3), 675–690.

Shao, W., Tian, X., 2015. Adaptive soft sensor for quality prediction of chemical processes based on selective ensemble of local partial least squares models. Chem. Eng. Res. Des. 95, 113–132.
Shephard, G.S., Thiel, P.G., Stockenstrom, S., Sydenham, E.W., 1996. Worldwide survey of fumonisin contamination of corn and corn-based products. J. AOAC Int. 79 (3), 671–687.
Shrout, T.R., Zhang, S.J., 2007. Lead-free piezoelectric ceramics: alternatives for PZT? J. Electrocer. 19 (3), 113–126.
Siqueira, J.R., Abouzar, M.H., Backer, M., Zucolotto, V., Poghossian, A., Oliveira, O.N., Schoning, M.J., 2009. Carbon nanotubes in nanostructured films: potential application as amperometric and potentiometric field-effect (bio-) chemical sensors. Phys. Status Solidi A 206 (3), 462–467.
Sivapalasingam, S., Friedman, C.R., Cohen, L., Tauxe, R.V., 2004. Fresh produce: a growing cause of outbreaks of foodborne illness in the United States, 1973 through 1997. J. Food Protect. 67 (3), 2342–2353.
Skotheim, T.A., Reynolds, J.R., 2006. Handbook of Conducting Polymers, Conjugated Polymers: Theory, Synthesis, Properties and Characterization. CRC Press, New York.
Sliwinska, M., Wisniewska, P., Dymerski, T., Namiesnik, J., Wardencki, W., 2014. Food analysis using artificial senses. J. Agric. Food Chem. 62 (3), 1423–1448.
Someya, T., Kato, Y., Sekitani, T., Iba, S., Noguchi, Y., Murase, Y., Kawaguchi, H., Sakurai, T., 2005. Conformable, flexible, large-area networks of pressure and thermal sensors with organic transistor active matrixes. Proceed. Natl. Acad. Sci. USA 102 (3), 12321–12325.
Song, E., Choi, J.W., 2015. Multi-analyte detection of chemical species using a conducting polymer nanowire-based sensor array platform. Sensor. Actuat. B 215, 99–106.
Song, X.L., Xu, S.F., Chen, L.X., Wei, Y.Q., Xiong, H., 2014. Recent advances in molecularly imprinted polymers in food analysis. J. Appl. Polym. Sci. 131 (16).
Steffens, C., Franceschi, E., Corazza, F.C., Herrmann, P.S.P., Oliveira, J.V., 2010. Gas sensors development using supercritical fluid technology to detect the ripeness of bananas. J. Food Eng. 101 (3), 365–369.
Stewart, M.E., Anderton, C.R., Thompson, L.B., Maria, J., Gray, S.K., Rogers, J.A., Nuzzo, R.G., 2008. Nanostructured plasmonic sensors. Chem. Rev. 108 (3), 494–521.
Su, Z.Q., Ding, J.W., Wei, G., 2014. Electrospinning: a facile technique for fabricating polymeric nanofibers doped with carbon nanotubes and metallic nanoparticles for sensor applications. RSC Adv. 4 (3), 52598–52610.
Sundramoorthy, A.K., Gunasekaran, S., 2014. Applications of graphene in quality assurance and safety of food. Trends Anal. Chem. 60, 36–53.
S'vorc, L., Tomcik, P., Svitkova, J., Rievaj, M., Bustin, D., 2012. Voltammetric determination of caffeine in beverage samples on bare boron-doped diamond electrode. Food Chem. 135 (3), 1198–1204.
Taylor, D.M., Macdonald, A.G., 1987. Ac admittance of the metal-insulator-electrolyte interface. J. Phys. D 20 (3), 1277–1283.
Tiemann, M., 2007. Porous metal oxides as gas sensors. Chem. Eur. J. 13 (3), 8376–8388.
Torres, C.M., Pico, Y., Manes, J., 1996. Determination of pesticide residues in fruit and vegetables. J. Chromatogr. A 754 (3), 301–331.
Trojanowicz, M., Hitchman, M.L., 1996. Determination of pesticides using electrochemical biosensors. Trends Anal. Chem. 15 (3), 38–45.
Valeur, B., Leray, I., 2000. Design principles of fluorescent molecular sensors for cation recognition. Coord. Chem. Rev. 205, 3–40.
van den Bogaard, A.E., Stobberingh, E.E., 1999. Antibiotic usage in animals: impact on bacterial resistance and public health. Drugs 58 (3), 589–607.

van Lenteren, J.C., 2000. A greenhouse without pesticides: fact or fantasy? Crop Prot. 19 (3), 375–384.
Van Tassel, P.R., 2012. Polyelectrolyte adsorption and layer-by-layer assembly: electrochemical control. Curr. Opin. Colloid Interface Sci. 17 (3), 106–113.
Varshney, M., Li, Y., Srinivasan, B., Tung, S., 2007. A label-free, microfluidics and interdigitated array microelectrode-based impedance biosensor in combination with nanoparticles immunoseparation for detection of *Escherichia coli* O157 : H7 in food samples. Sensor. Actuat. B 128 (3), 99–107.
Verpoorte, E.M.J., Vanderschoot, B.H., Jeanneret, S., Manz, A., Widmer, H.M., Derooij, N.F., 1994. 3-dimensional micro-flow manifolds for miniaturized chemical-analysis systems. J. Micromech. Microeng. 4 (3), 246–256.
Vitas, A.I., Aguado, V., Garcia-Jalon, I., 2004. Occurrence of Listeria monocytogenes in fresh and processed foods in Navarra (Spain). Int. J. Food Microbiol. 90 (3), 349–356.
Vlasov, Y., Legin, A., Rudnitskaya, A., Di Natale, C., D'Amico, A., 2005. Nonspecific sensor arrays ("electronic tongue") for chemical analysis of liquids (IUPAC Technical Report). Pure Appl. Chem. 77 (3), 1965–1983.
Vlasov, Y.G., Ermolenko, Y.E., Legin, A.V., Rudnitskaya, A.M., Kolodnikov, V.V., 2010. Chemical sensors and their systems. J. Anal. Chem. 65 (3), 880–898.
von der Kammer, F., Ferguson, P.L., Holden, P.A., Masion, A., Rogers, K.R., Klaine, S.J., Koelmans, A.A., Horne, N., Unrine, J.M., 2012. Analysis of engineered nanomaterials in complex matrices (environment and biota): General considerations and conceptual case studies. Environ. Toxicol. Chem. 31 (3), 32–49.
Vonau, W., Guth, U., 2006. pH monitoring: a review. J. Solid State Electrochem. 10 (3), 746–752.
Waggoner, P.S., Craighead, H.G., 2007. Micro- and nanomechanical sensors for environmental, chemical, and biological detection. Lab Chip 7 (3), 1238–1255.
Wang, J., 2000. Analytical Electrochemistry. John Wiley & Sons, New York.
Wang, J., 2005a. Carbon-nanotube based electrochemical biosensors: a review. Electroanalysis 17 (3), 7–14.
Wang, J., 2005b. Nanomaterial-based amplified transduction of biomolecular interactions. Small 1 (3), 1036–1043.
Wang, J., Lin, Y.H., 2008. Functionalized carbon nanotubes and nanofibers for biosensing applications. Trends Anal. Chem. 27 (3), 619–626.
Wang, Q., Moser, J.E., Gratzel, M., 2005a. Electrochemical impedance spectroscopic analysis of dye-sensitized solar cells. J. Phys. Chem. B 109 (3), 14945–14953.
Wang, X.L., Sato, T., Xing, B.S., Tao, S., 2005b. Health risks of heavy metals to the general public in Tianjin, China via consumption of vegetables and fish. Sci. Total Environ. 350 (3), 28–37.
Wang, N., Zhang, N.Q., Wang, M.H., 2006a. Wireless sensors in agriculture and food industry: recent development and future perspective. Comput. Electron. Agric. 50 (3), 1–14.
Wang, X., Zhou, J., Song, J., Liu, J., Xu, N., Wang, Z.L., 2006b. Piezoelectric field effect transistor and nanoforce sensor based on a single ZnO nanowire. Nano Lett. 6 (3), 2768–2772.
Wang, C., Yin, L., Zhang, L., Xiang, D., Gao, R., 2010. Metal oxide gas sensors: sensitivity and influencing factors. Sensors 10 (3), 2088–2106.
Wei, Z.B., Wang, J., Zhang, X., 2013. Monitoring of quality and storage time of unsealed pasteurized milk by voltammetric electronic tongue. Electrochim. Acta 88, 231–239.
Whitener, K.E., Sheehan, P.E., 2014. Graphene synthesis. Diamond Relat. Mater. 46, 25–34.

Whitesides, G.M., 2006. The origins and the future of microfluidics. Nature 442 (3), 368–373.
Wilson, D.M., Hoyt, S., Janata, J., Booksh, K., Obando, L., 2001. Chemical sensors: portable handheld field instrument. IEEE Sensors J. 1 (3), 256–274.
Winter, C.K., Davis, S.F., 2006. Organic foods. J. Food Sci. 71 (9), R117–R124.
Wittenberg, N.J., Haynes, C.L., 2009. Using nanoparticles to push the limits of detection. Nanomed. Nanobiotechnol. 1 (3), 237–254.
Wohltjen, H., Snow, A.W., 1998. Colloidal metal-insulator-metal ensemble chemiresistor sensor. Anal. Chem. 70 (3), 2856–2859.
Wright, J.S., Torres, R.D., Peters, B., Hope, D.T., Tovo, L.L., 2015. In-line chemical sensor deployment in a tritum plant. Fusion Sci. Technol. 67 (3), 639–642.
Xie, Y.F., Li, Y., Niu, L., Wang, H.Y., Qian, H., Yao, W.R., 2012. A novel surface-enhanced Raman scattering sensor to detect prohibited colorants in food by graphene/silver nanocomposite. Talanta 100, 32–37.
Yang, G.J., Huang, J.L., Meng, W.J., Shen, M., Jiao, X.A., 2009. A reusable capacitive immunosensor for detection of *Salmonella* spp. based on grafted ethylene diamine and self-assembled gold nanoparticle monolayers. Anal. Chim. Acta 647 (3), 159–166.
Yoon, J., Cao, X., Zhou, Q., Ma, L.Q., 2006. Accumulation of Pb, Cu, and Zn in native plants growing on a contaminated Florida site. Sci. Total Environ. 368 (3), 456–464.
Zakaria, A., Shakaff, A.Y.M., Masnan, M.J., Ahmad, M.N., Adom, A.H., Jaafar, M.N., Ghani, S.A., Abdullah, A.H., Aziz, A.H.A., Kamarudin, L.M., Subari, N., Fikri, N.A., 2011. A biomimetic sensor for the classification of honeys of different floral origin and the detection of adulteration. Sensors 11 (3), 7799–7822.
Zampa, M.F., de Brito, A.C.F., Kitagawa, I.L., Constantino, C.J.L., Oliveira, O.N., da Cunha, H.N., Zucolotto, V., dos Santos, J.R., Eiras, C., 2007. Natural gum-assisted phthalocyanine immobilization in electroactive nanocomposites: physicochemical characterization and sensing applications. Biomacromolecules 8 (3), 3408–3413.
Zaragoza, P., Fuentes, A., Ruiz-Rico, M., Vivancos, J.-L., Fernandez-Segovia, I., Ros-Lis, J.V., Barat, J.M., Martinez-Manez, R., 2015. Development of a colorimetric sensor array for squid spoilage assessment. Food Chem. 175, 315–321.
Zhang, C.L., Yu, S.H., 2014. Nanoparticles meet electrospinning: recent advances and future prospects. Chem. Soc. Rev. 43 (3), 4423–4448.
Zhang, S., Xie, C., Bai, Z., Hu, M., Li, H., Zeng, D., 2009. Spoiling and formaldehyde-containing detections in octopus with an e-nose. Food Chem. 113 (3), 1346–1350.

7

PREVENTION OF FOOD SPOILAGE USING NANOSCALE SENSORS

Krystian Mistewicz, Marian Nowak
Silesian University of Technology, Institute of Physics, Center for Science and Education, Katowice, Poland

1 Introduction

The prevention of food spoilage is one of the key tasks in the food industry, because it minimalizes a risk to consumers' health, and reduces economic losses experienced by manufacturers and retailers, as well as, wastage of food, which affects the world's food supply (Adley, 2014). It has crucial significance for improvement of the agricultural and food products shelf life and increases safety of manufacturing, processing, and shipping. Recently, we have witnessed significant interest in nanotechnology development in agriculture and food industry, including particularly food spoilage prevention. The counteraction of growth of undesirable bacteria or pathogens can be performed by use of some additives, for example, nanomaterials with antimicrobial properties (Sondi and Salopek-Sondi, 2004), as well as, by nanocoatings for hygienic food contact surfaces (Cushen et al., 2012). Simultaneously, nanoscale sensors are used to determine quality and freshness of food. The real-time monitoring of humidity, excess gases, amines, and microorganisms in production, packaging, and storage processes allows producers to avoid the undesired components and protect against unwanted physical and chemical food changes. It should be emphasized that nanomaterials can be used as reactive particles incorporated into packaging in a form of nanosensors that provide quality and security control of products (Kuswandi et al., 2011).

In this paper we present shortly the advances in nanobiosensors (section 2) and gas nanosensors (section 3) that are applicable in the food industry. Owing to the limited volume of this paper, many application of carbon nanostructures (Terranova et al., 2012) as nanobiosensors and gas nanosensors could not be presented.

Nanobiosensors. http://dx.doi.org/10.1016/B978-0-12-804301-1.00007-2

The special emphasis in this paper is placed on humidity nanosensors constructed from antimony sulfoiodide (SbSI) gel and SbSI single nanowires (section 4). SbSI is a semiconducting ferroelectrics. Properties of this material in the form of nanostructures with high surface-to-volume ratio make its electrical responses extremely sensitive to species adsorbed on the surface (Nowak, 2010). The surface area of the sonochemically prepared SbSI gel was estimated as about 75 m^2/g (Nowak et al., 2008). The indirect allowed energy band gap of SbSI nanomaterial equals $E_{gIa}(T = 283\ K) = 1.77(1)$ eV and $E_{gIa}(T = 303\ K) = 1.757(6)$ eV (Nowak et al., 2013). The Curie temperature $T_C = 293.0(2)$ K of SbSI nanomaterial was determined (Nowak et al., 2014a). Such value of the Currie temperature is characteristic for SbSI single crystals (Sasaki, 1964; Ishikawa, 1980; Toyoda, 1986; Palaniappan et al., 1987). However, the determined maximum value of dielectric constant $\varepsilon(T_c = 292(1)\ K) = 1.6 \times 10^4$ reported for SbSI gel (Szperlich et al., 2009) is much lower than the $\varepsilon(T_c = 291\ K) = 6.2 \times 10^4$ measured along the polar axis of the best SbSI crystals (Ishikawa et al., 1984). Also the calculated Curie constant $C_W = 1.41(14) \times 10^4$ K of SbSI xerogel is much lower than the value of 2.8×10^4 K reported for SbSI single crystals (Ishikawa et al., 1984). Taking it into consideration, one should remember that SbSI is a highly anisotropic material (Nowak, 2010). The permittivity of SbSI single crystal of the polar direction (along *c* axis) is approximately 2000 larger than for the perpendicular direction (Merz and Nitsche, 1964). It should be also emphasized that SbSI xerogel has very large porosity so the determined average value of dielectric constant does not represent the exact value for SbSI nanowires.

In vitro cytotoxicity studies have shown that L-ascorbic acid treated fraction of SbSI demonstrated significant toxicity (Gladkovskaya et al., 2015). L-Ascorbic acid increases SbSI amorphia and promotes surface defects, hence it creates good conditions for the byproducts trapping and contaminating with the residing toxicions (S^{2-}, Cl^-, Sb^{3+}). The tested particles showed fairly high ability to kill quickly expanding cells, and therefore can be used in antitumor treatments (Gladkovskaya et al., 2015).

Different phenomena can be applied for the gas sensing. Development of new type of nanosensors is one of the strategic branch of nanotechnology contribution to food industry, since demand for easy to apply, portable devices for use by nonspecialists is still increasing. Therefore, the conductive (Nowak et al., 2014b), photoconductive (Nowak et al., 2014a; Mistewicz et al., 2014), capacitive (Nowak et al., 2014b), and impedance (Starczewska et al., 2012) humidity nanosensors constructed from SbSI gel and SbSI single nanowires are reviewed in sections 4.1, 4.2, 4.3, and 4.4, respectively. SbSI nanosensors are competitive with the standard analytical techniques in terms of cost, performance, sensitivity, reliability, and power consumption.

2 Nanobiosensors

Microbial management during the food processing operations is crucial for preventing contamination and for improving the product safety, quality and production hygiene (Adley, 2014). In the past microbiological testing was relied on traditional growth based methods that used nutrient media. The time required to get results using these techniques is long and forward processing decisions and confirming manufacturing processes are static; results that may take days are now deemed to be inadequate. Therefore, methods that are fast, inexpensive and require little expertise or training to correctly interpret are highly desirable (Duncan, 2011).

Currently, the attention of scientific community has been focused on nanobiosensors with potential to be used for detection of toxins and pathogens rapidly and efficiently. Ko and coworkers have developed a colorimetric nanobiosensor based on the chromogenic effect of latex microspheres hybridization with gold nanoparticles (Ko et al., 2010). Au nanoparticles were chosen as signal indicators, because of their ability to attach to various functional ligands, and form their assemblies and complexes with various conjugates. Au nanoparticles functionalized with antibody attachable molecules have the potential for use in indicator-analyte displacement reaction. Therefore, the indicator-analyte displacement reaction can be applied for the detection of toxins. In Ko et al. (2010) a toxin analog, 2,4-dinitrophenol–bovine serum albumin (DNP–BSA) was attached to gold nanoparticles. It was allowed to hybridize the DNP–BSA with an anti-DNP antibody on latex microspheres. Due to this, latex microspheres had changed their color to pinkish-red (Fig. 7.1). The bright red color corresponds to case of no contamination, and white color indicates that the contamination threshold was exceeded. Not only the detection but also the quantification of DNP–glycine (GLY) toxin was done observing a competition between the analog-conjugated-gold nanoparticles and the toxin molecules. A visible color change from pinkish-red to white was observed as an effect of gold nanoparticles displacement from host latex microspheres in the presence of GLY molecules (Ko et al., 2010). The rapid, straightforward, and inexpensive DNP–GLY detection method can be probably applied to a variety of material compositions and particle sizes. However, future experiments are needed to extend this

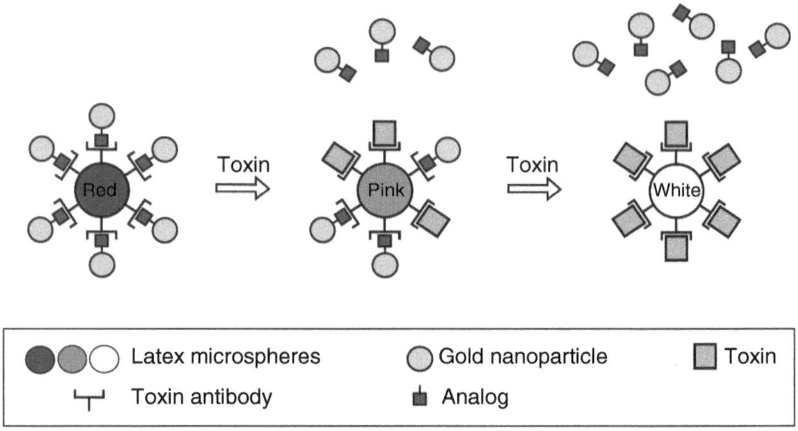

Figure 7.1. Indicator-analyte displacement between analog-conjugated gold nanoparticles and toxin molecules.

novel and promising technique to detect various toxins, allergens, pathogens, and other contaminants.

The immunosensor logic gates converting binding events between an antigen and an antibody into a measurable electrical signal were studied in (Liu et al., 2007). The polyaniline nanowires were used as the transducer for detection of *B. cereus* and *Escherichia coli* pathogens. According to (Liu et al., 2007), conductive polyaniline nanowire based immunosensors are relatively inexpensive to fabricate and easy to operate, which makes them an ideal candidate for multiarray biosensor architecture that can achieve reliable detection of pathogens. Liu and coworkers presented fabrication, characterization, and modeling of fundamental building blocks that implemented forward error-correction (FEC) on the polyaniline-based immunosensor. These building blocks included specialized AND and OR logic gates that operated based on specific binding between antigen and antibodies. The multiarray forward error-correcting biosensor was composed of four different pads: sample application, conjugate, capture, and absorption pads. The operation principle of this biosensor can be described as follows. Initially, the gap between the electrodes in the capture pad is open. After the sample is applied to the application pad, the solution containing the antigen flows to the conjugate pad, dissolves with the polyaniline-labeled antibody (Ab-P) and forms an antigen–antibody–polyaniline complex. Due to capillary action the complex is transported into the capture pad containing the immobilized antibodies. A structure similar to sandwich is formed, when second antibody–antigen reaction occurs. The polyaniline in the sandwich forms a molecular wire and bridges the two electrodes. The polymer structures extend out to bridge adjacent cells. It leads to an impedance variation, which can be sensed as an electric current change across the electrodes.

Etefagh and coworkers observed the considerable changes in the electrical resistance of the biosensor constructed from copper oxide nanoparticles and nanolayers, in the presence of *Aspergillus niger* fungi (Etefagh et al., 2013). CuO nanoparticles were synthesized by the sol-gel method. CuO nanolayers were deposited on glass substrates using a spray pyrolysis technique. It is well known, the activation of fungi in bread is accompanied by the production of some gaseous products, such as carbon dioxide. Therefore, the food decay in early stage can be recognized by detection of these gases. In order to eliminate undesired influence of moisture and carbon monoxide on electrical properties of investigated biosensor, Etefagh and coworkers used silica gel and calcium carbonate as adsorbers.

According to Haes et al. (2004) the platforms, based on the tunability of the localized surface plasmon resonance (LSPR) of Au and Ag nanoparticles, can be applied as a signal transduction mechanism for biosensing applications in solution. A rapid test for biomolecular reactivity was performed by measure of change in optimal scattering wavelength induced by interparticle coupling as the signal transduction method. For this experiment 4 mL of antihuman IgG/60 nm gold particles and 4 mL of human IgG/60 nm gold particles were mixed in a glass vial under stirring. Illumination of the sample vial was done by using a white light beam. The scattered light was recorded with a digital camera. The complementary proteins bound to each other causing the gold nanoparticles to be brought into closer proximity. The electromagnetic coupling of Au particles resulted in noticeable red shift of the LSPR. Due to nanoparticle aggregation color of the scattered light changed from green to yellow to orange in the space of about 15 min.

The innovative nanocantilevers-based devices were considered as biosensors (Sozer and Kokini, 2008) for detection of contaminant chemicals, toxins, and antibiotic residues in food products. The sensing principle is based on their ability to detect biological-binding interactions, such as between antigen and antibody, enzyme and substrate or cofactor and receptor and ligand, through physical and/or electromechanical signaling. For instance, pathogen detection relies on their ability to vibrate at various frequencies in dependence on the biomass of the pathogenic organisms. In Gfeller et al. (2005) change in resonance frequency as a function of the increasing mass on a cantilever array was proposed for detection of *E. coli* cells growth. *E. coli* cells were deposited on top of agarose coated cantilever. At first, the thin nutritive layer (about 200 nm thick) stayed in equilibrium with the humid environment. The bacterial cells growth led to assimilation of water, protein, salts, and carbohydrates out of the nutritive layer. It absorbed water to recover the equilibrium with the humid environment. This compensation influenced of additional mass load onto the cantilever.

One of the recent advances in food technology is development of kinetically programmable, and cost-efficient time-temperature indicator (TTI) protocol constructed from plasmonic nanocrystals (Zhang et al., 2013). It has been tailored for advanced quality control as well as for assessment of perishable products. Properties of the plasmonic nanocrystals are highly dependent on the temperature history during the custody chain (Zhang et al., 2013). This nontoxic, ecofriendly, and substantially low-cost plasmonic time-temperature indicator gives a sharp contrast red–green color change (Zhang et al., 2013). The color change is widely tunable by simply varying the concentrations of the relevant reagents (Zhang et al., 2013). It is another advantage of this nanomaterial, guaranteeing a largely flexible programmability.

Food-allergen management is perceived as one of the most important public health issues related to food safety. According to (Pilolli et al., 2013), even a little intake of allergen can trigger unpredictable, highly variable reactions, depending on the dose and the sensitivity of affected individual, thus compelling the allergic consumer to avoid allergen-containing food totally. The biosensors for food-allergen management can be classified into three groups related with transduction-based method: optical, electrochemical, and electromechanical. The operation principles of most optical biosensors are based upon measurement of changes in the surface properties of a sensor chip when the analyte is bound to a sensing layer on the device by sorption or complex formation of the receptor-target couple. A wide range of possible different chemical and physical phenomena (Table 7.1) can be applied for optical food allergens detection, including absorption, fluorescence, and surface-plasmon resonance (SPR). Electrochemical biosensors comprise following types: potentiometric (zero-current devices that measure the potential across an interface), impedance (small sinusoidal voltage is applied at a particular frequency and the resulting alternating current is measured), amperometric, and voltammetric (both based on the measurement of current as a function of applied electrode-solution voltage). The group of electromechanical biosensors consists mainly of quartz-crystal microbalances (QCMs) and micro-electromechanical systems (MEMS). Table 7.2 presents detailed summary of biosensors for

Table 7.1 Optical Biosensors for Detection of Food Allergens

Transduction	Analyte	Receptor	Assay	Matrixes	Run time	Nanomaterial	LOD	References
SPR (direct label-free detection)	α-lactalbumin	Polyclonal-Ab	Direct	Milk, colostrum, whey protein concentrates, infant formulae		None	0.94 ng/mL	Indyk (2009)
SPR (direct label-free detection)	Bovine Ig G caprine IgG	Polyclonal-Ab	Direct	Milk	<4 min	None	2.6 ng/mL 3.1 ng/mL	Crosson et al. (2010)
i-SPR (direct label-free detection)	Peanut, milk, lupine, soy, egg, hazelnut, almond, cashew nut, brazil nut, pine nut, pecan, macadamia nut, pistachio nut	Monoclonal and polyclonal Ab	Direct	Allergenic food cookies dark chocolate	8 min	None	low mg/kg range for cookies and chocolates	Rebe Raz et al. (2010)
SPR (direct label-free detection)	α-lactalbumin β-lactoglobulin BSA Lactoferrin IgG	Affinity-purified polyclonal-Ab	Direct	Raw and processed milk whey fractions milk-derived product	NA	None	NA	Billakanti et al. (2010)
Optical fiber SPR (direct label-free detection)	Ara h1	Polyclonal Ab	Sandwich	Chocolate	20 min	Magnetic nanobead	0.09 μg/mL	Pollet et al. (2011)
IMS-SERS	Ovalbumin	Polyclonal Ab	Direct	Whole milk	20 min	Magnetic microbeads + silver dendrites	1 μg/mL	He et al. (2011)
Colorimetric (indirect detection, enzyme label)	Lectin, Ara h3, Gliadin, Ana o3, Mitochondrion DNA, 16S rRNA	DNA	Sandwich	Soybean, cashew, peanut wheat, beef, chicken, fish, shrimp, cereal bar, chocolate chips, wheat biscuit, dark chocolate, fried mud carps	30 min (after PCR)	None	Cashew: 0.5 pg	Wang et al. (2011a)

(*Continued*)

Table 7.1 Optical Biosensors for Detection of Food Allergens (*cont.*)

Transduction	Analyte	Receptor	Assay	Matrixes	Run time	Nanomaterial	LOD	References
Colorimetric (indirect detection, enzyme label)	Sin a1, ITS(18S-26S), Jug r2, Oleosin, Mtd, Pru du1, Avenin, 2S Albumin mRNA	DNA	Sandwich	Walnut, almond, hazelnut, seed, sesame, celery, oat, mustard, lupine, wheat, biscuits, chocolate, omelet, wafers, bread sticks	30 min (after PCR)	None	Sesame 0.5 pg	Wang et al. (2011b)
Optical density variation (indirect detection(enzyme label)	Cor a1, Ar h2, Le	DNA	Sandwich	Seeds, cookies, cereals, chocolates, pastas, soups, and snacks, dairy products, jams, ice cream, frozen ready meals	2.5 h incubation time	None	1 μg/g	Tortajada-Genaro et al. (2012)
F-SLW fluorimetry (indirect detection, NPs label)	Soy protein	Polyclonal Ab	Competitive	Soy milk, soy yoghourt, fruit, soy juice	incubation time: 2 h ± 1.5 h	Nile blue-doped silica NPs (257 nm)	0.05 mg/L	Godoy-Navajas et al. (2011)

Abbreviations: SPR, Surface-plasmon resonance; IMS-SERS, Immuno-magnetic separation–surface-enhanced Raman spectroscopy; F-SLW, Front-surface long-wavelength; Ab, Antibody; NP, Nanoparticle; NA, Not available.

Reprinted from Pilolli et al., 2013 Advances in biosensor development based on integrating nanotechnology and applied to food-allergen management, Trend. Anal. Chem., 47, 12–26, Copyright (2013), with permission from Elsevier.

Table 7.2 Electrochemical and Electromechanical Biosensors for Detection of Food Allergens

Transduction	Analyte	Receptor	Assay	Matrixes	Nanomaterial	LOD	References
DPV (direct detection in presence of redox probe)	Casein	Polyclonal Ab	Direct	Cheese	AuNPs + Poly(L-Arg)/MWCNTs	50 ng/mL in standard solutions	Cao et al. (2011)
DPV (direct detection in presence of redox probe)	β-lactoglobulin	Ab	Direct	Cake, cheese snacks, sweet biscuits	Graphene	0.85 pg/mL in standard solutions	Wang et al. (2010b)
DPV (indirect detection, alkaline phosphatase label)	Hazelnut genes: Cor a 1.03, Cor a 1.04	DNA	Sandwich	Hazelnut, Chocolate, Soy milk, Chocolate cream, Biscuit, Breakfast cereal, Lecithin supplement, Ketchup, Peanut butter. Snack, Peanut	None	Cor a 1.03: 03 nM Cor a 1.04: 0.1 nM	Bettazzi et al. (2008)
RT-CA (indirect detection, alkaline phosphatase label)	Hazelnut gene: Cor a 1.04	DNA	Sandwich	—	Magnetic microbeads	0.2 nM	Berti et al. (2009)
SWV (indirect detection, dissolution of ZnO-QDs)	Chicken ovomucoid	Con A	Direct	—	ZnO QDs (3 nm)	1 ng/mL	Yang et al. (2011)
f-EIS (direct detection in presence of redox probe)	Peanut Ara h1	Monoclonal Ab	Direct		None	0.3 nM	Huang et al. (2008)
f-EIS (direct detection in presence of redox probe)	Peanut Ara h1	Monoclonal Ab	Direct		None	0.08 nM	Huang and Suni (2008)
f-EIS (direct detection in presence of redox probe)	Peanut Ara h1	DNA	Indirect	Peanut milk beverage	None	0.35 fM in standard solutions	Sun et al. (2012)

(*Continued*)

Table 7.2 Electrochemical and Electromechanical Biosensors for Detection of Food Allergens (*cont.*)

Transduction	Analyte	Receptor	Assay	Matrixes	Nanomaterial	LOD	References
Faradaic single frequency impedance measurement (direct detection in presence of redox probe)	Peanut protein Ara h1	Polyclonal Ab	Direct	Canned soup	None	NA	Singh and Suni (2010)
nf-EIS (direct detection)	Peanut protein Ara h1	Polyclonal Ab	Direct		3 nm Au coated nanoporous membrane	NA	Singh et al. (2010)
QCM	Shrimp	Polyclonal Ab	Direct	spiked water samples	AuNPs (10 nm)	0.333 ng/mL	Xiulan et al. (2010)
QCM	Gliadin	Polyclonal Ab	Direct	Wheat, barley, oat, rice, foxtail millet, corn, buckwheat, soybean, pancake mix, custard mix, baby rice, buckwheat, rye biscuit, digestive biscuit, plain crackers, red date wheat crispy biscuit, cream cracker, crisp flakes of rice and wheat	AuNPs (25 nm)	Standard solutions: 8 ppb food matrix: 1 ppm	Chu et al. (2012)

Abbreviations: DPV, Differential pulse voltammetry; RT-CA, Real-time chronoamperometry; SWV, Square-wave voltammetry; f-EIS or nf-EIS, Faradaic or non-Faradaic electrochemical-impedance spectroscopy; QCM, Quartz-crystal microbalance.

detection of food allergens that are relied on electrochemical and electromechanical transduction method.

3 Gas Nanosensors

Traditional methods of gas detecting such as mass spectrometry, chromatography, and IR spectroscopy are usually time consuming, expensive, complex, and large in size. Due to this disadvantages, they are not appropriate to apply in food industry. On the contrary, gas sensors can work in a rapid, continuous, easy, and nondestructive way. Broadly, gas sensors (Comini et al., 2009), according to the type of transducer, can be divided into the following types: electrical-based, capacitive-type, FET, solid-state electrochemical, optical, thermometric, acoustic wave, and cantilever-based. As observed in past decades, recent progress in nanotechnology, coupled with efforts toward size miniaturization and sensor power reduction, has prompted intense development of gas sensors based on nanomaterials.

It is well known that excess moisture and oxygen are major contributors to food spoilage. The exclusion of oxygen is preferred in order to inhibit growth of aerobic spoilage organisms, whereas, carbon dioxide is typically used in food packs to decrease bacterial growth rates (Bültzingslöwen et al., 2002). The presence of both carbon dioxide and nitrogen can reduce the rate of respiring fruits as well as vegetables and assists to retard oxygen-based spoilage. The traditional way to analyze vapor or gas content inside a package is based on destruction of the package and random testing during a production run, which is time consuming and costly. Those drawbacks are avoided in noninvasive methods that use nanosensors to continually and easily monitor the gas content of a package headspace. Therefore in the past decade many efforts have been made to evolve new types of nanomaterials with excellent sensing properties.

The photoactivated indicator ink containing TiO_2 nanoparticles, a blue, solvent-soluble, colored ion-paired methylene blue dye, glycerol and the polymer zein, was proposed for in-package oxygen detection (Mills and Hazafy, 2008). This indicator loses its color rapidly upon exposure to UVA light and remains colorless in an oxygen-free atmosphere, returning to its original blue color upon exposure to air. In the latter step the rate of color recovery is proportional to the level of ambient oxygen and the same film can be UV-activated repeatedly. Discussed ink made of TiO_2 nanoparticles provides an easy, visual method to identify modified atmosphere packages (MAPs) with possible compromised seal integrity (Mills and Hazafy, 2008).

In Bültzingslöwen et al. (2002) a noninvasive method of measuring carbon dioxide content in MAP has been described. It is based upon lifetime analysis of luminescent dyes standardized by fluorophore-encapsulated polymer nanobeads. The sensor exhibits high stability, resolution of better than 1%, and limit of detection of 0.08%.

Zhang and Zhang presented that the resistance sensor constructed from SnO_2–ZnO nanocomposite can be used successfully for trimethylamine (TMA) detecting (Zhang and Zhang, 2008). TMA is one of the most important organic amines, which is produced in the process of metabolism by animal organs and proteins. It is also reported as the toxic gas that is emitted during the decay of fishes and seashells after their death. Decrease of freshness of fishes causes increase of concentration of TMA, too. The SnO_2–ZnO nanocomposite showed the advantages of high selectivity, excellent sensitivity, strong stability, and prompt response/recovery characteristics in detecting TMA at 330°C.

Another novel promising method for monitoring changes in food quality is use of wireless passive antennas made of nanopatterned Au-nanoparticle plasmonic array on silk substrates (Tao et al., 2012). The dielectric changes of the sensed food influence resonant responses of the antenna. The shifts in the impedance and resonant frequency is used to transduce (and monitor) the variation of dielectric properties associated with density/firmness and material composition. This includes parameters such as moisture, gas emission, and salt content that correlate to changes of food quality during the spoilage process.

The moisture content monitoring can be easily and in an invasive way realized using carbon-coated Cu nanoparticles dispersed in a tenside film (Luechinger et al., 2007). Exposure to water or solvent vapor causes swelling of the polymer matrix and results in larger degrees of internanoparticle separation. These changes cause sensor strips to reflect or absorb different colors of light. Discussed polymer/Cu nanoparticle hybrid films revealed high sensitivity with optical shifts in the visible light range of up to 50 nm for a change of 1% in relative humidity.

4 SbSI Nanosensors of Humidity

The nanocrystalline SbSI can be prepared sonically in different liquids—ethanol (Nowak et al., 2008), methanol (Starczewska et al., 2009), isopropanol (Kwolek et al., 2015), and water (Nowak et al., 2014b; Gladkovskaya et al., 2015)—from the constituents, the elements antimony, sulfur, and iodine. In a standard procedure performed at room temperature and ambient pressure, 10 ml of ethanol was poured over 1.70955 g Sb, 1.78320 g I_2, and 0.45005 g S (Nowak and Jesionek, 2012). Such mixture corresponds to the stoichiometric ratio of the constituents in SbSI. The used container (eg, 54 mL Pyrex glass cylinder of 20 mm inside diameter) was closed before sonication to prevent volatilization of the reagents. The sonication was performed in VCX-750 ultrasonic reactor (Sonics and Materials, Inc., frequency 20 kHz, with 750 W peak electrical power, and 565 W/cm^2 power density) with cup-horn (630-0503 Sonics and Materials, Inc.). The use of other types of ultrasonic reactors was mentioned (Nowak et al., 2008; Starczewska et al., 2009; Nowak and Jesionek, 2012; Nowak et al., 2014b; Kwolek et al., 2015; Gladkovskaya et al., 2015). To maintain constant temperature of 323 K, water in VCX-750 reactor was continuously pumped through thermostated reservoir.

During the sonication a sol was formed. It was observed that the color of the slurry changed gradually from red (before sonication) into olive, green, yellow, and then into red–orange, indicating the growth process of SbSI nanowires (Nowak et al., 2008). The sonochemical reaction was continued so as to complete the gelation of SbSI. To extract the eventual remaining substrates the SbSI gel was many times [eg, eight times (Nowak, 2010)] rinsed with pure ethanol and centrifuged. The details of the procedure and the used equipment were published in Nowak (2010) and Nowak et al. (2014b). At the end the centrifuged precipitates were covered by colorless alcohol. The liquid was evaporated in air at 338 K, so the so-called SbSI xerogel was obtained.

The sonochemically prepared SbSI gel consists of a large quantity of nanowires with uniform shape and high crystallinity. The typical scanning electron microscopy (SEM) micrograph of SbSI sonochemically prepared in ethanol is shown in Fig. 7.2a. This material exists as a porous gel composed of nanowires with average lateral dimensions of 20–50 nm and average lengths up to several micrometers. The powder X-ray diffraction (XRD) patterns indicated that the product was well crystallized. The intensities and positions of the peaks were in good agreement with literature values for $Pna2_1$ orthorhombic phase of bulk SbSI (Iwata et al., 1966; Kikuchi et al., 1967).

The high-resolution transmission electron microscopy (HRTEM) image of a single SbSI nanowire (Fig. 7.2b) exhibited clear lattice fringes parallel to the nanowire axis. The interplanar spacing in Fig. 7.2b is about 0.654(6) nm, which corresponds to the interplanar spacing 0.65203 nm of (110) planes in the $Pna2_1$ structure of paraelectric bulk SbSI (Iwata et al., 1966; Kikuchi et al., 1967).

Figure 7.2. Preparation of sensors constructed of SbSI nanowires. (a) SEM micrograph of SbSI gel sonochemically prepared in ethanol; (b) typical HRTEM image of a single nanowire from SbSI gel; (c) SEM micrograph of a single SbSI nanowire ultrasonically welded to Au/Ti microelectrodes on Si/SiO_2 substrate; (d) AFM image of a single SbSI nanowire ultrasonically welded with Au/Ti microelectrodes on Si/SiO_2 substrate. (Part b) Reprinted from Nowak et al., 2014b. Fabrication and characterization of SbSI gel for humidity sensors. Sens. Actuat. A 210, 119–130. Copyright (2014), with permission from Elsevier; (part d) reprinted from Mistewicz, K. et al., 2013. Quantum effects in conductivity and photoconductivity of a single SbSI nanowire. Acta Phys. Pol. A 124, 827–829. Copyright (2013), with permission from Institute of Physics of the Polish Academy of Sciences.

According to the results of X-ray photoelectron spectroscopy (XPS), at least the surface layer of sonochemically prepared SbSI nanowires is of *p*-type (Nowak et al., 2009).

Characterization of SbSI gel was accomplished also using other techniques, such as energy dispersive X-ray (EDX) analysis (Nowak et al., 2008), atomic force microscopy (AFM) (Mistewicz et al., 2013), transmission electron microscopy (TEM) (Nowak et al., 2008), selected area electron diffraction (SAED) (Nowak

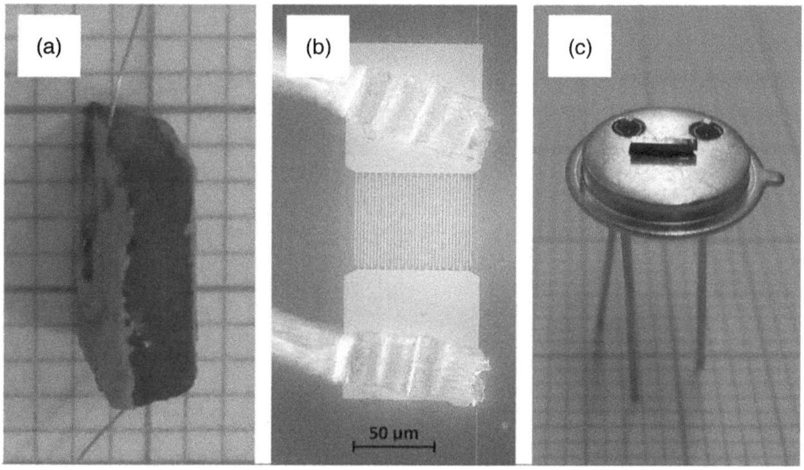

Figure 7.3. Images of SbSI humidity sensors: (a) composed of SbSI xerogel, (b) single SbSI nanowires nanowelded to Au/Ti interdigitated microelectrodes, and (c) standardized metal semiconductor package TO-5 with SbSI sensor.

et al., 2008), infrared (IR) absorbance (Starczewska et al., 2008), and diffuse reflection spectroscopy (DRS) (Nowak et al., 2008).

Humidity nanosensors were constructed from SbSI xerogel (Starczewska et al., 2012) as well as from single SbSI nanowires (Mistewicz et al., 2014). Fig. 7.3a presents the rectangular sample of SbSI xerogel. This sample was cut from the sonochemically fabricated material. The sample dimensions were following: 1.73, 1.65, and 1.50 mm. A silver paste (SPI Supplies) was used to paint electrodes on the largest opposite surfaces of the sample. Copper wire was applied to perform electrical connection from these electrodes as described in (Nowak and Jesionek, 2012). In (Starczewska et al., 2012) such a sample was used as a humidity sensor.

Fig. 7.3b shows a Si/SiO_2 substrate with Au/Ti interdigitated microelectrodes (with 1 µm spacing) to which single SbSI nanowires (Figs. 7.2c,d) were welded ultrasonically, according to the technology described in (Mistewicz et al., 2013, 2014, 2016). Fig. 7.3c presents standardized metal semiconductor package TO-5 with such prepared SbSI sensor.

4.1 Conductive SbSI Sensors of Humidity

The influence of humidity on electric current flowing through unilluminated SbSI gel and SbSI single nanowires in N_2 is shown in Fig. 7.4. Details of the measurements and the used equipment have been published in (Nowak et al., 2014b; Mistewicz et al., 2014).

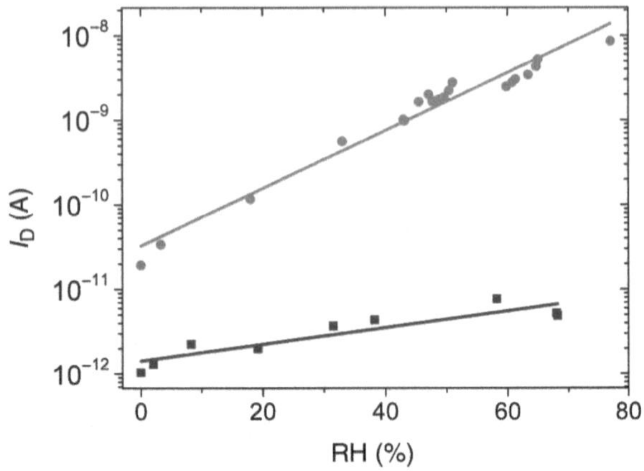

Figure 7.4. Influence of humidity on electric current flowing through unilluminated SbSI gel (●) and single SbSI nanowires (■) in N_2 ($T = 280$ K; $p = 4 \times 10^4$ Pa; *solid lines* represent the best fitted dependence described by Eq. (7.2); Values of the fitted parameters are given in Table 7.3.

A proton-transfer process was established (Starczewska et al., 2012) as the dominant conduction mechanism through the adsorbed water layers on SbSI nanowires. It was suggested (Nowak et al., 2009) that acceptors in sonochemically prepared SbSI gel exist due to the iodine vacancies in SbSI lattice. At low humidity, H_2O molecules are very easily chemically adsorbed on the surface of SbSI by a dissociation mechanism

$$H_2O + V_{I^-} + h^+ \rightarrow V_{I^-}OH^- + H^+ \qquad (7.1)$$

where V_I^- is the ionized iodine vacancy, $V_I^-OH^-$ is a hydroxyl group occupying an iodine lattice site, H^+ represents the free proton and h^+ is a hole. As pointed out in (Àlvarez et al., 2002, two mechanisms of the influence of electrical conduction by H^+ ions generated by this reaction can exist. In one of them, the H^+ ions jump through interstitials sites, and/or they are transported through channels in crystalline structure into the interior of grains. Second, the H^+ ions could be transported by jumps between adjacent OH^- groups. Defects existing in grains boundaries and/or near sample surfaces can trap such groups (Àlvarez et al., 2002). Results of measurements performed at low RH suggest water tightly bound with SbSI. According to (Starczewska et al., 2012), mobile protons generated by the dissociation of water molecules are responsible for the observed impedance characteristics. It should be emphasized that at low RH the electrical response pattern is stable after some days. It is known (Spanier et al., 2006) that surface hydroxylation is predominant in oxides, for example, $BaTiO_3$. This phenomenon is observed above all in the case of oxides prepared by wet chemical methods. It is characteristic for this technology that chemisorbed OH^- is stable in an ultrahigh-vacuum environment at elevated temperatures (>670 K) (Spanier et al., 2006). The presence of OH on the surface of SbSI nanowires has been detected in infrared spectrum of SbSI gel after high-temperature treatment in ultrahigh-vacuum (Starczewska et al., 2008). It should be noticed that the OH adsorbates enhance ferroelectricity at the positively polarized surface (Spanier et al., 2006). This phenomenon maintains a characteristic ferroelectric displacement pattern (Spanier et al., 2006).

The response at high humidity is determined by superficially adsorbed water. With increasing RH, water can be physisorbed on SbSI nanowire via hydrogen bonding. The adsorbed H_2O molecule associates with the neighboring water molecules and forms clusters. These clusters are created on nanowires surfaces as well as at the nanowire-boundaries. Then the nanowire's conductance increases a lot according to Grotthuss's chain reaction (Jacobson et al., 2004), where proton transfer occurs among the hydronium, and an ion-conductive layer forms on the surface of SbSI nanowires. The concentration of H^+ ion decreases with the decrease of water vapor, which causes the increase of resistance.

The results presented in Fig. 7.4 were fitted well with the following dependence:

$$I_{\mathrm{D}}(\mathrm{RH}) = I_{\mathrm{Do}} e^{\alpha_{\mathrm{H}} * \mathrm{RH}} \tag{7.2}$$

where $I_{\mathrm{D}}(\mathrm{RH})$ is electric current for relative humidity (RH), I_{Do} is the preexponential factor that describes electric current in dry N_2, and the constant α_{H} is related to the sensitivity of electrical conductivity on humidity. Values of fitted parameters are given in Table 7.3.

Table 7.3 Values of Parameters of Eq. (7.2) Least-Square Fitted to the Data I_D(RH) Presented in Fig. 7.4 for Gel and Single Nanowires of SbSI

Sensor	α_H [%$^{-1}$]	I_{D0} [pA]
SbSI single nanowires	0.0226(39)	1.43(23)
SbSI gel	0.0787(34)	32.5(54)

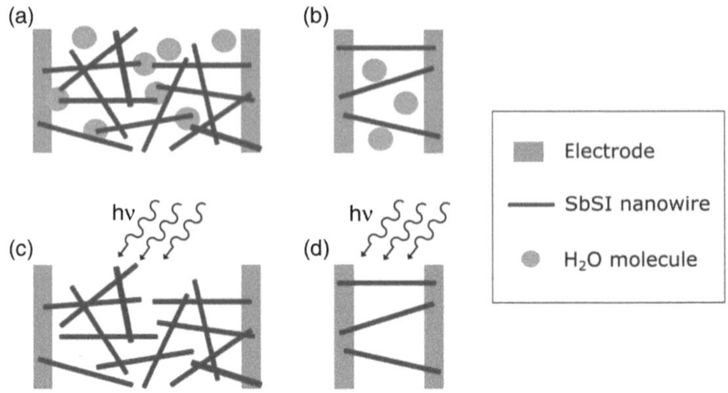

Figure 7.5. Schematic diagrams of unilluminated (a,b) and illuminated (c,d) sensors containing SbSI gel (a,c) and a few single SbSI nanowires (b,d) in moist environment.

Many effects may contribute to the exponential RH dependence of conductance. For example, the exponential dependence of the conductivity with the humidity can be explained by the change of permittivity arising from water adsorption (Ha et al., 2002). In another approach, jumps over the potential barrier in order to move from one to the next equilibrium position evoke exponential dependence on humidity (Haspel et al., 2013). The exponential behavior can also be attributed to the fact that the Debye screening length is much larger than NW radius so that the whole NW volume is affected by the gating of water molecules on the surface (Kiasari et al., 2012). The RH sensitivity of SbSI nanowires can also be related to the number of water-molecule adsorption sites, which increases by the presence of defect lattice sites (Traversa, 1995).

However, it should be taken into serious consideration that in the case of SbSI nanowires welded to the microelectrodes the dark current increases twice with increase of humidity from dry to wet environment, while in the case of SbSI gel the I_D increases exponentially by nearly three orders of magnitude with the same increase of humidity (Fig. 7.4). It demonstrates that the conductance of assemblies of SbSI nanowires in wet N_2 is mainly caused by H_2O clusters agglomerated on the nanowire-boundaries (Fig. 7.5a).

SbSI gel samples synthesized sonochemically are made up of assemblies of single crystal nanowires with random orientation. Electrical transport in such systems generally results both from bulk and surface properties of the nanowires and from properties of the interfaces between them. The latter are characterized by interface states leading to a potential barrier between the nanowires. The authors of (Comini et al. 2005; Comini, 2006) suggested that in DC configuration with multiple wires that also the potential barriers, that can appear due to wire-to-wire contact, could be important for a sensing mechanism. The resulting charge exchange between adsorbed molecules and the material modifies the energy barrier for grain-to-grain current percolation and, in turn, the electrical conductance of the layers (Comini et al., 2005).

Different mechanisms of water adsorption are well known (Agmon, 1995; Henderson, 2002) but a probable mechanism for the ferroelectric SbSI was proposed in (Nowak et al., 2014a). The perfectly aligned electric dipoles in a ferroelectric terminate at the opposite surfaces with unpaired positive and negative bound charges. Due to this spontaneous polarization $\vec{P_S}$, the charge density on their end-faces can reach 10^{18} charges per square meter (Gundel et al., 1989). The charge screening of $\vec{P_S}$ by free charge carriers and adsorbed species is concentrated near the interphase boundary of a single-domain section of the ferroelectric phase. Configuration of the band edges at the surfaces of monodomain ferroelectric is presented in Fig. 7.6a. Electrostatic force dominates the charge compensation and facilitates water adsorption. Both van der Waals and dipole–dipole interactions of water with ferroelectric SbSI could lead to water adsorption. The polarization-dependences of physisorption on ferroelectric surfaces have been reported in Li et al. (2008) and Tiwari and Donn (2009). It has also been proved (Garbarz et al., 2000) that SbSI in ferroelectric phase adsorbs water vapor much more intensively than paraelectric one.

To establish the model describing the exponential increase of SbSI conductivity with ongoing water adsorption additional investigations are needed, for example, examination of humidity influence on photoconductivity (Nowak et al., 2014a).

4.2 Photoconductive SbSI Sensors of Humidity

Fig. 7.7 shows typical DC photoconductivity current (I_{PC}) responses on illumination of SbSI single nanowires. The switching on and switching off illumination was performed in vacuum and in nitrogen of different RH. Details of the measurements and the used equipment were published in Mistewicz et al. (2014). Shape of responses presented in Fig. 7.7 depends on the value of relative humidity. After switching illumination on, the electric photoconductivity current increases fast, attains maximum, and then decreases slowly to a stationary value. Obviously, the rise of I_{PC} after switching on illumination is caused by photogeneration of excess carriers in the semiconducting SbSI. Switching the illumination off affects monotonous decrease of electric conductivity current due to recombination of excess carriers. However, the first

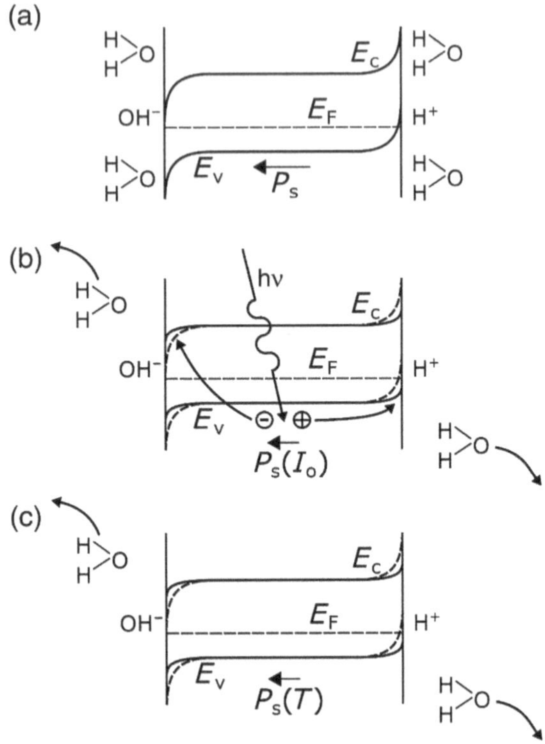

Figure 7.6. Schematic band diagrams of the surface space–charge region at p-type ferroelectric semiconductor under different conditions: (a) Unilluminated ($I_0 = 0$) sample in thermal equilibrium at temperature T_1; (b) super-bandgap illumination ($I_0 > 0$) with carriers separation under an electric field; (c) unilluminated ($I_0 = 0$) sample in thermal equilibrium at temperature $T_2 > T_1$; $\vec{P_S}$, spontaneous polarization; ⊕ and ⊖, photogenerated holes and electrons, E_c and E_v, bottom of conduction energy band and top of valence energy band, respectively, hν, photons; detailed description in the text. Reprinted from Nowak et al. (2014a). Copyright (2014), with permission from Elsevier.

pulse in the transient characteristic exhibits a complex shape that builds to a peak, relaxes to a minimum, and recovers to a saturated level (Fig. 7.7d). It can be recognized as a so-called hook anomaly, observed for infrared detectors (Sclar, 1984). Spontaneous random voltage spiking often occurs in IR extrinsic detectors. The hook anomaly spikes exhibit fast rise times, narrow time widths and longer tails that are usually system bandwidth limited. The height distribution of the spikes and the spike rate are strongly dependent, above some threshold, on the voltage applied to the detector. In addition, they also depend on temperature and the signal and background flux levels (Sclar, 1984). In the case of

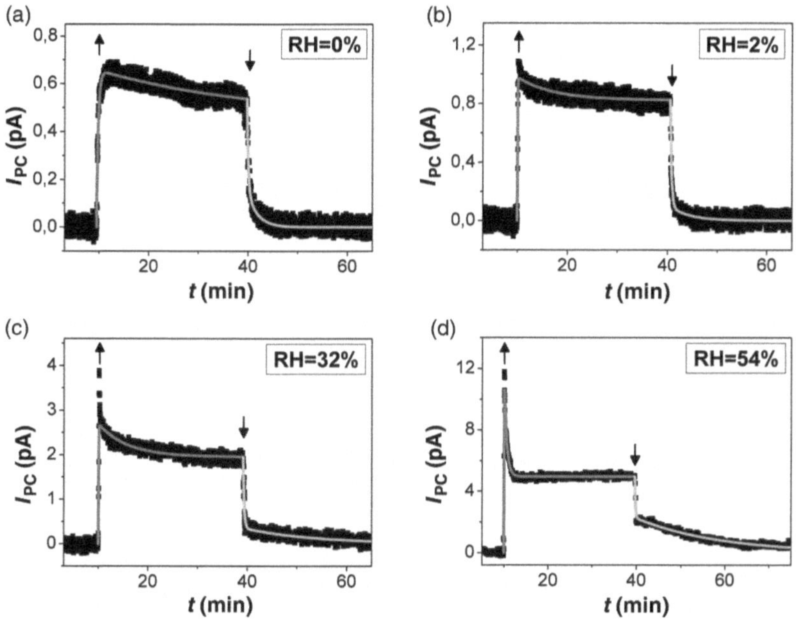

Figure 7.7. Hook anomaly in photoconductivity current responses on switching on (↑) and switching off (↓) illumination of SbSI single nanowires: (a) In vacuum ($p = 10^{-2}$ Pa) and (b–d) in moist N_2 ($p = 4 \times 10^4$ Pa). Characteristics were obtained for $\lambda = 488$ nm; $I_0 = 2 \times 10^{22}$ photon/(m^{2}s); $E = 1.5 \times 10^6$ V/m; $T = 280$ K. *Solid curves* represent the best fitted dependences described by Eq. (7.3). Reprinted from Mistewicz et al., 2014. SbSI single nanowires as humidity sensors. Acta Phys. Pol. A, 126, 1113–1114, Copyright (2014), with permission from Institute of Physics of the Polish Academy of Sciences.

humidity sensors constructed from SbSI single nanowires, hook anomaly pulse depends on relative humidity.

The transient characteristics of photoconductivity of humidity sensing SbSI single nanowires were fitted (Mistewicz et al., 2014) by the following empirical relations:

$$I_{PC}(t) = I_{PC1}\left[1 - A \cdot \exp\left(-\frac{t}{\tau_{11}}\right)\right] + I_{S1}\exp\left(-\frac{t}{\tau_{12}}\right) \quad \text{for} \quad t_{on} < t < t_{off} \tag{7.3a}$$

$$I_{PC}(t) = I_{PC2}\exp\left(-\frac{t}{\tau_{21}}\right) + I_{S2}\exp\left(-\frac{t}{\tau_{22}}\right) \quad \text{for} \quad t > t_{off} \tag{7.3b}$$

where I_{PC1} is the stationary value of photoconductivity current flowing under constant illumination; A, I_{S1}, I_{PC2}, I_{S2} are preexponential

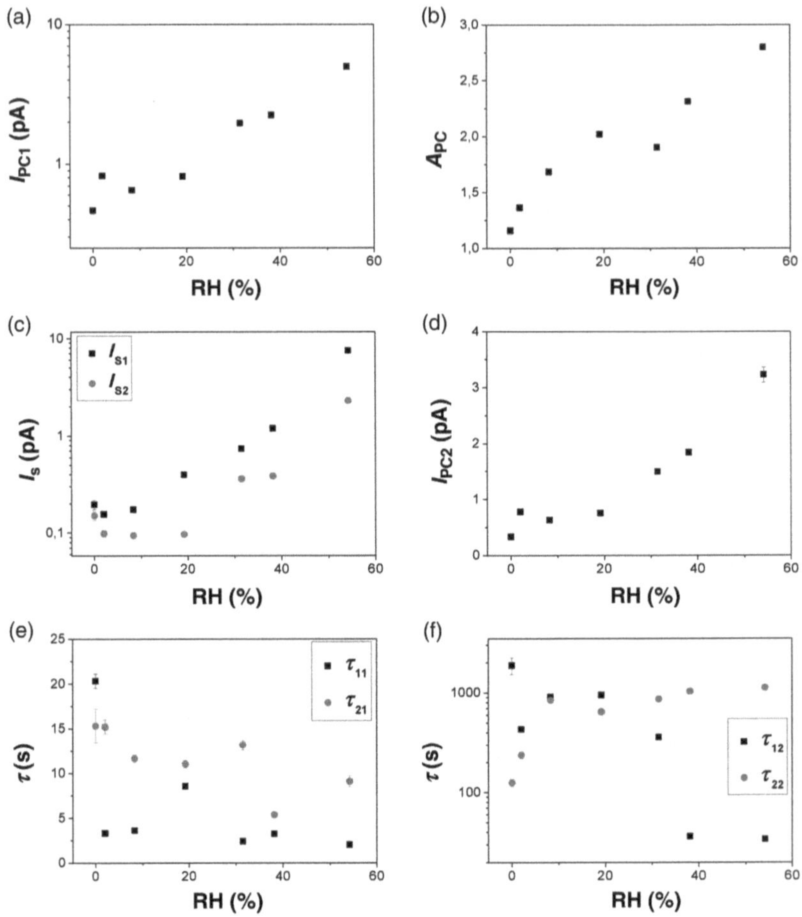

Figure 7.8. Influence of humidity on parameters of Eq. (7.3) fitted to the photoconductivity transient characteristics of SbSI single nanowires presented in Fig. 7.7 ($T = 280$ K, $E = 1.5 \times 10^6$ V/m, $\lambda = 488$ nm, $I_L = 2 \times 10^{22}$ photon/m^{2}s).

factors; τ_{11}, τ_{12}, τ_{21}, and τ_{22} are time constants. The fitting of the time dependence of photocurrent with superposition of two exponentials is very good (Fig. 7.7). The values of the fitted parameters are presented in Fig. 7.8. It should be noted that the value of I_{PC1} parameter (Fig. 7.8a) increases monotonically over 10 times in the relative humidity range from 0% to 54%.

It should be emphasized that the final settled value of photocurrent (I_{PC1}) is positive in the whole range of RH (Fig. 7.8a). Different influence of humidity on transient photoconductivity characteristics was observed (Nowak et al., 2014a) in the case of SbSI gel, made up of large quantity of nanowires (Fig. 7.9). It was found

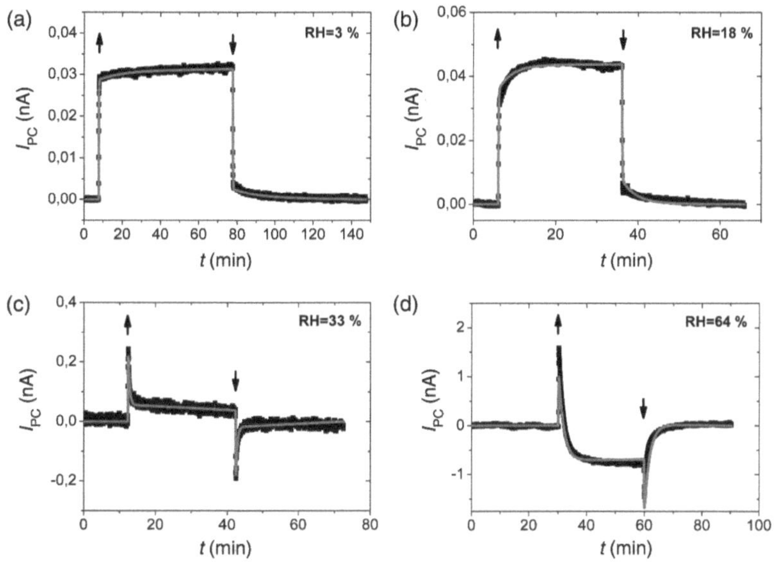

Figure 7.9. Responses on switching on (↑) and switching off (↓) illumination of SbSI gel in N_2 of different humidities ($\lambda = 488$ nm; $I_0 = 3.1(8) \times 10^{21}$ photon/(m^2s); $E = 3000$ V/m; $T = 280$ K; $p = 4 \times 10^4$ Pa; solid curves represent the best fitted dependencies described by Eq. (7.6); values of the fitted parameters are presented in Figure 7.12. Reprinted from Nowak et al., 2014. Transient characteristics and negative photoconductivity of SbSI humidity sensor, Sens. Actuat. A, 210, 32–40., Copyright (2014), with permission from Elsevier.

that, while the value of photocurrent under constant illumination is positive in the case of small and medium RH, it is negative for RH higher than some critical value $RH_C = 39.8\%$ (Nowak et al., 2014a).

At low RH the increase of I_{PC} after illumination of SbSI is caused by photogeneration of excess carriers (Fig. 7.9a,b). When illumination is switched off, the excess carriers recombine and the electric conductivity current decreases monotonically (Figs. 7.9a,b). This is the well-known kinetics of photoconductivity in semiconductors. In the case of SbSI, the kinetics change with the increase of humidity (Fig. 7.9). At medium and high RH (Figs. 7.9c,d) the I_{PC} increases fast after illumination of SbSI, attains maximum, and then decreases slowly to a stationary value. Under constant illumination the photoconductivity current is positive at medium RH. However, it is negative for higher humidity (Fig. 7.9d). It denotes that in the case of large RH, the electric current in illuminated SbSI gel is smaller than in the unilluminated one. This phenomenon can be recognized as so-called negative photoconductivity.

After switching off illumination for medium and high RH (Fig. 7.9c,d) the electric current first drops fast, reaches minimum,

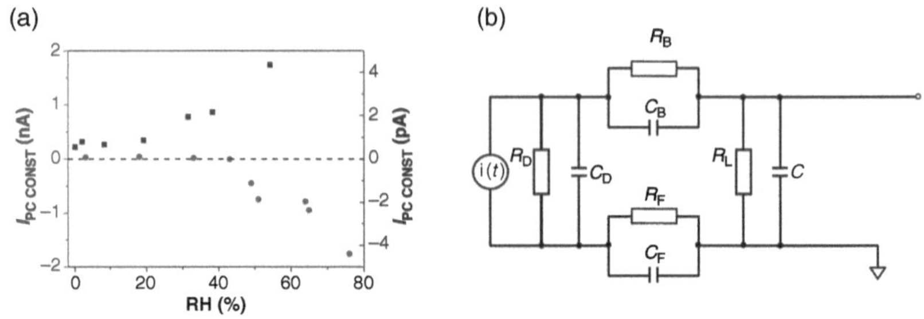

Figure 7.10. (a) Influence of humidity on value of photoconductivity current flowing under constant illumination in the case of SbSI gel (●) and single nanowires of SbSI (■) in N_2 ($\lambda = 488$ nm; $I_0 = 3.1(8) \times 10^{21}$ photon/(m²s); $E = 3000$ V/m; $T = 280$ K; $p = 4 \times 10^4$ Pa). (b) Circuit representation of a photoconductive detector of light.

and then increases with time to the intensity of current that was measured before the illumination of the sample. It should be noticed that the minimum values of I_{PC} in these cases are negative.

Fig. 7.10a shows influence of humidity on photoconductivity current $I_{PCconst}$ flowing under constant illumination of SbSI gel and single nanowires of SbSI in N_2. In the latter case $I_{PCconst}$ increases with the increase of RH as it was observed in the case of electric current I_D flowing through unilluminated SbSI single nanowires and SbSI gel in N_2 (Fig. 7.4). However, in the case of SbSI gel, the nonmonotonic dependence of $I_{PCconst}$ on RH (Fig. 7.10a) is quite different from the exponential changes of electric current flowing through the unilluminated sample of SbSI gel (Fig. 7.4). The positive photoconductivity observed in relatively dry N_2, that is, the increase of electric current under illumination of SbSI gel (Fig. 7.9a,b) is caused by the photogeneration of excess electron-hole pairs. However, we observe the decrease of conductance of SbSI gel upon illumination with increasing of ambient humidity (Fig. 7.9c,d). In the case of very large ambient humidity even the negative photoconductivity is observed (Fig. 7.9d); that is, illumination can cause the decrease of current to values smaller than dark current.

Usually, negative photoconductivity, in bulk materials (Bube, 1992; Papaioannou et al., 1989; Akimov et al., 2000) as well as in nanostructures (Huang et al., 2010), is attributed to two-center recombination effect: interaction between an upper electron trapping level and lower recombination center/acceptor level could give rise to a decrease in thermal current by illumination. Therefore, coexistence of trapping and recombination centers in energy gap of bulk SbSI single crystals has been proposed as the reason of negative photoconductivity reported in

(Grekov et al., 1971; Fridkin, 1980). However, the negative photoconductivity of Ce_2O nanowires (Fu et al., 2007), Co-doped ZnO nanobelts (Peng et al., 2008), and polymeric solid-state bridges (Kim and Choi, 2010) in ambient air or water vapor atmosphere has been attributed to the photodesorption of water molecules from nanoobjects' surface. The latter phenomenon is very probable in the investigated case of the influence of humidity on photoconductivity of SbSI gel.

Illumination of SbSI gel can affect desorption of water molecules and the observed decrease of conductivity current (Fig. 7.9). After illumination switching off, water molecules can be adsorbed again and electrical conductivity of SbSI gel increases (Fig. 7.9). The photodesorption of water from SbSI can be determined by two mechanisms. The first one involves a photonic interaction of light with SbSI. The second involves a thermal action of light. We describe the photonic approach: upon illumination, excess electrons and holes, that are generated by photons, selectively accumulate to positive or negative domain surfaces to compensate the polarization charge (Fig. 7.6b). The influence of photogenerated nonequilibrium carriers on screening of spontaneous polarization and the formation of interphase boundaries in bulk SbSI has been analyzed in (Fridkin, 1980; Gerzanich et al., 1982). It was found that illumination of both the poly- and mono-domain SbSI crystals gives rise to a sharp decrease in the value of the remanent polarization as compared to the spontaneous polarization (Gerzanich et al., 1982). The illumination induced changes in the equilibrium potential distribution (observed, eg, due to surface photovoltaic effect (Adamowicz, 1990; Kronik and Shapira, 2001) or photocapacitance (Matys et al., 2012) evoke some charge transfer and reduce the physisorption. The water is desorbed from SbSI surface. When the illumination is switched off, the excess electrons and holes recombine, the equilibrium polarization field is restored, and water readsorbs on the surface.

The thermal approach to the photodesorption of water from SbSI surface is based on pyroelectricity: in the case of ferroelectric, the temperature change creates an electric polarization change inside the crystal (compare Figs. 7.6a,c). This causes the decrease of surface band-bending and physisorption. Thermal conductivity of the sample is very small (Nowak et al., 2014a). Therefore, temperature of the illuminated nanowires can be much higher than temperature of the sample support and it is rather difficult to measure.

There was no significant hysteresis in the influence of illumination intensity on responses of DC photoconductivity of SbSI gel (Nowak et al., 2014a), proving that the water could be repeatedly

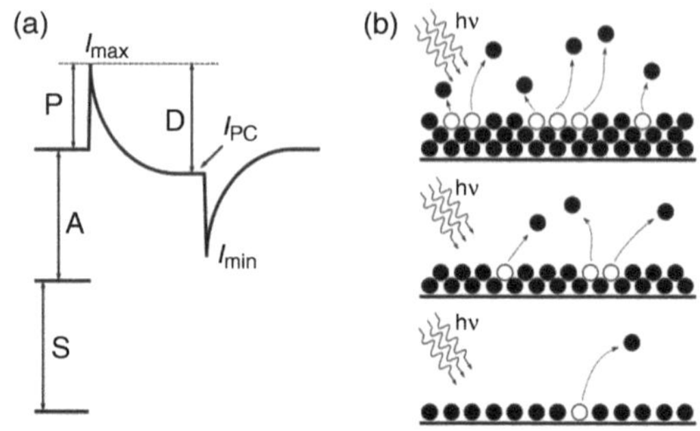

Figure 7.11. Diagrams showing (a) influence of different processes on conductivity and photoconductivity currents in SbSI nanowires and (b) desorption of water from illuminated SbSI surface. *S*, thermal generation of free carriers in a semiconductor; *A*, increase of conductivity evoked by the adsorbed water molecules; *P*, photogeneration of excess carriers; *D*, decrease of conductivity evoked by desorption of water molecules.

adsorbed and removed. Physisorbed H_2O is weakly bonded and desorbs easily from SbSI gel.

In the case of a competition between the photo-generated current enhancement and the current reduction due to water photodesorption (Fig. 7.11), the total current can be given as:

$$I_T = I_S + I_A + I_P - I_{DA} \quad (7.4)$$

where I_S represents the bias current flowing through a sample in vacuum, I_A is the current enhancement due to additional conductance caused by water adsorption, $I_D = I_S + I_A$ is humidity dependent dark current flowing through an unilluminated sample in water vapor atmosphere, I_P is the current enhancement caused by photogeneration of excess carriers in a semiconductor, I_{DA} is the current reduction caused by water photodesorption, and $I_{PC} = I_T - I_D$ is the photoconductivity current.

At relative humidity lower than some critical value (RH_c), the photo-generated current enhancement is higher than the current reduction due to water desorption. However, if RH is higher than RH_c, the I_P is less than the I_{DA}, and the negative photoconductivity occurs.

The change in the charge transport upon exposure of SbSI gel to water vapor may be due to a surface band bending induced by ionosorbed charged surface species. The surface band bending induces shift of the energy levels (relative to the Fermi level), change of the charge carrier population, and induces barrier changes at grain boundaries. The results of conductivity changes produced by H_2O adsorption on SnO_2 have been correlated to water-induced band-bending effects (Thiel and Madey, 1987; Batzill et al., 2006; Batzill and Diebold, 2007). Therefore, to describe the photoconductivity in SbSI gel one can apply the presented in (Sclar, 1984) photodetector model (Fig. 7.10b) that takes explicit account of the detector constants. In this model the voltage drop across semiconductor is augmented by voltage barriers at the respective contacts. The barriers tend to be blocking to charge flow.

Traditional ways of looking at barriers and photoconductivity phenomena have been summarized in Bube (1992). The electrical conductance of a semiconductor usually increases upon exposure to light irradiation with photon energy higher than its band gap due to (1) increasing the density of free carriers throughout the material (this is treated by the standard photoconductivity model), (2) decreasing the intergrain barrier height by changing the charge in the intergrain states, and (3) increasing the possibility for tunneling through the intergrain barriers by decreasing the depletion layer widths in the adjacent grains. In many cases these phenomena occur simultaneously.

The barriers are represented by opposed diodes that are shunted by resistances R_B and R_F to represent the back and forward resistances of the respective contacts. This detector model with diodes replaced by capacitances is shown in Fig. 7.10b. The product of $\tau_D=R_D C_D$ represents the detector dielectric relaxation time. The products $\tau_B=R_B C_B$ and $\tau_F=R_F C_F$ characterize the response properties of the two contacts. The registered time response τ of the detector is given by

$$\frac{1}{\tau}=\frac{1}{\tau_P}+\frac{1}{\tau_D}+\frac{1}{\tau_B}+\frac{1}{\tau_F}+\frac{1}{\tau_L} \tag{7.5}$$

where $\tau_L=R_L C$ represents the time response of the measurement circuit for a given choice of a load resistor R_L and capacitance C associated with the measurement circuit, τ_p is time constant describing the main phenomenon observed in the sample, that is, it represents effective carrier lifetime observed in photoconductivity or time constant of desorption of water molecules.

So, the time dependences of photoconductivity current in cases of illumination changes of a detector are described by the following function (Sclar, 1984):

$$I_{PC}(t)=I_{PCmax}\left[1-\exp\left(-\frac{t}{\tau}\right)\right]+I_{PCmax}\cdot\alpha\cdot\exp(-\beta t)\cdot\sinh(\gamma t) \quad \text{for}\, t_{on}<t<t_{off} \tag{7.6a}$$

$$I_{PC}(t) = I_{PCconst} - \left\{ I_{PCmax}\left[1 - \exp\left(-\frac{t}{\tau}\right)\right] + I_{PCmax} \cdot \alpha \cdot \exp(-\beta t) \cdot \sinh(\gamma t) \right\} \quad \text{for } t > t_{off} \tag{7.6b}$$

where t_{on} and t_{off} represent time of switching on and switching off illumination, respectively, $I_{PCconst}$ represents magnitude of saturated photoconductivity current flowing just before illumination switching off, I_{PCmax} is a parameter dependent on illumination intensity, optical reflectivity, absorption coefficient of the light, carriers mobility, bias voltage, and the relatively difficult to establish geometrical parameters of the sample composed of a large number of chaotically arranged nanowires. α, β, and γ are given by complicated expressions involving the parameters of pertinent circuit components (Fig. 7.10b)

$$\alpha = \frac{4aC_BR_BC_FR_F}{(C_BR_B + C_FR_F)^2}\left\{\left(1 - \frac{\tau}{C_FR_F}\right)\left(1 - \frac{\tau}{C_BR_B}\right) + \frac{C_D}{C_FC_B}\left(1 - \frac{\tau}{C_DR_D}\right)\left[C_F\left(1 - \frac{\tau}{C_FR_F}\right) + C_B\left(1 - \frac{\tau}{C_BR_B}\right)\right]\right\} \tag{7.7a}$$

$$\beta = \frac{C_BR_B + C_FR_F}{2C_BR_BC_FR_F(1+a)}\left\{1 + \frac{a\left[\frac{1}{\tau} + \frac{\frac{C_F}{C_FR_F} + \frac{C_B}{C_BR_B}}{C_FC_B}\right]}{\frac{1}{C_FR_F} + \frac{1}{C_BR_B}}\right\} \tag{7.7b}$$

$$\gamma = \beta\left[1 - \frac{1 + \frac{R_B + R_F}{R_B}}{C_BR_BC_FR_F(1+a)\beta^2}\right] \tag{7.7c}$$

$$a = \frac{C_F + C_B}{C_FC_B}C_D \tag{7.7d}$$

Least-square fitting of the time dependences of photoconductivity current in cases of illumination changes of SbSI gel in moist

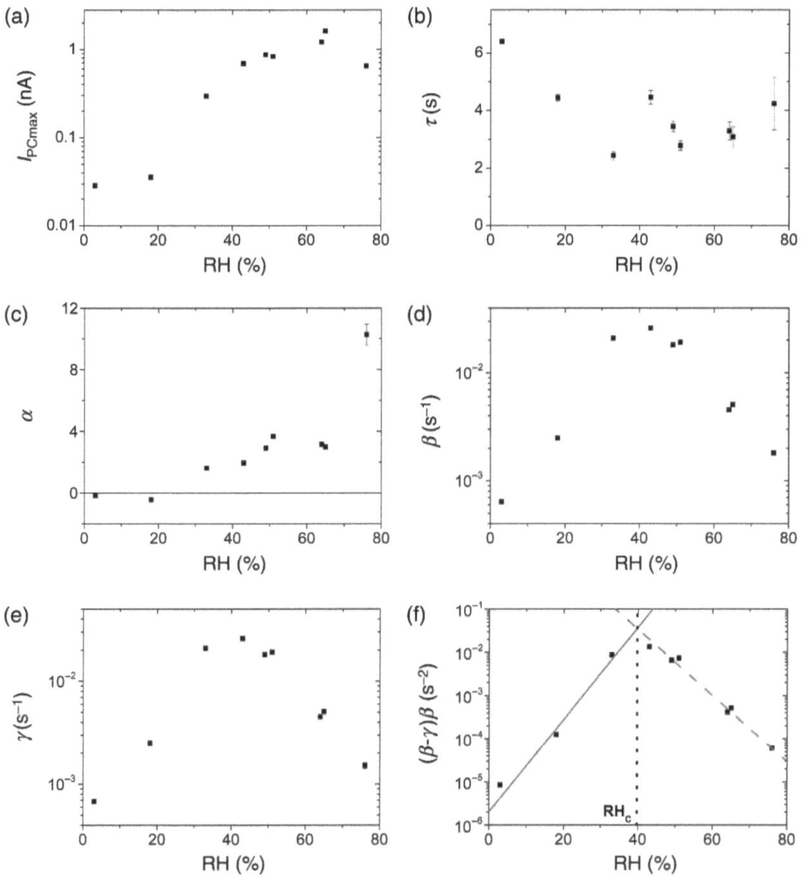

Figure 7.12. Influence of humidity on parameters of Eq. (7.6) fitted to the photoconductivity current responses on switching on and switching off illumination of SbSI gel in N_2 (λ = 488 nm; I_0 = 3.1(8) × 10^{21} photon/(m^2s); E = 3000 V/m; T = 280 K; p = 4 × 10^4 Pa; *solid and dash lines* present the best fitted exponential dependence Eq. (7.9); values of the fitted parameters are given in the Table 7.4). Reprinted from Nowak et al., 2014. Transient characteristics and negative photoconductivity of SbSI humidity sensor, Sens. Actuat. A, 210, 32–40., Copyright (2014), with permission from Elsevier.

N_2 (presented, eg, in Fig.7. 9) allowed to determine the values of I_0, τ, α, β, and γ for different humidities (Fig. 7.12). The nonlinear fitting has been performed using Origin 8.5 program. Fig. 7.12 presents also combination of β and γ parameters that contains less parameters of the equivalent electric circuit

$$(\beta-\gamma)\beta=\frac{2R_B+R_F}{R_B^2R_F[C_FC_B+(C_F+C_B)C_D]} \tag{7.8}$$

Table 7.4 Parameters of Eq. (7.9) Fitted to Values of $(\beta-\gamma)\beta$ Presented in Fig. (7.12f) for SbSI Gel in N_2 ($RH_c = 39.8\%$ for $\lambda = 488$ nm, $I_0 = 3.1(8) \times 10^{21}$ photon/(m^2s), $E = 3000$ V/m, $T = 280$ K, $p = 4 \times 10^4$ Pa)

	Fitted Parameters	
Range of RH	A_1, s^{-2}	B_1, $(\%)^{-1}$
$RH < RH_c$	$2(200) \times 10^{-6}$	39(27)
$RH > RH_c$	0.25(4)	−0.18(1)

Reprinted from Nowak et al., 2014 Transient characteristics and negative photoconductivity of SbSI humidity sensor, Sensor. Actuator. A, 210, 32–40. Copyright (2014), with permission from Elsevier.

Unfortunately, even the last parameter is an essentially intractable function of the detector constants. Of course the model should not possess too many components, because the fitting of many variables will be unreliable. Therefore, the unambiguous quantitative influence of humidity on properties of SbSI nanowires, contacts between them, and contacts between SbSI and metal electrodes is very difficult to establish. Only the influence of humidity on I_{PCmax}, τ, α, β, γ, and $(\beta-\gamma)\beta$ can be established (Fig. 7.12).

The following conclusions can be derived from the results presented in Fig. 7.12. Increase of humidity for RH≤20% evokes small increase of I_{PCmax} and small decrease of τ. Simultaneously, values of α are negative and little decrease with the increase of RH. The slope of I_{PCmax}(RH) is much greater for RH in the range from 20% to 66%, while the value of τ is practically independent of RH. In the same range of RH, values of α are positive and increase strongly with the increase of humidity. The absolute values of β and γ increase with the increase of RH, attain maxima for $RH_c = 39.8\%$, and then decrease. The influence of humidity on values of $(\beta-\gamma)\beta$ can be fitted with exponential function:

$$(\beta-\gamma)\beta = A_1 e^{B_1 * RH} \tag{7.9}$$

where A_1 is the preexponential factor that describes $(\beta-\gamma)\beta$ in dry N_2, and the constant B_1 is related to the sensitivity of $(\beta-\gamma)\beta$ on humidity. The power coefficient B_1 is positive for $RH < RH_c$ and negative for $RH > RH_c$ (Table 7.4). It should be emphasized that the value of $(\beta-\gamma)\beta$ changes by about four orders of magnitude in each of the mentioned ranges of humidity. This is a great advantage of transient photoconductivity humidity sensors made of SbSI gel.

Illumination intensity affects values of I_{PCmax}, τ, α, β, γ, and $(\beta-\gamma)\beta$. The values of I_{PCmax} can be least-squares fitted with the semi-empirical power equation (Nowak et al., 2014a):

$$I_{PCmax} = A_2 I_0^{\delta} \tag{7.10}$$

where $A_2 = 1.5(33) \times 10^{-26}$ nA and $\delta = 1.20(4)$. It should be noted that usually values of power coefficients γ and A_2 depend on photon energy (Nowak et al., 2013). Probably, they depend on humidity, too. It should be established in future investigations.

Values of τ can be least-squares fitted with the linear function (Nowak et al., 2014a):

$$\tau = A_3 + B_3 I_0 \tag{7.11}$$

where $A_3 = 1.65(4)$ s and $B_3 = 5.6(2)\cdot10^{-22}$ s^2m^2/photon. One can notice that carrier lifetimes in semiconductors are usually very small. So, the transients of photoconduction are expected to be very quick, and photogenerated electron holes rapidly disappear as soon as the illumination is off. Probably, the determined time constant τ is longer due to the boundary formation time (that is, so-called Maxwellian relaxation time proportional to ε and to the reciprocal of electric conductivity), as well as the time controlled by slow physisorption rate of H_2O.

The influence of illumination intensity on values of $(\beta-\gamma)\beta$ was fitted with the linear function (Nowak et al., 2014a):

$$(\beta-\gamma)\beta = A_4 + B_4 I_0 \tag{7.12}$$

where $A_4 = 7.2(16) \times 10^{-4}$ s^{-2} and $B_4 = 1.8(1) \times 10^{-24}$ $s^{-1}m^2$/photon.

Comparison of the results observed in the case of SbSI single nanowires nanowelded to metal electrodes (Mistewicz et al., 2014) with the results observed in the case of SbSI gel (Nowak et al., 2014a) leads to the following conclusion. The desorption of H_2O molecules from contacts between illuminated SbSI nanowires (Figs. 7.5a,c), and contacts between SbSI nanowires and metal electrodes is the more important factor that determines sensitivity of the gel humidity sensors. In Nowak et al. (2014a) it was shown that negative photoconductivity observed for SbSI gel is caused by light-induced desorption of H_2O molecules from SbSI nanowires surfaces, contacts between the nanowires, or contacts between SbSI and metal electrodes (Figs. 7.5a,c).

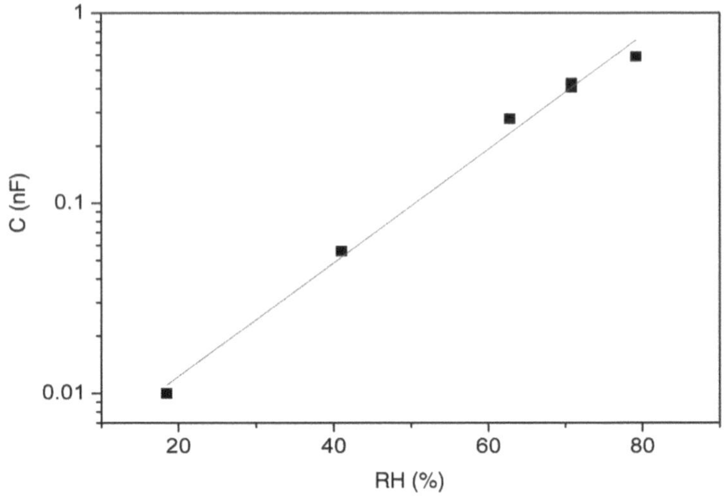

Figure 7.13. Humidity dependence of the capacity of SbSI gel sonochemically prepared in water (E = 415 V/m; f = 1 kHz; $p = 10^5$ Pa; T = 280(1) K; humidity 73%; *solid line* represents the best fitted dependence described by Eq. (7.13); Values of the fitted parameters are given in the text). Reprinted from Nowak et al., 2014 Fabrication and characterization of SbSI gel for humidity sensors, Sens. Actuat. A, 210, 119–130, Copyright (2014), with permission from Elsevier.

4.3 Capacitive SbSI Sensors of Humidity

The strong influence of humidity on electric capacity of SbSI gel in N_2 is shown in Fig. 7.13. Details of the measurements and the used equipment have been published in Nowak et al. (2014a). When the humidity increases, the capacity of SbSI gel increases by nearly two orders of magnitude following an exponential dependence (Fig. 7.13). Such dependence was reported also for barium perovskites (Viviani et al., 2001) and vanadyl phthalocyanine (VOPc) (Aziz et al., 2012). The increase in the capacitance can be attributed to the enhanced dielectric constant of the active layer due to adsorbed water vapor.

The data presented in Fig. 7.13 was fitted well with the following dependence:

$$C(RH) = C_0 e^{B_2 * \mathrm{RH}} \tag{7.13}$$

where C(RH) is capacity for relative humidity RH, C_0 = 3.09(55) pF is the preexponential factor that describes capacity in dry N_2, and the constant B_2 = 0.069(3) $(\%)^{-1}$ is related to the sensitivity of capacity on humidity.

There are many examples of the strong interaction of the ferroelectric materials [eg, $BaTiO_3$ (Wang et al., 2010a)] with the dipoles of adsorbed and absorbed water molecules. Due to strong dipole moments of H_2O molecules, the water adsorption/desorption affects dielectric properties of ferroelectric sensors. As it was reported and discussed in Garbarz et al. (2000), the ferroelectric state of bulk SbSI adsorbs water vapor much more intensively than a paraelectric one. The important question is whether the strong dipole of absorbed water acts as a spectator or as an actor in all ferroelectric applications. The spontaneous polarizability of SbSI is high near the Curie temperature (Gerzanich et al., 1982). It can influence the polarizability of the water clusters adsorbed between nanowires. As dielectric constant of H_2O is approximately 80, the capacitance of porous material increases with RH (Agarwal and Sharma, 2002).

4.4 Impedance SbSI Sensors of Humidity

One expects that resistance-humidity properties of SbSI xerogel are affected by semiconducting of its nanowires and surface adsorption of water as well as grain-boundary potential barriers and ferroelectric-paraelectric phase transition. Therefore, in Starczewska et al. (2012) the impedance spectroscopy was used for investigations of the influence of humidity on electrical characteristics of this material.

Fig. 7.14 presents the influence of humidity on Nyquist plots of SbSI gel for constant low and high temperature. Details of the measurements and the used equipment has been published in (Starczewska et al., 2012). In the case of high humidity the Nyquist characteristics consists of at least two covering semicircles. The sizes of the flattened semicircles in Nyquist plots of SbSI gel decrease with the temperature increase and increase with the humidity decrease (Fig. 7.14).

In Starczewska et al. (2012) different equivalent circuits (eg, Maxwell circuit, Voigt circuit, and their combinations) were analyzed as appropriate for complex nonlinear curve fitting of the experimental data. It was done using LEVM program (Macdonald, 1992). The best fitting was achieved for equivalent circuits shown in Fig. 7.14. Both circuits contain the parallel connected resistance R_1, capacitance C_1, and constant phase element CPE. Every ferroelectric and semiconducting SbSI nanowire can be represented by parallel connection of capacitance and resistance (Nowak and Jesionek, 2012). Capacitance of constant phase element is described by $C(\omega) = A(i\omega)^{n-1}$, where parameter A is a constant and the exponent n (that varies between 0 and 1) represents hopping charges or the screening of dipoles (Biju and Jain, 2008). CPE behaves as ideal resistor for $n = 0$ and as ideal capacitor for $n = 1$. Macroscopic

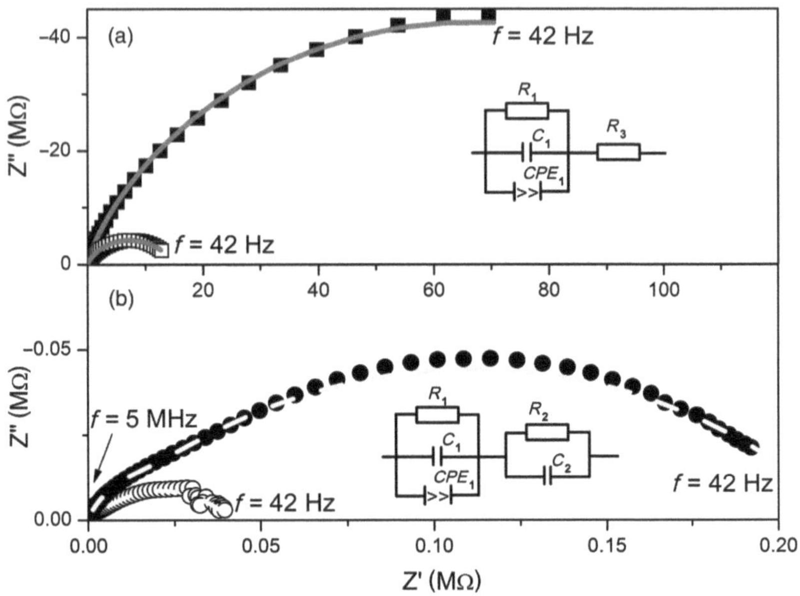

Figure 7.14. Nyquist plots for SbSI gel at various temperatures and humidities. (a) RH = 20%, (b) RH = 80%, ■, T = 281 K, □, T = 304K, •, T = 281 K, ○, T = 300 K). *Signs* represent experimental data. *Curves* show the best fitted dependencies calculated for equivalent circuits presented in insets. Fig. 7.15 presents values of the fitted parameters. Reprinted from Starczewska et al., 2012, Influence of humidity on impedance of SbSI gel, Sens. Actuat. A, 183, 34–42, Copyright (2012), with permission from Elsevier.

heterogeneities (that is, surface roughness, grain sizes, porosity, etc.) present at the sample electrodes can be represented by CPEs (Barsoukov and Macdonald, 2005). Also the ion adsorption on roughened electrodes can be described by CPEs (Barsoukov and Macdonald, 2005). Authors of Metikos-Hukovic et al. (1999) suggested that the CPE could be related to random mixtures of a conductor and an insulator in the effective medium approximation. According to Ricciardi et al. (2009) the CPEs can be applied in the case of activation energy distribution in the reduction process. CPEs can also describe contributions from dynamic disorders evoked, for example, by diffusion (Metikos-Hukovic et al., 1999). In Starczewska et al. (2012) some additional explanations were given: the CPEs represent time-dependent phenomena. Such phenomena occur when dipoles fluctuate due to interaction with their surroundings.

Model presented by Starczewska et al. (2012) describes SbSI gel as a simple RC circuit. However, this description is also appropriate for the case of a random distribution of elements possessing various properties. Authors of Castela and Simões (2003a) presented that in the case of a multilayer structure with uniform composition and properties, a complicated equivalent circuit of a finite series of parallel RC's networks can be transformed into a simple circuit. Such a simple circuit is composed of a resistance in series with a network of parallel other resistance and one capacitance. However, in the case of most real systems, the impedance response is better described by a resistance in parallel with a CPE (Castela and Simões, 2003b).

The equivalent circuits shown in Figs. 7.14a,b fit better the Nyquist characteristics of SbSI gel in cases of low and large RH, respectively. The resistance R_3 in series in circuit, presented in Fig. 7.14a, is replaced by capacitance C_2 parallel with resistance R_2 in the circuit shown in Fig. 7.14b. The R_2C_2 component in the last equivalent circuit is needed for fitting the responses registered at higher frequencies with the increase of RH.

Fig. 7.15 presents values of parameters determined from similar fittings performed for various humidities and temperatures of SbSI gel (Starczewska et al., 2012). Increase of humidity causes decrease of the resistance R_1 (Fig. 7.15). Due to small diameters of SbSI nanowires the ratio of surface-to-volume is large. Therefore, one can expect that electrical conductance of SbSI gel is mainly determined by surface conduction mechanism. As in cases of other semiconductors, the R_1 can decrease with increasing temperature. However, the resistance of SbSI gel also decreases with the increase of the number of adsorbed H_2O layers. Additionally, the resistance of SbSI gel decreases with the increase of ion mobility in the adsorbed water.

One can observe different slopes of the $R_1(T)$ dependencies for humidities lower and higher than 35%. Similar results were presented in Faia et al. (2004). The resistance R_3 is practically independent of temperature. However, it seems that R_3 increases with increasing humidity (Fig. 7.15). Surprisingly the value of R_2 (the component of the equivalent circuit presented in Fig. 7.14b) is much greater than R_3 present in equivalent circuit for small RH (Fig. 7.14a). Probably the R_3 also exists in series with the R_2C_2 component in the equivalent circuit for large RH. However, due to its small value, in comparison with the R_2C_2 resistance, the R_3 is undetectable in the used method of investigations. The resistance R_2 decreases with the increase of humidity (Fig. 7.15) while the capacitance C_2 is practically independent of humidity (Fig. 7.15).

Value of C_1 depends very little on temperature (Fig. 7.15). For RH higher than 35% the coefficient n describing CPE practically equals 0.5 (Fig. 7.15). Therefore, CPE is related with diffusion processes in this range of humidity. The exponent n increases with

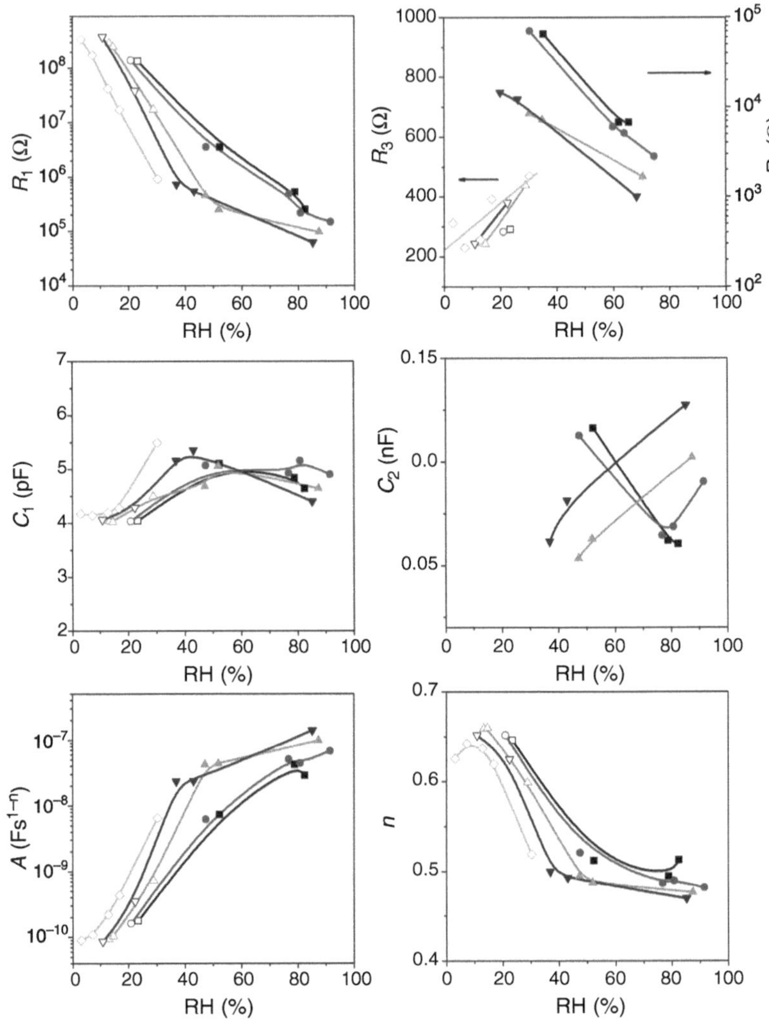

Figure 7.15. Influence of temperature on humidity dependence of resistance and capacitance parameters of equivalent circuits. As presented in Figure 7.14 (■, $T = 279$ K; ●, $T = 281$ K; ▲, $T = 288$ K; ▼, $T = 293$ K; ◆, $T = 311$ K) (*open and full signs*, models presented in Fig. 7.14a,b). Reprinted from Starczewska et al. 2012, Influence of humidity on impedance of SbSI gel, Sensor. Actuator. A, 183, 34–42, Copyright (2012), with permission from Elsevier.

decreasing humidity (Fig. 7.15) and attains the value of about 0.65 for RH = 10%. Probably, it is due to changes of the adsorbed H_2O that fills up the volume between the ends of SbSI nanowires (Starczewska et al., 2012). The value of *n* seems to be little dependent on temperature (Fig. 7.15). The parameter A describing CPE increases two orders of magnitude with the increase of humidity for RH smaller than 35%. In the case of humidity higher than 35%, the increase of parameter A with increasing RH is smaller.

One can explain the different behavior of frequency dependence of impedance of SbSI gel in the cases of low and high humidity, after taking into account chemisorbed and physisorbed water molecules. The tightly bound (water molecules trapped at iodine vacancies) and loosely associated water exists in SbSI gel in different ratios according to the humidity level.

5 Conclusions

Research and production of nanobiosensors and gas nanosensors applicable to prevention of food spoilage have relatively short history. However, there are many types of useful sensors. Among them there are humidity sensors constructed of antimony sulfoiodide (SbSI) nanowires, for example, the conductive, photoconductive, capacitive, as well as impedance SbSI sensors. The current flowing through ferroelectric and semiconducting SbSI nanowires is mainly controlled by the surface effect, which is sensitive to the environment. The high surface-to-volume ratio of SbSI nanowires makes them attractive for gas sensing applications. The not only quantitatively but also qualitatively different transient responses of current on illumination changes proved the strong interaction of this nanomaterial with the water dipoles.

The sonochemical synthesis is a simple, convenient, fast, mild, efficient, and environmentally friendly route for producing SbSI nanowires in great amount from elements: antimony, sulfur, and iodine in a single step at relatively low temperature. It is highly favorable for future mass production of multiple nanowires sensors. The produced nanowires can be used in the form of xerogel or can be transferred from the xerogel to separate devices, for example, realizing the resistors, photoresistors, and the FET-based gas sensors. The presented results show the special usefulness of SbSI xerogel as humidity sensor. When the humidity increases to 80%, the dark current increases by nearly three orders of magnitude and the capacity increases by nearly two orders of magnitude. Single nanowire semiconductor sensors have smaller power requirements than multiple nanowires sensors. However, the last sensors (eg, fabricated from SbSI gel) are simpler for fabrication and need minimal processing steps. This advantage is highly favorable for future mass production of the multiple nanowires sensors.

Small-sized SbSI nanosensors of humidity can find a lot of applications in prevention of food spoilage. However, the recognition of the response of this new type of sensors for different gasses is necessary to understand the adsorption processes in ferroelectric nanostructures. Probably, the presented transient photoconductivity method of humidity sensing is more general: alcohols, aromatics, amines, and phosphonates should all be adsorbed and photodesorbed from SbSI. Such detailed investigations should be provided in near future.

The key issue in the interpretation of experimental data is a model equivalent circuit that represents the sample in a realistic way. SbSI gel is a complicated object in which nanowires are connected with each other in parallel and in series. Simultaneously, interfaces (ie, nanowires–nanowires and nanowires–electrodes) are present in samples made of SbSI gel. In humid environment films of H_2O covering the nanowires as well as water affecting the mentioned interfaces make the electrical description of the sample more complicated. From a practical point of view, the models should not possess too many components. Obviously, the fitting of many variables becomes unreliable. Despite the difficulties in physical sense interpretation of the components, the equivalent circuits fit well the impedance properties of the investigated samples. Therefore, the equivalent circuits allow one to describe the humidity sensor properties of the SbSl gel for different temperatures.

Acknowledgments

The authors are indebted to their colleagues of the Institute of Physics of the Silesian University of Technology, Katowice (Poland), for their aid in carrying out the investigations on SbSI sensors.

This paper was partially supported by the Silesian University of Technology, Gliwice, Poland), projects no. BK-243/RIF/2016 and BKM-500/RIF/2016.

References

Adamowicz, B., 1990. Investigation of electron processes at the p- and n-type Si(111) real surface by the surface photovoltage method. Surf. Sci. 232, 1–8.

Adley, C.C., 2014. Past, present, and future of sensors in food production. Foods. 3, 491–510.

Agarwal, S., Sharma, G.L., 2002. Humidity sensing properties of (Ba, Sr) TiO_3 thin films grown by hydrothermal–electrochemical method. Sens. Actuat. B. 85, 205–211.

Agmon, N., 1995. The Grotthuss mechanism. Chem. Phys. Lett. 244, 456–462.

Akimov, B.A., Bogoyavlenskiy, V.A., Ryabova, L.I., Vasil'kov, V.N., 2000. Experimental study of negative photoconductivity in n-PbTe(Ga) epitaxial films. Phys. Rev. B. 61, 16045–16051.

Àlvarez, R., Guerrero, F., Garcia-Belmonte, G., Bisquert, J., 2002. Study of the humidity effect in the electrical response of the KSbMoO6 ionic conductive ceramic at low temperature. Mater. Sci. Eng. B 90, 291–295.

Aziz, F., Sayyad, M.H., Sulaiman, K., Majlis, B.H., Karimov, K.S., Ahmad, Z., Sugandi, G., 2012. Influence of humidity conditions on the capacitive and resistive response of an Al/VOPc/Pt co-planar humidity sensor. Meas. Sci. Technol. 23, 014001.

Barsoukov, E., Macdonald, J.R. (Eds.), 2005. Impedance Spectroscopy, Theory, Experiment, and Applications, second ed, Wiley-Interscience, New York.

Batzill, M., Diebold, U., 2007. Surface studies of gas sensing metal oxides. Phys. Chem. Chem. Phys. 9, 2307–2318.

Batzill, M., Bergermayer, W., Tanaka, I., Diebold, U., 2006. Tuning the chemical functionality of a gas sensitive material: water adsorption on SnO_2 (101). Surf. Sci. 600, L29–L32.

Berti, F., Laschi, S., Palchetti, I., Rossier, J.S., Reymond, F., Mascini, M., Marrazza, G., 2009. Microfluidic-based electrochemical genosensor coupled to magnetic beads for hybridization detection. Talanta 77, 971–978.

Bettazzi, F., Lucarelli, F., Palchetti, I., Berti, F., Marrazza, G., Mascini, M., 2008. Disposable electrochemical DNA-array for PCR amplified detection of hazelnut allergens in foodstuffs. Anal. Chim. Acta 614, 93–102.

Biju, K.P., Jain, M.K., 2008. Sol–gel derived TiO_2:ZrO_2 multilayer thin films for humidity sensing application. Sens. Actuat. B 128, 407–413.

Billakanti, J.M., Fee, C.J., Lane, F.R., Kash, A.S., Fredericks, R., 2010. Simultaneous, quantitative detection of five whey proteins in multiple samples by surface plasmon resonance. Int. Dairy J. 20, 96–105.

Bube, R.H., 1992. Photoelectronic Properties of Semiconductor. Cambridge University Press, Cambridge.

Bültzingslöwen, C., McEvoy, A.K., McDonagh, C., MacCraith, B.D., Klimant, I., Krause, C., Wolfbeis, O.S., 2002. Sol–gel based optical carbon dioxide sensor employing dual luminophore referencing for application in food packaging technology. Analyst 127, 1478–1483.

Cao, Q., Zhao, H., Yang, Y., He, Y., Ding, N., Wang, J., Wu, Z., Xiang, K., Wang, G., 2011. Electrochemical immunosensor for casein based on gold nanoparticles and poly(L-Arginine)/multiwalled carbon nanotubes composite film functionalized interface. Biosens. Bioelectron. 26, 3469–3474.

Castela, A.S., Simões, A.M., 2003a. An impedance model for the estimation of water absorption in organic coatings. Part I: A linear dielectric mixture equation. Corrosion Sci. 45, 1631–1646.

Castela, A.S., Simões, A.M., 2003b. An impedance model for the estimation of water absorption in organic coatings. Part II: A complex equation of mixture. Corrosion Sci. 45, 1647–1660.

Chu, P.T., Lin, C.S., Chen, W.J., Chen, C.F., Wen, H.W., 2012. Detection of gliadin in foods using a quartz crystal microbalance biosensor that incorporates gold nanoparticles. J. Agric. Food Chem. 60, 6483–6492.

Comini, E., 2006. Metal oxide nano-crystals for gas sensing. Anal. Chim. Acta 568, 28–40.

Comini, E., Faglia, G., Sberveglieri, G., Calestani, D., Zanotti, L., Zha, M., 2005. Tin oxide nanobelts electrical and sensing properties. Sens. Actuator. B. 111, 2–6.

Comini, E., Faglia, G., Sberveglieri, G., 2009. Solid State Gas Sensing. Springer, New York.

Crosson, C., Thomas, D., Rossi, C., 2010. Quantification of immunoglobulin g in bovine and caprine milk using a surface plasmon resonance-based immunosensor. J. Agric. Food Chem. 58, 3259–3264.

Cushen, M., Kerry, J., Morris, M., Cruz-Romero, M., Cummins, E., 2012. Nanotechnologies in the food industry: recent developments, risks, and regulation. Trends Food Sci. Technol. 24, 30–46.

Duncan, T.V., 2011. Applications of nanotechnology in food packaging and food safety: barrier materials, antimicrobials and sensors. J. Colloid. Interf. Sci. 363, 1–24.

Etefagh, R., Azhir, E., Shahtahmasebi, N., 2013. Synthesis of CuO nanoparticles and fabrication of nanostructural layer biosensors for detecting *Aspergillus niger* fungi. Sci. Iran. F. 20 (3), 1055–1058.

Faia, P.M., Furtado, C.S., Ferreira, A.J., 2004. Humidity sensing properties of a thick-film titania prepared by a slow spinning process. Sensor. Actuat. B. 101, 183–190.

Fridkin, V.M., 1980. Ferroelectric Semiconductors. Consultants Bureau, New York.

Fu, X.Q., Wang, C., Feng, P., Wang, T.H., 2007. Anomalous photoconductivity of CeO_2 nanowires in air. Appl. Phys. Lett. 91, 073104.

Garbarz, B., Bak, W., Starzyk, F., 2000. The IR properties of SbSI and SbSI:Cl ferroelectric crystal needles and μ–crystals. Ferroelectrics 239, 9–16.

Gerzanich, E.I., Lyakhovitskaya, V.A., Fridkin, V.M., Popovkin, B.A., 1982. SbSI and other ferroelectric $A^{V}B^{VI}C^{VII}$ materials. In: Kaldis, E. (Ed.), Current Topics in Materials Science. Amsterdam, North-Holland, pp. 55–190.

Gfeller, K.Y., Nugaeva, N., Hegner, M., 2005. Micromechanical oscillators as rapid biosensor for the detection of active growth of *Escherichia coli*. Biosens. Bioelectron. 21, 528–533.

Gladkovskaya, O., Rybina, I., Gun'ko, Yu.K., Erxleben, A., O'Connor, G.M., Yu, Rochev, 2015. Water-based ultrasonic synthesis of SbSI nanoneedles. Mater. Lett. 160, 113–116.

Godoy-Navajas, J., Aguilar Caballos, M.P., Gómez-Hens, A., 2011. Heterogeneous immunoassay for soy protein determination using nile blue-doped silica nanoparticles as labels and front-surface long-wavelength fluorimetry. Anal. Chim. Acta 701, 194–199.

Grekov, A.A., Rodin, A.I., Fridkin, V.M., 1971. Induced impurity photoconductivity of the ferroelectric semiconductor SbSI. Fiz. Tekh. Poluprovod 5, 1287–1292.

Gundel, H., Riege, H., Wilson, E.J.N., Handerek, J., Zioutas, K., 1989. Fast polarization changes in ferroelectrics and their application in accelerators. Nucl. Instrum. Methods Phys. Res. Section A 280, 1–6.

Ha, D.H., Nham, H., Yoo, K.-H., So, H.-M., Lee, H.-Y., Kawai, T., 2002. Humidity effects on the conductance of the assembly of DNA molecules. Chem. Phys. Lett. 355, 405–409.

Haes, A.J., Stuart, D.A., Nie, S., Van Duyne, R.P., 2004. Using solution-phase nanoparticles, surface-confined nanoparticle arrays and single nanoparticles as biological sensing platforms. J. Fluoresc. 14, 355–367.

Haspel, H., Bugris, V., Kukovecz, A., 2013. Water sorption induced dielectric changes in titanate nanowires. J. Phys. Chem. C 117, 16686–16697.

He, L., Haynes, C.L., Diez-Gonzalez, F., Labuza, T.P., 2011. Rapid detection of a foreign protein in milk using IMS–SERS. J. Raman Spectrosc. 42, 1428–1434.

Henderson, M.A., 2002. The interaction of water with solid surfaces: fundamental aspects revisited. Surf. Sci. Rep. 46, 1–308.

Huang, Y., Bell, M.C., Suni, I.I., 2008. Impedance biosensor for peanut protein Ara h 1. Anal. Chem. 80, 9157–9161.

Huang, Y., Suni, I.I., 2008. Degenerate Si as an Electrode Material for Electrochemical Biosensors. J. Electrochem. Soc. 155, J350–J354.

Huang, H.M., Chen, R.S., Chen, H.Y., Liu, T.W., Kuo, C.C., Chen, C.P., Hsu, H.C., Chen, L.C., Chen, K.H., Yang, Y.J., 2010. Photoconductivity in single AlN nanowires by subband gap excitation. Appl. Phys. Lett. 96, 062104.

Indyk, H.E., 2009. Development and application of an optical biosensor immunoassay for α-lactalbumin in bovine. Int. Dairy J. 19, 36–42.

Ishikawa, K., 1980. Electric field effect on the absorption edge in SbSI, Japan. J. Appl. Phys. 19, 1301–1309.
Ishikawa, K., Tomoda, W., Toyoda, K., 1984. Crystal growth of SbSI from the vapor phase. J. Cryst. Growth 69, 399–403.
Iwata, Y., Fukui, S., Koyano, N., Shibuya, I., 1966. Refined crystal structure and type of phase transition in ferroelectric SbSI. J. Phys. Soc. Jpn. 21, 1846–1847.
Jacobson, P.A., Rosa, L.G., Othon, C.M., Kraemer, K.L., Sorokin, A.V., Ducharme, S., Dowben, P.A., 2004. Water absorption and dielectric changes in crystalline poly vinylidene fluoride-trifluoroethylene. copolymer films. Appl. Phys. Lett. 84, 88–90.
Kiasari, N.M., Soltanian, S., Gholamkhass, B., Servati, P., 2012. Room temperature ultra-sensitive resistive humidity sensor based on single zinc oxide nanowire. Sens. Actuator. A 182, 101–105.
Kikuchi, A., Oka, Y., Sawaguchi, E., 1967. Crystal structure determination of SbSI. J. Phys. Soc. Jpn. 23, 337–354.
Kim, D., Choi, J., 2010. Water absorption dependent photoconductivity of a freestanding polymer bridge. J. Phys. Condens. Matter 22, 242201.
Ko, S., Gunasekaran, S., Yu, J., 2010. Self-indicating nanobiosensor for detection of 2,4-dinitrophenol. Food Control 21, 155–161.
Kronik, L., Shapira, Y., 2001. Surface photovoltage spectroscopy of semiconductor structures: at the crossroads of physics, chemistry, and electrical engineering. Surf. Interf. Anal. 31, 954–965.
Kuswandi, B., Wicaksono, Y., Jayus, Abdullah, A., Heng, L.Y., Ahmad, M., 2011. Smart packaging: sensors for monitoring of food quality and safety. Sens. Instrumen. Food Qual. 5, 137–146.
Kwolek, P., Pilarczyk, K., Tokarski, T., Mech, J., Irzmański, J., Szaciłowski, K., 2015. Photoelectrochemistry of n-type antimony sulfoiodide nanowires. Nanotechnology 26, 105710.
Li, D., Zhao, M.H., Garra, J., Kolpak, A.M., Rappe, A.M., Bonnell, D.A., Vohs, J.M., 2008. Direct in situ determination of the polarization dependence of physisorption on ferroelectric surfaces. Nat. Mater. 7, 473–477.
Liu, Y., Chakrabartty, S., Alocilja, E.C., 2007. Fundamental building blocks for molecular biowire based forward error-correcting biosensors. Nanotechnology 18, 424017.
Luechinger, N.A., Loher, S., Athanassiou, E.K., Grass, R.N., Stark, W.J., 2007. Highly sensitive optical detection of humidity on polymer/metal nanoparticle hybrid films. Langmuir 23, 3473–3477.
Macdonald, J.R., 1992. Complex Nonlinear Least-Squares Immitance Fitting Program LEVM. Available from http://www.physics.unc.edu/~macd/
Matys, M., Adamowicz, B., Hashizume, T., 2012. Determination of the deep donor-like interface state density distribution in metal/Al_2O_3/n-GaN structures from the photocapacitance–light intensity measurement. Appl. Phys. Lett. 101, 231608.
Merz, W.J., Nitsche, R., 1964. Ferroelectricity in SbSI and other compounds of the group 15-16-17 elements. Izvest. Akad. Nauk SSSR Ser. Fiz. 28, 681–682.
Metikos-Hukovic, M., Omanovic, S., Jukic, A., 1999. Impedance spectroscopy of semiconducting films on tin electrodes. Electrochim. Act. 45, 977–986.
Mills, A., Hazafy, D., 2008. A solvent-based intelligence ink for oxygen. Analyst 133, 213–218.
Mistewicz, K., Nowak, M., Wrzalik, R., Śleziona, J., Wieczorek, J., Guiseppi-Elie, A., 2016. Ultrasonic processing of SbSI nanowires for their application to gas sensors. Ultrasonics 69, 67–73.
Mistewicz, K., Nowak, M., Wrzalik, R., Jesionek, M., Szperlich, P., Paszkiewicz, R., Guiseppi-Elie, A., 2013. Quantum effects in electrical conductivity and photoconductivity of single SbSI nanowire. Acta Phys. Pol. A 124, 827–829.

Mistewicz, K., Nowak, M., Szperlich, P., Jesionek, M., Paszkiewicz, R., 2014. SbSI Single Nanowires as humidity sensors. Acta Phys. Pol. A 126, 1113–1114.

Nowak, M., 2010. Photoferroelectric Nanowires. In: Lupu, N. (Ed.), Nanowires science and technology. Intech, Croatia, pp. 269–308.

Nowak, M., Jesionek, M., 2012. Carbon nanotubes filled with ternary chalcohalides. In: Peng, X. (Ed.), Nanowires: recent advances. Intech, Croatia, pp. 263–293.

Nowak, M., Szperlich, P., Bober, Ł., Szala, J., Moskal, G., Stróż, D., 2008. Sonochemical preparation of SbSI gel. Ultrason. Sonochem. 15, 709–716.

Nowak, M., Talik, E., Szperlich, P., Stróż, D., 2009. XPS analysis of sonochemically prepared SbSI ethanogel. Appl. Surf. Sci. 255, 7689–7694.

Nowak, M., Bober, Ł., Borkowski, B., Kępińska, M., Szperlich, P., Stróż, D., Sozańska, M., 2013. Quantum efficiency coefficient for photogeneration of carriers in SbSI nanowires. Opt. Mater. 35, 2208–2216.

Nowak, M., Mistewicz, K., Nowrot, A., Szperlich, P., Jesionek, M., Starczewska, A., 2014a. Transient characteristics and negative photoconductivity of SbSI humidity sensor. Sensor. Actuator. A 210, 32–40.

Nowak, M., Nowrot, A., Szperlich, P., Jesionek, M., Kępińska, M., Starczewska, A., Mistewicz, K., Stróż, D., Szala, J., Rzychoń, T., Talik, E., Wrzalik, R., 2014b. Fabrication and characterization of SbSI gel for humidity sensors. Sens. Actuator. A 210, 119–130.

Palaniappan, L., Gnanam, F.D., Ramasamy, P., 1987. Electrical conductivity of antimony sulphoiodide. Mater. Sci. Eng. 85, L1–L3.

Papaioannou, G.J., Nowak, M., Euthymiou, P.C., 1989. Influence of illumination intensity on negative photoconductivity of Si ion-implanted GaAs:Cr. J. Appl. Phys. 65, 4864–4868.

Peng, L., Zhai, J.L., Wang, D.J., Wang, P., Zhang, Y., Pang, S., Xie, T.F., 2008. Anomalous photoconductivity of cobalt-doped zinc oxide nanobelts in air. Chem. Phys. Lett. 456, 231–235.

Pilolli, R., Monaci, L., Visconti, A., 2013. Advances in biosensor development based on integrating nanotechnology and applied to food-allergen management. Trend. Anal. Chem. 47, 12–26.

Pollet, J., Delport, F., Janssen, K.P.F., Tran, D.T., Wouters, J., Verbiest, T., Lammertyn, J., 2011. Fast and accurate peanut allergen detection with nanobead enhanced optical fiber SPR biosensor. Talanta 83, 1436–1441.

Rebe Raz, S., Liu, H., Norde, W., Bremer, M.G., 2010. Food allergens profiling with an imaging surface plasmon resonance-based biosensor. Anal. Chem. 82, 8485–8491.

Ricciardi, S., Ruiz-Morales, J.C., Nuñez, P., 2009. Origin and quantitative analysis of the constant phase element of a platinum SOFC cathode using the state-space model. Sol. Stat. Ionics 180, 1083–1090.

Sasaki, Y., 1964. Photoconductivity of ferroelectric photoconductor SbSI. J. Appl. Phys 3, 558–559.

Sclar, N., 1984. Properties of doped silicon and germanium infrared detectors. Prog. Quant. Electron. 9, 149–257.

Singh, R., Suni, I.I., 2010. Minimizing nonspecific adsorption in protein biosensors that utilize electrochemical impedance spectroscopy. J. Electrochem. Soc. 157, J334–J337.

Singh, R., Sharma, P.P., Baltus, R.E., Suni, I.I., 2010. Nanopore immunosensor for peanut protein Ara h1. Sens. Actuat. B 145, 98–103.

Sondi, I., Salopek-Sondi, B., 2004. Silver nanoparticles as antimicrobial agent: a case study on *E. coli* as a model for Gram-negative bacteria. J. Colloid. Interf. Sci. 275, 177–182.

Sozer, N., Kokini, J.L., 2008. Nanotechnology and its applications in the food sector. Trends Biotechnol. 27, 82–89.

Spanier, J.E., Kolpak, A.M., Urban, J.J., Grinberg, I., Ouyang, L., Yun, W.S., Rappe, A.M., Park, H., 2006. Ferroelectric phase transition in individual single-crystalline $BaTiO_3$ nanowires. Nano Lett. 6, 735–739.
Starczewska, A., Wrzalik, R., Nowak, M., Szperlich, Bober, Ł., Szala, J., Stróż, D., Czechowicz, D., 2008. Infrared spectroscopy of ferroelectric nanowires of antimony sulfoiodide. Infrared Phys. Technol. 51, 307–315.
Starczewska, A., Wrzalik, R., Nowak, M., Szperlich, P., Jesionek, M., Moskal, G., Rzychoń, T., Szala, J., Stróż, D., Maślanka, P., 2009. Influence of the solvent on ultrasonically produced SbSI nanowires. Ultrason. Sonochem. 16, 537–545.
Starczewska, A., Nowak, M., Szperlich, P., Toroń, B., Mistewicz, K., Stróż, D., Szala, J., 2012. Influence of humidity on impedance of SbSI gel. Sens. Actuator. A 183, 34–42.
Sun, X., Guan, L., Shan, X., Zhang, Y., Li, Z., 2012. Electrochemical detection of peanut allergen Ara h 1 using a sensitive DNA biosensor based on stem-loop probe. J. Agric. Food Chem. 60, 10979–10984.
Szperlich, P., Nowak, M., Bober, Ł., Szala, J., Stróż, D., 2009. Ferroelectric properties of ultrasonochemically prepared SbSI ethanogel. Ultrason. Sonochem. 16, 398–401.
Tao, H., Brenckle, M.A., Yang, M., Zhang, J., Liu, M., Siebert, S.M., Averitt, R.D., Mannoor, M.S., McAlpine, M.C., Rogers, J.A., Kaplan, D.L., Omenetto, F.G., 2012. Silk-based conformal, adhesive, edible food sensors. Adv. Mater. 24, 1067–1072.
Terranova, M., Orlanducci, S., Rossi, M. (Eds.), 2012. Carbon Nanomaterials for Gas Adsorption. CRC Press, Boca Raton.
Thiel, P.A., Madey, T.E., 1987. The interaction of water with surfaces: fundamental aspects. Surf. Sci. Rep. 7, 211–385.
Tiwari, D., Donn, S., 2009. Photochemistry on a polarisable semi-conductor: what do we understand today? J. Mater. Sci. 44, 5063–5079.
Tortajada-Genaro, L.A., Santiago-Felipe, S., Morais, S., Gabaldón, J.A., Puchades, R., Maquieira, Á., 2012. Multiplex DNA detection of food allergens on a digital versatile disk. J. Agric. Food Chem. 60, 36–43.
Toyoda, K., 1986. Electrical properties of SbSI crystals in the vicinity of the ferroelectric Curie point. Ferroelectrics 69, 201–215.
Traversa, E., 1995. Ceramic sensors for humidity detection: the state-of-the-art and future developments. Sens. Actuat. B 23, 135–156.
Viviani, M., Buscaglia, M.T., Buscaglia, V., Leoni, M., Nanni, P., 2001. Barium perovskites as humidity sensing materials. J. Europ. Ceramic Soc. 21, 1981–1984.
Wang, H.Y., Xiong, X.M., Zhang, J.X., 2010a. Moisture effect on the dielectric and structure of $BaTiO_3$-based devices. Microelectron. Reliab. 50, 887–890.
Wang, J., Munir, A., Li, Z., Zhou, H.S., 2010b. Aptamer-Au NPs conjugates-accumulated methylene blue for the sensitive electrochemical immunoassay of protein. Talanta 81, 63–67.
Wang, W., Han, J., Wu, Y., Yuan, F., Chen, Y., Ge, Y., 2011a. Simultaneous detection of eight food allergens using optical thin-film biosensor chips. J. Agric. Food Chem. 59, 6889–6894.
Wang, W., Li, Y., Zhao, F., Chen, Y., Ge, Y., 2011b. Optical thin-film biochips for multiplex detection of eight allergens in food. Food Res. Int. 44, 3229–3234.
Xiulan, S., Yinzhi, Z., Jingdong, S., Liyan, S., He, Q., Weijuan, Z., 2010. A quartz crystal microbalance-based Immunosensor for Shrimp Allergen Determination in Food. Eur. Food Res. Technol. 231, 563–570.
Yang, C., Gu, B., Xu, C., Xu, X., 2011. Self-assembled ZnO quantum dot bioconjugates for direct electrochemical determination of allergen. J. Electroanal. Chem. 660, 97–100.

Zhang, W.-H., Zhang, W.-D., 2008. Fabrication of SnO_2–ZnO nanocomposite sensor for selective sensing of trimethylamine and the freshness of fishes. Sens. Actuator. B 134, 403–408.

Zhang, C., Yin, A.-X., Jiang, R., Rong, J., Dong, L., Zhao, T., Sun, L.-D., Wang, J., Chen, X., Yan, C.-H., 2013. A general time-temperature indicator for perishable products based on kinetically programmable Ag overgrowth on Au nanorods. ACS Nano 7, 4561–4568.

8

BIOSENSOR TECHNOLOGIES FOR ANALYSES OF FOOD CONTAMINANTS

Elif Burcu Bahadır*, Mustafa Kemal Sezgintürk**

**Namık Kemal University, Scientific and Technological Research Center, Tekirdağ, Turkey; **Namık Kemal University, Faculty of Science, Chemistry Department, Biochemistry Division, Tekirdağ, Turkey*

1 Introduction

The control of food safety in the production process and end products is a crucial step. The safe production and marketing of harmless food ensures the health of the consumer. The detection and monitoring of contaminants, such as, chemical compounds, toxins, and pathogens in foods are important for human health. Traditional methods, such as, liquid- and gas chromatography with different detectors are very sensitive and selective; but they are expensive and require several days and a lot of pretreatment steps for sample preparation. Due to these challenges, simpler and cheaper analysis methods are required and in this sense biosensors have received great attention. Biosensors are ideal devices for monitoring contaminants. They are a subgroup of chemical sensors in which a biological recognition element is used for analyte detection. They are known as integrated receptor–transducer devices and the response depends on the interaction between biorecognition element and target analyte. The main advantages of biosensors are low cost, fast, and portable detection and they enable real-time and in situ monitoring of food contaminants. Publications on biosensors used in food analysis can be found in literature. However, most of biosensors developed for food contaminant detection are not commercially available. Commercial biosensors will be useful for food quality and food safety. And they will provide unique solutions for food analysis in terms of specificity and rapidity. In the coming years the use of commercial biosensors is expected to spread in the food industry.

Nanobiosensors. http://dx.doi.org/10.1016/B978-0-12-804301-1.00008-4

This chapter provides an overview on the biosensor technology for the food contaminant detection and it focuses on analysis of food security (detection of compound contaminants, allergens, toxins, pathogens, and additives, etc.). In this context, biosensors that are used for contaminant detection are summarized by using tables and the selected biosensors are explained in detail after tables. An overview on biosensors used in the field of food is offered to readers (Van Dorst et al., 2010).

2 Biosensors

A biosensor is known as an analytical device, which incorporates biologically derived recognition molecules, such as, enzyme, antibody, cell, aptamer, or single-stranded DNA (ssDNA) with a transducer to convert a biological signal into an electrical signal. A biosensor consists of three essential elements, a bioreceptor, a transducer, and a signal-processing system. The bioreceptor is immobilized on the electrode surface that is able to detect the specific target analyte. The transducer is known as a converter. Electrochemical, optic, piezoelectric, and thermometric transducers are the common types. Fig. 8.1 shows a schematic diagram of a biosensor.

The fundamental of electrochemical biosensors is the chemical reaction between biomolecule and analyte, which either generates an amperometric current, a potential, a potentiometric charge accumulation; or alters the conductive properties or the impedance of a medium between the electrodes (Bahadır and Sezgintürk, 2015a).

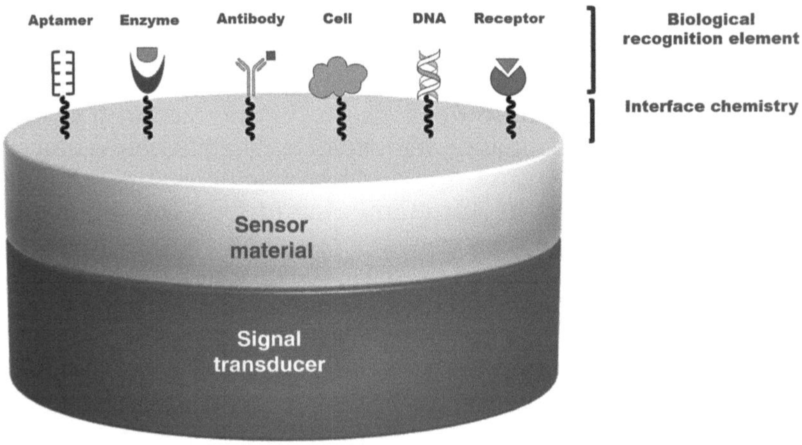

Figure 8.1. A schematic diagram of a biosensor.

An optic biosensor consists of a light source; an optic fiber, both transmitting the light and acting as the substrate; a biorecognition element, which is usually immobilized to the surface of the end of face of the fiber; and a detector to measure the output light signal. When the recognition molecule interacts with target analyte, it undergoes physiochemical transformations that alter its optic features. This transduction mechanism produces optic signals that are proportional to analyte concentration (Ligler and Taitt, 2002).

The piezoelectric biosensor detection principle is based on small changes in mass, which occur in response to the binding of the analyte to the biorecognition element on the piezoelectric crystal. After an increase in mass, a corresponding frequency shift occurs. Fig. 8.2 shows transducer types, which are used for the fabrication of these biosensors (Velusamy et al., 2010).

3 Application of Biosensors for Food Contaminants Detection

Food safety and quality is closely linked to human health. Food contaminants can originate from a vast of different sources, such as, the inappropriate use of veterinary drugs and pesticides, the formation of mycotoxins and marine toxins, and bacterial contamination. Contaminants, such as, pathogens, toxins, pesticides, and antibiotics contribute to problems (Ricci et al., 2007). The detection and monitoring of these molecules in food can be done by using conventional analytical methods, such as, ultrahigh liquid chromatography/mass spectrometry/mass spectrometry (LC/MS/MS) and gas chromatography/mass spectrometry (GC/MS); but they are time consuming, expensive, and require specialized persons. However, biosensors are cheap, fast, and portable detectors, which provide in situ and real-time analysis, without difficult sample preparation (Van Dorst et al., 2010). Food contaminants are summarized in Fig. 8.3. In this chapter, we summarized biosensors that are used for food-contaminant detection.

3.1 Biosensors for Xenobiotic Compounds in Food

3.1.1 Biosensors for Additives in Food

The food industry requires simple, cheap, and rapid analytical methods for the control of chemicals used as preservatives, dyes, and in foodstuffs. In this part of the chapter, biosensors used for the detection of additives in food are reviewed and are summarized in Table 8.1.

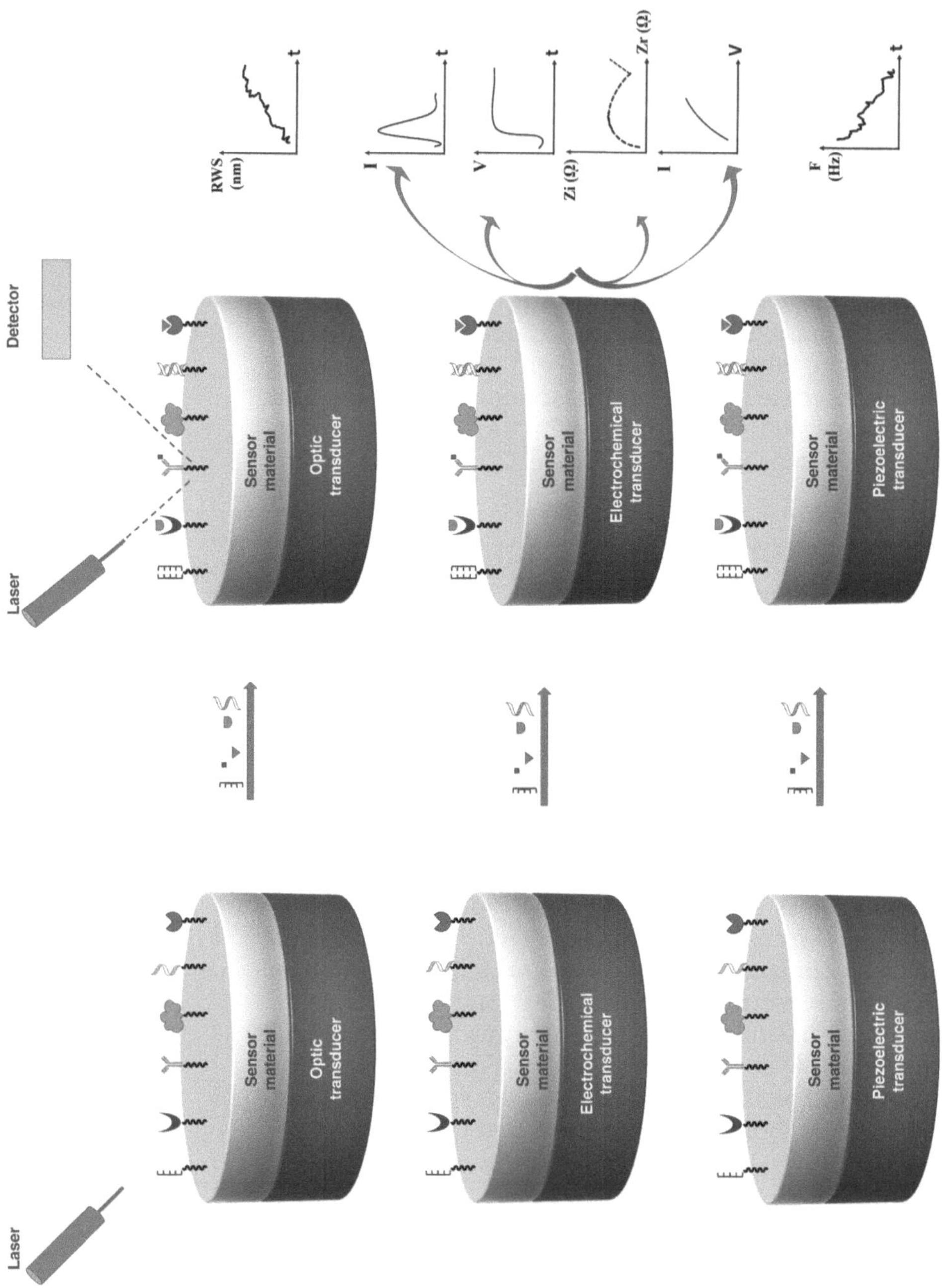

Figure 8.2. Transducer types of biosensors.

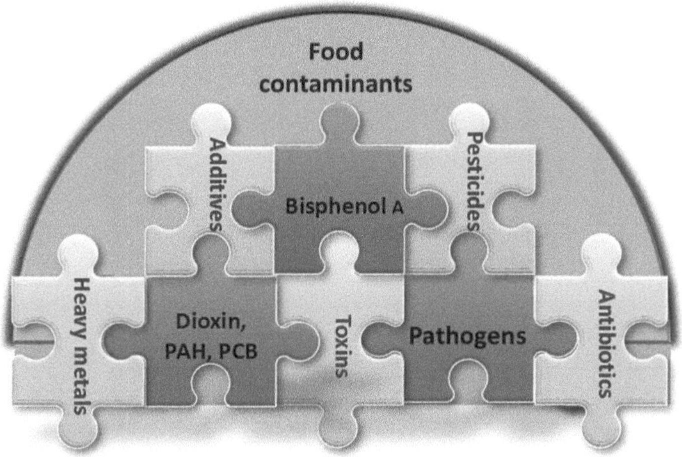

Figure 8.3. Food contaminants.

Glutamate is one of the 22 amino acids and is known as monosodium glutamate (monosodium salt) or MSG, which is employed as a flavor or taste enhancer in food (Basu et al., 2006; Pasco et al., 1999; Batra et al., 2014). They developed amperometric biosensors for MSG detection. Pasco et al. (1999) and Batra et al. (2014) used glutamate dehydrogenase and glutamate oxidase and Basu et al. (2006) used both enzymes for the construction of their biosensors. They found a linear relationship between the current and the glutamate level. They used oxygen electrode, carbon-paste wax electrode, and gold electrode as transducer elements. Due to the high conductivity of gold, it enhances electron transfer and because of this Batra et al. found the lowest detection limit. Basu et al. (2006) and Batra et al. (2014) used their biosensors in foodstuffs, such as, soy sauce, tomato sauce, chicken Thai soup, and tomato soup. They found a good correlation between these values and the values obtained by standard enzymic colorimetric method. Both of these biosensors have a half-life of over 60 days. Isa and Ghani (2009) developed chemical MSG sensor. They used 2B graphite pencil as an electrode and coated with chitosan for glutamate analysis in foods by using flow-injection analysis. They compared their sensor with high-liquid chromatography (HPLC) method and their sensor was better than the standard method in terms of reliability, simplicity, and cost.

Propyl gallate (PG) is a synthetic phenolic antioxidant in foodstuffs and can be found in dehydrated foodstuffs, edible fats, and oils. Morales et al. (2005a,b) developed amperometric PG biosensors. In both biosensors they used tyrosinase as a biorecognition element.

Table 8.1 Biosensors for Detection of Additives in Literature

Target Molecule	Immobilization Steps	Biorecognition Element	Transducer Type	Linear Range	Limit of Detection	Reference
Monosodium glutamate	Oxygen electrode/polypropylene film/L-glutamate oxidase (L-GLOD) + L-glutamate dehydrogenase (L-GLDH)	Enzyme	Electrochemical	0.02–1.2 mg/L	0.1 mg/L	Basu et al. (2006)
Propyl gallate (PG)	Graphite–teflon–tyrosinase-composite electrodes	Enzyme	Electrochemical	10^{-5}–2×10^{-4} M	9.10^{-7} M	Morales et al. (2005b)
Benzoic acid	GCE/$CaCO_3$ nanomaterials/tyrosinase	Enzyme	Electrochemical	5.6×10^{-7}–9.2×10^{-5} M	8×10^{-8} M	Shan et al. (2008)
Nitrite	GCE/HRP/[ZnCrABTS]–catalase/[ZnAlCl]	Enzyme	Electrochemical	4–50 μM	4 μM	Chen et al. (2008)
Glutamate	Graphite electrode/PVC membrane–chitosan	Chemical	Electrochemical	0.01–1 mM	0.008 mM	Isa and Ghani (2009)
Sudan II	Pencil graphite electrode/dsDNA	DNA	Electrochemical	0.5–4 μg/mL	0.4 μg/mL	Ensafi et al. (2012)
Tyramine	Pt/phosphate doped-ppy film/tyrosinase	Enzyme	Electrochemical	$4–80 \times 10^{-6}$ M	5.7×10^{-7} M	Apetrei and Apetrei (2013)
Sulfite	Au/carboxylated gold coated magnetic nanoparticles (Fe_3O_4@AuNPs)/sulfite oxidase (SOx)	Enzyme	Electrochemical	0.5–1000 μM	0.15 μM	Rawal et al. (2012)
Sulfite	Carbon-paste electrode/MWCNTs	Chemical	Electrochemical	1.6–32 mg/L	1 mg/L	Silva et al. (2015)
Glutamate	Carbon-paste wax electrodes/glutamate dehydrogenase	Enzyme	Electrochemical	2–40 mM	0.3 mM	Pasco et al. (1999)
PG	Graphite–teflon–tyrosinase-composite electrode	Enzyme	Electrochemical	$8–200 \times 10^{-6}$ M	1.3×10^{-6} M	Morales et al. (2005a,b)
2-Tert-butylhydroquinone	GCE/Au–SnO_2/gold nanowires–SWCNTs	Chemical	Electrochemical	0.05–230 μM	0.058 μM	Du et al. (2014)
Glutamate	Au/ppy nanoparticles–polyaniline film (ppyNPs–PANI)/glutamate oxidase (GluOx)	Enzyme	Electrochemical	0.02–400 μM	0.1 nM	Batra et al. (2014)

Benzoic acid	Oxygen electrode/mushroom tissue + glutaraldehyde (GA)/phenol oxidase	Enzyme	Electrochemical	25–100 μM	25 μM	Sezgintürk et al. (2005a)
Sulfite	Platinium foil/SOx	Enzyme	Electrochemical	0.01–1 mM	—	Zhao et al. (2006)
Sulfite	ITO/Prussian blue nanoparticles/ppy-composite/SOx	Enzyme	Electrochemical	0.5–1000 μM	0.12 μM	Rawal and Pundir (2012)
Sulfite	Aluminium electrode/PANI/SOx	Enzyme	Electrochemical	0.006–5 mM	0.002 mM	Bahmani et al. (2010)
Sulfite	Oxygen electrode/Teflon membrane/*M. vulgaris* tissue homogenate/GA	Cell	Electrochemical	0.2–1.8 μM	0.2 μM	Sezgintürk and Dinçkaya (2005)
Benzoic acid	Pt foil/PANI/GA/PPO	Enzyme	Electrochemical	0.3–40 μM	0.3 μM	Li et al. (2010)
Malic acid	SPCE/AuNPs/tetrathiafulvalene (TTF)/malate quinone oxidoreductase (MQO)/BSA	Enzyme	Electrochemical	2–19.6 μM	2 μM	Molinero-Abad et al. (2014)
Nitrite	GCE/polydiphenylamine (PDDA)–Pt composite	Chemical	Electrochemical	2 μM–9.5 mM	1 μM	Unnikrishnan et al. (2013)
L-Glutamate	Filter paper/nafion (NF) sol–gel silicate–3,3′5,5′-tetramethylbenzidine (TMB)/chitosan–GluOx–HRP	Enzyme	Optic	0.01–10 mM	5 μM	Muslima et al. (2012)
Sulfite	SPCE/redox active osmium polymer/SOx	Enzyme	Electrochemical	1–100 μM	0.5 μM	Spricigo et al. (2010)
Sulfite	GCE/mercury film/SOx + gelatin	Enzyme	Electrochemical	2×10^{-4}–2.8×10^{-3} M	2×10^{-4} M	Dinçkaya et al. (2007)
Amaranth Ponceau 4R	GCE/CNT–ppy	Chemical Chemical	Electrochemical Electrochemical	4×10^{-9}–5×10^{-7} M 8×10^{-9}–10^{-6} M	5×10^{-10} M 10^{-9} M	Wang et al. (2015b)
2-Phenylethanol	Oxygen electrode/polyelectrolyte complex gel membrane + sodium alginate + cellulose sulfate + poly(methylene)-*co*-guanidine/*Gluconobacter oxydans*	Cell	Electrochemical	0.02–0.7 mM	1 μM	Schenkmayerová et al. (2015)

The principle of their biosensors was based on electrochemical oxidation of PG at the graphite–Teflon-composite electrode.

Benzoic acid is a preservative that is added to foods and beverages to prevent microbial growth or undesirable chemical changes. However, higher level of benzoic acid is harmful to human health (Shan et al., 2008; Li et al., 2010). Shan et al. (2008), Sezgintürk et al. (2005a) and Li et al. (2010) developed amperometric biosensors for benzoic acid detection. Shan et al. (2008), Sezgintürk et al. (2005a) and Li et al. (2010) used tyrosinase, mushroom tissue, and polyphenol oxidase as a biorecognition element, respectively. The biosensor principle of Shan et al. (2008) was based on inhibition of tyrosinase with benzoic acid. Other biosensors' principle was based on the inhibitory effect of benzoic acid to polyphenol-oxidase enzyme. The mushroom tissue is rich in terms of polyphenol-oxidase enzyme. Benzoic acid represents competitive inhibitory kinetics to polyphenol oxidase. The principle of the method was based on the decrease in the differentiation of oxygen concentration that had been caused by the inhibition of polyphenol oxidase by benzoic acid. Researchers determined the benzoic acid concentration in real beverage sample by using biosensor and HPLC method. The relative errors were acceptable; also the results demonstrated the practical usage of the biosensor.

Nitrite is widely used as a preservative in cured and processed meats because it is a color- and flavor enhancer (Unnikrishnan et al., 2013). Chen et al. (2008) developed an amperometric biosensor for nitrite determination. For the construction of the biosensor they used catalase and horseradish peroxidase (HRP). Catalase and HRP were immobilized on the outer layer and on the layer of the electrode surface. In the presence of H_2O_2, a cathodic current was expected because of the enzymatic reduction of H_2O_2. But HRP is theoretically suppressed by the H_2O_2 decomposition by catalase. Nitrite inhibits this catalase activity and thus causes a decrease in catalase activity whereas nitrite has no effect on HRP activity.

Sulfite is used as a preservative agent in food, beverages, and pharmaceuticals to prevent oxidation and bacterial growth and reduce enzymatic reactions during production and storage (Rawal et al., 2012; Zhao et al., 2006; Rawal and Pundir, 2012; Bahmani et al., 2010; Spricigo et al., 2010; Dinçkaya et al., 2007). They developed an amperometric biosensor by using sulfite oxidase (SOx) as a biorecognition element. The principle of using these biosensors was based on the enzymatic oxidation of sulfite in the presence of SOx. Hydrogen peroxide was formed by oxidation of sulfite and the determination of the response current was based on oxidation of H_2O_2 at the electrode (Bahmani et al., 2010;

Dinçkaya et al., 2007). Rawal et al. (2012) and Rawal and Pundir (2012) used carboxylated gold-coated magnetic nanoparticles and Prussian-blue nanoparticles and the detection limits were lower than the other biosensors. The lower LOD was originated from high conductivity of nanoparticles. They enhance electron transfer between the electrolyte solution and electrode. Sezgintürk and Dinçkaya (2005) used *Mycena vulgaris* tissue homogenate as SOx source. For developing their biosensor, a dissolved oxygen probe was used. The principle of the biosensor was based on monitoring of the decrease of oxygen level in biological material and it was correlated with SOx activity. They monitored the consumed oxygen concentration related to the sulfite level.

Sudan dyes (Sudan I, II, III, and IV) are industrial dyes and are used as color pigments in chemical industries, such as, oils, fats, and plastics. Ensafi et al. (2012) developed a genosensor for the determination of Sudan II. Their study was composed of two sections; in the first part, they used Sudan II as a DNA intercalate on double-stranded (dsDNA)-modified pencil graphite electrode and monitored the changes of guanine and adenine oxidation signals before and after the intercalation. The changes between guanine- and adenine-oxidation responses were proportional to the Sudan II level. In the second part for the determination of Sudan II, adsorptive-stripping voltammetry was used on pretreated pencil-graphite electrode for the determination of Sudan II. The peak current was linearly dependent on the Sudan II level.

2-Tert-butylhydroquinone (TBHQ) is an antioxidant and frequently used to prevent oxidative degradation of fats and oils (Du et al., 2014). They developed an electrochemical sensor based on nanocomposites-modified glassy carbon electrode for the detection of TBHQ. The synergistic effects between Au, SnO_2, and graphenes–single-walled carbon nanotubes (SWCNTs) facilitated the electron-transfer reaction and increased the electroactive surface area. The sensor had advantages, such as, low cost, robustness, and high sensitivity for detecting TBHQ in real samples. Considering these advantages, the sensor holds great promise for applications in food industry.

Amaranth (E123) and Ponceau 4R (E124) are dyes used as additives in food and they are in the strict restriction list (Wang et al., 2015b). They developed a sensor, which was based on new carbon nanotube (CNT) and polypyrrole (ppy)-composite modified electrode for the simultaneous determination of the isomers of Amaranth and Ponceau 4R. They applied their sensor for the simultaneous detection of Ponceau 4R and Amaranth in fruit drink samples and also this method provided a useful tool for their simultaneous determination.

Phenyl acetic acid (PA) is used as a flavor in food industry. PA is produced either by the hydrolysis of benzyl cyanide using strong, toxic, and hazardous acids or by the transformation of L-phenylalanine via the Ehrlich pathway of yeasts to 2-phenylethanol (2-PE) and PA using acetic acid bacteria (Schenkmayerová et al., 2015). Schenkmayerová et al. (2015) developed a microbial biosensor for the detection of 2-PE by using *G. oxydans* as a biorecognition element. They immobilized the bacteria onto modified Clark-oxygen electrode. The change of oxygen concentration depended on 2-PE concentration. PA and phenyl acetaldehyde did not interfere with the biosensor response. The results were compatible with the gas-chromatography method.

3.1.2 Biosensors for Antibiotics in Food

Most of antibiotics originate from natural sources, as well as, semisynthetic or synthetic molecules. They have been used for treatment of bacterial diseases in humans and animals. When antibiotics are used in excess, it promotes the development of bacterial resistance and leads to a decrease in their efficiency in treating human or animal diseases. Monitoring of antibiotics is very important for safety of food products for human consumption (Fernández et al., 2010; Wang et al., 2014b; Mishra et al., 2014). The detection of antibiotics and drugs is important for food quality and safety. In new technologies, there is a lack of rapid and sensitive screening of drug residues in foods. Biosensors for antibiotic detection in food are summarized in Table 8.2.

Chloramphenicol (CAP) is a broad-spectrum antibiotic, which inhibits the activity of both Gram-positive and Gram-negative bacteria (Yadav et al., 2014). Fernández et al. (2010) developed an optic immunosensor for CAP detection. They used a novel portable six-channel SPR biosensor based on the plasmon of gold-diffraction grating surface. Their optic immunosensor could analyze simultaneously on-site antibiotic residues from three different families of fluoroquinolones, sulfonamides, and phenicols in whole milk samples. The chips were covalently biofunctionalized with haptenized proteins by means of a previously formed mixed self-assembled monolayer (SAM) (two types of mercapto-alkyl reagents containing polyethyleneglycol units). The samples or standards were mixed with specific polyclonal antibodies and applied to the sensor device. Milk samples were analyzed directly without pretreatment procedures. Alibolandia et al. (2015) and Yadav et al. (2014) developed an amperometric aptasensor. Alibolandia et al. (2015) used aptamer–CdTe quantum dots and graphene oxide and Yadav et al., 2014 immobilized aptamer onto the poly(4-amino-3-hydroxynapthalenesulfonicacid)-modified

Table 8.2 Biosensors for Detection of Antibiotics in Literature

Target Molecule	Immobilization Steps	Biorecognition Element	Transducer Type	Linear Range	Limit of Detection	Reference
Enroflaxin Sulfapyridine Chloamphenicol (CAP)	Gold grating surface/polyethylene glycol + alkylthiol reagents/haptanized proteins/ethanolamine antiantibiotic solutions	Antibody Antibody Antibody	Optic Optic Optic	— — —	0.30 μg/L 0.29 μg/L 0.26 μg/L	Fernández et al. (2010)
Tetracycline (TC)	Direct competitive enzyme linked aptamer assay (dc-ELAA)/streptavidin (SA)/biotin aptamer/TC–HRP	Aptamer	Optic	0.1–1000 ng/mL	0.0978 ng/mL	Wang et al. (2015d)
Oxytetracycline (OTC)	ELAA/biotinylated aptamer/OTC–BSA/OVA/streptavidin–HRP	Aptamer	Optic	—	12.3 μg/L	Kim et al. (2014a)
CAP	GO/aptamer–CdTeQDs	Aptamer	Electrochemical	0.1–1000 nM	0.2 ppb	Alibolandia et al. (2015)
TC	Au/AP solution/aptamer	Aptamer	Electrochemical	5–5.10^3 ng/mL	1 ng/mL	Chen et al. (2014)
TC	Indirect competitive enzyme linked aptamer assay (IC-ELAA)/TC-BSA/aptamer-biotin/streptavidin–HRP	Antibody	Optic	0.1–100 ng/mL	9.6×10^{-3} ng/mL	Wang et al. (2014b)
OTC	GCE/zinc cation-exchanged montmorillonite (Zn–Mt/GCE)	Chemical	Electrochemical	0.8–40 μM	10 μM	Sun et al. (2014)
Ciprofloxacin	GCE/poly(alizarin red)/electrodeposited graphene (PAR/EGR)	Chemical	Electrochemical	4×10^{-8}–1.2×10^{-4} M	0.01 μM	Zhang et al. (2014a)
TC	GCE/MWCNTs–COOH/anti-TC aptamer/ethanolamine	Aptamer	Electrochemical	10^{-8} – 5×10^{-5} M	5×10^{-9} M	Zhou et al. (2012)
CAP	Edge plane pyrolitic graphite/poly(4-amino-3-hydroxynapthalene sulfonic acid)(*p*-AHNSA)/mercaptoethanol/aptamer	Aptamer	Electrochemical	0.1–2500 nM	0.02 nM	Yadav et al. (2014)
Benzyl penicillin	Pt/penicillinase	Enzyme	Electrochemical	2.6×10^{-7}–6.6×10^{-7} M	7.9×10^{-8} M	Prado et al. (2015)
Neomycin	Paper strip/MWCNTs/antineomycin	Antibody	Electrochemical	0.2–125 ng/mL	0.04 ng/mL	Wu et al. (2012b)
Penicillin G Ampicillin	Au/PPy/penicillin G–ampicillin–BSA conjugate	Antibody Antibody	Piezoelectric Piezoelectric	2.5–250 ng/mL 2.5–500 ng/mL	0.8 ng/mL 3.9 ng/mL	Karaseva and Ermolaeva (2014)
Streptomycin (SRT)	Au chip/11–MAU/EDC/NHS/anti-SRT	Antibody	Piezoelectric	0.3–50 ng/mL	0.3 ng/mL	Mishra et al. (2014)
CAP	Nitrocellulose membrane/anti-CAP/BSA/anti-mouse IgG–HRP conjugate/3′,3′,5′,5′-tetramethylbenzidine (TMB)	Antibody	Visual	—	0.1 ng/mL	Xing et al. (2015)

edge plane pyrolytic graphite for construction of their aptasensors. Due to the high conductivity of graphene, the aptasensors have lower detection limits. In order to assess the practical feasibility of the aptasensor, Alibolandia et al. (2015) tried to determine CAP in milk samples. The aptasensor was evaluated using the standard-addition method of calibration. The results proved that the aptasensor could be used to sense CAP in milk samples.

Xing et al. (2015) developed a multicomponent lateral-flow assay strip on an antibody–antigen interaction for the rapid and simultaneous detection of Pb (II), microcystin–leucine–arginine, CAP, testosterone, and chlorothalonil. These analytes were immobilized in the different test lines on a nitrocellulose (NC) membrane. The related antibodies of these analytes interacted with the interested analytes and no crosscontamination was observed. Samples/standards were dropped in the conjugation pad, which included colloidal gold-labelled primary-antibody composites. Then the reaction solution was dropped to the strip. The solution moved up the NC membrane with the aid of the absorbent pad. The color intensity of the test region was inversely proportional to the level of analyte in the sample.

Penicillin antibiotics have been widely used for treating and preventing infectious diseases of animals. Wu et al. (2014) developed an amperometric enzyme biosensor for penicillin analysis. The biosensor of Wu et al. (2014) was composed of layer-by-layer film containing single-graphene nanosheets preadsorbed with hematein, ionic liquids (ILs), and penicillinase. The penicillinase catalyzed the hydrolysis of penicillin to penicilloic acid, where H^+ measured amperometrically. To assess their biosensor for real-sample analysis, they used milk. For the milk samples, they found relative standard deviations (RSD) and detection limit less than 4.5% and 10^{-9} M, respectively. Thus, they decided that their biosensor can be successfully used for the monitoring of penicillin residues in milk.

Tetracycline (TC) is a member of the broad-spectrum tetracycline group of antibiotics. It has been widely used as veterinary drug and as feed additive. Chen et al. (2014) and Zhou et al. (2012) developed an aptasensor for TC determination. Chen et al. (2014) found a linear relationship between the *log* concentration of TC and the charge transfer resistance (ΔR_{ct}) and Zhou et al. (2012) reported that the current increased linearly with the increasing TC concentrations. Wang et al. (2014b) developed a biosensor based on an indirect competitive enzyme-linked aptamer assay. A 76mer-ssDNA aptamer was used as a recognition element. They used their aptasensor for detecting TC in honey and they demonstrated that their biosensor had a low detection limit and a wide linear range.

Oxytetracycline (OTC) is a member of tetracycline group and a common antibacterial agent used for the control of animal diseases (Kim et al., 2014a). They developed an optic biosensor using ssDNA aptamers for the detection of OTC. Their method was based on an indirect competitive enzyme-linked aptamer assay using a biotin-labelled aptamer. The OTC3 aptamer was used as an aptasensor probe. The OTC3 had high specificity for OTC. They compared their aptasensor with HPLC method and results were in good agreement. Sun et al. (2014) determined OTC by using direct voltammetric method at a Zn–Mt/GCE. The Zn–Mt was used to improve the direct electrochemical activity of OTC. OTC exhibited very sensitive oxidation peak on the Zn–Mt/GCE. The oxidation peak current of OTC on Zn–Mt film modified GCE was linear with its concentration.

Ciprofloxacin (CPFX) is a second-generation fluoroquinolone antibiotic used for humans and animals treatment (Zhang et al., 2014a). They modified their electrode with PAR/EGR-composite film and detected CPFX in the presence of ascorbic acid, uric acid, and dopamine. Their biosensor principle was based on electrocatalytic oxidation of CPFX on PAR/EGR. The oxidation peak current was directly proportional to the level of CPFX.

Aminoglycoside antibiotics are veterinary drugs for treating infectious diseases. Wu et al. (2012b) developed paper strip immunosensor for neomycin (an aminoglycoside antibiotic) determination. For the construction of their biosensor, they used filter paper, SWCNTs, and antineomycin antibodies. They used amperometry for the determination of neomycin concentration. Mishra et al. (2014) fabricated a biosensor for the detection of streptomycin (an aminoglycoside antibiotic) residues in milk samples using flow injection analysis-electrochemical quartz crystal nanobalance technique. Monoclonal antibody specific to streptomycin was attached onto the thiol-modified gold–quartz crystal chip. They compared their biosensor with the enzyme-linked immunosorbent assay (ELISA) kit and a good correlation was found.

3.1.3 Biosensors for Bisphenol A in Food

Bisphenol A (BPA) is one kind of the endocrine disrupting compounds (EDCs) and interferes with the normal function of the endocrine system causing adverse effects for humans. It affects not only the endocrine system, but also thyroid function, central nervous system, endocrine pancreas, immune system, and the reproduction system. BPA is a monomer used in the synthesis of epoxy resins and polycarbonate plastics, such as, feeding bottles, water bottles, food packaging, and cans. Therefore, a vast amount of BPA can migrate into food through leaching from final products

(Wang et al., 2015e; Alkasir et al., 2010). The detection of BPA with traditional methods, such as, LC/MS/MS, GC/MS, and capillary electrophoresis can detect BPA at a low level but they are time consuming, cannot be applied on-site, and require pretreatment procedures. Electrochemical sensors and biosensors are enabled for rapid and on-site detection of BPA (Wang et al., 2015e; Alkasir et al., 2010; Portaccio et al., 2010; Yin et al., 2010; Wu et al., 2012a). Biosensors for BPA detection in food are summarized in Table 8.3.

Yin et al. (2010), Alkasir et al. (2010), Pan et al. (2015), Wang et al. (2014a), and Wu et al. (2012a), developed an amperometric biosensor for BPA monitoring. All of them used tyrosinase as biorecognition element. The principle of BPA detection was based on an electrochemical reaction. BPA was hydroxylated to the corresponding phenol in the presence of tyrosinase. The tyrosinase catalyzed the oxidation of the obtained polyhydric phenol to the corresponding *o*-quinone. Finally, the obtained *o*-quinone reduced electrochemically to the corresponding polyhydric phenol on the electrode surface (Wu et al., 2012a). Yin et al. (2010) fabricated their biosensor by attaching tyrosinase on multiwalled carbon nanotubes (MWNCTs)–cobalt phthalocyanine (CoPc)–silk fibroin (SF)-composite modified GCE. In MCWNTs–CoPc–SF-composite film, SF provided a proper environment for retaining of tyrosinase activity, MWNCTs were conductive materials and increased the electron transfer and CoPc represented a good electrocatalytic activity to electrooxidation of BPA. Alkasir et al. (2010) used magnetic nickel nanoparticles (NiNPs) as an enzyme immobilization material. They compared the analytical response of NiNPs biosensor with those based on iron oxide (Fe_3O_4) and AuNPs. Pan et al. (2015) constructed their biosensor by using graphene–gold nanocomposite material as immobilization matrix of tyrosinase. Wang et al. (2014a) immobilized tyrosinase onto AuNPs and thioctic acid amide SAMs modified gold electrode. AuNPs enhanced the electrochemical response. Wu et al. (2012a), immobilized enzyme onto the electrode surface by using nanographene material. Nanographene-based tyrosinase biosensor was fast and sensitive and had a lower background current. Due to the high electrical conductivity of MWCNTs, AuNPs, magnetic nanoparticles, and graphene; they provide low detection limits. Pan et al. (2015) used AuNPs and graphene together and achieved the properties of conductivity, adsorption, and biocompatibility.

Apart from tyrosinase biosensor for BPA detection, laccase was used as the recognition element. Portaccio et al. (2013) developed a biosensor based on a bionanocomposite (laccase–thionine–carbon black)-modified screen-printed electrode. Thionine was used as an electrochemical mediator coupled with a

Table 8.3 Biosensors for Detection of BPA in Literature

Target Molecule	Immobilization Steps	Biorecognition Element	Transducer Type	Linear Range	Limit of Detection	Reference
BPA	Carbon-paste electrode (CPE)/ionic liquid/ZnO–CNTs nanocomposite	Chemical	Electrochemical	0.002–700 μM	9 nM	Najafia et al. (2014)
BPA	Nitrocellulose membrane/AuNPs–anti-BPA conjugate/BSA	Antibody	Visual	0.05–10 ng/mL	0.076 ng/mL	Mei et al. (2013)
BPA	GCE/copper centered metal organic framework (Cu-MOF)/ Tyrosinase /Chi	Enzyme	Electrochemical	5×10^{-8}–3×10^{-6} M	13 nM	Wang et al. (2015e)
BPA	GCE/grapheme hypercrosslinked resin MN202 composite	Chemical	Electrochemical	5×10^{-9}–2×10^{-5} M	1.2×10^{-9} M	Zhou et al. (2014)
BPA	CPE/graphite powder + thionine + tyrosinase	Enzyme	Electrochemical	0.15–45 μM	0.15 μM	Portaccio et al. (2010)
BPA	GCE/MWCNTs–cobalt phthalocyanine (CoPc)–silk fibrion (SF) composite/tyrosinase	Enzyme	Electrochemical	5×10^{-8}–3×10^{-6} M	3×10^{-8} M	Yin et al. (2010)
BPA	GCE/jugBPA/anti-BPA/BPA/peroxidase-labelled anti-BPA	Antibody	Electrochemical	—	2 pg/mL	Wang et al. (2014c)
BPA	GCE/MWCNTs/grapheme oxide nanoribbons (GONRs)/citosan	Chemical	Electrochemical	0.005–150 μg/L	1 ng/mL	Xin et al. (2015)
BPA	Ecetylene black paste electrode/chi-GR-composite film	Chemical	Electrochemical	80 nM–1 μM 1–20 μM	8 nM	Deng et al. (2014)
BPA	SPE/AuNPs/tyrosinase/GA/Ni or AuNPs magnetic nanoparticles SPE/AuNPs/tyrosinase/GA/Fe_3O_4 magnetic nanoparticles SPE/AuNPs/tyrosinase/GA/AuNPs magnetic NPs	Enzyme Enzyme Enzyme	Electrochemical Electrochemical Electrochemical	9.1×10^{-7}–4.8×10^{-5} M 2.2×10^{-8}–4.10^{-5} M 4.2×10^{-8}–3.6×10^{-5} M	7.1×10^{-9} M 8.3×10^{-9} M 10^{-8} M	Alkasir et al. (2010)
BPA	GCE/GR–gold nanocomposite material–tyrosinase-Chi	Enzyme	Electrochemical	0.025–3 μM	1 nM	Pan et al. (2015)
BPA	Au/thioctic acidamide (T-NH_2)/AuNPs/tyrosinase	Enzyme	Electrochemical	3.99×10^{-7}–2.34×10^{-4} mol/L	1.33×10^{-7} mol/L	Wang et al. (2014a)
BPA	Fiber optic sensor/APTES/GA/BPA–BSA/anti-flourescence labelled monoclonal antibody	Antibody	Optic	0.5–100 μg/L	0.06 μg/L	Long et al. (2014)
BPA	SCPE/laccase–thionine–carbon black/NF	Enzyme	Electrochemical	0.5–50 μM	0.2 μM	Portaccio et al. (2013)
BPA	SiO_2/APTES/gold nanoclusters (AuNCs)/BPA-MIPs	Chemical	Optic	0–13.1 μM	0.1 μM	Wu et al. (2012a)
BPA	GCE/nanographene–tyrosinase–chitosan film	Enzyme	Electrochemical	100–2000 nmol/L	33 nmol/L	Wu et al. (2012a)
BPA	GCE/AuNPs/reduced graphene oxide (rGO)–(4-ferrocenylethylene) phenylamine (Fc-NH_2)	Chemical	Electrochemical	0.005–10 μM	2 nM	Huang et al. (2015)

nanostructured carbon black. Thionine has an electrocatalytic activity towards the reduction of BPA/laccase oxidation products. They tested their biosensor using tomato juice coming from metal cans coated with epoxy resins, which are known to contain and leak BPA. No appreciable current signals were obtained. The same samples were analyzed by using HPLC and BPA concentration found to be smaller than the detection limit. To assess the accuracy of their biosensor, the samples were spiked with known amounts of BPA. The results exhibited good recovery values and a low matrix effect.

3.1.4 Biosensors for Other Heavy Metals in Food

Heavy metal contamination is a critical threat to human health because these substances are nonbiodegradable. There are many biosensors and sensors for heavy metals detection. Biosensors for heavy metal detection in food are summarized in Table 8.4.

Chouteau et al. (2005) and Berezhetskyy et al. (2008) developed conductometric biosensors by using alkaline-phosphatase enzyme as the recognition element. In both of the biosensors, heavy metals act as enzyme inhibitors. Therefore, in the presence of alkaline phosphatase and heavy metals, the reaction induces a change in pH and in conductivity. The change in conductivity was detected with a conductometric transducer (Berezhetskyy et al., 2008).

Moyo et al. (2014) developed an enzyme biosensor for Pb^{2+} and Cu^{2+} detection. They immobilized HRP on maize tassel MWCNTs through electrostatic interactions. When heavy metals are introduced into the solution, the catalytic activity and the reduction current of HRP decreases. The decrease in the reduction current was proportional to the concentration of either Pb^{2+} or Cu^{2+} in aqueous solution. Syshchyk et al. (2015) developed an enzyme-biosensor system based on porous silicon photoluminescence for detection of glucose, urea, and heavy metals. The basis of this biosensor was based on the variations in medium pH changes in photoluminescence of porous silicon. The conversion of a biochemical signal into an optic one was applied for direct determination of glucose and urea, as well as, for inhibitory analysis of heavy metal ions. The enzymes urease and GOx were used as a bioselective material. The presence of heavy metal ions (Cu^{2+}, Pb^{2+}, and Cd^{2+}) inhibited the enzymatic reactions, which caused a decrease in signal.

Ghica et al. (2013) and Ayenimo and Adeloju (2015) developed electrochemical enzyme biosensors for heavy metals and they used GOx as the recognition element. For the construction of the biosensor, they used cobalt- or copper hexacyanoferrate-modified carbon

Table 8.4 Biosensors for Detection of Heavy Metals in Literature

Target Molecule	Immobilization Steps	Biorecognition Element	Transducer Type	Linear Range	Limit of Detection	Reference
Cd^{2+}	Sensor chip/Si_3N_4 layer/BSA membranes/*Chlorella vulgaris* microalgae + GA/alkaline phosphatase + AChE	Enzyme	Electrochemical	—	10 ppb	Chouteau et al. (2005)
Zn^{2+}				—	10 ppb	
Hg^{2+}	Graphene oxide/carbon dots labelled oligodeoxyribonucleotide (ODN-CDs)	Aptamer	Optic	5–2000 nM	2.6 nM	Cui et al. (2015)
Hg^{2+}	Graphene quantum dots (GQDs)	Chemical	Optic	120–2000 nM	0.439 nM	Li et al. (2015)
Hg^{2+}	Imidazol fluorophore (PIPT)/CdTe@SiO_2 QDs	Chemical	Optic	—	6.5 nM	Ma et al. (2015)
Hg^{2+}	Au/hairpin-shaped capture probe A/MCH/methylene blue (MB)/nicking endonuclease (NEase)	Enzyme	Electrochemical	$10–10^{-5}x10^{-7}$ M	8.7×10^{-11} M	Chen et al. (2015)
Cd^{2+}	Au/AP + BSA + GA + glycerol	Enzyme	Electrochemical	—	0.5 ppm	Berezhetskyy et al. (2008)
Zn^{2+}		Enzyme	Electrochemical	—	2 ppm	
Co^{2+}		Enzyme	Electrochemical	—	2 ppm	
Ni^{2+}		Enzyme	Electrochemical	—	5 ppm	
Pb^{2+}		Enzyme	Electrochemical	—	40 ppm	
Pb^{2+}	GCE/maize tassel–MWCNTs/HRP/NF layer	Enzyme	Electrochemical	0.092–0.55 mg/L	2.5 µg/L	Moyo et al. (2014)
Cu^{2+}			Electrochemical	0.068–2 mg/L	4.2 µg/L	
Pb^{2+}	SPCE/GR/polyaniline/polystyerene	Chemical	Electrochemical	10–500 µg/L	3.30 µg/L	Promphet et al. (2015)
Cd^{2+}			Electrochemical	10–500 µg/L	4.43 µg/L	
Cd^{2+}	Au/poly(allylamine hydrochloride)-coated AuNPs/BSA/GA + *Spirulina* cells	Cell	Electrochemical	—	10^{-20} M	Tekaya et al. (2014)
Hg^{2+}				—	10^{-20} M	
Cd^{2+}	Plastic-clad silica optic fiber/silver metal + ITO/pyrrole/chitosan layer	Chemical	Optic	0–200 µg/L	0.129 nM	Verma and Gupta (2015)
Pb^{2+}				0–200 µg/L	0.158 nM	
Hg^{2+}				0–200 µg/L	0.293 nM	
Hg^{2+}	Au/2-mercaptoethanol/thiolated oligonucleotide/thymine/Hg^{2+}/thymine	DNA	Electrochemical	0.2–50 nM	0.1 nM	Tortolini et al. (2014)

(*Continued*)

Table 8.4 Biosensors for Detection of Heavy Metals in Literature (*cont.*)

Target Molecule	Immobilization Steps	Biorecognition Element	Transducer Type	Linear Range	Limit of Detection	Reference
Hg^{2+}	Au/Cu_2O + nanochitosan (Cu_2O@NCs)–ssDNA	DNA	Electrochemical	1–100 nM	0.15 nM	Liu et al. (2015b)
Pb^{2+}	Au/thiolated ssDNA/MCH/MB	DNA	Electrochemical	0.05–1 μM	34.7 nM	Jarczewska et al. (2015)
Cu^{2+}	Silicon wafer/porous silica layer/urease + GOx	Enzyme	Optic	10 nM–1 mM	10 nM	Syshchyk et al. (2015)
Pb^{2+}		Enzyme	Optic	10 nM–1 mM	10 nM	
Cd^{2+}		Enzyme	Optic	10 nM–1 mM	10 nM	
Cd^{2+}	Carbon film electrode/cobalt hexacyanoferrate films/GOx/GA/BSA	Enzyme	Electrochemical	1.5–6 μM	2.4 μM	Ghica et al. (2013)
Co^{2+}		Enzyme	Electrochemical	5–35 μM	2.1 μM	
Cu^{2+}	Carbon film electrode/cobalt hexacyanoferrate films/GOx/GA/BSA	Enzyme	Electrochemical	0.2–3 μM	0.2 μM	
Ni^{2+}		Enzyme	Electrochemical	20–120 μM	3.3 μM	
Cd^{2+}		Enzyme	Electrochemical	1.5–6 μM	1.2 μM	
Co^{2+}		Enzyme	Electrochemical	2–40 μM	0.9 μM	
Cu^{2+}		Enzyme	Electrochemical	0.2–3 μM	0.5 μM	
Ni^{2+}		Enzyme	Electrochemical	20–120 μM	4.8 μM	
Cu^{2+}	Pt/ppy–GOx	Enzyme	Electrochemical	0.079–16 μM	0.079 μM	Ayenimo and Adeloju (2015)
Hg^{2+}		Enzyme	Electrochemical	0.025–5 μM	0.025 μM	
Cd^{2+}		Enzyme	Electrochemical	0.1–15 μM	0.024 μM	
Pb^{2+}		Enzyme	Electrochemical	0.04–62 μM	0.044 μM	
Pb^{2+}	Micro well plate/DNAzyme/invertase–AuNPs–complementary DNA conjugates	Enzyme	Electrochemical	1 pM–8 nM	1 pM	Zhang et al. (2015)
As^{3+}	CPE/*Porphyridium cruentum* +graphite powder	Cell	Electrochemical	2.5–20 μg/L	2.5 μg/L	Zaib et al. (2014)
Cd^{2+}	Ion selective electrode (ISE)/nylon membrane/sol–gel [ethanol–tetra ethyl orthosilicate (TEOS)]/*Bacillus badius* cells	Cell	Electrochemical	10 μg/L–1 mg/L	1 μg/L	Verma et al. (2011)

film electrodes, glucose oxidase (GOx) attached by crosslinking with glutaraldehyde, and platinum electrode with ultrathin ppy–GOx, respectively. Both of the biosensors' measurement principles were based on the inhibition of enzyme. Ayenimo and Adeloju (2015) found lower detection limits because the platinum electrode was more conductive than the carbon-film electrode.

Tekaya et al. (2014) designed a bienzymatic biosensor by immobilizing *Arthrospira platensis* cells on gold interdigitated transducers. Phosphatase and esterase activities were inhibited by heavy metals and by pesticides, respectively and the inhibition activities were monitored at different concentrations of pesticides and heavy metals. They reported that the results showed that there was no synergistic effect between the two families of pollutants. Zaib et al., 2014 developed a novel biosensor based on carbon-paste electrode modified with *P. cruentum* biomass and used for As (III) detection in contaminated water. The ability of the biosensor in the presence of different interfering metal (Na^+, K^+, Ca^{2+}, and Mg^{2+}) ions was examined. The usability of *P. cruentum*-modified biosensor was tried by using Fe^{3+}, Mn^{2+}, Cd^{2+}, Cu^{2+}, Ni^{2+}, Hg^{2+}, and Pb^{2+} metals as interferents.

Tortolini et al. (2014) and Liu et al. (2015b) developed a DNA biosensor for Hg (II) detection. The biosensor of Tortolini et al. (2014) was based on the formation of the complex thymine–Hg–thymine (T–Hg–T). Also, Hg^{2+} tended to bind two thymines, generating a T–Hg–T complex with a formation constant higher than that one of the coupling adenine–thymine, which was used for a selective, fast, and cheap Hg^{2+} monitoring. In the presence of the Hg^{2+} in sample, it led to form of T–Hg–T complex causing an increase of the faradic current, which was detected by square-wave voltammetry. Liu et al. (2015b) used cuprous oxide (Cu_2O)- and nanochitosan composites and a thymine-rich ssDNA for the construction of a biosensor and used electrochemical impedance spectroscopy as a detection method. The Cu_2O@NCs modified with DNA showed significant sensitivity in detecting Hg^{2+} compared with pristine Cu_2O and nanochitosan. This behavior led to a high difference in charge-transfer resistance during Hg^{2+} detection.

Li et al., 2015 and Ma et al., 2015 developed optic biosensors for Hg (II) detection. They used graphene quantum dots and CdTe@SiO_2 quantum dots for the construction of their biosensors respectively. Li et al. (2015) found lower detection limit because graphene has high conductivity. Promphet et al. (2015) fabricated a graphene/polyaniline/polystyrene (G/PANI/PS) nanoporous fiber modified screen-printed carbon electrode (SPCE) for simultaneous determination of lead (Pb^{2+}) and cadmium (Cd^{2+}). Square-wave anodic-stripping voltammetry was employed for the simultaneous determination of Pb^{2+} and Cd^{2+} in the presence of bismuth (Bi^{3+}) on G/PANI/PS nanoporous fiber-modified SPCE. Under optimal conditions, the anodic current was directly proportional to metal ion concentration.

3.1.5 Biosensors for Pesticides in Food

Extensive use of pesticides causes a tremendous release of amounts of their residues in the environment, such as, soil, water, and it also affects food safety. Several techniques, such as, HPLC and LC/MS–GC/MS are used for pesticide detection. When compared to these techniques biosensor technologies are preferable because of their minimum cost, high sensitivity, fast response, and easy operation (Zhang et al., 2014b). Biosensors for pesticides detection in food are summarized in Table 8.5.

Enzyme-based biosensors have been mostly used for pesticide detection (Table 8.5). Acetylcholine esterase (AChE) inhibition-based electrochemical biosensors are important for the analysis of pesticides, due to their simplicity, rapidity, specificity, and simple sample preparation (Zhang et al., 2014b). When AChE is immobilized on the electrode surface, it hydrolyzes acetylthiocholine to thiocholine; thus an irreversible oxidation peak is observed. The increment in the pesticide concentration decreases the oxidation current of thiocholine. Zhai et al. (2013) and Zhou et al. (2013) developed an amperometric enzyme biosensor for carbofuran detection. Zhai et al. (2013) electrodeposited chitosan–Prussian blue–MWCNTs–hollow-gold nanospheres film onto the gold-electrode surface, AChE and NF were immobilized onto the film. Zhou et al. (2013) used SnO_2 nanoparticles, carboxylic graphene, NF, and chitosan to immobilize AChE on GCE. Zhou et al. (2013) found lower limit of detection (LOD) than Zhai et al. (2013), because the conductivity of graphene is better than that of MWCNTs.

Zhang et al. (2014b) developed an enzyme biosensor for paraoxon- and carbaryl detection and they used organophosphate hydrolase and AChE as biorecognition element. Organophosphate hydrolase (OPH) was recognized as an alternative recognition enzyme the for direct detection of organophosphate. They prepared their biosensor by using electrostatically layer-by-layer assembly technique and bienzyme biosensing system was incorporated electrostatically. Sinha et al. (2010), Tuoro et al. (2011), and Kesik et al. (2014) developed an amperometric enzyme biosensor by using AChE. Zinc oxide, graphite powder, and MWCNTs–poly(benzenamine) (graphite electrode) were used for the construction of their the biosensors, respectively. Using graphite powder and graphite electrode decreased LOD as a result of their high conductivity.

Oliveira et al. (2013) developed an enzyme biosensor for carbamate-pesticide monitoring. For the construction their biosensor

Table 8.5 Biosensors for Detection of Pesticides in Literature

Target Molecule	Immobilization Steps	Biorecognition Element	Transducer Type	Linear Range	Limit of Detection	Reference
Chlorpyrifos	GCE/NF–carboxylic graphene (CGR)–	Enzyme	Electrochemical	10^{-13}–10^{-8} M	5.3×10^{-14} M	Liu et al.
Carbaryl	AgNPS/AChE	Enzyme	Electrochemical	10^{-12}–10^{-8} M	5.45×10^{-13} M	(2014a,b)
Dichlorvos	Boron-doped diamond (BDD) electrode/ honeycomb-like hierarchically ion liquids–AuNPs–porous carbon composite/AChE	Enzyme	Electrochemical	4.5×10^{-13}–4.5×10^{-9} M	2.99×10^{-13} M	Wei and Wang (2015)
Paraoxon	GCE/MWCNT–polyethyleneimine (PEI)/ MWCNT–DNA/OPH + AChE	Enzyme	Electrochemical	0.5–40 and 10–80 μM	0.5 μM	Zhang et al. (2014b)
Carbaryl		Enzyme	Electrochemical	0.5–50 and 10–100 μM	10 μM	
Glyphosate	Strip/hydrogel-agarose + rhodamine	Enzyme	Visual	0–0.5 mg/L	0.1 mg/L	Liu et al.
Malathion	B-covered gold nanoparticles	Enzyme	Visual	0–0.5 mg/L	—	(2014a,b)
Phoxim	(RB-AuNPs)–hydrogel/	Enzyme	Visual	0–0.5 mg/L	—	
Dimethoate	AChE + acetylthiocholine (ATC)	Enzyme	Electrochemical	0–0.5 mg/L	0.02 mg/L	
Dichlorvos		Enzyme	Electrochemical	0–0.5 mg/L	0.01 mg/L	
Chlorpyrifos		Enzyme	Electrochemical	0–0.5 mg/L	0.1 mg/L	
Dichlorvos	CdTeQDs + AChE + ChOx + acetylcholine (ACh)	Enzyme	Optic	4.49–6.78 nM	4.49 nM	Meng et al. (2013)
Lindane	Stainless steel/*Streptomyces* strain M7 (SM7)	Cell	Electrochemical	0–750 μg/L	120 μg/L	Rodriguez et al. (2015)
Malathion	Au/Chi-PB–MWCNTs–hollow gold	Enzyme	Electrochemical	0.05–75 nM	0.05 nM	Zhai et al.
Chlorpyrifos	nanospheres nanocomposite fikm/AChE/NF	Enzyme	Electrochemical	0.05–75 nM	0.05 nM	(2013)
Monocrotophos		Enzyme	Electrochemical	0.1–50 nM	0.1 nM	
Carbofuran		Enzyme	Electrochemical	5–80 nM	2.5 nM	

(*Continued*)

Table 8.5 Biosensors for Detection of Pesticides in Literature (*cont.*)

Target Molecule	Immobilization Steps	Biorecognition Element	Transducer Type	Linear Range	Limit of Detection	Reference
Methyl parathion Carbofuran	GCE/NF–CGR–SnO_2-NPs/AChE/Chi/NF	Enzyme Enzyme	Electrochemical Electrochemical	10^{-13}–10^{-10}; 10^{-10}–10^{-8} M 10^{-12}–10^{-10}; 10^{-10}–10^{-8} M	5×10^{-14} M 5×10^{-13} M	Zhou et al. (2013)
Paraoxon	SPE/sol–gel ZnO matrix/AChE/	Enzyme	Electrochemical	0.035–1.38 ppm	0.035 ppm	Sinha et al. (2010)
Paraoxon Dichlorvos	CPE/AChE	Enzyme Enzyme	Electrochemical Electrochemical	0.86–22.93 ppb 4.2–33.16 ppb	0.86 ppb 4.2 ppb	Tuoro et al. (2011)
Paraoxon Parathion Chlorfenvinphos	Graphite electrode/MWCNTs/EDC/NHS/poly(benzenamine) [poly(SNS-NH_2)]/AChE	Enzyme Enzyme Enzyme	Electrochemical Electrochemical Electrochemical	0.1–10 μg/L 0.1–7 μg/L 0.1–12.5 μg/L	2.46 ng/L 0.542 ng/L 4.9 ng/L	Kesik et al. (2014)
Carbaryl Monocrotophos	GCE/ionic liquid functionalized graphene (IL-GR)/gelatin/AChE/GA	Enzyme Enzyme	Electrochemical Electrochemical	10^{-14}–10^{-8} M 10^{-13}–5×10^{-8} M	5.3×10^{-15} M 4.6×10^{-14} M	Zheng et al. (2015)
CPO Malaoxon	SPE/PVA polymeric matrix/AChE	Enzyme	Electrochemical	5×10^{-10}–5×10^{-12} M 1.01×10^{-10}–9.17×10^{-11} M	5×10^{-12} M 2.17×10^{-10} M	Mishra et al. (2015)
Carbofuran Methyl parathion	GCE/NiNPs–CGR–NF/AChE-Chi/NF GCE/NiNPs–CGR–NF/AChE-Chi/NF	Enzyme Enzyme	Electrochemical Electrochemical	10^{-12}–10^{-10}; 10^{-10}–10^{-8} M 10^{-13}–10^{-10}; 10^{-10}–10^{-8} M	5×10^{-13} M 5×10^{-14} M	Yang et al. (2013)
Methamidophos Monocrotophos	Boron doped diamond (BDD)/Pt–carbon aerogels (Pt–CAs)/AChE	Enzyme	Electrochemical	10^{-11}–10^{-6} M	3.1×10^{-13} M	Liu and Wei (2014)

Malaoxon Omethoate Methidation–oxon	SPE/tetramethoxysilane (TMOS)/PEG600/ hydroxyethyl cellulose (HEC)	Enzyme	Electrochemical	7.5×10^{-5}– 2.5×10^{-6} M 5×10^{-6}–$5x10^{-8}$ M 5×10^{-7}–10^{-8} M	10^{-9} M 10^{-7} M 10^{-8} M	Najwa et al. (2014)
Fenvalerate	GCE/Chi/GA/antifenvalerate/ovalbumin	Antibody	Electrochemical	10^{-13}–10^{-1} mg/L	0.80 µg/L	Wang et al. (2013)
Carbofuran	Graphene doped carbon-paste electrode (GPE)/PB/laccase	Enzyme	Electrochemical	4.98×10^{-7}–5.88×10^{-6} M	0.022 mg/kg	Oliveira et al. (2013)
Carbaryl		Enzyme	Electrochemical	7.44×10^{-8}–8.47×10^{-7} M	0.001 mg/kg	
Formetanate		Enzyme	Electrochemical	2.49×10^{-7}–4.76×10^{-6} M	0.013 mg/kg	
Pirimicarb		Enzyme	Electrochemical	2.99×10^{-7}–5.66×10^{-6} M	0.007 mg/kg	
Ziram		Enzyme	Electrochemical	2.49×10^{-8}–5.66×10^{-7} M	0.002 mg/kg	
Ethion	Potentiometric strips/MWCNTs–polyvinyl chloride (MWCNTs–PVC) composite/α-cyclodextrin/BChE	Enzyme	Electrochemical	0–330 ng/mL	22 ng/mL	Khaled et al. (2014)
Carbaryl	CPE/Chi–AuNPs–Tyr–laccase	Enzyme	Electrochemical	9.9×10^{-8}–2.91×10^{-6} M	1.98×10^{-8} M	Oliveira et al. (2014)
Formetanate hydrochloride		Enzyme	Electrochemical	9.99×10^{-7}–3.21×10^{-5} M	2.15×10^{-7} M	
Propoxur		Enzyme	Electrochemical	4.99×10^{-7}–1.92×10^{-5} M	1.87×10^{-7} M	
		Enzyme	Electrochemical	9.99×10^{-8}–	1.68×10^{-9} M	
Ziram		Enzyme	Electrochemical	3.38×10^{-7} M		
Paraoxon–methyl	Nonconducting pyroceramic substrate/ gold film/PAH-coated AuNPs/BSA–*Spirulina* cells	Cell	Conductometric	—	10^{-18} M	Tekaya et al. (2013)
Parathion–methyl		Cell	Conductometric	—	10^{-20} M	
Triazine		Cell	Conductometric	—	10^{-20} M	
Diuron		Cell	Conductometric	—	10^{-12} M	

they immobilized *Trametes versicolor* laccase on graphene-doped carbon-paste electrode modified with Prussian blue films. Their biosensor's principle was based on the laccase-activity inhibition and 4-aminophenol was used as a redox mediator. They tested their biosensor in tomatoes and potatoes. The accuracy of the biosensor was controlled by the standard-additions method and they found acceptable recoveries.

Wei and Wang (2015) and Liu and Wei (2014) used boron-doped diamond electrode for organophosphorous-pesticide detection. Liu and Wei (2014) prepared their electrode with Platinum–carbon aerogels (Pt–CAs) composite. Pt-CAs composite increased the reactive area of the electrode surface, thus the interfacial resistance decreased and the electron transfer rate enhanced. Wei and Wang (2015) used honeycomb-like hierarchically ILs ([BSmim] HSO_4)–AuNPs–porous carbon-composite for the construction of their biosensor. Both of them found that the inhibition rate of organophosphorous pesticides were linearly proportional to their concentration.

Lindane is a pesticide potentially harmful to environment and human health. Rodriguez et al. (2015) used *Streptomyces* strain M7 (SM7), which was capable of growing in the presence of organochlorine pesticides (OCPs). SM7 used lindane as a carbon source. Electrochemical impedance spectroscopy (EIS) was used as the transduction method. Bacterial activity was measured by the impedance changes produced in the cells due to lindane removal from the culture medium. This proposed label-free impedimetric biosensor detected lindane without sample pretreatment.

Fenvalerate is a pyrethroid insecticide for several insect pests including Diptera larvae, Hemitera, and Lepidoptera. It is commonly used to prevent and control of pests in cotton, vegetables, corn, tea, fruit trees, and other crops. Wang et al. (2013) developed an impedimetric immunosensor for fenvalerate detection. To construct their biosensor, the electrode was modified with chitosan and then fenvalerate antibodies crosslinked on the GCE with glutaraldehyde. When interaction between antifenvalerate antibodies and fenvalerate occurred, the electron transfer resistance increased. The change of resistance was proportional to the logarithm of the fenvalerate concentration. For real sample analysis they used tea and their results were consistent with GC–MS results.

3.1.6 Biosensors for Other Xenobiotic Compounds in Food

Dioxins, dioxin-like polychlorinated biphenyls (DL-PCBs), polychlorinated biphenyls (PCBs), and polycyclic aromatic hydrocarbons (PAHs) are other xenobiotic compounds in food. The biosensors used for the detection of other xenobiotic compounds are summarized in Table 8.6.

Table 8.6 Biosensors for Detection of Other Contaminants in Literature

Target Molecule	Immobilization Steps	Biorecognition Element	Transducer Type	Linear Range	Limit of Detection	Reference
Dioxin-like PCBs	Gold chip/EDC/NHS/PCB–BSA/ethanolamine/anti-DL-PCBs	Antibody	Optic	100–600 ng/mL	—	Tsutsumi et al. (2008)
Dioxin	Gold chip/NHS–EDC/anti-dioxin/dioxin–HRP/anti-HRP antibody	Antibody	Optic	—	0.1 ng/mL	Shimomura et al. (2001)
Benzo(a)pyrene (BaP)	GCE/NF–GR/hemin/ssDNA	DNA	Electrochemical	$2–22 \times 10^{-8}$ M	1.12×10^{-8} M	Ni et al. (2014)
BaP 2-Hydroxybiphenyl (HBP)	Gold chip/polydimethylsiloxane (PDMS)/BaP–BSA + BPA–HBP/BaP–monoclonal antibody	Antibody Antibody	Optic Optic	10 ppt–300 ppb 10 ppt–1000 ppb	10 ppt	Gobi and Miura (2004)
BaP	GCE/PB/AuNPs/cysteine/anti-BaP/BSA	Antibody	Electrochemical	0.2–100 ng/mL	0.08 ng/mL	Zhang et al. (2014c)
BaP	SPGE/mercaptohexanol (MCH)/thiolated ssDNA + AuNPs/streptavidin–alkaline phosphatase (AP)/1-naphthylphosphate/1-naphthol	DNA	Electrochemical	—	—	Lanzone et al. (2013)
Naphthalene Phenanthrene	Glass microscope slide/polyvinylactate (PVAc)/FITC–dextran/ssDNA	DNA DNA	Optic Optic	0–10 mg/L	—	Doong et al. (2005)
BaP	Gold coated sensor chip/cystemaine dihyrochloride (CA)/1-pyrenebutyric acid (PBA)/NHS–EDC/BaP/anti-BaP	Antibody	Piezoelectric	—	5 μM	Boujday et al. (2010)
Aroclor	SPCE/rabbit IgG/antiaroclor/aroclor + alkaline phosphatase	Antibody	Electrochemical	0.01–50 ppm	—	Laschı et al. (2003)
Delor 103 2,4,4′-trichlorobiphenyl	Petri dishes/prepolymerized TMOS with biphenyl gels/*Pseudomonas* species P2	Cell Cell	Optic Optic	0.5–5 mg/L 0.5–5 mg/L	0.5 mg/L 0.2 mg/L	Gavlasova et al. (2008)

Dioxins and DL-PCBs are hazardous, toxic, prevalent, and persistent chemical compounds. These compounds originate from the environment and enter into the food chain (Chobtang et al., 2011). Tsutsumi et al. (2008) developed an optic biosensor immunoassay (BIA) for DL-PCBs and they used a surface plasmon resonance (SPR) sensor (Biacore 3000) as the transducer. Their immunosensor measured 2,3',4,4',5-pentachlorobiphenyl (PCB 118) specifically. For real-sample analysis they used fish because DL-PCB isomer is found mostly in fish.

PCBs are highly persistent organic compounds and they are slightly soluble in water. They enter into the food chain and affect adversely the health of human and animals (Gavlasova et al., 2008). Laschı et al. (2003) developed an immunosensor for aroclor detection. Anti-PCB antibodies were directly attached onto disposable SPCE. A competition between the PCBs presents in the sample and a fixed concentration of an alkaline phosphatase-labelled PCB occurred aroclor was determined by electrochemical detection. Gavlasova et al. (2008) developed whole-cell optic biosensor for semiquantitative detection of PCBs. Their biosensor was based on the formation of yellow meta-ring fission products by silica entrapped cells. For the construction of the whole cell biosensor, *Pseudomonas* spp. P2 immobilized in a silica matrix which contained biphenyl. They compared the results of their biosensor with GC-ECD (Gas Chromatography – Electron Capture Detector) results and they found no differences between the results.

PAHs are the compounds originating from the incomplete combustion of organic substances, such as fossil fuels. PAHs have four or more ring structures, and their metabolites are mutagenic and toxic and they are harmful to human health. BaP is the basic PAH compound that has five structural rings, and causes cancer (Ni et al., 2014, Boujday et al., 2010). Ni et al. (2014), Zhang et al. (2014c), and Lanzone et al. (2013) developed an electrochemical biosensor for BaP detection. Antibody and ssDNA was used as biorecognition element. Ni et al. (2014) used hemin, graphene, and NF for the construction of their biosensor. They found that the hemin/NF–graphene/GCE was a viable platform for the immobilization of DNA. This DNA biosensor was treated separately in BaP, H_2O_2, and in their mixture. In amperometric measurements there was an apparent oxidation peak after the electrode was immersed in H_2O_2. In the presence of H_2O_2, hemin mimicked cytochrome P450 to metabolize BaP, and as a result its metabolites were determined. The DNA damage was induced by this metabolite. Also in this study they demonstrated that this model system could successfully metabolize the BaP PAH molecule into its ultimate carcinogenic form, which facilitated the explanation of the mechanism of the DNA damage induced by BaP. Subsequently, in their indirect-amperometric method for the analysis of BaP, an aqueous solution was developed based on the linear changes of anodic current of guanine as a function of the BaP concentration (Ni et al., 2014). Zhang et al. (2014c) developed an amperometric immonosensor for BaP detection. For the construction of their immunosensor, Prussian blue was firstly electrodeposited onto a GCE and it was used as a mediator. After that, the nanogold layer was formed on the Prussian blue film electrochemically. Then, L-cysteine and nanogold particles were self-assembled layer-by-layer. Finally, anti-BaP was adsorbed onto the bilayer nanogold films for the construction of their immunosensor. They found that the current response of the immunosensor was inversely proportional to the concentration of BaP.

3.2 Biosensor for Toxins in Food

Mycotoxins are toxic secondary metabolites produced by several species of fungi, such as, *Aspergillus, Fusarium,* and *Penicillium.* They occur during the growth of the plant or after harvesting. The temperature and humidity of storage is important for the growth of mycotoxins. The consumption of plants that include mycotoxins introduces these mycotoxins into the human food chain (Panini et al., 2011). Methods, such as, HPLC, LC–tandem MS, immunochromatographic strips, molecularly imprinted polymer solid-phase extraction, and ELISA have been proposed. However, these laboratory techniques require either sophisticated instrumentation or highly qualified personnel to be employed. Therefore, the development of a convenient, rapid, and low cost analytical method for mycotoxin detection is important (Liu et al., 2014a). Biosensors for toxin detection in food are summarized in Table 8.7.

Zearalenone (ZEN) is a nonsteroidal estrogenic mycotoxin biosynthesized by a variety of *Fusarium* fungi. ZEN is found mostly feed and food causing severe risk to animals and humans (Liu et al., 2014a). They developed a label-free amperometric immunosensor for the detection of ZEN. Their immunosensor was based on mesoporous carbon (MC) and trimetallic nanorattles [Au-core and AgPt-shell structure (Au@AgPt)]. The Au@AgPt nanorattles were loaded onto the MC by physical adsorption. Panini et al. (2011) fabricated a microfluidic immunoassay for ZEN detection. Their immunoassay was based on a competitive direct immunoassay and anti-ZEN primary antibodies were immobilized on magnetic microspheres 3-aminopropyl-modified gold electrode. Their detection limits were between 1.7 ng/L to 0.41 μg/L. Due to

Table 8.7 Biosensors for Detection of Mycotoxins in Literature

Target Molecule	Immobilization Steps	Biorecognition Element	Transducer Type	Linear Range	Limit of Detection	Reference
Zearalenone (ZEN)	GCE/mesoporous carbon/Au@AgPt nanorattles/anti-ZEN/BSA	Antibody	Electrochemical	0.005–15 ng/mL	1.7 pg/mL	Liu et al. (2014a)
Aflatoxin M1 (AFM1)	Ag wire electrode/MUA/EDC/NHS/anti-AFM1/BSA	Antibody	Electrochemical	39–1000 pg/mL	39 pg/mL	Bacher et al. (2012)
Aflatoxin B1 (AFB1)	GCE/Chi/AuNP/anti-AFB1–AFB1–anti-AFB1–catechol–Au–Fe_3O_4	Antibody	Electrochemical	0.6–110 ng/mL	0.2 ng/mL	Masoomia et al. (2013)
Sterigmatocystin	Au/MWCNTs/NHS/EDC/aflatoxin-detoxifizyem (ADTZ)	Enzyme	Electrochemical	4.16×10^{-5}–1.33×10^{-3} mg/mL	13 μM	Yao et al. (2006)
AFB1	Gold-coated quartz crystal/4-aminothiophenol (4-ATP)/anti-AFB1/BSA	Antibody	Piezoelectric	0.1–4 ng/mL	0.012 ng/mL	Chauhan et al. (2015)
Ochratoxin A (OTA)	Glass/APTES/succinic anhydride/EDC/NHS/cysteamine (cys)/Gold nanorods (GNR)/OTA aptamer	Aptamer	Optic	0.1 nM–10 μM	0.1 nM	Park et al. (2014a)
OTA	Silica optic fiber/APTES/GA/ethylene diamine/dethiobiotin–EDC–NHS/magnetic bead–signal probe–OTA	Aptamer	Optic	6–500 nM	3 nM	Wang et al. (2015c)
AFB1	GNRs/MUA/EDC/NHS/anti-AFB1–BSA conjugates	Antibody	Optic	0.5–20 ng/mL	0.16 ng/mL	Xu et al. (2013)
Patulin	Au/NaCl–polyoxyethylenes orbitan–EDTA/NHS/EDC/Glutamine-binding protein–patulin/ethanolamine hydrochloride/antipatulin	Antibody	Optic	0–80 nM	0.1 nM	Pennacchio et al. (2014)
AFB1	Pt/Aflatoxin oxidase–sol gel/EDC–NHS/MWCNTs	Enzyme	Electrochemical	1–225 ng/mL	0.5 ng/mL	Li et al. (2011)
OTA	Chip dextran matrix/streptavidin/EDC/NHS/ethanolamine/biotin tagged aptamer	Aptamer	Optic	0.094–100 ng/mL	0.005 ng/mL	Zhu et al. (2015)
AFM1	SPE/anti-IgG/polyvinyl alcohol (PVA)/anti-AFM1/AFM1–HRP	Antibody	Electrochemical	30–160 ppt	25 ppt	Micheli et al. (2005)
OTA	ITO/chi-SWCNTs/rabbit-immunoglobulin (rIgGs)/BSA ITO/chi-MWCNTs/rIgGs/BSA	Antibody Antibody	Electrochemical Electrochemical	0.25–6 ng/dL 0.25–6 ng/dL	0.25 ng/dL 0.25 ng/dL	Kaushik et al. (2010)

AFB1	ITO/MWCNTs/anti-AFB1/BSA	Antibody	Electrochemical	0.25–1.375 ng/mL	0.08 ng/mL	Singh et al. (2013)
AFB1	Hemib/G–quadruplex HRP–mimicking DNAzyme–aptamer AFB1	Aptamer	Electrochemical	0.1–10 ng/mL	0.11 ng/mL	Shim et al. (2014)
Fumonisin (FB)	Nitrocellulose strip/HRP loaded goat antirabbit/antifumonisin	Antibody	Electrochemical	2.5–500 μgL/	2.5 μg/L	Mirasoli et al. (2012)
Patulin	Au–quartz wafer/antipatulin/BSA	Antibody	Piezoelectric	50–140 nM	—	Funari et al. (2014)
AFB1	SPE/PB/ChOx–AChE	Enzyme	Electrochemical	10–60 ppb	2 ppb	Rejeb et al. (2009)
FB	Nitrocellulose membrane (NC)/gold–anti-FB/FB/gold–anti-FB	Antibody	Visual	2.5–40 ng/mL	2.5 ng/mL	Li et al. (2012)
Deoxynivalenol (DON)	SPE/immunomegnetic beads (IMBs)/DON–*N*,*N*-2-disuccinimdyl carbonated (DSC)–carbamylated HAS (cHSA)/avidin–biotin–HRP–complex	Antibody	Electrochemical	100–4500 ng/mL	63 ng/mL	Romanazzo et al. (2010)
OTA	SPCE/magnetic beads/anti-OTA/OTA labeled HPR	Antibody	Electrochemical	0.01–20 ppb	0.008 ppb	Perrotta et al. (2012)
Sterigmatocystin	Au/Chi–SWCNTs/aflatoxin oxidase	Enzyme	Electrochemical	10–1480 ng/mL	3 ng/mL	Chen et al. (2010)
AFM1	SPCE/superparamegnetic nanoparticles/anti-AFM1/AFM1-HRP conjugate/5-methylphenazinium methyl sulfate (MPMS)	Enzyme	Electrochemical	0.01–0.1 ppb	0.01 ppb	Paniel et al. (2010)
Deoxynivalenol (DON)	NC membrane/gold labelled anti-DON/DON–cationic bovine serum albumin (CBSA)	Antibody	Visual	0–400 ng/mL	—	Xu et al. (2010)
AFM1	SPE/BSA–AFM1/gold conjugated anti-AFM1	Antibody	Electrochemical	15–1000 ng/mL	15 ng/mL	Vig et al. (2009)
AFB1	NRL array/Cy5-labeled anti-AFB1	Antibody	Optic	1.5–5.1 ng/g (corn) 0.6–1.4 ng/g (nut)	— —	Sapsford et al. (2006)
AFM1	Silicon wafers with SiO_2/Cr and Pt layers/Fe_3O_4/PANI thin film/AFM1–aptamer	Aptamer	Electrochemical	6–60 ng/L	1.98 ng/L	Nguyen et al. (2013)
OTA	SPE/PVA–anti-OTA	Antibody	Electrochemical	0.4–13.1 μg/kg	0.4 μg/kg	Alarcón et al. (2006)
OTA	ITO/PANI–stearic acid (SA)/OTA aptamer	Aptamer	Electrochemical	0–10 ng/mL 1–25 μg/mL	0.1 ng/mL	Panini et al. (2011)

(*Continued*)

Table 8.7 Biosensors for Detection of Mycotoxins in Literature (*cont.*)

Target Molecule	Immobilization Steps	Biorecognition Element	Transducer Type	Linear Range	Limit of Detection	Reference
AFB1	Au/gold binding protein–protein G/anti AFB1/BSA	Antibody	Optic	—	0.1 μg/mL	Park et al. (2014b)
OTA	Au/thiolated DNA_1/MCH/DNA_2/AuNPs/DNA_3/DNA_4/OTA	Aptamer	Electrochemical	2.5 pM–2.5 nM	0.75 pM	Yang et al. (2014)
OTA	Au/1,6-hexanedithiol/AuNPs/OTA–OVA/OTA/AP-labeled secondary antibody	Antibody	Electrochemical	10 pg/mL–10 ng/mL	8.2 pg/mL	Liu et al. (2009)
AFB1	GCE	Chemical	Electrochemical	0–012 mg/L	—	Gavrilova et al. (2014)
Staphylococcal enterotoxin B (SEB)	Optic fiber + gold layer/anti-SEB/BSA/SEB	Antibody	Optic	0–100 ng/mL	—	Slavik et al., 2002
SEB	CM5 sensor chip/EDC/NHS/anti-SEB	Antibody	Optic	—	1 ng/mL	Nedelkov et al. (2000)
Staphylococcal enterotoxin A (SEA)	Gold chip/16-mercapto-1-hexanol + 6-MCH/EDC/NHS/anti-SEA/glycine	Antibody	Optic	100–1000 ng/mL	100 ng/mL	Tsai and Li (2009)
Paralytic shellfish toxins (PSP)	Biacore Q/reusable sensor chips/EDC-NHS/jetfamine solution/ethanol amine/anti-PSP Biacore T100/reusable sensor chips/EDC–NHS/jetfamine solution/ethanolamine/anti-PSP	Antibody Antibody	Optic Optic	— —	— —	Haughey et al. (2011)
PSP	Microarray cartridge/anti-PSP/PSP/labelled antibody (antirabbit IgG)	Antibody	Optic	—	20 pg/mL	Meneely et al. (2013)
Cylindrospermopsin (CYN)	Carboxymethylated surface of CM5 chip/EDC–NHS/Jeffamine solution/ethanolamine/anti-CYN	Antibody	Optic	4.4–11.1 ng/mL	—	Elliott et al. (2013)
Okadaic acid (OA)	Crystal surface/polyethleneimine (PEI)/GA/OA–BSA/anti-OA	Antibody	Piezoelectric	0–25 μg/mL	1.9 μg/mL	Tang et al. (2002)

the high electron transfer rate of Au@AgPt nanorattles and its large surface area, suitable pore arrangement and high conductivity of MC, Liu et al. (2014a) found a lower detection limit.

The International Agency for Research on Cancer has categorized aflatoxin B1 as a group 1 human carcinogen and aflatoxins G1, G2, and B2 as group two possible human carcinogens. These toxins are carcinogenic, mutagenic, and teratogenic. Aflatoxin B1 (AFB1) enters the food chain with consumption of foods including aflatoxins (Li et al., 2011; Singh et al., 2013). Li et al. (2011) and Rejeb et al. (2009) developed amperometric enzyme biosensors for AFB1 detection. Aflatoxin oxidase (AFO) and AChE were used as biorecognition molecules. Li et al. (2011) buried AFO in sol–gel, attached to MWCNTs-modified Pt electrode. The covalent coupling between AFO and MWCNTs supported enzymatic activity of AFO and the oxidation of AFB1 occurred. The biosensor principle of Rejeb et al. (2009) was based on AChE inhibition by AFB1. In the presence of AChE, an amperometric choline-oxidase biosensor was used for monitoring the residual activity. For the construction of the biosensor, the choline oxidase (ChOx) immobilized onto screen-printed electrode modified with Prussian blue and detected the H_2O_2 concentration. Singh et al. (2013) and Masoomi et al. (2013) developed an amperometric immunosensor for AFB1 detection. Singh et al. (2013) deposited carboxylated multiwalled carbon nanotubes (c-MWCNTs) electrophoretically onto indium tin oxide (ITO) glass. The monoclonal AFB1 antibodies (anti-AFB1) immobilized onto c-MWCNTs/ITO electrode. The response current of the electrode increased when the concentration of the antigen (AFB1) was increased. The nonenzymatic sandwich-type amperometric immunosensor of Masoomi et al. (2013) was fabricated using a modified GCE/chitosan/AuNP/anti-AFB1, and a Fe_3O_4 magnetic core with a gold shell functionalized with 3-[(2-mercaptoethylimino)methyl]benzene-1,2-diol and labelled with AFB1. The differential-pulse voltammetry currents increased with the increase of AFB1 concentration. The storage stability of these biosensors was determined in 1 week and it was found that they retained 92% and 90% current response. These immunosensors were incubated in AFB1 and different possible interfering agents. The response of biosensors did not change and these biosensors indicated good selectivity for the measurement of AFB1.

Xu et al. (2013), Sapsford et al. (2006), and Park et al. (2014b) fabricated an optic biosensor for the monitoring of AFB1. Xu et al. (2013) used GNRs as a sensing material and it made the biosensor more stable. GNR–AFB1–BSA composites were aggregated after mixing with free antibodies resulting in significant changes in absorption intensity. At the same time, in the existence of AFB1 molecules in samples, competitive immune reaction between AFB1 molecules and antibodies occurred. In the bifunctional protein crosslinker-based SPR biosensor of Park et al. (2014b) used gold binding protein–protein G layer as a bifunctional crosslinker for rapid self-oriented immobilization of antibodies on gold substrates without any chemical treatment. The detection limits were 0.16 ng/mL and 1 μg/mL, respectively. As a result of high conductivity of GNRs, lower detection limits were obtained.

The consumption of milk and milk products by human population is quite high, particularly in infants and children, increasing exposure to aflatoxin M1 (AFM1). AFM1 is a stable molecule during the pasteurization, storage, and preparation of various milk products (Paniel et al., 2010). Paniel et al. (2010) and Micheli et al. (2005) developed an amperometric immunosensor for the detection of AFM1 in food products. Their sensors were based on a competitive immunoassay and HRP was used as a tag. Paniel et al. (2010) coated anti-AFM1 antibodies with magnetic nanoparticles and they were used to separate the bound and unbound fractions. The samples containing AFM1 were incubated with anti-AFM1 antibodies and HRP conjugate. Competition occurred between the AFM1 antigen and the anti-AFM1 antibodies conjugate. Micheli et al. (2005) immobilized the anti-AFM1 antibodies directly on the electrode surface, and a competition occurred between free AFM1 and AFM1–HRP conjugate. The detection limits of these biosensors were 10 ppt and 25 ppt, respectively. The magnetic nanoparticles provided lower LOD values. Both of these immunosensors had the vital advantage of measuring AFM1 concentration, centrifugation, dilution, and other pretreatment procedures. Bacher et al. (2012) and Vig et al. (2009) developed an impedimetric immunosensor for AFM1 detection. Bacher et al. (2012) used silver-wire electrode and 11-mercaptoundecanoic acid for SAM for the construction of their biosensor. Vig et al. (2009) used colloidal gold- and silver electrodeposition on screen-printed electrode for the construction of their biosensor. Both of the researchers used monoclonal antibody for AFM1 detection. The antigen–antibody interaction was calculated by monitoring impedance.

Sterigmatocystin (ST) is a precursor of AFB1 (Yao et al., 2006). Yao et al. (2006) and Chen et al. (2010) developed an amperometric enzyme biosensor for ST detection. Aflatoxin–detoxifizyme and AFO were used as biorecognition biomolecules in the fabrication of their biosensors, respectively. Yao et al. (2006) and Chen et al. (2010) used MWCNTs and chitosan–SWCNTs for the immobilization of enzymes.

Ochratoxin A (OTA) is a mycotoxin produced by several *Aspergillus* and *Penicillium* species, which occurs in growth period or during storage. Kaushik et al. (2010), Perotta et al. (2012), Alarcón et al. (2006) and Liu et al. (2009) developed an amperometric immunosensor for OTA detection. Kaushik et al. (2010) incorporated carboxylic group functionalized single-walled and multiwalled carbon nanotubes into chitosan to construct nanobiocomposite film onto indium–tin oxide-coated glass plate for coimmobilization of rabbit-immunoglobulin and bovine serum albumin. Using CNTs in biosensor construction resulted in increased electroactive surface area and enhanced electron transport. Perotta et al. (2012) used protein G-functionalized magnetic beads (MBs) for the construction of their biosensor. Magnetic beads provided a solid matrix for immobilization of anti-OTA antibodies and OTA and for their interaction. The principle of this immunosensor was based on a direct competitive assay between OTA antigen and OTA–HRP conjugate. HRP, in the presence of H_2O_2, catalyzes the oxidation of pyrocatechol to benzoquinone, whose back electrochemical reduction was detected on the electrode surface. Liu et al. (2009) fabricated their biosensor by using SAM on the gold electrode and 1,6-hexanedithiol. This layer increased the surface area and loading of OTA–ovalbumin complex and hence the sensitivity of the biosensor increased. For the construction of the biosensor, alkaline phosphatase was used as a tag. Electrochemical signal occurred from the oxidation of enzymatic product of 1-naphthyl phosphate and it was in a linear relationship with the OTA level. Park et al. (2014a), Wang et al. (2015c), and Zhu et al. (2015) developed an optic biosensor for OTA detection. Park et al. (2014a) and Wang et al. (2015c) used GNRs and MBs for the construction of their biosensor. Zhu et al. (2015) used streptavidin protein as a crosslinker, immobilized onto the surface of a sensor chip, and the biotin aptamer was captured through streptavidin–biotin interaction.

Patulin is a mycotoxin produced by the *Penicillium expansum, Aspergillus, Penicillium,* and *Paecilomyces* fungal species (Pennacchio et al., 2014). They developed a competitive immunoassay for patulin detection. The detection method is based on SPR and laser beam–induced interactions between probe and target molecules in the vicinity of gold surface of the biochip leading to the shift in resonance conditions and consequently to slight but easily detectable change of reflectivity. Funari et al. (2014) developed a piezoelectric biosensor for patulin detection. Oriented antibodies were tethered onto the gold surface of a quartz crystal microbalance through the photonics immobilization technique.

Fumonisins (FB), are produced by *Fusarium* species mostly affect corn and other agricultural products (Mirasoli et al., 2012). They developed an immunosensor, which consisted of a competitive lateral-flow immunoassay based on enzyme-catalyzed chemiluminescence detection and a portable charge-coupled device (CCD) camera used in a contact-imaging configuration. Li et al. (2012) developed a competitive lateral-flow immunoassay for the determination of total FBs (FB1, FB2, and FB3). It was used for fumonisins detection in maize samples. It consisted of a sample pad, conjugation pad, and test and control regions. The color intensity of the test region was in a linear relationship with fumonisins mixture (FB1:FB2:FB3; 12:4:1) concentration.

Deoxynivalenol (DON) is produced by *Fusarium* fungi and infects several cereal products, such as, wheat, corn, barley, oats, and rye intended for human and animal consumption (Xu et al., 2010). Romanazzo et al. (2010) developed a biosensor for DON detection. They used an enzyme-linked-immunomagnetic-electrochemical assay and immunomagnetic beads coupled with eight magnetized screen-printed electrodes as electrochemical transducers. Xu et al. (2010) developed a colloidal-gold lateral-flow immunoassay strip for the monitoring of DON concentration in wheat and maize samples. They also developed it by using anti-DON monoclonal antibody. This strip was based on visual detection.

Staphylococcus aureus is a major human pathogen, which produces numerous toxins, including the pyrogenic toxins—Staphylococcal enterotoxins (SEs) and toxic-shock syndrome toxin-1 (Nedelkov et al., 2000). Slavik et al. (2002) and Nedelkov et al. (2000) developed an optic biosensor for Stapylococcal enterotoxin B (SEB) detection. The biosensor of Slavik et al. (2002) was based on a spectral interrogation of surface plasmons in a miniature-sensing element based on a side-polished single-mode optic fiber with a thin metal overlayer. For specific detection of SEB, the SPR sensor was functionalized with a covalently crosslinked double-layer of antibodies against SEB. Nedelkov et al. (2000) used biomolecular interaction-analysis mass spectrometry for the detection of bacterial toxins in food products. In this research, two-step approach was used. In the first step, they used SPR to detect the binding of the toxin(s) to antibodies immobilized on a surface of a sensor chip. In the second step, they identified the bound toxin(s) by matrix-assisted laser desorption/ionization time-of-flight mass spectrometry. Tsai and Li (2009) developed an optic immunosensor for the detection of Staphylococcal enterotoxin A (SEA). The immunosensor surfaces were fabricated from various thiol mixtures of different concentration ratios of 16-mercapto-1-hexanol to 6-mercapto-1-hexanol. The surface was functionalized with a

1:20 mixed SAM and they reported that it exhibited the best result in terms of SEA detection sensitivity.

3.3 Biosensors for Pathogens in Food

There are five primary bacterial pathogens found in food, which include *Salmonella, Campylobacter, Escherichia coli O157, Listeria monocytogenes,* and *Clostridium perfringens.* The detection of pathogenic microorganism contamination in food and water has an important role for ensuring food safety, security and public health. Several different procedures such as polymerase chain reaction (PCR), ELISA, and flow cytometry have been used so far in literature (Sharma and Mutharasan, 2013; McGrath et al., 2012; Wang et al., 2015a). There are also a lot of biosensors that have been developed in literature. They are summarized in Table 8.8.

E. coli O157:H7 is one of the most dangerous foodborne pathogenic bacteria and it can cause severe, sometimes life-threatening illness. *Enterobacter sakazakii* is a pathogen, and it can cause life-threatening disease. Duo et al. (2013) developed an electrochemical immunosensor for *E. coli* O157:H7 and *E. sakazakii* detection using carbon screen-printed low-density arrays. MWCNTs/sodium alginate/carboxymethyl chitosan-composite films were immobilized on the screen-printed electrodes to increase the sensitivity of the working electrode. HRP-labelled antibodies of two bacteria were immobilized on different parts of screen-printed electrode. In order to detect the specificity of the immunosensor for *E. sakazakii* and *E. coli* O157:H7, they used five kinds of foodborne pathogenic bacteria (*E. coli, Salmonella pullorum, Shigella flexneri, Bacillus subtilis, Bacillus cereus,* and *S. aureus*). They reported that the crossreactivity at the electrode did not occur and this immunosensor was able to detect two pathogenic bacteria together without crosscontamination. The reproducibility of the immunosensor was tested by using 10-replicate determinations. The RSD of *E. sakazakii* and *E. coli* O157:H7 was 6.50% and 6.68%, respectively. These results represented that the immunosensor had acceptable precision and fabrication reproducibility.

Liu et al. (2014b) and Kim et al. (2014b) developed an optic immunosensor for *E.coli* and *Salmonella* detection. Liu et al. (2014b) used propidium iodide and Kim et al. (2014b) used quantum dots nanoparticles as the fluorescent dye. The fluorescence response of the biosensors increased with increasing cell concentration. Duan et al. (2014) developed a fluorescent aptasensor based on the AccuBlue dye for the detection of pathogenic bacteria. The sensor consisted of AccuBlue, an aptamer strand, and its complementary strand (cDNA). They fabricated two models for *Vibrio*

Table 8.8 Biosensors for Detection of Pathogenic Bacteria in Literature

Target Molecule	Immobilization Steps	Biorecognition Element	Transducer Type	Linear Range	Limit of Detection	References
E. coli	SPCE/MWCNTs–sodium alginate (SA)–carboxymethyl chitosan (CMC)-composite films/HRP-labeled antibodies	Antibody	Electrochemical	10^4–10^{10} cfu/mL	4.57×10^3 cfu/mL	Dou et al. (2013)
E. sakazakii	SPCE/MWCNTs–SA–CMC-composite films/HRP labeled antibodies	Antibody	Electrochemicall	10^4–10^{10} cfu/mL	3.27×10^3 cfu/mL	
E. coli	Fiber probe/anti-*E.coli*–propidium iodide	Antibody	Optic	10^4–10^7 cells/mL	10^4 cells/mL	Liu et al. (2014b)
Salmonella	Microfluidic chip/quantum dots nanoparticles–anti-*Salmonella*–super paramagnetic particles	Antibody	Optic	0–10^6 cfu/mL	10^3 cfu/mL	Kim et al. (2014b)
V. parahaemolyticus (VP)	VP-aptamer + t-aptamer + cDNA + AccuBlue	Aptamer	Optic	50–106 cfu/mL	35 cfu/mL	Duan et al. (2014)
S. typhimurium (ST)	ST-aptamer + t-aptamer + cDNA + AccuBlue	Aptamer	Optic	50–106 cfu/mL	25 cfu/mL	
Cronobacter (CB)	Au chip/hexa(ethylene glycol) undecanethiol (EG6)/poly(HEMA)/anti-CB	Antibody	Optic	—	10^4 cells/mL	Rodriguez-Emmenegger et al. (2011)
L. monocytogenes	Au/MPA/EDC/NHS/anti-*L. monocytogenes*/OVA/L. monocytogenes/polyclonal antibody–HRP conjugate	Antibody	Electrochemical	10^2–10^6 cfu/mL	10^2 cfu/mL	Cheng et al. (2014)
Salmonella enteritidis	Quantum dots/EDC/NHS/anti-*S. enteritidis* + anti-*S. aureus* + anti-*E. coli*	Antibody	Optic	—	10^2 cfu/mL	Wang et al. (2015a)
S. aureus		Antibody	Optic	—	10^2 cfu/mL	
E. coli		Antibody	Optic	—	10^2 cfu/mL	

parahaemolyticus and *Salmonella typhimurium* detection. One is the "signal on" model. In this model the aptamer was first bound to the target and cDNA and AccuBlue were added, the cDNA was hybridized with the free unreacted aptamer, and as a result a double-stranded DNA (dsDNA) duplex was formed. Thus this hybridization enabled AccuBlue to get into the dsDNA and exhibit significantly increased fluorescence intensity. The other is "signal off" model. In this model the aptamer was hybridized with cDNA. Addition of AccuBlue increased the fluorescence intensity. When the target was added, the aptamer was bound to the target and the cDNA detached from the dsDNA duplex and fluorescence intensity decreased.

L. monocytogenes (LM) is a Gram-positive pathogen that causes serious disease. It contaminates many food products, such as, meats, milk, cheese, ice cream, and raw vegetables. Cheng et al. (2014) developed a novel electrochemical immunosensor by SAM-modified gold electrodes for the detection of LM. Mouse monoclonal antibody of LM was immobilized on the SAM through a stable acyl-amino ester intermediate. Then, a direct sandwich format was developed, in which a polyclonal antibody for LM was conjugated to HRP as the enzyme label. In the presence of H_2O_2, HRP catalyzed the oxidation of thionine, whose back electrochemical reduction was detected on an gold electrode and the response current was proportional with LM concentration. *S. aureus, E. coli* O157:H7, *Shigella, Salmonella,* and *L. monocytogenes* were used to evaluate the specificity of the immunosensor system. There were 0.7%, 10.6%, 0.8%, and 27.5% crossreactions for *S. aureus, Salmonella, Shigella, E. coli* O157:H7, respectively. For real sample analysis they used milk. First, cultured *L. monocytogenes* cells were inoculated into milk at different concentrations. Then, the immunosensor was applied to detect the contaminated milk. They found that the concentration of *L. monocytogenes* in milk was consistent with the results obtained from the plate count method.

4 Commercial Biosensors for Food Contaminants

The food industry looks for cheap, easy analytical methods for checking the safety and quality control of foods. In a word, the determination of chemical and biological contaminants is essential for food safety and public health. There are many scientific publications on biosensors for food industries; however, only a few biosensors are commercially available and most of them determine glucose and lactic acid. Commercial biosensors provide

Table 8.9 Commercial Biosensors for the Food Industry (Bahadır and Sezgintürk, 2015b)

Company	Biosensor	Country
Oriental Electric	Fish deterioration tracking	China
Massachusetts Institute of Technology	Detection of *E. coli* O157:H7 in lettuce (Canary)	United States
Michigan State University's Electrochemical Biosensor	Detection of *E. coli* O157:H7 and *Salmonella* in meat products in United States	United States
Georgia Research Tech Institute	Detection of *Salmonella* and *Campylobacter* in pork industry	United States
Naval Research Laboratory	Detection of *Staphylococcal enterotoxin B* and *Botulinum toxin A* in tomatoes, sweet corn, beans, and mushrooms	United States
Universitat Autonoma de Barcelona in collaboration with CSIC	Detection of atrazine traces	Spain
Molecular Circuitry, Inc.	*E. coli* O157, *Salmonella, Listeria,* and *Campylobacter*	United States
Research International	Proteins, toxins, virus, bacteria, spores, and fungi (simultaneous analysis)	United States
Universal Sensors	Ethanol, methanol, glucose, sucrose, lactose, L-aas, glutamine, ascorbic acid, and oxalate	United States
Texas Instruments, Inc.	Peanut Allergens, antibiotics	United States
Yellow Springs Instruments Co.	Glucose, sucrose, lactose, L-lactate, galactose, L-glutamate, ethanol, H_2O_2, starch, glutamine, choline	United States
Affinity Sensors	*S. aureus* and cholera toxin	United Kingdom
Ambri Limited	Pathogens, such as, *Salmonella, Enterococcus*	United States
Biacore AB	Water soluble vitamins, chemical veterinary residues, and mycotoxins	Sweden
Biomerieux	Microorganisms	France
Biosensor Systems Design	Microorganisms and toxic substances	United States
Biosensores S.L.	Toxic substances	Spain
IVA Co Ltd	Heavy metals	Rusia
Motorola	Microorganisms and GMOs	Japan
Iventus Bio Tec	Ascorbic acid	Germany

important solutions for food analysis in terms of specificity and time saving (Bahadır and Sezgintürk, 2015b). The commercial biosensors for food contaminants are summarized in Table 8.9.

5 Conclusions and Future Perspectives

This chapter has highlighted the applications of biosensors for microbiological and chemical contaminants in food. Examples of electrochemical, optic, and piezoelectric techniques have been described and the three types of biosensors that appear in the literature have been taken into consideration. A range of molecules with biorecognition properties has been used as sensing elements in biosensors. There are many different ways to combine a new biosensor for food contaminant detection.

The global market of food analysis requires a reliable, cheap method for evaluating food quality and safety. Biosensors are suitable for improving food diagnostics in terms of quality control. They play a vital role as powerful analytical devices in food industry, providing fast, cost-effective, high sensitivity, and specificity in measurements. Among these developed biosensors for food contaminant analysis, only a few of them are commercially available and they are primarily used to determine glucose and lactic acid. Commercial biosensors will be useful for analysis of food quality control parameters, such as, additives, toxins, and pathogens detection. In the coming years the use of commercial biosensors is expected to spread in the food industry.

Acknowledgment

This chapter was supported by The Scientific and Technological Council of Turkey (TÜBİTAK, Project number: 113 Z 678).

References

Alarcón, S.H., Palleschi, G., Compagnone, D., Pascale, M., Visconti, A., Barna-Vetr, I., 2006. Monoclonal antibody–based electrochemical immunosensor for the determination of ochratoxin A in wheat. Talanta 69, 1031–1037.

Alibolandia, M., Hadizadeh, F., Vajhedin, F., Abnous, K., Ramezani, M., 2015. Design and fabrication of an aptasensor for chloramphenicol based on energy transfer of CdTe quantum dots to graphene oxide sheet. Mater. Sci. Eng. C 48, 611–619.

Alkasir, R.S.J., Ganesana, M., Won, Y., Stanciu, L., Andreescu, S., 2010. Enzyme functionalized nanoparticles for electrochemical biosensors: a comparative study with applications for the detection of bisphenol A. Biosens. Bioelectron. 26, 43–49.

Apetrei, I.M., Apetrei, C., 2013. Amperometric biosensor based on polypyrrole and tyrosinase for the detection of tyramine in food samples. Sens. Actuat. B 178, 40–46.
Ayenimo, J.G., Adeloju, S.B., 2015. Inhibitive potentiometric detection of trace metals with ultrathin polypyrrole glucose oxidase biosensor. Talanta 137, 62–70.
Bacher, G., Pal, S., Kanungo, L., Bhand, S., 2012. A label-free silver wire based impedimetric immunosensor for detection of aflatoxin M1 in milk. Sens. Actuat. B 168, 223–230.
Bahadır, E.B., Sezgintürk, M.K., 2015a. Electrochemical biosensors for hormone analyses. Biosens. Bioelectron. 68, 62–67.
Bahadır, E.B., Sezgintürk, M.K., 2015b. Applications of commercial biosensors in clinical, food, environmental, and biothreat/biowarfare analyses. Anal. Biochem. 478, 107–120.
Bahmani, B., Moztarzadeh, F., Rabiee, M., Tahriri, M., 2010. Development of an electrochemical sulfite biosensor by immobilization of sulfite oxidase on conducting polyaniline film. Synt. Met. 160, 2653–2657.
Basu, A.K., Chattopadhyay, P., Roychudhuri, U., Chakraborty, R., 2006. Biosens. Bioelectron. 21, 1968–1972.
Batra, B., Kumari, S., Pundir, S.S., 2014. Construction of glutamate biosensor based on covalent immobilization of glutmate oxidase on polypyrrole nanoparticles/polyaniline modified gold electrode. Enzyme Microb. Technol. 57, 69–77.
Berezhetskyy, A.L., Sosovska, O.F., Durrieu, C., Chovelon, J.M., Dzyadevych, S.V., Tran-Minh, C., 2008. Alkaline phosphatase conductometric biosensor for heavy-metal ions determination. ITBM-RBM 29, 136–140.
Boujday, S., Nasri, S., Salmain, M., Pradier, C., 2010. Surface IR immunosensors for label-free detection of benzo[a]pyrene. Biosens. Bioelectron. 26, 1750–1754.
Chauhan, R., Solanki, P.R., Singh, J., Mukherjee, I., Basu, T., Malhotra, B.D., 2015. A novel electrochemical piezoelectric label free immunosensor for aflatoxin B1 detection in groundnut. Food Control 52, 60–70.
Chen, J., Liu, D., Li, S., Yao, D., 2010. Development of an amperometric enzyme electrode biosensor for sterigmatocystin detection. Enzyme Microb. Technol. 47, 119–126.
Chen, H., Mousty, C., Chen, L., Cosnier, S., 2008. A new approach for nitrite determination based on a HRP/catalase biosensor. Mater. Sci. Eng. C 28, 726–730.
Chen, D., Yao, D., Xie, C., Liu, D., 2014. Development of an aptasensor for electrochemical detection of tetracycline. Food Control 42, 109–115.
Chen, D.M., Gao, Z.F., Jia, J., Li, N.B., Luo, H.Q., 2015. A sensitive and selective electrochemical biosensor for detection of mercury(II) ions based on nicking endonuclease-assisted signal amplification. Sens. Actuat. B 210, 290–296.
Cheng, C., Peng, Y., Bai, J., Zhang, X., Liu, Y., Fan, X., Ning, B., Gao, Z., 2014. Rapid detection of *Listeria monocytogenes* in milk by self-assembled electrochemical immunosensor. Sens. Actuat. B 190, 900–906.
Chobtang, J., Boer, I.J.M., Hoogenboom, R.L.A.P., Haasnoot, W., Kijlstra, A., Meerburg, B.G., 2011. The need and potential of biosensors to detect dioxins and dioxin-like polychlorinated biphenyls along the milk, eggs, and meat food chain. Sensors 11, 11692–11716.
Chouteau, C., Dzyadevych, S., Durrieu, C., Chovelon, J., 2005. A bi-enzymatic whole cell conductometric biosensor for heavy metal ions and pesticides detection in water samples. Biosens. Bioelectron. 21, 273–281.
Cui, X., Zhu, L., Wu, J., Hou, Y., Wang, P., Wang, Z., Yang, M., 2015. A fluorescent biosensor based on carbon dots-labeled oligodeoxyribonucleotide and graphene oxide for mercury (II) detection. Biosens. Bioelectron. 63, 506–512.

Deng, P., Xu, Z., Kuang, Y., 2014. Electrochemical determination of bisphenol A in plastic bottled drinking water and canned beverages using a molecularly imprinted chitosan–graphene composite film modified electrode. Food Chem. 157, 490–497.
Dinçkaya, E., Sezgintürk, M.K., Akyılmaz, E., Ertaş, F.N., 2007. Sulfite determination using sulfite oxidase biosensor based glassy carbon electrode coated with thin mercury film. Food Chem. 101, 1540–1544.
Doong, R., Shih, H., Lee, S., 2005. Sol–gel-derived array DNA biosensor for the detection of polycyclic aromatic hydrocarbons in water and biological samples. Sens. Actuat. B 111–112, 323–330.
Dou, W., Tang, W., Zhao, G., 2013. A disposable electrochemical immunosensor arrays using 4-channel screen-printed carbon electrode for simultaneous detection of *Escherichia coli* O157:H7 and *Enterobacter sakazakii*. Electrochim. Acta 97, 79–85.
Du, Y., Gao, X., Ye, X., Zheng, Z., Feng, Q., Wang, C., Wu, K., 2014. Composition and architecture-engineered Au–SnO_2/GNs-SWCNTs nanocomposites as ultrasensitive and robust electrochemical sensor for antioxidant additives in foods. Sens. Actuat. B 203, 926–934.
Duan, N., Wu, S., Ma, X., Xia, Y., Wang, Z., 2014. A universal fluorescent aptasensor based on AccuBlue dye for the detection of pathogenic bacteria. Anal. Biochem. 454, 1–6.
Elliott, C.T., Redshaw, C.H., George, S.E., Campbell, K., 2013. First development and characterisation of polyclonal and monoclonal antibodies to the emerging fresh water toxin cylindrospermopsin. Harmful Algae 24, 10–19.
Ensafi, A.A., Rezaei, B., Amini, M., Heydari-Bafrooei, E., 2012. A novel sensitive DNA–biosensor for detection of a carcinogen, Sudan II, using electrochemically treated pencil graphite electrode by voltammetric methods. Talanta 88, 244–251.
Fernández, F., Hegnerová, K., Piliarik, M., Sanchez-Baeza, F., Homola, J., Marco, M., 2010. A label-free and portable multichannel surface plasmon resonance immunosensor for on-site analysis of antibiotics in milk samples. Biosens. Bioelectron. 26, 1231–1238.
Funari, R., Ventura, B.D., Carrieri, R., Morra, L., Lahoz, E., Gesuele, F., Altucci, C., Velotta, R., 2014. Detection of parathion and patulin by quartz-crystal microbalance functionalized by the photonics immobilization technique. Biosens. Bioelectron. 67, 224–229.
Gavlasova, P., Kuncova, G., Kochankova, L., Mackova, M., 2008. Whole cell biosensor for polychlorinated biphenyl analysis based on optical detection. Int. Biodeter. 62, 304–312.
Gavrilova, M.A., Slepchenko, G.B., Mikheeva, E.V., Derybina, V.I., 2014. Voltammetric determination of aflatoxin B1. Procedia Chem. 10, 114–119.
Ghica, M.E., Carvalho, R.C., Amine, A., Brett, C.M.A., 2013. Glucose oxidase enzyme inhibition sensors for heavy metals at carbon film electrodes modified with cobalt or copper hexacyanoferrate. Sens. Actuat. B 178, 270–278.
Gobi, K.V., Miura, N., 2004. Highly sensitive and interference-free simultaneous detection of two polycyclic aromatic hydrocarbons at parts-per-trillion levels using a surface plasmon resonance immunosensor. Sens. Actuat. B 103, 265–271.
Haughey, S.A., Campbell, K., Yakes, B.J., Prezioso, S.M.F, DeGrasse, S.L., Kawatsu, K., Elliott, C.T., 2011. Comparison of biosensor platforms for surface plasmon resonance based detection of paralytic shellfish toxins. Talanta 85, 519–526.
Huang, N., Liu, M., Li, H., Zhang, Y., Yao, S., 2015. Synergetic signal amplification based on electrochemical reduced graphene oxide-ferrocene derivative hybrid and gold nanoparticles as an ultra-sensitive detection platform for bisphenol A. Anal. Chim. Acta. 853, 249–257.

Isa, I.M., Ghani, S.A., 2009. A nonplasticized chitosan based solid-state electrode for flow injection analysis of glutamate in food samples. Food Chem. 112, 756–759.

Jarczewska, M., Kierzkowska, E., Ziółkowski, R., Górski, L., Malinowska, E., 2015. Electrochemical oligonucleotide-based biosensor for the determination of lead ion. Bioelectrochemistry 101, 35–41.

Karaseva, N.A., Ermolaeva, T.N., 2014. Piezoelectric immunosensors for the detection of individual antibiotics and the total content of penicillin antibiotics in foodstuffs. Talanta 120, 312–317.

Kaushik, A., Solanki, P.R., Pandey, M.K., Kaneto, K., Ahmad, S., Malhotra, B.D., 2010. Carbon nanotubes–chitosan nanobiocomposite for immunosensor. Thin Solid Films 519, 1160–1166.

Kesik, M., Kanik, F.E., Turan, J., Kolb, M., Timur, S., Bahadir, M., Toppare, L., 2014. An acetylcholinesterase biosensor based on a conducting polymer using multiwalled carbon nanotubes for amperometric detection of organophosphorous pesticides. Sens. Actuat. B 205, 39–49.

Khaled, E., Kamel, M.S., Hassan, H.N.A., Abdel-Gawad, H., Aboul-Enein, H.Y., 2014. Performance of a portable biosensor for the analysis of ethion residues. Talanta 119, 467–472.

Kim, C., Lee, L., Min, J., Lim, M., Jeong, S., 2014a. An indirect competitive assay-based aptasensor for detection of oxytetracycline in milk. Biosens. Bioelectron. 51, 426–430.

Kim, G., Moon, J., Moh, C., Lim, J., 2014b. A microfluidic nano-biosensor for the detection of pathogenic *Salmonella*. Biosens. Bioelectron. 67, 243–247.

Lanzone, V., Scarano, S., Del Carlo, M., Corrado, F., Esposito, M., Minunni, M., Compagnone, D., 2013. Sensing benzo[a]pyrene-DNA adducts formation via decrease of hybridization reaction. Sens. Actuat. B 179, 187–193.

Laschı, S., Mascını, M., Scortıchını, G., Fraänek, M., Mascını, M., 2003. Polychlorinated biphenyls (PCBs) detection in food samples using an electrochemical immunosensor. J. Agric. Food Chem. 26, 1816–1822.

Li, S., Tan, Y., Wang, P., Ka, J., 2010. Inhibition of benzoic acid on the polyaniline–polyphenol oxidase biosensor. Sens. Actuat. B144, 18–22.

Li, S., Chen, J., Cao, H., Yao, D., Liu, D., 2011. Amperometric biosensor for aflatoxin B1based on aflatoxin-oxidase immobilized on multiwalled carbon nanotubes. Food Control 22, 43–49.

Li, Y., Zhou, Y., Lu, S., Guo, D., Ren, H., Meng, X., Zhi, B., Lin, C., Wang, Z., Li, X., Liu, Z., 2012. Development of a one-step test strip for rapid screening of fumonisins B1, B2, and B3 in maize. Food Control 24, 72–77.

Li, Z., Wang, Y., Ni, Y., Kokot, S., 2015. A rapid and label-free dual detection of Hg (II) and cysteine with the use of fluorescence switching of graphene quantum dots. Sens. Actuat. B 207, 490–497.

Ligler, F.S., Taitt, C.A.R., 2002. Optical Biosensors: Present and Future. Elsevier, Amsterdam.

Liu, Y., Wei, M., 2014. Development of acetylcholinesterase biosensor based on platinum carbon aerogels composite for determination of organophosphorus pesticides. Food Control 36, 49–54.

Liu, X., Deng, Y., Jin, X., Chen, L., Jiang, J., Shen, G., Yu, R., 2009. Ultrasensitive electrochemical immunosensor for ochratoxin A using gold colloid-mediated hapten immobilization. Anal. Biochem. 389, 63–68.

Liu, L., Chao, Y., Cao, W., Wang, Y., Luo, C., Pang, X., Fan, D., Wei, Q., 2014a. A label-free amperometric immunosensor for detection of zearalenone based on trimetallic Au-core/AgPt-shell nanorattles and mesoporous carbon. Anal. Chim. Acta 847, 29–36.

Liu, T., Zhao, Y., Zhang, Z., Zhang, P., Li, J., Yang, R., Yang, C., Zho, L., 2014b. A fiber-optic biosensor for specific identification of dead *Escherichia coli* O157:H7. Sens. Actuat. B 196, 161–167.

Liu, Q., Jiang, X., Zhang, Y., Zhen, L., Jing, W., Liu, S., Sui, G., 2015a. A novel test strip for organophosphorus detection. Sens. Actuat. B 210, 803–810.
Liu, S., Kang, M., Yan, F., Peng, D., Yang, Y., He, L., Wang, M., Fang, S., Zhang, Z., 2015b. Electrochemical DNA biosensor based on microspheres of cuprous oxide and nano-chitosan for Hg(II) detection. Electrochim. Acta 160, 64–73.
Long, F., Zhu, A., Zhou, X., Wang, H., Zhao, Z., Liu, L., Shi, H., 2014. Highly sensitive and selective optofluidics-based immunosensor for rapid assessment of bisphenol A leaching risk. Biosens. Bioelectron. 55, 19–25.
Ma, F., Sun, M., Zhang, K., Wang, S., 2015. A ratiometric fluorescence sensor for highly selective and sensitive detection of mercuric ion. Sens. Actuat. B 209, 377–383.
Masoomia, L., Sadeghi, O., Banitaba, M.H., Shahrjerdi, A., Davarani, S.S.H., 2013. A nonenzymatic nanomagnetic electro-immunosensor for determination of Aflatoxin B1 as a model antigen. Sens. Actuat. B 177, 1122–1127.
McGrath, T.F., Elliott, C.T., Fodey, T.L., 2012. Biosensors for the analysis of microbiological and chemical contaminants in food. Anal. Bioanal. Chem. 403, 75–92.
Mei, Z., Qu, W., Deng, Y., Chu, H., Cao, J., Xue, F., Zheng, L., El-Nezamic, H.S., Wu, Y., Chen, W., 2013. One-step signal amplified lateralflow strip biosensor for ultrasensitive and on-site detection of bisphenol A (BPA) in aqueous samples. Biosens. Bioelectron. 49, 457–461.
Meneely, J.P., Campbell, K., Greef, C., Lochhead, M.J., Elliott, C.T., 2013. Development and validation of an ultrasensitive fluorescence planar waveguide biosensor for the detection of paralytic shellfish toxins in marine algae. Biosens. Bioelectron. 41, 691–697.
Meng, X., Wei, J., Ren, X., Ren, J., Tang, F., 2013. A simple and sensitive fluorescence biosensor for detection of organophosphorus pesticides using H_2O_2-sensitive quantum dots/bi-enzyme. Biosens. Bioelectron. 47, 402–407.
Micheli, L., Grecco, R., Badea, M., Moscone, D., Palleschi, G., 2005. An electrochemical immunosensor for aflatoxin M1 determination in milk using screen-printed electrodes. Biosens. Bioelectron. 21, 588–596.
Mirasoli, M., Buragina, A., Dolci, L.S., Simoni, P., Anfossi, L., Giraudi, G., Roda, A., 2012. Chemiluminescence-based biosensor for fumonisins quantitative detection in maize samples. Biosens. Bioelectron. 32, 283–287.
Mishra, G.K., Sharma, A., Bhand, S., 2014. Ultrasensitive detection of streptomycin usingflow injection analysis-electrochemical quartz crystal nanobalance (FIA-EQCN) biosensor. Biosens. Bioelectron. 67, 532–539.
Mishra, R.K., Alonso, G.A., Istambouli, G., Bhand, S., Marty, J., 2015. Automated flow-based biosensor for quantification of binary organophosphates mixture in milk using artificial neural network. Sens. Actuat. B 208, 228–237.
Molinero-Abad, B., Alonso-Lomillo, M.A., Domínguez-Renedo, O., Arcos-Martínez, M.J., 2014. Malate quinone oxidoreductase biosensors based on tetrathiafulvalene and gold nanoparticles modified screen-printed carbon electrodes for malic acid determination in wine. Sens. Actuat. B 202, 971–975.
Morales, M.D., González, M.C., Serra, B., Reviejo, A.J., Pingarrón, J.M., 2005a. Composite amperometric tyrosinase biosensors for the determination of the additive propyl gallate in a reversed micellar medium. Sens. Actuat. B 106, 572–579.
Morales, M.D., González, M.C., Reviejo, A.J., Pingarrón, J.M., 2005b. A composite amperometric tyrosinase biosensor for the determination of the additive propyl gallate in foodstuffs. Microchem. J. 80, 71–78.
Moyo, M., Okonkwoa, J.O., Agyei, N.M., 2014. An amperometric biosensor based on horseradish peroxidase immobilized onto maize tassel-multiwalled carbon nanotubes modified glassy carbon electrode for determination of heavy metal ions in aqueous solution. Enzyme Microb. Technol. 56, 28–34.

Muslima, N.Z., Ahmad, M., Heng, L.Y., Saad, B., 2012. Optical biosensor test strip for the screening and direct determination of L-glutamate in food samples. Sens. Actuat. B 161, 493–497.

Najafia, M., Khalilzadeh, M.A., Karimi-Maleh, H., 2014. A new strategy for determination of bisphenol A in the presence of Sudan I using a ZnO/CNTs/ionic liquid paste electrode in food samples. Food Chem. 158, 125–131.

Najwa, B.O., Idriss, B., Georges, I., Ihya, A., Elhabib, A., Régis, R., Thierry, N., 2014. Enzymatic biosensor associated with molecularly imprinted polymers for sensitive and selective detection of organophosphorus insecticides in olive oil. Sci. Innov. 2 (6–1), 1–6.

Nedelkov, D., Rasooly, A., Nelson, R.W., 2000. Multitoxin biosensor–mass spectrometry analysis: a new approach for rapid, real-time, sensitive analysis of staphylococcal toxins in food. Int. J. Food Microbiol. 60, 1–13.

Nguyen, B.H., Tran, L.D., Do, Q.P., nguyen, H.L., Tran, N.H., nguyen, P.X., 2013. Label-free detection of aflatoxin M1 with electrochemical Fe_3O_4/polyaniline-based aptasensor. Mater. Sci. Eng. C 33, 2229–2234.

Ni, Y., Wang, P., Song, H., Lin, X., Kokot, S., 2014. Electrochemical detection of benzo(a)pyrene and related DNA damage using DNA/hemin/nafion–graphene biosensor. Anal. Chim. Acta. 821, 34–40.

Oliveira, T.M.B.F., Barroso, M.F., Morais, S., Araújo, M., Freire, C., Lima-Neto, P., Correia, A.N., Oliveira, M.B.P.P., Delerue-Matos, C., 2013. Laccase–Prussian blue film–graphene doped carbon paste modified electrode for carbamate pesticides quantification. Biosens. Bioelectron. 47, 292–299.

Oliveira, T.M.B.F., Barroso, M.F., Morais, S., Araújo, M., Freire, C., Lima-Neto, P., Correia, A.N., Oliveira, M.B.P.P., Delerue-Matos, C., 2014. Sensitive bi-enzymatic biosensor based on polyphenoloxidases–gold nanoparticles–chitosan hybridfilm–graphene doped carbon paste electrode for carbamates detection. Bioelectrochemistry 98, 20–29.

Pan, D., Gu, Y., Lan, H., Sun, Y., Gao, H., 2015. Functional graphene-gold nano-composite fabricated electrochemical biosensor for direct and rapid detection of bisphenol A. Anal. Chim. Acta 853, 297–302.

Paniel, N., Radoi, A., Marty, J., 2010. Development of an electrochemical biosensor for the detection of aflatoxin M1 in milk. Sensors 10, 9439–9448.

Panini, N.V., Salinas, E., Messina, G.A., Raba, J., 2011. Modified paramagnetic beads in a microfluidic system for the determination of zearalenone in feedstuffs samples. Food Chem. 125, 791–796.

Park, J., Byun, J., Mun, H., Shim, W., Shin, Y., Li, T., Kim, M., 2014a. A regeneratable, label-free, localized surface plasmon resonance (LSPR) aptasensor for the detection of ochratoxin A. Biosens. Bioelectron. 59, 321–327.

Park, J.H., Kim, Y., Kim, I., Ko, S., 2014b. Rapid detection of aflatoxin B1 by a bifunctional protein crosslinkerbased surface plasmon resonance biosensor. Food Control 36, 183–190.

Pasco, N., Jeffries, C., Davies, Q., Downard, A.J., Roddick-Lanzilotta, A.D., Gorton, L., 1999. Characterisation of a thermophilic L-glutamate dehydrogenase biosensor for amperometric determination of L-glutamate by flow injection analysis. Biosens. Bioelectron. 14, 171–178.

Pennacchio, A., Ruggiero, G., Staiano, M., Piccialli, G., Oliviero, G., Lewkowicz, A., Synak, A., Bojarski, P., D'Auria, S., 2014. A surface plasmon resonance based biochip for the detection of patulin toxin. Opt. Mater. 36, 1670–1675.

Perrotta, P.R., Arévalo, F.J., Vettorazzi, N.R., Zón, M.A., Fernández, H., 2012. Development of a very sensitive electrochemical magneto immunosensor for the direct determination of ochratoxin A in red wine. Sens. Actuat. B 162, 327–333.

Portaccio, M., Tuoro, D.D., Arduini, F., Lepore, M., Mita, D.G., Diano, N., Mita, L., Moscone, D., 2010. A thionine-modified carbon paste amperometric biosensor for catechol and bisphenol A determination. Biosens. Bioelectron. 25, 2003–2008.

Portaccio, M., Tuoro, D.D., Arduini, F., Moscone, D., Cammarota, M., Mita, D.G., Lepore, M., 2013. Laccase biosensor based on screen-printed electrode modified with thionine–carbon black nanocomposite, for bisphenol A detection. Electrochim. Acta 109, 340–347.

Prado, T., Foguel, M.V., Gonçalves, L.M., Sotomayor, M.P.T., 2015. β-Lactamase-based biosensor for the electrochemical determination of benzylpenicillin in milk. Sens. Actuat. B 210, 254–258.

Promphet, N., Rattanarat, P., Rangkupan, R., Chailapakul, O., Rodthongkum, N., 2015. An electrochemical sensor based on graphene/polyaniline/polystyrene nanoporous fibers modified electrode for simultaneous determination of lead and cadmium. Sens. Actuat. B 207, 526–534.

Rawal, R., Pundir, C.S., 2012. Development of an amperometric sulfite biosensor based on SO_x/PBNPs/PPY modified ITO electrode. Int. J. Biol. Macromol. 51, 449–455.

Rawal, R., Chawla, S., Pundir, C.S., 2012. An electrochemical sulfite biosensor based on gold-coated magnetic nanoparticles modified gold electrode. Biosens. Bioelectron. 31, 144–150.

Rejeb, I.B., Arduini, F., Arvinte, A., Amine, A., Gargouri, M., Micheli, L., Bala, C., Moscone, D., Palleschi, G., 2009. Development of a bio-electrochemical assay for AFB1detection in olive oil. Biosens. Bioelectron. 24, 1962–1968.

Ricci, F., Volpe, G., Micheli, L., Palleschi, G., 2007. A review on novel developments and applications of immunosensors in food analysis. Anal. Chim. Acta. 605, 111–129.

Rodriguez, M.L.L., Benimeli, C., Madride, R.E., Giacomelli, C.E., 2015. A simple *Streptomyces* spore-based impedimetric biosensor to detect lindane pesticide. Sens. Actuat. B 207, 447–454.

Romanazzo, D., Ricci, F., Volpe, G., Elliott, C.T., Vesco, S., Kroeger, K., Moscone, D., Stroka, J., Egmond, H.V., Vehniäinene, M., Palleschi, G., 2010. Development of a recombinant Fab-fragment based electrochemical immunosensor for deoxynivalenol detection in food samples. Biosens. Bioelectron. 25, 2615–2621.

Rodriguez-Emmenegger, C., Avramenko, O.A., Brynda, E., Skvor, J., Alles, A.B., 2011. Poly(HEMA) brushes emerging as a new platform for direct detection of food pathogen in milk samples. Biosens. Bioelectron. 26, 4545–4551.

Sapsford, K.E., Taitt, C.R., Fertig, S., Moore, M.H., Lassman, M.E., Maragos, C.M., Shriver-Lake, L.C., 2006. Indirect competitive immunoassay for detection of aflatoxin B1 in corn and nut products using the array biosensor. Biosens. Bioelectron. 21, 2298–2305.

Schenkmayerová, A., Bertóková, A., Šefcovicová, J., Štefuca, V., Bucko, M., Vikartovská, A., Gemeiner, P., Tkác, J., Katrlík, J., 2015. Whole-cell *Gluconobacter oxydans* biosensor for 2-phenylethanol biooxidation monitoring. Anal. Chim. Acta. 854, 140–144.

Sezgintürk, M.K., Dinçkaya, E., 2005. Direct determination of sulfite in food samples by a biosensor based on plant tissue homogenate. Talanta 65, 998–1002.

Sezgintürk, M.K., Göktug, T., Dinçkaya, E., 2005a. Detection of benzoic acid by an amperometric inhibitor biosensor based on mushroom tissue homogenate. Food Technol. Biotechnol. 43 (4), 329–334.

Shan, D., Li, Q., Xue, H., Cosnier, S., 2008. A highly reversible and sensitive tyrosinase inhibition-based amperometric biosensor for benzoic acid monitoring. Sens. Actuat. B 134, 1016–1021.

Sharma, H., Mutharasan, R., 2013. Review of biosensors for foodborne pathogens and toxins. Sens. Actuat. B 183, 535–549.
Shim, W., Mun, H., Joung, H., Ofori, J.A., Chung, D., Kim, M., 2014. Chemiluminescence competitive aptamer assay for the detection of aflatoxin B1 in corn samples. Food Control 36, 30–35.
Shimomura, M., Nomura, Y., Zhang, W., Sakino, M., Lee, K., Ikebukuro, K., Karube, I., 2001. Simple and rapid detection method using surface plasmon resonance for dioxins, polychlorinated biphenylx, and atrazine. Anal. Chim. Acta. 434, 223–230.
Silva, E.M., Takeuchi, R.M., Santos, A.L., 2015. Carbon nanotubes for voltammetric determination of sulfite in some beverages. Food Chem. 173, 763–769.
Singh, C., Srivastava, S., Ali, M.A., Gupta, T.K., Sumana, G., Srivastava, A., Mathur, R.B., Malhotra, B.D., 2013. Carboxylated multiwalled carbon nanotubes based biosensor for aflatoxin detection. Sens. Actuat. B 185, 258–264.
Sinha, R., Ganesana, M., Andreescu, S., Stanciu, L., 2010. AChE biosensor based on zinc oxide sol–gel for the detection of pesticides. Anal. Chim. Acta. 661, 195–199.
Slavik, R., Homola, J., Brynda, E., 2002. A miniature fiber optic surface plasmon resonance sensor for fast detection of *Staphylococcal enterotoxin* B. Biosens. Bioelectron. 17, 591–595.
Spricigo, R., Richter, C., Leimkühler, S., Gorton, L., Scheller, F.W., Wollenberger, U., 2010. Sulfite biosensor based on osmium redox polymer wired sulfite oxidase. Colloids Surf. A 354, 314–331.
Sun, J., Gan, T., Zhu, H., Shi, Z., Liu, Y., 2014. Direct electrochemical sensing for oxytetracycline in food using a zinc cation-exchanged montmorillonite. Appl. Clay Sci. 101, 598–603.
Syshchyk, O., Skryshevsky, V.A., Soldatkin, O.O., Soldatkin, A.P., 2015. Enzyme biosensor systems based on porous silicon photoluminescence for detection of glucose, urea, and heavy metals. Biosens. Bioelectron. 66, 89–94.
Tang, A.X.J., Pravda, M., Guilbault, G.G., Piletsky, S., Turne, A.P.F., 2002. Immunosensor for okadaic acid using quartz crystal microbalance. Anal. Chim. Acta. 471, 33–40.
Tekaya, N., Saiapina, O., Ouada, H.B., Lagarde, F., Ouada, H.B., Jaffrezic-Renault, N., 2013. Ultra-sensitive conductometric detection of pesticides based on inhibition of esterase activity in *Arthrospira platensis*. Environ. Pollut. 178, 182–188.
Tekaya, N., Saiapina, O., Ouada, H.B., Lagarde, F., Namour, P., Ouada, H.B., Jaffrezic-Renault, N., 2014. Bi-enzymatic conductometric biosensor for detection of heavy metal ions and pesticides in water samples based on enzymatic inhibition in *Arthrospira platensis*. J. Environ. Protect. 5, 441–453.
Tortolini, C., Bollella, P., Antonelli, M.L., Antiochia, R., Mazzei, F., Favero, G., 2014. DNA-based biosensors for Hg^{2+} determination by polythymine–methylene blue modified electrodes. Biosens. Bioelectron. 67, 524–531.
Tsai, W., Li, I., 2009. SPR-based immunosensor for determining *Staphylococcal enterotoxin* A. Sens. Actuat. B 136, 8–12.
Tsutsumi, T., Miyoshi, N., Sasaki, K., Maitani, T., 2008. Biosensor immunoassay for the screening of dioxin-like polychlorinated biphenyls in retail fish. Anal. Chim. Acta. 617, 177–183.
Tuoro, D.D., Portaccio, M., Lepore, M., Arduini, F., Moscone, D., Bencivenga, U., Mita, D.G., 2011. An acetylcholinesterase biosensor for determination of low concentrations of Paraoxon and Dichlorvos. N. Biotechnol. 29, 132–138.
Unnikrishnan, B., Ru, P., Chen, S., Mani, V., 2013. Nitrite determination at electrochemically synthesized polydiphenylamine-Pt composite modified glassy carbon electrode. Sens. Actuat. B 177, 887–892.

Van Dorst, B., Mehta, J.M., Bekaert, K., Rouah-Martin, E., Coen, W.D., Dubruel, P., Blust, R., Robbens, J., 2010. Recent advances in recognition elements of food and environmental biosensors: a review. Biosens. Bioelectron. 26, 1178–1194.
Velusamy, V., Arshak, K., Korostynska, O., Oliwa, K., Adley, C., 2010. An overview of foodborne pathogen detection: in the perspective of biosensors. Biotechnol. Adv. 28, 232–254.
Verma, R., Gupta, B.D., 2015. Detection of heavy metal ions in contaminated water by surface plasmon resonance–based optical fiber sensor using conducting polymer and chitosan. Food Chem. 166, 568–575.
Verma, N., Kumar, S., Kaur, H., 2011. Whole cell based disposable biosensor for cadmium detection in milk. Adv. Appl. Sci. Res. 2 (6), 354–363.
Vig, A., Radoi, A., Munoz-Berbel, X., Gyemant, G., Marty, J., 2009. Impedimetric aflatoxin M1 immunosensor based on colloidal gold and silver electrodeposition. Sens. Actuat. B 138, 214–220.
Wang, M., Kang, H., Xu, D., Wang, C., Liu, S., Hu, X., 2013. Label-free impedimetric immunosensor for sensitive detection of fenvalerate in tea. Food Chem. 141, 84–90.
Wang, N., Zhao, H., Ji, X., Li, X., Wang, B., 2014a. Gold nanoparticles-enhanced bisphenol A electrochemical biosensor based on tyrosinase immobilized onto self-assembled monolayers-modified gold electrode. Chinese Chem. Lett. 25, 720–722.
Wang, S., Yong, W., Liu, J., Zhang, L., Chen, Q., Dong, Y., 2014b. Development of an indirect competitive assay-based aptasensor for highly sensitive detection of tetracycline residue in honey. Biosens. Bioelectron. 57, 192–198.
Wang, X., Reisberg, S., Serradji, N., Anquetin, G., Pham, M.C., Wu, W., Dong, C.Z., Piro, B., 2014c. E-assay concept: detection of bisphenol A with a label-free electrochemical competitive immunoassay. Biosens. Bioelectron. 53, 214–219.
Wang, B., Wang, Q., Cai, Z., Ma, M., 2015a. Simultaneous, rapid and sensitive detection of three food-borne pathogenic bacteria using multicolor quantum dot probes based on multiplexfluoroimmunoassay in food samples. LWT Food Sci. Technol. 61, 368–376.
Wang, M., Gao, Y., Sun, Q., Zhao, J., 2015b. Ultrasensitive and simultaneous determination of the isomers of Amaranth and Ponceau 4R in foods based on new carbon nanotube/polypyrrole composites. Food Chem. 172, 873–879.
Wang, R., Xiang, Y., Zhou, X., Liu, L., Shi, H., 2015c. A reusable aptamer-based evanescent wave all-fiber biosensor for highly sensitive detection of Ochratoxin A. Biosens. Bioelectron. 66, 11–18.
Wang, S., Liu, J., Yong, W., Chen, Q., Zhang, L., Dong, Y., Su, H., Tan, T., 2015d. A direct competitive assay-based aptasensor for sensitive determination of tetracycline residue in Honey. Talanta 131, 562–569.
Wang, X., Lu, X., Wu, L., Chen, J., 2015e. 3D metal-organic framework as highly efficient biosensing platform for ultrasensitive and rapid detection of bisphenol A. Biosens. Bioelectron. 65, 295–301.
Wei, M., Wang, J., 2015. A novel acetylcholinesterase biosensor based on ionic liquids-AuNPs-porous carbon composite matrix for detection of organophosphate pesticides. Sens. Actuat. B 211, 290–296.
Wu, L., Deng, D., Jin, J., Lu, X., Chen, J., 2012a. Nanographene-based tyrosinase biosensor for rapid detection of bisphenol A. Biosens. Bioelectron. 35, 193–199.
Wu, X., Kuang, H., Hao, C., Xing, C., Wang, L., Xu, C., 2012b. Paper supported immunosensor for detection of antibiotics. Biosens. Bioelectron. 33, 309–312.
Wu, Y., Tang, L., Huang, L., Han, Z., Wang, J., Pan, H., 2014. A low detection limit penicillin biosensor based on single graphene nanosheets preadsorbed with hematein/ionic liquids/penicillinase. Mater. Sci. Eng. C 39, 92–99.

Xin, X., Sun, S., Li, H., Wang, M., Jia, R., 2015. Electrochemical bisphenol A sensor based on core–shell multiwalled carbon nanotubes/graphene oxide nanoribbons. Sens. Actuat. B 209, 275–280.

Xing, C., Liu, L., Song, S., Feng, M., Kuang, H., Xu, C., 2015. Ultrasensitive immunochromatographic assay for the simultaneous detection offive chemicals in drinking water. Biosens. Bioelectron. 15 (66), 445–453.

Xu, Y., Huang, Z., He, Q., Deng, S., Li, L., Li, Y., 2010. Development of an immunochromatographic strip test for the rapid detection of deoxynivalenol in wheat and maize. Food Chem. 119, 834–839.

Xu, X., Liu, X., Li, Y., Ying, Y., 2013. A simple and rapid optical biosensor for detection of aflatoxin B1 based on competitive dispersion of gold nanorods. Biosens. Bioelectron. 47, 361–367.

Yadav, Y.K., Agrawal, B., Chandra, P., Goyal, R.N., 2014. In vitro chloramphenicol detection in a *Haemophilus* influenza model using an aptamer-polymer based electrochemical biosensor. Biosens. Bioelectron. 55, 337–342.

Yang, X., Qian, J., Jiang, L., Yan, Y., Wang, K., Liu, Q., Wang, K., 2014. Ultrasensitive electrochemical aptasensor for ochratoxin a based on two-level cascaded signal amplification strategy. Bioelectrochemistry 96, 7–13.

Yang, L., Wang, G., Liu, Y., Wang, M., 2013. Development of a biosensor based on immobilization of acetylcholinesterase on NiO nanoparticles–carboxylic graphene–nafion modified electrode for detection of pesticides. Talanta 113, 135–141.

Yao, D., Cao, H., Wen, S., Liu, D., Bai, Y., Zheng, W., 2006. A novel biosensor for sterigmatocystin constructed by multiwalled carbon nanotubes (MWNT) modified with aflatoxin–detoxifizyme (ADTZ). Bioelectrochemistry 68, 126–133.

Yin, H., Zhou, Y., Xu, J., Ai, S., Cui, L., Zhu, L., 2010. Amperometric biosensor based on tyrosinase immobilized onto multiwalled carbon nanotubes-cobalt phthalocyanine-silk fibroin film and its application to determine bisphenol A. Anal. Chim. Acta. 659, 144–150.

Zaib, M., Saeed, A., Hussain, I., Athar, M.M., Iqbal, M., 2014. Voltammetric detection of As(III) with *Porphyridium cruentum*–based modified carbon paste electrode biosensor. Biosens. Bioelectron. 62, 242–248.

Zhai, C., Sun, X., Zhao, W., Gong, Z., Wang, X., 2013. Acetylcholinesterase biosensor based on chitosan/Prussian blue/multiwall carbon nanotubes/hollow gold nanospheres nanocomposite film by one-step electrodeposition. Biosens. Bioelectron. 42, 124–130.

Zhang, X., Wei, Y., Ding, Y., 2014a. Electrocatalytic oxidation and voltammetric determination of ciprofloxacin employing poly(alizarin red)/graphene composite film in the presence of ascorbic acid, uric acid, and dopamine. Anal. Chim. Acta. 835, 29–36.

Zhang, Y., Arugula, M.A., Wales, M., Wild, J., Simonian, A.L., 2014b. A novel layer-by-layer assembled multienzyme/CNT biosensor for discriminative detection between organophosphorus and nonorganophosphrus pesticides. Biosens. Bioelectron. 67, 287–295.

Zhang, Y., Su, Q., Xu, J., Zhang, Y., Chen, S., 2014c. Detecting of benzo[a]pyrene using a label-free amperometric immunosensor. Int. J. Electrochem. Sci. 9, 3736–3745.

Zhang, J., Tang, Y., Teng, L., Lu, M., Tang, D., 2015. Low-cost and highly efficient DNA biosensor for heavy metal ion using specific DNAzyme-modified microplate and portable glucometer-based detection mode. Biosens. Bioelectron. 68, 232–238.

Zhao, M., Hibbert, D.B., Gooding, J.J., 2006. Determination of sulfite in beer samples using an amperometric fill and flow channel biosensor employing sulfite oxidase. Anal. Chim. Acta. 556, 195–200.

Zheng, Y., Liu, Z., Jing, Y., Li, J., Zhan, H., 2015. An acetylcholinesterase biosensor based on ionic liquid functionalized graphene–gelatin-modified electrode for sensitive detection of pesticides. Sens. Actuat. B 210, 389–397.
Zhou, L., Li, D., Gaia, L., Wang, J., Li, Y., 2012. Electrochemical aptasensor for the detection of tetracycline with multiwalled carbon nanotubes amplification. Sens. Actuat. B 162, 201–208.
Zhou, Q., Yang, L., Wang, G., Yang, Y., 2013. Acetylcholinesterase biosensor based on SnO_2 nanoparticles–carboxylic graphene–nafion modified electrode for detection of pesticides. Biosens. Bioelectron. 49, 25–31.
Zhou, W., Sun, C., Zhou, Y., Yang, X., Yang, W., 2014. A facial electrochemical approach to determinate bisphenol A based on graphene-hypercrosslinked resin MN202 composite. Food Chem. 158, 81–87.
Zhu, Z., Feng, M., Zuo, L., Zhu, Z., Wang, F., Chen, L., Li, J., Shan, G., Luo, S., 2015. An aptamer-based surface plasmon resonance biosensor for the detection of ochratoxin A in wine and peanut oil. Biosens. Bioelectron. 65, 320–326.

9

ANALYTICAL AND ADVANCED METHODS-BASED DETERMINATION OF MELAMINE IN FOOD PRODUCTS

Kobun Rovina, Shafiquzzaman Siddiquee

University Malaysia Sabah, Biotechnology Research Institute, Sabah, Malaysia

1 Introduction

Melamine (1,3,5-triazine-2,4,6-triamine, or $C_3H_6N_6$) contains a important amount of nitrogen, which accounts for about 66% of its mass. It is a triazine analog together with three amino groups (Fig. 9.1). Melamine tends to hydrolyze under strong acid or alkali conditions resulting in the formation of cyanuric acid (2,4,6-trihydroxy-1,3,5-triazine), ammeline (4,6-diamino-2-hydroxy-1,3,5-triazine) and ammelide (6-amino-2,4-dihydroxy-1,3,5-triazine) (Mauer et al., 2009; Mecker et al., 2012). It is a synthetic organic nitrogenous compound that is widely used in the production of melamine–formaldehyde polymer resins, which are a component in many plastics, adhesives, glues, dyes, fertilizers, and laminated products such as plywood, cement, cleansers, and fire retardant paint (Finete et al., 2014; Andersen et al., 2008). However, the exposure to melamine may be particularly dangerous to human health. Melamine has low acute oral toxicity, but its high concentrations can induce renal pathology and even death in babies and children (Chan et al., 2008; Guan et al., 2009). Melamine can be hydrolyzed to cyanuric acid in vitro. Cyanuric acid will in turn associate with melamine to form an insoluble melamine–cyanurate complex which form stones in the urinary system, probably leading to acute renal failure by obstruction (Sun et al., 2010a; Kobayashi et al., 2010). Besides, it will also cause a variety of renal toxic effects such as nephrolithiasis, chronic kidney inflammation and bladder carcinoma (Hau et al., 2009).

Nanobiosensors. http://dx.doi.org/10.1016/B978-0-12-804301-1.00009-6

Figure 9.1. Structure of melamine.

Melamine is artificially added to milk products or pet food to increase the apparently protein content. Consequently, the addition of melamine to milk fraudulently increases the nitrogen concentration in milk resulting in a "false" appearance of a higher level of protein by the Kjeldahl method (Finete et al., 2013). The conventional assessing methods such as either the Kjeldahl or Dumas tests cannot distinguish nitrogen origin, when melamine was illegally adulterated into raw milk and dairy food to artificially enhance the apparent crude protein level. The toxicity of melamine caught the attention of physicians as a result of a recent spate of renal injury after exposure to melamine-tainted milk in China. Besides, it can cause proliferative lesions of the urinary tract through the irritative stimulation of calculi. It is harmful to the health of people as well as animals (Ogasawara et al., 1995).

Melamine has been reported and banned in food products and additives due to serious cases (Sun et al., 2010a). Melamine has low oral acute toxicity; however, it can form insoluble crystal with its hydrolyzate in urinary system. Long-term intake of melamine-tainted milk products can cause urinary calculus, renal dysfunction, and even death, especially in infants and children (Kobayashi et al., 2010). Moreover, a variety of toxic effects from melamine, including nephrolithiasis, chronic kidney inflammation, and bladder carcinoma, have been studied in animals. Hau et al. (2009) have reviewed the epidemiology, clinical features, and investigative findings concerning the only outbreak of melamine poisoning in humans. The melamine ingestion caused multiple toxic effects, including chronic kidney inflammation due to the acid crystals formed of melamine and cyanurate and their deposition in the renal tubules.

In the Sanlu Milk Scandal in China that occurred in 2008, the intentional addition of melamine in dairy products created a global hotspot of concern over the use of melamine. They reported that the crude protein level can increase more than 4% from the addition of 1% of melamine, owing to its high nitrogen content. The Chinese government promulgated an interim control limit for melamine at 1 mg kg^{-1} for infant formula and at 2.5 mg kg^{-1} for other milk products in Oct. 2008, and this interim limit became official for all foodstuffs in Apr. 2011 (Wu and Zhang, 2013). Some Chinese reports said large amounts of melamine had been illegally added to infant formula and some dairy products (Qiao et al., 2010; Tyan et al., 2009; Xin and Stone, 2008). Ingelfinger (2008) has reported that approximately 294,000 children have been hospitalized and 6 have died after ingestion of an infant milk formula contaminated with melamine. The World Health Organization, US Food and Drug Administration (FDA), the European community, and other countries and regions have established regulation of Maximum Residue Limits (MRLs) of melamine in all food products. Many countries have adopted the standard acceptable level of 1 mg kg^{-1} for infant formula milk powder and 2.5 mg kg^{-1} for other milk products. It has provided a sufficient margin of safety for dietary exposure relative to the Tolerable Daily Intake (TDI) of melamine, which is 0.2 mg kg^{-1} body weight (WHO, 2009). Beyond the limits, the Codex Alimentarius Commission allowed a maximum limit of 0.15 mg kg^{-1} melamine in liquid infant formula (FAO and WHO, 2011).

It is vital to monitor and control the harmful effects of melamine. Recently, many analytical methods have been developed for the determination and quantification of melamine and its analogs in food. Most of the published reports have detected the use of melamine using mass spectrometry (MS)–based methods, including surface desorption atmospheric pressure chemical ionization (DAPCI) mass spectrometry (DAPCI-MS) (Yang et al., 2009), extractive electrospray ionization mass spectrometry (EESI-MS) (Zhu et al., 2009a,b) and liquid chromatography-tandem mass spectrometry (LC-MS/MS) (Smoker and Krynitsky, 2008; Turnipseed et al., 2008) have addressed as confirmatory and screening for melamine, due to their high selectivity and lower detection limits. Many of these methods are widely applied for determination of melamine due to their sensitivity and high accuracy. However, limitations of MS techniques rely on its high cost, complexity, and difficulty of on-site applicability. To overcome these limitations, methods based on the spectroscopic properties of melamine, most of after reactions are proposed for melamine screening, particularly in milk samples. Other techniques are included such as gas chromatography (GC) (Yokley et al., 2000), high performance liquid chromatography (HPLC) (Ehling et al., 2007), micellar liquid chromatography (MLC) (Beltran-Martinavarro et al., 2014; Batista et al., 2014) high-performance liquid chromatography/mass spectrum (HPLC/MS) (Kim et al., 2008), surface enhanced raman spectroscopy (Lin et al., 2008) and capillary zone electrophoresis/mass spectrum (CE/MS) (Cook et al., 2005). Despite the high sensitivity and specificity that can be achieved by GC, HPLC, and CE methods, they require expensive instrumentation, extensive sample preparation, are time consuming, and must be monitored by skilled personnel. Additionally, the immunoassay method can offer more advantages compared to analytical instrumental methods because of high throughput and rapid turnaround time. One of the most famous immunoassay methods is enzyme-linked immunosorbent assay (ELISA), which is more efficient and can be used for the analysis of large samples. However, it is essential that labor-intensive operations including incubation, washing, and enzymatic reactions during the signal generation process. Thus, a simple, rapid, low-cost, convenient, high-sensitivity method is required to develop for detecting melamine in dairy products.

Nowadays, as a kind of novel technique, sensors, especially chemo- and bio-sensors, have been rapidly developed and received considerable attention. Developments in chemistry, physics, biochemistry, and molecular biology lead to the sensor techniques due to their simplicity, selectivity, sensitivity, low cost, and, in some cases, the advantages of miniaturization, portability, and integration into automated devices already in use for detection of melamine (Farre et al., 2009). In this regard, sensing techniques are an alternative method for determination of the melamine level in dairy products.

This chapter has discussed the current state of advanced (sensors) detection techniques in melamine and its compounds. Other sections deal with the current and past melamine contamination incidents together with modern instrument analytical methods for determination of the presence of melamine and its analogs. Analytical determinations of melamine in food products are significantly important due to its toxicity in humans and animals, and other countries around the world have very strictly controlled those foods likely to be contaminated during harvest or storage.

2 Melamine Structure and Application

Melamine (2, 4, 6-triamino-1, 3, 5-triazine; $C_3H_6N_6$; MW: 126.12) is a trivial polar molecular compound and a kind of triazine analogs that contains three amino groups commonly formed of white crystals content high nitrogen level (Liu et al., 2012a). Remarkably, melamine can hydrolyze via deamination reactions and formed of cyanuric acid (2,4,6-trihydroxy-1,3,5-triazine), ammeline (4,6-diamino-2-hydroxy-1,3,5-triazine), and ammelide (6-amino-2,4-dihydroxy-l,3,5-triazine) throughout strong acidic and alkaline conditions. Therefore, melamine and related triazines are able to form self-assembling in high molecular weight complexes, via organized by intramolecular networks of hydrogen bonds and π–π aromatic ring stacking (Whitesides et al., 1991).

Cyanuric acid is commonly an oxytriazine melamine analog that is involved as a by-product in melamine synthesis and is an accepted component of feed-grade biuret, a ruminant feed additive. Its derivatives are regulated in the USA as components of sanitizing solutions for use on food-processing equipment, utensils, and other items that come in contact with food. It is used in drinking water for disinfection purposes by using sodium dichloroisocyanurate that rapidly dechlorinates to cyanurate (Matte et al., 1990). Ammelide and ammeline are monoamino- and diaminooxytriazine analogs that are produced as by-products of melamine synthesis. Ammeline is usually used in lubricating greases (USFDA, 2007). It is found by heating dicyandiamide with aqueous ammonia at 160–170°C and synthesized by heating melamine with concentrated sulfuric acid (H_2SO_4) for a short time at 190°C. Ammelide decomposed at 170°C with water to form carbon dioxide (CO_2) and ammonia (NH_3). It converted into cyanuric acid by oxidizing agents or by boiling with acids/alkaline (Bann and Miller, 1958). However, the uses of ammelide information are not available.

In 2007, an approximately 1.2 million tons of melamine has been produced in the whole world; the predominant producers are located in China and Western Europe (Bizzari and Yokose, 2008). Melamine can be produced from three different starting materials, which are urea, dicyandiamide, and hydrogen cyanide. Synthetical chemical production of melamine commonly used such as plastics, adhesives, laminates, paints, permanent-press fabrics, flame retardants, textile finishes, tarnish inhibitors, paper coatings, and fertilizer mixtures (Zhang et al., 2008; Liang et al., 2009). Melamine is also used in the production of pesticides, when metabolized by two strains of soil bacteria, *Pseudomonas* strain A and *Klebsiella terragena*, giving rise through successive deamination reactions to ammeline, ammelide, and cyanuric acid, with further breakdown to biuret, urea, and, ultimately, ammonia and carbon dioxide (Shelton et al., 1997).

In 2007, melamine was used in degradation product that is detected on vegetable crops after spray application (Patakioutas et al., 2007). Trace amounts of melamine may occur in food from the uses of various fertilizers. Besides, melamine and its triazine compounds are used as a nitrogen source in slow-release urea-based fertilizer mixtures. The levels of melamine occurrence in food are unknown while applied various fertilizers. In addition, the previous study has been tested melamine for nitrogen release characteristics and response on the application of grass, the melamine–urea combination showed significant nitrogen plant uptake (Mosdell et al., 1987). Melamine-containing resins are mostly used in the manufacturing of particleboard. In a review of the wood product industry in Finland reported that almost 30,000 tons of urea-melamine resins are used annually for the production of particleboard and veneer products alone. Thus, it is important noted that legitimate uses of these melamine and its analogs may vary from one country to other countries (Antikainen et al., 2004).

3 Toxicology of Melamine and Its Metabolite

Melamine will be found at low levels in food and feed mostly from its legal use as an antiseptic material in laminates and plastics and as a trace contaminant in nitrogen supplements used in animal feed. Melamine and cyanuric acid are quickly absorbed and excreted in a crystal form in the urine of monogastric animals, as they are not metabolized by the body. The absorption, metabolism, distribution, excretion, and toxicokinetics of melamine have been systemically investigated in various mammalian species such as mice, dairy goats, and pigs (Baynes et al., 2010). Hau et al. (2009) have reported that direct contact with melamine results in skin irritation, and inhalation of melamine causes respiratory tract irritation. Oral ingestion affects the digestive tract, resulting in nausea, vomiting, and diarrhea. In severe cases, hypertension, edema, or oliguria may be observed, due to the interaction of melamine with isomers to form crystals.

Melamine has been reported to have low acute toxicity, with lethal doses from oral intake in mice and rats exceeding 3,000 mg kg^{-1} body weight. Although melamine has low toxicity, it may lead to kidney stones eventual renal failure, and ultimately death, when it forms an insoluble compound with the analog cyanuric acid as shown in Fig. 9.2. The combination of melamine and cyanuric acid causes the formation of crystals, which can lead to acute renal failure in animals, especially cats, within 48 h after ingestion. Pathogenesis may be through formation of high-molecular weight network complexes by self-association of melamine–cyanuric acid, which has poor aqueous solubility and precipitates in renal tubules, causing urinary system damage and ultimately death. However, some subchronic studies in rats and mice have been reported lethal doses of up to 18,000 mg of melamine kg^{-1} feed are related to the excretory organs (Hau et al., 2009).

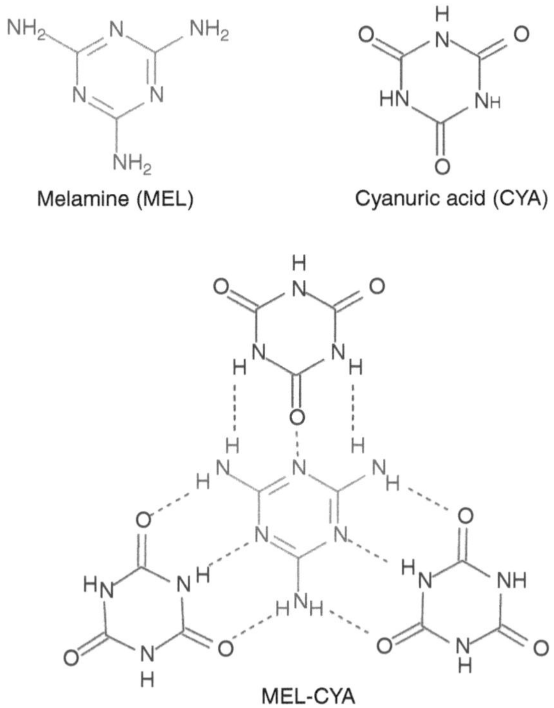

Figure 9.2. The chemical structures of melamine, cyanuric acid, and MEL-CYA (Sun et al., 2010a).

4 Melamine Contamination Cases

Melamine has been added to cattle feed as a nonprotein nitrogen (NPN) since 1958. However, it was discontinued in 1978 because it was found to be completely hydrolyzed in ruminants (Newton and Utley, 1978). Brown et al. (2007) have reported incidents of pet illness and acute renal failure occurring in several Asian countries due to consumption of pet food products manufactured at a single factory in Thailand in Mar. 2004. Subsequently, consumers and other veterinarians have been reported on more illnesses and deaths in pets that are potentially associated with the pet food consumed. In 2007 and 2008, there were two reported outbreaks of food borne illnesses associated with melamine in products such as milk, infant formula, frozen yogurt, pet food, biscuits, candy, and coffee drinks (Andersen et al., 2008).

The contamination of food with melamine became a hot issue in Sep. 2008 when the World Health Organization (WHO) reported that more than 47,000 infants and young children in China had been hospitalized due to urinary problems, renal tube blockages, kidney stones, and the death of several infants due to the consumption of melamine-contaminated infant formula and other related dairy products. The contamination of milk products was further investigated by the Administration of Quality Supervision, Inspection and Quarantine (AQSIQ), which revealed that 22 commercially available brands from China contained melamine in their milk products. Melamine and its compounds were found at levels ranging from 0.8% to 26% in wheat gluten and rice protein extracts imported from China, which are used to manufacture pet food in the USA. The State Council of China announced that melamine levels in most food products ranged from 0.09 to 619 mg kg^{-1}, with products from the Shijangzhuang Sanlu Company containing the highest levels of melamine, up to 2563 mg kg^{-1} (WHO, 2008b). Other melamine-contaminated food products that came from China included liquid milk, confectionaries, frozen yogurt, beverages, and biscuits. Other countries have also reported the presence of melamine in other food products such as nondairy creamer, ammonium bicarbonate, dried whole egg, and fresh hen eggs. The melamine content in eggs was probably due to tainted feed and other milk-containing products made in China.

5 Tolerable Daily Intake (TDI) and Risk Assessment of Melamine

The increasing food safety concern due to melamine-contaminated infant formula in 2008 has caused national authorities and other countries to assess total daily intake (TDI) in order to protect the general population, including infants, where the dietary exposure of melamine and the risk caused by it will be assessed (Wu and Zhang, 2013). WHO (2004) has defined the TDI as the estimated maximum amount of an agent to which individuals in the population may be exposed daily over their lifetimes without appreciable health risk. The intensified food safety concern over melamine in infant formula has prompted national authorities and countries to assess its TDI for protection of general population including young children. Dietary exposure to melamine and the risk from melamine-tainted infant formula have been assessed. Therefore, many countries worldwide have currently established a tolerance level for melamine in order to ensure a safe food supply and protect the health of its citizens.

The US Food and Drug Administration (FDA) is the first organization that has provided the estimation of TDI of melamine due to reports of pet foods contaminated with melamine. The initial TDI is established by the FDA to be 0.63 mg kg^{-1} for melamine.

This initial value is based on the analysis of data that is taken from the selected animal toxicity assay that is conducted by the National Toxicology Program (NTP) of the US Department of Health and Human Services (NTP, 1983). The World Health Organization has allowed a TDI for melamine intake of 0.2 mg kg^{-1} of body mass and the maximum residue limits (MRLs) for melamine have been established in many countries in the world. The MRLs for infant formula in China and United States was 1.0 mg kg^{-1} and 2.5 mg kg^{-1} for milk and other milk products. In Europe, the Food Safety Authority has permitted a limit of up to 2.5 mg kg^{-1} for all products that have milk content greater than 15% (WHO, 2009). The Malaysia Ministry of Health has established MRLs for melamine content, which are 1.0 mg kg^{-1} in baby food products and 2.5 mg kg^{-1} for adult food products (Wu and Zhang, 2013), while other countries such as New Zealand, Nigeria, and Canada have also set up their individual MRLs for melamine content in food.

6 Modern Instrument Analytical Methods

6.1 Capillary Electrophoresis

The capillary electrophoresis method has separated melamine by applying a strong electric field for separation of molecules. The compounds are separated based on the difference in their hydrophobicity, size, and charge (Sun et al., 2010b). Mostly capillary electrophoresis is a cost-effective and also potential analytical method used for small molecules separation. It has several advantages requiring few samples, consuming fewer reagents, and having a fast separation speed. The pretreatment steps for samples involving melamine in milk and other matrices are simple. However, when coupled with UV light, the sensitivity and reproducibility of the detection is reduced.

Chang et al. (2010) have described the use of 11-mercaptoundecanoic acid-capped AuNPs (MUA–AuNPs) for selective extraction of melamine prior to analysis by capillary electrophoresis with UV detection. The highest degree of melamine-induced aggregation of MUA–AuNPs was occurred at pH 5.0, indicating that the NP aggregation is mainly because of hydrogen bonding between the carboxylate groups of MUA and the amine groups of melamine. Under the optimum extraction and separation conditions, the limit of detection at a signal-to-noise ratio of 3 was estimated to be 77 pM for melamine, with linear range of 1–1000 nM. However, Xia et al. (2010) have developed a simple, rapid, and accurate method for the simultaneous determination of melamine and related compounds by capillary zone electrophoresis with diode array detection (CZE–DAD) and successfully applied in egg, dairy products, and pet feed. The optimized method showed good performance with linear coefficient of 0.9964, precision (≤4%), accuracy (82.2–101.6%), and high sensitivity.

6.2 Mass Spectrometry

MS methods are well known for being able to easily determine contaminants due to the specificity and high sensitivity. Currently, several LC-MS and LC-MS/MS techniques are employed in labeling melamine and improving the precision of developing methods (Chu et al., 2010; Desmarchelier et al., 2009). The FDA (2009) has proposed the use of LC-MS/MS as short-term method for determination of melamine and cyanuric acid residues in foods at the levels of 0.25 mg kg^{-1} for infant formula and other dairy-containing food products or ingredients. Besides, matrix-assisted laser desorption mass spectrometry (MALDI-MS) is a sensitive methods but require extensive, time consuming, and sample pretreatment costly, which are included the extraction, purification, and derivatization of samples (Sun et al., 2010b).

Zhao et al. (2015) have established a method for sensitive, fast, and quantitative determination of melamine in milk and dairy products using high-field asymmetric ion mobility spectrometry (FAIMS). The solid-phase extraction (SPE) technology has been used for purification after the sample is extracted by organic solvents, and followed by the analysis of FAIMS. Under optimum condition, the results showed that the SPE–FAIMS technique determines the amount melamine in samples with a low concentration at 0.1 mg kg^{-1}. The ion intensity has a linear relationship with melamine concentration in the range from 0.3 to 25 mg L^{-1}, with a good linearity of 0.9975. The limits of detection (LOD) and limits of quantification (LOQ) were found to be 0.1 mg kg^{-1} and 0.3 mg kg^{-1} in milk and dairy products, respectively, with relative standard deviation (RSD) is less than 8.0%. The results demonstrated that FAIMS has great potential as a powerful tool for food analysis and safety inspection.

Rajapandiyan et al. (2015) have addressed a simple, rapid and massive determination of melamine in liquid milk and powder samples by surface-enhanced Raman spectroscopy (SERS) technique. The proposed method used silver nanoparticles (AgNPs) decorated on a cylindrical support for detecting at low amount of sample (5 μL) without drying procedures. The linear range of melamine in water has 5–400 parts per billion (ppb) with detection limit of 2.5 ppb (S/N = 3). Mirzajani and Ahmadi (2015) have presented a feasible protocol for the preparation of melamine conjugated Fe_3O_4 magnetic nanoparticles. Magnetic nanocomposite has developed as a novel and sensitive solid phase for removal, preconcentration, and spectrophotometric determination of trace amount of malachite green (MG). The calibration curve is showed a linear range of 0.1–8 μg mL^{-1} with detection limit of 0.05 μg mL^{-1}, and the RSD found less than 1.66%. Furthermore, Kailasa and Wu (2015) have developed electrospray ionization mass spectrometry (ESI–MS) with simple, rapid, and sensitive method for determination of melamine without any sample preparation was demonstrated. The ESI–MS/MS on the protonated melamine at m/z 127 is produced the characteristic fragment ions at m/z 110, 85, 68, 60. The developed method has a good linearity in the range of 0.5–10.0 μg mL^{-1}, with detection limit of 0.1 μg mL^{-1} and RSD of 8.9% ($n = 5$). Together with the existing methods on MS and NP-based sensing methodologies for melamine detection, the developed method has provided an efficient platform for detection of melamine in dairy products without any sample preparation steps or postseparation procedures (Kailasa and Wu, 2015).

Meng et al. (2015) have developed a new method based on quadrupole time-of-flight mass spectrometry (QTOF-MS) by screening and confirmation for infant milk powder samples. They used matrix-matched internal standard calibration curves of UPLC–MS/MS are linear, with correlation coefficient is 0.994. The limit of detection at 5.0 μg kg^{-1} with recovery rates from 64.8% to 103%. The proposed method showed high sensitivity that able to use as an effective tool for rapid screening and highly sensitive multiresidue determination of melamine and their metabolites in milk powder samples.

6.3 Chromatography Techniques

High performance liquid chromatography (HPLC) has recognized to be one of the most powerful techniques for the analysis of biological matrices and pharmaceutical preparations due to its high efficiency and good reproducibility. This method is able to simultaneously detect at the presence of melamine and its analogs level in rice, wheat, and corn flours. HPLC methods with ultraviolet detection (HPLC-UV) (Deabes and El-Habib, 2012; Filazi et al., 2012; Rambla-Alegre et al., 2010; Venkatasami and Sowa, 2010; Yan et al., 2012; Zheng et al., 2012) and electrokinetic capillary chromatography with amperometric detection (Wang et al., 2010) have been applied for determination of melamine in milk samples. The determination of melamine using HPLC-UV may be highlighted because this technique has good reproducibility, sensitivity, and high accuracy in food samples (WHO, 2009).

However, UV detection is susceptible for interfering compounds that absorb in the same spectral region of melamine, around 240 nm (Lutter et al., 2011). Besides, erroneous quantification can occur if there is insufficient care to the chromatographic conditions and optimizations of samples are not conducted (Sun et al., 2010b). Therefore, most of the HPLC-UV methods have complex sample pretreatments, such as solid phase or liquid-liquid extraction, derivatization, and preconcentration involving several steps.

Fashi et al. (2015) have studied the present application of electro-membrane microextraction technique combined with HPLC-UV detection for fast extraction-preconcentration and determination of melamine in dairy products. They have accounted based on the extraction of charged melamine molecules are migrated from 6.5 mL feed solution through a liquid membrane immobilized in the pores of a hollow fiber into 20 μL of the acceptor solution. The best performance was achieved by using 2-nitrophenyloctylether, tris-(2-ethylhexyl) phosphate (10%), and di-(2-ethylhexyl) phosphate (10%) as the extraction solvent; application of a potential of 90V as the driving force and by adjusting the pH of feed and receiving phases at 5.5 and 1.0, respectively. The working range was 7–8000 ng mL^{-1}, with detection limit between 2.0 and 5.8 ng mL^{-1}.

A new and sensitive precolumn derivatization method has developed for the analysis of melamine leached from tableware by HPLC with fluorescence (HPLC-Fluo) detection (Zhang et al., 2014). The HPLC sensitivity is significantly enhanced by introducing 10-methyl-acridone-2-sulfonyl chloride (MASC) with excellent fluorescence property into the melamine molecule. Meanwhile, derivatization greatly increases the hydrophobicity of melamine. Thus, the common reversed phase column can be used for the HPLC analysis of high hydrophilic melamine. The detection limit obtained by the method of 0.40 μg L^{-1}. This is the first time that HPLC-Fluo detection has applied for the analysis of melamine. The method is successfully applied for the analysis of melamine leached from tableware. The results have indicated that the leaching of melamine from tableware is obvious when hot water or milk is added (Zhang et al., 2014). Furthermore, experimental studies of the natural photoluminescence of melamine in aqueous solutions showed that its fluorescence intensity (at 250/365 nm) is appropriated for analytical purposes (Finete et al., 2014). The exploitation of such melamine property provided on the basis of development of a new, simple, precise, and accurate method by high performance liquid chromatography with HPLC-Fluo to determine melamine in kitchen plasticware following aqueous extraction using a microwave oven. The detection limit has found at 0.027 μg mL^{-1} with a linear range up to 10 μg mL^{-1}. Different samples of kitchen plasticware have analyzed by the developed and optimized method; showed melamine concentrations in the aqueous extract up to 17 μg mL^{-1}, which corresponded to 86.0 mg kg^{-1} in these utensils.

Finete et al. (2015) have vital aimed a feasibility and applicability of the developed method for melamine determination in plasticware using HPLC-Fluo detection for the analysis of trace levels of melamine in UHT bovine milk. HPLC-Fluo method has developed based on the natural fluorescence emission of melamine (250/365 nm), and validated for melamine determination as an adulterant of bovine UHT whole milk. The clean up treatment of milk samples by deproteinization was optimized and never found any matrix effect. A linear range has obtained from 0.05 to 10.0 μg mL^{-1}, with linear coefficient (r^2) of 0.9998. The limits of detection and quantification of melamine are 0.0081 and 0.027 μg mL^{-1}, respectively. Extraction of milk samples fortified melamine with three concentration levels, two of which corresponded and established by WHO for melamine in foods; overall mean recovery of 95.4 ± 1.2% (n = 9). This recovery value has satisfied the performance criteria established by the Codex Alimentarius via analytical methods for determination of food residues, proved the usefulness and effectiveness of the proposed method.

Zhang et al. (2014) have established a new method based on the strong-cation-exchange restricted access material that synthesized by atom transfer radical polymerization (ATRP). Poly(3-sulfopropyl methacrylate-*co*-ethylene dimethacrylate) [p(SPM/EDMA)] has grafted on the silica by surface-initiated ATRP first for synthesizing process. The Sil-*g*-p(SPM/EDMA)-*g*-pGMMA has incorporated with protein exclusion and cation exchange, showing the property of cation-exchange restricted access material. The application of Sil-*g*-p(SPM/EDMA)-*g*-pGMMA is used for determination of melamine and cyromazine in bovine milk using the online solid-phase extraction and HPLC method. Besides, the Sil-*g*-p(SPM/EDMA)-*g*-pGMMA has successfully applied for the sample pretreatment and an HILIC column as analytical column. It can be used for the cation extraction from biological samples by direct HPLC injection and found good accuracy, precision, and low detection limits.

A novel dispersive micro solid-phase extraction (DMSPE) cleanup method based on the PCX sorbent is applied to the analysis of melamine and cyromazine residues in milk and milk powder, and ultra-HPLC high-resolution mass spectrometry (UHPLC–HRMS) used as instrument detection (Chen et al., 2015). The melamine and cyromazine are first extracted and then adsorbed to the PCX powder.

Subsequently, the analytes in PCX sorbent are eluted with ammonium hydroxide/acetonitrile (2.5:97.5 v/v) through a simple unit device equipped with syringe and nylon syringe filter. All the samples are analyzed by UHPLC–HRMS/MS on a Waters Acquity BEH HILIC column with the presence of formic acid and ammonium formate in water/acetonitrile as the mobile phase with gradient elution. The method is provided a significant improvement for determination of melamine and cyromazine in milk and milk powder in terms of efficient, rapid, economical, and miniaturized sample preparation methods, which yielded fewer matrix effects compared with SPE method. The established cleanup method is expected to be widely applied for the sample preparation of alkaline contaminants at trace levels in the future (Chen et al., 2015).

Micellar liquid chromatography-method has been developed for determination of melamine in swine kidney reported by Beltran-Martinavarro et al. (2014). Melamine is eluted from the HPLC column without interferences in <8.0 min. Mobile phase is a 0.11 M sodium dodecyl sulfate −7.5% propanol at pH 3 solution running under isocratic mode through a C18 column at 1 mL min^{-1} and absorbance detection at 210 nm. This method is validated according to the Food and Drug Administration guidelines: sensitivity, calibration range (0.3–30 mg g^{-1}), linearity ($r^2 > 0.9997$), precision (<7.6%), accuracy (8.3–3.6%), recovery (82.1–92.4%) and robustness (<5.1%). In similar studies Rambla-Alegre et al. (2010) have eluted melamine in nearly 9.3 min without overlapping the protein band or other endogeneous compounds. Validation has performed the following analytical parameters, which are: linearity (0.02–100 ppm; $r^2 = 0.999$), limit of detection (5 ppb), intra- and interday precision (R.S.D. < 7.6% and <9.7%, respectively) and robustness (R.S.D. < 7.4% for retention time and <5.0% for area). Recoveries for several milk samples are found in the range of 85–109%.

Silica, C8, and C18 monolithic column are the only option for separation on a sequential injection chromatography (SIC) system due to its low-pressure operation. The separation of molecules poorly retained by this stationary phase is a difficult task. Batista et al. (2014) have mentioned the first use of micellar chromatography on an SIC system, with the applicability used by the determination of melamine in milk. The mobile phase is composed by aqueous sodium dodecyl sulfate (SDS) solution and propanol (92.5:7.5). The adsorption of SDS monomers on the C18 stationary phase modified its chromatographic properties, enabling satisfactory melamine separation from the milk components. The sample pretreatment procedure is online implemented by dilution of the sample with SDS using a multiposition valve. A linear response has been observed within 2.0 and 6.0 mg L^{-1} of melamine with a detection limit is 0.6 mg L^{-1} and coefficients of variation at 2.9% (n = 6). The procedure is more suitable for fast determination of melamine below the limits detection established by FDA and WHO, with minimized reagent consumption. Results for different milk samples agreed with those obtained by high-performance liquid chromatography at the 95% confidence level (Batista et al., 2014).

The gas chromatography (GC) method is a common confirmation method used to analyze melamine and its related compounds. The FDA initially developed a GC-MS screening method in which melamine in sample extracts has derivatized with trimethylsilyl (TMS) agent prior to analysis (FDA, 2008). This method has been further improved by the application of gas chromatography coupled with tandem mass spectrometry (GC–MS/MS) and the use of 2,6-diamino-4-chloropyrimidine as an internal standard (IS) with an LOD which could be reached 2 μg kg^{-1} (Miao et al., 2009). However, instrumentation is expensive and sample pretreatment time-consuming in these methods.

6.4 Enzyme-Linked Immunosorbent Assays (ELISA)

Enzyme-linked immunosorbent assays (ELISA) are screening methods commonly used for large amounts of samples due to easy sample preparation and low cost. This method is sufficiently sensitive, fast, and highly selective for the determination of melamine (Garber and Brewer, 2010; Yin et al., 2010). Zhou et al. (2012) have developed a monoclonal antibody (mAb) for an inhibition enzyme-linked immunosorbent for monitoring the level of melamine in milk samples. The standard curve was linear in the range from 0.03 to 9 ng mL^{-1} with a detection limit of 0.01 ng mL^{-1}. The sensitivity of the assay was 0.35 ng mL^{-1}. The obtained results have showed a potential method as a tool for the rapid and reliable monitoring of melamine in liquid milk and milk powder samples (Zhou et al., 2012). Besides, fluorescence polarization immunoassay (FPIA) is especially applied for milk and milk powders with an LOD of 9.3 ng mL^{-1} reported by Wang et al. (2011). This method is sensitivity compared to other methods; however, the problem of these methods are the possibility of false positives, which occurred because most of the antibodies showed cross-reactivity to structures of melamine-related compounds. Therefore, positive samples from ELISA methods need to be rechecked for final confirmation (Liu et al., 2012b). However, sample analyses should only be included in semiquantitative analyses and use a specific antibody that works against analytes. A summary of the modern instrument analytical methods that has been used for the determination of melamine is shown in Table 9.1.

Table 9.1 Summary of Published Methods for Determination of Melamine and Related Compounds in Food Products

Method	Matrix	Analyses	Detection Limit	References
LC-MS/MS	Human urine	Melamine, cyanuric acid	0.00066 mg kg^{-1}	Panuwet et al. (2012)
LC-MS/MS	Rat plasma, liver, kidney, spleen, bladder, and brain	Melamine	0.00560–0.0083l mg kg^{-1}	Wu et al. (2009)
LC-MS/MS	Milk-based products and other food and beverage products	Melamine	0.01–0.10 mg kg^{-1}	Ibáñez et al. (2009)
LC-MS/MS	Milk and dairy products	Melamine, cyromazine	10 mg kg^{-1}	Feng et al. (2008)
LC-MS/MS	Kidney tissue	Melamine, ammeline, ammelide, Cyanuric acid	0.092–0.140 mg kg^{-1}	Filigenzi et al. (2008)
LC-MS/MS	Catfish, trout, tilapia, salmon, and shrimp	Melamine	0.0032 mg kg^{-1}	Andersen et al. (2008)
LC-MS/MS	Porcine muscle	Melamine	0.0017 mg kg^{-1}	Filigenzi et al. (2007)
LC-MS/MS	Infant formula		4 ng g^{-1}	Tittlemier et al. (2009)
LC-MS	Catfish, pork, chicken, and pet food	Melamine, cyanuric acid	0.01 mg kg^{-1}	Varelis and Jeskelis (2008)
HPLC-MS/MS	Kidney tissue		1 μg mL^{-1}	Filigenzi et al. (2008)
HPLC-MS/MS	Muscle tissue		1.7 ng g^{-1}	Filigenzi et al. (2007)
HPLC-MS/MS	Porcine muscle tissue	Melamine	1.7 ng g^{-1}	Filigenzi et al. (2007)
HPLC-DAD	Milk and dairy products	Melamine	0.035–0.110 mg kg^{-1}	Filazi et al. (2012)
HPLC-DAD	Liquid milk	Melamine	0.018 mg kg^{-1}	Sun et al. (2010b)
HPLC-DAD, LC-MS/MS	Plant origin protein powder	Melamine	10 and 0.5 mg kg^{-1}	Ding et al. (2008)
HPLC-DAD	Rice protein concentrate feeds	Melamine, Ammeline, Ammelide, Cyanuric acid	55–113 mg kg^{-1}	Muñiz-Valencia et al. (2008)
HPLC-DAD	Pet food	Melamine	0.1 μg mL^{-1}	Kim et al. (2008)
HPLC-UV	Infant formula	Melamine	0.1 mg kg^{-1}	Venkatasami and Sowa (2010)
HPLC-UV	Dairy products	Melamine	0.003 mg kg^{-1}	Zheng et al. (2012)

Table 9.1 Summary of Published Methods for Determination of Melamine and Related Compounds in Food Products (*cont.*)

Method	Matrix	Analyses	Detection Limit	References
HPLC-UV	Cereal flours	Melamine, ammeline, ammelide, cyanuric acid	5–90 mg kg^{-1}	Ehling et al. (2007)
HPLC-Fluo	Bovine UHT whole milk	Melamine	0.0081 μg/mL^{-1}	Finete et al. (2015)
HPLC-Fluo	Kitchen plasticware	Melamine	0.0081 μg mL^{-1}	Finete et al. (2014)
GC-MS/MS	Milk and milk products	Melamine, ammeline, ammelide, cyanuric acid	0.002 mg kg^{-1}	Miao et al. (2009)
GC-MS	Milk products	Melamine	0.0003 mg kg^{-1}	Li et al. (2009)
GC-MS	Liquid milk and milk powder	Melamine	0.1 mg kg^{-1}	Xu et al. (2009)
GC-MS	Eggs	Melamine	0.01 mg kg^{-1}	Xia et al. (2009)
GC-MS	Muscle		10 μg kg^{-1}	Zhu et al. (2009b)
LC-GC-MS	Soil		2.5 ppb, 0.05 ppb	Yokley et al. (2000)
MALDI-MS	Urine		12.5 μg mL^{-1}	Tang et al. (2009)
MALDI-MS	Pet food	Cyanuric acid	1.0 mg kg^{-1}	Han et al. (2009)
NIR/FTIR	Infant formula powder		1 ppm	Mauer et al. (2009)
DAPCI-MS	Milk		0.08 mg kg^{-1}	Yang et al. (2009)
nanoEESI	Milk		0.2 mg L^{-1}	Li et al. (2009)
CE-DAD	Milk		0.047 μg mL^{-1}	Chen et al. (2009)
LTP-MS/MS	Milk powder		6 ppb	Huang et al. (2009)
LTP- MS/MS	Lamb tissues, serum		10 ng g^{-1}	Lv et al. (2010)
NMR	Milk product		0.69 mg kg^{-1}	Lachenmeier et al. (2009)
MLC	Liquid and powdered milk samples	Melamine	9 ng mL^{-1}	Beltran-Martinavarro et al. (2012)
MLC	Swine kidney	Melamine	5 ppb	Beltran-Martinavarro et al. (2014)
MLC	Milk	Melamine	0.6 mg L^{-1}	Batista et al. (2014)
MLC	Plasma and urine	Melamine	50 ppb	Marco-Peiro et al. (2012)

7 Advanced Methods for Determination of Melamine

7.1 Electrochemical Sensors

Electrochemical sensors are extensively employed because of their instrumental simplicity, moderate cost and portability. They can provide many advantages such as good sensitivity and selectivity, remarkable rapidity, and facility that are applicable to most areas of analytical chemistry (Chen and Chatterjee, 2013). Various electrochemical strategies have been explored for the development of electrochemical sensors. Cyclic voltammetry (CV) has received great interest as it can be used for the elucidation of electrode processes and redox mechanisms. Differential pulse voltammetry (DPV) and square-wave voltammetry (SWV) are particularly useful in the determination of trace amounts of electroactive compounds. Besides, electrochemical impedance spectroscopy (EIS) is a powerful tool for examining many chemical and physical processes in solution, as well as in solids. It may be providing information on the reaction parameters, corrosion rates, electrode surfaces porosity, coating, mass transport, and interfacial capacitance measurements (Li et al., 2015).

In recent years, electrochemical techniques have attracted more and more attention in analytical chemistry due to their wide applications. Moreover, compared to analytical methods, electrochemical sensors have provided a highly sensitive and selective approach for the detection of numerous biological and environmental compounds. Several studies have been conducted based on an electrochemical sensor for determination of melamine. Rovina et al. (2015) have developed a novel electrochemical sensor method based on the modification of gold electrode with chitosan, calcium oxide nanoparticles and an ionic liquid for the determination in the presence of melamine in milk products. Under optimal conditions, they found the modified gold electrode able to detect with different concentrations of melamine from 9.6×10^{-3} to 9.6×10^{-15} M, with a detection limit of 9.6×10^{-16} M. The developed melamine sensor method is a very simple and fast procedure for analyzing melamine level in milk and milk products.

Furthermore, William and Thiago (2014) have also developed a simple, rapid, inexpensive, and accurate analysis based on the electrochemical monitoring of the reduction of the melamine copper chloride ionic pair ($MELH^+[CuCl_2]^-$). In this method, the anodic dissolution of the copper electrode is initiated electrochemically to release copper ions, which formed the melamine chloride copper ionic pair at the electrode surface. After this process, the melamine chloride copper ionic pair can be reduced electrochemically at the surface of the electrode and quantified under the optimized conditions. This protocol eliminates the necessity for contamination of the sample with a copper salt. The quantification of melamine in milk has been performed using DPV, where the peak current response is linearly increased with different melamine concentrations ranging from 5 to 90 μmol L^{-1}, with detection limit to be 0.85 μmol L^{-1} (3 σ $slope^{-1}$). The accuracy of the method is successfully proven by testing in real samples (William and Thiago, 2014).

A sol-gel based electrochemical sensor has developed for the detection of melamine in real samples. Thin film of sol–gel polymers with specific binding sites of melamine has developed with glassy carbon electrode surface by electropolymerization method (Xu et al., 2014). Under the optimum conditions, the developed method showed a good recognition toward melamine referring to other molecules structural. With acceptable regeneration and stability, the method could detected with different range of melamine from 6.3×10^{-7} to 1.1×10^{-4} mol L^{-1}, with a detected limit of 6.8×10^{-8} mol L^{-1}. The result indicated that electrochemical sensor is potentially applied for melamine detection in real samples.

Zhao et al. (2012) have developed a sensitive and environment friendly electrochemical determination of melamine by DPV utilizing a glassy carbon (GC) electrode coated with the multiwall carbon nanotube/chitosan (MWCNT/CS) composite. A MWCNT/CS composite containing abundant groups (–OH, hydrogen bond) was prepared by electrostatic self-assembly with MWCNTs and CS. $[Fe(CN)6]^{3/4}$ was used as the electrochemical probe to characterize the behaviors of the MWCNT/CS-modified GC electrode by CV and EIS. Thus, the melamine detection was achieved through the interaction between melamine and MWCNT/CS with the detection limit of 3 nM. This method can be directly applied to the determination of melamine in real milk, and did not need many complicated pretreatments such as centrifugation, filtration, and the use of toxic solvent (eg, trichloroacetic acid and methanol).

7.2 Molecularly Imprinted Polymers (MIPs)

Molecularly imprinted polymers (MIPs) are used in an approach to synthesize a polymer matrix with selective binding sites and specific recognition sites complementary in shape, size, and functional groups to template molecules, thus showing specific binding behaviors to the target molecules (Wang et al., 2011). The templates can be imprinted during polymerization and extracted for creating three-dimensional binding sites for the purpose of retaining the memory of a specific molecule. The general synthetic procedures of MIPs have involved complexation in a solution of a template molecule with functional monomers through noncovalent bonds, followed by polymerization of these monomers around the template with the help of a cross-linker in the presence of an initiator. After completing the polymerization process, template molecules are removed by extensive washing steps to disrupt the interactions between the template and the monomer for freeing the binding sites. The template extraction directly influences the limit of detection and precision. Generally, efficient extraction of the original templates is not easily located at the interior area because of extremely low analyte diffusion rates in highly cross-linked MIPs (Gao et al., 2007; Peng-Ju et al., 2007).

MIPs technique has developed for the improvement of the selectivity of electrochemical method. The affinity matrix is first prepared by the polymerization of functional monomers and cross-linkers in the presence of template molecules. Then, the template molecules are removed from the polymer. Thus, MIPs are obtained with the specific cavities complementarily in shape, size, and functional groups from the template molecules. MIPs electrochemical sensors have received considerable attention for the determination of melamine because of their high selectivity and stability (Li et al., 2011; Xu et al., 2014; Wu et al., 2012).

Liu et al. (2015a) have developed MIPs electrochemical sensor based on carbon nanotubes-ionic liquid composite to carrier and enhance the number of imprinted cavities that able to improve the selectivity and sensitivity of the sensor (Fig. 9.3). The developed sensor method has showed a linear relationship with the concentration of melamine in the range of 0.4–9.2 μM, with a detection limit of 0.11 μM. The developed method is successfully applied for determination of melamine in milk products with good recovery rates. Li et al. (2011) have used melamine as the template molecule, where the MIP particles are synthesized and embedded in a solid paraffin carbon paste to prepare the MIP sensor. The detection signals are amplified because of enzymatic reaction to the H_2O_2 catalytic oxidation with detection limit is 0.005 μmol L^{-1}. Moreover, a novel MIP impedimetric sensor based on gold electrode modified with MIP poly (2-mercaptobenzimidazole) (PMBI) is promoted for melamine detection previously reported by Wu et al. (2012). Under optimal condition, a linear response curve is obtained from 1.0×10^{-8} to 5.0×10^{-5} M, with a detection limit of 3.0×10^{-9} M. The proposed method displays remarkable advantages, such as sensitivity, high accuracy, wider linear range, and low detection limit.

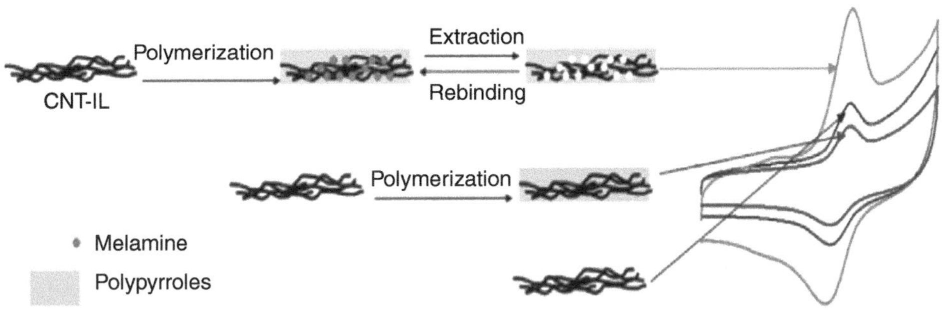

Figure 9.3. The preparation process of CNT-IL/MIP and CNT-IL/NIP, and the comparison of electrochemical response of melamine at different modified electrodes (Liu et al., 2015a).

Additionally, Liu et al. (2015b) have established the combination techniques of coupling molecularly imprinted solid-phase extraction (MISPE) with high-performance liquid chromatography (HPLC) for simultaneous measurement of melamine. The molecular recognition ability and the binding capability of the as-prepared polymers toward melamine is evaluated via static and dynamic binding tests, and found that the MD-MIPs have showed better affinity and selectivity for both templates when comparing to single-templated MIPs and nonimprinted polymers. An approach based on MISPE and HPLC have developed and optimized for melamine detection in powdered milk. The detection limit is 0.13 μg g^{-1} ($S/N = 3$), and RSD of intraday and interday determination for melamine range from 3.3% to 4.7% (Liu et al., 2015b). The MIPs are synthesized in an ethanol-water system, with melamine as the template and methacrylic acid as the organic functional monomer. A binding study exhibited that the MIP has showed excellent affinity and high selectivity for melamine and related compounds. Under optimized conditions, a detection limit is 2.59 μg kg^{-1}. The proposed MISPE–UPLC method is a highly selective and sensitive method for determination of melamine, and their metabolites are used in the control and quality assurance of milk (Ge et al., 2015). Besides, Zhu et al. (2015) have developed a novel molecularly imprinted stir bar (MIP-SB) with melamine as the template molecule. The developed method has determined melamine in powdered milk by combining MIP-SB sorptive extraction with HPLC. The liner range from 0.0631–12.6 ng mL^{-1} with good correlation coefficient of 0.9983, and the limit of detection is 0.0127 ng mL^{-1}. The method built is simple and suitable for the determination of melamine in milk products with satisfactory results, which is of great significance for quality control of milk products.

7.3 Aptamer-Based Sensors

Aptamer is a synthetic single-stranded nucleic acid or peptide molecules binding to specific target with high affinity and specificity. It is normally screened with a combinatorial technique called systematic evolution of ligands by exponential enrichment (SELEX) (Citartan et al., 2012). Aptamers offer several advantages over antibodies as they are spontaneously synthesized and possess appropriate storage properties and could be modified so easily. Owing to their inherent superiorities, it can widely be used as recognition elements in biosensor construction and entirely helpful in the detection of a wide range of substances from metal ions, organic molecules, peptides, and proteins to cells (Song et al., 2008).

The ability of pyrimidine to form hydrogen bonds with purines has been well documented, and the hydrogen bonding between thymine and 2,6-diamino-5-methylpyrimidin-4-one, which is a similar molecular structure as melamine has been reported by Chen et al. (2004). The single stranded oligonucleotide containing thymines are binders to melamine via hydrogen bonding. Based on this, recently several aptamer-based biosensors have developed for melamine detection. The selected aptamers are commonly the single stranded oligonucleotide of thymines with the difference on the length of the sequence.

Qi et al. (2010) have proved that single-strand DNA oligomers including polyAn, polyCn, polyGn, and polyTn could make AuNPs stable without aggregation owing to enough negative charges on their surface. The presence of melamine in the medium makes the polyTn-stabilized AuNPs aggregate, indicating that the specific interactions have occurred between melamine and polyTn, and the stability of a single base unit in polyTn decreases with the length increase of the polyTn. The novel specific recognition of melamine with thymine and this triple H-bonds motif in aqueous medium has been shown. Later, Huang et al. (2011) developed an aptamer-AuNPs-based colorimetric sensor using label-free and labeled AuNPs. In the label-free AuNPs procedure (as shown in Fig. 9.4), AuNPs are coated by the negative-charged citrate ions, which could inhibit AuNPs from aggregation in aqueous solution, while causing decline in the AuNPs aggregation in the presence of high salt concentrations. However, aptamers (poly–T10) could strongly adsorb on AuNPs and enhance the stability of AuNPs against the NaCl-induced aggregation. In the presence of melamine, aptamers will competitively bind with melamine by the stronger affinity, which will decrease the salt tolerance of AuNPs and result in the subsequent aggregation of AuNPs. The label-based AuNPs

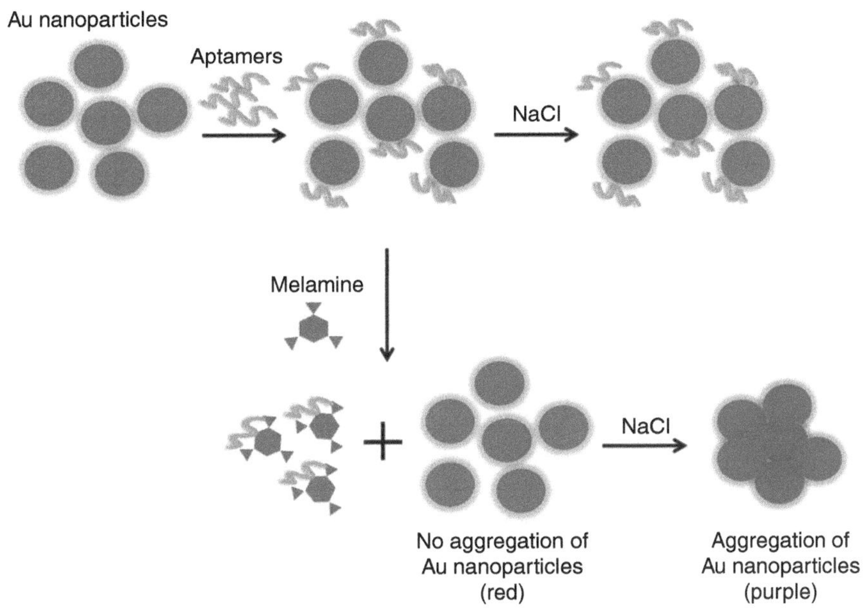

Figure 9.4. Aptamer-based colorimetric sensor for the detection of melamine (Huang et al., 2011).

procedure, the single-strand oligonucleotide labeled AuNPs are used for melamine determination, which is based on the combination of thymine with melamine (Huang et al., 2011). The HSssDNA (HS–$(CH_2)_6$-TAGCTATG GAATTCCTCGTAGGCATTTTTT) was first attached to the surface of AuNPs. The DNA-functionalized AuNPs can assemble when the oligonucleotides hybridize, which causes the changes of colors. When added to the melamine, the functionalized AuNPs are assembled upon binding of the oligonucleotide to melamine, which results in a red-to-blue color change. This assembly process is particularly reversible. The dissociation occurs upon thermal denaturation, leading to a blue-to-red color change. Both assays are high selectivity and high sensitivity with the detection limits are of 41.7 nM and 46.5 nM, respectively. The label-based method has provided better stability, a better accuracy, and a larger response range when compared with a label-free method.

Similarly, an AuNPs-based colorimetric apta-sensor for melamine has been developed using salt-induced aggregation of AuNPs (Yun et al., 2014). They used 50–TTTTTTTTTTTTTTTTTTTT-30(dT20) as melamine aptamer. The detection limits have found 1.5 mg L^{-1} by naked eye and 0.5 mg L^{-1} by ultraviolet–visible (UV–vis) spectrometer, respectively. The resonance scattering (RS) analysis is a new, rapid, simple, convenient, and sensitive analytical technique and applied to the analysis of trace nucleic acid, protein, drug, and others. The RS effects of nanoparticles have combined with a highly selective aptamer reaction to assay Hg^{2+} and K^{+} with satisfactory results (Cai et al., 2010). A RS spectral assay has developed using aptamer-modified AgNPs as probe (Liang et al., 2011). AgNPs are modified by a single-strand DNA to fabricate an Ag-aptamer probe for melamine. In the presence of NaCl, the probe is stable. The increased RS intensity is linear with melamine concentration in the range of 6.31–378.4 mg L^{-1} with a detection limit of 3.1 mg L^{-1}.

7.4 Optical Sensors

Optical sensors used as analytical tools are of vital importance in the field of chemo-/biosensors. By employing optical transduction techniques, they are provided chemical information ranging from analyte concentration and binding kinetics to microscopic imaging and molecular structure. Based on photonic attributes of optical sensors, a variety of signal transduction pathways are utilized including absorbance, transmission, fluorescence intensity, chemiluminescence, refractive index, polarization, and reflectivity (Luchansky and Bailey, 2012). In the following sections, we discuss various optical sensors for melamine analysis including colorimetric sensors, fluorescence sensors, chemiluminescence sensors, luminescence sensors, and surface plasmon resonance. Over the past decades, the optical sensors have been applied in a number of important areas including food safety, security, life science, environmental monitoring, and medicine (Bhatta et al., 2010; Md Muslim et al., 2012).

7.4.1 Colorimetric Sensors

Colorimetry, a kind of visual detection method, has drawn great attention due to its obvious advantages including simple instrument, easy operation, low cost while relatively high sensitivity. With the development of nanoscience and nanotechnology, more and more opportunities are provided on the application of nanomaterials in colorimetric sensing. Localized surface plasmon resonance (LSPR) is one of the most obvious properties of metal nanoparticles (NPs) such as gold nanoparticles (AuNPs) and silver nanoparticles (AgNPs). LSPR of the metal NPs is primarily connected with the NPs size, shape, composition, interparticles distance, small or large aggregates of metal NPs, and dielectric constant (refractive index) of the surrounding medium (Li et al., 2015). When the distances between the AuNPs/AgNPs become less than the average particle diameter, aggregation of the dispersed NPs occurred. Interparticle plasmon coupling of AuNPs/AgNPs obtained in aggregated NPs and the color of the NPs solution changed, which enabled the development of sensitive colorimetric sensor (Deng et al., 2012).

AuNPs have been widely used as colorimetric probes for several advantages of chemical and biosensing, such as simple synthesis process, unique and readily tuned optoelectronic properties, excellent biocompatibility, high extinction coefficient, and good suitability for multifunctionalization (Saha et al., 2011; Kuang et al., 2011). The AuNPs based colorimetric assays are of great interest because molecular recognition events can be easily transformed into color changes that arise from the interparticle plasmon coupling during AuNPs aggregation (red-to-purple or blue) or redispersion of an AuNPs aggregate (purple-to-red) (Zhao et al., 2008; Liu et al., 2011). The AuNPs-based colorimetric methods are definitely suitable, owing to their simplicity and visibility. Based on the mechanism of color changes of AuNPs mentioned earlier, there are a few reports of melamine detection that could be classified into two types: AuNPs aggregation-based colorimetric sensors and AuNPs nonaggregation-based colorimetric sensors (Chi et al., 2010; Kuang et al., 2011). One is owing to the aggregation of AuNPs; the other is in respect of the nonaggregation of AuNPs.

In the AuNPs aggregation-based colorimetric sensors, two outlines have been utilized for the determination of melamine using unmodified AuNPs and ligand-functionalized AuNPs as colorimetric indicators. The first outline employs the unmodified AuNPs as colorimetric probes that can be synthesized by the traditional Frens' method (Grabar et al., 1995). As illustrated in Fig. 9.5A, melamine can strongly bind with AuNPs through the interaction between amine groups of melamine and AuNPs, thus readily showed the stabilizing agents (citrates) from surfaces of AuNPs and caused the aggregation of AuNPs. Besides, the neighbor melamine-coated AuNPs could be cross-linked by NH...N hydrogen bonds between melamine molecules. Li et al. (2010) have validated that only three amine groups of melamine are responsible for the cross-linking interaction between melamine and AuNPs, which could induce the aggregation of AuNPs. The limit of detection for melamine is 0.4 ppm, and the sample pretreatment taken only 12 min at room temperature. This method is especially promising for on-site rapid detection of melamine contamination in foods such as eggs and animal feed. In a similar study done by Guo et al. (2010) and reported a label-free AuNPs based colorimetric kit for the detection of melamine employing 5 nm AuNPs. The assay was performed by naked eye in comparison with the standard colorimetric card without the aid of any instrument. Determination of 1 ppm level was achieved by visual detection, and the

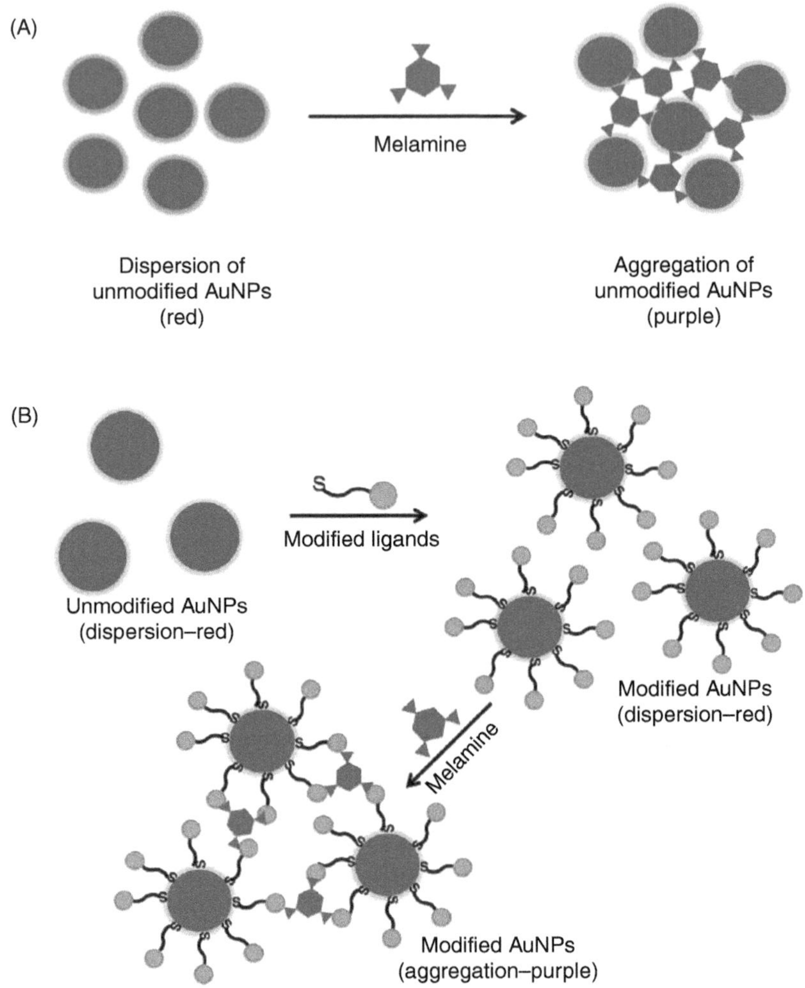

Figure 9.5. Schematic presentation of the colorimetric mechanism for melamine determination employing (A) unmodified AuNPs and (B) modified AuNPs (Li et al., 2015).

method was more suitable for rapid determination of melamine in most milk products, which could be used in the dairy industry and quality assurance departments, as well as by supermarket managers, customers, and so on. The method for melamine in liquid milk and infant formula, respectively, demonstrated a detection limit of 1.0 and 4.2 ppm by naked eyes, while the detection limit of 0.15 and 2.5 ppm by UV-vis spectrometer. Chi et al. (2010) have successfully explored a simple and effective colorimetric visualization of melamine in milk products employing 13 nm citrate-stabilized AuNPs with the aid of $NaHSO_4$ optimization. The NaHSO4-optimized AuNPs system was exhibited as having an excellent detection limit as low as ~25 ppb and the ability of $NaHSO_4$ to promote the ligand exchange between citrate and melamine at the surface of AuNPs, thus promoting the aggregation of AuNPs.

AuNPs-based colorimetric assay has stabilized and improved the sensitivity; it should be functionalized AuNPs with some suitable ligands. There have been a few reports for melamine sensing using ligand functionalized AuNPs as colorimetric recognition elements. As shown in Fig. 9.5B, the mechanism of this outline usually relies on the unique interaction between ligand and melamine allowing the visual sensing of melamine. Ai et al. (2009) have developed a MTT-stabilized AuNPs-based colorimetric sensor for visual detection of melamine. The first step was synthesis of a thiolfunctionalized cyanuric acid derivative 1-(2-mercaptoethyl)-1,3,5-triazinane-2,4,6-trione (MTT). Then, the MTT-stabilized AuNPs (12 nm) was prepared by ligand-exchange reaction using MTT and citrate-stabilized AuNPs. As a result it was initially demonstrated that the color change of AuNPs (from red to blue) was induced by the triple hydrogen-bonding recognition between melamine and MTT, which resulted in excellent selectivity for detection of melamine in milk products with complicated components. Detection limit found 2.5 ppb with visual detection within 1 min. This method was shown to be quite promising for on-site and real-time detection of melamine in raw milk, infant formula, and other milk products.

Kumar et al. (2014) have also reported a simple and sensitive colorimetric method by using AuNPs of 21-nm size synthesized by the citrate reduction method. The method is based on the principle that the melamine causes the aggregation of AuNPs of wine red color to change to blue/purple. Under optimized conditions, AuNPs are highly specific for melamine and can detect melamine down to a concentration of 0.05 mg L^{-1} (Kumar et al., 2014). Moreover, nanoparticle aggregates induced by synergistic effect of electrostatic interaction and hydrogen-bonding recognition between melamine and 3-mercaptopriopionic acid (MA), that conjugated on the surface of AuNPs are used for determination of melamine. MA molecules are conjugated on AuNP surfaces to form MA-AuNPs and detected as low as 0.4 μg mL^{-1}. The color change can be readily seen by the naked eye at 30 μg mL^{-1} melamine without the aid of any advanced instruments (Cai et al., 2014). The change in color can be visualized with the naked eye or a UV–vis spectrometer.

Recently, a simple and rapid field-portable colorimetric method for the detection of melamine in liquid milk was reported by Xin et al. (2015). Methanobactin (Mb) could reduce Au (III) to Au (0) and mediate the synthesis of AuNPs. The melamine interacted with oxazolone ring of Mb, which interrupted the formation of Au-NPs. This method could be used to detect melamine in liquid milk by naked eye observation with a detection limit of 5.56 106 M (0.7 mg kg^{-1}). Further, the plasmon absorbance of the formed Au-NPs have allowed the quantitative detection of melamine by UV–vis spectrometer with detection limit as low as 2.38×10^7 M (ie, 0.03 mg kg^{-1}). Wu et al. (2014) have developed a method for determination of melamine monomer's migratory quantity (MMMQ) based on the hydrogen bonding interaction between DNA thymine and melamine. The method has described a simple and rapid technique that focus on the measurement of MMMQ from melamine tableware by AuNPs and dynamic light scattering. With the presence of probe DNA (p-DNA), the AuNPs are more stable in NaCl solution, whereas they became aggregated when the p-DNA hybridizes with melamine (Fig. 9.6). The change in the hydrodynamic diameter of AuNPs could be detected by DLS technology. Under the optimal conditions, the average diameter increased linearly with the concentration of melamine over the range from 5.0 to 320.0 μg L^{-1} with a detection limit of 2.0 μg L^{-1} (3σ/slope). The DLS method has great potential in the analysis of the migration of melamine monomers.

Silver nanoparticles (AgNPs) have been proved to be effective catalytic materials for various applications due to their large surface-to-volume ratio and electronic properties, and also showed higher extinction coefficients than AuNPs at the same size (Li et al., 2014). In the metal nanoparticle-based colorimetric assays, AgNPs have been used in less extension than AuNPs; however, a number of colorimetric methods have been explored based on color change from yellow to brown during AgNPs dispersion and aggregation (Vilela et al., 2012). Ma et al. (2011) have developed a reliable assay for melamine sensing using dopamine-stabilized AgNPs as a colorimetric reader. Only dopamine was used as both a reducer to reduce Ag^+ and stabilizer to functionalize the generated AgNPs to synthesize monodispersed AgNPs with the color of bright yellow. Upon the mixture of dopamine with melamine, the aggregated-AgNPs correspondingly the color changed from bright yellow to brown. Han and Li (2010) have established a novel sensitive and low-cost colorimetric method using the stable ρ-nitroaniline-modified AgNPs, based on the melamine-induced aggregation and color changed yellow to blue of AgNPs owing to electron donor–acceptor interaction between melamine and *ρ*-nitroaniline at the AgNPs interface, with the detection limit as low as 0.1 ppm.

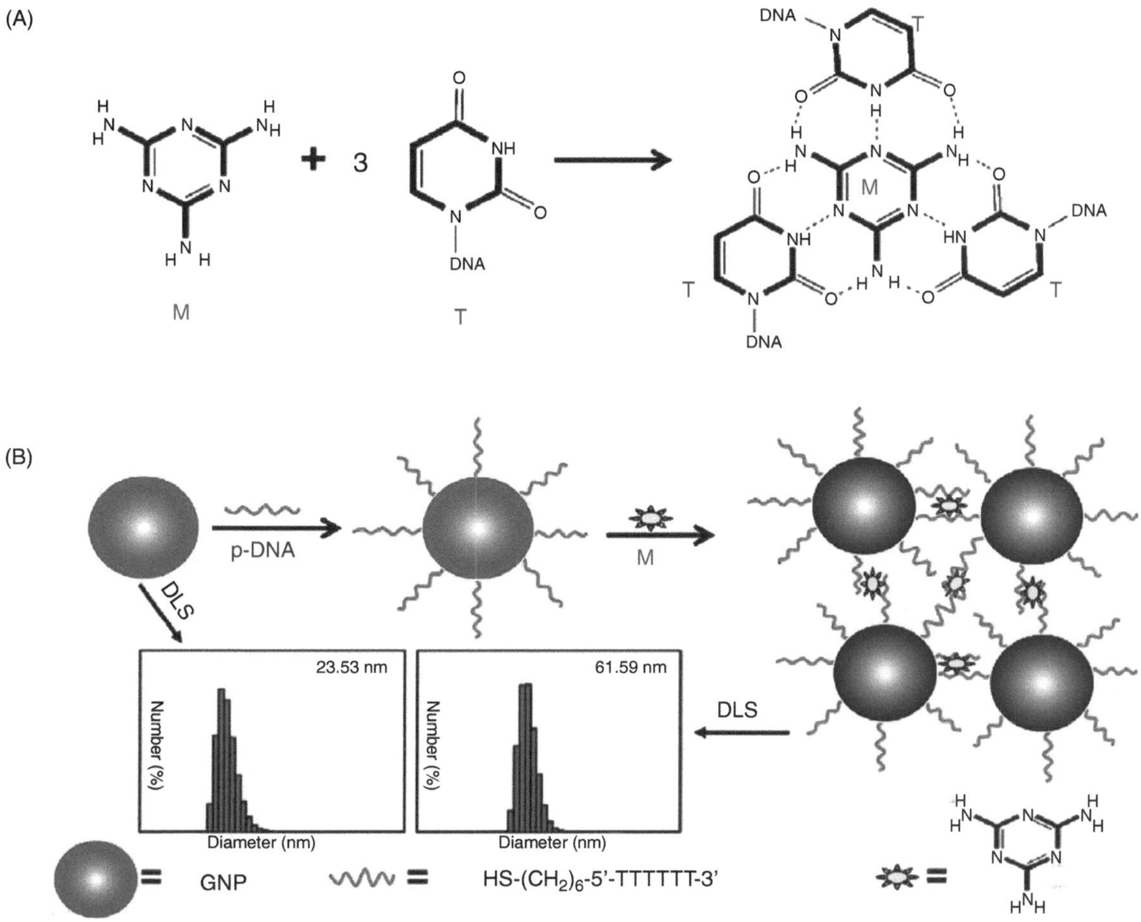

Figure 9.6. Illustration of the detection of melamine using GNPs probes and DLS. (A) Hydrogen-bonding recognition between melamine *(M)* and thymine *(T)*, and (B) the aggregation of the p-DNA modified GNPs recognized with melamine. From Wu et al. (2014).

Song et al. (2015) have developed a simple, rapid, and low-cost assay with high sensitivity for the melamine detection in milk samples using sulfanilic acid-modified silver nanoparticles (SAA-AgNPs). The interaction between exocyclic amine of melamine and SAA induces rapid aggregation of SAA-AgNPs accompanied by the naked eye. The extinction ratio ($A_{620\ nm}/A_{390\ nm}$) is correlated with the melamine concentration over the range of 0.1–3.1 μM with detection limit of 10.6 nM. The method is applied successfully for determination in the presence of melamine in pretreated milk products. Wang et al. (2015) have established a colorimetric detection method based upon an in situ formation of AgNPs through modified Tollens process at room temperature. The triazine ring

N atoms of melamine molecules are designed to complex the Ag^+ through electron donor–acceptor interaction. During the formation procedure of AgNPs, the melamine molecule is covalently bonded with the Ag^+ ions and adsorbed to the surface of AgNPs, resulting in the aggregation of the adjacent AgNPs with detectable declines of absorption signal. The method is detected as low as 10 nM, and discriminated melamine among sixteen other milk-relevant coexisting compounds.

Furthermore, the detection mechanism is based on the "oxidative etching-aggregation" of silver nanoparticles (AgNPs) by the cooperation effect of melamine and electron acceptors such as H_2O_2, ozone or $Fe(NO_3)_3$. It is an innovative and versatile functional colorimetric sensor for melamine and H_2O_2 are developed by Wang et al. (2012) with simplicity, excellent selectivity, and ultrasensitivity. The detection limits of the technique for melamine as low as 0.08, 0.16, and 3 nM when H_2O_2, ozone, and $Fe(NO_3)_3$ are used as an electron acceptor, respectively. This method is promising and opens up a new perspective for development of highly sensitive colorimetric sensors based on the special reactivity of AgNPs with nucleophilic reagents and electron acceptors (Wang et al., 2012).

Ma et al. (2009) have successfully developed a stable, highly sensitive, small-molecule sensor based on the relaxation of magnetic nanoparticles (MNPs) with an LOD 0.6 ng g^{-1}. The detection method has proved to be a powerful candidate as a sensor technology for rapid, sensitive detection of melamine and may find widespread applications. Xu et al. (2013) also successfully determined melamine in dairy products by the Fe_3O_4 MNPs-H_2O_2-ABTS detection system that accurate, reliable, sensitive, and convenient. The whole process is achieved in less than 1 h from the beginning of sampling to the final step of obtaining the results from the spectrophotometer.

7.4.2 Fluorescence Sensors

Fluorescence sensors for targets detection have produced great attention worldwide on account of their simplicity, high sensitivity, and easy operation. Fluorescent method has required an external light source (short-wavelength light) to initiate the electronic transitions in an atom or molecule, which then produces luminescence (a longer-wavelength light) (Perumal and Hashim, 2013). Recently, an environmental friendly procedure based on triton X-114 has been developed for melamine determination as an adulterant of protein content in milk (Carina et al., 2015). A linear response was detected from 1.0 to 6.0 mg L^{-1} concentration of melamine, with detection limit of 0.8 mg L^{-1}.

Fluorescence resonance energy transfer (FRET) is a nonradiative process whereby an excited state donor transfers energy to a proximal ground state acceptor through long-range dipole–dipole interactions, and the rate of energy transfer is extremely dependent on the extent of spectral overlap, the relative orientation of the transition dipoles, and especially the distance between the donor and acceptor molecules (typically in the range of 1–10 nm) (Chen et al., 2013). Guo et al. (2011) have exhibited a novel sensitive turn-on FRET-based fluorescent senor for melamine detection using fluorescein as donor and AuNPs as acceptor. Accordingly, the fluorescence of fluorescein could be significantly quenched via FRET between fluorescein and the AuNPs. Consequently, a rapid and sensitive detection of melamine was carried out within 15 min with the detection limit at 1 nM level. Besides, Cao et al. (2014a) have developed a rapid, convenient and highly sensitive method based on fluorescence resonance energy transfer between Rhodamine B and citrate-stabilized AuNPs. The Rhodamine B is electrostatically adsorbed to the surface of AuNPs and the fluorescence emission of Rhodamine B is quenched as shown in Fig. 9.7. Under the optimal conditions, the fluorescence recovery efficiency was linearly proportional with different concentrations of melamine in the range of 5–1000 μg L^{-1}, with detection limit of 0.18 μg L^{-1}. The developed method is effectively applied for melamine detection in milk and powdered infant formula.

Among fluorescence techniques, a number of fluorescence-based assays rely on fluorescence resonance energy transfer (FRET) spectroscopic technique, which occurs when the emission spectrum of the donor and the absorption spectrum of the acceptor are overlapped. FRET has been widely used in various research areas, such as nucleic acid detection, immunoassay, and ion detection. To date, there are only a few reports regarding the FRET detection of melamine. Most of these FRET-based melamine-sensing systems are primarily based on the use of fluorescence from organic dyes (Guo et al., 2011; Tang et al., 2013). However, most organic fluorophores suffer from photo bleaching and poor stability in the ambient environment, which results in irreproducible fluorescence signals for analysis. To overcome these shortcomings, some types of nanomaterials, such as QDs have been used as signaling probe. Dai et al. (2014) have been constructed a FRET system between amino-functionalized carbon dots (C-dots) and AuNPs. In this system, C-dots are treated as energy donors, while AuNPs are treated as energy acceptors. Under the optimized experimental conditions, melamine could be detected based on fluorescence intensity of C-dots, while detection limit was 36 nM.

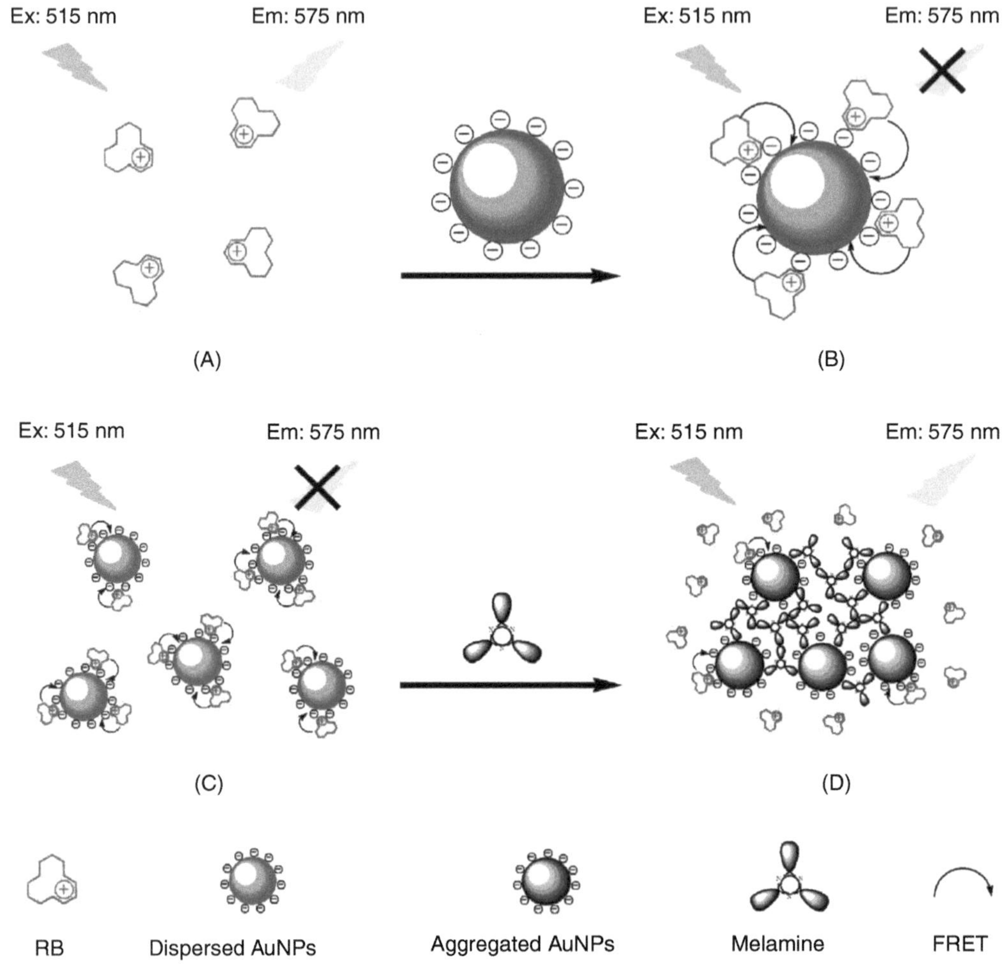

Figure 9.7. Schematic illustration for the detection of melamine based on FRET between the AuNPs and Rhodamine RB. From Cao et al. (2014a).

The developed method is applied for determination of melamine in milk samples with satisfactory results.

7.5 Quantum Dots

Quantum dots (QDs), are semiconductor nanoparticles with some advantages over organic and inorganic fluorophores: good optical properties, stability against photobleaching and chemical reaction, size-control luminescence, broad excitation bands and sharp emission bands, as well as high photoluminescence efficiency (Zhao et al., 2010; Ren and Yan, 2012). Quantum dots are widely used as fluorescent probes for analysis of various analytes. Typically, fluorescent QD are constructed from inorganic semiconductor core materials such as cadmium telluride (CdTe) or cadmium selenium (CdSe). Fluorescent QD has been used in determination of melamine. These methods are based on fluorescence resonance energy transfer between melamine and different QD including CdTe (Tang et al., 2013) and cadmium sulfide (Wang et al., 2012) or an inner filter effect of Au and Ag nanoparticles on the fluorescence of CdS (Cao et al., 2013), CdTe (Zhang et al., 2012), and ZnSe (Cao et al., 2014b).

A simple, sensitive, and precise room-temperature phosphorescence method has been developed based on l-cysteine-capped Mn-doped zinc sulfide (ZnS) quantum dots (QDs) as a probe for determination of melamine in dairy products (Demirhan et al., 2015). The phosphorescence intensity has been quenched with different concentrations of melamine in a linear range from 50 to 500 ng mL^{-1}, with a detection limit of 5.95 ng mL^{-1} (Demirhan et al., 2015). Xu and Lu (2015) have also developed a facile strategy to prepare mesoporous structured ratiometric fluorescence MIP sensor for highly sensitive and selective determination of melamine using CdTe QDs as target-sensitive dye and hematoporphyrin as reference dyes. Under optimum conditions, these mesoporous-structured ratiometric fluorescence MIP at QDs sensors showed detection limits as low as 38 nM, with recycling process is sustainable at least 10 times without obvious efficiency decrease. The feasibility of the developed method in real samples is successfully evaluated through the analysis of melamine in raw milk and milk powder samples with satisfactory recoveries of 92–101%. However, Li et al. (2013) have established the fluorescence quenching method based on thioglycolic acid (TGA)–capped CdTe QDs. In strong alkaline aqueous solution, the selectivity of the method has greatly improved due to most heavy metal ions showing no interference as they are in the precipitation form or in their anion form. Under optimum conditions, the fluorescence intensity change of TGA coated CdTe quantum dots have linearly proportional with different concentrations of melamine on the range of 1.0×10^{-11} to 1.0×10^{-5} mol L^{-1}, with a correlation coefficient of 0.9943 and a detection limit of 5×10^{-12} mol L^{-1}. The mechanism of fluorescence quenching of the QDs has been proposed based on the infrared spectroscopy information and electrophoresis experiments in the presence of melamine under alkaline condition.

A very practical and competitive sensing strategy for melamine has been developed based on the efficient inner filter effect (IFE) of Ag nanoparticles (AgNPs) on the fluorescence of eco-friendly ZnSe QDs (Cao et al., 2014b). The fluorescence emission of ZnSe QDs could be quenched significantly by AgNPs via IFE, which resulted in a fluorescence "off-state." The presence of melamine could induce the aggregation of AgNPs and modulate the IFE process as shown in Fig. 9.8. Hence, the IFE-decreased fluorescence could gradually recover in a melamine concentration-dependent way, thereby

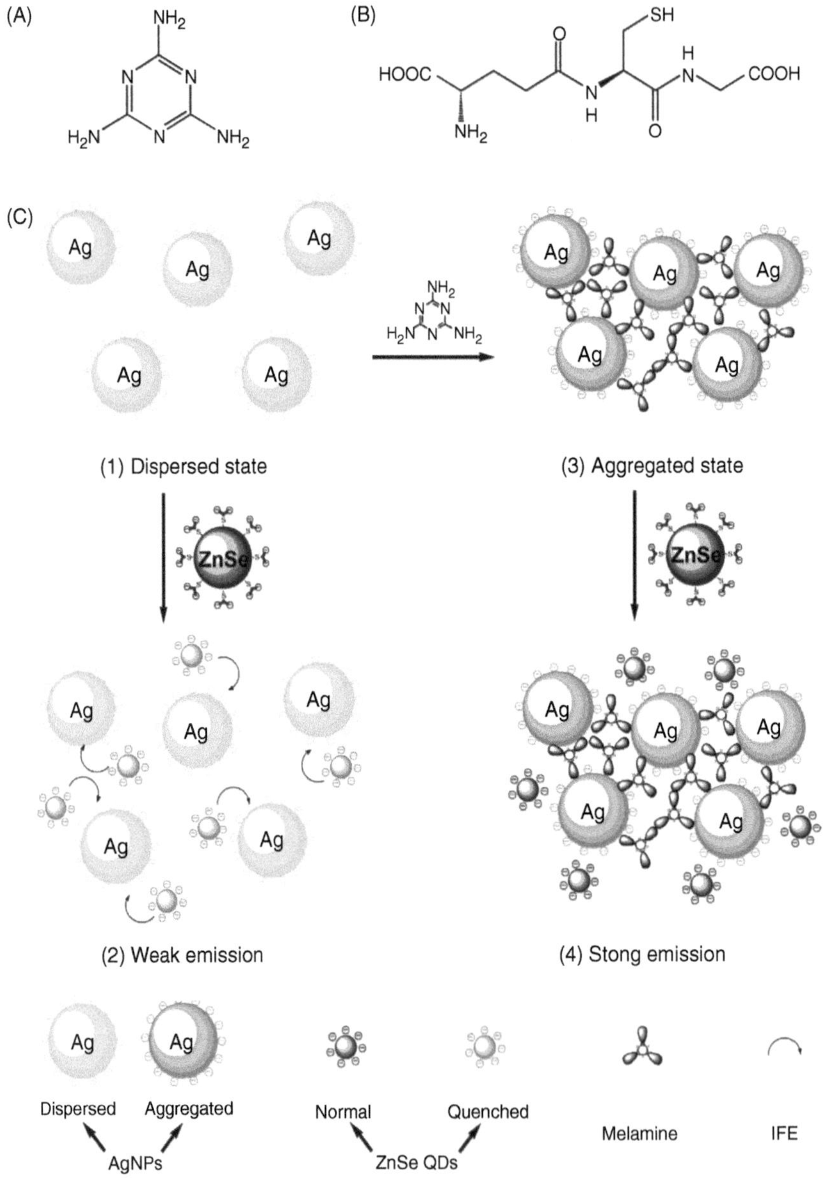

Figure 9.8. The molecular structures of (A) melamine and (B) l-GSH; (C) schematic illustration of the "turn-on" fluorescent detection of melamine based on the IFE of AgNPs on ZnSe QDs. From Cao et al. (2014b).

achieving the fluorescence "turn-on" sensing for melamine. The decent linear relationship for melamine has obtained in the range from 1.0 to 36 μg L^{-1}, and the detection limit is as low as 0.11 μg L^{-1}. The IFE-based strategy is successfully applied for melamine analysis in raw milk and egg samples and satisfactory results have been achieved.

Liu et al. (2012b) have developed QDs method based on the fluorescence changes of the water-soluble $CuInS_2$ QDs. The water-soluble $CuInS_2$ QDs capped by mercaptopropionic acid are directly synthesized by hydrothermal. The fluorescence emission of $CuInS_2$ QDs is quenched by the oxidation of the surface of the QDs with H_2O_2, and the quenched fluorescence of $CuInS_2$ QDs could be recovered upon the addition of small amounts of melamine. Under optimum conditions, there is a good linear relationship between the fluorescence intensity of $CuInS_2$ QDs and the concentration range of melamine from 1.0×10^{-8} to 1.0×10^{-5} mol L^{-1}, with a detection limit as low as 5 nM. Besides, Zhang et al. (2012) have established an inner filter effect (IFE) of gold nanoparticles (AuNPs) on the fluorescence of CdTe QDs. When thioglycolic acid-capped CdTe QDs has assorted with citrate-stabilized AuNPs, the fluorescence of CdTe QDs is significantly quenched via the IFE of AuNPs. The presence of melamine could have induced the aggregation and corresponding absorbance variation of AuNPs, which resulted in the recovery of IFE-decreased emission of CdTe QDs. The detection limit of melamine in raw milk is 0.02 mg L^{-1}. Cao et al. (2013) have effectively constructed a simple, rapid, and ultrasensitive analytical method for determination of melamine in raw milk based on inner filter effect (IFE) of AuNPs on the fluorescence of CdS QDs. With the presence of citrate-stabilized AuNPs, the fluorescence of l-cysteine (l-Cys)-capped CdS QDs is remarkably quenched by AuNPs via IFE. Under the optimum conditions, melamine in raw milk is detected on the range of 0.05–0.35 mg L^{-1}, with detection limit of 0.017 mg L^{-1}. The proposed method is a significant assessment of melamine in raw milk, which displays several advantages such as high sensitivity, low cost, and nontime-consuming compared to traditional methods.

7.6 Chemiluminescence Sensors (CL)

Chemiluminescence (CL) is referred to the phenomenon that the emission of light without incandescence due to a chemical reaction, in which a chemical species is excited to an electronically excited state and spontaneous generation of light occurs when it returns to its ground state (Li et al., 2014). It has low background scattering light interference and owns a variety of advantages

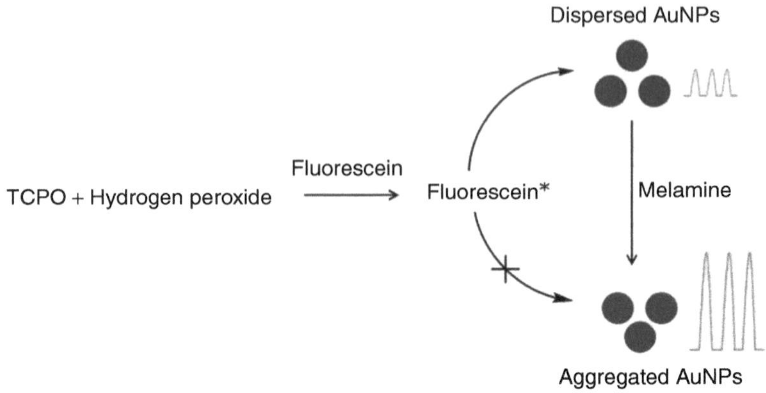

Figure 9.9. The interaction of bis(2,4,6-trichlorophenyl)oxalate-hydrogen peroxide-fluorescein with AuNPs in the presence of melamine. From Du et al. (2015).

such as high sensitivity, wide linear range, simple instrumentation, low cost, rapidity, safety and controllable emission rate (Lin et al., 2013). Du et al. (2015) have studied a turn-on chemiluminescence resonance energy transfer method fabricated for the determination of melamine by using bis(2,4,6-trichlorophenyl) oxalate-hydrogen peroxide-fluorescein chemiluminescence reaction as a donor and dispersed AuNPs as an acceptor (Fig. 9.9). Melamine could induce the aggregation of AuNPs, leading to a dramatic red-shift of the absorption band of dispersed AuNPs. The absorption band of the aggregated AuNPs does not overlap with the chemiluminescence spectrum of the reaction. In this method is able to detect melamine as low as 3×10^{-13} mol L^{-1}. The method is applied for determination of melamine in spiked milk samples, with recoveries within the range 94.1–104.2%.

Zeng et al. (2011) have studied the phenomenon of melamine that can obviously enhance the CLs signal of the Luminol-H_2O_2 system in basic medium; a simple, rapid, and sensitive flow injection chemiluminescence (FI-CL) method for determination of melamine has been developed. Under the optimum conditions, the linear range for determination of melamine is 0.2–80 μg mL^{-1}, with a detection limit at 0.12 μg mL^{-1}. The method is satisfactorily applied for determination of melamine in milk-based products, and satisfactory results are obtained without interferences from the sample matrix. Tang et al. (2011) have established a sensitive CLs method based on Luminol-$K_3Fe(CN)_6$ system in alkaline medium for determining melamine in urine and plasma samples. Under optimum conditions, the CLs intensity is a good linear relationship with melamine in the range of 9.0×10^{-9}–7.0×10^{-6} g mL^{-1} with a correlation coefficient of 0.9992. The detection limits are 3.54 ng mL^{-1} for urine sample and 6.58 ng mL^{-1} for plasma sample. In this method has provided more producible and stable approach for sensitive detection and quantification of melamine in urine and plasma samples (Tang et al., 2011). Moreover, CLs coupled with the flow injection (FI) system are offering high sensitivity, instrumental simplicity, sampling efficiency and reduced consumption, is an attractive analytic method. An ultrasensitive CLs method for determination of melamine at picogram level has been reported by Zhang et al. (2011) in a flow system. It is found that melamine can accelerate the electrons transferring rate of excited 3-aminophthalate with notable enhanced CLs intensity of luminol-hydrogen peroxide reaction. The increased CLs intensity was proportional to the concentration of melamine in the range from 2.5 to 250 pg mL^{-1} (R^2 = 0.9953), with a detection limit of 0.9 pg mL^{-1} and the RSD lower than 5.0%.

In addition, gold/silver alloy nanoparticles (Au/AgNPs) have been applied for the detraction of melamine in food product in the field of CL-based sensors. Manzoori et al. (2011) have developed an ultrasensitive CL method for the determination of low concentrations of melamine employing permanganate–formaldehyde system ($KMnO_4$–HCHO). In the absence of Au/AgNPs, a weak CL could be observed in that $KMnO_4$ as CL reagent can oxidize HCHO. While in the presence of small volume of Au/AgNPs solution, the CL of $KMnO_4$–HCHO could be obviously enhanced. However, upon the addition of melamine, the CL intensity of $KMnO_4$–HCHO could be significantly decreased induced by melamine-coated NPs formed through the interaction between the three primary amine groups of melamine and NPs. Furthermore, the CL intensity was proportional to melamine concentration, realizing the determination of melamine with a detection limit of 8 pg mL^{-1}.

7.6.1 Luminescence Sensors

Luminescence sensors are new type of luminescent nanomaterials and accounted metal nanoclusters (NCs), are generally composed of a few to roughly a hundred atoms, and the sizes are below 2 nm which approaches the Fermi wavelength of electrons, leading to the observation with different optical, electrical, and chemical properties as compared to nanoparticles (Shang et al., 2011). In recent years, metal NCs, especially Au and Ag NCs, have attracted a great deal of interest on account that they possess very strong, robust, size or scaffold-dependent tunable fluorescence, combine with good photo-stability and biocompatibility, and show very high emission rates (Zhang and Wang, 2014). Due to these excellent features, Au and Ag NCs have been recognized as ideal fluorescent probes in the field of fluorescence sensing for melamine.

Han et al. have synthesized oligonucleotide stabilized fluorescent silver nanoclusters (DNA–AgNCs) and applied in the melamine detection for the first time. In the study, DNA–AgNCs could exhibit an emission band centered at 665 nm when excited at 597 nm. In the presence of melamine, the fluorescence intensity of DNA–AgNCs could be gradually increased due to the microenvironment change of AgNCs induced by the interaction between the amine groups of melamine and thymine of DNA through strong hydrogen bonds. The method showed a good biocompatibility and high sensitivity with the detection limit at 10 nM level. Recently, an environment friendly and cost-effective turn-on fluorescent assay for melamine has been developed using bovine serum albumin stabilized gold nanoclusters (BSA–AuNCs) as the fluorescence probe (Dai et al., 2014). The fluorescence of AuNCs could be quenched by Hg^{2+} via high-affinity metallophilic Hg^{2+} and Au^{+} interactions (Xie et al., 2010). Moreover, melamine has been revealed greater coordination affinity toward Hg^{2+} (Du et al., 2013), which may result in the reduction of quenching ability of Hg^{2+} to AuNCs and accordingly the fluorescence enhancement of AuNCs. Based on the melamine-induced antiquenching ability of Hg^{2+} to fluorescent AuNCs, with the detection limit as low as 0.15 mM. The method can successfully be established and applied in real samples including raw milk and milk powder.

7.7 Surface Plasmon Resonance (SPR)

Surface plasmon resonance (SPR) uses surface plasmon waves (electromagnetic wave) to detect changes when the target analyte interacts with a biorecognition element on the sensor. In principle, when the SPR biosensor is exposed to any changes, it will induce changes in the refractive index, which is used to measure or observe the reaction (Perumal and Hashim, 2013). Liu et al. (2014) have successfully developed SPR assay of melamine based on the competitive reaction in which the assayed melamine with varying concentrations in solution competed with the pre-immobilized bovine serum albumin (BSA)-melamine. The BSA-melamine conjugate was immobilized on the carboxymethylated dextran-covered SPR chips, and then followed by the injection of the mixed solution of melamine. The linear range of the assay was 0.25–13 nM and 13–250 nM. The developed method is convenient and cost-effective, holding great promise for determination of small molecules with low concentrations.

Some of the recent advances in melamine detection techniques are summarized in Table 9.2 (Rovina and Siddiquee, 2015).

Table 9.2 List of Recent Advances in Melamine Detection Techniques

Method	LOD	References
MIP-capped CdTe quantum dots (MIP-CdTe QDs)	0.6 μM	Huy et al. (2014)
Chemosensor based on MIP film	5 nM	Pietrzyk et al. (2009)
Electrochemical sensor based on *para*-aminobenzoic acid	0.36 μM	Ting et al. (2011)
Electrochemical sensor based on modified carbon paste electrode with carbon nanotubes and a molybdenum (VI) complex (MC)	43.0 nM	Beitollahi and Sheikhshoaie (2011)
Electrochemical sensor based on multiwalled carbon nanotubes modified electrode.	2.0 nM	Zhu et al. (2010)
Surface plasmon resonance based on 6-hydrazino-1,3,5-triazine-2,4-diamine	0.5 μg mL^{-1}	Fodey et al. (2011)
Potentiometric sensor based on a MIP	6.0 μM	Liang et al. (2009)
Colorimetric sensor based on unmodified Gold NPs	0.4 μg mL^{-1}	Li et al. (2010)
Colorimetric sensor based on unmodified Gold NPs	25 ppb (5 min)	Chi et al. (2010)
Colorimetric sensor based on MTT-stabilized gold NPs	2.5 ppb (1 min)	Ai et al. (2009)
Colorimetric sensor based on silver NPs	0.1 μg g^{-1}	Han and Li (2010)
Colorimetric sensor based on multicolor QDs	0.6 ng g^{-1}	Peng et al. (2009)
Colorimetric sensor based on magnetic NPs	0.6 ng g^{-1}	Ma et al. (2009)
Colorimetric sensor based on Fe_3O_4 MNPs	3×10^{-6} M	Gao et al. (2007)
Glassy carbon electrode by chemical coupling of horseradish peroxidase	1.0×10^{-11} M	Xu et al. (2013)
Indium tin oxide electrode modified by gold NPs	0.13 ppb	Akter et al. (2013)
Graphite-epoxy composite electrode by using bismuthyl chloride	2.5×10^{-12} M	Zhu et al. (2011)
Screen printed carbon electrode	98.3 ppb	Liao et al. (2011)
Gold electrode modified by oligonucleotide (d(T)20) film	9.6×10^{-9} M	Cao et al. (2009)
Glassy carbon electrode by chemical coupling of horseradish peroxidase	1.0×10^{-11} M	Xu et al. (2013)
Indium tin oxide electrode modified by gold NPs	0.13 ppb	Akter et al. (2013)
Graphite-epoxy composite electrode by using bismuthyl chloride	2.5×10^{-12} M	Zhu et al. (2011)
Screen printed carbon electrode	98.3 ppb	Liao et al. (2011)
Gold electrode modified by oligonucleotide (d(T)20) film	9.6×10^{-9} M	Cao et al. (2009)
Optical biosensor based on polyclonal antibody	0.5 μg mL^{-1}	Fodey et al. (2011)

(*Continued*)

Table 9.2 List of Recent Advances in Melamine Detection Techniques (*cont.*)

Method	LOD	References
Chemiluminescenc based on Luminol-H_2O_2	0.12 μg mL^{-1}	Zeng et al. (2011)
Chemiluminescenc based on Luminol-$K_3Fe(CN)_6$	3.54 ng mL^{-1} (urine), 6.58 ng mL^{-1} (plasma)	Tang et al. (2011)
Planar wave guide fluorescence immunosensor	6.6 μg L^{-1}	Guo et al., 2014
Fluorescence based on a polyclonal antibody	9.3 ng mL^{-1}	Wang et al. (2011)
Fluorescence based on AuNPs	0.02 mg L^{-1}	Zhang et al. (2012)
Colorimetric aptasensor based on AuNPs	0.5 mg L^{-1}	Yun et al. (2014)
Aptamers (poly–T10) based on AuNPs	41.7 nM	Huang et al. (2011)
Aptamer probe based on AgNPs	3.1 mg L^{-1}	Liang et al. (2011)

8 Conclusions

Most substantial studies on analytical chemistry, analytical methods, toxicology, exposure, and establishment of TDI recommendation of melamine in infant formula and milk products have been conducted after the melamine issues in China. This chapter described varying methods for determination of the presence of melamine and its analogs in various matrices. The different methods each have their own advantages and disadvantages. The analytical methods are developed and used to analyze the present level of melamine in food samples. However, they require expensive instrumentation and extensive sample preparation, are time consuming, and need highly skilled personnel. Additionally, immunoassay methods are widely used to detect melamine but require labor-intensive operations including incubation, washing, and enzymatic reactions during the signal generation process.

Compared to the existing methods, advances in detection techniques are offering advantages due to their simplicity, low cost, rapidity, effectiveness, and high sensitivity. One of the famous methods is MIPs, which have high selectivity for target molecules, high affinity and robustness of recognition. Due to their stability, low cost, and ease in preparation, MIPs have gained wide acceptance as molecular recognition materials in developing sensors. Nanoparticles have been used for determining the presence of melamine because of unique functions such as biostability, biofunctionalization, and spectral that has not found in bulk-sized materials. Besides, it has high surface area, nontoxicity, good environmental acceptability, inexpensive, electrochemical activity, and high electron communication features. With recent developments in nanotechnology, new methods of designing colorimetric assays are emerging. The major advantage of colorimetric biosensor can be easily observed by the naked eye, and therefore, do not require any sophisticated instruments. Development of electrochemical sensors based on modified electrodes can be very simple, low cost, and offer accurateness, high stability, convenience, and a sensitive method for the determination of melamine. Electrochemical chemo-/biosensors undoubtedly will contribute in the near future to improvements in quality control in food processing, improving product quality and safety with minimal investment. The detailed study of melamine detection will open up new directions of research in thefield of sensors and sensing. The development of powerful analytical devices based on the combination of multiple techniques and the application of new materials will be established in the near future.

Acknowledgment

This work was supported by grants from the Ministry of Science, Technology and Innovation, MOSTI (No. SCF0090-IND-2013).

References

Ai, K., Liu, Y., Lu, L., 2009. Hydrogen-bonding recognition-induced color change of gold nanoparticles for visual detection of melamine in raw milk and infant formula. J. Am. Chem. Soc. 131, 9496–9497.

Akter, H., Shaikh, A.A., Taslima, R., Chowdhury, Rahman, M.S., Bakshi, P.K., Saleh Ahammad, A.J., 2013. Gold nanoparticle-modified indium tin oxide electrode for highly sensitive electrochemical detection of melamine. ECS Electrochem. Lett. 2, 13–15.

Andersen, W.C., Turnipseed, S.B., Karbiwnyk, C.M., Clark, S.B., Madson, M.R., Gieseker, C.M., 2008. Determination and confirmation of melamine residues in catfish, trout, tilapia, salmon, and shrimp by liquid chromatography with tandem mass spectrometry. J. Agric. Food. Chem. 56, 4340–4347.

Antikainen, R., Haapanen, R., Rekolainen, S., 2004. Flows of nitrogen and phosphorus in Finland: the forest industry and use of wood fuels. J. Clean. Prod. 12, 919–934.

Bann, B., Miller, S.A., 1958. Melamine and derivatives of melamine. Chem. Rev. 58, 131–172.

Batista, A.D., Nascimento, C.F., Melchert, W.R., Rocha, F.R.P., 2014. Expanding the separation capability of sequential injection chromatography: determination of melamine in milk exploiting micellar medium and on-line sample preparation. Microchem. J. 117, 106–110.

Baynes, R.E., Barlow, B., Mason, S.E., Riviere, J.E., 2010. Disposition of melamine residues in blood and milk from dairy goats exposed to an oral bolus of melamine. Food Chem. Toxicol. 48, 2542–2546.

Beitollahi, H., Sheikhshoaie, I., 2011. Electrocatalytic and simultaneous determination of isoproterenol, uric acid and folic acid at molybdenum (VI) complex-carbon nanotube paste electrode. Electrochim. Acta 56, 10259–10263.

Beltran-Martinavarro, B., Peris-Vicente, J., Marco-Peiro, S., Esteve-Romero, J., Rambla-Alegre, M., Carda-Broch, S., 2012. Use of micellar mobile phases for the chromatographic determination of melamine in dietetic supplements. Analyst 137, 269–274.

Beltran-Martinavarro, B., Peris-Vicente, J., Carda-Broch, S., Esteve-Romero, J., 2014. Development and validation of a micellar liquid chromatography based method to quantify melamine in swine kidney. Food Control 46, 168–173.

Bhatta, D., Stadden, E., Hashem, E., Sparrow, I.J.G., Emmerson, G.D., 2010. Multi-purpose optical biosensors for real-time detection of bacteria, viruses, and toxins. Sens. Actuat. B 149, 233–238.

Bizzari, S., Yokose, K., 2008. Melamine. Chemical Economics Handbook. SRI Consulting, Menlo Park, CA.

Brown, C.A., Jeong, K.S., Poppenga, R.H., Puschner, B., Miller, D.M., Ellis, A.E., 2007. Outbreaks of renal failure associated with melamine and cyanuric acid in dogs and cats in 2004 and 2007. J. Vet. Diagn. Invest. 19, 525–531.

Cai, W., Fan, Y., Jiang, Z., Yao, J., 2010. A highly sensitive and selective resonance scattering spectral assay for potassium ion based on aptamer and nanosilver aggregation reactions. Talanta 81, 1810–1815.

Cai, H.H., Yu, X., Dong, H., Cai, J., Yang, P.H., 2014. Visual and absorption spectroscopic detections of melamine with 3-mercaptopriopionic acid-functionalized gold nanoparticles: a synergistic strategy induced nanoparticle aggregates. J. Food Eng. 142, 163–169.

Cao, Q., Zhao, H., Zeng, L., Wang, J., Wang, R., Qiu, X., He, Y., 2009. Electrochemical determination of melamine using oligonucleotides modified gold electrodes. Talanta 80, 484–488.

Cao, X.Y., Shen, F., Zhang, M.W., Guo, J.J., Luo, Y., Li, X., Liu, H., Sun, C., Liu, J., 2013. Efficient inner filter effect of gold nanoparticles on the fluorescence of CdS quantum dots for sensitive detection of melamine in raw milk. Food Control 34, 221–229.

Cao, X., Shen, F., Zhang, M., Guo, J., Luo, Y., Xu, J., Li, Y., Sun, C., 2014a. Highly sensitive detection of melamine based on fluorescence resonance energy transfer between Rhodamine B and gold nanoparticles. Dyes Pigment. 111, 99–107.

Cao, X., Shen, F., Zhang, M., Sun, C., 2014b. Rapid and highly-sensitive melamine sensing based on the efficient inner filter effect of Ag nanoparticles on the fluorescence of eco-friendly ZnSe quantum dots. Sens. Actuat. B 202, 1175–1182.

Carina, F.N., Diogo, L.R., Fábio, R.P.R., 2015. A fast and environmental friendly analytical procedure for determination of melamine in milk exploiting fluorescence quenching. Food Chem. 169, 314–319.

Chan, E.Y.Y., Griffiths, S.M., Chan, C.W., 2008. Public-health risks of melamine in milk products. Lancet 372, 1444–1445.

Chang, C.W., Chu, S.P., Tseng, W.L., 2010. Selective extraction of melamine using 11-mercaptoundecanoic acid–capped gold nanoparticles followed by capillary electrophoresis. J. Chromatogr. A 1217, 7800–7806.

Chen, A.C., Chatterjee, S., 2013. Nanomaterials-based electrochemical sensors for biomedical applications. Chem. Soc. Rev. 42, 5425–5438.

Chen, D., Meena, M., Sharma, S.K., McLaughlin, L.W., 2004. Formation and stability of a Janus-Wedge type of DNA triplex. J. Am. Chem. Soc. 126, 70–71.
Chen, W., Jiang, Y., Ji, B.Q., Zhu, C.Q., Liu, L.Q., Peng, C.F., Jin, M.K., 2009. Automated and ultrasensitive detection of methyl-3-quinoxaline-2-carboxylic acid by using gold nanoparticles probes SIA-rt-PCR. Biosens. Bioelectron. 24, 2858–2863.
Chen, G., Song, F., Xiong, X., Peng, X., 2013. Fluorescent nanosensors based on fluorescence resonance energy transfer (FRET). Ind. Eng. Chem. Res. 52, 11228–11245.
Chen, D., Zhao, Y., Miao, H., Wu, Y., 2015. A novel dispersive micro solid phase extraction using PCX as the sorbent for the determination of melamine and cyromazine in milk and milk powder by UHPLC-HRMS/MS. Talanta 134, 144–152.
Chi, H., Liu, B., Guan, G., Zhang, Z., Han, M.Y., 2010. A simple, reliable and sensitive colorimetric visualization of melamine in milk by unmodified gold nanoparticles. Analyst 135, 1070–1075.
Chu, P.W.S., Chan, K.M., Cheung, S.T.C., Wong, Y.C., 2010. Review of analytical techniques used in proficiency-testing programs for melamine in animal feed and milk. TrAC—Trend. Anal. Chem. 29, 1014–1026.
Citartan, M., Gopinath, S.C.B., Tominaga, J., Tan, S.C., Tang, T.H., 2012. Assays for aptamer-based platforms. Biosens. Bioelectron. 34, 1–11.
Cook, H.A., Klampfl, C.W., Buchberger, W., 2005. Analysis of melamine resins by capillary zone electrophoresis with electrospray ionization-mass spectrometric detection. Electrophoresis 26, 1576–1583.
Dai, H.C., Shia, Y., Wang, Y., Suna, Y., Hua, J., Ni, P., Li, Z., 2014. A carbon-dot-based biosensor for melamine detection by fluorescence resonance energy transfer. Sens. Actuat. B 202, 201–208.
Deabes, M.M., El-Habib, R., 2012. Determination of melamine in infant milk formula, milk powder and bsaa fish samples by HPLC/DAD. J. Environ. Anal. Toxicol. 2, 2–4.
Demirhan, B., Demirhan, B., Kara, H.S., 2015. Room-temperature phosphorescence determination of melamine in dairy products using l-cysteine-capped Mn-doped zinc sulfide (ZnS) quantum dots. J. Dairy Sci. 98, 2992–3000.
Deng, H., Xu, Y., Liu, Y.H., Che, Z.J., Guo, H.L., Shan, S.X., Sun, Y., Liu, X.F., Huang, K.Y., Ma, X.W., Wu, Y., Liang, X.J., 2012. Determination of cyromazine in cucumber by molecularly imprinted solid phase extraction/high performance liquid chromatography. Anal. Chem. 84, 1253–1258.
Desmarchelier, A., Cuadra, M.G., Delatour, T., Mottier, P., 2009. Simultaneous quantitative determination of melamine and cyanuric acid in cow's milk and milk-based infant formula by liquid chromatography–electrospray ionization tandem mass spectrometry. J. Agric. Food. Chem. 57, 7186–7193.
Ding, T., Xu, J., Li, J., Shen, C., Wu, B., Chen, H., Li, S., 2008. Determination of melamine residue in plant origin protein powders using high performance liquid chromatography-diode array detection and high performance liquid chromatography-electrospray ionization tandem mass spectrometry. J. Chromatogr. 26, 6–9.
Du, J.J., Yin, S.Y., Jiang, L., Ma, B., Chen, X.D., 2013. A colorimetric logic gate based on free gold nanoparticles and the coordination strategy between melamine and mercury ions. Chem. Commun. 49, 4196–4198.
Du, J., Wang, Y., Zhang, W., 2015. Gold nanoparticles-based chemiluminescence resonance energy transfer for ultrasensitive detection of melamine. Spectrochim. Acta Part A Mol. Biomol. Spec. 149, 698–702.
Ehling, S., Tefera, S., Ho, I.P., 2007. High-performance liquid chromatographic method for the simultaneous detection of the adulteration of cereal flours

with melamine and related triazine by-products ammeline, ammelide, and cyanuric acid. Food Addit. Contam. 24, 1319–1325.
FAO and WHO, 2011. Codex Alimentarius Commission. Report of the fifth session of the codex committee on contaminants in foods, joint FAO/WHO food standards programme. Codex committee on contaminants in foods, 5th session, Netherlands.
Farre, M., Kantiani, L., Perez, S., Barcelo, D., 2009. Sensors and biosensors in support of EU Directives. TrAC. Trends Anal. Chem. 28, 170–185.
Fashi, A., Yaftian, M.R., Zamani, A., 2015. Determination of melamine in dairy products using electromembrane–LPME followed by HPLC. Food Chem. 188, 92–98.
Food and Drug Administration (FDA), 2008. Melamine Pet Food Recall in 2007, Guidance for industry safety testing of drug metabolites, Department of Health and Human Services.
Food and Drug Administration (FDA), 2009. Analytical Methods for Melamine and Triazine Analogs.
Feng, J.W., Cai, Q.R., Liu, X.C., Yu, Y.G., Peng, Y.F., Zhang, Y., 2008. Determination of melamine and cyromazine in milk and dairy products by LC-MS/MS. J. Food. Sci. Tech. 24, 1058–1060.
Filazi, A., Sireli, U.T., Ekici, H., Can, H.Y., Karagoz, A., 2012. Determination of melamine in milk and dairy products by high performance liquid chromatography. J. Dairy Sci. 95, 602–608.
Filigenzi, M.S., Tor, E.R., Poppenga, R.H., Aston, L.A., Puschner, B., 2007. The determination of melamine in muscle tissue by liquid chromatography/tandem mass spectrometry. Rapid Commun. Mass Spectrom. 21, 4027–4032.
Filigenzi, M.S., Puschner, B., Aston, L.R., Poppenga, R.H., 2008. Diagnostic determination of melamine and related compounds in kidney tissue by liquid chromatography/tandem mass spectrometry. J. Agric. Food. Chem. 56, 7593–7599.
Finete, V.L.M., Gouvêa, M.M., Marques, F.F.C., Netto, A.D.P., 2013. Is it possible to screen for milk or whey protein adulteration with melamine, urea and ammonium sulphate, combining Kjeldahl and classical spectrophotometric methods? Food Chem. 141, 3649–3655.
Finete, V.L.M., Gouvêa, M.M., Marques, F.F.C., Netto, A.D.P., 2014. Characterization of newfound natural luminescent properties of melamine, and development and validation of a method of high performance liquid chromatography with fluorescence detection for its determination in kitchen plastic ware. Talanta 123, 128–134.
Finete, V.L.M., Gouvêa, M.M., Marques, F.F.C., Netto, A.D.P., 2015. Validation of a method of high performance liquid chromatography with fluorescence detection for melamine determination in UHT whole bovine milk. Food Control 51, 402–407.
Fodey, T.L., Thompson, C.S., Traynor, I.M., Haughey, S.A., Kennedy, D.G., Crooks, S.R.H., 2011. Development of an optical biosensor based immunoassay to screen infant formula milk samples for adulteration with melamine. Anal. Chem. 83, 5012–5016.
Gao, L.Z., Zhuang, J., Nie, L., Zhang, J.B., Zhang, Y., 2007. Intrinsic peroxidase-like activity of ferromagnetic nanoparticles. Nat. Nanotechnol. 2, 577–583.
Garber, E.A.E., Brewer, V.A., 2010. Enzyme-linked immunosorbent assay detection of melamine in infant formula and wheat food products. J. Food Process. Preserv. 73, 701–707.
Ge, X., Wu, X., Wang, J., Liang, S., Sun, H., 2015. Highly sensitive determination of cyromazine, melamine, and their metabolites in milk by molecularly imprinted solid-phase extraction combined with ultra-performance liquid chromatography. J. Dairy Sci. 98, 2161–2171.

Grabar, K.C., Freeman, R.G., Hommer, M.B., Natan, M.J., 1995. Preparation and Characterization of Au Colloid Monolayers. Anal. Chem. 67, 735–743.
Guan, N., Fan, Q.F., Ding, J., Zhao, Y.M., Lu, J.Q., Ai, Y., Xu, G.B., Zhu, S.N., Yao, C., Jiang, L.N., Miao, J., Zhang, H., Zhao, D., Liu, X.Y., Yao, Y., 2009. Melamine-contaminated powdered formula and urolithiasis in young children. N. Engl. J. Med. 360, 1067–1074.
Guo, L.Q., Zhong, J.H., Wu, J.M., Fu, F.F., Chen, G.N., Zheng, X.Y., 2010. Visual detection of melamine in milk products by label-free gold nanoparticles. Talanta 82, 1654–1658.
Guo, L., Zhong, J., Wu, J., Fu, F.F., Chen, G., Chen, Y., Zheng, X., Lin, S., 2011. Sensitive turn-on fluorescent detection of melamine based on fluorescence resonance energy transfer. Analyst 136, 1659–1663.
Guo, H., Zhou, X., Zhang, Y., Song, B., Liu, L., 2014. Highly sensitive and rapid detection of melamine in milk products by planar waveguide fluorescence immunosensor (PWFI). Sens. Actuators B Chem. 194, 114–119.
Han, C.P., Li, H.B., 2010. Visual detection of melamine in infant formula at 0.1 ppm level based on silver nanoparticles. Analyst 135, 583–588.
Han, C., Wu, Y., Zhou, Y., Gong, Y., Liu, C., Wang, L., Shen, Y., 2009. Microwave-assisted extraction of melamine residues from pet food and analysis by ion-exchange LC–DAD. J. Chromatogr. 70, 927–931.
Hau, A.K., Kwan, T.H., Li, P.K., 2009. Melamine toxicity and the kidney. Clin. J. Am. Soc. Nephrol. 20, 245–250.
Huang, G.M., Ouyang, Z., Cooks, R.G., 2009. High-throughput trace melamine analysis in complex mixtures. Chem. Commun. Camb. Eng. 5, 556–558.
Huang, H., Li, L., Zhou, G., Liu, Z., Ma, Q., Feng, Y., Zeng, G., Tinnefeld, P., He, Z., 2011. Visual detection of melamine in milk samples based on label-free and labeled gold nanoparticles. Talanta 85, 1013–1019.
Huy, B.T., Seo, M.H., Zhang, X.F., Lee, Y., 2014. Selective optosensing of clenbuterol and melamine using molecularly imprinted polymer-capped CdTe quantum dots. Biosens. Bioelectron. 57, 310–316.
Ibáñez, M., Sancho, J.V., Hernández, F., 2009. Determination of melamine in milk-based products and other food and beverage products by ion-pair liquid chromatography–tandem mass spectrometry. Anal. Chim. Acta 649, 91–97.
Ingelfinger, J.R., 2008. Melamine and the global implications of food contamination. N. Engl. J. Med. 359, 2745–2748.
Kailasa, S.K., Wu, H.F., 2015. Electrospray ionization tandem mass spectrometry for rapid, sensitive and direct detection of melamine in dairy products. J. Ind. Eng. Chem. 21, 138–144.
Kim, B., Perkins, L.B., Bushway, R.J., Nesbit, S., Fan, T., Sheridan, R., 2008. Determination of melamine in pet food by enzyme immunoassay, high-performance liquid chromatography with diode array detection, and ultra-performance liquid chromatography with tandem mass spectrometry. J. AOAC Int. 91, 408–413.
Kobayashi, T., Okada, A., Fujii, Y., Niimi, K., Hamamoto, S., Yasui, T., Tozawa, K., Kohri, K., 2010. The mechanism of renal stone formation and renal failure induced by administration of melamine and cyanuric acid. Urol. Res. 38, 117–125.
Kuang, H., Chen, W., Yan, W., Xu, L., Zhu, Y., Liu, L., Chu, H., Peng, C., Wang, L., Kotov, N.A., Xu, C., 2011. Crown ether assembly of gold nanoparticles: melamine sensor. Biosens. Bioelectron. 26, 2032–2037.
Kumar, N., Seth, R., Kumar, H., 2014. Colorimetric detection of melamine in milk by citrate-stabilized gold nanoparticles. Anal. Biochem. 456, 43–49.
Lachenmeier, D.W., Humpfer, E., Fang, F., Schutz, B., Dvorteak, P., Sproll, C., Spraul, M., 2009. NMR-Spectroscopy for nontargeted screening and simultaneous quantification of health-relevant compounds in foods: the example of melamine. J. Agric. Food. Chem. 57, 7194–7199.

Li, J., Qi, H.Y., Shi, Y.P., 2009. Determination of metoprolol tartrate in tablets and human urine using flow-injection chemiluminescence method. J. Pharm. Biomed. Anal. 42, 384–388.
Li, L., Li, B.X., Cheng, D., Mao, L.H., 2010. Visual detection of melamine in raw milk using gold nanoparticles as colorimetric probe. Food Chem. 122, 895–900.
Li, J., Chen, Z., Li, Y., 2011. A strategy for constructing sensitive and renewable molecularly imprinted electrochemical sensors for melamine detection. Anal. Chim. Acta 706, 255–260.
Li, X., Li, J., Kuang, H., Feng, L., Yi, S., Xia, X., Huang, H., Chen, Y., Tang, C., Zeng, Y., 2013. An ultrasensitive method for the determination of melamine using cadmium telluride quantum dots as fluorescence probes. Anal. Chim. Acta. 802, 82–88.
Li, N., Liu, D.Q., Cui, H., 2014. Metal-nanoparticle-involved chemiluminescence and its applications in bioassays. Anal. Bioanal. Chem. 406 (23), 5561–5571.
Li, Y., Xu, J., Sun, C., 2015. Chemical sensors and biosensors for the detection of melamine. RSC Adv. 5, 1125–1147.
Liang, R.N., Zhang, R.M., Qin, W., 2009. Potentiometric sensor based on molecularly imprinted polymer for determination of melamine in milk. Sens. Actuators B: Chem. 141, 544–550.
Liang, X.S., Wei, H.P., Cui, Z.Q., Deng, J.Y., Zhang, Z.P., You, X.Y., Zhang, X.E., 2011. Colorimetric detection of melamine in complex matrices based on cysteamine modified gold nanoparticles. Analyst 136, 179–183.
Liao, C.W., Chen, Y.R., Chang, J.L., Zen, J.M., 2011. Single-run electrochemical determination of melamine in dairy products and pet foods. J. Agric. Food. Chem. 59, 9782–9787.
Lin, M., He, L., Awika, J., Yang, L., Ledoux, D.R., Li, H., Mustapha, A., 2008. Detection of melamine in gluten, chicken feed and processed foods using surface enhanced raman spectroscopy and HPLC. J. Food Sci. 73, 129–134.
Lin, Z., Chen, H., Lin, J.M., 2013. Peroxide induced ultra-weak chemiluminescence and its application in analytical chemistry. Analyst 138, 5182–5193.
Liu, D.B., Wang, Z., Jiang, X.Y., 2011. Gold nanoparticles for the colorimetric and fluorescent detection of ions and small organic molecules. Nanoscale 3, 1421–1433.
Liu, Y., Todd, E.E.D., Zhang, Q., Shi, J., Liu, X., 2012a. Recent developments in the detection of melamine. J. Zhejg. Uni. Sci. B. 13, 525–532.
Liu, S., Hu, J., Zhang, H., Su, X., 2012b. $CuInS_2$ quantum dots-based fluorescence turns off/on probe for detection of melamine. Talanta 101, 368–373.
Liu, S.S., Yi, X.Y., Wang, J.X., 2014. Sensitive detection of melamine at surface plasmon resonance chips pre-immobilized with bovine serum albumin-melamine conjugate. Chin. J. Anal. Chem. 42, 695–700.
Liu, B., Xiao, B., Cui, L., Wang, M., 2015a. Molecularly imprinted electrochemical sensor for the highly selective and sensitive determination of melamine. Mat. Sci. Eng. C 55, 457–461.
Liu, J., Song, H., Liu, J., Liu, Y., Li, L., Tang, H., Li, Y., 2015b. Preparation of molecularly imprinted polymer with double templates for rapid simultaneous determination of melamine and dicyandiamide in dairy products. Talanta 134, 761–767.
Luchansky, M.S., Bailey, R.C., 2012. High-Q optical sensors for chemical and biological analysis. Anal. Chem. 84 (2), 793–821.
Lutter, P., Savoy-Perroud, M.C., Campos-Gimenez, E., Meyer, L., Goldmann, T., Bertholet, M.C., Mottier, P., Desmarchelier, A., Monard, F., Perrin, C., Robert, F., Delatour, T., 2011. Screening and confirmatory methods for the

determination of melamine in cow's milk and milk-based powered infant formula: validation and proficiency-tests of ELISA, HPLC-UV, GC-MS and LC-MS/MS. Food Control 22, 903–913.

Lv, X.W., Wang, J., Wu, L., Qiu, J., Li, J.G., Wu, Z.L., Qin, Y.C., 2010. Tissue deposition and residue depletion in lambs exposed to melamine and cyanuric acid-contaminated diets. J. Agric. Food. Chem. 58, 943–948.

Ma, W., Chen, W., Qiao, R.R., Liu, C.Y., Yang, C.H., Li, Z.K., Xu, D.H., Peng, C.F., Jin, Z.Y., Xu, C.L., Zhu, S.F., Wang, L.B., 2009. Rapid and sensitive detection of microcystin by immunosensor based on nuclear magnetic resonance. Biosens. Bioelectron. 25, 240–243.

Ma, Y.R., Niu, H.Y., Zhang, X.L., Cai, Y.Q., 2011. One-step synthesis of silver/dopamine nanoparticles and visual detection of melamine in raw milk. Analyst 136, 4192–4196.

Manzoori, J.L., Amjadi, M., Hassanzadeh, J., 2011. Enhancement of the chemiluminescence of permanganate-formaldehyde system by gold/silver nanoalloys and its application to trace determination of melamine. Microchim. Acta 175, 47–54.

Marco-Peiro, S., Beltran-Martinavarro, B., Rambla-Alegre, M., Peris-Vicente, J., Esteve-Romero, J., 2012. Validation of an analytical methodology to quantify melamine in body fluids using micellar liquid chromatography. Talanta 88, 617–622.

Matte, D., Solastiouk, B., Merlin, A., Deglise, X., 1990. Kinetic study of the N-chlorination of cyanuric acid in the aqueous phase. Can. J. Chem. 68, 307–313.

Mauer, L.J., Chernyshova, A.A., Hiatt, A., Deering, A., Davis, R., 2009. Melamine detection in infant formula powder using near-and mid-infrared spectroscopy. J. Agric. Food. Chem. 57, 3974–3980.

Md Muslim, N.Z., Ahmad, M., Heng, L.Y., Saad, B., 2012. Optical biosensor test strip for the screening and direct determination of L-glutamate in food samples. Sens. Actuators B Chem. 161, 493–497.

Mecker, L.C., Tyner, K.M., Kauffman, J.F., Arzhantsev, S., Mans, D.J., Gryniewicz Ruzicka, C.M., 2012. Selective melamine detection in multiple sample matrices with a portable Raman instrument using surface enhanced Raman spectroscopy-active gold nanoparticles. Anal. Chim. Acta 733, 48–55.

Meng, Z., Shi, Z., Liang, S., Dong, X., Lv, Y., Sun, H., 2015. Rapid screening and quantification of cyromazine, melamine, ammelide, ammeline, cyanuric acid, and dicyandiamide in infant formula by ultra-performance liquid chromatography coupled with quadrupole time-of-flight mass spectrometry and triple quadrupole mass spectrometry. Food Control 55, 158–165.

Miao, H., Fan, S., Wu, Y.N., Zhang, L., Zhou, P.P., Li, J.G., Chen, H.J., Zhao, Y.F., 2009. Simultaneous determination of melamine, ammelide, ammeline, and cyanuric acid in milk and milk products by gas chromatography–tandem mass spectrometry. Biomed. Environ. Sci. 22, 87–94.

Mirzajani, R., Ahmadi, S., 2015. Melamine supported magnetic iron oxide nanoparticles (Fe_3O_4@Mel) for spectrophotometric determination of malachite green in water samples and fish tissues. J. Ind. Eng. Chem. 23, 171–178.

Mosdell, D.K., Daniel, W.H., Freeborg, R.P., 1987. Melamine and ammeline as nitrogen sources for turf grasses. Fertilizer Res. 11, 79–86.

Muñiz-Valencia, R., Ceballos-Magaña, S.G., Rosales-Martinez, D., Gonzalo-Lumbreras, R., Santos-Montes, A., Cubedo-Fernandez-Trapiella, A., Izquierdo-Hornillos, R.C., 2008. Method development and validation for melamine and its derivatives in rice concentrates by liquid chromatography: application to animal feed sample. Anal. Bioanal. Chem. 392, 523–531.

Newton, G.L., Utley, P.R., 1978. Melamine as a dietary nitrogen source for ruminants. J. Ani. Sci. 47, 1338–1344.
National Toxicology Program (NTP), 1983. Carcinogenesis bioassay of melamine, pp. 81–86.
Ogasawara, H., Imaida, K., Ishiwata, H., Toyoda, K., Kawanishi, T., Uneyama, C., Hayashi, S., Takahashi, M., Hayashi, Y., 1995. Urinary bladder carcinogenesis induced by melamine in F344 male rats: correlation between carcinogenicity and urolith formation. Carcinog. 16, 2773–2777.
Panuwet, P., Nguyen, J.V., Wade, E.L., D'Souza, P.E., Ryan, P.B., Barr, D.B., 2012. Quantification of melamine in human urine using cation-exchange based high performance liquid chromatography tandem mass spectrometry. J. Chromatogr. B. 887, 48–54.
Patakioutas, G., Savvas, D., Matakoulis, C., Sakellarides, T., Albanis, T., 2007. Fate of cyromazine and its metabolite melamine applied via nutrient solution to a closed-cycle cultivation of bean (*Phaseolus vulgaris* L.). J. Agric. Food Chem. 55, 9928–9935.
Peng, C.F., Li, Z.K., Zhua, Y.Y., Chen, W., Yuan, Y., Liu, L.Q., Li, Q.S., Xu, D.H., Qiao, R.R., Wang, L.B., Zhu, S.F., Jin, Z.Y., Xu, C.L., 2009. Simultaneous and sensitive determination of multiplex chemical residues based on multicolor quantum dot probes. Biosens. Bioelectron. 24, 3657–3662.
Peng-Ju, W., Jun, Y., Qing-De, S., Yun, G., Xiao-Lan, Z., Ji-Bao, C., 2007. Rapid removal of template from molecularly imprinted polymers by accelerated solvent extraction chin. J. Anal. Chem. 35, 484–488.
Perumal, V., Hashim, U., 2013. pH Measurement using in house fabricated interdigitated capacitive transducer. J. Eco. Fin. Adm. Sci. 12, 1–15.
Pietrzyk, A., Kutner, W., Chitta, R., Zandler, M.E., D'Souza, F., Sannicolo, F., Mussini, P.R., 2009. Melamine acoustic chemosensor based on molecularly imprinted polymer film. Anal. Chem. 81, 10061–10070.
Qi, W.J., Wu, D., Ling, J., Huang, C.Z., 2010. Visual and light scattering spectrometric detections of melamine with polythymine-stabilized gold nanoparticles through specific triple hydrogen-bonding recognition. Chem. Commun. 46, 4893–4895.
Qiao, G.H., Guo, T., Klein, K.K., 2010. Melamine in Chinese milk products and consumer confidence. Appetite 55, 190–195.
Rajapandiyan, P., Tang, W., Yang, J., 2015. Rapid detection of melamine in milk liquid and powder by surface-enhanced Raman scattering substrate array. Food Control 56, 155–160.
Rambla-Alegre, M., Peris-Vicente, J., Marco-Peiró, S., Beltrán-Martinavarro, B., Esteve-Romero, J., 2010. Development of an analytical methodology to quantify melamine in milk using micellar liquid chromatography and validation according to EU Regulation 2002/654/EC. Talanta 81, 894–900.
Ren, H.B., Yan, X.P., 2012. Ultrasonic assisted synthesis of adenosine triphosphate capped manganese-doped ZnS quantum dots for selective room temperature phosphorescence detection of arginine and methylated arginine in urine based on supramolecular Mg^{2+} adenosine triphosphate–arginine ternary system. Talanta 97, 16–22.
Rovina, K., Siddiquee, S., 2015. A review of recent advances in melamine detection techniques. J. Food Comp. Anal. 43, 25–38.
Rovina, K., Siddiquee, S., Wong, N.K., 2015. Development of melamine sensor based on ionic liquid/nanoparticles/chitosan with modified gold electrode for determination of melamine in milk product. Sens. BioSens. Res. 4, 16–22.
Saha, K., Agasti, S.S., Kim, C., Li, X.N., Rotello, V.M., 2011. Gold nanoparticles in chemical and biological sensing. Chem. Rev. 112, 2739–2779.

Shang, L., Dong, S.J., Nienhaus, G.U., 2011. Ultra-small fluorescent metal nanoclusters: synthesis and biological applications. Nano Today 6, 401–418.
Shelton, D.R., Karns, J.S., McCarty, G.W., Durham, D.R., 1997. Metabolism of melamine by klebsiella terragena. Appl. Environ. Microbiol. 63, 2832–2835.
Smoker M., Krynitsky, A.J., 2008. Interim Method for Determination of Melamine and Cyanuric Acid Residues in Foods using LC-MS/MS: Version 1.0. US FDA/CFSAN Laboratory Information Bulletin LIB No. 4422.
Song, S., Wang, L., Li, J., Zhao, J., Fan, C., 2008. Aptamer-based biosensors. TRAC-Trend. Anal. Chem. 27, 108–117.
Song, J., Wu, F., Wan, Y., Ma, L., 2015. Colorimetric detection of melamine in pretreated milk using silver nanoparticles functionalized with sulfanilic acid. Food Control 50, 356–361.
Sun, F.X., Ma, W., Xu, L.G., Zhu, Y.Y., Liu, L.Q., Peng, C.F., Wang, L.B., Kuang, H., Xu, C.L., 2010a. Analytical methods and recent developments in the detection of melamine. TrAC- Trend. Anal. Chem. 29, 1239–1249.
Sun, H.W., Wang, L.X., Ai, L.F., Liang, S.X., 2010b. A sensitive and validated method for determination of melamine residue in liquid milk by reversed phase high-performance liquid chromatography with solid-phase extraction. Food Control 21, 686–691.
Tang, H.W., Ng, K.M., Chui, S.S.Y., Che, C.M., Lam, C.W., Yuen, K.Y., Siu, T.S., Lan, L.C.L., 2009. Analysis of melamine cyanurate in urine using matrix-assisted laser desorption/ionization mass spectrometry. Anal. Chem. 81, 3676–3682.
Tang, X.S., Shi, X.Y., Tang, Y.H., Yue, Z.J., He, Q.Q., 2011. Chemiluminescence determination of melamine with Luminol-$K_3Fe(CN)_6$ system. J. Pharm. Anal. 2, 104–107.
Tang, G., Du, L., Su, X., 2013. Detection of melamine based on fluorescence resonance energy transfer between CdTe QDs and Rhodamine B. Food Chem. 141, 4060–4065.
Ting, Y., Deng, J., Lin, X., Ding, L., Li, Y., Li, H., Ting, X., 2011. Electrochemical sensor based on a poly(para-aminobenzoic acid) film modified glassy carbon electrode for the determination of melamine in milk. Electrochim. Acta 56, 4595–4602.
Tittlemier, S.A., Lau, B.P.Y., Menard, C., Corrigan, C., Sparling, M., Gaertner, D., Pepper, K., Feeley, M., 2009. Melamine in infant formula sold in Canada: occurrence and risk assessment. J. Agric. Food. Chem. 57, 5340–5344.
Turnipseed, S., Casey, C., Nochetto, C., Heller, D.N., 2008. Determination of Melamine and Cyanuric Acid Residues in Infant Formula using LC-MS/MS. Laboratory Information Bulletin LIB No. 4421, vol. 24.
Tyan, Y., Yang, M., Jong, S., Wang, C., Shiea, J., 2009. Melamine contamination. Anal. Bioanal. Chem. 395, 729–735.
US Food and Drug Administration (USFDA), 2007. Food and Drug Administration, Center for Food Safety and Nutrition.
Varelis, P., Jeskelis, R., 2008. Preparation of [$^{13}C_3$]-melamine and [$^{13}C_3$]-cyanuric acid and their application to the analysis of melamine and cyanuric acid in meat and pet food using liquid chromatography-tandem mass spectrometry. Food Addit. Contam. 25, 1208–1215.
Venkatasami, G., Sowa, R.J.J., 2010. A rapid, acetonitrile-free, HPLC method for determination of melamine in infant formula. Anal. Chim. Acta 665, 227–230.
Vilela, D., Gonzalez, M.C., Escarpa, A., 2012. Sensing colorimetric approaches based on gold and silver nanoparticles aggregation: chemical creativity behind the assay: areview. Anal. Chim. Acta 751, 24–43.
Wang, J., Jiang, L., Chu, Q., Ye, J., 2010. Residue analysis of melamine in milk products by micellar electrokinetic capillary chromatography with amperometric detection. Food Chem. 121, 215–219.

Wang, Q., Haughey, S.A., Sun, Y.M., Eremin, S.A., Li, Z.F., Liu, H., Xu, Z.L., Shen, Y.D., Lei, H.T., 2011. Development of a fluorescence polarization immunoassay for the detection of melamine in milk and milk powder. Anal. Bioanal. Chem. 399, 2275–2284.

Wang, G.L., Zhu, X.Y., Jiao, H.J., Dong, Y.M., Wu, X.M., Li, Z.J., 2012. Oxidative etching-aggregation of silver nanoparticles by melamine and electron acceptors: an innovative route toward ultrasensitive and versatile functional colorimetric sensors. Anal. Chim. Acta 747, 92–98.

Wang, H., Chen, D., Yu, L., Chang, M., Ci, L., 2015. One-step, room temperature, colorimetric melamine sensing using an in-situ formation of silver nanoparticles through modified Tollens process. Spectrochim. Acta Part A: Mol. Biomol. Spectrosc. 137, 281–285.

Whitesides, G.M., Mathias, J.P., Seto, C.T., 1991. Molecular self-assembly and nanochemistry: a chemical strategy for the synthesis of nanostructures. J. Sci. 254, 1312–1319.

World Health Oranization (WHO), 2004. The global burden of disease, Geneva, Switzerland

World Health Oranization (WHO), 2008. The World Health Report 2008—primary Health Care (Now More Than Ever), Geneva, Switzerland.

World Health Oranization (WHO), 2009. Toxicological and health aspects of melamine and cyanuric acid. Report of a WHO Expert Meeting In collaboration with FAO Supported by Health Canada, Geneva.

William, R.A., Thiago, R.L.C.P., 2014. Use of copper electrode for melamine quantification in milk. Electrochim. Acta 117, 379–384.

Wu, Y., Zhang, Y., 2013. Analytical chemistry, toxicology, epidemiology and health impact assessment of melamine in infant formula: recent progress and developments. Food Chem. Toxicol. 56, 325–335.

Wu, Y.T., Huang, C.M., Lina, C.C., Ho, W.A., Lin, L.C., Chiu, T.F., Tarn, D.C., Lin, C.H., Tsai, T.H., 2009. Determination of melamine in rat plasma, liver, kidney, spleen, bladder and brain by liquid chromatography–tandem mass spectrometry. J. Chromatogr. A. 1216, 7595–7601.

Wu, B., Wang, Z., Zhao, D., Lu, X., 2012. A novel molecularly imprinted impedimetric sensor for melamine determination. Talanta 101, 374–381.

Wu, L., Chen, K., Lu, Z., Li, T., Shao, K., Shao, F., Han, H., 2014. Hydrogen-bonding recognition-induced aggregation of gold nanoparticles for the determination of the migration of melamine monomers using dynamic light scattering. Anal. Chim. Acta 845, 92–97.

Xia, X., Ding, S.Y., Li, X.W., Gong, X., Zhang, S.X., Jiang, H.Y., Li, J.C., Shen, J.Z., 2009. Validation of a confirmatory method for the determination of melamine in egg by gas chromatography–mass spectrometry and ultra-performance liquid chromatography–tandem mass spectrometry. Anal. Chim. Acta 651, 196–200.

Xia, J., Zhou, N., Liu, Y., Chen, B., Wu, Y., Yao, S., 2010. Simultaneous determination of melamine and related compounds by capillary zone electrophoresis. Food Control 21, 912–918.

Xie, J.P., Zheng, Y.G., Ying, J.Y., 2010. Highly selective and ultrasensitive detection of Hg^{2+} based on fluorescence quenching of Au nanoclusters by Hg^{2+}–Au^{+} interactions. Chem. Commun. 46, 961–963.

Xin, H., Stone, R., 2008. Chinese probe unmasks high-tech adulteration with melamine. Sci. 322, 1310–1311.

Xin, J.Y., Zhang, L.X., Chen, D.D., Lin, K., Fan, H.C., Wang, Y., Xia, C.G., 2015. Colorimetric detection of melamine based on methanobactin-mediated synthesis of gold nanoparticles. Food Chem. 174, 473–479.

Xu, S., Lu, H., 2015. One-pot synthesis of mesoporous structured ratiometric fluorescence Molecularly imprinted sensor for highly sensitive detection of melamine from milk samples. Biosens. Bioelectron. 73, 160–166.
Xu, X.M., Ren, Y.P., Zhu, Y., Cai, Z.X., Han, J.L., Huang, B.F., Zhua, Y., 2009. Direct determination of melamine in dairy products by gas chromatography/mass spectrometry with coupled column separation. Anal. Chim. Acta 650, 39–43.
Xu, Q., Wei, H.P., Du, S., Li, H.B., Ji, Z.P., Hu, X.Y., 2013. Detection of subnanomolar melamine based on electrochemical accumulation coupled with enzyme colorimetric assay. J. Agric. Food Chem. 61, 1810–1817.
Xu, G., Zhang, H., Zhong, M., Zhang, T., Lu, X., Kan, X., 2014. Imprinted sol–gel electrochemical sensor for melamine direct recognition and detection. J. Electroanal. Chem. 713, 112–118.
Yan, H., Cheng, X., Sun, N., Cai, T., Wu, R., Han, K., 2012. Rapid and selective screening of melamine in bovine milk using molecularly imprinted matrix solid-phase dispersion coupled with liquid chromatography-ultraviolet detection. J. Chromatogr. B. 908, 137–142.
Yang, S.P., Ding, J.H., Zheng, J., Hu, B., Li, J.Q., Chen, H.W., Zhou, Z.Q., Qiao, X.L., 2009. Detection of melamine in milk products by surface desorption atmospheric pressure chemical ionization mass spectrometry. Anal. Chem. 81, 2426–2436.
Yin, W., Liu, J., Zhang, T., Li, W., Liu, W., Meng, M., He, F., Wan, Y., Feng, C., Wang, S., Lu, X., Xi, R.J., 2010. Preparation of monoclonal antibody for melamine and development of an indirect competitive ELISA for melamine detection in raw milk, milk powder, and animal feeds. J. Agr. Food Chem. 58, 8152–8157.
Yokley, R.A., Mayer, L.C., Rezaaiyan, R., Manuli, M.E., Cheung, M.W., 2000. Analytical method for the determination of cyromazine and melamine residues in soil using LC-UV and GC-MSD. J. Agric. Food Chem. 48, 3352–3358.
Yun, W., Li, H., Chen, S.Q., Tu, D.W., Xie, W.Y., Huang, Y., 2014. Aptamer-based rapid visual biosensing of melamine in whole milk. Eur. Food Res. Technol. 238, 989–995.
Zeng, H.J., Yang, R., Wang, Q.W., Li, J.J., Qu, L.B., 2011. Determination of melamine by flow injection analysis based on chemiluminescence system. Food Chem. 127, 842–846.
Zhang, L.B., Wang, E.K., 2014. Metal nanoclusters: new fluorescent probes for sensors and bioimaging. Nano Today 9, 132–157.
Zhang, J., Lewin, M., Pearce, E., Zammarano, M., Gilman, J.W., 2008. Flame retarding polyamide 6 with melamine cyanurate and layered silicates. Polym. Adv. Technol. 19, 928–936.
Zhang, J., Wu, M., Chen, D., Song, Z., 2011. Ultrasensitive determination of melamine in milk products and biological fluids by luminol-hydrogen peroxide chemiluminescence. J. Food Comp. Anal. 24, 1038–1042.
Zhang, M., Cao, X., Li, H., Guan, F., Guo, J., Shen, F., Luo, Y., Sun, C., Zhang, L., 2012. Sensitive fluorescent detection of melamine in raw milk based on the inner filter effect of Au nanoparticles on the fluorescence of CdTe quantum dots. Food Chem. 135, 1894–1900.
Zhang, S., Yu, Z., Hu, N., Sun, Y., Suo, Y., You, J., 2014. Sensitive determination of melamine leached from tableware by reversed phase high-performance liquid chromatography using 10-methyl-acridone-2-sulfonyl chloride as a precolumn fluorescent labeling reagent. Food Control 39, 25–29.
Zhao, W.A., Brook, M.A., Li, Y.F., 2008. Design of gold nanoparticle-based colorimetric biosensing assays. ChemBioChem 9, 2363–2371.

Zhao, W., Fung, Y., Waisum, O., Cheung, M.P.L., 2010. L-Cysteine-capped CdTe quantum dots as a fluorescence probe for determination of cardiolipin. Anal. Sci. 26, 879–884.

Zhao, T.K., Liu, L.H., Li, G.M., Dang, A.L., Li, T.H., 2012. Electrochemical determination of melamine with a glassy carbon electrode coated with a multi-wall carbon nanotube/chitosan composite. Electrochem. Soc. 159, 141–145.

Zhao, W., Wang, Y., Li, J., Li, L., Wang, Q., Han, K., Zhang, Y., Li, X., Li, P., Luo, J., Wang, X., 2015. Determination of melamine in milk and dairy products by microchip-based high-field asymmetric ion mobility spectrometry combined with solid-phase extraction. Food Chem. 188, 489–495.

Zheng, X.L., Yu, B.S., Li, K.X., Dai, Y.N., 2012. Determination of melamine in dairy products by HILIC-UV with NH_2 column. Food Control 23, 245–250.

Zhou, Y., Li, C.Y., Li, Y.S., Ren, H.L., Lu, S.Y., Tian, X.L., Hao, Y.M., Zhang, Y.Y., Shen, Q.F., Liu, Z.S., Meng, X.M., Zhang, J.H., 2012. Monoclonal antibody based inhibition ELISA as a new tool for the analysis of melamine in milk and pet food samples. Food Chem. 135, 2681–2686.

Zhu, L., Gamez, G., Chen, H., Chingin, K., Zenobi, R., 2009a. Rapid detection of melamine in untreated milk and wheat gluten by ultrasound-assisted extractive electrospray ionization mass spectrometry (EESI-MS). Chem. Commun. 5, 559–561.

Zhu, X., Wang, S., Liu, Q., Xu, Q., Xu, S., Chen, H.J., 2009b. Determination of residues of cyromazine and its metabolite, melamine, in animal-derived food by gas chromatography–mass spectrometry with derivatization. J. Agric. Food. Chem. 57, 11075–11080.

Zhu, H., Zhang, S.X., Li, M.X., Shao, Y.H., Zhu, Z.W., 2010. Electrochemical sensor for melamine based on its copper complex. Chem. Commun. 46, 2259–2261.

Zhu, Y., Zhang, Y., Li, J., Han, Y., Dong, G., Zhang, H., Al, E.T., 2011. Determination of melamine in fresh milk by electrochemistry with solid phase microextraction at bismuthyl chloride modified graphite epoxy composite electrode. Am. J. Anal. Chem. 2, 612–618.

Zhu, L., Xu, G., Wei, F., Yang, J., Hu, Q., 2015. Determination of melamine in powdered milk by molecularly imprinted stir bar sorptive extraction coupled with HPLC. J. Colloid Interface Sci. 454, 8–13.

10

NANOMATERIAL-BASED SENSOR PLATFORMS FOR FACILE DETECTION OF FOOD CONTAMINANTS

Rükan Genç

Mersin University, Functional Nanomaterials Lab, Chemical Engineering Department, Yenişehir/Mersin, Turkey

1 Introduction

Food contaminated with pathogenic microorganisms and associated toxins is posing a major public health concern with thousands of outbreaks worldwide, especially in developing countries, which can cause acute and chronic diseases, and in the worst case scenario epidemics and death (Dwivedi and Jaykus, 2011). Among the diverse number of organisms contaminating food, the majority of which are nonpathogenic, foodborne bacteria such as *Campylobacter, Salmonella, Listeria monocytogenes, Escherichia coli* (*E. coli*) O157:H7, *Staphylococcus aureus,* and *Bacillus cereus,* a number of *protozoa (Cryptosporidium, Giardia, Toxoplasma)* and viruses are considered to pose major problems for the food industry (Food and Drug Administration, 2012; Scallan et al., 2011; Wang et al., 2012a). In order to prevent public health and improve the food safety, a range of screening methods including bio/chemosensors and assays have been developed. However, due to biological diversity of foodborne pathogenic agents, today most efforts are given to solve this challenge by improving the existing detection methods in order to increase the sensitivity and selectivity to the desired pathogen or toxin without sophisticated instrumentation, and to prevent possible epidemics while maintaining economic wellness, as well (Farakos and Frank, 2014; Gilmartin and O'Kennedy, 2012; Pérez-López and Merkoçi, 2011).

Under the influence of new technological achievements, in the past decade, nanomaterial-based strategies for the detection of contaminants has been gaining more interest in the scientific communities and food industry for improved food

Nanobiosensors. http://dx.doi.org/10.1016/B978-0-12-804301-1.00010-2

safety. Nanotechnology, with the unique properties of the matter in "nano" size, such as enhanced optical, catalytic, electrochemical, and magnetic features, have the potential to impact on several aspects of the food science and technology during different phases of the product development (eg, product safety and quality control, product processing and storage) by arousing the possibility to develop new techniques to improve the food safety.

Biosensors are analytical devices that use specific biochemical reactions mediated by isolated enzymes, immunosystems, tissues, organelles, or whole cells to detect chemical compounds usually by electrical, thermal or optical signals (IUPAC, 2005). Broadly, a biosensor platform consists of a biorecognition element, signal-producing moiety (an enzyme or dye), and a transducer. Nanomaterials can be used for single purpose or one or more nanostructures carrying multiple functions might be used at once. For example, nanoparticles can be used for either separation and concentration of living organisms and small molecules even at trace amounts from a complex matrix of the food, or as a catalytic mediator facilitating the reaction; moreover, they can be a functional building block of the sensor platform leading an effective immobilization and their electro-activity or optic features may amplify the biological signal and improve the signal detection from food-relevant pathogens and contaminants leading enhanced sensitivity and decreased limit of detection to assist food treatment utilities and policy makers and, most important, would help to protect consumers.

In this chapter, considering the appeal of nanomaterial based technologies with increased number of research with an annual average number of 500 review paper/year reached in 2014 (www.scopus.com, keyword "nanomaterial"), we limited our discussion to scientific achievements in the past five years and focused on the recent advances in nanomaterial-based biosensors for detection of foodborne contaminants. For more information on the altered properties of nanomaterials and their potential in detection systems, the reader is kindly directed to the recent comprehensive reviews and books covering theoretical aspects of the topic: Genc (2014), Jianrong et al. (2004), Merkoçi (2009), Mody et al. (2010), Pérez-López and Merkoçi (2011), Staleva et al. (2009), West and Halas (2003), Zhu et al. (2014).

2 Current Approaches for Pathogen Detection from Contaminated Food

Most of the foods and agricultural products are potential source of bacteria transferred by cross-contamination across the food supply chain. According to USA Centers for Disease Control and Prevention (CDC), "When two or more people get the same illness from the same contaminated food or drink, the event is called a foodborne disease outbreak". CDC reports that are summarized in Table 10.1 exposes the fact that in the past 5 years (2010–2015) more than 50 foodborne outbreaks were investigated by the FDA (US Food and Drug Administration). Another detailed statistic on outbreaks between 2007 and 2012 caused by low water activity foods was reported by Farakos and Frank (2014). One other review worth mentioning is on major pathogens and foodborne illness in the United States (Scallan et al., 2011) in which the authors give an statistical occurrence of food-based illnesses by three categories: food-associated (1) illness, (2) hospitalization, and (3) death. In most of these events, people exposed to the pathogen or its toxins had been hospitalized due to severe clinical syndromes such as hemolytic uremic syndrome (*E. coli* O157:H7), listeriosis and weakened immune system (*L. monocytogenes*), diarrhea, fever, and abdominal cramps [*Salmonella Paratyphi* B variant L(+) tartrate(+)], kidney failure and a few deaths were also reported (www.cdc.gov/outbreaks). However, the estimates of annual occurrences in only the United States suggest a much more scary scenario, in which each year the consumption of contaminated foods with 31 different pathogens causes 37.2 million events of illness, and from these 228,744 people are hospitalized and 2,612 people die (estimates from data gathered in 2006).

The main cause of foodborne outbreaks is improper handling of food or poor hygienic practices in food service and food retail operations; prevention of such outbreaks could be achieved by following the regulations of good agricultural and manufacturing practices (www.fda.gov). There are several foodborne pathogen test kits that are approved by independent organizations including AOAC, FDA-BAM, and AFNOR; however, standard methods approved by ISO for food-pathogen detection are colony counting on an agar plate and polymerase chain reaction (PCR) tests (The Food Safety and Inspection Service (FSIS), 2015). These methods are laborious and may take from hours to several days to establish a significant result; in addition, the achievable sensitivity may not be sufficient to detect pathogens with virulence as low as 10–100 CFU/mL (Food and Drug Administration, 2012; Gilmartin and O'Kennedy, 2012). Many attempts have been made to develop new technologies to overcome these limitations and provide faster or even real-time pathogen monitoring systems, such as small lab-in-factory devices that can be handled by unskilled workers with no or less sample pretreatment steps.

Table 10.1 Foodborne Disease Outbreaks in the United States Between 2010 and 2015

Microorganism	Product Implied	Infected/Death(s)	Year
E. coli 0157:H7	Cheese	38/0	2010
	Beef	21/0	
	Romaine lettuce	60/0	2011
	Lebanon bologna	14/0	
	Hazelnuts	8/0	
	Ready-to-eat salads	33/0	2013
	Ground beef	12/0	2014
	Ready-to-eat salads	33/0	2013
E. coli 0121	Raw clover sprouts	19/0	2014
	Frozen food products	35/0	2013–2014
E. coli 0145	Shredded romaine lettuce	33/0	2010
	Multistate outbreak: source not idendified	9/1	2012
E. coli 026	Raw clover sprouts	29/0	2012
	Farm cantaloupes	147/33	2011
	Ricotta salata cheese	22/4	2012
	Farmstead cheeses	6/1	2013
	Soy products Sprouts	5/2	2014
L. monocytogenes	Cheese	5/1	2014
	Dairy products	8/1	
	Blue Bell ice cream	10/3	2015
	Commercially produced, prepackaged caramel apples	35/7	
	Shell eggs	1919/0	2010
	Turkish pine nuts	43/0	2011
	Bean sprouts	115/0	2014
Salmonella Enteritidis	Raw, frozen, stuffed Chicken Entrees produced by Aspen Foods	3/0	2015
	Raw, frozen, stuffed chicken Entrees produced by Barber Foods	7/0	

Table 10.1 Foodborne Disease Outbreaks in the United States Between 2010 and 2015 *(cont.)*

Microorganism	Product Implied	Infected/Death(s)	Year
S. Paratyphi B variant L(+) tartrate(+)	Ground beef	12/0	2014
	Frozen raw tuna	62/0	2015
	Kosher broiled chicken livers	190/0	2011
Salmonella Heidelberg	Ground turkey	136/0	2011
	Tyson Chicken	9/0	2014
	Chicken	634/0	
	Microbiology Laboratories	109/1	2010
S. Typhimurium	Peanut butter	42/0	2012
	Ground beef	22/0	2013
Salmonella Montevideo, and *Salmonella Mbandaka*	Tahini sesame paste	16/1	2013
Salmonella Braenderup	Mangoes	127/0	2012
	Nut butter	6/0	2014
S. Braenderup	Mangoes	127/0	201
Salmonella I 4,[5],12:i:-	Alfalfa sprouts	140/0	2009–2010
Salmonella Newport	Alfalfa sprouts	44/0	2010
Salmonella Chester, 2010	Cheesy Chicken Rice frozen entrée	44/0	2010
S. Typhi (Typhoid Fever), 2010	Frozen Mamey fruit pulp	9/0	2010
Salmonella Hartford and *Salmonella* Baildon, 2010	Restaurant Chain A	75/0	2010
S. Montevideo	Red and Black Pepper/Italian-Style Meats	272/0	2010
Salmonella Agona	Whole, fresh, imported papayas	106/0	2011
Salmonella Hadar	Turkey burgers	12/0	2011
Salmonella Panama	Del Monte cantaloupe	20/0	2011
Salmonella Bareilly and *Salmonella* Nchanga	Raw, scraped, ground tuna product	425/0	2012
Salmonella Stanley	Raw cashew cheese	17/0	2013
Salmonella Bredeney	Peanut butter	42/0	2013

Modified from FDA and CMC data.

2.1 Nanomaterials for Analyte Separation and Concentration

One of the major obstructions faced during pathogen detection occurs due to the complex nature of most food matrixes. Recovery of a detectable number of microorganism requires time-consuming sample preparation steps including separation and concentration of microbial targets with filtration and centrifugation cycles; these methods require cell culture-based enrichment steps in order to regain the loss during separation. In order to achieve sensitive and specific detection methodology, efforts are focused on simple, fast, and broadly applicable multiple sample processing systems.

In 1995, Winoto-mMorbach and Muller-Ruchholtz reported a large-scale magnetic nanoparticle preparation method using a phase-separation technique in a high-pressure homogenizer. The prepared nanospheres were further successfully conducted to the in vitro cell separation process by labeling the target cells with monoclonal antibody-functionalized nanoparticles recovered simply by external magnetic field (Winoto-Morbach and Müller-Ruchholtz, 1995). This study inspired further researches on cell and macromolecule recovery from complex media such as saliva, blood, and urea for biomedical uses. Food technology is no different than when an increasing number of research was published in the past 20 years on functionalization of magnetic nanoparticles for their use in pretreatment of the food to enhance capture efficiency (Fig. 10.1). However, the trend in recent years is to focus on new strategies with combinational use of magnetic nanoparticles with other nanoparticles and surfaces or synthesis new composites with additional catalytic and optic properties to facilitate signal transduction while eliminating the laborious mid-steps.

Zhao et al. (2009) reported a method that could detect foodborne pathogens, *Salmonella typhimurium, Shigella flexneri,* and *E. coli* O157:H7, combining magnetic particles with optically enhanced ultra-small-sized quantum dots and they achieved the detection of 10^{-3} CFU/mL in 2 h. Basically, an immunoaffinity assay was built by sandwich-like complex formation between two functional nanoparticles: pathogen specific antibody modified γ-Fe_2O_3 magnetic nanoparticles were used for cell recovery in the presence of magnetic field, quantum dots (QDs) were conducted as fluorescence labels with different emission wavelengths, a strong alternative of fluorophore dyes with poor energy lifetime and high photo-bleaching. When magnetically separated pathogen faces to the secondary antibody conjugated inorganic

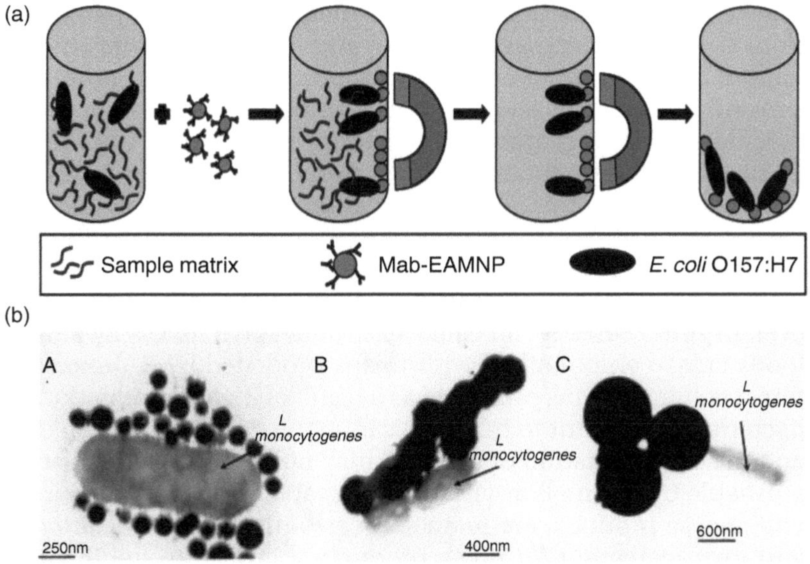

Figure 10.1. Schematic depiction of (a) immunomagnetic separation from food matrix and (b) transmission electron microscope image of *L. monocytogenes* separated by nanosized and submicron-sized magnetic particles (A) Reprinted from Cloutier et al. (2015); (B) reprinted from reference Shan et al. (2014), with permission from Elsevier.

nanoprob (quantum-dot), an immunoaffinity-based sandwich form, after second magnetic separation from unbound tag, presence of a specific pathogen can be measured by monitoring the fluorescence signal. This strategy not only facilitates the pathogen enrichment in the sample simultaneously and ease the inter-washing steps during labeling by use of magnetic nanoparticles, but also by simply applying ODs emitting the light at different wavelengths (620, 560, and 520 nm) as specific tag for different targets, one can easily conduct it to a multidetection system where ten or thousands of different foodborne pathogen specie can be screened in a food matrix at limit of detections reaching to single cell level.

In another study reported by Bhattacharya et al. (2011), antibody-labeled gold-magnetite nanocomposites were also implemented to an immune-affinity assay for *S. aureus* as model target. Authors designed a system that combines the magnetic futures of the magnetite with the surface-specific optic properties of nanosized gold nanoparticles by iterative deposition of gold nanoparticles on the previously decorated gold-seed stabilized magnetite nanoparticles. Through combination of these two important features of nanomaterials, authors attempt to achieve a dual role in one nanoparticle: first a separation module to recover pathogen from suspension by magnetic field, and second, an optically enhanced biorecognition element to capture *S. aureus* at ultra-low concentrations. They showed that *S. aureus* cells could immune-magnetically be captured at concentrations of 10^2–10^7 CFU/mL^{-1} in 30 mins, and by the designated method 100 CFU/mL can be detected by fluorometric and optical methods. Researchers also demonstrated the specificity of the assay by comparing the response from *E. coli* via confocal microscopy.

S. aureus is part of the common flora of healthy people, however, in some cases it can cause gastrointestinal illness by eating foods cross-contaminated with toxins produced by *S. aureus* by direct contact. Toxins produced as a result of high reproduction of bacteria are resistant to heat and cannot be destroyed by cooking and the contamination of the food may not be necessarily distinguishable by its smell or visually noticeable. Unpasteurized milk and cheese products are main sources of the bacteria *S. aureus*, and toxin detection for those products is critical to fullfill good food practice. Orlov et al. (2013) provided a methodology for detection of both *Staphylococcal enterotoxin* A (SEA) and toxic shock syndrome toxin (TSST) simultaneously in neat milk by using 3D-fiber structure located in micropipettes to filter the sample in situ and also to produce a high surface are for increased analyte recovery (Fig. 10.2). Magnetic nanoparticles of size below 50 nm were embedded through the microfibers of the membrane and used to capture and label the toxins. Signal detection was carried out analyzing nonlinear response of the superparamagnetic nanoparticles (SPIONs) in the presence of toxins at changing concentrations (Fig. 10.2a). Through developed technology, they reached detection limits of 0.1 and 4 ng/mL for TSST, 0.3 and 10 pg/mL for SEA, at both low (150 uL) and high volumes (30 mL), respectively (Orlov et al., 2013). This developed system, combining sandwich–type assay with immune-magnetic detection and filtering apparatus adaptable to continuous measurement in large sample volumes, is highly effective (Fig. 10.2b) and versatile, and can be adaped for real-time detection of various toxins at once from complex sample media without additional sample preparation, and can easily be applied to other industries.

Another example of the use of magnetic nanostructures in separation and detection is achieved by effectively combining magnetism with surface-enhanced Raman spectroscopy (SERS) to detect *E. coli* O157 in apple juice. Researchers synthesized immune-functional core-shell nanoclusters from magnetite covered with gold seeds to separate the pathogens from the media, and gold nanoparticles as SERS label by attaching them to Raman

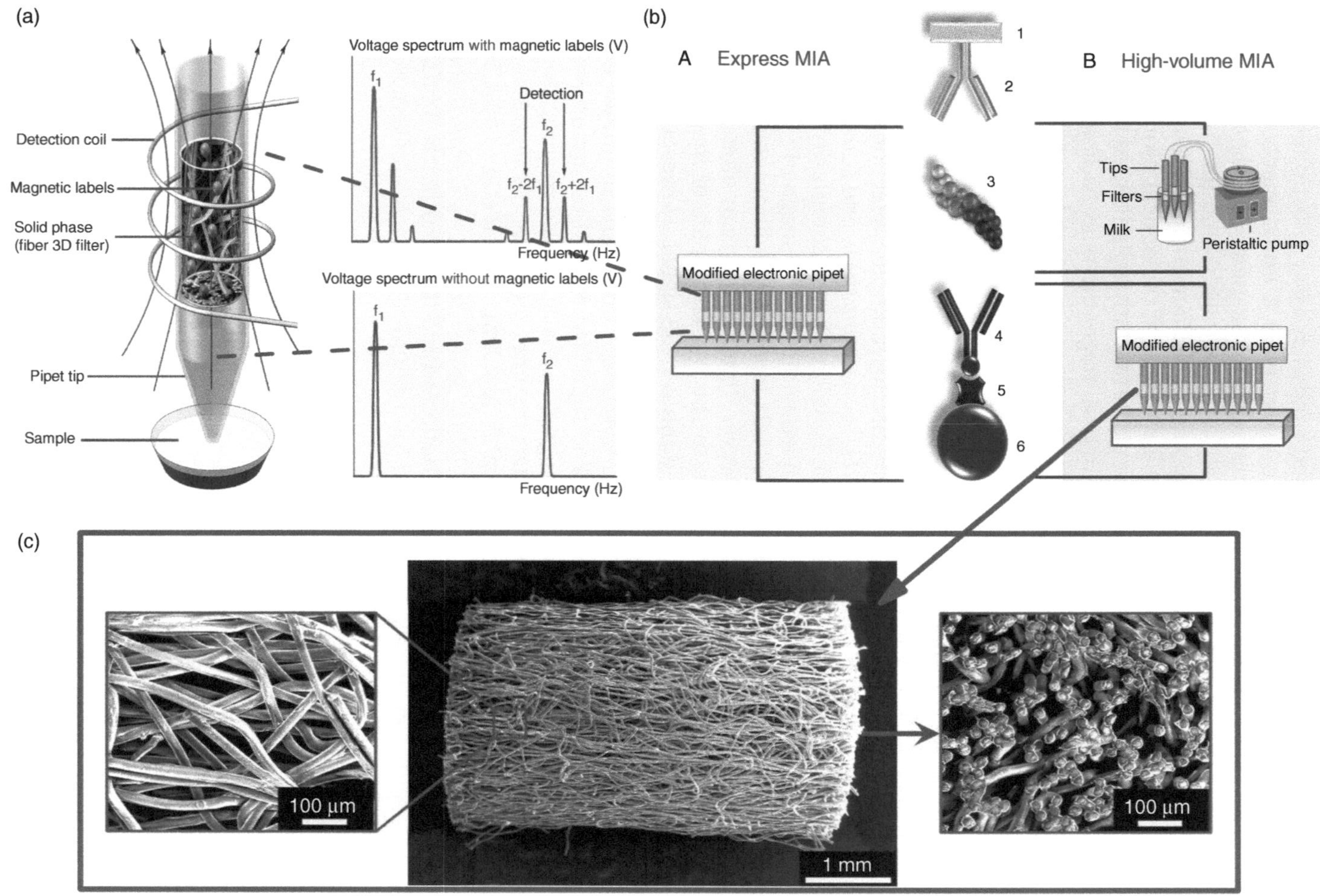

Figure 10.2. Schematic representation of (a) magnetic nanoparticles integrated in designed micropipette tip and the nonlinear-response-based detection; (b) designated 3D solid phase magnetic immunosandwich on (c) 3D fiber filters monitored with scanning electron microscope at different magnifications. Left to right: 500 × (lateral surface), 60 × (close view lateral surface), and 500 × (top surface fragment). Modified with permission from Orlov et al. (2013). Copyright (2015) American Chemical Society.

reporter molecules (RRM) and the detector antibody (dAb). Briefly, reported SERS-based sandwich assay constitutes three steps: (1) enrichment of the target from a liquid media, and further separation by magnetic field via magnetic particle concentrator; (2) sandwich complex formation with immune-functionalized gold-based raman tag; and (3) enhanced signal detection in raman spectra by taking advantage of the optical properties of nano-sized gold (Najafi et al., 2014). Although, formed sandwich between the Au-RRM-dAb and magnetite-gold-pathogen (recovery ranged 84–94%) results in limit of detection at concentrations as low as 10^2 CFU/mL. However, SERS technology has the potential to reach single molecule level, and that threshold can be caught by amendments focused on SERS reporter-nanoparticle couples.

In a recent study published by Cho et al. (2014a) magnetic nanoparticle-based immune-separation strategy was conducted to build fluorometric method for multiplex detection of pathogens including *E. coli* O157:H7, *S. typhimurium*, and *L. monocytogenes* from preinoculated foods with different complexity. Researchers performed a sandwich assay directly on magnetic beads by use of antibody—denatured BSA labelled with fluorophores for signal detection in situ. However, that strategy can often cause background problems due to the blackish color of magnetic beads that may interfere or diminish the signal. However, researchers cleverly solved the challenge by use of strategy consist of enzymatic cleavage of the fluorophore from particle surface by addition of proteinase *K*. The signal from free fluorophore was then read out following a magneto-separation and washing step to get rid of the unbound label and other interferences (Fig. 10.3) which resulted in a considerably low limit of detection (<5 CFU/mL) for the tested pathogens (*E. coli* O157:H7, *S. typhimurium*, and *L. monocytogenes*) in buffer, where it stayed below <5 CFU/mL even for preinoculated food samples; spinach, chicken, and milk in less than 2 h.

Sample pretreatment is critical for rapid screening of pathogens. Compared to the filtration and centrifugation-based separation techniques, sample enrichment using magnetic immunoseparation has more potential by eliminating laborsome steps to overcome the complexity of the food matrix and easiness of the separation procedure proceed simply by application of a magnetic field with a standard magnet that eliminates the use of specific equipments. Thus, implementation of immune/DNA-based magnetic separation techniques with other detection techniques, detection systems with less step and higher sensitivity can be achieved with enhanced the signal by enabling altered and specific sample recovery (Cho et al., 2014b).

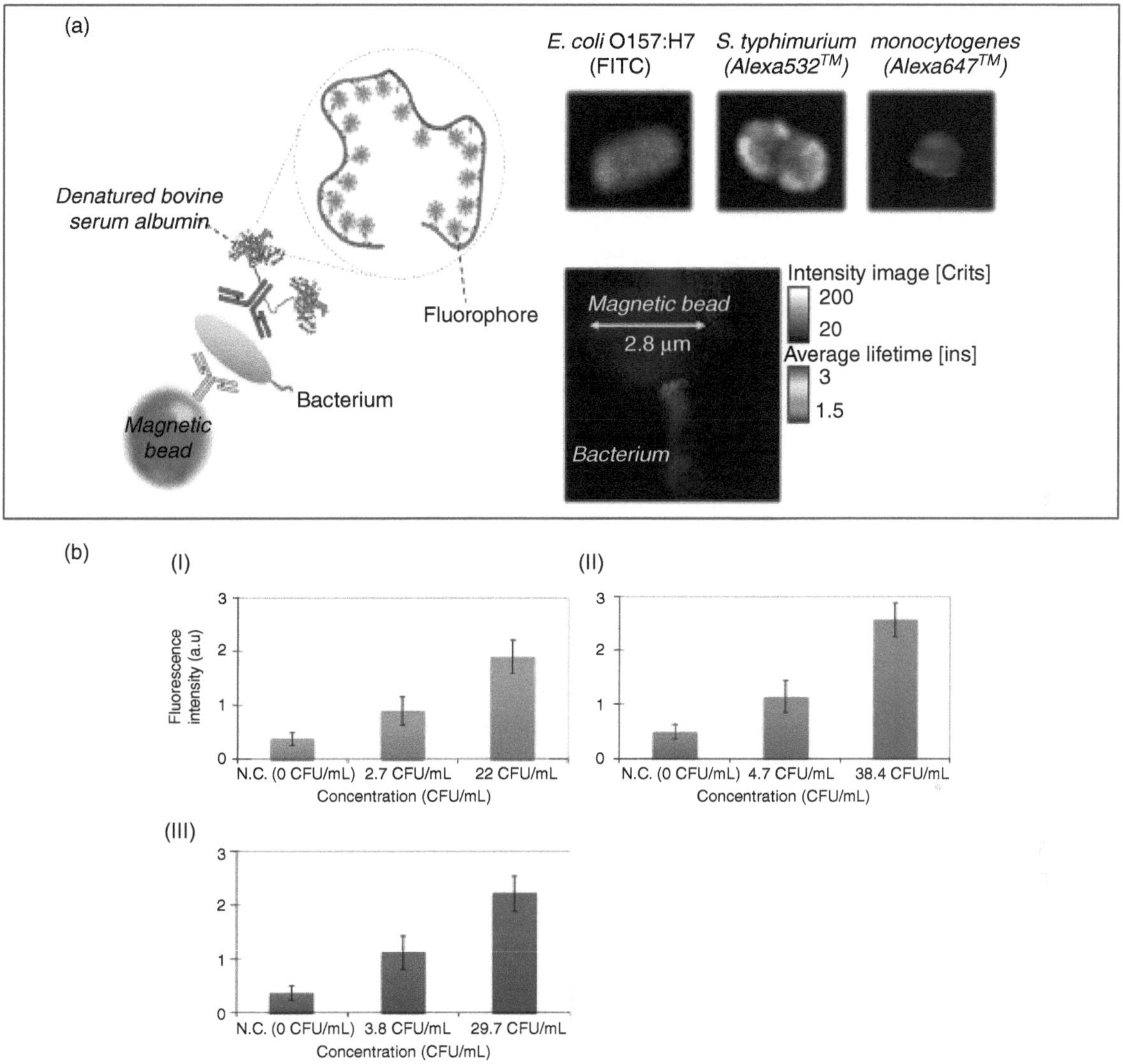

Figure 10.3. (a) Scheme depicting an in-situ fluorescent immunoassay proceed by affinity-based complex formation between pathogenic bacteria captured on immune-magnetic nanobeads and fluorophore-labeled antibody-dBSA conjugate (IFdBSA) (b) assay response for real sample analysis of (I) for *E. coli* O157:H7 (in spinach wash), (II) for *S. typhimurium* (in chicken meat), and (III) for *L. monocytogenes* (2% reduced milk). Reprinted from Cho et al. (2014a), with permission from Elsevier.

3 Nanomaterials as Recognition Element and Signal Enhancer

A widely reported class of biosensors is immunosensors which base on immunochemical reactions between an antigen (Ag) and an enzyme-labeled specific antibody (Ab) and the amount of target molecule (Ag) is quantified by measuring the enzymatic activity by electrochemically, optically or fluorometrically (Thévenot et al., 1999). The enzymatic activity is theoretically one enzyme for one target which is a determiner for the limits of the assay and often poses an obstacle in achieving distinguishable signal at low analyte concentrations. To overcome this, there are several approaches including the signal enhancement by multienzyme carriers (Genç et al., 2011) or colorimetrically-electrochemically enhanced inorganic nanomaterials which designed to be responsive in the presence of a specific analyte or mediator (Rong-Hwa et al., 2010; Zhang et al., 2015).

In this section, we aim to address recent developments on nanomaterial design for a final goal of amplification of the detectable signal (electrochemical, optical or fluorescence) from a single molecule or organism in comparison with single enzyme or dye-based analyte labeling strategies. Metal nanostructures with affiliated optical properties at the nanoscale, called surface plasmons, occur due to excitations of collective oscillations of charge densities make them valuable in design of molecular labels with analyte- switched optic properties (Bolduc and Masson, 2011; Kim et al., 2011; Staleva et al., 2009). There are plenty of inorganic and organic materials or their composites under study for this purpose; however, gold and silver, by far, attract considerable attention as they exploit surface plasmon-enhanced optical diffraction features (Fig. 10.4) which provide a quite unique opportunity for manipulation and transmission of light simply by changing their size, shape, chemical composition and environment (DeSantis et al., 2013; Ma et al., 2013; Saha et al., 2012; Tan et al., 2011). The geometry and size dependent properties of metallic nanoparticles have the capacity to direct the course of several fields including nano/biophotonics, imaging, data storage, bio/chemo sensing, and biomedicine (Genc, 2014).

Gold and silver nanoparticles, with shape, size, and surface tailorable properties, were two of the most studied metallic nanomaterials where their colloidal solutions show a wide scale of color change as response to the changes around their environment. By simply modifying their surface for a specific molecule, one can easily design a solution-based detection system. In such system, analyte adsorption to particle surface due to chemical

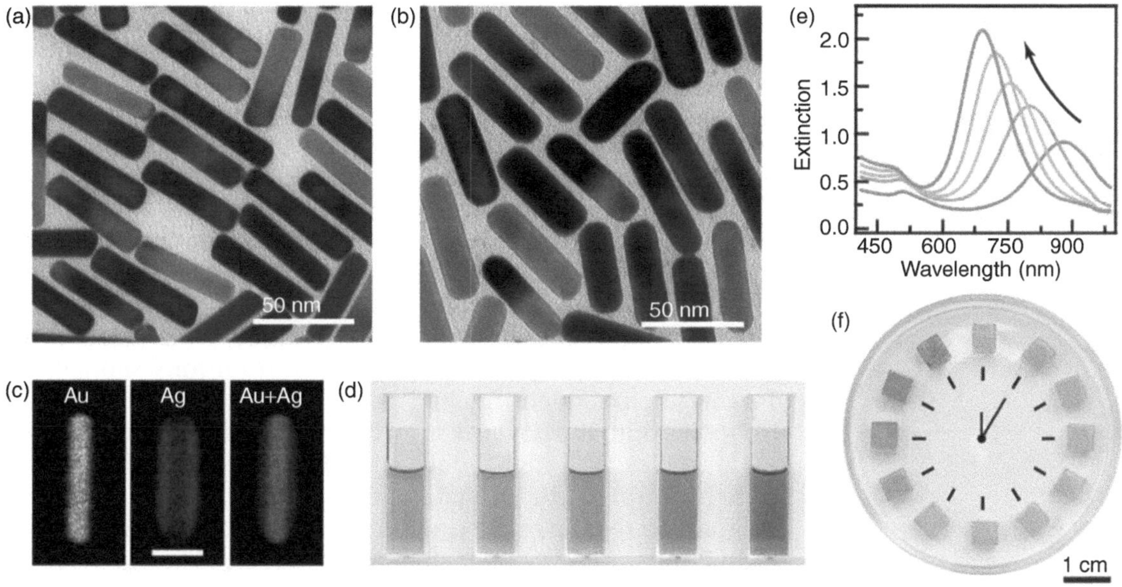

Figure 10.4. (a,b) Reaction condition dependent transformation of Au nanorods to Au/Ag nanorods depicted in transmission electron microscopy images of the Au and Au/Ag nanorods, (c) Energy-dispersive X-ray (EDX) mapping demonstrating the elemental composition of core/shell nanorods (scale bar = 20 nm), and (d) optical changes in color with (e) a continuous longitudinal plasmon band shift from 880 to 690 nm through the generation of Au/Ag nanorods, and (f) brilliant illustration of the relation between reaction time and the nanomaterials' changing plasmonic features by photographing hydrogel embedded nanostructures resembling a colourful clock. Adapted from Zhang et al. (2013) with permission. Copyright (2015) American Chemical Society.

attachments or bio-affinity results in a color shift from blue to red light band and a color change from red to purple (or blue) becomes visible to naked eye at relatively high analyte concentrations. This evident color change can easily be detected by simple spectrophotometric equipments or even with a modified smart-phone (Jiang et al., 2014). Proper design of metal-based colorimetric labels, presence of *E. coli* O157:H7 has been detected based on a catalytic oxidation reaction of peroxidase substrate by hydrogen peroxide using 4-mercaptophenylboronic acid-functioned Au@Pt nanoparticles which is adsorbed on the surface of *E. coli* O157:H7. In such procedure, Su et al. (2013) achived the detection of 7 colony forming unit of *E. coli* in 1 mL sample which they claimed that the color change is distinguishable with the naked eye.

Surface modification is the curial step for the design of such nanomaterials where a slightest change on the surface may result in particle aggregation which effects the color discrimination, and so the assays' sensitivity. Use of these nanoparticles for pathogen or toxin detection from such complex media like food makes the process even more difficult because the unspecific surface adsorption of other macromolecules (pathogens, enzymes, polymers, lipids ext.) could interfere with signal and may result in false positives. Thus, surface modifications providing highly specific recognition of the target or preseparation steps should be taken as underlined in earlier sections. For example, combination of the optic properties of gold nanoparticles with magnetism, Sung et al. reported that a recognition, separation, and signal detection could easily be achieved on a single nanoparticle: an antibody/gold nanoparticle/magnetic nanoparticle nanocomposite (antibody/AuNP/MNPs). To apply these particles for foodborne pathogen detection, *S. aureus* from milk sample, authors were embedded the gold nanoparticles and anti-*S. aureus* antibodies onto the bovine serum albumin (BSA) coated magnetic nanoparticles. Although, achieved LOD of 1.5×10^5 CFU is little bit high compared to the state of the art, reaching cell capture of 78% for *S. aureus* from the protein rich media of milk without additional separation steps or significant particle aggregation could be accepted as success for development of such colorimetric assay (Sung et al., 2013). Following an immune-magnetic separation step, (Cho and Irudayaraj, 2013) reached 3 CFU/mL of *E. coli* O157:H7 and 15. CFU/mL of *S. typhimurium* with an in situ ELISA system designed in a way that a multicomponent AuNP network was formed through layer by layer self-assembly of various AuNPs modified separately with primary, secondary, tertiary antibodies and finally a reporter enzyme. Although, brunch-like construction of gold nanoparticles and enzyme on the tips of this cluster have the possibility to collapse in, due to inter-molecular interactions, with amplified signal by means of the increased number of nanotags and enzymes, researchers achieved highly low limit of detections compared to the traditional ELISA.

Functional gold nanoparticles are reported as amplifier in electrochemical immunoassay for food pathogens which is essential for a sensitive measurement. As mentioned earlier, optical enhancement in metal nanoparticles are related to the size, shape and environment of the particle. As well as spherical nanoparticles, 1D gold nanorods (AuNR) also show interesting optical properties controlled by their length–diameter ratio and could be helpful for bacterial detection due to large surface area. Zhang et al. (2015) emphasized use of silica oxide (SiO_2)-AuNR modified with *E. coli*-specific antibody (dAb) and ferrocenecarboxylic acid (FCA) as signal-generating molecule to determine the quality of dairy products. A sandwich-based detection strategy was pursued on electrode surface by use of only 10 μL sample. Change in impedance due to the presence of FCA after washing of unbound label was read and correlated with the number of pathogen. Comparison studies with other methods reported, functionalized gold nanorod-based labels enables the detection of *E. coli* even at very low numbers, 60 CFU/mL, with a broad linear range between 1.0×10^2 to 5.0×10^4 CFU/mL. Following optimization studies, researchers achieved an enormous success in cell recovery from food samples including pure fresh milk, yogurt in shelf-life, and expired yogurt in the range of 95.1–106% with high selectivity among the control pathogens (Fig. 10.5) (Zhang et al., 2015). This study would have a great impact on further studies on the use of impedance immunosensors, where not only food pathogens but also related toxins can be detected. In a recent study parallel to

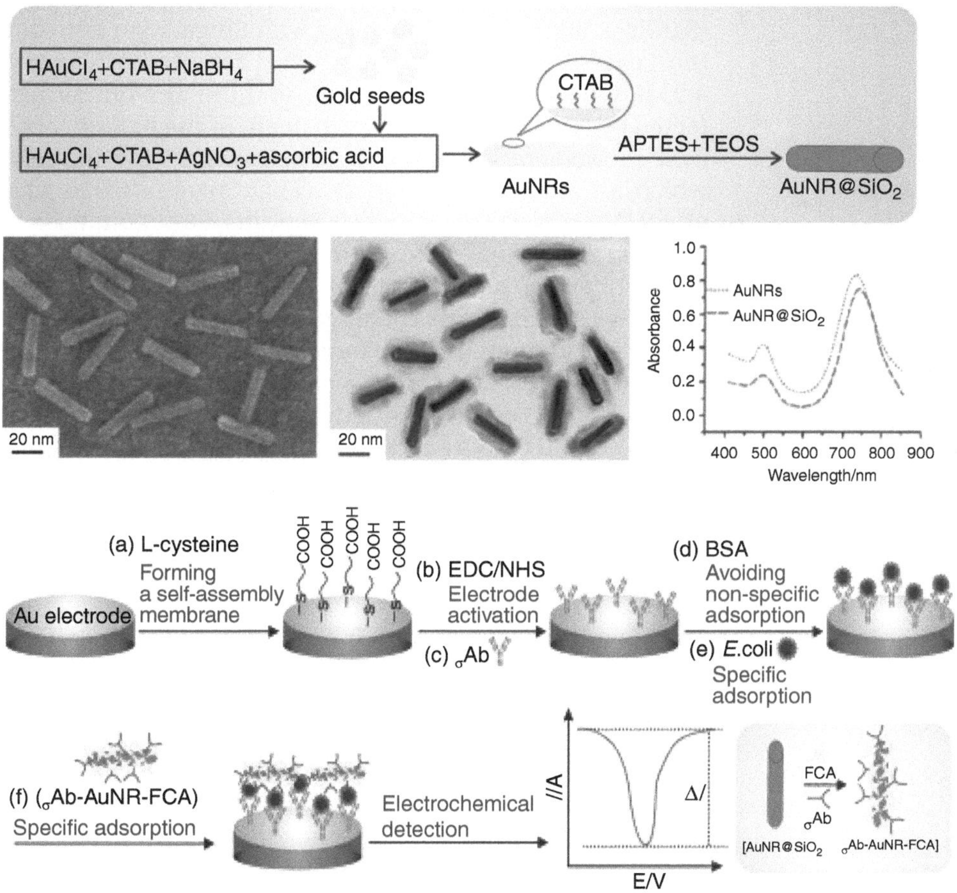

Figure 10.5. Synthesis of AuNRs and AuNR@SiO$_2$ nanorods with plasmonic properties and their use as signal enhancing tag in electrochemical immunosensor for detection of *E. coli*. Adapted from Zhang et al. (2015), with permission from Elsevier.

this, *Clostridium difficile* toxin A (TcdA) and toxin B (TcdB) were detected by conducting single domain antibody conjugated gold nanoparticles as signal enhancers. Increased concentration of toxin on electrode surface drives a relative electron transfer resistance, which had a proportional relation to analyte concentration in a range of 1–100 pg/mL. With this methodology, researchers reported a LOD of 0.61 and 0.60 pg/mL for TcdA and TcdB, respectively, which demonstrates a great potential for its future applications to the food industry (Zhu et al., 2015).

Oligonucleotides are of great interest since they can be designed with high specificity to a specific target. Shiga toxins are group of protein toxins secreted by *Shigella dysenteriae,* and some *E. coli* strains and digestion of the contaminated food with this toxin affects protein synthesis and may cause cell apoptosis as a result (Sandvig, 2001). Food poisoning with Shiga toxin at toxic levels in most cases leads to kidney failure and hemolytic uremic syndrome and if it gets to a level that lungs and the nervous system are affected by toxin, it often results in death of the patient. Stx 1 and stx 2 are the specific genes for Shiga toxin producing strains of *E. Coli* which are mainly classified as STEC; "Big Six": O26, O45, O103, O111, O121, and O145, plus O157 strains. Functionalization of gold nanoparticles (AuNPs) from thiol groups of single stranded oligonucleotides has been showed as a strong capture probe against to these two specific genes belongs to STEC strains. Hybridization of the oligo-fragment with the target leads a color change due to particle aggregation and can be evaluated optically by proportionate the shift observed in absorbed light to the amount of DNA (Quintela et al., 2015). Sun et al. (2015) attempted to design a surface enhanced fluorescence (SEF)–based DNA sensor for detection of the *E. coli* O157:H7 eaeA gene and they observed spacing distance controlled increment fluorescence intensity with the capability of analyt driven fluorescence turn "ON/OFF". Under optimal conditions, the DNA sensor produced a linear range from 10^{-17} to 10^{11} M with a correlation coefficient of 0.9947 and a detection limit of 3.33×10^{-18} M, and was also found to be specific in targeting eaeA. The DNA sensor demonstrated a new strategy of combining eaeA recognition and Au@Ag nanorods for fluorescence signal enhancement, and increased sensitivity in the detection of bacteria specific genes in trace amount from food samples.

Dark field imaging (DFI) is a distinguished technique which enables the visualization of unstained or transparent samples in both optic and electron microscopy by enhancing the contrast in such way that light illumination diminishes the dispersed light and only the scattered beams hit the sample. As a result, a bright image of specimen can be formed on a dark background without staining (Qian et al., 2010). Although, DFI is a simple and effective technique which have found wide use in microscopy, it requires high degrees of illumination which in some point could destroy or damage the biological samples. Therefore, combination of DFI with the enhanced optical properties of gold has been used to detect *E. coli* strains from water, milk, and fruit juice samples. For instance, using low-cost instrumentation like entry-level dark-field microscope setup, a the light source (tungsten lamp) and an automated image analysis algorithm for bacterial counting, a detection limit of 2–6 $\times$ 10^4 CFU/mL where higher than 70% cell recovery for targeted DH5α *E. coli* strain from a cocktail of control strains (*E. coli* BL21 and Rosetta) was achieved using antibody-conjugated gold nanoparticles of 90 nm as light scattering reporter in dark-field imaging (Xu et al., 2012).

Another interesting approach is application of inductively coupled plasma mass spectrometry (ICP-MS) to selectively detect enterohemorrhagic *E. coli* (EHEC) serotype O157:H7 which is one of the major causes of foodborne illness. ICP-MS is an analysis technique to determine trace amounts of metals and nonmetals from the liquid and solid materials by inductively heating gas to plasma state in which ions are produced, and separated ions are quantified with a mass spectrometer. Following an immuno-affinity–based separation of the strain of whim through gold nanoparticle (Au NP) labeling with a specific antibody; ICP-MS has been showed to reach a quite low LOD with 500 CFU/mL for the pathogen EHEC compared to the nonpathogenic controls, *E. coli* strain DH5α, ATCC35218, and ATCC25922. However, this procedure requires a high-tech instrument like ICP-MS, promising results gathered in 40 min with a degree of high specificity and low limit of detection that increases its potential for industrial use (Li et al., 2010).

Development of point of care detection systems are one of the immerged research fields which permits one-step and rapid analysis of several macromolecule and whole cell without expertise or expensive equipments (Choi et al., 2010). The lateral flow assay (LFA) is one of the cheapest and simplest formats of point of care detection where nanotechnology can be implemented easily to enhance the analyte signal. A basic LFA system is composed of series of capillary pads (absorption and conjugation), a membrane where capture molecule is embedded on, control and detection zones in which analyte complexion with the capturing probe can be monitored by use of a color tag. Instead of using enzymes and dyes as label, an optically enhanced nanoparticle (Fig. 10.6) can be more effective as molecular label since they are more stable by means of photo-bleaching and provides more binding side for target molecule (Choi et al., 2010; Rohrman et al., 2012).

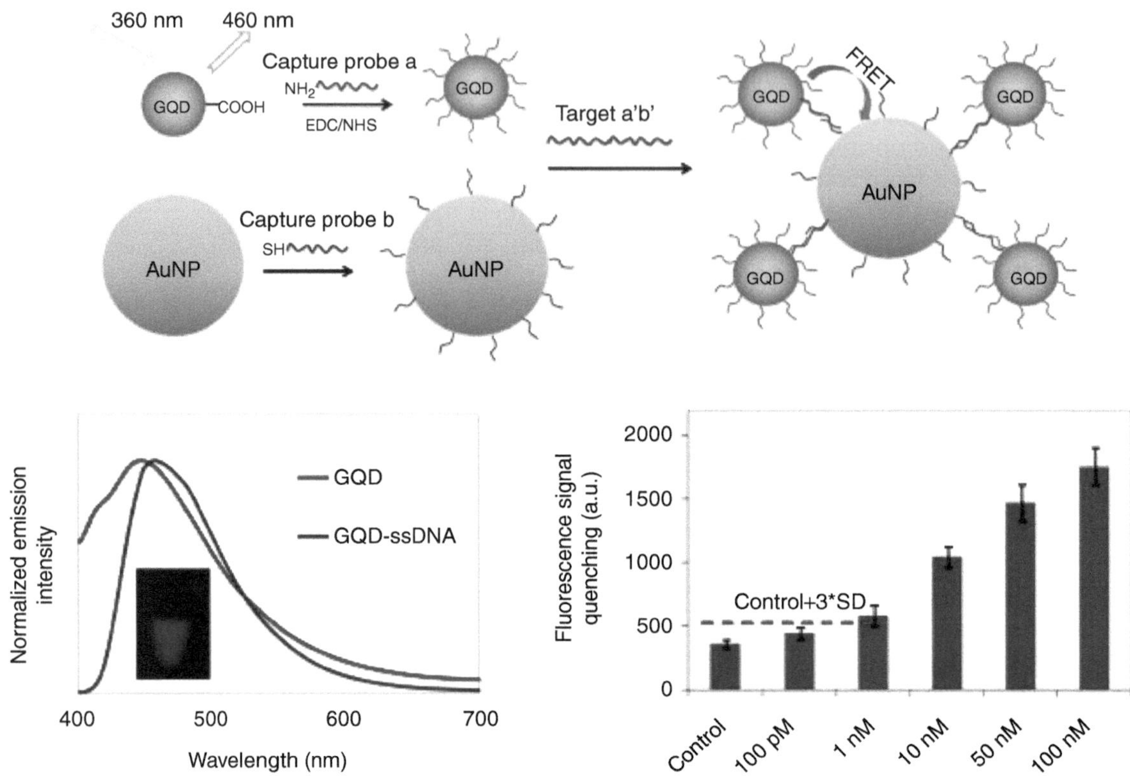

Figure 10.6. A lateral flow assay (LTF) introducing target modified gold nanoparticles as color tag. From Rohrman et al. (2012).

This simple design allows signal enhancement by use of gold nanoparticle (Au-NP)–based molecular label for colorimetric detection of any analyte with naked-eye distinguishable signal enhancement. A lateral flow assay (LFA) developed by Rong-Hwa et al. (2010) for detection of *S. enterotoxin* B (SEB) on a secondary antibody modified porous nitrocellulose membrane could be a good example to such system. In the presence of toxin, an affinity-based accumulation of toxin-loaded Au-NPs on membrane surface forms a red line as an indicator of the toxin which the reaction terminates in less than 5 min and the color change can be detected by naked eye even at toxin concentrations of nanomolar level with no sign of cross reaction. Researchers approved that additional use of silver nanoparticles also has positive effect with amplified signal to a range that toxin can be detected in as low as pg/mL level (Rong-Hwa et al., 2010). LFA has been also conducted to DNA-based detection system to detect *E. coli* O157:H7 through hybridization step allowing a sandwich-like complex formation between capture probe immobilized on detection zone, DNA fragment specific to enterohemorraghic *E. coli* and signaling probe labeled with gold nanoparticles (AuNPs). The signaling probe was designed by surface modification of gold nanoparticles with locked nucleic acid (LNA) nucleotides which are known with their advantage of increased hybridization specificity and a high-affinity targeting of complementary RNA and DNA (Vester and Wengel, 2004). Application of LNA as probe instead of a complementary DNA fragment has been demonstrated to be effectively increment signal intensity by three fold and allows the detection of the target in 0.4 nm range through a portable strip reader with the possibility to color discrimination by naked eye at target concentrations above 3 nm (Rastogi et al., 2012).

Adapting refurbished conventional glucose electrode strips to pathogen detection, Davis et al. (2013) convey a cost-efficient biosensor to detect *L. monocytogenes* in food samples. To do so, first gold nanoparticles were immobilized on screen-printed carbon electrode (SPCE) strips and then surface was modified with *L. monocytogenes* specific monoclonal antibodies. When enzyme-labeled secondary antibodies covered on AuNPs were used as signal element, 2 log CFU/g pathogen was effectively determined in wild blueberry samples. Further, researchers have tested the specificity of the sensor with two other pathogens (*E. coli* O157:H7 and *S. Typhimurium*) and SPCE biosensor from reused electrodes showed high sensitivity for *L. monocytogenes* over *E. coli* O157:H7 and *S. Typhimurium*. This study is a good demonstrator of the versatility of metallic nanoparticles by means of adoptability to different sensor platforms with the requirement of such small modifications.

In addition to colorimetric nanotags made up of gold and silver, super-tiny-sized quantum dots (QDots) below 10 nm with long fluorescence life time and high stability over traditional fluorescence dyes are in fact becoming the arising star of the new generation detection systems. Although synthesis methodologies for those inorganic nanocrystals are relatively laborious and require the use of toxic chemicals, quantum dots provide a bouquet of colors with size- and surface-chemistry-dependent fluorescence emission at a quite broad range, in addition to resulting optical prosperity which paves the way for new possibilities for multi-detection systems (Resch-genger et al., 2008; Gun'ko et al. 2010). For example, simple fluorometric strategies using quantum dots (QDs) as fluorescent probes are reported widely for immunodetection of whole cell pathogens from food sources such as indirect immunofluorescence detection of *Bacillus anthracis* (*B. anthracis*) (Liu et al. 2011), fluorescent carboxymethyl chitosan/CdS quantum dots (CMCS–CdS QDs) composites for specific detection of *S. aureus* (Wang et al., 2011), and immunological bead-free cell detection method using quantum dots (QDs) as reporter to detect *E. coli* O157:H7 and *Salmonella* in ground beef after immunomagnetic separation with magnetic nanoparticles (Wang et al., 2012b). By following these cost and time efficient strategies, new approaches were developed for rapid monitoring of pathogens and toxins from more complex samples.

One of the major physical phenomena which focus the attentions to QDots is the fluorescence resonance energy transfer (FRET). FRET is a physical response to energy transfer between a donor molecule and another nearby molecule acting as acceptor. When two chromospheres come close to a certain molecular distance below 10 nm, excited light absorbed by the donor triggers the transfer of excitation energy from the donor to the acceptor via nonradiative dipole–dipole coupling (Saini et al., 2006; Wegner et al., 2013). The so-called energy transfer results in decreased donor fluorescence and an increase in acceptor fluorescence intensity due to reduction of excited state lifetime (Wegner et al., 2013). This distinguishes property provided by FRET responsive molecules and materials can be used in various areas to probe intramolecular interactions between macromolecules and their interactions, in vitro/in vivo monitoring, nucleic acid analysis, signal transduction in chemo/bio-sensors and light harvesting in solar cells etc. At most of the sensing strategies, FRET-based detection systems mainly consist of a capturing moiety (eg, antibody, DNA, RNA fragments and receptors) and signal element, however, their use for whole cell detection is limited due to the molecular distance driven signal transduction. Thus, instead of whole cell pathogen probing, detecting specific molecules, such as antigens, gene, and protein biomarkers, is more suitable for this method. Graphene-based quantum dots are quite versatile materials that can act as donor, where their combination with AuNPs in close proximity can provide FRET effect as a result. Shi et al. (2015) showed *S. aureus* specific gene sequence detection by design of a FRET biosensor platform as shown in Fig. 10.6. They performed a classic sandwich assay in two steps: first, cohybridization of the target oligos with the capture oligos localized on GQDs; and second, reporter probe modified AuNPs were administrated. Hybridization of the reporter probe with the target captured on the quantum dot surface provided a bridge between donor and the acceptor leading a fluorescence quenching proportional to the target concentration. Authors claimed a limit of detection (LOD) reached to 1 nm for *S.aureus* gene and have shown that the assay has a good sequence selectivity for both single-base mismatched oligos and double-base mismatched oligos. Using two type of donor, green-emitting quantum-dots (gQDs) and red-emitting quantum-dots (rQDs), and amorphous carbon nanoparticles (CNPs) as acceptor, aptamer-based dual detection of *Vibrio parahaemolyticus* and *S. typhimurium* has been provided with such a strategy that one QD was modified *V. parahaemolyticus* recognizing aptamer while the other with *S. typhimurium* recognizing aptamer. Fluoresence de-quencing (CNPs behave as quencher of QDs) due to analyte presence results in a linear correlation with number of pathogens and signal increment ranging in a wide scale from 50 to 10^6 CFU/mL and low LOD of 25 and 35 CFU/mL was observed for *V. parahaemolyticus* and *S. typhimurium*, respectively. Method exhibited quite high selectivity to target pathogens over other bacteria strains (*S. dysenteriae, L. monocytogenes, S. aureus,* and *E. coli*). Moreover, real sample analysis with chicken and shrimp has resulted in a good agreement with plate-counting method at changing pathogen concentrations (Duan et al., 2014).

CdSe/ZnS core/shell quantum dots are one of the most used types of inorganic tags where they found a great amount of use in biosensors. A biosensor system for detection of pathogens has been developed by using dendron-shaped nanocrystals for detection of two types of model pathogen: a bacteria (*E. coli* O157:H7) and a virus (*hepatitis B*) on a microporous membrane modified with the capture antibody. Following a preincubation of the quantum dot-based nanotag with the capture antibodies, a sandwich assay has been performed through the porous membrane and target concentration was measured with the photoluminescence formed as a result of the formed complex of capture probe-Ag-reporter probe. Optimizing several parameters including membrane pore size, operation time and pH, detectable level of 2.3 CFU/mL for *E. coli* O157:H7 and 5 ng/mL for hepatitis B surface Antigen (HBsAg) has been reached in 30 min (Liu et al., 2007). In this same context, Jackeray et al. (2011) also conducted CdSe/ZnS quantum dots (QDs) modified with *Salmonella typhi* (*S. typhi*) specific immunoglobulin G (IgG) and immobilized them on polycarbonate (PC) membrane to determine water contamination with bacteria via on site monitoring with fluorescent micrographs. With designed system, a detection sensitivity of 100 cell/mL was obtained which is 100-fold higher compared to the detectable limits achieved by conventional dye (FITC)-based assay (Jackeray et al. 2011).

As mentioned before, size and material chemistry related fluorescence properties of quantum dots can be integrated into multidetection systems. Zhao et al. (2009) combined sample enrichment strategy using magnetic materials with semiconductor-material-based quantum dots possessing fluorescence intensity at different emission wavelengths (620, 560, and 520 nm) in order to detect foodborne pathogens, *S. typhimurium*, *S. flexneri*, and *E. coli* O157:H7, simultaneously. Silica-coated γ-Fe_2O_3 magnetic nanoparticles were modified with antibodies by glutaraldehyde treatment and used for immunoseparation of bacteria from food sample. QDs with different emission wavelengths (620, 560, and 520 nm) were functionalized separately with targeting antibodies to establish specific tags for each bacteria. Pretreated bacteria samples were incubated with the tags, and each complex was visualized using fluorescence microscope and demonstrated specific attachment of antibody-conjugated QDs on bacterial cell wall. Multimodal foodborne pathogen bacteria detection with the reported assay using fluorescence measurements at changing pathogen concentrations, simultaneously, resulted in a detection limit of 10^{-3} CFU/mL in a food matrix (Zhao et al., 2009).

Not only by use of antigens and genetic material as capturing moiety in an assay, but strategies targeting the cell specific ligands and reactions can also be used to recognized pathogens; for example, a quantum dot-ferrichrome bioprobe has been designed to specifically detect *Pseudomonas fluorescens* (*P. fluorescens*), taking into account the fact that ferrichrome is strong chelator of iron (III) and the resulting complex results in the formation of cell clusters. Once the cell clusters formed due to specific recognition of the chelate by cell membrane receptors, cells were easily identified by fluorescence microscope and changes in fluorescence intensity and spectrophotometric data were conducted to proportionate the signal to the cell number. However, while the intelligent use of quantum dot-ferrichrome bioprobe provided a single-cell range detection of a specific pathogen, reported assay lacks the applicability to detect other pathogens until specific siderophores exist for the pathogen of interest (Wu et al., 2009). In another example, colistin, which is a polypeptide antibiotic effective on most of the Gram-negative bacteria by integrating to the cell wall (Soon et al., 2012) has been used to detect *E. coli* as a model for Gram-negative bacteria. By combining colistin with quantum dot (CdSe/ZnS), Carrillo-Carrión et al. (2011) intended to achieve an intensely fluorescent nanotag (Colis–QDs) with the ability to adhere to bacterial cell wall. The presence of colistin molecules on QDs eases the accumulation of Colis–QDs on bacterial cell wall due to the amphiphilic character and specific affinity of the molecule to the cell wall lipids. Fluorescence intensity-based single-step detection of the *E. coli* using the assay provided a detection limit of 28 cell/mL for *E. coli* in an extremely short analysis time (<15 min).

Nanomaterial-based detection strategies are not limited to metalic nanoparticles, several other organic and inorganic materials and their combinations have been synthesized for signal enhancement applications and modified with diverse number of capture molecule to achieve altered recognition capacity. A recent example is the use of lectin concanavalin A (Con A) to enable cell adhesion of *E. coli* O157:H7 to the mannose-conjugated magnetic nanoparticles for cell recovery and enrichment. Separated cells have been introduced to silica nanoparticles doped with two fluorescent dyes (FSiNPs) that are designed for two-color flow cytometry (FCM) assay. By optimizing the system for eliminating false negatives and background signal, assay provided *E. coli* O157:H7 detection at levels of 7 cells/mL and 6 cells/mL from PB buffer and precontaminated water, respectively, in less than 4 h, including the enrichment step. Fluorescent dye embedded to silica nanoparticles has amplified the FCM signal due to high concentration dye absorption to a single nanoparticle, which increases the number of fluorescent dye per cell (He et al., 2011a). Another approach introducing signal probe functionalized mesoporous silica nanoparticles (MSN) in chemiluminescence immunoassay has been developed to determine *S. enterotoxin* B (SEB) in milk, water, and serum samples. Horseradish peroxidase enzyme (HRP) was used as a chemically active colorometric label, and chemically attached to nanoparticle surface together with anti-SEB antibody. Chen et al. (2012) performed the assay simultaneously with two monoclonal antibodies that can be recognized by different cellular epitopes of a single bacteria and get a linear response within the range of 0.01–1.00 ng/mL ($r = 0.9967$), and the detection limit (DL) of 4 pg/mL (3σ). This approach was successfully demonstrated as a simple, cost-effective, sensitive method to detect SEB in complex media like milk, water, and serum.

4 Nanomaterials-Based Signal Transducers and Sensor Platforms

Nanomaterial-enriched sensor platforms with enhanced conductivity and reactivity have been shown to alter the signal transduction, selectivity, and stability compared to the bare electrodes of sensors. Being superconductive, carbon and graphene-based nanostructures are leading materials for electrode modification, where they provide huge surface area with several active centers for capture probe, and besides, they form a conductive channel to canalize the electron transfer through the electrode. Viswanathan et al. (2009) conducted these intrinsic advantages by immobilizing an enzyme [acetylcholinesterase enzyme (AChE)] on vertically assembled single-wall carbon nanotubes, SWCNTs, which chemically localized onto the gold electrode for insecticide detection from vegetable crops (Fig. 10.7). Further modification of the functionalized electron surface with capture oligo pioneered a ligand specific enzyme inhibition pathway that increased the sensitivity and amplified signal. Using the aforementioned strategy, authors obtained a limit of detection as low as 1×10^{-3} nm for both food contaminants: methyl parathion and chlorpyrifos. Another example worth reporting is on glass lab-on-a-chip microfluidic sensor system performed by enzymatic inhibition of acetylcholinesterase. In the designated sensor, enzymes were deposited on magnetic particles, which were used to introduce a controlled amount of enzyme through the microchannels and helped to tailor surface immobilization of enzyme to the capture region by an externally controlled magnetic field. Under optimized conditions, Authors achieved detection of carbofuran, a toxic carbamate pesticide, at nanomolar level (Llopis et al., 2009).

One-dimensional semiconductive nanowires and nanotubes are powerful tools for sensor and photovoltaic studies due to their highly interesting shape-altered properties (eg, ultra-sensitivity, surface area, and crystal structure–enhanced electrochemical activity, ease of surface modification) in which titanium oxide is known to be the most abundant one with unique properties, such as electron mobility, optical band gap, and surface reactivity. Introducing these properties to develop an immunosensor for *L. monocytogenes*, a rapid detection system has been explored by mounting hydrothermally synthesized TiO_2 nanowire bundles on gold microelectrodes with mask welding (Wang et al., 2008). Once *L. monocytogenes* were captured by monoclonal antibodies immobilized onto TiO_2 nanowire bundles, an impedance jump occurs. Correlating the difference in impedance with the number of bacteria, comparison studies with standard immunoassay Dot Blot analysis (LOD = 2.2×10^5 CFU/mL) have revealed that the presence of titania nanowires increased the sensitivity of sensor, and by developed method researchers show the possibility to detect 1.0×10^2 CFU *L. monocytogenes* per mL without interference of other pathogens or molecules. In a good agreement with the aforementioned studies, Yang et al. (2009) developed a reusable electrochemical impedance spectroscopy-based sensor platform for the detection of *Salmonella* spp. by controlled deposition of gold nanoparticles on glassy carbon electrode surface with the help of grafted ethylene diamine. Authors reported that with the designed sensor a detection limit of 1.0×10^2 CFU/mL from highly complex food matrixes like preprocessed commercial pork can easily be achieved.

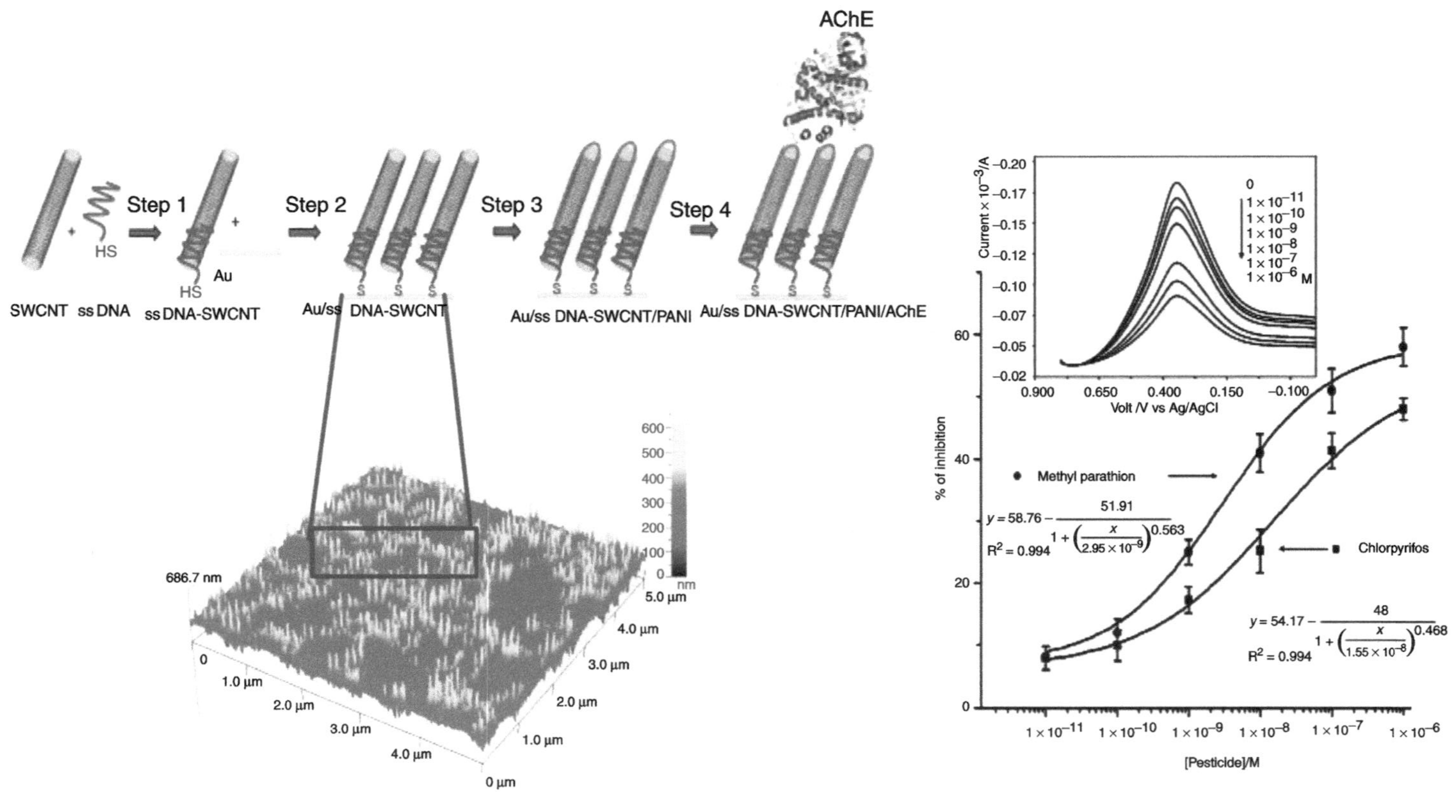

Figure 10.7. ***S. aureus* detection strategy using FRET-based genosensor where target hybridization with the capture oligo forms a close proximity, resulting in energy transfer from QDs to AuNPs and affects the QDs emission intensity which is proportional to the target concentration.** Reprinted from reference Shi et al. (2015), with permission from Elsevier.

One of the challenges of electrode surface modification with charged and complex structures like DNA fragments is to find strategies to overcome uncontrollable accumulation of capture probe to the surface which results in tight packing of molecules or activity lose, heterogenity through the surface due to uncontrolled cluster formation which results in poor reproducibility between the measurements, and possibility to insulating layer formation which may interfere with the signal. Aiming to overcome these challenges, Wang et al. (2009) provided an intelligent solution to the problem in which not the surface of the material but, as represented in Fig. 10.8, the nanopores (73 ± 7 nm diameter) of the aluminum anodized oxide membranes were used to immobilize capture probe oligo against *E. coli* O157:H7. A novel chemical functionalization procedure (polymerase-extending (PE) method) to immobilize single strand probe to the inner cavities of the nanopores without mentioned limitations has been introduced. Hybridization occurred by principle of complementary base pairing in the presence of Taq DNA polymerase and deoxyribonucleotide triphosphates (dNTPs) at optimal reaction conditions (Fig. 10.8) inside the nanopores. With this clever strategy, researchers eliminated limitations of traditional biofunctionalization methods (Wang et al., 2009). Basically, detection method consists of the extension of the probe strand, first with PE method and real-time hybridization of the probe with target which results in an immediate change on electrode surface. This signal difference due to hybridization is measured by cyclic voltammeter (CV) and electrochemical impedance spectroscopy (EIS). Researchers have performed series of optimization studies with this system and at optimum assay conditions; they achieved sensitivity down to 0.5 nm target DNA.

Marques et al. (2009) introduced another interesting approach for DNA immobilization on an electrode surface by embedding gold nanoparticles in a lab-made graphite-epoxy composite that they call "nano-AuGEC" As summarized in Fig. 10.9, the main idea is to obtain controlled assemblies of DNA-based capturing probes on an electrode surface tailored by the number of active binding sides offered by gold nanocomposites and providing enough spacing for DNA hybridization. They for the first time used PCR for double-tagging of DNA: one tag, —SH, is for capture probe (ssDNA), assembly on gold NPs through S = Au bonds, and the other (—DIG) is to label counter DNA fragment, a synthetic DNA of *Salmonella* sp. as a model to food pathogens, for enzyme labeling after hybridization. Group succeeded amperometric detection of synthetic DNA at femtomole levels and provide surface controlled immobilization of targeting probe with amplified transduction.

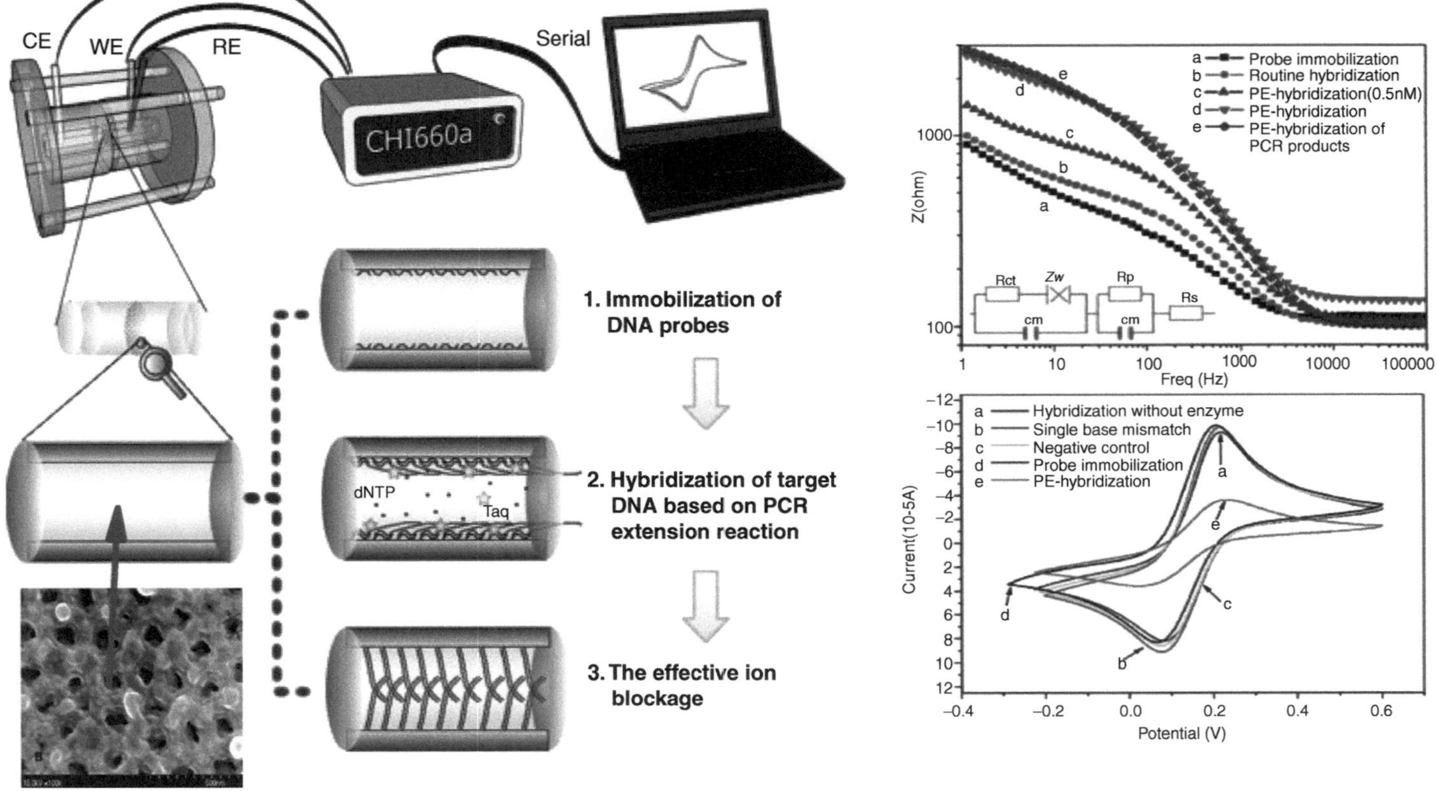

Figure 10.8. Schematic representation of a nanotube-based enzyme inhibition assay for detection of two different insecticides on vegetable crops and corresponding AFM image of vertically aligned single wall carbon nanotubes (SWCNTs) on gold electrode as conductive transduction platform.
Adapted from Viswanathan et al. (2009), with permission from Elsevier.

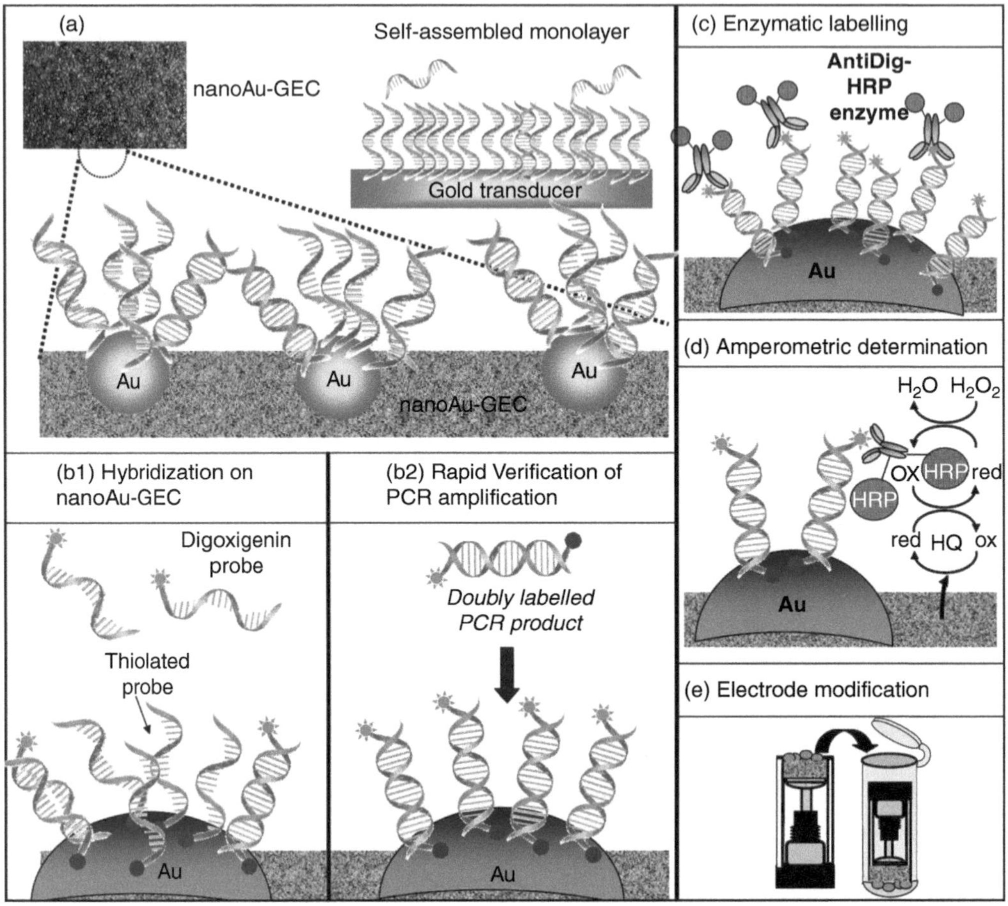

Figure 10.9. Representation of a novel in situ polymerase-extending hybridization method explored on the inner nanopores of the aluminum anodized oxide membranes and electrode design novel. Electrochemical detection of *E. coli* O157:H7 specific target DNA can be realized by implementing the designated electrode to both cyclic voltammetry and impedimetric analysis. Abbreviations: WE: working electrode; RE: reference electrode; CE: counter electrode. Adapted from Wang et al. (2009), with permission from Elsevier.

High-quality surface-enhanced Raman scattering (SERS) spectra is another detection technique in which signal enhancement over one million–sfold can be reached by two principle mechanisms: an electromagnetic enhancement and a chemical charge transfer effect, in which electromagnetic enhancement has higher dominance. Under incident laser light, the presence of analyte in close proximity to the SERS-active metal surface effects the excitation of surface plasmons; thus, the performance is directly related to the surface properties, such as nature and characteristics of the metal surfaces and roughness (Dao et al., 2015). Nanotechnology with the capability to produce surfaces with controlled size, shape, inter-particle distance, edges, and roughness, compared with the thousands of different combinations of inorganic and organic compounds, raises unlimited possibilities to obtain the best SERS platform. Among common metals employed in SERS (gold, silver, and copper), silver have been reported as the most effective signal enhancer and applied to wide range of research fields including biomedical, pharmaceutical, environmental, and food analysis. Wu et al. (2012), reported SERS-based detection of aflatoxins by fabricating arrays of silver nanorods (AgNR) via oblique angle deposition (OAD), which have been shown to enhance SERS signal more than 10^8-fold with low batch-to-batch difference. The concentration-dependent SERS detection was provided by peak analysis and compared with the Raman spectra calculated by density function theory (DFT). Studies revealed detection of toxins in femtomolar range. One other report worthy of mention is the study conducted by He et al., 2011b by taking advantage of one-dimensional nanostructures. Researchers prepared a SERS platform by growing one-directional zinc oxide nano-necklace (ZnO NN) arrays on r-plane sapphires using chemical vapor deposition. Designed SERS-active substrate exhibited 10^4-fold signal amplification and LOD of 0,1 μM for melamine as model food contaminant when the surface was covered with 45 nm gold film. On the other hand, using a glycopeptidic antibiotic, vancomycin, modified silver nanorods (AgNR) as SERS platform, six different species of bacterial pathogen from mungbean sprouts with LOD of 10^2 CFU/mL have been detected selectively using both bench-top and handheld Raman spectrometer (Wu et al., 2013).

Acetamiprid is an odorless insecticide that is common contaminant of agricultural products. In order to prevent end-users from acetamiprid exposure, Fei et al. (2015) developed a nanomaterial-based label-free electrochemical impedimetric sensor by following the earlier studies using nanoparticles, such as Au NPs (Li et al. 2013), carbon nanotubes (Zelada-Guillén et al., 2009), and graphene-based nanomaterials containing graphene oxide (GOx)/ C_3N_4 nanocomposites (Li et al., 2014), Au NPs decorated graphene sheets (Hu et al. 2011a), and 3,4,9,10-perylene tetracarboxylic acid functionalized graphene sheets (Hu et al. 2011b). In such a system, acetamiprid specific aptamer was used as capture probe and immobilized on different nanomaterials to evaluate their capacity by means of level of amplified electrochemical detection signals by increasing the immobilized number of aptamer. Comparison studies taking account of the electron transfer resistance profiles of three different electrode surfaces (pure Au NPs, multiwalled carbon nanotube-reduced GOx nanoribbon (MWCNT-rGONR), and gold-immersed multiwalled carbon nanotube-reduced GOx nanoribbon composites (Au/MWCNT-rGONR) initiated by formation of aptamer-nanostructure-target complex revealed that Au/MWCNT-rGONR was the most sensitive signal aptasensing platform with femtomol range limit of detection for acetamiprid (Fei et al., 2015).

A lateral flow system–based direct-charge transfer (DCT) biosensor has been fabricated on a conductive polymer nanowire modified with an antibody against the food pathogen (*B. cereus*) acting as both sensing platform and signal transducer. The so designed biosensor was constructed by polyaniline nanowire–based membrane (the molecular electrical transducer), cellulose membrane (sample application and absorption pads), fiberglass membrane grade G6 (tconjugate pad), and nitrocellulose membrane (capture pad) assembled on an etched copper platform. In the presence of the target, a switch ON/OFF system proceeds and an electrical circuit is thus formed. Direct electron charge flow generates a resistance signal, which is proportionate to the concentration of the target. This highly specific and sensitive detection methodology with LOD of 10 CFU *B. cereus*/mL performed in as fast as 6 min, could be quite inspirational for future attempts to create simpler and cheaper detection devices (Pal et al., 2007).

Another nanomaterial-based sensor platform worth mention is the platform developed by Li et al. (2015) for detection of *E. coli* O157:H7 using all the featured properties of nanomaterials described in this chapter. As schematized in Fig. 10.10, in the proposed approach GOx has been conducted as carrier for immobilized thionine (Thi) due to its large surface area and altered conductivity. On the other hand, optically enhanced Au/SiO_2 nanocomposites (Au-SiO_2) were combined with GOx-Thi as signal amplifier and hemin intercalated into the G-quadruplex forming hemin/G-quadruplex structure was used as bioelectrocatalytic HRP-mimicking DNAzyme. As clearly schematized in Fig. 10.11, platform was tested for detection of *E. coli* O157:H7 eaeA gen and trials revealed that sensor can efficiently respond to 0.01 nm eaeA gen (S/N = 3) with a quite wide linear detection range from 0.02 to 50.0 nm. As suggested by all of recent literature briefly touched upon, this study is another declaration of the fact that implementation of nanostructures with changing, size, shape, and chemical composition alters the signal transduction with increased detection sensitivity and selectivity while decreasing the limit of detection to single cell or femtomole levels which definitely would be inspirational for future attempts on point-of-care detection devices for food safety and evaluation and.

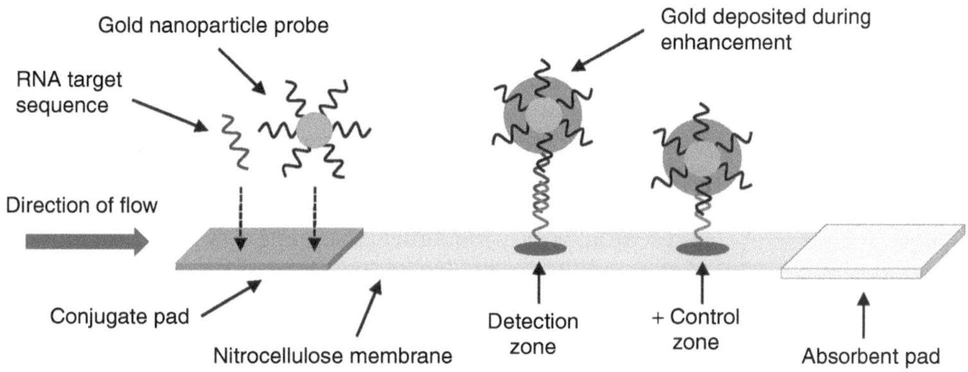

Figure 10.10. Double-tag DNA-based detection of Salmonella sp. on gold nanoparticles modified electrodes. From Marques et al. (2009).

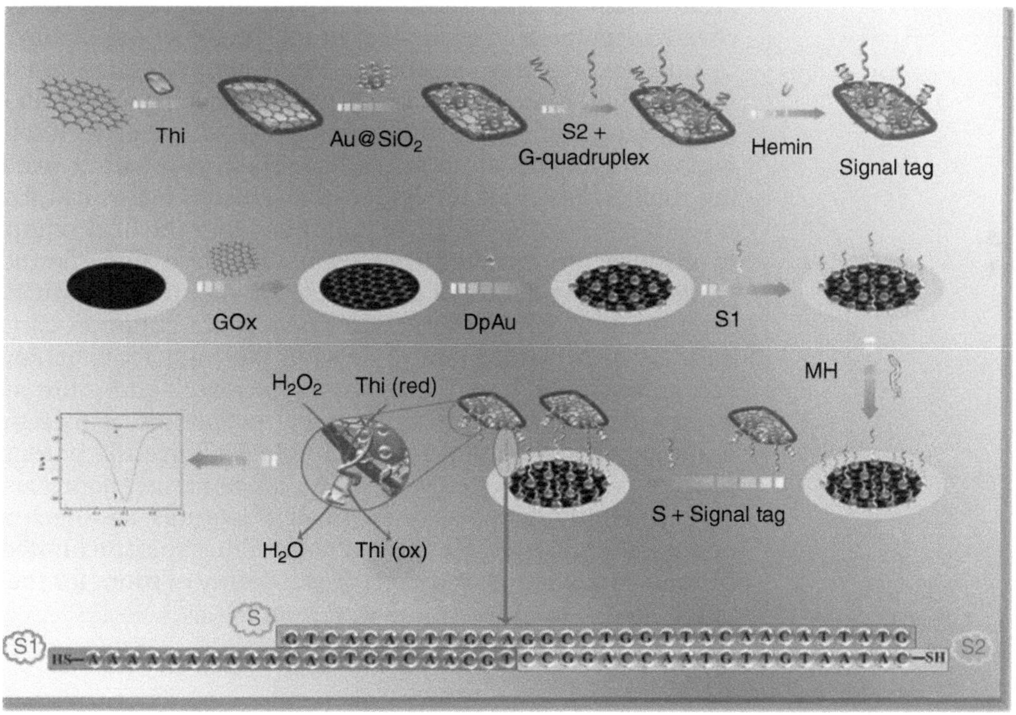

Figure 10.11. Scheme of pathogen detection using thionine (Thi) embedded GOxnanosheet modified electrode in which gold–silica nanocomposites were acted as signal enhancer by controlled immobilization of bioelectrocatalytic HRP-mimicking DNAzyme hemin/G-quadruplex structure as signal probe. From Li et al. (2015).

5 Conclusions and Future Perspectives

Nanotechnology is a new industry that uses altered properties of the matter in nanosize and manipulates it for a particular purpose in combination with other technologies. It is already approved by scientific community as a golden key for development of detection systems for medical and pharmaceutical sectors, but increased number of research worldwide indicates that in very near future, a new storm of possibilities may also arouse for on-field use of nanomaterial-based foodborne pathogen detection systems in food industry.

In this chapter, for this end, several aspects of recent advances in nanotechnology for detection of foodborne pathogens and associated toxins in the past 5 years have been addressed by briefly covering updated outbreak statics and traditional methods. This includes inorganic nanoparticles and their composites with other materials showing shape, size, and chemical-composition-tailored characteristics have been addressed by means of their use in different stages of detection. In addition to covering the technical aspects of the use of nanotechnology in food industry focused applications, sample pretreatment, signal transduction and enhancement have been covered separately in terms of their role on the sensor platform.

Current literature emphasized that nanomaterial-based sensors have the potential to reach single molecule level with ultra-high selectivity; on the other hand, like all other technologies, nanotechnological approaches have their own handicaps, limiting their widespread implementation to the real world. Lack of reproducibility, the need for upgrades in analytical equipment to achieve the most efficiency possible, and uncertainties regarding the regulation of large-scale nanomaterial production to avoid unwanted environmental impacts are common concerns, and it is highly important to develop the capacity to achieve the full potential of nanomaterials. In broader terms, the superb electrochemical and physiochemical properties of materials in nanoscale coupled with new technologies have the potential to provide smaller devices with multiplex analysis, bioprobes with altered selectivity, and so on. With all its features, nanotechnology opens a door to a wonderland of possibilities that are limited only by one's imagination and there is still plenty of room for their use in the field of food safety and governance.

References

Bhattacharya, D., Chakraborty, S.P., Pramanik, A., Baksi, A., Roy, S., Maiti, T.K., Ghosh, S.K., Pramanik, P., 2011. Detection of total count of *Staphylococcus aureus* using antitoxin antibody labelled gold magnetite nanocomposites: a novel tool for capture, detection, and bacterial separation. J. Mater. Chem. 21, 17273–17282.

Bolduc, O.R., Masson, J.-F., 2011. Advances in surface plasmon resonance sensing with nanoparticles and thin films: nanomaterials, surface chemistry, and hybrid plasmonic techniques. Anal. Chem. 83, 8057–8062.
Carrillo-Carrión, C., Simonet, B.M., Valcárcel, M., 2011. Colistin-functionalized CdSe/ZnS quantum dots as fluorescent probe for the rapid detection of *Escherichia coli*. Biosens. Bioelectron. 26, 4368–4374.
Chen, L., Zhang, Z., Zhang, X., Fu, A., Xue, P., Yan, R., 2012. A novel chemiluminescence immunoassay of *Staphylococcal enterotoxin* B using HRP-functionalized mesoporous silica nanoparticle as label. Food Chem. 135, 208–212.
Cho, I.-H., Irudayaraj, J., 2013. In situ immuno-gold nanoparticle network ELISA biosensors for pathogen detection. Int. J. Food Microbiol. 164, 70–75.
Cho, I.-H., Mauer, L., Irudayaraj, J., 2014a. In-situ fluorescent immunomagnetic multiplex detection of foodborne pathogens in very low numbers. Biosens. Bioelectron. 57, 143–148.
Cho, I.-H., Radadia, A.D., Farrokhzad, K., Ximenes, E., Bae, E., Singh, A.K., Oliver, H., Ladisch, M., Bhunia, A., Applegate, B., Mauer, L., Bashir, R., Irudayaraj, J., 2014b. Nano/micro and spectroscopic approaches to food pathogen detection. Annu. Rev. Anal. Chem. 7, 65–88.
Choi, D.H., Lee, S.K., Oh, Y.K., Bae, B.W., Lee, S.D., Kim, S., Shin, Y.-B., Kim, M.-G., 2010. A dual gold nanoparticle conjugate-based lateral flow assay (LFA) method for the analysis of troponin I. Biosens. Bioelectron. 25, 1999–2002.
Cloutier, B.C., Cloutier, A.K., Alocilja, E.C., 2015. Optimization of electrically active magnetic nanoparticles as accurate and efficient microbial extraction tools. Biosensors 5 (1), 69–84.
Dao, T.C., Luong, T.Q.N., Cao, T.A., Nguyen, N.H., Kieu, N.M., Luong, T.T., Le, V.V., 2015. Trace detection of herbicides by SERS technique, using SERS-active substrates fabricated from different silver nanostructures deposited on silicon. Adv. Nat. Sci. Nanosci. Nanotechnol. 6, 035012.
Davis, D., Guo, X., Musavi, L., Lin, C.-S., Chen, S.-H., Wu, V.C.H., 2013. Gold nanoparticle-modified carbon electrode biosensor for the detection of *Listeria monocytogenes*. Ind. Biotechnol. 9 (1), 31–36.
DeSantis, C.J., Weiner, R.G., Radmilovic, A., Bower, M.M., Skrabalak, S.E., 2013. Seeding bimetallic nanostructures as a new class of plasmonic colloids. J. Phys. Chem. Lett. 4, 3072–3082.
Duan, N., Wu, S., Dai, S., Miao, T., Chen, J., Wang, Z., 2014. Simultaneous detection of pathogenic bacteria using an aptamer-based biosensor and dual fluorescence resonance energy transfer from quantum dots to carbon nanoparticles. Microchim. Acta 182, 917–923.
Dwivedi, H.P., Jaykus, L.-A., 2011. Detection of pathogens in foods: the current state-of-the-art and future directions. Crit. Rev. Microbiol. 37, 40–63.
Farakos, S.M.S.F., Frank, J., 2014. The Microbiological Safety of Low Water Activity Foods and Spices. Springer, New York.
Fei, A., Liu, Q., Huan, J., Qian, J., Dong, X., Qiu, B., Mao, H., Wang, K., 2015. Label-free impedimetric aptasensor for detection of femtomole level acetamiprid using gold nanoparticles decorated multiwalled carbon nanotube-reduced graphene oxide nanoribbon composites. Biosens. Bioelectron. 70, 122–129.
Food and Drug Administration, 2012. Bad Bug Book: Foodborne Pathogenic Microorganisms and Natural Toxins, second ed. http://www.fda.gov/downloads/Food/FoodborneIllnessContaminants/UCM297627.pdf.
Genc, R., 2014. Biosensors: Recent Advances and Mathematical Challenges, In: Osma, J.F., Stoytcheva, M. (Eds.), OmniaScience Monographs. OmniaScience.
Genç, R., Murphy, D., Fragoso, A., Ortiz, M., O'Sullivan, C.K., 2011. Signal-enhancing thermosensitive liposomes for highly sensitive immunosensor development. Anal. Chem. 83, 563–570.

Gilmartin, N., O'Kennedy, R., 2012. Nanobiotechnologies for the detection and reduction of pathogens. Enzyme and microbial technology 50 (2), 87–95.

Gun'ko, Y.K., Moloney, M.M., Gallagher, S., Govan, J., Hanley, C., 2010. New quantum dot sensors. Nanotechnology 7679, 76791V.

He, X., Zhou, L., He, D., Wang, K., Cao, J., 2011a. Rapid and ultrasensitive *E. coli* O157:H7 quantitation by combination of ligand magnetic nanoparticles enrichment with fluorescent nanoparticles-based two-color flow cytometry. Analyst 136, 4183.

He, L., Shi, J., Sun, X., Lin, M., Yu, P., Li, H., 2011b. Gold-coated zinc oxide nanonecklaces as a SERS substrate. J. Nanosci. Nanotechnol. 11, 3509–3515.

Hu, Y., Hua, S., Li, F., Jiang, Y., Bai, X., Li, D., Niu, L., 2011a. Green-synthesized gold nanoparticles decorated graphene sheets for label-free electrochemical impedance DNA hybridization biosensing. Biosens. Bioelectron. 26, 4355–4361.

Hu, Y., Li, F., Bai, X., Li, D., Hua, S., Wang, K., Niu, L., 2011b. Label-free electrochemical impedance sensing of DNA hybridization based on functionalized graphene sheets. Chem. Commun. (Camb) 47, 1743–1745.

IUPAC, 2005. IUPAC Compendium of Chemical Terminology 2005, second ed., Compendium of Chemical Terminology, (the "Gold Book"). Scientific Publications, Oxford.

Jackeray, R., Abid, C.K.V.Z., Singh, G., Jain, S., Chattopadhyaya, S., Sapra, S., Shrivastav, T.G., Singh, H., 2011. Selective capturing and detection of *Salmonella typhi* on polycarbonate membrane using bioconjugated quantum dots. Talanta 84, 952–962.

Jiang, J., Wang, X., Chao, R., Ren, Y., Hu, C., Xu, Z., Liu, G.L., 2014. Smartphone-based portable bacteria pre-concentrating microfluidic sensor and impedance sensing system. Sens. Actuat. B-Chem. 193, 653–659.

Jianrong, C., Yuqing, M., Nongyue, H., Xiaohua, W., Sijiao, L., 2004. Nanotechnology and biosensors. Biotechnol. Adv. 22, 505–518.

Kim, D.-K., Yoo, S.M., Park, T.J., Yoshikawa, H., Tamiya, E., Park, J.Y., Lee, S.Y., 2011. Plasmonic properties of the multispot copper-capped nanoparticle array chip and its application to optical biosensors for pathogen detection of multiplex DNAs. Anal. Chem. 83, 6215–6222.

Li, F., Zhao, Q., Wang, C., Lu, X., Li, X.-F., Le, X.C., 2010. Detection of *Escherichia coli* O157:H7 using gold nanoparticle labeling and inductively coupled plasma mass spectrometry. Anal. Chem. 82, 3399–3403.

Li, F., Yang, L., Chen, M., Li, P., Tang, B., 2013. A selective amperometric sensing platform for lead-based on target-induced strand release. Analyst 138, 461–466.

Li, X.-R., Li, X.-L., Xu, M.-C., Xu, J.-J., Chen, H.-Y., 2014. Gold nanodendrities on graphene oxide nanosheets for oxygen reduction reaction. J. Mater. Chem. A 2, 1697–1703.

Li, Y., Deng, J., Fang, L., Yu, K., Huang, H., Jiang, L., Liang, W., Zheng, J., 2015. A novel electrochemical DNA biosensor based on HRP-mimicking hemin/G-quadruplex wrapped GOx nanocomposites as tag for detection of *Escherichia coli* O157:H7. Biosens. Bioelectron. 63, 1–6.

Liu, Y., Brandon, R., Cate, M., Peng, X., Stony, R., Johnson, M., 2007. Detection of pathogens using luminescent CdSe/ZnS dendron nanocrystals and a porous membrane immunofilter. Anal. Chem. 79, 8796–8802.

Liu, X.-H., Luo, J.-P., Tian, Q., Liu, C.-X., Cai, X.-X., 2011. Detection of *Bacillus anthracis* using fluorescence immunoassay with quantum dots labels. Chin. J. Anal. Chem. 39, 163–167.

Llopis, X., Pumera, M., Alegret, S., Merkoçi, A., 2009. Lab-on-a-chip for ultrasensitive detection of carbofuran by enzymatic inhibition with replacement of enzyme using magnetic beads. Lab Chip 9, 213–218.

Ma, H., Chi, H., Wu, J., Wang, M., Li, J., Hoshina, H., Saiki, S., Seko, N., 2013. A novel avenue to gold nanostructured microtubes using functionalized fiber as the ligand, the reductant, and the template. ACS Appl. Mater. Interfaces. 5 (17), 8761–8765.

Marques, P.R.B. de O., Lermo, A., Campoy, S., Yamanaka, H., Barbé, J., Alegret, S., Pividori, M.I., 2009. Double-tagging polymerase chain reaction with a thiolated primer and electrochemical genosensing based on gold nanocomposite sensor for food safety. Anal. Chem. 81, 1332–1339.

Merkoçi, A., 2009. Biosensing Using Nanomaterials, Structure. John Wiley & Sons, Hoboken, NJ.

Mody, V.V., Siwale, R., Singh, A., Mody, H.R., 2010. Introduction to metallic nanoparticles. J. Pharm. Bioallied Sci. 2, 282–289.

Najafi, R., Mukherjee, S., Hudson, J., Sharma, A., Banerjee, P., 2014. Development of a rapid capture-cum-detection method for *Escherichia coli* O157 from apple juice comprising nano-immunomagnetic separation in tandem with surface enhanced Raman scattering. Int. J. Food Microbiol. 189, 89–97.

Orlov, A.V., Khodakova, J.A., Nikitin, M.P., Shepelyakovskaya, A.O., Brovko, F.A., Laman, A.G., Grishin, E.V., Nikitin, P.I., 2013. Magnetic immunoassay for detection of staphylococcal toxins in complex media. Anal. Chem. 85, 1154–1163.

Pal, S., Alocilja, E.C., Downes, F.P., 2007. Nanowire labeled direct-charge transfer biosensor for detecting *Bacillus* species. Biosens. Bioelectron. 22, 2329–2336.

Pérez-López, B., Merkoçi, A., 2011. Nanomaterials-based biosensors for food analysis applications. Trends Food Sci. Technol. 22 (11), 625–639.

Qian, W., Huang, X., Kang, B., El-Sayed, M.A., 2010. Dark-field light scattering imaging of living cancer cell component from birth through division using bioconjugated gold nanoprobes. J. Biomed. Opt. 15, 046025.

Quintela, I.A., de los Reyes, B.G., Lin, C.-S., Wu, V.C.H., 2015. Simultaneous direct detection of Shiga-toxin producing *Escherichia coli* (STEC) strains by optical biosensing with oligonucleotide-functionalized gold nanoparticles. Nanoscale 7, 2417–2426.

Rastogi, S.K., Gibson, C.M., Branen, J.R., Aston, D.E., Branen, A.L., Hrdlicka, P.J., 2012. DNA detection on lateral flow test strips: enhanced signal sensitivity using LNA-conjugated gold nanoparticles. Chem. Commun. (Camb). 48, 7714–7716.

Resch-genger, U., Grabolle, M., Cavaliere-jaricot, S., Nitschke, R., Nann, T., 2008. Quantum dots versus organic dyes as fluorescent labels. Nat. Methods 5, 763–775.

Rohrman, B.A., Leautaud, V., Molyneux, E., Richards-Kortum, R.R., 2012. A lateral flow assay for quantitative detection of amplified HIV-1 RNA. PLoS One 7, e45611.

Rong-Hwa, S., Shiao-Shek, T., Der-Jiang, C., Yao-Wen, H., 2010. Gold nanoparticle-based lateral flow assay for detection of *Staphylococcal enterotoxin* B. Food Chem. 118, 462–466.

Saha, K., Agasti, S.S., Kim, C., Li, X., Rotello, V.M., 2012. Gold nanoparticles in chemical and biological sensing. Chem. Rev. 112, 2739–2779.

Saini, S., Singh, H., Bagchi, B., 2006. Fluorescence resonance energy transfer (FRET) in chemistry and biology: non-Förster distance dependence of the FRET rate. J. Chem. Sci. 118, 23–35.

Sandvig, K., 2001. Shiga toxins. Toxicon 39, 1629–1635.

Scallan, E., Hoekstra, R.M., Angulo, F.J., Tauxe, R.V., Widdowson, M.-A., Roy, S.L., Jones, J.L., Griffi, P.M., 2011. Foodborne illness acquired in the United States: major pathogens. Emerg. Infect. Dis. 17, 7–15.

Shan, S., Zhong, Z., Lai, W., Xiong, Y., Cui, X., Liu, D., 2014. Immunomagnetic nanobeads based on a streptavidin-biotin system for the highly efficient and specific separation of *Listeria monocytogenes*. Food Control 45, 138–142.

Shi, J., Chan, C., Pang, Y., Ye, W., Tian, F., Lyu, J., Zhang, Y., Yang, M., 2015. A fluorescence resonance energy transfer (FRET) biosensor-based on graphene quantum dots (GQDs) and gold nanoparticles (AuNPs) for the detection of mecA gene sequence of *Staphylococcus aureus*. Biosens. Bioelectron. 67, 595–600.

Soon, R.L., Li, J., Boyce, J.D., Harper, M., Adler, B., Larson, I., Nation, R.L., 2012. Cell surface hydrophobicity of colistin-susceptible versus resistant *Acinetobacter baumannii* determined by contact angles: methodological considerations and implications. J. Appl. Microbiol. 113, 940–951.

Staleva, H., Skrabalak, S.E., Carey, C.R., Kosel, T., Xia, Y., Hartland, G.V., Kealley, C.S., Cortie, M.B., Maaroof, A.I., Wu, X., Sun, Y., Pelton, M., Wang, H., Zou, S., Chem, P., Phys, C., Okamoto, H., Rycenga, M., Hou, K.K., Cobley, C.M., Andrea, G., Camargo, P.H.C., Rodríguez-fernández, J., Funston, A.M., Álvarez-puebla, R.A., Liz-marzán, L.M., Zijlstra, P., Chon, J.W.M., Gu, M., Schwartz, A.G., 2009. Nanophotonics: plasmonics and metal nanoparticles. Phys. Chem. Chem. Phys. 11, 5866.

Su, H., Zhao, H., Qiao, F., Chen, L., Duan, R., Ai, S., 2013. Colorimetric detection of *Escherichia coli* O157:H7 using functionalized Au@Pt nanoparticles as peroxidase mimetics. Analyst 138, 3026–3031.

Sun, J., Ji, J., Sun, Y., Abdalhai, M.H., Zhang, Y., Sun, X., 2015. DNA biosensor-based on fluorescence detection of *E. coli* O157:H7 by Au@Ag nanorods. Biosens. Bioelectron. 70, 239–245.

Sung, Y.J., Suk, H.-J., Sung, H.Y., Li, T., Poo, H., Kim, M.-G., 2013. Novel antibody/gold nanoparticle/magnetic nanoparticle nanocomposites for immunomagnetic separation and rapid colorimetric detection of *Staphylococcus aureus* in milk. Biosens. Bioelectron. 43, 432–439.

Tan, Y., Zang, X., Gu, J., Liu, D., Zhu, S., Su, H., Feng, C., Liu, Q., Lau, W.M., Moon, W.J., Zhang, D., 2011. Morphological effects on surface-enhanced raman scattering from silver butterfly wing scales synthesized via photoreduction. Langmuir 27, 11742–11746.

The Food Safety and Inspection Service (FSIS), 2015. Foodborne Pathogen Test Kits Validated by Independent Organizations. http://www.fsis.usda.gov/wps/wcm/connect/f97532f4-9c28-4ecc-9aee-0e1e6cde1a89/Validated-Test-Kit-Spreadsheet.pdf?MOD=AJPERES

Thévenot, D., Toth, K., Durst, R., Wilson, G., 1999. Electrochemical biosensors: recommended definitions and classification. Pure Appl. Chem., De Gruyter 71 (12), 2333–2348.

Vester, B., Wengel, J., 2004. LNA (locked nucleic acid): high-affinity targeting of complementary RNA and DNA. Biochemistry 43, 13233–13241.

Viswanathan, S., Radecka, H., Radecki, J., 2009. Electrochemical biosensor for pesticides based on acetylcholinesterase immobilized on polyaniline deposited on vertically assembled carbon nanotubes wrapped with ssDNA. Biosens. Bioelectron. 24, 2772–2777.

Wang, R., Dong, W., Ruan, C., Kanayeva, D., Tian, R., Lassiter, K., Li, Y., 2008. TiO2 nanowire bundle microelectrode based impedance immunosensor for rapid and sensitive detection of *Listeria monocytogenes*. Nano Lett. 8, 2625–2631.

Wang, L., Liu, Q., Hu, Z., Zhang, Y., Wu, C., Yang, M., Wang, P., 2009. A novel electrochemical biosensor based on dynamic polymerase-extending hybridization for *E. coli* O157:H7 DNA detection. Talanta 78, 647–652.

Wang, X., Du, Y., Li, Y., Li, D., Sun, R., 2011. Fluorescent identification and detection of *Staphylococcus aureus* with carboxymethyl chitosan/CdS quantum dots bioconjugates. J. Biomater. Sci. Polym. Ed. 22, 1881–1893.

Wang, Y., Ye, Z., Ying, Y., 2012a. New trends in impedimetric biosensors for the detection of foodborne pathogenic bacteria. Sensors (Basel) 12, 3449–3471.

Wang, L., Wu, C.-S., Fan, X., Mustapha, A., 2012b. Detection of *Escherichia coli* O157:H7 and *Salmonella* in ground beef by a bead-free quantum dot-facilitated isolation method. Int. J. Food Microbiol. 156, 3–87.

Wegner, K.D., Lanh, P.T., Jennings, T., Oh, E., Jain, V., Fairclough, S.M., Smith, J.M., Giovanelli, E., Lequeux, N., Pons, T., Hildebrandt, N., 2013. Influence of luminescence quantum yield, surface coating, and functionalization of quantum dots on the sensitivity of time-resolved FRET bioassays. ACS Appl. Mater. Interf. 5, 2881–2892.

West, J.L., Halas, N.J., 2003. Engineered nanomaterials for biophotonics applications: improving sensing, imaging, and therapeutics. Annu. Rev. Biomed. Eng. 5, 285–292.

Winoto-Morbach, S., Müller-Ruchholtz, W., 1995. Preparation and functionality testing of biodegradable and biocompatible magnetic albumin nanospheres for cell labelling and immunomagnetic separation. Eur. J. 41(1), 55–61.

Wu, S.-M., Zhang, Z.-L., Wang, X.-D., Zhang, M.-X., Peng, J., Xie, Z.-X., Pang, D.-W., 2009. Quantum dot: ferrichrome bioprobes for recognition of *Pseudomonas fluorescens*. J. Phys. Chem. C 113, 9169–9174.

Wu, X., Gao, S., Wang, J.-S., Wang, H., Huang, Y.-W., Zhao, Y., 2012. The surface-enhanced Raman spectra of aflatoxins: spectral analysis, density functional theory calculation, detection, and differentiation. Analyst 137, 4226–4234.

Wu, X., Xu, C., Tripp, R.A., Huang, Y., Zhao, Y., 2013. Detection and differentiation of foodborne pathogenic bacteria in mungbean sprouts using field deployable label-free SERS devices. Analyst 138, 3005–3012.

Xu, X., Chen, Y., Wei, H., Xia, B., Liu, F., Li, N., 2012. Counting bacteria using functionalized gold nanoparticles as the light-scattering reporter. Anal. Chem. 84, 9721–9728.

Yang, G.-J., Huang, J.-L., Meng, W.-J., Shen, M., Jiao, X.-A., 2009. A reusable capacitive immunosensor for detection of *Salmonella* spp. based on grafted ethylene diamine and self-assembled gold nanoparticle monolayers. Anal. Chim. Acta 647, 159–166.

Zelada-Guillén, G.A., Riu, J., Düzgün, A., Rius, F.X., 2009. Immediate detection of living bacteria at ultralow concentrations using a carbon nanotube–based potentiometric aptasensor. Angew. Chem. Int. Ed. Engl. 48, 7334–7337.

Zhang, C., Yin, A.-X., Jiang, R., Rong, J., Dong, L., Zhao, T., Sun, L.-D., Wang, J., Chen, X., Yan, C.-H., 2013. Time–temperature indicator for perishable products based on kinetically programmable Ag overgrowth on Au nanorods. ACS Nano 7 (5), 4561–4568.

Zhang, X., Zhang, F., Zhang, H., Shen, J., Han, E., Dong, X., 2015. Functionalized gold nanorod-based labels for amplified electrochemical immunoassay of *E. coli* as indicator bacteria relevant to the quality of dairy product. Talanta 132, 600–605.

Zhao, Y., Ye, M., Chao, Q., Jia, N., Ge, Y., Shen, H., 2009. Simultaneous detection of multifood-borne pathogenic bacteria based on functionalized quantum dots coupled with immunomagnetic separation in food samples. J. Agric. Food Chem. 57, 517–524.

Zhu, K., Dietrich, R., Didier, A., Doyscher, D., Märtlbauer, E., 2014. Recent developments in antibody-based assays for the detection of bacterial toxins. Toxins (Basel). 6, 1325–1348.

Zhu, Z., Shi, L., Feng, H., Zhou, H.S., 2015. Single domain antibody coated gold nanoparticles as enhancer for *Clostridium difficile* toxin detection by electrochemical impedance immunosensors. Bioelectrochemistry 101, 153–158.

11

EVANESCENT FIELD EFFECT–BASED NANOBIOSENSORS FOR AGRO-ENVIRONMENTAL AND FOOD SAFETY

Nóra Adányi*, Krisztina Majer-Baranyi*, András Székács**

**National Agricultural Research and Innovation Center, Food Science Research Institute, Budapest, Hungary; **National Agricultural Research and Innovation Center, Agro-Environmental Research Institute, Budapest, Hungary*

1 Introduction

Nanotechnology has a multiple and rapidly expanding role in food industry. Nanomaterials are used in food production or inspection with their unique, nanosized dispersity, or as novel methodologies utilizing nanostructures in analytical or technological processes. Nanotechnology is of particular importance in the development of biosensors: sensitivity and other attributes of biosensors can be substantially improved by using nanomaterials in biosensor construction. In the past decades, nanostructures, including nanotubes, nanoparticles, nanorods, nanofibers as well as thin films have been intensively investigated for improved physicochemical properties and biosensor applications (Bata-Vidács et al., 2013). A great interest has focused on the development of selective and sensitive sensors for the detection of very low levels of chemical and biological substances and for the measurement of molecular interactions in situ and in real time. These requirements can be achieved with optical sensors, particularly by those based on evanescent wave principles at physical phase boundaries capable to allow direct monitoring of nuance modifications in optical properties. At boundaries between thin layers of different materials with negligible absorption, radiation is partially reflected and transmitted. Evanescent waves or fields are created at such adequate phase separation surfaces, when waves (not necessarily limited to light, but including broader range electromagnetic or other waves) cause rapidly decaying interactions based on interference

Nanobiosensors. http://dx.doi.org/10.1016/B978-0-12-804301-1.00011-4

or quantum mechanical processes. These evanescent fields are not spatially limited in their interaction to the waveguide surface, but are allowed to penetrate to the sample overlayer, and therefore they are also affected by surface characteristics and conditions, and as such they are sensitive to changes in the electronic polarization distribution within the evanescent fields. As a result, the evanescent signal is modified by ongoing surface processes, possibly leading to substantial changes in these minute signals, which in turn, create a possibility to detect surface changes with outstandingly high sensitivity and in real time. Wave interference is typically generated by nanosized waveguide gratings created on the waveguide surface, but can also be monitored otherwise (Mach–Zehnder or slab interferometry, etc.). Changes at the interface between the waveguide and the medium can arise in several different ways: new atoms may replace existing atoms and change polarizability, the density of the material within the evanescent field may increase or decrease, an adlayer of particles with different polarizability property can be built up or removed (Erdélyi et al., 2008). Among the large variety of optical methods used for biosensing the enfilade of evanescent wave optical devices can measure the interaction between complementary molecules in real time, without any need of labels (Lechuga, 2000a,b). These direct label-free sensing methods provide important information in some fields of application, for example, dynamic/kinetic information based on concentration changes detected in chemical or biological binding processes. All commercial methods use evanescent field techniques to monitor the affinity reaction via variations in the refractive index (RI) by changes in ligand concentration. The actual limit of detection (LOD), defined as the lowest concentration discernible above the mean of the blank plus 3 times its standard deviation (blank + 3 SD), is continuously pushed forward, and LODs of 0.1–1 pg/mm^2 are in reach of highly sensitive transducers (Brecht and Gauglitz, 1995), and for quantitative analysis limits of quantification (LOQs) also display a similar favorable shift.

In this chapter we discuss the measuring principles of the methods of demonstrated or potential utility in optical waveguide grating based biosensing. The most prevailing label-free optical biosensors apply surface plasmon resonance (SPR) based instrumentation, the measurement method and applications of which are discussed in detail in other chapters of this book. Therefore, our review focuses on optical waveguide grating based measurement principles other than SPR, as well as their possibilities in application for environmental and food analysis to promote agro-environmental and food safety in compliance with the "farm to fork" concept (Fan et al., 2008; Schmitt et al., 2008; Estevez et al., 2012).

2 Label-Free Optical Biosensor Techniques Based on Evanescent Field Effect

The measuring principles of the evanescent field based label-free optical biosensor techniques can be classified with respect to the optical effects created on the basis of reflectance, interference, or fluorescence spectroscopy, taking advantage of the evanescent field principle in case of a stripe or film waveguides, as well as mode coupling and being or not being in resonance, using polarized or nonpolarized light to observe phase shifts (Brecht and Gauglitz, 1995). Effects involved include phase changes, coupling effects between waveguide structures, SPR, grating couplers, dual beam, and mode beat interferometry. Besides reflectometric techniques, ellipsometry can also be applied, making use of light of varying polarization states and exhibiting high sensitivity, or simpler techniques as multiple reflection interference (white light interference) at thin films. Most techniques discussed are assumed to be governed by changes in optical density (Brecht and Gauglitz, 1995). Most important measurement setups include different optical waveguides, interferometer based sensors, RIfS, total internal reflection fluorescence (TIRF), total internal reflection ellipsometry (TIRE), ring resonator.

2.1 Optical Waveguide-Based Sensor Structures

In optical waveguide sensors light is incident or coupled into a special waveguide at a specific angle, in which it propagates with total internal reflection (TIR) to an appropriate photodetector. The entire metal (gold) layer used in SPR may serve as a waveguide for the plasmon generated. In other waveguide-based sensors a waveguide layer, coming in direct contact with the liquid phase studied, is created. In all these cases (including SPR) surface processes at the liquid-waveguide interface affect the resonance conditions of the evanescent field. Planar waveguides (PWGs) are films made of high RI material embedded between lower RI materials. When light wave introduced into the high RI film arrives at the boundary at an angle greater than the critical angle of TIR the light wave is confined inside the waveguide. The laser illuminates the waveguide at varying angles, and light is coupled into the waveguide only at specific angles determined by the effective RI of the guided mode. The basic physics of the evanescent field effect defines the types of application (eg, monitoring interfacial processes), yet detection may be achieved by different techniques (eg, measurement of in- or outcoupling angles for grating couplers, Mach–Zehnder interferometry). Characteristic changes may arise in several different ways: new atoms of differing polarizability may replace existing atoms, the density of the material within the evanescent field may increase or decrease; an adlayer of particles with polarizability different from that of the surrounding medium may be built up (or removed) at the interface between the waveguide and the medium (Horváth et al., 2002; Erdélyi et al., 2008). SPR, normal and reverse symmetry waveguides, metal-clad waveguide (MCWG) sensors, and RMs fall under this group of sensors. The measuring principles of these sensors are summarized in Fig. 11.1.

2.1.1 Normal Symmetry Waveguide Sensors

The simplest waveguide design consists of an optically dense slab with RI higher than that of both the (glass) substrate and the medium above the waveguide (Fig. 11.1a). Typically the system

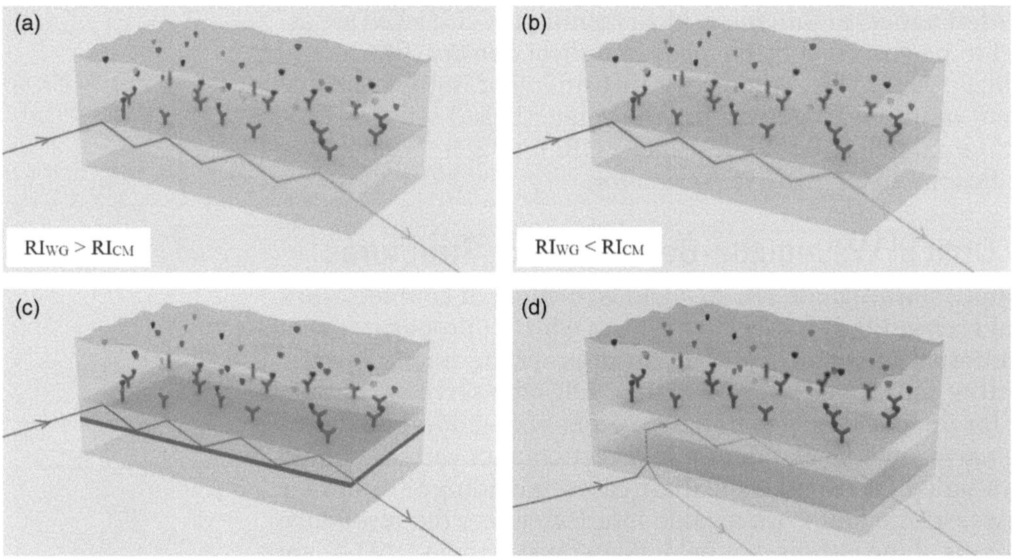

Figure 11.1. Measurement principles of optical waveguide based sensor setups. (a) Conventional normal symmetry waveguide sensor. Laser light is guided into the high refractive index (RI) waveguide layer, where it propagates by total internal reflection (TIR), and loses intensity as evanescent field is generated at the waveguide-cover medium boundary. The RI of the waveguide (RI_{WG}) is higher than that of the cover medium (RI_{CM}). (b) Reverse symmetry waveguide sensor. The setup is the same as in the normal symmetry sensor, but the RI of the waveguide (RI_{WG}) is lower than that of the cover medium (RI_{CM}), which creates an evanescent field of reversed symmetry capable to penetrate deeper into the cover medium. (c) Metal-clad waveguide (MCWG) sensor. A fully reflective metal layer is positioned below the waveguide layer causing TIR at the film-cover boundary and normal metallic reflection at the film-metal boundary, and therefore, a stronger evanescent field. (d) Resonant mirror sensor. Unlike in the previous ones, light does not descend directly into the waveguide, but is directed through the lower substrate layer, and is coupled into the waveguide through the leaky spacer layer, where it undergoes TIR and creates an evanescent field. In turn, light intensity of the reflected light as a function of the incidence angle is detected.

thus consists of a thin, high RI slab (eg, titanium dioxide, $n\sim2$), supported on optical glass ($n\sim1.5$) and being in contact with an aqueous cover medium ($n\sim1.4$). With such conventional (normal symmetry) arrangement, the evanescent field of red or green laser light guided within the waveguide typically extends into the cover medium at depths of 100–200 nm (Horváth et al., 2002; Erdélyi et al., 2008). In such normal arrangements, the resonance angle at which the evanescent field occurs is detected from the intensity of light arriving to the photodetectors at the end of the waveguide.

2.1.2 Reverse Symmetry Waveguide

By modifying the optical characteristics of the waveguide layer and the sensor structure, the evanescent field can be pushed deeper into the medium solution, and due to this deeper penetration light interaction with components in the cover medium can be monitored at a thicker adjacency of the waveguide surface, which allows measurement of light interaction with large biomolecules or even cells. In such a reverse symmetry waveguide nanoporous silica is used as a lower cladding layer, as it has an RI of 1.193, much lower than that of water. As a result, more light can be concentrated near the sensing surface to enhance the sensor sensitivity (Fig. 11.1b) (Fan et al., 2008). Using conventional waveguide construction the evanescent field decays exponentially from the boundary of the waveguide into the external medium by normal symmetry, and light propagating in waveguide sensors penetrates the cover medium to a depth of 100–200 nm with high intensity. With this model the molecular scale analytes with layer a thickness of 1–20 nm (eg, protein absorption, affinity binding) can be measured with high sensitivity, however, it was found to depend highly on the waveguide design (Horváth et al., 2002). Opposed to the conventional waveguide geometry; in the reverse symmetry configuration, the support with very low RI drives the guided modes into the cover medium. With this arrangement, very large evanescent fields (extending ~10–200 μm away from the waveguide) can be achieved due to the inverted field symmetry (Horváth et al., 2002).

2.1.3 Metal-Clad Waveguide Sensors

The structure of the MCWG is similar to the conventional waveguide but with an extra layer of metal introduced below the waveguide. Therefore, the MCWG is a planar four-layered waveguide structure comprising a substrate, a thin metal-cladding, a nanoporous dielectric waveguide film and a cover medium (Fig. 11.1c). The setup is advantageous in sensing of biomolecules due to the high surface area of the nanopores in the waveguide layer and a sharp dip in the reflection spectrum due to characteristics of the MCWG mode. A nanostructured porous anodic Al_2O_3/Al film fabricated on a glass substrate was successfully applied as MCWG sensor (Hotta et al., 2010). Similarly to the conventional dielectric waveguide, light is guided in the waveguide film of the MCWG, however, while the guided light undergoes incomplete TIR (due to slight partial transmission and scattering) at both the film/cover- and the film/substrate-boundary in the conventional waveguides, TIR in the MCWG takes place only at the film/cover-boundary, and normal reflectance (with lesser light loss) occurs at the film/metal-boundary (Fan et al., 2008).

In the MCWG, the light is guided in a low RI waveguide layer separated from the high RI substrate by the metal layer, which also serves sensitivity improvement through intensifying light-matter interaction by pushing light into the solution layer. There are two types of MCWGs, the dip and peak types (Skivesen et al., 2005). When the imaginary part of the metal dielectric constant is high and the metal layer is thin (~1–5 nm), the MCWG is operated as a peak type and a characteristic peak emerges in the reflection. On the other hand, when the imaginary part of the metal dielectric constant is small and the metal thickness is somewhat thicker (~10–50 nm), the MCWG is operated as a dip type, in which the reflection features a characteristic resonant dip (Fan et al., 2008). The MCWG has much longer penetration depth than RM, SPR, and regular waveguide (Zourob et al., 2003). A comparison between the RM and the MCWG shows the MCWG is about 3 times more sensitive to the bulk RI change in the solution, indicating a much larger fraction of light in the evanescent field (Zourob and Goddard, 2005). Skivesen et al. (2005) optimized metal-clad waveguides achieving high sensitivity in the determination of adlayer thickness and changes in RI. In a detailed mathematical model, both the angular shift and the width of the resonances in the optical signal were considered using the Fresnel reflection coefficients, and came to the conclusion that MCWGs allow a sixfold improvement in analytical sensitivity compared to SPR sensors.

2.1.4 Resonant Mirror-Based Sensors

The RM technique is based on the use of a thin layer (~100 nm) of a high RI dielectric material on a thicker layer (~1 mm) of low RI material (all set up on a glass substrate or prism) (Fig. 11.1d). In this technique, however, light is not incident to the waveguide, but is directed through the lower substrate layer, and is coupled into the waveguide through the leaky spacer layer, and it undergoes multiple TIR at the top interface, allowing the evanescent wave to penetrate to the sample overlayer. A substantial part of the incident light is reflected from the spacer layer, and the higher evanescence field is generated, the more the light undergoes a phase change upon reflection; and by monitoring the angle at which this occurs, it is possible to detect changes within the evanescent field as the resonance angle is sensitive to the RI change near the waveguide, which can be detected at the output side of the RM. RM supports both the transverse electric (TE) and the transverse magnetic (TM) modes with different resonant angles (Cush et al., 1993; Goddard et al., 1994; Fan et al., 2008; Conroy et al., 2009).

2.2 Reflectance-Based Sensors

As seen, the element of (partial) reflectance measurement was already involved in the RM technique, other optical methods are fully based on the detection of changes in optical characteristics in the light beam reflected from the surface studied. Characteristics detected include fluorescence, polarization orientation, and phase shift. Changes in these characteristics in the reflected light due to molecular interactions of the evanescent field created at the waveguide-covering medium interface are detected.

2.2.1 Total Internal Reflectance Fluorescence

The TIRF measuring method exploits the principle of TIR occurring as a result of an incident light beam crossing a sensor interface of an RI higher than the medium ($RI_{WG} > RI_{CM}$) at incidence above the critical angle (Hecht, 1998). The evanescent field is affected by compounds immobilized on the sensor surface in the medium: physicochemical excitation of these molecules not only modifies the electromagnetic evanescent field, but also reduces background signal from molecules in the solution (Fig. 11.2a), and also affect the intensity and the penetration depth of the evanescent wave. In contrast, penetration depth is not affected by changes in the incident polarization. Field intensity decreases exponentially with the distance from the sensor surface and practically vanishes by a distance of ~100 nm. Relying on excitation of surface-bound labels by the evanescent wave generated by laser light at the interface, this technique provides possibility for studying the changes of surface-bound fluorescent species (Oheim and Schapper, 2005; Morgan and Mitchell, 2006). The exciting light is most commonly coupled into the sample with a prism. The same phenomenon is utilized in planar waveguide fluorescence immunosensor (PWGFIs). The TIRF optical system can be equipped either with PWGs (Schmitt et al., 2008) or with fiber optics. Preininger et al. (2000) introduced a fiber optic TIRF instrumentation based on an optical three-layer system with a fiber core, polymer

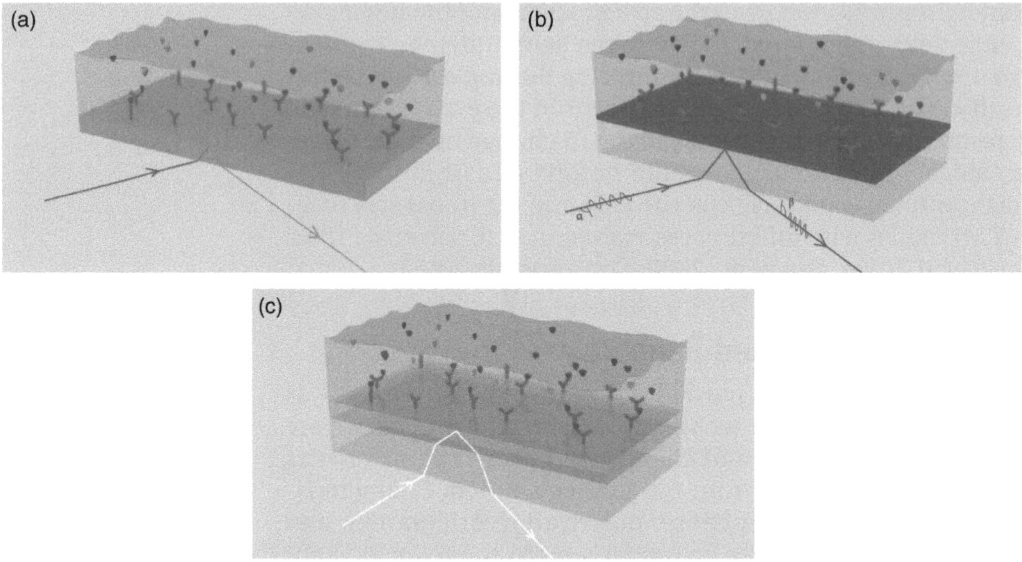

Figure 11.2. Measurement principles of optical reflectance based sensor setups. (a) Total internal reflectance fluorescence (TIRF) sensor. Laser light is directed onto the high refractive index (RI) waveguide layer, and reflected fluorescent light as a function of the incidence angle is detected. (b) Total internal reflection ellipsometry (TIRE) sensor. The setup is similar as in TIRF but polarity of the reflected light as a function of the incidence angle is detected. (c) Reflectometric interference spectroscopy (RIfS) sensor. White light is directed onto the high RI waveguide layer, where it undergoes partial reflection, and phase shift as a function of the incidence angle is detected.

layer, and sensing medium. In this model the midlayer polymer film is optically denser (higher RI) than the two other media, so the light can undergo multiple reflections in the guiding polymer film as long as the condition of TIR exists. Tedeschi et al. (2003) modified the fiber optic TIRF system by using a HeNe laser emitting at 543 nm as the optical source and a photon counter (photomultiplier tube and discriminator) as detector. The laser beam is modulated by a chopper (at a frequency of 315 Hz) and is coupled to a 200-mm delivery fiber. Light exiting from the delivery fiber is coupled to the measurement fiber by means of two objectives.

2.2.2 *Total Internal Reflection Ellipsometry*

Ellipsometry employs linearly polarized light reflected at oblique incidence from a surface, resulting in elliptically polarized reflected light. Changes in the shape and orientation of the reflected light are dependent on the direction of incident light and reflective surface properties (Conroy et al., 2009). The method of TIRE (also called plasmon enhanced ellipsometry) combines the advantages of the spectroscopic ellipsometry and the Kretchmann-type SPR geometry of TIR (Fig. 11.2b) (Westphal and

Bornmann, 2002; Poksinski and Arwin, 2003; Nabok et al., 2005). The TIRE system is based on a rotating polarizer multiwavelength ellipsometer equipped with a cylindrical lens or prism to couple light to the internal interface and a flow cell designed for TIR measurements. Light from the light source passes through the rotating polarizer and is then refracted into a glass prism in optical contact with a glass slide with a thin Au layer (about 50 nm) on the side adjacent to the prism. After reflection the light passes through the analyzer (a fixed polarizer) and the polarization is determined by analyzing the signals from a diode array in the detector unit (Poksinski and Arwin, 2003; Le et al., 2010).

2.2.3 Reflectometric Interference Spectroscopy

RIfS is a label-free detection method based on the multiple reflection of white light at transparent thin interface layers. If the optical path length through these layers is less than the coherence wavelength, the different partial beams interfere and form an interference pattern depending on the wavelength (Proll et al., 2009; Conroy et al., 2009). The product of physical thickness and RI (optical thickness) of the sensitive layer determines the position of the interference spectrum, its shift is proportional to the change in optical thickness. For use of RIfS as an optical probe an interference layer with appropriate RI must be used as a substrate for the biosensing layer. For simplicity, perpendicular irradiance is chosen in practice (Brecht and Gauglitz, 1995). Interactions at solid surfaces can be observed time resolved, aiming at thermodynamic and kinetic constants of these processes (Fig. 11.2c). RIfS is temperature dependent, however, it has a greater depth of penetration into the solution than SPR (Mohrle et al., 2006).

2.3 Grating-Based Biosensors

Grating is a region on top of or below a waveguide prepared usually of dielectric and transparent materials. The use of optical grating at sensor layer interfaces enables coupling light into the waveguide at combinations of incident angles and light frequency (resonance). Under noncoupling conditions, light propagates the same way as it would without any grating. Under specific coupling conditions, however, light completely alters its propagation route: it is guided into the waveguide, and undergoes TIR until it reaches the waveguide edge or an additional coupling element, which will couple it out. The setup of certain grating based biosensors utilizing in- and outcoupling are summarized in Fig. 11.3a.

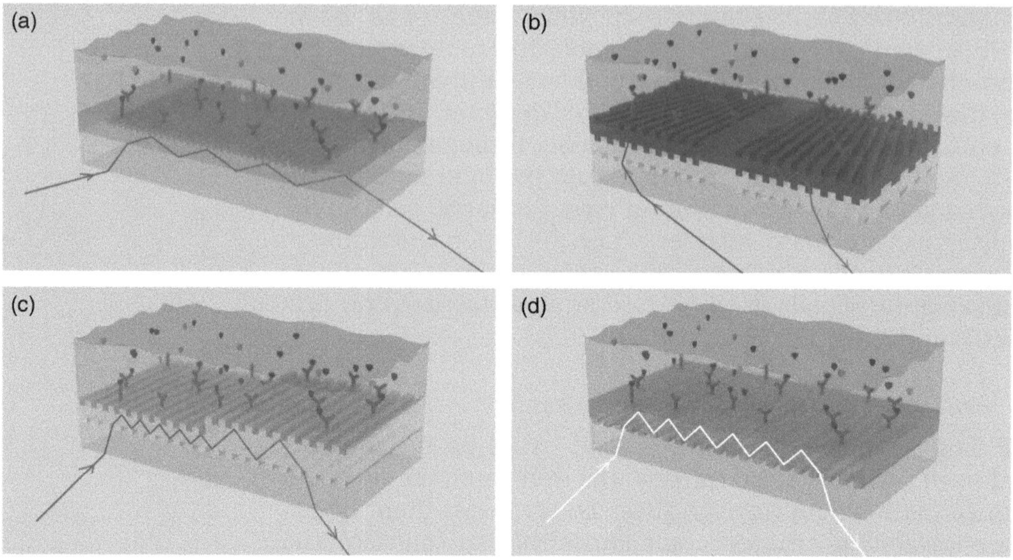

Figure 11.3. Measurement principles of optical grating based sensor setups. Laser light is directed through the lower substrate layer, and is coupled into the waveguide, where it propagates by total internal reflection (TIR). (a) Optical waveguide lightmode spectroscopy (OWLS) sensor. The incoupled light reaching the end of the waveguide layer as a function of the incidence angle is detected. (b) Light pointer based on chirped grating couplers (CGS). The incoupled light is outcoupled from the waveguide layer by the nonparallel outcoupling grating, and outcoupled light as a function of the incidence angle is detected. (c) Wavelength-interrogated optical sensor (WIOS). Similar to the light pointer, but light is outcoupled by parallel outcoupling gratings of differing layer thickness from the incoupling region. (d) Resonant waveguide grating (RWG) sensor setup is created in microplate based high throughput systems, where the bottom of each well of the microplate is a RWG sensor. Broadband light is coupled in and out of the waveguide. Wavelength shift in the outcoupled light due to interactions at the waveguide-covering medium interface is detected.

2.3.1 Incoupling Mode Sensors

The grating incoupler has several unique features, which makes it very powerful and versatile in optical evanescent wave techniques. In particular, both the TM and TE modes can be measured, and by tuning the waveguide parameters (eg, thickness and/or RI of high RI slab), the number of excitable modes can be selected. Both TM and TE represent orthogonal states of polarization and will propagate in a waveguide with different velocity, since their sensitivity toward RI changes outside the waveguide is different. Thus, the two modes experience a phase difference increasing with propagation length. Other parameters are the RI in the superstate and the observation wavelength. Polarization sensitive detection allows determination of this phase difference and its change, respectively (Brecht and Gauglitz, 1995; Conroy et al., 2009).

2.3.1.1 Optical Waveguide Lightmode Spectroscopy

The setup of the OWLS sensor is similar to that of the conventional waveguide, but with an optical grating crated on the waveguide-cover medium interface (Fig. 11.3a). OWLS is based on the determination of the actual resonance angle (corresponding to given surface characteristics) with high accuracy and precision, when a linearly polarized laser light is diffracted by a grating and coupled into a thin waveguide (Adányi et al., 2006). The incoupling is a resonance phenomenon that occurs at a defined angle of incidence that depends on the RI of the medium covering the surface of the waveguide. In the waveguide layer, light is guided by TIR to the edges where it is detected by photodiodes. By varying the angle of incidence of the light, the mode spectrum is obtained from which the effective RIs are calculated for both electrical and magnetic field waves (Luppa et al., 2001; Vörös et al., 2002; Jiang et al., 2008; Conroy et al., 2009).

To excite the guided mode the light should be coupled into the waveguide either at the cross section (butt coupling) or from the surface with a prism or grating. Detecting the guided modes we get either information about the RI of the material that covers the waveguide film, or the RI and thickness of a deposited thin layer can be determined. In the first case the optical grating coupler waveguide is used as refractometer. In the second case, it can be used as a sensor for a specific molecule if the sensor surface is coated with a monomolecular chemoresponsive layer that makes the target molecule selectively adsorb on the surface. Both the RI and the thickness of the adsorbed layer can be determined and the process of adsorption monitored in situ. In a typical grating coupler waveguide technique the sensor chip consists of a high RI waveguide film deposited on a lower index substrate and having an optical grating for in- or outcoupling light to or from the waveguide. A cuvette is attached to the top of the sensor surface that allows a cover material to be brought into contact with the waveguide. This cover material can be a solvent from which adsorption takes place and an adsorbed layer is built up on the waveguide surface (Erdélyi et al., 2008).

2.3.1.2 Tunable Wavelength Interrogated Sensor Technology

Tunable wavelength interrogated sensor technology (TWIST) creates an evanescent signal by modifying the wavelength of the incident laser (as opposed to the angle of incidence in other techniques, eg, OWLS). TWIST utilizes a tunable wavelength laser at a fixed angle coupled into a silicon oxynitride film waveguide; and intensity of the incoupled light, separately for the TE and TM components, is recorded as a function of the laser wavelength applied. Resonance conditions are dependent on the RI at the grating surface, and are shifted by modifications of the surface. (Tiefenthaler, 1992; Grego et al., 2012).

2.3.2 Outcoupling Mode Sensors

To investigate compact reading systems for biosensing there is the need for integrated sensing mechanisms without mechanical movement the outcoupling grating systems are coming to the front. As by using grating couplers there are three parameters to monitor the change of the effective RI: the coupling angle, the grating period, and the wavelength. The simplest method of outcoupling mode sensors, often termed as reflection mode grating coupler (RMGC), is a conventional normal symmetry waveguide sensor with a single in- and outcoupling grade and uses a charge-coupled device (CCD) camera for detection of the outcoupled light (Piehler et al., 1997).

2.3.2.1 Light Pointer Based on Chirped Grating Couplers

Chirped grating coupler (CGS)-based light pointers utilize unique nonparallel in- and outcoupling gratings (Fig. 11.3b). Light pointers based on CGSs measure RI changes at the surface of a waveguide chip using an optical arrangement. The key role is played by CGCs: grating couplers with a grating periodicity linearly varying with the position along the line grating. The CGCs utilize light incident to the sensing surface at a fixed angle providing uniform illumination over an extended sensor area. The asymmetrical geometry of the grating creates incoupling at one specific position and outcoupling at another one. Outcoupled light is detected by a CCD camera or a position-sensitive detector.

In the case of an infinite sensing layer on the input grating coupler, the effective waveguide RI will only be dependent on the sensing layer RI if we assume that the thickness of the waveguide layer is constant. As biochemical components are bound at or detached from the surface of the waveguide due to molecular interactions (resulting in changes in surface-bound mass) the coupling position is shifted, which can be monitored in real time by the beam position (Dübendorfer and Kunz, 1998; Erdélyi et al., 2008).

2.3.2.2 Wavelength-Interrogated Optical Sensor

In the wavelength-interrogated optical sensor (WIOS), parallel in- and outcoupling gratings are used, but with differing layer thickness at the in- and outcoupling regions (Fig. 11.3c). The resonance condition of the grating coupler is interrogated via spectral tuning of the light source and monitoring of the corresponding intensity of the outcoupled light. WIOS detects the RI changes in the evanescent wave of a straight waveguide grating using a wavelength sweep. The method employs a special semiconductor laser source, vertical cavity surface-emitting laser (VCSEL). VCSEL is constructed with laser diodes with a monolithic laser resonator, and produces low beam divergence laser. Light from VCSEL, emitting at around 763 nm, is incident on the first grating. The waveguide mode is excited and propagates into the waveguiding layer. The second grating couples out the guided light (at a different angle) and directs it on the detector. In the WIOS configuration, if fluorescent labels excited at the laser wavelength are used, the evanescent wave excites fluorescent light that is partially coupled back into the waveguide. The fluorescent light is also outcoupled with the second grating (Kehl et al., 2015; Erdélyi et al., 2008). This type of outcoupling is applied in resonant double grating waveguide structure (DGWS) sensors (Sharon et al., 1997; Kappel et al., 2004; Muriano et al., 2012) and the bidiffractive grating biosensor (BDG) (O'Brien et al., 2000) as well.

2.3.2.3 Resonant Waveguide Grating Sensors

Optical biosensors based on sensing changes in refraction index are suitable for studying biomolecular interactions, but are limited in throughput. Based on microplates, a high throughput system was developed, where the optical sensors in each well of the microplate are resonant waveguide grating (RWG) sensors and consist of a substrate with an optical grating and a high RI waveguide coating (Fig. 11.3d). When illuminated with broadband light at a fixed angle of incidence, these sensors reflect only a narrow band of wavelengths. The light coupled into the waveguide film propagates parallel to the surface in the plane of the waveguide film and creates an evanescent wave in the liquid adjacent to the interface with a penetration depth of ~150 nm. Binding of compounds to the immobilized target induces a change in the effective index of refraction of the waveguide, and this is manifest as a shift in the wavelength of light reflected from the sensor. The magnitude of this wavelength shift is proportional to the amount of analyte that binds to the immobilized target (Daghestani and Day, 2010; Erdélyi et al., 2008).

2.4 Interferometer-Based Biosensors

Interferometric biosensors, as other evanescent field-based sensors, comprise a sensor chip with an optical waveguide, which is in contact with the analyte. The configuration of integrated planar optical waveguide interferometer biosensors is based on a PWG with structured optical elements in order to generate an interferometric signal. For this purpose, the incident light (laser) is split into two beams: the object beam passing through the sensor in contact with the medium to be analyzed, and a reference beam propagating through the same waveguide structure unaffected by the presence of the analyte(s). When biomolecules adsorb on the sensing area, the phase shift or velocity of the guided mode changes as the light propagating in the waveguide interacts with components in the medium. Thus, the principle of these sensors is similar to all evanescent field based sensors, that is, the field is being modified by substances in close proximity to the sensor surface region, resulting in this case in a phase shift in the object beam compared to the reference beam. To improve analytical sensitivity and LOD, the optical phase difference is detected through an interference signal between the object beam and the reference beam created at the sensor output. Such sensor architecture not only outstandingly improves the analytical features of the label-free and nondestructive detection, but it also allows minimized reagent consumption, analysis time, and thus, cost efficacy, offering possibilities for multiplexing, miniaturization and automation (Lukosz et al., 1997; Schneider et al., 1997; Brandenburg et al., 2000; Prieto et al., 2003; Ymeti et al., 2005; Schmitt et al., 2007; Jiang et al., 2008; Kozma et al., 2014).

2.4.1 Mach–Zehnder Interferometer

Mach–Zehnder interferometers (MZIs) employ an optical waveguide split into two arms, and after a certain distance they are recombined again allowing difference measurement between the simultaneously used two channels. The sensor arm will be exposed to a variation of the RI due to immunoreactions in the sensor channel. During this distance, light traveling in the sensing arm will experience a phase shift in comparison with guided light in the reference arm. Coherent, single frequency, single polarization light from a laser enters the single-mode input waveguide and is split equally at a Y-junction. One branch has a window over the top of it allowing the evanescent field of that branch to interact with the sample while the reference arm is protected from the sample with a thick cladding layer. The two branches recombine at the output, resulting in interference, and a photodetector measures the intensity. Generally, the waveguide structure must be single polarization and single mode so that multimodal and cross-polarization interference does not appear at the output. Typically, increasing the length of the sensing window increases the sensing signal (Fan et al., 2008; Kozma et al., 2014).

Liu et al. (2013) demonstrated a highly sensitive label-free MZI biosensor based on silicon nitride slot waveguide. Unlike the conventional MZI sensors, the sensing arm of the sensor consists of a slot waveguide while the reference arm consists of a strip waveguide. Thanks to the slot waveguide's property to provide high optical intensity in a subwavelength-size low RI region (slot region), which allows high light-analyte interaction, higher sensitivity can be obtained as compared to conventional waveguides using the slot waveguide as sensing region. Lu et al. (2011) studied an interferometer biosensor with improved long-time stability and simple operation by using a novel prism-chamber assembly and an inexpensive waveguide made by sputtering a tapered nanometric layer of Ta_2O_5 on a single-mode glass waveguide.

Harris et al. (2015) investigated a refractometry measurement based on a tapered bend-insensitive fiber MZI operating in both guided and leaky mode. The novel approach allows for RI measurement by monitoring power changes in leaky outer-cladding modes, and simultaneous phase based temperature detection using inner-cladding modes, which remain guided.

2.4.2 Young's Interferometer

In a YI, the coherent, single mode, single polarization light at the input is split into a reference arm and a sensing arm. Instead of recombining the arms as in the MZI, the optical output of the two arms combines to form interference fringes on a detector screen, such as a CCD. The spatial intensity distribution represents a phase shift due to a RI change, and the distance between the two arms and the distance between the YI output and the screen, respectively. A phase change on the reference arm causes a shift in the position of the interference fringes can be analyzed quickly with a fast Fourier transform of the spatial intensity (Brandenburg and Henninger, 1994; Fan et al., 2008; Kozma et al., 2014). Schmitt et al. (2008) described the development of a label free biosensor based on a free-space Young's interferometer (YI) configuration with a planar optical waveguide chip as sensor element. Two gratings located in the waveguide substrate are used for light coupling. The interference pattern resulting from an overlap of two coherent light beams is generated by means of a double slit. With this arrangement a very stable sensor performance has been realized.

2.4.3 Hartman Interferometer

A third and very similar interferometry configuration that has been used for biosensing is the Hartman interferometer (HI). In this configuration, functionalization molecules are patterned in strips on top of a PWG (Schneider et al., 1997). Light is coupled

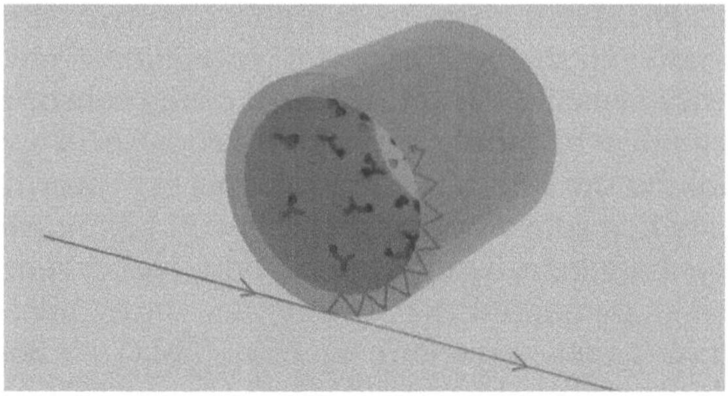

Figure 11.4. Measurement principle of the optical ring resonator sensor setup. The incident laser light directed tangentially onto a microcapillary waveguide is coupled into the high refraction index capillary wall waveguide, where it propagates by total internal reflection and creates an evanescent field on the internal capillary surface. In turn, light intensity of the reflected light is detected.

into and out of the PWG using gratings. Integrated optics is placed near the output of the chip to create interference between pairs of functionalized strips.

2.5 Optical Ring Resonator–Based Biosensors

The optical ring resonator is constituted in the circular cross-section of the microcapillary fluidics (Fig. 11.4). In a ring resonator, the light propagates in the form of whispering gallery modes (WGMs) or circulating waveguide modes, which result from TIR of light along the curved boundary between the high and low RI media. The WGM has the evanescent field present at the ring resonator surface and responds to the binding of biomolecules. The capillary wall acts as a waveguide for repeatedly circulating light coupled in from a tapered fiber cable.

2.5.1 Optofluidic Ring Resonators

The principle of the optofluidic ring resonator (OFRR) is rather simple: when a tangential laser is passed at the external surface of a microsized glass capillary, the circular capillary wall forms a ring resonator supporting WGMs, as an evanescent field that extends even beyond the inner wall of the OFRR capillary. In other words, optical characteristics of the light external to the OFRR capillary are affected by molecular processes taking place on the internal capillary surface as well. Thus, when analyte molecules bind to biorecognition elements (antibodies, receptors, etc.) immobilized at the inner surface, the effective RI of the OFRR changes, and in turn, the resonant wavelength spectrally shifts. Thus, by directly or indirectly monitoring the WGM spectral shift, it is possible to obtain both quantitative and kinetic information about the binding of molecules near the surface. Ring resonator biosensors have been presented in different configurations as microfabricated ring-, disk- or microtoroid-shaped resonators on a chip, standalone dielectric microspheres and capillary based OFRRs (White et al., 2007; Fan et al., 2008; Yang et al., 2008; Gounaridis et al., 2015).

2.5.2 Fluorescent Core Microcapillary Sensors

Microcapillaries are efficiently applied for sensing not only in OFRRs, but also in fluorescent core microcapillary (FCM) structures, in which fluorescent coating is being applied on the inner surface of microcapillaries, and the fluorescent layer thus created is excited with a diode laser or light emitting diode (LED). The FCM makes use of the high RI fluorescent layer, and WGMs develop due to the difference in RI between the fluorescent layer and the glass wall. The main advantage of the technique is its robustness, but an element of limited reproducibility is that the thickness of the coating is hard to control, affecting sensor sensitivity (Manchee et al., 2011; Rowland et al., 2013; Lane et al., 2014).

2.6 Optical Fiber-Based Biosensors

Recently the fiber optic-based biosensors are coming into the front, as fibers are inexpensive material for optical sensor design providing efficient signal delivery. Fiber Bragg gratings (FBGs) are to be applied as versatile sensor with a wide range of applications. An FBG is constructed in a short segment of optical fiber that reflects particular wavelengths of light and transmits all others. This is achieved by creating a periodic variation in the RI of the fiber core, which generates a wavelength-specific dielectric mirror. Illuminating the fiber with two focused intersecting laser beams, RI perturbations can be written into the fiber core that have periodicities on the order of the wavelength. The resulting structure functions as a band rejection filter, reflecting a narrow band of light at the Bragg wavelength (Riant, 2003; Bosch et al., 2007; Fan et al., 2008). To expose the evanescent field from the FBG, several strategies have been introduced. One possibility is to create a surface grating on the side of the fiber, requiring bending and polishing until the fiber core is exposed, and then a grating is physically patterned on that surface. Biosensing was also carried out with a D-shape fiber patterned with a FBG or with chemically etching the optical fiber down to its core (Schroeder et al., 2001; Lowder et al., 2007; Saini et al., 2007).

Long-period gratings manufactured with periodicity on the order of 100 μm to 1 mm, three or four orders of magnitude larger than traditional FBGs. In a long-period grating, core modes couple into the cladding modes of the fiber. In the cladding of the fiber, there is an increased field intensity, the spectral position of the reflection peak becomes more sensitive to the RI change in the surrounding medium, so they sense the RI changes around the cladding without etching or complex structural design (Chen et al., 2006; Fan et al., 2008).

2.7 Lab-on-a-Chip Applications

Microfluidics or lab-on-a-chip (LOC) are miniaturized devices integrating all or major steps of a chemical or biochemical analysis in a single device or chip. This includes sample preparation, amplification and purification, sample/fluid manipulation, and finally sample detection and analysis, all merged on a single chip. Most current devices are simple planar microfluidic devices that do not incorporate the detection and after the reaction has taken place, the readout must be done with complex instrumentation in laboratory settings. Integrated optics-based biosensors have become the most suitable technology for LOC integration due to their ability for miniaturization, their extreme sensitivity, robustness, reliability, and their potential for multiplexing and mass production at low cost. Photonic biosensors are well-established technologies for the sensitive monitoring of molecular interactions. They could afford the requirements for the "on-chip" detection in LOC platforms due to their outstanding characteristics of sensitivity, label-free and real-time detection. The detection principle of most optical biosensors is based on the evanescent field detection providing well developed technology for fabricating new devices and well characterized detection methods (Estevez et al., 2012; Jokerst et al., 2012; Ríos et al., 2012).

3 Nanobiosensor Applications for Agro-Environmental and Food Safety

Environmental and food safety as a key aspect in food production and processing has come forward into the forefront during the past decades. In harmony with the food safety "farm to fork" concept throughout the agri-food sector, agro-environmental safety has become recognized as a major factor affecting food safety, and along with stricter food safety measures analysis of produce and food contaminants gained increased importance. Biosensors—especially by the adaptation of nanotechnology—represent a cutting-edge frontier in environmental and food analysis. Thus, due to their excellent analytical features, integrated optical sensors including interferometers, resonators, coupling-based devices, such as grating couplers and chemically induced field profile change sensors have been proposed to be applied for monitoring chemical contaminants (Lambeck, 2006) and MZI sensors to be applied in water quality management (De Graaf et al., 2012). With the long expansion of immunoanalysis, antibodies have long been the most popular affinity-based recognition elements used, just like in other areas of bioanalysis, in biosensor development as well. In addition, however, great attention has been focused recently on the utilization of novel recognition elements including binding proteins, molecularly imprinted polymers or aptamers. These techniques often provide improved characteristics, like specificity, sensitivity, stability, applicability, or cost-efficacy (Alocilja and Radke, 2003; Wagacha and Muthomi, 2008; Van Dorst et al., 2010; Arora et al., 2011; Scognamiglio et al., 2014; Warriner et al., 2014). Yet, antibody-based immunosensors remain to provide different strategies for immunosensing both in direct or indirect (competitive) formats depending on the design by which the structure of the bioactive components are built up on the sensor surface (Fig. 11.5a). The proper antibodies used as capture elements immobilized on the sensor surface, organic microcontaminants, proteins, or microbes are capable to selectively bind to such bioactivated surfaces. Alternatively, the surface of the immunosensor can be sensitized with antigens or antigen-conjugates for different contaminants. The analyte (antigen) content in the sample inhibits, in a competitive manner, binding of the antibodies added to the bioactivated surface. Sensitivity can be enhanced if the antigen and the proper antiserum is preincubated prior to being injected onto the sensor surface, and then detection is carried out as a binding inhibition test, resulting in a further decrease in LOD relative to the direct, immobilized antibody method. The inhibition assay format is very prosperous especially for the detection of low molecular weight environmental analytes. The signal measured by the immunosensor could be further amplified by nanoparticles coupled to the antibody used.

Considering these factors, there is a wide variety of biosensor applications, and the emerging technologies of lab-on-a-chip microdevices and nanosensors offer further opportunities for the development of new biosensors with much better performance also for their use in environmental, food, and biothreat/biowarfare analysis. Analytical characteristics of the evanescent field effect–based sensors can be attenuated by the use of nanostructures, that is, improvements in sensitivity, LOD, miniaturization,

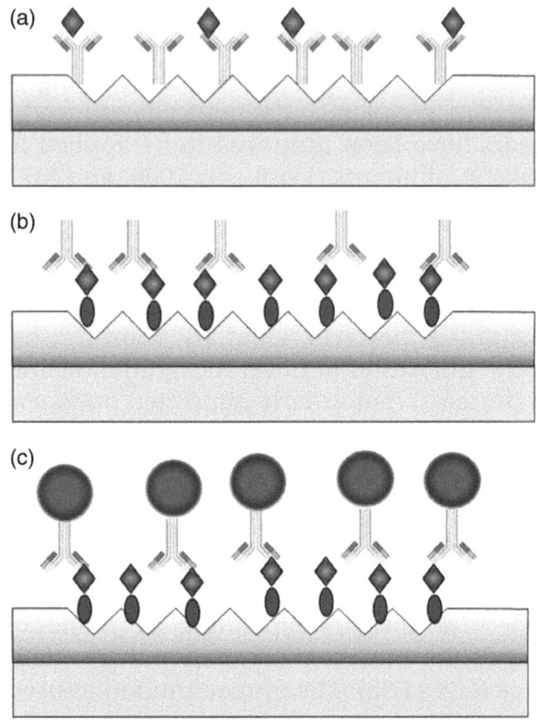

Figure 11.5. Immunosensor setups for direct and competitive detection of analytes. In the direct format (a) antibodies are immobilized on the sensor surface allowing direct capture of the analyte (antigen) from the cover medium. In the competitive format (b) antigen or its conjugate is immobilized on the sensor surface, and binding of analyte-specific antibodies, inhibited by analyte content of the sample added together or preincubated with the analyte-specific antibodies, is detected. The competitive format can be amplified (c) by labeling the antibodies with nanoparticles. (● nanoparticle, antibody, ◆ antigen, antigen-conjugate).

or reusability are achieved (Jianrong et al., 2004; Fan et al., 2008; Yang et al., 2008; Bahadir and Sezgintürk, 2015).

In this section nanobiosensors based on evanescent field effect developed for the determination of the most important biological and chemical contaminants are summarized. The survey collects the appropriate biosensing developments for the determination of hazardous residues (pesticides, drug residues used in animal husbandry), mycotoxins and microbial contamination. The summary solely covers applications of the evanescent field effect–based optical biosensor techniques discussed earlier, and makes an attempt to demonstrate the variegation of the immunosensor developments.

3.1 Nanobiosensors for Pesticide and Drug Residues and Their Indicator Proteins

As a chemical family, the amino-s-triazines, including atrazine and simazine, are a group of herbicides with a wide range of former uses until the restriction or ban in the European Union and worldwide on most of them. Atrazine used to be a worldwide market-leading herbicide active ingredient in corn (in the European Union until its ban in 2004 due to its ubiquitous and unpreventable water contamination), whereas simazine used to have the most crops and sites authorized for use. Most of their uses became classified as restricted due to the ground and surface water contaminating tendency of these compounds, and practically only terbutylazine remained authorized in the European Union from this group. In parallel, monitoring these compounds not only in food or feed commodities, but also in environmental matrices, including surface water and soil, became of key importance in environmental and food analysis. Several sensorics based analytical methods are suggested for the determination of the triazine derivatives. Schipper et al. (1997) introduced an MZI sensor for atrazine allowed a fast, semi-quantitative, competitive, slope direction measurement. This fast method (~10 min) with a limiting sensitivity around 0.1 ng mL^{-1} (the EU limit for given pesticide residues) is especially suitable for prescreening applications. Prescreening of water samples can reduce the number of samples to be analyzed with classical complex and time-consuming instrumental analytical techniques like high-performance liquid chromatography coupled with mass spectroscopy (HPLC–MS) or gas chromatography coupled with mass spectroscopy (GC–MS). Skládal et al. (1999) tested a competitive affinity RM biosensor for the determination of atrazine in soil. For methanolic extracts of soil the same sensor chip was repeatedly applicable for as high as 120 assays. Detection of atrazine concentrations from 0.1 ng mL^{-1} to 10 μg mL^{-1} was evaluated, but the practical LOD in soil extracts was found to be 1 ng mL^{-1} due to matrix effects. Biosensor formats using label-free detection with different optical transducers (RMGC, MZI, SPR, RIfS) were compared for the detection of simazine (Brecht and Gauglitz, 1997). All determinations were devised in the competitive mode: antibodies were preincubated with the sample allowing inhibition of the antibody binding sites by the analyte, and subsequently the concentration of the free antibodies was quantified by binding to a transducer modified with a derivative of the analyte. Test cycles below 15 min were achieved, but sensor performance was limited by drift and noise of the transducers. LODs were found comparable for all transducer formats and reached 0.05–0.15 ng mL^{-1} under laboratory conditions. With RMGC a test midpoint of 0.4 ng mL^{-1} and a working range of 0.2–1.5 ng mL^{-1} was estimated. The MZI technique provided a calibration curve midpoint of 0.3 ng mL^{-1} and a LOD of 0.1 ng mL^{-1}. When using SPR based sensor, test midpoint of 0.5 ng mL^{-1} was found with an estimated LOD of 0.2 ng mL^{-1}, whereas with RIfS apparatus a test midpoint of 0.28 ng mL^{-1} of the target active ingredient was found. Field trials with real-world samples were performed with the SPR and RIfS detection–based immunosensors, and the performance of the devices was to some extent dependent on the sample matrix. Results from ground water or surface water with limited amounts of dissolved organic carbon showed acceptable agreement with HPLC or GC-MS determinations (Brecht and Gauglitz, 1997). Different integrated optical channel waveguide MZIs were characterized and applied as label-free immunoprobes with the advantage that referencing the output signal allowed the elimination of signal fluctuations due to coupling and light-source instabilities, resulting in an increase of the signal-to-noise ratio by a factor of up to 10. Using an improved device, a simazine immunosensor test has been carried out with a test midpoint of 0.3 ng mL^{-1} and a LOD of approximately 0.1 ng mL^{-1} (Drapp et al., 1997). An RMGC was characterized and the potential of the device was investigated for simazine detection with indirect format. Using Fab fragments of an antisimazine polyclonal antibody (pAb) a test midpoint of 0.6 ng mL^{-1} and LOD of approximately 0.25 μg L^{-1} was achieved (Piehler et al., 1997). A TIRF-based biosensor was used to discriminate a mixture of atrazine and simazine with the use of cross-reactive antibodies for these two analytes. The multivariate calibration of the sensor signal was performed using artificial neural networks, as the relationship between the sensor signals and the concentration of the analytes was found to be highly nonlinear. The examination of the procedure showed that a 2-step measurement using only one antibody per step allowed a quantification of both analytes in a mixture with LODs of 0.2 and 0.3 ng mL^{-1} for atrazine and simazine, respectively. Although these LODs are above the level set in the EU directive for single pesticide residues in drinking water, the biosensor proposed has several advantages compared with common chromatographic procedures: it allows fully automated use, it can be regenerated allowing several hundreds of measurements without user input, and even remote control for online monitoring in the field is possible with the measurement of a sample taking only 30 min in total (Reder et al., 2003).

Nanobiosensor techniques were developed as analytical possibilities for determination of other pesticides as well. Mosiello et al. (1997) developed a fiber optic TIRF sensor using the red fluorophore Cy.5, based on a compact, inexpensive solid-state laser source and photodetector for the quantitative detection of herbicide active ingredient 2,4-dichlorophenoxy-acetic acid (2,4-D). The working range of the assay with the monoclonal antibody (mAb) based sensor was found to be 0.07–2.2 ng mL^{-1} of 2,4-D.

The determination of the selective postemergent general use acetanilide herbicide propanil, banned in the European Union in 2002 but still in wide use in the United States and worldwide, is becoming of increasing importance in environmental and drinking water samples. Optimization of a TIRF-based immunosensor resulted in a LOD of 0.0006 ng mL^{-1} and an LOQ of 0.0045 ng mL^{-1}. Such highly improved sensitivity for detection of single pesticide residues was facilitated by the application of the immunosensor in the ultrasensitive, fully automated and robust biosensor River Analyser (RIANA) developed in the scope of an EU project (1995–1999) and further developed into the Automated Water Analyser Computer Supported System (AWACSS) (Tschmelak et al., 2005a,b), and it indicates that biosensors can compete with traditional instrumental analytical methods in the field of water analysis (Tschmelak et al., 2004a).

OWLS-based immunosensors were applied both in direct and in indirect format for the detection of the herbicide active ingredient trifluralin. This compound used to be commonly used as a preemergence herbicide until its ban in the European Union in 2008 due to its endocrine disrupting effects and decrease in use in the United States due to its classification as a possible human carcinogen. The immobilized pAb based direct OWLS immunosensor allowed the detection of trifluralin only above 100 ng mL^{-1} due to the small molecular size of the target analyte. The immobilized antigen-conjugate based OWLS system, however, allowed the detection of trifluralin in the concentration range of 2×10^{-7} to 3×10^{-5} ng mL^{-1}. Results of trifluralin determinations in spiked and field surface water samples were compared to those obtained in parallel enzyme-linked immunosorbent assay (ELISA) tests and in GC–MS analyses, and indicated an increase of six orders of magnitude in LOD (Székács et al., 2003).

Yang et al. (2008) developed a novel label-free OFRR biosensor for detection of a long-banned organophosphorus insecticide active ingredient of outstanding ecotoxicity, parathion-methyl. The linear working range was obtained from 0.01 to 250 ng mL^{-1} with a response time of 0.1–0.5 min and a LOD of 0.0094 ng mL^{-1}, suggesting that the OFRR biosensor offers good opportunities for direct detection of toxic contamination in field samples.

A TIRE sensor was developed for the detection of the old type pesticide of general toxicity, pentachlorophenol (PCP) (Basova et al., 2010). Due to its broad cytotoxic activity, PCP used to be registered a herbicide, insecticide, fungicide, algaecide, and disinfectant active ingredient, but its use has gradually been restricted particularly because of side effects due to its wide cellular toxicity. Pesticide uses on PCP have been banned in the European Union and European countries as early as at the end of the 1980s, and it has gradually lost utility in areas other than wood preservation, with this latter also coming to an end in 2015, when PCP was globally banned worldwide by the United Nations. Analysis of a restricted on banned substance is always of importance both for monitoring and for official inspectory purposes. In the TIRE sensor, thin films of eight octa-tosylamido-substituted zinc phthalocyanine derivatives ($ZnPcR_8$) were investigated as active layers of optical sensors. Using TIRE detection technique, response characteristics of the $ZnPcR_8$ films to PCP were studied in the concentration range of 0.1–20 ng mL^{-1}, and also validated for natural river and wastewater samples. It was demonstrated that the specific binding of PCP molecules caused substantial shifts of the spectrum. The LOD for PCP was found to be 0.7 ng mL^{-1} (Basova et al., 2010).

Environmental and food safety is affected not only by pesticide residues as contaminants of agricultural origin, but also by residues of veterinary drugs or even pharmaceuticals. Moreover, technical contaminants are also of safety concerns. Therefore, such compounds also emerge as important target analytes. A direct-binding OWLS based immunosensor detecting sulfamethazine (SMZ), a sulfonamide antibiotic with wide range of antimicrobial activity, commonly used in animal and fish feedstuffs for prophylactic and therapeutic purpose, was developed. Anti-SMZ antibodies were covalently anchored in situ on the surface of the sensor chip activated by 3-aminopropyltriethoxysilane and glutaraldehyde. The LOD was found to be 2.78 ng mL^{-1}, specific sensor responses increased logarithmically with SMZ concentrations up to 2.8 mg mL^{-1}, and the immunosensor was favorably reusable for SMZ screening (Kim et al., 2008a). Chloramphenicol antibiotic was measured in standard solutions with an OWLS immunosensor. Immobilization of antichloramphenicol antibodies has been performed onto the surface of OWLS chips with glutaraldehyde as crosslinker, and the signal measured was proportional to the analyte concentration in the range of 32 ng mL^{-1} to 320 μg mL^{-1} (Adányi et al., 2006). Another highly toxic and ubiquitous pollutant antibiotic sulfadimidine is reported to be detected with a portable and reusable optofluidics based fiber optic biosensor (Hao et al., 2015). The indirect competitive immunosensor using Cy5.5-labeled antibody quantitatively detects sulphadimidine in the concentration range of 0.17–10.73 ng·mL^{-1} with a LOD of

0.05 ng·mL^{-1}, and allows regeneration more than 300 times without losing sensitivity.

Biosensors gained importance also in hormone detection. A TIRF based competitive immunosensor, utilizing a surface bound progesterone hapten and antiprogesterone mAb as a biological recognition element, was developed for progesterone testing in daily milk. The assay time was minimized in about 5 min, and the method allowed a LOD of 0.04 ng mL^{-1} and a LOQ of 0.34 ng mL^{-1}. An outstanding advantage of the method is that through its rapid performance without the need for any sample-preparatory steps it allows automatization and in-line integration into the milking process machinery for continuous monitoring the estrous cycle of dairy cattle (Käppel et al., 2007). Highly sensitive detection methods based on RIfS and TIRF with immunochemistry to determine testosterone content in aqueous samples were also developed (Tschmelak et al., 2004d, 2006). The TIRF-based biosensor setup enabled rapid and ultrasensitive analyses in real-world samples of testosterone with a remarkable LOD of 0.2 ng L^{-1}. All real-world samples, even those containing the antigen only in the sub-ng L^{-1} range (eg, 0.9 ng L^{-1}), could be determined with recovery rates between 70% and 120%.

An immunosensor approach based on nonlinear two-photon fluorescence spectroscopy coupled with a DGWS as a resonant Ta_2O_5 sensing chip, has been used for the detection of methylboldenone, an androgenic anabolic steroid used illegally as a growth promoter. This synthetic steroid is detected by the biosensor down to 0.1 ng mL^{-1} concentration. The detection relies on a direct competitive format, using a boldenone-rhodamine conjugate as fluorescent competitor and a specific antimethylboldenone antibody. The developed immunosensor presents great potential as a robust device for its implementation on the detection of small illegal and toxic contaminants. A significant increase in the detectability of the analyte with two-photon fluorescence in comparison with one-photon fluorescent immunoassay measurements was achieved with an IC_{50} of 4.6 ± 0.13 ng mL^{-1} and an LOD of 0.1 ± 0.004 ng mL^{-1} (Muriano et al., 2012).

A TIRF-based indirect competitive PWGFI immunosensor was devised for sensitive, rapid, and simple quantification of melamine, originated from plasticware, in buffer and milk products. The chip carrying immobilized melamine-BSA conjugate was reusable and highly resistive to nonspecific binding of proteins. The regeneration of the sensor allows more than 100 assay cycles within 20 min for each assay cycle. For the standard curve, the LOD was 6.6 ng mL^{-1} and the quantitative detection range was 26.6–517.5 ng mL^{-1}. The results meet the requirements for the maximum amount of melamine allowed in infant formula (1.0 mg kg^{-1}) and all other food products (2.5 mg kg^{-1}) set by the World Health Organization. Minor cross-reactivity effects were noted to occur among closely resembling pollutant molecules. Recoveries obtained in the melamine-fortified blank samples were between 89.8 and 103.2% with relative standard deviation lower than 8.1% (Guo et al., 2014).

Bisphenol A (BPA) is another common food contaminant used in food packaging materials and is a ubiquitous environmental pollutant with endocrine disrupting potential. As a result, BPA has restrictions in certain applications and a ban in the European Union and Canada on use in baby bottles. In turn, BPA is required to be monitored in food and environmental samples. A reusable PWG immunosensor was developed for the detection of BPA in water samples with high sensitivity (Xiao-hong et al., 2014a). In the competitive immunosensor format, anti-BPA antibodies labeled with a fluorophore moiety bind to the sensor surface if BPA content in the sample does not inhibit the process, and the incident light coupled into the PWG and propagating there by TIR excites the fluorophore label on the antibodies on the sensor surface. The method was shown to allow an LOD of 30 ng mL^{-1}, a detection range of 124 ng mL^{-1}–9.60 μg mL^{-1} and a test midpoint of 1.09 ± 0.25 μg mL^{-1}. The rapid method (20 min per assay cycle) allows regenerable use with more than 300 assay cycles on real water samples.

A growing concern has been expressed in recent years regarding the impact of endocrine disrupting chemicals (EDCs), thought to mimic natural hormones and thus, to cause adverse effects on human and wildlife growth, development and reproduction. A large number of chemicals have been reported as EDCs (as also seen earlier) affecting water quality, and the fact that both their chemical characteristics and possible hormonal receptors can be multivarious renders the analytical detection of these compounds difficult. In turn, EDCs are often detected by mechanism-based bioanalytical procedures or analyte-specific immunoanalytical methods. A common pathway in endocrine effects is the disruption of vitellogenesis, normal production of the yolk-related protein, vitellogenin (Vtg) involved in oviproduction. Vtg has, therefore, become a widely used biomarker for assessing exposure to environmental EDCs.

A label-free carp (*Cyprinus carpio*) Vtg sensor has a strong potential for on-site monitoring on the possible contamination of edible fish with endocrine disruptors as a sum parameter in an inland carp farm. A direct binding OWLS based immunosensor was developed for carp Vtg detection. A linear relationship was found between analyte concentration and sensor response, together with the LOD of 0.00675 nM. The reusability of the immunosensor was reasonably good, as presumed from the coefficient of variability (CV) of 6.02% for nine repetitive measurements. When a female and male carp sera induced with 17β-estradiol injection were analyzed, biomarker induction was even identifiable at 2000-fold serum dilution (Kim et al., 2008b). Another biosensor application was also developed for the detection of Vtg concentration in seawater flatfish (*Paralichthys olivaceus*) (Kim et al., 2007a). Very similar development was performed for Vtg determination in carp, where purified pAb raised in rabbits against lipovitellin (Lpv) purified from carp ovary was used (Adányi et al., 2013). The serum showed fully cross-reactivity with Vtg and was applied both in direct and indirect measurement formats. The direct OWLS immunosensor and the sandwich ELISA assay developed showed linear measuring range in 600–12,000 and 200–3,000 ng mL^{-1} Lpv concentrations in standard solutions, respectively. The competitive immunosensor allowed a sensitive detection range for Lpv between 3 and 100 ng mL^{-1}. Vtg content in blood and liver samples from male and female carp were determined by indirect OWLS method. Vtg levels in the liver of female carp was lower than determined in the blood samples, therefore, blood occurs to be a more proper tissue to monitor elevated Vtg in male carp due to possible exposure to water-contaminating endocrine disrupting chemicals in fish breeding or fishponds. To monitor the EDC pollution of the specific environmental regions, an OWLS-based immunosensor was developed for frog (*Bombina orientalis*) Vtg determination in biological samples as a biomarker for exogenous estrogen compounds (Majer-Baranyi et al., 2015a). Antibody against Vtg was produced with purified Lpv from the homogenized ovary of oriental fire-bellied toads (*B. orientalis*). When measuring Vtg in direct manner, the linear measuring range for Vtg was 0.1–10 μg mL^{-1}, while for the competitive measurement, the range for Vtg was 0.5–50 ng mL^{-1}. Heart, liver, and gonad samples from male animals were spiked with Vtg and were analyzed using the newly developed method, and female toads and spawn samples were tested and compared to the calibration curve obtained by the spiked samples.

3.2 Nanobiosensors for Mycotoxins and Toxins in Foodstuff

Mycotoxins are toxic secondary metabolites produced by fungi, and these biochemical compounds may contaminate various agricultural commodities either before harvest or under postharvest conditions, and can be present in a wide range of food and feed commodities including cereal grains, oil seeds, spices, dried fruits, apple juice, wine, and meat products from animals fed with contaminated meal. Tropical conditions such as high temperatures and moisture, monsoons, unseasonal rains during harvest, and flash floods typically lead to fungal proliferation and production of mycotoxins (Bhat and Vasanthi, 2003). Some mycotoxins such as aflatoxins are considered by the US Food and Drug Administration (US FDA) to be unavoidable contaminant of food, therefore the goal has been to minimize contamination. Food-borne mycotoxins likely to be of greatest significance in Africa and other tropical developing countries are fumonisins and aflatoxins (WHO, 2006). Mycotoxins attract worldwide attention because of the significant life quality and economic losses associated with their impact on human health, animal productivity and trade (WHO, 2006; Wu 2006; Wagacha and Muthomi, 2008). Food-borne toxins are widespread and dangerous for human and animal health. It is very important to control the level of T-2 mycotoxin, which has a high toxic effect and may serve as biological weapons (Alocilja and Radke, 2003; Wagacha and Muthomi 2008; Starodub et al., 2012).

As aflatoxins produced by *Aspergillus* species, particularly aflatoxin B1 (AFB1) pose substantial health risks to society, their routine monitoring and control in foodstuffs is a definite requirement. A TIRF-based indirect competitive immunosensor was developed for sensitive and rapid quantification of AFB1 in buffer, corn, and nut products. Methanol extracts of AFB1-spiked samples were incubated with Cy5-anti-AFB1 mAbs, and the sample/antibody mix was injected over a waveguide surface containing immobilized AFB1. AFB1 content in the sample binds to the mAbs, preventing its binding to the surface haptens, and in result, the fluorescence signal decreases. The method allowed an LOD of 0.3 ng mL^{-1} for AFB1 in buffer, as well as 1.5–5.1 and 0.6–1.4 ng g^{-1} in different corn and nut products, respectively (Sapsford et al., 2006). The OWLS technique has been applied for the detection of AFB1 and OTA in both competitive and in direct immunosensors. When using a noncompetitive method, sensor responses were obtained first only at analyte concentrations of 5–10 ng mL^{-1}. In both cases, the responses were very unstable, but the sensitive detection range of the competitive detection method was between 0.5 and 10 ng mL^{-1}. Determination from barley and wheat flour samples well correlated with those with ELISA: the regression coefficient between the two methods for ochratoxin and aflatoxin was determined as 0.96 and 0.89, respectively (Adányi et al., 2007).

Aflatoxin M1 (AFM1) contamination of milk occurs after the metabolism of AFB1 by the cow. Even though the transformation from AFB1 to AFM1 turns AFB1 about 1000 times less toxic, AFM1 still belongs to the most toxic mycotoxin class. An OWLS-based immunosensing method was investigated for the detection of AFM1 using indirect (competitive) immunoassay method. Comparing the different sample preparation methods, it was concluded, that the best results were obtained when 100-fold diluted filtered or centrifuged samples were measured. In the case of filtered milk samples, AFM1 could be detected in the range of 0.001–0.1 ng mL^{-1} with an LOD of 0.0005 ng mL^{-1}. As for centrifuged milk samples, the LOD was as low as 0.0001 ng mL^{-1}, and the dynamic measuring range was found between 0.005–0.01 ng mL^{-1} (Szalontai et al., 2014).

An integrated analytical system was demonstrated for chemiluminescence detection of ochratoxin A (OTA), produced by some *Aspergillus* species. The method conjugates an indirect competitive ELISA strategy developed in polydimethylsiloxane microfluidics with integrated microfabricated hydrogenated amorphous silicon photodiodes. An LOD of 0.85 ng mL^{-1} was obtained for OTA detection in a phosphate buffered saline solution using a straight-channel configuration. Comparable LODs were obtained for beer extracts, but only a higher LOD of OTA of 28 ng mL^{-1} was obtained for red wine extracts. A two-channel U-shape microfluidic device was also developed to perform the simultaneous analysis of a reference solution and of an OTA contaminated solution, in which the measurement errors were efficiently reduced of one order of magnitude in the LOD of OTA in beer and red wine extracts. Based on the results, the miniaturized system can be developed into one stage of a future highly sensitive, portable, and fully integrated "toxin-chip" for applications for monitoring food safety (Novo et al., 2013).

Contamination by deoxynivalenol (DON), a trichothecene mycotoxin produced by *Fusarium* species, occurs in cereals worldwide; therefore, efforts have been made toward the development of rapid and sensitive methods for the detection of this compound. An OWLS technique based label-free method was developed for detection of DON in both competitive and in direct immunoassay formats using DON-specific pAbs. The direct method was found to result in an unstable sensor response and sensitivity insufficient to determine DON in different grains. In contrast, a competitive immunosensor format provided reproducible quantitative detection in the sub-ng L^{-1} range. The sensitive detection range of the competitive detection method was between 0.01 and 50 ng mL^{-1}. After the establishment of the indirect method, spiked wheat flour samples were investigated. Results obtained with spiked samples showed that OWLS detection has a potential for quick determination of DON in wheat samples (Majer-Baranyi et al., 2011). A biolayer interferometry (BLI) based measuring method was developed to detect DON without the need for labeling. A BLI assay of DON in wheat flour can be done in 7 min, and the determined LOD was 0.1 mg kg^{-1}. To monitor DON in wheat rapidly, a biosensor using the principle of BLI was developed. The signal from the sensor was substantially amplified through the use of a primary antibody–colloidal gold conjugate. The amplification was much greater in the presence of wheat matrix than in buffered solution, suggesting matrix components may have contributed to the enhancement. The LOD of the method was 0.09 mg kg^{-1}, while the LOQ was 0.35 mg kg^{-1}. Recovery from wheat spiked over the range from 0.2 to 5 mg kg^{-1} averaged 103% (RSD = 12%). The quantitative assay compared favorably (r^2 = 0.9698) with a reference chromatographic method for 40 naturally contaminated wheats (Maragos, 2011, 2012). PWG-based indirect competitive immunoassay was performed recently as a novel screening method for simultaneous analysis of multiple mycotoxins. Through light propagation in the waveguide, fluorescence from labeled antibodies in antigen antibody complex bound to the surface of a biochip was measured. The LOD of DON detected by PWG-based system was approximately 400 mg kg^{-1}. However, a further study demonstrated that, for wheat samples containing 1250–2000 mg kg^{-1} DON (still within the regulatory limits), the results obtained from the PWG based system were only 71% of those values detected by GC–MS. While the method is suitable for simultaneous analysis of mycotoxins, PWG has lower sensitivity compared to some other analytical techniques (Tittlemier et al., 2013; Ran et al., 2013).

As *Fusarium* mycotoxin zearalenone (ZON) is also a widely occurring estrogenic pollutant of fungal origin in agricultural plants and commodities, routine monitoring of this substance is prompted. An OWLS method was developed and again the competitive immunosensor was found to be superior in sensitivity as compared to the corresponding noncompetitive sensor, and the former surpassed in detection sensitivity, to a 50,000-fold extent, the corresponding ELISA. In their common detectable concentration range (1–100 ng mL^{-1}, 0.1–10 μg kg^{-1}) the two methods, OWLS and ELISA, detected closely similar toxin contents (Székács et al., 2009).

Methods for detection of *Fusarium* mycotoxins T-2 and ZON in sub-ppb concentrations were introduced by using TIRE in both direct and competitive immunosensors. A direct immunosensor of ZON provided sensitivity from 0.1 ng mL^{-1} up to 100 μg mL^{-1}, while for competitive immunosensor the LOD was found to be 0.01 ng mL^{-1}. MAbs proved to be 10-fold more selective to

T-2 mycotoxin than pAbs, although pAbs provided a wider range of linear response (0.1–150 ng mL^{-1}) to T-2 mycotoxin (Nabok et al., 2009). The investigation of grain products samples revealed the presence of high concentrations of T-2 (>600 ng mL^{-1}) in mold products (Nabok et al., 2007, 2009, 2011; Starodub et al., 2012). TIRE was also applied to the detection of T-2. MAbs proved to be 10-fold more selective to T-2 mycotoxin than pAbs, although pAbs provided a wider range of linear response (0.1–150 ng/mL^{-1}) to T-2 mycotoxin. Despite its high sensitivity, this biosensing technique requires expensive equipment, and in situ measurements are not possible (Prieto-Simón et al., 2007; Nabok et al., 2007, 2009).

Mycotoxin contamination is of particular interest, within food safety, in the safety of spice products as well. Spice plants and herbs may suffer unintended, accidental, or occasional chemical or microbiological contamination during their cultivation, processing, mixing, or storage. Moreover, for various chemical micro-contaminants and microbiological pollutants, not only accidental, but deliberate contamination of spices may occur, either aiming for unethical and illegal market gains by food fraud or adultery, or for societally malignant purposes. Although such intentional food fraud or biothreat type contamination is not typical for mycotoxins, possible use of T-2 mycotoxin in prepence of malice has been emphasized in the literature (Starodub et al., 2012), and cases of intentional release to the market of food (spice) products known by the producer or distributor to be contaminated with mycotoxins due to natural (eg, climate conditions) or technological (eg, improper harvest or storage) causes is also considered unethical deliberate contamination that puts the consumer at risk. Optical biosensors for mycotoxins or microorganisms have a great potential in quality assurance of spice production and provision of spice safety. In the scope of the EU-project SPICED within the 7th Framework Programme for Research of the EU, Adányi et al. (2015a,b) developed an OWLS-based immunosensor for monitoring composition of herbs, and similar OWLS immunosensor are being developed for monitoring mycotoxins in spice paprika (Majer-Baranyi et al., 2015b).

Staphylococcal enterotoxin B (SEB) is an exotoxin excreted by the *Staphylococcus aureus* bacterium. *Staphylococcus* species thrive and produce toxins in unrefrigerated meats, dairy, and bakery products. A capillary biosensor applying the waveguiding properties of the capillary was demonstrated to integrate the signal over an increased surface area without simultaneously increasing the background noise from the detector. A dose-response study was also conducted using the protein toxin SEB and antigen concentrations ranging from 10 pg mL^{-1} to 1 ng mL^{-1}. The LOD was 30 pg mL^{-1} and the range of quantitation for the curve was between 30 and 400 pg mL^{-1} (Ligler et al., 2002).

Marine dinoflagellates of the genera *Alexandrium* are well-known producers of the potent neurotoxic paralytic shellfish toxin, saxitoxin that upon entering the food web ultimately present a serious risk to public health in addition to causing extensive economic losses. Direct coastal monitoring of *Alexandrium* spp. can provide early warning of potential shellfish contamination and risks to consumers by using a PWGFI. The disposable PWG was constructed on a transparent substrate onto which an array of a toxin-protein conjugate is deposited, assembled in a cartridge allowing the introduction of sample, and detection reagents. High affinity antibody to paralytic shellfish toxins was used in the competitive assay format with a detection signal generated via a fluorescently labeled secondary antibody. Assay speed was optimized to enable measurement within 15 min. A rapid, portable sample preparation technique was developed for *Alexandrium* spp. in seawater The assay was validated and the LOD and detection capability (CCβ), defined as the smallest content of the substance that may be detected, identified and/or quantified in a sample with a given error probability, were determined as 12 and 20 pg mL^{-1} respectively, with an intra-assay CV of 11.3% at the CCβ and an average recovery of 106% (Meneely et al., 2013).

Yessotoxin (YTX), a group of lipophilic sulfated polyether compounds, is produced by a number of planktonic algal species. Association measurements were done in a RM biosensor. Aminosilane surfaces were used, on which 3′,5′-cyclic-nucleotide-specific phosphodiesterase (PDE) from bovine brain was immobilized. Over this anchored ligand different amounts of YTX were added and typical association curve of pseudo-first-order kinetic profiles were observed. The PDE–YTX association was suitable for determination of the toxin concentration in shellfish samples. The assay had sufficient sensitivity and can be used on simple shellfish extracts with an LOD of about 0.5 μM (570 ng mL^{-1}) (Pazos et al., 2004).

Microcystins are a major group of cyanobacterial heptapeptide toxins found in freshwater and brackish environments. An automated online optical biosensor was developed for detection of microcystin-LR (containing leucine and arginine as variable L-amino acids). The competitive immunosensor was devised with covalently anchored microcystin-conjugate and fluorescence-labeled antimicrocystin mAb, and allowed a quantification range of 0.2–4 ng mL^{-1}, an LOD of 0.09 ng mL^{-1}, validated by HPLC, as well as continuous real time determination and high specificity (Shi et al., 2013). An element of novelty of the optimized fluorescence

PWG immunosensor is that it applies a single chain variable fragment (scFv) of an avian antibody for recognition, allowing a detection range of 0.21–5.9 ng mL^{-1} and a practical LOD of 0.19 ng mL^{-1} for free microcystin-LR in solution, as well as high reproducibility, low reagent volumes, inexpensive cartridge format and assay time of less than 20 min (Murphy et al., 2015).

3.3 Nanotechnology-Based Microbial Sensors

Selective discrimination of bacterial strain types and detection of actual cell concentrations is of great importance in all areas of microbial research including toxicity monitoring, clinical diagnostics, food safety, and environmental science (Xi et al., 2011). Novel sensitive sensing molecules (antibodies, lectins) and immobilization methods, enhanced sensitivity and selectivity are needed for real-time detection of contaminating microbes present in different samples. The four major food-borne pathogens are *Salmonella, Listeria monocytogenes, Campylobacter*, and *Escherichia coli* O157:H7, while accordingly the most studied and developed methods are for the selective determination of *E. coli* strains, especially the *E. coli* O157:H7. This *E. coli* strain causes a severe food-borne disease through its secreted powerful poison, called verotoxin that binds to receptors in human kidney, brain, and gut cells, and leads to the death of the affected cells. *E. coli* O157:H7 is transmitted primarily through consumption of contaminated foods, such as raw or undercooked ground meat products, raw milk, and contaminated raw vegetables and sprouts. Different types of immunosensors are reported for selective determination of *E. coli* cells (Alocilja and Radke, 2003; Starodub et al., 2012; Bata-Vidács et al., 2013).

A fiber optic biosensor was devised for the detection of pathogenic *E. coli* O157:H7 in spiked ground beef samples (DeMarco and Lim, 2002). The portable evanescent wave–based biosensor used a 635-nm laser diode and optical fiber probes. The evanescent signal generated had a width of approximately 1000 nm over the surface of the fiber optics. Optical emission upon excitation of fluorescent molecules of cyanine 5 dye-labeled anti-*E. coli* O157:H7 pAbs was coupled into the evanescent field, and the fluorescent signal was detected. The method allowed high accuracy: 100% correct identification of positive samples was achieved both with silica and polystyrene waveguides (at 9.0×10^3 and 5.2×10^2 cfu g^{-1} for 25 g ground beef samples, respectively). An integrating waveguide biosensor was described for the detection of waterborne *E. coli* O157, based on a fluorescent sandwich immunosensor performed inside a glass capillary waveguide. The genomic DNA of captured

E. coli O157 cells was extracted and quantitative real-time PCR subsequently performed to assess biosensor-capture efficiency. In vitro microbial growth in capillary waveguide was also documented. The biosensor was shown to allow for quantitative detection of as few as 10 cells per capillary (0.075 mL volume) and can be used in conjunction with cell amplification, PCR, and microarray technologies to positively identify a pathogen (Zhu et al., 2005). Horváth et al. (2003) tested a grating-coupled planar optical waveguide sensor by monitoring the adhesion of *E. coli* K12 cells to the sensor surface. Nonspecific detection of *E coli* was demonstrated with reverse symmetry waveguides. The LOD is estimated to be 60 cells mm^{-2}, a few 100 times better than that in the RM application detected. OWLS technique was also used for the direct measurement of *E. coli* by immobilizing anti-*E. coli* IgG onto the surface of the amino silanized waveguide sensor by covalent coupling with glutaraldehyde. The measuring range was found to be between $3–10^4$ and $3–10^7$ cfu mL^{-1} (Adányi et al., 2006).

An HI using an evanescent wave was described with a single planar wave of polarized light. Using this technique *Salmonella typhimurium* was possible to directly detect in the range of 5×10^8 to 5×10^{10} cfu mL^{-1} with a detection time of 5 min (Schneider et al., 1997). In situ immobilization was studied over an optical grating coupler (OGC, OWLS) sensor chip and measuring method was developed for anti-*Salmonella* antibody, which was raised against the common structural antigen of *Salmonella* spp., followed by label-free direct binding strategy determination of *S. typhimurium*. The sensitivity of the current OGC immunosensor was found reasonably good with LOD of 1.3×10^3 cfu mL^{-1}. The sensor prepared was quite specific to *Salmonella* binding and was also repeatable even after 10 cycles of measurement with a CV of 2.87%. The selectivity of the OGC immunosensor for *Salmonella* detection was tested for other hazardous bacteria of hygienic importance in the food industry, as of *L. monocytogenes*, *E. coli*, and *Pseudomonas aeruginosa* at the same concentration range, where the sensor response was 0.25, 0.57, and 2.32%, respectively, compared to the surface mass change for the *S. typhimurium* injection of the same concentration (Kim et al., 2007b). The feasibility of rapid detecting *S. typhimurium* in food products using a microfluidic based nanobiosensor for immunomagnetic separation and quantum dot (QD) labeling was investigated. For selective detection of *Salmonella*, anti-*Salmonella* pAbs were covalently immobilized onto the QD surface. To separate and concentrate the cells from the sample, superparamagnetic particles and a microfluidic chip were used. A portable fluorometer was used to measure the fluorescence signal from the QD nanoparticles attached to *Salmonella* in the samples. The sensitivity for detection of pathogenic *Salmonella* was evaluated using serially diluted *S. typhimurium* in borate buffer and chicken extract. The LOD of the sensor was 10^3 cfu mL^{-1} *Salmonella* in both borate buffer and food extract (Kim et al., 2015).

An RM-based biosensor was developed for detecting *S. aureus*, as one of the first optosensoric application to detect whole cells. The cells were detected at concentrations in the range of 8×10^6 to 8×10^7 cells mL^{-1} in buffer and 4×10^3 to 1.6×10^6 cells mL^{-1} in milk at a detection time of 5 min on the basis of their binding capacity to human immunoglobulin G immobilized on the aminosilane-derivatized surface of the sensor with or without a hydrogel coating of carboxymethyl-dextran. The use of human immunoglobulin G labeled with colloidal gold resulted in a 1000-fold increase in assay sensitivity (Watts et al., 1994).

A birefringent waveguide biosensor system with an integrated microfluidic channel was used to detect pathogenic bacteria, such as *L. monocytogenes*, in food, especially in dairy products. The detection sensitivity was computationally and experimentally optimized by altering the thickness and length of the TiO_2 layer applied, which is the sensing surface in the waveguide platform. The LOD of *L. monocytogenes* was detected as 10^6 cfu mL^{-1}, and the specificity of detection of *L. monocytogenes* was confirmed by comparison with the detection of *E. coli* O157. The concentration of *L. monocytogenes* in a milk buffer was measured to elucidate the feasibility of this method for pathogen detection in food samples (Sim et al., 2012).

3.4 Multianalyte Nanobiosensor Instrumentation for Environmental and Food Analysis

Several applications are known, in which, thanks to the technical investigations delineated before, different substrates can be analyzed in the same time by using multichannel instrumentation. To monitor the quality of water samples, an integrated evanescent wave multichannel immunosensor (EWMI) system was developed for the parallel detection of 2.4-D, microcystin-LR, bisphenol A, and melamine (Xiao-hong et al., 2014b). With the use of the evanescent field, the fluorescent dyes near the chip surface are excited, and the generated fluorescence signals quantitatively relate to the concentrations of analyte. Incident light is coupled into the planar optical waveguide chip via a beveled angle to form eight individual reflection spots for multianalyte biosensing. Fluorescence emissions are collected by eight multimode fibers placed in parallel underneath the chip. The method was indicated to be capable to detect the four target analytes simultaneously and specifically detected within an analysis time of about 20 min for each assay cycle. LODs for 2,4-D, microcystin-LR, bisphenol A and melamine were 2.17 mg mL^{-1}, 30 μg mL^{-1}, 25 μg mL^{-1}, and 7.5 mg mL^{-1}, respectively. The regeneration of the PWG chip allowed more than 300 assays after each other (Xiao-hong et al., 2014b).

The RIANA analyzer system with analyte detection based on TIRF technology (laser diode 635 nm, 12 mW) was developed, and the analyte recognition was based on competitive, indirect measuring system. The automated and robust RIANA biosensor was developed further to the Automated Water Analyser Computer Supported System (AWACSS) (Tschmelak et al., 2005a,b). In these systems, analyte derivatives were immobilized onto the transducer surface as bioactive recognition elements and analyte-specific antibodies labeled with fluorescent markers were incubated with the analyte samples. To verify the reproducibility, precision and robustness of the TIRF immunosensor, two common statistical methods were used to evaluate the LOD and LOQ for simultaneous multianalyte calibration with atrazine, BPA, and estrone in Milli-Q water. Using the automated biosensor, a LOD below 0.020 ng mL^{-1} was achieved without sample pretreatment and preconcentration for each of the analytes in a simultaneous multianalyte calibration. Calibration of isoproturon herbicide (in single analyte mode) was performed in the same measuring device, and resulted in a LOD of 0.016 ng mL^{-1}, and a LOQ of 0.091 ng mL^{-1} (Tschmelak et al., 2004b). Moreover the RIANA instrumentation was applied as immunosensor for seven different sulfonamide derivatives (sulfadiazine, sulfamethoxazole, sulfadimidine, sulfamethizole, sulfadimethoxine, sulfathiazole, and sulfamethoxypyridazine). The results of the immunosensor were compared using a mathematical model. In an automated biosensor system, LODs below 0.01 ng mL^{-1} and LOQs below 0.1 ng mL^{-1} were achieved without sample preconcentration for sulfonamide compounds. Unstrained spiked and unspiked real water samples with complex matrices (drinking, ground, and surface water) were measured (Tschmelak et al., 2004c).

A portable and reusable optofluidics-based fiber optic biosensor reported to detect sulphadimidine (see earlier) was indicated to simultaneously detect atrazine, melamine and BPA as well, and was suggested for use for in dairy products (Hao et al., 2015). The multianalyte sensor of superior selectivity also showed excellent regeneration capability.

A small, portable, and easily operated biosensor called RAPTOR was investigated for rapid immunoassays in the field. This fiber optic biosensor is designed to detect up to four different analytes at one time performing sandwich fluoroimmunosensoric determinations on four optical waveguides simultaneously, with assay results from all four fibers available in less than 12 min. With a slight modification, the multianalyte detection capability of RAPTOR was doubled by using the same four optical waveguides. Three different sets of capture antibodies were immobilized onto the four waveguides in such a combination that any one of eight analytes was identified in a sample by the pattern of the response. The following eight antigens could be clearly detected at the concentrations tested: *Staphylococcal* enterotoxin type-B (SEB), *Francisella tularensis* LVS, *Bacillus anthracis Sterne*, ricin, *Yersinia pestis* F1 antigen, MS2 bacteriophage, cholera toxin, and *Salmonella typhimurium* (Anderson et al., 2004).

A BDG-based potential field deployable instrument was created, which was capable of multiple agent analysis for a number of agents. This technique employs colloidal gold conjugates to enhance the attenuation of the guided light beams prior to coupling out of the waveguide. Four separate antigens were evaluated, as *S. aureus* enterotoxin B (SEB), ricin, *Clostridium botulinum* toxin, and *F. tularensis* was detected with a target sensitivity of <1 ng mL^{-1}, 10 ng mL^{-1}, <100 ng mL^{-1}, and 10^5 ng mL^{-1}, respectively (O'Brien et al., 2000).

4 Conclusions

The earlier survey of theoretical and practical aspects of nanobiosensor applications intended to underline the excellent feasibility of the wide-ranging use of the physical phenomenon, evanescent field effect in sensorics based spectrometric detectors. This well-studied process makes it possible to study the chemical, biochemical, or even microbiological processes at molecular level without labeling. Different physical or chemical methods are available for binding or conjugating biomolecules responsible for the selectivity of the novel nanodesigned biosensors. The earlier detailed view on nanosensor technologies in conjunction with the various types of bioactive components applied provides a broad range of nanobiosensors. These biologically active components in the sensors and the detection methods utilized can be varied according to the actual demand toward the highly selective, quick analytical procedures. Utilization of the sensors extends from single use disposable or in situ portable devices to high-tech research instrumentation. Beyond the most developed clinical biosensors, applications have reached the field of equal importance, monitoring of components affecting agro-environmental and food safety.

The panoramic view of the state of art of the relevant evanescent field-based sensor techniques and the review of some of the most revealing applications in environmental and food analysis presented will hopefully set future nanobiosensor development forward.

Abbreviations

2,4-D	2,4-Dichlorophenoxyacetic acid
AFB1	Aflatoxin B1 mycotoxin
AFM1	Aflatoxin M1 mycotoxin
AWACSS	Automated Water Analyser Computer Supported System
BDG	Bidiffractive grating biosensor
BLI	Biolayer interferometry
BPA	Bisphenol A
CCD	Charge-coupled device
CCβ	Detection capability
CGS	Chirped grating couplers
CV	Coefficient of variability
DON	Deoxynivalenol mycotoxin
EDC	Endocrine disrupting chemical
ELISA	Enzyme-linked immunosorbent assay
EWMI	Evanescent wave multichannel immunosensor
FBG	Fiber Bragg grating
FCM	Fluorescent core microcapillary
GC–MS	Gas chromatography coupled with mass spectroscopy
HI	Hartman interferometer
HPLC–MS	High-performance liquid chromatography coupled with mass spectroscopy
LED	Light emitting diode
LOC	Lab-on-a-chip
LOD	Limit of detection
LOQ	Limit of quantification
Lpv	Lipovitellin
mAb	Monoclonal antibody
MCWG	Metal-clad waveguide sensors
MZI	Mach–Zehnder interferometer
OFRR	Optofluidic ring resonator
OTA	Ochratoxin A mycotoxin
OWLS	Optical waveguide lightmode spectroscopy
pAb	Polyclonal antibody
PCP	Pentachlorophenol
PWG	Planar waveguide
PWGFI	Planar waveguide fluorescence immunosensor
QD	Quantum dot
RI	Refractive index
RIANA	River Analyser
RIfS	Reflectometric interference spectroscopy
RM	Resonant mirror
RMGC	Reflection mode grating coupler
RWG	Resonant waveguide grating
scFv	Single chain variable fragment

SEB	Staphylococcal enterotoxin B
SMZ	Sulfamethazine
SPR	Surface plasmon resonance
TE	Transverse electric
TIR	Total internal reflection
TIRE	Total internal reflection ellipsometry
TIRF	Total internal reflection fluorescence spectroscopy
TM	Transverse magnetic
TWIST	Tunable wavelength interrogated sensor technology
VCSEL	Vertical cavity surface-emitting laser
Vtg	Vitellogenin
WGM	Whispering gallery mode
WIOS	Wavelength-interrogated optical sensing
YI	Young's interferometer
YTX	Yessotoxin
ZON	Zearalenone mycotoxin

Acknowledgments

Research was funded by project NUKR.SFPP 984637 in the NATO SPS Programme and received partial funding from the EU-project SPICED (Grant Agreement: 312631) with the financial support from the 7th Framework Programme of the European Union. This publication reflects the views only of the authors, and the European Commission cannot be held responsible for any use which may be made of the information contained therein.

References

Adányi, N., Váradi, M., Kim, N., Szendrő, I., 2006. Development of new immunosensors for determination of contaminants in food. Curr. Appl. Phys. 6, 279–286.

Adányi, N., Levkovets, I.A., Rodriguez, G.S., Ronald, A., Váradi, M., Szendrő, I., 2007. Development of immunosensor based on OWLS technique for determining aflatoxin B1 and ochratoxin A. Biosens. Bioelectron. 22, 797–802.

Adányi, N., Majer-Baranyi, K., Nagy, A., Németh, Gy., Szendrő, I., Székács, A., 2013. Optical waveguide light-mode spectroscopy immunosensor for detection of carp vitellogenin. Sensor. Actuat. B: Chem. 176, 932–939.

Adányi, N., Majer-Baranyi, K., Szendrő, I., Székács, A., Label-free immunosensors with OWLS detection for food safety, Euro Food Chem XVIII Conference, Madrid, Spain, October 13–16, 2015.

Adányi, N., Majer-Baranyi, K., Wang, B., Szendrő, I., Székács, A., Development of immunosensors based on optical waveguide lightmode spectroscopy (OWLS) technique for determining active substance in herbs, Ninth International Conference on Instrumental Methods of Analysis-Modern Trends and Applications, Kalamata, Greece, September 20–24, 2015.

Alocilja, E.C., Radke, S.M., 2003. Market analysis of biosensors for food safety. Biosens. Bioelectron. 18, 841–846.

Anderson, G.P., Lingerfelt, B.M., Taitt, C.R., 2004. Eight analyte detection using a four-channel optical biosensor. Sens. Lett. 2, 1–7.

Arora, P., Sindhu, A., Dilbaghi, N., Chaudhury, A., 2011. Biosensors as innovative tools for the detection of food borne pathogens. Biosens. Bioelectron. 28, 1–12.

Bahadir, E.B., Sezgintürk, M.K., 2015. Applications of commercial biosensors in clinical, food, environmental, and biothreat/biowarfare analyses. Anal. Biochem. 478, 107–120.

Basova, T., Hassan, A., Yuksel, F., Gürek, A.G., Ahsen, V., 2010. Optical detection of pentachlorophenol in water using thin films of octa-tosylamido substituted zinc phthalocyanine. Sensor. Actuat. B: Chem. 150, 523–528.

Bata-Vidács, I., Adányi, N., Beczner, J., Farkas, J., Székács, A., 2013. Nanotechnology and Microbial Food Safety. In: Mendez-Vilas, A. (Ed.), Microbial Pathogens and Strategies for Combating Them: Science, Technology and Education. Formatex, Zubaran, Spain, pp. 155–159.

Bhat, R.V., Vasanthi, S., 2003. Mycotoxin food safety risks in developing countries. In: Unnevehr, L.J. (Ed.), Food Safety in Food Security and Food Trade. International Food Policy Research Institute (IFPRI), Vision 2020 for Food, Agriculture and Environment, Washington, pp. 1–2.

Bosch, M.E., Sánchez, A.J.R., Rojas, F.S., Ojeda, C.B., 2007. Recent development in optical fiber biosensors. Sensors 7, 797–895.

Brandenburg, A., Henninger, R., 1994. Integrated optical Young interferometer. Appl. Optics 33, 5941–5947.

Brandenburg, A., Krauter, R., Künzel, C., Stefan, M., Schulte, H., 2000. Interferometric sensor for detection of surface-bound bioreactions. Appl. Optics 39, 6396–6405.

Brecht, A., Gauglitz, G., 1995. Optical probes and transducers. Biosens. Bioelectron. 10, 923–936.

Brecht, A., Gauglitz, G., 1997. Label free optical immunoprobes for pesticide detection. Anal. Chim. Acta 347, 219–233.

Chen, N., Yun, B., Cui, Y., 2006. Cladding mode resonances of etch-eroded fiber Bragg grating for ambient refractive index sensing. Appl. Phys. Lett. 88, 133902.

Conroy, P.J., Hearty, S., Leonard, P., O'Kennedy, R.J., 2009. Antibody production, design and use for biosensor-based applications. Semin. Cell. Dev. Biol. 20, 10–26.

Cush, R., Cronin, J.M., Stewart, W.J., Maule, C.H., Molloy, J., Goddard, N.J., 1993. The resonant mirror: a novel optical biosensor for direct sensing of biomolecular interactions part I: principle of operation and associated instrumentation. Biosens. Bioelectron. 8, 347–354.

Daghestani, H.N., Day, B.W., 2010. Theory and applications of surface plasmon resonance, resonant mirror, resonant waveguide grating, and dual polarization interferometry. Biosens. Sens. 10, 9630–9646.

De Graaf, B.R., Williamson, F., Koerkamp, M.K., Verhoef, J.W., Wuestman, R., Bajema, B., Trietsch, E., Van Delft, W., 2012. Implementation of an innovative sensor technology for effective online water quality monitoring in the distribution network. Water Pract. Technol. 7, 099.

DeMarco, D.R., Lim, D.V., 2002. Detection of *Escherichia coli* O157:H7 in 10- and 25-gram ground beef samples with an evanescent-wave biosensor with silica and polystyrene waveguides. J. Food Prot. 4, 595–725.

Drapp, B., Piehler, J., Brecht, A., Gauglitz, G., Luff, B.J., Wilkinson, J.S., Ingenhoff, J., 1997. Integrated optical Mach-Zehnder interferometers as simazine immunoprobes. Sensor. Actuat. B: Chem. 39, 277–282.

Dübendorfer, J., Kunz, R.E., 1998. Compact integrated optical immunosensor using replicated chirped grating coupler sensor chips. Appl. Optics 37, 1890–1894.

Erdélyi, K., Frutos, A.G., Ramsden, J.J., Szendrő, I., Voirin, G., 2008. Grating-based optical biosensors. In: Marks, R.S., Cullen, D.C., Karube, I., Lowe, C.R., Weetall, H.H. (Eds.), Handbook of Biosensors and Biochips. John Wiley & Sons Ltd, New York, pp. 569–586.

Estevez, M.C., Alvarez, M., Lechuga, L.M., 2012. Integrated optical devices for lab-on-a-chip biosensing applications. Laser Photon. Rev. 6, 463–487.
Fan, X., White, I.M., Shopova, S.I., Zhu, H., Suter, J.D., Sun, Y., 2008. Sensitive optical biosensors for unlabeled targets: a review. Anal. Chim. Acta 620, 8–26.
Goddard, N.J., Pollard-Knight, D., Maule, C.H., 1994. Real-time biomolecular interaction analysis using the resonant mirror sensor. Analyst 119, 583–588.
Gounaridis, L., Groumas, P., Schreuder, E., Heideman, R., Katopodis, V., Kouloumentas, C., Avramopoulos, H., 2015. Design of grating couplers and MMI couplers on the TriPleX platform enabling ultra-compact photonic-based biosensors. Sensor. Actuat. B: Chem. 209, 1057–1063.
Grego, S., Gilchrist, K.H., Carlson, J.B., Stoner, B.R., 2012. A compact and multichannel optical biosensor based on a wavelength interrogated input grating coupler. Sensor. Actuat. B: Chem. 161, 721–727.
Guo, H., Zhou, X., Zhang, Y., Song, B., Liu, L., Zhang, J., Shi, H., 2014. Highly sensitive and rapid detection of melamine in milk products by planar waveguide fluorescence immunosensor (PWFI). Sensor. Actuat. B: Chem. 194, 114–119.
Hao, X.-J., Zhou, X.-J., Zhang, Y., Long, F., Song, L., Shi, H.-C., 2015. Portable and reusable optofluidics-based biosensing platform for ultrasensitive detection of sulfadimidine in dairy products. Sensors 15, 8302–8313.
Harris, J., Lu, P., Larocque, H., Chen, L., Bao, X., 2015. In-fiber Mach–Zehnder interferometric refractive index sensors with guided and leaky modes. Sensor. Actuat. B: Chem. 206, 246–251.
Hecht, E., 1998. Total internal reflection and the evanescent wave. Optics, 121–127.
Horváth, R., Lindvold, L.R., Larsen, N.B., 2002. Reverse-symmetry waveguides: theory and fabrication. Appl. Phys. B. 74, 383–393.
Horváth, R., Pedersen, H.C., Skivesen, N., Selmeczi, D., Larsen, N.B., 2003. Optical waveguide sensor for on-line monitoring of bacteria. Opt. Lett. 28, 1233–1235.
Hotta, K., Yamaguchi, A., Teramae, N., 2010. Properties of a metal clad waveguide sensor based on a nanoporous-metal-oxide/metal multilayer film. Anal. Chem. 82, 6066–6073.
Jiang, X., Li, D., Xu, X., Ying, Y., Li, Y., Ye, Z., Wang, J., 2008. Immunosensors for detection of pesticide residues. Biosens. Bioelectron. 23, 1577–1587.
Jianrong, C., Yuqing, M., Nongyue, H., Xiaohua, W., Sijiao, L., 2004. Nanotechnology and biosensors. Biotechnol. Adv. 22, 505–518.
Jokerst, J.C., Emorya, J.M., Henry, C.S., 2012. Advances in microfluidics for environmental analysis. Analyst 137, 24–34.
Kappel, C., Selle, A., Bader, M.A., Marowsky, G., 2004. Resonant double-grating waveguide structures as inverted Fabry-Perot interferometers. J. Opt. Soc. Am. B. 21, 1127–1136.
Käppel, N.D., Pröll, F., Gauglitz, G., 2007. Development of a TIRF-based biosensor for sensitive detection of progesterone in bovine milk. Biosens. Bioelectron. 22, 2295–2300.
Kehl, F., Bischof, D., Michler, M., Keka, M., Stanley, S., 2015. Design of a label-free, distributed Bragg grating resonator based dielectric waveguide biosensor. Photonics 2, 124–138.
Kim, N., Hyung-Seok, R., Woo-Yeon, K., 2007a. Flatfish vitellogenin detection using optical waveguide lightmode spectroscopy-based immunosensor. J. Microbiol. Biotechnol. 17, 1445–1451.
Kim, N., Park, I.-S., Kim, W.-Y., 2007b. *Salmonella* detection with a direct-binding optical grating coupler immunosensor. Sensor. Actuat. B-Chem. 121, 606–615.
Kim, N., Kim, D.-K., Kim, W.-Y., 2008a. Sulfamethazine detection with direct-binding optical waveguide lightmode spectroscopy-based immunosensor. Food Chem. 108, 768–773.

Kim, N., Kim, D.K., Cho, Y.J., Moon, D.K., Kim, W.Y., 2008b. Carp vitellogenin detection by an optical waveguide lightmode spectroscopy biosensor. Biosens. Bioelectron. 24, 391–396.

Kim, G., Moon, J.-H., Moh, C.-Y., Lim, J.-G., 2015. A microfluidic nano-biosensor for the detection of pathogenic Salmonella. Biosens. Bioelectron. 67, 243–247.

Kozma, P., Kehl, F., Ehrentreich-Förster, E., Stamm, C., Bier, F.F., 2014. Integrated planar optical waveguide interferometer biosensors: a comparative review. Biosens. Bioelectron. 58, 287–307.

Lambeck, P.V., 2006. Integrated optical sensors for the chemical domain. Meas. Sci. Technol. 17, 93–116.

Lane, S., Chan, J., Thiessen, T., Meldrum, A., 2014. Whispering gallery mode structure and refractometric sensitivity of fluorescent capillary-type sensors. Sensor. Actuat. B: Chem. 190, 752–759.

Le, N.C.H., Gubala, V., Gandhiraman, R.P., Coyle, C., Daniels, S., Williams, D.E., 2010. Total internal reflection ellipsometry as a label-free assessment method for optimization of the reactive surface of bioassay devices based on a functionalized cycloolefin polymer. Anal. Bioanal. Chem. 398, 1927–1936.

Lechuga, L.M., 2000a. Optical sensors based on evanescent field sensing part 1: surface plasmon resonance sensors. Quim. Anal. 19, 54–60.

Lechuga, L.M., 2000b. Optical sensors based on evanescent field sensing part II: integrated optical sensors. Quim. Anal. 19, 61–67.

Ligler, F.S., Breimer, M., Golden, J.P., Nivens, D.A., Dodson, J.P., Green, T.M., Haders, D.P., Sadik, O.A., 2002. Integrating waveguide biosensor. Anal. Chem. 74, 713–719.

Liu, Q., Tu, X., Kim, K.W., Kee, J.S., Shin, Y., Han, K., Yoon, Y.-J., Lo, G.-Q., Park, M.K., 2013. Highly sensitive Mach–Zehnder interferometer biosensor based on silicon nitride slot waveguide. Sensor. Actuat. B: Chem. 188, 681–688.

Lowder, T.L., Gordon, J.D., Schultz, S.M., Selfridge, R.H., 2007. Volatile organic compound sensing using a surface relief d-shaped fiber bragg gratings and a polydimethylsiloxane layer. Opt. Lett. 32, 2523–2525.

Lu, D.F., Qi, Z.-M., Liu, R.-P., 2011. An interferometric biosensor composed of a prism-chamber assembly and a composite waveguide with a Ta_2O_5 nanometric layer. Sensor. Actuat. B: Chem. 157, 575–580.

Lukosz, W., Stamm, Ch., Moser, H.R., Ryf, R., Diibendorfer, J., 1997. Difference interferometer with new phase-measurement method as integrated-optical refractometer, humidity sensor and biosensor. Sensor. Actuat. B: Chem. 38-39, 316–323.

Luppa, P.B., Sokoll, L.J., Chan, D.W., 2001. Immunosensors: principles and applications to clinical chemistry. Clin. Chim. Acta 314, 1–26.

Majer-Baranyi, K., Székács, A., Szendrő, I., Kiss, A., Adányi, N., 2011. Optical waveguide lightmode spectroscopy technique-based immunosensor development for deoxynivalenol determination in wheat samples. Eur. Food Res. Technol. 233, 1041–1047.

Majer-Baranyi, K., Adányi, N., Nagy, A., Bukovskaya, O., Szendrő, I., Székács, A., 2015a. Label-free immunosensor for monitoring vitellogenin as a biomarker for exogenous oestrogen compounds in amphibian species. Int. J. Environ. An. Ch. 95, 481–493.

Majer-Baranyi, K., Zalán, Zs, Mörtl, M., Székács, A., Szendrő, I., Adányi, N., Optical waveguide lightmode spectroscopy technique-based immunosensor development for aflatoxin B1 determination in spice paprika samples, Ninth International Conference on "Instrumental Methods of Analysis-Modern Trends and Applications," Kalamata, Greece, September 20–24, 2015.

Manchee, C.P.K., Zamora, V., Silverstone, J., Veinot, J.G.C., Meldrum, A., 2011. Refractometric sensing with fluorescent-core microcapillaries. Opt. Exp. 19, 21540–21551.

Maragos, C.M., 2011. Detection of deoxynivalenol using biolayer interferometry. Mycotox. Res. 27, 157–165.
Maragos, C.M., 2012. Signal amplification using colloidal gold in a biolayer interferometry-based immunosensor for the mycotoxin deoxynivalenol. Food Addit. Contam. 29, 1108–1117.
Meneely, J.P., Campbell, K., Greef, C., Lochhead, M.J., Elliott, C.T., 2013. Development and validation of an ultrasensitive fluorescence planar waveguide biosensor for the detection of paralytic shellfish toxins in marine algae. Biosens. Bioelectron. 41, 691–697.
Mohrle, B.P., Kohler, K., Jaehrling, J., Brock, R., Gauglitz, G., 2006. Label-free characterization of cell adhesion using reflectometric interference spectroscopy (RIfS). Anal. Bioanal. Chem. 384, 407–413.
Morgan, C.G., Mitchell, A.C., 2006. Total internal reflection fluorescence imaging using an upconverting cover slip for multicolour evanescent excitation. J. Microsc. 222, 48–57.
Mosiello, L., Nencini, L., Segre, L., Spanň, M., 1997. A fibre-optic immunosensor for 2,4-dichlorophenoxyacetic acid detection. Sensor. Actuat. B: Chem. 39, 353–359.
Muriano, A., Thayil, K.N.A., Salvador, J.-P., Loza-Alvarez, P., Soria, S., Galve, R., Marco, M.-P., 2012. Two-photon fluorescent immunosensor for androgenic hormones using resonant grating waveguide structures. Sensor. Actuat. B: Chem. 174, 394–401.
Murphy, C., Stack, E., Krivelo, S., McPartlin, D.A., Byrne, B., Greef, C., Lochhead, M.J., Husar, G., Devlin, S., Elliott, C.T., O'Kennedy, R.J., 2015. Detection of the cyanobacterial toxin, microcystin-LR, using a novel recombinant antibody-based optical-planar waveguide platform. Biosens. Bioelectron. 67, 708–714.
Nabok, A.V., Tsargorodskaya, A., Hassan, A.K., Starodub, N.F., 2005. Total internal reflection ellipsometry and SPR detection of low molecular weight environmental toxins. Appl. Surf. Sci. 246, 381–386.
Nabok, A.V., Tsargorodskaya, A., Holloway, A., Starodub, N.F., Gojster, O., 2007. Registration of T-2 mycotoxin with total internal reflection ellipsometry and QCM impedance methods. Biosens. Bioelectron. 22, 885–890.
Nabok, A.V., Tsargorodskaya, A., Mustafa, M.K., Székács, A., Székács, I., Starodub, N.F., 2009. Detection of low molecular weight toxins using optical phase detection techniques. Procedia Chem. 1, 1491–1494.
Nabok, A., Tsargorodskaya, A., Mustafa, M.K., Székács, I., Starodub, N.F., Székács, A., 2011. Detection of low molecular weight toxins using an optical phase method of ellipsometry. Sensor. Actuat. B: Chem. 154, 232–237.
Novo, P., Moulas, G., Prazeres, D.M.F., Chu, V., Conde, J.P., 2013. Detection of ochratoxin A in wine and beer by chemiluminescence-based ELISA in microfluidics with integrated photodiodes. Sensor. Actuat. B: Chem. 176, 232–240.
O'Brien, T., Johnson, III, L.H., Aldrich, J.L., Allen, S.G., Liang, L.-T., Plummer, A.L., Krak, S.J., Boiarski, A.A., 2000. The development of immunoassays to four biological threat agents in a bidiffractive grating biosensor. Biosens. Bioelectron. 14, 815–828.
Oheim, M., Schapper, F., 2005. Non-linear evanescent-field imaging. J. Phys. D Appl. Phys. 38, 185–197.
Pazos, M.-J., Alfonso, A., Vieytes, M.R., Yasumoto, T., Vieites, J.M., Botana, L.M., 2004. Resonant mirror biosensor detection method based on yessotoxin–phosphodiesterase interactions. Anal. Biochem. 335, 112–118.
Piehler, J., Brandenburg, A., Brecht, A., Wagner, E., Gauglitz, G., 1997. Characterization of grating couplers for affinity-based pesticide sensing. Appl. Opt. 36, 6554–6562.

Poksinski, M., Arwin, H., 2003. In situ monitoring of metal surfaces exposed to milk using total internal reflection ellipsometry. Sensor. Actuat. B: Chem. 94, 247–252.
Preininger, C., Mencaglia, A., Baldini, F., 2000. Polymer-coated optical fibres for application in a direct evanescent wave immunoassay. Anal. Chim. Acta 403, 67–76.
Prieto, F., Sepúlveda, B., Calle, A., Llobera, A., Domínguez, C., Lechuga, L.M., 2003. Integrated Mach–Zehnder interferometer based on ARROW structures for biosensor applications. Sensor. Actuat. B: Chem 92, 151–158.
Prieto-Simón, B., Noguer, T., Campŕs, M., 2007. Emerging biotools for assessment of mycotoxins in the past decade. Trac-Trend. Anal. Chem. 26, 689–702.
Proll, G., Markovic, G., Steinle, L., Gauglitz, G., 2009. Reflectometric interference spectroscopy. In: Rasooly, A., Herold, K.E. (Eds.), Biosensors and Biodetection: Methods in Molecular Biology. Humana Press, New York, pp. 167–178.
Ran, R., Wang, C., Han, Z., Wu, A., Zhang, D., Shi, J., 2013. Determination of deoxynivalenol (DON) and its derivatives: current status of analytical methods. Food Control 34, 138–148.
Reder, S., Dieterle, F., Jansen, H., Alcock, S., Gauglitz, G., 2003. Multi-analyte assay for triazines using cross-reactive antibodies and neural networks. Biosens. Bioelectron. 19, 447–455.
Riant, I., 2003. Fiber Bragg gratings for optical telecommunications. Physique 4, 41–49.
Ríos, Á., Zougagh, M., Avila, M., 2012. Miniaturization through lab-on-a-chip: utopia or reality for routine laboratories? Anal. Chim. Acta 740, 1–11.
Rowland, K.J., Francois, A., Hoffman, P., Monro, T., 2013. Fluorescent polymer coated capillaries as optofluidic refractometric sensors. Opt. Exp. 21, 11492–11505.
Saini, S.S., Stanford, C., Lee, S.M., Park, J., DeShong, P., Bentley, W.E., Dagenais, M., 2007. Monolayer detection of biochemical agents using etched-core fiber bragg grating sensors. IEEE Photon. Technol. Lett. 19, 1341–1343.
Sapsford, K.E., Taitt, C.R., Fertig, S., Moore, M.H., Lassman, M.A., Maragos, C.M., Shriver-Lake, L.C., 2006. Indirect competitive immunoassay for detection of aflatoxin B1 in corn and nut products using the array biosensor. Biosens. Bioelectron. 21, 2298–2305.
Schipper, E.F., Bergevoet, A.J.H., Kooyman, R.P.H., Greve, J., 1997. New detection method for atrazine pesticides with the optical waveguide Mach-Zehnder immunosensor. Anal. Chim. Acta 341, 171–176.
Schmitt, K., Schirmer, B., Hoffmann, C., Brandenburg, A., Meyrueis, P., 2007. Interferometric biosensor based on planar optical waveguide sensor chips for label-free detection of surface bound bioreactions. Biosens. Bioelectron. 22, 2591–2597.
Schmitt, K., Oehse, K., Sulz, G., Hoffmann, C., 2008. Evanescent field sensors based on tantalum pentoxide waveguides. Sensors 8, 711–738.
Schneider, B.H., Edwards, J.G., Hartman, N.F., 1997. Hartman interferometer: versatile integrated optic sensor for label-free, real-time quantification of nucleic acids, proteins, and pathogens. Clin. Chem. 43, 1757–1763.
Schroeder, K., Ecke, W., Mueller, R., Willsch, R., Andreev, A., 2001. A fibre Bragg grating refractometer. Meas. Sci. Technol. 12, 757–764.
Scognamiglio, V., Arduini, F., Palleschi, G., Rea, G., 2014. Biosensing technology for sustainable food safety. Trac-Trend. Anal. Chem. 62, 1–10.
Sharon, A., Rosenblatt, D., Friesem, A.A., 1997. Resonant grating-waveguide structures for visible and near-infrared radiation. J. Opt. Soc. Am. A 14, 2985–2993.
Shi, H.C., Song, B.D., Long, F., Zhou, X.H., He, M., Lv, Q., Yang, H.Y., 2013. Automated online optical biosensing system for continuous real-time

determination of microcystin-LR with high sensitivity and specificity: early warning for cyanotoxin risk in drinking water sources. Environ. Sci. Technol. 47, 4341–4434.
Sim, J.H., Kwak, Y.H., Choi, C.H., Paek, S.-H., Park, S.S., Seo, S., 2012. A birefringent waveguide biosensor platform for label-free live cell detection of *Listeria monocytogenes*. Sensor. Actuat. B: Chem. 173, 752–759.
Skivesen, N., Horvath, R., Pedersen, H.C., 2005. Optimization of metal-clad waveguide sensors. Sensor. Actuat. B: Chem. 106, 668–676.
Skládal, P., Deng, A., Kolář, V., 1999. Resonant mirror-based optical immunosensor: application for the measurement of atrazine in soil. Anal. Chim. Acta 399, 29–36.
Starodub, N.F., Ogorodniichuk, Y.A., Sitnik, Y.A., Slishik, N.F., 2012. Biosensors for the control of some toxins, viral and microbial infections to prevent actions of bioterrorists. In: Nikolelis, D.P. (Ed.), Portable Chemical Sensors. NATO Science for Peace and Security Series A: Chemistry and Biology. Springer, The Netherlands, pp. 95–117.
Szalontai, H., Kiss, A., Adányi, N., 2014. Determination of Aflatoxin m1 in milk samples by an OWLS-based immunosensor. Acta Aliment. Hung. 43, 148–155.
Székács, A., Trummer, N., Adányi, N., Váradi, M., Szendrő, I., 2003. Development of a non-labeled immunosensor for the herbicide trifluralin via optical waveguide lightmode spectroscopic detection. Anal. Chim. Acta 487, 31–42.
Székács, A., Adányi, N., Székács, I., Majer-Baranyi, K., Szendrő, I., 2009. Optical waveguide lightmode spectroscopy immunosensors for environmental monitoring. Appl. Optics. 48, B151–B158.
Tedeschi, L., Domenici, C., Ahluwalia, A., Baldini, F., Mencaglia, A., 2003. Antibody immobilisation on fibre optic TIRF sensors. Biosens. Bioelectron. 19, 85–93.
Tiefenthaler, K., 1992. Integrated optical couplers as chemical waveguide sensors. Adv. Biosens. 2, 261–289.
Tittlemier, S.A., Roscoe, M., Drul, D., Blagden, R., Kobialka, C., Chan, J., Gaba, D., 2013. Single laboratory evaluation of a planar waveguide-based system for a simple simultaneous analysis of four mycotoxins in wheat. Mycotox. Res. 29, 55–62.
Tschmelak, J., Proll, G., Gauglitz, G., 2004a. Immunosensor for estrone with an equal limit of detection as common analytical methods. Anal. Bioanal. Chem. 378, 744–745.
Tschmelak, J., Proll, G., Gauglitz, G., 2004b. Verification of performance with the automated direct optical TIRF immunosensor (River Analyser) in single and multi-analyte assays with real water samples. Biosens. Bioelectron. 20, 743–752.
Tschmelak, J., Kumpf, M., Proll, G., Gauglitz, G., 2004c. Biosensor for seven sulphonamides in drinking, ground, and surface water with difficult matrices. Anal. Lett. 37, 1701–1718.
Tschmelak, J., Proll, G., Gauglitz, G., 2004d. Sub-nanogram per litre detection of the emerging contaminant progesterone with a fully automated immunosensor based on evanescent field techniques. Anal. Chim. Acta 519, 143–146.
Tschmelak, J., Proll, G., Riedt, J., Kaiser, J., Kraemmer, P., Bárzaga, L., Wilkinson, J.S., Hua, P., Hole, J.P., Nudd, R., Jackson, M., Abuknesha, R., Barceló, D., Rodriguez-Mozaz, S., De Alda, M.J., Sacher, F., Stien, J., Slobodník, J., Oswald, P., Kozmenko, H., Korenková, E., Tóthová, L., Krascsenits, Z., Gauglitz, G., 2005a. Automated Water Analyser Computer Supported System (AWACSS) Part I: project objectives, basic technology, immunoassay development, software design and networking. Biosens. Bioelectron. 20, 1499–1508.
Tschmelak, J., Proll, G., Riedt, J., Kaiser, J., Kraemmer, P., Bárzaga, L., Wilkinson, J.S., Hua, P., Hole, J.P., Nudd, R., Jackson, M., Abuknesha, R., Barceló, D., Rodriguez-Mozaz, S., De Alda, M.J., Sacher, F., Stien, J., Slobodník, J., Oswald, P., Kozmenko, H., Korenková, E., Tóthová, L., Krascsenits, Z., Gauglitz, G., 2005b. Automated water analyser computer supported system (AWACSS) Part

II: intelligent, remote-controlled, cost-effective, on-line, water-monitoring measurement system. Biosens. Bioelectron. 20, 1509–1519.

Tschmelak, J., Kumpf, M., Käppel, N., Proll, G., Gauglitz, G., 2006. Total internal reflectance fluorescence (TIRF) biosensor for environmental monitoring of testosterone with commercially available immunochemistry: antibody characterization, assay development and real sample measurements. Talanta 69, 343–350.

Van Dorst, B., Mehta, J., Bekaert, K., Rouah-Martin, E., De Coen, W., Dubruel, P., Blust, R., Robbens, J., 2010. Recent advances in recognition elements of food and environmental biosensors: a review. Biosens. Bioelectron. 26, 1178–1194.

Vörös, J., Ramsden, J.J., Csúcs, G., Szendrő, I., De Paul, S.M., Textor, M., Spencer, N.D., 2002. Optical grating coupler biosensors. Biomaterials 23, 3699–3710.

Wagacha, J.M., Muthomi, J.W., 2008. Mycotoxin problem in Africa: current status, implications to food safety and health and possible management strategies. Int. J. Food Microbiol. 124, 1–12.

Warriner, K., Subrayal, M., Reddy, S.M., Namvar, A., Neethirajan, S., 2014. Developments in nanoparticles for use in biosensors to assess food safety and quality. Trends Food Sci. Technol. 40, 183–199.

Watts, H.J., Lowe, C.R., Pollard-Knight, D.V., 1994. Optical biosensor for monitoring microbial cells. Anal. Chem. 66, 2465–2470.

Westphal, P., Bornmann, A., 2002. Biomolecular detection by surface plasmon enhanced ellipsometry. Sensor. Actuat. B: Chem. 84, 278–282.

White, I.M., Zhu, H., Suter, J.D., Hanumegowda, N.M., Oveys, H., Zourob, M., Fan, X., 2007. Refractometric sensors for lab-on-a-chip based on optical ring resonators. IEEE Sens. J. 7, 28–35.

World Health Organization (WHO), 2006. Mycotoxins in African foods: implications to food safety and health. AFRO Food Saf. News., World Health Org. Food Saf. (FOS) 2, 1–2.

Wu, F., 2006. Mycotoxin reduction in Bt corn: potential economic, health, and regulatory impacts. Transgen. Res. 15, 277–289.

Xi, F., Gao, J., Wang, J., Wang, Z., 2011. Discrimination and detection of bacteria with a label-free impedemetric biosensor based on self-assembled lectin monolayer. J. Electroanal. Chem. 656, 252–257.

Xiao-hong, Z., Lan-Hua, L., Wei-Qi, X., Bao-Dong, S., Jian-Wu, S., Miao, H., Han-chang, S., 2014a. A reusable evanescent wave immunosensor for highly sensitive detection of bisphenol A in water samples. Sci. Rep. 4, 4572.

Xiao-hong, Z., Bao-dong, S., Han-chang, S., Lan-hua, L., Hong-li, G., Miao, H., 2014b. An evanescent wave multi-channel immunosensor system for the highly sensitive detection of small analytes in water samples. Sensor. Actuat. B: Chem. 198, 150–156.

Yang, G., White, I.M., Fan, X., 2008. An opto-fluidic ring resonator biosensor for the detection of organophosphorus pesticides. Sensor. Actuat. B: Chem. 133, 105–112.

Ymeti, A., Kanger, J.S., Greve, J., Besselink, G.A.J., Lambeck, P.V., Wijn, R., Heideman, R.G., 2005. Integration of microfluidics with a four-channel integrated optical Young interferometer immunosensor. Biosens. Bioelectron. 20, 1417–1421.

Zhu, P., Shelton, D.R., Karns, J.S., Sundaram, A., Li, S., Amstutz, P., Tang, C.-M., 2005. Detection of water-borne *E. coli* O157 using the integrating waveguide biosensor. Biosens. Bioelectron. 21, 678–683.

Zourob, M., Goddard, N.J., 2005. Metal clad leaky waveguides for chemical and biosensing applications. Biosens. Bioelectron. 20, 1718–1727.

Zourob, M., Mohr, S., Brown, B.J.T., Fielden, P.R., McDonnell, M., Goddard, N.J., 2003. The development of a metal clad leaky waveguide sensor for the detection of particles. Sensor. Actuat. B: Chem. 90, 296–307.

12

MICRO- AND NANOTECHNOLOGY-BASED APPROACHES TO DETECT PATHOGENIC AGENTS IN FOOD

Nathaniel C. Cady*, Vincenzina Fusco, Giuseppe Maruccio†, Elisabetta Primiceri†, Carl A. Batt‡**

**SUNY Polytechnic Institute, Colleges of Nanoscale Science & Engineering, Albany, NY, United States; **National Research Council of Italy, Institute of Sciences of Food Production, CNR-ISPA, Bari, Italy; †University of Salento, Department of Mathematics and Physics "Ennio De Giorgi," CNR NANOTEC—Institute of Nanotechnology, Lecce, Italy; ‡Cornell University, Department of Food Science, Ithaca, NY, United States*

1 Introduction

Although great progress has been made in food safety diagnostics, which has resulted in improved food safety and security awareness worldwide, foodborne diseases due to the consumption of food containing harmful parasites, viruses, and bacteria or chemical compounds, are still widespread. The degree of severity of foodborne illnesses may be different depending on several factors with symptoms ranging from mild to severe and life-threatening. Moreover, some foodborne diseases are being associated with chronic sequelae (Porter et al., 2013a; Gibney et al., 2014; Keithlin et al., 2014a, 2014b, 2015), such as, inflammatory bowel disease (IBD) (Ford et al., 2014; Merger and Croitoru, 1998; Doorduyn et al., 2008), irritable bowel syndrome (IBS) (Doorduyn et al., 2008; Haagsma et al., 2010; Ford et al., 2014), Guillain–Barré syndrome (Poropatich et al., 2010; Ford et al., 2014), and reactive arthritis (Locht et al., 2002; Porter et al., 2013b; Ford et al., 2014). The World Health Organization (2014) recently estimated that 2.2 million people, including many children, annually die due to waterborne and foodborne diarrheal diseases (food safety fact sheet, http://www.who.int/mediacentre/factsheets/fs399/en).

Nanobiosensors. http://dx.doi.org/10.1016/B978-0-12-804301-1.00012-6

Annual estimates of foodborne disease range from 48 million cases in the United States (http://www.cdc.gov/foodborneburden), to 4 million episodes of domestically acquired foodborne illness in Canada (http://www.phac-aspc.gc.ca/efwd-emoha/efbi-emoa-eng.php), 4.1 million cases in Australia (Kirk et al., 2014), 1 million in the United Kingdom (FSA, 2011), and more than 320,000 in the European Union (EFSA, 2015). Of these illnesses, *Salmonella* spp., *Listeria monocytogenes*, *Campylobacter* spp., *Staphylococcus aureus*, *Toxoplasma gondii*, and *Clostridium perfringens* are the most common pathogens causing foodborne illness, death (Thomas et al., 2013), and considerable economic losses (McLinden et al., 2014). Factors, such as, climate change, the globalization of both the food market, and consumer lifestyle habits (including a shift toward healthy, minimally processed foods) impact these trends. Further exacerbating these trends is an increase in human population with demographic shift toward an aging of population, which in turn is composed by a growing number of immunocompromised people who are more susceptible to foodborne diseases. The epidemiology of foodborne disease is changing in parallel, due to the emergence, reemergence, and worldwide spread of some pathogenic microorganisms. This is the case, for example, of the etiological agent of several outbreaks in neonatal intensive care units, discovered as novel species in 1980 (Farmer et al., 1980), and subsequently reclassified as *Cronobacter sakazakii* (Iversen et al., 2007), whose emergence has been so concerning that its absence in powdered infant formula and dried dietary foods for special medical purposes intended for under 6 months of age infants, has been regulated (European Commission Regulation EC No. 2073/2005, amended by EU regulation No. 1441/2007). Additionally, new variants of established pathogens are reemerging due to the acquisition of further virulence factors. These include, but are not limited to *L. monocytogenes*, *Campylobacter* spp., *Salmonella* spp., methicillin-resistant *S. aureus* (MRSA), and multidrug-resistant Gram-negative bacteria, such as, *Klebsiella pneumoniae* and *Escherichia coli*. *L. monocytogenes*, worldwide recognized as one of the most concerning pathogen, in dairy products specially made from raw unpasteurized milk, continues to be a public health burden being able to cause listeriosis with a 20% case-fatality rate that increases up to 75% in pregnant women, elderly, and immunocompromised adults (Quero et al., 2014). Indeed, food operators have been imposed to ensure the conformity of their productions with the safety criteria [Regulation (EC) No. 2073/2005 and its amendments] in risky foods, such as, ready-to-eat foods. National Surveillance Programs of this pathogen have been adopted (http://www.cdc.gov/nationalsurveillance/

listeria-surveillance.html; http://www.hc-sc.gc.ca/fn-an/legislation/pol/policy_listeria_monocytogenes_2011-eng.php; https://sites.anses.fr/en/minisite/listeria/ecdc-fwd-surveillance-report-seven-priority-diseases-2010-2012). Campylobacteriosis, mainly caused due to the consumption of contaminated meat (principally poultry) is one of the most widespread infectious diseases, whose incidence and prevalence has highly increased in both developed and developing countries in the past decade, becoming endemic in some regions of Africa, Asia, and the Middle East (Kaakoush et al., 2015). Salmonellosis is another global concern for public health, causing an estimated tens of millions of human cases, up to a hundred thousand (http://www.who.int/mediacentre/factsheets/fs139/en). *S. aureus* is among the top five pathogens causing foodborne disease in the United States (http://www.cdc.gov/foodborneburden/2011-foodborne-estimates.html), while in the European Union, 364 staphylococcal food poisoning cases, that is, 6.4% of all outbreaks recorded, causing 2,532 cases, 288 hospitalizations, and 1 death have been documented (EFSA, 2015). This pathogenic bacterium produces a wide range of virulence factors and exotoxins. Staphylococcal enterotoxins (SEs), produced by *S. aureus* and few other species of coagulase positive Staphylococci (CPS), are the etiological agents of the staphylococcal food poisoning (Hennekinne et al., 2012; Fusco et al., 2011). CPS- and SEs diagnostics is crucial to assess the compliance of the foodstuffs with the European regulatory requirements (EC regulations No. 2073/2005, 1441/2007, etc.) (Hennekinne et al., 2010). Indeed, given the risk of staphylococcal food poisoning and the heat stability of SEs, which allow their persistence in a foodstuff even if the relevant vegetative cells have been inactivated; it is necessary to enumerate CPS at a specific step of the production process when the number of Staphylococci is expected to be the highest (process-hygiene criteria) and to screen samples for the presence of SEs above the well-specified *M* values (food-safety criteria). Hospital and community-associated infections in humans due to MRSA are a public-health concern worldwide, with animals being used as foods and their derived products considered as potential sources of MRSA (Alba et al., 2015; Riva et al., 2015; Traversa et al., 2015).

Antibiotic-resistant pathogenic bacteria, whose advent and spreading is being mainly caused by unwise use of antibiotics in human- and animal medicine, are detrimental to public health and result in an economic burden. It has been estimated that in the United States, drug-resistant infections involve more than 2 million people annually, provoking at least 23,000 deaths as a result. In the European Union, more than 25,000 people die yearly due to antibiotic-resistant bacteria, with the antimicrobial resistance accounting to at least €1500 million for extra health care costs and lost productivity.

1.1 Why Food Operators Should be so Concerned About Microbial Food Safety

Apart from legal constraints (Regulation 178/2002), food operators should consider that foodborne pathogenic microorganisms and/or their metabolites may contaminate a food matrix at every step of the food chain (from farm to fork) and, depending on a wide range of biotic and abiotic factors, may proliferate and thus become responsible for foodborne infections, toxic responses, or intoxications. Food micro- and macrostructure, as well as, environmental factors affect storage and preparation resulting in ineffective or sublethal elimination of contaminating bacteria. The resulting microbiota may include viable, dead, stressed, or injured microbial cells (Fusco et al., 2011, 2012; Quero et al., 2014; Fusco and Quero, 2012, 2014). Enumerating the viable, dead, stressed, or injured microbial cells in a food matrix is of utmost importance, especially if those organisms are pathogenic (Fusco et al., 2011, 2012; Quero et al., 2014; Fusco and Quero, 2012, 2014). Some microorganisms enter a viable but nonculturable state due to environmental stresses, a form of injury, which is the result of a number of physiological changes (Stiles and Witter, 1965). Standard microbiological enumeration using selective media leads to a great underestimation of their numbers (Speck, 1970). The problem is that under some conditions these injured cells may recover, grow, and present as a health challenge to an individual consuming the food. Molecular methods that are not dependent on the growth of the targeted microorganism can circumvent the underestimation of injured cells; but the methods need to have some means to differentiate injured from truly dead cells (Norton and Batt, 1999). Typically this involves the use of target, such as mRNA, whose presence is correlated with viability or DNA intercalating dyes, such as, EMA (ethidium monoazide) and PMA (propidium monoazide), which are able to penetrate damaged cell walls/membranes, bind to the DNA through a photoactivation-mediated process, that makes the resulting complex not amplifiable by polymerase chain reaction (PCR) (reviewed by Fusco and Quero, 2014).

Apart from public health risk, economic losses due to the recall of products in cases of sanitary alarms (https://webgate.ec.europa.eu/rasff-window/portal; http://www.inspection.gc.ca; http://www.fda.gov/Safety/Recalls, http://www.foodstandards.gov.au) may occur. Ethnic foods, whose popularity is increasing worldwide due to several factors, such as, high immigration rates, increased international travels, cosmopolitan eating habits, and globalization of food supplies are frequently being involved in foodborne outbreaks and recalls due to contamination with undesired chemicals, including allergens, toxins, and biologic agents or pathogens (eg, viruses, bacteria, and parasites), thus prompting the need to assess, control, and prevent the emerging microbial and toxicological risks associated with these foods (Fusco et al., 2015). Given the globalization of the food supply and the great potential of pathogenic bacteria as biothreat agents, a foodborne pathogen, as well as, a food safety incident can rapidly and easily spread to a broader geographical area, tainting the global agro-food security, and increasing the risk of crossborder transmission of foodborne diseases.

The threat of bioterrorism has raised the risk of biologic warfare including the deliberate contamination of food and beverages at any step of food/beverage chain. Food and waterborne pathogens and their toxins have been listed as Category B agents of second highest priority in terms of public health and national security (Heymann, 2001; Wein and Liu, 2005; FDA, 2007; Dembek and Noah, 2011; Fusco and Quero, 2014).

The high incidence of foodborne pathogen contamination, as well as, the rising threat of biologic attacks necessitate the development of reliable, sensitive, multiplexed, high-throughput, portable, and automated platforms to detect foodborne illness-causing agents and their toxins in foods (Fusco and Quero, 2014). A plethora of platforms addressing one or more but not all the aforementioned requirements have been developed (Cady et al., 2011; Okeke et al., 2011; Yoon and Kim, 2012; Wang et al., 2012a,b,c; Cho et al., 2014b; Sin et al., 2014; Abu-Salah et al., 2015; McNerney, 2015). The current trend is to make use of micro- and nanofabrication technologies to achieve the miniaturization of these platforms into lab-on-a-chip (LOC) devices, also referred to as biomedicalnano- and microelectromechanical systems (BioNEMS or BioMEMS), for an integrated and automated sample preparation and (quantitative) detection.

Herein, the latest advances in the micro- and nanotechnologies applied to the foodborne pathogen diagnostics are summarized pinpointing drawbacks and advantages of the most promising.

2 Food Pathogen Detection Using Micro- and Nanotechnologies

Micro- and nanoscale technologies have been leveraged from their roots in the semiconductor industry, where they are used to manufacture computer chips and electronics, to a wide range of other applications, including sensors, biosensors, biomedical devices, and optics. Microtechnologies include devices and system components on the size scale of 1–100 μm (10^{-6}–10^{-4} m) while nanotechnologies are typically defined as those within the size scale of 1–100 nm (10^{-9}–10^{-7} m). In this text, we define micro- and nanotechnologies as those including both chip-based technologies (devices fabricated in a small, chip-like format), as well as, those using micro- or nanoscale materials (nanoparticles, microparticles, quantum dots, etc.). The primary reasons for utilizing micro- and nanotechnologies for biosensing, especially for detection of food pathogens, is that these technologies can provide enhanced sensitivity (Billington et al., 2014), utilize small sample size and reduced reagent volume (Cady et al., 2003, 2011; Liu et al., 2007; Lui et al., 2009), can be highly multiplexed (Shu et al., 2014), can reduce detection / processing time (Law et al., 2014), and can offer improved portability for use outside of the laboratory (Lui et al., 2009). This combination of features makes micro- and nanoscale biosensors ideal for detection of pathogens in the field (such as, the farm or food processing site), while also improving time of detection. This has the potential to reduce the likelihood of disease outbreak and can facilitate rapid corrective actions once contamination is detected. Furthermore, detection of multiple pathogens in a single device or single assay can streamline surveillance. For instance, rather than performing 10 separate enrichment and detection assays for 10 individual organisms, these steps can be combined into a single assay or single device (eg, LOC devices) enabling parallel detection of all 10 organisms. The small size of these devices also provides the added benefit of using smaller volumes of sample and analytical reagents. This can significantly reduce the cost of operation (especially when considering costly reagents), as well as, reduce the burden of collecting and processing large samples. The small size format can also have unique consequences over larger, laboratory-scale devices, such as, faster reaction times, higher signal-to-noise values, and in some cases, higher sensitivity/lower limits of detection.

2.1 General Pathogen Detection Schemes

Pathogen detection using micro- and nanoscale technologies typically focuses on quantitative detection of (1) nucleic acids (DNA or RNA) from target organisms, (2) detection of whole/intact target organisms, or (3) other targets including antigens, toxins, or metabolites that are unique to the target organisms. Conventional detection methods largely rely on total target-cell enumeration on selective growth media, with more current approaches focusing on nucleic acid- or antigen based-detection methods. Growth-based detection schemes involve the separation or enrichment of bacteria from food matrices, followed by growth on selective media, and/or differential growth media. Such growth media represses growth of other organisms, while encouraging growth of the target organism. This prevents false-positive results that could occur from the growth of nonpathogenic organisms within the food matrix. Another conventional approach is to measure the metabolic activity or selective growth of target organisms using a panel of media/reagents. One of the most widely used formats is the BiomerieuxAPI Strip (http://www.biomerieux-diagnostics.com/), which is essentially a panel of multiple small-volume reservoirs into which bacteria can be inoculated. Each reservoir contains growth media and/or biochemical reagents that enable visual confirmation of various metabolic activities or growth limitations. These panels provide an initial screening for bacterial identification based on the growth preferences or metabolic activity differences between species or strains of bacteria. However, this assay format is still time consuming (often taking ≥12 h to achieve results).

The primary reason that micro- and nanotechnologies do not utilize enrichment and growth-based techniques is that they are typically intended for high throughput or rapid detection, both of which are hampered by time-consuming growth-based methods. As mentioned earlier, most growth-based methods take between 12 and 24 h (or longer) to receive results. The one downfall of using nonenrichment-based techniques is that they can suffer from a lack of sensitivity, which limits their application where the analysis must meet the zero-tolerance threshold. Thus, some micro- and nanotechnologies may still rely on some degree of growth-based enrichment to achieve the necessary limit of detection.

The first detection, we discuss in detail, is the nucleic acid-based detection. Nucleic acid-based detection of food pathogens is not a new or novel approach. In fact, the most popular nucleic acid-based approach, the PCR, has been employed for food detection since the 1990s (Bsat et al., 1994). Nucleic acid-based approaches offer a high degree of selectivity, because each bacterial species, and indeed each individual subtype or strain will have its own unique DNA (and RNA) sequences. This, combined with modern next generation sequencing (NGS), enables both sequence discovery and assay development to target unique sequences for detection and discrimination. Micro- and nanotechnologies have further enabled miniaturized, chip-based formats for performing nucleic-acid analysis, which greatly enhances portability, and in some cases improves multiplexing capability. For instance, so-called LOC systems have been developed for PCR-based detection of a variety of food pathogens (Cady et al., 2011; Tachibana et al., 2015). In these types of detection formats, total nucleic acids are extracted from a sample, followed by microliter-scale PCR reactions, typically in a chip-based format. The small size of the reaction chamber enables extremely rapid heat transfer, which can reduce the PCR cycling time and even improve the efficiency of PCR amplification. In order to quantify successful PCR amplification, most microchip-based PCR systems employ real-time PCR assays. These technologies have been further enhanced by adding multiplexing capability via multiplex real-time PCR (Shu et al., 2014) and through microarray-based methods (Tortajada-Genaro et al., 2015).

For antigen-based detection there are many different approaches and device/assay formats that can be employed. In its most basic format, an antigen-based method targets whole cells, toxins, or target analytes with an antigen-binding agent [antibody (Ab), aptamer, etc.]. While antibodies are still the most common type of antigen-binding agent employed, DNA, RNA, and protein-based aptamers are growing in popularity. Aptamers are generated by a process known as the "systematic evolution of ligands by exponential enrichment" or SELEX (Potyrailo et al., 1998). In this approach, libraries of DNA, RNA, or polypeptides are selected for their ability to bind to target antigens, and then enriched by multiple washing and reselection procedures, such that unique, high-affinity sequences are identified. Once a high-affinity sequence is identified, it can be integrated with a wide range of antigen-based detection approaches (Ozalp et al., 2014; Lee et al., 2015; Wu et al., 2015). DNA and RNA aptamers are the most commonly used aptamers for biosensing applications, mainly due to their resistance to protease degradation, their ability to withstand thermal denaturation and refolding cycles. One of the other advantages of using nucleic acid-based aptamers is the relative ease of producing large quantities of these aptamers (using modern DNA and RNA synthesis methods). This is in contrast to protein-based aptamers, which can be difficult to produce in cell-free protein-expression systems, or via traditional in vivo protein-expression systems.

For pathogen detection, antigen-binding agents can be tethered to a surface, in order to remove the target from the bulk sample, or can be attached to a nano- or microparticle, which enables detection or enhanced removal from the sample. The antigen-binding event can be coupled with a unique transduction mechanism for direct detection (such as, changes in optic or electric properties) or the binding event can be used for immobilization and performance of a secondary detection assay (such as, a secondary labeling step to produce fluorescence, chemiluminescence, etc.). For instance, whole cells can be immobilized on a chip and directly detected through micro-optic methods (St John et al., 1998), or via microscale separation methods, followed by the more conventional enzyme linked immunosorbent assay (ELISA) described by Wang et al. (2013c). In the case of St John et al. (1998), micrometer-wide lines of anti-*E. coli* antibodies were deposited onto silicon surfaces for an optic diffraction-based detection assay. When *E. coli* were presented to the Ab-modified surface, cells were attached to the Ab stripes, which acted as a diffraction grating and resulted in a measurable increase in the intensity of diffracted laser light from the surface. For ELISA-based methods, such as described by Wang et al. (2013a,b,c), cells could be captured on Ab-modified magnetic beads (immunomagnetic separation) and then subjected to labeling with enzyme-linked secondary antibodies to generate an optic readout. Both of these methods rely on capture of whole cells, which differs starkly from methods that rely on detection of molecular targets.

In summary, nucleic acid- and antigen-based detection methods are most commonly employed for micro- and nanotechnology based biosensors of food pathogens. While there are limited examples of enrichment and growth-based approaches that utilize LOC microtechnologies (Heo and Hua 2009), these methods are typically slow, have reduced selectivity and specificity, and are therefore outside the context of this text. Other emerging technologies include the use of bacteriophage for immobilization or detection (Billington et al., 2014). While phage are nanoscale in size (typically 100 nm or smaller) they, too, fall beyond the scope of this text. In the proceeding sections, specific examples of nano- and microscale technologies for food pathogen detection are provided, along with explanation of the detection principles.

2.2 Immunomagnetic Separation and Detection

One of the most widely used separation methods employing microtechnology is immunomagnetic separation (IMS). In this approach, microscale magnetic beads are coated with antibodies that target specific pathogens. These beads are then mixed with a sample where they bind to the target pathogens and can then be removed and concentrated by applying an external magnetic field. In this way, the target pathogens are removed from the sample and can be detected using techniques, such as, fluorescent detection (Cho et al., 2014a) or even flow cytometry (Seo et al., 1998). IMS has been used for detection of a wide range of pathogens in many

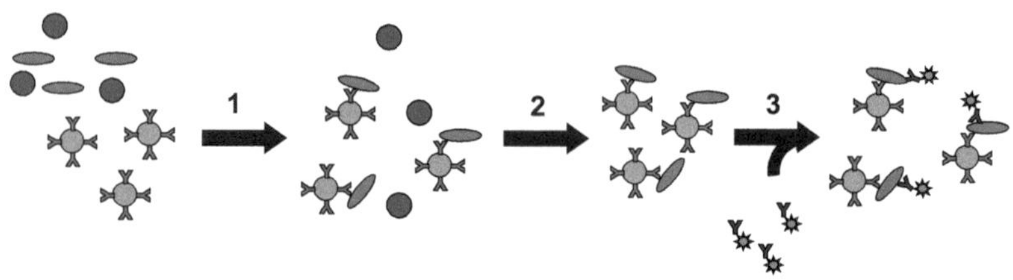

Figure 12.1. Immunomagnetic separation and detection scheme. Immunomagnetic beads are mixed with a mixed sample containing target *(ovals)* and nontarget *(circles)* organisms (Step 1). After mixing, magnetic separation and washing (Step 2) only target organisms are bound to the beads. This can be followed by labeling with secondary antibodies/labels (Step 3).

different sample matrices. One of the primary reasons for using IMS is that immunomagnetic micro- or nanobeads can be directly added to the food matrix (often with some dilution) and then be captured via magnetic separation (Ogden et al., 2001) (Fig. 12.1). This obviates the need for complicated filtration and concentration methods, which are more typically employed for enrichment and growth-based approaches. This has enabled detection of bacterial pathogens in complex samples, such as, ground beef, apple juice, and milk (Bennett et al., 1996; Gehring and Tu 2005; Seo et al., 1998). Once target pathogens are captured and then concentrated (by IMS) they can then be detected by conventional culturing on selective media, or by a variety of modern methods, for example, postcapture fluorescent labeling (Zhu et al., 2011; Wang et al., 2013c), postcapture PCR detection (Wang et al., 2013b), postcapture electrochemical detection (Gehring and Tu 2005), postcapture flow cytometry (Seo et al., 1998), or even postcapture labeling with fluorescent nanomaterials/quantum dots (Su and Li 2004).

Most commercially available immunomagnetic beads are in the 1–100 μm size range, but more recently research groups have demonstrated that nanoscale magnetic particles can also be used for IMS. For instance, Liu et al. (2015) demonstrated nanoparticle-based magnetic separation followed by immunochromatographic detection of *Vibrio parahaemolyticus*. Varshney et al. (2005) also demonstrated nanomagnetic particle-based separation and Ab-based detection of *E. coli* from ground beef. Other examples include immunomagnetic nanoparticles for *E. coli* enrichment and immunochromatographic detection (Qi et al., 2011), as well as, immunomagnetic nanoparticle concentration of *L. monocytogenes* for PCR-based detection (Yang et al., 2007). One advantage of using nanoparticles versus microparticles for IMS, is that with nanoparticles, one can achieve higher degrees of binding (ie, more particles bound to each target cell). Since microparticles are of the same size scale as bacterial cells, typically only one or two microparticles can bind to a target cell at one time. When using nanoparticles at a high concentration, hundreds of nanoparticles can attach to a single cell. This can enhance the total magnetic activity of the target cell postattachment. This can also enable the targeting of multiple antigens to a single cell. This is particularly interesting for scenarios in which cells present multiple types of antigens that could be targeted. For instance, using nanoparticles, one could target both cellular appendages (flagella, pili, etc.), as well as, cell surface markers (proteins, LPS, polysaccharides, etc.) without having steric interference between the particle-based tags.

2.3 Pathogen Detection Using Nanoscale Materials

Nanoscale materials have unique properties that are recently being utilized for pathogen detection technologies. These include optic and electric properties that are much different in nanoscale versus bulk forms of the same material. A good example of this is semiconductor quantum dots, which are crystalline nanoparticles that are highly fluorescent, are resistant to photo bleaching, and have a large Stokes shift (difference between the maximum optic excitation and emission wavelengths) (Cady et al., 2007). Quantum dots are synthesized from semiconducting materials, such as, cadmium selenide and lead selenide, which in bulk form are nonfluorescent, but in their nanoscale form are susceptible to quantum effects and become highly fluorescent materials. Quantum dots are superior to dye-based fluorophores as they present large Stokes shifts (difference between their peak fluorescent excitation wavelength and their peak emission wavelength), which increases multiplexing capability (by simplifying the optic components of the detection system); and their increased fluorescence intensity and resistance to photo bleaching enables increased detection sensitivity. Quantum dots have been used for direct fluorescent staining of multiple foodborne pathogens (Bruno, 2014), immunochromotogarphic detection of *Campylobacter jejuni* (Xu et al., 2013), and detection of *Salmonella* spp. in ground beef (Wang et al., 2012b). Quantum dots have also been used in place of traditional dye-based fluorophores in nucleic acid-based detection schemes. For instance, so-called quantum-dot molecular beacons were synthesized for detection of unique DNA sequences for *Salmonella typhimurium*, *L. monocytogenes*, and *Bacillus thuringiensis* (Cady et al., 2007; Burris et al., 2013). In this approach, short-DNA hairpins were attached to a quantum dot at one end and a fluorescent quencher (or gold particle) at the opposite end. In this configuration, the fluorophore does not emit light (does not

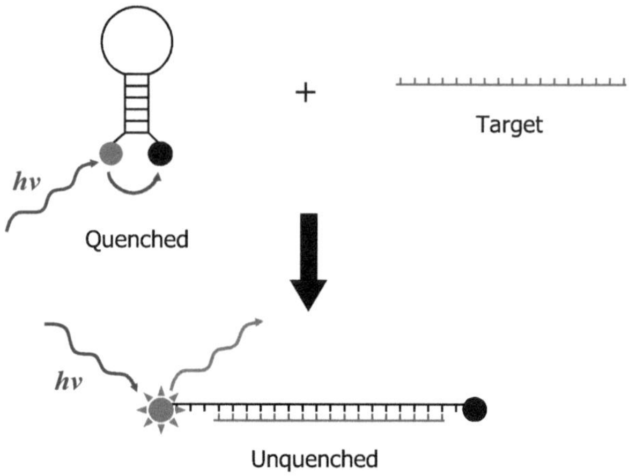

Figure 12.2. Example of a molecular beacon, used for fluorescence-based detection of target nucleic acids. The DNA hairpin, containing a fluorophore and a quencher molecule, is nonfluorescent (quenched), whereas the unfolded hairpin bound to its target is fluorescent (unquenched).

fluoresce) because of a phenomenon known as fluorescent resonant energy transfer (FRET). Optimal energy transfer (and subsequent quenching) occurs when the fluorophore and quencher are in very close proximity (typically around 5 nm distance). When the DNA hairpin is exposed to target DNA, it can unravel and hybridize to the target DNA, physically separating the fluorophore and quencher, resulting in a sharp increase in fluorescence. This results in a high signal-to-noise assay, because there is no fluorescent signal produced until the target DNA is present. Further, it can be combined with techniques, such as PCR, in which the molecular beacon is used as a "probe" to detect amplified DNA that is a product of the PCR reaction. Similar to the 5'-nuclease (TaqMan) assay, this adds another level of specificity to the detection assay, reducing the likelihood of false positive results (Fig. 12.2).

Other nanoscale materials that have been used for food pathogen detection include the carbon-based materials, graphene, fullerene, and carbon nanotubes. Graphene is an atomically thin, sheet-like nanomaterial consisting of π-bonded carbon, which has superior conductivity to traditional metal conductors. Fullerenes, also known as C60 or carbon nanospheres, are spheres of single-layer π-bonded carbon that also have unique electric properties. Carbon nanotubes are essentially extended, cylindrical forms of graphene, resembling a rolled tube of π-bonded carbon sheets. All of these materials are being widely explored because of their extremely high conductivity, high strength, and high sensitivity to surface-based interactions. Some of the challenges in using these materials are that they are highly inert. Thus, to use them in sensor configurations where target-binding events are needed, unique attachment/crosslinking schemes must be employed. For instance, to covalently attach affinity-based substrates (antibodies and aptamers) to a carbon nanotube or graphene surface, the carbon material must be exposed to strong acids or high-energy plasma to break carbon—carbon bonds and produce carbonyl or other reactive groups that can be used for chemical crosslinking reactions. Other approaches include the use of hydrophobic interactions between the affinity molecule and the highly nonpolar surfaces of these materials. This can include using nonpolar amino acids (eg, phenylananine) as a linker portion, by grafting affinity molecules onto molecules containing aromatic groups, or by fully saturated carbon chains. One other challenge for using carbon-based devices is the lack of controlled growth and deposition of these materials onto sensor substrates. While some progress has been made in this area, it is still quite difficult to grow a large, uniform film of graphene on a flat substrate (such as, a silicon wafer or glass surface). Likewise, precise localization of carbon nanotubes and fullerenes onto a surface can be challenging. Typically carbon nanotubes are deposited en masse, resulting in poorly ordered structures, which can diminish from the unique electric properties of well-aligned single nanotubes.

Graphene-based electric sensors have been used to detect *E. coli* via impedance measurements (Wang et al., 2013a). In this measurement approach, graphene was embedded in a filter material and was then further modified by addition of gold nanoparticles. This formed a high surface area electrode surface that could then be used to detect *E. coli* cells via affinity-based binding and subsequent electric impedance measurement. As with other impedance-based measurements, the added capacitance and resistance of the bacterial cells changes the frequency-dependent impedance of the electrode (in this case the graphene-modified paper). Fullerenes have also been used for electrochemical antigen-based detection of *E. coli* (Li et al., 2013). In this case, a fullerene-based electrode was modified with Au–SiO_2 nanoparticles, which were subsequently modified with anti-*E. coli* antibodies for affinity-based targeting. Interestingly, this sensor utilized a sandwich-binding approach in which successfully bound *E. coli* cells could be further modified with antibodies carrying a glucose oxidase enzyme. This, in turn, enabled the sensor to be used in an aperometric (current-generating) format. This unique approach of using multiple nanomaterials for bacterial sensing resulted in a drastic increase in sensitivity, with a detection limit of just 15 CFU mL^{-1}.

Beyond using nanoscale carbon-based materials for micro- and nanobacterial sensors, other carbon-based electric sensors have been developed. These include screen-printed carbon electrodes, such as, those used by Afonso et al. (2013) for detection of *Salmonella*. Screen-printing is essentially a low resolution form of pattern transfer that enables the printing of liquid-based materials through a previously patterned porous "screen" material. The screen has regions in which the pores are blocked (the masking region), as well as, regions where the pores are open. When the screen is placed against a flat surface, a liquid-based ink can be pressed through the open pores, transferring onto the flat substrate. This same technology is used for high-volume, rapid printing of textiles. For microscale electrodes, screen-printing is a fast and easy way of generating electrodes in a high throughput manner. This aids in reducing the final cost of the sensor and can enable the precise patterning of materials (such as, carbon paste) that would be difficult to achieve with other methods.

Nanoscale electric wires (nanowires) and three-dimensional field-effect transistors (FETs) are relatively recent technologies that are being employed for biosensing applications. A FET is a semiconductor device that is sensitive to the amount of charge (or electric field) at its surface. In electric devices, FETs are configured with two terminals (a source and drain) that are connected within an electric circuit. A third portion of the device, the "gate," lies between the source and drain, and can be perturbed by a metal electrode, or by adjusting the charge/electric field environment in its proximity. For biosensing applications, this is useful because many biologic molecules and even whole cells have an inherent electric charge. For instance, due to the phosphate backbone in nucleic acids, these molecules have a highly negative charge. Thus, FETs can be highly sensitive to the amount of DNA present in a solution, DNA binding, or hybridization events at the surface of the gate. Typically, a change in electric field at the gate of the FET results in a change in current flow between the source and the drain of the device. In some cases, enhancement of the electric field can increase current (by generating more charge carriers within the gate region) or can decrease current (by pushing away charge carriers in the gate region).

Although FETs have been used in semiconductor devices for many years, they are only recently gaining popularity for biosensing applications. One FET configuration that is particularly popular is the so-called finFET. As the local charge environment around nanowires or finFETs changes, conductivity through these devices is either greatly enhanced or diminished, thus resulting in a sensitive transduction mechanism. This is due to the three-dimensional nature of the finFET versus a traditional FET having a planar (flat) configuration. finFETs (sometimes referred to as a type of nanowire) have a three-dimensional gate that protrudes out of the surface plane of the device. This gives the device a very high surface area-to-volume ratio, which in turn, makes it more sensitive to surface-based interactions. Similar to finFETs, nanowire devices are being developed for biosensing applications. Nanowire sensors differ from finFETs primarily by the fabrication processes used to produce them. Unlike finFETs, which are fabricated using traditional semiconductor processing, nanowire devices are typically fabricated using solution-phase or gas-phase chemical growth of nanowires, followed by selective placement of the nanowires onto electric contacts. While this can produce extremely high sensitivity devices, it can introduce challenges into the fabrication process. To date there are no examples of finFET-based sensors for food pathogen detection, but nanowire-based sensors have been used for electric detection of whole cells, including those of the food pathogen *Bacillus cereus* (Pal et al., 2007) and for the detection of pathogen DNA (Jeong et al., 2015) (Fig. 12.3).

Additional examples of nanomaterial-based pathogen detection include those using colorimetric and Raman spectroscopy-based transduction mechanisms. Many types of metallic nanoparticles have unique optic properties, including distinct

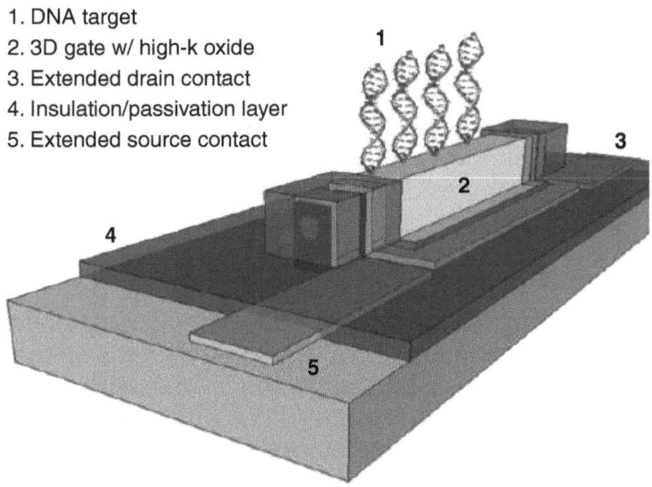

Figure 12.3. Conceptual schematic of a nanowire/finFET-based DNA biosensor. The narrow, three-dimensional gate region *(2)* is highly sensitive to small changes in electric charge. Thus, addition of DNA to the surface, or DNA hybridization events *(1)* can drastically affect the measured source *(5)* to drain *(3)* current through the device. This results in a high sensitivity DNA detection platform.

shifts in optic absorbance based on particle size or aggregation. These properties have been leveraged for gold nanoparticle-based colorimetric detection of *S. aureus* in milk (Sung et al., 2013) and more generalized gold nanoparticle colorimetric assays for *S. aureus* (Wang et al., 2012a). In these assays, gold particles are modified with affinity reagents (antibodies or aptamers), which enable binding to bacterial targets. As the gold particles attach to the target cells, they form clusters, which result in a shift in their optic properties. Typically, aggregation of gold particles causes a shift in their absorbance spectrum from shorter wavelengths to longer wavelengths (also known as a red-shift).

Gold and silver nanoparticles can also be used to enhance the signal intensity for Raman spectroscopy, which is also known as surface-enhanced Raman spectroscopy (SERS). Unlike infrared or absorption spectroscopy, Raman spectroscopy utilizes the inelastic scattering of photons off chemicals or materials. The Raman spectrum provides much more information content as compared to the normal fluorescence spectra of the fluorophore itself, making it possible to multiplex many targets into a single SERS-based assay (Lowe et al., 2010). In the SERS configuration, the Raman signal exhibits a significant increase in signal intensity when brought in close proximity to a noble metal, such as, gold or silver, increasing sensitivity. This is due to the generation of surface plasmons on the metallic particle, which greatly enhance the Raman spectral signal. This SERS-based technique has been used for detecting multiple pathogens, including *E. coli* in ground beef (Guven et al., 2011; Najafi et al., 2014; Cho et al., 2015). Key to all of these applications is the ability to carefully control the geometry and size of nanoparticles and then optimizing their performance for a specific application, for example, SERS-enhancing nanoparticles (Alvarez-Puebla and Liz-Marzán, 2010).

2.4 Lab-on-a-Chip–Based Detection

The integration of biosensors for pathogen detection with microfluidic components for sample pretreatment opens a huge possibility of progress and innovation in the field of LOC and micrototal analysis systems (μTAS). The rationale is to avoid the need of off-chip procedures and specialized personnel in order to increase the throughput and reduce the assay cost. This is a key advantage of miniaturization in a chip format, which is already impacting multiple fields, from medical diagnostics to food and environmental monitoring. In particular, applications for food control span from food safety (including the detection of pathogens, their viability status, as well as, the released toxins (Yang et al., 2010; Han et al., 2013) to food authenticity (eg, for the identification/differentiation of fish species in commercial products) (Rasmussen and Morrissey 2008). However, LOC systems enable determination of other analytes of food significance, such as, pesticides contaminations (Drechsel et al., 2015), vitamins (Lok et al., 2012), antioxidants (Kovachev et al., 2010), and allergens (Chiriacò et al., 2015). Beyond miniaturization and automation, other concepts and methodologies exploited in LOC systems often come from the traditional approaches, while fabrication methods come from the microelectronics industry (optic lithography) or were adapted for handling the polymeric materials employed for the microfluidic network (soft lithography). Without the intention of providing a comprehensive overview of the whole literature, we review a few technology demonstrators in the following sections.

To achieve an on-chip detection of foodborne pathogens two main approaches have been developed: (1) methods based on nucleic acids amplification, and (2) methods based on the detection of whole cell without the need of cell lysis.

LOC based on nucleic acid detection for the rapid quantification of pathogens and their viability has been demonstrated by multiple groups. Zhao et al. (2006) reported a microsystem for the analysis of *E. coli* in water and food by evaluating the GroEL-mRNA content after heating at 47°C as a viability indicator. They carried out an asymmetric RT-PCR and employed an electrochemical transduction with gold nanoparticles and silver electrodeposition for the detection of the amplicons. Although both heat shock stimulation of GroEL-mRNA production and RT-PCR were performed in a commercial thermal cycler, this approach can be fully miniaturized in a chip format.

Recently several isothermal amplification techniques have been also developed on LOC platform to avoid the need of microheaters for thermal cycling, which allow a faster reaction if compared with conventional PCR methods. These methods mimic in vivo strategies for nucleic acid synthesis by taking advantages from the use of accessory proteins such as helicases. One of the most widely used isothermal amplification approach is the so-called loop-mediated isothermal amplification (LAMP) that requires a set of four or six primers and a DNA polymerase with strand displacement activity. The amplicons have a stem-loop structure with inverted repeats of the target sequences. This technique is very efficient, allowing the synthesis of large quantities of DNA in 5–10 min. This technology has widely been applied for developing point-of-care test for microorganism detection. In particular, Sun et al. (2015) realized a platform able to perform on-chip sample preparation, amplification, and real-time detection. Specifically,

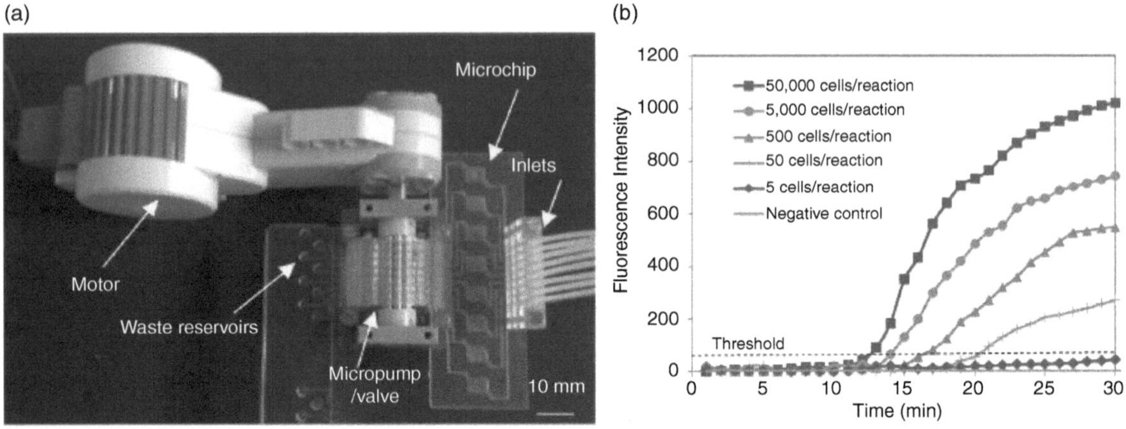

Figure 12.4. (a) Experimental setup for on-chip isothermal amplification and genomic detection of *Salmonella* spp. and (b) real-time response for various assays performed at different cell concentrations. Reproduced with permission from Sun et al. (2015). Copyright 2015, The Royal Society of Chemistry.

they implemented a LAMP, which works at a constant low temperature with no need of performing accurate, power-consuming thermal cycles. As a proof-of-concept, they applied this device for the determination of *Salmonella* spp. in food samples resulting in assay duration smaller than 40 min and a detection limit around 10 cells μL^{-1}. These performances have to be compared to the corresponding values for traditional culture assays, which can take days. Fig. 12.4a shows a picture of the setup, which includes the microchip, inlets, waste reservoirs, micropumps, valves, and a motor. The analysis was performed on spiked samples prepared from pork, using magnetic beads to remove contaminants, while the readout was in real-time and based on fluorescence. The platform had multiplexing capacity and was able to analyze up to eight samples in parallel. In Fig. 12.4b, the results of the assay are reported for different cell concentrations; 20 min are enough to reach the selected signal threshold at 50 cells μL^{-1}.

A device performing a similar task was optimized by Morant-Minana and Elizalde (2015). In this case, however, the authors employed a voltametric transduction instead of a fluorescence one for the genomic detection of *Campylobacter* spp. Since the bacterial motility is important in determining its virulence, flagellin genes were addressed in this study after inoculating the pathogen in a chicken sample. The hybridization sensor was fabricated on a thermoplastic polymer (cyclo-olefin polymer) and consisted of a three-electrode configuration with immobilized thiolated probes and $Fe(CN)_6^{3-/4-}$ used as redox couple. Bajwa et al. (2013) instead made use of impedance spectroscopy for miniaturizing a

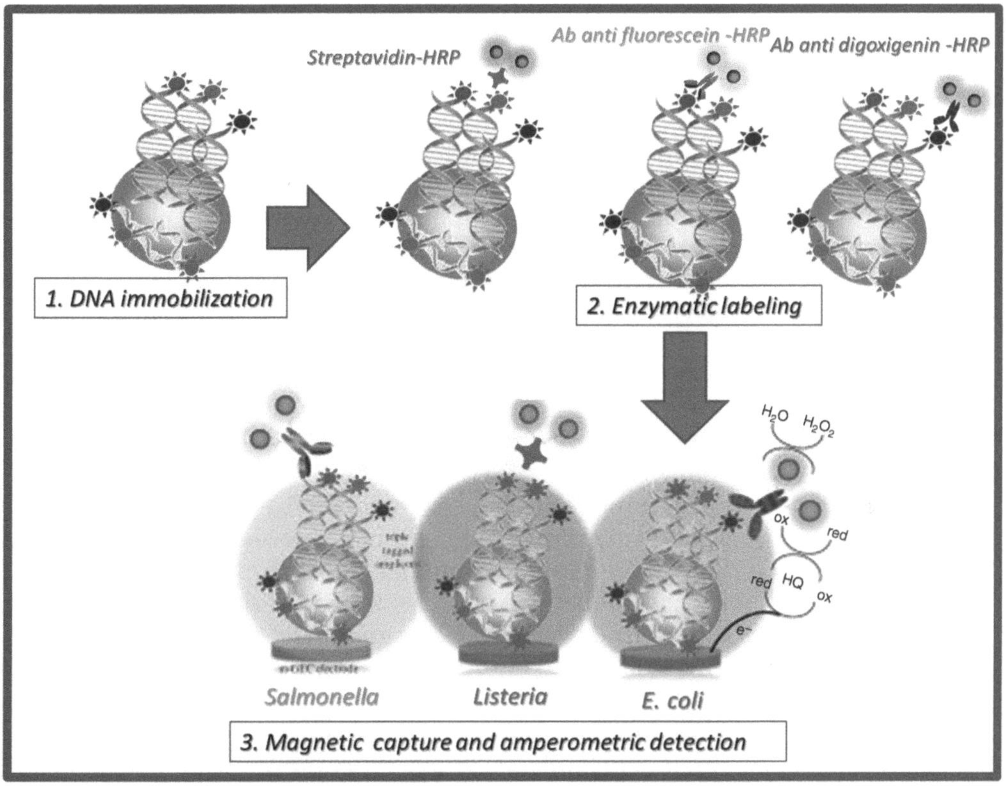

Figure 12.5. Schematic representation of the methods proposed by Brandão et al. (2014) for the simultaneous detection of *Salmonella enterica*, *L. monocytogenes*, and *E. coli* based on triple-tagging multiplex PCR and electrochemical magneto-genosensing on silica magnetic particles. Reproduced with permission from Brandão et al. (2015).

plate-count technique in order to determine the presence and location of bacterial colonies thanks to an array of microelectrodes. In general, electric-readout schemes present significant advantages in terms of portability because the control electronics and measurements requirements are more suitable for field use.

Multiple-analyte detection is one of the most important improvements that is possible by exploiting the possibility offered by miniaturization and LOC platforms. In this respect Brandão et al. (2015) have proposed a new platform for the multiplexed detection of different foodborne pathogens (Fig. 12.5) by taking advantage from microfluidic devices with different reaction chambers for the readout.

They developed a triple-tagging multiplex PCR coupled with a magnetic actuation and an electrochemical detection for the identification of three of the most important pathogens in food, in particular, *S. typhimurium*, *L. monocytogenes*, and *E. coli*. They used PCR specific tagging primers for each bacterial strain targeted; probes specific for *S. typhimurium* were tagged with fluorescein, probes for *L. monocytogenes* with biotin, and those for *E. coli* with digoxigenin. Amplified- and tagged sequences were bound on silica magnetic particles that are used for immobilization onto electrodes thanks to a magnetic actuation exploiting the properties of a magneto-graphite epoxy composite electrode (m-GEC). At the end, amperometric detection has been achieved thanks to the use of specific electrochemical reporters (Ab Anti-Fluorescein–HRP, Streptavidin–HRP, and Ab Anti-Digoxigenin–HRP).

All the systems previously described require the external contribution of an operator to perform some steps of the process but the final goal of a LOC platform would be to obtain a fully automatic assay for "sample-to-results" analysis. In this respect, centrifugal microfluidic has the potential to really accelerate the development of new analytical devices because fluidic handling and sample processing can be achieved without external pumping, but just with a rotational motor to create the necessary forces to drive the flow. Kim et al. (2014) have described a centrifugal microfluidic device (Fig. 12.6), for the detection of *Salmonella* in milk samples. The reported device integrates all the functions required for analytic processes, from DNA extraction to final detection. In particular they have integrated cell lysis and DNA extraction, isothermal recombinase polymerase amplification (RPA), and detection onto a single platform. A laser diode has been used for valve actuation, cell lysis, and temperature control in the isothermal amplification step. Lateral flow strips allow a direct visual observation of the results of the analysis. The entire analysis requires only 30 min with a detection limit of 10 CFU mL^{-1} in water and 10^2 CFU mL^{-1} in real samples (Kim et al., 2014).

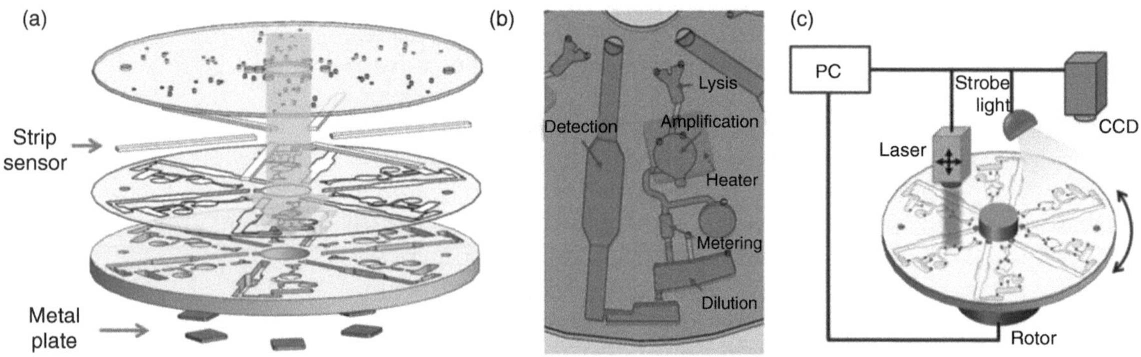

Figure 12.6. (a) Schematic view of the lab-on-a-disc showing the multilayer structure. (b) Top view of the functional units for cell lysis, isothermal amplification, metering, dilution, and detection. (c) Schematic illustration of the experimental setup. Reproduced with permission from Kim et al. (2014). Copyright 2014, American Chemical Society.

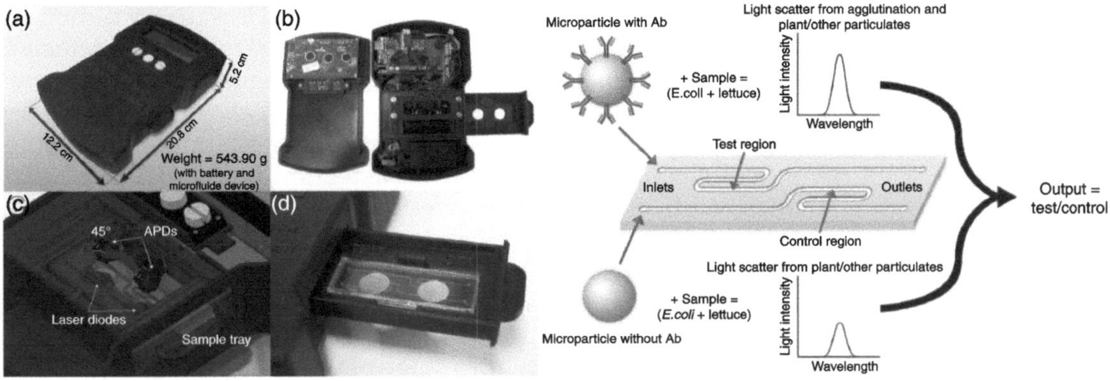

Figure 12.7. Images of the portable system and microfluidic network implemented by You et al. (2011). Reproduced with permission from You et al. (2011). Copyright 2011, Elsevier B.V.

A different approach for foodborne pathogens is based on whole-cell identification. Several strategies can be used for detection of bacterial cells, such as, the compact and portable system reported by You et al. (2011). Their platform (Fig. 12.7) was battery-powered and included microfluidic channels, readout, and control electronics (microprocessor, differential operation amplifier, and display). Transduction was based on light scattering from agglutinated Ab coated-latex particles and the analysis was performed with the support of Mie-scatter simulations. The proof-of-concept demonstration was carried out by detecting *E. coli* K12 and O157:H7 in iceberg lettuce. A further advantage is the very short assay time (≤6 min), while the reported detection limit was 10 CFU mL^{-1}. The authors' efforts also concentrated on reducing the off-chip steps for sample processing and enrichment, which required the use of commercial filters and syringe.

Due to its high sensitivity and easy setup, electrochemical impedance spectroscopy (EIS) has been extensively applied for biosensor fabrication to quantify foodborne bacteria. Impedance can be used to measure the release of ionic metabolites induced by bacterial metabolism and cell growth; distinguishing viable and dead cells (Silley and Forsythe, 1996; Wawerla et al., 1999). More recently, impedimetric biosensors into a chip format offered new opportunities in the field of bacterial detection in particular in term of selectivity and specificity. In fact, microfabricated systems allow the functionalization of the electrode surface with molecules able to capture specific bacterial cells. In this case, impedance spectroscopy can be used to monitor not only the electric properties of the medium but also the electric parameters (such as, capacitance and electron transfer resistance) of the interface between electrode and solution. Thus, every biorecognition event happening at the electrode surface can be detected as an increase of impedance values. For example, Radhakrishnan et al. (2013) were able to detect *L. monocytogenes* by EIS with very high sensitivity using a mouse monoclonal Ab immobilized onto a gold electrode. They reported detection limits of 5 CFU mL^{-1} and 4 CFU mL^{-1} for ideal solutions and filtered tomato extract, respectively.

A step forward has been done by Jiang et al. (2014) who moved toward an integrated platform by presenting a low-cost, miniaturized, and sensitive bacteria sensor based on electric impedance spectroscopy coupled with a smartphone. The proposed platform is composed of interdigitated electrodes on a microhole silicon substrate and a microfluidic chamber based on a nanoporous filter paper, which is used to preconcentrate bacteria in sample solutions, which lowered the detection limit to 10 bacterial cells mL^{-1} (Jiang et al., 2014).

Other electrochemical methods can be used for a sensitive detection of foodborne bacteria. To improve the sensor performance, there are two main solutions exploited by research groups; the use of nanomaterials for electrode and the discovery of new functionalization strategies to obtain a better electric response.

In this respect, DNA nanostructures have attracted great interest for their unique capacity to self-assemble in a predictable form and the possibility of easy coupling with surface and other functionalities. For example, Giovanni et al. (2015) have developed a DNA nanopyramid to anchor antibodies to the electrode surface in order to perform electrochemical ultrasensitive immunological assay for *E. coli* detection. They also use a secondary Ab conjugated with ferrocene for a voltammetric readout achieving a very high sensitivity of 1.2 CFU mL^{-1} by exploiting the great advantages of DNA nanopyramids: (1) the vertex for antibodies binding has a stable height so Abs can protrude easily to capture bacterial cells; and (2) the hollow structure enhances the electron transfer with redox molecules in solution, in comparison to other anchoring functionality (Giovanni et al., 2015) (Fig. 12.8).

Despite significant advances in nanoscale LOC devices, there are significant challenges for the direct detection of pathogens in foods. Interfacing the large, complicated, and different food samples with the microscale fluid interrogation volumes requires unique solutions. These approaches need to take a large sample (typically 1–25 g) diluted and in many cases liquefied at least 10-fold, and deliver it to a very small volume in the device. New tools for sample pretreatment and pathogens concentration are strongly needed to achieve an efficient detection.

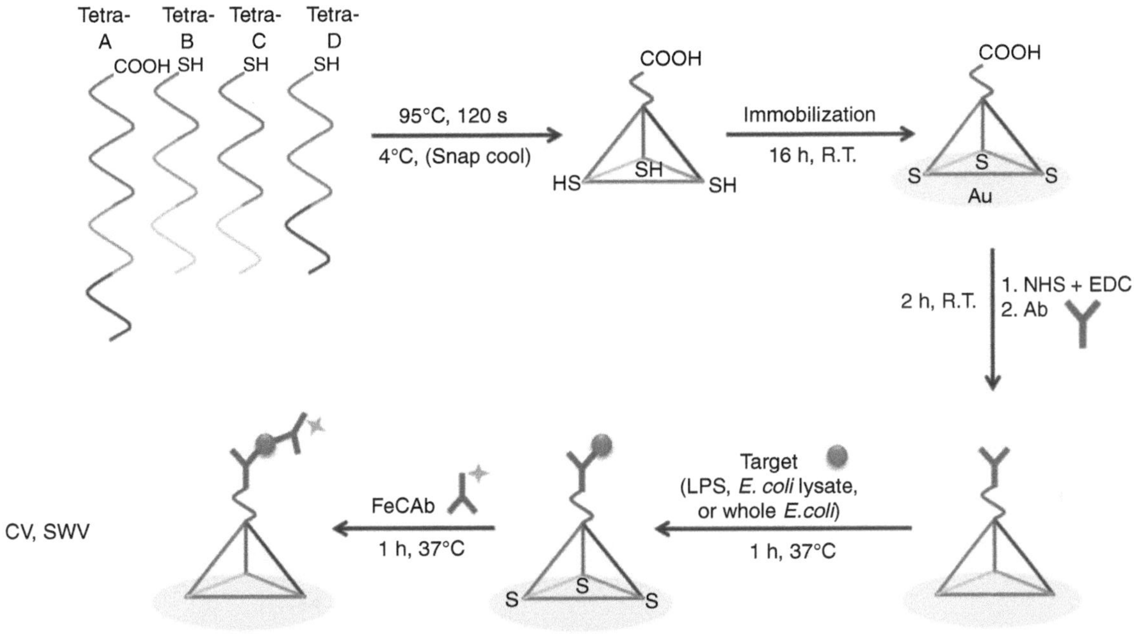

Figure 12.8. Schematic assembling methods of DNA nanopyramids and Ab functionalization for bacteria detection. Reproduced with permission from Giovanni et al. (2015). Copyright 2015, Wiley-VCH Verlag GmbH & C.

In this respect, Zhang et al. (2015) proposed a microfluidic method for selective capture of microorganisms as an effective and rapid way based on aptamers targeting living bacterial cells, such as *E. coli*. They developed the best aptamer sequences by using a whole-cell based aptamers selection and integrated them into the microfluidic device as recognition and capture elements for selective capture of microorganisms. The results demonstrated that such a device was able to capture *E. coli* selectively and distinguish different species of bacteria. Moreover it can be integrated with very different detection strategies in order to achieve a complete analysis on chip (Zhang et al., 2015).

Other approaches can be used for bacterial concentration in microfluidic devices by exploiting various physical mechanisms, such as, gravitational force, electric force, dielectrophoretic force, magnetic force, and acoustic-primary radiation force. Compared to biochemical techniques based on molecular recognition, physical approaches do not require any recovery step that can potentially alter viability or properties of isolated bacterial cells, which are viable and free in suspension enabling easy coupling with downstream on chip analysis.

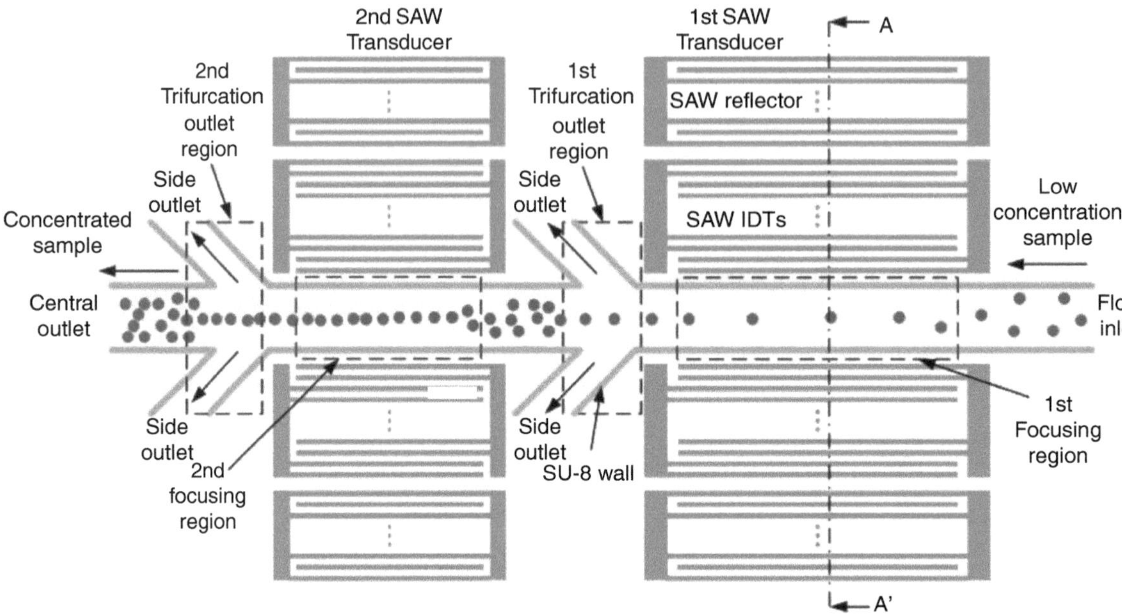

Figure 12.9. Schematic of a SAW-based SU-8 microfluidic concentrator proposed by Mu et al. (2015). The device has two focusing regions followed by two trifurcation outlet regions when viewed from the top. Reproduced with permission from Mu et al. (2015). Copyright 2014, Elsevier.

Mu et al. (2015) have recently developed a surface acoustic wave (SAW) microfluidic device to concentrate bacterial cells in suspensions. A typical SAW device is composed by couples of interdigitated electrodes on a piezoelectric substrate used for applying an electric field to the piezoelectric material that respond to this perturbation by generating a mechanical deformation, so when an alternating electric signal is applied, an acoustic wave propagating parallel to the surface is generated. This acoustic wave can be used for particles/microorganism manipulation inside microchannels (Fig. 12.9).

The device is composed of SU-8 microfluidic network of channels aligned with SAW reflectors and is based on a two-stage concentration design. In each concentration module bacterial cells are acoustically focused to the middle of the channel. So, when the microchannel is split, focused cells go to the next concentration step while clear liquid (free of bacteria) is drained from the two lateral outlets. This device has been applied to concentrate *E. coli* DH5 leading to an overall concentration factor of 59 and a particle recovery efficiency of 81.0 ± 17.2%.

Recently Moral-Zamora et al. (2015) has reported one of the first examples of LOC platform with a concentration and detection

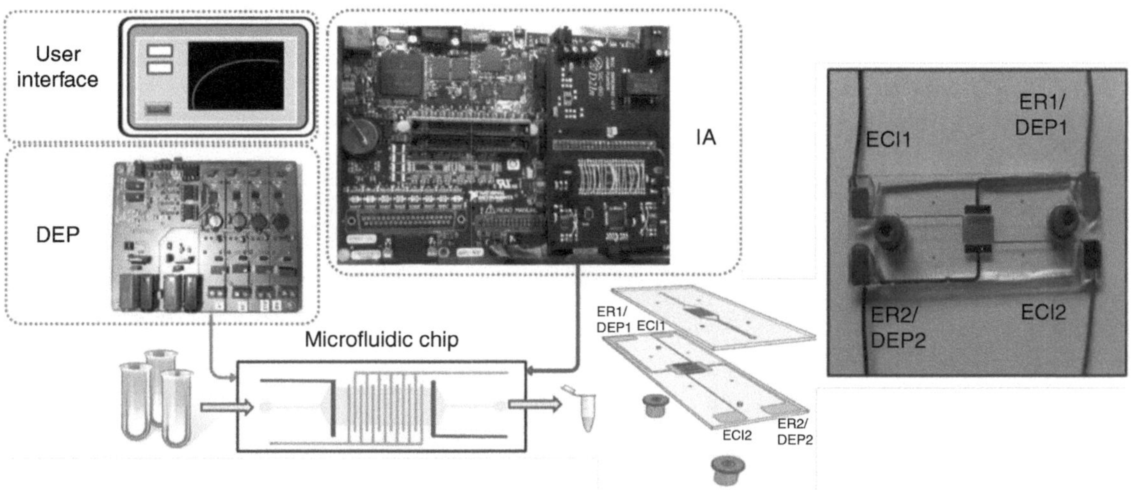

Figure 12.10. Overview of the combined DEP and impedance-analyzer system and structure of the prototype. From, with permission, Moral-Zamora et al. (2015). Copyright 2015, Wiley-VCH Verlag GmbH & C.

of bacteria integrated on the same device. They described a prototype consisting of a dielectrophoretic (DEP) actuator, to preconcentrate the bacterial cells, and an impedance sensor, to measure bacteria concentration by monitoring changes in conductivity of the medium. Dielectrophoresis is a separation technique that allows the separation of particles on the basis of their polarizablity. It is realized by applying a nonuniform electric field, in these conditions all particles can be polarized and manipulated by the dielectric force. The intensity of this force depends strongly on the medium and particles' properties, such as, electric characteristics, shape, and size. So by changing intensity and frequency of the electric field, it is possible to selectively manipulate specific particles. In the work of Moral-Zamora et al. (2015) the electrodes for DEP and impedance measurement are included in a single microfluidic chamber. The platform has been applied to the identification of *E. coli* (Fig. 12.10).

In a different approach, but still exploiting microfluidics, Clime et al. (2015) optimized a tool that was able to remove the debris particles from beef samples and extract the pathogens without the need of membranes. It is based on the concepts of hydrodynamic focusing and inertial lateral migration (Fig. 12.11). Both the inlets and the outlets were designed with a three-channel geometry. On the inlet side, the central channel was employed to inject the sample while the lateral channels provided clean PBS buffer leading to hydrodynamic focusing of the sample beam. Within the central separation channel, the debris particles migrated to the border.

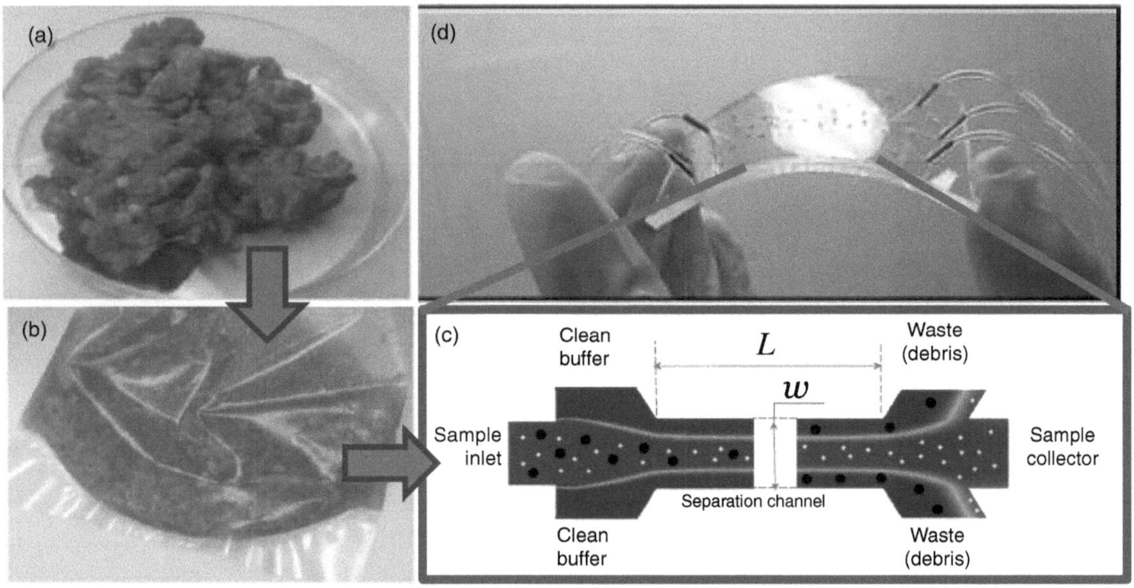

Figure 12.11. Sketch of off-chip and on-chip sample preparation/treatment procedures as implemented by Clime and coworkers. Reprinted with permission from Clime et al., Biomed. Microdev. 2015, 17. Copyright 2014 Springer-Verlag Berlin Heidelberg.

Then, on the outlet side, laminar flow allows the separation of the debris from the sample to be analyzed. Numerical simulation by finite-element methods and preliminary tests using fluorescent particles facilitated the optimization of the device. The separation performance was found to depend on the flow rate. Although the first steps (sample homogenization and suspension filtration) were performed off-chip, this work provides a useful contribution for a full integration of the assay in a chip format. The limitations in exploiting all the advantages of microfluidics come from the solid nature of the sample matrix, but can be overcome when analyzing liquid food samples (eg, wine, milk, oil, or other beverages).

Paramagnetic beads coated with a target capture molecules, such as, an Ab, lectin, or aptamer have been perhaps the best-developed sample preparation method (Haukanes and Kvam, 1993). IMS and detection can be also performed on chip. Among the various literature reports, it is worth mentioning here the work of Berenguel-Alonso et al. (2014), which employed magnetic beads (MB) to enhance an MB-based amplification reaction for the detection of *E. coli* O157:H7. Permanent magnets were integrated on a rotating disk at various distances from the rotation axis, intercepting the reaction chamber at different locations to drive the magnetic mixing (Fig. 12.12). This process favored an enzymatic

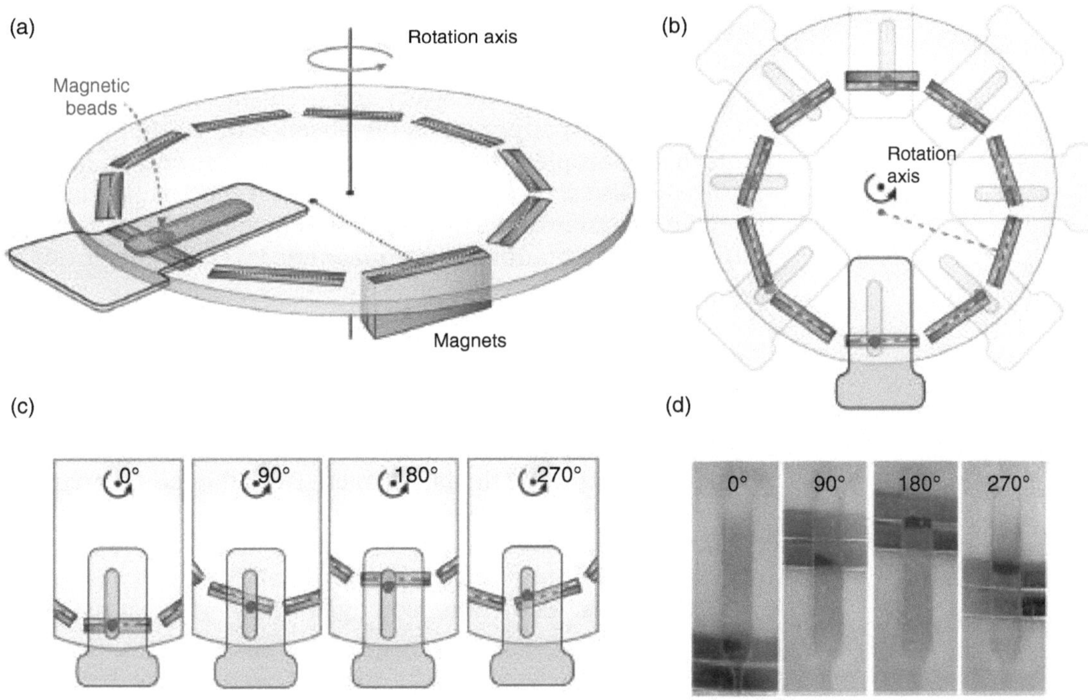

Figure 12.12. Experimental setup ferromagnetic-beads driven mixing and reaction enhancement as reported by Berebuel-Alonso et al. (2014). Reproduced with permission from Berebuel-Alonso et al. (2014). Copyright 2014, Springer-Verlag Berlin Heidelberg.

reaction resulting in the production of a fluorescent compound detected in the immunoassay. Superparamagnetic nanoparticles can also be employed for this scope. Moreover, after antigen–Ab selective binding, the formed complexes can be also exploited for implementing a magnetoresistive transduction. Magnetoresistance-based sensors (MR) were employed as transducers to detect antigens in a sandwich assay by using an Ab functionalized with magnetic labels. Typically, MR transducers are based on the Giant Magnetoresistive Effect (GMR) observed in multilayer structures composed of ferromagnetic and nonmagnetic layers. The effect consists of a change in the electric resistance depending on the direction of magnetization of ferromagnetic layers. In biosensors, the magnetization direction can be controlled, not by applying an external magnetic field, but by the presence of the magnetic labels. In this way, it is possible to detect analytes thanks to the changes in magnetoresistance induced by the presence of magnetic labels of the analytes after their binding to the capture probe on the transducer surface.

For example, Mujika et al. (2009) detected *E. coli* O157:H7 by means of giant magnetoresistive multilayer sensors integrated in a SU-8 three-dimensional microfluidic module fabricated by lithography of the two separate layers and their successive bonding. An advantage of this approach is the absence of magnetic contaminants in the samples, resulting in a low baseline and an improved sensitivity. Also in this case, the readout is electric, because it is based on a change of the junction resistance when the magnetization relative orientations of the magnetic layers change from parallel to antiparallel and vice versa.

Combining the use of magnetic beads, filtration, and ultimately an integrated real-time PCR detection component, Lui et al. (2009) were able to carry out an sample to detection process on a single platform. Challenges still exist in the extension of this type of platform to the wide variety of food samples that complicate the physical separation of the target agent from the food matrix.

In conclusion, LOC platforms are driving a revolution in the technologies for addressing food safety issues. Despite consumer health, the economic impact on the related industry is also relevant, as well as, security aspects against bioterrorism risks.

3 Summarizing Remarks

Recent advances in micro- and nanotechnologies have served as the basis for the development of a wide range of diagnostic platforms and diagnostic assays that extend from sample to assay readout. A wide range of approaches, to detect food pathogens using micro- and nanotechnologies, have been presented. These include the use of nanomaterials as enhancers for analytical assays, microchip-based sensors for small volume, rapid diagnostic assays, and unique LOC systems that incorporate multiple sample processing and analytical steps onto a single platform. One of the great advantages of using micro- and nanotechnologies for food-pathogen detection is the promise of more rapid and portable detection platforms. This is particularly useful in both agricultural and food processing scenarios, where rapid time to detection is paramount for preventing contamination or disease outbreak. Incorporation of multiple laboratory steps into a single, automated platform also reduces the likelihood of human error, and sets the stage for remote (operator-free) detection of pathogens in many environments. While there are multiple scientific and engineering challenges that must be overcome to achieve commercialization and wide acceptance of these technologies, there has been significant progress over the past 15–20 years of micro- and nanoscale biosensor development. This is especially apparent with LOC systems, which in some cases have already seen commercial development and application. For some of the emerging technologies, such as, those that use nanomaterials to enhance detection-assay sensitivity, the road to commercialization may be longer. However, it will be exciting to observe the next decade of technology development and the broad implementation of these devices and technologies in agricultural and food processing settings. We also anticipate the eventual use of such devices by individuals for pathogen detection at home. Additionally, these technologies have much broader applications for medical diagnostics and prognostics. There are many parallel efforts to develop similar technologies to be used in clinical settings and for personalized medicine and personalized diagnostics. What we hope to see in the future is more crosstalk between the food science and medical fields, within the context of pathogen detection. Although the sample matrices can be quite different, the core technology for many of these biosensing platforms is the same, and thus technology sharing between these fields is paramount.

References

Abu-Salah, K.M., Zourob, M.M., Mouffouk, F., Alrokayan, S.A., Alaamery, M.A., Ansari, A.A., 2015. DNA-based nanobiosensors as an emerging platform for detection of disease. Sensors 15, 14539–14568.

Afonso, A.S., Perez-Lopez, B., Faria, R.C., Mattoso, L.H., Hernandez-Herrero, M., Roig-Sagues, A.X., Maltez-da Costa, M., Merkoci, A., 2013. Electrochemical detection of *Salmonella* using gold nanoparticles. Biosens. Bioelectron. 40, 121–126.

Alba, P., Feltrin, F., Cordaro, G., Porrero, M.C., Kraushaar, B., Argudín, M.A., Nykäsenoja, S., Monaco, M., Stegger, M., Aarestrup, F.M., Butaye, P., Franco, A., Battisti, A., 2015. Livestock-associated methicillin resistant and methicillin susceptible *Staphylococcus aureus* sequence type (CC)1 in European farmed animals: high genetic relatedness of isolates from italian cattle herds and humans. PLoS ONE 10, e0137143.

Alvarez-Puebla, R.A., Liz-Marzán, L.M., 2010. SERS-based diagnosis and biodetection. Small 6, 604–610.

Bajwa, A., Tan, S.T., Mehta, R., Bahreyni, B., 2013. Rapid detection of viable microorganisms based on a plate count technique using arrayed microelectrodes. Sensors 13, 8188–8198.

Bennett, A.R., MacPhee, S., Betts, R.P., 1996. The isolation and detection of *Escherichiacoli* O157 by use of immunomagnetic separation and immunoassay procedures. Lett. Appl. Microbiol. 22, 237–243.

Berenguel-Alonso, M., Granados, X., Faraudo, J., Alonso-Chamarro, J., Puyol, M., 2014. Magnetic actuator for the control and mixing of magnetic bead-based reactions on-chip. Anal. Bioanalytic. Chem. 406, 6607–6616.

Billington, C., Hudson, J.A., D'Sa, E., 2014. Prevention of bacterial foodborne disease using nanobiotechnology. Nanotechnol. Sci. Appl. 7, 73–83.

Brandão, D., Liébana, S., Campoy, S., Cortés, M.P., Alegret, S., Pividori, M.I., 2015. Simultaneous electrochemical magnetogenosensing of foodborne bacteria based on triple-tagging multiplex amplification. Biosens. Bioelectron. 74, 652–659.

Bruno, J.G., 2014. Application of DNA aptamers and quantum dots to lateral flow test strips for detection of foodborne pathogens with improved sensitivity versus colloidal gold. Pathogens 3, 341–355.

Bsat, N., Wiedmann, M., Czajka, J., Barany, F., Piani, M., Batt, C., 1994. Food safety applications of nucleic acid-based assays. Food Technol. 48, 142–145.

Burris, K.P., Wu, T.C., Vasudev, M., Stroscio, M.A., Millwood, R.J., Stewart, Jr., C.N., 2013. Mega-nano detection of foodborne pathogens and transgenes using molecular beacon and semiconductor quantum dot technologies. IEEE Trans. Nanobiosci. 12, 233–238.

Cady, N.C., Stelick, S., Batt, C.A., 2003. Nucleic acid purification using microfabricated silicon structures. Biosens. Bioelectron. 19, 59–66.

Cady, N.C., Strickland, A.D., Batt, C.A., 2007. Optimized linkage and quenching strategies for quantum dot molecular beacons. Mol. Cell. Probes 2, 116–124.

Cady, N.C., Stelick, S.J., Batt, C.A., 2011. PCR-based detection of *Bacillus anthracis* using an integrated microfluidic platform. Int. J. Biomed. Nanosci. Nanotechnol. 211, 152–166.

Chiriacò, M.S., de Feo, F., Primiceri, E., Monteduro, A.G., de Benedetto, G.E., Pennetta, A., Rinaldi, R., Maruccio, G., 2015. Portable gliadin-immunochip for contamination control on the food production chain. Talanta 142, 57–63.

Cho, I.H., Mauer, L., Irudayaraj, J., 2014a. In-situ fluorescent immunomagnetic multiplex detection of foodborne pathogens in very low numbers. Biosens. Bioelectron. 57, 143–148.

Cho, I.H., Radadia, A.D., Farrokhzad, K., Ximenes, E., Bae, E., Singh, A.K., Oliver, H., Ladisch, M., Bhunia, A., Applegate, B., Mauer, L., Bashir, R., Irudayaraj, J., 2014b. Nano/micro and spectroscopic approaches to food pathogen detection. Ann. Rev. Anal. Chem. 7, 65–88.

Cho, I.H., Bhandari, P., Patel, P., Irudayaraj, J., 2015. Membrane filter-assisted surface enhanced Raman spectroscopy for the rapid detection of *E. coli* O157:H7 in ground beef. Biosens. Bioelectron. 64, 171–176.

Clime, L., Hoa, X.Y.D., Corneau, N., Morton, K.J., Luebbert, C., Mounier, M., Brassard, D., Geissler, M., Bidawid, S., Farber, J., Veres, T., 2015. Microfluidic filtration and extraction of pathogens from food samples by hydrodynamic focusing and inertial lateral migration. Biomed. Microdev. 17 (1).

Dembek, Z.F., Noah, D.L., 2011. Food- and water-borne pathogens. Encyclopedia of Bioterrorism Defense, 1–4.

Doorduyn, Y., van Pelt, W., Siezen, C.L., van der Horst, F., van Duynhoven, Y.T., Hoebee, B., et al., 2008. Novel insight in the association between Salmonellosis or Campylobacteriosis and chronic illness, and the role of host genetics in susceptibility to these diseases. Epidemiol. Infect. 136, 1225–1234.

Drechsel, L., Schulz, M., von Stetten, F., Moldovan, C., Zengerle, R., Paust, N., 2015. Electrochemical pesticide detection with AutoDip: a portable platform for automation of crude sample analyses. Lab Chip 15, 704–710.

EFSA (European Food Safety Authority) and ECDC (European Centre for Disease Prevention and Control), 2015. The European Union summary report on trends and sources of zoonoses, zoonotic agents and foodborne outbreaks in 2013. EFSA J. 13, 3991.

Farmer, III, J.J., Asbury, M.A., Hickman, F.W., Brenner, D.J., 1980. The Enterobacteriaceae study group. *Enterobacter sakazakii*: a new species of "Enterobacteriaceae" isolated from clinical specimens. Int. J. Syst. Bacteriol. 30, 569–584.

FDA, 2007. Food protection plan. Food Drug Admin.

Ford, L., Kirk, M., Glass, K., Hall, G., 2014. Sequelae of foodborne illness caused by 5 pathogens, Australia, Circa 2010. Emerg. Infect. Dis. 20, 1865–1871.

Food Standard Agency, 2011. Foodborne disease strategy 2010-15 an FSA programme for the reduction of foodborne disease in the UK. Version 1.0. Published May 2011, following public consultation.
Fusco, V., Quero, G.M., 2012. Nucleic acid-based methods to identify, detect and type pathogenic bacteria occurring in milk and dairy products. In: Eissa, A.A. (Ed), Structure and function of food engineering. InTech.
Fusco, V., Quero, G.M., 2014. Culture-dependent and -independent nucleic acid-based methods used in the microbial safety assessment of milk and dairy products. Comp. Rev. Food Sci. Food Saf. 13, 493–537.
Fusco, V., Quero, G.M., Morea, M., Blaiotta, G., Visconti, A., 2011. Rapid and reliable identification of *Staphylococcus aureus* harbouring the *enterotoxin gene cluster* (*egc*) and quantitative detection in raw milk by real time PCR. Int. J. Food Microbiol. 144, 528–537.
Fusco, V., Riccardi, M., Quero, G.M., 2012. Thin agar layer- versus most probable number-PCR to enumerate viable and stressed *Escherichia coli* O157:H7 and application in a traditional raw milk pasta filata cheese. Int. J. Food Microbiol. 159, 1–8.
Fusco, V., den Besten, H.M., Logrieco, A.F., Rodriguez, F.P., Skandamis, P.N., Stessl, B., Teixeira, P., 2015. Food safety aspects on ethnic foods: toxicological and microbial risks. Curr. Opin. Food Sci. (6), 24–32.
Gehring, A.G., Tu, S.I., 2005. Enzyme-linked immunomagnetic electrochemical detection of live *Escherichia coli* 0157: H7 in apple juice. J. Food Protect. 68, 146–149.
Gibney, K.B., O'Toole, J., Sinclair, M., Leder, K., 2014. Disease burden of selected gastrointestinal pathogens in Australia, 2010. Int. J. Infect. Dis. 28, 176–185.
Giovanni, M., Setyawati, M.I., Tay, C.Y., Qian, H., Kuan, W.S., Leong, D.T., 2015. Electrochemical Quantifi cation of *Escherichia coli* with DNA nanostructure. Adv. Funct. Mater. 25, 3840–3846.
Guven, B.N., Basaran-Akgul, E., Temur, U., Boyaci, I.H., 2011. SERS-based sandwich immunoassay using antibody coated magnetic nanoparticles for *Escherichia coli* enumeration. Analyst 136, 740–748.
Haagsma, J.A., Siersema, P.D., de Wit, N.J., Havelaar, A.H., 2010. Disease burden of post-infectious irritable bowel syndrome in The Netherlands. Epidemiol. Infect. 138, 1650–1656.
Han, J.H., Kim, H.J., Sudheendra, L., Gee, S.J., Hammock, B.D., Kennedy, I.M., 2013. Photonic crystal lab-on-a-chip for detecting staphylococcal enterotoxin B at low attomolar concentration. Anal. Chem. 85, 3104–3109.
Haukanes, B.I., Kvam, C., 1993. Application of magnetic beads in bioassays. Nature Biotechnol. 11, 60–63.
Hennekinne, J.A., Ostyn, A., Guillier, F., Herbin, S., Prufer, A.L., Dragacci, S., 2010. How should staphylococcal food poisoning outbreaks be characterized? Toxins 2, 2106–2116.
Hennekinne, J.A., De Buyser, M.L., Dragacci, S., 2012. *Staphylococcus aureus* and its food poisoning toxins: characterization and outbreak investigation. FEMS Microbiol. Rev. 36, 815–836.
Heo, J., Hua, S.Z., 2009. An overview of recent strategies in pathogen sensing. Sensors 9, 4483–4502.
Heymann, D.L., 2001. Strengthening global preparedness for defense against infectious disease threats. Committee on Foreign Relations, United States Senate hearing on the threat of bioterrorism and the spread of infectious diseases.
Iversen, C., Lehner, A., Mullane, N., Bidlas, E., Cleenwerck, I., Marugg, J., Fanning, S., Stephan, R., Joosten, H., 2007. The taxonomy of *Enterobacter sakazakii*:

proposal of a new genus *Cronobacter* gen. nov. and description of *Cronobacter sakazakii* comb. nov. *Cronobacter sakazakii* subsp. *sakazakii*, comb. nov., *Cronobacter sakazakii* subsp. *malonaticus* subsp. nov., *Cronobacter turicensis* sp. nov., *Cronobacter muytjensii* sp. nov., *Cronobacter dublinensis* sp. nov. and *Cronobacter genomospecies* I. BMC Evol. Biol. 7, 64.

Jeong, J., Kim, H., Lee, J.B., 2015. Enzymatic polymerization on DNA modified gold nanowire for label-free detection of pathogen DNA. Int. J. Mol. Sci. 16, 13653–13660.

Jiang, J., Wang, X., Chao, R., Ren, Y., Hu, C., Xu, Z., Liu, G.L., 2014. Smartphone based portable bacteria pre-concentrating microfluidicsensor and impedance sensing. Sens. Actuat. B 193, 653–659.

Kaakoush, N.O., Castaño-Rodríguez, N., Mitchell, H.M., Man, S.M., 2015. Global epidemiology of *Campylobacter* infection. Clin. Microbiol. Rev. 28, 687–720.

Keithlin, J., Sargeant, J., Thomas, M.K., Fazil, A., 2014a. Chronic sequelae of *E. coli* O157: systematic review and meta-analysis of the proportion of *E. coli* O157 cases that develop chronic sequelae. Foodborne Pathog. Dis. 11, 79–95.

Keithlin, J., Sargeant, J., Thomas, M.K., Fazil, A., 2014b. Systematic review and meta-analysis of the proportion of *Campylobacter* cases that develop chronic sequelae. BMC Pub. Health 14, 1203.

Keithlin, J., Sargeant, J.M., Thomas, M.K., Fazil, A., 2015. Systematic review and meta-analysis of the proportion of non-typhoidal *Salmonella* cases that develop chronic sequelae. Epidemiol. Infect. 143, 1333–1351.

Kim, T.H., Park, J., Kim, C.J., Cho, Y.K., 2014. Fully integrated lab-on-a-disc for nucleic acid analysis of foodborne pathogens. Anal. Chem. 86, 3841–3848.

Kirk, M., Glass, K., Ford, L., Brown, K., Hall, G., 2014. Foodborne illness in Australia: annual incidence circa 2010. National Centre for Epidemiology and Population Health, Australian National University. Commonwealth of Australia, 2014.

Kovachev, N., Canals, A., Escarpa, A., 2010. Fast and selective microfluidic chips for electrochemical antioxidant sensing in complex samples. Anal. Chem. 82, 2925–2931.

Law, J.W., Ab Mutalib, N.S., Chan, K.G., Lee, L.H., 2014. Rapid methods for the detection of foodborne bacterial pathogens: principles, applications, advantages and limitations. Front. Microbiol. 5, 770.

Lee, S.H., Ahn, J.Y., Lee, K.A., Um, H.J., Sekhon, S.S., Sun Park, T., Min, J., Kim, Y.H., 2015. Analytical bioconjugates, aptamers, enable specific quantitative detection of *Listeria monocytogenes*. Biosens. Bioelectron. 68, 272–280.

Li, Y., Fang, L., Cheng, P., Deng, J., Jiang, L., Huang, H., Zheng, J., 2013. An electrochemical immunosensor for sensitive detection of *Escherichia coli* O157:H7 using C60 based biocompatible platform and enzyme functionalized Pt nanochains tracing tag. Biosens. Bioelectron. 49, 485–491.

Liu, Y., Cady, N.C., Batt, C.A., 2007. A plastic microchip for nucleic acid purification. Biomed. Microdev. 9, 769–776.

Liu, Y., Zhang, Z., Wang, Y., Zhao, Y., Lu, Y., Xu, X., Yan, J., Pan, Y., 2015. A highly sensitive and flexible magnetic nanoprobe labeled immunochromatographic assay platform for pathogen *Vibrio parahaemolyticus*. Int. J. Food Microbiol. 211, 109–116.

Locht, H., Mølbak, K., Krogfelt, K.A., 2002. High frequency of reactive joint symptoms after an outbreak of *Salmonella enteritidis*. J. Rheumatol. 29, 767–771.

Lok, K.S., Muttalib, S., Lee, P.P.F., Kwok, Y.C., Nguyen, N.T., 2012. Rapid determination of vitamin B-12 concentration with a chemiluminescence lab on a chip. Lab Chip 12, 2353–2361.

Lowe, A.J., Huh, Y.S., Strickland, A.D., Erickson, D., Batt, C., 2010. Multiplex single nucleotide polymorphism genotyping using ligase detection reaction coupled to surface enhanced Raman spectroscopy. Anal. Chem. 82, 5810–5814.
Lui, C., Cady, N.C., Batt, C.A., 2009. Nucleic acid-based detection of bacterial pathogens using integrated microfluidic platform systems. Sensors 9, 3713–3744.
McLinden, T., Sargeant, J.M., Thomas, K., Papadopoulos, A., Fazil, A., 2014. Component costs of foodborne illness: a scoping review. BMC Pub. Health 14, 509.
McNerney, R., 2015. Diagnostics for developing countries. Diagnostics 5, 200–209.
Merger, M., Croitoru, K., 1998. Infections in the immunopathogenesis of chronic inflammatory bowel disease. Semin. Immunol. 10, 69–78.
Moral-Zamora, B., Punter-Villagrassa, J., Oliva-Branas, A.M., Á lvarez-Azpeitia, J.M., Colomer-Farrarons, J., Samitier, J., Homs-Corbera, A., Miribel-Catalál, P.L., 2015. Combined dielectrophoretic and impedance system for on-chipcontrolled bacteria concentration:Application to *Escherichia coli*. Electrophoresis 36, 1130–1141, 2015.
Morant-Minana, M.C., Elizalde, J., 2015. Microscale electrodes integrated on COP for real sample *Campylobacter* spp. detection. Biosens. Bioelectron. 70, 491–497.
Mu, C., Zhang, Z., Min Lin, Dai, Z., Cao, X., 2015. Development of a highly effective multi-stage surface acoustic wave SU-8 microfluidic concentrator. Sens. Actuat. B 215, 77–85.
Mujika, M., Arana, S., Castano, E., Tijero, M., Vilares, R., Ruano-Lopez, J.M., Cruz, A., Sainz, L., Berganza, J., 2009. Magnetoresistive immunosensor for the detection of *Escherichia coli* O157:H7 including a microfluidic network. Biosens. Bioelectron. 24, 1253–1258.
Najafi, R., Mukherjee, S., Hudson, Jr., J., Sharma, A., Banerjee, P., 2014. Development of a rapid capture-cum-detection method for *Escherichia coli* O157 from apple juice comprising nano-immunomagnetic separation in tandem with surface enhanced Raman scattering. Int. J. Food Microbiol. 189, 89–97.
Norton, D.M., Batt, C.A., 1999. Detection of viable *Listeria monocytogenes* with a 5′ nuclease PCR assay. Appl. Environ. Microbiol. 65, 2122–2127.
Ogden, I.D., Hepburn, N.F., MacRae, M., 2001. The optimization of isolation media used in immunomagnetic separation methods for the detection of *Escherichiacoli* O157 in foods. J. Appl. Microbiol. 91, 373–379.
Okeke, I.N., Peeling, R.W., Goossens, H., Auckenthaler, R., Olmsted, S.S., de Lavison, J.F., Zimmer, B.L., Perkins, M.D., Nordqvist, K., 2011. Diagnostics as essential tools for containing antibacterial resistance. Drug Res. Updates 14, 95–106.
Ozalp, V.C., Bayramoglu, G., Kavruk, M., Keskin, B.B., Oktem, H.A., Arica, M.Y., 2014. Pathogen detection by core-shell type aptamer-magnetic preconcentration coupled to real-time PCR. Anal. Biochem. 447, 119–125.
Pal, S., Alocilja, E.C., Downes, F.P., 2007. Nanowire labeled direct-charge transfer biosensor for detecting *Bacillus* species. Biosens. Bioelectron. 22, 2329–2336.
Poropatich, K.O., Fischer Walker, C.L., Black, R.E., 2010. Quantifying the association between *Campylobacter* infection and Guillain-Barré syndrome: a systematic review. J. Health Nutr. 28, 545–552.
Porter, C.K., Choi, D., Cash, B., Pimentel, M., Murray, J., May, L., Riddle, M.S., 2013a. Pathogen-specific risk of chronic gastrointestinal disorders following bacterial causes of foodborne illness. BMC Gastroenterol. 13, 46.
Porter, C.K., Choi, D., Riddle, M.S., 2013b. Pathogen-specific risk of reactive arthritis from bacterial causes of foodborne illness. J. Rheumatol. 40, 712–714.

Potyrailo, R.A., Conrad, R.C., Ellington, A.D., Hieftje, G.M., 1998. Adapting selected nucleic acid ligands (aptamers) to biosensors. Anal. Chem. 70, 3419–3425.

Qi, H., Zhong, Z., Zhou, H.X., Deng, C.Y., Zhu, H., Li, J.F., Wang, X.L., Li, F.R., 2011. A rapid and highly sensitive protocol for the detection of *Escherichia coli* O157:H7 based on immunochromatography assay combined with the enrichment technique of immunomagnetic nanoparticles. Int. J. Nanomed. 6, 3033–3039.

Quero, G.M., Santovito, E., Visconti, A., Fusco, V., 2014. Quantitative detection of *Listeria monocytogenes* in raw milk and soft cheeses: culture-independent versus liquid- and solid-based culture-dependent real time PCR approaches. LWT Food Sci. Technol. 58, 11–20.

Radhakrishnan, R., Jahne, M., Rogers, S., Suni, I., 2013. Detection of *Listeria monocytogenes* by electrochemical impedance spectroscopy. Electroanalysis 25 (9), 2231–2237.

Rasmussen, R.S., Morrissey, M.T., 2008. DNA-based methods for the identification of commercial fish and seafood species. Comp. Rev. Food Sci. Food Saf. 7, 280–295.

Riva, A., Borghi, E., Cirasola, D., Colmegna, S., Borgo, F., Amato, E., Pontello, M.M., Morace, G., 2015. Methicillin-Resistant *Staphylococcus aureus* in raw milk: prevalence, SCCmec typing, enterotoxin characterization, and antimicrobial resistance patterns. J. Food Prot. 78, 1142–1146.

Seo, K.H., Brackett, R.E., Frank, J.F., 1998. Rapid detection of *Escherichia coli* O157:H7 using immuno-magnetic flow cytometry in ground beef, apple juice, and milk. Int. J. Food Microbiol. 44, 115–123.

Shu, B., Zhang, C., Xing, D., 2014. Segmented continuous-flow multiplex polymerase chain reaction microfluidics for high-throughput and rapid foodborne pathogen detection. Anal. Chim. Acta 826, 51–60.

Silley, P., Forsythe, S., 1996. Impedance microbiology—a rapid change for microbiologists. J. Appl. Bacteriol. 80, 233–243.

Sin, M.L., Mach, K.E., Wong, P.K., Liao, J.C., 2014. Advances and challenges in biosensor-based diagnosis of infectious diseases. Exp. Rev. Mol. Diagn. 14, 225–244.

Speck, M., 1970. Selective culture of spoilage and indicator organisms. J. Milk Food Technol. 33, 163–167.

St John, P.M., Davis, R., Cady, N., Czajka, J., Batt, C.A., Craighead, H.G., 1998. Diffraction-based cell detection using a microcontact printed antibody grating. Anal. Chem. 70, 1108–1111.

Stiles, M., Witter, L., 1965. Thermal inactivation, heat injury, and recovery of *Staphylococcus aureus*. J. Dairy Sci. 48, 677–681.

Su, X.L., Li, Y., 2004. Quantum dot biolabeling coupled with immunomagnetic separation for detection of *Escherichia coli* O157:H7. Anal. Chem. 76, 4806–4810.

Sun, Y., Quyen, T.L., Hung, T.Q., Chin, W.H., Wolff, A., Bang, D.D., 2015. A lab-on-a-chip system with integrated sample preparation and loop-mediated isothermal amplification for rapid and quantitative detection of *Salmonella* spp. in food samples. Lab Chip 15, 1898–1904.

Sung, Y.J., Suk, H.J., Sung, H.Y., Li, T., Poo, H., Kim, M.G., 2013. Novel antibody/gold nanoparticle/magnetic nanoparticle nanocomposites for immunomagnetic separation and rapid colorimetric detection of *Staphylococcus aureus* in milk. Biosens. Bioelectron. 43, 432–439.

Tachibana, H., Saito, M., Shibuya, S., Tsuji, K., Miyagawa, N., Yamanaka, K., Tamiya, E., 2015. On-chip quantitative detection of pathogen genes by autonomous microfluidic PCR platform. Biosens. Bioelectron. 74, 725–730.

Thomas, M.K., Murray, R., Flockhart, L., Pintar, K., Pollari, F., Fazil, A., Nesbitt, A., Marshall, B., 2013. Foodborne Path. Dis. 10, 639–648.

Tortajada-Genaro, L.A., Rodrigo, A., Hevia, E., Mena, S., Ninoles, R., Maquieira, A., 2015. Microarray on digital versatile disc for identification and genotyping of *Salmonella* and *Campylobacter* in meat products. Anal. Bioanal. Chem.

Traversa, A., Gariano, G.R., Gallina, S., Bianchi, D.M., Orusa, R., Domenis, L., Cavallerio, P., Fossati, L., Serra, R., Decastelli, L., 2015. Methicillin resistance in *Staphylococcus aureus* strains isolated from food and wild animal carcasses in Italy. Food Microbiol. 52, 154–158.

Varshney, M., Yang, L., Su, X.L., Li, Y., 2005. Magnetic nanoparticle-antibody conjugates for the separation of *Escherichia coli* O157:H7 in ground beef. J. Food Prot. 68, 1804–1811.

Wang, J., Gao, J., Liu, D., Han, D., Wang, Z., 2012a. Phenylboronic acid functionalized gold nanoparticles for highly sensitive detection of *Staphylococcus aureus*. Nanoscale 4, 451–454.

Wang, L., Wu, C.S., Fan, X., Mustapha, A., 2012b. Detection of *Escherichia coli* O157:H7 and *Salmonella* in ground beef by a bead-free quantum dot-facilitated isolation method. Int. J. Food Microbiol. 156, 83–87.

Wang, Y., Ye, Z., Ying, Y., 2012c. New trends in impedimetric biosensors for the detection of foodborne pathogenic bacteria. Sensors 12, 3449–3471.

Wang, Y., Ping, J., Ye, Z., Wu, J., Ying, Y., 2013a. Impedimetric immunosensor based on gold nanoparticles modified graphene paper for label-free detection of *Escherichia coli* O157:H7. Biosens. Bioelectron. 49, 492–498.

Wang, Z., Wang, J., Yue, T., Yuan, Y., Cai, R., Niu, C., 2013b. Immunomagnetic separation combined with polymerase chain reaction for the detection of *Alicyclobacillus acidoterrestris* in apple juice. PLoS ONE 8, e82376.

Wang, Z., Yue, T., Yuan, Y., Cai, R., Niu, C., Guo, C., 2013c. Development and evaluation of an immunomagnetic separation-ELISA for the detection of *Alicyclobacillus* spp. in apple juice. Int. J. Food Microbiol. 166, 28–33.

Wawerla, M., Stolle, A., Schalch, B., Eisgruber, H., 1999. Impedance microbiology: applications in food hygiene. J. Food Prot. 62, 1488–1496.

Wein, L.M., Liu, Y., 2005. Analyzing a bioterror attack on the food supply: the case of botulinum toxin in milk. Proc. Natl. Acad. Sci. USA 102, 9984–9989.

Wu, W., Zhao, S., Mao, Y., Fang, Z., Lu, X., Zeng, L., 2015. A sensitive lateral flow biosensor for *Escherichia coli* O157:H7 detection based on aptamer mediated strand displacement amplification. Anal. Chim. Acta 861, 62–68.

Xu, F., Di, X., Ming, X., Xu, H., Li, B., Li, P., Aguilar, Z.P., Cheng, T., Wu, X., Wei, H., 2013. Quantum dot-based immunochromatography test strip for rapid detection of *Campylobacter jejuni*. J. Nanosci. Nanotechnol. 13, 4552–4559.

Yang, H., Qu, L., Wimbrow, A.N., Jiang, X., Sun, Y., 2007. Rapid detection of *Listeria monocytogenes* by nanoparticle-based immunomagnetic separation and real-time PCR. Int. J. Food Microbiol. 118, 132–138.

Yang, M.H., Sun, S., Bruck, H.A., Kostov, Y., Rasooly, A., 2010. Lab-on-a-chip for label free biological semiconductor analysis of Staphylococcal enterotoxin B. Lab Chip 10, 2534–2540.

Yoon, J.Y., Kim, B., 2012. Lab-on-a-chip pathogen sensors for food safety. Sensors 12, 10713–10741.

You, D.J., Geshell, K.J., Yoon, J.Y., 2011. Direct and sensitive detection of foodborne pathogens within fresh produce samples using a field-deployable handheld device. Biosens. Bioelectron. 28, 399–406.

Zhang, C., Lv, X., Han, X., Man, Y., Saeed, Y., Qing, H., Deng, Y., 2015. Whole-cell based aptamer selection for selective capture of microorganisms using microfluidic devices. Anal. Methods 7, 6339.

Zhao, W.T., Yao, S.J., Hsing, I.M., 2006. A microsystem compatible strategy for viable *Escherichia coli* detection. Biosens. Bioelectron. 21, 1163–1170.

Zhu, P., Shelton, D.R., Li, S., Adams, D.L., Karns, J.S., Amstutz, P., Tang, C.M., 2011. Detection of E. coli O157:H7 by immunomagnetic separation coupled with fluorescence immunoassay. Biosens. Bioelectron. 30, 337–341.

13

CONTAMINANT SENSORS: NANOSENSORS, AN EFFICIENT ALARM FOR FOOD PATHOGEN DETECTION

Cheunjit Prakitchaiwattana*, Rachatida Det-udom**
**Chulalongkorn University, Faculty of Science, Department of Food Technology, Bangkok, Thailand; **Chulalongkorn University, Faculty of Science, Food Research and Testing Laboratory (FRTL), Bangkok, Thailand*

1 Introduction

1.1 Demand in Microbiological Method Developments

Pathogenic bacteria causing food poisoning is a major threat to human health due to their rapid growth, ability to survive in food, and the possibility of cross-contamination. In food manufacturing systems, foods may be contaminated by microbiological hazards/pathogens at any step in the food chain from primary production to processing to transportation to retail to consumer's storage and to meal preparation. Generally, prevention of contamination from pathogens is mainly dependent on the application of hygienic practices throughout the food supply (Todd, 1997). Nowadays, food regulatory agencies have established strict control programs to avoid pathogen contamination through the food chain. These include food-safety-related programs such as good agricultural practices (GAP), good manufacturing practices (GMP), hazard analysis and critical control point (HACCP), and the other food codes recommending specific practices for particular foods. These programs can significantly reduce pathogens in food and have been mandatorily and/or voluntarily adopted worldwide by food industries. However, to improve efficiency in food safety control systems, monitoring of contaminated pathogens, their growth and the toxins generated along the food chain is also needed. Thus, an analytical method to detect pathogens is one of the scientific

Nanobiosensors. http://dx.doi.org/10.1016/B978-0-12-804301-1.00013-8

fields that has been extensively developed in order to be used as an efficient alarm for food safety monitoring along with a quality assurance system. The method, which is considered efficient for a quality assurance system, should be able to rapidly process a large number of samples with a simple operating procedure.

1.2 Rapid/Alternative Methods

Traditionally, microbiological analysis comprises various operating procedures including collection and preparation of food samples, culturing and isolation of the microorganisms in/on appropriate media, enumeration and/or detection, as well as identification of isolates to genus, species, or strain levels. Importantly, the standard methods for analyzing pathogen have been established based on these methodologies since the techniques of microbiological culture can provide great sensitivity and specificity.

However, the disadvantages of the conventional methods are that they are laborious, time-consuming, labor intensive, and a large amount of materials is required for each analysis. Due to an increasing interest in food safety for consumers, modern food-processing technologies and quality assurance demand rapid, cost-effective, and more reliable methods for monitoring microbiological safety along the food chain, rather than the use of conventional methods. According to these requirements, detection methods that offer rapid, efficient, high sample throughput and on-site analytical strategies have been developed through recent advances in immunology, molecular biology, computer technology, biotechnology, and engineering. These methods represent a new generation of technologies for microbiological analysis of foods. With evolution of technology, various rapid methods have been developed and the time required to detect and identify microorganisms has been significantly reduced. In addition, this progress helps reduce cost of the analysis and allows shorter holding time of sensitive food products (Hobson et al., 1996; Cox and Fleet, 1997; Hall, 2002; Velusamy et al., 2010). There are many specialized equipments and products commercially available for food industries. Furthermore, some of the alternative methods have already been adopted as standard methods by competent authorities of various countries. Current rapid/alternative methods developed based on different concepts widely used in analysis of foodborne pathogens are described below with summarized concepts, advantages, and drawbacks.

1.2.1 Biochemical Identification Test Kits

In identification of isolates to genus, species, or strain level, colonies isolated from food samples are generally subjected to several specific tests. Since bacterial species differ in their colony morphologies and metabolic activities, species can be differentiated by using selective and/or differential media. For more definitive identification of bacteria, series of biochemical tests are used to differentiate closely related organisms. These various tests are designed to identify various metabolic properties of different bacterial species. In the past, food microbiologists used a range of media in plates and test tubes for biochemical identification of isolates. To reduce operating time, labor, and materials, manual miniaturized identification systems were first introduced to the microbiological laboratory. These systems offered several advantages over conventional testing. For example, the use of the kits considerably reduced time and labor involved in preparing and utilizing conventional media, as well as volume of media in tubes and plates. Additionally, preparation cost and storage space required for standard reagent and media as well as identification time were also decreased (Fung, 1969; Cox and Fleet, 1997). The commercial biochemical test kits, which are chromogenic substrates, carbon compound substrates, and enzyme substrates in a form of strip test and the others, have been manufactured and used extensively. Furthermore, the test kits have continuously been improved by the introduction of automation. Various commercial systems have been widely accepted as automated system for microbial identification. Some automated systems transform identification process by reducing the time required to identify rapidly growing bacteria to as little as 2–4 h in contrast to one or more days as previously required by the conventional methodologies (Stager and Davis, 1992; Cox and Fleet, 1997). However, all systems used to identify bacteria have limitations. Importantly, biochemical properties do not accurately reflect the entire extent of the genomic complexity of a given species. Furthermore, phenotypic properties are sometimes unstable and the results may also be influenced by changes in the conditions such as growth substrate, temperature, and pH (Rosselló-Mora and Amann, 2001). In addition, the other drawbacks on commercial systems are the construction of databases, and panel configurations of the biochemical tests. For example, panel configurations are seldom modified after the tests are produced. However, a number of newly described taxa may substantially increase after a period of time (Janda and Abbott, 2002).

1.2.2 Membrane Filter Techniques

The membrane filter technique was introduced as an effective method in testing fluid samples for microbiological contamination. The use of membrane filter for isolation and enumeration of microorganisms in food or beverage samples were well studied and reported (Sharpe, 1985; McDaniels et al., 1987; Ciebin et al., 1995; Rompré et al., 2002; Hachich et al., 2012). Also, the filtration technique, called standard membrane filter (MF), is effectively used and accepted by the government laboratory to monitor drinking water [American Public Health Association (APHA), 2005, 2007; Hachich et al., 2012]. Membrane filter technology offers a number of analytical advantages over traditional methods. It involves fewer preparation steps and reduces detection time. The preparation steps, particularly for pathogen analysis such as enrichment culturing or preincubation in nutrient medium, can be omitted since large volume of sample is filtered, the sample size is increased, and limit of detection is improved. The filters, with bacterial cells bound to the surfaces, can be removed from one medium to the others without disturbing the individuality of the developed cells or colonies. This condition is beneficial when an enrichment culturing is required to recover injured cells prior to culture on selective medium. After sample filtration, the membranes are transferred to an appropriate medium for incubation. However, colony development generally takes the same amount of time as conventional growth on agar media (Ciebin et al., 1995; Cox and Fleet, 1997; Rompré et al., 2002; Hachich et al., 2012). Alternatively, the micro-colonies grown on the membrane surfaces can be examined with a stereomicroscope, which can be done within 6–12 h after incubation in some cases (Sharpe, 1995). The main application includes the traditional use of a filter to provide colony counts, direct membrane-plating method (Sharpe, 1995), direct epifluorescent filter techniques (DEFT) (Pettipher, 1983), and the hydrophobic grid membrane filter (HGMF) (Sharpe and Michaud, 1974) and their modified products (Hachich et al., 2012). The HGMF and modified products were initially developed for total aerobic plate count but, by incubation of the membrane on selective media, the application was expanded to hygiene indicator bacteria and also several pathogens (McDaniels et al., 1987; Ciebin et al., 1995; Rompré et al., 2002; Hachich et al., 2012). This technology is commercially available and widely used in food laboratories. The major limitations of the membrane filter technique are fouling and clogging of the membrane, which restricts the flow of sample during filtration. Thus, this approach is limited to testing of liquid foods such as water and beverages, which have fewer particles to clog the membrane pores.

1.2.3 Dry Rehydratable Film

This technique comes from an availability of prepared selective and differential media in a form of dry films. Undiluted or diluted test suspension is added to the film and gelling agent is used to solidify the suspension. The films are then incubated for colony development, which can be noticed from the color change due to the incorporation of specific chromogenic agent in medium formulation. Actually, the films are selective and differential media are developed relying on structural or metabolic properties of the selected species. Compared to conventional methods, the advantage of rehydratable film is that the colonies present on the film can be identified without being subjected to a confirmation test (Ellis and Meldrum, 2002; Kinneberg and Lindberg, 2002). Currently, there are several dry rehydratable films manufactured, and widely used for analysis of microbial quality and some pathogens in foods. Also, performances of some systems have been approved by a large number of government agencies.

1.2.4 Impedance Microbiology

Impedance microbiology is the detection and quantification of bacterial growth by measuring the electrical impedance change when ionic molecules are formed during the metabolism of microorganisms (Yang and Bashir, 2008). The ionic molecules are released due to two mechanisms. First, the conductivity of culture media can be increased by the conversion of nonionized glucose to two molecules of lactic acid, as well as the conversion of lactic acid and oxygen to produce carbonic acid. The second mechanism is a result of the ion exchange occurring at the cell membrane where ions such as sodium and potassium are transported through double lipid layer of the membrane and the conductivity of the medium is changed (Owicki and Wallace Parce, 1992). The electrical properties are simply measured by inserting a pair of electrodes into the medium and recording the flow of current between them. To detect the change in impedance properties of the medium, bacteria population must reach approximately 6–7 log cells/mL. The time required to initiate this change is called the detection time (DT). Under controlled conditions, DT is inversely related to the initial population of viable microorganisms in the medium (Firstenberg-Eden, 1985; Cox and Fleet, 1997). To detect bacteria, media have to be formulated to provide selectivity for specific bacteria, as well as able to produce changes in impedance.

Commercial instruments to detect bacteria through impedance change have been developed by several manufacturers. The main advantages of impedance testing are convenience, ability to detect viable and nonviable cells, rapidness, high sensitivity, and small sample size required. The drawbacks of the method are the interference from food matrices, need for a specific selective medium, and issues of sample contamination (Wawerla et al., 1999; Flint et al., 2015)

1.2.5 Bioluminescence

The method for detection of microorganisms in foods by bioluminescence assay is based on two concepts well documented by Cox and Fleet (1997). The first concept is the determination of total viable population of microorganisms by measurement of ATP using bioluminescent reaction called ATP-bioluminescence (Bülte and Reuter, 1985; Larocco, 1986; Griffiths, 1993). Second, specific microorganisms are detected by measuring their bioluminescence produced by the lux genes, called bacterial bioluminescence.

For ATP-bioluminescence, several limitations such as the lack of specificity restrict its application to the detection of bacteria in foods. For example, total ATP is measured without differentiation between target and nontarget cells. However, a number of rapid ATP-bioluminescence methods are developed to evaluate microbial growth and its metabolisms. Also, the equipment for ATP-bioluminescence assay has been commercialized for routine analysis in the food industries to monitor hygiene and test for general microbial load and sterility (Barrichello and Allil, 1997; Hawronskyj and Holah, 1997; Colquhoun et al., 1998; Paez et al., 2003; Costa et al., 2006).

For the other bioluminescence concept, bacterial bioluminescence is the biochemical emission of visible light by living organisms. The preeminent bioluminescent bacteria are marine bacteria in the genera *Vibrio* and *Photobacterium*. Luciferase is the enzyme from bacteria that catalyzes light emission and the substrates of this enzyme are reduced flavin mononucleotide (FMNH2), oxygen and a long-chain aliphatic aldehyde. The gene specific for luciferase and genes required for light production are clustered in an operon called the lux operon (Meighen, 1993). Bacteriophage is further used in development of the lux gene incorporation method. The lux gene of bioluminescence bacteria has been cloned, incorporated into the DNA of the bacteriophage, and transferred to the DNA of nonbioluminescence bacteria in order to make them bioluminescent (Sutherland et al., 1994; Baker et al., 1992). The light can be detected in vivo in the living animal using a sensitive detection system. Therefore, this assay is ideally for bioluminescence imaging protocols. This method offers numerous applications including detection of foodborne pathogens (Stewart et al., 1991; Baker et al., 1992; Kyriakides and Patel, 1995; Gahan, 2012). Through this technology, phages specifically recognize, infect, and replicate within defined host cells. The host range is determined by the structural proteins of the phage. Compared to the use of antibodies, bacteriophage has similar binding abilities, but it is more easily produced in vitro, and is cost effective, stable, and can replace antibodies in various available detection technologies (Smartt and Ripp, 2011). Thus, this assay can potentially be used as a rapid detection method with high specificity and developed for various applications including pathogen analysis.

1.2.6 Immunoassay

An immunoassay is a biochemical test based on specific interaction of antigens and antibodies to form detectable complexes (Sipponen et al., 1976; Uotila et al., 1981). Quantifying antigen or antibody can be achieved by labeling of antigen or antibody. Immunoassay is developed by the use of enzyme, radioisotope, magnetic or fluorescence as an antigen or antibody label. Other techniques include agglutination, nephelometry, turbidity, and Western blot (Debnath et al., 2010). Application of immunoassay widely used in the microbiological analysis of foods are enzyme-linked immunosorbent assay (ELISA), immunomagnetic separation (IMS), and agglutination assay/immunoprecipitation/immunochromatography. The ELISA is the most commonly used antibody assay for foodborne pathogen detection. Principally, ELISA is an immunoassay using solid support to immobilize antigens or antibodies with the antibody–enzyme conjugates to produce a detectable signal and indicate the antigen–antibody complex. ELISA have been commercially developed and applied for detection of various pathogens and their toxins in foods (Candlish, 1991).

Furthermore, antibodies coupled to magnetic particles or beads are also used in immunomagnetic separation (IMS) technology to isolate pathogens from preenrichment media (Oggel et al., 1990). Specific antibodies are attached to the surface of beads, which are later bound with their antigens. Coated beads are mixed with the food sample in preenrichment culture. After applying magnetic field, target cells/antigen species bound to the beads, are separated from the culture fluid. Captured cells/antigens can be identified or further tested using the other assays (Feng, 1997). This method has been continuously improved by advance technologies compared to the other methods.

Immunoprecipitation is a technique of protein precipitating caused by a specific antibody. For agglutination assay or immunochromatography, the microbes, after reacting with specific antibodies will be cross-linked and a visually observable network or cell clump formed. The agglutination assay is simply conducted by mixing antigen with antibody on the surface of microscope slide. For immunochromatography, antibody is coupled to colored latex beads or colloidal particles and mixed with the enriched sample to obtain results (Feldsine et al., 1997). Immunoprecipitation has been commercialized and applied to a wide range of analyses of pathogenic bacteria and their toxins.

1.2.7 Nucleic Acid-Based Protocols

Several nucleic acid–based detection systems are developed for foodborne pathogens. Most of them are designed for rapid results with excellent detection limits and specificity (Bsat et al., 1994; Olsen et al., 1995; Naravaneni and Jamil, 2005; Nocker et al., 2009; Josefsen et al., 2010; Chen et al., 2011; Mandal et al., 2011; Zhou et al., 2013). Nucleic acid–based assays for detection of pathogens such as polymerase chain reaction (PCR), real-time PCR, nucleic acid hybridization probe, DNA fingerprinting methods, Pulsed-Field Gel Electrophoresis (PFGE), PCR-ribotyping, DNA microarray analysis, Reverse Transcriptase PCR (RT-PCR) and nucleic acid sequence-based amplification (NASBA) have been developed. An average 24–36 h is required to extract nucleic acids from bacterial colony isolates (Olsen et al., 1995; Naravaneni and Jamil, 2005). Also, amplification of target genes used in several identification tests based on PCR is time consuming. To eliminate these steps and improve specificity, methods such as using propidium monoazide as a sample treatment for selective detection of live bacteria combining with microarray technology (Nocker et al., 2009), real-time PCR (Josefsen et al., 2010), and loop-mediated isothermal amplification (Chen et al., 2011) have been developed.

Rapid molecular determination techniques are commonly used in screening, but increasing numbers of the methods are also recognized as confirmatory. Furthermore, the sensitive, specific and rapid molecular methods for diagnosis of foodborne bacteria targeted on their DNA and RNA are developed with the ability to distinguish their viability status. The molecular methods designed for bacterial-RNA involve both mRNA and rRNA. Since mRNA has a very short half-life, it could be used as a marker of bacterial cell viability (Keer and Birch, 2003). However, the use of 16S rRNA, which was able to resist different bacterial stress treatments, was demonstrated as an excellent detection method, even though 16S rRNA has longer half-life compared to mRNA (Hu et al., 2013a; Tolker-Nielsen et al., 1997; Keer and Birch, 2003). The most used methods for DNA study are Southern blotting, hybridization, isothermal amplification, PCR-based techniques and, recently DNA-arrays.

The only molecular technique which is cultural independent is PCR-denaturing gradient gel electrophoresis (PCR-DGGE) and PCR-temperature gradient gel electrophoresis (PCR-TGGE). These techniques can be used to identify microbial ecology in food samples by directly amplifying target bacterial DNA/RNA without cultivation step. The PCR amplicons are separated by DGGE or TGGE which could differentiate DNA with small difference in nucleic base sequences. The DNA/RNA fragment on the gel could be further examined. These methods have been used for detection of foodborne pathogens (Muyzer and Smalla, 1998). However, in amplification of mixed DNA/RNA from food sample, only DNA of the main populations (3–4 strains depending on microbial species) would be amplified (Prakitchaiwattana et al., 2004). Thus, the true ecology of bacterial contamination in food might not be detected, which would be the limitation of this technology.

2 Foodborne Pathogens and Their Significant Characteristics Used for Detection

To date, many approaches are proposed and implemented throughout the whole food chain to address safety concerns. These approaches mostly focus on significant foodborne pathogens specified by regulatory authorities and/or standard organization agencies. Brief information about each pathogen including its key properties for determination and current status is described as follows.

2.1 *Salmonella*

This Gram-negative, motile, rod-shaped *Salmonella* is one of the well-known foodborne pathogens causing illness called salmonellosis. Cell envelope of *Salmonella* spp. consists of a complex lipopolysaccharide structure, which may function as an endotoxin. Once ingested, *Salmonella* viable cell are colonized in ileum and colon, by binding on the surface of epithelial cell, and using type III secretion system (T3SS) to inject effector proteins directly into the host cell cytoplasm (Velge et al., 2012). Current methods and key characteristics used in the current detection of *Salmonella* include (1) nucleic acid-based methods utilizing the specific characteristics of *invA* gene encoding T3SS proteins (Malorny et al., 2003), the conserved region in chromosomal replication origin (oriC) (Jeong et al., 2013), and tetrathionate respiration *ttrRSBCA* gene (Malorny et al., 2004), for example; (2) immunoassay-based methods utilizing the specific characteristics of the immune response of antisera with O-polysaccharide repeat (O-antigen) and flagellin protein (H-antigen), such as IgG-layer protein. *Salmonella* are commonly found in meat, poultry, and their products. However, most of salmonellosis cases are caused by improper food sanitization (CFSAN, 2005). *Salmonella* causing severe disease includes

Salmonella thyphi, and *Salmonella thyphimurium.* Based on the Commission Regulation (EC) No 2073/2005 (EC, 2005), *Salmonella* should not be detected in 25 g ($n = 5$, $c = 0$) in ready-to-eat foods containing raw egg placed on the market during their shelf life.

2.2 *Shigella*

Shigella is Gram-negative, nonmotile, rod-shaped bacteria closely related to *Salmonella.* This bacteria cause illness called shigellosis. The pathogen enters human body via ingestion, survives acidic condition of the stomach, and colonizes at large intestine where colonic epithelium and inflammatory colitis invasion begin. Encoded by invasion plasmid antigens (ipa) genes, and induced by membranous cell, epithelial cell, and macrophages of the host's lymphoid follicle (Schroeder and Hilbi, 2008), the A, B, C, and H antigens are used as key property for detection. The other leading characteristics used in detection include invasion-associated loci (*ial* gene) region, the transcriptional regulator invasive virulence gene (*virA* and *virB* genes), mxi-spa locus in the virulence plasmid of *S. flexneri,* and elongation factor Tu (*tuf* gene). *Shigella* is differentiated from *Escherichia coli* by its pathogenicity, physiology (failure to ferment lactose or decarboxylate lysine) and serology (Hale and Keusch, 1996). The common sources of *Shigella* are raw, undercooked, low pH, or unsanitary food such as ground meat, raw seafood, sushi, salad, coleslaw, unpasteurized juice (Warren et al., 2006a). In order to monitor the safety of ready to eat food placed on market, *Shigella* should not be detected in 25 grams. However, testing of more or less food may be specified during outbreak investigations or when sampling is based on the other purposes (Food Safety Authority of Ireland (FSAI), 2014). Also, *E. coli* and thermotolerant coliforms are usually used as index for *Shigella* in drinking water (WHO chronicle, 1996).

2.3 *Escherichia Coli/E. coli* O157:H7

Diverse strains of *Escherichia coli* are both commensal and pathogenic, belonging to the Enterobacteriaceae family. This bacterial species is generally found in the environment, food, and intestines of warm-blooded animals. Therefore, the *E. coli* present in food product/food manufacturing process indicates poor hygiene. Some *E. coli* strains can cause serious foodborne disease. The pathogenic *E. coli* is serotyped into six groups based on their virulence properties. *E. coli* O157:H7 is one of the important serotypes causing illness by producing *Shiga* toxin encoded by *stx1* and *stx2* gene (CFSAN, 2005). O and H antigens are also used as key properties for *E. coli* strain identification, including *wzx* and *wzy* in the O-antigen gene cluster of O145 strain, *rfbE* in the O-antigen gene cluster of O157 strain, *fliC* loci encoding flagellin protein, or O and H antigen-antibody reaction. Foods likely to be contaminated with *E.coli* are ground beef, unpasteurized milk, raw cheese, raw fruit, and vegetable, as well as food with pH < 4.6, for example, yogurt, mayonnaise, fermented sausages (CFSAN, 2005). Normally, *E. coli* is used as sanitary index for food manufacturing so that 3-class sampling plans are usually specified. For example, based on process hygiene criteria established by the Commission Regulation (EC) No 2073/2005, *E. coli* tested at manufacturing process should be ≤ 100 cfu/g ($n = 5$, $c = 2$, $m = 100$ cfu/g, $M = 1{,}000$ cfu/g) for satisfactory result in unpasteurized fruit juice and cheese (EC, 2005). Based on the WHO standard on drinking water, *E. coli* must not be detected in any 100 mL of drinking water sample (WHO chronicle, 1996).

2.4 *Listeria Monocytogenes*

Listeria monocytogenes is a facultative anaerobe and one of the most virulent foodborne pathogens causing listeriosis. Multiple proteins related to infection cycle, including internalin A and B (inlA and inlB), hemolysin (encoded from *hlyA* gene), PI-PLC, PC-PLC, and *actA* (Jadhav et al., 2012) are the key characters used for detection. *L. monocytogenes* can grow in low temperature condition, so high-risk foods include ready-to-eat meals, smoked seafood, raw milk and dairy products. Based on the Codex Guidelines on the Application of General Principles of Food Hygiene to the Control of *L. monocytogenes* in Foods, *L. monocytogenes* should not be detected in 25 g ($n = 5$, $c = 0$) of ready-to-eat food in which growth can occur. This microbiological criterion is used for ready-to-eat foods from the end of manufacture or port of entry to the point of sale (CODEX Alimentarius, 2007).

2.5 *Vibrios*

Vibrio species is a group of bacteria in which several members are significant foodborne pathogens including *V. parahaemolyticus* and *V. cholerae*. These bacteria are normally found in warm estuary and marine water and can be transmitted to humans by open wound infection and ingestion. Hence, consuming raw seafood may result in risk from *Vibrio* infection. Once infected, *Vibrio* causes gastroenteritis disease in human but pathogenicity varies between strains. *V. parahaemolyticus* pathogenic strain has two major virulence factor genes, which encode proteins TDH and

TRH (Raghunath, 2015). While TDH is pore-forming toxin, TRH is heat labile. These two proteins are widely used as key properties for detection. Furthermore, *V. cholerae* O1 and O139 strains produce cholerae toxin and toxin coregulated pilus encoded from the network of genes including *ctxA*, and also *ToxRS* operon (Bina et al., 2003; Fedio et al., 2007). These virulence-associated regulators are the major characters used for *V. cholerae* detection. The *ToxRS* regulator is also found in *V. parahaemolyticus* which is a crucial element for this bacteria detection. The symptoms of vibriosis patients also differ from strain to strain. *V. parahaemolyticus* causes nonbloody diarrhea while *V. cholerae* is causative agent of cholera. To ensure safety, ICMSF recommends *V. parahaemolyticus* should be found ≤2 log cfu/g in live bivalve mollusk ($n = 10$, $c = 1$, $m = 100$, $M = 10{,}000$) (ICMSF, 2011).

2.6 *Staphylococcus Aureus*

Staphylococcus aureus is normal flora bacteria generally found on skin, hair, nose, and throat of human and warm blood animals. When contaminated in food, *S. aureus* can produce short, extracellular, water-soluble, heat-resistant proteins causing gastrointestinal illness. These proteins are called *Staphylococcal* enterotoxins (SEs) including SEA, SEB, SEC1, SEC2, SEC3, SED, SEE, SEG, SEH, and SEI. The enterotoxins are causative agents of *Staphylococcal* food poisoning (CFSAN, 2005). *S. aureus* also produces thermonuclease considered as important virulence factor. The thermonuclease encoding gene, *nuc*, is widely used as unique marker for *S. aureus* determination (Hu et al., 2013b). In the other approaches based on cultural plating method, the acriflavine resistance ability and its encoding gene are utilized for detection. Furthermore, the acriflavine resistance-encoding gene is also used for *S. aureus* determination in nucleic acid-based assay (Trnčíková et al., 2009). Additionally, the 23s rRNA conserved regions is widely employed in *S. aureus* determination and typing (Dolzani et al., 1994; Jeong et al., 2013). High-risk foods include uncooked food, salad, cold-cut meat product, pudding, pastries, and sandwiches. Microbiological limits for *S. aureus*, for example; ≤100 cfu/g in Cooked crustaceans and molluscan shellfish ($n = 5$, $c = 2$, $m = 100$ cfu/g, $M = 1{,}000$ cfu/g) (Ababouch et al., 2005), ≤10^4 cfu/g in all fish (USFDA, 2011), ≤10 cfu/g in frozen milk-based product and fresh cheese ($n = 5$, $c = 2$, $m = 10$ cfu/g, $M = 100$ cfu/g) (Institute of Medicine (US) and National Research Council (US), 2003). *Staphylococcal* enterotoxins should not be detected in 25 g ($n = 5$, $c = 0$) in cheese, milk powder, and whey powder placed on the market during their shelf life (EC, 2005).

2.7 *Bacillus Cereus*

Bacillus cereus is Gram-positive, rod shaped, spore-forming bacteria which have been recognized as one of the major foodborne pathogens. The pathogenicity which is associated with its toxin-producing ability varies among strains. The pathogenicity of *B. cereus* is associated with the production of tissue-destructive exoenzymes including hemolysins, phospholipases, emesis-inducing toxin, and pore-forming enterotoxins such as hemolysin BL (HBL), nonhemolytic enterotoxin (NHE), and cytotoxin K (Kotiranta et al., 2000; Liu, 2007). One of the main factors affect toxin formation is growth condition. For instance, *B. cereus* will produce enterotoxins in gastrointestinal track of host, but it will produce the emetic toxin in contaminated food (Bottone, 2010). Thus, the important genetic characters used in *B. cereus* assay are enterotoxin producing genes (*hlb, nhe, cytK, entFM*) and emetic toxin–producing gene (CER and ces) (Liu, 2007; Kim et al., 2012). A variety of food, especially dried foods such as cereal, flour-based ready-to-eat-foods, and catered foods left at room temperature for a long period of time are potential source of *B. cereus* contamination (CFSAN, 2005). Microbiological limits of *B. cereus* are for example: ≤10 cfu/g in breakfast cereal, ≤50 cfu/g in powdered infant formulae, ≤100 cfu/g in dried fruit and coleslaw, and ≤1,000 cfu/g in herbs (Institute of Medicine (US) and National Research Council (US), 2003).

2.8 *Clostridium Botulinum*

Clostridium botulinum is the cause of a life-threatening foodborne illness called botulism due to the neurotoxin production that grows in food. The botulinum toxins can be grouped based on their antigenic specificity into 7 types. Among them, types A, B, E, and F cause human botulism with the estimated median lethal dose for human of less than 100 ng by ingestion (Sharma et al., 2005). Normally, one strain of *C. botulinum* only produces one type of toxin, but dual type toxin-producing strains are recently reported (CFSAN, 2005). The new, rapid approach for *C. botulinum* detection in food samples are mostly based on virulence factor and toxin-encoding genes. These include nontoxic nonhemagglutinin component gene (*ntnh*), botulinum neurotoxins gene cluster (*bonts* gene), 16s rRNA genes (*rrs* genes), RNA polymerase subunit B (*rpoB* gene), hemagglutinin cluster gene (*HA*) and the *bonts* gene-associated orf-X cluster (*orf-Xs* gene) (Lindström et al., 2001; Vanhomwegen et al., 2013; Grenda et al., 2014). The toxins produced from these genes are also key targets for toxin detection in food products (Sharma et al., 2005; Rivera et al., 2006; Brunt et al., 2010). The high-risk foods are underprocessed low-acid canned food, vacuum-sealed meat and fish product, and herb-infused oil and -cheese.

2.9 *Clostridium Perfringens*

Clostridium perfringens is another *Clostridium* pathogenic strain associated with foods. Different from *C. butulinum*, this bacterium causes the most common food poisoning symptoms such as diarrhea, abdominal cramp, and nausea. The strains containing *C. perfringens* enterotoxin (CPE) gene are considered foodborne pathogens. Also, an ability to form heat-resistant spore contributes to its pathogenicity. *C. perfringens* infection begins as a result of ingesting vegetative cells contaminated in food and caused toxin formation in gastrointestinal tract of the host. Spores are then produced along with CPE, which is released after lysis of the sporulating cells (Grass et al., 2013). The α, β, ε, and ι toxin-encoding gene (*cpa or plc, cpb, cpe,* and *iA)* are the key characters used for alternative methods and approaches for *C. perfringens* determination (Tansuphasiri, 2001; Shimizu et al., 2009; Banerjee and Bhunia, 2010; Yang et al., 2010; Kaneko et al., 2011; Chon et al., 2012; Guran and Oksuztepe, 2013). The potential sources of *C. perfringens* are meat products, stew, casseroles, and gravy. Microbiological limits of this pathogen are, for example: <10 cfu/g in coleslaw, salad, dried fruit, and vegetable, and <100 cfu/g in ready-to-eat food (Institute of Medicine (US) and National Research Council (US), 2003).

2.10 Campylobacter

Camphylobacter species, Gram-negative, heat-sensitive, spiral shape bacteria, cause common gastroenteritis. At least 16 species are currently reported as foodborne pathogens. Some *Camphylobacter* can produce cytolethal distending toxin (CDT), including *C. jejuni* and *C. coli,* which are the most frequently found strains causing outbreaks worldwide (CFSAN, 2005; Dasti et al., 2010). The *hipO* gene encoding hippuricase enzyme is used to identify *C. jejuni* while *glyA* gene-encoding serine hydroxymethyltransferase enzyme is used as specific characteristics to detect the presence of *C. coli* (Vondrakova et al., 2014). The other key properties used for detection are the difference between *gyrB* and *rnpB* gene sequence (Zoder et al., 2011), and igG antigen-antibody specific reaction (Sista et al., 2006). Foods that may be contaminated with *Campylobacter* are raw and undercooked poultry, unpasteurized milk, and contaminated water. In managing the risk from *Campylobacter* in meat industries, certain guidelines for authority to measure the effectiveness of the slaughter and grinding process are proposed by Food Safety and Inspection Service, USDA (FSIS, 2015). Samples are taken as part of a moving window and the results are used to determine if a manufacturer is meeting the performance standard on a continuous basis. The maximum number of positive samples allowed in one window is specified. For example, *Campylobacter* positive samples shall be less than 7.7% of the total number of samples taken for a window in raw chicken part, and less than 1.9% in comminuted chicken.

Current rapid methods based on different concepts for qualitative and quantitative determination of these pathogens, their detection limits and status, implying advantages and drawbacks for further development, are described in Table 13.1.

3 Potential of Nanotechnology in Microbiological Analysis Development

The main conventional or current techniques for detection of pathogen contamination in foods include immunology-based methods (ie, enzyme-linked immunosorbent assays (ELISA)) and nucleic acid–based protocols (ie, polymerase chain reaction (PCR)-based assays). These assays may take from several hours up to a few days to yield an answer. Due to this inadequacy, many researchers have recently been focused on the development of rapid methods which are simpler, quicker, more sensitive, and possibly more cost-effective than the conventional for detection and quantification of pathogens in foods.

With remarkable advances in nanoscience over the past decade, nanotechnology-enabled sensors and systems have been increasingly used to develop rapid methods for food contaminant detection. The uniqueness in physical, chemical, and biological properties that emerges at nanoscale can be adapted to solve problems in foodborne pathogen detection. A major advantage of nanotechnology-based detection methods is that the surface-to-volume ratio of nanostructures is much larger than that of macroscopic-sized materials. Thus, it is possible to detect trace amounts of analytes with significantly increased sensitivity (NIN, 2009). Nanotechnology-enabled sensors provide great promise as practical and cost-effective techniques for monitoring the presence of chemical and biological contaminants in foods during harvesting, processing, storage, and retail.

Table 13.1 Technological Concepts, Details, and Status Used in Current Foodborne Pathogen Determination

Pathogen	Technological Concept	Key Characters Used for Detection	Detection Limit	Status	References
Salmonella spp.	1. Cultural-based method				
	SMS test	Motility and Lysine decarboxylation activity	$\geq 10^0$ cfu/25 g in food	Commercialize, AFNOR approved	Biomerieux (2014) AFNOR AES-10-04-05-04
	RAPID' Salmonella	C-8 esterase activity and motility			BIO-RAD (2015) AFNOR *BRD 07-11 12-05*
	2. Nucleic acid–based method: DNA hybridization				
	TaqMan qualitative rt-PCR	*InvA* mRNA	$\geq 10^1$ cfu/25 g in food with enrichment	New approach	González-Escalona et al. (2009)
		TtrRSBCA locus	$\geq 10^0$ cfu/mL in pure culture; $\geq 10^0$ cfu/25 g in food		Malorny et al. (2004)
	Multiplex rt-PCR	*OriC* gene of *S.* Typhimurium	$\geq 10^2$ cfu/g in food		Jeong et al. (2013)
	Multiplex PCR	*InvA*gene	$\geq 10^0$ cfu/mL in food homogenate		Vantarakis et al. (2000)
	3. Immunoassay				
	RapidCheck SELECT	D1 and SE antigen	$\geq 10^0$ cfu/25 g in food	Commercialized, AFNOR approved	Giannella (1996)
	Thermo Scientific Oxoid Salmonella Precis	C-8 esterase activity and IgG antigen	$\geq 10^{-1}$ cfu/25 g in food		Thermo Fisher Scientific (2015) AFNOR *UNI 03-06 12-07*
	Multiplex fiber optic biosensor	IgG2α-type antigen	$\geq 10^3$ cfu/mL pure culture; $\geq 10^1$ cfu/g in food	New approach	Ohk and Bhunia (2013)
	Membrane dot ELISA	*o*-antigen of serotype D	N/A	New, validated method	Tawe (2014)

Shigella	1. Nucleic acid based method				
	Enriched, multiplex-PCR	*ial* and *ipa* gene	≥10^4 cfu pure culture	New approach, compare to CMMEF	Jiménez et al. (2010)
	Multiplex PCR	VirA gene	≥10^3 cfu in food ≥10^1 cfu/mL in food homogenate with enrichment	New approach	Vantarakis et al. (2000)
		ipaH, mxiC, and *virB* genes	≥10^0 cfu/25 g food with enrichment		Binet et al. (2014)
	Duplex-PCR	*ipaH* gene	≥10^3 cfu/mL pure culture; ≥10^0 cfu/10 g food with enrichment		Senachai et al. (2013)
	Nested PCR	*ial* gene	≥10^3 cfu/g in food ≥10^0 cfu/g in food	New approach, compared to NMKL no. 151	Lindqvist (1999)
	rt-PCR: CRENAME	*tuf* gene	≥10^0cfu/100 mL water sample	New approach, compared to US-EPA Method 1604	Maheux et al. (2011)
	2. Immunoassay				
	Membrane dot ELISA	*o*-antigen serotype A to D	N/A	New, validated method	Tawe (2014)
	Flow-through immunocapture (FTI)	*ipaH* genes	≥ 10^0 cfu/25 g food	New approach	Warren et al. (2006b)

(*Continued*)

Table 13.1 Technological Concepts, Details, and Status Used in Current Foodborne Pathogen Determination (*cont.*)

Pathogen	Technological Concept	Key Characters Used for Detection	Detection Limit	Status	References
E. coli	1. Immunoassay				
	PANI-Glu-antibody fabrication biosensor	polyclonal *E.coli* antigen	$\geq 10^1$ cfu/mL pure culture	New approach	Muhammad-Tahir and Alocilja (2003)
	2. Nucleic acid–based method				
	PCR	*wzx* gene in O104	N/A	EU-RL recommended method	Tawe (2014)
	rt-PCR	*wzx* gene in O104, *fliC* gene in H4, *wzy* gene in O121 serotype			
	3. Bioluminescence assay				
	Hand-held fluorescence detector	β-D-glucuronidase	$\geq 10^0$ cfu/mL water sample	New approach	Wildeboer et al. (2010)
	4. Other property				
	Soleris *E. coli* method	pH reduction via *E.coli* metabolism	$\geq 10^1$ cfu/g in food	Commercialized, validated by AOAC	Foti et al. (2012)
	Mie scattering detection	Hydrophobicity of *E.coli* cell membrane	$\geq 10^1$ cfu/g in food	New approach	Liang et al. (2014)
E. coli O157:H7:	1. Nucleic acid–based method				
	3M Molecular Detection Assay *E.coli* O157	Distinct regions *E. coli* O157:H7 genome	$\geq 10^0$ cfu/25 g in food	Commercialized, AFNOR approved	FSIS (2015)
	Multiplex RT-PCR	*stx1* (*stx2, wzy,* and *eae* genes)	$\geq 10^0$ cfu/g food	New approach	Fratamico and DebRoy (2010)
	MPN quantitative real-time PCR	*fliC H7* and *rfbE* gene	$\geq 10^0$ cfu/g in food	New approach, compared with MPN cultural method	Russo et al. (2014)

E. coli O157:H7:	2. Immunoassay				
	IMS Pathatrix System-cytometric bead array	O- and H- antigen	$\geq 10^{-1}$ cfu/g in food	New approach	Leach et al. (2010)
	Enzyme-linked immunomagnet-icchemiluminescence		$\geq 4 \times 10^{2}$ cfu/g in food with enrichment		Gehring et al. (2006)
	PANI-Glu-antibody fabrication biosensor	O antigen	$\geq 10^{1}$ cfu/mL pure culture	New approach, compared with USFAD-BAM, 2001	Muhammad-Tahir and Alocilja (2003)
	Au/PANI/Glu/antibody sensor	Monoclonal anti-*E. coli* antibody	$\geq 10^{2}$ cfu/mL pure culture	New approach, comparing with related study	Chowdhury et al. (2012)
	Multiplex fiber optic biosensor	Anti-*E. coli* O157 antibodies	$\geq 10^{3}$ cfu/mL pure culture; $\geq 10^{1}$ cfu/g in food	New approach	Ohk and Bhunia (2013)
	IMS with biosensor	Monoclonal anti-*E.coli* O157:H7 antibodies	$\geq 10^{0}$ cfu/mL pure culture		Setterington and Alocilja (2012)
	3. Electrochemical impedance				
	MEMS biosensor	Polyclonal IgG *E.coli* antigen	$\geq 3 \times 10^{3}$ cfu/mL pure culture	New approach	Shibajyoti (2011)
	AMP modified biosensor	Lipid compound in *E.coli* cell membrane	$\geq 10^{3}$ cfu/mL pure culture		Li et al. (2014)

(*Continued*)

Table 13.1 Technological Concepts, Details, and Status Used in Current Foodborne Pathogen Determination (*cont.*)

Pathogen	Technological Concept	Key Characters Used for Detection	Detection Limit	Status	References
L. monocytogenes	1. Nucleic acid–based method				
	Coamplication rt-PCR	*hlyA* gene	$\geq 10^1$ cfu/25 g in food	New method, validated with ISO 11290-1:1996/ 1: 2004	Rodríguez-Lázaro et al. (2005); Gianfranceschi et al. (2014)
	LAMP method		$\geq 10^0$ cfu/food portion with enrichment step	New approach, comparing with MPN cultural method	Shan et al. (2012)
	MPN quantitative real-time PCR				Russo et al. (2014)
	Prodect-List59 probe	*inlA* gene	$\geq 10^0$ cfu/g in food	New approach, comparing with cultural method	Ingianni et al. (2001)
	rt-PCR		$\geq 10^0$ cfu/mL pure culture; $\geq 10^1$ cfu/g in food		Heo et al. (2014)
	Macrophage-based and rt-PCR	*L. monocytogenes* invasion property + *hlyA* gene	$\geq 10^1$ cfu/g in food	New approach	Day and Basavanna (2015)
	2. Immunoassay				
	Aptamer functionalized fiber-optic biosensor	internalin A protein	$\geq 10^0$ cfu/g in food with enrichment step	New approach	Ohk et al. (2010)
	Multiplex fiber-optic biosensor	MAb-C11E9 antibodies	$\geq 10^1$ cfu/g in food with enrichment step		Ohk and Bhunia (2013)
	3. Other technology				
	Cell-based biosensor	Virulence factor causing Ped-2E9 cell cytotoxicity	$\geq 10^3$ cfu/g in food with enrichment step	New approach	Banerjee and Bhunia (2010)
	SPR sensor with Lm phage	ActA protein	$\geq 2 \times 10^6$ cfu/mL pure culture		Nanduri et al. (2007)

Vibrio	1. Nucleic acid–based method				
	rt-PCR	*ctxA* gene of *V. cholerae*	$\geq 10^0$ cfu/g in food	New approach, comparing with cultural method	Fedio et al. (2007)
		toxR gene of *V. parahaemolyticus*	$\geq 10^1$ cfu/mL pure culture	New approach	Wang et al. (2013)
	rt-PCR (with bacteriophage)	*Vibrio* phage MAR10	$\geq 8.5 \times 10^3$cfu/mL pure culture		Argentina (2014)
	Multiplex rt-PCR	*toxR* gene of *V. parahaemolyticus*	$\geq 10^1$ cfu/g in food	New approach, comparing with commercial kit	Jeong et al. (2013)
	Direct-PCR with APW enrichment	*toxR* and *tdh* gene of *V. parahaemolyticus*	N/A	New approach, comparing with MPN cultural method	Blanco-Abad et al. (2009)
	2. Bioluminescence				
	Fluorogenic assay	Trypsinlike activity of *V. parahaemolyticus*	$\geq 10^1$ cfu/g in food	New approach, comparing with MPN cultural method	Miyamoto et al. (1990)
	3. Immunoassay				
	Surface Immuno-functionalization	Antivibrio polyclonal antibody	$\geq 7.0 \times 10^3$ cfu/g pure culture	new approach	Laczka et al. (2014)

(*Continued*)

Table 13.1 Technological Concepts, Details, and Status Used in Current Foodborne Pathogen Determination (*cont.*)

Pathogen	Technological Concept	Key Characters Used for Detection	Detection Limit	Status	References
S. aureus	1. Nucleic acid–based method				
	Multiplex PCR	*sea, seb, sec, sed,* and *see* gene	N/A	New approach	El-Jakee et al. (2013)
	Multiplex rt-PCR	23s rRNA	$\geq 10^1$ cfu/g in food		Jeong et al. (2013)
	Aptamer-GNPs and light-scattering detection system	SA17 and SA61 aptamer	N/A	New approach	Chang et al. (2013)
	rt-PCR with enrichment step	Gene encoding acriflavine resistance proteins	$\geq 10^1$ cfu/mL pure culture; $\geq 10^0$ cfu/g in food sample with enrichment	New validated method	Trnčková et al. (2009)
	rt-PCR	Thermonuclease (*nuc*) gene	$\geq 10^1$ cfu/g in food	New approach	Hein et al. (2001)
	2. Immunoassay				
	Immuno-PCR method (toxin detection)	SEA and TSST toxin	≥ 10 pg/mL in culture supernatant	New approach	Maerle et al. (2014)
	3. Cultural-based method				
	TEMPO MPN procedure	pH reduction via *S. aureus* metabolism	$\geq 10^1$ cfu/food portion	Validated commercialized method	Costa et al. (2008)
	RAPID staph	Tellurite reduction and lecithinase activity	$\geq 10^1$ cfu/g in food		Lauer et al. (2007)
	4. Other technology				
	Cell-based biosensor (toxin detection)	α-Hemolys in *S. aureus*	≥ 10 μg/mL toxin in food matrix	New approach	Banerjee and Bhunia (2010)

B. cereus	1. Nucleic acid–based method				
	PFGE	DNA fingerprint	N/A	New approach	Kaminska et al. (2015)
	Electric DNA arrays	*hblA, hblC, hblD, nheA, nheB, nheC, cytK-2,* and *ces* gene	$\geq 10^8$ cfu in pure culture		Liu (2007)
	PCR	*hblA* gene	N/A	New approach, comparing with commercial kit	Mäntynen and Lindström (1998)
	Multiplex PCR with IAC	*Hbl A, nhe A,* and *cyt K* gene	$\geq 10^1$ cfu/g in food	New approach	Kalyan Kumar et al. (2010)
	Multiplex PCR	*EntFM, hblC, nheA, cytK, ces,* and *CER* gene	N/A		Kim et al. (2012)
	2. Immunoassay				
	IMS with biosensor	Polyclonal antibacillus antibodies	$\geq 4 \times 10^1$ cfu/mL pure culture	New approach	Setterington and Alocilja (2012)
	Sandwich immunoassay—direct-charge transfer biosensor	IgG antigen	$\geq 10^1$ cfu/g in food		Pal et al. (2008)
	Cell-based biosensor	Enterotoxin production	10 μg/mL toxin in food matrix		Banerjee and Bhunia (2010)
		Virulence potential causing Ped-2E9 cell cytotoxicity	$\geq 10^3$ cfu/g in food		

(*Continued*)

Table 13.1 Technological Concepts, Details, and Status Used in Current Foodborne Pathogen Determination (*cont.*)

Pathogen	Technological Concept	Key Characters Used for Detection	Detection Limit	Status	References
C. botulinum	1. Nucleic acid–based method				
	1.1 Cell detection				
	rt-PCR	*ntnh* gene	≥10^{-1} spore/g food sample	New approach	Guran and Oksuztepe (2013)
	Multiplex PCR	*BoNT/C, BoNT/D, BoNT/CD,* and *BoNT/DC* gene	N/A	New approach	Anniballi et al. (2013)
		BoNT gene	≥10^{-2} spore/g in food with enrichment	New approach	Lindström et al. (2001)
	Quadruplexrt-PCR assay	*BoNT/A, BoNT/B, BoNT/E,* and *BoNT/F* gene	N/A	New approach	Satterfield et al. (2010)
	High-density DNA resequencing microarray	*rrs, rpoB, BoNT, ntnh, botR, ha, orf,* and *p47* gene cluster	N/A	New approach	Vanhomwegen et al. (2013)
	Gene disc array	*BoNT/A, BoNT/B, BoNT/E* and *BoNT/F* gene	≥10^{3} cfu/mL pure culture	Validated method	Fach et al. (2011)
	1.2 Toxin detection				
	RNA aptasensor	*BoNT/A* toxin	≥20.3 ng/mL in food matrix	New approach	Banerjee and Bhunia (2010)
	2. Immunoassay (nurotoxin detection)				
	Electrochemiluminescence (ECL) assays	*BoNT/A, BoNT/B, BoNT/E* and *BoNT/F* protein	≥50 pg/mL	New approach	Rivera et al. (2006)
	Affinity immunochromatography column assays		≥5 ng/mL in food matrix	New approach	Brunt et al. (2010)
	Lateral-flow assays	*BoNT/A, BoNT/B,* and *BoNT/E* protein	≥10 ng/mL in food	New approach	Sharma et al. (2005)

C. perfringens	1. Nucleic acid–based method				
	IMS-PCR	Immunogens + *cpa* and *cpe* genes	$\geq 10^1$ cfu/g in food	New approach	Yang et al. (2010)
	Duplex PCR assay	*plc* and *cpe* gene	N/A		Tansuphasiri (2001)
	Multiplex PCR	*cpa, cpb, etx, iA,* and *cpe* genes	N/A		Guran and Oksuztepe (2013)
	rt-PCR	*cpa* gene	$\geq 1.8 \times 10^3$ cfu/g in food	New approach, comparing with cultural method	Chon et al. (2012)
	Nested PCR	*cpe* gene	3 cfu/g food sample	New approach	Kaneko et al. (2011)
	Fish with filter cultivation	16S rRNA region	$\geq 10^2$ cfu/g in food sample		Shimizu et al. (2009)
	2. Other method				
	Cell-based biosensor	phospholipase C	≥ 10 μg/mL in food matrix	New approach	Banerjee and Bhunia (2010)
Campylobacter	1. Nucleic acid–based method				
	Multiplex rt-PCR	*hipO* gene (*C. jejuni*); *glyA* gene (*C. coli*); *pepT* gene (*C. lari*)	$\geq 10^3$ cfu/mL in food	New approach	Vondrakova et al. (2014)
	Lyophilized Multiplex rt-PCR	gyrB and rnpB gene	$\geq 10^1$ cfu/mL in food	Validated commercialized method	Zoder et al. (2011)
	2. Immunoassay				
	Spreeta dual-channel SPR sensor	IgG antigen	$\geq 10^2$ cfu/mL both pure culture and in food	New approach	Sista et al. (2006)

This section provides an overview of recent progress relating to the use of nanotechnology-based novel techniques in detection of pathogens in food. Efficiency and potential of the developed methods are compared with the standard or current methods. The techniques—immunology-based, nucleic acid-based, and cellular-based protocols—are elaborated in this part.

3.1 Nanotechnology and Nanomaterials-Based Immunoassay

Methods based on antigen–antibody (protein-based) bindings are widely used for detection of foodborne pathogens. A number of antibody types and formats, including in-house synthesized and/or commercially produced, are available for immune detection such as conventional and heavy-chain antibodies, as well as polyclonal, monoclonal, and recombinant antibodies. In spite of short assay time compared to that of the traditional culture techniques, antibody-based detection still lacks the ability to detect microorganisms in real time and/or to monitor online. However, if the quantity of the pathogen is high enough, the immunoassay can be used to provide real-time information. Problems that may arise with immunoassay-based methods are low sensitivity, low affinity of antibody to pathogens or other analytes being measured, and potential interference from contaminants (Meng and Doyle, 2002). To overcome these problems, the integration of traditional immunoassay-based methods with another technology that could help increase method sensitivity such as nanotechnology/nanomaterials has gained much attention. In addition, the key advantage of immunoassay-based technology is the compatibility with nanotechnology-based approach to enhance pathogen detection (Gilmartin and O'Kennedy, 2012). Therefore, a number of nanotechnology-based immunoassays have been developed and used to detect major foodborne pathogens, as mentioned later.

3.1.1 Surface Plasmon Resonance (SPR)–Based Immunoassay

Immunology-based method is coupled with immunosensors based on surface plasmon resonance (SPR) for pathogen or analyte detection in complex biological media with high specificity and sensitivity. The method was developed based on the detection of antigen binding to antibody molecules immobilized on the sensor surface (Disley et al., 1998; Sakai et al., 1998; Toyama et al., 1998).

SPR is an optical phenomenon occurring when p-polarized light strikes an electrically conducting metal layer under total internal reflection conditions. Plasmons, converted from free electrons in the metal surface during SPR process, are changed when the composition of the medium changes resulting in the change in the angle of incident light (SPR angle) at which the resonance occurs. Apart from the composition of the medium, the SPR angle also depends on the properties of metal film, the wavelength of incident light, and the refractive index of media on either side of the metal film. Since the SPR angle is highly sensitive to any change in the refractive index of the metal–dielectric interface, the occurrence of binding between antigen and antibody can be monitored by measuring the SPR angle and the minimum intensity of reflected light (Salamon et al., 1997; Poltronieri et al., 2014). Furthermore, the sensitivity of an SPR immunosensor for the detection of low concentration antigens can also be improved by controlling the orientation of the antibodies immobilized on the sensor surface. The binding activity of an antibody immobilized on a solid surface is usually less than that in the aqueous phase. Therefore, the development of an immobilization method for antibodies is required to construct a highly oriented layer (Choi et al., 2001; Oh et al., 2004a,b; Jyoung et al., 2006).

Immunosensor for detection of pathogens using SPR is described by several research groups. For example, Oh et al. (2004a) developed SPR-based immunosensor using protein G for the detection of *Salmonella* Typhimurium. A protein G layer was constructed by chemically binding to self-assembled monolayer of 11-mercaptoundecanoic acid on gold (Au) surface. SPR spectroscopy was used to determine the formation of modified protein G layer on Au film as well as the binding of antibody and antigen. The SPR-based immunosensor could detect *S.* Typhimurium from pure culture at the detection limit of 10^2 cfu/mL. In the same year, Oh et al. (2004b) also applied a similar technique on detection of *Salmonella paratyphi*. Free amine (–NH2) of protein G was substituted into thiol (–SH) using 2-iminothiolane so that self-assembled protein G was fabricated on Au surface. This technique was able to detect *Salmonella* paratyphi from pure culture with the same detection limit as *S.* Typhimurium (Oh et al., 2004a) at 10^2 cfu/mL. However, when this technique was used to detect *Vibrio cholera* strains, it showed specific binding capacity on *V. cholera* with detection limit of 10^5 cells/mL (Jyoung et al., 2006), which is not as sensitive as the detection of *S. paratyphi* (Oh et al., 2004b). Additionally, several factors such as sample preparation directly affected the limit of detection of the SPR. Also, the binding of cell to the antibody-immobilized substrate was diffusion-limited. Therefore, the limit of detection could be further improved by optimization of sample preparation methods (Taylor et al., 2006).

SPR was further developed by modifying with nanomaterial-based approach to enhance the detection of pathogens and/or their toxins. Zhu et al. (2009) proposed a technique called localized surface plasmon resonance (LSPR)-based hybrid nanoparticles for detection of *Staphylococcus aureus* enterotoxin B. This LSPR-based nanobiosensor was designed on a basis of the hybrid gold–silver (Au–Ag) nanoparticles array. To make the hybrid nanoparticles, the Ag nanoparticles on glass substrate were coated with an Au thin film. The generated surface plasmons from the array carried the biological interaction message into the corresponding spectra. The hybrid Au–Ag nanoparticles could be used to detect *S. aureus* enterotoxin B at the level of nanogram per milliliter (ng/mL).

Recently, Singh et al. (2015) proposed the use of graphene for enhancing sensitivity of SPR immunosensors. The application of graphene benefited not only in amplifying SPR signal but also in controlling of the immobilization of biotinylated cholera toxin antigen on copper coordinated nitrilotriacetic acid. Through this system, the detection limit of an immunosensor for anticholera toxin antibodies was at 4 pg/mL. Hence, this type of nanomaterial-based modification could enhance detection sensitivity of some pathogens and their toxins to as low as 10^2 cells/mL or ng/mL compared to standard or current methods. Although this nanobiosensor demonstrates potential in monitoring, detection, and identification of foodborne pathogens, the improvement of the method specificity in differentiating antibody and antigen binding properties of each pathogen is still required. Also, method efficiency in multiplex detection or in various types of food should be further investigated and validated. Therefore, several significant challenges need to be addressed before SPR-based immunoassay can be used in detection of pathogens.

3.1.2 Quantum Dots–Based Immunoassay

Quantum dots (QDs) are nanosized crystalline particles that emit fluorescence with size-dependent emission wavelengths (Alivisatos et al., 2005; Wang et al., 2006) due to quantum confinement. The color of emitted light is a function of the particle size subject to change from red to blue when the QDs particle size becomes smaller. Compared to organic dye molecules, QDs have their own optical and electronic advantages, such as high fluorescence quantum yields, narrow emission bands, high stoke shifts, and stability against photobleaching. Especially, the controlled size enables the multiplex simultaneous detection of different targets, and thus significantly improves the speed of the detection (Wang et al., 2015).

QDs have been found useful in detection of foodborne pathogens as demonstrated in several reports. For example, Kuo et al. (2008) developed water-soluble cadmium-telluride (CdTe) QDs coated with denatured bovine serum albumin and conjugated the coated QDs with anti-*E. coli* antibodies by cross-linking. The antibody-conjugated CdTe QDs were then used to detect *E. coli* O157:H7 in comparison with the detection of *Listeria monocytogenes* using fluorescence microscopy. The result showed that anti-*E. coli* antibody, conjugated to CdTe QDs, still maintained its bioactivity and biorecognition specificity, which indicated the success in detection of *E. coli* O157:H7. Furthermore, Su and Li (2004) proposed the use of QDs as a fluorescence marker coupled with immunomagnetic separation for detection of *E. coli* O157:H7. Magnetic beads coated with anti-*E. coli* O157 antibodies were used to selectively capture the target bacteria, and biotin-conjugated anti-*E. coli* antibodies were later added to create sandwich immuno complexes. After magnetic separation, the immuno complexes were labeled with QDs by biotin-streptavidin conjugation followed by fluorescence measurement. *E. coli* K12 and *Salmonella* Typhimurium were used as controls for testing of method specificity. This technique could specifically detect pure culture of *E. coli* O157:H7 with the detection limit of 10^3 cfu/mL by using less than 2 h of detection time.

Yang and Li (2006) used QDs for simultaneous detection of *E. coli* O157:H7 and *S.* Typhimurium. Highly fluorescent semiconductor QDs with different emission wavelengths of 525 and 705 nm were conjugated to anti-*E. coli* O157 and anti-*Salmonella* antibodies, respectively. Since different particle sizes of QDs can be excited by a single wavelength of light, resulting in different emissions that can be measured simultaneously, multiple detection is achievable. The lower detection limit in this type of detection was 10^4 cfu/mL compared to that of single detection at 10^3 cfu/mL in the work carried out by Su and Li (2004). Furthermore, Zhao et al. (2009) used antibody-conjugated semiconductor QDs as fluorescence markers to simultaneously detect three foodborne pathogenic bacteria from pure culture. QDs with different emission wavelengths of 620, 560, and 520 nm were immobilized with anti-*S.* Typhimurium, anti-*S. flexneri*, and anti-*E. coli* O157:H7 antibodies, respectively. Target bacteria formed a complex with magnetic beads called "bead-cell." The bead–cell complexes then reacted with QD-antibody conjugates to form "sandwich" complexes. The intensities of the different emissions were measured and used for characterization and quantification of each bacterial strain. As a result, simultaneous bacterial detection and quantification was achieved with the detection limit of 10^3 cfu/mL.

Recently, Wang et al. (2015) developed the multicolor QD probes based on multiplex fluoroimmunoassay to detect three key pathogenic bacteria from both pure cultures and spiked in several food samples. The multicolor QDs, with different emission wavelengths at 504 nm, 557 nm, and 604 nm, were conjugated with anti-*S. enteritidis*, anti-*S. aureus*, and anti-*E. coli* antibodies, respectively. Through immunological recognition, target bacteria were selectively and rapidly captured by QD-antibody conjugates with a better detection limit of 10^2 cfu/mL in pure culture compared to the previously mentioned QD-based techniques. When this multicolor QD probes were used on spiked bacteria in food, the detection limit and range were different depending on type of food. Generally the detection limit of spiked bacteria in food was poorer than that of pure culture at a range of 10^2–10^3 cfu/mL. In the same year, Kim et al. (2015) reported the use of QDs based immunoassay for detection of *S.* Typhimurium from pure culture and spiked in food matrix of chicken extracts. *S.* Typhimurium was detected using a microfluidic device to demonstrate the advantages of using magnetic beads together with QDs as a fluorescent label. *Salmonella* cells were separated and concentrated using anti-*Salmonella* antibody–coated magnetic beads. The cells were labeled with antibody-conjugated QDs to form "sandwich" complexes. The fluorescence signals from the complexes were measured using a fluorometer to detect and quantify *c*. The detection limit of the method was 10^3 cfu/mL in both pure culture and food sample. In conclusion, the results obtained from various techniques used on different kinds of food demonstrated the feasibility, applicability, and anti-interference ability of these combined methods. Although these techniques may be used in a highly sensitive quantitative analysis, accuracy, precision, and specificity of the methods need to be validated in comparison with the standard methods.

3.1.3 Magnetic Nanoparticle Probes–Based Immunoassay

Magnetic nanoparticles have been used in rapid detection of foodborne pathogens. The magnetic nanoparticles are coated with biospecific materials to facilitate the binding of analytes such as antibodies, bacteria, proteins, DNA, and so forth. The bound analytes are separated from the sample by using magnetic field. For magnetic nanoparticle immunoassay, most research is focused on adjusting various detection techniques for pathogens from pure culture and food matrices. For example, Cheng et al. (2009) developed a method for detection of *E. coli* O157:H7 using biofunctional magnetic nanoparticles (BMNPs) coupled with adenosine triphosphate (ATP) bioluminescence. The BMNPs were made by immobilizing anti-*E. coli* antibody on the magnetic nanoparticles, and used to capture the *E. coli*. Isolated from the sample solution by magnetic field, the target *E. coli* was then quantified using an ATP bioluminescence. When BMNPs was applied to detect *E. coli* spiked into pasteurized milk, the detection limit was 20 cfu/mL and detection time was 1 h.

In the same year, Ravindranath et al. (2009) developed magnetic nanoparticles functionalized with anti-*E. coli* O157:H7 or anti-*S.* Typhimurium antibodies for isolation of both strains from a mixture of bacteria and from food samples. The pathogens were then detected by Fourier transform infrared method. The magnetic nanoparticles showed highly selective to both bacterial strains in the tested food matrices, which were diluted milk and spinach extract, with detection limits ranged from 10^4–10^5 cfu/mL and the detection time shorter than 30 min. Furthermore, Guven et al. (2011) developed a method using a combination technique between immunomagnetic separation (IMS) and surface-enhanced Raman scattering (SERS) for detect and quantify *E. coli*. Biotin-labeled anti-*E. coli* antibodies were immobilized onto gold-coated magnetic nanoparticles and used to capture *E. coli* cells. Raman labels were fabricated using gold nanoparticles coated with 5,5-dithiobis-(2-nitrobenzoic acid) (DTNB) and afterward with a molecular recognizer. The DTNB-labeled nanoparticles were later interacted with the complex between the magnetic nanoparticle and anti-*E. coli* antibodies. The limit of detection and limit of quantification for this method were 8 cfu/mL, and 24 cfu/mL, respectively. The method was also highly selective for *E. coli* detection since there was no significant response when tested with *Enterobacter aerogenes*, *Enterobacter dissolvens*, and *Salmonella enteriditis* as nontarget bacteria. The method was further used to detect *E. coli* in water samples and the obtained results were in line with the standard plate count method. Interestingly, this method is rapid with total analysis time around 1 h.

Najafi et al. (2014) also developed a similar technique using IMS and SERS to detect *E. coli* O157 from liquid media and apple juice. The anti-*E. coli* O157 antibody was immobilized on magnetite–gold magnetic nanoparticles (MNPs). The capture efficiency for *E. coli* O157 using MNP was about 84–94%. Cross reactivity was not detected with background nontarget bacteria including *Listeria monocytogenes*, *Bacillus cereus*, *Enterococcus faecalis*, *Staphylococcus epidermidis*, *Micrococcus luteus*, *Streptococcus mutans*, *Citrobacter freundii*, *Enterobacter cloacae*, *Escherichia vulneris*, *Hafnia alvei*, *Klebsiella oxytoca*, *Pseudomonas aeruginosa*, and *Salmonella enteritidis*. For detection of the target pathogen, the SERS nanoprobes were constructed by conjugating gold nanoparticles with Raman reporter molecules and the detector antibody (dAb). The Raman label dAb was interacted with the complex between MNPs and anti-*E. coli* O157 antibody. This immunoassay could be used to separate and detect *E. coli* O157 in apple juice at the detection limit of 100 cfu/mL within an hour of analysis time. With a similar concept, Nguyen et al. (2014) demonstrated combined effect of magnetic silica nanotubes (MSNTs) for bacterial separation and impedimetric sensor for the label-free detection of *S.* Typhimurium. The outer surface of MSNTs was functionalized with positive charges for concentration and isolation of bacteria from sample. Antibody specific to *S.* Typhimurium was immobilized on the interdigitated microelectrode of an impedimetric sensor. The complex between target bacteria and MSNT was conjugated with antibody immobilized impedimetric sensor. The MSNT could significantly enhance the performance of impedimetric sensor by generating highly discriminated impedance signals correlated with different bacterial concentrations. The total analysis time was around 30 min.

Cho and Irudayaraj (2013) applied magnetic nanoparticle technology to enhance ELISA sensor efficiency. An in situ immunogold nanoparticles (AuNP) network-based ELISA biosensor was developed with a sample concentration technique of immunomagnetic separation to detect foodborne pathogens by using *E. coli* O157 and *S.* Typhimurium as a model. The sensor was optimized by formation of immuno-AuNP network onto the antigen-binding site at the outer membrane layer of bacteria. The method was also validated using microtiter immunoassay. The proposed biosensor was able to detect bacterial cells at a low concentration of 3 cells/mL for *E. coli* O157:H7 and *S.* Typhimurium in buffer. In liquid food samples, the limits of detection were at 3 and 15 cfu/mL for *E. coli* O157:H7 and *S.* Typhimurium, respectively. Additionally, the total analysis time was less than 2 h. To conclude, this technique demonstrated high sensitivity and specificity for both bacteria in liquid food samples. After that, Cho et al. (2015) also modified the magnetic nanoparticles based technique by using membrane filtration and silver intensification step together with SERS to detect *E. coli* O157:H7 from pure culture and inoculation in ground beef. Nontarget bacteria, *Salmonella* Typhimurium, *Salmonella enteritidis,* and *Listeria monocytogenes* were used to test the method specificity. The target bacteria was captured and separated by two different antibodies bearing nanoparticle complexes, which were magnetic nanoparticles coated with monoclonal antibody, and gold nanoparticles conjugated with polyclonal antibody and Raman reporter of 4-mercapto benzoic acid. After centrifugal filtration, the remaining bacteria–nanoparticle complexes on the 0.45 micron pore filter membrane were located by silver intensification before subsequent detection by SERS. This technique could detect *E. coli* O157:H7 as low as 10 cfu/mL within 1 h in pure culture and 3 h in ground beef. The magnetic nanoparticle based immunoassay and its modifications reviewed above demonstrate the ability of the methods to monitor target bacteria with improved analytical sensitivity compared to the current techniques and present an opportunity to develop rapid detection method for routine food analysis.

3.1.4 Carbon Nanotube–Based Immunoassay

Carbon nanotubes (CNTs) are one-dimensional cylindrical nanostructure with a diameter of a few nanometers and a length of up to several millimeters. CNTs can be generally classified into two major groups, which are single-wall carbon nanotubes (SWCNTs) and multiwall carbon nanotubes (MWCNTs) (Wang, 2005). CNTs have been investigated for potential use as biosensor and bacterial filter due to their unique electronic, chemical, mechanical, and structural properties (Niyogi et al., 2002). CNTs have been proved to be good absorbent for various bacteria of different sizes and shapes, including foodborne pathogens such as *E. coli, S. aureus, B. subtilis,* and so forth (Deng et al., 2008; Upadhyayula et al., 2009; Lee et al., 2012; Deokar et al., 2013). For example, Yang and Bashir (2008) proposed the use of optical CNT immunosensor to improve the detection of *Staphylococcal* enterotoxin B (SEB). Anti-SEB antibodies were immobilized onto CNT surface by electrostatic adsorption. Then, the complex between anti-SEB antibody and CNTs was bound to a polycarbonate layer to form a sandwich link. ELISA assay was used to detect the SEB. With the CNTs, the sensitivity of this immunosensor was increased at least 6-fold. Also, Villamizar et al. (2008) demonstrated the use of SWCNTs based method to detect *Salmonella infantis.* The technique was based on a field effect transistor (FET) in which SWCNT network acted as the conductor channel. Anti-*Salmonella* antibodies (rabbit polyclonal antibody for *Salmonella* "O" and "H" antigens) were adsorbed onto the SWCNTs. Afterward, Tween 20 was used to protect SWCNTs from the nonspecific binding of other bacteria or proteins. This method could specifically detect *S. infantis* from pure culture with the presence of the other competitive bacteria such as *Streptococcus pyogenes* and *Shigella sonnei.* The FET devices could detect *S. infantis* at the concentration of 100 cfu/mL in 1 h. Recently, Liu et al. (2014) designed multifunctional MWCNTs based method. The MWCNTs functioned as immunomagnetic-fluorescent sensors for detection of *Vibrio alginolyticus* while *E. coli, S. aureus, B. subtilis,* and *P. aeruginosa* strains were employed as nontarget bacteria. The developed MWCNTs could accurately detect *V. alginolyticus* in pure culture with the detection limit of 2.8×10^2 cfu/mL without cross-reaction with the other four strains tested. The device was also used to determine trace amount of spiked *V. alginolyticus* in shrimp and water from fishpond and showed great potential for rapid and sensitive detection as well as quantification similar to its use in pure culture. Although CNTs-based method is effective, the major disadvantage is that CNTs are toxic and difficult to manipulate. In addition, the other problem on CNT-based sensor is the low sensitivity of the system while it is used on the analytes with no affinity to CNTs. Therefore, several significant challenges need to be addressed before CNTs can be widely accepted as detection method for foodborne pathogens.

3.1.5 Other Nanotechnology and Nanomaterial-Based Immunoassay

Apart from the nanotechnology and nanomaterial based immunoassay as categorized above, another device and their application have also been developed and further evaluated for detection of pathogens in food. For instance, Zhao et al. (2007) demonstrated a method called disposable electrochemical immunosensor to detect *Vibrio parahaemolyticus* (VP). This method was developed using the screen-printed electrode coated with agarose/nanogold membrane, and horseradish peroxidase labeled with anti-VP antibody. *E. coli* O157:H7, *Salmonella pullorum* and *Staphylococcus* were used as nontarget strains for testing of method specificity. Through this device, *V. parahaemolyticus* from pure culture was detected based on the inhibition of the enzymatic activity on the oxidation reaction of thionine by H_2O_2 due to the formation of immunocomplex on the surface of the immunosensor. The immunosensor also exhibited high specificity with a detection limit of 7.4×10^4 cfu/mL. In the same year, Kim et al. (2007) developed screening methods of label-free optical grating coupler (OGC) immunosensor with direct binding to detect *Salmonella* along with various immobilization methods for an anti-*Salmonella* antibody over the amino-, epoxide- and thiol-terminal silanes. Anti-*Salmonella* antibody and peroxidase-labeled anti-*Salmonella* antibody, immunized against *Salmonella* common structural antigen from goat serum, were used in this study. The limit of detection for the OGC immunosensor was around 1.3×10^3 cfu/mL. The method was relatively specific to target pathogen, *S.* Typhimurium, when tested with the other nontarget organisms such as *L. monocytogenes*, *E. coli*, and *P. aeruginosa*.

Furthermore, Sim et al. (2012) demonstrated the application of a nanomaterial based immunoassay called birefringent waveguide biosensor platform to detect label-free live cell of *Listeria monocytogenes*. The device consisted of a waveguide sensor platform for determining the immune reaction, and a birefringence measurement system for measuring the phase difference between two orthogonal polarizations, emerged from the change of refractive index due to the immune reaction in the microfluidic channel of the sensor. Fabrication of the waveguide sensor was compatible with conventional integrated circuit processes. Sensitivity of waveguide platform was optimized by testing different thickness of the overlaid titanium dioxide (TiO_2). Additionally, sensitivity of waveguide composition was also influenced by the method for controlling the length of TiO_2 film. To conclude, the investigated method was able to detect immune reactions for pathogenic bacteria including *Listeria monocytogenes*. Furthermore, Joung et al. (2013) proposed the electrochemical immunoassay based on nanosized pores embedded in an insulating membrane which helped increase capture efficiency of antibodies. The method was developed using a nanoporous membrane-based impedimetric immunosensor for label-free detection of pathogenic bacteria in whole milk by functionalizing a commercially available alumina nanoporous membrane with hyaluronic acid. By using *Escherichia coli* O157:H7 as a model pathogen, the modified device could effectively reduce the nonspecific binding of biomolecules and other cells, as well as allow successful immobilization of antibodies. The antibody–pathogen interactions caused changes in impedance of electrolytes through nanopores. The impedance spectra were then analyzed by normalized impedance change. The detection limit for *E. coli* O157:H7 from pure culture was found to be as low as 10 cfu/mL and for those spiked in whole milk was as low as 83.7 cfu/mL. The device was specific to *E. coli* O157:H7 when tested with the presence of *Staphylococcus aureus*, *Bacillus cereus*, and nonpathogenic *E.coli* DH5 as nontarget bacteria.

Bok et al. (2013) developed a fluorescence biosensor for detection of *Clostridium botulinum* neurotoxin type A (BoNT/A). The ultra-bright, photostable dye-doped nanoparticle tags was used in combination with thin films of nanoporous organosilicate to achieve high intrinsic signal amplification. The combination of optical effects could enhance the fluorescence and consequently improve the ratio between signal and noise significantly. BoNT/A was detected in phosphate buffer saline at the level as low as 21.3 fg/mL in 4 h of analysis time. For application on food, the detection limit of BoNT/A was around 150 fg/mL in both orange juice and tap water. This result demonstrated improvement of detection limit comparing to that of the other ELISA-based methods. Furthermore, immunosensor based on electrodeposition of gold nanoparticles (AuNPs) is one of the techniques that have been increasingly developed to potentially use for detection of food-borne pathogens. Hu et al. (2014) reported the combined use of the electrodeposited AuNPs and streptavidin-biotin affinity system as enzyme immunosensor to detect *Salmonella pullorum* and *Salmonella gallinarum*. The electron communication was improved and the electrode conductivity was also strengthened as a result of the enhancement of electron transfer by electrodeposition of AuNPs. In addition, the number of the conjugated biomolecules was significantly increased by this electrodeposited AuNPs and the streptavidin-biotin system. Through this system, *S. pullorum* and *S. gallinarum* were detected at 1.95×10^2 cfu/mL. Similarly, Wang et al. (2014) also applied the concept of electrodeposited AuNPs and ionic liquids for immunosensor to detect *S. pullorum*. The immunosensor was constructed by modifying screen-printed carbon electrode with electrodeposited AuNPs, horseradish peroxidase labeled with anti-*S. pullorum* antibody, and ionic liquids (ILs). AuNPs were first electrodeposited on the electrode surface to raise the amount of bound antibodies. The electrodeposited AuNPs were then modified with ILs to protect the antibodies from the test environment inactivation, as well as to sustain their biological activity and the electrode stability. With combination of electrodeposited AuNPs and ILs, reproducibility, stability, specificity, and sensitivity of the method were enhanced.

3.2 Nanotechnology and Nanomaterials-Based Nucleic Acid Assay

Various techniques based on nucleic acid assay including nucleic acid probe hybridization (ie, enzyme-linked immunosorbent assay), polymerase chain reaction (PCR) (ie, multiplex PCR), and other nucleic acid–based molecular biology have been explored to improve the rapidness and automation of foodborne pathogen detection. This type of assay is relatively time-saving with the analysis time of less than 24 h. (Bennett, 2005; Carroll, 2008; Wellinghausen et al., 2009). However, disadvantages of the method are low level of automation, expensive instrument and complicated sample preparation, for example. As a result, nanotechnology such as electrochemical DNA/cDNA biosensors has been greatly developed to improve the efficiency among different DNA hybridization assay methods. Also, various kinds of modified electrodes have been invented for the ssDNA immobilization (Teles and Fonseca, 2008a,b; Xu et al., 2009). Among them, nanotechnology/nanomaterials have been widely utilized due to the advantages such as high surface area, desirable electronic properties and electrocatalytic activity. The numerous benefits of nanotechnology/nanomaterials enable the improved design in probes for rapid identification of several pathogens.

Furthermore, the other technique under the nucleic acid assay is based on PCR protocol. The integration of nanotechnology/nanomaterials to this most conventional nucleic acid detection methods can assist in enhancement of sensitivity through polymerase-mediated amplification. Researchers have been focused on achieving higher sensitivity through either short PCR rounds or assay-intrinsic signal enhancement. The modified device could ideally facilitate the detection of pathogens with the ability to discriminate single nucleotide.

3.2.1 Nanotechnology and Nanomaterials-Based Nucleic Acid Probe Hybridization Assay

Nucleic acid hybridization techniques are widely used in detection and identification of pathogens. It has also been used extensively in a variety of applications such as the development of optical, electrochemical, gravimetric, surface plasmon resonance (SPR)-based, and electric biosensors (Piunno et al., 1995; Minunni et al., 2003; Kelley et al., 1999; Drummond et al., 2003). Unlike protein-based detection, nucleic acid recognition sites can be synthesized and regenerated for multiple uses. The technique employs a labeled or label-free segment of single-stranded nucleic acids, ssDNA or ssRNA, in annealing and complementing DNA/RNA sequence. Probe and target sequences have to share a significant amount of similarity (Ramachandran et al., 2004). After hybridization occurs, the probe-target binding is detected by the transducer such as optical, electrochemical and SPR-based, and so forth.

Several studies integrated nanotechnology to nucleic acid probe to facilitate the detection of pathogens. In early stage, the incorporation of nanotechnology was limited to clinical diagnosis as well described by Kaittanis et al. (2010). Colorimetric polynucleotide detection method based on mercaptoalkyloligonucleotide-modified gold nanoparticle probes was first reported by Elghanian et al. (1997). Later, the development of real-time multicolor DNA detection with chemoresponsive diffraction gratings and nanoparticle probes was published (Bailey et al., 2003). The work demonstrated the unique interaction between gold nanoparticles (AuNPs) and DNA which led to a shift in the SPR peak. Apart from contributing to rapid and highly specific detection, the ssDNA carrying nanoparticles also facilitated real-time detection with micropatterned diffraction gratings and AuNPs. The bound nanoparticle probe with diffraction grating caused changes in local dielectric environment and subsequently resulted in the diffracted pattern of incident light. In addition, AuNPs were also used for the colorimetric detection of DNA-binding molecules allowing the optical identification of the DNA-bound molecules through color changes. Consequently, feasibility of visual profiling on pathogens was increased by this nanosystem.

During the last decade, DNA hybridization techniques have been developed and also applied in detection and identification of foodborne pathogens. Qi et al. (2012) developed an electrochemical DNA biosensor with chitosan-Co_3O_4 nanorod-graphene composite to detect *Staphylococcus aureus*. Co_3O_4 nanorods, graphene (GR), and chitosan (CTS) were used to fabricate nanocomposite materials. The nanomaterials were further modified on carbon ionic liquid electrode (CILE) and eventually used as the substrate electrode to construct electrochemical DNA biosensor. The ssDNA probe was immobilized on the CTS–Co_3O_4–GR/CILE surface by electrostatic attraction. Methylene blue (MB) was used as an electrochemical indicator to monitor the hybridization events via the decrease or increase of the reduction peaks. The prepared electrochemical DNA sensor was applied to detect *nuc* gene sequence fragment from the *S. aureus*. The method helped increase the quantity of ssDNA adsorbed on the electrode surface and subsequently stimulate electrochemical response of MB due to the exceptional ability in electron transfer, large surface of GR, good film-forming ability of CTS, and high conductivity of CILE. Interestingly, this device could sensitively and selectively detect the target DNA with the detection limit as low as 4.3×10^{-13} M and with the discrimination ability to the one-base and three-base mismatched ssDNA sequences. Liu et al. (2014) developed bioactive paper-based platform consisting of AuNPs and AuNP–thiolated DNA probe conjugates for detection of viable *Listeria minocytogenes* by sensing hlyA mRNA gene. The mRNA of the target bacteria was amplified and applied to the paper-based platform for a visual test using sandwich hybridization assays. When the RNA products migrated along the platform by capillary action, the AuNPs accumulated at the designated area. Subsequently, the results were visually detected without any complicated instrument for qualitative analysis or with the aid of image software for quantification. Under optimized conditions, genomic RNA from *L. monocytogenes* could be detected at the level of 0.5 pg/μL. The method was also able to specifically detect *L. monocytogenes* from actual samples at the concentration as low as 20 cfu/mL.

In the last few years, aptamer-based biochemical detection has been widely examined and reported as a reliable analytical method for pathogens. Aptamers are ssDNA, ssRNA, or modified nucleic acids which can recognize target molecules such as toxins, proteins, and cells through specific binding and folding into three-dimensional structures (Morris et al., 1998; Blank et al., 2001). Aptamers with randomized sequences can be obtained from a combinatorial oligonucleotide library via an in vitro selection process called systematic evolution of ligands by exponential enrichment (SELEX). Compared to antibodies, aptamers have added advantages such as ease of synthesis, higher specificity, stability with wider range of targets, inexpensiveness, and ability to be modified with functional groups. Application of aptamers is not only used for therapeutics and diagnostics (Jayasena, 1999; Que-Gewirth and Sullenger, 2007) but also extended to the detection and identification of foodborne pathogens. Based on aptamer specificity, the system could be further adapted for quantification as a method developed by Lee et al. (2012, 2015). In 2012, they identified an RNase-resistant RNA aptamer by a cell-based subtractive SELEX technology. The RNA aptamers specifically bound to the cell surface of *E. coli* O157:H7, and were hence able to distinguish pathogenic strains from the others. An RNA library generated by in vitro transcription of synthetic DNA templates was incubated with the *E. coli* K12 strain for aptamer selection and characterization. The RNAs binding to the *E. coli* K12 strain were discarded. The remaining RNAs were then used for the selection of O157:H7-specific aptamers. After six rounds of subtractive cell-SELEX, it was found that the selected aptamer specifically bound to the O157:H7 serotype, without binding to the K12 strain. The technique was confirmed by aptamer-immobilized ELISA, real-time PCR analysis, or an aptamer-linked precipitation experiment. The obtained RNA aptamer was able to separate the virulent serotype from non-pathogenic strain by binding to an O157:H7-specific lipopolysaccharide, including the O antigen. As a result, the isolated RNA aptamer could potentially be mounted on a biosensor system as a ligand to directly detect *E. coli* O157:H7.

Lian et al. (2015) proposed aptamer/graphene interdigitated gold electrode epiezoelectric sensor for detecting *Staphylococcus aureus*. The device was constructed by immobilizing *S. aureus* aptamers as sensing elements on the graphene surface of interdigital gold electrodes (IDE). 4-Mercapto-benzenediazonium tetrafluoroborate was used as a cross-linking agent to chemically bind graphene to IDE. Subsequently, the modified IDE was connected to a series electrodepiezo electric quartz crystal (SPQC). After the specific binding to *S. aureus* occurred, the aptamer were dropped from the surface of the graphene. The electric parameters of the electrode surface was shifted and resulted in the change of oscillator frequency of the SPQC. This detection was completed within

60 min with 41 cfu/mL of detection limit. Additionally, Lee et al. (2015) also developed an aptamer-based sandwich assay (ABSA) platform that demonstrated a potential for use in pathogen detection. Aptamers selected by whole-cell SELEX with high affinity and specificity was used to identify live *L. monocytogenes*. ABSA was carried out in a three-step assay including immobilizing the capture aptamer onto a 96-well microplate, adding samples to the well, and finally detecting *L. monocytogenes* via fluorescent-labeled aptamer. The aptamer pair, being used as a very specific conjugate for sandwich complex, significantly increased the sensitivity of ABSA platform. The proposed ABSA platform can also distinguish *L. monocytogenes* from the other *Listeria* species and the other bacterial genera. These improvements in sensitivity and specificity not only facilitated the detection of *L. monocytogenes* at extremely low concentrations, but also permitted the development of routine assay platform in a 96-well plate format for multivalent diagnostics.

Through similar methodology, Abbaspour et al. (2015) also reported a highly sensitive and selective dual-aptamer-based sandwich assay for the detection of *S. aureus*. In this bioassay, the primary aptamer was immobilized on magnetic beads (MBs) coated with streptavidin, which served as a capture probe. When a sample containing *S. aureus* was added, bacterium was captured by specific interaction between aptamer and bacterium. The secondary aptamer, conjugated to silver nanoparticles (Apt-AgNP), was later added to complete the sandwich and sensitively reported the detection of the target. After resuspension and magnetic separation of MBs carrying the affinity complex, the supernatant containing AgNP ions was transferred onto the surface of an electrode for differential pulse stripping voltammetric measurement. The method showed a detection limit as low as 1.0 cfu/mL. When this method was used for quantification of *S. aureus* in actual water samples, the results demonstrated an acceptable consistency with a plate count method. In conclusion, the proposed sandwich assay benefited from an increase in specificity, as a result of MBs which were functioned as carriers of affinity ligands for solution-phase recognition with fast magnetic separation. The secondary aptamer conjugated to silver nanoparticles also enhanced sensitivity of the assay. The electrochemical detection method further showed many advantages in terms of simplicity, reduction of analysis time, low cost, and improved detection limit.

3.2.2 Nanotechnology and Nanomaterials-Based Polymerase Chain Reaction (PCR) Assay

Nucleic acid probe hybridization assay is not the only nucleic acid–based method which is used in the detection and identification of foodborne pathogens, but polymerase chain reaction (PCR) is also extensively employed. The PCR assays are able to amplify single DNA target and multiple targets (multiplex PCR) for specific detection of pathogens. For detection of viable cells, reverse transcriptase PCR was further developed to detect mRNA transcripts from the specific genes of target bacteria (Klein and Juneja, 1997; Hill et al., 1998). Both conventional and real-time PCR can be applied to detect and identify bacterial pathogens. Furthermore, the variation in genomic targets that can be selected for use is highlighted by several key references as summarized in Byrne et al. (2015). The key advantage of PCR-based assays is the ability to simultaneously detect multiple pathogens in a single sample using multiplex PCR. Additionally, PCR assays can be constructed to differentiate between viable and nonviable cells by nanotechnology modification.

Nanomagnetic beads as magnetic primers have been used in new strategies for DNA electrochemical sensor to improve sensitivity and selectivity. For example, Lermo et al. (2007) used magnetic primers for in situ amplification of *Salmonella* genome on magnetic beads and further separated the amplified product with magneto electrodes based on graphite–epoxy composite. The PCR assay was done by immobilization of the biotinylated capture probe on the modified magnetic bead, together with hybridization of IS200 sequence specific for *Salmonella* and the digoxigenin probe. Enzymatic labeling was subsequently performed using the anti-digoxigenin horseradish peroxidase antibody. After in situ amplification of the DNA template on magnetic beads, the doubly labeled PCR product was thus rapidly detected by the electrochemical magneto biosensing strategy. The amperometric change due to the doubly labeled duplex was then evaluated. The results showed the direct DNA amplification successfully performed on magnetic beads by using the magnetic primer. In addition, the magneto electrode was able to detect changes at single nucleotide polymorphism, when stringent hybridization conditions were used. In conclusion, this strategy could be used for the rapid and sensitive detection of PCR amplified samples.

The nanomagnetic beads based separation combined with multiplex PCR assay was also demonstrated by Yang et al. (2013). The technique was proposed for simultaneous detection of *Salmonella* Typhimurium, *E. coli* O157:H7, and *Listeria monocytogenes*, and also tested in food products. Magnetic nanobeads based immunomagnetic separation was used to separate the target bacterial cells while multiplex PCR was used to amplify the target genes. Initially, the target bacteria were also treated with propidium monoazide prior to DNA extraction to eliminate the interference from nonviable cells. The result showed the detection limit of the assay was about 100 cfu/mL for all three bacterial strains in pure culture and around 1,000 cfu/g for all strains spiked in food samples of lettuce, tomato, and ground beef. Although this assay is highly innovative compared to the other nucleic acid–based methods, it requires bacterial isolation, lysis, and/or isolation of bacterial DNA which limits the application in actual circumstances during field work.

3.3 Nanotechnology and Nanomaterials-Based Whole Cells Detection Assay

Large numbers of research are directed towards capture and detection of the analytes, such as nucleic acids, protein, and enzymes, and so forth. However, in term of real time or online analysis, whole cell methods may provide information, especially for food safety, more extended than that of cell component approaches, for example, cell morphology, cell number, spatial distribution within a sample, correlation of genetic identity, biochemical activities, and cell behavior (Brehm-Stecher, 2008). Ultimately, these types of information may be useful in determining pathogenicity and metabolic status, which indicate bacterial cells' ability to cause disease. The information can be used to identify rational options to manage the risk associated with the specific cell such as food product disposal, reprocessing, or recall (Brehm-Stecher, 2008).

Until now, several whole cell detection methods were developed with an integration of nanotechnology and/or nanomaterials for characterization of pathogens in food samples. Among them, Surface-Enhanced Raman Spectroscopy (SERS) is a method with high selectivity and sensitivity resulting from the presence of surface enhancement mechanisms only at the surface. SERS have been evaluated as a possible method to detect and characterize different pathogens by their spectra characteristic. Sundaram et al. (2013) evaluated the use of SERS coupled with biopolymer encapsulated silver nanosubstrates for identification of foodborne pathogens. Principal component analysis was used to develop classification models to differentiate pathogens based on spectra characteristic. The encapsulated biopolymer, to be used as SERS substrate, was fabricated using silver nitrate, polyvinyl alcohol solution, and trisodium citrate. Prepared as fresh culture to maintain viability, foodborne pathogen isolates from chicken rinse including Gram-negative *Salmonella* Typhimurium and *E. coli*, and Gram-positive *Listeria innocua* and *Staphylococcus aureus* were used. After suspended in sterile deionized water, each bacteria culture was placed on the nanosubstrate individually and the SERS spectral data were recorded over the Raman shift between 400 and 1800 cm^{-1}. Comparing between the SERS spectra from Gram-negative bacteria and those of the Gram-positive, the result indicated that all of them had similar cell wall and cell membrane structures. In contrast, their major differences were the nucleic acid and amino acid structure between 1200 and 1700 cm^{-1} and at the fingerprint region between 400 and 700 cm^{-1}. The result demonstrated that silver biopolymer nanoparticles could be a potential SERS substrate for pathogen detection.

Based on the same technology, Chen et al. (2014) evaluated the SERS technique for detection of *E. coli*. SERS active filters with silver nanorod (AgNR) array were employed for cell characterization in comparison with standard glass AgNR substrate. The fabrication technique of AgNR array film was oblique angle deposition. AgNR was then deposited on the smooth surface of Anodisc and GA-8 filters with the pore size of 0.2 μm. The obtained AgNR filters were used for preconcentration of bacteria and subsequently detected by on-chip SERS. It was found that AgNR filters could improve bacterial SERS signal and limit of detection by at least two orders of magnitude compared to the standard glass. The other nanotechnology to enhance fluorescent labeling for detection of bacterial cells was also reported. For example, Liu et al. (2015) synthesized gold nanoclusters with bright red fluorescence for visualization enhancement during detection of *E. coli*. To significantly improve the red fluorescence intensity, the lysozyme functionalized fluorescence gold nanoclusters (lysozyme-AuNCs) were fabricated and specifically attached onto the surface of *E. coli* from pure culture. The red fluorescence intensity of lysozyme-AuNCs was linearly enhanced at a range of 2.4×10^4 to 6.0×10^6 cfu/mL with the detection limit of 2.0×10^4 cfu/mL. The visualization fluorescence technique could enable the rapid and real-time detection of bacteria, or apply to other functional proteins such as antibody, enzyme, and peptide, or use in the fields of biochemistry and biomedicine. In summary, although the method has potential to be selectively quantify and provide useful information based on SERS technology, process in cell separation and preparation, which may take several hours, is still required. Thus, further development can be done by combining rapid cell capture or concentrating technique to reduce assay time.

Nanobased methods developed based on different concepts in analysis of foodborne pathogens, some analytical concepts of each methods are depicted in Fig. 13.1. In addition, efficiency and potentials of the nanobased assays are also summarized and illustrated in Fig. 13.2.

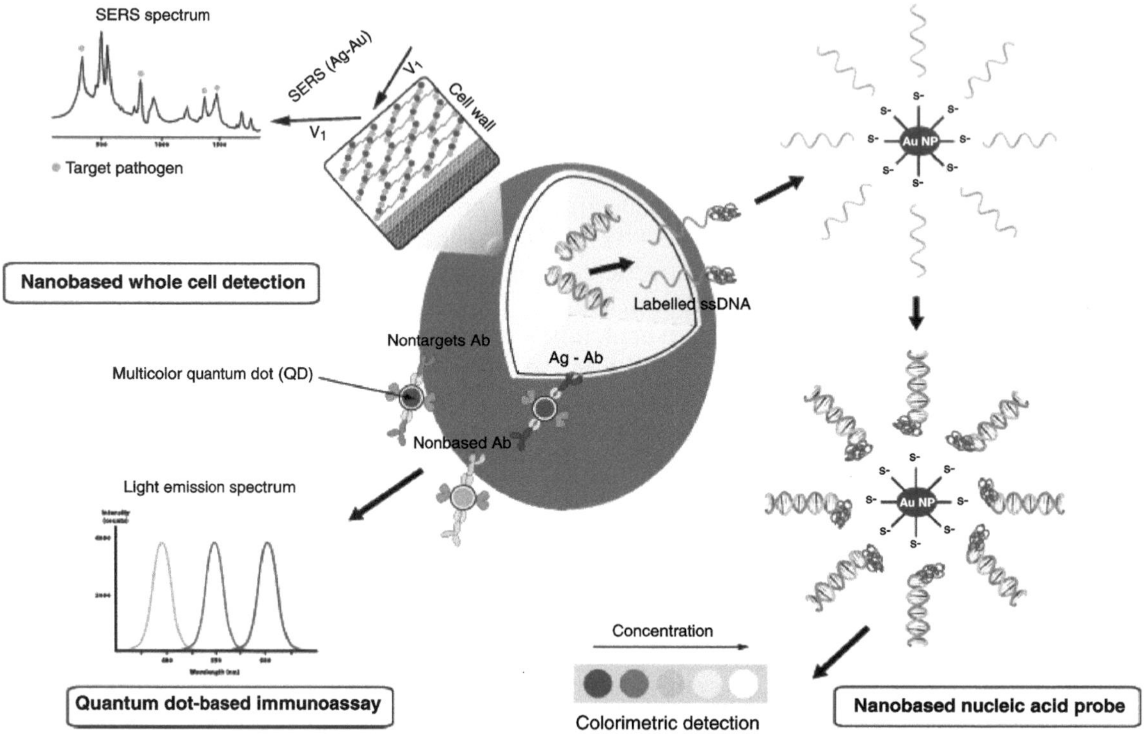

Figure 13.1. Nanotechnology-based strategies for the identification of foodborne pathogens. Some examples of nanotechnology, quantum dot-based immunoassay with antibody labeled; multicolor quantum dot bound to target antigen, quantifying by light emission intensity. Nanobased nucleic acid probe with AuNP probe hybridized to labeled ssDNA quantifying by colorimetric detection. Nanobased whole cell detection using SERS with Ag-Au coated surface to determination of cell wall properties through SERS spectrum. *Ag-Ab*, Antigen-Antibody; *Ag-Au*, Silver-gold coated surface; *AuNP*, gold nanoparticle; ssDNA, single strand DNA.

4 Conclusions and Outlook

For any newly developed analytical method, speed, sensitivity, specificity, and cost effectiveness are key concerns. In food analysis, authority often requires evidence demonstrating the absence of pathogenic organisms in food product, at the level causing adverse health effect to consumers. For example, the method must be able to reliably detect single cells in 25 g samples. This requirement poses an additional challenge for new assays. The concept and evolution of analytical methods for foodborne pathogen detection described in the first part is used as a basis for discussion on benefits, and limitations of each means. These limitations lead to further development with advanced technology of nanotechnology/nanomaterials. The key properties of each significant pathogen in the conventional detection method are also utilized in the development of nanobased methods.

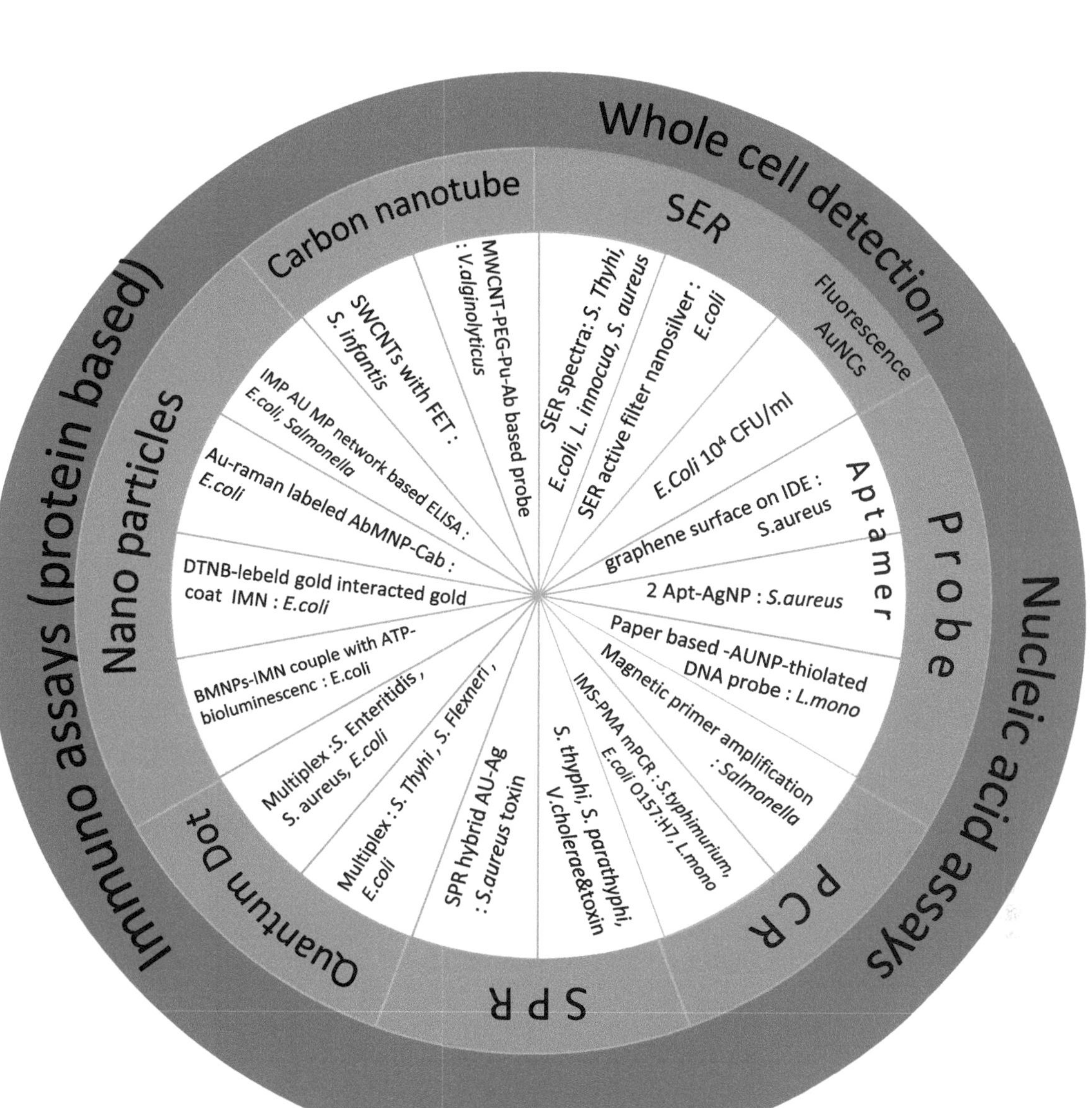

Figure 13.2. Summary of nanobased methods in the analysis of foodborne pathogens in terms of sensitivity and speed. *EC, Escherichia coli; LM, Listeria monocytogenes; Sal, Salmonella spp.; SI, Salmonella infantis; ST, Salmonella typhi; STm, Salmonella Typhimurium; SA, Staphylococcus aureus; VA, Vibrio alginolyticus.*

Nanotechnology-based methods have been combined with the traditional detection method to enhance the sensitivity and speed of detection. The recent researches demonstrate potential of nano-based methods to be used for real time monitoring. Currently, disadvantages are also found when the analytical techniques such as immunology-based, nucleic acid–based, and whole cell–based detections are combined with the nanotechnology, for example,

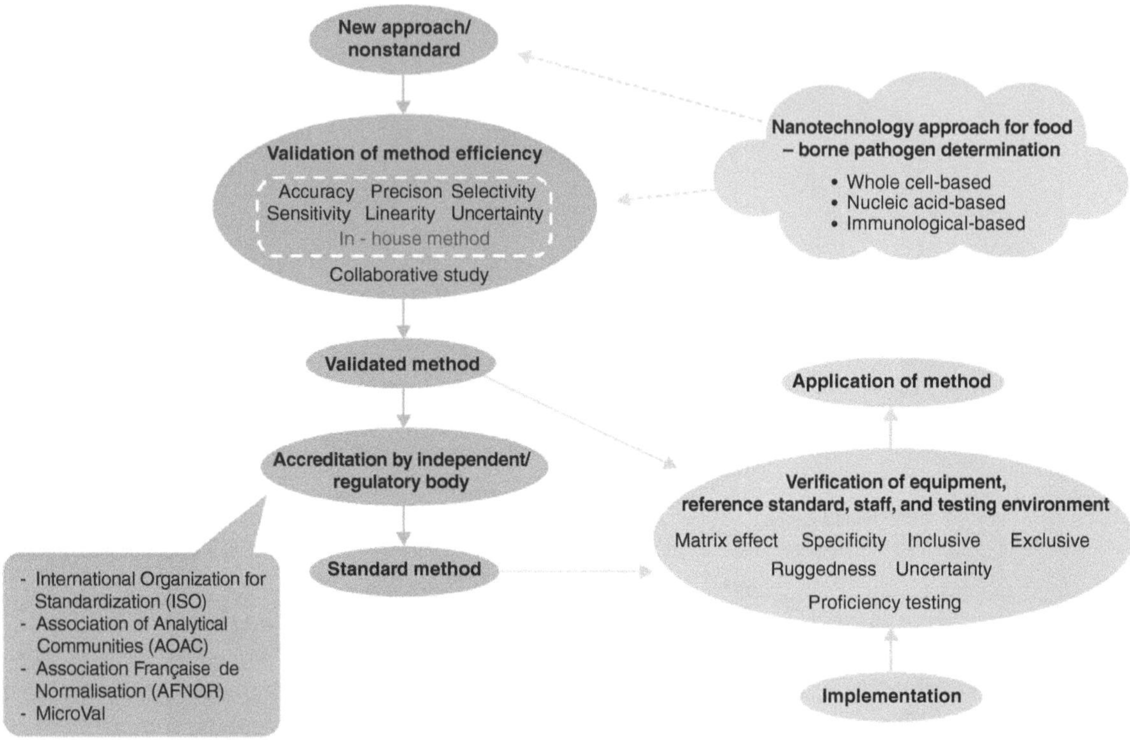

Figure 13.3. Method validation process for accreditation of newly developed analytical methods and status of developed nanobased methods based on this review.

method accreditation, safety concern of some techniques, and high expenses, in relation to the current available method.

In accreditation of newly developed nanotechnology-based methods, the reviewed methods are still at the initial stage of submitting the assay for consideration by the independent or regulatory body. To determine the reliability and applicability in the food safety control system of the new approach as shown in Fig. 13.3, the method validation or verification must be done (Joint FAO/IAEA Expert Consultation, 1997). Validation process starts with defining performance characteristics, comparing these characteristics with a reference method, and statistical evaluating to determine equivalence. In contrast, verification study is to ensure that the method when performed by another laboratory is still reliable and meets all the defined performance characteristics (Simone and Robouch, 2014). During validation for regulatory use, reference methods namely standard method, "gold standard," or official method, such as those specified in the national regulation (*FDA's Bacteriological Analytical Manual* [USFDA-BAM],

Association Française de Normalisation [*AFNOR*], British Standards Institution [BSI], *Japanese Industrial Standards Committee* [JISC]) or international standard bodies (The *Association of Analytical Communities* [AOAC] and The International Organization for Standardization [ISO]) are used to compare with the novel assays before implementation (The FDA Foods Program Science and Research Steering Committee [SRSC], 2011). Thus, the nanobased protocols as reviewed in this chapter are mainly at the stage of partial validation in which accuracy, sensitivity and specificity are evaluated.

Based on the review, several significant challenges need to be addressed before applications in detection of foodborne pathogens. In SPR-based immunoassay, the efficiency of the detection methods in multiplexed pathogens/toxins and for the use in food samples should be further investigated and/or validated. For quantum dots based immunoassay, this technique has potential to be highly sensitive method for quantitation. For further application in foods, accuracy, precision, and specificity need to be compared with the standard methods. For the magnetic nanoparticle based immunoassay and their modifications, the methods demonstrate the ability to monitor target bacteria with improved sensitivity compared to the current techniques. Even though CNTs-based method is effective, the major disadvantage is that CNTs are toxic and difficult to manipulate. In addition, CNT-based sensing systems have low sensitivity for analytes that have no affinity to CNTs. Therefore, further research is required to overcome certain drawbacks of the CNTs-based system prior to the use in actual food.

In nanotechnology-based nucleic acid assay, ability to be synthesized and regenerated of nucleic acid recognition sites for multiple uses is the key advantage compared to the protein-based detection. Several concepts are employed in this approach including (1) nucleic acid probe and (2) polymerase chain reaction (PCR). The nucleic acid probe and aptamer technology shows ideal sensitivity as low as single cell detection. PCR modified by nanotechnology can differentiate between viable and nonviable cells with high sensitivity. Furthermore, both nanobased nucleic acid probe and nanobased PCR assays as reviewed in this chapter have been developed by testing in several food matrices showing high potential in further application. Although the methods are highly innovative, all nucleic acid–based methods except for aptamer technique require bacterial isolation, lysis and/or isolation of bacterial DNA, which can limit their application in actual circumstances during field work.

Key advantage of whole cells detection–based nanotechnology is the potential in providing extended information that is only available within the context of an intact cell relative to cell component approaches. However, the whole cell methods have not been widely studied compared to the other nanobased methods. Based on the review in this chapter, although the whole cell method has potential to be used in selective quantification and provide useful information based on SERS technology, process in cell separation and preparation, which may take several hours, is still required. Thus, further development can be done by combining rapid cell capture or concentrating technique to reduce assay time.

For final remarks, the developed method should be cost-effective. Ideally, the method should not require expensive reagents or equipment and should be simple to perform under different circumstances, with minimal preprocessing of sample. The nanobased method may not be ideal since they generally require expensive reagent and equipment. The use of the method is still limited to clinical laboratories, which are a larger market compared to food-testing laboratories. However, if in-depth studies and researches are further conducted to establish the nanobased method in the field of food microbiology, especially, to detect pathogens at the beginning of food processing such as the use in multiple bacteria detection and real-time monitoring, these applications will have clear advantages over the current methods. In case the nanobased method is applicable for the use as such, the cost of the analysis may be a minor concern compared to the greater benefits obtained in food analysis.

References

Ababouch, L., Gandini, G., Ryder, J., 2005. Annex 4: Microbiological standards (European Unoin): existing and upcoming standards. Food and Agriculture Organization of the United Nations, Rome.

Abbaspour, A., Norouz-Sarvestani, F., Noori, A., Soltani, N., 2015. Aptamer-conjugated silver nanoparticles for electrochemical dual-aptamer-based sandwich detection of *Staphylococcus aureus*. Biosens. Bioelectron. 68, 149–155.

Alivisatos, A.P., Gu, W., Larabell, C., 2005. Quantum dots as cellular probes. Ann. Rev. Biomed. Eng. 7, 55–76.

American Public Health Association, (APHA), 2005. Membrane filter technique for members of the coliform group. American Public Health Association, Washington, DC.

American Public Health Association (APHA), 2007. Samples. American Public Health Association, Washington, DC.

Anniballi, F., Auricchio, B., Woudstra, C., Fach, P., Fiore, A., Skarin, H., Bano, L., Segerman, B., Knutsson, R., De Medici, D., 2013. Multiplex real-time PCR for detecting and typing *Clostridium botulinum* group III organisms and their mosaic variants. Biosecur. Bioterror. 11, S207–S214.

Argentina, A., 2014. Rapid Detection of *Vibrio parahaemolyticus* Using Bacteriophage and Real-time PCR. Ontario, Canada, MSc Thesis, The University of Guelph.
Bailey, R.C., Nam, J.-M., Mirkin, C.A., Hupp, J.T., 2003. Real-time multicolor DNA detection with chemoresponsive diffraction gratings and nanoparticle probes. J. Am. Chem. Soc. 125, 13541–13547.
Baker, J.M., Griffiths, M.W., Collins-Thompson, D.L., 1992. Bacterial bioluminescence: applications in food microbiology. J. Food Protect. 55, 62–70.
Banerjee, P., Bhunia, A.K., 2010. Cell-based biosensor for rapid screening of pathogens and toxins. Biosens. Bioelectron. 26, 99–106.
Barrichello, A., Allil, M., 1997. Bioluminescência: uma nova ferramenta para tornar o controle microbiológico mais rápido, fácil e preciso. Rev. Inst. Lat. Cândido Tostes 52, 71–79.
Bennett, R.W., 2005. *Staphylococcal enterotoxin* and its rapid identification in foods by enzyme-linked immunosorbent assay-based methodology. J. Food Protect. 68, 1264–1270.
Bina, J., Zhu, J., Dziejman, M., Faruque, S., Calderwood, S., Mekalanos, J., 2003. ToxR regulon of *Vibrio cholerae* and its expression in vibrios shed by cholera patients. Proc. Natl. Acad. Sci. USA 100, 2801–2806.
Binet, R., Deer, D.M., Uhlfelder, S.J., 2014. Rapid detection of *Shigella* and enteroinvasive *Escherichia coli* in produce enrichments by a conventional multiplex PCR assay. Food Microbiol. 40, 48–54.
Biomerieux 2014. AFNOR certification validation study simple method for *Salmonella* (SMS®) for the detection of *Salmonella* spp. in food, feed, and environmental samples: summary report–V1. Association Française de Normalisation (AFNOR)
BIO-RAD, 2015. RAPID' Salmonella. AFNOR certification reference No. BRD 07/11-12/15. Available from : http://nf-validation.afnor.org/en/wp-content/uploads/sites/2/2014/03/BRD-07-11-12-05_en.pdf
Blanco-Abad, V., Ansede-Bermejo, J., Rodriguez-Castro, A., Martinez-Urtaza, J., 2009. Evaluation of different procedures for the optimized detection of *Vibrio parahaemolyticus* in mussels and environmental samples. Int. J. Food Microbiol. 129, 229–236.
Blank, M., Weinschenk, T., Priemer, M., Schluesener, H., 2001. Systematic evolution of a DNA aptamer binding to rat brain tumor microvessels: selective targeting of endothelial regulatory protein pigpen. J. Biol. Chem. 276, 16464–16468.
Bok, S., Korampally, V., Darr, C.M., Folk, W.R., Polo-Parada, L., Gangopadhyay, K., Gangopadhyay, S., 2013. Femtogram-level detection of *Clostridium botulinum* neurotoxin type A by sandwich immunoassay using nanoporous substrate and ultra-bright fluorescent suprananoparticles. Biosens. Bioelectron. 41, 409–416.
Bottone, E.J., 2010. *Bacillus cereus*, a volatile human pathogen. Clin. Microbiol. Rev. 23, 382–398.
Brehm-Stecher, B.F., 2008. Methods for whole cell detection of microorganisms. In: Camesano, T.A., and Mello, C.M. (Eds.), Microbial Surfaces: Structure, Interactions, and Reactivity, vol. 984, ACS Symposium Series, pp. 29–51.
Brunt, J., Webb, M.D., Peck, M.W., 2010. Rapid affinity immunochromatography column-based tests for sensitive detection of *Clostridium botulinum* neurotoxins and *Escherichia coli* O157. Appl. Environ. Microbiol. 76, 4143–4150.
Bsat, N., Wiedmann, M., Czajka, J., Barany, F., Piani, M., Batt, C.A., 1994. Food safety applications of nucleic acid-based assays: polymerase and ligase chain reaction can help detect foodborne pathogens and track isolates to their source of infection. Food Technol. 48, 142–145.

Bülte, M., Reuter, G., 1985. The bioluminescence technique as a rapid method for the determination of the microflora of meat. Int. J. Food Microbiol. 2, 371–381.
Byrne, B., Gilmartin, N., Lakshmanan, R.S., O'Kennedy, R., 2015. Antibodies, enzymes, and nucleic acid sensors for high throughput screening of microbes and toxins in food. In: Bhunia, A.K., Kim, M.S., Taitt, C.R. (Eds.), High Throughput Screening for Food Safety Assessment. Woodhead Publishing.
Candlish, A.A.G., 1991. Immunological methods in food microbiology. Food Microbiol. 8, 1–14.
Carroll, K.C., 2008. Rapid diagnostics for methicillin-resistant *Staphylococcus aureus*: current status. Mol. Diag. Ther. 12, 15–24.
Center for Food Safety and Applied Nutrition (CFSAN), 2005. Bad Bug Book. Center for Food Safety & Applied Nutrition, Foodborne Pathogenic Microorganisms and Natural Toxins Handbook. Available from: http://www.cfsan.fda.gov/~mow/intro.html
Chang, Y.-C., Yang, C.-Y., Sun, R.-L., Cheng, Y.-F., Kao, W.-C., Yang, P.-C., 2013. Rapid single-cell detection of *Staphylococcus aureus* by aptamer-conjugated gold nanoparticles. Sci. Report. 3, 1863.
Chen, S., Wang, F., Beaulieu, J.C., Stein, R.E., GE, B., 2011. Rapid detection of viable *Salmonellae* in produce by coupling propidium monoazide with loop-mediated isothermal amplification. Appl. Environ. Microbiol. 77, 4008–4016.
Chen, J., Wu, X., Huang, Y.-W., Zhao, Y., 2014. Detection of *E. coli* using SERS active filters with silver nanorod array. Sens. Actuat. B: Chem. 191, 485–490.
Cheng, Y., Liu, Y., Huang, J., LI, K., Zhang, W., Xian, Y., Jin, L., 2009. Combining biofunctional magnetic nanoparticles and ATP bioluminescence for rapid detection of *Escherichia coli*. Talanta 77, 1332–1336.
Cho, I.-H., Irudayaraj, J., 2013. In-situ immuno-gold nanoparticle network ELISA biosensors for pathogen detection. Int. J. Food Microbiol. 164, 70–75.
Cho, I.-H., Bhandari, P., Patel, P., Irudayaraj, J., 2015. Membrane filter-assisted surface enhanced Raman spectroscopy for the rapid detection of *E. coli* O157:H7 in ground beef. Biosens. Bioelectron. 64, 171–176.
Choi, J.W., Park, J.H., Lee, W., OH, B.K., Min, J., Lee, W.H., 2001. Fluorescence immunoassay of HDL and LDL using protein A LB film. J. Microbiol. Biotech. 11, 979–985.
Chon, J.W., Park, J.S., Hyeon, J.Y., Park, C., Song, K.Y., Hong, K.W., Hwang, I.G., Kwak, H.S., Seo, K.S., 2012. Development of real-time PCR for the detection of *Clostridium perfringens* in meats and vegetables. J. Microbiol. Biotech. 22, 5.
Chowdhury, A.D., De, A., Chaudhuri, C.R., Bandyopadhyay, K., Sen, P., 2012. Label-free polyaniline based impedimetric biosensor for detection of *E. coli* O157:H7 bacteria. Sens. Actuat. B: Chem. 171–172, 916–923.
Ciebin, B.W., Brodsky, M.H., Eddington, R., Horsnell, G., Choney, A., Palmateer, G., Ley, A., Joshi, R., Shears, G., 1995. Comparative evaluation of modified m-FC and m-TEC media for membrane filter enumeration of *Escherichia coli* in water. Appl. Environ. Microbiol. 61, 3940–3942.
CODEX Alimentarius, 2007. Guidelines on the application of general principles of food hygiene to the control of *Listeria monocytogenes* in foods. CODEX alimentarius. Available from : http://www.fao.org/input/download/standards/10740/CXG_061e.pdf
Colquhoun, K.O., Timms, S., Fricker, C.R., 1998. A simple method for the comparison of commercially available ATP hygiene-monitoring systems. J. Food Protect. 61, 499–501.
Costa, P.D., Andrade, N.J., Brandão, S.C.C., Passos, F.J.V., Soares, N.D.F.F., 2006. ATP-bioluminescence assay as an alternative for hygiene-monitoring

procedures of stainless-steel milk contact surfaces. Braz. J. Microbiol. 37, 345–349.
Costa, A., Cagnès, S., Vivier, C., Fillet, R., Chatellier, S., 2008. Performance of the TEMPO® STA method in comparison with BAM and alternative methods for enumeration of *Staphylococcus aureus* in Food. Available from: http://www.biomerieux-usa.com/sites/subsidiary_us/files/doc/tempo_sta_-_iafp08_22.07.08_ac-1.pdf
Cox, J., Fleet, G.H., 1997. New directions in the microbiological analysis of foods. In: Hocking, A.D. (Ed.), Foodborne Micriorganisms of Public Health Significance. sixth ed. Australian Institute of Food Science & Technology, Sydney, pp. 71–138.
Dasti, J.I., Tareen, A.M., Lugert, R., Zautner, A.E., Gross, U., 2010. *Campylobacter jejuni*: a brief overview on pathogenicity-associated factors and disease-mediating mechanisms. Int. J. Med. Microbiol. 300, 205–211.
Day, J.B., Basavanna, U., 2015. Real-time PCR detection of *Listeria monocytogenes* in infant formula and lettuce following macrophage-based isolation and enrichment. J. Appl. Microbiol. 118, 233–244.
Debnath, M., Prasad, G.B.K.S., Bisen, P.S., 2010. Molecular Diagnostics: Promises and Possibilities. Springer, Netherlands.
Deng, S., Upadhyayula, V.K.K., Smith, G.B., Mitchell, M.C., 2008. Adsorption equilibrium and kinetics of microorganisms on single-wall carbon nanotubes. IEEE Sens. J. 8, 954–962.
Deokar, A.R., Lin, L.-Y., Chang, C.-C., Ling, Y.-C., 2013. Single-walled carbon nanotube coated antibacterial paper: preparation and mechanistic study. J. Mater. Chem. B 1, 2639–2646.
Disley, D.M., Cullen, D.C., You, H.-X., Lowe, C.R., 1998. Covalent coupling of immunoglobulin G to self-assembled monolayers as a method for immobilizing the interfacial-recognition layer of a surface plasmon resonance immunosensor. Biosens. Bioelectron. 13, 1213–1225.
Dolzani, L., Tonin, E., Lagatolla, C., Monti-Bragadin, C., 1994. Typing of *Staphylococcus aureus* by amplification of the 16S–23S rRNA intergenic spacer sequences. FEMS Microbiol. Lett. 119, 167–173.
Drummond, T.G., Hill, M.G., Barton, J.K., 2003. Electrochemical DNA sensors. Nat. Biotech. 21, 1192–1199.
El-Jakee, J., Marouf, S.A., Ata, N.S., Abdel-Rahman, E.H., Abd El-Moez, S.I., Samy, A.A., El-Sayed, W.E., 2013. Rapid method for detection of *Staphylococcus aureus* enterotoxins in food. Global Veterinaria 11, 7.
Elghanian, R., Storhoff, J.J., Mucic, R.C., Letsinger, R.L., Mirkin, C.A., 1997. Selective colorimetric detection of polynucleotides based on the distance-dependent optical properties of gold nanoparticles. Science 277, 1078–1081.
Ellis, P., Meldrum, R., 2002. Comparison of the compact dry TC and 3M petrifilm ACP dry sheet media methods with the spiral plate method for the examination of randomly selected foods for obtaining aerobic colony counts. J. Food Protect. 65, 423–425.
European Commission (EC), 2005. Commission Regulation (EC) No 2073/2005 of November 15, 2005 on microbiological criteria for foodstuffs. Available from : http://eur-lex.europa.eu/legal-content/EN/ALL/?uri=CELEX%3A32005R2073
Fach, P., Fenicia, L., Knutsson, R., Wielinga, P.R., Anniballi, F., Delibato, E., Auricchio, B., Woudstra, C., Ågren, J., Segerman, B., de Medici, D., van Rotterdam, B.J., 2011. An innovative molecular detection tool for tracking and tracing *Clostridium botulinum* types A, B, E, F and other botulinum neurotoxin producing *Clostridia* based on the GeneDisc cycler. Int. J. Food Microbiol. 145 (Suppl. 1), S145–S151.

Fedio, W., Blackstone, G.M., Kikuta-Oshima, L., Wendakoon, C., Mcgrath, T.H., Depaola, A., 2007. Rapid detection of the *Vibrio cholerae ctx* Gene in food enrichments using real-time polymerase chain reaction. J. AOAC Int. 90, 1278–1283.

Feldsine, P.T., Forgey, R.L., Falbo-Nelson, M.T., Brunelle, S.L., 1997. *Escherichia coli* O157:H7 visual immunoprecipitate assay: a comparative validation study. J. AOAC Int. 80, 43–48.

Feng, P., 1997. Impact of molecular biology on the detection of foodborne pathogens. Mol. Biotechnol. 7, 267–278.

Firstenberg-Eden, R., 1985. Electrical impedance method for determining microbial quality of foods. In: Habermehl, K.O. (Ed.), Rapid Methods and Automation in Microbiology and Immunology. Springer, Berlin Heidelberg.

Flint, S., Naila, A., Bashir, R., 2015. Impedance microbiology and microbial screening strategy for detecting pathogens in food. In: Bhunia, A.K., Kim, M.S., Taitt, C.R. (Eds.), High Throughput Screening for Food Safety Assessment. Woodhead Publishing.

Food Safety Authority of Ireland (FSAI), 2014. Guidance note no.3: guidelines for the interpretation of results of microbiological testing of ready-to-eat foods placed on the market (revision 1). Food Safety Authority of Ireland. Available from : https://www.fsai.ie/publications_GN3_microbiological_limits/

Foti, D., Romano, L., Alles, S., Mozola, M.A., 2012. Validation of the Soleris® *E. coli* method for detection and semi-quantitative determination of *Escherichia coli* in foods. J. AOAC Int. 95, 786–794.

Fratamico, P., DebRoy, C., 2010. Detection of *Escherichia coli* O157:H7 in food using real-time multiplex PCR assays targeting the *stx 1, stx 2, wzy* O157, and the *fliC* h7 or *eae* Genes. Food Anal. Method. 3, 330–337.

Fung, D.Y.C., 1969. Rapid methods for determining *Staphylococcal* toxins and *Salmonellae*-associated with poultry products. PhD dissertation, Iowa State University.

Gahan, C.G., 2012. The bacterial lux reporter system: applications in bacterial localization studies. Curr. Gene Ther. 12, 12–19.

Gehring, A.G., Albin, D.M., Irwin, P.L., Reed, S.A., TU, S.-I., 2006. Comparison of enzyme-linked immunomagnetic chemiluminescence with U.S. Food and Drug Administration's Bacteriological Analytical Manual method for the detection of *Escherichia coli* O157:H7. J. Microbiol. Method. 67, 527–533.

Gianfranceschi, M.V., Rodriguez-Lazaro, D., Hernandez, M., González-García, P., Comin, D., Gattuso, A., Delibato, E., Sonnessa, M., Pasquali, F., Prencipe, V., Sreter-Lancz, Z., Saiz-Abajo, M.-J., Pérez-De-Juan, J., Butrón, J., Kozačinski, L., Tomic, D.H., Zdolec, N., Johannessen, G.S., Jakočiūnė, D., Olsen, J.E., De Santis, P., Lovari, S., Bertasi, B., Pavoni, E., Paiusco, A., De Cesare, A., Manfreda, G., De Medici, D., 2014. European validation of a real-time PCR-based method for detection of *Listeria monocytogenes* in soft cheese. Int. J. Food Microbiol. 184, 128–133.

Giannella, R., 1996. Salmonella. In: Baron, S. et al., (Ed.), Barons Medical Microbiology. University of Texas, Medical Branch, Texas.

Gilmartin, N., O'Kennedy, R., 2012. Nanobiotechnologies for the detection and reduction of pathogens. Enzyme Microb. Technol. 50, 87–95.

González-Escalona, N., Hammack, T.S., Russell, M., Jacobson, A.P., DE Jesús, A.J., Brown, E.W., Lampel, K.A., 2009. Detection of live *Salmonella* sp. cells in produce by a TaqMan-based quantitative reverse transcriptase real-time PCR targeting invA mRNA. Appl. Environ. Microbiol. 75, 3714–3720.

Grass, J.E., Gould, L.H., Mahon, B.E., 2013. Epidemiology of foodborne disease outbreaks caused by *Clostridium perfringens*, United States, 1998–2010. Foodborne Pathog. Dis. 10, 131–136.

Grenda, T., Kukier, E., Goldsztejn, M., Kwiatek, K., 2014. In-house validation of PCR-based procedure for specific detection of *Clostridium botulinum* types C and D. Bull. Vet. Inst. Pulawy 58 (3), 393–398.
Griffiths, M.W., 1993. Applications of bioluminescence in the dairy industry. J. Dairy Sci. 76, 3118–3125.
Guran, H.S., Oksuztepe, G., 2013. Detection and typing of Clostridium perfringens from retail chicken meat parts. Lett. Appl. Microb. 57, 77–82.
Guven, B., Basaran-Akgul, N., Temur, E., Tamer, U., Boyac, I.H., 2011. SERS-based sandwich immunoassay using antibody coated magnetic nanoparticles for *Escherichia coli* enumeration. Analyst 136, 740–748.
Hachich, E.M., Di Bari, M., Christ, A.P.G., Lamparelli, C.C., Ramos, S.S., Sato, M.I.Z., 2012. Comparison of thermotolerant coliforms and *Escherichia coli* densities in freshwater bodies. Braz. J. Microb. 43, 675–681.
Hale, T.L., Keusch, G.T., 1996. Shigella. In: Baron, S. (Ed.), Medical Microbiology. fourth ed. University of Texas Medical Branch at Galveston, Galveston, TX, Available from: http://www.ncbi.nlm.nih.gov/books/NBK8038.
Hall, R.H., 2002. Biosensor technologies for detecting microbiological foodborne hazards. Microb. Infect. 4, 425–432.
Hawronskyj, J.-M., Holah, J., 1997. ATP: a universal hygiene monitor. Trends Food Sci. Technol. 8, 79–84.
Hein, I., Lehner, A., Rieck, P., Klein, K., Brandl, E., Wagner, M., 2001. Comparison of different approaches to quantify *Staphylococcus aureus* cells by real-time quantitative PCR and application of this technique for examination of cheese. Appl. Environ. Microbiol. 67, 3122–3126.
Heo, E.J., Song, B.R., Park, H.J., Kim, Y.J., Moon, J.S., Wee, S.H., Kim, J.-S., Yoon, Y., 2014. Rapid detection of *Listeria monocytogenes* by real-time PCR in processed meat and dairy products. J. Food Protect. 77, 453–458.
Hill, W.E., Datta, A.R., Feng, P., Lampel, K.A., Payne, W.L., 1998. Identificationof foodborne bacterial pathogens by gene probes. AOAC International, Gaithershurg, MD.
Hobson, N.S., Tothill, I., Turner, A.P.F., 1996. Microbial detection. Biosens. Bioelectron. 11, 455–477.
Hu, C., Dou, W., Zhao, G., 2014. Enzyme immunosensor based on gold nanoparticles electroposition and Streptavidin-biotin system for detection of *S. pullorum* and *S. gallinarum*. Electrochim. Acta. 117, 239–245.
Hu, X., Xu, B., Yang, Y., Liu, D., Yang, M., Wang, J., Shen, H., Zhou, X., Ma, X., 2013a. A high throughput multiplex PCR assay for simultaneous detection of seven aminoglycoside-resistance genes in *Enterobacteriaceae*. BMC Microbiol. 13, 58–158.
Hu, Y., Meng, J., Shi, C., Hervin, K., Fratamico, P.M., Shi, X., 2013b. Characterization and comparative analysis of a second thermonuclease from *Staphylococcus aureus*. Microbiol. Res. 168, 174–182.
International Commission on Microbiological Specification for Food (ICMSF), 2011. Microorganisms in Foods 8: Use of Data for Assessing Process Control and Product Acceptance. Springer, US.
Ingianni, A., Floris, M., Palomba, P., Madeddu, M.A., Quartuccio, M., Pompei, R., 2001. Rapid detection of *Listeria monocytogenes* in foods, by a combination of PCR and DNA probe. Mole. Cell. Probes 15, 275–280.
Institute of medicine (US) and National Research Council (US) Committee on the Review of the use of scientific criteria and performance standards for safe food, 2003. Food and Drug Administration and Environmental Protection Agency Guidance Levels for Seafood, National Academies Press, Washington, DC. Available from: http://www.ncbi.nlm.nih.gov/books/NBK221561/

Jadhav, S., Bhave, M., Palombo, E.A., 2012. Methods used for the detection and subtyping of *Listeria monocytogenes*. J. Microb. Method. 88, 327–341.

Janda, J.M., Abbott, S.L., 2002. Bacterial identification for publication: when is enough, enough? J. Clin. Microb. 40, 1887–1891.

Jayasena, S.D., 1999. Aptamers: an emerging class of molecules that rival antibodies in diagnostics. Clin. Chem. 45, 1628–1650.

Jeong, Y., Jung, H., Hong, J.-H., 2013. Multiplex real-time polymerase chain reaction for rapid detection of *Staphylococcus aureus*, *Vibrio parahaemolyticus*, and *Salmonella typhimurium* in milk and kimbap. J. Korean Soc. Appl. Biol. Chem. 56, 715–721.

Jiménez, K.B., Mccoy, C.B., Achí, R., 2010. Detection of *Shigella* in lettuce by the use of a rapid molecular assay with increased sensitivity. Braz. J. Microb. 41, 993–1000.

Joint FAO/IAEA Expert Consultation,1997. Validation of analytical methods for food control. In: Nations, F.A.A.O.O.T.U. (Ed.), Vienna, Austria: Food and Agriculture Organization.

Josefsen, M.H., Löfström, C., Hansen, T.B., Christensen, L.S., Olsen, J.E., Hoorfar, J., 2010. Rapid quantification of viable *Campylobacter* bacteria on chicken carcasses, using real-time PCR and propidium monoazide treatment, as a tool for quantitative risk assessment. Appl. Environ. Microbiol. 76, 5097–5104.

Joung, C.-K., Kim, H.-N., Lim, M.-C., Jeon, T.-J., Kim, H.-Y., Kim, Y.-R., 2013. A nanoporous membrane-based impedimetric immunosensor for label-free detection of pathogenic bacteria in whole milk. Biosens. Bioelectron. 44, 210–215.

Jyoung, J.-Y., Hong, S., Lee, W., Choi, J.-W., 2006. Immunosensor for the detection of *Vibrio cholerae* O1 using surface plasmon resonance. Biosens. Bioelectron. 21, 2315–2319.

Kaittanis, C., Santra, S., Perez, J.M., 2010. Emerging nanotechnology-based strategies for the identification of microbial pathogenesis. Adv. Drug Deliv. Rev. 62, 408–423.

Kalyan Kumar, T.D., Murali, H.S., Batra, H.V., 2010. Multiplex PCR assay for the detection of enterotoxic *Bacillus cereus* group strains and its application in food matrices. Ind. J. Microb. 50, 165–171.

Kaminska, P.S., Fiedoruk, K., Jankowska, D., Mahillon, J., Nowosad, K., Drewicka, E., Zambrzycka, M., Swiecicka, I., 2015. One-day pulsed-field gel electrophoresis protocol for rapid determination of emetic *Bacillus cereus* isolates. Electrophoresis 36, 1051–1054.

Kaneko, I., Miyamoto, K., Mimura, K., Yumine, N., Utsunomiya, H., Akimoto, S., Mcclane, B.A., 2011. Detection of enterotoxigenic *Clostridium perfringens* in meat samples by using molecular methods. Appl. Environ. Microbiol. 77, 7526–7532.

Keer, J.T., Birch, L., 2003. Molecular methods for the assessment of bacterial viability. J. Microb. Method. 53, 175–183.

Kelley, S.O., Boon, E.M., Barton, J.K., Jackson, N.M., Hill, M.G., 1999. Single-base mismatch detection based on charge transduction through DNA. Nucleic Acids Res. 27, 4830–4837.

Kim, N., Park, I.-S., Kim, W.-Y., 2007. *Salmonella* detection with a direct-binding optical grating coupler immunosensor. Sens. Actuat. B-Chem. 121, 606–615.

Kim, J.-M., Forghani, F., Kim, J.-B., Park, Y.-B., Park, M.-S., Wang, J., Park, J., OH, D.-H., 2012. Improved multiplex PCR assay for simultaneous detection of *Bacillus cereus* emetic and enterotoxic strains. Food Sci. Biotechnol. 21, 1439–1444.

Kim, G., Moon, J.-H., Moh, C.-Y., Lim, J.-G., 2015. A microfluidic nano-biosensor for the detection of pathogenic *Salmonella*. Biosens. Bioelectron. 67, 243–247.

Kinneberg, K.M., Lindberg, K.G., 2002. Dry rehydratable film method for rapid enumeration of Coliforms in foods (3M ™) Petrifilm™ Rapid Coliform Count plate: collaborative study. J. AOAC Int. 85, 56–71.

Klein, P.G., Juneja, V.K., 1997. Sensitive detection of viable *Listeria monocytogenes* by reverse transcription-PCR. Appl. Environ. Microbiol. 63, 4441–4448.

Kotiranta, A., Lounatmaa, K., Haapasalo, M., 2000. Epidemiology and pathogenesis of *Bacillus cereus* infections. Microb. Infect. 2, 189–198.

Kuo, Y.-C., Wang, Q., Ruengruglikit, C., Yu, H., Huang, Q., 2008. Antibody-conjugated CdTe quantum dots for *Escherichia coli* detection. J. Phys. Chem. C 112, 4818–4824.

Kyriakides, A.L., Patel, P.D., 1995. Luminescence techniques for microbiological analysis of foods. In: Patel, P.D. (Ed.), Rapid Analysis Techniques in Food Microbiology. Springer, US.

Laczka, O.F., Labbate, M., Seymour, J.R., Bourne, D.G., Fielder, S.S., Doblin, M.A., 2014. Surface immuno-functionalisation for the capture and detection of *Vibrio* species in the marine environment: a new management tool for industrial facilities. PLoS ONE 9, e108387.

LaRocco, K.A., Littel, K.J., Pierson, M.D., 1986. The bioluminescent ATP assay for determining the microbial quality of foods. In: Pierson, M.D., Stern, N.J. (Eds.), Foodborne Microorganisms and their Toxins: Developing Methodology. Marcel Dekker, New York, pp. 145–175.

Lauer, W.F., Gary, S.M.T., Patel, A., 2007. Validation of RAPID' Staph for Enumeration of Staphylococcus aureus in Foods. J. AOAC Int. 90 (2), 414–426.

Leach, K.M., Stroot, J.M., Lim, D.V., 2010. Same-day detection of *Escherichia coli* O157:H7 from spinach by using electrochemiluminescent and cytometric bead array biosensors. Appl. Environ. Microbiol. 76, 8044–8052.

Lee, J.-Y., Jin, J.-H., Kim, J.H., Kim, M.J., Lee, C.J., Min, N.K., 2012. Plasma-activated carbon nanotube-based high sensitivity immunosensors for monitoring *Legionella pneumophila* by direct detection of maltose binding protein peptidoglycan-associated lipoprotein (MBP-PAL). Biotech. Bioeng. 109, 1471–1478.

Lee, S.-H., Ahn, J.-Y., Lee, K.-A., Um, H.-J., Sekhon, S.S., Sun Park, T., Min, J., Kim, Y.-H., 2015. Analytical bioconjugates, aptamers, enable specific quantitative detection of *Listeria monocytogenes*. Biosens. Bioelectron. 68, 272–280.

Lermo, A., Campoy, S., Barbé, J., Hernández, S., Alegret, S., Pividori, M.I., 2007. In situ DNA amplification with magnetic primers for the electrochemical detection of food pathogens. Biosens. Bioelectron. 22, 2010–2017.

Li, Y., Afrasiabi, R., Fathi, F., Wang, N., Xiang, C., Love, R., She, Z., Kraatz, H.-B., 2014. Impedance based detection of pathogenic *E. coli* O157:H7 using a ferrocene-antimicrobial peptide modified biosensor. Biosens. Bioelectron. 58, 193–199.

Lian, Y., He, F., Wang, H., Tong, F., 2015. A new aptamer/graphene interdigitated gold electrode piezoelectric sensor for rapid and specific detection of *Staphylococcus aureus*. Biosens. Bioelectron. 65, 314–319.

Liang, P.-S., Park, T.S., Yoon, J.-Y., 2014. Rapid and reagentless detection of microbial contamination within meat utilizing a smartphone-based biosensor. Sci. Report 4, 5953.

Lindqvist, R., 1999. Detection of *Shigella* spp. in food with a nested PCR method: sensitivity and performance compared with a conventional culture method. J. Appl. Microbiol. 86, 971–978.

Lindström, M., Keto, R., Markkula, A., Nevas, M., Hielm, S., Korkeala, H., 2001. Multiplex PCR assay for detection and identification of *Clostridium botulinum* types A, B, E, and F in food and fecal material. Appl. Environ. Microbiol. 67, 5694–5699.

Liu, Y., 2007. Electric DNA arrays for determination of pathogenic *Bacillus cereus [Elektronisk resurs]*. KTH, Stockholm.
Liu, H., Zhan, F., Liu, F., Zhu, M., Zhou, X., Xing, D., 2014. Visual and sensitive detection of viable pathogenic bacteria by sensing of RNA markers in gold nanoparticles based paper platform. Biosens. Bioelectron. 62, 38–46.
Liu, J., Lu, L., Xu, S., Wang, L., 2015. One-pot synthesis of gold nanoclusters with bright red fluorescence and good biorecognition abilities for visualization fluorescence enhancement detection of *E. coli*. Talanta 134, 54–59.
Maerle, A.V., Ryazantsev, D.Y., Dmitrenko, O.A., Petrova, E.E., Komaleva, R.L., Sergeev, I.V., Trofimov, D.Y., Zavriev, S.K., 2014. Detection of *Staphylococcus aureus* toxins using immuno-PCR. Russ. J. Bioorg. Chem. 40, 526–531.
Maheux, A.F., Bissonnette, L., Boissinot, M., Bernier, J.-L.T., Huppé, V., Picard, F.J., Bérubé, È., Bergeron, M.G., 2011. Rapid concentration and molecular enrichment approach for sensitive detection of *Escherichia coli* and *Shigella* species in potable water samples. Appl. Environ. Microbiol. 77, 6199–6207.
Malorny, B., Bunge, C., Helmuth, R., 2003. Evaluation of *Salmonella* spp.: Specific Primer-Sets for Validation within the Food PCR Project. Federal Institute for Health Protection of Consumers and Veterinary Medicine, National Reference Laboratory for Salmonella, Berlin, Germany.
Malorny, B., Paccassoni, E., Fach, P., Bunge, C., Martin, A., Helmuth, R., 2004. Diagnostic real-time PCR for detection of *Salmonella* in food. Appl. Environ. Microbiol. 70, 7046–7052.
Mandal, P., Biswas, A., Choi, K., Pal, U.K., 2011. Methods for rapid detection of foodborne pathogens: an overview. Am. J. Food Technol. 6, 87–102.
Mäntynen, V., Lindström, K., 1998. A rapid PCR-based DNA test for enterotoxic *Bacillus cereus*. Appl. Environ. Microbiol. 64, 1634–1639.
McDaniels, A.E., Bordner, R.H., Menkedick, J.R., Weber, C.I., 1987. Comparison of the hydrophobic-grid membrane filter procedure and standard methods for coliform analysis of water. Appl. Environ. Microbiol. 53, 1003–1009.
Meighen, E.A., 1993. Bacterial bioluminescence: organization, regulation, and application of the *lux* genes. FASEB J. 7, 1016–1022.
Meng, J., Doyle, M.P., 2002. Introduction: microbiological food safety. Microb. Infect. 4, 395–397.
Minunni, M., Tombelli, S., Scielzi, R., Mannelli, I., Mascini, M., Gaudiano, C., 2003. Detection of β-thalassemia by a DNA piezoelectric biosensor coupled with polymerase chain reaction. Anal. Chim. Acta 481, 55–64.
Miyamoto, T., Miwa, H., Hatano, S., 1990. Improved fluorogenic assay for rapid detection of *Vibrio parahaemolyticus* in foods. Appl. Environ. Microbiol. 56, 1480–1484.
Morris, K.N., Jensen, K.B., Julin, C.M., Weil, M., Gold, L., 1998. High-affinity ligands from in vitro selection: complex targets. Proc. Natl. Acad. Sci. 95, 2902–2907.
Muhammad-Tahir, Z., Alocilja, E.C., 2003. Fabrication of a disposable biosensor for *Escherichia coli* O157:H7 detection. IEEE Sens. J. 3, 345–351.
Muyzer, G., Smalla, K., 1998. Application of denaturing gradient gel electrophoresis (DGGE) and temperature gradient gel electrophoresis (TGGE) in microbial ecology. Antonie van Leeuwenhoek 73, 127–141.
Najafi, R., Mukherjee, S., Hudson, Jr., J., Sharma, A., Banerjee, P., 2014. Development of a rapid capture-cum-detection method for *Escherichia coli* O157 from apple juice comprising nano-immunomagnetic separation in tandem with surface enhanced Raman scattering. Int. J. Food Microbiol. 189, 89–97.
Nanduri, V., Bhunia, A.K., Tu, S.-I., Paoli, G.C., Brewster, J.D., 2007. SPR biosensor for the detection of *L. monocytogenes* using phage-displayed antibody. Biosens. Bioelectron. 23, 248–252.

Naravaneni, R., Jamil, K., 2005. Rapid detection of food-borne pathogens by using molecular techniques. J. Med. Microbiol. 54, 51–54.
Nguyen, P.-D., Tran, T.B., Nguyen, D.T.X., Min, J., 2014. Magnetic silica nanotube-assisted impedimetric immunosensor for the separation and label-free detection of *Salmonella typhimurium*. Sens. Actuat. B-Chem. 197, 314–320.
NIN 2009. Nanotechnology-enabled sensing. Report of the National Nanotechnology Initiative Workshop. 5–7 May 2009.
Niyogi, S., Hamon, M.A., Hu, H., Zhao, B., Bhowmik, P., Sen, R., Itkis, M.E., Haddon, R.C., 2002. Chemistry of single-walled carbon nanotubes. Accounts Chem. Res. 35, 1105–1113.
Nocker, A., Mazza, A., Masson, L., Camper, A.K., Brousseau, R., 2009. Selective detection of live bacteria combining propidium monoazide sample treatment with microarray technology. J. Microbiol. Method 76, 253–261.
Oggel, J.J., Nundy, D.C., Randall, C.J., 1990. Modified 1-2 Test™ system as a rapid screening method for detection of *Salmonella* in foods and feeds. J. Food Protect. 53, 656–658.
Oh, B.-K., Kim, Y.-K., Park, K.W., Lee, W.H., Choi, J.-W., 2004a. Surface plasmon resonance immunosensor for the detection of *Salmonella typhimurium*. Biosens. Bioelectron. 19, 1497–1504.
Oh, B.-K., Lee, W., Kim, Y.-K., Lee, W.H., Choi, J.-W., 2004b. Surface plasmon resonance immunosensor using self-assembled protein G for the detection of *Salmonella paratyphi*. J. Biotechnol. 111, 1–8.
Ohk, S.-H., Bhunia, A.K., 2013. Multiplex fiber optic biosensor for detection of *Listeria monocytogenes, Escherichia coli* O157:H7 and *Salmonella enterica* from ready-to-eat meat samples. Food Microbiol. 33, 166–171.
Ohk, S.H., Koo, O.K., Sen, T., Yamamoto, C.M., Bhunia, A.K., 2010. Antibody–aptamer functionalized fibre-optic biosensor for specific detection of *Listeria monocytogenes* from food. J. Appl. Microbiol. 109, 808–817.
Olsen, J.E., Aabo, S., Hill, W., Notermans, S., Wernars, K., Granum, P.E., Popovic, T., Rasmussen, H.N., Olsvik, Ø., 1995. Probes and polymerase chain reaction for detection of food-borne bacterial pathogens. Int. J. Food Microbiol. 28, 1–78.
Owicki, J.C., Wallace Parce, J., 1992. Biosensors based on the energy metabolism of living cells: the physical chemistry and cell biology of extracellular acidification. Biosens. Bioelectron. 7, 255–272.
Paez, R., Taverna, M., Charlon, V., Cuatrim, A., Etchverry, F., Costa, L.H., 2003. Application of ATP-bioluminescence technique for assessing cleanliness of milking equipment, bulk tank, and milk transport tanker. Food Protect. Trend. 23, 308–314.
Pal, S., Ying, W., Alocilja, E.C., Downes, F.P., 2008. Sensitivity and specificity performance of a direct-charge transfer biosensor for detecting *Bacillus cereus* in selected food matrices. Biosys. Eng. 99, 461–468.
Pettipher, G.L., 1983. The Direct Epifluorescent Filter Technique for the Rapid Enumeration of Micro-organisms. Research Studies Press Ltd., Letchworth, Herts.
Piunno, P.A.E., Krull, U.J., Hudson, R.H.E., Damha, M.J., Cohen, H., 1995. Fiber-optic DNA sensor for fluorometric nucleic acid determination. Anal. Chem. 67, 2635–2643.
Poltronieri, P., Mezzolla, V., Primiceri, E., Maruccio, G., 2014. Biosensors for the detection of food pathogens. Foods 3, 511.
Prakitchaiwattana, C.J., Fleet, G.H., Heard, G.M., 2004. Application and evaluation of denaturing gradient gel electrophoresis to analyze the yeast ecology of wine grapes.
Qi, X., Gao, H., Zhang, Y., Wang, X., Chen, Y., Sun, W., 2012. Electrochemical DNA biosensor with chitosan-Co3O4 nanorod-graphene composite for

the sensitive detection of *Staphylococcus aureus nuc* gene sequence. Bioelectrochemistry 88, 42–47.

Que-Gewirth, N.S., Sullenger, B.A., 2007. Gene therapy progress and prospects: RNA aptamers. Gene Ther. 14, 283–291.

Raghunath, P., 2015. Roles of thermostable direct hemolysin (TDH) and TDH-related hemolysin (TRH) in *Vibrio parahaemolyticus*. Frontiers Microbiol. 5, 805.

Ramachandran, A., Flinchbaugh, J., Ayoubi, P., Olah, G.A., Malayer, J.R., 2004. Target discrimination by surface-immobilized molecular beacons designed to detect *Francisella tularensis*. Biosens. Bioelectron. 19, 727–736.

Ravindranath, S.P., Mauer, L.J., Deb-Roy, C., Irudayaraj, J., 2009. Biofunctionalized magnetic nanoparticle integrated mid-infrared pathogen sensor for food matrixes. Anal. Chem. 81, 2840–2846.

Rivera, V.R., Gamez, F.J., Keener, W.K., White, J.A., Poli, M.A., 2006. Rapid detection of *Clostridium botulinum* toxins A, B, E, and F in clinical samples, selected food matrices, and buffer using paramagnetic bead-based electrochemiluminescence detection. Anal. Biochem. 353, 248–256.

Rodríguez-Lázaro, D., Pla, M., Scortti, M., Monzó, H.J., Vázquez-Boland, J.A., 2005. A novel real-time PCR for *Listeria monocytogenes* that monitors analytical performance via an internal amplification control. Appl. Environ. Microbiol. 71, 9008–9012.

Rompré, A., Servais, P., Baudart, J., De-Roubin, M.-R., Laurent, P., 2002. Detection and enumeration of coliforms in drinking water: current methods and emerging approaches. J. Microbiol. Method. 49, 31–54.

Rosselló-Mora, R., Amann, R., 2001. The species concept for prokaryotes. FEMS Microbiol. Rev. 25 (1), 39–67.

Russo, P., Botticella, G., Capozzi, V., Massa, S., Spano, G., Beneduce, L., 2014. A fast, reliable, and sensitive method for detection and quantification of *Listeria monocytogenes* and *Escherichia coli* O157:H7 in ready-to-eat fresh-cut products by MPN-qPCR. BioMed Res. Int. 2014, 608296.

Sakai, G., Ogata, K., Uda, T., Miura, N., Yamazoe, N., 1998. A surface plasmon resonance-based immunosensor for highly sensitive detection of morphine. Sens. Actuat. B-Chem. 49, 5–12.

Salamon, Z., Macleod, H.A., Tollin, G., 1997. Surface plasmon resonance spectroscopy as a tool for investigating the biochemical and biophysical properties of membrane protein systems. II: Applications to biological systems. Biochimica et Biophysica Acta (BBA) 1331, 131–152.

Satterfield, B.A., Stewart, A.F., Lew, C.S., Pickett, D.O., Cohen, M.N., Moore, E.A., Luedtke, P.F., Apos, Neill, K.L., Robison, R.A., 2010. A quadruplex real-time PCR assay for rapid detection and differentiation of the *Clostridium botulinum* toxin genes A, B, E, and F. J. Med. Microbiol. 59, 55–64.

Schroeder, G.N., Hilbi, H., 2008. Molecular pathogenesis of *Shigella* spp.: controlling host cell signaling, invasion, and death by type III secretion. Clin. Microbiol. Rev. 21, 134–156.

Senachai, P., Chomvarin, C., Wongboot, W., Boonyanugomol, W., Tangkanakul, W., 2013. Duplex PCR for detection of *Salmonella* and *Shigella* spp in cockle samples. SE Asian J. Trop. Med. Publ. Health 44, 866–874.

Setterington, E.B., Alocilja, E.C., 2012. Electrochemical biosensor for rapid and sensitive detection of magnetically extracted bacterial pathogens. Biosensors 2, 15.

Shan, X., Zhang, Y., Zhang, Z., Chen, M., Su, Y., Yuan, Y., Alam, M.J., Yan, H., Shi, L., 2012. Rapid detection of food-borne *Listeria monocytogenes* by real-time quantitative loop-mediated isothermal amplification. Food Sci. Biotech. 21, 101–106.

Sharma, S.K., Eblen, B.S., Bull, R.L., Burr, D.H., Whiting, R.C., 2005. Evaluation of lateral-flow *Clostridium botulinum* neurotoxin detection kits for food analysis. Appl. Environ. Microbiol. 71, 3935–3941.
Sharpe, A.N., 1985. Membrane filter techniques for food microbiological analysis. In: Habermehl, K.O. (Ed.), Rapid Methods and Automation in Microbiology and Immunology. Springer, Berlin Heidelberg.
Sharpe, A.N., 1995. Development and evaluation of membrane filtration techniques in microbial analysis. In: Patel, P.D. (Ed.), Rapid Analysis Techniques in Food Microbiology. Springer, US.
Sharpe, A.N., Michaud, G.L., 1974. Hydrophobic grid-membrane filters: new approach to microbiological enumeration. Appl. Microbiol. 28, 223–225.
Shibajyoti, G.D., 2011. MEMS biosensor for detection of *Escherichia coli* O157:H7 in food products. MSc, University of Missouri, Columbia.
Shimizu, S., Ootsubo, M., Kubosawa, Y., Fuchizawa, I., Kawai, Y., Yamazaki, K., 2009. Fluorescent in situ hybridization in combination with filter cultivation (FISHFC) method for specific detection and enumeration of viable *Clostridium perfringens*. Food Microbiol. 26, 425–431.
Sim, J.H., Kwak, Y.H., Choi, C.H., Paek, S.-H., Park, S.S., Seo, S., 2012. A birefringent waveguide biosensor platform for label-free live cell detection of *Listeria monocytogenes*. Sens. Actuat. B-Chem. 173, 752–759.
Simone, G., Robouch, P., 2014. EURL-FA Guide: Protocol for verification studies of singlelaboratory/in-house validated methods. In: Additives, E.U.R.L.F.F., third ed., European Commission.
Singh, M., Holzinger, M., Tabrizian, M., Winters, S., Berner, N.C., Cosnier, S., Duesberg, G.S., 2015. Noncovalently functionalized monolayer graphene for sensitivity enhancement of surface plasmon resonance immunosensors. J. Am. Chem. Soc. 137, 2800–2803.
Sipponen, P., Ruoslahti, E., Vuento, M., Engvall, E., Stenman, U., 1976. CEA and CEA-like activity in gastric cancer. Acta Hepato-Gastroenterologica 23, 276–279.
Sista, S., Wei, D., Oyarzabal, O., Simonian, A., 2006. Sensitive surface plasmon resonance biosensor for the near-real time detection of *Campylobacter jejuni*. Meeting Abstracts, MA2005-01, 1468.
Smartt, A., Ripp, S., 2011. Bacteriophage reporter technology for sensing and detecting microbial targets. Anal. Bioanal. Chem. 400, 991–1007.
Stager, C.E., Davis, J.R., 1992. Automated systems for identification of microorganisms. Clin. Microbiol. Rev. 5, 302–327.
Stewart, G.S.A.B., Denyer, S.P., Lewington, J., 1991. Microbiology illuminated: gene engineering and bioluminescence. Trends Food Sci. Tech. 2, 7–10.
Su, X.-L., Li, Y., 2004. Quantum dot biolabeling coupled with immunomagnetic separation for detection of *Escherichia coli* O157:H7. Anal. Chem. 76, 4806–4810.
Sundaram, J., Park, B., Kwon, Y., Lawrence, K.C., 2013. Surface-enhanced Raman scattering (SERS) with biopolymer encapsulated silver nanosubstrates for rapid detection of foodborne pathogens. Int. J. Food Microbiol. 167, 67–73.
Sutherland, A.D., Bell, C., Limond, A., Deakin, J., Hunter, E.A., 1994. The biotrace method for estimating bacterial numbers in milk by bioluminescence. Int. J. Dairy Tech. 47, 117–120.
Tansuphasiri, U., 2001. Development of duplex PCR assay for rapid detection of enterotoxigenic isolates of *Clostridium perfringens*. SE Asian J. Trop. Med. Publ. Health 32, 9.
Tawe, J., 2014. Validation of PCR Assays for Setection of *Shiga* Toxin-Producing *E. coli* O104:H4 and O121 in Food. Uppsala University School of Engineering, Uppsala, Available from: https://www.diva-portal.org/smash/get/diva2:705306/FULLTEXT01.pdf.

Taylor, A.D., Ladd, J., Yu, Q., Chen, S., Homola, J., Jiang, S., 2006. Quantitative and simultaneous detection of four foodborne bacterial pathogens with a multichannel SPR sensor. Biosens. Bioelectron. 22, 752–758.

Teles, F.R.R., Fonseca, L.P., 2008a. Applications of polymers for biomolecule immobilization in electrochemical biosensors. Mater. Sci. Eng. C 28, 1530–1543.

Teles, F.R.R., Fonseca, L.P., 2008b. Trends in DNA biosensors. Talanta 77, 606–623.

The FDA Foods Program Science and Research Steering Committee [SRSC], 2011. Guidelines for the validation of analytical methods for the detection of microbial pathogens in foods. In: D.O.H.A.H, Food and Drug Administration.

The Food Safety and Inspection Service (FSIS), 2015. Foodborne Pathogen Test Kits Validated by Independent Organizations. Available from: http://www.fsis.usda.gov/wps/wcm/connect/fsis-content/internet/main/topics/regulatory-compliance/new-technologies/new-technologies

Thermo Fisher Scientific, 2015. Salmonella Precis. AFNOR certification reference No. UNI 03/06-12/07. Available from: http://nf-validation.afnor.org/en/wp-content/uploads/sites/2/2014/03/UNI-03-06-12-07_en.pdf

Todd, E.C., 1997. Epidemiology of foodborne diseases: a worldwide review. World Health Stat. Q. 50, 30–50.

Tolker-Nielsen, T., Larsen, M.H., Kyed, H., Molin, S., 1997. Effects of stress treatments on the detection of *Salmonella typhimurium* by in situ hybridization. Int. J. Food Microbiol. 35, 251–258.

Toyama, S., Shoji, A., Yoshida, Y., Yamauchi, S., Ikariyama, Y., 1998. Surface design of SPR-based immunosensor for the effective binding of antigen or antibody in the evanescent field using mixed polymer matrix. Sens. Actuat. B-Chem. 52, 65–71.

Trnčíková, T., Hrušková, V., Oravcová, K., Pangallo, D., Kaclíková, E., 2009. Rapid and sensitive detection of *Staphylococcus aureus* in food using selective enrichment and real-time PCR targeting a new gene marker. Food Anal. Method 2, 241–250.

Uotila, M., Ruoslahti, E., Engvall, E., 1981. Two-site sandwich enzyme immunoassay with monoclonal antibodies to human alpha-fetoprotein. J. Immunol. Method. 42, 11–15.

Upadhyayula, V.K.K., Deng, S., Mitchell, M.C., Smith, G.B., 2009. Application of carbon nanotube technology for removal of contaminants in drinking water: a review. Sci. Total Environ. 408, 1–13.

U.S. Food and Drug Administration (USFDA), 2011. Fish and Fishery Products Hazards and Controls Guidance, fourth ed. U.S. Department of Health and Human Services FDA, Maryland.

Vanhomwegen, J., Berthet, N., Mazuet, C., Guigon, G., Vallaeys, T., Stamboliyska, R., Dubois, P., Kennedy, G.C., Cole, S.T., Caro, V., Manuguerra, J.-C., Popoff, M.-R., 2013. Application of high-density DNA resequencing microarray for detection and characterization of botulinum neurotoxin-producing *Clostridia*. PLoS ONE 8, e67510.

Vantarakis, A., Komninou, G., Venieri, D., Papapetropoulou, M., 2000. Development of a multiplex PCR detection of *Salmonella* spp. and *Shigella* spp. in mussels. Lett. Appl. Microb. 31, 105–109.

Velge, P., Wiedemann, A., Rosselin, M., Abed, N., Boumart, Z., Chaussé, A.M., Grépinet, O., Namdari, F., Roche, S.M., Rossignol, A., Virlogeux-Payant, I., 2012. Multiplicity of *Salmonella* entry mechanisms, a new paradigm for *Salmonella* pathogenesis. Microbiol. Open. 1, 243–258.

Velusamy, V., Arshak, K., Korostynska, O., Oliwa, K., Adley, C., 2010. An overview of foodborne pathogen detection: in the perspective of biosensors. Biotechnol. Adv. 28, 232–254.

Villamizar, R.A., Maroto, A., Rius, F.X., Inza, I., Figueras, M.J., 2008. Fast detection of *Salmonella Infantis* with carbon nanotube field effect transistors. Biosens. Bioelectron. 24, 279–283.

Vondrakova, L., Pazlarova, J., Demnerova, K., 2014. Detection, identification, and quantification of *Campylobacter jejuni, coli,* and *lari* in food matrices all at once using multiplex qPCR. Gut Path. 6, 1–9.

Wang, J., 2005. Carbon-nanotube-based electrochemical biosensors: a review. Electroanalysis 17, 7–14.

Wang, Q., Kuo, Y., Wang, Y., Shin, G., Ruengruglikit, C., Huang, Q., 2006. Luminescent properties of water-soluble denatured bovine serum albumin-coated CdTe quantum dots. J. Phys. Chem. B 110, 16860–16866.

Wang, D., Fang, Z., Xie, C., Liu, Y., 2013. Construction of method for rapid detection of *Vibrio parahaemolyticus* using the quantitative real-time PCR based on the ToxR gene. Adv. J. Food Sci. Tech. 5, 9.

Wang, D., Dou, W., Zhao, G., Chen, Y., 2014. Immunosensor based on electrodeposition of gold-nanoparticles and ionic liquid composite for detection of *Salmonella pullorum.* J. Microb. Method. 106, 110–118.

Wang, B., Wang, Q., Cai, Z., Ma, M., 2015. Simultaneous, rapid and sensitive detection of three food-borne pathogenic bacteria using multicolor quantum dot probes based on multiplex fluoroimmunoassay in food samples. LWT – Food Sci. Technol. 61, 368–376.

Warren, B., Parish, M., Schneider, K., 2006a. *Shigella* as a foodborne pathogen and current methods for detection in food. Crit. Rev. Food Sci. Nutr. 46, 551–567.

Warren, B.R., Yuk, H.-G., Schneider, K.R., 2006b. Detection of *Shigella sonnei* in selected foods by flow-through immunocapture followed by real-time polymerase chain reaction or isolation on MacConkey ager. J. Rapid Method. Autom. Microbiol. 14, 309–324.

Wawerla, M., Stolle, A., Schalch, B., Eisgruber, H., 1999. Impedance microbiology: applications in food hygiene. J. Food Protect. 62, 1488–1496.

Wellinghausen, N., Siegel, D., Gebert, S., Winter, J., 2009. Rapid detection of *Staphylococcus aureus* bacteremia and methicillin resistance by real-time PCR in whole blood samples. Eur. J. Clin. Microbiol. Infect. Dis. 28, 1001–1005.

World Health Organization (WHO) chronicle, 1996. Guidelines for drinking-water quality. In: Health Criteria and Other Supporting Information, second ed. Geneva: World Health Organization (WHO) Chronicle; 1996.

Wildeboer, D., Amirat, L., Price, R.G., Abuknesha, R.A., 2010. Rapid detection of *Escherichia coli* in water using a hand-held fluorescence detector. Water Res. 44, 2621–2628.

Xu, K., Huang, J., Ye, Z., Ying, Y., Li, Y., 2009. Recent development of nano-materials used in DNA biosensors. Sens. Actuat. B-Chem. 9, 24.

Yang, L., Bashir, R., 2008. Electrical/electrochemical impedance for rapid detection of foodborne pathogenic bacteria. Biotechnol. Adv. 26, 135–150.

Yang, L., Li, Y., 2006. Simultaneous detection of *Escherichia coli* O157:H7 and *Salmonella typhimurium* using quantum dots as fluorescence labels. Analyst 131, 394–401.

Yang, Z.-Y., Shim, W.-B., Kim, K.-Y., Chung, D.-H., 2010. Rapid detection of enterotoxigenic *Clostridium perfringens* in meat samples using immunomagnetic separation polymerase chain reaction (IMS–PCR). J. Agric. Food Chem. 58, 7135–7140.

Yang, Y., Xu, F., Xu, H., Aguilar, Z.P., Niu, R., Yuan, Y., Sun, J., You, X., Lai, W., Xiong, Y., Wan, C., Wei, H., 2013. Magnetic nano-beads based separation combined with propidium monoazide treatment and multiplex PCR assay for simultaneous detection of viable *Salmonella typhimurium, Escherichia coli* O157:H7, and *Listeria monocytogenes* in food products. Food Microbiol. 34, 418–424.

Zhao, G., Xing, F., Deng, S., 2007. A disposable amperometric enzyme immunosensor for rapid detection of *Vibrio parahaemolyticus* in food based on agarose/Nano-Au membrane and screen-printed electrode. Electrochem. Comm. 9, 1263–1268.

Zhao, Y., Ye, M., Chao, Q., Jia, N., Ge, Y., Shen, H., 2009. Simultaneous detection of multifood-borne pathogenic bacteria based on functionalized quantum dots coupled with immunomagnetic separation in food samples. J. Agric. Food Chem. 57, 517–524.

Zhou, B., Xiao, J., Liu, S., Yang, J., Wang, Y., Nie, F., Zhou, Q., Li, Y., Zhao, G., 2013. Simultaneous detection of six food-borne pathogens by multiplex PCR with a GeXP analyzer. Food Control 32, 198–204.

Zhu, S., Du, C., Fu, Y., 2009. Localized surface plasmon resonance-based hybrid Au–Ag nanoparticles for detection of *Staphylococcus aureus* enterotoxin B. Optic. Mater. 31, 1608–1613.

Zoder, P., Petrauskene, O.V., Bolchakova, E., Brzoska, P.M., Brevnov, M.G., Schumaker, M., Nunez, A., Balachandran, P., Furtado, M.R., 2011. Design and validation of a lyophilized multiplex real-time PCR assay for *Campylobacter* pathogen detection and speciation in fast PCR conditions. Available from https://www.thermofisher.com/content/dam/LifeTech/global/applied-sciences/pdfs/food-safety/foodsafety_lifelink_doc_lineitem_22.pdf.

14

CONTAMINANT SENSORS: NANOTECHNOLOGY-BASED CONTAMINANT SENSORS

Niha Mohan Kulshreshtha*, Divya Shrivastava*, Prakash Singh Bisen,†**

**Jaipur National University, School of Life Sciences, Jaipur, Rajasthan, India; **Jiwaji University, School of Studies in Biotechnology, Gwalior, Madhya Pradesh, India; †Tropilite Foods Pvt. Ltd., R&D Division, Gwalior, Madhya Pradesh, India*

1 Introduction

Food diagnostics is an emerging field that applies "modern" methods of detection of microbes, bacteria, chemicals, biotoxins, heavy metals, and prions in all steps of the food chain from raw materials to end products. Nanobiosensor technology has been a major driver used for the analysis of foods. Throughout the world, food production, preparation, and distribution have become increasingly complex, and raw materials are often sourced globally. Changes in food processing techniques, food distribution, and the emergence of new food pathogens have changed the epidemiology of foodborne diseases. Foodborne microorganisms are continuously changing due to their inherent ability to evolve and their amazing capacity to adapt to different forms of stress (Bisen et al., 2012, 2014b). New primary production technologies and food manufacturing practices are introduced all the time; food consumption patterns and the demographic structure of many countries continue to change. Approximately one-third of all food manufactured in the world is lost due to spoilage (Gustavsson et al., 2011).

1.1 Why Contaminant Sensing is Required in the Food Industry

Food diagnostics is a relatively new and emerging area fuelled in large part by the ever-increasing demand for food safety.

Nanobiosensors. http://dx.doi.org/10.1016/B978-0-12-804301-1.00014-X

In addition, nanosensor-based approaches are also essential in the areas of food authentication, detection of foodborne pathogenic microorganisms, and screening for food allergens and adulteration (Debnath et al., 2010). The contaminant can spread rapidly with the quick and efficient distribution systems at multiple locations limiting the reaction time (Hall, 2002). New, flexible tools are required for evaluating and managing new food safety challenges with the use of food safety management tools, most important HACCP (hazard analysis critical control point), and the consequent application of hygienic measures, based on good manufacturing/hygienic practice (GMP/GHP) (Debnath et al., 2010; Bisen et al., 2012, 2014b). However, the lack of reliable data is often limiting the usefulness of this approach and therefore data collection is one of the priorities for future food safety strategies. The safety and quality of food can be tested in source laboratories but the cost and time involved limits the utility (Bisen et al., 2012, 2014b). Also, this type of analysis cannot account for mishandling during transportation or storage, after the product has left the source. Therefore, there is increasing demand for real-time sensors that can sense the quality and safety of the foods on site (Warriner et al., 2014). An outline of the types of contaminants that can enter food at different stages is given in Fig. 14.1.

1.1.1 Ensuring Safe Storage and Transportation within the Sell-By Period

Packaged food products are usually marked with expiry dates that indicate the preferred use by period. However, these periods are determined in standardized conditions assuming that the recommended conditions of storage and transportation have been maintained. In some cases, these conditions might be compromised in the absence of direct physical examination, which will lead to false indications about the suitability of food for consumption. The packing material in some cases may get damaged and

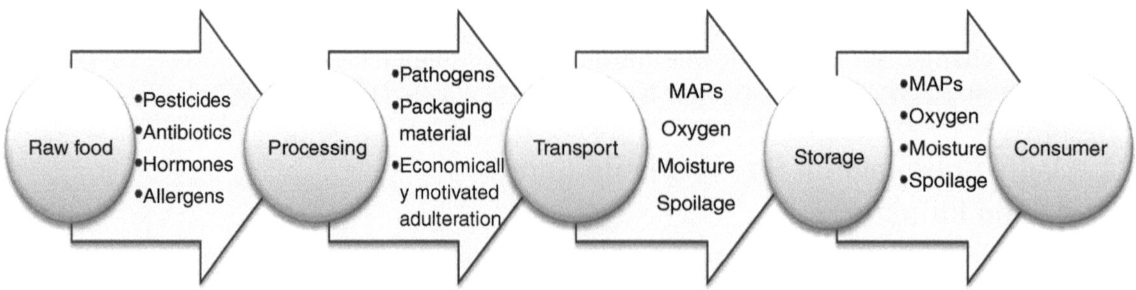

Figure 14.1. Types of food contaminants at various stages from source to consumer.

spoil the food, leading to inadvertent leakage of oxygen or moisture, restricting the use of batch analysis at the source (Luechinger et al., 2007). The majority of foodborne disease outbreaks result from such unintentional contamination of a product as a result of inappropriate processing, handling, and/or packaging. Intentional contamination of our food supply with biological or chemical agents also is a significant threat to national security. Bioterrorist attacks on our food supply could be accomplished with selected bacterial pathogens or toxins. A convenient method would be to use noninvasive sensing mechanism on each and every packet, which would directly inform the consumer about the compromised quality. The contaminants that can be sensed include gaseous indicators of leakage (oxygen, moisture) or spoilage (amines, volatile organic compounds, etc.) and the biological agents. A key step in establishing an effective food-safety program in a forward-deployed theater is to have adequate laboratory diagnostics that can identify and characterize rapidly any agent that could cause the quality of food.

1.1.2 Reducing Potential Health Risks

Food is a rich source of nutrients that attracts microbiological growth. The natural micro flora of food or beverage can consist of three main components: those associated with the raw material, those acquired during processing, and those surviving preservation and storage. They can be further subdivided into harmless organisms, producing either desirable or undesirable flavor changes in the food, and pathogens, forming dangerous enterotoxins (Bisen, 2014b). During processing, foods are subjected to highly complex and rapidly changing environments in which the microorganism may evade inactivation and detection in spite of hygienic manufacturing practices (Hall, 2002). Further, microbial sensing technologies are continuously subject to challenges in the form of novel combinations of types of food and pathogenic microorganisms. In recent times, there has been an increase in the demand for minimally processed food, which further raises expectations of pathogen detection methods as the pathogen elimination points are greatly reduced. Traditional method of pathogen detection requires an enrichment step to bring the target in the detectable range of biosensor. This step increases the time required for analysis and sophisticated containment facilities are required to carry out such analysis. Moreover, not all pathogens are cultivable in similar laboratory conditions. Further detection methods based on traditional culture based batch analysis do not ensure complete food safety (Swaminathan and Feng, 1994). Traditional methods also suffer from the drawbacks of being time consuming and labor intensive (Baeumner, 2003). The costs incurred on production companies upon potential product recalls due to microbial contamination have increased the interest in on-site pathogen testing (Nugen and Baeumner, 2008). In addition, there has been an increased awareness among people about the possible inadvertent or deliberate contamination of food products, therefore decentralized sensing has become very crucial (Ravichandran, 2010; Vaseashta, 2006). Unique problems encountered during contaminant sensing in food matrices include light scattering, sample opacity, and numerous other interferences (Singh et al., 2009).

1.1.3 Detection of Banned Dyes and Adulterants

In addition to the pathogenic contamination, which might inadvertently enter food, certain dyes, and adulterants are deliberately added in order to gain economic benefits. Food adulteration has caused serious illnesses among infants, adults, and animals (Kumar et al., 2015; Xin and Stone, 2008; Trivedi et al., 2009; Kobayashi et al., 2010). To counteract such incidences of "economically motivated adulteration" (EMA), there is a requirement of rigorous monitoring of food products. Owing to the health issues related to their consumption, many countries have banned the use of these dyes in foods. However, cases have been reported in which Sudan I dye in concentrations as high as 4000 mg/kg has been found in hot chilli products (ASTA, 2005; Botek et al., 2007). This incident led to prohibition of the import of hot chilli and its products by member states of European Commission in the absence of an accompanying analytical report validating absence of Sudan I and IV Sudan, Red 7B and Rhodamine B, curcuma, curry, sumac, and palm oil (RASFF, 2004, 2005; EFSA, 2005).

1.1.4 Detection of Pharmacological Residues Such as Antibiotics and Hormones

Antibiotics and hormones are routinely used in livestock for economic benefits by increasing meat supply and milk production. There is a growing concern regarding the presence of antibiotic/antibiotic residues in food products due to the evolution of MDR pathogenic strains in humans (Craig et al., 2013). Similarly, unmonitored use of hormones may lead to their residual levels in finished products, which might have unfavorable effects on consumers.

1.1.5 Detection of Pesticides

Pesticides are routinely used in agriculture to reduce crop losses due to microbiological or insect pests. However, lack of awareness among farmers leads to indiscriminate use of pesticides, which might or might not be required for a particular crop. Also, if pesticides are used above a prescribed limit, they may enter the crop plant and cause health concern. Therefore, pesticide levels in the food products needs to be monitored. There are three potential sources of pesticide residues in food grains, arising from (1) application of pesticides to protect the growing crop, (2) contamination of the environment by highly stable pesticides previously applied for other purposes, and (3) application of insecticides to protect the harvested crop during storage and handling.

1.1.6 Detection of Contamination Caused by Packaging Material

Plastic packaging, containers, and cling films often have instructions on how to use them safely to keep chemical migration to a minimum. More than 30 types of plastics have been used as packaging materials including polyethylene, polypropylene, polycarbonates, and polyvinyl chlorides. In certain cases, the material such as melamine used in packaging might enter the food and adversely affect the quality of food.

1.1.7 Detection of Allergens

Certain people are allergic to some specific food items and need to avoid them in order to prevent allergies. For common allergens, such as gluten, gliadin, and so forth, foods free of these allergens are specially processed for the allergenic population. As these components can trigger allergic reactions even in minute quantities, extensive testing of food is required to ensure that the processed food is free of allergen.

2 Biosensors

A biosensor is a device that can report the presence or activity of analytes using a biomolecular component providing specificity to the sensor by binding or interacting with the analyte and is able to cause a detectable change in mass, fluorescence, electric charge, or refractive index, and a transducer element, able to transform this interaction into a suitable electronic signal (Fritz, 2008; Bisen, 2014a). They are coupled together in one of the four possible ways, such as membrane entrapment, physical adsorption, and covalent bonding. The ideal characteristics of an efficient sensing system include speed, sensitivity, accuracy, real-time detection with feasible cost among others (Fig. 14.2).

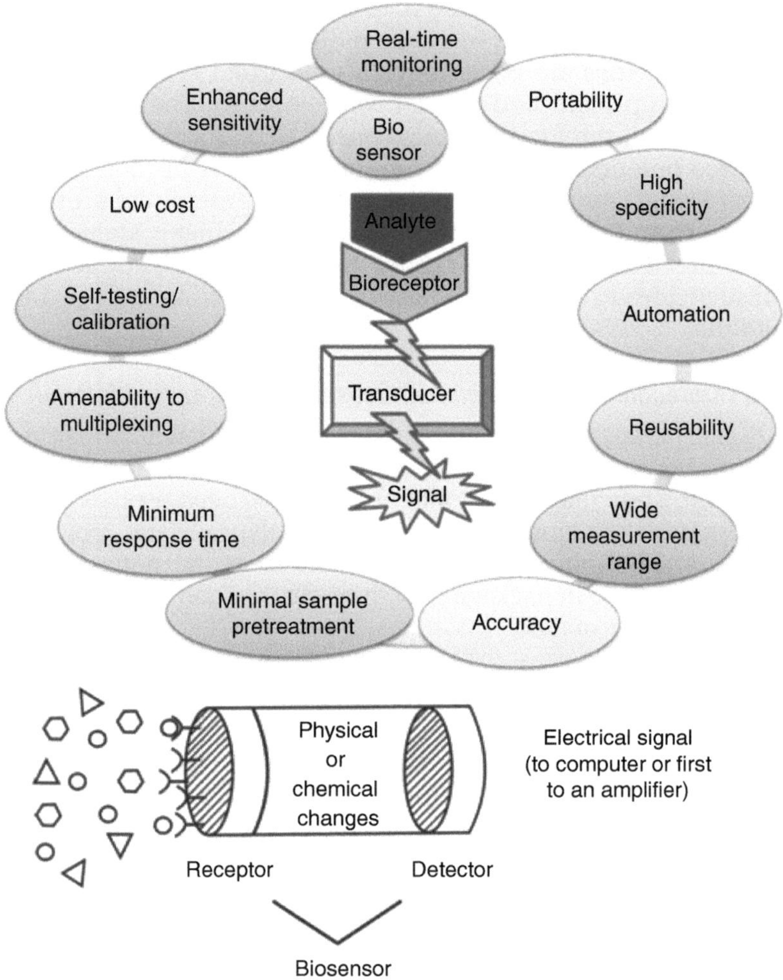

Figure 14.2. Utility and specific properties of biosensors.

Most of the biosensor formats, which were initially developed for health sector, are now beginning to be exploited in the food industry to assess safety and quality of food products (Warriner et al., 2014; Bisen, 2014a). However, only a few of them have been commercialized for food industry due to difference in sample sizes, types of matrices, and so forth. The "bio" and "sensor" elements can be coupled together in one of four possible ways:membrane entrapment, physical adsorption, covalent bonding, and cross-linking. Thus, a typical biosensor consists of mainly three parts: (1) biological material, (2) transducer, and (3) signal processors (Fig. 14.2).

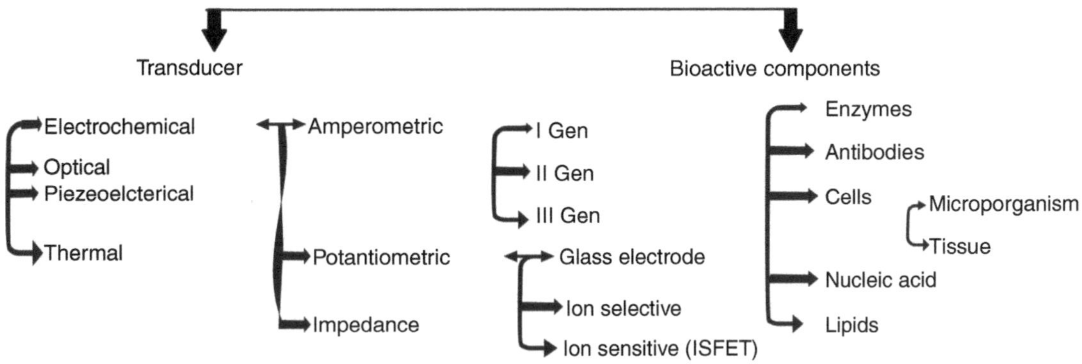

Figure 14.3. Types of transducers and bioactive components.

2.1 Classification of Biosensors

Biosensors may be classified according to components or immobilization techniques used such as transducers and the bioactive compounds. The bioelement is very specific to the analyte to which it is sensitive. Depending on the transducing mechanism used, the biosensors can be of many types such as: (1) resonant biosensors, (2) optical detection biosensors, (3) thermal detection biosensors, (4) ion-sensitive FET (ISFET) biosensors, and (5) electrochemical biosensors (Bisen, 2014a). The electrochemical biosensors based on the parameter measured can be further classified as (1) conductometric, (2) amperometric, and (3) potentiometric. Therefore, biosensors can be divided into different types based on the type of detection (Fig. 14.3).

2.2 Properties of Biosensors

Some of the notable properties of a good biosensor are specificity, linearity, response time, simplicity, continuous monitoring ability, reproducibility, portability, and cost effectiveness (Table 14.1).

2.3 Biosensor Components

2.3.1 *Bioactive Components*

The biological part of the biosensor specifically reacts with the analyte of interest, sparking a signal that is detectable by the attached transducer. The bioactive components may be a purified enzyme, antibodies, cells, nucleic acids, and/or lipids.

Table 14.1 Various Biosensor Transducers, Principles, and Applications

Transducer System	Principle	Applications
Enzyme electrode	Amperometric	Enzyme substrate and immunological system
Conductometer	Conductance	Enzyme substrate
Piezoelectric crystal	Mass change	Volatile gases and vapors
Thermistor	Calorimetric	Enzyme, prganelle, whole cell, or tissue sensors for substrate, products, gases, pollutants, antibiotics, vitamins, and so forth
Optoelectronic/wave guide and fiber optic device	Optical pH	Enzyme substrates and immunological systems
Ion-sensitive electrode (ISE)	Potentiometric	Ions in biological media, enzyme electrodes, enzyme immunosensors
Field effect transistor (FET)	Potentiometric	Ions, gases, enzyme substrates, and immunological analytes.

2.3.1.1 Sensing DevicesOptical

The biosensor is based on optical diffraction or electrochemo luminescence properties and the output transduced signal measured is light. Optical diffraction-based devices use a silicon wafer coated with a protein via covalent bonds. The resulting signal can be measured or can be further amplified before measuring for improved sensitivity and allowing multiple analytes to be detected by using different monitoring wavelengths. Use of immobilized luciferase greatly reduces the cost of analyses.

2.3.1.1.1 Calorimetric Biosensors Many enzyme catalyzed reactions are exothermic, generating heat and the temperature changes. In these cases, one can carry out measurements in temperature-controlled small packed bed columns with immobilized enzymes and determine the temperature at the entry and exit of these columns by incorporating suitable thermistors (Fig. 14.4).

Calorimetric transducing devices measure the heat released upon a biochemical reaction occurring at the sensor surface. On the basis of the type of heat transfer, they can be categorized as follows:

1. *Isothermal calorimeters*: maintain isothermal conditions and the amount of energy required for cooling or heating is measured.

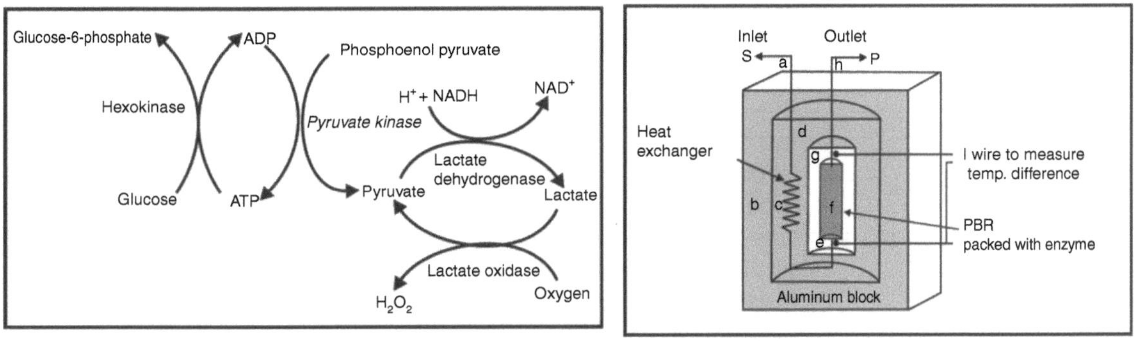

Figure 14.4. Coupling of many heat-increasing enzymes in exothermic reaction and schematic diagram of a calorimetric biosensor.

2. *Heat conduction calorimeters*: measure the difference in temperature between the highly conducting reaction vessel and the surrounding heat sink.
3. *Isoperibol calorimeter*: measures the temperature difference between the thermally insulated (adiabatic) reaction vessel and the surrounding isothermal jacket.

In spite of the progress in the field of biosensors, the demand for development of improved biosensors that are more sensitive in terms of concentration and volume of analyte and which can simultaneously sense broader range of analytes continues to increase (Fritz, 2008). These sensors are characterized by a reaction between analyte and biosensor component to produce a color change (Su et al., 2013). Nanoparticles can also be modified by an ionic ally to couple desired protein. Another approach is aggregation of colloidal nanoparticles with subsequent colorimetric changes, which can be perceived even by an untrained person (Lim et al., 2012).

2.3.1.1.2 Fluorescence Fluorescence analysis using organic dyes has been used traditionally to detect pathogens. They are based on excitation of analyte using suitable laser light followed by detection of emitted fluorescence where fluorescent nanoparticles are used to enhance the signal while providing photostability. Quantum dots labeled with binding molecules can also be used as a fluorescent tracer (Penn et al., 2003).

2.3.1.1.3 Surface Plasmon Resonance Surface Plasmon resonance transducers measure minute changes in the angle of reflectance of plasmon waves on dielectric interface. When a plane polarized light passes through a prism whose one side is coated with a metal, at an appropriate angle, total internal reflection is observed, which induces formation of a charge wave at the interface (called Plasmon waves) moving up to a few microns. The incident angle of the total internal reflection is measured using a photodetector. The angle of reflectance changes when the refractive index of the surface changes, for example, in response to binding of analyte (Warriner et al., 2014). SPR offers advantages such as being label free and rapid (Cho et al., 2014). The disadvantages include slow diffusion driven mass transfer, low sensitivity due to small refractive index, and insufficient depth of layer influenced by SPR (Wang et al., 2010). To this end antibody nanoparticle conjugates have been used to enhance the signal in sandwich assays (Wang et al., 2010).

2.3.1.1.4 Antigen antibody The binding between an antigen and its corresponding antibody is very specific. This property of antibody is exploited while designing biosensors based on antibodies. Antibodies are usually covalently bonded on the surface of the transducer by conjugation of amino, aldehyde, carboxyl, or sulfhydryl groups (Shaikh and Patil, 2012). The binding reaction between the antibody and antigen can be monitored as a time-dependent change of fluorescence signal, which is proportional to the reaction ratio of antibody to analyte (Bisen, 2014a). Though antibodies have similar limitations with enzymes, immunosensors offer advantages of rapid and on site measurements over traditional immunoassays (Shaikh and Patil, 2012). Immunosensors usually make use of optical or acoustic transducers (Fig. 14.5).

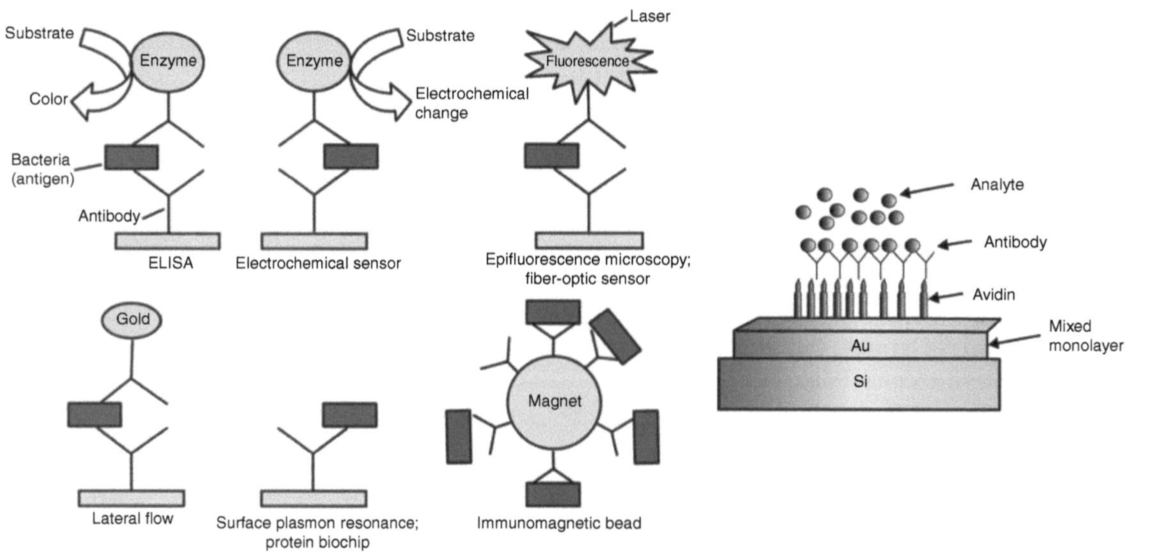

Figure 14.5. Immunosensor design.

2.3.1.2 Acoustic/Piezoelectric

In this mode, sensing molecules are attached to a piezoelectric surface (a mass to frequency transducer) in which interactions between the analyte and the sensing molecules set up mechanical vibrations that can be translated into an electrical signal proportional to the analyte, such as quartz crystal.

Acoustic transducers detect a change in mass density, viscoelastic, electric, or dielectric properties of chemically interactive membrane placed in contact with a piezoelectric material. Following are two commonly used acoustic transducers:

2.3.1.2.1 Bulk acoustic wave sensor (BAW) BAW measures the change in resonance frequency of a resonator (eg, quartz crystal resonator in quartz crystal microbalance), which may be due to mass or viscoelastic changes at the sensor surface. Nonspecific binding effects might interfere with the detection.

2.3.1.2.2 Surface acoustic wave sensor (SAW) SAW measures the change in surface acoustic waves (not the bulk), which may be due to surface effects such as mass, viscosity, pressure, magnetic fields, strain, temperature, and irradiation with UV rays (Arugula and Simonian, 2014).

2.3.1.3 Resonant

An acoustic wave transducer is coupled with an antibody (bioelement) in this mode. The mass of the membrane changes when the analyte molecule (or antigen) gets attached to the membrane. The resulting change in the mass subsequently changes the resonant frequency of the transducer, which is then measured.

2.3.1.4 Thermal-detection

They are constructed by combining immobilized enzyme molecules with temperature sensors, which are based on biological reactions, namely absorption or production of heat, which in turn changes the temperature of the medium in which the reaction takes place. The use of sophisticated and expensive instrumentation is the major drawback of this technique.

2.3.1.5 Ion-Sensitive

The ISFET can be constructed by covering the sensor electrode with a polymer layer. This type of biosensor is also called an enzyme field effect transistor (ENFET) and is primarily used for pH detection (Fig. 14.6). The technology is based on semiconductor device with ion-sensitive surface, which interacts with the ions and the potential change is subsequently measured.

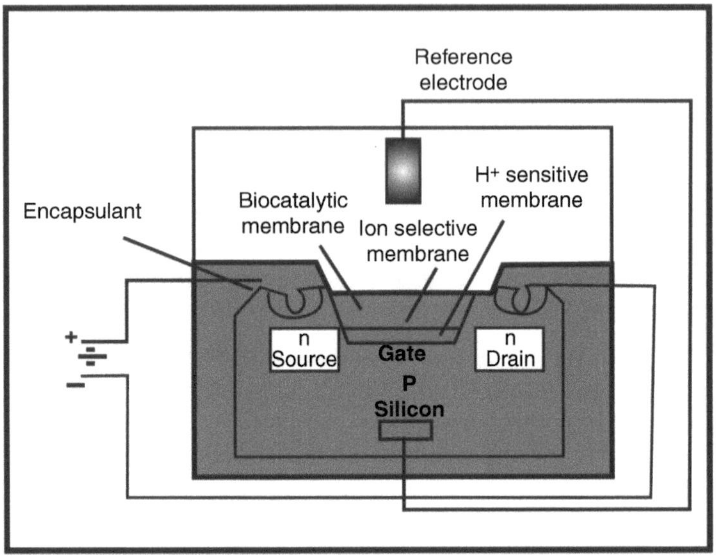

Figure 14.6. Schematic diagram of the section across the width of an ENFET.

2.3.1.6 Electrochemical

The chemical reactions produce or consume ions or electrons that can be sensed and used as measuring parameter in this class of biosensors. The sensing molecule reacts specifically with compounds to be detected, sparking an electrical signal proportional to the concentration of the analyte (Fig. 14.7).

High sensitivity, selectivity, and ability to operate in turbid solutions are advantages of electrochemical biosensors. Electrochemical biosensors are mainly used for the detection of hybridized

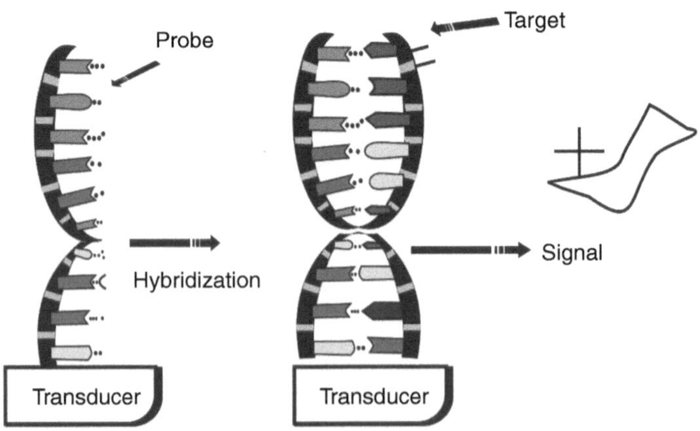

Figure 14.7. Electrochemical DNA biosensor.

DNA, DNA-binding drugs, glucose concentration, and so forth. Electrochemical biosensors can be classified based on measuring electrical parameters as: (1) conductometric, (2) amperometric, and (3) potentiometric.

2.3.1.6.1 Conductometric The measured parameter is the electrical conductance/resistance of the solution. This change is measured and calibrated to a proper scale but having relatively low sensitivity.

2.3.1.6.2 *Amperometric* (measurement of the current resulting from a redox reaction) Amperometric detection is based on measuring the oxidation or reduction of an electroactive compound at a working electrode (sensor) (Fig. 14.8). Enzyme-electropolymer-based amperometric biosensors has also been developed as an innovative platform for time-temperature integrators (Reyes-De-Corcuera et al., 2005)

In vivo sensing: Amperometric biosensors have been also applied for in vivo sensing since their size may be reduced. By the appropriate casting of membranes onto the biosensors tip, high selectivity and biocompatibility may be achieved (Fig. 14.8).

2.3.1.6.3 Potentiometric The measured parameter in this type of sensor is oxidation or reduction potential of an electrochemical reaction. The voltage at which these reactions occur indicates a particular reaction and particular species (Fig. 14.9).

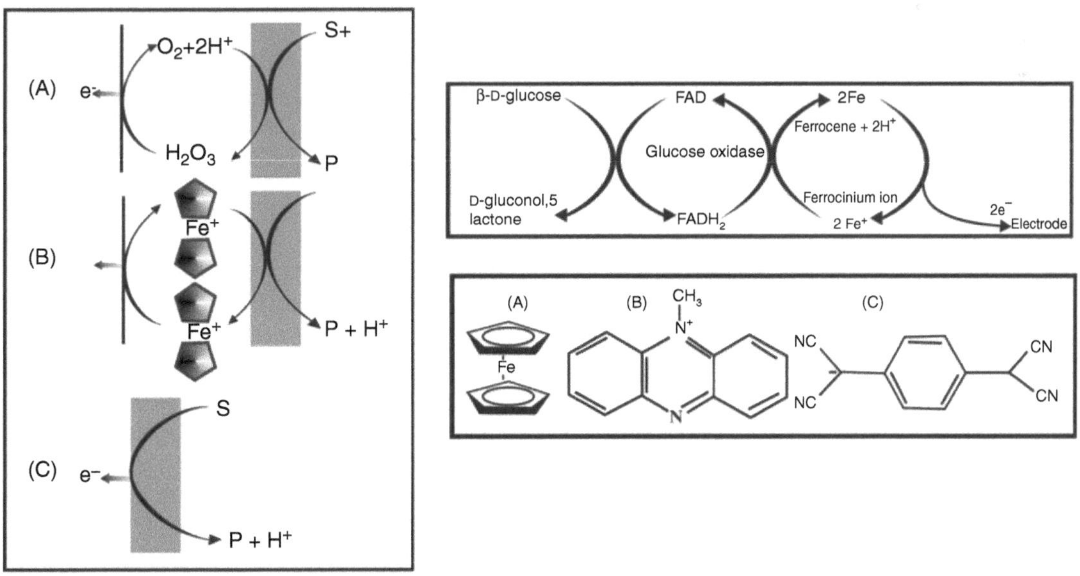

Figure 14.8. Amperometric biosensors for flavor-oxidase enzymes illustrating the three generations in the development of biosensors.

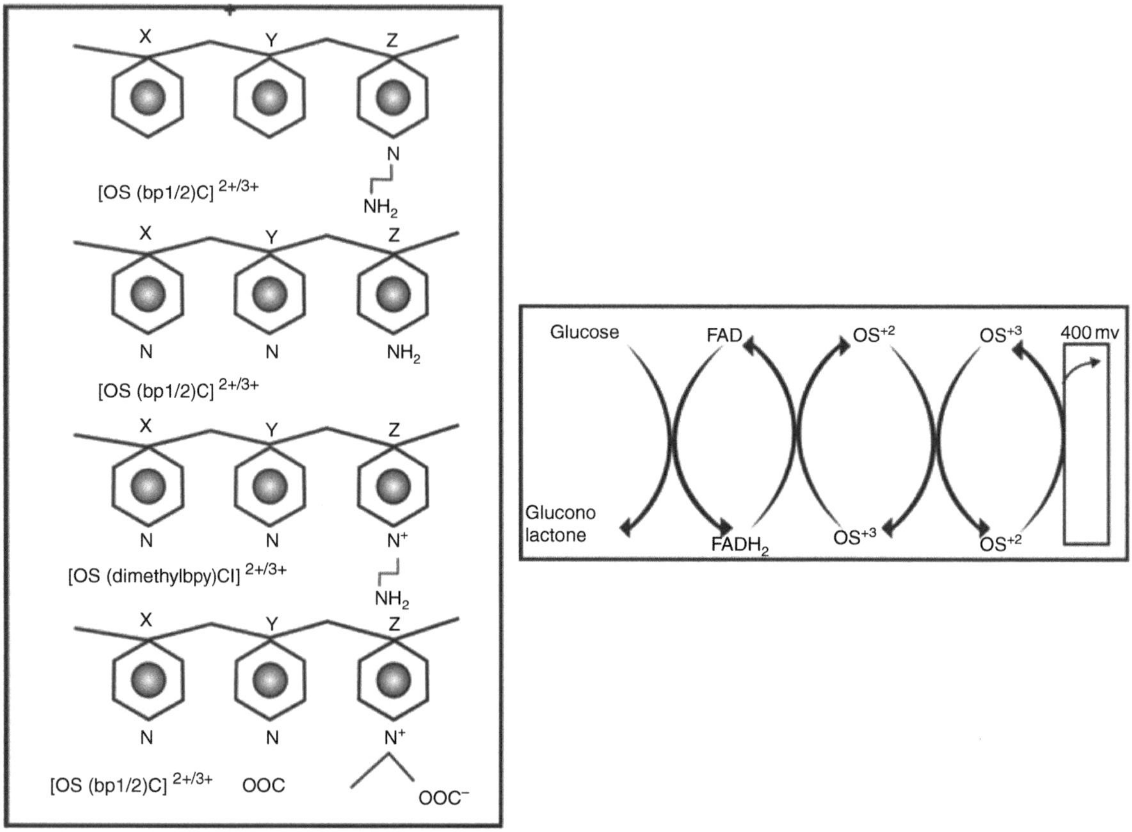

Figure 14.9. Structure of enzyme "wiring" polymers. Variation of X, Y, and Z allow the properties of the polymer to be customized and the redox cycles occurring at a three-dimensional redox epoxy-wired enzyme electrode. The wired enzyme is a flavin (FAD)-containing oxidase enzyme.

Ion selective electrodes utilized in biosensing devices can be classified as follows:

1. *Normal pH electrodes*: These are normal hydrated glass membrane electrodes, which sense cations in a concentration dependent manner by producing a transverse electric potential arising due to competition for binding sites.
2. *Glass pH electrode for gases*: These electrodes are composed of gas permeable membrane selective for specific gases like carbon dioxide, hydrogen sulfide, or ammonia. The measurement is based on pH differences between electrode and the membrane arising due to diffusion of these gases through the membrane.
3. *Solid-state electrode*: In these electrodes, the glass membrane is replaced by conductive membrane made of silver compounds. These electrodes have been used in the determination of cyanide and iodide ions.

Recently ion selective electrodes have been used in the production of ISFETs, which have been conjugated with enzymes to develop biosensors.

3 Nanosensors

Nanosensor can be defined as any sensing device which incorporates a nanoscale component, that is, a component having dimensions within 1–100 nm at least in one dimension (Jianrong et al., 2004). Nanomaterials can be used in sensors both to enhance sensing as well as transduction.

3.1 Nanomaterials: Ideal Properties for Use in Sensors

The physical, chemical, and biological properties of nanomaterials are remarkably different from their macroscopic counterparts (Ravichandran, 2010). These properties impart unique characteristics to the nanomaterials which can be exploited for biosensor design to improve the specificity, accurary, linearity, response time, simplicity, continuous monitoring ability, reproducibility, portability, and cost (Li et al., 2011)

3.1.1 Size

The size of nanomaterials is comparable to biological macromolecules. Therefore, these materials are well suited to detect analytes present in the microenvironments which are not accessible to the macro sensors (Wei et al., 2009). Due to their nanoscale size, nanoparticles are ideal candidates for direct localized detection. A number of biosensors are now available which do not require preprocessing of the sample, and therefore can be used for a direct estimation of analyte. Small size of the sensor element also enables miniaturization of the sensor increasing the portability for onsite testing. The detection is, therefore, possible not only at the source, but also at the distribution/consumer level. With the help of microfluidic systems, detections are possible utilizing small sample volumes in addition to also effectively reducing the cost and time required for analysis. The fabrication cost of the electrode is also reduced due to reduction in volume of expensive materials (Wei et al., 2009). Owing to the small size, the multiple nanosensors may be equipped in single instrument to provide multianalyte detection or to obtain spatial distribution of a single analyte (Wei et al., 2009). The favorable size and optical properties allow nanosensors to be incorporated into food packaging for noninvasive, visual analysis of food quality during storage, transportation, and consumption (Duncan, 2011).

3.1.2 *Electrical Properties*

There is enhanced electron transfer due to high electrical conductivity; therefore, the sensitivity of the sensor is greatly increased for example, gold nanoparticles (Majdalawieh et al., 2014). Electrochemical measurements are considerably quick in nanoelectrodes due to radial (nonplanar) diffusion compared to macroelectrodes operating via planar diffusion (Wei et al., 2009).

3.1.3 *Optical Properties*

The optical properties of nanoparticles are considerably different from the respective bulk materials owing to the quantum effect. There is higher energy difference between the conduction and the highest valence band due to the compact packing density of electrons in nanoparticles resulting release of higher energy upon return of electron from excited state to ground state for the spectral shift. The semiconductor properties are also affected leading to changes in electrochemical phenomena like surface Plasmon resonance (Warriner et al., 2014). Quantum dots, a type of nanoparticles have found a new application as a substitute for enzyme labeling. Quantum dots are able to absorb more photons compared to organic fluorescent dyes, provide more brightness, and release the energy in the form of photo luminescence omitting the requirement of an additional assay step (Pisanic Ii et al., 2014). In addition, they are photostable and provide high signal to noise ratio (Syed, 2014). As the emission peak is dependent upon the size of nanocrystal, different sizes of quantum dots can be used to obtain different color labels for multiplexing.

3.1.4 *Chemical Properties*

The chemical reactivity is also enhanced compared to macromaterials, due to small size enabling fast detection in both liquid and gaseous phases (Vaseashta, 2006).

3.1.5 *Magnetic Properties*

Magnetic nanoparticles labeled with a suitable capture reagent, for example, antibodies can be used to concentrate the analyte, a technique called IMS (immunomagnetic separation). Some nanoparticles behave as ferromagnetic particles when an external magnetic field is applied because of having a lower number of unpaired electrons than the bulk material. These properties are useful in generating magneto-resistive nanosensors. A large number of analytical approaches are being used in food industry requiring a sample preconcentration to achieve detectable signal.

3.1.6 Semiconductor Property

A large number of enzymes used in biosensors are oxidoreductases producing an electroactive product detected amperometrically. These enzymes require mediators to prevent interference from other electroactive constituents in the sample. Use of semiconductor nanoparticles allow direct wiring of enzyme redox centers to the electrode by passing the requirement of exogenous mediator and reducing response time of the sensor (Palanisamy et al., 2012; Hu et al., 2012).

3.1.7 Enhanced Sensing Surface Area

Nanoparticles have a high surface area volume ratio; large surface areas work more efficiently for immobilization of bioaffinity agent. The higher surface area increases the number of bioaffinity agents that can be incorporated per unit volume and therefore increase the sensitivity of the device. Due to large surface area of nanoelectrodes, nanosensors can be used to detect analytes in poorly conducting media even in the absence of suitable electrolyte (Wei et al., 2009) particularly useful for detecting analytes in complex and variable food matrices (Singh et al., 2009).

3.1.8 High Accuracy

Any minute changes in the sensor environment are transduced accurately for reliable detection of the analyte due to enhanced sensing surface.

3.1.9 Quick Response

Use of nanoparticles allows direct wiring of enzyme redox centers to the electrode by passing the requirement of exogenous mediator and reduce response time of the sensor (Palanisamy et al., 2012; Hu et al., 2012). These properties of nanomaterials have improved the design and efficiency of biosensors to meet the expectations of growing market demands (Fig. 14.10).

3.2 Types of Nanosensors

Nanomaterials can improve sensor design at various levels. They can improve sensing abilities as well as transduction properties. Likewise, nanosensors can be classified into following types:

3.2.1 Colorimetric Metal Nanoparticle Detectors

Colorimeteric sensors offer simplicity and ease of operation and detection. Certain nanoparticles can undergo easily visible color changes in response to presence of analyte and this property can

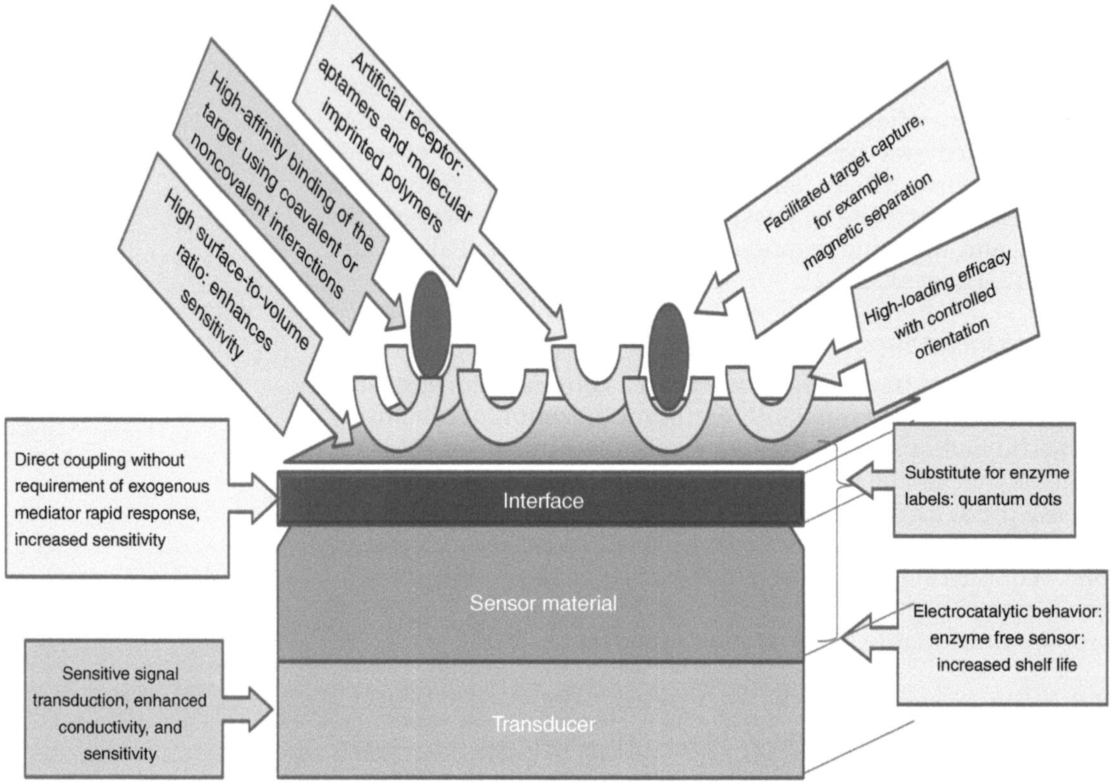

Figure 14.10. Design and efficiency of biosensors with nano sensors and nanomaterials.

be utilized to develop nanoparticle-based colorimetric assay (Ai et al., 2009; Kuang et al., 2011). The formation of nanoparticles can also be controlled by presence of analyte to produce a visible effect proportional to the concentration of analyte (Cao et al., 2010).

3.2.2 Carbon Nanotube Biosensor

CNT-based nanomaterials have the unique capability to alter their properties in presence of some chemical species but are otherwise chemically inert. They are morphologically flexible, biocompatible, and their size is comparable to biomolecules, making them an ideal candidate for biosensor design and applications having high surface area: weight ratio; conducting properties; quick, accurate, and reversible measurements; easy derivatization; and ability to generate electro-chemiluminiscence in aqueous solutions (Vaseashta, 2006). Carbon nanotubes can be used as a substitute owing to their superior mechanical properties for silicon-based chips in nanoelectromechanical systems (Sapmaz et al., 2003).

3.2.3 Electronic Nose

The quality of certain food items is ascertained primarily by their aroma. Traditionally the aroma quality is analyzed by a trained panel who decodes the complexity and heterogeinity of the aroma and rates them on the basis of experience. This type of analysis might not be reproducible and may vary from person to person (Peris and Escuder-Gilabert, 2009). The other alternative is analytical analysis of the headspace using gas chromatography. This method has higher reliability but has high operating cost and suffers from the disadvantage of long analysis time per analysis (García et al., 2006). Electronic noses, which mimic the human nose in sensing the aroma and rate it according to the pattern of different volatile components, bypass the problem of reliability in expert panel-based methods and cost of analytical methods. An electronic nose (Persaud and Dodd, 1982; Gan et al., 2005) comprises three components: a sensory component, which is usually an array sensor enabled to sense different types of aroma imparting volatile compounds; a unit to collect the signals generated from the sensory unit; and, finally, a software to recognize the pattern of VOC's to ascertain food quality (García et al., 2006) (Fig. 14.11).

The sensory component of electronic noses can be surface acoustic wave sensor, metal oxide semiconductor sensor, or quartz resonator. They can be used for the analysis of VOC arising from both liquid and solid food samples (Schaller et al., 1998;

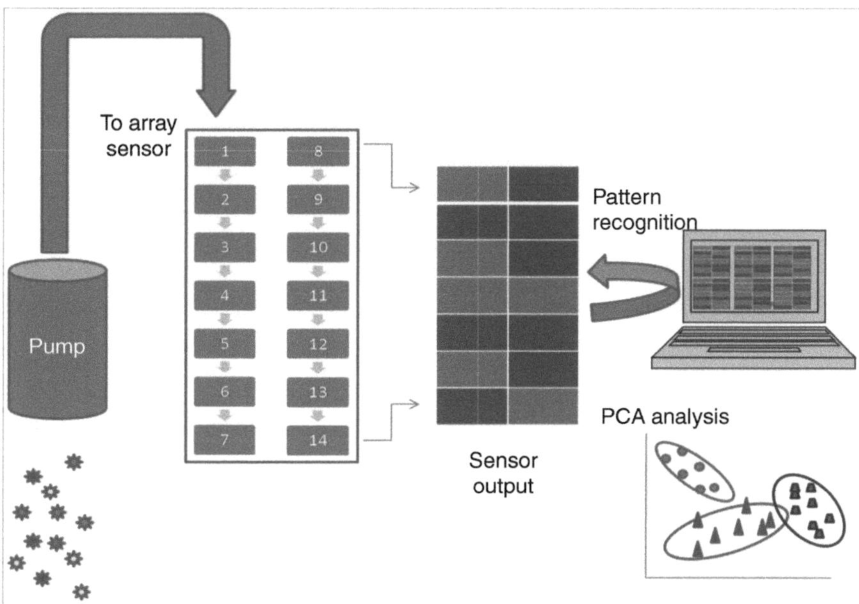

Figure 14.11. Diagrammatic view of e-nose (electronic nose).

Frías et al., 2002). Electronic noses offer several advantages including portability, low detection time, low cost per analysis, and high reliability, but the presence of high concentration of interfering substances, for example, ethanol and water in wines may reduce their sensitivity (Francis and Cynkar, 2000). Electronic noses have been incorporated directly in packaging material to quickly sense the presence of pathogens by their metabolic products (Ravichandran, 2010). Some notable examples of utility of e-nose in ascertaining food quality are listed in Table 14.2.

3.2.4 Electronic Tongue

An electronic tongue is similar to an electronic nose and works for nonvolatile analytes. E-tongues can be used to analyze the quality of foodstuffs like wine, beer, and tea. The quality of these foodstuffs is determined by detecting their bitterness, sourness, or astringency, which is mainly attributed to the presence of polyphenols. These can be incorporated directly in packaging material to quickly sense the "taste," mimicking the physico-chemical interaction of food molecules with taste buds present on tongue (Ravichandran, 2010).

Table 14.2 Selected Applications of e-nose in the Food Industry

Food Item	Type of Analysis	References
Wines	Quality	Penza and Cassano (2004); Buratti et al. (2004)
Cheese	Detection of mould	Ampuero and Bosset (2003)
Milk	Quantification of off flavors; identification of single strains of disinfectant–resistant bacteria in mixed cultures	Ampuero and Bosset (2003)
Meat products	Quality changes	Vestergaard et al. (2007)
Vegetable oils	Flavor analysis	Gan et al. (2005)
RBD palm olein	Storage stability	Gan et al. (2005)
	Lard adulteration	Man et al. (2005)
Wheat	Age and insect damage	Zhang and Wang (2007)
Sesame oil	Maize oil adulteration	Hai and Wang (2006)
Virgin olive oils	Adulteration	Oliveros et al. (2002)
Porcine meat loaf	Sensory quality	Hansen et al. (2002)

As in the case of electronic nose, multiple analytes are tested in a single run to get a blueprint pattern of sample composition, thereby reducing the detection time to a large extent, which is very crucial in food industry (Ravichandran, 2010). E-tongue sensors based on amperometric transducer are of four types metal, conducting polymer, phtalocyanine film, and biosensors. Similarly the detection mode can be fixed potential or pulse sweeping potential.

Metal microchip based e-tongue is composed of capillary electrophoresis coupled with screen-printed electrode. E-tongues based biosensor are composed of an enzyme, solid electrode, and biochemical transducer. The conductivity of conducting polymer sensors might vary with the analyte (Scampicchio et al., 2008), and they are sensitive to humidity but they offer advantages such as rapid adsorption desorption and partial selectivity. Electrodes are composed of coordination compounds containing a transition metal with phthalocyanine film (Baldwin et al., 2011) and used to sense bitterness in olive oils (Apetrei et al., 2004). The performance of metal sensor, biosensor, or conductive polymer based e-tongue can be enhanced by utilizing nanodimensions to increase surface-to-volume ratio to reduce the detection limits. Miniaturized sensor arrays are now being developed. Often, a lipid membrane is used for recognition, which translates the relevant substances into electric potential across membrane. The amount of lipids present in the membrane can be varied to optimize detection limit (Iiyama et al., 2009).

3.2.5 *Nanocantilevers*

Nanocantilever biosensors can respond to mechanical bending caused by a slight change in temperature, pH, DNA hybridization, interaction between antigen and antibody, adsorption of pathogen, or formation of self-assembled monolayers at the sensor surface (Fritz, 2008). The detection is based on physical and chemical signaling stimulated by biological interactions such as those between antigen–antibody, enzyme–substrate, and ligand–receptor. They are made up of silicon-based materials such as silicon nitride or silicon dioxide forming the bottom layer and a gold-made reflective top layer (Fritz, 2008). They are used for detection of chemical contaminants, toxins, and antibiotic residues in food (Ravichandran, 2010). Vibrational frequency can also be customized as per biomass of pathogenic microorganism to simultaneously detect multiple pathogens (Ravichandran, 2010).

Cantilever sensors consist of a cantilever beam sensitive to their environment having bending in order of nanometer and the sensors are called as nanomechanical sensor (Fritz, 2008). A position sensitive detector (PSD) is used to transform the mechanical bending into positional change of laser spot. Alternatively, a

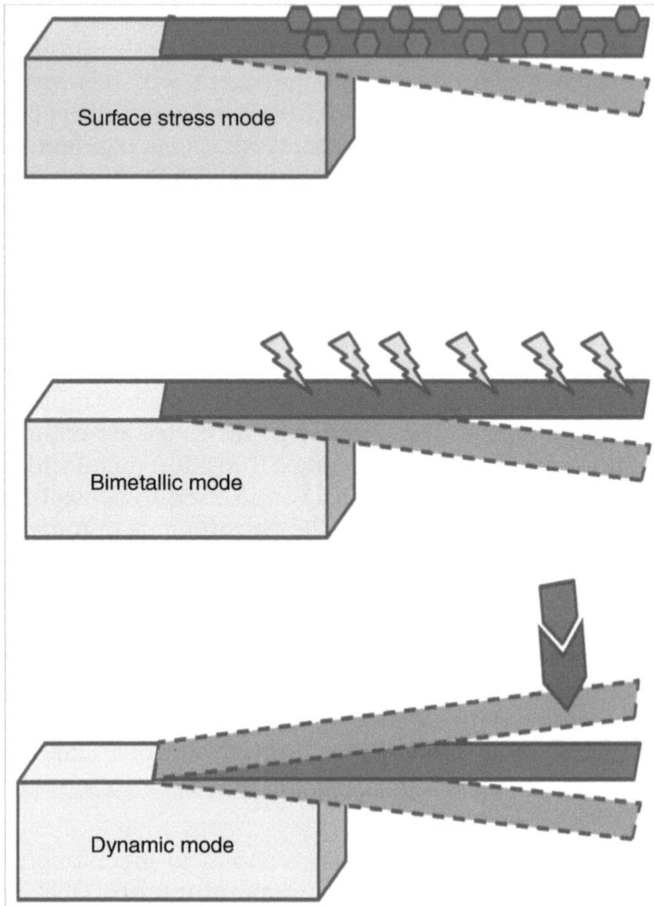

Figure 14.12. Different modes of nanocantelevers.

peizoresistive readout may be used for measuring change in resistance upon mechanical bending (Fritz, 2008). Nanocantilevers can be operated in different modes: surface stress mode, bimetallic mode or dynamic mode (Fig. 14.12).

Surface stress mode is the most commonly used in which molecules bind to the top and bottom surface of the cantilever asymmetrically, causing cantilever bending. The two surface layers may be functionalized for enhanced asymmetric binding using silane chemistry by layering amino or mercapto silane, which can be further cross-linked to desired receptor molecule via their end groups. Alternatively, electrostatic binding of positively charged molecules to the negatively charged silicon dioxide may be achieved (Fritz, 2008). The gold surface may be functionalized utilizing thiol chemistry by exploiting the affinity of gold to sulfur groups. Thus thiol-labeled DNA or cysteine-rich proteins may be bound to this surface acting as receptor for the complementary DNA or specific antibody (Fritz, 2008). The layers can also be covered with inert coating like thiolated polyethylene glycol to prevent adsorption (Fritz, 2008). The bending of beam due to differential thermal expansion of the two surfaces is sensed in bimetallic mode. Temperature changes as little as 10^{-5} K can be sensed using this mode by careful optimization (Ziegler, 2004). Dynamic mode is based on the property of cantilevers to behave as harmonic oscillators, which can be excited at their resonant frequency, depending on the effective mass of the oscillator. When the analyte is bound to the surface of the cantilever, its effective mass increases, decreasing the resonant frequency and the mass change can be measured even up to single molecule (Yang et al., 2000).

There are several advantages of cantilever biosensors including their amenability to microfabrication. These sensors are label free, allowing them to detect unmodified molecules and, therefore, can be used for real-time monitoring. They can be applied to a wide variety of molecules by appropriate functionalization with small sample volumes. However, the requirement of sophisticated functionalization and limited theoretical description and low repeatability in liquid samples limit their use (Craig et al., 2013).

3.2.6 Optical

Fiber optic nanoprobe: Nanoprobes consist of an optical fiber of nanometer radius whose outer wall is coated with silver/aluminum or gold and the tip is functionalized by silanization to enable it to covalently bind with antibody. Fluorescent analyte is detected by the probe, which is mounted on an inverted microscope coupled to a photomultiplier tube (PMT) detector (Vo-Dinh et al., 2001). Cells are maintained in a viable state by heating microscope stage to 37°C. The optical nanoprobe sensor can be used to manipulate subcellular locations owing to their small size.

3.2.7 Nanoelectromechanical System (NEMS)

Nanoelectromechanical systems (NEMS) are the successors of microelectromechanical systems (MEMS). They work by converting electrical signal into nanoscale mechanical motion and mechanical motion into electrical signal. The devices offer advantages like rapid response, portability, and low cost because of nanoscale moving parts. They can be used as active cells by devices and can be used to control the environment during storage. They are able to detect chemical or biochemical signals using advanced transducers (Ravichandran, 2010). The MEMS technology for detection of trans food content has been commercialized by Polychromix, Wilmington, MA, USA (Ritter, 2005).

3.2.8 Aptamer Sensor

Aptamers are single-stranded oligonucleotide sequences or peptide molecules with the capability to recognize various specific target molecules ranging from small ions to large proteins with high affinity and specificity. They are the novel biomolecular recognition elements, which can be utilized as receptors in biosensors (among other applications) due to small size, cost efficiency, and design flexibility. Aptamer-based biosensors are able to detect analytes which were, up to now (using antibody technology) very difficult to measure or even detect, like toxic or nonimmunogenic substances (Figs. 14.13 and 14.14). Immunobiosensor with EC, optical (SPR fluorescence) or mass-sensitive (piezoelectric microbalance) transducers are gaining extensive research interest. Aptamer has a wide range of molecular and therapeutic targets,

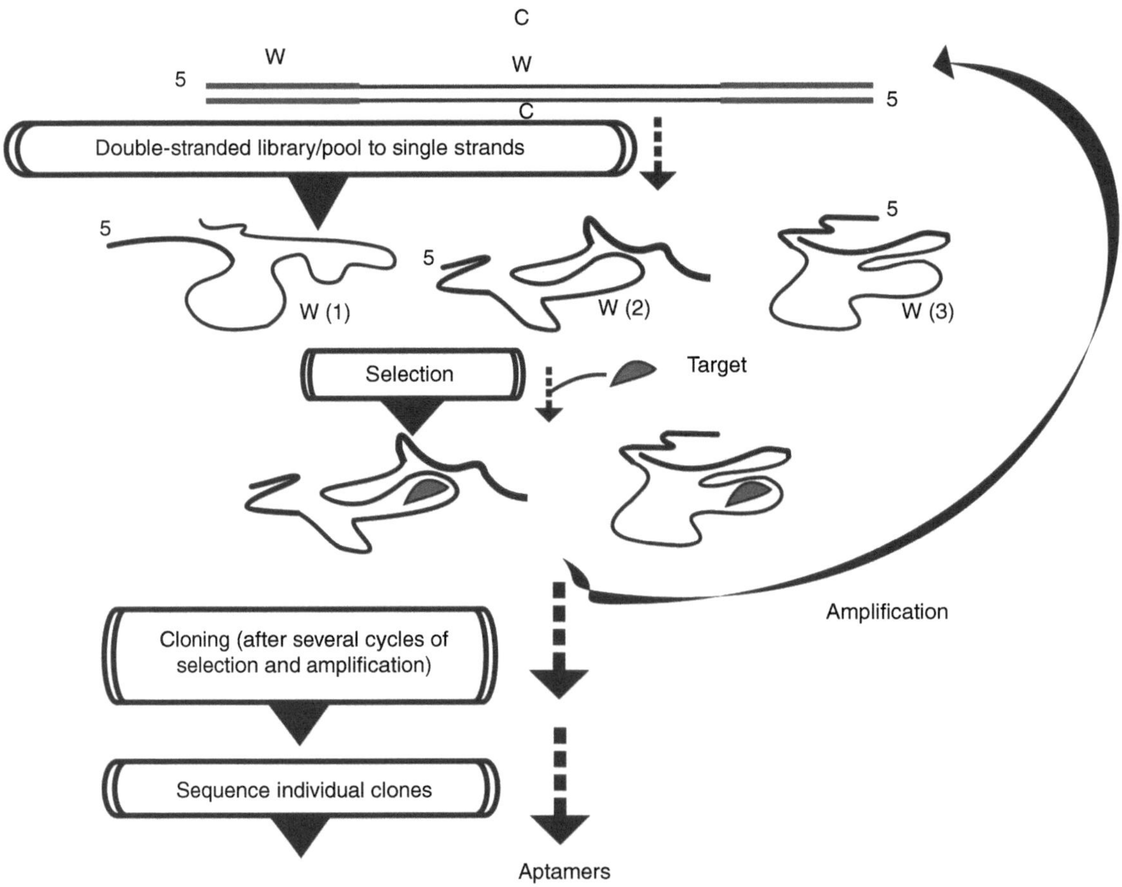

Figure 14.13. The approach can be used to isolate high-affinity aptamers for a wide variety of protein and small molecules.

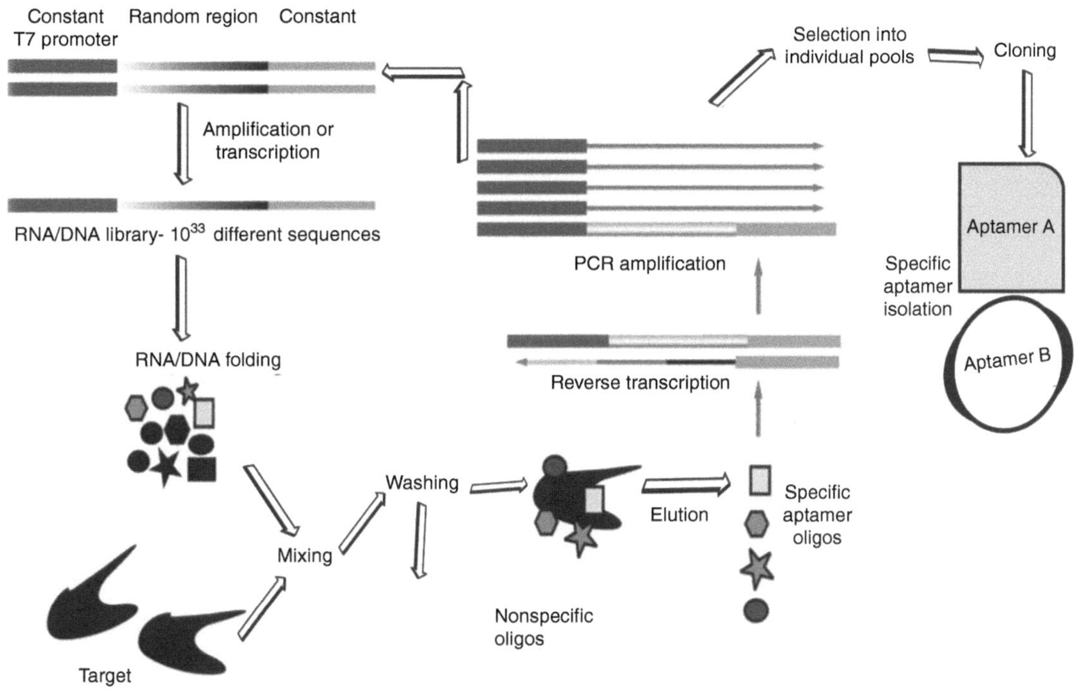

Figure 14.14. Aptamers can be used to select specific regions.

including amino acids, any class of proteins (enzymes, membrane proteins, viral proteins, cytokines, growth factors, and immunoglobulins), drugs, metal ions, other small bio/organic/inorganic small molecules, and even whole cells. Moreover, combinatorial chemical synthesis offers a wide variety of methods for aptamer sequence modifications such as the terminal tagging chemical groups (Selvakumar and Thakur, 2012). A major limitation of aptamer sensors, is the potential of biodegradation or bio-fouling when used in complex biological matrices containing high levels of nucleases which can be overcome by encapsulating the aptamer sensor in suitable polymer (Li et al., 2015).

Aptamers resemble antibodies but they undergo structural modification after binding with their target. Aptamers are developed by systematic evolution of ligands by exponential enrichment (SELEX) (Tuerk and Gold, 1990).

3.2.9 Surface-Enhanced Raman Scattering–Based Sensor

SERS nanosensors are based on the principle of Raman spectroscopy where the Raman scattering signal of analytes adsorbed on metal surfaces is amplified 100–1000-folds. In Raman spectroscopy, the molecules are identified on the basis of their characteristic spectral fingerprint arising due to the unique pattern of molecular vibrations (Duncan, 2011). In order to overcome the major limitation of Raman scattering, the roughened metal surfaces can be used in the proximity of analyte molecules to enhance the signal called the surface-enhanced Raman scattering. The effect arises out of interaction of localized electric fields generated due to photoexitation of surface plasmon of the metal and the molecular electronic states of the analyte. Rough surfaces having lots of curvatures works well for SERS. Thus metallic nanostructures are particularly useful for SERS-based sensing as the interaction is dependent on orientation of the analyte with respect to the surface. Nanomaterials like fractal-like or patterned gold nanostructures (Lin et al., 2008), AuNPs, AgNPs, and Au–Ag core–shell nanoparticles (Ravindranath et al., 2011) have been utilized for the detection. SERS-based nanosensors have been reported for detection of chemicals like melamine (Liu et al., 2010), malachite green, crystal violet (He et al., 2008a), perchlorate (Gu et al., 2009), bacteria (He et al., 2008b; Liu et al., 2008; Wang et al., 2010; Ravindranath et al., 2011, Wang et al., 2011, Craig et al., 2013) and viruses (Fan et al., 2010; Zheng and He, 2014). SERS can be used for rapid screening of samples in conjuction with HPLC analysis to eliminate false-positive samples (Duncan, 2011). The advantages include high sensitivity, minimal sample preparation requirements, high throughput screening, label-free detection, short analyses time, and real-time analysis, but sophisticated instrumentation is required for analysis (Craig et al., 2013).

3.2.10 Liposome Nanovesicles

Liposomes are spherical molecules with an aquesous interior surrounded by a phospholipid bilayer (Fig. 14.15). Nanoliposomes offer advantages of being biocompatible and biodegradable in addition to being small in size. They can be used to improve solubility and bioavailability of bioactive agent and prevent their undesirable interactions with other molecules (Mozafari, 2010). Nanovesicles have been used frequently for encapsulation and delivery in food industry. Regarding biosensing applications, these vesicles can be conveniently used in immunoassays to store a large number of fluorescent molecules, amplifying the signal greatly, thus increasing the sensitivity of the assay (Monroe, 1990). For immunoassay, these polar headgroups of constituting phospholipids can be conjugated to the antibody of interest. This conjugation can be covalent, biotin-streptavidin-based or Fc-binding protein (protein G/protein A) based (Chen et al., 2005). Initially

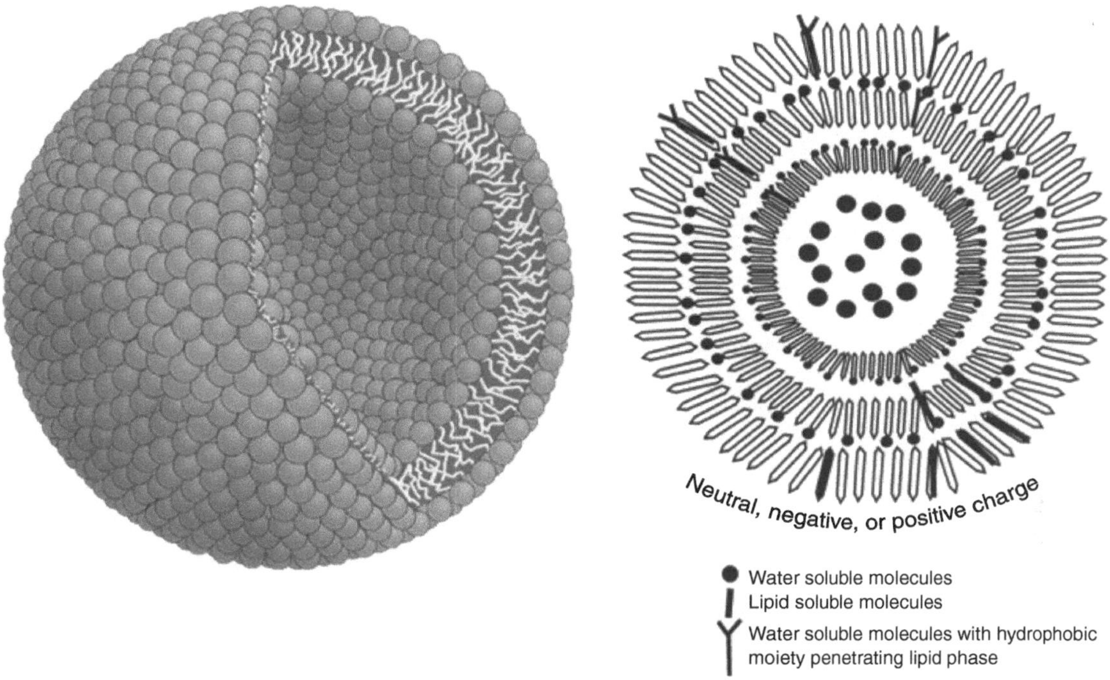

Figure 14.15. Diagrammatic representation of a liposome and inner view showing three bilayer phospholipid membranes alternating with aqueous spaces.

used in solid phase analysis, liposomes have now been used in continuous immunuassays called flow-injection immunoassays (Locascio-Brown et al., 1990).

3.2.11 Fluorescence Enzyme–Linked Immuno Sorbent Assay

A variant of the conventional ELISA method is the fluorescence enzyme–linked immuno sorbent assay (F-ELISA) technique, which has been utilized to detect very low concentrations of proteins in blood. It is called Fluorescence ELISA or F-ELISA as it involves use of an enzymatic reporter, which generates a fluorescent product. The method involves production of microscopic beads with protein-specific antibodies coated on their surface. The beads are placed isolated from each other in femtolitre volume reaction chambers. As the immunocomplex is captured, it is labeled with the fluorogenic enzymatic reporter to generate a fluorescent signal that can be detected using fluorescnence imaging. Quantitative data can be obtained by counting the number of fluorescent wells relative to the total number of wells (Rissin et al., 2010).

3.2.12 Nanowire Biosensor

Nanowires can be defined as unidimensional nanostructures with large aspect ratios useful for manipulating electrons, photons, plasmons, phonons, and atoms (Guo et al., 2013). Nanowires have high surface-to-volume ratio, minimum power consumption, and the potential to be packed in high-density nanoscale devices. They can be used for efficient transport of electrons for optical excitation. Owing to their high surface-to-volume ratio and quantum confinement effect, minor perturbations are sufficient to strongly affect their electrical properties enabling single molecule detection. 1-D structure can be used direct due to limited loss of signal intensities, label free readout, and are particularly suitable for rapid and real-time monitoring (Wanekaya et al., 2006). Optical detection is aided with nanowires because of their high mechanical flexibility, sidewall smoothness, and diameter uniformity (Guo et al., 2013).

3.2.13 Ion Channel Switch

Ion channel switch biosensors utilize electrical transduction due to flux of ions through the channel. The interactions can be measured at the membrane surface to which the ion channel is tethered. Anlaytes such as bacteria, protein, DNA, and drugs can be quantitatively analyzed using this sensor (Woodhouse et al., 1999). Ion channel switch biosensor has been used for rapid screening of influenza A virus in clinical samples (Oh et al., 2008).

3.2.14 Viral

Pathogen detection using biosensor involves specific surface functionalization using molecular probes specific for the target. Antibodies and nucleic acid probes have been routinely utilized for this purpose. However, these probes are susceptible to environmental conditions and are sometimes cross-reactive, costly, and technical expertise is required for the operation. Bacteriophages have been exploited as alternative probes for pathogen detection due to their specificity, selectivity to the host, and ease of amplification. Additionally, as phages can propagate only in live bacteria, they reduce the chances of false positives (Brovko et al., 2012). Once the phage is bound to the host, the host cell can be detected by labeling the phage using fluorescent dyes which can be visualized using fluorescent microscopy or detected using flow cytometery. Phages can be utilized for the replacement of antibodies in detection involving SPR. Genetically engineered phages with modified surface properties to display specific molecules (phage display) or to emit detectable signal (reporter phages) have been utilized for probing. Phage amplification and phage-induced lysis-based methods have also been developed (Cho et al., 2014). However, drawbacks such as induction of lysis of host, drying effect resulting in loss of capturing ability, limit their use (Singh et al., 2013).

3.2.15 PEBBLE: Probes Encapsulated by Biologically Localized Embedding

Spherical optical nanosensors or PEBBLEs range in size from 20 to 600 nm in diameter and consist of fluorescent indicators entrapped in a matrix consisting of polyacrylamide/polydecylmethacrylate or sol–gel (silica, ormosil) nanoparticle (Clark et al., 1999). They encapsulate the sensing element within an inert matrix, thereby protecting it. Owing to their small size, specificity, and sensitivity comparable to macroscopic ion selective electrodes, they can be used for sensing ions in cellular environment with much better response time and detection limit (Buck et al., 2004). PEBBLEs have been developed for intracellular detection of small analytes like protons, calcium, magnesium, zinc, potassium, sodium, chloride, and hydroxyl ions as well as oxygen and glucose.

3.2.16 Nanoshell Biosensor

Metal nanoshells are composed of a core of dielectric nanoparticle like silica encapsulated by an ultrathin metal shell usually made up of gold. These nanoshells have unique optical properties that are amenable to fine-tuning. By manipulating the size and composition of each layer of nanoshell, they can be made to absorb or scatter light in different regions of electromagnetic spectrum. These particles can be easily conjugated with antibodies and other probes and are also effective substrates for SERS, making them an ideal candidate for biosensor design (Hirsch et al., 2006). The restriction of "free-electron-like" metal substrates such as Au, Ag, and Cu nanostructures in SERS can be overcome by using a shell-isolated nanoparticle-enhanced Raman spectroscopy (SHINERS) technique. This technique has a potential to make SERS applicable to surface with any composition and morphology by utilizing Au-core silica-shell nanoparticles (Au@SiO(2) NPs) (Li et al., 2013).

3.2.17 Array Biosensors

Array biosensors are composed of a solid substrate harboring a two-dimensional array of recognition sites (biochips). Biochips can have a separate detection system as in array plate biochips or an integrated detection system such as integrated circuit microsensor chip as in the case of detector array biochips (Vo-Dinh et al., 2001). Biochips offer advantages of multiplexed detection and can be conveniently used for medical diagnosis.

3.2.18 Microfluidic Devices/Lab on a Chip

Microfluidic devices are nanosensors in miniature format. The main advantage includes high sensitivity, greatly reduced working volumes, and scope of miniaturization to increase portability for on-site monitoring (Ravichandran, 2010). Silicon-based microfluidic devices are called laboratory on chip technology and has been used to test food additives like benzoate, sorbate, beta hydroxyl benzoic acid esters, glutamate by carrying out electrophoresis in chip set up (Bodor et al., 2001).

4 Detection Using Nanosensors

The nanosensors have shown great promise for the fabrication of novel biosensors with faster response and higher sensitivity than that of planar sensor configurations, due to their small dimensions combined with a dramatically increased contact surface and strong binding with biological and chemical reagents. Such nanosensors have important applications in food industry.

4.1 Detection of Leakage or Spoilage

Production, processing, and shipment of food products could be made more secure through the use of nanosensors for pathogen and contaminant detection. Nanomaterials are being developed with enhanced mechanical and thermal properties to ensure better protection of foods from exterior mechanical, thermal, chemical, or microbiological effects.

4.1.1 Oxygen

To increase the shelf life, a variety of food items are packaged under vacuum or under modified gaseous atmosphere (Valdés et al., 2009). Leakage of oxygen inside food packaging causes its deterioration owing to lipid oxidation and microbial growth. It may also cause discoloration, change in texture, odor, and flavor (Bowles and Lu, 2014). The traditional method of measurement of oxygen in package is destructive in nature where package is punctured to extract atmosphere of the package and loaded into electrochemical fuel cell for analysis (Bültzingslöwen et al., 2002).

A photoactivated indicator ink developed by Mills (2005) provides users with a method to directly detect modified atmosphere packages (MAPs) which might be damaged during transport or storage. The ink contains a redox active dye—methylene blue—and TiO_2 or SnO_2 nanoparticles, which change color in response to oxygen and allow users to visually analyze the package qualitatively before buying. An oxygen-sensing device has been developed with methylene blue TiO_2 nanocomposite deposited on glass slide, followed by spin-coating of triethanolamine embedded in hydroxy ethyl cellulose under UV irradiation, resulting in bleaching of blue color due to reduction of methylene blue to leuco form. In the presence of molecular oxygen, the dye is reoxidized and blue color reappears (Lee et al. 2005; Gutiérrez-Tauste et al., 2007; Mills and Hazafy, 2008, 2009).

Luminescent nanostructures such as polystyrene-block-vinyl pyrrolidone nanobeads have also been utilized to sense oxygen. The detection is based on quenching of luminescence of the luminescent nanobeads (stained with metal ligand complex). Although not directly applied in food industry but being used in monitoring bioreactors (Borisov and Klimant, 2009).

4.1.2 Moisture

A noninvasive method for detecting moisture level in package was devised by employing a matrix of copper nanoparticles covered by silicon tenside film and a wetting agent. The strip color is based on the change of color of internanoparticle separation due to swelling of polymer matrix in humid environment (Luechinger et al., 2007).

4.1.3 Carbon Dioxide

Carbon dioxide is usually added in packaging to reduce bacterial growth (Bültzingslöwen et al., 2002). To monitor the MAPs, CO_2 is traditionally analyzed using infrared absorption spectrometry. But the analysis leads to destruction of packaging and also does not guarantee complete quality control (Bültzingslöwen et al., 2002). A nondestructive method based on immobilized fluorescent pH indicator using modified silica matrix has been described (Bültzingslöwen et al., 2002). Analysis of the lifetime of luminescent dye was carried out using polymer nanobeads, which encapsulate a fluorophore. This CO_2 sensor has a wide detection range of 0.8–100% having low (0.6%) cross-sensitivity with oxygen.

4.1.4 Gaseous Amine

Gaseous amines such as trimethylamine and others are widely used as indicators of fish and meat spoilage. A nanosensor composed of nanofibrils of perylene-based fluorophores has been used to detect gaseous amines even at very low level (Che et al., 2008; Che and Zhang, 2009). The same could also be determined with SnO_2 composites and TiO_2 microrods whose conductance was dependent on changes in gaseous amine levels (Zhang and Zhang, 2008). The detection limit is found to be low 1 μg mL^{-1} and the sensor is selective and stable with prompt response and recovery.

A quartz crystal microbalance (QCM)-based sensor was developed to detect Trimethylamine (TMA) by depositing a polyaniline TiO_2 nanocomposite on electrode (Zheng et al., 2008). Surfactant-modified ZnO/polyvinyl pyrrolidone composite films are spin-coated on Al electrodes and utilize for sensing TMA by Tang et al. (2006) with a detection limit as little as 0.4 μg mL^{-1}.

4.1.5 Hypoxanthine

Hypoxanthine, a major catabolic product of ATP, imparts a bitter taste in fish and is an indicator of fish freshness. In flatfish, it can be used as measure of duration of icing (Valdés et al., 2009). An amperometric biosensor using gold nanoparticles based on activity of xanthine oxidase was developed by Cubukcu et al. (2007) for determination of hypoxanthine in tuna fish samples. Similar approach was used by Agui et al. (2006) for the amperometric determination of hypoxanthine in stored samples of sardines and chicken meat.

4.1.6 Volatile Organics

A sensor series composed of SnO nanobelts or ZnO-TiO_2 nanocomposites has been utilized for simultaneous detection of volatile organic compounds like acetone, ethanol and carbon monoxide (Comini et al., 2005, 2006; Barreca et al., 2007).

4.1.7 Ethylene Gas

Ethylene is a well-known indicator of fruit ripening, softening, and senescence. Continuous monitoring of ethylene allows storage and distribution centers to manage their produce according to its status. WO_3–SnO_2 nanocomposites have been used for the detection of ethylene gas and consequent ripening of the fruit (Pimtong-Ngam et al., 2007). A WO_3–SnO_2 binary oxide sensor is inserted in the middle of a quartz tube. Ethylene gas is detected using conductivity measurements at 300°C. There is a linear response between 2 and 8 μg mL^{-1} with sensor.

4.2 Detection of Pathogens and Their Products

4.2.1 Pathogens

New technology allows government regulatory agencies to identify a bacterial pathogen and trace it back to its source more rapidly. The majority of foodborne disease outbreaks result from unintentional contamination of a product as a result of inappropriate processing or handling (WHO, 2007). According to data from Centers for Disease Control and Prevention, one-sixth of the US population becomes ill due to foodborne diseases (CDC, 2011). The conventional method to detect microbial contamination includes culture dependent and biochemical identification, immunological detection, and PCR-based detection (Velusamy et al., 2010). Culture-based methods are accurate and reliable but are time consuming and labor intensive. They usually take 7–10 days for confirmatory results (Chu et al., 2008; Velusamy et al., 2010). Immunological methods on the other hand are highly sensitive and faster but have narrow detection range (Craig et al., 2013). PCR-based methods are also very sensitive but might give errors arising from nontarget amplification and require pathogen enrichment. Therefore, rapid, sensitive, and portable pathogen detection is the need of the hour (Craig et al., 2013). SERS-based approach captures pathogens of interest using selective biomolecules, imparting specificity to the sensor. They can be easily used to differentiate between species and strains (Velusamy et al. 2010, 2012). Although there has been significant progress in biosensor design, simultaneous detection of multiple pathogens is often accompanied by compromises in either precision or sensitivity (Ravindranath et al. 2011).

4.2.1.1 *E. coli*

E. coli O157:H7 is major foodborne pathogen which causes an estimated 70,000 infections annually in the USA due to consumption of contaminated raw clover sprouts, ground beef, salads, frozen food samples, organic spinach, and spring-mix blends (CDC, 2015). The infections due to production of microbial toxins damage the intestinal lining, causing stomach cramps and anaemia called haemolytic uremic syndrome and hemorrhagic colitis. Sensing techniques that enable culture enrichment or amplification are of great interest (Valdés et al., 2009) as the minimum dose of *E. coli* O157:H7 that can cause infection is fewer than 100 organisms (Tuttle et al., 1999).

The quartz crystal microbalance (QCM) DNA sensor method involves immobilization of a thiolated, *E. coli* O157:H7 specific ssDNA (eaeA gene) on the surface of QCM sensor through self-assembly (Mao et al., 2006). The hybridization of the ssDNA probe to the complementary target DNA (in presence of pathogen) resulted in mass change, which could be sensed in the form of variation in resonant frequency of the QCM. Streptavidin conjugated Fe_3O_4 nanoparticles are used to amplify the frequency change to enhance the mass. *E. coli* O157:H7 eaeA gene fragments are amplified using asymmetric PCR with biotin labeled primers detected to sense upto 2.67×10^2 colony forming unit (cfu) mL^{-1}. Functionalized –COOH multiwall carbon nanotubes have been utilized in association with Nafion modified glassy carbon electrode (GCE) for detection of coliforms amperometrically (Cheng et al., 2008). A suspension of Nafion (perfluorosulfonated cation-exchanger polymer) solution/MWCNT is poured on the carbon electrode and the solvent is allowed to evaporate to generate the modified GCE. Galactosidase enzyme, characteristic of coliforms when reacts with *p*-aminophenol β-galactopyranoside substrate added in the bacterial solution, hydrolyzes it to *p*-aminophenol detected by the MWCNT/Nafion modified GCE. The method having a detection limit of 10 cfu mL^{-1} for the detection of *E. coli* in river water is normally recommended in the food industry due to rapidity and sensitivity.

Zhang et al. (2007) recommended bismuth nanofilm to modify a glassy carbon electrode for *E. coli* detection for medical, environmental and food applications. The assay is based on the presence of the enzyme β-D-glucuronidase in *E. coli*, which catalyzes the hydrolysis of externally added substrate 4-nitrophenyl β-D-glucuronidase to 4-nitrophenol (4-NP). The reaction is detected in the form of a reduction peak at –0.53 V. The assay is completed in just 3 h and could detect upto 100 cfu mL^{-1} *E. coli* cells.

Immobilized self-assembled peptide nanotubes (PNTs) on carbon paste electrode have been developed to detect *E. coli* O157:H7 electrochemically (Cho et al., 2008). FITC labeled anti-*E. coli* antibodies are then attached to the electrode. The antigen–antibody interaction is sensed by cyclic voltammetry using redox probe; the device is highly portable allowing easy field sampling (Cho et al., 2008).

Impedance-based biosensor for rapid detection of *E. coli* O157:H7 in ground beef has been recommended in which a gold microelectrode is coupled with magenetic (Fe/Fe_2O_3), antibody coated nanoparticles using biotin streptavidin chemistry (Varshney and Li, 2007). Once the target is captured onto the magnetic nanoparticles (via antigen–antibody interaction), the nanoparticles are concentrated on an interdigitated array microelectrode for detection. The method did not require a redox probe, however the LOD (limit of detection) is quite high (3×10^5 cfu mL^{-1}). A further improvement on this type of sensor is to embed labeled magnetic nanoparticles in custom made microfluidic flow cell to achieve lower LOD (Varshney et al., 2005; Varshney et al., 2007).

Temur et al. (2010) employed a SERS-based sandwich immunoassay along with gold nanoparticles to detect *E. coli* with a limit of detection of 5 cfu mL^{-1}. In addition, the method has proved to be highly specific for detection of *E. coli* against *Enterobacter aerogenes* and *Enterobacter dissolvens*. Combined with a SERS-based sandwich immunoassay along with antibody labeled magnetic nanoparticles was employed to enumerate *E. coli* with a linear correlation from 10^1–10^4 cfu mL^{-1} and a LOD of 8 cfu mL^{-1} and analysis time less than 70 min (Guven et al., 2011).

Detection of *E. coli* using optical method has also been reported by Ravindranath et al. (2009). Abdalhai et al. (2015) have developed an electrochemical genosensor for *E. coli* O157:H7 in beef sample. Cho et al. (2015) proposed a simple and rapid technique to detect low levels of *E. coli* O157:H7 in pure culture and ground beef-based on SERS. They utilized two antibody-conjugated nanoparticle complexes (gold and magnetic) to capture and separate target bacteria. The bacteria complexed with the nanoparticles were then localized using silver intensification and analyzed by SERS having high sensitivity (up to 10 cfu mL^{-1}) and short detection time (1–3 h).

4.2.1.2 Salmonella

Salmonella is the second most frequently occurring pathogen causing foodborne illnesses in the USA (CDC, 2015). Infection of *Salmonella enteritidis* and *S. typhimurium* causes Salmonellosis—an enteric disease. *Salmonella infantis* has also been reported to cause foodborne outbreaks associated with the egg and especially the chicken meat. Villamizar et al. (2008) developed a network of single-walled carbon nanotubes SWCNTs to construct a field effect transistor sensor for detecting *S. infantis*. SWCNTs were conjugated with anti-*Salmonella* antibodies. The interaction between antigen–antibody resulted in a decrease in electric current. The detection limit of this biosensor was 100 cfu mL^{-1} of the pathogen in 1 h.

4.2.1.3 *Mycobacterium avium* spp. paratuberculosis

Mycobacterium avium spp. paratuberculosis has been quantitatively detected from contaminated milk by observing effects of magnetic particle agglomeration (Kaittanis et al., 2007).

4.2.1.4 Brucella

Brucella has been detected in blood serum of infected cattle with synthesized magnetic nanoparticles (Fornara et al., 2008).

4.2.1.5 *Listeria monocytogenes*

An IMS-based sensor utilizing magnetic nanoparticles has been developed by immobilizing the specific antibodies on the surface of the functionalized iron oxide nanoparticles. The sensor is utilized for detection of *L. monocytogenes* in artificially contaminated milk samples (Yang et al., 2007). Electrochemical detection of *L. monocytogenes* has also been reported (Wang et al., 2009).

4.2.1.6 *Vibrio parahaemolyticus*

Outbreaks involving *V. parahaemolyticus* are prevalent in areas where raw and undercooked shellfish is consumed in daily diet. An amperometric immunosensor has been described for the detection of *V. parahaemolyticus* (Zhao et al., 2007). The sensor was constructed by employing a composite of agarose doped gold nanoparticles to coat electrode. This membrane is then used to immobilize HRP-labeled anti-*V. parahaemolyticus* antibody. The reduction of cathodic peak current in relation to the presence of *V. parahaemolyticus* was used to sense the pathogen.

4.2.1.7 *Bacillus cereus*

A direct-charge transfer (DCT) biosensor based on antigen–antibody interaction and direct electron flow to generate a resistance signal has been employed to detect the foodborne pathogen, *B. cereus* (Pal et al., 2007). It utilizes *B. cereus* specific antibodies as sensing element and polyaniline nanowire as electrical transducer with a sensitivity of 10–100 cfu mL^{-1} for pure cultures of *B. cereus*.

4.2.1.8 Simultaneous Detection of Multiple Pathogens

Simultaneous detection of multiple pathogens is of particular interest as it saves time and effort to check for presence of different pathogens in addition to saving the sample. Several multipathogen detection nanosensors have been developed.

A SERS-based nanosensor has been described to detect multiple pathogens like *E. coli*, *Salmonella*, or *Listeria* in several matrices in real time (Liu et al., 2008; Weidemaier et al., 2015). The assay bypasses the requirement of wash steps required for complex samples. In addition, the method allows continuous monitoring of pathogen through enrichment process. The SERS-labeled immunoassay reagents are added in the enrichment vessel, and the real-time signal is monitored through the wall of the vessel, which also enables a bio-contained food safety testing (Liu et al., 2008; Weidemaier et al., 2015).

A sensitive multiplex fluorescence immunoassay using multicolor quantum dot probes labeled with the specific antibodies to simultaneously detect three pathogens *S. enteritidis*, *Staphylococcus aureus*, and *E. coli* have been developed to test multiple food samples (Wang et al., 2015). Simultaneous detection of different strains of *Shiga*-toxin producing *E. coli* (STEC) have been demonstrated using optical biosensing with gold nanoparticles functionalized with oligonucleotide (Quintela et al., 2015). SERS-based strategy has been utilized to detect *E. coli, E. coli* O157:H7, *Staphylococcus epidermidis, E. coli* DH 5α, *S. aureus,* and *Salmonella typhimurium* individually and in mixture (Chu et al., 2008). They generated AgNP substrate by oblique angle deposition method. SERS bands common to all bacterial species and specific to different species could be identified. PCA was applied to discriminate between Gram stain types, species, and strains. The reliability was improved by applying linear discrimination analysis. *E. coli, S. typhimurium,* and *S. aureus* at a concentration of 10 cfu mL^{-1} were detected using SERS probes composed of silver nanospheres formed by assembly of silver nanocrystals. Fan et al. (2011) utilized SERS-based strategy involving silver nanosubstrates to differentiate *E. coli* O157:H7, *Listeria monocytogenes, S. epidermidis,* and *Enterococcus faecelis* and compared two approaches, one involving gold-coated microscope slide and second involving internal deposition of AgNPs inside cells. It was concluded that the latter strategy was able to enhance the Raman signals to a large extent compared to the former by detecting and discriminating even single cell (Fan et al., 2011).

Wang et al. (2011) developed a SERS-based immunomagnetic sensor for multipathogen detection in food matrix where *Salmonella enterica* serovar Typhimurium and *S. aureus* were detected in spinach wash water and peanut butter sample using silica-coated magnetic probes labeled with pathogen specific antibodies. Gold nanoparticles integrated with Raman reporters and labeled with antibodies against same pathogen were utilized for the sandwich assay and obtained LOD of 10^3 cfu mL^{-1}. Ravindranath et al. (2011) developed a SERS-based method utilizing different nanoprobes labeled with Raman reporter molecules in addition to pathogen specific antiligands. When the specific pathogen is present, these probes quickly bind to the pathogen surface and filter out on Millipore nanoporous membrane of 450 nm pore size, while the unbound probe is washed out. The probes attached to the captured bacteria can then be used to directly carry out Raman spectra analysis without the requirement of additional processing steps. This strategy was successfully used to detect *S. typhimurium, S. aureus,* and *E. coli* O157:H7 by incorporating gold nanoparticles functionalizd with *S. typhimurium* aptamers, silver NP functionalized with anti-*S. aureus* antibodies and core-shell NPs functionalized with anti-*E. coli* O157:H7 antibodies. The nanoparticles were coupled with Raman reporters and the pathogen was subsequently detected using confocal Raman approach and an LOD ranging from 10^2–10^3 cfu mL^{-1} was achieved (Ravindranath et al., 2011).

4.3 Toxins

4.3.1 *Microcystin-LR*

Traditionally, Microcystin-LR toxin produced by cyanobacterium *Microcystis aeruginosa* in water has been detected using ELISA but now this method can be replaced with a 10 times more sensitive method using carbon nanotubes coated with anti-MCLR antibodies (Wang et al., 2009). The binding of the toxin on the surface initiates a change in electrochemical conduction, which can be detected up to 0.6 nM sensitivity (Wang et al., 2009).

4.3.2 *Aflatoxin in Milk*

A highly sensitive immunosensor that detects aflatoxin B17 in milk has been developed. This sensor is based on piezoelectric AuNP (Jin et al., 2009).

4.3.3 *Ochratoxin A*

Ochratoxin-A foodborne fungal contaminant can be detected electrochemically by utilizing chitosan nanocomposite along with cerium oxide nanoparticle immuosensor (Kaushik et al., 2009). A plasmonic-based optical biosensor for Ochratoxin-A has been developed by Todescato et al. (2014); tested in different matrices like dried milk, juices, and wheat mix. A sensitive Ochratoxin-A sensor based on aptamer has been developed by employing Ochratoxin-specific aptamer in conjunction with complementary DNA and double-strand specific fluorescent dye pico green to detect the toxin (Lv et al., 2014).

4.3.4 *Staphylococcal Enterotoxin B*

Nanowire transistors (Mishra et al., 2008) and carbon nanotubes-based optical immunodetection (Yang et al., 2008) is being used for enterotoxin B detection.

4.3.5 *Cholera Toxin*

Carbon nanotubes-based technology has been developed for detection of cholera toxin with very high sensitivity (Viswanathan et al., 2006).

4.3.6 *Botulinum Toxin Serotype A*

Botulism has been of major concern due to its possible bioterrorism applications. Quantum dots method labeled with the respective antibody has been developed for higher LOD (Warner et al., 2009).

4.4 Banned Dyes and Adulterants

4.4.1 Urea

Urea is added in milk to increase solid not fat value (SNF) and nitrogen content. This misleads the conventional testing of protein content. More than 70 mg dL^{-1} urea in milk can lead to indigestion, urinary tract obstruction, renal failure, gastrointestinal bleeding, and cancer (Nikoleli et al., 2010). Urease functionalized AuNPs-based bio-sensors (Nair et al., 2013) and unmodified gold nanoparticles-based aptamer sensors (Kumar et al., 2015) have been developed for detection of urea.

4.4.2 Melamine

Melamine in dairy products can lead to urinary calculi, renal failure and may result in death in infants (Cheng et al., 2010; Lin et al., 2008). Melamine adulteration has lead to major outbreaks related with contaminated infant formula milk in China in 2008 and contaminated pet food in United States in 2007, leading to market recalls and huge economic losses. Melamine levels need to be monitored (Smoker and Krynitsky, 2008; Turnipseed et al., 2008) in order to maintain the safe level in infant formula (up to 1 mg L^{-1}) and food products (up to 2.5 mg L^{-1}). Nanosensor-based approaches have been utilized to monitor melamine levels in food products replacing time consuming traditional chromatographic detection. An optical sensor has been developed which is based on the property of cyanuric acid to selectively bind melamine. Presence of melamine causes concentration dependent aggregation AuNPs fuctionalized with cyanuric acid resulting in a visible color change from red to blue. This sensor is highly sensitive and can detect melamine up to a concentration as low as 2.5 ppb (Ai et al., 2009). Another approach was based on formation of AuNP formation in presence of reductant. In presence of melamine, the reductant is unable to cause formation of AuNP with no change in color to red (Cao et al., 2010). Melamine has also been detected colorimetrically by using AuNPs and ether modified thiols (Kuang et al., 2011). A visual method based on 3-mercaptopriopionic acid MA molecules conjugated on AuNP surfaces to form MA-modified AuNPs (MA-AuNPs), acting as nanoprobes has been used for the detection of melamine in infant formula (Cai et al., 2014).

Li et al. (2014) have overcome the problems of random spreading and dilution of liquids by fabricating a super hydrophobic-oleophobic mesh-like surface from Ag nanowires for SERS-based detection of melamine and Sudan I in water and toluene. Spiked milk samples were also successfully analyzed using this sensor.

4.4.3 Food Colorants

Ponceau 4R and Allura Red in soft drinks (Zhang et al., 2010) and Sudan 1 in ketchup or chilli powder (Mo et al., 2010) has been detected using carbon nanotubes. The colorant specific oxidation peak undergoes intensity changes on the basis of intensity of analyte. Sunset yellow and chrysoidine have been detected using a SERS-based method utilizing a substrate composed of SiO2@Au nanoshells with a detection limit of 1 ppm and 0.5 ppm respectively (Xie et al., 2014).

4.5 Pharmacological Residues

A SERS-based technique utilizing dendritic AgNPs has been reported for detection of ciprofloxacin, enrofloxacin, and chloramphenicol (He et al., 2009) replacing time-consuming traditional chromatographic antibiotic detection. The nanosubstrate is composed of AgNPs deposited on a gold-coated glass slide. The three antibiotics could be resolved by PCA and LOQ by around 20 ppb. Chloremphenicol and crystal violet, used as an illegal antifungal agent in aquaculture, has also been detected using Klarite and Q-SERS substrates (Lai et al., 2011b) with an LOD of 20–50 ppb. An immuno-chromatographic test strip based on Ru(phen)3(2+)-doped silica fluorescent nanoparticle (FN) has been developed for detection of enrofloxacin (ENR) residues in chicken meat (Huang et al., 2013) and for furazolidone, enrofloxacin, and malachite green (an industrial dye) in tilapia fillets using SERS (Zhang et al., 2012). Clenbuterol, a drug used illegally to promote growth in food animals can impair lung and heart functions in humans. A competitive SERS immnunoassay has been developed with 4,4'-dipyridyl, and clenbuterol antibody labeled AuNPs attached to a substrate for competitive combination to detect this drug in pig urine (Zhu et al., 2011). Sulpha drugs such as sulfamethazine, sulfamerazine, and sulfamethoxazole are used commonly to promote growth of farm animals and to control disease outbreaks. These have been detected using Klarite SERS-active substrate with an LOD of 10 ng mL^{-1} (Lai et al., 2011a). Leanness of swine carcass is improved using drug raptopamine, which is banned in China and European Union due to the potential health risk though legal in USA (Zhai et al., 2011). Overdose of Theophyllin, a drug widely used for airway obstruction treatment can be lethal (Liu et al., 2011b; Mitenko et al., 1973). Theophyllin is found in certain food products such as tea, coffee, and chocolate and has been detected using AgNPs coated with biomimetic receptors with an LOD of 10 μM in spiked green tea samples (Liu et al., 2011b). Phenolic estrogens have been detected indirectly in infant formula using SERRS (surface-enhanced resonance Raman scattering) with an LOD of 0.1 ppb (Han et al., 2011).

4.6 Pesticides

There is a growing concern over the presence of trace amounts of pesticide residues in products derived from the crops. These residues might lead to long-term heath effects (Craig et al., 2013). Therefore, to assure safety of consumers, monitoring of pesticides in agricultural produce on a routine basis is very important. Nanosensor-based rapid, sensitive, and portable methods are being developed to replace routinely used time-consuming chromatographic method for pesticide analysis (Vongsvivut et al., 2010).

4.6.1 Parathion and Paraoxon

Parathion is a moderately toxic organophosphorus insecticide. Wang and Li (2008) developed a ZrO_2/Au nanocomposite film electrode for determination of Parathion with the LOD of 3 ng mL^{-1}. The sensor requires a preabsorption of parathion on electrode and detection was based on square wave voltametry.

Paraxon is a metabolite of organophosphorous insecticide parathion and has neurotoxic property. Constantine et al. (2003) developed an optical nanosensor for detecting paraoxon. The in-solution detection was carried out in a film assembly composed of layers of alternately arranged film of chitosan, organophosphorous hydrolase, and thioglycolic acid capped CdSe quantum dots. Exposure to organophosphorous compound led to its hydrolysis aided by OP hydrolase resulted in altered conformation and subsequent decrease luminescence of quantum dots. A similar approach based on CdSe–ZnS quantum dots was developed by Ji et al. (2005). However, real samples need to be tested to validate the effectiveness (Valdés et al., 2009). Vamvakaki and Chaniotakis (2007) used inhibition of activity of enzyme acetyl cholinesterase to detect Paraxon. The enzyme was encapsulated into liposome along with a pH sensitive fluorophore pyramine. Upon addition of substrate thioacetlycholine, hydrolysis resulted in decrease in pH, which could be sensed by pyramine. The change in pH reduction in presence of pesticide could be monitored fluorometrically.

4.6.2 Amitrole

Amitrole is a herbicide with a maximum detectable residue limit of 0.01 mg kg^{-1} by employing a carbon paste electrode modified with Fe(II)-phthalocyanine (FePh) nanoparticles to detect amitrole using square wave voltametry (Siswana et al., 2006). Crystal violet and malachite green are banned antimicrobials, which are used in aquaculture to improve production. If not used within prescribed limits, they can be detected in fish grown in contaminated water (Zheng and He, 2014).

4.6.3 Fonofos (Organophosphorous Pesticide)

A highly sensitive SERS-based technique for detecting fonofos pesticide utilizing metal colloids and AgNPs has been developed by Vongsvivut et al. (2010).

4.6.4 Sulphur-Containing Pesticides

A SERS-based approach have been used to identify sulphur containing pesticides in grapes, apple, mango, pear, and peach peels (Liu et al., 2011a). Silver-coated AuNPs have been used with strong absorption of plasmonic resonance to achieve much higher enhancement of Raman scattering than the individual AuNPs or AgNPs in this technique.

4.6.5 Tricyclazole

Quantification of tricyclazole was based on SERS and a portable Raman instrument is used to carry out the analysis in paddy rice (Tang et al., 2012). Pesticides that enter through water are detected using nanoscale liposome-based detector (Vamvakaki and Chaniotakis, 2007).

4.7 Contamination Caused by Packaging Material

Melamine used in packaging material may also enter food under inappropriate storage/ handling conditions. As already described earlier, melamine contamination can be detected by using AuNPs (Ai et al., 2009; Cao et al., 2010; Kuang et al., 2011).

4.8 Allergens

Gliadin (inflammation causing protein in patients suffering from celiac disease) and gluten can be detected using nanostructured silver island films in close proximity to rhodamine labeled antigliadin antibodies. The metal-enhanced fluorescence is detected using enhanced fluorescence linked immuno-sorbent assay (Staiano et al., 2009) while peanut allergens by Speroni et al. (2010).

5 Nanobiosensors in Food Technology Market

The nanotechnology market in future will be driven by enhanced sensitivity and high throughput detection. The food and medical industries will offer maximum opportunities in this regard in areas of quality management and control during production, processing, packaging, storage, and transportation. The growth in

Table 14.3 Commercialized Nanobiosensors for the Food Industry

Contaminant	Broad Category	Method of Detection	Commercial Name	References
Trans fat content	NEMS	Digital transform spectrometer	Polychromix	Ritter (2005)
Pathogens in food and water	Cantilever	Sensing ligand receptor interactions	European funded project: BioFinger	Jain (2008)
Flavor analysis of vegetable oils	Surface acoustic wave electronic nose	SAW vapor print	Z-Nose	Gan et al. (2005)

revenue generation is expected to be concomitant with advancements in nanotechnology. Nanosensors incorporating improved nanomaterials showing better characteristics will be the major thrust in the market. Factors such as improved design, reduced complexity, and wireless architechture will also play a critical role. As these nanosensors begin to prove their worth, their demand will be enhanced as evident from the example of nanobots—the nanosensors containing tiny transducer chips for storage. The total market share for nano sensor in 2012 was $1.2 billion and some important companies involved with nanosensor research and development are (1) Dow Corning, (2) Samsung, (3) Boeing, (4) Lockheed Martin, (5) IBM, (6) Motorola, (7) Agilent, (8) startups (Nanomix and Ambri). Few nanobiosensors that have been commercialized are listed in Table 14.3.

6 Safety and Challenges

Toxicity of nanoparticles is dependent upon a large number of factors including the size, structure, shape, elemental constituents, mass concentration, solubility, surface area, surface charge, and surface chemistry (Tiede et al., 2008). Enhanced bioavailability of certain nutrients or food additives might have unfavorable effects (Ravichandran, 2010).

Nanomaterials used in food packaging are usually not inhaled or ingested by the end users, although there is a risk that they may enter food. If this happens, their use should be based on acceptable daily intake (ADI) (Ravichandran, 2010). In addition, the effect of these nanomaterials on the normal microbial flora of gut and mouth needs to be assessed beforehand (Sozer and Kokini, 2009). A second possible exposure is via the inadvertent release of nanomaterials in the environment such as disposal of smart surfaces. This type of exposure may have unforeseen, untested effects (Ravichandran, 2010). Therefore, the long-term fate and disposal of smart surfaces should be considered prior to use. Presently there are no worldwide-accepted rules or regulations regarding nanotechnology-based food products. Manufacturers have to follow general existing regulations for food products; however, the guidelines will vary from country to country. The USFDA does not specifically cover nanoparticles. The Institute of Food Science and Technology (UK), European Commission reiterates the requirement of modification of existing laws for example, ingredient labeling to allow consumer to take informed decisions. TA Swiss (Swiss center for technology assessment) allows use if additives appear on a positive list of tested items identifiable by an E-number, which is again not nanomaterial-specific (Ravichandran, 2010). The list of generally accepted as safe (GRAS) additives should be reexamined when used at a nanoscale level (Ravichandran, 2010). Ultimate success of any nanomaterial-based product depends on consumer acceptance. Public needs to be groomed to increase acceptability of these 'Atomically modified foods." They should be allowed to take informed decisions by incorporating the nanomaterial component in the labeling (Ravichandran, 2010).

7 Future Prospects

The most promising breakthrough of the development of online or on-site, sensitive, low-cost, rapid methods for routine use are expected to be made in the area of nanosensor technology. Many prototypes for food diagnostic application in the food and drink industry are currently being developed. They have high potential for automation and allow the construction of simple and portable equipment for fast analysis. These properties will open up many applications within quality and process control, quality and safety control of raw materials, and for HACCP monitoring. New technology must have high sensitivity, high specificity, high precision (repeatability), at the same time rapid, robust, and cheap. There is currently no method that will fulfill all requirements. A variety of nanosensor platforms are available for sensing applications in food industry that offer specific, sensitive, and cost-effective detection of contaminants. However, these technologies have to prove their worth for real food samples. The major challenge among the detection systems used is the refinement of sample preparation steps so that even an unskilled person can perform the analysis. An ideal sensor would be one in which the sample preparation module is integrated with the on-site sensing system. Thus, versatile, multifunctional sensing systems, which can work in a variety of complex food matrices are the need of the hour. Although multiplexing has been initiated with the e-nose technology for detection and analysis of complex aroma compounds, further work is required to delineate sample processing steps for different types of food matrices so that the same sensor could work in different types of food samples. Efforts are also likely to be made to reduce the material cost in nanosensor-based systems and to integrate sensors in packaging material for direct consumer analysis. Future applications might include protein quality and the detection of allergens, genetically modified proteins, BSE prions, pathogens, and biocide residues.

8 Conclusions

Nanomaterials have no doubt been the major drivers in the field of biosensing in food industry catering to the specific needs of the industry including safety testing, real-time monitoring of food products during storage and transportation and on-site testing to ensure consumer satisfaction. With their widely acclaimed qualities such as appropriate and easily alterable size, unique electrical, optical, chemical, magnetic, and semiconducting properties and enhanced sensing surface area that can be easily functionalized, they surpass the routinely used biosensors in terms of sensitivity and accuracy, quick response, and direct localized detection. A range of nanosensor platforms is now available with widely differing detection principles. The simplest nanosensors are colorimetric metal nanoparticle detectors, which produce a visible reaction upon reaction with analyte. Carbon nanotubes have superior mechanical properties and have found application in nanoelectromechanical system (NEMS). Electronic noses and electronic tongues mimic the human nose and tongue by decoding patterns of volatile and nonvolatile compounds respectively. Nanocantilevers work by generating mechanical bending of nanodevices in response to the presence of analyte. Optical nanosensors can be used to probe subcellular locations using fluorescence phenomenon and a fiber-optic nanoprobe. Aptamer sensors can be used to design biological recognition molecules similar to antibodies based on custom requirements. SERS utilize Raman scattering signal of analytes adsorbed on nanomaterial surfaces. Liposome nanovesicles have been used to amplify fluorescent signals in immunosensing applications. PEBBLE can be used for intramolecular detection of small analytes. Microfluidic devices/lab-on-a-chip devices offer advantages of working with small sample volumes and rapid detection. Nanosensors have been applied for various applications including detection of pathogens, toxins, spoilage, banned chemicals, pharmacological residues, pesticides, packaging material residues, and allergens. In future, further advancements are likely to be directed at decreasing the cost of the sensor, increasing the rapidity for on site testing, automated analysis, as well as streamlining the sample preparation methods for diverse food matrices. Prospective applications include development of nanosensors for detecting genetically modified proteins, BSE prions, and biocide residues.

Acknowledgments

NMK would like to acknowledge Dr Purnima Dhall, SERB Scientist, Dayal Singh College, University of Delhi, for sharing her academic resources.

References

Abdalhai, M.H., Fernandes, A., Xia, X., Musa, A., Ji, J., Sun, X., 2015. Electrochemical genosensor to detect pathogenic bacteria (*E. Coli* O157:H7) as applied in real food samples (fresh beef) to improve food safety and quality control. J. Agri. Food Chem. 63, 5017–5025.

Agui, L., Manso, J., Yañez, P., Pingarron, J., 2006. Amperometric biosensor for hypoxanthine based on immobilized xanthine oxidase on nanocrystal gold–carbon paste electrodes. Sens. Actuat. B 113, 272–280.

Ai, K., Liu, Y., Lu, L., 2009. Hydrogen-bonding recognition-induced color change of gold nanoparticles for visual detection of melamine in raw milk and infant formula. Am. Chem. Soc. 131, 9496–9497.

Ampuero, S., Bosset, J.O., 2003. The electronic nose applied to dairy products: a review. Sens. Actuat. B 94, 1–12.

Apetrei, C., Rodrıguez-Méndez, M.L., Parra, V., Gutierrez, F., De Saja, J.A., 2004. Array of voltammetric sensors for the discrimination of bitter solutions. Sens. Actuat. B 103, 145–152.

Arugula, M.A., Simonian, A., 2014. Novel trends in affinity biosensors: current challenges and perspectives. Meas. Sci. Technol. 25, 032001.

ASTA, 2005. ASTA White Paper on Sudan and Related Dyes. Available from: http://www.astaspice.org/pubs/SudanWhitePaper.pdf.

Baeumner, A.J., 2003. Biosensors for environmental pollutants and food contaminants. Anal. Bioanal. Chem. 377, 434–445.

Baldwin, E.A., Bai, J., Plotto, A., Dea, S., 2011. Electronic noses and tongues: applications for the food and pharmaceutical industries. Sensors 11, 4744–4766.

Barreca, D., Comini, E., Ferrucci, A.P., Gasparotto, A., Maccato, C., Maragno, C., Sberveglieri, G., Tondello, E., 2007. First example of ZnO−TiO2 nanocomposites by chemical vapor deposition: structure, morphology, composition, and gas sensing performances. Chem. Mater. 19, 5642–5649.

Bisen, P.S., 2014a. Biosensor. Laboratory Protocols in Applied Life Sciences. CRC Press, Boca Raton, pp. 1173–1212.

Bisen, P.S., 2014b. Microbes in food and dairy producton. Microbes in Practice. IK International, New Delhi, pp. 1089–1157.
Bisen, P.S., Debnath, M., Prasad, G.B.K.S., 2012. Food diagnostics, food preservation, and foodborne microbial disease. Microbes: Concepts and Applications. Wiley-Blackwell, Hoboken, New Jersey, pp. 649–684.
Bodor, R., Madajová, V., Kaniansky, D., Masár, M., Jöhnck, M., Stanislawski, B., 2001. Isotachophoresis and isotachophoresis-zone electrophoresis separations of inorganic anions present in water samples on a planar chip with column-coupling separation channels and conductivity detection. J. Chromatogr. A 916, 155–165.
Borisov, S.M., Klimant, I., 2009. Luminescent nanobeads for optical sensing and imaging of dissolved oxygen. Microchim. Acta 164, 7–15.
Botek, P., PouStka, J., Hajslova, J., 2007. Determination of banned dyes in spices by liquid chromatography–mass spectrometry. Czech J. Food Sci. 25, 17–24.
Bowles, M., Lu, J., 2014. Removing the blinders: a literature review on the potential of nanoscale technologies for the management of supply chains. Technol. Forecast. Soc. 82, 190–198.
Brovko, L.Y., Anany, H., Griffiths, M.W., 2012. Bacteriophages for detection and control of bacterial pathogens in food and food-processing environment. Adv. Food Nutr. Res. 67, 241–288.
Buck, S.M., Xu, H., Brasuel, M., Philbert, M.A., Kopelman, R., 2004. Nanoscale probes encapsulated by biologically localized embedding (PEBBLEs) for ion sensing and imaging in live cells. Talanta 63, 41–59.
Bültzingslöwen, C.V., McEvoy, A.K., McDonagh, C., MacCraith, B.D., Klimant, I., Krause, C., Wolfbeis, O.S., 2002. Sol–gel based optical carbon dioxide sensor employing dual luminophore referencing for application in food packaging technology. Analyst 127, 1478–1483.
Buratti, S., Benedetti, S., Scampicchio, M., Pangerod, E., 2004. Characterization and classification of Italian Barbera wines by using an electronic nose and an amperometric electronic tongue. Anal. Chim. Acta. 525, 133–139.
Cai, H., Yu, X., Dong, H., Cai, J., Yang, P., 2014. Visual and absorption spectroscopic detections of melamine with 3-mercaptopriopionic acid-functionalized gold nanoparticles: a synergistic strategy induced nanoparticle aggregates. J. Food Eng. 142, 163–169.
Cao, Q.A., Zhao, H., He, Y.J., Li, X.J., Zeng, L.X., Ding, N., Wang, J.A., Yang, J., Wang, G.W., 2010. Hydrogen-bonding-induced colorimetric detection of melamine by nonaggregation-based Au-NPs as a probe. Biosens. Bioelectron. 25, 2680–2685.
CDC, 2011. Estimates: Findings. Centers for Disease Control and Prevention, Atlanta, GA, http://www.cdc.gov/foodborneburden/index.html.
CDC, 2015. Reports of Selected *E. coli* Outbreak Investigations. Atlanta, GA, Centers for Disease Control and Prevention. Available from: http://www. cdc. gov/ ecoli/outbreaks.html.
Che, Y., Zhang, L., 2009. Enhanced fluorescence sensing of amine vapor based on ultrathin nanofibers. Chem. Commun. 34, 5106–5108.
Che, Y., Yang, X., Loser, S., Zang, L., 2008. Expedient vapor probing of organic amines using fluorescent nanofibers fabricated from an n-type organic semiconductor. Nano Lett. 8, 2219–2223.
Chen, C.S., Baeumner, A.J., Durst, R.A., 2005. Protein G-liposomal nanovesicles as universal reagents for immunoassays. Talanta 67, 205–211.
Cheng, Y., Liu, Y., Huang, J., Xian, Y., Zhang, W., Zhang, Z., Jin, L., 2008. Rapid amperometric detection of coliforms based on MWNTs/Nafion composite film modified glass carbon electrode. Talanta 75, 167–171.

Cheng, Y., Dong, Y., Wu, J., Yang, X., Bai, H., Zheng, H., Ren, D., Jhou, Y., Li, M., 2010. Screening melamine adulterant in milk powder with laser Raman spectrometry. J. Food Compos. Anal. 23, 199–202.

Cho, E.C., Choi, J.W., Lee, M., Koo, K.K., 2008. Fabrication of an electrochemical immunosensor with self-assembled peptide nanotubes. Colloid Surf. A 313, 95–99.

Cho, I.H., Radadia, A., Farrokhzad, K., Ximenes, E., Bae, E., Singh, A.K., Oliver, H., Ladisch, M., Bhunia, A.K., Applegate, B., 2014. Nano/micro and spectroscopic approaches to food pathogen detection. Annu. Rev. Anal. Chem. 7, 65–88.

Cho, I.H., Bhandari, P., Patel, P., Irudayaraj, J., 2015. Membrane filter-assisted surface-enhanced Raman spectroscopy for the rapid detection of *E. coli* O157:H7 in ground beef. Biosens. Bioelectron. 64, 171–176.

Chu, H., Huang, Y., Zhao, Y., 2008. Silver nanorod arrays as a surface-enhanced Raman scattering substrate for foodborne pathogenic bacteria detection. Appl. Spectrosc. 62, 922–931.

Clark, H.A., Hoyer, M., Philbert, M.A., Kopelman, R., 1999. Optical nanosensors for chemical analysis inside single living cells. 1. Fabrication, characterization, and methods for intracellular delivery of PEBBLE sensors. Anal. Chem. 71, 4831–4836.

Comini, E., Faglia, G., Sberveglieri, G., Calestani, D., Zanotti, L., Zha, M., 2005. Tin oxide nanobelts electrical and sensing properties. Sens. Actuat. B 111, 2–6.

Comini, E., Faglia, G., Sberveglieri, G., Zanotti, L., 2006. Oxide nanobelts as conductometric gas sensors. Mater. Manuf. Process. 21, 229–232.

Constantine, C.A., Gattás-Asfura, K.M., Mello, S.V., Crespo, G., Rastogi, V., Cheng, T.C., DeFrank, J.J., Leblanc, R.M., 2003. Layer-by-layer films of chitosan, organophosphorus hydrolase, and thioglycolic acid-capped CdSe quantum dots for the detection of paraoxon. J. Phys. Chem. B 107, 13762–13764.

Craig, A.P., Franca, A.S., Irudayaraj, J., 2013. Surface-enhanced Raman spectroscopy applied to food safety. Annu. Rev. Food Sci. Technol. 4, 369–380.

Cubukcu, M., Timur, S., Anik, U., 2007. Examination of performance of glassy carbon paste electrode modified with gold nanoparticle and xanthine oxidase for xanthine and hypoxanthine detection. Talanta 74, 434–439.

Debnath, M., Prasad, G.B.K.S., Bisen, P.S., 2010. Food diagnostics. Molecular Diagnostics: Promises and Possibilities. Springer, Dordrecht, pp. 467–482.

Duncan, T.V., 2011. Applications of nanotechnology in food packaging and food safety: barrier materials, antimicrobials and sensors. J. Colloid Interface Sci. 363, 1–24.

EFSA, 2005. Review the toxicology of a number of dyes illegally present in food in the EU. EFSA J. 263, 1–71.

Fan, C., Hu, Z., Riley, L.K., Purdy, G.A., Mustapha, A., Lin, M., 2010. Detecting food- and waterborne viruses by surface-enhanced Raman spectroscopy. J. Food Sci. 75, M302–M307.

Fan, C., Hu, Z., Mustapha, A., Lin, M., 2011. Rapid detection of food- and waterborne bacteria using surface-enhanced Raman spectroscopy coupled with silver nanosubstrates. Appl. Microbiol. Biotechnol. 92, 1053–1061.

Fornara, A., Johansson, P., Petersson, K., Gustafsson, S., Qin, J., Olsson, E., Ilver, D., Krozer, A., Muhammed, M., Johansson, C., 2008. Tailored magnetic nanoparticles for direct and sensitive detection of biomolecules in biological samples. Nano lett. 8, 3423–3428.

Francis, L., Cynkar, W., 2000. Investigation of an "electronic nose". Aust. NZ Wine Ind. 15, 54–56.

Frías, S., Conde, J.E., Rodríguez, M.A., Dohnal, V., Pérez-Trujillo, J.P., 2002. Metallic content of wines from the Canary Islands (Spain). Application of artificial neural networks to the data analysis. Food/Nahrung 46, 370–375.

Fritz, J., 2008. Cantilever biosensors. Analyst 133, 855–863.
Gan, H.L., Man, Y.C., Tan, C.P., NorAini, I., Nazimah, S.A.H., 2005. Characterization of vegetable oils by surface acoustic-wave-sensing electronic nose. Food Chem. 89, 507–518.
García, M., Aleixandre, M., Gutiérrez, J., Horrillo, M.C., 2006. Electronic nose for wine discrimination. Sens. Actuat. B 113, 911–916.
Gu, B., Ruan, C., Wang, W., 2009. Perchlorate detection at nanomolar concentrations by surface-enhanced Raman scattering. Appl. Spectrosc. 63, 98–102.
Guo, X., Ying, Y., Tong, L., 2013. Photonic nanowires: from subwavelength waveguides to optical sensors. Acc. Chem. Res. 47, 656–666s.
Gustavsson, J., Cederberg C., van Otterdijk R., Meybeck A., 2011. FAO: Global Food Losses and Food Waste: Extent, Causes, and Prevention. Rome, Food and Agriculture Organization, pp. 1–23.
Gutiérrez-Tauste, D., Domènech, X., Casañ-Pastor, N., Ayllón, J.A., 2007. Characterization of methylene blue/TiO 2 hybrid thin films prepared by the liquid phase deposition (LPD) method: application for fabrication of light-activated colorimetric oxygen indicators. J. Photochem. Photobiol. A 187, 45–52.
Guven, B., Basaran-Akgul, N., Temur, E., Tamer, U., Boyacı, I˙.H., 2011. SERS-based sandwich immunoassay using antibody-coated magnetic nanoparticles for *Escherichia coli* enumeration. Analyst 136, 740–748.
Hai, Z., Wang, J., 2006. Electronic nose and data analysis for detection of maize oil adulteration in sesame oil. Sens. Actuat. B 119, 449–455.
Hall, R., 2002. Biosensor technologies for detecting microbiological foodborne hazards. Microbes Infect. 4, 425–432.
Han, X.X., Pienpinijtham, P., Zhao, B., Ozaki, Y., 2011. Coupling reaction-based ultrasensitive detection of phenolic estrogens using surface-enhanced resonance Raman scattering. Anal. Chem. 83, 8582–8588.
Hansen, T., Boholt, K., Gammelgaard, E., Byrne, D.V., 2002. Sensory and electronic nose analysis of porcine meat loaf in relation to ingredient quality. Sensors, Proceedings of IEEE, vol. 1, pp. 736–740.
He, L., Kim, N.J., Li, H., Hu, Z., Lin, M., 2008a. Use of a fractal-like gold nanostructure in surface-enhanced Raman spectroscopy for detection of selected food contaminants. J. Agric. Food Chem. 56, 9843–9847.
He, L., Liu, Y., Lin, M., Mustafa, A., Wang, Y., 2008b. Detecting single *Bacillus* spores by surface-enhanced Raman spectroscopy. Sens. Instrum. Food Qual. 2, 247–253.
He, L., Lin, M., Li, H., Kim, N.J., 2009. Surface-enhanced Raman spectroscopy coupled with dendritic silver nanosubstrate for detection of restricted antibiotics. J. Raman Spectrosc. 41, 739–744.
Hirsch, L.R., Gobin, A.M., Lowery, A.R., Tam, F., Drezek, R.A., Halas, N.J., West, J.L., 2006. Metal nanoshells. Ann. Biomed. Eng. 34, 15–22.
Hu, Y.F., Zhang, Z.H., Zhang, H.B., Luo, L.J., Yao, S.Z., 2012. Sensitive and selective imprinted electrochemical sensor for pnitrophenol based on ZnO nanoparticles/carbon nanotubes doped chitosan film. Thin Solid Films 520, 5314–5321.
Huang, X., Aguilar, Z.P., Li, H., Lai, W., Wei, H., Xu, H., Xiong, Y., 2013. Fluorescent Ru(phen)3(2+)-doped silica nanoparticles-based ICTS sensor for quantitative detection of enrofloxacin residues in chicken meat. Anal. Chem. 85, 5120–5128.
Iiyama, S., Ezaki, S., Toko, K., 2009. Sensitivity-improvement of taste sensor by change of lipid concentration in membrane. Sens. Actuat. B Chem. 141, 343–348.
Jain, K.K., 2008. The Handbook of Nanomedicine. Springer, Dordrecht, Netherlands.

Ji, X., Zheng, J., Xu, J., Rastogi, V.K., Cheng, T.C., DeFrank, J.J., Leblanc, R.M., 2005. (CdSe) ZnS quantum dots and organophosphorus hydrolase bioconjugate as biosensors for detection of paraoxon. J. Phys. Chem. B 109, 3793–3799.

Jianrong, C., Yuqing, M., Nongyue, H., Xiaohua, W., Sijiao, L., 2004. Nanotechnology and biosensors. Biotechnol. Adv. 22, 505–518.

Jin, X., Jin, X., Chen, L., Jiang, J., Shen, G., Yu, R., 2009. Piezoelectric immunosensor with gold nanoparticles enhanced competitive immunoreaction technique for quantification of aflatoxin B1. Biosens. Bioelectron. 24, 2580–2585.

Kaittanis, C., Naser, S.A., Perez, J.M., 2007. One-step, nanoparticle-mediated bacterial detection with magnetic relaxation. Nano Lett. 7, 380–383.

Kaushik, A., Solanki, P.R., Pandey, M.K., Ahmad, S., Malhotra, B.D., 2009. Cerium oxide-chitosan based nanobiocomposite for foodborne mycotoxin detection. Appl. Phys. Lett. 95, 173703-1-173703-3.

Kobayashi, T., Okada, A., Fujii, Y., Niimi, K., Hamamoto, S., Yasui, T., Tozawa, K., Kohri, K., 2010. The mechanism of renal stone formation and renal failure induced by administration of melamine and cyanuric acid. Urol. Res. 38, 117–125.

Kuang, H., Chen, W., Yan, W., Xu, L., Zhu, Y., Liu, L., Chu, H., Peng, C., Wang, L., Kotov, N.A., Xu, C., 2011. Crown ether assembly of gold nanoparticles: melamine sensor. Biosens. Bioelectron. 26, 2032–2037.

Kumar, P., Lambadi, P.R., Navani, N.K., 2015. Nonenzymatic detection of urea using unmodified gold nanoparticles based aptasensor. Biosens. Bioelectron. 72, 340–347.

Lai, K., Zhai, F., Zhang, Y., Wang, X., Rasco, B.A., Huang, Y., 2011a. Application of surface enhanced Raman spectroscopy for analyses of restricted sulfa drugs. Sens. Instrum. Food Qual. Saf. 5, 91–96.

Lai, K., Zhang, Y., Du, R., Zhai, F., Rasco, B.A., Huang, Y., 2011b. Determination of chloramphenicol and crystal violet with surface enhanced Raman spectroscopy. Sens. Instrum. Food Qual. Saf. 5, 19–24.

Lee, S.K., Sheridan, M., Mills, A., 2005. Novel UV-activated F indicator. Chem. Mater. 17, 2744–2751.

Li, S., Singh, J., Li, H., Banerjee, I.A., 2011. Fabrication and Evaluation of Nanoparticle-Based Biosensors in Biosensor Nanomaterials. Wiley-Blackwell, Hoboken, New Jersey.

Li, J.F., Tian, X.D., Li, S.B., Anema, J.R., Yang, Z.L., Ding, Y., Wu, Z.F., Zeng, Y.M., Chen, Q.Z., Ren, B., Wang, Z.L., Tian, Z.Q., 2013. Surface analysis using shell-isolated nanoparticle-enhanced Raman spectroscopy. Nat. Protoc. 8, 52–65.

Li, X., Lee, H.K., Phang, I.Y., Lee, C.K., Ling, X.Y., 2014. Superhydrophobic-oleophobic Ag nanowire platform: an analyte-concentrating and quantitative aqueous and organic toxin surface-enhanced Raman scattering sensor. Anal. Chem. 86, 10437–10444.

Li, Z., Muhandiramlage, T.P., Keogh, J.P., Hall, Jr., H.K., Aspinwall, C.A., 2015. Aptamer-functionalized porous phospholipid nanoshells for direct measurement of Hg^{2+} in urine. Anal. Bioanal. Chem. 407, 953–960.

Lim, S., Koo, O.K., You, Y.S., Lee, Y.E., Kim, M.S., Chang, P.S., Kang, D.H., Yu, J.H., Choi, Y.J., Gunasekaran, S., 2012. Enhancing nanoparticle-based visible detection by controlling the extent of aggregation. Sci. Rep. 2, 456.

Lin, M., He, L., Awika, J., Yang, L., Ledoux, D.R., Mustapha, A., 2008. Detection of melamine in gluten, chicken feed, and processed foods using surface enhanced Raman spectroscopy and HPLC. J. Food Sci. 73, T129–T134.

Liu, Y., Chen, Y.R., Nou, X., Kim, M.S., Chao, K., 2008. Label-free SERS for rapid species identification of *Escherichia coli, Listeria monocytogenes*, and *Salmonella typhimurium* bacteria. Spectroscopy 23, 48.

Liu, B., Lin, M., Li, H., 2010. Potential of SERS for rapid detection of melamine and cyanuric acid extracted from milk. Sens. Instrum. Food Qual. Saf. 4, 13–19.
Liu, B., Han, G., Zhang, Z., Liu, R., Jiang, C., Wang, S., Han, M.Y., 2011a. Shell thickness-dependent Raman enhancement for rapid identification and detection of pesticide residues at fruit peels. Anal. Chem. 84, 255–261.
Liu, P., Liu, R., Guan, G., Jiang, C., Wang, S., Zhang, Z., 2011b. Surface-enhanced Raman scattering sensor for theophylline determination by molecular imprinting on silver nanoparticles. Analyst 136, 4152–4158.
Locascio-Brown, L., Plant, A.L., Horvath, V., Durst, R.A., 1990. Liposome flow injection immunoassay: implications for sensitivity, dynamic range, and antibody regeneration. Anal. Chem. 62, 2587–2593.
Luechinger, N.A., Loher, S., Athanassiou, E.K., Grass, R.N., Stark, W.J., 2007. Highly sensitive optical detection of humidity on polymer/metal nanoparticle hybrid films. Langmuir 23, 3473–3477.
Lv, Z., Chen, A., Liu, J., Guan, Z., Zhou, Y., Xu, S., Li, C., 2014. A simple and sensitive approach for Ochratoxin A detection using a label-free fluorescent aptasensor. PloS One 9, e85968.
Majdalawieh, A., Kanan, M.C., El-Kadri, O., Kanan, S.M., 2014. Recent advances in gold and silver nanoparticles: synthesis and applications. J. Nanosci. Nanotechnol. 14, 4757–4780.
Man, Chen Y.B., Syahariza, Z.A., Mirghani, M.E.S., Jinap, S., Bakar, J., 2005. Analysis of potential lard adulteration in chocolate and chocolate products using Fourier transform infrared spectroscopy. Food Chem. 90, 815–819.
Mao, X., Yang, L., Su, X.L., Li, Y., 2006. A nanoparticle amplification based quartz crystal microbalance DNA sensor for detection of *Escherichia coli* O157:H7. Biosens. Bioelectron. 21, 1178–1185.
Mills, A., 2005. Oxygen indicators and intelligent inks for packaging food. Chem. Soc. Rev. 34, 1003–1011.
Mills, A., Hazafy, D., 2008. A solvent-based intelligence ink for oxygen. Analyst 133, 213–218.
Mills, A., Hazafy, D., 2009. Nanocrystalline SnO 2-based, UVB-activated, colorimetric oxygen indicator. Sens. Actuat. B 136, 344–349.
Mishra, N.N., Maki, W.C., Cameron, E., Nelson, R., Winterrowd, P., Rastogi, S.K., Filanoski, B., Maki, G.K., 2008. Ultra-sensitive detection of bacterial toxin with silicon nanowire transistor. Lab Chip 8, 868–871.
Mitenko, P.A., Richard, I., Ogilvie, M.D., 1973. Rational intravenous doses of theophylline. N. Engl. J. Med. 289, 600–603.
Mo, Z., Zhang, Y., Zhao, F., Xiao, F., Guo, G., Zeng, B., 2010. Sensitive voltammetric determination of Sudan I in food samples by using gemini surfactant–ionic liquid–multiwalled carbon nanotube composite film modified glassy carbon electrodes. Food Chem. 121, 233–237.
Monroe, D., 1990. Novel liposome ixnniunoassays for detecting antigens, antibodies, and haptens. J. Liposome Res. 1, 339–377.
Mozafari, M.R., 2010. Nanoliposomes: preparation and analysis. Liposomes. Humana Press, New York, pp. 29–50.
Nair, L.V., Philips, D.S., Jayasree, R.S., Ajayaghosh, A., 2013. A near-infrared fluorescent nanosensor (AuC@ Urease) for the selective detection of blood urea. Small 9, 2673–2677.
Nikoleli, G.P., Nikolelis, D.P., Methenitis, C., 2010. Construction of a simple optical sensor based on air stable lipid film with incorporated urease for the rapid detection of urea in milk. Anal. Chim. Acta 675, 58–63.
Nugen, S.R., Baeumner, A.J., 2008. Trends and opportunities in food pathogen detection. Anal. Bioanal. Chem. 391, 451–454.

Oh, S.Y., Cornell, B., Smith, D., Higgins, G., Burrell, C.J., Kok, T.W., 2008. Rapid detection of influenza A virus in clinical samples using an ion channel switch biosensor. Biosens. Bioelectron. 23, 1161–1165.

Oliveros, M.C.C., Pavon, J.L.P., Pinto, C.G.B., Laespada, M.E.F., Cordero, B.M., Forina, M., 2002. Electronic nose based on metal oxide semiconductor sensors as a fast alternative for the detection of adulteration of virgin olive oils. Analytica Chimica Acta 459, 219–228.

Pal, S., Alocilja, E.C., Downes, F.P., 2007. Nanowire labeled direct-charge transfer biosensor for detecting *Bacillus* species. Biosens. Bioelectron. 22, 2329–2336.

Palanisamy, S., Cheemalapati, S., Chen, S.M., 2012. An enzymatic biofuel cell based on electrochemically reduced graphene oxide and multiwalled carbon nanotubes/zinc oxide modified electrode. Int. J. Electrochem. Sci. 7, 11477–11487.

Penn, S.G., He, L., Natan, M.J., 2003. Nanoparticles for bioanalysis. Curr. Opin. Chem. Biol. 7, 609–615.

Penza, M., Cassano, G., 2004. Chemometric characterization of Italian wines by thin-film multisensors array and artificial neural networks. Food Chem. 86, 283–296.

Peris, M., Escuder-Gilabert, L., 2009. A 21st century technique for food control: electronic noses. Anal. Chim. Acta 638, 1–15.

Persaud, K., Dodd, G.H., 1982. Analysis of discrimination mechanisms of the mammalian olfactory system using a model nose. Nature 299, 352–355.

Pimtong-Ngam, Y., Jiemsirilers, S., Supothina, S., 2007. Preparation of tungsten oxide–tin oxide nanocomposites and their ethylene sensing characteristics. Sens. Actuat. A 139, 7–11.

Pisanic Ii, T.R., Zhang, Y., Wang, T.H., 2014. Quantum dots in diagnostics and detection: principles and paradigms. Analyst 139, 2968–2981.

Quintela, I.A., Benildo, G., Lin, C.S., Wu, V.C., 2015. Simultaneous direct detection of *Shiga*-toxin producing *Escherichia coli* (STEC) strains by optical biosensing with oligonucleotide-functionalized gold nanoparticles. Nanoscale 7, 2417–2426.

RASFF, 2004. Rapid Alert System for Food and Feed (RASFF). Annual Report on the Functioning of the RASFF, 2004. Available from: http://ec.europa.eu/food/food/rapidalert/report2004_en.pdf.

RASFF, 2005. Rapid Alert System for Food and Feed (RASFF). Annual Report, 2005. Available from: http://ec.europa.eu/food/food/rapidalert/report2005_ en.pdf.

Ravichandran, R., 2010. Nanotechnology applications in food and food processing: innovative green approaches, opportunities and uncertainties for global market. Int. J. Green Nanotechnol. Phys. Chem. 1, P72–P96.

Ravindranath, S.P., Mauer, L.J., Deb-Roy, C., Irudayaraj, J., 2009. Biofunctionalized magnetic nanoparticle integrated mid-infrared pathogen sensor for food matrixes. Anal. Chem. 81, 2840–2846.

Ravindranath, S.P., Wang, Y., Irudayaraj, J., 2011. SERS driven cross-platform based multiplex pathogen detection. Sens. Actuat. B 152, 183–190.

Reyes-De-Corcuera, J.I., Cavalieri, R.P., Powers, J.R., Tang, J., Kang, D.H., 2005. Enzyme-electropolymer-based amperometric biosensors: an innovative platform for time-temperature integrators. J. Agric. Food. Chem. 53, 8866–8873.

Rissin, D.M., Kan, C.W., Campbell, T.G., Howes, S.C., Fournier, D.R., Song, L., Piech, T., Patel, P.P., Chang, L., Rivnak, A.J., Ferrell, E.P., Randall, J.D., Provuncher, G.K., Walt, D.R., Duffy, D.C., 2010. Single-molecule enzyme-linked immunosorbent assay detects serum proteins at subfemtomolar concentrations. Nat. Biotechnol. 28, 595–599.

Ritter, S.K., 2005. An eye on food. Chem. Eng. News 83, 28–34.
Sapmaz, S., Blanter, Y.M., Gurevich, L., Van der Zant, H.S.J., 2003. Carbon nanotubes as nanoelectromechanical systems. Phys. Rev. B 67, 235414.
Scampicchio, M., Ballabio, D., Arecchi, A., Cosio, S.M., Mannino, S., 2008. Amperometric electronic tongue for food analysis. Microchim. Acta 163, 11–21.
Schaller, E., Bosset, J.O., Escher, F., 1998. "Electronic noses" and their application to food. LWT—Food Sci. Technol. 31, 305–316.
Selvakumar, L.S., Thakur, M.S., 2012. Nano RNA aptamer wire for analysis of vitamin B_{12}. Anal. Biochem. 427, 151–157.
Shaikh, S., Patil, M.A., 2012. Analysis, designing, and working principal of optical fiber (OF) biosensors. Int. J. Eng. Sci. 1, 52–57.
Singh, A.K., Senapati, D., Wang, S., Griffin, J., Neely, A., Candice, P., Naylor, K.M., Varisli, B., Kalluri, J.R., Ray, P.C., 2009. Gold nanorod–based selective identification of *Escherichia coli* bacteria using two-photon Rayleigh scattering spectroscopy. ACS Nano. 3, 1906–1912.
Singh, A., Poshtiban, S., Evoy, S., 2013. Recent advances in bacteriophage-based biosensors for foodborne pathogen detection. Sensors 13, 1763–1786.
Siswana, M., Ozoemena, K., Nyokong, T., 2006. Electrocatalytic behavior of carbon paste electrode modified with iron(II) phthalocyanine (FePc) nanoparticles towards the detection of amitrole. Talanta 69, 1136–1142.
Smoker, M., Krynitsky, A.J., 2008. Interim method for determination of melamine and cyanuric acid residues in foods using LC-MS/MS: version 1.0. U.S. Food Drug Adm. Lab. Inf. Bull. LIB 4422. pp.1–27.
Sozer, N., Kokini, J.L., 2009. Nanotechnology and its applications in the food sector. Trends Biotechnol. 27, 82–89.
Speroni, F., Elviri, L., Careri, M., Mangia, A., 2010. Magnetic particles functionalized with PAMAM-dendrimers and antibodies: a new system for an ELISA method able to detect Ara h3/4 peanut allergen in foods. Anal. Bioanal. Chem. 397, 3035–3042.
Staiano, M., Matveeva, E.G., Rossi, M., Crescenzo, R., Gryczynski, Z., Gryczynski, I., Iozzino, L., Akopova, I., D'Auria, S., 2009. Nanostructured silver-based surfaces: new emergent methodologies for an easy detection of analytes. ACS Appl. Mater. Interf. 1, 2909–2916.
Su, H., Zhao, H., Qiao, F., Chen, L., Duan, R., Ai, S., 2013. Colorimetric detection of *Escherichia coli* O157:H7 using functionalized Au@Pt nanoparticles as peroxidase mimetics. Analyst 138, 3026–3031.
Swaminathan, B., Feng, P., 1994. Rapid detection of foodborne pathogenic bacteria. Annu. Rev. Microbiol. 48, 401–426.
Syed, M.A., 2014. Advances in nanodiagnostic techniques for microbial agents. Biosens. Bioelectron. 51, 391–400.
Tang, H., Yan, M., Ma, X., Zhang, H., Wang, M., Yang, D., 2006. Gas sensing behavior of polyvinylpyrrolidone-modified ZnO nanoparticles for trimethylamine. Sens. Actuat. B 113, 324–328.
Tang, H., Fang, D., Li, Q., Cao, P., Geng, J., Sui, T., Wang, X., Iqbal, J., Du, Y., 2012. Determination of tricyclazole content in paddy rice by surface enhanced Raman spectroscopy. J. Food Sci. 77, T105–T109.
Temur, E., Boyacı, I˙.H., Tamer, U., Unsal, H., Aydogan, N., 2010. A highly sensitive detection platform based on surface-enhanced Raman scattering for *Escherichia coli* enumeration. Anal. Bioanal. Chem. 397, 1595–1604.
Tiede, K., Boxall, A.B., Tear, S.P., Lewis, J., David, H., Hassellöv, M., 2008. Detection and characterization of engineered nanoparticles in food and the environment. Food Addit. Contam. 25, 795–821.

Todescato, F., Antognoli, A., Meneghello, A., Cretaio, E., Signorini, R., Bozio, R., 2014. Sensitive detection of Ochratoxin A in food and drinks using metal-enhanced fluorescence. Biosens. Bioelectron. 57, 125–132.

Trivedi, U.B., Lakshminarayana, D., Kothari, I.L., Patel, N.G., Kapse, H.N., Makhija, K.K., Patel, P.B., Panchal, C.J., 2009. Potentiometric biosensor for urea determination in milk. Sens. Actuat. B 140, 260–266.

Tuerk, C., Gold, L., 1990. Systematic evolution of ligands by exponential enrichment: RNA ligands to bacteriophage T4 DNA polymerase. Science 249, 505–510.

Turnipseed, S., Casey, C., Nochetto, C., Heller, D.N., 2008. Determination of melamine and cyanuric acid residues in infant formula using LC-MS/MS. US Food Drug Adm. Lab. Inf. Bull., 24.

Tuttle, J., Gomez, T., Doyle, M.P., Wells, J.G., Zhao, T., Tauxe, R.V., Griffin, P.M., 1999. Lessons from a large outbreak of *Escherichia coli* O157:H7 infections: insights into the infectious dose and method of widespread contamination of hamburger patties. Epidemiol. Infect. 122, 185–192.

Valdés, M.G., González, A.C.V., Calzón, J.A.G., Díaz-García, M.E., 2009. Analytical nanotechnology for food analysis. Microchim. Acta 166, 1–19.

Vamvakaki, V., Chaniotakis, N.A., 2007. Pesticide detection with a liposome-based nano-biosensor. Biosens. Bioelectron. 22, 2848–2853.

Varshney, M., Li, Y., 2007. Interdigitated array microelectrode based impedance biosensor coupled with magnetic nanoparticle– antibody conjugates for detection of *Escherichia coli* O157:H7 in food samples. Biosens. Bioelectron. 22, 2408–2414.

Varshney, M., Yang, L., Su, X.L., Li, Y., 2005. Magnetic nanoparticle-antibody conjugates for the separation of *Escherichia coli* O157:H7 in ground beef. J. Food Protect. 68, 1804–1811.

Varshney, M., Li, Y., Srinivasan, B., Tung, S., 2007. A label-free, microfluidics and interdigitated array microelectrode-based impedance biosensor in combination with nanoparticles immunoseparation for detection of *Escherichia coli* O157:H7 in food samples. Sens. Actuat. B 128, 99–107.

Vaseashta, A., 2006. Carbon nanotubes–based devices and sensors. Mater. Manuf. Proc. 21, 710–716.

Velusamy, V., Arshak, K., Korostynska, O., Oliwa, K., Adley, C., 2010. An overview of foodborne pathogen detection: in the perspective of biosensors. Biotechnol. Adv. 28, 232–254.

Velusamy, V., Arshak, K., Korostynka, O., Vaseashta, A., Adley, C., 2012. Real-time detection of foodborne pathogens. Technological Innovations in Sensing and Detection of Chemical, Biological, Radiological, Nuclear Threats, and Ecological Terrorism. Springer, Dordrecht, Netherlands, pp. 149–158.

Vestergaard, M., Madsen, N.T., Bligaard, H.B., Bredahl, L., Rasmussen, P.T., Andersen, H.R., 2007. Consequences of two or four months of finishing feeding of culled dry dairy cows on carcass characteristics and technological and sensory meat quality. Meat Sci. 76, 635–643.

Villamizar, R.A., Maroto, A., Rius, F.X., Inza, I., Figueras, M.J., 2008. Fast detection of *Salmonella infantis* with carbon nanotube field effect transistors. Biosens. Bioelectron. 24, 279–283.

Viswanathan, S., Wu, L.C., Huang, M.R., Ho, J.A.A., 2006. Electrochemical immunosensor for cholera toxin using liposomes and poly (3, 4-ethylenedioxythiophene)-coated carbon nanotubes. Anal. Chem. 78, 1115–1121.

Vo-Dinh, T., Cullum, B.M., Stokes, D.L., 2001. Nanosensors and biochips: frontiers in biomolecular diagnostics. Sens. Actuat. B 74, 2–11.

Vongsvivut, J., Robertson, E.G., McNaughton, D., 2010. Surface-enhanced Raman spectroscopic analysis of fonofos pesticide adsorbed on silver and gold nanoparticles. J. Raman Spectrosc. 41, 1137–1148.

Wanekaya, A.K., Chen, W., Myung, N.V., Mulchandani, A., 2006. Nanowire-based electrochemical biosensors. Electroanalysis 18, 533–550.

Wang, M., Li, Z., 2008. Nano-composite ZrO_2/Au film electrode for voltammetric detection of parathion. Sens. Actuat. B 133, 607–612.

Wang, L., Chen, W., Xu, D., Shim, B.S., Zhu, Y., Sun, F., Liu, L., Peng, C., Jin, J., Xu, C., Kotov, N.A., 2009. Simple, rapid, sensitive, and versatile SWNT– paper sensor for environmental toxin detection competitive with ELISA. Nano Lett. 9, 4147–4152.

Wang, Y., Lee, K., Irudayaraj, J., 2010. Silver nanosphere SERS probes for sensitive identification of pathogens. J. Phys. Chem. C 114, 16122–16128.

Wang, Y., Ravindranath, S., Irudayaraj, J., 2011. Separation and detection of multiple pathogens in a food matrix by magnetic SERS nanoprobes. Anal. Bioanal. Chem. 399, 1271–1278.

Wang, B., Wang, Q., Cai, Z., Ma, M., 2015. Simultaneous, rapid and sensitive detection of three foodborne pathogenic bacteria using multicolor quantum dot probes based on multiplex fluoroimmunoassay in food samples. LWT—Food Sci. Technol. 61, 368–376.

Warner, M.G., Grate, J.W., Tyler, A., Ozanich, R.M., Miller, K.D., Lou, J., Marks, J.D., Bruckner-Lea, C.J., 2009. Quantum dot immunoassays in renewable surface column and 96-well plate formats for the fluorescence detection of botulinum neurotoxin using high-affinity antibodies. Biosens. Bioelectron. 25, 179–184.

Warriner, K., Reddy, S.M., Namvar, A., Neethirajan, S., 2014. Developments in nanoparticles for use in biosensors to assess food safety and quality. Trends Food Sci. Technol. 40, 183–199.

Wei, D., Bailey, M.J., Andrew, P., Ryhänen, T., 2009. Electrochemical biosensors at the nanoscale. Lab Chip 9, 2123–2131.

Weidemaier, K., Carruthers, E., Curry, A., Kuroda, M., Fallows, E., Thomas, J., Sherman, D., Muldoon, M., 2015. Real-time pathogen monitoring during enrichment: a novel nanotechnology-based approach to food safety testing. Int. J. Food Microbiol. 198, 19–27.

WHO., 2007. Food Safety and Foodborne Illness. Fact Sheet No. 237. Geneva, Switz.: World Health Organ. Available from: http://www.who.int/mediacentre/factsheets/ fs237/en/.

Woodhouse, G., King, L., Wieczorek, L., Osman, P., Cornell, B., 1999. The ion channel switch biosensor. J. Mol. Recogn. 12, 328–334.

Xie, Y., Chen, T., Cheng, Y., Wang, H., Qian, H., Yao, W., 2014. SiO_2@ Au nanoshells-based SERS method for detection of sunset yellow and chrysoidine. Spectrochim. Acta A 132, 355–360.

Xin, H., Stone, R., 2008. Tainted milk scandal: Chinese probe unmasks high-tech adulteration with melamine. Science 322, 1310–1311.

Yang, J., Ono, T., Esashi, M., 2000. Mechanical behavior of ultrathin microcantilever. Sens. Actuat. A 82, 102–107.

Yang, H., Qu, L., Wimbrow, A.N., Jiang, X., Sun, Y., 2007. Rapid detection of *Listeria monocytogenes* by nanoparticle-based immunomagnetic separation and real-time PCR. Int. J. Food Microbiol. 118, 132–138.

Yang, M., Kostov, Y., Rasooly, A., 2008. Carbon nanotubes based optical immunodetection of *Staphylococcal* Enterotoxin B (SEB) in food. Int. J. Food Microbiol. 127, 78–83.

Zhai, F., Huang, Y., Li, C., Wang, X., Lai, K., 2011. Rapid determination of raptopamine in swine urine using surface-enhanced Raman spectroscopy. J. Agric. Food Chem. 59, 10023–10027.

Zhang, H., Wang, J., 2007. Detection of age and insect damage incurred by wheat, with an electronic nose. J. Stored Product. Res. 43, 489–495.

Zhang, W., Zhang, W., 2008. Fabrication of SnO_2–ZnO nanocomposite sensor for selective sensing of trimethylamine and the freshness of fishes. Sens. Actuat. B 134, 403–408.

Zhang, W., Tang, H., Geng, P., Wang, Q., Jin, L., Wu, Z., 2007. Amperometric method for rapid detection of *Escherichia coli* by flow injection analysis using a bismuth nano-film modified glassy carbon electrode. Electrochem. Commun. 9, 833–838.

Zhang, Y., Zhang, X., Lu, X., Yang, J., Wu, K., 2010. Multi-wall carbon nanotube film-based electrochemical sensor for rapid detection of Ponceau 4R and Allura Red. Food Chem. 122, 909–913.

Zhang, Y., Huang, Y., Zhai, F., Du, R., Lai, K., 2012. Analyses of enrofloxacin, furazolidone and malachite green in fish products with surface-enhanced Raman spectroscopy. Food Chem. 135, 845–850.

Zhao, G., Xing, F., Deng, S., 2007. A disposable amperometric enzyme immunosensor for rapid detection of *Vibrio parahaemolyticus* in foodbased on agarose/Nano-Au membrane and screen-printed electrode. Electrochem. Commun. 9, 1263–1268.

Zheng, J., He, L., 2014. Surface-enhanced Raman spectroscopy for the chemical analysis of food. Compr. Rev. Food Sci. Food Saf. 13, 317–328.

Zheng, J., Li, G., Ma, X., Wang, Y., Wu, G., Cheng, Y., 2008. Polyaniline–TiO_2 nano-composite-based trimethylamine QCM sensor and its thermal behavior studies. Sens. Actuat. B 133, 374–380.

Zhu, G., Hu, Y., Gao, J., Zhong, L., 2011. Highly sensitive detection of clenbuterol using competitive surface enhanced Raman scattering immunoassay. Anal. Chim. Acta. 697, 61–66.

Ziegler, C., 2004. Cantilever-based biosensors. Anal. Bioanal. Chem. 379, 946–959.

15

NANOCOMPOSITE BIOSENSORS FOR POINT-OF-CARE—EVALUATION OF FOOD QUALITY AND SAFETY

Anisha A. D'Souza, Durga Kumari, Rinti Banerjee
Nanomedicine Lab, Department of Biosciences and Bioengineering, Indian Institute of Technology, Mumbai, India

1 Introduction

Increased outbreaks in foodborne illness and the need to meet the demands of growing populations have been the driving forces for ensuring that food production undergoes quality control for evaluation of its fitness for human consumption (Sekhon, 2014). Food safety is crucial at every step right from its production to its preservation, transportation, and consumption. Consumers expect optimal quality, good stability, and a safe food product at a reasonable price. Monetary and legal aspects related to food products as a consequence of food contamination are of grievous concern to food producers accompanied with loss of consumer confidence (Odumeru, 2002). With these concerns, food safety and nutritional quality have always been a priority issue for producers, food safety regulators, the food industry, and consumers (Karlsson, 2004). Foodborne illness could also be as serious as botulism, listeriosis, or hemolytic uremia syndrome, which often need hospitalization. Hence, an accurate and quick monitoring of food quality should be feasible during various stages of food production until consumption.

Conventional analytical methods mandate high numbers of pathogens or analytes for visual or identifiable changes in food. Low levels of analytes in food generate low signals with great variations, which are difficult to detect with confidence (Rodríguez-Lorenzo et al., 2012). Moreover, many analytical techniques are destructive and affect the seal of the packaged food (Duncan, 2011). Until recently, most of the biosensors developed

Nanobiosensors. http://dx.doi.org/10.1016/B978-0-12-804301-1.00015-1

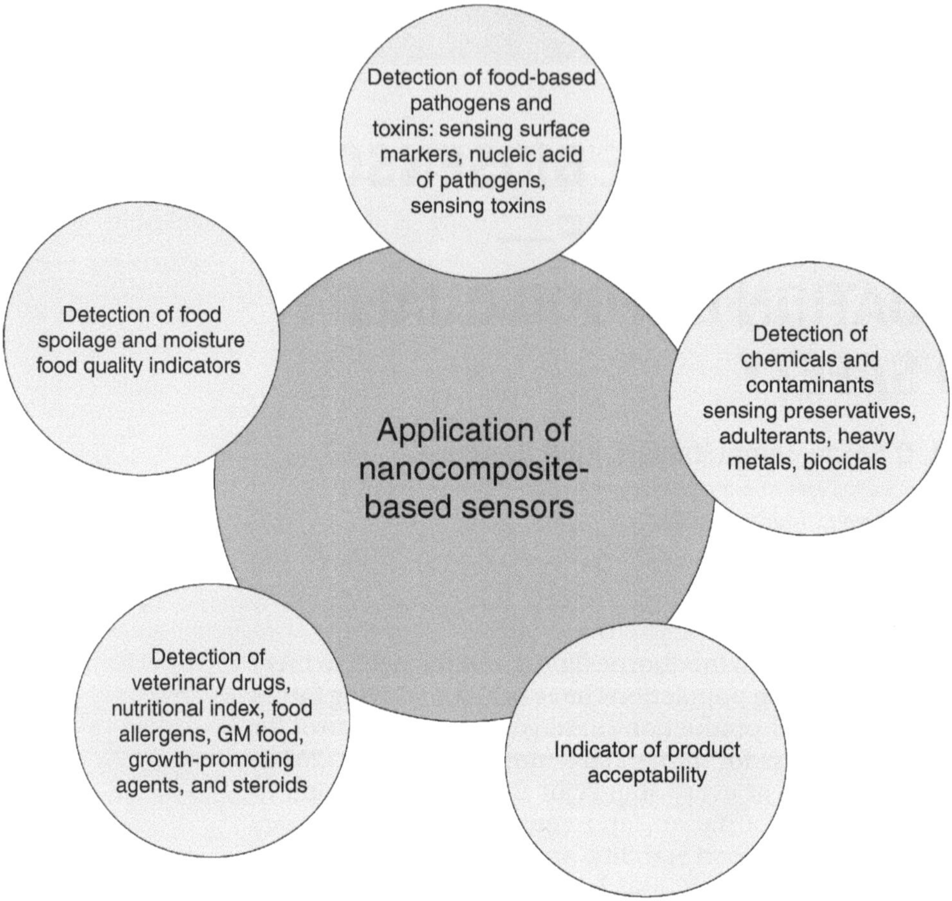

Figure 15.1. Nanocomposite based biosensors in food quality and safety.

were marketed for medical applications. Moreover, food matrices offer many limitations due to their opacity, coloring matter, light scattering and complexity of food materials. Advances in the area of analytical methods and scientific tools, with better detection methods in the complex matrix of foods, thus dictate the need for development of advanced biosensors for the food industry.

Nanocomposite-based biosensors provide new solutions to tackle critical analytical evaluation and detection problems of contaminants and achieving quality and safety in the food industry. The unique traits of nanoscale composites with improved stability make them outstanding for sensor applications (Fig. 15.1). Their use has enabled improved sensitivity and portability with unparalleled performance and real-time sensing in complex matrices. The current chapter features the various nanocomposites employed in improving the conventional analytical and sensing applications in the food industry. Additionally, monitoring of food freshness, pathogen contamination and toxins in food products, after being dispatched until they are consumed requires point-of-care sensing technology. The chapter describes specific demands of nanocomposite-based nanosensors in addressing crucial segments related to food safety and quality, such as determining contamination, pesticide residues, antibiotics, shelf life and organoleptic assessment at point of care. The chapter also discusses the existing and future prospects of nanotechnology-based biosensors along with their commercial potential.

2 Biosensors and the Food Industry

As per the International Union of Pure and Applied Chemistry, biosensors are devices that contain an amalgamation of receptors with transducers for selective quantification using incorporated biological recognition elements (Thévenot et al., 2001). The receptor of a biosensor recognizes and interacts with the analyte. Receptors could be enzymes, antigens (Ag), nucleic acids, or organelles performing an affinity, biochemical, or catalytic reaction. Binding or catalysis between receptors and substrate transforms the physical, chemical, or biologic information to some form of energy, such as, electrochemical, optic, fluorescence, mass, or calorimetric signals, which is measured by the transducer. The transducer thus provides quantifiable physicochemical signals proportional to the analyte concentration (Thakur and Ragavan, 2013).

Biosensors are widely used to determine contamination, raw material and product composition. They also keep tab on quality check. Besides government's mandatory regulations, quality check also encompasses nutritive value; degree of food spoilage, especially of meat and fish; degradation during storage; processing, distribution, and presentation fit for human consumption; peak freshness, with minimized opportunity for foodborne diseases; and health hazards on food consumption. Monitoring organoleptic characteristics, such as, smell, odor, and taste of food is also an important parameter for determining food quality. Rapid, nondestructive detection of foodborne pathogens, pesticide residues and allergens with high analytical sensitivity are the major needs of the food industry (Murugaboopathi et al., 2013). Additionally, on-site monitoring of fermentation, antibiotic residues, and toxins in food is also required in the food industry.

Noninvasive sensing technology at point-of-care settings may be beneficial for food testing, especially if the analytes can

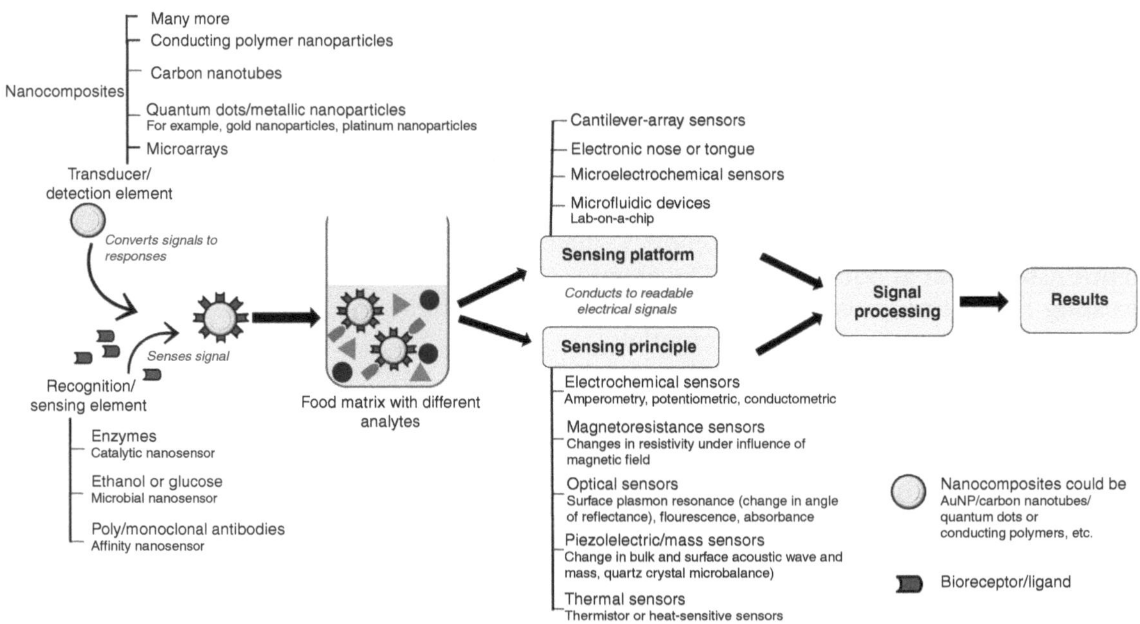

Figure 15.2. Role of nanotechnology in sensors for food safety and quality assessment.

be determined just by visual observation of the product. The demand for quick and specific biosensors for analysis of food has thus been continuously rising and expanding to address the different aspects of food quality and safety. Friendly applicability and simplified communication with end users are the ultimate striving goals for assuring quality of food products.

3 Nanocomposite-Based Biosensors

Nanotechnology is "the understanding and control of matter that involves manipulation of structures at dimensions of 1–100 nm" (Sargent, 2014). Application of nanotechnology to sensors or biosensors in the food industry is known as "nanosensors" or "nanobiosensors," respectively. Nanosensors are relatively a new concept in the food industry compared to their application in pharmaceuticals. Nanocomposites are utilized in several ways to improve the efficiency of nanosensors. They essentially consist of nanoparticles (NPs) attached to any of the various recognition labels followed by detection through any of the sensing approaches illustrated in Fig. 15.2. Nanotechnology in sensing, permits miniaturization of sensors which facilitates their integration with normal food production and packaging machinery and equipment, or can be used offline as point-of-care devices (Leonard et al., 2003). The demand for analyte samples and reagents used in nanosensors is less than that of conventional methods of detection. Additionally, the time factor involved for detection is significantly reduced from hours to minutes compared to many traditional laboratory analytical techniques.

Nanomaterials differ in properties from those of the bulk macroscopic material due to their existence and interaction at the atomic levels. The small size of nanomaterials (1–100 nm) results in large surface areas with high surface reactivity, high electric conductivity, enhanced magnetic properties and improved quantum confinement effects, such as, luminescence and catalytic properties, making them very sensitive (Kaittanis et al., 2010) with high analyte-capturing efficiencies (Duncan, 2011). The slightest modification of the nanostructures with respect to their size, self-assembly, surface composition, and shape alters and tunes their electronic, conductive, fluorescence, light-scattering, binding properties and the absorption-emission spectra. They can also permit engineering with different molecules and combinations of individual bioaffinity agents and act as transducers to enable ultrasensitive detection, as well as, multiplex sensing assays for different pathogens and protein-based toxins (Tallury et al., 2010; Xu and Ying, 2011). For instance, in one study, combinations of metallic platinum nanocomposites and single-walled carbon nanotubes coated on glassy carbon provided enhanced sensitivity for detection of hydrogen peroxide and glucose over individual nanocomposites by voltammetry within 3 s. The presence of nanocomposites ensured that the electric contact of the glass electrode and the biosensor loaded with glucose oxidase enzyme was more sensitive than that of an unmodified glass electrode loaded with the same enzyme (Hrapovic et al., 2004).

The unique properties of nanomaterials can thus be easily exploited for improving sensitivity, robustness, and reproducibility of biosensors, diagnosing within a shorter span of time, and at the same time being relatively economical (Nath et al., 2008). Nanoproperties change before and after binding with analytes. A simple example would be illumination with different-colored fluorescence on interaction with different pathogens. They permit real-time and online qualitative and semiquantitative detection. They can be also called "naked eye chemosensors." Such online sensors provide greater control over product properties during manufacturing, thus improving product quality and cost reduction (Ravindranath et al., 2009). Incorporation of nanoparticles (NPs) into composites also permits intimate contact with analytes like pathogens, as NPs often exhibit size similar to biologic species, such as, bacilli and viruses.

Gold nanoparticles (AuNPs) exhibit length-dependent absorption properties; permitting simultaneous detection of organisms. Hybrid nanocomposites of iron oxide–gold nanorods and iron oxide nanodumbbells have tunable optic and magnetic properties. Iron oxide–gold nanorods upon binding with pathogens exhibit different aspect ratios and thus differ in plasmon properties and also facilitate magnetic separation. This functionality was explored for multiplex optic detection (Wang and Irudayaraj, 2010). Nanowires conjugated to fluorescent antibodies (Ab) can detect pathogenic microorganisms. For instance, if the food contains traces of *Salmonella*, *Salmonella* Abs conjugated to fluorescent antibodies are coated on the nanowire. The Ab-coated nanowire upon interaction with *Salmonella* would alter the fluorescence of the probes, making them visible (Dingman, 2008). The nanocomposite-based nanosensors thus permit visual identification of pathogens at point-of-care without significant lag time, and serve as a simple technique for consumers to know the quality and safety of food products.

Coupling of metallic NPs to affinity ligands can lead to sensing of genetic strands, bacteria, or toxins using any of the sensing principles based on change in fluorescence, magnetic, Raman scattering, and electrochemical changes, as depicted in Fig. 15.2 (Alivisatos et al., 2005). The high sensitivity of nanosensors permits real-time detection and quantitation at molecular and single-cell levels (Tallury et al., 2010). Online real-time analysis has significantly reduced the time for pathogen detection. Earlier analytical techniques needed over 24 h for permitting sufficient growth of pathogens to be detected (Mahalik and Nambiar, 2010).

NPs exhibiting magnetic, fluorescence and conductive materials have been exploited for detection, imaging and tracking through various probes. For instance, quantum dots (QDs) permit labeling and visualization, replacing classical fluorophores. Metallic NPs are conducting in nature and thus wiring of the biologic unit facilitates a rapid transfer of electrons. Hence, electrodes are often modified with a monolayer of AuNPs conjugated to bioelements. In one study, glucose oxidase conjugated to silver/gold/platinum NPs electrodeposited on metal electrodes enhanced transfer of electrons between enzyme and electrode with improved stability (Zhu et al., 2007). The role of conducting polymer nanowires and semiconductors as nanomaterials in sensor applications has already been reviewed (Willner et al., 2007).

Application of nanotechnology in nanosensors is, however, not restricted only to recognition labels but could range from recognition element to transducer or running systems. Platforms technology of nanotechnology in microfluidics, microelectromechanical

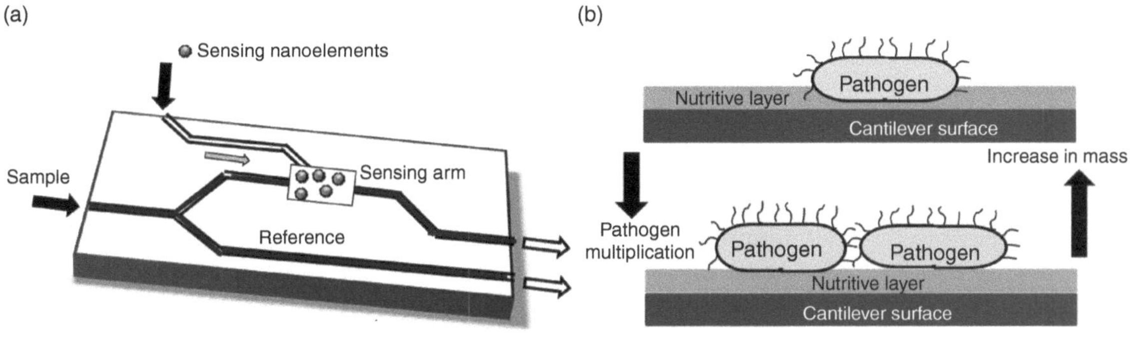

Figure 15.3. Schematic representative of (a) simplified microfluidic system (b) mass sensors used in nanocantilever. Increase in bacterial growth increases bacterial mass and depletes nutritive layer deflected by nanocantilever due to change in mass.

systems, cantilevers, and microarrays have further widened the realm of quality analysis (Sozer and Kokini, 2012) (Fig. 15.3).

Microfluidic systems based on silicon have been popular as laboratory-on-a-chip technology. They are multifunctional wherein sampling, separation, reaction, and detection can be performed on the same device. *Nanocantilever*-based sensors have been used to detect various molecular interactions, such as, Ag–Ab, enzyme–substrate–cofactor, chemical interactions, receptor–ligand and nucleic-acid hybridization (Hall, 2002; Kumar, 2007). Bio-Finger is a portable nanosensor consisting of nanocantilevers made of piezoresistive material for detection of pathogens, chemicals, and toxins in food based on ligand–receptor interactions. Any binding or interaction between an Ag and Ab induces mechanical stress on the nanocantilever, which is coated with antibodies, and changes the electric resistance of the material. Multiple such nanocantilevers can be arranged in an array to form a single chip. The total analytical time for detecting the analyte is 15–20 min. The cost of Bio-Finger is expected to be $10, cheaper than the existing diagnostic techniques. Arrays have an upper hand in obtaining sensitive and simultaneous detection of analytes. Reference sensors can also be incorporated into the array, minimizing the noise.

4 Application of Nanosensors in the Food Industry

4.1 Detection of Foodborne Pathogens and Toxins

Food contamination caused by pathogens (bacteria, fungi, protozoa, viruses, prions, and worms) and toxins remains a major concern globally. Conventional techniques for identification of pathogens include microscopic examination, growth patterns of culture upon culturing in differential media, their biochemical tests, and immunoassays (Tallury et al., 2010).

These techniques are laborious, with lag times greater than 24 h for pathogen amplification. The process of sample preparation remains a determinant for specificity and sensitivity of many assays (Kaittanis et al., 2010). Lately, molecular diagnostics, such as, identification of DNA segments and proteins of the pathogen's genome and virulent strains using polymerase chain reaction (PCR), ligase chain reaction, and checkerboard hybridization have been employed (Tallury et al., 2010). However, these methods mandate the need of undamaged microbial DNA for detection. Identification of circulating Abs and specific cell-wall epitopes made of carbohydrate or proteins can also be achieved by immunoassays, phage typing, and flow cytometry; but are limited to homogeneous, purified, and simple matrix samples (Kaittanis et al., 2010).

4.1.1 Sensing the Surface Markers

RuBpy dye [Tris(2,2′-bipyridyl) dichlororuthenium (II) hexahydrate] labeled with Ab-conjugated silica fluorescent NPs (60 nm in diameter), which were conjugated to anti-*Escherichia coli* Ab, enabled single *E. coli* O157:H7 detection obtained from processed beef samples within 20 min. The fluorescent signal of the dye varies on the biorecognition of *E. coli*. Sensitivity of the method was similar to that of conventional plate counting but with a shorter readout time. The technique also permits high-throughput quantitation of 1–400 bacterial cells using microtiter plate reader (Zhao et al., 2004) (Fig. 15.4). Silicon/gold nanorod nanoarrays functionalized with anti-*Salmonella* Abs bearing organic-dye particles become visible in the presence of *Salmonella*. The high aspect ratio of silicon nanorods upon interaction of anti-*Salmonella* Abs with *Salmonella* Ags shows increased fluorescence intensity of dye molecules (Fu et al., 2008).

Electrostatic interactions of AuNPs functionalized with fluorescent π-conjugated anionic polymers like *p*-phenylene ethylene results in quenching of fluorescence. However, in cases of bacterial contamination, fluorescence emission increases due to displacement of the anionic polymer with the negative cell surface of bacteria, including *E. coli*, *Lactococcus lactis*, *Streptomyces coelicolor*, and *Bacillus subtilis*. Linear-discriminant analysis generates a characteristic signature plot specific for each microorganism. The method has been successful for detection of 1×10^9 CFU (ie, $OD_{600} = 1$) (Phillips et al., 2008). Different strains of norovirus, parvovirus, corona virus, adenovirus, simian rotovirus, herpes virus, and Sendai virus at a count of 100 particles in food and drinking water have been discriminated using AuNPs (Fan et al., 2010).

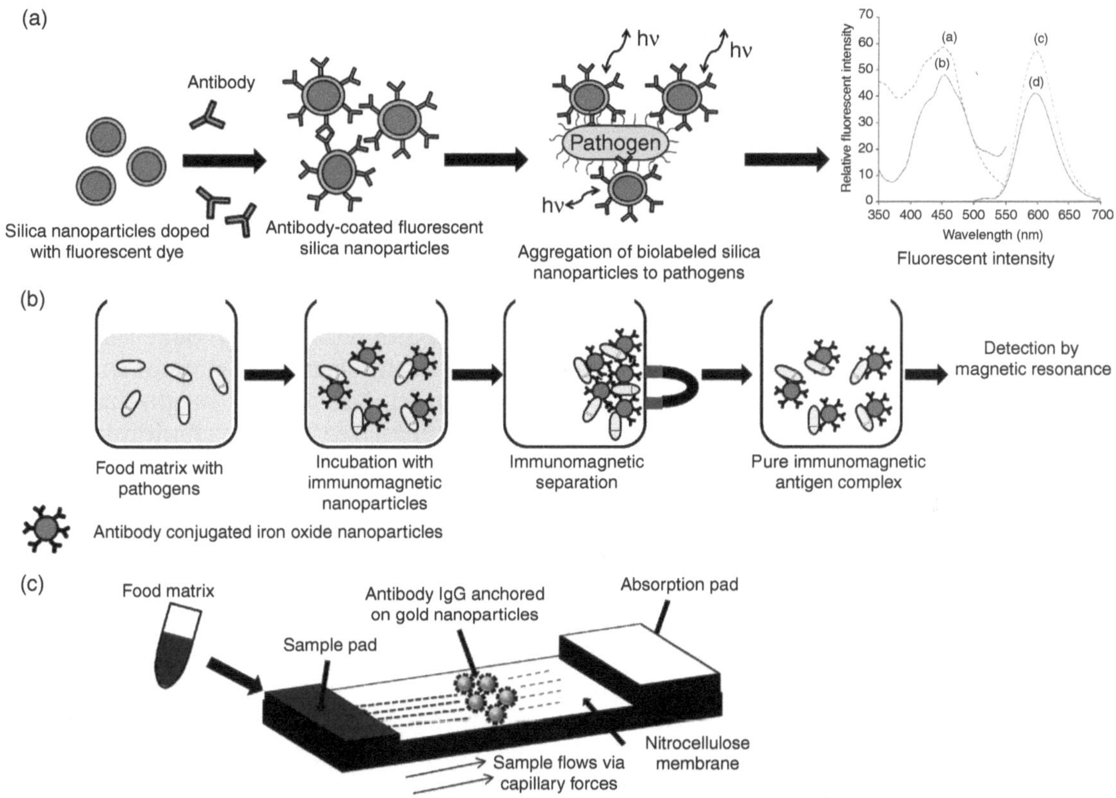

Figure 15.4. (a) *E. coli* Ab conjugated silica fluorescent NPs for detection of *E. coli* using fluorescence intensity; (b) quantitation and detection of pathogens using immunomagnetic NPs; (c) immunochromatographic separation of pathogens on nitrocellulose membrane

Surface-enhanced Raman scattering (SERS) based on nanostructured gold and silver nanorods as substrates, can detect and differentiate strains of bacillus species and simultaneously screen *E. coli*, *Listeria monocytogenes*, and *Salmonella typhimurium* within 10 s of collection time (He et al., 2008; Chu et al., 2008). Principal component analysis is used to analyze the Raman scatter and classify the data according to the presence of different species, strains, and Gram-types. Yet another advantage is the differentiation between viable and nonviable cells. Nonviable cells have significantly reduced SERS (Chu et al., 2008). Surface plasmon resonance (SPR) effect of AuNPs improved the performance and sensitivity 10-fold compared to conventional methods (Chen et al., 2004).

Interaction of magnetic NPs with pathogens undergoes spin–spin relaxation time (T_2) of adjacent water molecules (Chemla et al., 2000). Change in relaxation times (ΔT_2) can be detected by benchtop-magnetic relaxometer (20 MHz, 0.47 T) or superconducting-quantum interference device (SQUID) or by magnetic resonance imaging (60 MHz, 1.5 T) (Zheng et al., 2009b). The higher the relaxivity of iron-oxide NPs, the higher is the sensitivity of detection. *Mycobacterium avium* spp. *paratuberculosis* (MAP) was quantified in whole milk using spherical iron-oxide NPs and dextran-coated iron oxide nanorods (R^2 300 M/m per s) within 5–30 min without any interference (Kaittanis et al., 2007; Nath et al., 2008). In the presence of MAP, conjugation of iron-oxide NPs changed the spin–spin ΔT_2 of water protons. Low bacterial concentration caused assembly of NPs on bacterial surfaces, and hence high ΔT_2 was observed. While at high bacterial concentrations, ΔT_2 decreased due to the dispersed state of the NPs. In the presence of a magnetic field, NPs acquire magnetization. On switching off of the magnetic field, free NPs relax rapidly by Brownian rotation and thus do not contribute to any signals. Abs bound to pathogens are immobilized and undergo Neel relaxation; that is, the oscillations vary in the presence of a target with slow decay of magnetic flux, dissipating magnetic flux detectable by SQUID, as opposed to, the change in Brownian motion observed in free NPs (Chemla et al., 2000). Magnetic NPs conjugated to Ab and immobilized with SQUID on amylar film detected *L. monocytogenes* above a count of 6×10^6 (Grossman et al., 2004). However, the requirement of high infrastructural costs limits the application of these magnetic NP-based sensors for the food industry.

Nanosensors commonly use immunoassays with Ab immobilized on NPs coupled with immunocapture for detection (Warriner et al., 2014). Immunomagnetic separation is performed to concentrate the analytes or pathogens above the signal-to-noise ratio. Antibodies specific to analyte or pathogen are conjugated to magnetic particles followed by separation with a magnet prior to detection (Fig. 15.4b). *L. monocytogenes* have been captured by immunomagnetic NPs during a 2 h time duration. The immunomagnetic NPs consisted of magnetic NPs (diameter of 30 nm) functionalized with *L. monocytogenes* Ab via biotin–streptavidin linkages. The separated complex was injected into a microfluidic chip and detected using an impedimetric immunosensor at 102 Hz. Resistance of complexes dispersed in mannitol solution decreased the impedance. The method was sensitive in simple matrices, such as, buffers with a sensitivity of 10^3 CFU/mL. The sensitivity decreased in matrices like milk, lettuce-rinse samples, and beef homogenates with a limit of detection of 10^4–10^5 CFU/mL within 3 h due to interference of high protein/fat content. The method exhibited high specificity in the presence of other foodborne bacteria like *E. coli*, *Staphylococcus aureus*, and *Salmonella tyhphimurium* (Kanayeva et al., 2012). AuNPs conjugated to secondary enzyme-labeled

L. monocytogenes Ab detected *L. monocytogenes* with high specificity over other enteric bacteria at 2 log CFU/g in wild-blueberry samples. The NPs were immobilized on screen-printed carbon electrodes for amperometric sensing (Davis et al., 2013).

Magnetic iron-oxide NPs conjugated to vancomycin quantified *S. aureus* using a magnetic resonance system. Vancomycin interacted with the peptidoglycan of the bacterial cell wall. The method could detect almost 10 bacteria present in 10 μL of sample within 15 min. The magnetic NPs form clusters upon identification of the analyte and the self-assembled cluster complexes decrease the nuclear spin of surrounding water protons (T_2). Detection of T_2 was possible by a handheld portable device consisting of a microcoil array for nuclear magnetic-resonance measurements, a microfluidic network for handling sample and mixing with NPs, and a permanent magnet to generate a polarizing magnetic field (Lee et al., 2008). The combination of gold-microelectrode array in a microfluidic device along with magnetic NPs conjugated with Ab helps in detection of pathogens, for example, *E. coli* in ground beef. The bacterial cells were concentrated by immunocapture by a factor of 10^4–10^5 in a volume of 400 pL; and the metabolism was detected by impedance. Detection time was comparatively reduced with bacilli in a dilute sample (Gomez-Sjoberg et al., 2005). *E. coli* could be detected with nanocantilever coated with agarose-nutritive media for bacilli. As the mass of the cantilever array increased, the resonance frequency changed. The cantilever could detect a mass of approximately 100 active *E. coli* cells within 1 h (Fig. 15.3b). Selectivity of the method could be improved by adding antibiotics to the nutritive media (Gfeller et al., 2005). Likewise, magnetic NPs conjugated to anti-H5N1 Ab aggregated in the presence of avian flu virus. High transition-temperature superconducting quantum interference senses the immunomagnetic reduction of the reagents and allows quantitation of H5N1 in wash-free samples (Yang et al., 2008). Diagnostic kits containing AuNPs have been studied for detecting influenza virus in poultry. These diagnostic kits have AuNPs conjugated with specific protein of the avian influenza virus (Emami et al., 2012). Literature has ample such examples for detection of pathogens (Table 15.1).

An immunochromatographic assay for rapid detection of *S. aureus* has been developed by Huang et al. (Huang, 2006; Huang et al., 2007). On a defined zone of nitrocellulose membrane, IgG antiprotein A (cell wall protein of *S. aureus*) conjugated to AuNPs was mobilized. Two precooked foods naturally contaminated with bacterium and 11 processed foods artificially spiked with *S. aureus* were allowed to flow past the IgG antiprotein by capillary action. A red color was seen for the specific reaction with antiprotein A at

Table 15.1 Nanosensors Used for Detection of Pathogens

Analyte	Nanomaterial	Detection Method	Detection Limits	References
Bacilli				
Campylobacter jejuni	Lectin immobilized on gold surface of quartz crystal electrode	QCM	Reusable, limit of detection 10^3 cells within 30 min	Safina et al. (2008)
E. coli O157:H7	Core-shell Cu/AuNPs with anti-*E. coli* Ab	Anodic-stripping voltammetry	Detection limit is 30 CFU/mL within a time span of 2 h	Zhang et al. (2009)
	Magnetic nanobeads	Impedance spectroscopy	45 log CFU/g in freshly ground beef, assay time 35 min	Varshney et al. (2005)
	Ab immobilized gold-coated on QCM	QCM with dissipation monitoring	Log–log linear working range 10^7–10^9 cells/mL	Poitras and Tufenkji (2009)
	Anti-*E.coli* Ab immobilized on PEG alkanethiol monolayers on SPR	Sandwich assay SPR immunosensor	Detection limit 10^6 CFU/mL, highly specific against *Salmonella enteritidis*	Subramanian et al. (2006)
	Sugar attached to iron-oxide NPs	Fluorescent staining	Can isolate up to 88% of *E. coli* within 45 min	El-Boubbou et al. (2007)
	Ab coated on iron-oxide NPs	Optic detection using FTIR/portable mid-IR	10^4–10^5 CFU/mL in 2% milk and spinach extract	Ravindranath et al. (2009)
S. aureus, *Vibrio parahaemolyticus*, and *S. typhimurium*	Multicolor upconversion NPs (rare-earth ions) coupled with aptamers attached to magnetic NPs	Magnetic NPs used for immuneseparation with aptamers followed by visible luminescence	25 CFU/mL for *S. aureus* 10 CFU/mL for *V. parahaemolyticus* 15 CFU/mL for *S. typhimurium*	Wu et al. (2014)
S. typhimurium and *S. aureus*	AuNPs labeled with Raman reporter molecules	Raman scattering	10^3 CFU/mL in spinach wash and peanut butter emulsion	Zhuyi et al. (2011)

Table 15.1 Nanosensors Used for Detection of Pathogens (*cont.*)

Analyte	Nanomaterial	Detection Method	Detection Limits	References
S. typhimurium	Protein G-coated 11-mercaptoundecanoic acid monolayer on gold surface	SPR	10^2–10^9 CFU/mL	Oh et al. (2004)
S. typhimurium, *Shigella flexneri*, and *E. coli* O157:H7	Combination of magnetic NPs and Ab-conjugated semiconductor QD	Magnetic NPs immobilized with Ab and quantified by fluorescence	10^{-3} CFU/mL within 2 h in food matrix protein in high concentration decreased sensitivity of the method	Zhao et al. (2009)
Salmonella	Anti-*Salmonella* Ab immobilized to QD	Fluorescence at 415 nm UV	4×10^3 CFU/mL in food extracts	Kim et al. (2013)
S. aureus	AuNPs immobilized with anti-*S. aureus* Ab	Magnetic beads enhance capture of *S. aureus* from milk matrix florescence of gold quantified	Assay time 40 min, low sensitivity with detection limit of 5 log CFU/mL	Sung et al. (2013)
Viruses				
Cauliflower mosaic virus	Lead sulfide NPs linked to cauliflower mosaic virus 35S oligonucleotide	Sensitive differential pulse anodic stripping voltammetric	Detection limit was 4.38×10^{-12} M/L	Sun et al. (2008)
Tomato ring-spot virus and grapevine fan leaf virus	Mesoporous silicon	Change in electric parameters after adsorption	Effective virus size for detection was 50 nm	Vashpanov et al. (2008)
Spores				
Bacillus anthracis spores	Polyaniline-coated magnetic NPs (100 nm)	Detection mechanism depend on capillary flow of captured spores	Limit of detection 4.2×10^2 spores/mL within 16 min	Pal and Alocilja (2009)
Bacillus thuringiensis spores	60-base aptamer conjugated to zinc-sulfide capped CdSe QD	Detects at 655 nm with no photo bleaching of QDs	Semiquantitative, specific and detects 1000 CFU/mL	Ikanovic et al. (2007)

<25 CFU/g. The same technique could identify 28 different strains of *S. aureus* and 23 non-*S. aureus* strains (Fig. 15.4c).

NanoBioluminescence Detection Spray developed by AgroMicron contains a luminescent protein engineered to detect *Salmonella* and *E. coli.* The spray is directly applied to food or beverage and the proteins react with the pathogens, producing a visual glow.

4.1.2 Sensing the Nucleic Acids of Pathogens

AuNPs cause the free electrons in the conduction band to form plasmonic bands. Any minor change in the dielectric constant locally either by adsorption or by aggregation shifts the plasmonic bands. Interaction between nucleic acid and AuNPs causes a shift in the peak of SPR (Sun and Xia, 2002). For detection of anthrax, AuNPs are used in probes. Hybridization between the target gene and the AuNP-based probes alters the dielectric constant and hence diffraction of light. This method is simple, specific, and sensitive with a color-selective nature and was found to detect anthrax lethal factor as low as 40 fM concentration (Bailey et al., 2003).

Methicillin-resistant strains of *S. aureus* possess the mecA gene, which can be detected using DNA-modified AuNPs and does not require target or signal amplification (Storhoff et al., 2004). AuNPs capped with oligonucleotideprobes (alkanethiol-31 nucleotide probes) have been studied in detecting the invA gene of *Salmonella*. On amplification of the target gene by PCR, high sensitivity was observed in comparison with agarose gel electrophoresis. Detection limit of the probe was 21.78 ng/μL (Majdinasab et al., 2012). AuNPs assembled via 2-aminoethanethiol on the surface of gold chip were immobilized with gold-binding polypeptides (GBP)–protein A–human IgG conjugated to anti-*Salmonella* Ab. The surface modification intensified the binding signal intensity threefold and caused a 10-fold increase in sensitivity in comparison with the bare chip (Ko et al., 2009). The detection techniques of nucleic acids and those involving SPR techniques currently are limited to only clean, isolated samples, and evaluation of the same in mixed matrices of food analyte is warranted.

4.1.3 Sensing Toxins

Toxins disrupt the phospholipid bilayer of host cell plasma membranes by generating pores within the layer. Fungal toxins (mycotoxins), cyanobacterial toxins in freshwater, and marine toxins in shellfish are retained in food samples even after the death of the pathogen (Kayal and Charbit, 2006). Regulatory bodies have set maximum permitted levels for mycotoxins in food products

(European Commission Regulation, 2006). Detecting low levels of mycotoxins is a challenging issue in complex food matrices (Berthiller et al., 2014).

AuNPs coated with thiolated lactose derivatives mimic the distal portion of GM1 ganglioside, an extracellular matrix of intestinal epithelial cells, and conjugates with cholera toxins. Interaction of toxin and NPs occurs within 10 min with a red shift of the plasmonic band and allows a detection limit of 54 nM (Schofield et al., 2007). Similarly, globotriose-coated AuNPs resemble the globotriaosylceramide present on renal epithelia and intestinal microvilli. The B subunit of Shiga-like toxin interacts with this bio-artificial substrate and permits visual detection of the interaction (Chien et al., 2008).

In another study, multiplexed sandwich fluoroimmunoassays have been used for simultaneous examination of four toxins—cholera, ricin, Staphylococcal enterotoxin B, and Shiga-like toxin 1 using cadmium–selenide–zinc sulfide (CdSe/ZnS QDs) at four different wavelengths after binding to their specific Ab (Goldman et al., 2003). Staphylococcal enterotoxin B isolated from food samples by immunomagnetic separation using magnetic NPs coated with anti-Staphylococcal enterotoxin Ab could be quantified by fluorescence nanotransduction (Branen et al., 2007). Fluorescent quantum dots (QDs) have also quantitated Staphylococcal enterotoxin B, ricin, cholera, and Shiga-like toxin. Europium quantum dots depicted sensitivity of 10 pg/mL, 100 times more than conventional enzyme-linked immunosorbent assays (Tang et al., 2009b). Ab-labeled luminescent QDs have also been studied for detecting botulinium toxin serotype A with picomolar sensitivity (Warner et al., 2009).

Staphylococcal enterotoxin B reacts with polyclonal Ab-coated AuNPs and then with fixed polyclonal Ab on porous nitrocellulose membranes placed in a lateral-flow assay device. Toxin–Ab reaction reproducibly led to the formation of a red line at the detection zone within 5 min with increased sensitivity up to 10 pg/mL on silver enhancement (Rong-Hwa et al., 2010). Likewise, AuNPs conjugated to IgG-rabbit anti-Staphylococcal enterotoxin B immobilized on a polycarbonate surface, enabled detection of Staphylococcal enterotoxin B. The food analytes were mainly tomatoes, baby foods, and mushrooms spiked with enterotoxin, centrifuged, and passed through a cation exchanger, carboxymethyl cellulose analytical column to remove the interfering food materials. After incubation of these eluted samples with the membrane for 45 min, the membrane was exposed to HRP-conjugated anti-Staphylococcal enterotoxin B IgG and detected by chemiluminescence. Detection limit was 0.01 ng/mL (Yang et al., 2009).

Aflatoxin (toxins of Aspergillus) detection has been studied with AuNPs bearing aflatoxin Ab up to levels of 0.5 ng/mL in grains within 15 min. Immobilization of these NPs on immunochromatography strips facilitated portability with performance comparable to that of high-pressure liquid chromatography (HPLC) (Shim et al., 2007). Toxins in milk interact with aflatoxin M1 Ab anchored on iron-oxide NPs, isolating the toxins and allowing quantification by enzyme-linked immunosorbent assay (Radoi et al., 2008). Aflatoxin B1 in artificially contaminated milk up to concentrations of 0.01 ng/mL could be detected using piezoelectric immunosensor based on AuNPs (Jin et al., 2009). Nanoiron-oxide NPs core coated with AuNPs as shell, detected aflatoxin B2 in nuts when used as labels in a lateral-flow immunodipstick with similar results as obtained by HPLC but with a shorter readout time of 15 min. Samples for analysis were prepared by extraction of ground peanuts, hazelnuts, pistachios, and almonds with an admixture of methanol and water followed by filtration (Tang et al., 2009a). Immunomagnetic NPs conjugated to polydopamine-extracted aflatoxin from liquid food samples with 59% recovery in case of aflatoxin AF and 89% of aflatoxin B1 in red wine. It thus helps in concentrating the analyte, which can then be analyzed (McCullum et al., 2014).

Antimicrocystin-LR-coated single-walled carbon nanotubes bind to microcystin-LR (toxin released by cyanobacteria) in drinking water up to levels of 0.6 nM. The method is an improvement over the sampling time involved in the traditional method of quantitation by enzyme-linked immunosorbent assay (Wang et al., 2009a).

Recently, antideoxynivalenol conjugated to poly(maleic anhydride-alt-1-octadecene)–coated QDs were studied for detection of deoxynivalenol (trichothecenemycotoxin) in diluted extracts of maize with detection limit of 500 μg/L (Speranskaya et al., 2014). More such examples are depicted in Table 15.2.

4.2 Detection of Chemicals and Contaminants

4.2.1 Sensing Preservatives

Food preservatives like benzoates and taste-intensifying additives, such as, glutamates have been checked for their isolation and isotachophoresis zonal electrophoresis detection on a microfluidic poly(methylmethacrylate) chip. The method is very sensitive, with low limits of detection (μM/L) with no interference from food (Bodor et al., 2001). Concentrations of benzoic acid and its salt added as food preservatives are dictated by law, as well as, controlled by regulatory agencies (European Union Law, 1995). Nanomaterial made of glassy carbon electrodes can

Table 15.2 Nanosensors Used for Detection of Toxins

Analyte	Nanomaterial	Detection Method	Detection Limits	References
Aflatoxin B1	Anti-aflatoxin B1 monoclonal Ab–bovine serum albumin immobilized on nanostructured samarium nanorods onto indium–SnO_2–glass substrate	Amperometry immunosensor	57.82 pg/mL with a response time of 5 s	Singh et al. (2013)
Aflatoxin B1	Aflatoxin-oxidase embedded in sol–gel linked to multiwalled carbon nanotubes	Chronoamperometry	1.6 nM with an average response time of 44 s	Li et al. (2011)
Cholera toxin	Anti-cholera-toxin Ab adsorbed on AuNPs modified gold electrode	Change in capacitance	0.09 attoM	Loyprasert et al. (2010)
Clostridium botulinum neurotoxin type A	Dye-doped NP (22 nm) conjugated Ab against nanoporous organosilicate	Fluorescence biosensortoxin	145.8 fg/mL in orange juice and 164.2 fg/mL in tap water	Bok et al. (2013)
Microcystin-LR	Anti-microcystin-LR Ab on QDs	Square wave-stripping voltammetry detection technique	99 ng/L	Yu et al. (2009)
Microcystin-LR	Anti-microcystin-LR Ab on AgNPs used as electrodes	Label-free capacitance immunosensors	10 fM	Dawan et al. (2011)
Microcystin-LR	Anti-microcystin-LR Ab immobilized on gold electrode coated with L-cysteine-modified AuNPs	Label-free amperometric immunosensor	20 ng/L	Tong et al. (2011)
Ochratoxin-A	Anti-ochratoxin A polyclonal Ab bound to AuNPs onto working electrode	Differential-pulse voltammetry	0.2 ng/L	Bonel et al. (2010)
Ochratoxin-A	Rabbit Ab immobilized on nanocrystalline TiO_2-chitosan	Electrochemical impedance spectroscopy	Sensitivity is 7.5 mM, 4 times higher than only chitosan	Khan and Dhayal (2008)

(Continued)

Table 15.2 Nanosensors Used for Detection of Toxins (*cont.*)

Analyte	Nanomaterial	Detection Method	Detection Limits	References
Ochratoxin-A	Rabbit IgG Ab-bovine serum albumin immobilized onto nanostructured cerium oxide film fabricated onto an indium–SnO_2	Differential-pulse voltammetry	2.5 ng/L	Kaushik et al. (2009a)
Ochratoxin-A	Rabbit IgG Ab-bovine serum albumin conjugate immobilized on fumed silica NPs and chitosan-nanobiocomposite film	Differential-pulse voltammetry	0.3 ng/dL	Kaushik et al. (2009b)
Palytoxin (marine toxin)	Carbon nanotubes conjugated to anti-Palytoxin Ab	Electrochemiluminescent method	2.2 μg/kg of mussel meat, twofold higher than LC–MS/MS	Zamolo et al. (2012)
Sterigmatocystin	Aflatoxin-detoxifizyme covalently linked to multiwalled carbon nanotubes	Differential-pulse voltammetry	0.13 mM	Yao et al. (2006)
Sterigmatocystin	Aflatoxin-oxidase immobilized onto chitosan–single-walled carbon nanotubes–modified Au electrode	Amperometry	3 ng/mL	Chen et al. (2010)

detect benzoic acid in yogurt and other nonalcoholic beverages. Benzoic acid inhibits the biocatalytic activity of tyrosinase and polyphenol oxidase. Concentrations of benzoic acid up to 0.03 μM could be proportionally detected by catalytic inhibition of tyrosinase entrapped in titania gel modified with multiwalled carbon nanotubes and Nafion in nonalcoholic beverages. The results obtained were consistent with those obtained with HPLC method but with a better sensitivity of 1.06 μA/μM (Kochana et al., 2012). Amperometric quantitation of monosodium glutamate in food seasonings, sauces, and soups using glutamate oxidase anchored to a screen-printed electrode was in the range of 1–20 mg/dL detected within 2 min (Basu et al., 2006).

4.2.2 Sensing the Adulterants

Melamine is an adulterant often found in pet foods, raw milk, and infant foods to elevate the protein content. Cyanuric acid-functionalized AuNPs bind to melamine selectively and cause aggregation of AuNPs, which change in color from red to blue. Limits of detection are 2.5 ppb and the results are easily visible to the naked eye (Ai et al., 2009). Based on a similar principle, chemical reductants and crown ether-modified thiols specific to melamine binding have been included in sensing technology, which prevents formation of AuNPs. This results in a change of color (Kuang et al., 2011). SERS-based detection using patterned gold nanostructures (Bluray Disc surfaces) have also detected melamine up to levels of 2 ppm from contaminated noodles, wheat gluten, cakes and chicken feed (Lin et al., 2008; Liu and Ma, 2010).

Adulteration of urea in milk is an emerging concern. Blends of urease enzyme and polyvinyl pyrrolidone electrospun to form beaded nanofibers of dimensions 7–100 nm were used for urea biosensing. The implantable patch had been originally used for biosensing urea in kidney failure patients, which can also be extrapolated for urea contamination in milk (Gouma, 2010).

Water poisoning with cyanide could be detected up to levels of 2 nM by quenching of fluorescence exhibited by AuNPs (Liu et al., 2010). Detection of acrylamide in water is also possible by voltammetry, using single-walled carbon nanotubes coated on glassy carbon electrodes (Krajewska et al., 2008). Perchlorate, a food and water pollutant, has also been detected using 2-dimethylaminoethanethiol-modified AuNPs based on the SERS principle at 0.1 μg/L (Gu et al., 2009).

Addition of colorants like Ponceau 4R and Sudan I are illegal as they are carcinogenic adulterants (Duncan, 2011). Carbon nanotube-based colorimetric assays have been developed for detection of such food colorants often used in soft drinks and beverages (Zhang et al., 2010). Graphene-modified glassy carbon electrodes detected Sudan IV in ketchup and chili powder in the range of 2×10^{-7} M–8×10^{-5} M (Mo et al., 2010). Multiwalled carbon nanotubes with ionic liquid nanocomposites containing carbon–ceramic electrodes have been used for electrochemical detection of dyes used in food and beverages, such as, tartrazine and sunset yellow (Majidi et al., 2013).

4.2.3 Sensing Heavy Metals

Mercury (Hg) in elemental, inorganic, and organic forms is toxic to humans. Techniques for removal of mercury, cadmium (Cd), silver (Ag), and titanium (Ti) have been developed using superparamagnetic iron-oxide NPs functionalized with dimercaptosuccinic

Table 15.3 Nanosensors in Detection of Heavy Metals

Analyte	Nanomaterial	Detection Method	Detection Limits	References
Cd (II), Co (II), and Sb (III)	Mesoporous silicates conjugated to chalcone isothiocyanate	Color change observed on binding of metals; unbound—pale yellow, Co complex—cyan, Sb complex—yellow, Cd—colorless	Metal ions could be detected from 5 ppb to 1 ppm	El-Nahass et al. (2015)
Hg (II)	AuNPs functionalized with bovine serum albumin conjugated to rhodamine 6G;	Fluorescence response	0.1–2.5 μM	Chai et al. (2010)
	AuNPs functionalized with thioglycolic acid and rhodamine B	Rhodamine B fluorescence is quenched by AuNPs due to FRET, but Hg (II) disrupts the pair	4×10^{-10} M/L consistent with atomic absorption spectroscopy	Zheng et al. (2009a)
Hg (II), Ag (I), and Pb (II)	Fluorescent dye-labeled aptamer is adsorbed on multiwalled carbon nanotube	Fluorescent detection; metals disturb aptamer binding with nanotubes leading to multicolors	Limit of detection of 15 nM for Hg (II), 18 nM for Ag (I), and 20 nM for Pb (II) in water	Wang and Si (2013)

acid from water (Yantasee et al., 2007). AuNPs functionalized with Hg-specific oligonucleotide consisting of thymine can detect Hg (II) up to concentrations of 0.5 nM (Zhu et al., 2009). Monolithic nanostructured cages of cubic Fm3m (C10) monoliths aid in naked-eye detection of toxic metal ions like Hg (II), lead (II), and Cd (Balaji et al., 2006). Detailed reviews of optic sensors for detecting Hg are also reported (El-Safty and Shenashen, 2012). Iron-oxide NPs immobilized with alkaline phosphatase measure the amperometric reaction between alkaline phosphatase and ascorbic acid 2-phosphate on contact with Hg, Cu, Ag, and Pb in water or other complex matrices (Loh et al., 2008). Different nanosensors used for detection of heavy metals are listed in Table 15.3.

4.2.4 Sensing Biocidals

Organophosphates widely used as pesticides and insecticides are retained in plants, water, and fruits due to their high aqueous solubility. They irreversibly inhibit acetyl cholinesterase enzyme, essential for central nervous system functioning. High consumption causes buildup of the neurotransmitter and interferes with muscular responses and acts as a neurotoxin (Mulchandani et al., 2001). Conventionally, they are detected by chromatography, which is time-consuming and needs trained personnel for extensive preparation steps with no real-time detection. Organophosphorous pesticides can be selectively determined by Ab immobilized on magnetic NPs coated with 3-aminopropyl triethoxysilane based on enzyme-linked immune sorbent assay with high sensitivity (Hu et al., 2010). Acetyl cholinesterase immobilized on copolymer of polypyrrole and polyaniline followed by doping with multiwalled carbon nanotubes could detect thiocholine up to 1.0 ng/mL with good reproducibility. Polypyrrole is used as a polymer matrix as it can be easily electropolymerized in neutral pH solutions at lower oxidation potentials conducive for biologic molecules, while carbon nanotubes provide effective wiring and an anionic dopant. Copolymer of polypyrrole with aniline provided a porous matrix for entrapment of acetyl cholinesterase (Du et al., 2010b). A disposable screen-printed electrode consisting of multiwalled carbon nanotubes with *Drosophila melanogaster* acetyl cholinesterase and Prussian blue detected dichlorvos and carbofuran (organophosphate and carbamate pesticides) up to 0.5 μg/L in vegetable and water samples (Tang et al., 2011).

Residues of herbicide 2,4-D up to 250 pg/mL could be detected using cadmium telluride (CdTe) QD as label. The QD capped with mercaptopropionic acid was coupled to alkaline phosphatase using EDC coupling followed by conjugation to 2,4-D. Competitive binding between conjugated 2,4-D–alkaline phosphatase–CdTe QD and free 2,4-D Anti-2,4-D-IgG Ab immobilized in an immunoreactor column enabled fluoroimmunoassay detection (Vinayaka et al., 2009). Carbon nanotubes and metallic nanomaterial-based electrochemical sensors for detection of pesticide have been reviewed recently (Xiang et al., 2011). Some more examples are cited in Table 15.4.

Crystal violet and malachite green, fungicides, and antimicrobials banned by the US Food and Drug Administration (FDA), are common adulterants found in fish grown in contaminated waters. Patterned gold nanostructures have been able to detect them up to ~0.2 ppb levels based on the SER principle (He et al., 2008).

4.3 Detection of Food Spoilage and Moisture: Food Quality Indicators

Shelf life of food products refers to the period for which they can be used while maintaining the food quality. Extrinsic factors, such

Table 15.4 Nanosensors in Detection of Pesticides

Analyte	Nanomaterial	Detection Method	Detection Limits	References
2,4-Dinitrophenol	AuNPs functionalized with 2,4-dinitrophenol–bovine serum albumin and hybridized with latex microspheres	Colorimetric; competitive binding between conjugated AuNPs and toxin molecules for binding anti-DNP Ab	Rapid detection	Ko et al. (2010)
Chlorpyrifos and malathion	AuNPs	Color change due to pesticide adsorption on AuNP	Chlorpyrifos up to 20 ppb and malathion up to 100 ppb	Lisha and Pradeep (2009)
DDT	Anti-DDT Ab conjugated to AuNPs, nitrocellulose membrane strip immobilized with 2,2-bis-(*p*-chlorophenyl) acetic acid–bovine serum albumin conjugate	Dipstick-based competitive immunoassay; dipping the strip in immunocomplex solution binds to the immobilized Ag with red color formed in detection zone proportional to DDT concentration	27 ng/mL	Lisa et al. (2009)
Methyl parathion	Methyl-parathion degrading enzyme immobilized through CdTe QDs coated on a glassy carbon electrode with multiwalled carbon nanotube and AuNPs	Voltammetry; the enzyme selectively hydrolyzes the P–S bond of pesticide and not any other pesticides; real-time reusable monitoring	1.0 ng/mL	Du et al. (2010a)
Monocrotophos	Glassy carbon electrode modified with CdTe QDs and AuNPs immobilized to acetyl cholinesterase through chitosan microspheres	Voltammetry; QDs and AuNPs synergized and helped in amplifying detection sensitivity	0.3 ng/mL	Du et al. (2008)
N-methylcarbamate pesticide carbofuran	AuNPs labeled to an anti-carbofuran mAb and distributed on porous fiberglass-conjugated pad	Ovalbumin–carbofuran and goat antimouse IgG immobilized on nitrocellulose membrane and works as the test line and control line	0.25 mg/L in water within 10 min	Zhou et al. (2004)
Paraoxon	Paraoxon Ab loaded on AuNPs immobilized on an electrode using Nafion membrane	Voltammetry; electrochemical immunosensor	12 μg/mL	Hu et al. (2003)

as, gaseous atmosphere, storage temperature, and relative humidity coupled with intrinsic properties of food determine the shelf stability and perishability of food. Shelf-stable foods have lower water activity and/or low pH, whereas perishable foods have the opposite intrinsic properties (McMeekin and Ross, 1996). Stale food results in unacceptable flavor, odor, accumulation of gases, release of exudates, physicochemical changes in appearance, consistency, and color.

Stale meat or fish releases gaseous amines like ammonia due to floral growth or any chemical change of the product. Biogenic amines are also indicators of food spoilage (Ruiz-Capillas and Moral, 2004); for example, histamine, putrescine, and tyramine consumption have all been linked to hypertension and migraine (Linares et al., 2012). Xanthines are breakdown products of purines and are used for predicting shelf life of fish and meat (Devi et al., 2012). Xanthine oxidase conjugated to chitosan–ZnO NPs dispersed on a multiwalled carbon nanotube scaffold within matrix of polyaniline was studied as a xanthine sensor. Addition of ZnO NPs to carbon nanotubes promotes better and faster transport of electrons between the active sites of biomolecules (ie, xanthine oxidase and electrode with good chemical stability). Chitosan permitted formation of bioadhesive films for immobilization of xanthine oxidase. The sensor displayed a detection level of 0.1 mM without any interference from matrices (Devi et al., 2012). Conjugation of xanthine oxidase to AgNPs further enhanced the stability of enzymes on storage for 60 days at room temperature (Devi et al., 2013).

Moisture content and gases, especially oxygen in excess, are mainly responsible for food spoilage. ZnO- and TiO_2 NPs have been studied for detection of ethanol and oxygen, respectively (Robert, 2014). Photoactivated indicator ink, consisting of TiO_2- or SnO_2 NPs and redox-active methylene blue dye, detects even trace quantities of oxygen retained during packaging or during leak cases (Mills and Hazafy, 2008). Electrospun nanofibers and nanowebs made of carbon, metal-oxide semiconductors, polymers (nylon-6, polyacrylic acid, and polyvinyl alcohol), and ceramics have been employed as sensors for detection of ammonia, hydrogen sulfide, nitrogen dioxide, carbon dioxide, carbon monoxide, moisture, formaldehyde, and other volatile organic gases (Ding et al., 2010). Adsorption of gases changes the oxidation–reduction potential of oxide semiconductors and affects the electric conductance (Shevade et al., 2003). Polyacrylic-acid nanofibers coated with QCM (Quartz crystal microbalance) exhibit a sensitivity of 130 ppb for the detection of ammonia (Ding et al., 2005). Perylene-based nanofibrils containing fluorophores detect parts-per-trillion level

Table 15.5 Nanosensors in Gas Detection

Gaseous Analyte	Nanomaterial	Observation	References
Carbon dioxide	Fluorophore-encapsulated polymer nanobeads	0.8–100% with 1% resolution and minimal cross-sensitivity of 0.6%	Von Bültzingslöwen et al. (2002)
Ethylene gas	WO_3–SnO_2 nanocomposites (35 nm and 100 nm)	2–8 ppm; WO_3 increased the sensitivity	Pimtong-Ngam et al. (2007)
Gaseous amines	SnO_2 NPs–TiO_2 microrods composites	Conductance changes	Zhang and Zhang (2008)

of gaseous amines (Che and Zang, 2009). A tenside (detergent) film dispersed with carbon-coated CuNPs is also sensitive to humidity. Moisture causes the polymer to swell, resulting in separation of NPs, and hence absorption and reflection of different colors. On dissipation of moisture, the sensors are reusable (Luechinger et al., 2007). Some more examples are given in Table 15.5.

A research team at the Canadian Wheat Board Center for Grain Research has developed sensors for conducting polymeric NPs made of anilineboronic acid-phosphate complex to monitor the quality of grains (Neethirajan et al., 2009). The nanosensors particularly used for carbon dioxide detection essentially consist of anilineboronic acid-phosphate complexes between the gold-interdigitated array microelectrodes and indium-doped SnO_2-coated glass slides and electrolyte layer, followed by a layer of polytetrafluoroethylene as a selective layer. The conducting polymer responds to any volatile gases, insects, fungi, and analytes; detects the source and type of spoilage; and generates a change in electric resistance in the presence of moisture. The sensors have a low power requirement too, and hence can be placed at pest hide-outs (Neethirajan et al., 2010; Neethirajan et al., 2013).

4.4 Indicators of Product Acceptability

Standardization of sensory properties (color, smell, taste, and texture) conventionally relies on sensory analysis of trained expertise and thus is subject to human variation, is nonautomatic, and is expensive. Chemical fingerprinting of flavor and aroma is extremely complex and difficult due to the heterogeneity of different compounds, contributing to unique experiences of flavor (smell and taste) (Gardner and Bartlett, 1994). Gas chromatography cannot detect volatile compounds below their threshold limits and needs a time run up to 100 min for a single analysis

(García et al., 2006). Finally, odors perceived are of isolated peaks and not of integral sensation. Electronic nose and electronic tongue are designed similar to the olfactory and taste buds on the tongue and identify variations rather than quantification of any analytes.

Electronic NOSE (Natural Olfactory Sensor Emulator), also known as e-nose, can detect hydrophobic volatile gases having molecular weights in the range of 30–300 Da produced from food products (Reineccius, 1996). E-nose consists of three integrated sections—a sample loading part followed by a sensor and finally a data processing unit. The sample loading part transfers a sample to the sensors. Adsorption of volatiles to the sensor arrays changes the resistance of the sensors proportional to presence of the volatiles and senses specific chemical odors. The sensor arrays are composed of metal-oxide semiconductors, piezoelectric crystals conducting organic polymers, or quartz semiconductors. Polymers with intrinsic conductance, such as, polypyrroles may also be blended with varying sensitivity to analytes. Nonconductive polymers, such as, polyethylene oxide or ethyl cellulose admixed with carbon black may be used (Shevade et al., 2003). Adsorption of volatile gases with high vapor pressure (eg, fatty acids, alcohols formed during bacterial or yeast fermentation) induces oxidation–reduction of oxide semiconductors, changing the dimensions of the composites and affecting electric conductance (Shevade et al., 2003). The data-processing system processes the signals by intelligent pattern recognition algorithms, including discrimination factorial analysis for recognizing unknown samples, principal component analysis, and partial least squares for quantifying bitterness.

Alcohols, fatty acids, esters, and ketones yield strong responses, whereas carbon dioxide, water, and nitrogen dioxide have lower responses. The high surface–volume ratio of nanocarriers favors gas adsorption on sensor arrays and hence increased selectivity and sensitivity of sensors, detectable in a short span of time. It also minimizes the device size and power consumption and simplifies device circuitry. For example, electronic noses can easily detect *Ganoderma boninense*, a plant pathogen responsible for basal stem rot on tree trunks and food products (Abdullah et al., 2011; Durán and Marcato, 2013).

As described earlier, an electronic nose based on polymeric NPs of polyanilineboronic acid has been demonstrated for detecting ongoing spoilage during storage of grains (Neethirajan et al., 2013), as well as, discriminating strains of species producing mycotoxins and mold from healthy grains (Magan and Evans, 2000). A number of reviews on the same with details can be referred (Cabanes et al., 2009; Sahgal et al., 2007).

Cyranose 320 (Cyrano Sciences, Pasadena, CA) is a commercial handheld electronic nose consisting of nanocomposites of conducting polymers, for example, poly(vinyl butyral) and poly(ethylene oxide), with carbon-black nanotubes with advanced pattern recognition algorithms. For identification, the instrument has to be trained to associate and store smell prints of each volatile with distinct responses (Gouma and Sberveglieri, 2004). Cyranose 320 detects *Salmonella*, *E. coli*, and *S. aureus* in beef and sausages at concentrations from 50 ppb onward (Balasubramanian et al., 2005; Abdallah et al., 2013). It gives simple results, such as, "Accept" or "Reject," "Mixture Identified," or "Contaminated." Sensitivity and specificity of an electronic nose detecting *E. coli* O157:H7 and non-O157 strains ranged between 41.7% and 50% for nonnormalized data (Younts et al., 2003). Libra Nose is also a metalloporphyrin-coated quartz microbalance sensor array-based electronic nose for detection of alcohols, amines, ketones, and aromatic compounds (Di Natale et al., 2003).

Metalloporyphyrins mimic the mammalian olfactory receptors, which are also metalloproteins in nature. Chemical selectivity and sensitivity of sensors depend on metal and peripheral constituents of the complex (Brunink et al., 1996). Microfabricated beams of silicon (cantilevers) coated with polymer are also nanomechanical sensors for detecting volatile gases. Interaction of volatile gases results in swelling of the polymer, and the surface stresses cause bending of the cantilever or change in resonating frequency for cantilevers in static or dynamic mode (Gouma and Sberveglieri, 2004). Nano-ZnO film gas sensors evaluated quality of Chinese vinegar. ZnO NPs are sensitive to hydrogen sulfide, nitrogen dioxide, chlorine, ozone, and alcohols at room temperature. This is a unique feature compared to headspace gas chromatography, which requires heating to achieve equilibration of volatile gas in the headspace (Zhang et al., 2006). Few nanosensors can detect gases released from food products that are not suitable for consumption due to spoilage (Valdés et al., 2009; Ghasemi-Varnamkhasti et al., 2011; Śliwińska et al., 2014).

Quality of food can also be differentiated using nanosensors (Durán and Marcato, 2013). An electronic tongue responds to soluble portions of food using an array of sensors combined with pattern-recognition analysis. The pattern obtained is matched with the existing database of tastes. Electronic tongues are composed of layer-by-layer ultrathin films of molecular building lipid/polymer membranes with 10,000-fold increased sensitivity than that of the normal human tongue. Molecular assemblies of polyhydroxyl groups and association of sodium ions at particular pH decide the tastes of particular foods. In the presence of analyte, the building block self-assembles and forms receptors (Hou et al., 2012). The electrode charge density of the lipid/polymer membrane and ion distribution changes and gives a particular membrane electric potential, or SPR imaging measures molecules adsorbed by receptors for food.

Receptors for bitterness have been immobilized on polypyrrole-modified carbon nanotubes. Upon binding, polypyrrole reorients, which is recorded as an electric signal. Bitterness up to 1 fM could be detected in beverages (Song et al., 2012). Discrimination of degradation products in orange juice has been efficiently detected using microelectrodes of polyanilinenano fibers used as taste sensors (Medeiros et al., 2009). The electronic nose and tongue replace the human sensory test for quality determination, which often fails on routine and repeated analysis.

A number of glucose-sensitive sensors have been developed for determining glucose content in commercial beverages (eg, AuNPs coated with glucose-sensitive enzymes) (Ozdemir et al., 2010). *Pseudomonas fluoresens* immobilized on carbon nanotubes with epoxy composite could detect 0.5–4.0 mM of glucose and xenobiotics in anaerobic conditions with high sensitivity (Kirgoz et al., 2007).

4.5 Others

4.5.1 Detection of Veterinary Drugs

Animals treated with veterinary therapeutic products are exposed to wide range of chemicals legally or illegally. Majorly, they include antimicrobials used in treatment of infectious diseases or to enhance growth rates or feed efficiency. Residues of these chemicals, especially antibiotics in milk and poultry livestock, including penicillin, tetracycline, macrolides, and β-lactam, pose as toxic hazards, lead to allergic responses, or to the development of antibiotic-resistant strains in the poultry, as well as, in humans consuming them (Ferrini et al., 2008). Chloramphenicol molecularly imprinted poly(ethylene glycol dimethacrylate-*N*-methacryloyl-L-histidinemethylester) NPs attached onto the surface of SPR detected 40 ng/kg of chloramphenicol in a honey sample. The method was verified to be selective with thiamphenicol and florphenicol with similar chemical structures (Kara et al., 2013). Anti-penicillin G immobilized on gold electrode modified with a self-assembled nanolayer of thioctic acid detected 3×10^{-15} M of penicillin G in milk by impedimetry (Thavarungkul et al., 2007). Tobramycin could be detected by QCM nanosensor of gold surface containing a self-assembled monolayer of mercaptane up to a concentration of 5.7×10^{-12} M in chicken egg white and milk (Yola et al., 2014).

4.5.2 Detection of Nutritional Components

Lysine is used as marker for nutritional damage caused in food products due to heat treatment, for example, ultraheated temperature (UHT) treated milk. The estimation of lysine in processed food products, based on advanced "Maillard reaction," gives an indication of the protein damage due to food processing. Lysine has a free ε-amino group that can easily react with other reducing sugars present in food as ingredients. This reaction gives a specific color and flavor to food when heated to a high temperature, which acts as an indicator (Luong et al., 1997). A detection system based on biotin-labeled Ab, coupled with an enhanced chemiluminescence or fluorescence, detects protein present in analytes (Luong et al., 2008).

A quick, reliable, and sensitive immunodipstick colloidal gold coated with vitamin B12–IgY Ab has been developed for vitamin B12 measurement in food samples. The visible detection limit of vitamin B12 is 1 ng/mL. This immunodipstick probe is simple and effective with no requirement of trained operators. Vitamin B12 is essential for normal functioning of the neural system and brain. This is of special note for vegetarians, as plant products are deficit in vitamin B12 (Selva Kumar and Thakur, 2011).

Certain food products (eg, poultry, meat, cheese, and shrimp) are characterized by high amounts of cholesterol. They can also be quantified using carbon nanotubes. These carbon nanotubes have polymer enzyme coatings. Carbon nanotubes increase the electrocatalytic effect due to enhanced surface coverage area for measurement of electrocatalytic response using voltammetry. Carbon nanotubes promote electron transfer reactions in redox reactions (Yang et al., 2011b; Wisitsoraat et al., 2009). An online database has also reported infusing nanocapsules of plant steroids replacing meat cholesterol (Mulvaney, 2011). Some of the other nanobased systems to determine nutritional content are given in Table 15.6.

4.5.3 Detection of Growth-Promoting Agents and Steroids

Plant growth-promoting agents and steroids are used in food production and livestock reproduction, respectively, to gain maximum profits with minimum investments. However, consumption of these steroidal hormones puts humans at risk, while growth-promoting agents accumulate along the food chain in fatty tissues, and it is difficult to rid consumers' systems of the substances for a prolonged time. For example, tests are conducted to estimate progesterone in milk and other food products originating from farm animals. These diagnostics are primarily immunoassay-based techniques (Tothill, 2001). Improved versions of immunosensors

Table 15.6 Nanosensors for Detecting Nutritional Components

Analyte	Nanomaterial	Detection	Observation	References
Glucose	Graphite rod electrode modified by AuNPs containing glucose oxidase	Amperometric enzyme electrode	Linearity range 0.1–10 mM, detection limit 0.08 mM	German et al. (2010)
Glucose	Polyamide nanofibrous membrane coated on glassy carbon electrode containing glucose oxidase	Amperometric enzyme electrode using ferrocene as mediator	Linearity range 1–10 mM with detection limit of 6×10^{-6} M	Scampicchio et al. (2010)
Inulin	Fructose dehydrogenase and inulinase immobilized on a AuNPs–cysteamine self-assembled monolayer, modified gold electrode	Inulinase hydrolyzes inulin forming fructose detected by fructose dehydrogenase by electrochemical oxidation of tetrathiafulvalene used as mediator	Lifetime of sensor is 35 days with detection limit of 6.6×10^{-7} M/L	Manso et al. (2008)
Oxalic acid	Multiwalled carbon nanotubes modified glassy carbon electrode	Voltammetry; oxidation of oxalic acid was greatly improved over bare glassy carbon electrode	Detection limit 0.012 mM	Zheng et al. (2009b)
Phenolic compounds	Tyrosinase immobilized by drop-coating on glassy carbon electrode by polyamidic nanofibrous membrane	Amperometry	Response time of 16 s with a detection limit of 0.05 μM	Arecchi et al. (2010)
Polyphenols	Tyrosinase immobilized on carbon electrodes with electrodeposited AuNPs	Voltammetry	Lifetime of 18 days; rapidly detects phenol, catechol, caffeic acid, chlorogenic acid, gallic acid, and proto-catechu aldehyde in wines	Carralero Sanz et al. (2005)
Polyphenols	Laccase biosensor on a multilayer material of AuNPs, fullerenols, and *Trametes versicolor* laccase assembled layer by layer onto gold electrode surface	Amperometry	Limit of detection of 0.006 mM/L in wine expressed as polyphenol index 5.0–50 mg/L	Lanzellotto et al. (2014)

that are based on nanotechnology are required for on-site and real-time monitoring of food products, which are still in the nascent stage.

A detailed explanation of various growth-promoting agents in agricultural crops, such as, pesticides and insecticides is already covered in Section 4.2.4. Nanobiosensors composed of an atomic-force microscopy tip functionalized with acetolactate synthase enzyme detected the herbicide metsulfuron-methyl, an acetolactate synthase inhibitor, through force curves (Sekhon, 2014). Misuse of clenbuterol, used for treatment of asthma, for increasing the fat and protein in pigs has led to development of a florescence-based immunoassay. This assay, consisting of CdSe/CdS QDs and Fe_3O_4/AuNPs for Ag–Ab attachments, can estimate clenbuterol in the linear range of 0.5–20,000 pg/mL in pig urine (Wang et al., 2009b).

4.5.4 Detection of Food Allergens

Food allergens are food components that on consumption trigger allergic reactions that could range from minor skin rashes to severe, life-threatening conditions. Examples of major potential allergens include peanuts, wheat protein gluten, milk, egg, nuts, shellfish, and soybeans. The European Community has identified 14 allergenic food ingredients as food allergens and has fixed regulations for labeling them (Pilolli et al., 2013). Food allergy is described as an immune response that could be genetic or environmentally induced, and is IgE mediated or non-IgE mediated (cellular), involving oral, gastrointestinal, and respiratory tracts.

Evaluations of IgE-based hypersensitivity due to peanut protein Ara h1 using AuNPs localized plasmon array have been studied up to a detection limit of 2 nM (Olkhov et al., 2012).

Soy protein, a major food allergen of soymilk, soy yogurt, fruit, and soy juice, has been detected by silica NPs doped with Nile blue (particle size 257 nm). Nile blue-doped silica NPs prepared by reverse-micelle microemulsion and linked to antisoy protein Ab detected soy protein up to 0.05 mg/L with an incubation time of 2 h (Godoy-Navajas et al., 2011).

Egg-white protein (ovalbumin) added to whole milk can be detected up to 1 ppm within 20 min by combined technology of immunomagnetic separation of ovalbumin by magnetic microbeads followed by SERS using silver dendrites for quantitation (He et al., 2011). SPR-based techniques are coupled with sensors for quantification of α-lactalbumin present in marketed milk and whey protein isolates (Indyk, 2009). Adulteration of bovine and caprine milk with colostrum has been reported using an IgG-based SPR immunosensor with a short analytical real-time monitoring time of within 4 min during milking (Pilolli et al., 2013).

Table 15.7 Nanosensors Used in Detecting Food Allergens

Analyte	Nanomaterial	Detection	Observation	References
Casein	Ab-coated AuNPs + Poly(L-Arg)/multiwalled carbon nanotube	Differential-pulse voltammetry	50 ng/mL in cheese	Cao et al. (2011)
Chicken ovomucoid	Concanavalin A conjugated zinc oxide QDs sized 3 nm	Square-wave voltammetry	1 ng/mL in standard solutions	Yang et al. (2011a)
Gliadin	Ab-coated AuNPs (25 nm)	QCM	8 ppb in standard solutions, food matrix wheat, barley, oat, rice, corn, and soybean pancake mix—1 ppm	Chu et al. (2012)
Peanut protein Ara h1	Ab-coated 3 nm Au-coated nanoporous membrane	Electrochemical impedance spectroscopy	—	Singh et al. (2010)
Shrimp	Ab-coated AuNPs (10 nm)	QCM	0.33 μg/mL in spiked water samples	Jiang et al. (2013)

Quantitative determination of bovine serum albumin, immunoglobulin G, α-lactalbumin, β-lactoglobulin, and lactoferrin simultaneously in milk, whey fractions, and milk-derived products is possible using SPR (Billakanti et al., 2010). Imaging SPR with Ab array provided a complete allergen profile of hazelnut content in cookies and chocolate products (Rebe Raz et al., 2010). Some other examples are also given in Table 15.7.

4.5.5 Detection of Genetically Modified Food

Recent developments in genetic engineering and its application in crop production to meet the demands of ever-increasing global population have put many in a dilemma about its use in food production (Sicherer and Sampson, 2010). Considering regulatory aspects related to food products of genetically modified (GM) organism origin, a thorough assessment is thus required for risks related to consumption of these products. Sociopsychological aspects affect the acceptability of GM foods. Protein analysis by using immunosensors accounts for one of the growing markets in the food industry. SDI Europe Ltd. has developed an immunoassay that can detect GM food ingredients using dipstick devices. Detection of GM food products that are consumed by either humans or animals are based on PCR techniques that analyze specific DNA sequences. These developments have led to miniaturization of the instruments that were previously used in laboratories and are now available for on-site analysis of samples using DNA chips. These tests can also be modified into DNA biosensor devices (Tothill, 2001). SERS-barcoded nanosensors designed by encapsulation of AuNPs with silica conjugated to oligonucleotides detected *B. thuringiensis* (Bt) gene-transformed rice expressing insecticidal proteins up to 0.1 pg/mL (Chen et al., 2012).

5 Advanced Packaging Quality

A turning point in packaging is nanoreinforced packaging and active packaging providing an extra tensile strength of food packets. Nanosilver in combination with antibacterial agents may prove to be more potent to combat multiple bacterial growths with a smaller amount of dose required. TiO_2 has been suggested to improve upon the UV protection barrier, as well as, reduce bacterial contamination in food packaging when used as NPs (Ranjan et al., 2014). Several antimicrobial packaging and oxygen-scavenging food packages have been developed (Saxl, 2011). Imperm (Eastman Chemical Co., Kingsport, TN, USA), Aegis, Durethan, and others use a nanocomposite of nylon and nanoclay with enhanced barrier properties that can help keep the oxygen out and the carbonation in, especially in polyethylene bottles. Beer in nanocomposite bottles can keep for as long as 30 weeks. The same has also been used in soda, milk, and water bottles, as well as, in military food packages (Gruère, 2012). Milk packages made of these materials improved durability of milk up to 9 days, and grains stayed dry. Nanowax for fruits also provided better durability and transparency (Sekhon, 2014).

Aegis OX (Honeywell, Inc.) has oxygen scavengers and nanocomposite clay particles within the polyethylene bottle material for fruit juice, soft drink, and beer servings. Durethan KU2-2601 (Bayer AG), composed of polyamide and nanoclay with layered silicate barriers, with enhanced gloss and stiffness, in addition to barrier properties, is commonly used in paperboard juice containers (Ranjan et al., 2014). Nanolok, a polymeric resin composed of nanodispersed silicate, generated superabsorbent-hydrogel nanocomposites. The McDonald's burger containers contain an ecosynthetic adhesive made up of 5–150 nm starch nanospheres, which require less water for adhesion with reduced drying time (Kumari and Yadav, 2013).

Incorporation of enzymes, such as, cholesterol reductase or lactase increases the nutritional value of food. SiO_2 NPs immobilized with glutamate dehydrogenase and lactate dehydrogenase have shown enhanced enzyme activity (Ranjan et al., 2014).

6 Smart Food Packaging Sensors

Smart- or intelligent packaging indicates the quality of food within a package through use of intelligent nanosensors devices that are either incorporated into packaging or attached to packaging (ie, nanocomposite smart packaging). For instance, incorporation of oxygen scavengers or indicators within food packaging material would serve as indicators for food spoilage by sensing gaseous analytes and thus provide real-time food freshness (Lange et al., 2002; García et al., 2006). Food-packaging distributors also use nanotechnology-based biosensors for product tracking (nanotagging) by using radio-frequency identification tags to prevent counterfeiting. However, radio frequency identification tags are expensive, with the cheapest tag costing about 50 cents. Optic fibers also find wider application with calorimetric and florescent dyes that change florescence and absorbance with a change in pH or oxidation state.

Motivations to innovate new technology related to the food sector are consumer demands for quality food products globally and the need for faster transport of food. Purposes of packaging are to retard food deterioration, increase in shelf life and provide protection from environmental influences, such as, light and moisture. There are materials that change color on exposure to visible- or UV light, indicating the quality of the food product. This technology when coupled with information and communication technology can automatically send data to users or monitoring sections about the condition of food products. This combination can revolutionize the world of the packaging industry both at the consumer level and at the industry level. For example, substances that are produced due to food spoilage processes (lactic and acetic acids, alcohols, and amines) can be detected by the use of smart-packaging materials made of luminescent nanosized ZnO and organic luminophores. These nanophotonics assemblies can work as novel packaging materials. The principle involved behind this novel packaging is that it detects the change in properties (optic, electric, or mechanical) of sensors. These changes can be analyzed by visual inspection or by external devices that convert the packaging response into electric signals followed by data interpretation using computer systems. This combination of smart packaging with communication technology is cost-effective and utilizes photonics principles at the nanoscale level. This innovative technique of detection is termed nanophotonics with detection limits of 2×10^{-3} M/L for ZnO and 10^{-5} M/L for organic luminophore (Sarapulova et al., 2015).

Timestrip has developed an iStrip based on AuNPs that shows red color at temperatures above freezing. Freezing causes irreversible agglomeration of the AuNPs with loss of red color. Thus accidental exposure of chilled foods to room temperature could be determined using this iStrip. Various TiO_2- and SnO_2-based colorimetric oxygen indicators have been developed that bleach the color of redox dyes on exposure to UV irradiation. The color of the films developed depends on the extent of exposure to oxygen (Ranjan et al., 2014).

Nanobarcodes consisting of synthetic DNA barcodes present in film packages detect the food pathogens by fluorescence of a fluorescent probe under UV (Brody et al., 2008). GuardIN Fresh, a Nanomech subsidiary (Fayetteville, AR, USA), commercializes a nanotechnology-based scavenging system that scavenges ethylene gas responsible for hastened ripening of perishable products (Sekhon, 2014). An electronic tongue has also been used in food packaging to detect gases released due to spoilage of food. The gases cause the sensor strip to change its color in the presence of gases, indicating whether the food is fresh or stale (Saxl, 2011).

A research group at Netherlands is working on "Release on Command" food packaging with a nanobased bioswitch that releases preservatives when the food is about to spoil (Sekhon, 2014). Such sensing mechanisms would not only improve the food quality but also reduce food wastage.

7 Regulatory Aspects

Stringent regulatory guidelines must to ensure "safe" and "quality" food to consumers. There has been a lack of regulatory guidelines on the use of NPs in food packaging and storage at the US Food and Drug Administration. This needs to be taken seriously, as lack of guidance will hamper both the growth and the application of nanotechnology in the food industry (Servin et al., 2015).

Nanocomposites, the earliest commercial application in the food industry, include use of plastic nanocomposites applicable for packaging of food and beverages. Commercial biosensors based on nanocomposites are concise, simple, and easy to use for point-of-care biosensing. There are diagnostic assays that are fabricated using nanosensors for point-of-care assessment of Ab in milk samples, but they are not commercially available for field testing. Research is ongoing to make them available for testing on farm animals using nonserological milk samples (Vyas et al., 2015). Cellulosic nanowhisker production technology has been developed to replace the use of plastic and fiberglass, by making biocomposite from wheat straw. Nanobarcode technology has gained momentum in the past few years because of its ability to form a large number of combinations. But this technology is not yet well exploited (Dasgupta et al., 2015). Inorganic materials like SiO_2, MgO, and TiO_2 are being used as coating materials for used as oxygen barriers, and Ag for "active coating" for antibacterial activity (Chaudhry and Castle, 2011).

8 Future Aspects and Challenges

Nanocomposite technology is an emerging material science. Its wider application in the food industry extends from food production to food processing and ultimately to food consumption. Tracing food products, quality assessment, safety, and maintaining a nutritional index have all made the food industry take a deeper look into technology involved in innovative diagnostics and biosensing techniques. On-site and point-of-care detection for a large number of increasing analytes requires quick, sensitive, and accurate measurements. This has inspired many researchers in the food industry to develop nanotechnology-based sensors as diagnostic tools. Nanotechnology-based biosensors using nanocomposites will reduce the load of laboratory testing and provide on-site applications. There is an urgent need for more available technology in the food industry for easy, affordable, sensitive, and broad-spectrum analysis of food products, both at industrial level and at consumer level. This chapter has specifically emphasized developments in the field of nanotechnology-based biosensors being incorporated in food industry for detection of plant-based pathogens, toxins, GM food products, nutritional index, and regulatory aspects attached to them.

A major challenge with nanobiosensors is their use with real samples, as many biosensors fail when they come in contact with real food products because of the complexity of samples. Replacement and recalibration of nanobiosensors are another drawback that needs to be considered. Nanobiosensor-based on-site tests help in reduction of time for which products are kept in inventory, rapid response time and corporate liability costs (Alocilja and Radke, 2003).

Most of the research is still in its nascent stage for detection of analytes in real food without disrupting its final form. A nondestructive technology based on information directly available on the final packed food product would be very desirable. SPR-based sensors are a promising technique that does not require labeling with molecules and can easily assay crude samples without purification. The need for nanobiosensors has multiplied many fold in the recent past due to increased consumer awareness and regulatory action being imposed about food composition and safety.

Abbreviations

Ab Antibody
AuNP Gold nanoparticle
CFU Colony-forming unit
NP Nanoparticle
QD Quantum dot
SERS Surface-enhanced Raman scattering
SPR Surface plasmon resonance
SQUID Superconducting quantum interference device
QCM Quartz crystal microbalance

References

Abdallah, S.A., Al-Shatti, L.A., Alhajraf, A.F., Al-Hammad, N., Al-Awadi, B., 2013. The detection of foodborne bacteria on beef: the application of the electronic nose. Springerplus 2, 687.

Abdullah, A.H., Adom, A.H., Md. Shakaff, A.Y., Ahmad, M.N., Saad, M.A., Tan, E.S., Fikri, N.A., Markom, M.A., Zakaria, A., 2011. Electronic nose system for ganoderma detection. Sens. Lett. 9, 353–358.

Ai, K., Liu, Y., Lu, L., 2009. Hydrogen-bonding recognition-induced color change of gold nanoparticles for visual detection of melamine in raw milk and infant formula. J. Am. Chem. Soc. 131, 9496–9497.

Alivisatos, A.P., Gu, W., Larabell, C., 2005. Quantum dots as cellular probes. Annu. Rev. Biomed. Eng. 7, 55–76.

Alocilja, E.C., Radke, S.M., 2003. Market analysis of biosensors for food safety. Biosens. Bioelectron. 18, 841–846.

Arecchi, A., Scampicchio, M., Drusch, S., Mannino, S., 2010. Nanofibrous membrane based tyrosinase-biosensor for the detection of phenolic compounds. Anal. Chim. Acta 659, 133–136.

Bailey, R.C., Nam, J.M., Mirkin, C.A., Hupp, J.T., 2003. Real-time multicolor DNA detection with chemoresponsive diffraction gratings and nanoparticle probes. J. Am. Chem. Soc. 125, 13541–13547.

Balaji, T., El-Safty, S.A., Matsunaga, H., Hanaoka, T., Mizukami, F., 2006. Optical sensors based on nanostructured cage materials for the detection of toxic metal ions. Angew. Chem. Int. Ed. Engl. 45, 7202–7208.

Balasubramanian, S., Panigrahi, S., Logue, C.M., Marchello, M., Sherwood, J.S., 2005. Identification of salmonella-inoculated beef using a portable electronic nose system. J. Rapid Meth. Aut. Mic. 13, 71–95.

Basu, A.K., Chattopadhyay, P., Roychudhuri, U., Chakraborty, R., 2006. Development of biosensor based on immobilized L-glutamate oxidase for determination of monosodium glutamate in food. Indian J. Exp. Biol. 44, 392–398.
Berthiller, F., Burdaspal, P.A., Crews, C., Iha, M.H., Krska, R., Lattanzio, V.M.T., Macdonald, S., Malone, R.J., Maragos, C., Solfrizzo, M., Stroka, J., Whitaker, T.B., 2014. Developments in mycotoxin analysis: an update for 2012–2013. World Mycotox. J. 7, 3–33.
Billakanti, J.M., Fee, C.J., Lane, F.R., Kash, A.S., Fredericks, R., 2010. Simultaneous, quantitative detection of five whey proteins in multiple samples by surface plasmon resonance. Int. Dairy J. 20, 96–105.
Bodor, R., Žúborová, M., Ölvecká, E., Madajová, V., Masár, M., Kaniansky, D., Stanislawski, B., 2001. Isotachophoresis and isotachophoresis-zone electrophoresis of food additives on a chip with column-coupling separation channels. J. Sep. Sci. 24, 802–809.
Bok, S., Korampally, V., Darr, C.M., Folk, W.R., Polo-Parada, L., Gangopadhyay, K., Gangopadhyay, S., 2013. Femtogram-level detection of *Clostridium botulinum* neurotoxin type A by sandwich immunoassay using nanoporous substrate and ultra-bright fluorescent suprananoparticles. Biosens. Bioelectron. 41, 409–416.
Bonel, L., Vidal, J.C., Duato, P., Castillo, J.R., 2010. Ochratoxin A nanostructured electrochemical immunosensors based on polyclonal antibodies and gold nanoparticles coupled to the antigen. Anal. Method. 2, 335–341.
Branen, J.R., Hass, M.J., Douthit, E.R., Maki, W.C., Branen, A.L., 2007. Detection of *Escherichia coli* O157, *Salmonella enterica* serovar *Typhimurium*, and staphylococcal enterotoxin B in a single sample using enzymatic bio-nanotransduction. J. Food Prot. 70, 841–850.
Brody, A.L., Bugusu, B., Han, J.H., Sand, C.K., McHugh, T.H., 2008. Scientific status summary: innovative food packaging solutions. J. Food Sci. 73, R107–R116.
Brunink, J.A.J., Di Natale, C., Bungaro, F., Davide, F.A.M., D'amico, A., Paolesse, R., Boschi, T., Faccio, M., Ferri, G., 1996. The application of metalloporphyrins as coating material for quartz microbalance-based chemical sensors. Anal. Chim. Acta 325, 53–64.
Cabanes, F.J., Sahgal, N., Bragulat, M.R., Magan, N., 2009. Early discrimination of fungal species responsible of ochratoxin A contamination of wine and other grape products using an electronic nose. Mycotox. Res. 25, 187–192.
Cao, Q., Zhao, H., Yang, Y., He, Y., Ding, N., Wang, J., Wu, Z., Xiang, K., Wang, G., 2011. Electrochemical immunosensor for casein based on gold nanoparticles and poly(L-arginine)/multi-walled carbon nanotubes composite film functionalized interface. Biosens. Bioelectron. 26, 3469–3474.
Carralero Sanz, V., Mena, M.L., González-Cortés, A., Yáñez-Sedeño, P., Pingarrón, J.M., 2005. Development of a tyrosinase biosensor based on gold nanoparticles-modified glassy carbon electrodes: application to the measurement of a bioelectrochemical polyphenols index in wines. Anal. Chim. Acta 528, 1–8.
Chai, F., Wang, T., Li, L., Liu, H., Zhang, L., Su, Z., Wang, C., 2010. Fluorescent gold nanoprobes for the sensitive and selective detection for Hg^{2+}. Nanoscale Res. Lett. 5, 1856–1860.
Chaudhry, Q., Castle, L., 2011. Food applications of nanotechnologies: an overview of opportunities and challenges for developing countries. Trends Food Sci. Technol. 22, 595–603.
Che, Y., Zang, L., 2009. Enhanced fluorescence sensing of amine vapor based on ultrathin nanofibers. Chem. Commun., 5106–5108.
Chemla, Y.R., Grossman, H.L., Poon, Y., Mcdermott, R., Stevens, R., Alper, M.D., Clarke, J., 2000. Ultrasensitive magnetic biosensor for homogeneous immunoassay. Proc. Natl. Acad. Sci. USA 97, 14268–14272.

Chen, S.J., Chien, F.C., Lin, G.Y., Lee, K.C., 2004. Enhancement of the resolution of surface plasmon resonance biosensors by control of the size and distribution of nanoparticles. Opt. Lett. 29, 1390–1392.

Chen, J., Liu, D., Li, S., Yao, D., 2010. Development of an amperometric enzyme electrode biosensor for sterigmatocystin detection. Enzyme Microb. Technol. 47, 119–126.

Chen, K., Han, H., Luo, Z., Wang, Y., Wang, X., 2012. A practicable detection system for genetically modified rice by SERS-barcoded nanosensors. Biosens. Bioelectron. 34, 118–124.

Chien, Y.Y., Jan, M.D., Adak, A.K., Tzeng, H.C., Lin, Y.P., Chen, Y.J., Wang, K.T., Chen, C.T., Chen, C.C., Lin, C.C., 2008. Globotriose-functionalized gold nanoparticles as multivalent probes for Shiga-like toxin. ChemBioChem 9, 1100–1109.

Chu, H., Huang, Y., Zhao, Y., 2008. Silver nanorod arrays as a surface-enhanced Raman scattering substrate for foodborne pathogenic bacteria detection. Appl. Spectrosc. 62, 922–931.

Chu, P.T., Lin, C.S., Chen, W.J., Chen, C.F., Wen, H.W., 2012. Detection of gliadin in foods using a quartz crystal microbalance biosensor that incorporates gold nanoparticles. J. Agric. Food Chem. 60, 6483–6492.

Dasgupta, N., Ranjan, S., Mundekkad, D., Ramalingam, C., Shanker, R., Kumar, A., 2015. Nanotechnology in agro-food: from field to plate. Food Res. Int. 69, 381–400.

Davis, D., Guo, X., Musavi, L., Lin, C.S., Chen, S.H., Wu, V.C.H., 2013. Gold nanoparticle-modified carbon electrode biosensor for the detection of *Listeria monocytogenes*. Ind. Biotechnol. 9, 31–36.

Dawan, S., Kanatharana, P., Wongkittisuksa, B., Limbut, W., Numnuam, A., Limsakul, C., Thavarungkul, P., 2011. Label-free capacitive immunosensors for ultra-trace detection based on the increase of immobilized antibodies on silver nanoparticles. Anal. Chim. Acta 699, 232–241.

Devi, R., Yadav, S., Pundir, C.S., 2012. Amperometric determination of xanthine in fish meat by zinc oxide nanoparticle/chitosan/multiwalled carbon nanotube/polyaniline composite film bound xanthine oxidase. Analyst 137, 754–759.

Devi, R., Batra, B., Lata, S., Yadav, S., Pundir, C.S., 2013. A method for determination of xanthine in meat by amperometric biosensor based on silver nanoparticles/cysteine modified Au electrode. Process Biochem. 48, 242–249.

Di Natale, C., Macagnano, A., Martinelli, E., Paolesse, R., D'Arcangelo, G., Roscioni, C., Finazzi-Agro, A., D'Amico, A., 2003. Lung cancer identification by the analysis of breath by means of an array of non-selective gas sensors. Biosens. Bioelectron. 18, 1209–1218.

Ding, B., Yamazaki, M., Shiratori, S., 2005. Electrospun fibrous polyacrylic acid membrane-based gas sensors. Sens. Actuat. B 106, 477–483.

Ding, B., Wang, M., Wang, X., Yu, J., Sun, G., 2010. Electrospun nanomaterials for ultrasensitive sensors. Mater. Today 13, 16–27.

Dingman, J., 2008. Nanotechnology: its impact on food safety. J. Environ. Health 70, 47–50.

Du, D., Chen, S., Song, D., Li, H., Chen, X., 2008. Development of acetylcholinesterase biosensor based on CdTe quantum dots/gold nanoparticles modified chitosan microspheres interface. Biosens. Bioelectron. 24, 475–479.

Du, D., Chen, W., Zhang, W., Liu, D., Li, H., Lin, Y., 2010a. Covalent coupling of organophosphorus hydrolase loaded quantum dots to carbon nanotube/Au nanocomposite for enhanced detection of methyl parathion. Biosens. Bioelectron. 25, 1370–1375.

Du, D., Ye, X., Cai, J., Liu, J., Zhang, A., 2010b. Acetylcholinesterase biosensor design based on carbon nanotube-encapsulated polypyrrole and polyaniline copolymer for amperometric detection of organophosphates. Biosens. Bioelectron. 25, 2503–2508.

Duncan, T.V., 2011. Applications of nanotechnology in food packaging and food safety: barrier materials, antimicrobials and sensors. J. Colloid Interface Sci. 363, 1–24.

Durán, N., Marcato, P.D., 2013. Nanobiotechnology perspectives: role of nanotechnology in the food industry: a review. Int. J. Food Sci. Technol. 48, 1127–1134.

El-Boubbou, K., Gruden, C., Huang, X., 2007. Magnetic glyco-nanoparticles: a unique tool for rapid pathogen detection, decontamination, and strain differentiation. J. Am. Chem. Soc. 129, 13392–13393.

El-Nahass, M.N., Fayed, T.A., Shaaban, M.H., Hassan, F.M., 2015. Chalcone isothiocyanate-mesoporous silicates: selective anchoring and toxic metal ions detection. Sens. Actuat. B 210, 56–68.

El-Safty, S.A., Shenashen, M.A., 2012. Mercury-ion optical sensors. Trends Anal. Chem. 38, 98–115.

Emami, T., Madani, R., Rezayat, S.M., Golchinfar, F., Sarkar, S., 2012. Applying of gold nanoparticle to avoid diffusion of the conserved peptide of avian influenza nonstructural protein from membrane in Western blot. J. Appl. Poultry Res. 21, 563–566.

European Commission Regulation, 2006. Setting maximum levels for certain contaminants in foodstuffs (EC No. 1881/2006). Official J. European Union (L364), 5–24.

European Union Law, 1995. Additives other than authorised colours and sweeteners. E.P.A.C. Directive 95/2/EC.

Fan, C., Hu, Z., Riley, L.K., Purdy, G.A., Mustapha, A., Lin, M., 2010. Detecting food- and waterborne viruses by surface-enhanced Raman spectroscopy. J. Food Sci. 75, M302–M307.

Ferrini, A.M., Mannoni, V., Carpico, G., Pellegrini, G.E., 2008. Detection and identification of beta-lactam residues in milk using a hybrid biosensor. J. Agric. Food Chem. 56, 784–788.

Fu, J., Park, B., Siragusa, G., Jones, L., Tripp, R., Zhao, Y., Cho, Y.J., 2008. An Au/Si hetero-nanorod-based biosensor for *Salmonella* detection. Nanotechnology 19, 155502.

García, M., Aleixandre, M., Gutiérrez, J., Horrillo, M.C., 2006. Electronic nose for wine discrimination. Sens. Actuat. B 113, 911–916.

Gardner, J.W., Bartlett, P.N., 1994. A brief history of electronic noses. Sens. Actuat. B 18, 210–211.

German, N., Ramanaviciene, A., Voronovic, J., Ramanavicius, A., 2010. Glucose biosensor based on graphite electrodes modified with glucose oxidase and colloidal gold nanoparticles. Microchim. Acta 168, 221–229.

Gfeller, K.Y., Nugaeva, N., Hegner, M., 2005. Rapid biosensor for detection of antibiotic-selective growth of *Escherichia coli*. Appl. Environ. Microb. 71, 2626–2631.

Ghasemi-Varnamkhasti, M., Mohtasebi, S.S., Rodríguez-Méndez, M.L., Siadat, M., Ahmadi, H., Razavi, S.H., 2011. Electronic and bioelectronic tongues, two promising analytical tools for the quality evaluation of non-alcoholic beer. Trends Food Sci. Tech. 22, 245–248.

Godoy-Navajas, J., Aguilar Caballos, M.P., Gomez-Hens, A., 2011. Heterogeneous immunoassay for soy protein determination using Nile blue-doped silica nanoparticles as labels and front-surface long-wavelength fluorimetry. Anal. Chim. Acta 701, 194–199.

Goldman, E.R., Clapp, A.R., Anderson, G.P., Uyeda, H.T., Mauro, J.M., Medintz, I.L., Mattoussi, H., 2003. Multiplexed toxin analysis using four colors of quantum dot fluororeagents. Anal. Chem. 76, 684–688.

Gomez-Sjoberg, R., Morisette, D.T., Bashir, R., 2005. Impedance microbiology-on-a-chip: microfluidic bioprocessor for rapid detection of bacterial metabolism. J. Microelectromech. S. 14, 829–838.

Gouma, P.I.P., 2010. Electrospinning—a novel nanomanufacturing technique for hybrid nanofibers and their non-woven mats. In: Gouma, P.I.P. (Ed.), Nanomaterials for Chemical Sensors and Biotechnology. Pan Stanford Publishing Pte Ltd, Singapore.

Gouma, P., Sberveglieri, G., 2004. Novel materials and applications of electronic noses and tongues. MRS Bull. 29, 697–702.

Grossman, H.L., Myers, W.R., Vreeland, V.J., Bruehl, R., Alper, M.D., Bertozzi, C.R., Clarke, J., 2004. Detection of bacteria in suspension by using a superconducting quantum interference device. Proc. Natl. Acad. Sci. USA 101, 129–134.

Gruère, G.P., 2012. Implications of nanotechnology growth in food and agriculture in OECD countries. Food Policy 37, 191–198.

Gu, B., Ruan, C., Wang, W., 2009. Perchlorate detection at nanomolar concentrations by surface-enhanced Raman scattering. Appl. Spectrosc. 63, 98–102.

Hall, R.H., 2002. Biosensor technologies for detecting microbiological foodborne hazards. Microbes Infect. 4, 425–432.

He, L., Kim, N.J., Li, H., Hu, Z., Lin, M., 2008. Use of a fractal-like gold nanostructure in surface-enhanced Raman spectroscopy for detection of selected food contaminants. J. Agric. Food Chem. 56, 9843–9847.

He, L., Haynes, C.L., Diez-Gonzalez, F., Labuza, T.P., 2011. Rapid detection of a foreign protein in milk using IMS–SERS. J. Raman Spectrosc. 42, 1428–1434.

Hou, Y., Genua, M., Tada Batista, D., Calemczuk, R., Buhot, A., Fornarelli, P., Koubachi, J., Bonnaffe, D., Saesen, E., Laguri, C., Lortat-Jacob, H., Livache, T., 2012. Continuous evolution profiles for electronic-tongue-based analysis. Angew. Chem. Int. Ed. Engl. 51, 10394–10398.

Hrapovic, S., Liu, Y., Male, K.B., Luong, J.H., 2004. Electrochemical biosensing platforms using platinum nanoparticles and carbon nanotubes. Anal. Chem. 76, 1083–1088.

Hu, S.Q., Xie, J.W., Xu, Q.H., Rong, K.T., Shen, G.L., Yu, R.Q., 2003. A label-free electrochemical immunosensor based on gold nanoparticles for detection of paraoxon. Talanta 61, 769–777.

Hu, Y., Shen, G., Zhu, H., Jiang, G., 2010. A class-specific enzyme-linked immunosorbent assay based on magnetic particles for multiresidue organophosphorus pesticides. J. Agric. Food Chem. 58, 2801–2806.

Huang, S.H., 2006. Gold nanoparticle-based immunochromatographic test for identification of *Staphylococcus aureus* from clinical specimens. Clin. Chim. Acta 373, 139–143.

Huang, S.H., Wei, H.C., Lee, Y.C., 2007. One-step immunochromatographic assay for the detection of *Staphylococcus aureus*. Food Control 18, 893–897.

Ikanovic, M., Rudzinski, W.E., Bruno, J.G., Allman, A., Carrillo, M.P., Dwarakanath, S., Bhahdigadi, S., Rao, P., Kiel, J.L., Andrews, C.J., 2007. Fluorescence assay based on aptamer-quantum dot binding to *Bacillus thuringiensis* spores. J. Fluoresc. 17, 193–199.

Indyk, H., 2009. Development and application of an optical biosensor immunoassay for α-lactalbumin in bovine milk. Int. Dairy J. 19, 36–42.

Jiang, D., Ji, J., An, L., Sun, X., Zhang, Y., Zhang, G., Tang, L., 2013. Mast cell-based electrochemical biosensor for quantification of the major shrimp allergen Pen a 1 (tropomyosin). Biosens. Bioelectron. 50, 150–156.

Jin, X., Jin, X., Chen, L., Jiang, J., Shen, G., Yu, R., 2009. Piezoelectric immunosensor with gold nanoparticles enhanced competitive immunoreaction technique for quantification of aflatoxin B1. Biosens. Bioelectron. 24, 2580–2585.
Kaittanis, C., Naser, S.A., Perez, J.M., 2007. One-step, nanoparticle-mediated bacterial detection with magnetic relaxation. Nano Lett. 7, 380–383.
Kaittanis, C., Santra, S., Perez, J.M., 2010. Emerging nanotechnology-based strategies for the identification of microbial pathogenesis. Adv. Drug Deliv. Rev. 62, 408–423.
Kanayeva, D.A., Wang, R., Rhoads, D., Erf, G.F., Slavik, M.F., Tung, S., Li, Y., 2012. Efficient separation and sensitive detection of *Listeria monocytogenes* using an impedance immunosensor based on magnetic nanoparticles, a microfluidic chip, and an interdigitated microelectrode. J. Food Prot. 75, 1951–1959.
Kara, M., Uzun, L., Kolayli, S., Denizli, A., 2013. Combining molecular imprinted nanoparticles with surface plasmon resonance nanosensor for chloramphenicol detection in honey. J. Appl. Polym. Sci. 129, 2273–2279.
Karlsson, R., 2004. SPR for molecular interaction analysis: a review of emerging application areas. J. Mol. Recognit. 17, 151–161.
Kaushik, A., Solanki, P.R., Ansari, A.A., Ahmad, S., Malhotra, B.D., 2009a. A nanostructured cerium oxide film-based immunosensor for mycotoxin detection. Nanotechnology 20, 055105.
Kaushik, A., Solanki, P.R., Sood, K.N., Ahmad, S., Malhotra, B.D., 2009b. Fumed silica nanoparticles–chitosan nanobiocomposite for ochratoxin-A detection. Electrochem. Commun. 11, 1919–1923.
Kayal, S., Charbit, A., 2006. Listeriolysin O: a key protein of *Listeria monocytogenes* with multiple functions. FEMS Microbiol. Rev. 30, 514–529.
Khan, R., Dhayal, M., 2008. Nanocrystalline bioactive TiO_2–chitosan impedimetric immunosensor for ochratoxin-A. Electrochem. Commun. 10, 492–495.
Kim, G., Park, S.B., Moon, J.H., Lee, S., 2013. Detection of pathogenic salmonella with nanobiosensors. Anal. Methods 5, 5717–5723.
Kirgoz, Ü.A., Timur, S., Odaci, D., Pérez, B., Alegret, S., Merkoçi, A., 2007. Carbon nanotube composite as novel platform for microbial biosensor. Electroanalysis 19, 893–898.
Ko, S., Park, T.J., Kim, H.S., Kim, J.H., Cho, Y.J., 2009. Directed self-assembly of gold binding polypeptide-protein A fusion proteins for development of gold nanoparticle-based SPR immunosensors. Biosens. Bioelectron. 24, 2592–2597.
Ko, S., Gunasekaran, S., Yu, J., 2010. Self-indicating nanobiosensor for detection of 2,4-dinitrophenol. Food Control 21, 155–161.
Kochana, J., Kozak, J., Skrobisz, A., Woźniakiewicz, M., 2012. Tyrosinase biosensor for benzoic acid inhibition-based determination with the use of a flow-batch monosegmented sequential injection system. Talanta 96, 147–152.
Krajewska, A., Radecki, J., Radecka, H., 2008. A voltammetric biosensor based on glassy carbon electrodes modified with single-walled carbon nanotubes/hemoglobin for detection of acrylamide in water extracts from potato crisps. Sensors 8, 5832–5844.
Kuang, H., Chen, W., Yan, W., Xu, L., Zhu, Y., Liu, L., Chu, H., Peng, C., Wang, L., Kotov, N.A., Xu, C., 2011. Crown ether assembly of gold nanoparticles: melamine sensor. Biosens. Bioelectron. 26, 2032–2037.
Kumar, C.S.S.R., 2007. Nanomaterials for Biosensors. Wiley-VCH, Weinheim, Germany.
Kumari, A., Yadav, S.K., 2013. Nanotechnology in agri-food sector. Crit. Rev. Food Sci. Nutr. 54, 975–984.
Lange, D., Hagleitner, C., Hierlemann, A., Brand, O., Baltes, H., 2002. Complementary metal oxide semiconductor cantilever arrays on a single chip: mass-sensitive detection of volatile organic compounds. Anal. Chem. 74, 3084–3095.

Lanzellotto, C., Favero, G., Antonelli, M.L., Tortolini, C., Cannistraro, S., Coppari, E., Mazzei, F., 2014. Nanostructured enzymatic biosensor based on fullerene and gold nanoparticles: preparation, characterization and analytical applications. Biosens. Bioelectron. 55, 430–437.
Lee, H., Sun, E., Ham, D., Weissleder, R., 2008. Chip-NMR biosensor for detection and molecular analysis of cells. Nat. Med. 14, 869–874.
Leonard, P., Hearty, S., Brennan, J., Dunne, L., Quinn, J., Chakraborty, T., O'Kennedy, R., 2003. Advances in biosensors for detection of pathogens in food and water. Enzyme Microb. Tech. 32, 3–13.
Li, S.C., Chen, J.H., Cao, H., Yao, D.S., Liu, D.L., 2011. Amperometric biosensor for aflatoxin B1 based on aflatoxin-oxidase immobilized on multiwalled carbon nanotubes. Food Control 22, 43–49.
Lin, M., He, L., Awika, J., Yang, L., Ledoux, D.R., Li, H., Mustapha, A., 2008. Detection of melamine in gluten, chicken feed, and processed foods using surface enhanced Raman spectroscopy and HPLC. J. Food Sci. 73, T129–T134.
Linares, D.M., Del Río, B., Ladero, V., Martínez, N., Fernández, M., Martín, M.C., Álvarez, M.A., 2012. Factors influencing biogenic amines accumulation in dairy products. Front. Microbiol. 3, 180.
Lisa, M., Chouhan, R.S., Vinayaka, A.C., Manonmani, H.K., Thakur, M.S., 2009. Gold nanoparticles based dipstick immunoassay for the rapid detection of dichlorodiphenyltrichloroethane: an organochlorine pesticide. Biosens. Bioelectron. 25, 224–227.
Lisha, K.P., Pradeep, T., 2009. Enhanced visual detection of pesticides using gold nanoparticles. J. Environ. Sci. Health B 44, 697–705.
Liu, B., Ma, L., 2010. Research on multi-colony diploid genetic algorithm for production logistics scheduling optimization. In: Proceedings—2010 International Conference on Web Information Systems and Mining (WISM), pp. 13–17.
Liu, Y., Ai, K., Cheng, X., Huo, L., Lu, L., 2010. Gold-nanocluster-based fluorescent sensors for highly sensitive and selective detection of cyanide in water. Adv. Funct. Mater. 20, 951–956.
Loh, K.S., Lee, Y., Musa, A., Salmah, A., Zamri, I., 2008. Use of Fe_3O_4 nanoparticles for enhancement of biosensor response to the herbicide 2,4-dichlorophenoxyacetic acid. Sensors 8, 5775–5791.
Loyprasert, S., Hedstrom, M., Thavarungkul, P., Kanatharana, P., Mattiasson, B., 2010. Sub-attomolar detection of cholera toxin using a label-free capacitive immunosensor. Biosens. Bioelectron. 25, 1977–1983.
Luechinger, N.A., Loher, S., Athanassiou, E.K., Grass, R.N., Stark, W.J., 2007. Highly sensitive optical detection of humidity on polymer/metal nanoparticle hybrid films. Langmuir 23, 3473–3477.
Luong, J.H.T., Bouvrette, P., Male, K.B., 1997. Developments and applications of biosensors in food analysis. Trends Biotechnol. 15, 369–377.
Luong, J.H.T., Male, K.B., Glennon, J.D., 2008. Biosensor technology: technology push versus market pull. Biotechnol. Adv. 26, 492–500.
Magan, N., Evans, P., 2000. Volatiles in grain as an indicator of fungal spoilage, order descriptors for classifying spoiled grain and the potential for early detection using electronic nose technology: a review. J. Stored Prod. Protect. 36, 319–340.
Mahalik, N.P., Nambiar, A.N., 2010. Trends in food packaging and manufacturing systems and technology. Trends Food Sci. Technol. 21, 117–128.
Majdinasab, M., Aminlari, M., Sheikhi, M.H., Niakousari, M., Shekarforoosh, S., 2012. Detection of inv A gene of *Salmonella* by DNA-gold nanoparticles biosensor and its comparison with PCR. J. Exp. Nanosci. 8, 223–239.

Majidi, M., Fadakar Bajeh Baj, R., Naseri, A., 2013. Carbon nanotube–ionic liquid (CNT–IL) nanocamposite modified sol-gel derived carbon-ceramic electrode for simultaneous determination of sunset yellow and tartrazine in food samples. Food Anal. Method. 6, 1388–1397.
Manso, J., Mena, M.A., Yanez-Sedeno, P., Pingarron, J.M., 2008. Bienzyme amperometric biosensor using gold nanoparticle-modified electrodes for the determination of inulin in foods. Anal. Biochem. 375, 345–353.
McCullum, C., Tchounwou, P., Ding, L.S., Liao, X., Liu, Y.M., 2014. Extraction of aflatoxins from liquid foodstuff samples with polydopamine-coated superparamagnetic nanoparticles for HPLC-MS/MS analysis. J. Agric. Food Chem. 62, 4261–4267.
McMeekin, T.A., Ross, T., 1996. Shelf life prediction: status and future possibilities. Int. J. Food Microbiol. 33, 65–83.
Medeiros, E.S., Gregório, R., Martinez, R.A., Mattoso, L.H.C., 2009. A taste sensor array based on polyaniline nanofibers for orange juice quality assessment. Sens. Lett. 7, 24–30.
Mills, A., Hazafy, D., 2008. A solvent-based intelligence ink for oxygen. Analyst 133, 213–218.
Mo, Z., Zhang, Y., Zhao, F., Xiao, F., Guo, G., Zeng, B., 2010. Sensitive voltammetric determination of Sudan I in food samples by using gemini surfactant-ionic liquid-multiwalled carbon nanotube composite film modified glassy carbon electrodes. Food Chem. 121, 233–237.
Mulchandani, A., Chen, W., Mulchandani, P., Wang, J., Rogers, K.R., 2001. Biosensors for direct determination of organophosphate pesticides. Biosens. Bioelectron. 16, 225–230.
Mulvaney, D., 2011. Green Food: An A-to-Z Guide. SAGE Publications, Thousand Oaks, CA, USA.
Murugaboopathi, G., Parthasarathy, V., Chellaram, C., Anand, T.P., Vinurajkumar, S., 2013. Applications of biosensors in food industry. Biosci. Biotech. Res. Asia 10, 711–714.
Nath, S., Kaittanis, C., Tinkham, A., Perez, J.M., 2008. Dextran-coated gold nanoparticles for the assessment of antimicrobial susceptibility. Anal. Chem. 80, 1033–1038.
Neethirajan, S., Jayas, D.S., Sadistap, S., 2009. Carbon dioxide (CO_2) sensors for the agri-food industry—a review. Food Bioprocess Tech. 2, 115–121.
Neethirajan, S., Freund, M.S., Jayas, D.S., Shafai, C., Thomson, D.J., White, N.D.G., 2010. Development of carbon dioxide (CO_2) sensor for grain quality monitoring. Biosyst. Eng. 106, 395–404.
Neethirajan, S., Freund, M.S., Jayas, D.J., 2013. United States Patent 8454819 B2. Poly(aniline boronic acid) polymers and methods of use.
Odumeru, J., 2002. Inside microbiology: current microbial concerns in the dairy industry. Food Safety Magazine 8, 20–23.
Oh, B.K., Kim, Y.K., Park, K.W., Lee, W.H., Choi, J.W., 2004. Surface plasmon resonance immunosensor for the detection of *Salmonella typhimurium*. Biosens. Bioelectron. 19, 1497–1504.
Olkhov, R.V., Parker, R., Shaw, A.M., 2012. Whole blood screening of antibodies using label-free nanoparticle biophotonic array platform. Biosens. Bioelectron. 36, 1–5.
Ozdemir, C., Yeni, F., Odaci, D., Timur, S., 2010. Electrochemical glucose biosensing by pyranose oxidase immobilized in gold nanoparticle-polyaniline/AgCl/gelatin nanocomposite matrix. Food Chem. 119, 380–385.
Pal, S., Alocilja, E.C., 2009. Electrically active polyaniline coated magnetic (EAPM) nanoparticle as novel transducer in biosensor for detection of *Bacillus anthracis* spores in food samples. Biosens. Bioelectron. 24, 1437–1444.

Phillips, R.L., Miranda, O.R., You, C.C., Rotello, V.M., Bunz, U.H., 2008. Rapid and efficient identification of bacteria using gold-nanoparticle-poly(para-phenyleneethynylene) constructs. Angew. Chem. Int. Ed. Engl. 47, 2590–2594.
Pilolli, R., Monaci, L., Visconti, A., 2013. Advances in biosensor development based on integrating nanotechnology and applied to food-allergen management. Trends Anal. Chem.; TrAC 47, 12–26.
Pimtong-Ngam, Y., Jiemsirilers, S., Supothina, S., 2007. Preparation of tungsten oxide-tin oxide nanocomposites and their ethylene sensing characteristics. Sens. Actuat. A 139, 7–11.
Poitras, C., Tufenkji, N., 2009. A QCM-D-based biosensor for *E. coli* O157:H7 highlighting the relevance of the dissipation slope as a transduction signal. Biosens. Bioelectron. 24, 2137–2142.
Radoi, A., Targa, M., Prieto-Simon, B., Marty, J.L., 2008. Enzyme-linked immunosorbent assay (ELISA) based on superparamagnetic nanoparticles for aflatoxin M1 detection. Talanta 77, 138–143.
Ranjan, S., Dasgupta, N., Chakraborty, A., Melvin Samuel, S., Ramalingam, C., Shanker, R., Kumar, A., 2014. Nanoscience and nanotechnologies in food industries: opportunities and research trends. J. Nanopart. Res. 16, 1–23.
Ravindranath, S.P., Mauer, L.J., Deb-Roy, C., Irudayaraj, J., 2009. Biofunctionalized magnetic nanoparticle integrated mid-infrared pathogen sensor for food matrixes. Anal. Chem. 81, 2840–2846.
Rebe Raz, S., Liu, H., Norde, W., Bremer, M.G., 2010. Food allergens profiling with an imaging surface plasmon resonance-based biosensor. Anal. Chem. 82, 8485–8491.
Reineccius, G.A., 1996. Instrumental means of monitoring the flavor quality of foods. In: Lee, T.C., Kim, H.J. (Eds.), Chemical Markers for Processed and Stored Foods. American Chemical Society, University of Michigan, USA.
Robert, B., 2014. Nanomaterials for gas sensing: a review of recent research. Sensor Rev. 34, 1–8.
Rodríguez-Lorenzo, L., De La Rica, R., Álvarez-Puebla, R.A., Liz-Marzán, L.M., Stevens, M.M., 2012. Plasmonic nanosensors with inverse sensitivity by means of enzyme-guided crystal growth. Nat. Mater. 11, 604–607.
Rong-Hwa, S., Shiao-Shek, T., Der-Jiang, C., Yao-Wen, H., 2010. Gold nanoparticle-based lateral flow assay for detection of staphylococcal enterotoxin B. Food Chem. 118, 462–466.
Ruiz-Capillas, C., Moral, A., 2004. Free amino acids and biogenic amines in red and white muscle of tuna stored in controlled atmospheres. Amino Acids 26, 125–132.
Safina, G., Van Lier, M., Danielsson, B., 2008. Flow-injection assay of the pathogenic bacteria using lectin-based quartz crystal microbalance biosensor. Talanta 77, 468–472.
Sahgal, N., Needham, R., Cabanes, F.J., Magan, N., 2007. Potential for detection and discrimination between mycotoxigenic and non-toxigenic spoilage moulds using volatile production patterns: a review. Food Addit. Contam. 24, 1161–1168.
Sarapulova, O., Sherstiuk, V., Shvalagin, V., Kukhta, A., 2015. Photonics and nanophotonics and information and communication technologies in modern food packaging. Nanoscale Res. Lett. 10, 229.
Sargent, J.F., 2014. The National Nanotechnology Initiative: Overview, Reauthorization, and Appropriations Issues. Congressional Research Service. Report for Congress for Members and Committees of Congress. Arlington, VA, USA.
Saxl, O., 2011. Nanotech in food, beverage and related packaging. Applications and Markets to 2015Ion Publishers, UK.
Scampicchio, M., Arecchi, A., Lawrence, N.S., Mannino, S., 2010. Nylon nanofibrous membrane for mediated glucose biosensing. Sens. Actuat. B 145, 394–397.

Schofield, C.L., Field, R.A., Russell, D.A., 2007. Glyconanoparticles for the colorimetric detection of cholera toxin. Anal. Chem. 79, 1356–1361.
Sekhon, B.S., 2014. Nanotechnology in agri-food production: an overview. Nanotechnol. Sci. Appl. 7, 31–53.
Selva Kumar, L.S., Thakur, M.S., 2011. Competitive immunoassay for analysis of vitamin B(12). Anal. Biochem. 418, 238–246.
Servin, A., Elmer, W., Mukherjee, A., De La Torre-Roche, R., Hamdi, H., White, J., Bindraban, P., Dimkpa, C., 2015. A review of the use of engineered nanomaterials to suppress plant disease and enhance crop yield. J. Nanopart. Res. 17, 1–21.
Shevade, A.V., Ryan, M.A., Homer, M.L., Manfreda, A.M., Zhou, H., Manatt, K.S., 2003. Molecular modeling of polymer composite–analyte interactions in electronic nose sensors. Sens. Actuat. B 93, 84–91.
Shim, W.B., Yang, Z.Y., Kim, J.S., Kim, J.Y., Kang, S.J., Woo, G.J., Chung, Y.C., Eremin, S.A., Chung, D.H., 2007. Development of immunochromatography strip-test using nanocolloidal gold-antibody probe for the rapid detection of aflatoxin B1 in grain and feed samples. J. Microbiol. Biotechnol. 17, 1629–1637.
Sicherer, S.H., Sampson, H.A., 2010. Food allergy. J. Allergy Clin. Immunol. 125, S116–S125.
Singh, R., Sharma, P.P., Baltus, R.E., Suni, I.I., 2010. Nanopore immunosensor for peanut protein Ara h1. Sens. Actuat. B 145, 98–103.
Singh, J., Roychoudhury, A., Srivastava, M., Solanki, P.R., Lee, D.W., Lee, S.H., Malhotra, B.D., 2013. A highly efficient rare earth metal oxide nanorods based platform for aflatoxin detection. J. Mater. Chem. B 1, 4493–4503.
Śliwińska, M., Wiśniewska, P., Dymerski, T., Namieśnik, J., Wardencki, W., 2014. Food analysis using artificial senses. J. Agric. Food Chem. 62, 1423–1448.
Song, H.S., Kwon, O.S., Lee, S.H., Park, S.J., Kim, U.K., Jang, J., Park, T.H., 2012. Human taste receptor-functionalized field effect transistor as a human-like nanobioelectronic tongue. Nano Lett. 13, 172–178.
Sozer, N., Kokini, J.L., 2012. The applications of nanotechnology. In: Picó, Y. (Ed.), Chemical Analysis of Food: Techniques and Applications. Academic Press, Boston, MA.
Speranskaya, E.S., Beloglazova, N.V., Lenain, P., De Saeger, S., Wang, Z., Zhang, S., Hens, Z., Knopp, D., Niessner, R., Potapkin, D.V., Goryacheva, I.Y., 2014. Polymer-coated fluorescent CdSe-based quantum dots for application in immunoassay. Biosens. Bioelectron. 53, 225–231.
Storhoff, J.J., Lucas, A.D., Garimella, V., Bao, Y.P., Muller, U.R., 2004. Homogeneous detection of unamplified genomic DNA sequences based on colorimetric scatter of gold nanoparticle probes. Nat. Biotechnol. 22, 883–887.
Subramanian, A., Irudayaraj, J., Ryan, T., 2006. A mixed self-assembled monolayer-based surface plasmon immunosensor for detection of *E. coli* O157:H7. Biosens. Bioelectron. 21, 998–1006.
Sun, Y., Xia, Y., 2002. Shape-controlled synthesis of gold and silver nanoparticles. Science 298, 2176–2179.
Sun, W., Zhong, J., Qin, P., Jiao, K., 2008. Electrochemical biosensor for the detection of cauliflower mosaic virus 35 S gene sequences using lead sulfide nanoparticles as oligonucleotide labels. Anal. Biochem. 377, 115–119.
Sung, Y.J., Suk, H.J., Sung, H.Y., Li, T., Poo, H., Kim, M.G., 2013. Novel antibody/gold nanoparticle/magnetic nanoparticle nanocomposites for immunomagnetic separation and rapid colorimetric detection of *Staphylococcus aureus* in milk. Biosens. Bioelectron. 43, 432–439.
Tallury, P., Malhotra, A., Byrne, L.M., Santra, S., 2010. Nanobioimaging and sensing of infectious diseases. Adv. Drug Deliv. Rev. 62, 424–437.
Tang, D., Sauceda, J.C., Lin, Z., Ott, S., Basova, E., Goryacheva, I., Biselli, S., Lin, J., Niessner, R., Knopp, D., 2009a. Magnetic nanogold microspheres-based

lateral-flow immunodipstick for rapid detection of aflatoxin B2 in food. Biosens. Bioelectron. 25, 514–518.

Tang, S., Moayeri, M., Chen, Z., Harma, H., Zhao, J., Hu, H., Purcell, R.H., Leppla, S.H., Hewlett, I.K., 2009b. Detection of anthrax toxin by an ultrasensitive immunoassay using europium nanoparticles. Clin. Vaccine Immunol. 16, 408–413.

Tang, Z., Chen, H., Song, S., Fan, C., Zhang, D., Wu, A., 2011. Disposable screen-printed electrode coupled with recombinant *Drosophila melanogaster* acetylcholinesterase and multiwalled carbon nanotubes for rapid detection of pesticides. J. AOAC Int. 94, 307–312.

Thakur, M.S., Ragavan, K.V., 2013. Biosensors in food processing. J. Food Sci. Technol. 50, 625–641.

Thavarungkul, P., Dawan, S., Kanatharana, P., Asawatreratanakul, P., 2007. Detecting penicillin G in milk with impedimetric label-free immunosensor. Biosens. Bioelectron. 23, 688–694.

Thévenot, D.R., Toth, K., Durst, R.A., Wilson, G.S., 2001. Electrochemical biosensors: recommended definitions and classification. Biosens. Bioelectron. 16, 121–131.

Tong, P., Tang, S., He, Y., Shao, Y., Zhang, L., Chen, G., 2011. Label-free immunosensing of microcystin-LR using a gold electrode modified with gold nanoparticles. Microchim. Acta 173, 299–305.

Tothill, I.E., 2001. Biosensors developments and potential applications in the agricultural diagnosis sector. Comput. Electron. Agr. 30, 205–218.

Valdés, M., Valdés González, A., García Calzón, J., Díaz-García, M., 2009. Analytical nanotechnology for food analysis. Microchim. Acta 166, 1–19.

Varshney, M., Yang, L., Su, X.L., Li, Y., 2005. Magnetic nanoparticle-antibody conjugates for the separation of *Escherichia coli* O157:H7 in ground beef. J. Food Prot. 68, 1804–1811.

Vashpanov, Y., Young Son, J., Dal Kwack, K., 2008. Mesoporous silicon with modified surface for plant viruses and their protein particle sensing. Sensors 8, 6225–6234.

Vinayaka, A.C., Basheer, S., Thakur, M.S., 2009. Bioconjugation of CdTe quantum dot for the detection of 2,4-dichlorophenoxyacetic acid by competitive fluoroimmunoassay based biosensor. Biosens. Bioelectron. 24, 1615–1620.

Von Bültzingslöwen, C., McEvoy, A.K., McDonagh, C., MacCraith, B.D., Klimant, I., Krause, C., Wolfbeis, O.S., 2002. Sol-gel based optical carbon dioxide sensor employing dual luminophore referencing for application in food packaging technology. Analyst 127, 1478–1483.

Vyas, S.S., Jadhav, S.V., Majee, S.B., Shastri, J.S., Patravale, V.B., 2015. Development of immunochromatographic strip test using fluorescent, micellar silica nanosensors for rapid detection of *B. abortus* antibodies in milk samples. Biosens. Bioelectron. 70, 254–260.

Wang, C., Irudayaraj, J., 2010. Multifunctional magnetic-optical nanoparticle probes for simultaneous detection, separation, and thermal ablation of multiple pathogens. Small 6, 283–289.

Wang, S.E., Si, S., 2013. Aptamer biosensing platform based on carbon nanotube long-range energy transfer for sensitive, selective and multicolor fluorescent heavy metal ion analysis. Anal. Method. 5, 2947–2953.

Wang, L., Chen, W., Xu, D., Shim, B.S., Zhu, Y., Sun, F., Liu, L., Peng, C., Jin, Z., Xu, C., Kotov, N.A., 2009a. Simple, rapid, sensitive, and versatile SWNT-paper sensor for environmental toxin detection competitive with ELISA. Nano Lett. 9, 4147–4152.

Wang, X., Tao, G., Meng, Y., 2009b. A novel CdSe/CdS quantum dot-based competitive fluoroimmunoassay for the detection of clenbuterol residue in

pig urine using magnetic core/shell Fe_3O_4/Au nanoparticles as a solid carrier. Anal. Sci. 25, 1409–1413.
Warner, M.G., Grate, J.W., Tyler, A., Ozanich, R.M., Miller, K.D., Lou, J., Marks, J.D., Bruckner-Lea, C.J., 2009. Quantum dot immunoassays in renewable surface column and 96-well plate formats for the fluorescence detection of botulinum neurotoxin using high-affinity antibodies. Biosens. Bioelectron. 25, 179–184.
Warriner, K., Reddy, S.M., Namvar, A., Neethirajan, S., 2014. Developments in nanoparticles for use in biosensors to assess food safety and quality. Trends Food Sci. Technol. 40, 183–199.
Willner, I., Baron, R., Willner, B., 2007. Integrated nanoparticle-biomolecule systems for biosensing and bioelectronics. Biosens. Bioelectron. 22, 1841–1852.
Wisitsoraat, A., Karuwan, C., Wong-Ek, K., Phokharatkul, D., Sritongkham, P., Tuantranont, A., 2009. High sensitivity electrochemical cholesterol sensor utilizing a vertically aligned carbon nanotube electrode with electropolymerized enzyme immobilization. Sensors 9, 8658–8668.
Wu, S., Duan, N., Shi, Z., Fang, C., Wang, Z., 2014. Simultaneous aptasensor for multiplex pathogenic bacteria detection based on multicolor upconversion nanoparticles labels. Anal. Chem. 86, 3100–3107.
Xiang, L., Zhao, C., Wang, J., 2011. Nanomaterials-based electrochemical sensors and biosensors for pesticide detection. Sens. Lett. 9, 1184–1189.
Xu, X., Ying, Y., 2011. Microbial biosensors for environmental monitoring and food analysis. Food Rev. Int. 27, 300–329.
Yang, S.Y., Chieh, J.J., Wang, W.C., Yu, C.Y., Lan, C.B., Chen, J.H., Horng, H.E., Hong, C.Y., Yang, H.C., Huang, W., 2008. Ultra-highly sensitive and wash-free bio-detection of H5N1 virus by immunomagnetic reduction assays. J. Virol. Methods 153, 250–252.
Yang, M., Kostov, Y., Bruck, H.A., Rasooly, A., 2009. Gold nanoparticle-based enhanced chemiluminescence immunosensor for detection of staphylococcal enterotoxin B (SEB) in food. Int. J. Food Microbiol. 133, 265–271.
Yang, C., Gu, B., Xu, C., Xu, X., 2011a. Self-assembled ZnO quantum dot bioconjugates for direct electrochemical determination of allergen. J. Electroanal. Chem. 660, 97–100.
Yang, J.Y., Li, Y., Chen, S.M., Lin, K.C., 2011b. Fabrication of a cholesterol biosensor based on cholesterol oxidase and multiwall carbon nanotube hybrid composites. Int. J. Electrochem. Sci. 6, 2223–2224.
Yantasee, W., Warner, C.L., Sangvanich, T., Addleman, R.S., Carter, T.G., Wiacek, R.J., Fryxell, G.E., Timchalk, C., Warner, M.G., 2007. Removal of heavy metals from aqueous systems with thiol functionalized superparamagnetic nanoparticles. Environ. Sci. Technol. 41, 5114–5119.
Yao, D.S., Cao, H., Wen, S., Liu, D.L., Bai, Y., Zheng, W.J., 2006. A novel biosensor for sterigmatocystin constructed by multi-walled carbon nanotubes (MWNT) modified with aflatoxin-detoxifizyme (ADTZ). Bioelectrochemistry 68, 126–133.
Yola, M.L., Uzun, L., Ozaltin, N., Denizli, A., 2014. Development of molecular imprinted nanosensor for determination of tobramycin in pharmaceuticals and foods. Talanta 120, 318–324.
Younts, S., Alocilja, E., Osburn, W., Marquie, S., Gray, J., Grooms, D., 2003. Experimental use of a gas sensor-based instrument for differentiation of *Escherichia coli* O157:H7 from non-O157:H7 *Escherichia coli* field isolates. J. Food Prot. 66, 1455–1458.
Yu, H.W., Lee, J., Kim, S., Nguyen, G.H., Kim, I.S., 2009. Electrochemical immunoassay using quantum dot/antibody probe for identification of cyanobacterial hepatotoxin microcystin-LR. Anal. Bioanal. Chem. 394, 2173–2181.

Zamolo, V.A., Valenti, G., Venturelli, E., Chaloin, O., Marcaccio, M., Boscolo, S., Castagnola, V., Sosa, S., Berti, F., Fontanive, G., Poli, M., Tubaro, A., Bianco, A., Paolucci, F., Prato, M., 2012. Highly sensitive electrochemiluminescent nanobiosensor for the detection of palytoxin. ACS Nano 6, 7989–7997.

Zhang, W.H., Zhang, W.D., 2008. Fabrication of SnO_2-ZnO nanocomposite sensor for selective sensing of trimethylamine and the freshness of fishes. Sens. Actuat. B 134, 403–408.

Zhang, Q., Zhang, S., Xie, C., Zeng, D., Fan, C., Li, D., Bai, Z., 2006. Characterization of Chinese vinegars by electronic nose. Sens. Actuat. B 119, 538–546.

Zhang, D., Carr, D.J., Alocilja, E.C., 2009. Fluorescent bio-barcode DNA assay for the detection of *Salmonella enterica* serovar Enteritidis. Biosens. Bioelectron. 24, 1377–1381.

Zhang, Y., Zhang, X., Lu, X., Yang, J., Wu, K., 2010. Multi-wall carbon nanotube film-based electrochemical sensor for rapid detection of Ponceau 4R and Allura Red. Food Chem. 122, 909–913.

Zhao, X., Hilliard, L.R., Mechery, S.J., Wang, Y., Bagwe, R.P., Jin, S., Tan, W., 2004. A rapid bioassay for single bacterial cell quantitation using bioconjugated nanoparticles. Proc. Natl. Acad. Sci. USA 101, 15027–15032.

Zhao, Y., Ye, M., Chao, Q., Jia, N., Ge, Y., Shen, H., 2009. Simultaneous detection of multifood-borne pathogenic bacteria based on functionalized quantum dots coupled with immunomagnetic separation in food samples. J. Agric. Food Chem. 57, 517–524.

Zheng, A., Chen, J., Wu, G., Wei, H., He, C., Kai, X., Wu, G., Chen, Y., 2009a. Optimization of a sensitive method for the "switch-on" determination of mercury(II) in waters using rhodamine B capped gold nanoparticles as a fluorescence sensor. Microchim. Acta 164, 17–27.

Zheng, Y., Yang, C., Pu, W., Zhang, J., 2009b. Determination of oxalic acid in spinach with carbon nanotubes-modified electrode. Food Chem. 114, 1523–1528.

Zhou, P., Lu, Y., Zhu, J., Hong, J., Li, B., Zhou, J., Gong, D., Montoya, A., 2004. Nanocolloidal gold-based immunoassay for the detection of the *N*-methylcarbamate pesticide carbofuran. J. Agric. Food Chem. 52, 4355–4359.

Zhu, Y., Zhu, H., Yang, X., Xu, L., Li, C., 2007. Sensitive biosensors based on (dendrimer encapsulated Pt nanoparticles)/enzyme multilayers. Electroanalysis 19, 698–703.

Zhu, Z., Su, Y., Li, J., Li, D., Zhang, J., Song, S., Zhao, Y., Li, G., Fan, C., 2009. Highly sensitive electrochemical sensor for mercury(II) ions by using a mercury-specific oligonucleotide probe and gold nanoparticle-based amplification. Anal. Chem. 81, 7660–7666.

Zhuyi, W., Liyi, S., Fengqing, W., Shuai, Y., Yin, Z., Meihong, Z., 2011. The sol–gel template synthesis of porous TiO_2 for a high performance humidity sensor. Nanotechnology 22, 275502.

16

PLASMONIC NANOPARTICLES AND QUANTUM DOTS IN THE IDENTIFICATION OF INORGANIC AND ORGANIC CONTAMINANTS IN FOOD SAMPLES

Suresh Kumar Kailasa*, Jigneshkumar V. Rohit*, Rakesh Kumar Singhal**

**S. V. National Institute of Technology, Department of Applied Chemistry, Surat, Gujarat, India; **Bhabha Atomic Research Center, Analytical Chemistry Division, Mumbai, Maharashtra, India*

1 Introduction

The first demand in food science is to ensure food safety. Today's global movement of food materials presents researchers with a new challenge which is due to the threat of worldwide contamination of food samples. Chemical contaminants (inorganic and organic) may occur in our food from various sources. Inorganic contaminants (metal ions and anions) are the most abundant group of chemical elements on the earth's crust, and they are found in all foods. Some of these metals (iron, calcium, potassium, and zinc) are present in nature and are considered essential when speaking of human diet at least, within certain specific tolerances. On the other hand, metals, such as lead, cadmium, arsenic, and mercury, may be detected in foods and other commodities as contaminants and pose serious risks to the human health because of different factors, including the known bioaccumulation (Oldring et al., 2014). It is well known that inorganic contaminants (metal ions and anions) enter the environment due to increasing industrial activities. They are nonbiodegradable and can accumulate in the food chain, which poses a severe threat to the environment and human health. Similarly, organic chemicals including agrochemicals (residues of pesticides and veterinary drugs) and other

Nanobiosensors. http://dx.doi.org/10.1016/B978-0-12-804301-1.00016-3

inorganic chemicals are seriously contaminated with foods either by direct or indirect use, resulting in food-borne illness to the biological systems (Jackson 2009).

The continuously increasing volume of international trade in agricultural commodities requires the harmonization of quality and safety provisions to ensure the free movement of goods and the protection of the consumers. Therefore, selective and sensitive identification of chemical species is essentially required to monitor the food quality and food safety. In this connection, various analytical techniques including atomic absorption spectroscopy, inductively coupled plasma/mass spectrometry, inductively coupled plasma/atomic emission spectrometry, fluorescence, and ultraviolet (UV)–visible spectroscopic techniques have been used for the identification of inorganic contaminants in various food samples (Bings et al., 2006; Li et al., 2013a). Pesticides have traditionally been detected using capillary electrophoresis, high-performance liquid chromatography, and chromatographic techniques coupled with mass spectrometry, which offer powerful trace analysis with high reproducibility and very low detection limits (Chai and Tan, 2009; Qu et al., 2010). Although these techniques are highly sensitive and selective, they require tedious sample preparation and preconcentration procedures, expensive instruments, and professional personnel. Importantly, they cannot be used as portable devices for on-site detection.

With the rapid development of nanotechnology, versatile classes of nanomaterials with unique physical and chemical properties have been synthesized and empowered them with huge potential applications in multidisciplinary research areas (Kailasa and Wu, 2014, 2015; Kailasa et al., 2014). They have been proposed as chemical probes in assaying of various species including inorganic, organic, and biomolecules due to the high sensitivity of their optical response to their immediate surrounding environment (Scida et al., 2011; Turcheniuk et al., 2015). Chemical and nanofabrication have shown great impact to create new nanoobjects with unique physicochemical properties that strongly useful for the improvement of analytical instrument sensitivity with minimized sample preparations. Among them, noble metal nanoparticles (Au and Ag NPs), with sizes close to or smaller than the typical wavelength of light in the visible range, are of special interest and collectively called as plasmonic nanoparticles. The plasmon refers to coherent oscillations of conduction electrons in metals (and other free-electron materials) in response to incoming light. Such oscillations can be strongly confined to a metal-dielectric interface (known as surface plasmons) and therefore they can be strongly influenced by boundary conditions. Commonly employed plasmonic materials include the noble metals gold and silver, but more recently copper, aluminum, highly doped oxides, chalcogenides, and graphene also in same range. Nanostructures of these materials show dramatically distinct plasmonic performances due to differences in carrier concentration, carrier mobility, and ohmic loss. When the dimensions of such plasmonic materials are at the nanometer scale, interesting phenomena occur, in which light can be concentrated and manipulated by objects that are significantly smaller than its wavelength (Liz-Marzan et al., 2014). The unique properties of plasmonic nanostructures, such as the large electromagnetic field enhancements, high photothermal conversion efficiencies, and rich spectral responses, empower them with huge potential in a variety of applications, ranging from clinical therapy and diagnostics to optoelectronic control (Kociak and Stephan, 2014).

Quantum dots (QDs) are semiconducting materials that are, in general, synthesized with II–VI or III–V column elements of the periodic table. These are atom-like properties and can be referred to as artificial atoms with nanometer size ranging from 1 to 10 nm. It is well known that electron-hole pairs can be created upon the photoexcitation of QDs, causing recombination of energy levels, which yields multicolor emission of light. These unique photophysical properties makes them as suitable candidates for sensing of a variety of species at minimal volume samples (Silvi and Credi, 2015). The great potential of employing plasmonic nanostructures and QDs in analytical science is another important driving force underlying the development of nanoplasmonics- and QDs- based analytical methods for assays of inorganic and organic chemical species with high selectivity and sensitivity. The outstanding properties of plasmonic and QDs nanostructures endow them to be effective candidates for chemical and biological sensing, plasmon-enabled photothermal therapy, plasmon-assisted laser welding, plasmon-assisted photoacoustic imaging, plasmon-enhanced, fluorescence enhanced or quenched mechanisms.

In this chapter, we have paid particular attention to some representative works in the area of plasmonic nanoparticles- and QDs-based sensors for the detection of inorganic (metal ions) and organic contaminants in food samples. The assay of inorganic species (metal ions) and organic contaminants are emphasized using metallic NPs (Au and Ag NPs) and QDs as colorimetric and fluorescent probes. This chapter is mainly categorized into absorbance-based assay and fluorescent analysis, according to the measure techniques of the transduced signals. One thing that should be mentioned is that this is not a comprehensive review but rather a quick guide to the use of plasmonic NPs (Au and Ag NPs) and QDs as probes for assays of inorganic (metal ions) and organic contaminants in food samples. Only a small part of the published papers are included in this chapter, mainly on the basis of Au and Ag NPs and QDs-based UV-visible and fluorescence spectrometric methods of identification. We apologize to the authors of the many excellent reports that could not be cited here because of our space limitations.

2 Identification of Metal Ions by Plasmonic Nanoparticles and Quantum Dots as Probes

Inorganic contaminants including metal ions and anions are the most abundant group of chemical elements on the earth's crust and can be found in all foods. Some of them are essential to the diet, within certain specific tolerances, while others are present as contaminants and pose a risk to the human health. The knowledge of the risk by inorganic contamination in foodstuffs is an argument of great importance. Along the production chain, foods may come in contact with inorganic species at different stages of the production process: parts of industrial plants, storage tanks, tools, and mainly primary packaging. Identification of trace levels of inorganic contaminants in food samples involves sample preparation approaches that include decomposition of food samples and extraction before their identification by using various analytical techniques. Even though, a variety of analytical instruments, such as atomic absorption spectrometry, flame atomic absorption spectrometry, X-ray photoelectron spectroscopy, inductively coupled plasma (ICP) atomic emission spectrometry and ICP-mass spectrometry have been used for the analysis of inorganic species, unfortunately they require tedious sample preparation procedures and typical detection limits in the order of micrometer to nanometer, which are unsuitable for the accurate detection of inorganic contaminants in trace amounts in food samples (Kailasa and Wu, 2012; Aragay et al., 2011). To overcome from the above difficulties, plasmonic NPs (Au and Ag NPs), and QDs-based optical sensors have emerged as the platform of choice for the detection and discrimination of inorganic species (metal ions and anions) as they (1) enable sterile sensing without sample contamination, (2) can be miniaturized and multiplexed within lab-on-chip devices, and (3) offer readouts in the visible region at minimal volume of samples. In the past few years, the sensitivity and selectivity of UV-visible and fluorescence spectrometric techniques have greatly improved by the integration of plasmonic NPs and QDs in sample preparations, which shows great importance to improve detection limits and to allow an adequate miniaturization of the sensing devices. In this section, we devote our focus mostly into that borderline to show the benefits of a merger that involves plasmonic NPs and QDs nanostructured materials for the detection of inorganic contaminants in food samples.

Darbha et al. (2007) developed gold nanoparticle (Au NP)-based miniaturized battery-operated ultrasensitive nanomaterial surface energy transfer (NSET) probe for screening Hg^{2+} ion levels in water and fish. In this approach, Au NP surface acts as an acceptor and organic dye acts as a donor) that provides high sensitivity for the detection of metal ions because of their unique property of superquenching chromophores through both energy-transfer and electron-transfer processes. It was observed that the fluorescence intensity of fluorophore was completely quenched (nearly 100%) when the fluorophore was statically adsorbed on the particle (static quenching) via efficient nonradiative energy transfer. As a result, rhodamine B molecules are released from the Au NP surface, yielding a very distinct fluorescence signal enhancement (800-fold) within a few seconds by the addition of Hg^{2+} ion (130 ppb). Using this probe, authors observed that swordfish contain mercury at levels a little higher than the EPA limit of 0.55 ppm, whereas the amounts of mercury in other fish and shellfish, for example, lobster, crab, catfish, and tilapia, are lower than the EPA limit. It is well known that the surface plasmon resonance (SPR) is due to the collective oscillation of free conduction electrons in the noble metal in resonance with the incident electromagnetic radiation. The monodispersed Au NPs (~20 nm) shows a strong SPR peak in between 510 and 535 nm and Au NPs dispersed solution exhibits red or wine red color. The characteristic SPR peak can be red-shifted by the aggregation of Au NPs induced by the addition of analytes, resulting in a change in color, which provides direct, visual, and rapid detection of analytes with minimized sample preparations and volumes. For example, Guo et al. (2014a) described the use of unmodified Au NPs as probes for simple and label-free colorimetric detection of Cd^{2+} ion in rice samples. Authors observed that the unmodified Au NPs are easily aggregate in a high concentration of NaCl, but salt-induced aggregation is prevented by the addition of glutathione (GSH). However, the aggregation of unmodified Au NPs was induced by the addition of Cd^{2+} ion via spherical shaped complex between Cd^{2+} and GSH, yielding to weaken the stability of Au NPs, which results a shift in SPR peak from 520 to 640 nm, resulting in color change from red to blue that can be observed with naked eye. This method was successfully applied to detect Cd^{2+} ion in rice samples in several provinces of

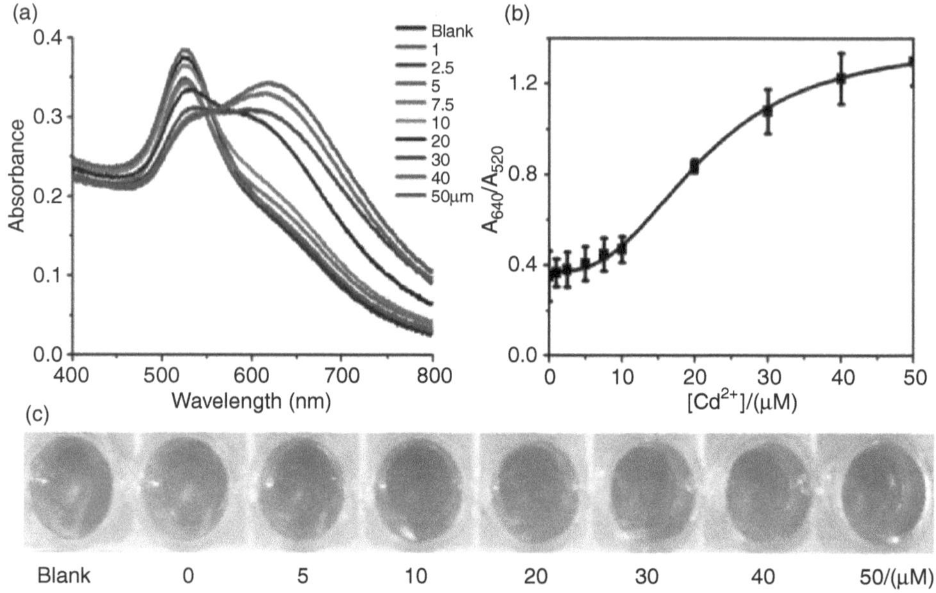

Figure 16.1. Detection of Cd^{2+} in rice samples. (a) UV-vis absorption spectra of Au NPs in the presence of different concentrations of Cd^{2+} in the rice digestion. (b) The plot of A_{640}/A_{520} of Au NPs versus the concentrations of Cd^{2+} in the rice digestion. (c) Photographs of Au NPs in the presence of different concentrations of Cd^{2+} in the rice digestion (the concentration of Au NPs is about 2 nM, 10 mM pH 6.0 phosphate sodium buffer, the concentration of NaCl is 40 mM, the concentration of GSH is 1 μM, the rice digestion is diluted to 500 times, the error bars are obtained based on three samples detection). Reprinted from Guo et al., 2014a, Anal. Chem. 86, 8530–8534, with permission.

China (Fig. 16.1). On the basis of this principle, developed various noble metal (Au and) NPs-based colorimetric sensors for monitoring various metal ions such as Fe^{3+}, Ba^{2+}, Cr^{3+}, Pb^{2+}, Al^{3+}, and Cd^{2+} ions in drinking water and biological samples (Mehta et al., 2014a,b, 2015a,b; Modi et al., 2014). These functionalized NPs act as selective colorimetric probes for detection of the above metal ions over various metal ions and exhibit the low limit of detection (LOD) values. Yu et al. (2009) illustrated the bifunctionalization of Au NPs with polythymine oligonucleotide T_{33} and citrate for the colorimetric sensing of Hg^{2+} ion in drinking water. It was observed that T_{33}-Au NPs exhibited poor sensitivity toward Hg^{2+} ion, however, the sensitivity of the probe was increased to 100-fold by the addition of Mn^{2+} ion, offering rapid analysis time (5 min) with detection limit of 10 nM for Hg^{2+} ion in drinking water. Sener et al. (2014) illustrated a colorimetric assay for the detection of Hg^{2+} ion in drinking water samples. The assay was based on the aggregation of citrate-capped Au NPs in the presence of Hg^{2+} ion and lysine. Hg^{2+} ions are spontaneously interacted with the surfaces of Au NPs, which facilitates to form a mercury layer. As a result, the color of the Au NPs was immediately changed from red to purple or gray after addition of lysine.

Iodide is one of the essential micronutrients for normal human growth. It plays a key role for healthy development especially in pregnant women and children. However, it can cause thyroid disease and mental retardation if there is an improper balance in the human body. In this connection, Maity et al. (2015) synthesized water soluble *p*-sulfonatocalix[4]arene functionalized Au NPs, used as a colorimetric probe for the detection of iodide in water and iodized edible salt samples. The characteristic SPR peak of functionalized Au NPs at 524 nm was red-shifted to 660 nm by the addition of iodide, indicating the aggregation of Au NPs. The aggregation mechanism is based on the chemisorptions of iodide on the surface of functionalized Au NPs, resulting in spontaneous oxidation of the anion to form iodine. The formed iodine was effectively absorbed on the surfaces of Au NPs via van der Waals interactions, which induces the aggregation of Au NPs.

3 Identification of Pesticides and Organic Contaminants by Plasmonic Nanoparticles as Probes

3.1 Analysis of Pesticides in Food Samples

Pesticides are chemicals that can be used to control and/or eliminate plant or animal pests and diseases. They are broadly classified into four types such as herbicides, insecticides, fungicides, and rodenticides; however, other types according to their mode of action and chemical structure again they classified as arsenic, carbamates, organophosphates, pyrethroids, and nitrophenol derivatives (World Health Organization, 2009; Aragay et al., 2011). It is well known that pesticides are widely used in agriculture sector to improve crop yield and to control insects and fungi. Due to their uses, drinking water and food are contaminated by the pesticide, which shows a potential dose-related chronic and acute toxicity against living organisms. In this connection, analysis of pesticides in fruits, vegetables, and other food products is a major analytical area with growing importance nanoscience and technology. For example, Sun et al. (2011) developed a simple label-free colorimetric sensing method for the detection of organophosphate nerve agents and pesticide using lipoic acid (LA) capped Au NPs as a probe. In this assay, acetylthiocholine (ATCh), and cationic thiocholine (TCh) were produced by the catalytic reaction of acetylcholine esterase (AChE), which results the aggregation of lipoic acid capped Au NPs induced by ATCh and TCh, which results a color change from red to blue and a red-shift toward longer wavelength. This specific feature allows them to use as a probe for sensing of organopesticides (OPs) since the function of AChE is inhibited by OPs. It was observed that the ultratrace OPs are completely inactivated AChE via nonreversible phosphorylation, which blocks the generation of TCh that can keep Au NPs in well dispersion state (Fig. 16.2). This sensor enables detection of ultratrace OPs in spiked apple juice samples with reduced sample preparations.

Similarly, Li et al. (2011a) described the use of citrate-coated Au NPs as a colorimetric probe for the rapid colorimetric assay

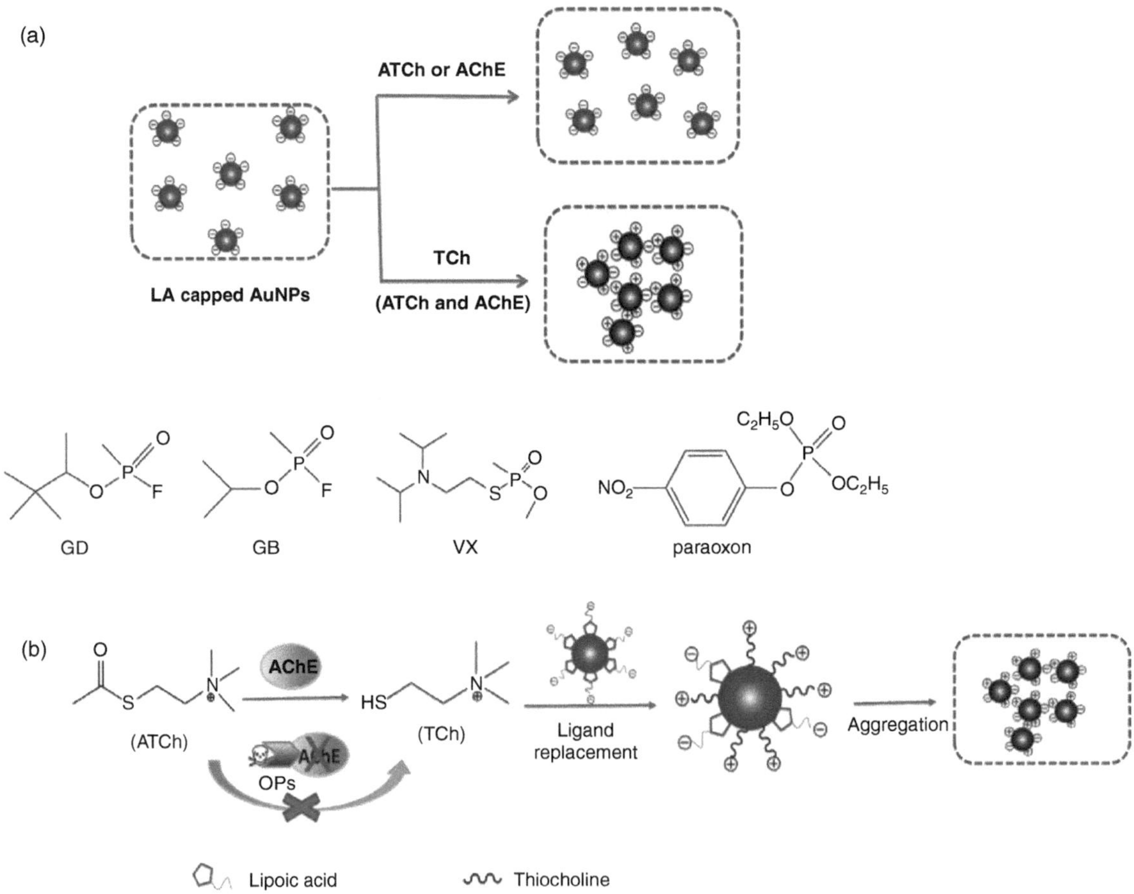

Figure 16.2. (a) Thiocholine-induced aggregation of the LA capped capped Au NPs and chemical structures of investigated OPs, in which two are G-type [Sarin (GB) and Soman (GD)], one is V-type (VX), and one is a pesticide (paraoxon). (b) Illustration of signaling for Ops based on the Au NPs probe. All operations were carried out under chemical safety principles. Reprinted from Sun et al., 2011, Biosens. Bioelectron. 28, 152–157, with permission.

of organophosphorus pesticides in Chinese cabbage. The sensing mechanism is based on catalytic hydrolysis of ATCh into TCh by AChE, resulting in the aggregation of Au NPs and the color change from claret-red to purple or even gray, and a new SPR peak at strong absorption peak at 675 nm. In PBS pH 8.0, the absorbance of Au NPs at 522 nm shows a linear relationship with mathamidophos concentration in the range 0.02–1.42 μg/mL, offering a detection limit of 1.40 ng/mL. Our group described the use of ascorbic acid (AA) functionalized Au NPs as a probe for the rapid, selective and sensitive colorimetric detection of dichlorvos in apple juice and wheat samples (D'souza et al., 2014). Upon the addition of dichlorvos, the SPR band of AA–Au NPs at 525 nm was red-shifted toward longer wavelength 620 nm, confirming the aggregation of AA–Au NPs induced by dichlorvos via hydrogen-bonding, which yields a color change from cherry red to purple. This method was successfully applied to detect dichlorvos in water, apple, and wheat samples.

Surface modified Au NPs were successfully used as sensors for the selective and sensitive colorimetric detection of pesticides in food samples. The AA was used as reducing and capping agent for the preparation of stable dispersed Au NPs in micellar media and used as a colorimetric probe for detection of thiram in tomato, cucumber, or watermelon samples (Rastegarzadeh and Abdali, 2013), Xu et al. (2011) described the use of citrate-capped Au NPs as a probe for the detection of acetamiprid in vegetable samples. Recently, Liu et al. (2015) synthesized different sizes of Au NPs (17.26, 24.13, and 30.35 nm) using citrate as a reducing and capping agent. The different size Au NPs were coated on the chip surface and their SPR signal intensity studied in the presence of atrazine. The maximum SPR signal was observed in chips modified with 30.35 nm Au NPs and exhibiting 1.91 and 7.17 times higher than chips modified with 24.13 and 17.26 nm Au NPs. The developed sensor is free from the interferences with the other pesticides. Su et al. (2012) prepared fluorescence Au NPs using gallic acid (GA) as a reducing and stabilizing agent. It was observed that the prepared GA–Au NPs are well dispersed with an average size of 37 nm and emitted fluorescence emission at 318 nm when the excitation wavelength is at 296 nm. The fluorescence intensity of GA–Au NPs was quenched by the addition of cyromazine via strong hydrogen-bonding between GA on the surfaces of Au NPs and cyromazine, resulting in a color change from purple to yellow–green. The developed probe showed high selectivity toward cyromazine in the presence of metal ions and pesticides and successfully applied to detect cyromazine in milk and pet food samples with good accuracy and precision. Fu et al. (2013)

developed highly sensitive colorimetric method for the detection of paraoxon in apple juice. Authors performed the colorimetric experiments in 96-well plate by adding 180 μL of the probe solution followed by the successive addition of 5 μL CuO NPs suspension, 5 μL of sodium ascorbate (0.02 M), and 5 μL of ATCh solutions. Then, 5 μL of AChE (0.05 mg/mL) solution was added with or without paraoxon. The color changes were investigated in the presence and absence of CuO NPs. It was noticed that the addition of CuO NPs played a key role in the direct inducement of Au NPs aggregation by the AChE–ATCl system and CH_3COOH, indicating that CuO NPs acted as the resource of the catalyst. CuO NPs were dissolved in enzymatically generated CH_3COOH from the AChE–ATCl system, yielding to release Cu^+ in the presence of reductant. The produced Cu^+ ion is typically catalyzed the conjugation between azide and terminal alkyne groups on surfaces of Au NPs, which results the aggregation of Au NPs. Therefore, Cu (I)-catalyzed click chemistry was greatly allowed to improve the colorimetric sensitivity for the detection of paraoxon based on the indirect modulation of click chemistry-induced Au NPs aggregation by the AChE–ATCl system. This probe was successfully applied to detect paraoxon in apple juice samples. Martí et al. (2013) functionalized Au NPs with novel reagent (thioctic acid capped Au NPs functionalized with pyridines or 2-[2-(dimethylamino)phenyl]) and used as a probe for the detection of organophosphorus pesticides. Author confirms that pyridine moiety on the Au NPs play key role to trigger an aggregation process via electrostatic interactions that results a bathochromic shift of the SPR absorption band and a change in the color from red to blue. Nair et al. (2003) developed nanomaterials-based colorimetric methods for detection of endosulfan, and chlorpyrifos and malathion (Lisha et al., 2009) in water samples. These probes were successfully applied for on-site monitoring of endosulfan and chlorpyrifos in drinking water samples with good accuracy and precision. Zheng et al. (2013b) developed a novel colorimetric method for the rapid and sensitive detection of glyphosate in water samples using cysteamine-capped Au NPs as a probe. The functionalized Au NPs were aggregated by the addition of glyphosate via electrostatic interactions, resulting in a color change from red to blue, which allows quantification by UV-visible spectrophotometry and monitoring with the naked eye.

Apart from the Au NPs, Ag NPs have great appeal in the development of colorimetric assays of various chemical species because of their unique optical properties and color changes from yellow to brown/orange/red/green based on their dispersed and aggregated systems (Vilela et al., 2012). The colorimetric assays of various pesticides using plasmonic sensors have received significant attention in nanoanalytical chemistry. The plasmonic properties of Ag NPs have played key role to improve the selectivity and sensitivity of UV-visible spectrometry for the ultratrace identification of target analytes in various matrices. For example, Menon et al. illustrated the use of *p*-sulfonato-calix[4]resorcinarene functionalized Ag NPs as a probe for colorimetric assay of dimethoate in water and wastewater samples (Menon et al., 2013). The assay was based on the red-shift of characteristic SPR peak of Ag NPs from 420 to 519 nm, yielding a color change from yellow to red, which is indicative of the aggregation of Ag NPs induced by dimethoate. The aggregation of Ag NPs was due to the selective interaction between *p*-sulfonato-calix[4]resorcinarene on the surfaces of Ag NPs and dimethoate through host–guest interactions such as electrostatic forces, cation–π interactions or π–π interactions. The developed probe is highly selective and shows good recoveries (98%) for the determination of dimethoate from water samples. In order to tune the optical properties and sensing applications of Ag NPs, our group successfully functionalized Ag NPs with 5-sulfo anthranilic acid dithiocarbamate (Rohit and Kailasa, 2014a), cyclen dithiocarbamate (Rohit and Kailasa, 2014b), dopamine dithiocarbamate (Rohit et al., 2014c) and 4-aminobenzenethiol (Patel et al., 2015) for the detection of tricyclazole fungicide, thiram and paraquat, mancozeb, and carbendazim fungicide in various food samples. In these methods, new SPR peaks were generated at longer wavelengths by the addition of the above pesticides in Ag NPs, resulting in a color change from yellow to orange or red or green, which can be directly observed with the naked eye. These sensing mechanisms are based on the electrostatic or hydrogen bonding or π–π interactions between functional ligands on the surfaces of Ag NPs and pesticides (Fig. 16.3). These probes have shown dynamic linear responses with the concentration of target analytes at micromolar concentrations, with micromolar and nanomolar detection limits. The common interfering chemical species do not interfere with the detection of target analytes, and enable to detect pesticides in rice, fruit, and vegetable samples with high precision and accuracy.

The development of dual nanoprobes for the colorimetric and fluorescence detection of chemical species in complex matrices has attracted considerable attention because of their simplicity, selectivity, and sensitivity. Cui et al. (2011) fabricated rhodamine B on the surfaces of Ag NPs and used it as dual probe for colorimetric and fluorescence assay of fenamithion in vegetable samples. The authors confirmed that the characteristic SPR peak was red-shifted to a longer wavelength by the addition of fenamithion, indicating the smaller individual Ag NPs were dissipated

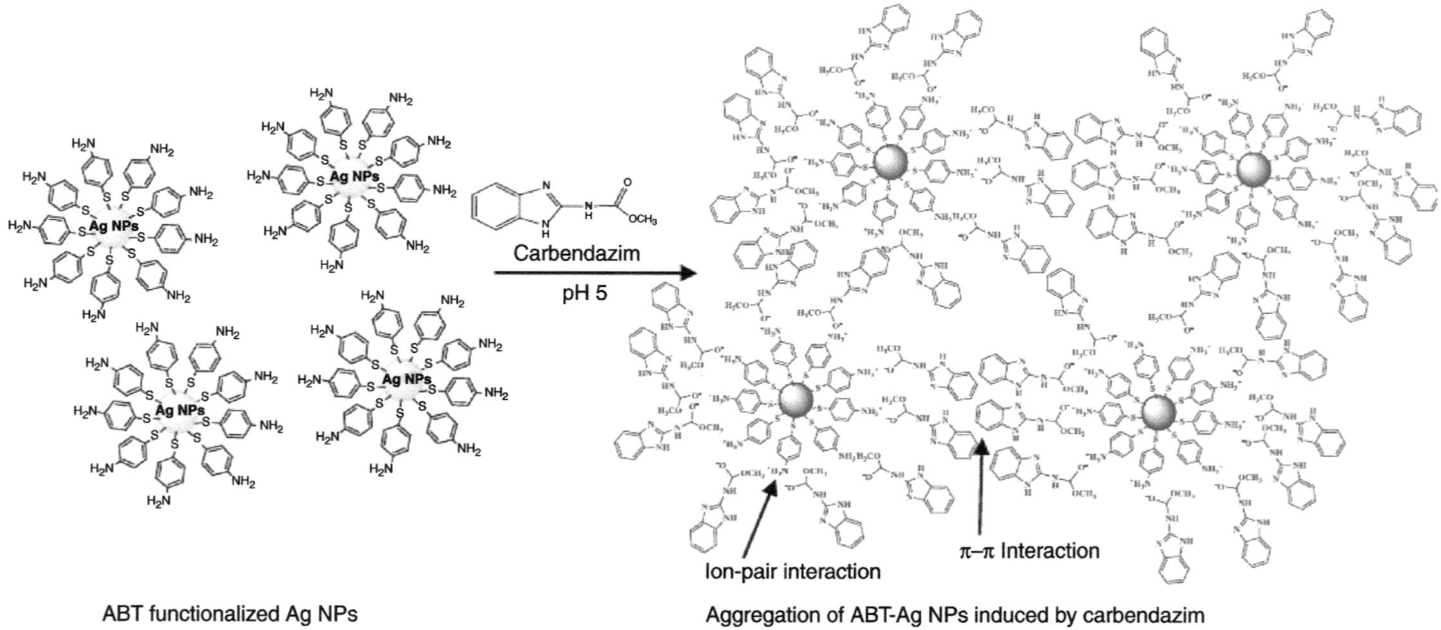

Figure 16.3. Schematic representation for the colorimetric detection of carbendazim using ABT–Ag NPs as a colorimetric probe. Reprinted from Patel et al., 2015, Sensor Actuat. B Chem. 206, 684–691, with permission.

from the original RB-Ag NPs and fenamithion was attached on the surfaces of Ag NPs via analyte-induced etching mechanism. As a result, the color of RB-Ag NPs was gradually turned into red and the intensity of the absorption peak at 531 (RB) nm was increased with increasing concentration of fenamithion, indicating that RB molecules were displaced from the surface of the Ag NPs, which induces the fluorescence recovery of RB in the Ag NPs. Vasimalai and John (2013) prepared chitosan (Chi) capped Ag NPs and used as a novel luminescent sensor for molecular recognition of malathion in water and fruit samples. The prepared Chi–Ag NPs showed the absorption maximum at 394 nm and emission maximum at 536 nm. It was observed that the absorption spectra of Chi–Ag NPs did not show any significant changes by the addition of malathion at μM concentration. However, the emission intensity of Chi–Ag NPs was quenched and the emission maximum was shifted toward longer wavelength in the presence of pM concentration of malathion. Based on the decrease in emission intensity, a good linearity was observed for the emission intensity against malathion concentration in the range of 1×10^{-9}–10×10^{-12} M, offering LOD at 94 fM/L (S/N = 3). This probe was successfully applied to detect malathion in fruits and water samples and the results agreed with HPLC. Liu et al. (2012) functionalized Au NPs with RB and used it as a dual probe (colorimetric and fluorometric) for the detection of organophosphorus and carbamate pesticides in environmental samples. The positively charged amino groups of RB can readily absorb onto the negatively charged citrate-Au NPs via electrostatic interaction, which results in the fluorescence quenching of RB by Au NPs. However, the fluorescence of RB was slowly recovered by the addition of both ATC and AChE into RB-Au NPs solution, which yields a color change from red to purple. The target analytes (organophosphorus and carbamate pesticides) inhibit the AChE activity, resulting in the prevention of thiocoline production, which causes it to restore the fluorescence of RB. This probe was successfully applied to detect several kinds of pesticides including carbaryl, diazinon, malathion, and phorate and allowed the detection of pesticides in agricultural products and river water. The type of nanomaterial, sensing mechanism, and LOD of plasmonic NPs- and QDs-based inorganic and pesticides sensors reviewed in this section are listed in Table 16.1.

Nanomaterials integrated lateral flow immunoassay (LFIA) technique has proven to be as a powerful analytical tool for online monitoring of chemical and biochemical species in food samples with high selectivity and sensitivity. For example, Zhou et al. (2004) coated nanocolloidal gold particles with an anticarbofuran monoclonal antibody (Mab) and dispensed on the conjugated pad of a

Table 16.1 Plasmonic NPs-Based Assays of Inorganic and Pesticides in Food Samples

Nanomaterials	Mechanism	Analyte	Limit of Detection	Food Sample	References
Au NP	Fluorescence enhancement	Hg^{2+}	$< 0.55^{a}$	Swordfish, lobster, crab, catfish, and tilapia	Darbha et al. (2007)
Au NPs	Red-shift in SPR peak (A_{640}/A_{520})	Cd^{2+}	5×10^{-8b}	Rice samples	Guo et al. (2014a)
Au NPs	Red-shift in SPR peak (A_{700}/A_{520})	Fe^{3+}	14.8×10^{-8b}	Water and urine samples	Mehta et al. (2014a)
Au NPs	Red-shift in SPR peak (A_{640}/A_{520})	Pb^{2+}	5.0×10^{-8b}	Water samples	Mehta et al. (2014b)
Ag NPs	Red-shift in SPR peak (A_{435}/A_{405}) and (A_{430}/A_{405})	Cr^{3+} and Ba^{2+}	6.45×10^{-8} and 8.02×10^{-8b}	Drinking, tap, and river water samples	Modi et al. (2014)
Au NPs	Red-shift in SPR peak (A_{660}/A_{523})	Al^{3+}	3.8×10^{-8b}	Drinking, tap, canal, and river water samples	Mehta et al. (2015a)
Au NPs	Red-shift in SPR peak (A_{685}/A_{523})	Cd^{2+}	6.3×10^{-8b}	Drinking, tap, canal, river, and industrial waste water samples	Mehta et al. (2015b)
Au NPs	Red-shift in SPR peak (A_{685}/A_{523})	Hg^{2+}	1.0×10^{-10b}	Drinking, tap, canal, and river water samples	Yu et al. (2009)
Au NPs	Red-shift in SPR peak (A_{725}/A_{525})	Hg^{2+}	2.9×10^{-9b}	Drinking and tap water samples	Sener et al. (2014)
Au NPs	Red-shift in SPR peak (A_{660}/A_{524})	I^{-}	8.0×10^{-8}	Water samples and iodized edible salt	Maity et al. (2015)
Au NPs	Red-shift in SPR peak (A_{625}/A_{525})	Organo phosphorus pesticides	2.28×10^{-12b}	Apple juice matrix	Sun et al. (2011)
Au NPs	Red-shift in SPR peak (A_{675}/A_{522})	Methamidophos	1.4^{c}	Chinese cabbage	Li et al. (2011a)
Au NPs	Red-shift in SPR peak (A_{620}/A_{525})	Dichlorvos	42.9×10^{-6b}	Apple and wheat samples	D'souza et al. (2014)
Au NPs	SPR peak decreasing	Thiram	1.7×10^{-7b}	Tomato, cucumber, or watermelon	Rastegarzadeh and Abdali (2013)
Au NPs	Red-shift in SPR peak (A_{700}/A_{520})	Acetamiprid	4.4×10^{-7b}	Vegetables (greens, plant eggs, and cucumber)	Xu et al. (2011)

Table 16.1 Plasmonic NPs-Based Assays of Inorganic and Pesticides in Food Samples (*cont.*)

Nanomaterials	Mechanism	Analyte	Limit of Detection	Food Sample	References
Au NPs	SPR signal enhancement	Atrazine	1.0[c]	Maize	Liu et al. (2015)
Au NPs	Fluorescence quenching	Cyromazine	0.07×10^{-6}[b]	Milk and pet food samples	Su et al. (2012)
Au NPs	Red-shift in SPR peak (A_{625}/A_{530})	Paraoxon	0.001[c]	Apple juice	Fu et al. (2013)
Au NPs	Red-shift in SPR peak (A_{658}/A_{526})	Organo-phosphorus pesticides	81[a]	Water and biological samples	Martí et al. (2013)
Au NPs	Red-shift in SPR peak (A_{643}/A_{524})	Endosulfan	2.0[a]	Drinking water	Nair et al. (2003)
Au NPs	Red-shift in SPR peak	Chlorpyrifos and malathion	0.01[a] and 0.04[a]	Drinking water	Lisha et al. (2009)
Au NPs	Red-shift in SPR peak (A_{650}/A_{524})	Glyphosate	5.88×10^{-8}[b]	Environmental water	Zheng et al. (2013b)
Ag NPs	Red-shift in SPR peak (A_{519}/ A_{420})	Dimethoate	80×10^{-9}[b]	Wastewater	Menon et al. (2013)
Ag NPs	Red-shift in SPR peak (A_{550}/ A_{400})	Tricyclazole fungicide	1.8×10^{-7}[b]	Rice	Rohit and Kailasa (2014a)
Ag NPs	Red-shift in SPR peak (A_{530}/ A_{396}) and (A_{510}/ A_{396})	Thiram and paraquat in	2.81×10^{-6}[b] M (thiram) and $7.21 \times 10 \times 10^{-6}$[b] (paraquat)	Environmental water, potato, and wheat samples	Rohit and Kailasa (2014b)
Ag NPs	Red-shift in SPR peak (A_{620}/ A_{394})	Mancozeb	21.1×10^{-6}[b]	Apple, grape, and tomato	Rohit et al. (2014c)
Ag NPs	Red-shift in SPR peak (A_{510}/ A_{397})	Carbendazim fungicide	1.04×10^{-6}[b]	Apple and Carrot	Patel et al. (2015)
Ag NPs	Colorimetric and fluorescence	Fenamithion	10×10^{-9}[b]	Vegetables	Cui et al. (2011)
Ag NPs	Colorimetric and fluorescence	Malathion	94×10^{-15}[b]	Water, mangoes, and grapes	Vasimalai and John (2013)

(*Continued*)

Table 16.1 Plasmonic NPs-Based Assays of Inorganic and Pesticides in Food Samples (*cont.*)

Nanomaterials	Mechanism	Analyte	Limit of Detection	Food Sample	References
Au NPs	Colorimetric and fluorescence	Organophosphorus and carbamate pesticides	0.0001–0.001[c]	Tomatoes and cucumbers	Liu et al. (2012)
Nanocolloidal gold	Color change	Carbofuran	2.5×10^{-7c}	Drinking water	Zhou et al. (2004)
Carbon NPs	Color change	Thiabendazole	8.5×10^{-3c}	Fruit juice	Blazková et al. (2010)
Nanocolloidal gold	Color change	Chlorpyrifos	10×10^{-3c}	Rice and letture	Kim et al. (2011)
Nanocolloidal gold	Color change	Cyromazine and melamine	2.2×10^{-4c} 0.26 $\times 10^{-4c}$	Milk, muscle, liver of swine	Le et al. (2013)
Au NPs	Color change	Imidacloprid, chlorpyrifos-methyl and isocarbophos	50×10^{-3c}, 100 $\times 10^{-3c}$ and 100 $\times 10^{-3c}$	Cabbage and soil	Wang et al. (2014)

[a] ppm
[b] Moles/liter
[c] ng/L

porous glass fiber. At the same time, ovalbumin (OVA)-carbofuran and goat antimouse IgG were also dispensed on the nitrocellulose (NC) membrane and then investigated its application for the detection of carbofuan (Zhou et al., 2004). It was noticed that the carbofuran-containing sample was migrated through the NC membrane and reacted with the anticarbofuran Mab labeled with the colloidal gold, which facilitates the appearance of a single dark red line in the control region and two color lines in the control and test regions. Blazková et al. (2010) developed an immunostrip test by integrating carbon particles for the detection of thiabendazole in fruit juice. The assay is based on the changing of color, because intense black color was appeared during the flow of an antibody-carbon complex and thiabendazole along the strip, which can be useful for the detection and visualization of thiabendazole. The obtained results agreed with results obtained by enzyme-linked immunosorbent assay (ELISA). The developed strip acted as a potential probe for screening of thiabendazole in fruit juices at levels far below the maximum residue limits. Similarly, Kim et al. (2011) described the use of competitive antigen-coated colloidal gold in immunochromatographic assay (ICA) of the organophosphorus insecticide chlorpyrifos in rice and lettuce samples. The ICA test strip was constructed using membranes that can be divided into three zones such as detection zone, sample pad, and an absorbent pad. The authors assembled chlorpyrifos Hapten-OVA conjugate and then antimouse IgG separately coated on the membrane. The assay was based on the migration in between analyte in the sample and the analyte hapten immobilized on the test strip that can be acted as the binding sites of the antibody-colloidal gold (Ab-CG) conjugate, which is also migrating on the test strip. This principle was successfully used for the identification of chlorpyrifos by ICA. Le et al. (2013) developed a novel colloidal gold-based lateral flow immunoassay for the detection of cyromazine and melamine in milk, muscle, liver of swine. This method exhibited good recoveries ranged from 73.9% to 104.2% with <11.9% relative standard deviation (RSD) for CA and MA, respectively. The authors compared the results of LFIA with HPLC, indicating that the results of LFIA were in good agreement, which demonstrates the practical applicability of the developed assay in real samples. Recently, Wang et al. (2014) illustrated the use of nanocolloidal gold in LFIA techniques for the simultaneous detection of three pesticides (imidacloprid, chlorpyrifos-methyl, and isocarbophos) in cabbage and soil. The developed nanocolloid gold-based LFIA consists four parts (a sample pad, a conjugate pad, a nitrocellulose membrane, and an absorbent pad). The authors added three imidacloprid, chlorpyrifos-methyl, and isocarbophos separately into the sample loading zone of the LFIA, and the sample solution without any pesticide was set as the control. The respective immunoreactions were investigated by changing the color. Similarly, Liu et al. (2011) established a highly sensitive and specific magnetic LFIA method for the detection of paraoxon methyl via the use of Fe_3O_4 particle aggregates instead of individual nanoparticles commonly adopted in conventional LFIA. The assay is based on the controllable Fe_3O_4 particle aggregates, offering improved visual detection limit for the detection of paraoxon methyl in real samples.

3.2 Analysis of Organic Contaminants in Food Samples

Melamine (2,4,6-triamino-1,3,5-triazine) is used in various industries for products such as polymer, fertilizer, flame retardants, and other products. Even though it is widely used in chemical industries, it is not allowed to be used as an additive in food or related ingredients due to its severe effects on the kidneys and the formation of stones in 10,000 infants. The organic molecule sensing by the discrete molecular nanoreceptors relies on the fact that the recognition of target species by molecular nanoreceptor is based on the spectral change and color change that can be used to monitor the quantity of the chemical species in the food samples. To improve the sensitivity and performance of metallic nanoprobes, several groups have paid more attention to fabricate NPs surfaces with a wide variety of organic molecules that can enhance the sensitivity of UV-visible spectrometer for the quantification of organic contaminants. For examples, Ai et al. (2009) synthesized 1-(2-mercaptoethyl)-1,3,5-triazinane-2,4,6-trione (MTT) and then functionalized on the surfaces of Au NPs (12 nm) by ligand-exchange reaction. The MTT-Au NPs are well dispersed in water with an average size of 12 nm and shown the SPR peak at 519 nm. The absorption spectrum of MTT-Au NPs was drastically reduced and a new SPR peak was generated at 700 nm by the addition of melamine, resulting in a color change from red to blue (Fig. 16.4). The sensing mechanism is based on the hydrogen bonding between MTT-Au NPs and melamine, which induces the aggregation of Au NPs, and shows that the detection limit for melamine is 0.2 nM. A simple, rapid, field-portable colorimetric method was developed for the detection of melamine citrate-capped Au NPs as a probe (Guo et al., 2010). The proposed nanosensor was successfully detected melamine in liquid milk and infant formula within 30 min, which avoids tedious sample preparations and advanced instrumentation. Similarly, Au NPs were functionalized with cysteamine (Liang et al., 2011) and 18-crown-6 ether (Kuang et al., 2011) and used as colorimetric probes for detection of melamine in milk products. The authors constructed two calibration graphs by

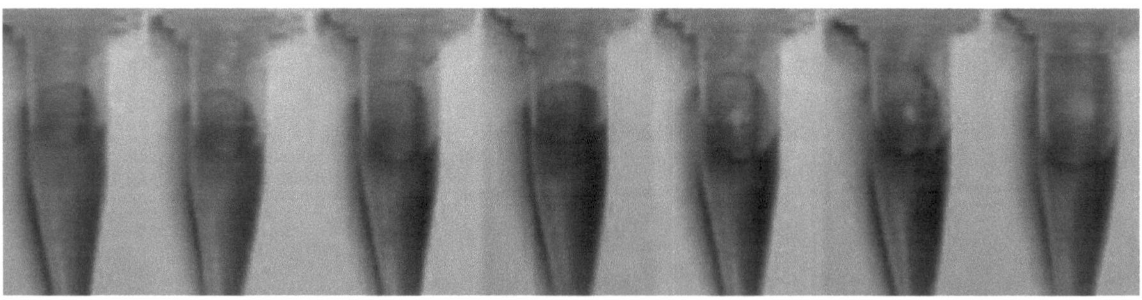

Figure 16.4. Visual color change of the MTT-stabilized gold nanoparticles upon addition of melamine with different concentrations under the optimized conditions (from left to right: 0, 20, 60, 100, 200, 280, 400 nM). Reprinted from Ai et al., 2009, J. Am. Chem. Soc. 131, 9496–9497, with permission.

taking readings at the absorption ratios of A_{720}/A_{520} and A_{620}/A_{520} (Liang et al., 2011). It was observed that the calibration plot at A_{720}/A_{520} gave a wider linear range, but was less sensitive than the calibration plot at ratio of A_{620}/A_{520}. The 18-crown-6 ether capped Au NPs were aggregated by the addition of melamine due to the formation of cavity complexes with amines and the detection limit was down to 6 ppb for the detection of melamine in milk products (Kuang et al., 2011).

Functionalized Au NPs have shown to be perfect candidates for colorimetric assays because of their ultrahigh extinction coefficients ($2.7 \times 10^8\ M^{-1}\ cm^{-1}$ for 13 nm Au NPs), more than 1000 times higher than those of organic dyes, allowing the detection limit to be as low as the nanomolar or picomolar level. As a result, Au NPs were successfully functionalized with various organic molecules such as citrate (Xing et al., 2013; Paul et al., 2015), cysteamine (Ma et al., 2013), ellagic acid (Zhang et al., 2013a) and 1,4-dithiothreitol (Xiao et al., 2015) and used as probes for colorimetric assay of melamine in milk products. In these methods, melamine was quantified by measuring the SPR peak of Au NPs in the presence of melamine. The characteristic SPR peak was gradually decreased and new peaks were progressively appeared with increasing concentration of melamine, resulting in a color change from red to purple or blue, which indicates that the aggregation of Au NPs induced by melamine. Interesting study was carried out on the sensitivity of the Au NPs based on their absorption ratios at A_{638}/A_{527} (Au NPs-I), A_{626}/A_{525} (Au NPs-II), and A_{680}/A_{526} (Au NPs-III). It was observed that the LOD for melamine is 2.37×10^{-8} M using Au NPs-II, which exhibited better sensitivity as compared to the LOD of the Au NPs-I (3.3×10^{-8} M) and Au NPs-III (8.9×10^{-8} M), respectively. All the earlier sensing methods mechanisms are based on the aggregation of Au NPs induced by melamine. Melamine contains amino group and ring nitrogen of hybrid aromatic that may strongly bind to the surface of Au NPs via ligand exchange with citrate ions or electrostatic or hydrogen bonding and finally results in the aggregation of Au NPs. Importantly, the coexisting organic and inorganic species did not interfere with the determination of melamine. These methods were successfully applied to detect melamine in milk and dairy products with high selectivity and sensitivity.

He et al. (2013) developed an unlabeled Au NPs-based sensor for the detection of tetracycline (TET) detection in aqueous solutions and milk. Au NPs were functionalized with an aptamer and poly(diallyldimethylammonium) (PDDA) and the color and the SPR peak of Au NPs were changed by the addition of TET, indicating that the formation of TET-aptamer complex. In the absence of

TET, aptamer was easily bound with PDDA to form complex that allows Au NPs in dispersion state. However, the interactions between aptamer and PDDA was disturbed by the addition of TET via aptamer-TEM complex, resulting in a color change from wine red to blue that can be used for the quantification of TET in milk. Similarly, Zhang et al. (2013b) developed 1-step synthetic method for the preparation of Au NPs as pyrocatechol violet (PCV) as a reducing and capping agent and used as a colorimetric probe for the detection of four antibiotics kanamycin monosulfate, neomycin sulfate, streptomycin sulfate, and bleomycin sulfate in milk. The Au NPs aggregation is based on the formation of hydrogen bonding between OH groups of PCV-Au NPs and —OH, $—NH_2$, —NH- groups of antibiotics, resulting in a color change and a red-shift in the SPR peak. The method shows good linearity between the absorption ratio (A_{670}/A_{530}) and the logarithm of KA concentration (1.0×10^{-8} to 5.0×10^{-7} M and 5.0×10^{-7} to 5.5×10^{-5} M), which allows to detect KA at 1.0×10^{-9} M. This method was free from the interferences and successfully detected KA in the various chemical species (glycerol, glucose, alanine, phenylalanine, K^+, Na^+, Mg^{2+}, Ca^{2+}, SO_4^{2-}, CO_3^{2-}, Cl^-, Br^-, and NO_3^-). Ma et al. (2011) described the use of dopamine stabilized Ag NPs as a probe for the colorimetric assay of melamine in milk. The sensing mechanism was based on the aggregation of dopamine-Ag NPs induced by melamine via Michael addition and Schiff base reactions between the quinone form of catechols and melamine, resulting in a color change from yellow to brown accompanied by a red-shift in their SPR peak, which can be observed with naked eye and then quantified by UV-visible spectrometry. Recently, Song et al. (2015) illustrated the use of sulfanilic acid-modified Ag NPs as a probe for the simple, rapid, and lower-cost assay of melamine in milk samples. The color and spectral changes were due the interaction between exocyclic amine of melamine and SAA–Ag NPs, which induces rapid aggregation of SAA–Ag NPs accompanied by a naked-eye visible color change, resulting in precise quantification of melamine that can be monitored by a simple UV-visible spectrometer. This probe provides simple platform for the detection of melamine with negligible interferences from metal ions, amino acids, and sugars, offering impressive detection limit of 10.6 nM for melamine in milk, which is much lower than the safety limits of several countries (8 μM for infant formula in China, 20 μM in both the United States and European Union and 1.2 μM in the CAC review for melamine in liquid infant formula). Two colorimetric procedures were established for the detection of cyanuric acid in milk and water samples using Ag NPs as a probe (Kappi et al., 2014). The first approach is based on the aggregation of melamine functionalized Ag NPs induced by cyanuric acid via hydrogen bonding, resulting in a red-shift in the SPR peak from 410 to 640 nm and a change in color from yellow to green, which can be observed with naked eye. In the second assay, the color of unmodified Ag NPs was changed from yellow to purple or orange red by the addition of stable cyanuric acid-melamine complexes, yielding generation of a new SPR peak at 525 nm and the color changes and spectral transitions of both assays with increasing cyanuric acid concentrations.

To improve the selectivity and sensitivity of chromatography, Li et al. (2011b) prepared functionalized nanocolloid gold with BSA antigen and goat anti-mouse IgG and then integrated with chromatography for assay of melamine in milk products and animal feed. This approach was successfully detected melamine within 3–10 min and minimized sample preparations. Zhou et al. (2011) developed Au NPs-based kit for colorimetric sensing of melamine in milk products. This method was based on the aggregation of Au NPs induced by melamine. The Au NPs-based kit acts as ecofriendly colorimetric probe for on-site detection of melamine in milk and dairy products without aid of any sophisticated instrument. Fluoroquinolones (FQs) are used as antibiotics for Gram-negative and Gram-positive bacteria and widely used for the treatment of respiratory diseases and gastroenteritis in food-origin animals such as cattle, pigs, and poultry, as well as for diseases in aquaculture. For example, Zhu et al. (2008) developed a rapid and sensitive colloidal gold immunochromatography test strip for the detection of 12 fluoroquinolones (enrofloxacin, ciprofloxacin, norfloxacin, flumequine, pefloxacin, ofloxacin, lomefloxacin, enoxacin, danofloxacin, amifloxacin, oxolinic acid, and marbofloxacin) chicken muscle and liver. The method exhibits several merits such as its simplicity, speed and avoided false-negative results, and successfully applied to detect fluoroquinolone residues in large numbers of chicken muscle and liver samples on site. Recently, Shi et al. (2015) described the use of unmodified Au NPs as a probe for the simple colorimetric assay of methylamphetamine in biological samples. The method was based on the aggregation of unmodified Au NPs induced by methylamphetamine, which results a red-shift in SPR peak and a change in color from red to blue, offers LOD 0.82 μM.

Lysozyme (Lys) is a ubiquitous alkaline protease, and ability to hydrolyze mucopolysaccharide into soluble glycopeptides. It also contains antimicrobial proteins that allowed the dissolution of bacteria by the cleavage of cell walls in pathogenic bacteria. In this connection, the detection of Lys in biological and food samples plays key roles in theory and practical applications in proteomics. For example, Chen et al. (2008) functionalized Au NPs with human serum albumin (HSA) and used as a colorimetric probe for the detection of Lys in egg white. The assay was based on the aggregation of HSA–Au NPs induced by Lys through electrostatic interactions and London-van der Waals attractive force, resulting in appearance of new peak at 600 nm and changing color from red to blue. Authors also studied matrix-assisted desorption/ionization mass spectrometry (MALDI-MS) for the confirmation of interaction between HSA–Au NPs and Lys. This probe was successfully applied to detect lysozyme in egg white. Recently, water-dispersible DIC-Au NPs were synthesized using dicoumarol (DIC) as a novel reducing and capping agent and used as a probe for the sensitive and selective detection of cysteine (Cys) and Lys in biofluids (Kasibabu et al., 2015). Authors studied various optimal experimental parameters (reagent, precursor concentrations, and reaction temperature) for the preparation of water-dispersed Au NPs with controlled size and color using DIC as a reducing and capping agent. Upon the addition of Cys and Lys, the intensity of the SPR peak at 525 nm was greatly decreased and generated a new SPR peaks at 666 and 570 nm for Cys and Lys, resulting in substantial color changes from cherry red to blue for Cys and to purple for Lys, respectively. It is noted that the high degree of the aggregation of DIC–Au NPs induced by Cys and Lys was observed by the addition of 0.4 M NaCl. The DIC-Au NPs acted as a selective colorimetric probe for the detection of Cys and Lys over other aminoacids, inorganic species and pesticides, which allows opening up a new way for the detection of Cys and Lys in biological samples. Zheng et al. (2013b) synthesized fluorescent triazolylcoumarin functionalized Au NPs and used as a novel, highly sensitive and selective dual-readout sensor (colorimetric and fluorometric) for the assay of Lys in egg white. It was noticed that the fluorescence of triazolylcoumarin molecules was quenched by Au NPs through the fluorescence resonance energy transfer (FRET), however, the fluorescence of triazolylcoumarin was restored by the addition of lysozyme, indicating that the strong interactions between Lys and Au NPs, which results in a change in color from red to purple. As a result, good recoveries were observed in the range 96.7–113.3% and 80.0–113.0%, allowing to distinguish Lys in complex biological samples. Similarly, Yuan et al. (2014b) developed a specific method for the colorimetric detection of *Staphylococcus aureus* based on aptamer recognition coupled to tyramine signal amplification. In this method, authors immobilized biotinylated aptamer specific on the surface of the wells of a microplate via biotin-avidin binding that can be trapped selectively *Staphylococcus aureus*. As a result, good linear relationship was observed between absorbance at 450 nm and the concentration of *S. aureus* (10–107 cfu/mL), offering the detection limit of 8 cfu/mL for *S. aureus*.

4 Identification of Pesticides and Organic Contaminants by Quantum Dots as Probes

In recent years, QDs are widely used as probes for the quantification of chemical species in complex matrices. The QDs are inorganic fluorophores and attracted great research interest in various fields including imaging, biolabeling, and chemical and biological assay (Gill et al., 2008). Zhang et al. (2010) described the use of mercapto propionic acid–capped CdTe QDs as fluorescent probes for a simple and on-site assay of organophosphorus pesticides in apples. The sensing mechanism is based on the surface coordination-originated fluorescence resonance energy transfer (FRET) of CdTe QDs in the presence of target analytes. It was observed that the strong green emission of CdTe QDs was effectively quenched by the addition of dithizone at basic media via a FRET mechanism between the emission of the CdTe QDs and the absorption of the surface dithizone–Cd complex, indicating that dithizone ligands on the surfaces of CdTe QDs are easily replaced by the hydrolyzate of organophosphorothioate, which facilitate to turn on the fluorescence of the system. The fluorescence turn-on chemosensor was successfully detected chlorpyrifos residues in apples at a limit of 5.5 ppb, which is under the maximum residue limit allowed by the US Environmental Protection Agency. Guo et al. (2014b) illustrated the advantage of inner filter effect (IFE) of Au NPs on CdTe QDs for the sensitive fluorescence detection of methamidophos in Chinese cabbage samples. It was observed that the fluorescence emission of CdTe QDs was remarkably decreased in the presence of Au NPs via IFE. The intensity of Au NPs absorption spectra was decreased due to the aggregation of Au NPs by AChE catalyzed hydrolysis of ATCh into TCh, which results in quenching the fluorescence of CdTe QDs. However, the Au NPs aggregation was prevented by inhibiting AChE activity with organophosphorus pesticides, preventing the aggregation of Au NPs and restoring the fluorescence of CdTe QDs. As a result, the response was linearly proportional to the concentration of methamidophos in the range of 0.06~0.78 mg/kg, offering the detection limit of 2 μg/kg (3σ). Recently, Yan et al. (2015) developed CdTe QDs-based fluorescence method for the detection of parathion-methyl (PM). This method is based on electron transfer (ET) between the hydrolysate of PM (*p*-nitrophenol) and cetyltrimethylammonium bromide (CTAB) on the surfaces of CdTe QDs. The CTAB molecules were effectively adsorbed on the surfaces of mercaptopropionic acid capped CdTe QDs, allowing to trap trace level PM molecules via strong hydrophobic interaction. As a result, the released *p*-nitrophenol is strongly quenched the fluorescence intensity of CdTe QDs/CTAB, which is proportional to PM concentration in the range of 25–3000 ng/mL. This probe was successfully applied to detect PM in environmental and agricultural samples (Yan et al., 2015). Similarly, Chen's group developed a rapid and sensitive competitive fluorescence-linked immunosorbent assay (cFLISA) for the detection of chlorpyrifos in drinking water using CdTe QDs-streptavidin as a probe (Chen et al., 2010). Authors functionalized CdTe QDs with 3-mercaptopropyl acid and streptavidin, which facilitates to improve the sensitivity of method, offering detection limit of 3.8 ng/mL. This method greatly improved the sensitivity 5.5-fold and reduced detection time by 1 h when compared with conventional ELISA, resulting in good recoveries ranged from 85.9% to 105.3% with coefficients of variation ranging from 6.3% to 13.5%. Garcia-Febrero et al. (2013) developed an electrochemical magneto immunosensor for the detection of paraquat in potato samples. The immunosensor was made with magnetic graphite–epoxy composite electrode and antigen biofunctionalized magnetic microparticles that allow to accurate detection of paraquat in potato extracts without additional clean-up or purification steps, offering the LOD (90% of the maximum signal) of 2.18 ± 2.08 μg/kg.

Li et al. (2013b) developed an ultrasensitive fluorescence method for the detection of melamine in milk and pet food samples using thioglycolic acid capped CdTe QDs at pH 11.0. In order to improve the sensitivity of the probe, authors investigated the effect of pH on the sensitivity, it was observed that the fluorescence intensity of CdTe QDs was increasing with increasing pH up to 11.0 and after the intensity was decreased, which is due to the formation of cadmium hydroxide product on the surface of QDs at higher pH. Under optimum conditions, the fluorescence intensity of TGA-coated CdTe QDs was linearly proportional to melamine concentration in the range of 1.0×10^{-11} to 1.0×10^{-5} mol/L with a correlation coefficient of 0.9943, offering detection limit of 5×10^{-12} mol/L. Lu et al. (2015) synthesized a novel fluorescent probe 7-(benzylamino)-9.9-dibutyl-9H-fluorene-2-carbaldehyde (BDFC) and functionalized on the surfaces of Au NPs that acts as a simple, sensitive and reliable "turn-on" fluorescence sensor for the detection of melamine in milk products based on the FRET efficiency between Au NPs and BDFC. It was noticed that the BDFC shows a very strong fluorescence emission peak at 550 nm, however, the fluorescence intensity was effectively quenched by the addition of Au NPs. Upon the addition of melamine into the above system, the aggregation of Au NPs induced by melamine, resulting in desorption of the probe from the surfaces of Au NPs, which restore the fluorescence of the probe. As a result, the detection of melamine was achieved by measuring the changes of the fluorescence intensity of the probe, offering lowers the detection limit down to 3.0 nmol/L. Similarly, Wu et al. (2015) introduced a new nanosensor for the detection of melamine in milk samples based on FRET between upconversion nanoparticles (UCNPs) and Au NPs. The fluorescence of UCNPs was quenched by the addition of Au NPs, however, the fluorescence of UCNPs was gradually recovered by the addition of melamine, indicating that melamine was released UCNPs on the surfaces of Au NPs by interacting itself with Au NPs surfaces via N–Au NPs bond, which induces the aggregation of Au NPs. The authors investigated various influencing factors including media pH (7.0), the concentration of Au NPs (1.23 nM) and incubation time (12 min) for sensitive detection of melamine in milk samples (Fig. 16.5). Miao et al. (2011) developed

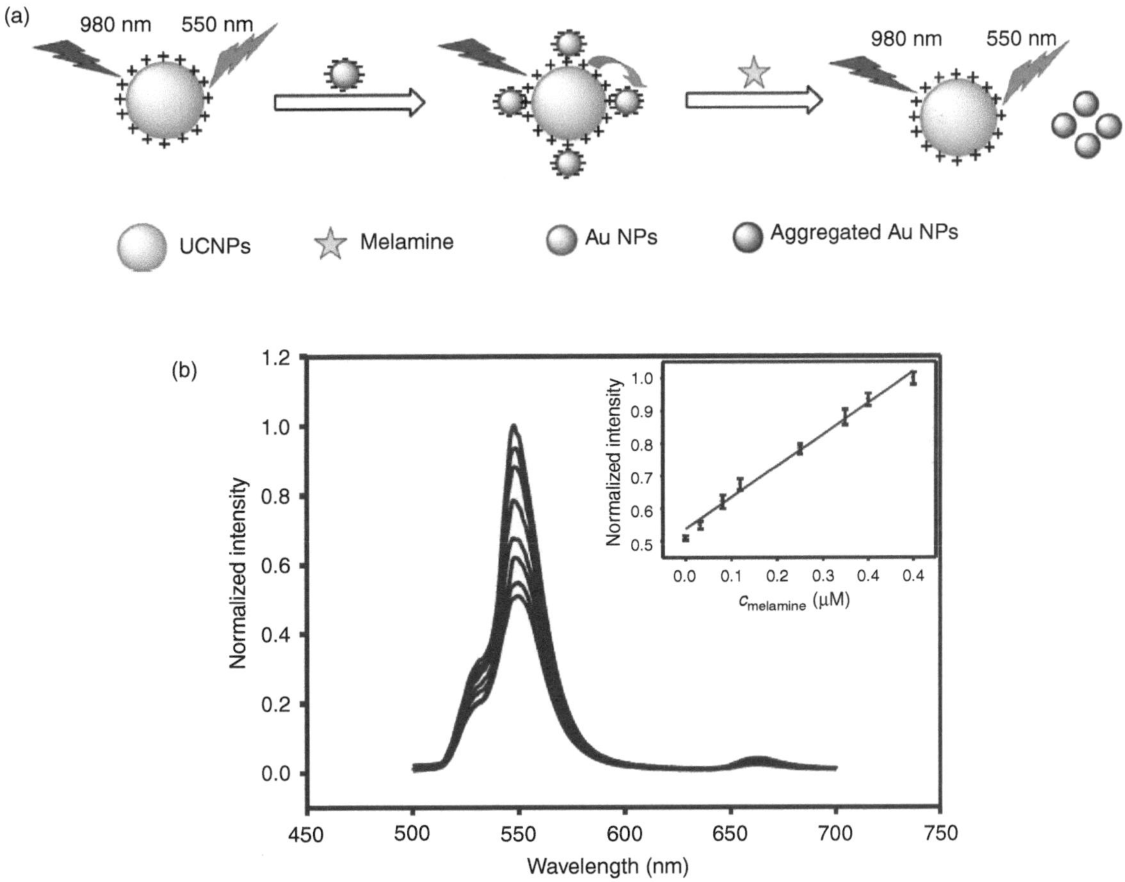

Figure 16.5. (a) Schematic illustration of the "turn-on" fluorescence assay for melamine. (b) Fluorescence spectra of UCNPs-Au NPs nanosensor in the presence of different concentration of melamine. The concentration of melamine (from bottom to top) is 0, 0.032, 0.08, 0.12, 0.25, 0.35, 0.4, 0.5 μM, respectively. [Au NPs]: 1.23 nM. [UCNPs]: 0.06 mg/mL. pH: 7.0. Inset: the enhanced fluorescence intensities as a function of the concentration of melamine. Reprinted from Wu et al., 2015, Talanta 136, 47–53, with permission.

a fluorescent assay for the detection of DNA specifically for *S. aureus* in milk samples. First, authors immobilized DNA on avid in modified wells of microplates, and then hybridized the capture DNA with one end of the target DNA. The recognition of the other end of the target DNA with a signal probe labeled with CdTe nanocrystals and Au-NPs at the 3′- and 5′-terminus, which fabricates the sandwich-type detection system for pathogenic bacteria. The feasibility of the method was investigated by detecting *S. aureus* in milk samples.

Recently, carbon nanodots (CDs) have proven to be a new class of fluorescent nanomaterials with a size of <10 nm due to their good water solubility, chemical inertness, easy preparation, and biocompatible nature. As a result, the application of CDs has been explored in various analytical techniques by using them as probes for the detection of various chemical species and bioimaging. For example, Yuan and coworkers developed a new "turn on" fluorescence nanosensor for the selective detection of Hg^{2+} ion in water samples using bis-(dithiocarbamato)copper(II) functionalized CDs ($CuDTC_2$-CDs) as a probe (Yuan et al., 2014a). The fluorescence of CDs was quenched by the formation of $CuDTC_2$-complexing by a combination of electron transfer and energy transfer mechanism, however the fluorescence was turn-on immediately by the addition of Hg^{2+} ion, indicating that the Hg^{2+} ion was effectively displaced Cu^{2+} ion in the $CuDTC_2$ complex, which facilitates shutting down the energy transfer pathway, offering the detection limit 4 ppb for Hg^{2+} ion. Similarly, Hou et al. (2015) prepared water soluble CDs using citric acid as carbon resource and L-tyrosine methyl ester as modification reagent by simple hydrothermal reaction. The strong and stable photoluminescence were observed in the prepared CDs with a quantum yield of 3.8%. Tyrosine methyl ester was oxidized in the presence of tyrosinase as catalyst of on the surface of CDs, yielding to generate quinone products, which quenches the fluorescence of CDs. Based on this mechanism, the authors introduced methyl parathion in the system, that can be decreased enzyme activity, resulting in a decrease in the fluorescence quenching rate. The quinine units are a kind of electron acceptor, which can cause the quenching the fluorescence of CDs. However, the fluorescence of CDs was effectively restored by inhibiting tyrosinase activity with methyl parathion, which can be constructed as a sensor for pesticides detection. The authors studied several optimum conditions including the pH of solution, the concentration of enzyme, the reaction temperature and the incubation time of methyl parathion and tyrosinase for the sensitive detection of methyl parathion. The probe was successfully applied to detect methyl parathion in cabbage, milk, and fruit juice samples. A simple, economical, and green hydrothermal method was developed for the preparation of water-soluble, high-fluorescent carbon QDs using aloe as a carbon source and used as a probe for the detection of tartrazine in steamed buns, honey and candy (Xu et al., 2015). The synthesized CDs were spherical shape with an average diameter of 5 nm and emit bright yellow photoluminescence with a quantum yield of approximately 10.37%. It was observed that the fluorescence intensity of CDs was greatly quenched by the addition of tartrazine via a static quenching process. The quenching of fluorescence intensity was made as it acts as a fluorescent probe for the detection of tartrazine and showed good linearity in the range of 0.25–32.50 μM. This method was effectively applied to monitor tartrazine in food samples, suggesting its great potential toward food routine analysis. The type of nanomaterial, sensing mechanism and LOD of plasmonic NPs- and QDs-based pesticides and organic contaminants sensors reviewed in this section are listed in Table 16.2.

5 Summary

Over the past 20 years plasmonic NPs- and QDs-based colorimetric assays have shown tremendous interest for on-site identification of various chemical species in complex samples with reduced sample preparations at minimal volume of samples. This chapter highlighted the application of plasmonic NPs and QDs in the recognition of pesticides and organic contaminants in food samples. Conclusions can be drawn that plasmonic NPs and QDs show great promise in the colorimetric assays of organic molecules in various food samples as they offer numerous advantages over conventional time-consuming offline measurements, and the unique optical and physicochemical properties brought by the intrinsic merits in recognition of chemical species by changing the color, which can be observed with naked eye. The sensing mechanisms of plasmonic NPs are based on the aggregation of Au/Ag NPs induced by target species via electrostatic or π–π interactions or hydrogen bonding. Similarly, QDs-based assays are based on the changes in emission spectra of QDs, which is due to either fluorescence quenching or fluorescence restoring by the addition of target analytes.

Undoubtedly, the rapid development of nanoscience provides significant improvement in the sensitivity and selectivity for the identification of chemical species with minimal volume samples. Among the nanomaterials, Au NPs have received considerable research interest in the chemical species identification by UV-visible and fluorescence spectrometry due to its virtues, including SPR,

Table 16.2 Plasmonic NPs- and QDs-Based Assays of Pesticides and Organic Contaminants in Food Samples

Nanomaterials	Capping Agent	Mechanism	Analyte	Limit of Detection	Food Sample	References
Au NPs	MTT	Red-shift (A_{700}/A_{519})	Melamine	0.2×10^{-9a}	Milk products	Ai et al. (2009)
Au NPs	Citrate	Red-shift (A_{660}/A_{520})	Melamine	0.2×10^{-6a}	Milk products	Guo et al. (2010)
Au NPs	Cysteamine	Red-shift (A_{720}/A_{520} and A_{620}/A_{520})	Melamine	1.0^{b}	Milk products	Liang et al. (2011)
Au NPs	18-crown-6 ether	Red-shift (A_{698}/A_{525})	Melamine	6.0^{c}	Milk products	Kuang et al. (2011)
Au NPs	Citrate	Red-shift (A_{640}/A_{520})	Melamine	$\sim70^{c}$	Milk products	Xing et al. (2013)
Au NPs	Citrate	Red-shift (A_{638}/A_{527})	Melamine	2.3×10^{-8a}	Milk products	Paul et al. (2015)
Au NPs	Cysteamine	Red-shift (A_{650}/A_{520})	Melamine	80×10^{-69}	Milk	Ma et al. (2013)
Au NPs	Ellagic acid	Disappearance of SPR peak	Melamine	1.6×10^{-9a}	Milk products	Zhang et al. (2013a)
Au NPs	1,4-dithiothreitol	Red-shift (A_{680}/A_{520})	Melamine	2.4×10^{-8a}	Milk	Xiao et al. (2015)
Au NPs	Aptamer and poly (diallyldimethyl-ammonium)	Red-shift (A_{680}/A_{520})	Tetracycline	45.8×10^{-6a}	Milk	He et al. (2013)
Au NPs	PCV	Red-shift (A_{670}/A_{530})	Four antibiotics	1.0×10^{-9a}	Milk	Zhang et al. (2013b)
Ag NPs	Dopamine	Red-shift (A_{600}/A_{410})	Melamine	10^{c}	Milk	Ma et al. (2011)
Ag NPs	Sulfanilic acid	Red-shift (A_{620}/A_{390})	Melamine	10.6×10^{-9a}	Milk products	Song et al. (2015)
Ag NPs	Melamine	Red-shift (A_{640}/A_{410})	Cyanuric acid	0.60^{b} and 0.25^{b}	Milk and water samples	Kappi et al. (2014)
Nanocolloid gold	BSA antigen and goat antimouse IgG	Color change	Melamine	0.002^{b}	Milk products and animal feed	Li et al. (2011a)
Au NPs	Citrate	Red-shift (A_{525}/A_{410})	Melamine	1^{b}	Milk products	Zhou et al. (2011)

Table 16.2 Plasmonic NPs- and QDs-Based Assays of Pesticides and Organic Contaminants in Food Samples (*cont.*)

Nanomaterials	Capping Agent	Mechanism	Analyte	Limit of Detection	Food Sample	References
Nanocolloid gold	Antigen and goat antimouse IgG	Color change	12 fluoroquinolones	5.0×10^{-7}[b]	Chicken muscle and liver	Zhu et al. (2008)
Au NPs	has	Red-shift (A_{600}/A_{525})	Lys	5.0×10^{-10}[a]	Egg white	Chen et al. (2008)
Au NPs	DIC	Red-shift (A_{575}/A_{525})	Lys	8.7×10^{-9}[a]	Egg white	Kasibabu et al. (2015)
Au NPs	Triazolylcoumarin	Colorimetric and fluorescence	Lys	23×10^{-6}[b]	Egg white	Zheng et al. (2013b)
CdTe QDs	Mercapto propionic acid	FRET	Organophosphorus pesticides	$\sim 0.1 \times 10^{-9}$[a]	Apples	Zhang et al. (2010)
Au NPs-CdTe QDs	—	Fluorescence quenching	Methamidophos	2.0[d]	Chinese cabbage	Guo et al. (2014b)
CdTe QDs	Cetyltrimethylammonium bromide	Fluorescence quenching via ET	Parathion-methyl	18×10^{-9}[a]	Rice	Yan et al. (2015)
CdTe QDs	3-mercaptopropyl acid and streptavidin	—	chlorpyrifos	3.8×10^{-6}[b]	Drinking water	Chen et al. (2010)
CdTe QDs	Thioglycolic acid		Melamine	5×10^{-12}[a]	milk powder and pet feeds	Li et al. (2013b)
Au NPs	BDFC	Fluorescence turn-on	Melamine	3.0×10^{-9}[a]	Milk and milk-based products	Lu et al. (2015)
Au NPs	Citrate	Fluorescence turn-on	Melamine	18.0×10^{-9}[a]	Milk samples	Wu et al. (2015)
CDs	—	Fluorescence turn-on	Hg^{2+}	4[c]	Tap, lake water	Yuan et al. (2014b)
CDs	L-tyrosine methyl ester	Fluorescence turn-on	Methyl parathion	4.8×10^{-11}[a]	Cabbage, milk and fruit juice	Hou et al. (2015)
CDs	—	Fluorescence quenching	Tartrazine	7.3×10^{-8}[a]	Steamed buns, and honey	Xu et al. (2015)

[a] Moles/liter
[b] mg/L
[c] ppb
[d] μg/kg

ease of synthesis and functionalization, and excellent stability. The functionalization of plasmonic NPs and QDs with specific molecules causes it to produce unique physicochemical properties on the surfaces of nanomaterials, allowing the preconcentration of the target analytes around the surface of nanomaterials, and to produce an effective adsorption of analytes on the surface of nanomaterials, which results in changes in their SPR peak and color that can be used for the quantification of chemical species. Although exciting progress in plasmonic NPs- and QDs-based sensors has been successfully made for the quantification of chemical species in various samples including food samples, they are still facing several challenges such as reproducibility and practical applications. It is expected that the development and integration of nanomaterials in spectrophotometric tools will continue growing in food analysis. In this regard, much work is still needed to fill the huge gap in the quantification of organic molecules in food samples by nanomaterials-based UV-visible and fluorescence spectrometric techniques. The integration of novel plasmonic NPs and QDs in new portable analytical instruments configurations would make technological advances for the routine food sample analysis, and also a revolutionary improvement in the recognition of all types of contaminants (inorganic, organic, pesticides, dyes, fungus, and bacteria) is expected to grow in foodomics in the near future.

References

Ai, K., Liu, Y., Lu, L., 2009. Hydrogen-bonding recognition-induced color change of gold nanoparticles for visual detection of melamine in raw milk and infant formula. J. Am. Chem. Soc. 131, 9496–9497.

Aragay, G., Pons, J., Merkoci, A., 2011. Recent trends in macro-, micro-, and nanomaterial-based tools and strategies for heavy-metal detection. Chem. Rev. 111, 3433–3458.

Bings, N.H., Bogaerts, A., Broekaert, J.A.C., 2006. Atomic spectroscopy. Anal. Chem. 78, 3917–3946.

Blazková, M., Rauch, P., Fukal, L., 2010. Strip-based immunoassay for rapid detection of thiabendazole. Biosens. Bioelectron. 25, 2122–2128.

Chai, M.K., Tan, G.H., 2009. Validation of a headspace solid-phase microextraction procedure with gas chromatography-electron capture detection of pesticide residues in fruits and vegetables. Food Chem. 117, 561–567.

Chen, Y.-M., Yu, C.-J., Cheng, T.-L., Tseng, W.-L., 2008. Colorimetric detection of lysozyme based on electrostatic interaction with human serum albumin-modified gold nanoparticles. Langmuir 24, 3654–3660.

Chen, Y., Ren, H.L., Liu, N., Sai, N., Liu, X., Liu, Z., Gao, Z., Ning, B., 2010. A fluoroimmunoassay based on quantum dot-streptavidin conjugate for the detection of chlorpyrifos. J. Agr. Food Chem. 58, 8895–8903.

Cui, Z., Han, C., Li, H., 2011. Dual-signal fenamithion probe by combining fluorescence with colorimetry based on Rhodamine B modified silver nanoparticles. Analyst 136, 1351–1356.

Darbha, G.K., Ray, A., Ray, P.C., 2007. Gold nanoparticle-based miniaturized nanomaterial surface energy transfer probe for rapid and ultrasensitive detection of mercury in soil, water, and fish. ACS Nano 1, 208–214.
D'souza, S.L., Pati, R.K., Kailasa, S.K., 2014. Ascorbic acid functionalized gold nanoparticles as a probe for colorimetric and visual read-out determination of dichlorvos in environmental samples. Anal. Methods 6, 9007–9014.
Fu, G., Chen, W., Yue, X., Jiang, X., 2013. Highly sensitive colorimetric detection of organophosphate pesticides using copper catalyzed click chemistry. Talanta 103, 110–115.
Garcia-Febrero, R., Valera, E., Muriano, A., Pividori, M.-I., Sanchez-Baeza, F., Marco, M.-P., 2013. An electrochemical magneto immunosensor (EMIS) for the determination of paraquat residues in potato samples. Anal. Bioanal. Chem. 405, 7841–7849.
Gill, R., Zayats, M., Willner, I., 2008. Semiconductor quantum dots for bioanalysis. Angew. Chem. Int. Ed. 47, 7602–7625.
Guo, L., Zhong, J., Wu, J., Fu, F.F., Chen, G., Zheng, X., Lin, S., 2010. Visual detection of melamine in milk products by label-free gold nanoparticles. Talanta 82, 1654–1658.
Guo, Y., Zhang, Y., Shao, H., Wang, Z., Wang, X., Jiang, X., 2014a. Label-free colorimetric detection of cadmium ions in rice samples using gold nanoparticles. Anal. Chem. 86, 8530–8534.
Guo, J., Li, H., Xue, M., Zhang, M., Cao, X., Luo, Y., Shen, F., Sun, C., 2014b. Highly sensitive detection of organophosphorus pesticides represented by methamidophos via inner filter effect of au nanoparticles on the fluorescence of CdTe quantum dots. *Food Anal. Methods* **7**, 1247–1255.
He, L., Luo, Y., Zhi, W., Zhou, P., 2013. Colorimetric sensing of tetracyclines in milk based on the assembly of cationic conjugated polymer-aggregated gold nanoparticles. Food Anal. Methods 6, 1704–1711.
Hou, J., Dong, J., Zhu, H., Teng, X., Ai, S., Mang, M., 2015. A simple and sensitive fluorescent sensor for methyl parathion based on L-tyrosine methyl ester functionalized carbon dots. Biosens. Bioelectron. 68, 20–26.
Jackson, L.S., 2009. Chemical food safety issues in the United States: past, present, and future. J. Agr. Food Chem. 57, 8161–8170.
Kailasa, S.K., Wu, H.F., 2012. Inorganic contaminants: sample preparation approaches. Pawliszyn, J., Le, X.C., Li, X.-F., Lee, H.K. (Eds.), Comprehensive Sampling and Sample Preparation, 3, Elsevier, Academic Press, Oxford, UK, pp. 743–782.
Kailasa, S.K., Wu, H.F., 2014. Recent developments in nanoparticle-based MALDI mass spectrometric analysis of phosphoproteomes. Microchim. Acta 181, 853–864.
Kailasa, S.K., Wu, H.F., 2015. Nanomaterials-based miniaturized extraction and preconcentration techniques coupled with matrix-assisted laser desorption/ ionization mass spectrometry for biomolecules assays. TrAC-Trend Anal. Chem. 65, 54–72.
Kailasa, S.K., Mehta, V.N., Wu, H.F., 2014. Recent developments of liquid-phase microextraction techniques directly combined with ESI- and MALDI-mass spectrometric techniques for organic and biomolecule assays. RSC Adv. 4, 16188–16205.
Kappi, F.A., Tsogas, G.Z., Giokas, D.L., Christodouleas, D.C., Vlessidis, A.G., 2014. Colorimetric and visual read-out determination of cyanuric acid exploiting the interaction between melamine and silver nanoparticles. Microchim. Acta 181, 623–629.
Kasibabu, B.S.B., Bhamore, J.R., D'souza, S.L., Kailasa, S.K., 2015. Dicoumarol assisted synthesis of water dispersible gold nanoparticles for colorimetric sensing of cysteine and lysozyme in biofluids. RSC Adv. 5, 39182–39191.

Kim, Y.A., Lee, E.H., Kim, K.O., Lee, Y.T., Hammock, B.D., Lee, H.S., 2011. Competitive immunochromatographic assay for the detection of the organophosphorus pesticide chlorpyrifos. Anal. Chim. Acta 693, 106–113.

Kociak, M., Stephan, O., 2014. Mapping plasmons at the nanometer scale in an electron microscope. Chem. Soc. Rev. 43, 3865–3883.

Kuang, H., Chen, W., Yan, W., Xu, L., Zhu, Y., Liu, L., Chu, H., Peng, C., Wang, L., Kotov, N.A., Xu, C., 2011. Crown ether assembly of gold nanoparticles: melamine sensor. Biosens. Bioelectron. 26, 2032–2037.

Le, T., Yan, P., Xu, J., Hao, Y., 2013. A novel colloidal gold-based lateral flow immunoassay for rapid simultaneous detection of cyromazine and melamine in foods of animal origin. Food Chem. 138, 1610–1615.

Li, H. S Guo, J., Pinga, H., Liua, L., Zhanga, M., Guana, F., Suna, C., Qian, Z., 2011a. Visual detection of organophosphorus pesticides represented by mathamidophos using Au nanoparticles as colorimetric probe. Talanta 87, 93–99.

Li, X., Luo, P., Tang, S., Beier, R.C., Wu, X., Yang, L., Li, Y., Xiao, X., 2011b. Development of an immunochromatographic strip test for rapid detection of melamine in raw milk, milk products and animal feed. J. Agr. Food Chem. 59, 6064–6070.

Li, M., Gou, H., Al-Ogaidi, I., Wu, N., 2013a. Nanostructured sensors for detection of heavy metals: a review. ACS Sustainable Chem. Eng. 1, 713–723.

Li, X., Lia, J., Kuanga, H., Feng, L., Yi, S., Xi, X., Huang, H., Chen, Y., Tang, C., Zeng, Y., 2013b. An ultrasensitive method for the determination of melamine using cadmium telluride quantum dots as fluorescence probes. Anal. Chim. Acta 802, 82–88.

Liang, X., Wei, H., Cui, Z., Deng, J., Zhang, Z., You, X., Zhang, X.-E., 2011. Colorimetric detection of melamine in complex matrices based on cysteamine-modified gold nanoparticles. Analyst 136, 179–183.

Lisha, K.P., Anshup, Pradeep, T., 2009. Enhanced visual detection of pesticides using gold nanoparticles. J. Environ. Sci. Heal. 44, 697–705.

Liu, C., Jia, Q., Yang, C., Qiao, R., Jing, L., Wang, L., Xu, C., Gao, M., 2011. Lateral flow immunochromatographic assay for sensitive pesticide detection by using Fe_3O_4 nanoparticle aggregates as color reagents. Anal. Chem. 83, 6778–6784.

Liu, D., Chen, W., Wei, J., Li, X., Wang, Z., Jiang, X., 2012. A highly sensitive, dual-readout assay based on gold nanoparticles for organophosphorus and carbamate pesticides. Anal. Chem. 84, 4185–4191.

Liu, X., Yang, Y., Mao, L., Li, Z., Zhou, C., Liub, X., Zheng, S., Hu, Y., 2015. SPR quantitative analysis of direct detection of atrazine traces on Au-nanoparticles: nanoparticles size effect. Sensor Actuat. B Chem. 218, 1–7.

Liz-Marzan, L.M., Murphy, C.J., Wang, J., 2014. Nanoplasmonics. Chem. Soc. Rev. 43, 3820–3822.

Lu, Q., Zhao, J., Xue, S., Yin, P., Zhang, Y., Yao, S., 2015. A "turn-on" fluorescent sensor for ultrasensitive detection of melamine based on a new fluorescence probe and AuNPs. Analyst 140, 1155–1160.

Ma, Y., Niu, H., Zhang, X., Cai, Y., 2011. One-step synthesis of silver/dopamine nanoparticles and visual detection of melamine in raw milk. Analyst 136, 4192–4196.

Ma, Y., Jiang, L., Mei, Y., Song, R., Tian, D., Huang, H., 2013. Colorimetric sensing strategy for mercury (II) and melamine utilizing cysteamine-modified gold nanoparticles. Analyst 138, 5338–5343.

Maity, D., Bhatt, M., Paul, P., 2015. Calix[4]arene functionalized gold nanoparticles for colorimetric and bare-eye detection of iodide in aqueous media and periodate aided enhancement in sensitivity. Microchim. Acta 182, 377–384.

Martí, A., Costero, A.M., Gaviña, P., Gil, S., Parra, M., Brotons-Gisbert, M., Sánchez-Royo, J.F., 2013. Functionalized gold nanoparticles as an approach to the direct colorimetric detection of dcnp nerve agent simulant. Eur. J. Org. Chem. 22, 4770–4779.

Mehta, V.N., Kailasa, S.K., Wu, H.F., 2014a. Sensitive and selective colorimetric sensing of Fe^{3+} ion by using *p*-amino salicylic acid dithiocarbamate functionalized gold nanoparticles. New J. Chem. 38, 1503–1511.

Mehta, V.N., Solanki, J.N., Kailasa, S.K., 2014b. Selective visual detection of Pb(II) ion via gold nanoparticles coated with a dithiocarbamate-modified 4′-aminobenzo-18-crown-6. Microchim. Acta 181, 1905–1915.

Mehta, V.N., Singhal, R.K., Kailasa, S.K., 2015a. A molecular assembly of piperidine carboxylic acid dithiocarbamate on gold nanoparticles for the selective and sensitive detection of Al^{3+} ion in water samples. RSC Adv. 5, 33468–33477.

Mehta, V.N., Solanki, J.N., Basu, H., Singhal, R.K., Kailasa, S.K., 2015b. Simple and sensitive colorimetric sensing of Cd^{2+}ion using chitosandithiocarbamate functionalized gold nanoparticles as a probe. Sensor Actuat. B Chem. 220, 850–858.

Menon, S.K., Modi, N.R., Pandya, A., Lodha, A., 2013. Ultrasensitive and specific detection of dimethoate using a *p*-sulphonato-calix[4]resorcinarene functionalized silver nanoprobe in aqueous solution. RSC Adv. 3, 10623–10627.

Miao, T., Wang, Z., Li, S., Wang, X., 2011. Sensitive fluorescent detection of *Staphylococcus aureus* using nanogold linked CdTe nanocrystals as signal amplification labels. Microchim. Acta 172, 431–437.

Modi, R.P., Mehta, V.N., Kailasa, S.K., 2014. Bifunctionalization of silver nanoparticles with 6-mercaptonicotinicacid and melamine for simultaneous colorimetric sensing of $Cr^{3+}$and Ba^{2+}ions. Sensor Actuat. B Chem. 195, 562–571.

Nair, A.S., Tom, R.T., Pradeep, T., 2003. Detection and extraction of endosulfan by metal nanoparticles. J. Environ. Monit. 5, 363–365.

Oldring, P.K.T., O'Mahony, C., Dixon, J., Vints, M., Mehegan, J., Dequatre, C., Castle, L., 2014. Development of a new modelling tool (FACET) to assess exposure to chemical migrants from food packaging. Food Add. Contam. Part A 31, 444–465.

Patel, G.M., Rohit, J.V., Singhal, R.K., Kailasa, S.K., 2015. Recognition of carbendazim fungicide in environmental samples by using 4-aminobenzenethiol functionalized silver nanoparticles as a colorimetric sensor. Sensor Actuat. B Chem. 206, 684–691.

Paul, I.E., Rajeshwari, A., Prathna, T.C., Raichur, A.M., Chandrasekarana, N., Mukherjee, A., 2015. Colorimetric detection of melamine based on the size effect of AuNPs. Anal. Methods 7, 1453–1462.

Qu, L.J., Zhang, H., Zhu, J.H., Yang, G.S., Aboul-Enein, H.Y., 2010. Rapid determination of organophosphorous pesticides in leeks by gas chromatography–triple quadrupole mass spectrometry. Food Chem. 122, 327–332.

Rastegarzadeh, S., Abdali, Sh., 2013. Colorimetric determination of thiram based on formation of gold nanoparticles using ascorbic acid. Talanta 104, 22–26.

Rohit, J.V., Kailasa, S.K., 2014a. 5-sulfo anthranilic acid dithiocarbamate functionalized silver nanoparticles as a colorimetric probe for the simple and selective detection of tricyclazole fungicide in rice samples. Anal. Methods 6, 5934–5941.

Rohit, J.V., Kailasa, S.K., 2014b. Cyclen dithiocarbamate-functionalized silver nanoparticles as a probe for colorimetric sensing of thiram and paraquat pesticides via host–guest chemistry. J. Nanopart. Res. 16, 1–16.

Rohit, J.V., Solanki, J.N., Kailasa, S.K., 2014c. Surface modification of silver nanoparticles with dopaminedithiocarbamate for selective colorimetric sensing of mancozeb inenvironmental samples. Sensor. Actuat. B Chem. 200, 219–226.
Scida, K., Stege, P.W., Haby, G., Messina, G.A., García, C.D., 2011. Recent applications of carbon-based nanomaterials in analytical chemistry: critical review. Anal. Chim. Acta 691, 6–17.
Sener, G., Uzun, L., Denizli, A., 2014. Lysine-promoted colorimetric response of gold nanoparticles: a simple assay for ultrasensitive mercury (II) detection. Anal. Chem. 86, 514–520.
Shi, Q., Shi, Y., Pan, Y., Yue, Z., Zhang, H., Yi, C., 2015. Colorimetric and bare eye determination of urinary methylamphetamine based on the use of aptamers and the salt-induced aggregation of unmodified gold nanoparticles. Microchim. Acta 182, 505–511.
Silvi, S., Credi, A., 2015. Luminescent sensors based on quantum dot–molecule conjugates. Chem. Soc. Rev. 44, 4275–4289.
Song, J., Wu, F., Yiqun, W., Ma, L., 2015. Colorimetric detection of melamine in pretreated milk using silver nanoparticles functionalized with sulfanilic acid. Food Cont. 50, 356–361.
Su, H., Chen, L., Sun, B., Ai, S., 2012. Fluorescence detection of cyromazine using gallic acid-reduced gold nanoparticles. Sensor Actuat. B Chem. 174, 458–464.
Sun, J., Guo, L., Bao, Y., Xie, J., 2011. A simple, label-free AuNPs-based colorimetric ultrasensitive detection of nerve agents and highly toxic organophosphate pesticide. Biosens. Bioelectron. 28, 152–157.
Turcheniuk, K., Boukherroub, R., Szunerits, S., 2015. Gold–graphene nanocomposites for sensing and biomedical applications. J. Mater. Chem. B 3, 4301–4324.
Vasimalai, N., John, S.A., 2013. Biopolymer capped silver nanoparticles as fluorophore for ultra sensitive and selective determination of malathion. Talanta 115, 24–31.
Vilela, D., González, M.C., Escarpa, A., 2012. Sensing colorimetric approaches based on gold and silver nanoparticles aggregation: chemical creativity behind the assay: a review. Anal. Chim. Acta 751, 24–43.
Wang, L., Cai, J., Wang, Y., Fang, Q., Wang, S., Cheng, Q., Du, D., Lin, Y., Liu, F., 2014. A bare-eye-based lateral flow immunoassay based on the use of gold nanoparticles for simultaneous detection of three pesticides. Microchim. Acta 181, 1565–1572.
World Health Organization, 2009. The WHO recommended classification of pesticides by hazard and guidelines to classification.
Wu, Q., Long, Q., Li, H., Zhang, Y., Yao, S., 2015. An upconversion fluorescence resonance energy transfer nanosensor for one step detection of melamine in raw milk. Talanta 136, 47–53.
Xiao, C., Zhang, X., Liu, J., Yang, A., Zhao, H., Li, X., He, Y., Yuan, Z., 2015. Sensitive colorimetric detection of melamine with 1,4-dithiothreitol modified gold nanoparticles. Anal. Methods 7, 924–929.
Xing, H.-B., Wu, Y.-G., Zhan, S.-S, Zhou, P., 2013. A rapid colorimetric detection of melamine in raw milk by unmodified gold nanoparticles. Food Anal. Methods 6, 1441–1447.
Xu, Q., Du, S., Gen-di., J., Li, H., Hu, X.Y., 2011. Determination of acetamiprid by a colorimetric method based on the aggregation of gold nanoparticles. Microchim. Acta 173, 323–329.
Xu, H., Yang, X., Li, G., Zhao, C., Liao, X., 2015. Green synthesis of fluorescent carbon dots for selective detection of tartrazine in food samples. J. Agr. Food Chem. 63, 6707–6714.

Yan, X., Li, H., Wang, X., Su, X., 2015. A novel fluorescence probing strategy for the determination of parathion-methyl. Talanta 131, 88–94.
Yu, C.-J., Cheng, T.-L., Tseng, W.-L., 2009. Effects of Mn^{2+} on oligonucleotide-gold nanoparticle hybrids for colorimetric sensing of Hg^{2+}: improving colorimetric sensitivity and accelerating color change. Biosens. Bioelectron. 25, 204–210.
Yuan, C, Liu, B., Liu, F., Han, M.-Y., Zhang, Z., 2014a. Fluorescence "turn on" detection of mercuric ion based on bis(dithiocarbamato)copper(ii) complex functionalized carbon nanodots. Anal. Chem. 86, 1123–1130.
Yuan, J., Yu, Y., Li, C., Ma, X., Xia, Y., Chen, J., Wang, Z., 2014b. Visual detection and microplate assay for *Staphylococcus aureus* based on aptamer recognition coupled to tyramine signal amplification. Microchim. Acta 181, 321–327.
Zhang, K., Mei, Q., Guan, G., Liu, B., Wang, S., Zhang, Z., 2010. Ligand replacement-induced fluorescence switch of quantum dots for ultrasensitive detection of organophosphorothioate pesticides. Anal. Chem. 82, 9579–9586.
Zhang, X., Wu, Z., Xue, Y., Zhang, Y., Zhao, H., He, Y., Li, X., Yuan, Z., 2013a. Colorimetric detection of melamine based on the interruption of the synthesis of gold nanoparticles. Anal. Methods 5, 1930–1934.
Zhang, X., Zhang, Y., Zhao, H., He, Y., Li, X., Yuan, Z., 2013b. Highly sensitive and selective colorimetric sensing of antibiotics in milk. Anal. Chim. Acta 778, 63–69.
Zheng, H., Qiu, S., Xu, K., Luo, L., Song, Y., Lin, Z., Guo, L., Qiu, B., Chen, G., 2013a. Colorimetric and fluorometric dual-readout sensor for lysozyme. Analyst 138, 6517–6522.
Zheng, J., Zhang, H., Qu, J., Zhu, Q., Chen, X., 2013b. Visual detection of glyphosate in environmental water samples using cysteamine-stabilized gold nanoparticles as colorimetric probe. Anal. Methods 5, 917–924.
Zhou, P., Lu, Y., Zhu, J., Hong, J., Li, B., Zhou, J., Gong, D., Montoya, A., 2004. Nanocolloidal gold-based immunoassay for the detection of the N-methylcarbamate pesticide carbofuran. J. Agr. Food. Chem. 52, 4355–4359.
Zhou, Q., Liu, N., Qie, Z., Wang, Y., Ning, B., Gao, Z., 2011. Development of gold nanoparticle-based rapid detection kit for melamine in milk products. J. Agr. Food Chem. 59, 12006–12011.
Zhu, Y., Li, L., Wang, Z., Chen, Y., Zhao, Z., Zhu, L., Wu, X., Wan, Y., He, F., Shen, J., 2008. Development of an immunochromatography strip for the rapid detection of 12 fluoroquinolones in chicken muscle and liver. J. Agr. Food Chem. 56, 5469–5474.

17

A TECHNIQUE COMES TO LIFE FOR SECURITY OF LIFE: THE FOOD CONTAMINANT SENSORS

Santanu Patra*, Ekta Roy*, Rashmi Madhuri*, Prashant K. Sharma**

**Indian School of Mines, Department of Applied Chemistry, Dhanbad, Jharkhand, India; **Indian School of Mines, Department of Applied Physics, Functional Nanomaterials Research Laboratory, Dhanbad, Jharkhand, India*

1 Introduction

In our daily routine of life, every morning we wake up and start our day with the newspaper and breakfast. Do we remember how many times we have read some news like: A toddler is hospitalized as a result of drinking contaminated fruit juice. Some preschooler dies because he or she eats a midday meal that is not thoroughly cooked and contaminated. An evening party group becomes ill with the same symptoms. A school cafeteria is unable to operate because some students had filed a complaint regarding their non-hygienic food supply, resulting in vomiting, diarrhea, and fever in several students. In each case, the illness or death was traced to something in the food supply.

According to oxford dictionary, food can be defined as "any nutritious substance that people or animals eat or drink or that plant can absorb in order to maintain life and growth".

This is the biological law of life, that we are alive till we are getting sufficient amount of food and water. Is the sufficient amount of food enough requirements for today's life? Our live depends on a reliable and safe food supply, which is free from any harmful contamination. So, the question is: What are the food contaminants? They are the actual culprits behind the scene. Food contamination refers to the presence of harmful chemicals and microorganisms in the food, which can cause consumer illness or food contamination and is the state of being impure or unfit

Nanobiosensors. http://dx.doi.org/10.1016/B978-0-12-804301-1.00017-5

for use due to the introduction of unwholesome or undesirable elements. The contamination occurs due to the introduction of unwholesome or undesirable elements into the food. Food can also be contaminated by insects, rodents, chemicals, microbes, or other foreign particles. They have threatened our lives in such an extent that before eating a food we are thinking about that. The impact of food contaminants on consumer health and well-being is either apparent in very short time span or it will take many years of processing prolonged exposure at low levels to show their impact, for example, cancer (Baggiani et al., 2007).

In short, anything added in the food product as extra, which is neither habitually found in the food nor the composition of food is called as contaminant.

In contamination, the addition of certain substances occurs in an unintended and unplanned manner, which sometimes may cause problems to human beings. For better understanding, the contamination can be divided into three major categories according to their source:

1. Physical contaminants
2. Chemical contaminants
3. Microbial contaminants

1.1 Physical Contaminants

Physical contaminants are certain types of substances that present in the food mixture and may or may not change or damage the property of food, however, their presence sometimes cause a health problem in the consumer's body. As for example, sometimes metal filings or broken pieces of glass have been found in our daily food. These types of materials do not spoil the quality or taste of foods, but they may cause injury if swallowed. The contamination occurs due to the presence of packaging materials, insects, and rodent droppings and can also be categorized as physical contaminants.

1.2 Chemical Contaminants

Chemical contamination mainly occurs to keep insects and other pests under control. For this, insecticides and herbicides are the most common chemicals used to reduce the losses due to insects and/or weeds. Both insecticides and herbicides come under same class known as pesticides. Sometimes the pesticide residues remain on food and enter into the food supply (Aragay et al., 2012). The US Department of Agriculture controls the use of all pesticides. To understand the effectiveness of the used chemicals, they must undergo certain tests and the main motive is to determine whether residues have the tendency to create health hazards. Water is considered to be the second source of food contamination from chemical sources (Holderbeke et al., 2014). For processing of every food product, water is the common solvent used. On this time, many poisonous substances can dissolve in and make the water highly toxic. In general, the terminology used to categorize anything as toxic substance is that they are harmful even if present in very low amount/concentration. Mercury, cadmium, lead, chloroform, benzene, and polychlorinated biphenyls (PCBs) are among the toxic substances that may get into water supplies.

1.3 Microbial Contaminants

The previously discussed physical and chemical contaminants do not have the ability to change the food itself but becomes seriously hazardous after their consumption with food (Scheu et al., 1998). If in a certain stage any undesired changes are found in the food itself, it may be considered that the food is spoiled. After food spoilage, consumption of food becomes unfit or undesirable. It is well understood that the presence of bacteria, fungi, and mold have the ability to cause desirable and undesirable changes in food. In general, the microorganisms and enzymatic compounds are more often involved in spoiling of food as coculprits.

The supply of fresh, tasty, authentic, and safe food products to consumers is the key fact to control the food industry. Moreover, the official authorities or government bodies should implement some very stern rules or food standard throughout the whole food chain, that is, starting from food cultivation to processing at the home or restaurant level to maintain the hygiene and quality of food products. Before reaching the consumer, the food products have to cross certain steps like processing, handling, and distribution and in their way, they can be contaminated by adulterants, various agrochemicals, microbial contaminants, veterinary drug residues, and environmental factors (Fig. 17.1). The use of packaging materials can protect foodstuffs from external contaminations but their storage and transportation ability sometimes may not be strictly followed by distributers. Due to the change of color or other sensory characteristics, the spoiled and unpleasant food products should be separated out and rejected by the consumers; otherwise it may cause food-borne illness.

Thus, for the minimization of risk or infection caused by contaminated food, fast and simple detection techniques should be

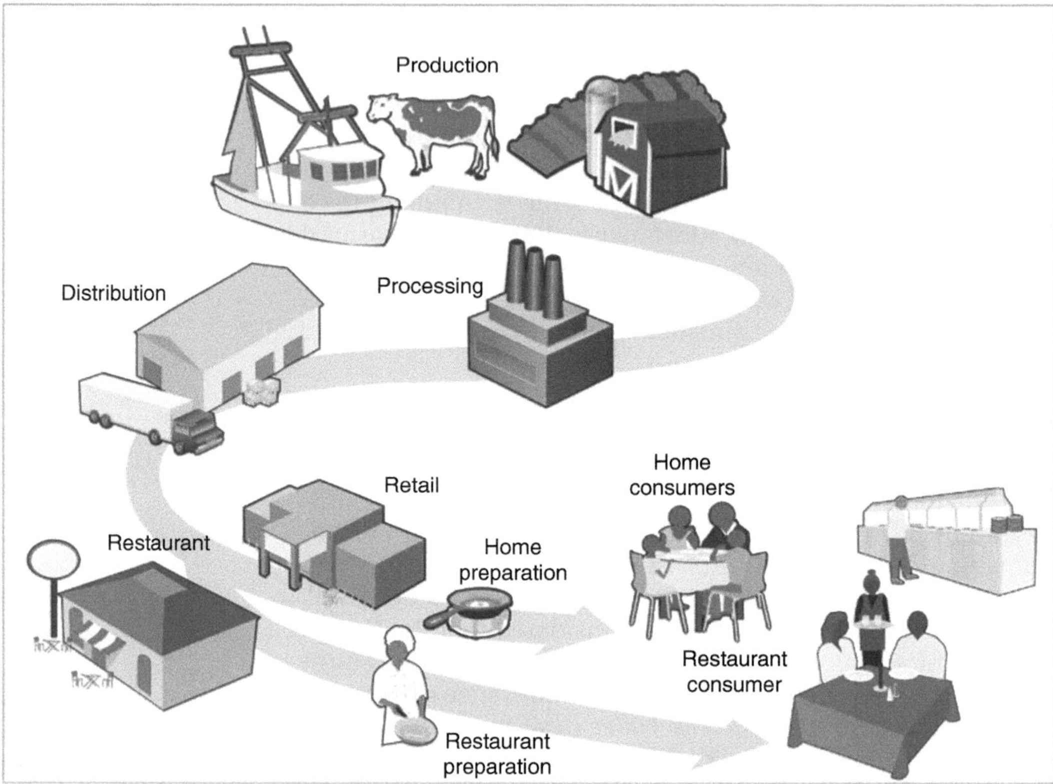

Figure 17.1. Food contamination chain. Food can be contaminated in its every stage from the production to processing to washing of cooking utensils.

used for the detection of pathogens and chemical residues in food. A variety of chemicals may come into our food supply at different stages of the food chain by means of intentional or unintentional addition. These chemicals include pesticide residues, food additives, environmental and industrial contaminants, mycotoxins, flavoring substances, and micronutrients (Pocas and Hogg, 2007). Based on the major impact on the food, herein, in this chapter we have classified them in three general categories:

1. contamination due to environmental and industrial exposure
2. food adulterant
3. food packaging materials

The contaminated food has the potential to make anyone sick and can cause symptoms like cramps, nausea, diarrhea, vomiting, nerve damage, allergies, and paralysis. Pathogenic bacteria, like allergens or bacterial toxins contaminated food has a relatively quick effect on the body. Therefore, the food industry needs suitable analytical methods for process and quality control. The determination of chemical and biological contaminants in foods is of paramount importance to the health of food because unlike the contamination of a physical nature, they cannot be displayed. The development of analytical methods and techniques in the environmental and food sciences has been developed parallel with growing consumer concerns it as relates to food safety and security (Kantianim et al., 2010). But the research areas have been complicated by the increasing mobility and globalization of foodstuffs and associated raw materials. They are directly and indirectly making contamination, from bad to worse, and growing on a global scale (Hamburg, 2011). Therefore, role of analytical instruments in the field of food contamination analysis and food industry is growing very fast day-by-day. To date, a variety of analytical techniques are commonly employed in food analysis. Typically, they include atomic spectroscopy (AS), high-performance liquid chromatography (HPLC), gas chromatography (GC), nuclear magnetic resonance (NMR), infrared (IR), mass spectrometry, enzyme-linked immunisorbent assay (ELISA), fluorescence, capillary electrophoresis (CE), supercritical fluid chromatography (SFC), solid phase extraction (SPE), supercritical fluid extraction (SFE), headspace (HS), and flow injection analysis (FIA) (Govindhan et al., 2014). Although thc listed techniques and instruments are very popular and sensitive, to decrease the complexity of food safety, food laboratories have been forced to replace their classical analytical methods with 21st-century techniques, which may facilitate the provision of adequate responses to this expanding global demand (Govindhan et al., 2014).

In recent years, sensors/biosensors are emerging to be a technology of choice in this respect as they are self-contained, compact, and low cost, with minimal power requirements. These devices have proven to be an extremely viable alternative to traditional analytical techniques such as chromatography with high sensitivity, as they are capable of providing specific quantitative analytical information via the utilization of chemical/biological recognition elements, which can translate interfaced contaminant inputs into useable output signals (Thévenot et al., 1999). The only drawback of these sensors for food applications is that they are not currently commercially available. The need today is fast and sensitive detection of food contamination at home level not like old one, where food samples are typically sent to laboratories, resulting in high costs and difficulties associated with real-time monitoring (Parellada et al., 1998). Despite the recent advances in food contaminant detection (Silva et al., 2013), there still exists many challenges and opportunities to improve the current sensor technology in order to have simple, rapid, versatile, and inexpensive tools for detection of food contaminants. The further integration of materials science, microfabrication, functional nanomaterials, and electrical engineering with sensor technologies will be required to advance wireless, real-time environmental and food monitoring and analysis (Govindhan et al., 2014).

> *Any sufficiently unconventional technology is indistinguishable from magic.*
>
> **Arthur C. Clarke**

This statement is found to be very true for nanotechnology-based molecular sensing. In that case, the detection limits "magically" and drops down to a very small value. For the detection of food quality control, combination of analytical chemistry and nanoscience are being applied extensively in recent days. The unique chemical, electrooptical, mechanical, and thermal properties of nanomaterials makes them useful materials for the preparation electronic noses and tongues, cantilevers, test-strips, surface-enhanced Raman scattering (SERS) nanoprobes, radio-frequency identification tags, immunonanodipsticks, biosensors/lab-on-chips, and so on, and applied for the detection of various contaminants in food samples as well as to speed up the analysis and improve the sensitivity of the food supervisory committees. Nanomaterials, like single-, double-, or multiwalled carbon nanotubes (SWCNTs, DWCNTs, MWCNTs); metal oxide nanoparticles; magnetic nanoparticles (MNPs); metal nanorods of ZnO/Au/Ag, and so on; graphene/graphene oxide; nanowires of metal oxide, polymer, silicon; quantum dots (QDs); cellulose derivative based (optical and/or photonic) nanofibers, and so on have been previously applied for the fabrication of sensing devices for food analysis to improve the performance of conventional methods without nanoparticles. The techniques possess several advantages in comparison to the conventional reported methods, namely, improved sensitivity, high selectivity, better transportability, data reproducibility, easy-to-operate, simultaneous analysis, error-free analysis of packed food materials, which made them a better replacement in food sensing area. The recent and significant advancements in the area of nanotechnology based food quality control are discussed in the following sections.

A sensor consists of mainly two parts, called the receptor and the transducer part. The receptor part of the sensor receives physical and chemical stimulus and transforms it into a form of energy. The transducer part can be capable of transducing this energy into a useful analytical signal (Kumar, 2007). In the different sections of this chapter, we have tried to incorporate the role of different types of sensors for food contaminant analysis.

2 Contamination due to Environmental and Industrial Exposure

In the food contamination chain, the very first member is production or producer (Fig. 17.1). The contamination of this level is mainly due to two aspects:

1. The soil in which vegetables/fruits grow
2. The water used for their impegration

For more production in the least time span, the farmers and industry people are using a large amount of pesticide, which is no doubt essential for a better crop but in a limited way. With the development of intensive agricultural practices, millions of tons of pesticides are widely used each year to control and/or eliminate plant or animal pests and diseases in the field of agriculture, medicine, industry, and related activities. Pesticides can be classified as herbicides, insecticides, fungicides, or other types according to their purpose. Arsenic, carbamates, organophosphates, pyrethroids, and nitrophenol derivatives most common chemical compound involved as pesticides (Fig. 17.2). According to literature the most hazardous pesticides are carbamates, coumarin, organochlorides, organophosphates, and arsenic and mercury derivatives (Aragay et al., 2012). All of these are known to be highly toxic to humans and affects their central nervous system (CNS). Due to their high toxicity, environmental agencies have set maximum values for their contamination levels in drinking and surface water (Richardson and Ternes, 2011). The pesticides either remain in the soil or enter surface waters and groundwater depending on their water solubility. The resulting compounds from pesticide degradation can remain in animals, vegetables, and water sources and they can become more concentrated as they move up the food chain. Some of the pesticides have limited persistence in the environment and their very low level in the agricultural product is very harmful to the human health. To avoid these, there is increasing interest in the development of systems to sense and/or remove those (Aragay et al., 2012).

Till now many methods has been reported for the detection of pesticides in various foods, vegetables, beverages, and water samples. Some of them are modified with nanoparticles (Aragay et al., 2012) and some are not (Sannino et al., 2004; Zawiyah et al., 2007; Sharif et al., 2006). But in recent days, owing to the properties given by the small size of nanomaterials, their large surface-to-volume ratios, their physicochemical properties, composition, and shape and also their unusual target binding characteristics makes them an advanced material for the fabrication of sensing devices. These nanoparticle-modified sensors

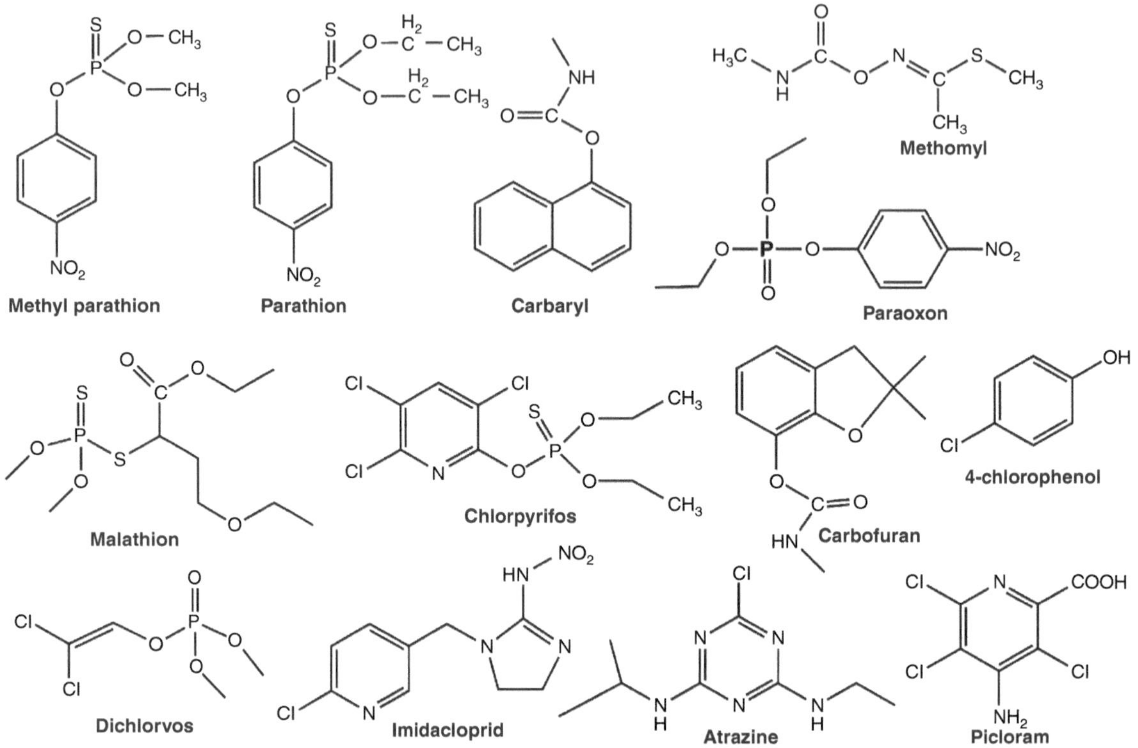

Figure 17.2. Chemical structure of some common pesticides.

can markedly improve the sensitivity and specificity of analyte detection. Based on the nature of nanoparticles, mainly electrochemical and optical detection techniques are for the sensing of pesticides. Herein, we have classified the type of sensors used for pesticides detection in the recent years.

2.1 Detection of Pesticides by Electrochemical Sensors

Owing to their simplicity, specificity, sensitivity, short analysis time, easy operation, and cost effectiveness, the electrochemical techniques are always the most popular ones for the fabrication of sensors (Roy et al., 2014a). The common nanoparticles used in these types of sensing devices must have high electron transfer rate or high electrocatalytic activity to obtain a lowest detection limit. Driven by these, great efforts have been made in the design, fabrication, and application of nanomaterials for electrochemical sensing devices (Roy et al., 2015a).

2.1.1 *Acetylcholine-Based Eelectrochemical Systems*

For the pesticide detection by electrochemical sensors, the acetylcholinase (AChE)-modified sensors were very popular. For the fabrication of ACchE-modified sensor, the acetylicholine was first immobilized onto the surface of working electrode. In the presence of AChE enzyme, the acetylthiocholine produced an electro-active species thiocholine. The product (thiocholine) could be used for the quantitative measurement of the enzyme activityas well as the biomarker for the biological effect of some pesticides (mainly organophosphate and carbamate) involved in the inhibition action. In Table 17.1 we have summarized the analytical data for the estimation of some common pesticide using acetylthiocholine modified electrochemical sensor in combination with nanomaterials.

The reaction mechanism on the surface of biosensor can be written as (Cai and Du, 2008):

$$\text{Acetylcholine} + H_2O \xrightarrow{\text{AChE}} \text{Thiocholine} + \text{HA}(\text{Acetic acid})$$
$$2\,\text{Thiocholine}(\text{Red}) \rightarrow \text{Thiocholine}(\text{Ox})(\text{dimeric}) + 2H^+ + 2e^-$$

Based on the previous mechanism, Cesarino et al. has reported the use of multiwalled carbon nanotubes (MWCNTs) and polyaniline (PANI) core–shell modified glassy carbon (GC) electrode (GCE) for the detection of carbaryl and methomyl with the limit of detection (LOD) 1.4 and 0.95 μmol L^{-1}, respectively (Cesarino et al., 2012). They applied these pesticide biosensors in the detection of carbaryl and methomyl pesticides in food samples like cabbage, broccoli, and apple using chronoamperometry (CA). In another paper, Cai and Du also reported a type of biosensor for the detection of carbaryl by modified screen-printed carbon electrode with the LOD 0.004 μg mL^{-1} and applied this for the detection of carbaryl in garlic samples (Cai and Du, 2008). Carbaryl and methomyl is known to be a widely used carbamate of family insecticides. The widespread use of carabryl in agriculture can be viewed in terms of their high insecticidal activity, low bioaccumulation, and relatively low environmental persistence due to rapid degradation in the environment (Cai and Du, 2008). Carbaryl often produced a highly toxic intermediary methyl isocyanate. In another report, Du et al. reported the detection of malathion, an organophosphate-based pesticide. They have used polypyrrole and MWCNTs with a LOD 1.0 ng mL^{-1} (Du et al., 2010a). In another paper, MWCNT and β-cyclodextrin-modified GCE was used for the detection of organophosphate pesticide dimethoate in garlic samples (Du et al., 2010b). Viswanathan et al. has reported an AChE-based electrochemical biosensor using SWCNTs for the detection of methyl parathion (MP) and chlorpyrifos (CPF) (Viswanathan et al., 2009).

Table 17.1 Detection of Pesticides by Electrochemical Acetylcholinase Biosensor

S. No.	Electrode Materials	Template(s)	Linear Range	LOD	Detection Technique	Real Samples	References
1.	GC/MWCNT/PANI/AChE	Carbaryl, Methomyl	0.4–8.0 μM L^{-1} for both	1.4, 0.95 μM L^{-1}	CV, SWV, CA	Cabbage, broccoli, apple	Cesarino et al. (2012)
2.	AChE-CAMC/SPCE	Carbaryl	0.01–0.5 and 2–20 μg mL^{-1}	0.004 μg mL^{-1}	CV, Amperometry	Garlic	Cai and Du (2008)
3.	Au–PtNPs/3-APTES/GC	Paraoxon ethyl, sarin, and aldicarb	150–200 nM, 40–50 nM, and 40–60 μM	150, 40, and 40 μM	CV, EIS, LSV	—	Upadhyay et al. (2009)
4.	AChE-PANI-PPy-MWCNTs/GCE	Malathion	0.01–0.5 μg mL^{-1} and 1–25 μg mL^{-1}	1.0 ng mL^{-1}	CV, Amperometry		Du et al. (2010a)
5.	MWCNTs-β-CD-CHIT/GCE	Dimethoate	0.01–2.44 and 2.44–10.00 μM	2 nM	CV, Amperometry	Garlic	Du et al. (2010b)
6.	Au/ssDNA–SWCNT/PANI/AChE	Methyl parathion and Chlorpyrifos	1.0×10^{-11} and 1.0×10^{-6} M	1×10^{-12} M	CV, EIS, SWV	Spiked river water samples	Viswanathan et al. (2009)
7.	AChE-Fe_3O_4 NPs/c-MWCNTs/ITO electrode	Malathion, Chlorpyrifos, Monocrotophos, and Endosulfan	0.1–70, 0.1–50, 0.1–70, and 0.1–100 nM L^{-1}	0.1 nM L^{-1} for all	CV, EIS, DPV, Amperometry	Cabbage, onion, spinach	Chauhan and Pundir (2012)
8.	AChE–Au–PDDA–PB/GCE	Monocrotophos	1.0–1000 and 1.0–10 ng mL^{-1}	0.8 pg mL^{-1}	CV, EIS, DPV, Amperometry	Garlic	Wu et al. (2011)
9.	AChE–AuNP–$CaCO_3$/silica sol–gel/AuE	Malathion and Chlorpyrifos	0.1–100 and 0.1–70 nM	0.1 nM	CV, EIS, Amperometry	River water samples	Chauhan et al. (2011)
10.	AChE–Au–PPy/GCE	Methyl parathion	0.005–0.12 and 0.5–4.5 μg mL^{-1}	2 ng mL^{-1}	CV	—	Gong et al. (2009)

11.	AChE/HSO_4^--AuNPs-porous carbon/BDD electrode	Dichlorvos	4.5×10^{-13}–4.5×10^{-9} M	2.99×10^{-13} M	CV, EIS, DPV	Lettuce leaves	Wei and Wang (2015)
12.	AChE–GNR–Nafion/GCE	Dichlorvos	5.0–100 ng mL^{-1} and 1.0–20 μg mL^{-1}	2.0 ng mL^{-1}	CV, Amperometry	River water samples	Wu et al. (2013)
13.	AChE/Fe_3O_4–CH/GCE	Carbofuran	5.0×10^{-9} M–9.0×10^{-8} M	3.6×10^{-9} M	CV, SWV	Cabbage	Jeyapragasam and Saraswathi (2014)
14.	BChE-CBNPs-SPE	Paraoxon	5–30 μg L^{-1}	5 μg L^{-1}	CV, EIS	—	Arduini et al. (2015)
15.	NF/AChE-CS/ZnO NPs-CGR-NF/GCE	Chlorpyrifos and Carbofuran	1.0×10^{-13}–1×10^{-8} M and 1.0×10^{-12}–1×10^{-8} M	5×10^{-14} M and 5.2×10^{-13} M	CV, DPV, Amperometry	Tap water and lake water	Wang et al. (2014)

GC, Glassy carbon electrode; MWCNT, multiwalled carbon nanotube; PANI, polyaniline; AChE, acetylcholinesterase; CV, cyclic voltammetry; SWV, square wave voltammetry; CA, Chronoamperometry; CAMC, cellulose acetate composite; SPCE, screen-printed carbon electrode; 3APTES, 3-aminopropyl triethoxy silane; EIS, electrochemical impedance spectra; LSV, linear sweep voltammetry; PPy, polypyrrole; β-CD, β-cyclodextrin; CHIT, chitosan; SWCNT, single-walled carbon nanotube; ITO, indium tin oxide; DPV, differential pulse voltammetry; PDDA, poly(dimethyldiallylammonium chloride); PB, Prussian blue; BDD, boron-doped diamond; GNR, graphene nanoribbon; CGR, carboxylic graphene; NF, nafion

MP and CPF are known to be the widely used organophosphorous with high toxicity. They obtained a detection limit of 1×10^{-12} M and were used for their detection in river water samples. Magnetic Fe_3O_4 and MWCNTs electrodeposited onto indium tin oxide (ITO)-coated glass plate was used for the successful determination of malathion, CPF, monocrotophos, and endosulfan by Chauhan and Pundir (2012). They obtained an LOD 0.1 nmol L^{-1} for all the pesticides. The biosensor was employed for the determination of pesticides in environmental and food samples (cabbage, onions, and spinach). Bimetals are known to be efficient materials in the electrochemical sensing due to their synergistic effect of the two individual metals and the uneven distribution of electrons occurs as a consequence of electronic interaction between the two different metals, which enhances its catalytic activity (Patra et al., 2015a; Roy et al., 2015b). Based on this Upadhyay et al. has reported the use of gold platinum (Au-Pt) bimetal for the detection of Paraoxon ethyl, sarin, and aldicarb with LODs 150 nM, 40 nM, and 40 μM, respectively (Upadhyay et al., 2009). Gold nanoparticles are widely used in the electrochemical sensing due to their intrinsic electrochemical characteristics. Wu et al. (2011) reported an electrochemical AChE biosensor using gold nanoparticles for the successful detection of monocrotophos (an organophosphate insecticide, persistent organic pollutant) in garlic samples. Similar work using porous gold-nanoparticle–$CaCO_3$ hybrid material modified Au electrode has been reported by Chauhan et al. (2011) for the detection of organophosphate pesticide malathion and CPF. Gold nanoparticles have also been used by Gong et al using gold-polypyrrole GCEs for the detection of MP with a detection limit as low as 2.0 ng mL^{-1} (Gong et al., 2009). Wei and Wang (2015) have reported a novel acetylcholinesterase biosensor based on ionic liquids-AuNPs-porous carbon composite matrix for detection of organophosphate pesticides dichlorvos. They also show its potential usefulness in real samples like lettuce leaf. Another report has come from Jeyapragasam and Saraswathi (2014) for the detection of carbofuran based on acetylcholinesterase immobilized onto iron oxide–chitosan nanocomposite. The sensor found its usefulness for the detection of this insecticide in some fruit and vegetable samples like lettuce, cabbage, green peppers, tomatoes, Chinese chives, and strawberries. Ding et al. (2014) have reported the assembly of carbon nanotubes on a nanoporous gold electrode for the designing of acetylcholinesterase electrochemical biosensor for malathion. Zinc oxide (ZnO) nanoparticles also have electrocatalytic activity and popularly used in electrochemical sensing of biomolecules (Patra et al., 2015b). Wang et al. (2014) used dispersed ZnO nanoparticles on carboxylic graphene for the preparation of sensors and used for the detection of CPF and carbofuran in some water samples.

Afar from MWCNTs, other carbon nanomaterials, like graphene are being employed for pesticide sensors owing to its remarkable electrochemical properties, show fascinating advantages in electrochemical biosensors. Compared to CNTs, graphene is reported to offer more advanced properties and is likely to exhibit fewer weaknesses (Brownson and Banks, 2010). Graphene is known to be an excellent electrical conductor, heterogeneous electron transfer occurs at the edges of the graphene or at defects in the basal plane. Thus, the high surface area of graphene translates to numerous defects and offers many electroactive sites. A graphene-based electrochemical biosensor has been reported by Wu et al. Here electrochemically reduced graphene oxide and Nafion nanocomposites were used for ultralow potential detection of organophosphate pesticide dichlorvos with an LOD 2 ng mL^{-1} (Wu et al., 2013).

Based on inhibition of butyrylcholinesterase, screen-printed biosensors modified with carbon black nanoparticles has been reported by Arduini et al. for the detection of paraoxon. For this, the electrode was first modified with a dispersion of carbon black nanoparticles (CBNPs) in a dimethylformamide-water mixture, and butyrylcholinesterase was then immobilized on the surface by cross-linking. An LOD of 5.0 μg L^{-1} was obtained by this method (Arduini et al., 2015).

2.1.1.1 Electrochemical-Based Immunosensors

Immunosensors are defined as the affinity ligand-based biosensor devices where the immunochemical reaction is coupled to a transducer. The fundamental principle mainly depends on the molecular recognition between antigens and antibodies to form a stable complex (Fig. 17.3). Electrochemical, optical, and microgravimetric immunosensors are previously reported. The

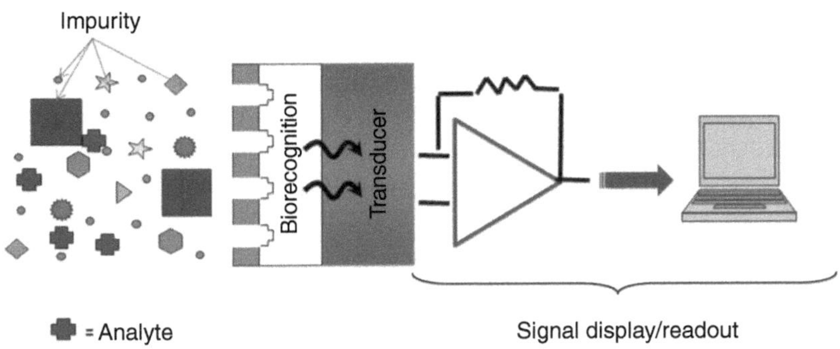

Figure 17.3. Schematic representation of an immunosensor.

advancement of immunoassay technology is a triumphant story especially for the clinical application and this continues to be an exciting area of research (Luppa et al., 2001).

Electrochemical immunosensors have appealed significant interest for many years because of their high sensitivity, low cost, and inherent miniaturization. Different types of nanomaterials, ranging from nanoparticles to graphene, opens up new doors for electrochemical immunoassays, in terms of improved electrochemical properties of the transducers and better conjugation with biomolecules and they sometimes act as electrochemical labels. The summary of immunosensors applied for pesticide detection was portrayed in Table 17.2.

Valera et al. (2010) reported immunosensors based on interdigitated μ-electrodes combined with AuNPs and specific antibodies to detect atrazine in red wine samples by conductometric measurements. The conductometric change occurs when secondary antibodies labeled with AuNPs are deposited onto the electrode gaps after the immunoassay. The inclusion of gold nanoparticle reduces the gap between interdigitated μ-electrodes and increases the electric field between them, generating an analytical readout signal. The main advantages lie in the short analysis time for assay completion; its take 5 h for 48 samples. The obtained LODs are far below the maximum residue level (50 $\mu g\ kg^{-1}$) established by the European Union for residues of atrazine in the wine grapes and other foodstuff products. Tang et al. (2008) has reported chitosan/gold nanoparticles for the detection of picloram herbicides. They obtained a much lower LOD 5.0 $ng\ mL^{-1}$. The prepared immunosensor was further used to detect picloram in rice, lettuce, and paddy field water. After this Chen et al. (2010a) has reported similar picloram detection method. Their system deals with ordered three-dimensional gold (Au) nanoclusters. They applied horseradish peroxidase (HRP)-labeled secondary antibody for enzyme-amplified amperometric measurement. They reported that the electrodeposited Au nanoclusters built direct electrical contact and immobilization interface with protein molecules without postmodification and positioning. Under the optimal conditions, a much lower LOD $5.0 \times 10^{-4}\ \mu g\ mL^{-1}$ was obtained. The authors highlighted that as compared to the chitosan/AuNP composite system devised by Tang et al. their system offers significantly greater specific surface area for molecular bioconjugation and better electron-exchange capability, thus facilitating a lower detection limit. Sharma et al. (2011) reported an electrochemical immunosensor for the detection of diuron herbicide involving the use of laser ablated gold electrodes fabricated on polystyrene substrate and modified with a Prussian blue-gold nanoparticle

Table 17.2 Detection of Pesticides by Other Electrochemical Sensors (MIP-Based, Immunosensor, and Normal Electrochemical Sensor)

S. No.	Name of the Sensor	Electrode Materials	Template(s)	Linear Range	LOD	Detection Technique	Real Samples	References
1.	Electrochemical sensor	Silver/nafion modified GCE	Methyl parathion and Parathion	0.3001–1.444 and 0.103–0.6179 μM (DPV) and 0.2–1.00 μM both (Amperometry)	0.0874 and 0.0835 μM (DPV) and 0.2815 and 0.2659 μM (Amperometry)	CV, DPV, Amperometry	Water	Kumaravel and Chandrasekaran (2010)
2.	Molecularly imprinted electrochemical sensor	MIP/rGO@Au/GCE	Carbofuran	5.0×10^{-8}–2.0×10^{-5} mol L^{-1}	2.0×10^{-8} mol L^{-1}	CV, EIS	Cabbage, cucumber	Tan et al. (2015)
3.	Electrochemical sensor	Au/Nafion/GCE	Methyl parathion	5.0×10^{-7}–1.2×10^{-4} M	1.0×10^{-7} M	CV, SWV	Water and vegetable samples	Kang et al. (2010)
4.	Molecularly imprinted electrochemical sensor	Imprinted PoPD-RGO electrode	Imidacloprid	7.5×10^{-7}–7.0×10^{-5} M	4.0×10^{-7} M	CV, LSV	Pear samples	Kong et al. (2013)
5.	Electrochemical sensor	GN-AuNRs/GCE	Methyl parathion	10.0–500.0 and 750.0–4000.0 ng mL^{-1}	0.82 ng/mL^{-1}	CV, EIS, SWV	Tap water, kiwi fruit	Zhu et al. (2014)
6.	Electrochemical sensor	NiOPc, hemin and graphene oxide/CPE	Carbofuran and diuron	5.0×10^{-6}–1.4×10^{-4} and 5.0×10^{-5}–1.0×10^{-3} mol L^{-1}	1.7×10^{-6} and 2.0×10^{-5} mol L^{-1}	CV	Carrot and soil	Wong and Sotomayor (2014)
7.	Electrochemical sensor	MWCNTs–CeO_2–Au/GCE	Methyl parathion	1.0×10^{-10}–1.0×10^{-7} M	3.02×10^{-11} M	DPV	River water and soil samples	Dong et al. (2013a)

(Continued)

Table 17.2 Detection of Pesticides by Other Electrochemical Sensors (MIP-Based, Immunosensor, and Normal Electrochemical Sensor) (*cont.*)

S. No.	Name of the Sensor	Electrode Materials	Template(s)	Linear Range	LOD	Detection Technique	Real Samples	References
8.	Electrochemical sensor	ZrO_2/CPE	Methyl parathion	5.0–3000.0 ng mL^{-1}	2.0 ng mL^{-1}	CV, SWV	Different water samples	Parham and Rahbar (2010)
9.	Electrochemical sensor	ZrO_2/Au film electrode	Parathion	20.0–140.0 ng mL^{-1}	3.0 ng mL^{-1}	CV, SWV	—	Wang and Li (2008)
10.	Electrochemical sensor	Agnf/TiO_2nf/GCE	Imidacloprid	1.0–5.0 μM	0.93 μM	Amperometry	Commercial samples and water samples	Kumaravel and Chandrasekaran (2011)
11.	Electrochemical sensor	PCz/CRGO/GCE	Imidacloprid	3.0–10.0 μM (both)	0.22 μM (CV), 0.44 μM (DPV)	CV, DPV	—	Lei et al. (2013)
12.	Electrochemical sensor	Au–TiO_2/Chit/GCE	Parathion	1.0–7.0 $\times 10^3$ ng mL^{-1}	0.5 ng mL^{-1}	CV, DPV	Onion, Lettuce, Cabbage	Qu et al. (2008)
13.	Electrochemical sensor	β-CDP/rGO/GCE	Imidacloprid	1.0×10^{-6}–1.5×10^{-4} mol L^{-1} (CV), 5.0×10^{-8}–1.5×10^{-5} mol L^{-1} and 2.0×10^{-5}–1.5×10^{-4} mol L^{-1} (DPV)	1.0×10^{-7} mol L^{-1} (CV), 2.0×10^{-8} mol L^{-1} (DPV)	CV, DPV	—	Chen et al. (2013)
14.	Electrochemical immunosensor	PB-GNPs modified LC-LAGE	Phenyl urea herbicide-Diuron	4.0 pM to 0.4 μM	4.0 pM	SWV	—	Sharma et al. (2011)
15.	Electrochemical immunosensor	Thin Au/Cr interdigitated μ-electrodes	Atrazine	0.32–2000.0 μg L^{-1}	0.16 μM	Conductometry	Red wine samples	Valera et al. (2010)

16.	Electrochemical immunosensor		Picloram	0.005–10.0 μg mL^{-1}	5.0 ng mL^{-1}	CA	Rice, lettuce, and paddy field water	Tang et al. (2008)
17.	Electrochemical immunosensor		Picloram	0.001–10.0 μg mL^{-1}	5.0×10^{-4} μg mL^{-1}	CA	Peach samples	Chen et al. (2010a)
18.	Electrochemical sensor	TiO_2-GO/SPE	4-chlorophenol	0.05–20.0 μM and 20.0–200.0 μM	0.02 μM	CV	Real water samples	Bai et al. (2014)
19.	Molecularly imprinted electrochemical sensor	FuAuNP–ATP–MIP–AuNP–MCNT/GCE	Methyl-parathion	0.1–1.1 ng mL^{-1} and 1.1–11.0 ng mL^{-1}	0.08 ng mL^{-1}	CV, LSSV	Water, apples, and cucumber	Wu et al. (2014)
20.	Molecularly imprinted electrochemical Sensor	PATP-AuNP's-GCE	Chlorpyrifos	0.5–10.0 μM	3.3×10^{-7} M	CV	Spiked tap water samples	Xie et al. (2010)

rGO, Reduced graphene oxide; PoPD, poly(*o*-phenylenediamine); NIOPc, nickel (II) 1,4,8,11,15,18,22,25-octabutoxy-29H,31H phthalocyanine; CPE, carbon paste electrode; Agnf, nanosilver Nafion; PCz, poly(carbazole); CRGO, chemically reduced graphene oxide; LAGE, laser ablated gold electrodes; PATP, electropolymerized polyaminothiophenol.

(PB-AuNP) film. The binding of available diuron-antibody on the conjugate coated electrode was detected using alkaline phosphatase rabbit anti-IgG (antiimmunoglobulin) antibody. The addition of 1-naphthyl phosphate substrate resulted in the production of electrochemically active product, 1-naphthol, which was monitored using square wave voltammetry technique. They also reported that PB-AuNP film has the advantage of a large surface-to-volume ratio, strong adhesion, and high surface activity, all of which give the assay greater sensitivity relative to the case of unmodified electrodes. The system is highly specific for diuron, with a sensitivity of 4.0 pM to 0.4 μM with LOD 4.0 pM.

2.1.1.2 MIP-Based Electrochemical Sensor

Although the enzyme and antibody-based sensors have certain good advantages like high selectivity but they have certain limitations like poor chemical/physical stability. This prevents their use in harsh environments (eg, acidic or basic media, organic solvents, high temperatures). Another disadvantage is the difficulty of obtaining antibodies that are specific to a particular pesticide. In addition, the natural antibodies are highly expensive and sometimes undergo mutation, thus varying the interaction capacity of the recognition sites. Artificial antibodies like molecularly imprinted polymers (MIPs) have attracted considerable attention for the detection of biomolecules due to their high selectivity and stability (Patra et al., 2015c). Generally MIPs are prepared by polymerization of selected monomers in the presence of template. Some noncovalent interaction occurs between the functional monomer and the template molecules in the prepolymer mixture (Fig. 17.4). After the polymerization the template molecule is extracted from the polymer matrix and that leaves certain cavities, which is selective and specific to the template molecule (Roy et al., 2015c). The summary of MIP-based sensors applied for pesticide detection is portrayed in Table 17.2.

Xie et al. has reported an MIP-based system for the detection of CPF. Polymerization was done by electropolymerized polyaminothiophenol (PATP) membranes at the surface of AuNP-modified GCEs (Xie et al., 2010). Before the elctropolymerization, the *p*-aminothiophenol (ATP) moiety first attached to the AuNPs by Au–S bonds at the surface of GCE. After this, CPF was further assembled itself on the monolayer of thiophenol moiety through the hydrogen bonding interaction. They also reported that the cyclic voltammogram obtained for CPF on the imprinted PATP-AuNP-GCE sensor is about 3.2-fold higher than the PATP-Au sensor. The detection limit is also lower in magnitude for imprinted PATP-Au sensor by 2-fold than the PATP-AuNP-GCE sensor. A linear relationship in the range of 0.5–10 μM CPF was obtained with

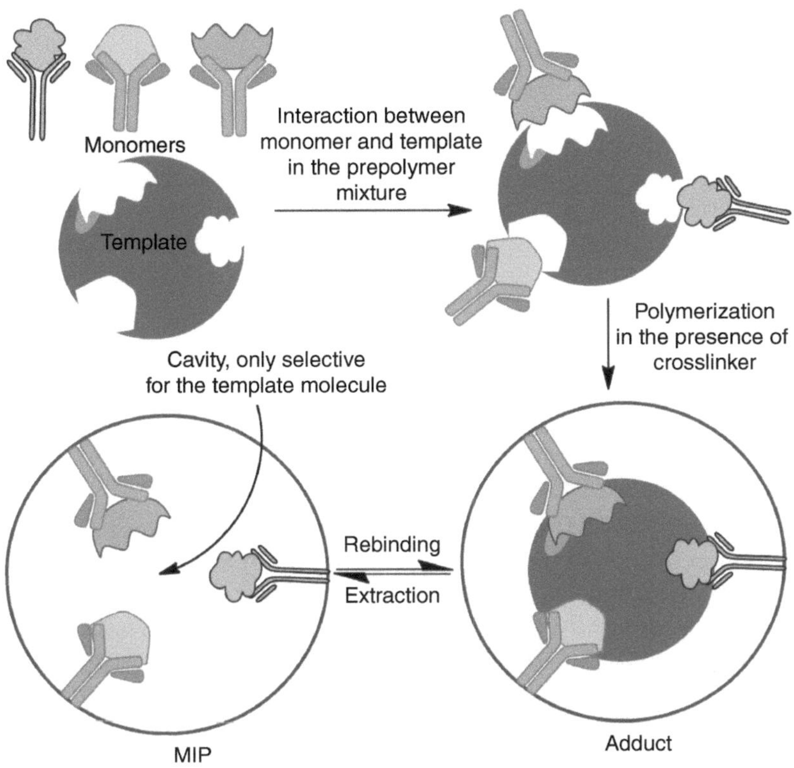

Figure 17.4. Schematic representation for the preparation of molecularly imprinted polymer (MIP).

a detection limit of 0.33 μM of CPF. The sensor was further used to detect CPF spiked tap water samples. Tan et al. (2015) reported the use of MIPs-based electrochemical sensing of carbofuran on reduced graphene oxide and gold nanoparticles modified electrode. Under the optimum conditions, the peak current of the sensor and carbofuran concentration showed a good linear relationship over the range from 5.0×10^{-8}–2.0×10^{-5} mol L^{-1}, with the detection limit 2.0×10^{-8} mol L^{-1}. The sensor was further used for the determination of the pesticide in vegetable samples like cabbage and cucumber with good recovery (97.5–110.6%). Another graphene oxide and MIP-based electrochemical sensor has been reported by Kong et al. (2013) and they used electropolmerized poly(*o*-phenylenediamine) membranes at reduced graphene oxide modified electrode for imidacloprid determination. They obtained a much lower LOD 4.0×10^{-7} M by linear sweep voltammetry technique (LSV). The real-time applicability of the sensor is determined by using it for the detection of imidacloprid in pear samples with a recovery range 91.3–96.6%. MP detection by molecularly imprinted electrochemical sensor based on nitrogen doped graphene sheets has been reported by Xue et al. They have reported the fabrication of MIP-based sensitive and selective electrochemical sensor using electropolymerization technique. Cyclic voltammetry technique was used for the electropolymerization onto an Au electrode using phosphate buffered saline (PBS) as supporting electrolyte, phenol as functional monomer and MP as template molecule (Xue et al., 2014). A much lower LOD 0.01 $\mu g\ mL^{-1}$ was obtained. The sensor was also applied for the detection of MP in water samples. Functionalized gold nanoparticle-decorated carbon nanotubes were also employed for the detection of MP (Wu et al., 2014). A much lower detection limit 0.01 $ng\ mL^{-1}$ was reported. Apple and cucumber was used for the real-time application.

2.1.1.3 Other Electrochemical Sensors

In addition to the aforementioned approaches described for the detection of pesticides, some of the other analytical methods have been applied to develop new, simple, low-cost, portable, and sensitive devices that do not require biological or synthetic receptors. The summary of other electrochemical sensors applied for pesticide detection was portrayed in Table 17.2. Kumaravel and Chandrasekaran (2010) have previously reported nanosilver/Nafion composite electrodes for electrochemical sensing of MP and parathion with an LOD of 0.0874 and 0.0835 μM. The water samples are used to prove the real-time applicability of the sensor. Wong et al. (2014) has reported the fabrication of electrochemical biomimetic sensor modified with hemin and graphene oxide for detection of carbofuran in food with recovery close to 100%. Carrot and tomato are used as real samples with an LOD of 9.0×10^{-9} mol L^{-1}. Kang et al. used gold nanoparticles and Nafion film modified GCEs for the MP detection in water and vegetable samples (Kang et al., 2010). Similar MP sensor has been reported by Zhu et al. They used graphene (GN)-Au nanorod nanoassembly modified GCE (Zhu et al., 2014). They also mentioned that the electrochemical synthesis of GN-AuNRs hybrid is environmentally friendly as it is not involved in the chemical reduction of graphene oxide (GO) and facile for just on the basis of electrostatic interaction between GO and AuNRs. The high conductivity, large surface area, high catalytic ability, and good adsorption capacity toward aromatic rings of graphene with the excellent electronic properties and adsorption capacity of AuNRs, the highly sensitive MP sensor was fabricated with the GN-AuNRs nanocomposite. The LOD was lower than many previously reported enzyme or nonenzyme-based sensors and was calculated to be 0.82 $ng\ mL^{-1}$. They have chosen kiwi fruit and water samples for the real-time application. Another GO-based sensor reported by Wong and Sotomayor (2014) described the use of an electrochemical sensor coupled to FIA system for the detection of pesticides. The proposed sensor was prepared using a carbon paste electrode modified with hemin and a nickel (II) complex. They have used carbofuran and diuron as model pesticides and applied it in the successful detection in carrot and soil samples. Ag/Cu nanoalloy was used by Sreedhar et al. (2015) for the detection of CPF in waterand soil samples. Graphene composite paste electrode was used as the working electrode. They have shown that peak currents increases and the overpotential decreases for CPF in Ag/Cu–graphene nanocomposite compared to Ag–graphene, Cu–graphene, graphene paste and carbon paste electrodes. This could be accredited to the synergistic combination of graphene and Ag/Cu alloy nanoparticles with good adsorption, large surface area, increased active sites leading to fast electron transfer rates and high electrical conductivity of nanocomposite. Fu et al. (2015) has used three-dimensional mono-6-thio-β-cyclodextrin covalently functionalized gold nanoparticle/SWCNT hybrids modified GCE for the electrochemical determination of MP using square wave anodic stripping voltammetry method. They obtained a lower detection limit of 0.1 nM and the proposed sensor was employed for the detection of the pesticide in environmental water samples with good recovery 92.3–109.6%. A highly sensitive electrochemical stripping analysis MP at MWCNTs-CeO_2-Au nanocomposite modified electrode was reported by Dong et al. (2013a). Furthermore, the proposed electrochemical sensor was successfully employed for the determination of MP in water and soil samples. Parham and Rahbar (2010) reported the ZrO_2-nanoparticles modified carbon paste electrode for the square wave voltammetric detection of MP. They also conclude that the ZrO_2 nanoparticles have a strong affinity toward the phosphate group on MP molecules, which provide sensitivity and selectivity of the modified carbon paste electrode. They found an LOD of 2.0 ng mL^{-1}. The electrode was successfully applied for the determination of MP in different water samples. Wang and Li (2008) also reported a nanocomposite of ZrO_2/Au film electrode for voltammetric detection of parathion. Kumaravel and Chandrasekaran (2011) have reported the electrochemical determination of imidacloprid using nanosilver-Nafion/nano-TiO_2 Nafion composite modified GCEs by amperometric method. Another imidacloprid sensor has been reported by Lei et al. (2013) and Chen et al. (2013) by using poly(carbazole)/chemically reduced graphene oxide modified GCEs and β-cyclodextrin polymer functionalized reduced-graphene oxide, respectively.

Gold-titanium dioxide (Au-TiO_2) nanocomposite was also used for electrochemical detection of trace organophosphates insecticides (Qu et al., 2008). Another TiO_2-based TiO_2–graphene nanoparticle was used for the detection of 4-chlorophenol (4-CP) by modified screen-printed electrode (Bai et al., 2014). At the selected experimental conditions, they found that an oxidation peak current increases with the increase of 4-CP concentrations over the range of 0.05–20.0 μM and 20.0–200.0 μM. The lower LOD was 0.02 μM.

2.1.2 Optical Method for the Pesticide Detection

Like the electrochemical sensor, the nanoparticles-based optical sensor has certain advantages and they have been applied for the detection pesticides.

2.1.2.1 Acetylcholinase-Based Optical Sensor

AChE mainly is used as a catalyst for the hydrolysis of the neurotransmitter acetylcholine (ACh) to inactive choline and the inhibition of AChE. This will lead to the accumulation of acetylcholine in the synaptic cleft, resulting in impeded neurotransmission. The AChE inhibition systems are monitored by determining the generated enzyme production. The traditional optical methods used for the measurement the amount of AChE and its inhibitors are spectrophotometric thiol assay by using Ellman reagent and the colorimetric detection of H_2O_2. But these systems suffer from the lack of sensitivity. To enhance the sensitivity, advanced techniques, including colorimetric and fluorescent assays and surface plasmon resonance (SPR) based on metallic/MNPs and QDs, have been developed (Xia et al., 2015). The optical sensors used for pesticide detection are summarized in Table 17.3.

Saa et al. previously reported this type of biosensor for the enzymatic growth of CdS QDs where the thiocholine molecules promote the decomposition of $S_2O_3^{2-}$ to H_2S, which, in the presence of Cd^{2+} ions, forms fluorescent nanocrystals of CdS (Saa et al., 2010). The formation of fluorescent nanoparticles is monitored by fluorescence spectroscopy. Thereafter, Zheng et al. (2011) reported an organophosphorous pesticide paraoxon/parathion biosensor based on fluorescence quenching. After the introduction of the pesticides into solution, they interact with the active centers of AChE, causing its activity to decrease. For this, there is a decrease in thiocholine production and by this way quenching rate of the QDs also decreases. The authors show the potential usefulness of the sensor in the real world by applying it to real fruit and vegetable samples containing picomolar concentrations of paraoxon and/or parathion. The combination of citrate-capped silver nanoparticles and AChE was also used for the detection of organophosphorus

Table 17.3 Detection of Pesticides by Optical Method

S.N.	Type of Sensor	Nanoparticle Used	Template(s)	Linear Range	LOD	Detection Technique	Real Samples	References
1.	Fluorescence immunosensor	CdS@ZnS QDs	Chlorpyriphos	—	2.9 nM	Fluorescence competitive LFIA	Rat plasma samples	Zou et al. (2010)
2.	Fluorescence immunosensor	Carbon NPs	Methiocarb	0.02–0.56 ng mL^{-1}	0.016 ng mL^{-1}	Colorimetric competitive LFIA	Surface water	Blazková et al. (2009)
3.	Fluorescence immunosensor	AuNPs	Carbofuran/ Triazophos		0.14/0.012 μM	Colorimetric competitive LFIA	Spiked, tap, surface, and groundwater	Guo et al. (2009)
4.	Fluorescence immunosensor	CdTe QDs	Chlorpyrifos		11.0 nM	FIA	Drinking water samples	Chen et al. (2010b)
5.	Enzymatic acetylcholinase biosensor	CdS QDs	BW284c51		50.0 nM	Fluorescence	—	Saa et al. (2010)
6.	Enzymatic acetylcholinase biosensor	CdTe QDs	Paraoxon/ Parathion	10^{-12}–10^{-6} M	1.05×10^{-11} M and 4.47×10^{-12} M	Fluorescence quenching	Vegetables and fruit	Zheng et al. (2011)
7.	Colorimetric sensor	AuNP	Acetamiprid	6.6×10^{-7}–6.6×10^{-5} M	4.4×10^{-8} M	UV–Vis spectroscopy	Vegetables	Xu et al. (2011)
8.	Acetylcholinase based colorimetric/ fluorometric sensor	RB-AuNP	Carbaryl, Diazinon, Malathion, and Phorate		0.1, 0.1, 0.3, and 1.0 μg L^{-1}	UV–Vis and Fluorescence	—	Liu et al. (2012)
9.	Acetylcholinase based optical sensor	citrate-capped Ag NP	Dipterex	0.25–37.5 ng mL^{-1}	0.18 ng mL^{-1}	UV–Vis spectroscopy	Spiked water samples	Li et al. (2014a)
10.	Fluorescent chemosensors	CdS nanoparticles	Dicofol	—	55.0 ppb	Fluorescence	—	Walia and Acharya (2014)

(*Continued*)

Table 17.3 Detection of Pesticides by Optical Method (*cont.*)

S.N.	Type of Sensor	Nanoparticle Used	Template(s)	Linear Range	LOD	Detection Technique	Real Samples	References
11.	Fluorescent chemosensors	CdTe QD, AuNP	Methamidophos	0.06–0.78 mg kg^{-1}	2.0 μg kg^{-1}	Fluorescence	Chinese cabbage	Guo et al. (2014)
12.	Colorimetric sensor	AuNP	Imidacloprid	5×10^{-7}–1.0×10^{-3} M	2.5×10^{-5} M	UV–Vis spectroscopy	—	Tan et al. (2014)
13.	Fluorescent chemosensors	AuNPs and RF-QDs	Acetamiprid	0.025–5.0 mg/mL^{-1}	16.8 mg L^{-1}	Fluorescence	Water and rice	Yan et al. (2014)
14.	Colorimetric sensor	core–shell Ag@ SiO2	Pyrethroids	1.0×10^{-6} M–2.0×10^{-3} M	1.0×10^{-6} M	UV–Vis spectroscopy	—	Li et al. (2011)
15.	Colorimetric sensor	AuNP	Nereistoxins	50–250 μg kg^{-1}	40.0 μg kg^{-1}	UV–Vis spectroscopy	Tea, kiwi fruit, rice and cabbage	Liu et al. (2015a)
16.	Colorimetric sensor	AuNP	Paraquat	—	0.2 μM	UV–Vis spectroscopy	—	Li et al. (2013a)
17.	Colorimetric sesnor	Ag nanoprobe	Dimethoate	10×10^{-8} M–100×10^{-8} M	0.8×10^{-9} M	UV–Vis spectroscopy	Industrial waste water	Menon et al. (2013)
18.	Host–guest based optical sensor	Molecular imprinted SiNPs	Atrazine		1.8 μM	Fluorescence	Lake water	Liu et al. (2011)
19.	Host–guest based optical sensor	AuNPs	Parathion	10^{-12} M–10^{-6} M	1.0×10^{-12} M	SERS		Wang et al. (2010)
20.	Host–guest based optical sensor	AuNPs	Atrazine		5.0 pM	SPR	Acetonitrile	Matsui et al. (2009)
21.	Host–guest based optical sensor	CdTe QDs	Fenamithion/ Acetamiprid		1.2×10^{-8} M/ 3.4×10^{-8} M	Fluorescence	Garlic samples	Qu et al. (2009)

QD, Quantum dot; LFIA, lateral flow immunoassay; FIA, fluoro immunoassay; RF-QD, ratiometric fluorescence quantum dots; SERS, surface enhanced Raman spectrometry; SPR, surface plasma resonance.

pesticide, dipterex (Li et al., 2014a). After the formation of thiocholine (TCh), AgNPs get aggregated and correspondingly the color of AgNPs in solution changes from bright yellow to pink, and the UV–Vis characteristic absorption peak of AgNPs at about 400 nm decreases. A new absorption band also appears at about 520 nm. They found a detection limit of 0.18 ng mL^{-1}. A highly sensitive, dual-readout assay based on gold nanoparticles for the detection of organophosphorus and carbamate pesticides was reported (Liu et al., 2012). Rhodamine B-covered gold nanoparticle was used for the colorimetric and fluorometric detection of the pesticides and the detection mechanism was based on inhibition activity of AChE. When AChE is pretreated with AChE inhibitors like organophosphorus compounds or arbamates pesticides, they can inhibit activity of AChE and prevent the generation of thiocholine (Fig. 17.5). As a consequence, the color of the rhodamine B-covered gold nanoparticle solutions remains red and the fluorescence of RB molecules becomes quenched (Fig. 17.5).

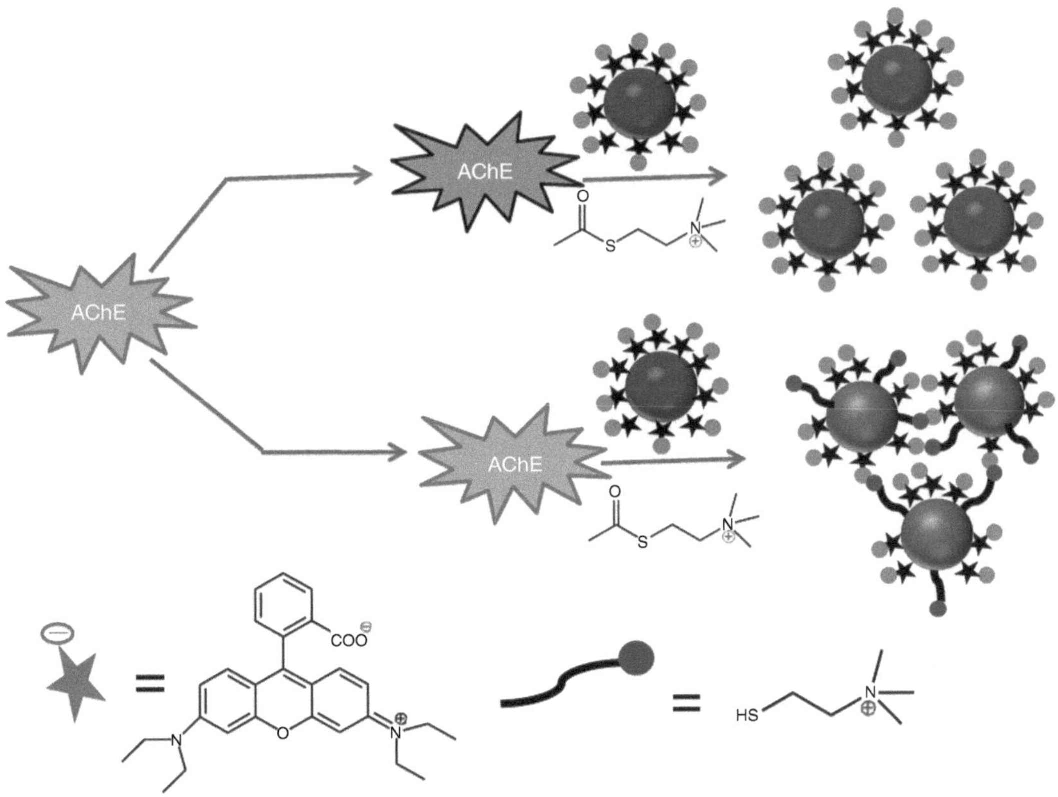

Figure 17.5. Design of the dual-readout (Colorimetric and Fluorometric) assay for pesticide detection. Modified with permission from Liu et al. (2012)

2.1.2.2 Optical Immunosensors

Taking the advantages of immunosensor technology, Zou et al. reported the use of a quantam dot–based lateral flow immunoassay (LFIA) for the detection biomarker of CPF trichloropyridinol (Zou et al., 2010). They obtained an LOD of 2.9 nM and applied it for the detection of the pesticide in rat plasma samples. A colorimetric LIFA method has been reported based on carbon nanoparticles by Blazkova et al. (2009). They used methiocarb as a template and obtained a detection limit of 0.02–0.56 ng mL^{-1}. A gold nanoparticle label has been used by Guo et al. for the detection of carbofuran and triazophos. The broad application of AuNPs in LFIA is related to the fact that these nanoparticles can easily be obtained and conjugated (Guo et al., 2009). The use of these nanoparticles also gives the advantage as final test result is observable by the naked eye. Chen et al. (2010b) labeled another CdTe QD-based immunoassay for the detection of pesticide CPF in drinking water. They treated CPF with 3-mercaptopropionic acid and bovine serum albumin (BSA) added to the resulting mixture. The obtained a CPF-BSA conjugate immobilized onto a microwell plate by simple adsorption method. A specific anti-CPF biotin was modified with biotin, mixed with streptavidin modified CdTe QDs. with the increase in CPF, the amount of QDs immobilized in the plate decreases, which converts into a falling fluorescence output signal. The system exhibits exceptional accuracy and low variability and applied in analysis of real drinking water samples.

2.1.2.3 MIP and Host–Guest Interaction-Based Optical Sensor

There is an increasing interest in the fabrication artificial antibody-based sensor discussed in the MIP-based electrochemical sensor section. Liu et al. (2011) has reported the use of fluorescent silica nanoparticle (SiNP) cores with combination of MIPs for the pesticide detection. The surface of SiNPs is functionalized with vinyl groups, which enables the polymerization of MIPs onto the nanoparticle surface. They have reported a lower LOD 1.8 μM and applied it for the determination of atrazine in lake water samples. Another MIP-based atrazine and parathion sensor was also reported (Wang et al., 2010; Matsui et al., 2009). Here gold nanoparticles were used to provide an enhancement of the local electromagnetic field by a coupling effect with the gold film. By this way, a very low detection limits (5.0 pM) were achieved using SPR technique and 1×10^{-12} M by SERS were achieved for atrazine and parathion respectively. In case of parathion detection they used mono-6-thio-β-cyclodextrin. In literature cyclodextrin is a good receptor for the host-guest interaction. Qu et al. (2009) has reported a simple, rapid, and sensitive detection system for acetamiprid and fenamithion based on *p*-sulfonatocalix [4] arene and CdTe QDs. They reported that the interaction between free QDs and fenamithion leads to a selective decrease in QD fluorescence. The sensor was further used for the pesticide detection in garlic samples.

2.1.2.4 Other Optical Sensor

Advances have been made to improve the sensitivity of optical pesticide sensors based on nanomaterials that do not require immunoassays, host-guest chemistry, or enzyme-inhibition strategies. Xu et al. (2011) has reported a method for the detection of acetamiprid based on the aggregation of gold nanoparticles. There is strong interaction between the cyano group of acetamiprid with gold nanoparticles resulting in the aggregation of gold nanoparticles and is accompanied by a color change from red to purple. The change in concentration of acetamiprid can be determined qualitatively and quantitatively by visually monitoring the color change or by using a spectrometer. They obtained a LOD of 4.4×10^{-8} M and applied the sensor for the detection in vegetable samples. Walia and Acharya (2014) reported the use of fluorescent cadmium sulfide nanoparticles for selective and sensitive detection of toxic pesticide dicofol. They reported that the existing chemosensor does not focus toward differentiating organophosphorus and organochlorine pesticides specifically. In this respect, they synthesized and characterized glutathione coated CdS nanoparticles. The LOD of dicofol is reported as 55 ± 11 ppb. Another fluorescent chemosensor for the detection of methamidophos was reported using inner filter effect of Au nanoparticles on the fluorescence of CdTe QD (Guo et al., 2014). They obtained an LOD of 2.0 μg kg^{-1} and applied the sensor for the successful detection in Chinese cabbage with a recovery range 95.7–112.1%. Another AuNPs-based imidacloprid colorimetric sensing device fabricated by Tan et al. (2014). They reported that gold nanoparticles had a cooperative effect on recognition of imidacloprid resulting in appreciable changes in color. The main advantage of their work over classical assays is that it can realize the prompt on-site and real-time detection of imidacloprid with high sensitivity (5.0×10^{-6} M) and selectively. They applied the proposed sensor for detecting imidacloprid in water samples. Yan et al. (2014) reported a fluorescent detection method for acetamiprid detection based on the inner filter effect of gold nanoparticles on ratiometric fluorescence QD (RF QD). The as-prepared RF QDs–based dual-emission nanosensor was fabricated by assembling green emissive QDs (λ_{em} = 539 nm) on the surface of red emissive QDs (λ_{em} = 661 nm)-doped silica microspheres. Acetamiprid can bind to the surface of AuNPs due to the presence of a cyano group that has good affinity with gold. This could make the aggregation of AuNPs and a color change from red to blue. For this, the inner filter effect of AuNPs on RF QDs becomes weakened and the PL intensity of RF QDs was recovered accordingly. They obtained a much lower LOD 16.8 mg L^{-1}. The real-time applicability of the sensor was shown by applying it for the determination of acetamiprid in water and rice samples with a recovery range 96–105%. Li et al. (2011) have reported an optical sensor in which sensing mechanism is based on the strong hydrogen-bond interactions between the amine groups of $Ag@SiO_2 - NH_2$ NPs and the trifluoromethylfluoryl group of the pyrethroid pesticide lambda-cyhalothrin (LC). The output signal comes from the color change after the addition of LC (from yellow to pink) of $Ag@SiO_2 - NH_2$ NPs. In case of other pyrethroids, the signal was considered negligible. Menon et al. (2013) has also reported an Ag nanoparticle-based colorimetric sensor and its detection in wastewater. They obtained an LOD of 0.8×10^{-9} M. Li et al. has used carboxylatopillar[5]arene-modified gold Nanoparticles for paraquat sensing. They obtained a detection limit 0.2 μM by UV–Vis spectroscopic method (Li et al., 2013a). A colorimetric and visual determination method of total nereistoxin-related insecticide has been reported by Liu et al. (2015a). Determination was based on the nereistoxin-induced aggregation of AuNPs. In this method, they convert nereistoxin related insecticides to nereistoxins in the pretreatment step that involves liquid–liquid extraction and hydrolysis for sample cleanup. After that, the interaction between nereistoxins and AuNPs provides a possibility to visually quantify nereistoxins due to a color change from red to blue. The interaction between nereistoxins and AuNPs is initiated by the electrostatic interactions and the strong Au–S covalent bonds. They report a detection range from 50–250.0 μg kg^{-1} and LOD 40.0 μg kg^{-1}. The as-prepared sensor was applied for the pesticide detection in tea, kiwi fruit, rice, and cabbage with an acceptable recovery range.

2.2 Contamination Due to Heavy Metal Ions

Heavy metal pollution is one of the most thoughtful environmental problems, which challenges global sustainability. Heavy metals enter the environment due to increasing industrial activities. Heavy metal ions such as Pb (II), Hg (II), As (V), and Cd (II) are highly toxic and carcinogenic even at a trace level (Duffus, 2002; Aragay et al., 2011). They are nonbiodegradable in nature and can accumulate in the food chain, which is a real threat to the environment and human health. It is therefore important to monitor the heavy metal concentrations in the environment, drinking water, food, and biological fluids. Conventional methods for heavy metal detection are based on atomic absorption spectroscopy (AAS), inductively coupled plasma/mass spectrometry, inductively coupled plasma/atomic emission spectrometry, ultraviolet–visible spectroscopy, and so on. The metal ion detection technique using nanomaterials for metal ion detection is summarized in Table 17.4.

Li et al. (2013b) has reported a review article on nanostructured sensors for detection of heavy metals. They reported that the conventional methods for the heavy metal detection are highly sensitive and selective but they require tedious sample preparation and preconcentration procedures, expensive instruments, and professional personnel. And this type of sensing device cannot be used as portable devices for on-site detection. A fast development of nanotechnology has provided new openings for improving the performance of sensors in terms of their sensitivity, LOD, selectivity, and reproducibility. In this chapter we only discuss the contribution made in the fabrication nanosensors for the detection of heavy metal ions in food samples during the past 6 years. Ping et al. (2014) has reported reduced graphene oxide modified disposable bismuth film electrode for the analysis of heavy metal ions in milk samples. Square wave anodic stripping voltammetry technique was used for the detection of cadmium and lead by modified screen-printed electrode. They obtained a good recovery of 84.9–90.7%, 84.6–91.6% for cadmium and lead, respectively. Similar graphene oxide (GO)-based cadmium (II) and lead (II) sensors have been reported by Yavuz et al. (2015), where GO was covalently attached to glassy carbon (GC) electrode (GC–O–GO) for fabricating nanosensors. Differential pulse anodic stripping voltammetry (DPASV) techniques were used for the detection of the metal ions in real samples like rice, soya, milk, and tap water samples. The sensor has a lowest LOD of 0.25 and 0.28 pM for Pb (II) and Cd (II), respectively. In another report by JianFeng et al. (2012), carbon composite electrode modified with bismuth oxide nanoparticles and the ionic liquid n-octylpyridinium hexafluorophosphate was fabricated and used for simultaneous determination cadmium and lead levels using square wave anodic stripping voltammetry. Under optimized conditions, they obtained a linear range of the composite electrode from 3.0 to 30.0 μg L^{-1} for both metal ions with the detection limit of 0.15 μg L^{-1} for cadmium (II) and 0.21 μg L^{-1} for lead (II). The sensor was applied in milk sample analyses and showed that the modified electrode was sensitive, reliable, and effective for the determination of trace heavy metals with a good recovery, holding great promise for routine analysis.

Table 17.4 Detection of Heavy Metal Ions in Food Samples

S. No.	Nanoparticles Used	Detection Technique	Metal Ions Detected	Linear Concentration Range	LOD	Real Samples Detected	References
1.	rGO	SWASV	Cd (II), Pb (II)	1.0–60.0 $\mu g\ L^{-1}$	0.5, 0.8 $\mu g\ L^{-1}$	Milk samples	Ping et al. (2014)
2.	GO	DPASV	Pb (II), Cd (II)	1.0×10^{-8}–1.0×10^{-12} M	0.25 and 0.28 pM	Rice, soya, milk, and tap water samples	Yavuz et al. (2015)
3.	Hydroxyapatite nanorods	FAAS	Cu (II), Zn (II), Pb (II)	Upto 250.0 $\mu g\ L^{-1}$ for Cu (II), 300.0 $\mu g\ L^{-1}$ for Zn (II) and 400.0 $\mu g\ L^{-1}$ for Pb (II)	0.72, 0.55, and 5.12 $\mu g\ L^{-1}$	Blood, urine, hair, tomatoes, fish muscle, chicken, spinach, rice, black tea	Mortada et al. (2015)
4.	Imprinted nanoparticles	FAAS	Zn (II)	0.5–120.0 $\mu g\ L^{-1}$	0.15 $\mu g\ L^{-1}$	Parsley, basin, red meat, milk, tomato	Behbahani et al. (2014a)
5.	Magnetic mesoporous silica nano-particles	FAAS	Pb (II)		1.3 $\mu g\ L^{-1}$	Bovine, buffalo, elk, sheep, goat muscle and liver	Aboufazeli et al. (2013)
6.	Bismuth oxide nanoparticles	SWASV	Cd (II), Pb (II)	3.0–30.0 $\mu g\ L^{-1}$	0.15 $\mu g\ L^{-1}$, 0.21 $\mu g\ L^{-1}$	Milk samples	JianFeng et al. (2012)
7.	Magnetic meso-porous silica	GFAAS	Cd (II)	10.0–200.0 $ng\ L^{-1}$	6.1 $ng\ L^{-1}$	Environmental water, human urine, and rice samples	Zhao et al. (2015)
8.	Magnetic nanoparticles	FAAS	Cd (II)	10.0–500.0 $ng\ mL^{-1}$	3.71 $ng\ mL^{-1}$	Green tea, rice, tobacco, carrot, lettuce, ginseng, spice, tap water, river water, sea water	Mirabi et al. (2015)
9.	MWCNT	FAAS	Cu (II), Pb (II), Fe (II), Ni (II), Zn (II)	1.0–2.6 $ng\ mL^{-1}$			Ghaedi et al. (2013)
10.	MWCNT	FAAS	Cu (II), Zn (II), Pb (II), Cd (II)		0.69, 0.48, 8.9, and 0.25 $ng\ mL^{-1}$		Soylak et al. (2011)
11.	MWCNT	AAS	Fe (III), Cu (II), Mn (II), Pb (II)	3.5–8.0 $\mu g\ L^{-1}$		Syzygium aromaticum, Zingiber officinale, rice flour	Ozcan et al. (2010)

12.	MWCNT	ICP-MS	V (V), Cr (VI), Cu (II), As (V), Pb (II)	0.005–10.0 μg L^{-1}	0.0021, 0.0038, 0.0035, 0.0013, 0.0036 μg L^{-1}	Herring, spinach, and water samples	Dai et al. (2012)
13.	MWCNT	GFAAS	V (V)	0.05–3.0 μg L^{-1}	0.012 μg L^{-1}	Black tea, coffee, tomato, cabbage, zucchini, apple, and chicken samples	Wadhwa et al. (2013)
14.	SBA-15 mesoporous silica	FAAS	Pb (II), Cu (II), Zn (II), Cd (II)	10.0–80.0, 1.0–40.0, 0.5–30.0, 1.0–20.0 ng mL^{-1}	4.5, 0.6, 0.2 and 0.5 ng mL^{-1}	Iranian black tea, Indian black tea, black pepper, hot pepper, tomato	Hajiaghababaei et al. (2013)
15.	Nanoporous carbon	FAAS	Cd (II), Cu (II)		0.04, 0.09 ng mL^{-1}	Wheat, tomato, lettuce, olive, apple, onion, celery, carrot, broccoli	Behbahani et al. (2015)
16.	Star-like polyaniline nanostructures	FAAS	Pb (II), Cu (II)	1.0–120.0 μg L^{-1} and 2.0–100.0 μg L^{-1}	0.4 and 0.9 μg L^{-1}	Water, Persian gulf shrimp, crab, various fish, potato, tomato, apple, mushroom	Behbahani et al. (2014b)
17.	SWCNT	FAAS	Cr (III), Fe (III)	0.25–5.0, 0.5–10.0 μg L^{-1}	2.12 and 4.08 μg L^{-1}	Herbal plants and food samples	Soylak and Unsal (2010)
18.	Graphene oxide/ nanogold	RRSET	Fluorine ion	6.0×10^{-8} and 1.3×10-5 mol L^{-1}	6.0×10^{-8} mol L^{-1}	Toothpaste	Liang et al. (2015)
19.	MWCNT	DPSV	Cd (II), Cu (II), Pb (II)	4.16–205.92 μg L^{-1}, 9.54–471.46 μg L^{-1}, and 6.69–204.16 μg L^{-1}	1.03, 2.12, 1.62 μg L^{-1}	Plant leaf (potato, tomato, mango) and their fruit/ vegetables with mushrooms, cashew nuts, sweet potato, hazelnuts, soybeans, and raisins	Roy et al. (2014b)

SWASV, Square wave anodic stripping voltammetry; DPASV, differential pulse anodic stripping voltammetry; FAAS, flame atomic absorption spectroscopy; GFAAS, graphite furnace atomic absorption spectroscopy; ICP-MS, inductively coupled plasma optical emission spectrometry; RRSET, resonance Rayleigh scattering-energy transfer.

AAS is the one of the popular techniques used for the detection of metal ions. The technique makes use of absorption spectrometry to determine the concentration of an analyte in a sample. A standard solution with known analyte content is required to establish the relation between the measured absorbance and the analyte concentration. This mainly relies on the Beer–Lambert Law. Two types of atomizers are most commonly used nowadays: flames and graphite tube atomizers. The flame atomizers-based AAS (FAAS) is the oldest and most commonly used. This technique is popularly applied in the detection of metal ions in recent years, also. Mortada et al. (2015) used this technique for the quantitative determination of copper (II), zinc (II), and lead (II) in biological and food samples after preconcentration with hydroxyapatite nanorods. These hydroxyapatite nanorods were prepared from recycled eggshells by using precipitation method. Their calibration curves were linear up to 250.0 μg L^{-1} for Cu (II), 300 μg L^{-1} for Zn (II), and 400 μg L^{-1} for Pb (II). The sensor is further used for the metal ion detection in blood, urine, hair, tomatoes, fish muscle, chicken, spinach, rice, and black tea. FAAS was used for the detection of single metal ion like zinc (II) by Behbahani et al. (2014a) using imprinted nanomaterials. They found a linear range 0.5–120 μg L^{-1} with a LOD 0.15 μg L^{-1} and applied this sensor to real sample detection in parsley, basin, red meat, milk, and tomato. Same FAAS sensors based on magnetic mesoporous silica nanoparticles was also reported for the detection of Pb (II) (Aboufazeli et al., 2013). To investigate the selectivity of this sensor toward Pb(II) ions detection, the effect of variety of ions on preconcentration and recovery of Pb(II) ions were also investigated by them and found a LOD of 1.3 μg L^{-1}. The real-time application was done by the determination in muscle and liver of bovine, buffalo, elk, sheep, and goat. MNPs-based FAAS sensor for Cd (II) detection was reported by Mirabi et al. Here, diphenylcarbazone/sodium dodecyl sulfate was immobilized on MNP Fe_3O_4 as a new sorbent SPE (Mirabi et al., 2015). MNPs have a relatively high adsorption as compared to the similar type of materials due to their large surface area. These Fe_3O_4 nanoparticles are highly monodispersed and magnetically separable from the sample solution. The detection limit of this method for cadmium was 3.71 ng mL^{-1} and the relative standard deviation (RSD) was 0.503%. The reported advantages of that method were its rapidity, easy preparation of sorbents, and a high concentration factor. Vegetable and water samples like green tea, rice, tobacco, carrot, lettuce, ginseng, spice, tap water, river water, and seawater was used for the detection of Cd (II) with a good recovery. MWCNTs have also been reported for the detection of metal ions by some earlier literature

(Ghaedi et al., 2013; Soylak et al., 2011; Ozcan et al., 2010). Ghaedi et al. (2013) reported multiwalled carbon nanotubes chemically functionalized with 2.3-silylpropylimino methylphenol and successfully applied for the SPE of some metal ions in food samples. In another report the same group reported an SPE based on the utilization of multiwalled carbon nanotubes in combination with [(E)-N1-(4-nitro-benzylidene)]-N2-[2-([E]-4-nitro-benzylideneamino)ethyl] ethane-1.2 diamine as a solid phase extractant. The detection limits of the four metal ions are 0.69, 0.48, 8.9, and 0.25 ng/mL^{-1}, respectively for Cu (II), Zn (II), Pb (II), and Cd (II), respectively (Soylak et al., 2011). Column SPEs of iron (III), copper (II), manganese (II), and lead (II) ions are reported from food and water samples on multiwalled carbon nanotubes. The validation of the fabricated sensor was monitored by applying it in the detection of metal ions from *Syzygium aromaticum*, *Zingiber officinale*, and rice flour (Ozcan et al., 2010). A SWCNTs disk was used for the determination of chromium and iron in food and herbal plant samples after SPE (Soylak and Unsal, 2010). They reported that analyte ions were quantitatively recovered at pH 8.0 on SWCNTs disk that contains 30 mg of nanotube. Nanoporous carbon has also been applied for trace detection of cadmium and copper ions in agricultural products (Behbahani et al., 2015). The nanoporous carbon modified with mercapto groups was used as a new SPE sorbent. Their experimental results demonstrate that simultaneous trace cadmium (II) and copper (II) ions can be quantitatively preconcentrated at pH 6.0 with recoveries >97% and the limits of detection obtained as 0.04 and 0.09 ng mL^{-1} for the ions of cadmium and copper, respectively. Finally, they applied the sorbent for successful trace determination of cadmium and copper ions in food samples (like wheat, tomato, lettuce, olive, apple, onion, celery, carrot, broccoli). A tetragonal star-like PANI nanostructures was used for efficient SPE and trace detection of Pb (II) and Cu (II) in agricultural products, seafoods, and water samples (Behbahani et al., 2014b). Due to their unique star like nanostructure of synthesized sorbent, the tendency of the sorbent toward selective extraction of lead and copper ion in the optimized pH is very interesting. They obtained a much lower LOD of 0.4 and 0.9 μg L^{-1} Pb (II) and Cu (II), respectively. The sensor was later applied for the detection of metal ions in water, Persian gulf shrimp, crab, various fish, potato, tomato, apple, and mushroom.

Ion imprinting technique is a useful method for the detection of metal ions due to their high selectivity, sensitivity, and specificity. An SBA-15 mesoporous silica functionalized with guanidin groups has been employed as solid phase extractant for preconcentration of Pb (II), Cu (II), Zn (II), and Cd (II) ions

(Hajiaghababaei et al., 2013). The detection limits of the sensor for the metal ion detection was found to be 4.5, 0.6, 0.2, and 0.5 ng mL^{-1} for Pb (II), Cu (II), Zn (II), and Cd (II), respectively. The sensor was applied for the detection in Iranian black tea, Indian black tea, black pepper, hot pepper, and tomato for the real-time application. An ion-imprinted magnetic mesoporous silica nanoparticles have been synthesized for selective magnetic SPE of trace Cd and its detection by graphite furnace AAS (GFAAS) (Zhao et al., 2015). They also reported that the direct determination of Cd (II) in real-world samples is difficult due to its low concentration and the complex matrix. Under the optimized conditions, the detection limit of the proposed method was calculated as 6.1 ng L^{-1}. This ion-imprinted sensor was further used for detection of Cd (II) in environmental water, human urine, and rice samples with recoveries of 89.3–116%. Another GFAAS sensor was reported by Wadhwa et al. (2013). They used MWCNTs as an SPE sorbent and successfully applied to the detection of vanadium (V) in microwave-digested food samples including black tea, coffee, tomato, cabbage, zucchini, apple, and chicken samples. Depending on mass spectroscopy, a Schiff base-chitosan grafted multiwalled carbon nanotubes was used as a novel SPE adsorbent for determination of heavy metals like V (V), Cr (VI), Cu (II), As (V), and Pb (II) (Dai et al., 2012). Here the adsorbent was synthesized by covalently grafting a Schiff base-chitosan (S-CHIT) onto the surfaces of oxidized MWCNTs. They applied the developed method to the determination of trace-metal ions in herring, spinach, river water, and tap water with good recoveries ranging from 91.0% to 105.0%. A fluorine ion sensor has also been reported based on graphene oxide/nanogold resonance Rayleigh scattering-energy transfer (RRSET) (Liang et al., 2015). After the addition of graphene oxide/nanogold (GO/NG) as RRS spectral probe with strong RRS peak at 370 nm, the color changed to gray. The RRS intensity also decreased with the increase of fluorine ion concentration due to the RRSET from GO/NG to the complex. This RRSET method-based sensor was successfully applied to the analysis of fluorine in toothpaste and water samples, with acceptable results. Our group also reported a MWCNT-based ion imprinted polymer for the simultaneous determination of heavy metals like Cd (II), Cu (II), and Pb (II) in biological samples (Roy et al., 2014b). In that report, we studied the contents of copper (II), cadmium (II), and lead (II) in various components of our biota (soil, water, and plants, that is, fruits and vegetables), collected from different areas like normal, industrial, coal-mine, river, and hill areas. A multitemplate-imprinted nanowire modified electrochemical sensor was fabricated for the accurate and rapid sensing of these metal ions. For the synthesis

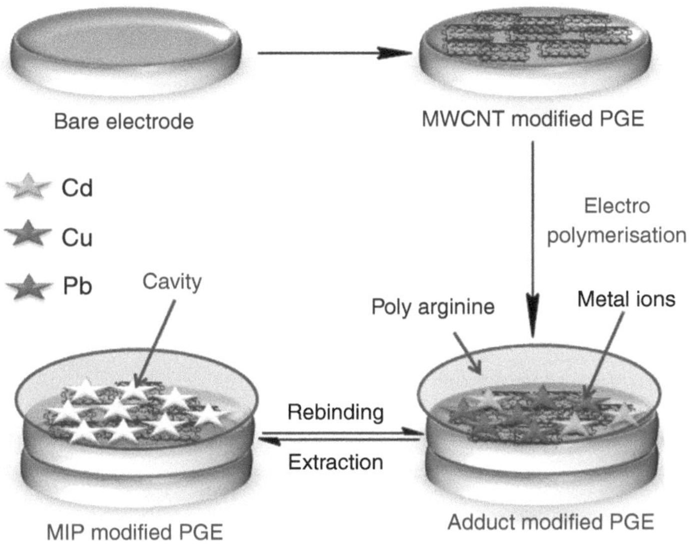

Figure 17.6. The schematic representation of the ion imprinted electrode fabrication. Modified with permission from Roy et al. (2014b)

of imprinted nanowire, MWCNTs were used as a core, on which a layer of conducting polyarginine was casted using an electropolymerization technique (Fig. 17.6). During the analysis of soil, water, and plant, it was observed that plants growing in a contaminated environment have a large accumulation of metal ions in their leaves and fruits. The paper reported that the imprinted nanowire is able to detect metal ions simultaneously or individually. The LODs were found to be 1.03, 2.12, 1.62 $\mu g\ L^{-1}$ for Cd (II), Cu (II), and Pb (II), respectively. The sensor was able to detect the metal ions in real samples like plant leaf (potato, tomato, and mango) and their fruits/vegetables with mushrooms, cashew nuts, sweet potato, hazelnuts, soybeans, and raisins.

2.3 Detection of Food-Pathogenic Bacteria

Pathogens are a class of species including bacteria, viruses, fungi, and protozoa. They are considered as a leading cause for loss of lives in the developing world and the rural areas of developed countries, due to lack of infrastructure and resources. These pathogens can be spread via plants, animals, and humans and by this manner infectious diseases can spread exponentially and lead to an illness. Early diagnosis method is the most effective method for preventing the spread of infectious diseases (Verma et al., 2015). But these methods are challenging because of

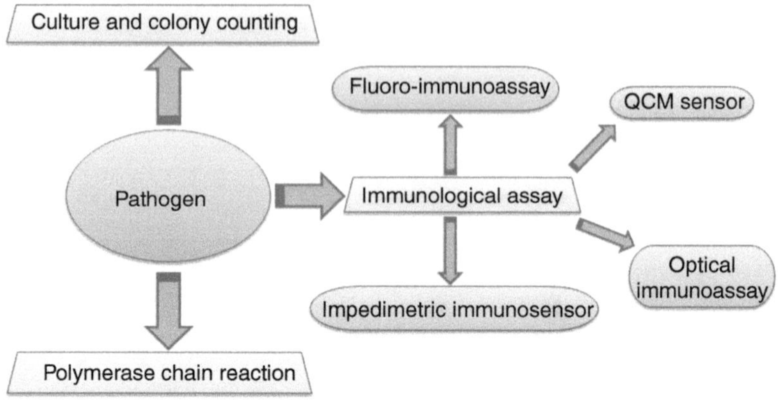

Figure 17.7. A list of techniques used so far for the detection of pathogens.

expensive equipment, specialized sample preparation, and slow data output. Modern biosensors have overcome these complications by miniaturizing devices and providing a simple rapid output that can be analyzed at the point-of care without specialized training.

Pathogenic bacteria detection technologies have been continuously improved. The conventional methods for detection of pathogens can unevenly be divided into three categories: culture and colony counting, immunological assays, and polymerase chain reaction (PCR)-based methods (Velusamy et al., 2010) (Fig. 17.7). These reported methods offer high sensitivity and specificity and also provide both quantitative and qualitative information. However, a major drawback is the need of high processing times. This clearly indicates a need for better solutions. The colony counting method involves the isolation and growth of suspicious pathogens, followed by their visual inspection (Verma et al., 2015). But very low levels of organisms can be identified due to the characteristic amplification during colony growth. Their slow turnaround times due to long incubation periods and the need of intensive labor make their use somewhat limited. In general, depending on the pathogen, initial results often takes at least 2 days, final results take some more time (Velusamy et al., 2010). Furthermore, in the colony counting method, a pathogen requires to be culturable. Another conventional method, immunological assays are very popular for pathogen detection owing to their adaptability for a wide variety of pathogens including bacteria and viruses. The ELISA method is an example of most popular immunological assay method. These assays trust on specific antibody recognition of antigens and other biomolecules to the target. The key advantage of immunological assays over colony counting is the reduced assay time although maintaining high specificity.

A fluoroimmunoassy method based on multicolor quantum dot probes has been reported for the detection of three pathogenic bacteria *Salmonella enteritidis, Staphylococcus aureus*, and *Escherichia coli* (Wang et al., 2015). In that report, the multicolor QDs having different emission wavelengths located at 504, 557, and 604 nm were conjugated with anti-*S. enteritidis*, anti-*S. aureus*, and anti-*E. coli* antibodies, respectively. The QD-antibody conjugates captures the target bacteria selectively and rapidly through immunological recognition. After this, the fluorescence intensities of the final complexes (measured at 504, 557, and 604 nm, respectively) have increased regularly with the increasing cell numbers of target bacteria. This method was further applied to real samples like soda, mineral water, apple juice, milk, tomato sauce, chicken sauce, pudding, cheese, and luncheon meat. ELISA is widely deployed in clinical laboratories as its capability of producing an optical response. Cho and Irudayaraj (2013) reported an in situ immuno-gold nanoparticle network ELISA biosensors for the detection of pathogens. The in situ network biosensor was able to detect pathogens at extremely low numbers: 3 cells mL^{-1} of *E. coli* O157:H7 and *Salmonella typhimurium* in buffer and 3 cfu mL^{-1} (colony forming unit mL^{-1}) of *E. coli* O157: H7 and 15 cfu mL^{-1} of *S. typhimurium* in real sample conditions within 2 h of inoculation. Immunomagnetic separation and rapid colorimetric detection technique is also reported (Sung et al., 2013). Here antibody/gold nanoparticle/magnetic nanoparticle nanocomposites were used for the detection of *S. aureus* in milk. They synthesized the nanocomposites by coating the MNPs with BSA then adsorbing the AuNPs and anti-*S. aureus* antibodies on their surface. Using the completed immunomagnetic nanostructures, *S. aureus* immunized in the milk sample and was captured and isolated from the medium using the permanent magnet. After this, captured cells were also applied to the selective filtration system to produce color signals that were used for the detection of the target pathogen. Based on colorimetric method, the sensor finds detection limits of this sensor were 1.5×10^3 and 1.5×10^5 cfu for *S. aureus* in PBS and the milk sample, respectively. Song et al. reported the fabrication of a lateral flow colloidal gold immunoassay strip for the simultaneous detection of *Shigella boydii* and *E. coli* O157:H7 in bread, milk, and jelly samples. For the rapid detections, double monoclonal antibodies (McAbs) were used and the sensitivity of the strip was reported as 10^6 cfu mL^{-1} for both *S. boydii* and *E. coli* O157:H7 without any cross-reaction in the presence of other related bacteria. They also reported that the preincubation of real samples (bread, milk, and jelly) can effectively improve the detection limit. In their study, when they preincubated the samples in broth for 10, 10, and 8 h for bread, milk, and jelly, respectively, the detection limit was significantly improved to 4 cfu mL^{-1} of the original bacterial content. Electrical impedimetric method is a rapid and label-free biosensor for pathogenic detection by monitoring the change in impedance caused by bacteria binding event. With this technique, Nguyen et al. (2014) has reported impedimetric immunosensor for the separation and label-free detection of *S. typhimurium*. They used magnetic silica nanotubes (MSNTs) bacterial separation and enhancement impedimetric signal. They functionalized the outer surface of MSNT positive charges for the successful concentration and the isolation of bacteria from a high-volume sample. These prepared MSNTs were very useful in antibody free bioseparation and in impedimetric signal enhancement. For real-time application, they used it for the successful determination in pure culture and the ground beef samples with a good detection limit. Another impedimetric immunosensor based on nanoporous membrane has been reported for label-free detection of pathogenic bacteria in whole milk (Joung et al., 2013). Their proposed immunosensor was successfully applied for the detection of *E. coli* O157:H7 in whole milk samples with the detection limit of 83.7 cfu mL^{-1} and 95% probability. Another electrochemical immunosensor for *E. coli* analysis was reported by Zhang et al. (2014). They fabricated a MWCNT/AuNPs hybrid nanostructures by electrodepositing AuNPs on MWCNTs surface via multipotential step technique and further used as a sensing electrode. The nanohybrid was used for the improvement of electrochemical performance for signal amplification, as well as the immobilization of *E. coli* capture antibody. They used this proposed strategy to evaluate *E. coli* in urban sludge. Dong et al. (2013b) reported another label-free electrochemical impedance immunosensor based on AuNPs/poly(amidoamine)-MWCNTs-chitosan nanocomposite modified GCEsfor the detection of *S. typhimurium* in milk. They used electrochemical impedance spectroscopy (EIS) and cyclic voltammetry (CV) technique to verify the stepwise assembly of the immunosensor. Due to the addition of MWCNTs, poly(amidoamine), and AuNPs, the sensitivity of the immunosensor was greatly enhanced.

Quartz crystal microbalance (QCM) is a reagent free technique, used for direct measurement of pathogens in food products. A QCM-based sensor was reported for sensitive and specific detection of *Salmonella enterica* (Ozalp et al., 2015). They combined aptamer-based magnetic separation system a QCM analysis for specific and real-time monitoring. Milk samples were used for the specific detection of *Salmonella* cells at 100 cfu mL^{-1}. Another QCM-based sensor based on gold nanoparticles was reported for the real-time and sensitive detection of *Salmonella*. The QCM bare gold sensor chip contain two sensing array and they modified them by covalently immobilizing the monoclonal capture antibody on the active spot and a mouse IgG antibody on the control spot by using a conventional amine coupling chemistry. They reported that after the binding of the *Salmonella* cells onto the immobilized anti-*Salmonella* antibody the sensor frequency gets altered. The sensor frequency mainly correlated to cells concentration in the buffer samples. They also reported QCM sensors with gold nanoparticles gave the highest sensitivity with a LOD ~10–20 cfu mL^{-1} compared to direct and sandwich assay. Platinum-coated MNPs clusters (Pt/MNCs) magnetophoretic chromatography was used to detect *E. coli* colorimetric in milk (Kwon et al., 2015. They functionalized half-fragments of monoclonal *E. coli* O157:H7 (EC) antibodies to Pt/MNCs and then were used to capture *E. coli* bacteria in milk. The sensor was good for real-time application with a lesser assay time. Najafi et al. (2014) reported a combined capture and detection method comprising of nanoimmunomagnetic separation (NIMS) and SERS for the detection of *E. coli*. They immobilized the capture antibodies on magnetite–gold (Fe_3O_4/Au) MNPs and then used for separation and concentration of the *E. coli* O157 cells from model liquid food matrix like apple juice. They reported a lowest detectable bacterial cell concentration in apple juice (10^2 cfu mL^{-1}) and a total analysis time of less than an hour. Many commercially available ELISA kits system are also available (like Assurance EIA, DuPont BAX, GeneQuence, Pathatrix from the company Biocontrol, DuPont, Neogen, Life Technologies, respectively). There are also some drawbacks associated with this technique, that is, requiring multiple steps, specialized training, and several hours of runtime (Ahmed et al., 2014; Lazcka et al., 2007).

Another type of sensor known as PCR sensor, where the amplification of target deoxyribonucleic acid (DNA) sequences offers PCR-based conventional methods a high degree of sensitivity that even capable of detecting single gene copies. In comparison to colony counting, the high sensitivity is achieved without an elongated incubation time. In this technique, bacteria do not need to be grown (Ahmed et al., 2014). The high specificity is achieved through the design of primers and probes, but there are certain limitations like the interference from nonpathogenic genetic material, which may lead to misleading results due to mismatch or nonspecific amplification (Ahmed et al., 2014; Velusamy et al., 2010). Possibly the main drawback of traditional PCR-based methods for pathogen detection is the failure to distinguish viable and nonviable cells because both contain the amplification target (Velusamy et al., 2010). To overcome this issue, reverse transcription PCR (RT-PCR)-based assays have been developed to target rapidly degrading messenger ribonucleic acid (mRNA) strands present during the cell's growth cycle (Velusamy et al., 2010; Yaron and Matthews, 2002).

3 Food Adulterant Detection

During past two decades, the world experienced some major food scandals. Among them, the Belgium dioxin scandal (1999), Irish pork crisis (2008), and the milk scandal of China (2008) gained the most media attention. Their international impacts come from the globalization and rapid distribution systems. It was reported at the end of May 1999, Belgium's major food scandal was due to the contamination of animal feed with about 40–50 kg of PCBs and almost 1.0 g of dioxins [polychlorinated dibenzo-p-dioxins (PCDDs) and polychlorinated dibenzofurans (PCDFs)]. The source of contamination was marked out of the stock of contaminated fat delivered to animal feed producers. Although they found the indication of contamination in Feb. 1999, the hypothesis of poisoning was only confirmed in Apr. 1999. Then the laboratory analyses revealed an exceptionally high level of dioxins in feed, meat, and eggs of the affected poultry. In the past, there were also several food contamination reported due to PCBs (Debacker et al., 2007; Covaci et al., 2008). Among them, the most notable ones are the Yusho and Yucheng rice oil poisonings from Japan 1968 (Tanabe et al., 1989) and Taiwan 1979 (Soong and Ling, 1997), respectively (Covaci et al., 2008). There are some other incidents which are less known to the public like the contamination of French cheese with technical oil from farm engines in 1976, the poultry poisonings due to PCB-contaminated plastic wire netting in France in the 1970s and the distribution of PCB-contaminated feed to pig farms in Montana, USA, 1979 (Lok and Powell, 2000). Other food contamination reported were poultry food chain contaminations in New York, USA (1971) where approximately 150,000 animals were victimized and Wilmington, USA (1972) where approximately 100,000 animals were victimized. Other reported contaminations include the Seveso incident in Italy, 1976 (Needham et al., 1999), the dumping of barrels with dioxin-contaminated residues in Lekkerkerk, The Netherlands (1980), and some other incidents that occurred between the 1996 and 2002 in Germany, Brazil, USA, and other countries, where feed ingredients (ball clay, lime

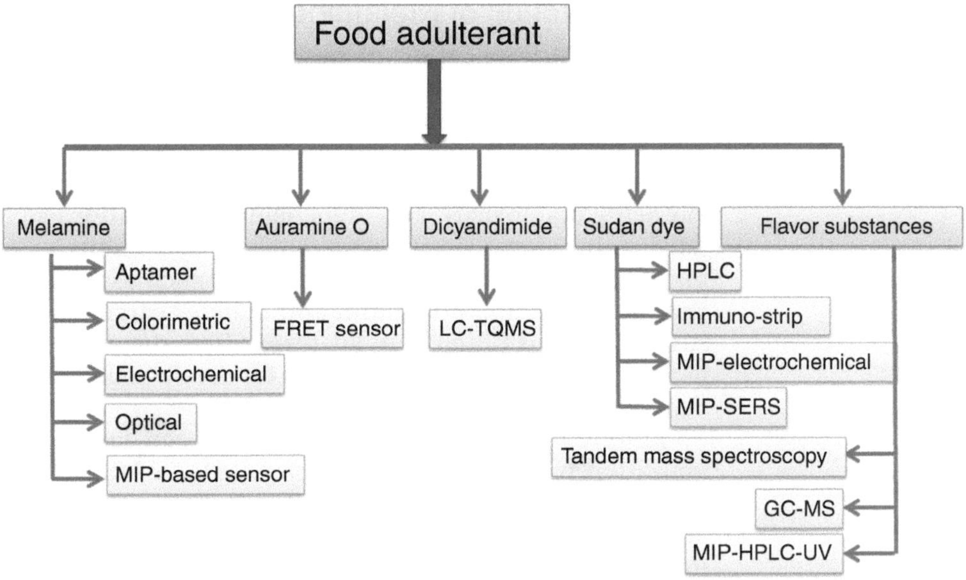

Figure 17.8. Various food adulterant and their detection techniques.

pulp, kaolinite, etc.) were contaminated with dioxins (Schecter et al., 2006). Another scandal, the Irish pork scandal of 2008, came to light in early Dec. 2008. At that time contaminated animal feed was supplied by one Irish manufacturer to 45 beef farms, nine pig farms, and one dairy farm across Ireland. This caused the contamination of pork with between 80 and 200 times higher than the European Union's recommended limit for dioxins and dioxin-like PCBs. For the analysis, pig fat samples were taken under the National Residues Programme for screening and they found a positive result for the presence nondioxin-like marker PCBs. They found 80–200 pg g^{-1} dioxin in pig fat where the EU legal limit was 1.0 pg g^{-1} (Banati, 2011). The H5N1 avian flu and the H1N1 swine flu in 2003 and 2009, respectively, dropped the use of poultry and pork meat. According to the WHO data (2009) there were 467 known human cases in 15 countries when 282 (60%) died due to avian flu. In swine flu, around 120 countries reported H1N1 cases. Among all, the Chinese milk scandal different because this was not an accident or unintentional contamination. The contaminated melamine (1.3.5-triazine-2.4.6-triamine, or $C_3H_6N_6$) contains a significant amount of nitrogen, which accounts for about 66% of its mass. The popular techniques used for sensing of food adulterant are shown in Fig. 17.8.

3.1 Detection of Melamine

The main application of melamine is in the production of melamine-formaldehyde polymer resins, which are a component in many plastics, adhesives, glues, fertilizer, and laminated products, such as plywood, cement, cleansers, and paint. The exposure to melamine is dangerous to human health and can cause low acute oral toxicity, and in its high concentrations can induce renal pathology and even death in babies and children. Besides, it may cause a variety of renal toxic effects such as nephrolithiasis, chronic kidney inflammation, and bladder carcinoma (Li et al., 2015). After this China scandal, researchers showed their interest in the formation of sensing devices for the detection of melamine in milk samples.

Various methods like CE, matrix-assisted laser desorption/ionization mass spectrophotometry (MALDI-MS), NMR, HPLC, micellar liquid chromatography (MLC), liquid chromatography-tandem mass spectrophotometry (LC–MS/MS), GC-mass spectrometry (GC–MS), low temperature plasma probe combined with tandem mass spectrophotometry (LTP-MS2), and chemiluminescence (CL) analysis has been reported so far for the detection of melamine (Rovina and Siddiquee, 2015). The CE method depends on the principle where a strong electric field is applied to separate molecules based on the difference in their hydrophobicity, size, and charge (Sun et al., 2010).

3.1.1 Aptamer-Based Sensor

Aptamer is known to be a synthetic single-stranded nucleic acid or peptide molecules, which can bind to a specific target with high affinity and specificity. Aptamers have more than a few advantages over antibodies as they are spontaneously synthesized, have appropriate storage properties, and could be modified without difficulty. Due to their inherent superiorities, they are widely used as recognition elements in biosensor construction for the detection of a wide range of substances. Qi et al. (2010) have reported a visual and light scattering spectrometric method for the detections of melamine with polythymine (polyTn)-stabilized gold nanoparticles through specific triple hydrogen-bonding recognition. Here single-stranded DNA oligomers including polyAn (polyadenine), polyCn (polycytosine), polyGn (polyguanine), and polyTn could make AuNPs stable without aggregation owing to enough negative charges on their surface (Qi et al., 2010). Due to the presence of melamine in the medium, the polyTn-stabilized AuNPs aggregate. This indicates that specific interactions occur between melamine and polyTn and the stability of a single base unit in polyTn decreases with the length increase of the polyTn. Later, Huang et al. (2011) reported a visual detection method for the detection of melamine in some milk samples based on label-free and labeled gold nanoparticles. In this procedure, the AuNPs are coated by the negative-charged citrate ions, which prevent the AuNPs from aggregation in aqueous solution. Whether in the presence of melamine, aptamers will bind to melamine with a stronger affinity and decrease the salt tolerance of AuNPs (Huang et al., 2011). This will result in the subsequent aggregation of AuNPs. This paper guarantees fast and reliable readout with naked eyes by making visual detection possible. After this Yun et al. (2014) also reported gold nanoparticles and aptamer-based visual biosensing platform for the detection of melamine. In that report, aptamer was used as a melamine recognition element and AuNPs as the color indicator (Yun et al., 2014). This provides a rapid and on-site detection of melamine by naked eyes or UV–Vis spectrometer. The main advantage lies in the analysis time and it is reported that the whole analytical process can be finished in 30 min without any assistance of any instrument. Under the optimized condition, the proposed method is very much useful to detect melamine in whole milk.

3.1.2 Colorimetric Sensor

Han and Li (2010) reported a visual detection method for melamine in infant formula based on silver nanoparticles. Here the stable *p*-nitroaniline-modified silver nanoparticles showed sensitivity and selectivity toward melamine (Han and Li, 2010). The presence of melamine induces the aggregation of nanoparticles due to electron donor–acceptor interaction between melamine and *p*-nitroaniline at the nanoparticles interface. This results in a shift of the surface plasmon band and a consequent color change of the AgNPs from yellow to blue. This color change was monitored using UV–Vis spectrophotometry. Kuang et al. reports 18-crown-6 ether functionalized AuNPs as an efficient platform for the detection of melamine by visually and spectroscopically. Under the optimal conditions, the paper reports a concentration range from 10 to 500 ppb with a LOD as 6.0 ppb for melamine (Kuang et al., 2011). Song et al. (2015) reports a colorimetric detection platform for melamine detection using sulfanilic acid functionalized silver nanoparticles. The formation of sulfanilic acid-AgNP conjugates dramatically improves the selectivity and only melamine reacts with that conjugate. The quantification of melamine is monitored by a simple UV–Vis spectrometer and obtained a detection limit of 10.6 nM. Xiao et al. (2015) reported a colorimetric sensing method for melamine detection based on

1,4-dithiothreitol functionalized gold nanoparticles. The hydrogen-bonding interaction between 1,4-dithiothreitoland melamine resulted in the aggregation of AuNPs, by which a color change of the AuNPs from wine red to blue and a red-shift of the SPR peak of the conjugates occur (Xiao et al., 2015).

3.1.3 Electrochemical Biosensor

Electrochemical sensors possess a large number of advantages over other techniques due to their fast response, low cost, simple operation, and high sensitivity. Zhu et al. (2011) report melamine detection based on electrochemical method for solid phase microextraction at bismuthyl chloride modified graphite epoxy composite electrode. The bismuthyl chloride layer functions as a solid phase microextractant and extracts melamine from an aqueous solution. The adsorption of melamine follows the Freundlich adsorption model and after the adsorption, a decrease in the reduction current of bismuthyl chloride observed. This process is used to determine melamine content with a LOD of 2.5×10^{-12} M and relative standard deviation of 2.7%. On that year Liao et al. also reported an electrochemical determination method for melamine in dairy product and pet food. A screen-printed carbon electrode was used for melamine detection by square wave voltammetry technique (Liao et al., 2011). Akter et al. reported an electrochemical biosensor by using ferricyanide as a redox indicator and an ITO electrode modified with gold nanoparticles (AuNPs) as base materials. A strong interaction between the $-NH_2$ group of melamine and the AuNPs was reported and after the attachment to the electrode, the cathodic peak current of ferricyanide slowly decreases, which was directly related to the concentration of melamine that had built up at the electrode (Akter et al., 2013).

3.1.4 Optical Biosensor

Over the past few decades' optical biosensors have been widely used in food safety. Fodey et al. reported an optical biosensor based on immunoassay for the detection of melamine. They reported that the antibody did not cross-react with any of the byproducts but some significant interference was observed with the insecticide cyromazine of which melamine is a metabolite. They found an LOD of <0.5 μg mL^{-1} in both infant formula and infant liquid milk (Fodey et al., 2011).

3.1.5 MIP-Based Sensor

Due to high selectivity and its numerous advantages MIPs are also applied for the detection of melamine over the years. Curcio et al. (2010) reported a molecularly imprinted SPE method to selectively extract melamine from two real samples, a food supplement and a freeze-dried meat sample. Ivanova-Mitseva et al. (2012) reported a cubic MIP nanoparticle with a fluorescent core for the detection of melamine and reported its use as a direct replacement for antibodies in biosensor applications. Later, an MIP-capped CdTe QD was reported by Huy et al. for the optosensing of melamine. Excellent selectivity of this CdTe QD based MIP was observed based on fluorescence quenching (Huy et al., 2014). A novel biosensor combining MIP and SERS for the determination of melamine in milk was developed (Hu et al., 2015). Here silver dendritic nanostructures functioned as SERS-active substrate for signal collection. They validated the use of MIPs to efficiently separate and enrich melamine from whole milk by means of static and kinetic adsorption tests. Simultaneous determination of melamine and dicyandiamide in dairy products by double imprinting technology was reported by Liu et al. They coupled molecularly imprinted SPE (MISPE) with HPLC for their detection (Liu et al., 2015b). The static and dynamic binding tests were also performed to prove better binding efficiency. The detection limit of the method 0.13 $\mu g\ g^{-1}$ for melamine and 0.07 $\mu g\ g^{-1}$ for dicyandiamide were obtained. Shrivastav et al. (2015) reported an MIP-based fiber optic SPR (MIP-SPR) sensor for melamine detection. The advantage of the sensor is its fast response, high selectivity and sensitivity, low cost. Another report for the melamine detection is reported by Zhu et al. (2015a), who reported the determination of melamine in powdered milk by molecularly imprinted stir bar sorptive extraction coupled with HPLC. As compared to non-imprinted stir bar, the sorptive capacity of molecularly imprinted stir bar was nearly 3 times higher. They obtained a much lower LOD, that is, 0.0127 $ng\ mL^{-1}$.

3.2 Detection of Auramine O

Other food adulterant detection methods have also been reported in the literature. Among them, auramine O (AO) is a common adulterant used to improve the appearance of medical herbs. This banned hazardous dye may also found in some herbs formulas, like medicinal electuaries and dietary supplements. AO possess a highly carcinogenic property and on accumulation, it will be a real threat for human body (Martelli et al., 1998). Therefore, the development of selective sensing device for AO administration is always in a demand. Zhu et al. (2015b) reported the preparation of fluorescent carbon dot for the detection of AO. Fluorescence resonance energy transfer process was used for the detection of

AO and the linear concentrations range of was found to be 0.18–16.0 μM with a LOD of 54.0 nM.

3.3 Detection of Dicyandimide

Another adulterant dicyandiamide is used as nitrification inhibitor for increasing the nitrogen utilization efficiency of grazed meadow. The dicyandiamide residues are considered as a contaminant. A core-shell nanoring aminofunctionalized magnetic polymer was synthesized and used for the detection of dicyandiamide by ultrafast liquid chromatography–tandem quadrupole mass spectrometry method combined with dispersive micro-SPE (Chen et al., 2014). The dicyandiamide had a good linearity in the range of 0.1–20.0 μg L^{-1}.

3.4 Detection of Sudan I

Sudan I is an industrial dye used to color oils, waxes, and polishes. But sometimes it is also added to foodstuffs and cosmetics for the color enhancement. Sudan I is classified as a carcinogenic and mutagenic compound by the International Agency for Research on Cancer. In the year of 2003, a French lab detected for the first time Sudan I in chilli products imported from India. After this, in 2005, the United Kingdom recalled more than 470 food products contaminated with Sudan I (Cheung et al., 2010). A large Sudan I adulteration incident spread worldwide in the same year. This created a tremendous panic and concern over food safety (Cheung et al., 2010). This led to the need for the development of a rapid Sudan I sensing system. HPLC–mass spectrometry (HPLC–MS) has been widely used for direct determination of Sudan dyes. But the application of this method becomes somewhat limited as it is a time-consuming and expensive technique. Further, some other techniques have also been applied for the detection of Sudan dyes. Wang et al. reported nanocolloidal gold-based immuno-dip strip assay for rapid detection this dye in chilli powder and tomato sauce (Wang et al., 2013). In this process, a semiquantitative dip strip assay was prepared by using nanocolloidal gold-labeled monoclonal antibody (8A10) for the rapid detection of Sudan red I in food samples. The protein–Sudan red I conjugate was coated onto a nitrocellulose membrane strip in a defined test line. The time required to perform this test was about 10 min and the method had a visual LOD of 10.0 ng g^{-1} in tomato sauce and chilli powder samples (Wang et al., 2013). Chen et al. (2011) reported a molecularly imprinted sol–gel polymer embedded with gold nanoparticles for the specific and selective detection of

Sudan I. Linear sweep voltammetric technique was used for the detection of Sudan I in the range of 0.1×10^{-7} to 1.0×10^{-5} M, with a detection limit of 2.0×10^{-9} M. Later, Li et al. (2014b) reported the extraction of Sudan dyes including Sudan I, II, III, and IV by magnetic titanium dioxide nanoparticles (Fe_3O_4@TiO_2) coated with sodium dodecylsulfate (SDS) as adsorbent. In the same year Xu et al. (2014) reported the preparation of MIP-nanofilm and used for the selective detection of Sudan dyes. The surface-initiated in situ polymerization technique was used for the preparation of MIP-SPR sensor chip by anchoring the MIP nanofilm of Sudan I on a gold chip. Gao et al. (2015) reported the determination of Sudan I in paprika powder by MIPs–thin layer chromatography–surface enhanced Raman spectroscopic biosensor. This is a reliable technique for the detection of Sudan dye adulterant in food samples.

3.5 Flavoring Substances Detection

The addition of herbs and plants as flavoring agents in food has a long tradition. But these natural products used for flavoring sometimes contain carcinogenic compounds. Estragole, safrole, methyl eugenol, *trans*-anethole, myristicin, and elemicin are used as bases for flavorings and they are present in many herbs. Coumarin is also used as a flavoring substance present at relatively high concentrations in a wide variety of plants in some essential oils, specifically cassia leaf, cinnamon leaf, bark, lavender, and peppermint oils. Pulegone and menthofuran can be detected in many mint products, such as candies and chewing gum. Furocoumarins, another natural plant constituents with phototoxic and photomutagenic properties are originated mainly in vegetables and fruits such as celery, carrots, parsnips, and citrus. The International Agency for Research in Cancer (IARC) classified some of them as carcinogenic to humans in class 1A. Bousova et al. (2011) reported a method for the detection of seven flavoring substances based on headspace solid-phase microextraction and tandem mass spectrometry. Lopez et al. (2015) reported straightforward methods to determine 14 flavoring substances by GC-MS. They applied their method for the detection of those flavoring substances in real samples like bakery products, chewing gum, breakfast cereals, nonalcoholic and alcoholic beverages, soup, sauces, meat, fish, dairy products, fruits, vegetables, nuts, and seeds (Lopez et al., 2015). One publication reported the preparation of molecularly imprinted stir bar based on magnetic field-induced self-assembly of multifunctional Fe_3O_4@PANI nanoparticles for the detection of vanillin (Wu et al., 2015). Vanillin is the popular flavoring agent used in wide range of commercial products such as beverages, foods, pharmaceutical products, and cosmetics. But the natural vanillin obtained from vanilla pods or bean can only fulfil 1% of the market demands. Therefore the available vanillin in the market is mostly obtained through chemical synthesis. It is reported vanillin has adverse effects on liver and kidney functions in babies. The results were measured by HPLC-UV method.

4 Food Packaging Materials

Among the intentional or unintentional addition in the food, packaging systems and other food-contact materials are also a source of contamination in food products and beverages. For this, in the past few decades a remarkable growth in the development of food packaging reported. A safety packaging makes food more convenient and protects from the attack of microorganisms, biological and chemical changes so the packaged foods can enjoy a longer shelf life. The synthetic polymers used for the packaging of food have high molecular weights and therefore a negligible biological availability. However, due to the use of lower molecular weight additives in these polymers, there is a possibility of human exposure to these lower molecular weight components (Pocas and Hogg, 2007). The main substances that may migrate from plastic materials to the food include monomers and starting substances, catalysts, solvents, and additives. Paper or board are used for carrying and storing the dry foods. These papers are composed of pulp from different vegetable sources and some additives like fillers, starch, and derivatives, wet strength sizing agents, retention aids, biocides, fluorescent whitening agents, and grease-proofing agents are used. Some polymers like polyethylene and waxes are used as a coating material in the paper. Tin metal that are used for beverages are internally coated with a polymeric layer, and thus the layer of food contact is not the metal but the lacquer. It mainly contains phenolic resins, often contains only small amounts of monomers, oligomers, and additives, and a large amount of other unknown or nondescribed components. Metal lids that have been used in glass jars are also a source of potential contaminants. The semicarbazide in baby food jars, resulting from degradation of azodicarbonamide and epoxidized soybean oil (ESBO), a plasticizer used in the plastisol gasket, were also considered as food contaminants (Pocas and Hogg, 2007).

But unlike most food additives, the exposure to chemicals migrating from packaging materials to the food is typically very low and always in the range from less than 10 ppb to 60 ppm in the food. The migrant species may not be the substance used in the production or conversion of the material, but sometimes an unknown reaction product. Some nonintentionally added substances like degradation products from additives or monomers and impurities, and substances originated in printing inks, adhesives, and solvents may also migrate into the food under certain conditions (Arvanitoyannis and Bosnea, 2004; Skjevrak et al., 2005). In this case the main issue is the uncertainty in the identity and/or in the biological properties of the migrating substances. Nonidentified and nondetected substances signify an uncharacterized contribution to the exposure of contaminants from food-contact materials.

5 Conclusions and Future Outlook

The detection of various intentional and unintentional food contaminants in food products based on nanomaterials has been presented in this chapter. The chapter comprises the efforts made by researchers in the past few years for the detection of pesticides, heavy metal ions, pathogen, various food adulterants, and some discussions about the contamination due to food packaging materials. Much development has been made so far for the preparation of nanosensor for their detection. But still there is a need for more research for their practical application. The growing industrial and agricultural waste is a major concern. The water and food largely are contaminated by this manner and enter into our food chain. Even today large food scandals are reported like popular noodles "Maggi" have come under a regulatory scanner in India. Samples collected from various parts of the country were found containing added monosodium glutamate (MSG) and lead in excess of the permissible limit and its sale is banned in many states. According to WHO, the contamination of lead account for 1,43,000 deaths per year with the highest burden in developing regions. Besides these large scandals, we have experienced some common scandals in our day-to-day life like the use of chalk powder; urea in sugar; water, formalin, detergent in milk; argemone seed in mustard seed; volatile oils in cloves; papaya seed in black pepper; and above all some artificial colors in all types of colored food. In the vegetable shop, some hormones are injected in their large amounts and these hormones are also harmful to our body. Besides many steps taken for the safest food, the food federation has more concerns about these scandals and has to improve technology to detect these adulterants or contaminants in a short time and accurately.

Acknowledgments

Authors are thankful to Department of Science and Technology, Government of India for sanction of the Fast Track Research Project for Young Scientists to Dr Rashmi Madhuri (Ref. No.: SB/FT/CS-155/2012) and Dr Prashant K. Sharma (Ref. No.: SR/FTP/PS-157/2011). Dr Sharma (FRS/34/2012-2013/APH) and Dr Madhuri (FRS/43/2013-2014/AC) are also thankful to Indian School of Mines, Dhanbad for grant of Major Research Project under the Faculty Research Scheme. We are also thankful to Board of Research in Nuclear Sciences (BRNS), Department of Atomic Energy, Government of India for major research project (34/14/21/2014-BRNS).

References

Aboufazeli, F., Zhad, H.R.L.Z., Sadeghi, O., Karimi, M., Najafi, E., 2013. Novel ion imprinted polymer magnetic mesoporous silica nano-particles for selective separation and determination of lead ions in food samples. Food Chem. 141, 3459–3465.

Ahmed, A., Rushworth, J.V., Hirst, N.A., Millner, P.A., 2014. Biosensors for whole-cell bacterial detection. Clin. Microbiol. Rev. 27, 631–646.

Akter, H., Shaikh, A.A., Chowdhury, T.R., Rahman, M.S., Bakshi, P.K., Saleh Ahammad, A.J., 2013. Gold nanoparticle-modified indium tin oxide electrode for highly sensitive electrochemical detection of melamine. ECS Electrochem. Lett. 2, 13–15.

Aragay, G., Pons, J., Merkoci, A., 2011. Recent trends in macro-, micro-, and nanomaterial-based tools and strategies for heavy-metal detection. Chem. Rev. 111, 3433–3458.

Aragay, G., Pino, F., Merkoci, A., 2012. Nanomaterials for sensing and destroying pesticides. Chem. Rev. 112, 5317–5338.

Arduini, F., Forchielli, M., Neagu, A.A.D., Cacciotti, I., Nanni, F., Moscone, D., Palleschi, G., 2015. Screen-printed biosensor modified with carbon black nanoparticles for the determination of paraoxon based on the inhibition of butyrylcholinesterase. Microchim. Acta 182, 643–651.

Arvanitoyannis, I.S., Bosnea, L., 2004. Migration of substances from food packaging materials to foods. Crit. Rev. Food Sci. Nutr. 44, 63–76.

Baggiani, C., Anfossi, L., Giovannoli, C., 2007. Solid phase extraction of food contaminants using molecular imprinted polymers. Anal. Chim. Acta 591, 29–39.

Bai, X., Huang, X., Zhang, X., Hua, Z., Wang, C., Qin, Q., Zhang, Q., 2014. TiO_2–graphene nanoparticle based electrochemical sensor for the bimodal-response detection of 4-chlorophenol. RSC Adv. 4, 13461–13468.

Banati, D., 2011. Consumer response to food scandals and scares. Trends Food Sci. Technol. 22, 56–60.

Behbahani, M., Salarian, M., Bagheri, A., Tabani, H., Omidi, F., Fakhari, A., 2014a. Synthesis, characterization and analytical application of Zn(II)-imprinted polymer as an efficient solid-phase extraction technique for trace determination of zinc ions in food samples. J. Food Compos. Anal. 34, 81–89.

Behbahani, M., Bide, Y., Salarian, M., Niknezhad, M., Bagheri, S., Bagheri, A., Nabid, M.R., 2014b. The use of tetragonal star-like polyaniline nanostructures for efficient solid phase extraction and trace detection of Pb(II) and Cu(II) in agricultural products, sea foods, and water samples. Food Chem. 158, 14–19.

Behbahani, M., Abolhasani, J., Amini, M.M., Sadeghi, O., Omidi, F., Bagheri, A., Salarian, M., 2015. Application of mercapto ordered carbohydrate-derived

porous carbons for trace detection of cadmium and copper ions in agricultural products. Food Chem. 173, 1207–1212.
Blazková, M., Micková-Holubová, B., Rauch, P., Fukal, L., 2009. Immunochromatographic colloidal carbon-based assay for detection of methiocarb in surface water. Biosens. Bioelectron. 25, 753–758.
Bousova, K., Mittendorf, K., Senyuva, H., 2011. A solid-phase microextraction GC/MS/MS method for rapid quantitative analysis of food and beverages for the presence of legally restricted biologically active flavorings. J. AOAC Int. 94, 1189–1199.
Brownson, D.A.C., Banks, C.E., 2010. Graphene electrochemistry: an overview of potential applications. Analyst 135, 2768–2778.
Cai, J., Du, D., 2008. A disposable sensor based on immobilization of acetylcholinesterase to multiwall carbon nanotube modified screen-printed electrode for determination of carbaryl. J. Appl. Electrochem. 38, 1217–1222.
Cesarino, I., Moraes, F.C., Lanza, M.R.V., Machado, S.A.S., 2012. Electrochemical detection of carbamate pesticides in fruit and vegetables with a biosensor based on acetylcholinesterase immobilised on a composite of polyaniline–carbon nanotubes. Food Chem. 135, 873–879.
Chauhan, N., Pundir, C.S., 2012. An amperometric acetylcholinesterase sensor based on Fe3O4 nanoparticle/multi-walled carbon nanotube-modified ITO coated glass plate for the detection of pesticides. Electrochim. Acta 67, 79–86.
Chauhan, N., Narang, J., Pundir, C.S., 2011. Immobilization of rat brain acetylcholinesterase on porous gold-nanoparticle – $CaCO_3$ hybrid material modified Au electrode for detection of organophosphorous insecticides. Int. J. Biol. Macromol. 49, 923–929.
Chen, L., Zeng, G., Zhang, Y., Tang, L., Huang, D., Liu, C., Pang, Y., Luo, J., 2010a. Trace detection of picloram using an electrochemical immunosensor based on three-dimensional gold nanoclusters. Anal. Biochem. 407, 172–179.
Chen, Y., Ren, H.l., Liu, N., Sai, N., Liu, X., Liu, Z., Gao, Z., Ning, B.a., 2010b. A fluoroimmunoassay based on quantum dot – streptavidin conjugate for the detection of chlorpyrifos. J. Agric. Food Chem. 58, 8895–8903.
Chen, S., Du, D., Huang, J., Zhang, A., Tu, H., Zhang, A., 2011. Rational design and application of molecularly imprinted sol–gel polymer for the electrochemically selective and sensitive determination of Sudan I. Talanta 84, 451–456.
Chen, M., Meng, Y., Zhang, W., Zhou, J., Xie, J., Diao, G., 2013. β-Cyclodextrin polymer functionalized reduced-graphene oxide: application for electrochemical determination imidacloprid. Electrochim. Acta 108, 1–9.
Chen, X.-H., Zhou, L.-X., Zhao, Y.-G., Pan, S.-D., Jin, M.-C., 2014. Application of nanoring amino-functionalized magnetic polymer dispersive micro-solid-phase extraction and ultra fast liquid chromatography–tandem mass spectrometry in dicyandiamide residue analysis of powdered milk. Talanta 119, 187–192.
Cheung, W., Shadi, I.T., Xu, Y., Goodacre, R., 2010. Quantitative analysis of the banned food dye Sudan-1 using surface enhanced Raman scattering with multivariate chemometrics. J. Phys. Chem. C 114, 7285–7290.
Cho, I.H., Irudayaraj, J., 2013. In situ immuno-gold nanoparticle network ELISA biosensors for pathogen detection. Int. J. Food Microbiol. 164, 70–75.
Covaci, A., Voorspoels, S., Schepens, P., Jorens, P., Blust, R., Neels, H., 2008. The Belgian PCB/dioxin crisis-8 years later: an overview. Environ. Toxicol. Pharmacol. 25, 164–170.
Curcio, M., Puoci, F., Cirillo, G., Iemma, F., Spizzirri, U.G., Picci, N., 2010. Selective determination of melamine in aqueous medium by molecularly imprinted solid phase extraction. J. Agric. Food Chem. 58, 11883–11887.

Dai, B., Cao, M., Fang, G., Liu, B., Dong, X., Pan, M., Wang, S., 2012. Schiff base-chitosan grafted multiwalled carbon nanotubes as a novel solid-phase extraction adsorbent for determination of heavy metal by ICP-MS. J. Hazard. Mater. 219–220, 103–110.

Debacker, N., Sasse, A., Wouwe, N.V., Goeyens, L., Sartor, F., Oyen, H.V., 2007. PCDD/F levels in plasma of a belgian population before and after the 1999 belgian PCB/DIOXIN incident. Chemosphere 67, S217–S223.

Ding, J., Zhang, H., Jia, F., Qin, W., Du, D., 2014. Assembly of carbon nanotubes on a nanoporous gold electrode for acetylcholinesterase biosensor design. Sens. Actuat. B 199, 284–290.

Dong, J., Wang, X., Qiao, F., Liu, P., Ai, S., 2013a. Highly sensitive electrochemical stripping analysis of methyl parathion at MWCNTs-CeO_2-Au nanocomposite modified electrode. Sens. Actuat. B 186, 774–780.

Dong, J., Zhao, H., Xu, M., Ma, Q., Ai, S., 2013b. A label-free electrochemical impedance immunosensor based on AuNPs/PAMAM-MWCNT-Chi nanocomposite modified glassy carbon electrode for detection of *Salmonella typhimurium* in milk. Food Chem. 141, 1980–1986.

Du, D., Ye, X., Cai, J., Liu, J., Zhang, A., 2010a. Acetylcholinesterase biosensor design based on carbon nanotube-encapsulated polypyrrole and polyaniline copolymer for amperometric detection of organophosphates. Biosens. Bioelectron. 25, 2503–2508.

Du, D., Wang, M., Cai, J., Zhang, A., 2010b. Sensitive acetylcholinesterase biosensor based on assembly of β-cyclodextrins onto multiwall carbon nanotubes for detection of organophosphates pesticide. Sens. Actuat. B 146, 337–341.

Duffus, J.H., 2002. Heavy metals: a meaningless term? (IUPAC technical report). Pure Appl. Chem. 74, 793–807.

Fodey, T.L., Thompson, C.S., Traynor, I.M., Haughey, S.A., Kennedy, D.G., Crooks, S.R.H., 2011. Development of an optical biosensor based immunoassay to screen infant formula milk samples for adulteration with melamine. Anal. Chem. 83, 5012–5016.

Fu, X.-C., Zhang, J., Tao, Y.-Y., Wu, J., Xie, C.-G., Kong, L.-T., 2015. Three-dimensional mono-6-thio-β-cyclodextrin covalently functionalized gold nanoparticle/single-wall carbon nanotube hybrids for highly sensitive and selective electrochemical determination of methyl parathion. Electrochim. Acta 153, 12–18.

Gao, F., Hu, Y., Chen, D., Li-Chan, E.C.Y., Grant, E., Lu, X., 2015. Determination of Sudan I in paprika powder by molecularly imprinted polymers–thin layer chromatography–surface enhanced Raman spectroscopic biosensor. Talanta 143, 344–352.

Ghaedi, M., Montazerozohori, M., Biyareh, M.N., Mortazavi, K., Soylak, M., 2013. Chemically bonded multiwalled carbon nanotubes as efficient material for solid phase extraction of some metal ions in food samples. Int. J. Environ. Anal. Chem. 93, 528–542.

Gong, J., Wang, L., Zhang, L., 2009. Electrochemical biosensing of methyl parathion pesticide based on acetylcholinesterase immobilized onto Au-polypyrrole interlaced network-like nanocomposite. Biosens. Bioelectron. 24, 2285–2288.

Govindhan, M., Adhikari, B.-R., Chen, A., 2014. Nanomaterials-based electrochemical detection of chemical contaminants. RSC Adv. 4, 63741–63760.

Guo, Y.-R., Liu, S.-Y., Gui, W.-J., Zhu, G.-N., 2009. Gold immunochromatographic assay for simultaneous detection of carbofuran and triazophos in water samples. Anal. Biochem. 389, 32–39.

Guo, J., Li, H., Xue, M., Zhang, M., Cao, X., Luo, Y., Shen, F., Sun, C., 2014. Highly sensitive detection of organophosphorus pesticides represented by methamidophos via inner filter effect of Au nanoparticles on the fluorescence of CdTe quantum dots. Food Anal. Methods 7, 1247–1255.
Hajiaghababaei, L., Tajmiri, T., Badiei, A., Ganjali, M.R., Khaniani, Y., Ziarani, G.M., 2013. Heavy metals determination in water and food samples after preconcentration by a new nanoporous adsorbent. Food Chem. 141, 1916–1922.
Hamburg, M.A., 2011. Advancing regulatory science. Science 331, 987.
Han, C.P., Li, H.B., 2010. Visual detection of melamine in infant formula at 0.1 ppm level based on silver nanoparticles. Analyst 135, 583–588.
Holderbeke, M.V., Geerts, L., Vanermen, G., Servaes, K., Sioen, I., De Henauw, S., Fierens, T., 2014. Determination of contamination pathways of phthalates in food products sold on the Belgian market. Environ. Res. 134, 345–352.
Hu, Y., Feng, S., Gao, F., Li-Chan, E.C.Y., Grant, E., Lu, X., 2015. Detection of melamine in milk using molecularly imprinted polymers–surface enhanced Raman spectroscopy. Food Chem. 176, 123–129.
Huang, H., Li, L., Zhou, G.H., Liu, Z.H., Ma, Q., Feng, Y.Q., Zeng, G.P., Tinnefeldc, P., He, Z.K., 2011. Visual detection of melamine in milk samples based on label-free and labeled gold nanoparticles. Talanta 85, 1013–1019.
Huy, B.T., Seo, M.-H., Zhang, X., Lee, Y.-I., 2014. Selective optosensing of clenbuterol and melamine using molecularly imprinted polymer-capped CdTe quantum dots. Biosens. Bioelectron. 57, 310–316.
Ivanova-Mitseva, P.K., Guerreiro, A., Piletska, E.V., Whitcombe, M.J., Zhou, Z., Mitsev, A., Davis, F., Piletsky, S.A., 2012. Cubic molecularly imprinted polymer nanoparticles with a fluorescent core. Angew. Chem. Int. Ed. 51, 5196–5199.
Jeyapragasam, T., Saraswathi, R., 2014. Electrochemical biosensing of carbofuran based on acetylcholinesterase immobilized onto iron oxide–chitosan nanocomposite. Sens. Actuat. B 191, 681–687.
JianFeng, P., Jian, W., YiBin, Y., 2012. Determination of trace heavy metals in milk using an ionic liquid and bismuth oxide nanoparticles modified carbon paste electrode. Chin. Sci. Bull. 57, 1781–1787.
Joung, C.-K., Kim, H.-N., Lim, M.-C., Jeon, T.-J., Kim, H.-Y., Kim, Y.-R., 2013. A nanoporous membrane-based impedimetric immunosensor for label-free detection of pathogenic bacteria in whole milk. Biosens. Bioelectron. 44, 210–215.
Kang, T.-F., Wang, F., Lu, L.-P., Zhang, Y., Liu, T.-S., 2010. Methyl parathion sensors based on gold nanoparticles and Nafion film modified glassy carbon electrodes. Sens. Actuat. B 145, 104–109.
Kantianim, L., Llorca, M., Sanchis, J., Farre, M., Barcelo, D., 2010. Emerging food contaminants: a review. Anal. Bioanal. Chem. 398, 2413–2427.
Kong, L., Jiang, X., Zeng, Y., Zhou, T., Shi, G., 2013. Molecularly imprinted sensor based on electropolmerized poly(o-phenylenediamine) membranes at reduced graphene oxide modified electrode for imidacloprid determination. Sens. Actuat. B 184, 424–431.
Kuang, H., Chen, W., Yan, W., Xu, L., Zhu, Y., Liu, L., Chu, H., Peng, C., Wang, L., Kotov, N.A., Xu, C., 2011. Crown ether assembly of gold nanoparticles: melamine sensor. Biosens. Bioelectron. 26, 2032–2037.
Kumar, C.S.S.R., 2007. Nanomaterials for Biosensors. John Wiley & Sons, New York.
Kumaravel, A., Chandrasekaran, M., 2010. A novel nanosilver/Nafion composite electrode for electrochemical sensing of methyl parathion and parathion. J. Electroanal. Chem. 638, 231–235.

Kumaravel, A., Chandrasekaran, M., 2011. Electrochemical determination of imidacloprid using nanosilver Nafion®/nanoTiO_2 Nafion® composite modified glassy carbon electrode. Sens. Actuat. B 158, 319–326.

Kwon, D., Lee, S., Ahn, M.M., Kang, I.S., Park, K.-H., Jeon, S., 2015. Colorimetric detection of pathogenic bacteria using platinum-coated magnetic nanoparticle clusters and magnetophoretic chromatography. Anal. Chim. Acta 883, 61–66.

Lazcka, O., Del Campo, F.J., Munoz, F.X., 2007. Pathogen detection: a perspective of traditional methods and biosensors. Biosens. Bioelectron. 22, 1205–1217.

Lei, W., Wu, Q., Si, W., Gu, Z., Zhang, Y., Deng, J., Hao, Q., 2013. Electrochemical determination of imidacloprid using poly(carbazole)/chemically reduced graphene oxide modified glassy carbon electrode. Sens. Actuat. B 183, 102–109.

Li, Y., Cui, Z., Li, D., Li, H., 2011. Colorimetric determination of pyrethroids based on core–shell Ag@SiO_2 nanoparticles. Sens. Actuat. B 155, 878–883.

Li, H., Chen, D.-X., Sun, Y.-L., Zheng, Y.B., Tan, L.-L., Weiss, P.S., Yang, Y.-W., 2013a. Viologen-mediated assembly of and sensing with carboxylatopillar[5]arene-modified gold nanoparticles. J. Am. Chem. Soc. 135, 1570–1576.

Li, M., Gou, H., Al-Ogaidi, I., Wu, N., 2013b. Nanostructured sensors for detection of heavy metals: a review. ACS Sustain. Chem. Eng. 1, 713–723.

Li, Z., Wang, Y., Nia, Y., Kokot, S., 2014a. Unmodified silver nanoparticles for rapid analysis of the organophosphorus pesticide, dipterex, often found in different waters. Sens. Actuat. B 193, 205–211.

Li, C., Chen, L., You, X., 2014b. Extraction of Sudan dyes from environmental water by hemimicelles-based magnetic titanium dioxide nanoparticles. Environ. Sci. Pollut. Res. 21, 12382–12389.

Li, Y., Xu, J., Sun, C., 2015. Chemical sensors and biosensors for the detection of melamine. RSC Adv. 5, 1125–1147.

Liang, A., Peng, J., Liu, Q., Wen, G., Lu, Z., Jiang, Z., 2015. Highly sensitive and selective determination of fluorine ion by graphene oxide/nanogold resonance Rayleigh scattering-energy transfer analytical platform. Food Chem. 181, 38–42.

Liao, C.-W., Chen, Y.-R., Chang, J.-L., Zen, J.-M., 2011. Single-run electrochemical determination of melamine in dairy products and pet foods. J. Agric. Food Chem. 59, 9782–9787.

Liu, R., Guan, G., Wang, S., Zhang, Z., 2011. Core-shell nanostructured molecular imprinting fluorescent chemosensor for selective detection of atrazine herbicide. Analyst 136, 184–190.

Liu, D., Chen, W., Wei, J., Li, X., Wang, Z., Jiang, X., 2012. A highly sensitive, dual-readout assay based on gold nanoparticles for organophosphorus and carbamate pesticides. Anal. Chem. 84, 4185–4191.

Liu, W., Zhang, D., Zhu, W., Zhang, S., Wang, Y., Yu, S., Liu, T., Zhang, X., Zhang, W., Wang, J., 2015a. Colorimetric and visual determination of total nereistoxin-related insecticides by exploiting a nereistoxin-driven aggregation of gold nanoparticles. Microchim. Acta 182, 401–408.

Liu, J., Song, H., Liu, J., Liu, Y., Li, L., Tang, H., Li, Y., 2015b. Preparation of molecularly imprinted polymer with double templates for rapid simultaneous determination of melamine and dicyandiamide in dairy products. Talanta 134, 761–767.

Lok, C., Powell, D., 2000. The Belgian dioxin crisis of the summer of 1999 a case study in crisis communications and management. February 1, 2000, Report #13, Department of Food Science, University of Guelph, p. 24. <http://www.foodsafetynetwork.ca>

Lopez, P., van Sisseren, M., De Marco, S., Jekel, A., de Nijs, M., Mol, H.G.J., 2015. A straightforward method to determine flavouring substances in food by GC–MS. Food Chem. 174, 407–416.

Luppa, P.B., Sokoll, L.J., Chan, D.W., 2001. Immunosensors—principles and applications to clinical chemistry. Clin. Chim. Acta 314, 1–26.
Martelli, A., Brambilla Campart, G., Canonero, R., Carrozzino, R., Mattioli, F., Robbiano, L., Cavanna, M., 1998. Evaluation of auramine genotoxicity in primary rat and human hepatocytes and in the intact rat. Mutat. Res. Genetic Toxicol. Environ. Mutagen. 414, 37–47.
Matsui, J., Takayose, M., Akamatsu, K., Nawafune, H., Tamaki, K., Sugimoto, N., 2009. Molecularly imprinted nanocomposites for highly sensitive SPR detection of a non-aqueous atrazine sample. Analyst 134, 80–86.
Menon, S.K., Modi, N.R., Pandya, A., Lodha, A., 2013. Ultrasensitive and specific detection of dimethoate using a p-sulphonato-calix[4]resorcinarene functionalized silver nanoprobe in aqueous solution. RSC Adv. 3, 10623–10627.
Mirabi, A., Dalirandeh, Z., Rad, A.S., 2015. Preparation of modified magnetic nanoparticles as a sorbent for the preconcentration and determination of cadmium ions in food and environmental water samples prior to flame atomic absorption spectrometry. J. Magn. Magn. Mater. 381, 138–144.
Mortada, W.I., Kenawy, I.M.M., Abdelghany, A.M., Ismail, A.M., Donia, A.F., Nabieh, K.A., 2015. Determination of Cu2 + , Zn2+ and Pb2+ in biological and food samples by FAAS after preconcentration with hydroxyapatite nanorods originated from eggshell. Mater. Sci. Eng. C 52, 288–296.
Najafi, R., Mukherjee, S., Hudson, Jr., J., Sharma, A., Banerjee, P., 2014. Development of a rapid capture-cum-detection method for *Escherichia coli* O157 from apple juice comprising nano-immunomagnetic separation in tandem with surface enhanced Raman scattering. Int. J. Food Microbiol. 189, 89–97.
Needham, L.L., Gethoux, P.M., Patterson, D.G., Brambilla, P., Smith, S.J., Sampson, E.J., Mocarelli, P., 1999. Exposure assessment: serum levels of TCDD in Seveso. Italy Environ. Res. 80, S200–S206.
Nguyen, P.-D., Tran, T.B., Nguyen, D.T.X., Min, J., 2014. Magnetic silica nanotube-assisted impedimetric immunosensor for the separation and label-free detection of *Salmonella typhimurium*. Sens. Actuat. B 197, 314–320.
Ozalp, V.C., Bayramoglu, G., Erdem, Z., Arica, M.Y., 2015. Pathogen detection in complex samples by quartz crystal microbalance sensor coupled to aptamer functionalized core–shell type magnetic separation. Anal. Chim. Acta 853, 533–540.
Ozcan, S.G., Satiroglu, N., Soylak, M., 2010. Column solid phase extraction of iron(III), copper(II), manganese(II) and lead(II) ions food and water samples on multi-walled carbon nanotubes. Food Chem. Toxicol. 48, 2401–2406.
Parellada, J., Narvaez, A., Lopez, M.A., Domınguez, E., Fernandez, J.J., Pavlov, V., Katakis, I., 1998. Amperometric immunosensors and enzyme electrodes for environmental applications. Anal. Chim. Acta 362, 47–57.
Parham, H., Rahbar, N., 2010. Square wave voltammetric determination of methyl parathion using ZrO_2-nanoparticles modified carbon paste electrode. J. Hazard. Mater. 177, 1077–1084.
Patra, S., Roy, E., Das, R., Karfa, P., Kumar, S., Madhuri, R., Sharma, P.K., 2015a. Bimetallic magnetic nanoparticle as a new platform for fabrication of pyridoxine and pyridoxal-5′-phosphate imprinted polymer modified high throughput electrochemical sensor. Biosens. Bioelectron. 73, 234–244.
Patra, S., Roy, E., Madhuri, R., Sharma, P.K., 2015b. Imprinted ZnO nanostructure-based electrochemical sensing of calcitonin: a clinical marker for medullary thyroid carcinoma. Anal. Chim. Acta 853, 271–284.
Patra, S., Roy, E., Madhuri, R., Sharma, P.K., 2015c. An imprinted Ag@CdS core shell nanoparticle based optical-electrochemical dual probe for trace level recognition of ferritin. Biosens. Bioelectron. 63, 301–310.

Ping, J., Wang, Y., Wu, J., Ying, Y., 2014. Development of an electrochemically reduced graphene oxide modified disposable bismuth film electrode and its application for stripping analysis of heavy metals in milk. Food Chem. 151, 65–71.
Pocas, M.d.F., Hogg, T., 2007. Exposure assessment of chemicals from packaging materials in foods: a review. Trends Food Sci. Technol. 18, 219–230.
Qi, W.J., Wu, D., Ling, J., Huang, C.Z., 2010. Visual and light scattering spectrometric detections of melamine with polythymine-stabilized gold nanoparticles through specific triple hydrogen-bonding recognition. Chem. Commun. 46, 4893–4895.
Qu, Y., Min, H., Wei, Y., Xiao, F., Shi, G., Li, X., Jin, L., 2008. Au–TiO2/Chit modified sensor for electrochemical detection of trace organophosphates insecticides. Talanta 76, 758–762.
Qu, F., Zhou, X., Xu, J., Li, H., Xie, G., 2009. Luminescence switching of CdTe quantum dots in presence of p-sulfonatocalix[4]arene to detect pesticides in aqueous solution. Talanta 78, 1359–1363.
Richardson, S.D., Ternes, T.A., 2011. Water analysis: emerging contaminants and current issues. Anal. Chem. 83, 4614–4648.
Rovina, K., Siddiquee, S., 2015. A review of recent advances in melamine detection techniques. J. Food Compos. Anal. 43, 25–38.
Roy, E., Patra, S., Madhuri, R., Sharma, P.K., 2014a. Gold nanoparticle mediated designing of non-hydrolytic sol-gel cross-linked metformin imprinted polymer network: a theoretical and experimental study. Talanta 120, 198–207.
Roy, E., Patra, S., Madhuri, R., Sharma, P.K., 2014b. Simultaneous determination of heavy metals in biological samples by a multiple-template imprinting technique: an electrochemical study. RSC Adv. 4, 56690–56700.
Roy, E., Patra, S., Kumar, D., Madhuri, R., Sharma, P.K., 2015a. Multifunctional magnetic reduced graphene oxide dendrites: synthesis, characterization and their applications. Biosens. Bioelectron. 68, 726–735.
Roy, E., Patra, S., Madhuri, R., Sharma, P.K., 2015b. Developing electrochemical sensor for point-of-care diagnostics of oxidative stress marker using imprinted bimetallic Fe/Pd nanoparticle. Talanta 132, 406–415.
Roy, E., Patra, S., Madhuri, R., Sharma, P.K., 2015c. Development of an imprinted polymeric sensor with dual sensing property for trace level estimation of zinc and arginine. Mater. Sci. Eng. C 49, 25–33.
Saa, L., Virel, A., Sanchez-Lopez, J., Pavlov, V., 2010. Analytical applications of enzymatic growth of quantum dots. Chem. Eur. J. 16, 6187–6192.
Sannino, A., Bolzoni, L., Bandini, M., 2004. Application of liquid chromatography with electrospray tandem mass spectrometry to the determination of a new generation of pesticides in processed fruits and vegetables. J. Chromatogr. A 1036, 161–169.
Schecter, A., Quynh, H.T., Papke, O., Tung, K.C., Constable, J.D., 2006. Agent Orange, dioxins, and other chemicals of concern in Vietnam: update 2006. J. Occup. Environ. Med. 48, 408–413.
Scheu, P.M., Berghof, K., Stahl, U., 1998. Detection of pathogenic and spoilage micro-organisms in food with the polymerase chain reaction. Food Microbiol. 15, 13–31.
Sharif, Z., Man, Y.B.C., Hamid, N.S.A., Keat, C.C., 2006. Determination of organochlorine and pyrethroid pesticides in fruit and vegetables using solid phase extraction clean-up cartridges. J. Chromatogr. A 1127, 254–261.
Sharma, P., Sablok, K., Bhalla, V., Suri, C.R., 2011. A novel disposable electrochemical immunosensor for phenyl urea herbicide diuron. Biosens. Bioelectron. 26, 4209–4212.
Shrivastav, A.M., Mishra, S.K., Gupta, B.D., 2015. Fiber optic SPR sensor for the detection of melamine using molecular imprinting. Sens. Actuat. B 212, 404–410.

Silva, L.M.d.C., Santos, V.P.S.d., Salgado, A.M., Pereira, K.S., 2013. Biosensors for contaminants monitoring in food and environment for human and environmental health. In: Rinken, T. (Ed.), State of the Art in Biosensors: Environmental and Medical Applications. InTech, pp. 151–167.

Skjevrak, I., Bred, C., Steffensen, I.-L., Mikalsen, A., Alexander, J., Fjeldal, P., Herikstad, H., 2005. Non-targeted multi-component analytical surveillance of plastic food contact materials: identification of substances not included in EU positive lists and their risk assessment. Food Addit. Contam. 22, 1012–1022.

Song, J., Wu, F., Wan, Y., Ma, L., 2015. Colorimetric detection of melamine in pretreated milk using silver nanoparticles functionalized with sulfanilic acid. Food Control 50, 356–361.

Soong, D.K., Ling, Y.C., 1997. Reassessment of PCDD/DFs and Co-PCBs toxicity in contaminated rice-bran oil responsible for the disease "Yu-Cheng". Chemosphere 34, 1579–1586.

Soylak, M., Unsal, Y.E., 2010. Chromium and iron determinations in food and herbal plant samples by atomic absorption spectrometry after solid phase extraction on single-walled carbon nanotubes (SWCNTs) disk. Food Chem. Toxicol. 48, 1511–1515.

Soylak, M., Yilmaz, E., Ghaedi, M., Montazerozohori, M., 2011. Solid phase extraction on multiwalled carbon nanotubes and flame atomic absorption spectrometry combination for determination of some metal ions in environmental and food samples. Toxicol. Environ. Chem. 93, 873–885.

Sreedhar, N.Y., Kumar, M.S., Krishnaveni, K., 2015. Sensitive determination of chlorpyrifos using Ag/Cu alloy nanoparticles and graphene composite paste electrode. Sens. Actuat. B 210, 475–482.

Sun, F.X., Ma, W., Xu, L.G., Zhu, Y.Y., Liu, L.Q., Peng, C.F., Wang, L.B., Kuang, H., Xu, C.L., 2010. Analytical methods and recent developments in the detection of melamine. TrAC Trends Anal. Chem. 29, 1239–1249.

Sung, Y.J., Suk, H.-J., Sung, H.Y., Li, T., Poo, H., Kim, M.-G., 2013. Novel antibody/gold nanoparticle/magnetic nanoparticle nanocomposites for immunomagnetic separation and rapid colorimetric detection of *Staphylococcus aureus* in milk. Biosens. Bioelectron. 43, 432–439.

Tan, S., Zhao, H., Tian, D., Wang, F., Liu, J., Li, H., 2014. Piperidine–calix [4] arene modified gold nanoparticles: Imidacloprid colorimetric sensing. Sens. Actuat. B 204, 522–527.

Tan, X., Hu, Q., Wu, J., Li, X., Li, P., Yu, H., Li, X., Lei, F., 2015. Electrochemical sensor based on molecularly imprinted polymer reduced graphene oxide and gold nanoparticles modified electrode for detection of carbofuran. Sens. Actuat. B 220, 216–221.

Tanabe, S., Kannan, N., Wakimoto, T., Tatsukawa, R., 1989. Isomer-specific determination and toxic evaluation of potentially hazardous coplanar PCBs, dibenzofurans and dioxins in the tissues of "Yusho" PCB poisoning victim and in the causal oil. Toxicol. Environ. Chem. 24, 215–231.

Tang, L., Zeng, G.-M., Shen, G.-L., Li, Y.-P., Zhang, Y., Huang, D.-L., 2008. Rapid detection of picloram in agricultural field samples using a disposable immunomembrane-based electrochemical sensor. Environ. Sci. Technol. 42, 1207–1212.

Thévenot, D.R., Toth, K., Durst, R.A., Wilson, G.S., 1999. Electrochemical biosensors: recommended definitions and classification. Pure Appl. Chem. 71, 2333–2348.

Upadhyay, S., Rao, G.R., Sharma, M.K., Bhattacharya, B.K., Rao, V.K., Vijayaraghavan, R., 2009. Immobilization of acetylcholineesterase-choline oxidase on a gold-platinum bimetallic nanoparticles modified glassy

carbon electrode for the sensitive detection of organophosphate pesticides, carbamates and nerve agents. Biosens. Bioelectron. 25, 832–838.
Valera, E., Ramón-Azcón, J., Barranco, A., Alfaro, B., Sánchez-Baeza, F., Marco, M.-P., Rodríguez, A., 2010. Determination of atrazine residues in red wine samples: a conductimetric solution. Food Chem. 122, 888–894.
Velusamy, V., Arshak, K., Korostynska, O., Oliwa, K., Adley, C., 2010. An overview of foodborne pathogen detection: in the perspective of biosensors. Biotechnol. Adv. 28, 232–254.
Verma, M.S., Rogowski, J.L., Jones, L., Gu, F.X., 2015. Colorimetric biosensing of pathogens using gold nanoparticles. Biotechnol. Adv. 33, 666–680.
Viswanathan, S., Radecka, H., Radecki, J., 2009. Electrochemical biosensor for pesticides based on acetylcholinesterase immobilized on polyaniline deposited on vertically assembled carbon nanotubes wrapped with ssDNA. Biosens. Bioelectron. 24, 2772–2777.
Wadhwa, S.K., Tuzen, M., Kazi, T.G., Soylak, M., 2013. Graphite furnace atomic absorption spectrometric detection of vanadium in water and food samples after solid phase extraction on multiwalled carbon nanotubes. Talanta 116, 205–209.
Walia, S., Acharya, A., 2014. Fluorescent cadmium sulfide nanoparticles for selective and sensitive detection of toxic pesticides in aqueous medium. J. Nanopart. Res. 16, 2778.
Wang, M., Li, Z., 2008. Nano-composite ZrO_2/Au film electrode for voltammetric detection of parathion. Sens. Actuat. B 133, 607–612.
Wang, J., Kong, L., Guo, Z., Xu, J.Y., Liu, J., 2010. Synthesis of novel decorated one-dimensional gold nanoparticle and its application in ultrasensitive detection of insecticide. J. Mater. Chem. 20, 5271–5279.
Wang, J., Wang, Z., Liu, J., Li, H., Li, Q.X., Li, J., Xu, T., 2013. Nanocolloidal gold-based immuno-dip strip assay for rapid detection of Sudan red I in food samples. Food Chem. 136, 1478–1483.
Wang, G., Tan, X., Zhou, Q., Liu, Y., Wang, M., Yang, L., 2014. Synthesis of highly dispersed zinc oxide nanoparticles on carboxylic graphene for development a sensitive acetylcholinesterase biosensor. Sens. Actuat. B 190, 730–736.
Wang, B., Wang, Q., Cai, Z., Ma, M., 2015. Simultaneous, rapid and sensitive detection of three food-borne pathogenic bacteria using multicolor quantum dot probes based on multiplex fluoroimmunoassay in food samples. LWT—Food Sci. Technol. 61, 368–376.
Wei, M., Wang, J., 2015. A novel acetylcholinesterase biosensor based on ionic liquids-AuNPs-porous carbon composite matrix for detection of organophosphate pesticides. Sens. Actuat. B 211, 290–296.
WHO, 2009. World Health Organization. Cumulative number of confirmed human cases of A via ninfluenzaA/(H5N1). <http://www.who.int/csr/disease/avian_influenza/country/en/>
Wong, A., Sotomayor, M.D.P.T., 2014. Determination of carbofuran and diuron in FIA system using electrochemical sensor modified with organometallic complexes and graphene oxide. J. Electroanal. Chem. 731, 163–171.
Wong, A., Materon, E.M., Sotomayor, M.D.P.T., 2014. Development of a biomimetic sensor modified with hemin and graphene oxide for monitoring of carbofuran in food. Electrochim. Acta 146, 830–837.
Wu, S., Lan, X., Zhao, W., Li, Y., Zhang, L., Wang, H., Han, M., Tao, S., 2011. Controlled immobilization of acetylcholinesterase on improved hydrophobic gold nanoparticle/Prussian blue modified surface for ultra-trace organophosphate pesticide detection. Biosens. Bioelectron. 27, 82–87.
Wu, S., Huang, F., Lan, X., Wang, X., Wang, J., Meng, C., 2013. Electrochemically reduced graphene oxide and Nafion nanocomposite for ultralow potential detection of organophosphate pesticide. Sens. Actuat. B 177, 724–729.

Wu, B., Hou, L., Du, M., Zhang, T., Wang, Z., Xue, Z., Lu, X., 2014. A molecularly imprinted electrochemical enzymeless sensor based on functionalized gold nanoparticle decorated carbon nanotubes for methyl-parathion detection. RSC Adv. 4, 53701–53710.

Wu, J., Yang, Z., Chen, N., Zhu, W., Hong, J., Huang, C., Zhou, X., 2015. Vanillin-molecularly targeted extraction of stir bar based on magnetic field induced self-assembly of multifunctional Fe_3O4@Polyaniline nanoparticles for detection of vanilla-flavor enhancers in infant milk powders. J. Coll. Interf. Sci. 442, 22–29.

Xia, N., Wang, Q., Liu, L., 2015. Nanomaterials-based optical techniques for the detection of acetylcholinesterase and pesticides. Sensors 15, 499–514.

Xiao, C., Zhang, X., Liu, J., Yang, A., Zhao, H., Li, X., He, Y., Yuan, Z., 2015. Sensitive colorimetric detection of melamine with 1,4-dithiothreitol modified gold nanoparticles. Anal. Methods 7, 924–929.

Xie, C., Li, H., Li, S., Wu, J., Zhang, Z., 2010. Surface molecular self-assembly for organophosphate pesticide imprinting in electropolymerized poly(p-aminothiophenol) membranes on a gold nanoparticle modified glassy carbon electrode. Anal. Chem. 82, 241–249.

Xu, Q., Du, S., Jin, G.-d., Li, H., Hu, X.Y., 2011. Determination of acetamiprid by a colorimetric method based on the aggregation of gold nanoparticles. Microchim. Acta 173, 323–329.

Xu, X.-Y., Tian, X.-G., Cai, L.-G., Xu, Z.-L., Lei, H.-T., Wang, H., Sun, Y.-M., 2014. Molecularly imprinted polymer based surface plasmon resonance sensors for detection of Sudan dyes. Anal. Methods 6, 3751–3757.

Xue, X., Wei, Q., Wu, D., Li, H., Zhang, Y., Fenga, R., Du, B., 2014. Determination of methyl parathion by a molecularly imprinted sensor based on nitrogen doped graphene sheets. Electrochim. Acta 116, 366–371.

Yan, X., Li, H., Li, Y., Su, Y., 2014. Visual and fluorescent detection of acetamiprid based on the inner filter effect of gold nanoparticles on ratiometric fluorescence quantum dots. Anal. Chim. Acta 852, 189–195.

Yaron, S., Matthews, K.R., 2002. A reverse transcriptase-polymerase chain reaction assay for detection of viable *Escherichia coli* O157:H7: investigation of specific target genes. J. Appl. Microbiol. 92, 633–640.

Yavuz, S., Erkal, A., Kariper, I.A., Solak, A.O., Jeon, S., Mülazımoğlu, I.E., Üstünd ağ, Z., 2015. Carbonaceous materials-12: a novel highly sensitive graphene oxide-based carbon electrode: preparation, characterization, and heavy metal analysis in food samples. Food Anal. Methods 9 (2), 322–331.

Yun, W., Li, H., Chen, S.Q., Tu, D.W., Xie, W.Y., Huang, Y., 2014. Aptamer-based rapid: visual biosensing of melamine in whole milk. Eur. Food Res. Technol. 238, 989–995.

Zawiyah, S., Che Man, Y.B., Nazimah, S.A.H., Chin, C.K., Tsukamoto, I., Hamanyza, A.H., Norhaizan, I., 2007. Determination of organochlorine and pyrethroid pesticides in fruit and vegetables using SAX/PSA clean-up column. Food Chem. 102, 98–103.

Zhang, X., Lu, W., Han, E., Wang, S., Shen, J., 2014. Hybrid nanostructure-based immunosensing for electrochemical assay of *Escherichia coli* as indicator bacteria relevant to the recycling of urban sludge. Electrochim. Acta 141, 384–390.

Zhao, B., He, M., Chen, B., Hu, B., 2015. Novel ion imprinted magnetic mesoporous silica for selective magnetic solid phase extraction of trace Cd followed by graphite furnace atomic absorption spectrometry detection. Spectrochim. Acta B 107, 115–124.

Zheng, Z., Zhou, Y., Li, X., Liu, S., Tang, Z., 2011. Highly-sensitive organophosphorous pesticide biosensors based on nanostructured films of acetylcholinesterase and CdTe quantum dots. Biosens. Bioelectron. 26, 3081–3085.

Zhu, Y., Zhang, Y., Li, J., Han, Y., Dong, G., Zhang, H., 2011. Determination of melamine in fresh milk by electrochemistry with solid phase microextraction at bismuthyl chloride modified graphite epoxy composite electrode. Am. J. Anal. Chem. 2, 612–618.

Zhu, W., Liu, W., Li, T., Yue, X., Liu, T., Zhang, W., Yu, S., Zhang, D., Wang, J., 2014. Determination of melamine in powdered milk by molecularly imprinted stir bar sorptive extraction coupled with HPLC. Electrochim. Acta 146, 419–428.

Zhu, L., Xu, G., Wei, F., Yang, J., Hu, Q., 2015a. Determination of melamine in powdered milk by molecularly imprinted stir bar sorptive extraction coupled with HPLC. J. Colloid Interface Sci. 454, 8–13.

Zhu, J.-H., Li, M.-M., Liu, S.-P., Liu, Z.-F., Li, Y.-F., Hu, X.-L., 2015b. Fluorescent carbon dots for auramine O determination and logic gate operation. Sens. Actuat. B 219, 261–267.

Zou, Z., Du, D., Wang, J., Smith, J.N., Timchalk, C., Li, Y., Lin, Y., 2010. Quantum dot-based immunochromatographic fluorescent biosensor for biomonitoring trichloropyridinol, a biomarker of exposure to chlorpyrifos. Anal. Chem. 82, 5125–5133.

18

NEW TRENDS IN THE FOOD INDUSTRY: APPLICATION OF NANOSENSORS IN FOOD PACKAGING

Thiago Caon*, Silvia Maria Martelli, Farayde Matta Fakhouri****

**Department of Pharmaceutical Sciences, Federal University of Santa Catarina, Florianópolis, Santa Catarina, Brazil; **Faculty of Engineering, Federal University of Grande Dourados, Campo Grande, Mato Grosso do Sul, Brazil*

1 Introduction

The main function of food packaging is to maintain the quality and safety of food during storage and transportation, and to extend the shelf life of food products by preventing unfavorable factors or conditions such as spoilage microorganisms, chemical contaminants, oxygen, moisture, light, and external force (Rhim et al., 2013). The quality of packaged food is directly related to the food and packaging material attributes and a minimum interaction between the food and the packaging is expected (Han, 2014). Traditional packaging systems have several limitations related to extending shelf life and maintaining the safety of food products. Therefore, advanced technologies, including active and intelligent packaging systems, have been developed in response to increasing consumer demands for minimally processed foods with fewer preservatives, more stringent regulatory requirements, market globalization, increased concerns for food safety, and more recently, the threat of food bioterrorism (Yam et al., 2005).

Intelligent or smart packaging is intended to monitor and provide information about the quality of the packaged food or its surrounded environment to predict or decide the safe shelf life. It alerts the consumer to contamination of pathogens, detect harmful chemicals (eg, pesticide residues from agriculture or other contaminants that can accumulate or pass through the food chain) or degradation

Nanobiosensors. http://dx.doi.org/10.1016/B978-0-12-804301-1.00018-7

products caused by food deterioration (Rhim et al., 2013). This technology may also be used to detect allergen proteins, preventing adverse reactions in consumers after taking food such as gluten, peanuts and tree nuts (Sanjib and Jatindra, 2014).

Chemical or biological sensors (specific recognition systems with physical or electrochemical signaling mechanisms) may be efficiently incorporated into food packaging to provide rapid, accurate "in situ" analysis, but these devices need to be suitable for printing technology for mass production, low-cost relative to the value of the food product, easy to use, reliable, simple, and reproducible in their range of operation, and environmentally gentle as well as food contact safe (Kuswandi et al., 2011). Another point is that sensors should meet ever increasing regulatory requirements (Vanderroost et al., 2014). In many situations, it is necessary to detect multiple analytes or pathogens simultaneously, which would not be possible with conventional sensors. With this in mind, sensors in nanoscale have been prioritized in recent researches on smart food packaging and also due to the other advantages provided by materials in these dimensions such as selectivity and sensitivity. Once nanoparticles provide a large surface area, a variety of biomolecules could be immobilized on these structures, increasing the number of sites able to interact with target analytes (Xu et al., 2009). Materials presenting specific electrical and optical properties may be selected, contributing to an application in "label-free" detection and in the development of sensors with enhanced sensitivities (able to detect low concentration of analyte) and improved response times (Gilmartin and O'Kennedy, 2012). Another benefit is that the quality of the food can be assessed in a nondestructive manner.

Given the importance of this subject for the food industry, after clarifying some key concepts related to intelligent packaging systems, this chapter presents the main types of sensors and nanosensors used in the food packaging field, a general classification system and their advantages. The objective here is not to demonstrate as these devices are constructed, or to provide an exhaustive literature review, but to present their limitations and benefits compared to traditional approaches as a way of encouraging researchers to develop novel solutions in order to improve food quality and safety.

2 Smart/Intelligent Packaging Systems

"Intelligent packaging" may be defined as an component or inherent property of a product, pack, or pack/product configuration, which provides intelligence enough to the function and application of the food product itself (Summers, 1992).

Another definition states that intelligent packaging is packaging whose function switches on and off in response to changing external/internal conditions, which usually includes a mechanism that can communicate the status of the product to the end user (Butler, 2001). Therefore, intelligent packaging can be stated as a system that senses and informs the condition of the product or monitors the condition of the packaged food to provide information about quality during transport, distribution, and storage (Yam et al., 2005). The term can be used in a broad sense, including features concerning product identity, authenticity, and traceability, tamper evidence and theft protection, as well as, safety and quality issues (Kuswandi et al., 2011). This is useful not only for quality control to ensure that consumers are able to purchase products, which are at their peak freshness and flavor, but also has the potential to improve food safety and reduce the frequency of foodborne illnesses (Duncan, 2011). Such technology would obviously benefit consumers, industry stakeholders, and food regulators (whole production chain).

Although distinctly different from the concept of active packaging, features of intelligent packaging can be used to check the effectiveness and integrity of active packaging systems (Hutton, 2003). While "active packaging" incorporates robust ways to control oxidation, microbial growth, hydrolysis and other degradation reactions, "smart/intelligent packaging" designs facilitate the real-time monitoring of food quality (Kerry et al., 2006) by detecting analytes through different approaches.

The most promising active packaging technologies are: antimicrobial packaging, oxygen scavengers, moisture regulators for fresh produce and dried foods, and ethylene absorbing systems for fruits and vegetables (Lalpuria et al., 2012). The addition of ethanol releasing sachets in high moisture bakery products, for instance, may suppress mold growth when low oxygen barrier packaging is used (Rooney, 1995). Metal nanoparticles, metal oxide nanomaterials, and carbon nanotubes have also been included into food packaging due to antimicrobial activity (Silvestre et al., 2011). It is noteworthy that a direct contact of these materials with food should be avoided because this latter affects the performance or function of the system and may cause migration problems (common to highly reactive metal nanoparticles). (Ahvenainen, 2003). One of the main advantages to include these technologies is the possibility of reducing packaging cost by complementing active packaging with cheaper traditional passive packaging (Rooney, 1995). Furthermore, consumers more often prefer or demand foods with active packaging systems because they may contain fewer additives, particularly preservatives, and have better quality and improved safety.

Intelligent packaging systems may include the following components (Kerry et al., 2006; Yam et al., 2005):

1. Indicators: time-temperature indicators (TTIs), integrity or gas indicators, freshness indicators (microbial or pathogen spoilage);
2. Barcodes and radio frequency identification tags (RFIDs);
3. Sensors: biosensors, gas sensors, fluorescence-based oxygen sensors.

Indicators are devices that indicate the presence or absence of a target compound or the degree of reaction between two or more compounds by means of a characteristic change, usually in color. The difference between *sensors* and indicators is that the latter do not have receptor and transducer components and instead communicate information through direct visual changes (Lee and Rahman, 2014). Although the production process of sensors usually is more complex, indicators are more susceptible to be affected by food matrix.

3 Sensor, Biosensor, and Nanosensor Concepts

Sensors may be defined as devices used to detect, locate, or quantify energy or matter, providing a signal for the detection or measurement of a physical or chemical property to which the device responds. Most sensors consist of a receptor and a transducer. Receptors transform physical or chemical information into a form of energy, whereas a transducer converts this energy into a useful, typically electrical, analytical signal. High development and production costs, strict industry specifications, safety considerations, and relatively limited demand (compared to the biomedical sector) from both industry and consumer, are the main obstacles to extensive commercial adoption of such systems within the food industry (Lee and Rahman, 2014).

Biosensors are compact analytical devices that are able to detect, record, and transmit information related to biochemical or biological reactions (Yam et al., 2005). In other words, biosensors are probes that combine biological materials and electronic components in order to obtain a measurable signal (Lee and Rahman, 2014). They typically comprise a bioreceptor (eg, enzymes, antibody/antigens, hormones, microbes, and nucleic acid) specific to the target analyte and a transducer responsible for the conversion of biological signals to a measurable response (Yam et al., 2005). The latter can assume several forms (eg, electrochemical, optical, acoustic, mass sensitive) depending on the parameters being measured. Electrochemical biosensors, for example, may be used to measure the current produced from oxidation-reduction reactions. Optical biosensors, in turn, are based on optical properties of analyte such as reflectance, bio- and chemiluminescence, UV–Vis absorption, and fluorescence. On the other hand, mass sensitive biosensors detect small changes caused by chemical interaction of analytes to small piezoelectric crystals (binding of certain analyte increases the mass of the crystal and subsequently change its oscillation frequency) (Morrison et al., 2007). An ideal biosensor would have the necessary sensitivity, high specificity, precision, and accuracy, in addition to being economical, small, portable, and capable of being used by semi-skilled operators. It should be stable under normal storage conditions and not be affected by physical parameters such as stirring, temperature, or pH (except in those situations where these parameters need to be monitored). Finally, such a sensor would offer good dynamic response (low measurement time) and would be biocompatible without any toxic or antigenic effects (Lee and Rahman, 2014). Risk assessments should be performed concomitantly with the development stage of these devices, including migration studies of materials from packages into food matrices.

Despite the development of several innovative platforms for the detection of pathogens and contaminants, most of these are incorporated within devices and require the extraction of a sample to determine the presence of the target analyte, making them impractical for use as biosensors in intelligent packaging. Other challenges for such systems are that ideally they should provide an easily distinguished response (ie, color change) and be cheap to manufacture (low cost of materials and devices), which are typically not characteristics of biosensors widely applied in clinical, biochemical, and environmental analyses (Kuswandi et al., 2011).

In this context, biotechnology and nanotechnology approaches have been combined to overcome these limitations, resulting in different nanosensors able to detect different pathogens and chemical or physical changes. The fact is that nanoparticles present particular optical and electrical properties and, due to large surface area, they may amplify the detection signal, which facilitates its use in food packaging.

Nanosensors, by definition, are nanotechnology-enabled sensors characterized by one of the following attributes: either the size of the sensor or its sensitivity is on the nanoscale or the spatial interaction distance between the sensor and the object is given in nanometers (Khanna, 2008). There are three major reasons why nanoscale technologies represent the next generation of biosensors: (1) they improve assay sensitivity, specificity, and limits of detection (LOD); (2) increase sample throughput; and (3) reduce assay complexity and cost (when the detection is performed "in situ"). Nanosensors often require less amount of analyte to register a response, which is largely due to the small area of the sensing surface (Driskell and Tripp, 2009). A bacterium is at least an order of magnitude greater in size than the nanomaterial, therefore a significant number of nanomaterials could bind to each bacterial cell and as a consequence signal amplification will occur (Gilmartin and O'Kennedy, 2012). Smaller sensing areas generally allow higher-density arrays to be fabricated, and this feature maximizes the number of analytes (eg, pathogens or biomarkers) that can be detected in a single test without increasing sample requirements. Assay complexity and cost can also be reduced by nanosensors, reducing or eliminating sample processing steps. On the other hand, it is important to note that several nanosensor devices self-described as "label free" often require a recognition element, such as an antibody for oligomers for selective sensor detection (Driskell and Tripp, 2009). Therefore, it is necessary to find a balance between complexity and selectivity so that the designed nanosensor has an affordable cost and meets the purpose for which it was created.

4 Classification Nanosensors

Nanosensors may be designed for controlling internal and external conditions of food products. With this in mind, they may be classified in nanosensors for detection of atmospheric impacts (external conditions) and nanosensors for microorganisms and chemical detection inside the package (internal condition) (Ileš et al., 2011).

4.1 Nanosensors for Detection of External Conditions

Decreasing dimensions, a great improvement in the development of computer technology, enables potential application of sensors in food, making it possible to monitor all atmospheric impacts on the product, from the factory to the final customer. For example, if micro- or nanosensors for temperature, moisture, or other measurement devices are added to freshly picked fruit, then while receiving fruit, salespersons would be able to read data to check freshness of this product and conditions of storage or transportation (Vergara et al., 2007). Based on this data, they could decide whether to take or refuse the goods in question. These types of sensors are in the development/research phase (Yang et al., 2006), and for data transmission cheap RFID, tags are used, which are very flexible, small and almost invisible, but still much greater than 1000 nm.

4.2 Nanosensors for Detection of Internal Conditions

Another two types of nanosensors in food products can be recognized on the nanoscale depending on the type of detection: nanosensors and nanobiosensors. Chemical sensors and biosensors are produced in a similar way by putting chemical or bacterial detector (reagents) on the probe, which in reaction with the targeted chemical element or bacteria causes a change on the nanosensor output (Ileš et al., 2011).

5 Radio Frequency Identification (RFID) Sensors

The increasing consumer demand for healthy and safe foods has led to the need for a well-structured and exact traceability system. Such a system can make consumers feel confident in their purchase by providing detailed information about where a product comes from, what its components are, and its processing and handling history. Consumers can get some specific information by entering the tracking code on the company's website. An efficient tracing system will also be beneficial to the producers as it makes it possible to promptly trace all the packages in the case of a product recall and it also provides information to help identify the cause of the problem (Lalpuria et al., 2012), improving quality management systems of food industry.

For this purpose, RFID tags may be efficiently employed. An RFID system includes a reader (or interrogator), a transponder (or a tag), and their associated antennas (Gerst et al., 2005). In a typical RFID system, a reader emits radio waves to capture data from an RFID tag, and the data is then passed onto a host computer (which may be connected to a local network or to the Internet) for analysis and decision making (Fig. 18.1) (Yam et al., 2005).

RFID tags are typically classified into two major types according to power supply: passive and active tags. Active tags such as transponders and beacons require a battery to power its microchip, present longer communication range (20–100 m), greater processing power, and a high operating frequency because they have their own power source. These devices are bulky and more

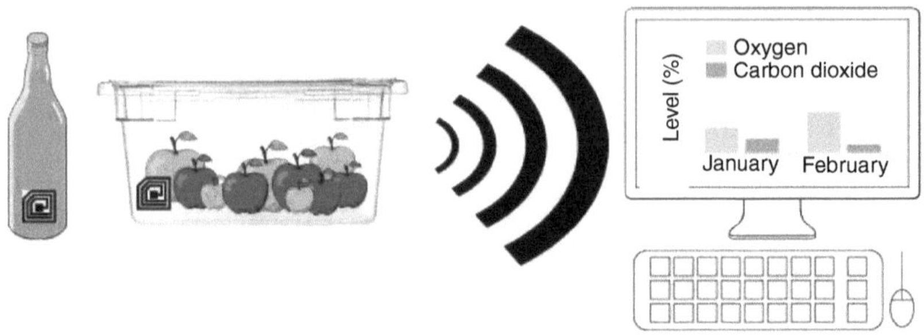

Figure 18.1. Smart packaging based on RFID technology for food quality control.

expensive due to the embedded battery. On the other hand, passive tags do not require any internal power source, reducing the size of these devices (desirable to be included in food packaging). They derive their transmission power from the signal received of an active reader. Therefore, passive RFID operation requires a stronger signal from the reader and the signal strength returned from the tag is constrained to low levels. The signal is transmitted only when they are close to the field of the reader. These tags have low cost with unlimited life, and that is the main reason to prioritize them in lower-priced mass-produced products (Kumar et al., 2009; Weinstein, 2005). A third category is also referred to by some research teams, the semipassive tags, which only use a battery to run the microchip's circuitry, but the transmitted signal uses the same method as that of passive tags. Unsurprisingly, semipassive tags have most of the limitations observed for the passive tag in terms of slow reading speed and a relatively short reading distance. External conditions can also be monitored using these tags, which are powered by a battery (Ruiz-Garcia et al., 2009).

There is another classification which distinguishes these devices in low-, intermediate-, and high-frequency RFIDs (Kumar et al., 2009). When tags are attached to metal on the surface or inside the package, data can be read inappropriately depending on the frequency range considered. Loss of radio signal induced by metal materials in the high magnetic field of a reader is much less frequent at low frequency RFID tags, which typically operate in a range between 125 and 135 kHz, compared to higher frequency signals. Water molecules can absorb microwave signals, resulting in signal loss or interference during data acquisition from microwave RFID tags. Since most foods contain high amounts of water in the composition, this signal interference requires further study to enable application of the technology in the food industry. This interference in signal strength by water is more evident for RFID tags operating at microwave frequency range (0.3–300 GHz). For this reason, RFID tags operating at lower frequencies (9–135 kHz) should primarily be used for such applications (Kumar et al., 2009) if the water content is not reduced during the food processing step.

Although RFID has been available for many years for tracking expensive items and livestock, its broad application in food packaging has only begun in recent years. Its use becomes even more interesting when they are integrated with temperature loggers, chemical sensors, or a biosensor to carry time–temperature history, chemical and microbial data given that various parameters may be monitored simultaneously, enhancing aspects associated to food safety (Siró, 2014).This remote and nondestructive approach of monitoring food products combining RFID with other sensors is especially relevant for the transport and storage of perishable products, such as fruit, vegetables, meats, and fish. These products need to be kept under strict conditions to ensure freshness and quality. Sensor-enabled RFID tags could help food distributors to optimize their supply chain by minimizing the amount of food that is spoiled before it reaches the retailer (Vanderroost et al., 2014). Flexible RFID tags, for instance, have been integrated with resistant sensors capable of detecting amines and devices able to monitor temperature and humidity as an approach to evaluate the freshness of fish. These devices have shown to be highly sensitive for detecting low concentrations of trimethylamine (close to 1 ppm), a marker associated to the freshness of fish, suggesting that the tag may be suited for detecting this compound in the headspace of food packaging (Smits et al., 2012). A prototype combining RFID system with metal oxide sensors has also been designed to monitor the ripening process of climacteric fruits during storage, transportation, and vending. A commercial off-the-shelf inductive coupling radio frequency transceiver operating at 13.56 MHz (fully compliant with the ISO15693 standard), micro–hot plate gas sensors, and readout electronics were integrated to monitor the conservation stage of these fruits (Vergara et al., 2007).

Cost analysis of this technology basically involves two points of view. The supporters of the technology argue that RFID significantly reduces costs, increases the transparency, and hence improve the visibility of the entire supply chain, leading one step further toward the achievement of the truly integrated and virtual supply chain. On the other hand, critics draw attention to the huge technical challenges such as integration with the existing information technology infrastructure and the even more substantial organizational changes required by the adoption of RFID such as the changes in the business processes, as well as the high cost of the RFID tags that hamper the implementation and use of RFID (Gerst et al., 2005).

It is expected, in next years, that the use of RFID on or in food packaging will increase because rapid cost reduction of the tags makes them affordable in these applications and that will be true for primary, secondary, and tertiary packaging. A key advance must be RFID labels being made with printed electronics instead of the silicon chip, which currently represents the main part of their cost. Later, the cost of applying a label will be eliminated by printing RFID directly onto packaging. Printing and similar laminar technology, which will replace the RFID silicon chip, can reduce tag price by 90% or more for very high production volumes (Harrop, 2012).

In this context, alternative materials should be used to replace transistors in the chip and nanotechnology may also be included to provide advantages or specific properties. Printable metallic nanoparticle inks have opened up new, cheaper ways to produce printed circuits on packaging. Once printed, sintering is necessary to partially melt and fuse adjacent nanoparticles together to form a conductive structure. A new method of electrical sintering of printed conductive silver nanoparticles to produce printed conductor structures such as antennas, circuits, and sensors on substrates offers a number of significant advantages over thermal oven sintering (Sharrock, 2012). This approach has been used to produce printed and low-cost copper acetate-based sensors. These devices operate by detecting hydrogen sulfide (H_2S) gas as a metabolic end product of sulfate-reducing bacteria, allowing monitor the quality of raw broiler meat. A printed planar coil antenna and printed capacitor were attached to the sensor to construct a wirelessly readable printed resonance circuit, which has been shown suitable for a commercial mass scale manufacturing process (Koskela et al., 2015). Unlike most other RFID-linked sensors, which require a battery to power the sensor and the active RFID that transmits the sensor data, the earlier mentioned type of printed sensor is passive, meaning that the only powered component in the system is the reader.

Single wall carbon nanotube patterns printed using inkjet represent another cheap solution in the field of wireless sensor systems, which also takes into account impedance changes for detecting the analyte. One or more of the different conductive inks used in producing a unique code might be made inherently sensitive to the target analyte in a way that affects its conductivity and thus the complex impedance of the code pattern when interrogated by a wireless reader. Alternatively, the analyte-responsive component might comprise an overlay or bridge between elements of the coding pattern sufficient to alter the complex impedance of the code (Sharrock, 2012).

In wireless sensors based on standard passive RFID tags, impedance changes, which vary according to the analyte concentration, usually are monitored by a sensing film deposited onto the resonant antenna of the RFID tag. While alcohols can be detected at low parts per million concentrations, the lower limit for ammonia so far achieved was calculated to be 20 ppb, demonstrating that analytes present sensitivity. Therefore, various overlay film combinations may be used to provide the most appropriate response and selectivity for the particular target analyte. Furthermore, the sensitivity is also dependent on the power of the RF emission from the reader and the distance between the reader and the RFID sensor, which also should be tailored. In some specific situations, it is not necessary to overlay a sensor film on the antenna to achieve sensor functionality from a passive RFID tag. For instance, milk can be easily monitored by attaching a standard disposable RFID tag to the outside of the milk carton. Changes in the solution dielectric constant as a function of age of the milk can be monitored by its effects, through the wall of the container, on the electromagnetic field penetration depth out of the plane of the sensor (Potyrailo et al., 2009).

Despite the historical importance of these sensors, the high cost of this technology has always been one of the limitations for its inclusion in lower price food products. Nanotechnology seems to open new opportunities to develop low-cost RFID tags, enabling rapid transfer of this technology in food packaging.

6 Gas Sensors

Excess oxygen leads to several types of deteriorative biochemical and chemical reactions in foods including fat oxidation, browning reactions, and pigment oxidation. Furthermore, most of the common food spoilage bacteria and fungi require oxygen for growth (Sandhya, 2010). In this context, the food industry has stored the products under vacuum or modified gaseous atmosphere to maintain quality and extend product shelf life by slowing these deteriorative reactions. Examination of the changes in residual oxygen level of modified atmosphere packaging or vacuum packed foods during the storage period can be an important indicator on the evolution of degradative processes in these products (Valdés et al., 2009), but many assays require destruction of the package. In processing facilities, packaged foods are tested randomly during a production run, typically one in every 300–400 packages (Mills, 2005). This process is time-consuming and costly, and yet does not ensure that unsampled packages meet quality and safety standards (Smolander et al., 1997). The ability to continually monitor the gas content of a package headspace would also provide a means to assess the safety and quality of the contained food long after it has left the production chain, suggesting that noninvasive leak detection and gas content methods would be invaluable (Duncan, 2011).

Gases produced from microorganism metabolism as result of food spoilage may be detected by several types of sensors, which are devices able to respond reversibly and quantitatively to the presence of a gaseous analyte by changing specific physical parameters of the sensor, and are monitored by an external device (Kerry et al., 2006). Systems presently available for gas detection include amperometric oxygen sensors, potentiometric carbon dioxide (CO_2)sensors, metal oxide semiconductor field effect transistors, organic conducting polymers, and piezo-electric crystal sensors (Kress-Rogers, 2001). Based on the type of transducer, they can be classified into two main groups: (1) optical (colorimetric, gas-induced fluorescence change) and (2) electrochemical (potentiometric, amperometric, and conductometric) (Puligundla et al., 2012). Many factors govern the feasibility of commercial success, including sensitivity, selectivity, response and recovery times, LOD, and power consumption (many of these parameters are common for other types of sensors here described).

Sensors based on conducting polymers (or electro-active conjugated polymers) consist of conducting particles embedded into an insulating polymer matrix. The resistance changes of the sensors produce a specific pattern corresponding to the gas under investigation (Arshak et al., 2007). Conducting polymers show unusual electrochemical properties that contribute to sensing, such as high electrical conductivity, low ionization potential, high electronic affinities, and optical properties, which are related to their conjugated π electron backbones (Retama et al., 2005; Wiziack et al., 2007). The conducting polymers belonging to polyene and polyaromatic such as polyaniline, polyacetylene, and polypyrrole have been widely studied (Ahuja et al., 2007). Electrochemically polymerized conducting polymers have a remarkable ability to switch between conducting oxidized (doped) and insulating reduced (undoped) states, which is the basis for several applications—the same principle is valid for those nanosensors used in pathogen detection (Rajesh et al., 2004). Nanosensors containing carbon black and polyaniline have been demonstrated to be able to detect three foodborne pathogens (*Bacillus cereus*, *Vibrio parahaemolyticus*, and *Salmonella* spp) by producing a specific response pattern for each microorganism (Arshak et al., 2007).

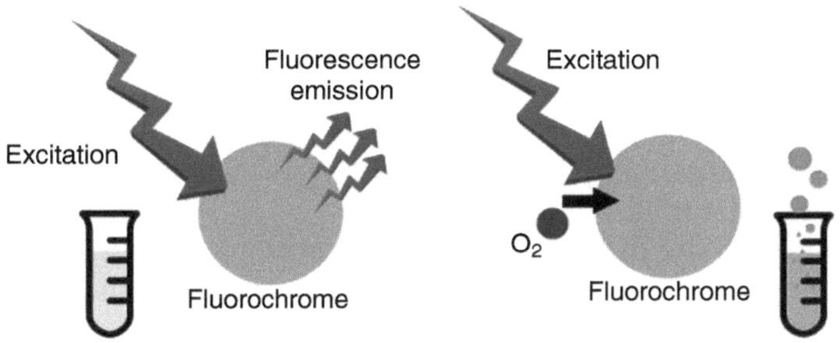

Figure 18.2. Quenching of fluorescence by oxygen.

Optical oxygen sensors are superior over conventional electrochemical sensors given that they do not consume oxygen, are not susceptible to stirring, are not affected by electromagnetic interference, have extremely high spatial resolution, are employed in hazardous areas, are capable of remote measurements and also for measuring gas and liquids (Koo et al., 2004; Lam and Kostov, 2010; Lee and Rahman, 2014).

Instruments and materials for optical oxygen sensing suitable for intelligent packaging applications are often composed of a solid-state material (usually a solid polymer matrix) based on the principle of luminescence quenching or absorbance changes as a result of direct interaction with the analyte (Papkovsky et al., 1995; Thompson and Lakowicz, 1993). When certain dye molecules absorb light at their absorption maxima, they display excited states and emit radiation at specific wavelengths (fluorescence or phosphorescence). Quenching of this excited luminescent dye can occur during collision with oxygen molecules (Eggins, 2002). Due to this energy transfer mechanism, luminescence intensity is decreased over time and the degree of quenching is proportional to the concentration of oxygen within the system (Fig. 18.2).

Fluorescence has attracted special attention among the many optical chemical sensors because it is highly sensitive, versatile, noninvasive, and usually presents low toxicity (Wolfbeis, 2008). Due to these advantages, a fluorescence quenching-based device for measuring oxygen in headspace as well as in dissolved liquids in transparent, semitransparent, and translucent sealed packages in the food, has been prioritized for commercial purposes. For oxygen sensing in a package, the system uses the O2xyDot, which is placed inside a bottle or package prior to filling and sealing. The device is monitored from the outside of the package through the use of a fiber optic pen, maintaining the integrity of package, which is extremely advantageous as discussed earlier this topic (Siró, 2014).

In contrast to conventional systems, the optochemical sensors are chemically inert, so it neither takes part in any chemical reactions nor consumes analyte, thereby helping to boost quality control through detection of product deterioration or microbial contamination by sensing gas analytes such as H_2S, CO_2, and amines (Lee and Rahman, 2014).

Similar to the oxygen detection systems, electrochemical sensors should not receive a higher priority in situations aiming to monitor the CO_2 level in food packaging. Although these sensors present a compact structure, ability of continuous monitoring, high selectivity, and low cost (Hong et al., 2005), they require high power for signal generation. In this situation, operating temperatures higher than 400°C should be provided, which are not suited for food packaging applications. Moreover, environmental factors such as temperature and humidity can affect the sensors performance, especially in the case of metal oxide sensors, by cross-sensitivity reactions (Puligundla et al., 2012). Based on this information, it becomes clear that the selection of the most appropriate sensor should consider not only the type of gas to be detected, but also if the sensor and its operating conditions are compatible with the spot where it should be added.

Simultaneous detection of different gases may be achieved by developing more complex systems able to integrate sensors that may operate through the same or different principles. Printable oxygen and CO_2 sensors have been developed for application in food packaging in different detection systems. The oxygen was measured by detecting the degree of quenching of a fluorescent ruthenium complex entrapped in a sol–gel matrix. To monitor CO_2 gas, a fluorescence-based and pH-sensitive 8-hydroxypyrene-1.3.6-trisulfonic acid indicator dye encapsulated in organically modified silica glass matrix was considered (McEvoy et al., 2003). It is worth noting that reduced levels of oxygen and increased levels of CO_2 in the atmosphere around the fresh produce are often recommended (Lalpuria et al., 2012).

On the other hand, a single nanosensor can be designed to detect more than one gas simultaneously. Nanobelts based on Tin oxide, for instance, have been used as conductometric gas sensors for a simultaneous detection of oxygen and ethanol (Comini et al., 2005). Taking into account the individual advantages of each semiconductor system, these materials can be synergistically combined for sensing different analytes in a single device. With this in mind, binary ZnO–TiO_2 nanocomposites have been prepared in view of high reactivity of TiO_2 and the large excitation binding energy of ZnO. ZnO systems have relatively low selectivity and high operating temperatures, but these disadvantages can be overcome by the introduction of TiO_2. These devices allowed the detection of different volatile organic compounds, which sensitivity may be controlled tailoring their morphological and structural features (Barreca et al., 2007).

In the past few years, an exponential increase in the number of publications dealing with nanomaterials for sensing purposes has been observed, in particular gas sensing. Small, simple, high sensitive, selective and reversible (ie, allowing multiple consecutive measurements) chemical sensors are required, but also low LOD and low operating temperatures (Vanderroost et al., 2014). This latter characteristic is extremely relevant for gas detection systems, enabling its application in food packaging.

In view of these advantages, UV-activated indicator inks based upon nanosized TiO_2 or SnO_2 particles and a redox-active dye have been developed for ion-package oxygen detection. Exposure to UV light led to activation (photobleaching) of the indicator and photoreduction of the marker. The rate of color change is proportional to the level of oxygen, providing to the consumers and retailers an easy, visual method to identify modified atmosphere packages with possible compromised seal integrity (Mills, 2005).

Unlike conventional gas sensors, which respond to analyte concentration, single-walled carbon nanotubes (SWNT) gas sensors respond to the surface coverage of analytes. As a result, they can offer very high detection limits for low vapor pressure analytes (Snow et al., 2005). Gas molecules adsorb onto the surfaces of the SWNTs, especially at defect points. Molecules with strong electron-donating or withdrawing capabilities lead to large changes in resistance, facilitating the detection (Cao and Rogers, 2009). Besides monitoring variations in the conductance of SWNT thin films, sensing can be accomplished by measuring the changes in capacitance (Snow et al., 2005). Different mechanisms for conductance and capacitance responses lead to different responses to analyte molecules with similar structures (Snow and Perkins, 2005). The ratio of the change in conductance to the change in capacitance can be used as a characteristic signature to distinguish different chemical vapors. A major disadvantage of SWNT gas sensors is the lack of specificity, which may be overcome by functionalization of SWNTs with specific receptors for target analytes. This approach has also been developed for selective detection of H_2S, CH_4, and CO_2 (Cao and Rogers, 2009).

Fluorescent nanofibers manufactured from an *n*-type organic semiconductor have been used to detect gaseous amines, acting as an indicator of fish and meat spoilage. These entangled nanofibers form a mesh-like and highly porous film upon deposition onto a substrate, which allows for maximal exposure to the gaseous analyte molecules, expedient diffusion of the molecules throughout the mesh-like film, and increased adsorption and accumulation of the gaseous molecules within the porous matrix. Because of this porous structure, a fast response (milliseconds) for the sensing is observed for nanofibril-film-based sensors, which is more advantageous over other polymer-film-based fluorescent sensors (Che et al., 2008). Similarly, SnO_2–ZnO nanocomposite sensors were produced for a selective sensing of trimethylamine, showing strong stability, high selectivity, and sensitivity even at high temperatures (Zhang and Zhang, 2008). Due to this advantage, this device may be included in food packaging where processing conditions require high temperatures.

Generally speaking, it is possible to observe that the research targeted by novel materials was primarily responsible for technological advances in nanosensors used for the detection of gases, which highlights the importance of developing multicenter studies. Combining different materials in single device, both selectivity and sensitivity to a particular gas may be tailored. Porous materials provide advantages in terms of speed of detection by facilitating the gas diffusion. On the other direction, thermally stable materials allow the processability at high temperatures with remarkable interest for the detection of gases and incorporation into food packaging.

7 Sensors for Food Pathogens and Contaminants

The Centers for Disease Control (CDC) estimate that each year approximately 48 million Americans (1 out of 6) get sick, 128,000 are hospitalized, and 3,000 die of foodborne illnesses. The main five pathogens contributing to domestically acquired foodborne illnesses are Norovirus, *Salmonella* nontyphoidal, *Clostridium perfringes*, *Campylobacter* spp., and *Staphylococcus aureus*. Among the main pathogens resulting in hospitalization, it can be mentioned Norovirus, *Salmonella* nontyphidal, *Campylobacter* spp., *Toxoplasma gondii*, and *Escherichia coli* (CDC, 2011). In this context, the ability to determine whether foods are contaminated by various bacteria, fungi, or viruses able to cause foodborne illnesses represents an important research objective as an approach for reducing hospitalization costs.

The conventional methods for detection of foodborne bacterial pathogens in food products are based on culturing the microorganisms on agar plates followed by standard biochemical probes; however, these procedures are time consuming and also depend on the ability of the pathogens to grow in various culture media (preenrichment, selective enrichment, selective plating, and identification media) (Law et al., 2014). The complexity of food matrices, heterogeneous distribution of level of pathogens and stress suffered by microorganisms during processing steps usually result in long duration enrichment procedures to increase the pathogen populations (López-Campos et al., 2012). All these methods are laborious given that they require the preparation of culture media, inoculation of plates, and colony counting (Law et al., 2014). Therefore, detection methods that are fast, inexpensive, and require little expertise or training to correctly interpret are especially desirable for nonlaboratory settings (Duncan, 2011).

In recent years, various rapid methods presenting high sensitivity and specificity have been developed to overcome the limitations of conventional methods for the detection and identification of foodborne pathogens (Zhao et al., 2014). Once a single pathogen present in food is enough to cause infection, therefore sensitivity is a criterion that should be prioritized in nanosensor production. Rapid methods are more time-efficient, labor-saving, and able to reduce human errors (Law et al., 2014). Rapid detection methods are typically categorized into nucleic acid-based, biosensor-based, and immunological-based methods (Zhao et al., 2014).

Immunological assays, which typically involve certain biotechnology effort, have been the most convenient biological detection methods due to highly specific antibody-antigen interactions (Duncan, 2011). In view of benefits in terms of selectivity, various biosensor prototypes have been designed, but a small number of them are commercially available. Toxin Guard, for instance, is a visual diagnostic system based on antibody-antigen reactions on polymer packaging films that indicate the presence of pathogenic bacteria (*Salmonella*, *Campylobacter*, *E. coli* O157, and *Listeria*), parasites (*Cyclospora*), pesticides, and genetic modification markers in food packaging. Immobilized antibodies are attached to a thin flexible polymer film layer and specific changes in its shape or color suggest the presence of target pathogens. The Food Sentinel Systems is another commercial device used in food packaging, which combines an immunochemical method to a uniquely designed commercial universal product code (UPC) bar system. A specific-pathogen antibody is applied to a membrane forming part of the barcode system, which, when exposed to contaminants such as *Salmonella* spp., *E. coli* 0157:H7, or *Listeria monocytogenes*, forms a localized dark bar on the membrane (Lee and Rahman, 2014).

Nanomaterial-based microbial sensors generally utilize this same approach (antibody-antigen interactions). These materials present particular optical and electrical properties, and the possibility of surface functionalization by using different chemical groups offer significant improvements in terms of selectivity, speed, and sensitivity compared to chemical or biological methods based on macroscale materials (Duncan, 2011). The surface functionalization allows the inclusion of specific properties to the system or maximizing/optimizing these, amplifying the detection signal (eg, inclusion of fluorescent or colorimetric markers) or providing higher selectivity. Given that bacteria are at least an order of magnitude greater in size than nanomaterials, a significant number of nanomaterials would be able to bind to each bacterial cell and, as a consequence, signal amplification will occur (Gilmartin and O'Kennedy, 2012).

The complexity of a detection method depends greatly on the matrix that contains the target pathogen. Matrices vary greatly in physical properties, chemical composition, and microbial (Zourob et al., 2008). Once food products are considered complex matrices, dealing with sample opacity, light scattering, color and other numerous interferences remains the major challenge for an efficient sensing. Although some all-optical methods have been devised, most detection approaches in real food systems require isolation of the target organism from the surrounding environment to ensure that signal-to-noise ratios are sufficiently large enough to be monitored (Duncan, 2011).

The immunomagnetic separation (IMS) technique, which uses magnetic particles coated with selective antibodies in combination with a magnet, may be used for this purpose (Fig. 18.3). It is able to selectively separate the target pathogen from the food matrix (Zourob et al., 2008). Magnetic beads may be able to separate the pathogen from interfering matrix components and other nontarget microorganisms. The properties of magnetic particles are modulated depending on their application. If particles are intended for pathogen IMS or detection, both size and composition should be optimized. Particles with sizes smaller than those of the pathogen to be detected are preferred over larger particles. Particles presenting a reduced size are characterized by a sufficiently high magnetic moment, exactly because of signal amplification effect mentioned earlier. This aspect highlights the importance of using magnetic particles in a nanoscale size (Zourob et al., 2008). Multiplex magnetic capture systems based on amine-functionalized nanoparticles have also been considered for a simultaneous detection of *E. coli*, *Salmonella*, *S. aureus*, and *B. cereus*, achieving capture efficiency values of 97, 55, 66.5, and 97.4%, respectively (Guven et al., 2011).

Unlike optical methods (colorimetric or fluorimetric), electrochemical detection has the advantage of being unaffected by light

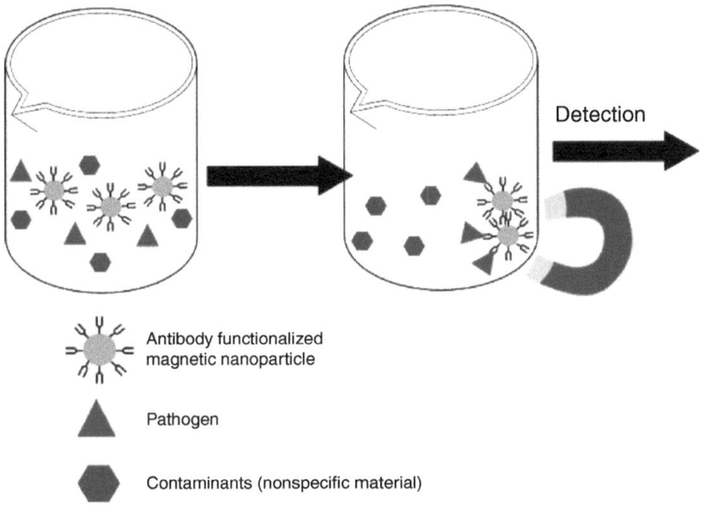

Figure 18.3. Schematic representation of nanomagnetic separation.

scattering and adsorption from the various food components, which also may be an alternative when complex matrices are found (Siró, 2014). Metallic nanoparticles have applications in both optical and electrochemical bioassays, resulting in an increased sensitivity due to their high surface (Wang et al., 2010a). Based on the fact that aggregated gold nanoparticles are blue while dispersed solutions from them are red, colorimetric assays have been designed for sensing *Salmonella typhimurium* (Wang et al., 2010b). In an attempt to increase the selectivity of these nanoparticles, bioreceptors can be immobilized on the surface through different coupling chemistry. With this in mind, and taking advantage of fluorescence detection in terms of sensitivity, a highly amplified fluorescence barcode assay based on gold nanoparticles functionalized with a large number of oligonucleotides strands and a corresponding recognition agent has been proposed for the rapid detection of *Salmonella enteritidis* (Cederquist and Kelley, 2012). An additional deposition of silver on these gold nanoparticles can also be performed to further increase the detection limit. In a sandwich complex using *Salmonella*-specific antibodies and antibody-coated Au-nanoparticles, for instance, a silver layer was deposited around the nanoparticles resulting in signal amplification of the chemiluminescence based assay and a detection limit of 5 cfu/mL (Wang et al., 2011). A similar approach, also involving gold nanoparticles coated by silver, was used for sensing *Campylobacter jejuni*, another foodborne pathogen (Cao et al., 2011). In view of their electrical properties, gold nanoparticles were also exploited to develop piezoelectric biosensors for "real-time" detection of

E. coli O157:H7, achieving detection limit close to 100 cfu/mL (Chen et al., 2008). The main advantage of piezoelectric biosensors, compared to optical devices, is its flexibility, facilitating immobilization and amplification procedures (Tombelli, 2012).

Multiplex detection of pathogens may be achieved by using devices such as quantum dots (QDs), which are colloidal semiconducting fluorescent nanoparticles (2–10 nm) consisting of a semiconductor material core (normally cadmium mixed with selenium or tellurium), which has been coated with an additional semiconductor shell (usually zinc sulphide). Compared to traditional dyes (eg, fluorescein isothiocyanate FITC), QDs display brighter fluorescence intensities for longer, and signal amplification has resulted in lower detection thresholds (Gilmartin and O'Kennedy, 2012). These devices may have a role in transduction or serve as a scaffold for selective recognition elements such as antibodies, aptamers, or nucleic acids (Siró, 2014). The simultaneous detection of several pathogens is due to their broad Stokes shift, which permits the simultaneous excitation of many different QDs at a single excitation wavelength (Gilmartin and O'Kennedy, 2012). Simultaneous detection of *E. coli* and *S. enteritidis*, for instance, was demonstrated by coupling IMS with QDs labeling. QDs having different emission wavelengths were conjugated with anti-*E. coli* and anti-*Salmonella* antibodies. The comparable detection limits of these multiplex assays to those observed in the single pathogen detection assays earlier show great possibilities for cost-effective, rapid multiplex devices (Dudak and Boyaci, 2009). Despite the apparent potential of QDs, some challenges still need to be overcome to lead to its widespread use. High concentration of proteins in food matrix, for instance, decreases the sensitivity of QDs. Quenching of the fluorescence signal also occurred when QDs were employed in combination with magnetic beads for bacterial capture (Zhao et al., 2009). Furthermore, toxicity issues have been reported for cadmium selenide-based QDs (Sharrock, 2012).

In addition to the QDs, DNA-based nanostructures may also be included in multiplexed molecular detection schemes. The functionality of these systems is based on a capture-mediated approach in which the nanoparticle-immobilized single-stranded DNA hybridizes with a target (complementary sequence). After that, the hybridization event is converted into an optical, electrical, or coupled response by labeling oligonucleotides with fluorescent dyes and/or by immobilizing them onto a surface.

Besides the use for detecting foodborne pathogens, these multiplexed devices may have a potential application for genetically modified organisms (GMOs) in food products, contributing to ensure the quality of these products in both domestic and international markets. In recent years, the safety of GMOs has aroused controversy in global trade due to the risks to the environment, ecological problems, and human health. Due to concerns, GMO risks, various protein- and DNA-based techniques have been purposed. The presence of inserted exogenous genes of GMOs has been often monitored by classic polymerase chain reaction (PCR) assays; but the occurrence of nonspecific amplification and false positive results has encouraged the investigation of alternative methods such as DNA-based nanostructures (Zhao et al., 2011).

Several types of DNA-based nanostructures for molecular sensing may be considered: (1) linear DNA-mediated nanostructures; (2) branched DNA nanostructures (fluorescence nanobarcodes, fluorescence DNA nanotags, ABC monomers, and DNA tile arrays); and (3) functional tertiary structures of DNA. When only small quantities of target are present in certain sample, signal amplification may be needed to enable the detection. Gold nanoparticles, QDs, and carbon nanotubes are excellent candidates for this purpose (Lee et al., 2010; Siró, 2014). For sensing different pathogens simultaneously or a same pathogen presenting different mutations in its DNA, nanostructures based on branched DNA molecules may be efficiently employed. These branched DNA molecules are obtained from rationally designed single-stranded oligomers, leading to the production of complex DNA nanostructures. A DNA with various branch points allows for not only multiple fluorescent labels per target, but different color ratios that may correspond to different target sequences (Lee et al., 2010). Antibodies or molecular probes and fluorescent dye molecules may be efficiently attached in these sites. Multicolor fluorescence-intensity-encoded "nanobarcodes" could be successfully manufactured by precisely controlling the dye type and number. Different ratios of different fluorescent dyes enable decoding of the nanobarcodes, independent of the dye positions (each color pattern correspond to a specific molecular probe). Using only three fluorescent dyes that can be distinguished optically, up to 1000 different coded structures can be obtained and linked to specific antibodies or molecular probes (Sharrock, 2012). In the preparation of DNA nanobarcodes, molecular probes are linked to the unprotected reactive ends of dendrimer-like DNA nanostructures. Molecular probes (eg,, DNA or RNA probes or even antibodies) can be easily attached to DNA nanobarcodes by using different DNA-manipulation enzyme tools (Li et al., 2005). Consequently, the resultant DNA nanobarcodes contain molecular recognition elements able to detect various pathogens simultaneously.

Another novel or recently resurrected approach to sensor label construction involves the use of conducting polymers. Typically, such conducting particles can be embedded into an insulating polymer matrix and changes in the resistance can be used for sensing. Polydiacetylene (PDA) has been considered an ideal candidate for sensing applications due to its unique chromatic/optical properties (Siró, 2014). Moreover, this polymer may be easily synthesized (Pires et al., 2010). Diacetylenic lipid monomers are readily polymerized in monolayers by UV irradiation to form a conjugated PDA backbone that changes color from deep blue to red in response to various stimuli (eg, increasing temperature or changes in pH) because of conformational changes in the conjugated backbone (Tieke et al., 1977). Another attractive feature of PDA systems in the context of sensing applications has been the fluorescence properties; blue phase PDA is nonfluorescent while the red-phase configuration exhibits strong fluorescence with minimal bleaching (Jelinek et al., 2015). Basically, the PDA molecules can be assembled in two different ways for biological assays: can be immobilized as a thin film (also known as Langmuir film) on a solid support or as vesicles (or liposomes) in a solution. Although the production of vesicles is simpler and more reproducible compared to that of films, and they also mimic the cell membrane more closely, thin films have been attracting attention due to their more general applicability (they can be used on different supports, such as glass and plastics). PDA thin films on a solid support also have more advantages over other arrangements in terms of chemical and physical stability (Pires et al., 2010). On the other hand, liquid-phase sensors possess strong optical properties because of the high concentration of PDA supramolecules in solution (Park et al., 2009). Based on this information, this polymer may be efficiently incorporated into an intelligent packaging systems (Pires et al., 2010). PDA films and vesicles are promising devices for the detection of bacteria and bacterial toxins. Recognition elements that are bound to specific markers on the surface of bacterial cells have been typically incorporated into the PDA matrices, providing chromatic signals. The incorporation of ganglioside G_{Tb1} into a PDA film, for instance, was able to detect the botulinum toxin (Charych et al., 1996). Mannose derivatives have been selected in various investigations as a ligand for *E. coli* given that the adsorption of this pathogen on the film surface involve interactions with mannose-containing receptors (Su et al., 2005). Multiplex pathogen-detection platforms based on PDA have also been considered. Additional immobilization properties in PDA liposomes were achieved by including an "interlinker", the ethylenediamine, allowing the simultaneous quantitative and qualitative analysis of six different pathogens: *Cryptosporidium parvum*, *Giardia lamblia*, *E. coli* O-157, *S. typhimurium*, *Shigella flexneri*, and *Encephalitozoon intestinalis* (Park et al., 2009). Recent studies have also demonstrated that blue-to-red and fluorescence transformations in phospholipid/PDA vesicles are possible even in the absence of specific recognition elements, which would reduce production costs significanlty. In this situation, bacteria or their secreted substances should present certain tropism by membrane, inducing the changes. Chromatic transformation phenomena are explained by fluidity changes and surface perturbations within the lipid domains, inducing structural/chromatic transitions of the adjacent PDA matrix through interfacial processes (interface between phospholipid and PDA). Based on this fact, the generated chromatic responses after interaction with the analyte may be tailored by controlling the phospholipid/PDA ratio (Silbert et al., 2006).

Although various technologies for bacterial detection based on nanosensors have been developed, few of them have achieved commercial status. Polymer-based biosensores seem to be closer to mass commercial production since they usually exhibit greater affinity/compatibility with the packaging material. Detection sensitivity, robustness of the transfer process to the packaging materials, production cost, biosensor stability (usually unstable in organic solvents, high temperature or extreme pH values), complexity of the food matrices, and the consistency of the initial optical state still are the main challenges to be overcome. Decision-making should take into consideration a joint analysis of all these parameters. For instance, QDs may be selected to a specific detection due to high sensitivity; however, they may be inappropriate for complex matrices (protein content affect the performance). Physicochemical or biological properties of analyte is one of the main criteria that guides the type of nanosensor to be selected as well as production aspects.

8 Main Challenges and Directions for Nanosensor Applications

Nowadays, most of the nanosensor research results are obtained in laboratories under controlled conditions, which may be different from those real conditions. In the food industry, the sensors should be capable of working under extreme temperatures, pressures, pH, ionic strength, and other changing parameters. In the real world, sensors have to operate in the presence of multiple microorganisms and other substances, which could interfere with the sensor measurements.

In addition to the significant cost of the devices, at least until the popularization of the technology, high infrastructure costs are expected. In fact, the technological advancement in the nanoindustries will probably change the fabrication process of the entire packaging industry, which is even greater when biotechnological efforts should be included (eg, production of nanobiosensors) (Ranjan et al., 2014).

The stability of nanosensors in presence of other species should also be considered. Changes in fingerprint signature may occur when the analytes interacts with different unknown species, which makes detection and determination harder (Arora et al., 2006). When food composition or processing conditions are strongly modified, new performance testing of nanosensors should be performed to ensure an efficient operability of these devices, attesting their robustness and selectivity.

Effect of food enzymes or those resulting from the processing steps on the stability of biological recognition components (other enzymes, antigens and antibodies) should not be neglected. Immobilization and stabilization approaches of biosensors based on the enzyme acetylcholinesterase (AChE), for example, have been not effective in detecting pesticide residues when this recognition component is accessed by proteases. With this problem in mind, AChE has been effectively stabilized into liposomes by hydrophobic interactions between the enzyme and the nanostructure membrane, preventing access of these proteases. Furthermore, porins embedded into the lipid membrane were used to select the target molecule (Vamvakaki and Chaniotakis, 2007).

Critical issues of sampling may occur in both internal and external nanosensors. The scale of the sensors is extremely small compared to the product, or condition, it is measuring. This implies that the sensor measures only a small part of the food or environment. Autonomous wireless sensors networks arranged in various sites are often mentioned as possible solutions to this problem (Kowalczyk, 2013).

As already mentioned, nanosensors may be used in multianalyte detection. In these situations, the functionalization of devices may be required to achieve appropriate selectivity. A similar approach is considered for complex food matrices. In this context, the main technical challenge during the development of novel nanosensors, especially immune, is to provide an effective standardized immobilization of biological materials such as antigens on the transducer material and optimization of the functional parameters of the device. Protein molecules, for instance, can be absorbed on the metal or silicon surfaces, but the effectiveness of this process is often not high. The active centers are often blocked at the sorption, leading to significant loss of the level of the specific interaction on the transducer surface. The level of nonspecific interactions can be unchanged and then the signal-to-background ratio sharply decreases, and the sensitivity of the immune sensor falls sharply. To avoid these problems, the transducer surface may be pretreated with specific materials (eg, polymer layers). The inclusion of these intermediate layers of other molecules may cause selective exposure of sites of the sensitive biological structures, improving the performance of devices (Nickolaj, 2013).

As with any new technology, there is a significant resistance to create awareness and gain acceptance of the use of nanotechnology in the food industry. Currently, no specific regulation for controlling or limiting the production of nanosensors in the food industry exists, and this is mainly owing to a lack of knowledge about the risks. In short, the development of nanosensor research and applications in the food sector is starting to emerge; however, it is still an incipient field compared to the medical applications (Sozer and Kokini, 2009). The safety of nanosensors really poses a challenge to both government and industry. In an industrial scale, it is important to evaluate the release of nanomaterials into the environment and to estimate the subsequent levels of exposure to these materials (risks should be identified and quantified) (Ranjan et al., 2014).

The main reason for this evaluation is that properties of materials at the nanoscale can be substantially different from conventional bulk forms or individual molecules, as previously mentioned, which may result in unexpected hazards. Typically, research teams carry out a careful analysis of nature of these materials as well as their applications for defining a conceptual risk categorization; however, risks can be under- or overestimated by using this approach. Insoluble and biologically stable nanoparticles (hard nanomaterial) presenting highly reactive surfaces may migrate from the packaging into a food and, after ingestion, they may cross biological barriers, reaching parts of the body in which access is restricted (Sanjib and Jatindra, 2014). With this in mind, undesirable transfers of nanosensor fragments may affect the food safety for the consumer and should be monitored. Although numerous nanotechnology-derived food packaging materials are commercially available in several countries, limited data are currently available on the potential migration of nanomaterials from manufactured products. In some situations, migration of nanoscale materials may even induce sensory changes of foods. TiO_2 nanoparticles, for example, may cause rancidity resulting from lipid oxidation in foods (de Azeredo, 2013). In this context, migration assays of nanopackaging materials have been recommended by regulatory agencies before approving them for commercialization. Factors such as food composition, pH, temperature, exposure time, radius of nanoparticle, and dynamic viscosity of polymer may affect the migration of nanoparticles from packaging material. Therefore, extreme and controlled conditions should be considered for the assays in order to reproduce the real conditions (Kuorwel et al., 2015; Metak et al., 2015). Overall, migration assays have been performed using food simulants because it is very complicated, time-consuming, and expensive to monitor this parameter into real food products (Kuorwel et al., 2015).

The food matrix is very complex, which also limits the use of certain analytical tools. Caution must be taken to state that nanomaterials are not been migrating from food packaging into food matrix because shortfalls in analytical methods are often observed. Risks of a given nanomaterial may not be quantified if a previous validation of the analytical method was not carried out. A highly selective, accurate, and sensitive technique is demanded (Huang et al., 2015).

Although few toxicology studies have been conducted with nanosensors or materials used in the production of these devices, consumers appear to be more confident with nanotechnology-derived packaging than nanotechnology-engineered foods. In a study with 153 subjects comparing both these approaches, the results supported the hypothesis that nanotechnology inside a food is perceived as less acceptable than being on the outside (ie, in the food packaging) (Siegrist et al., 2007).

The challenges described earlier are the areas in which nanosensor researchers will probably be working on in the next years. In the future, it is expected the development and commercialization of more multiplexed hybrid systems with different combinations of transducer platforms and recognition elements (Kowalczyk, 2013). To deal with the diverse contaminants in food, it is also expected that different sensing platforms (optical, mechanical, and chemical) will be integrated into one sensing device. New receptor-less physical sensors will continue to be explored as viable methods. At the individual consumer level, it is also expected that electronic noses can be used as integrands in other devices, such as refrigerator temperature sensors, infant products, and cellular phones, to detect odors related to food quality (Farahi et al., 2012).

9 Concluding Remarks

Considering the high number of foodborne-disease outbreaks occurring in the world, there is a growing need for the development of technologies capable of detecting pathogens and also to enable consumers to know the quality of the products they are consuming. The incorporation of chemical and biological sensors into food packaging materials has resulted in what is known as smart or intelligent packaging, which improves detection and tracking of the food condition during storage and transport for safety and biosecurity purposes. External or internal changes in the product's environment may be monitored depending on the selected device. At an even more advanced stage, foods capable of changing of color, flavor, or nutritional properties according to a person's dietary needs, allergies, or taste preferences will be developed by the companies, contributing to a further increase in the level of customer satisfaction.

The benefits of these sensors are even greater in situations in which the unique electro-optical properties of nanoscale particles are considered. Nanosensors can produce highly amplified signals, which allow detecting low concentrations of target analyte or pathogen. In addition, multiplex schemes to detect various targets simultaneously may be designed. The complexity of food matrices remains the main technological challenge, particularly when the colorimetric response is monitored, but magnetic nanoparticles or composed of other materials are changing this scenario.

At the moment, the most common commercial sensors are RFIDs, which are better suited for supply chain quality assurance and for monitoring the external conditions (temperature, humidity). Printed nanosensor labels also are expected to be a reality in the near future, but the cost of this technology still needs to be reduced. Furthermore, there is an urgent need for regulation of nanomaterials before their incorporation into food and processing including packaging. This risk is even greater when novel materials are considered, requiring periodic risk assessments.

Acknowledgments

We would like to thank the Mind the Graph team for granting us free access to its database, which was used to prepare the illustrations.

References

Ahuja, T., Mir, I.A., Kumar, D., Rajesh, 2007. Biomolecular immobilization on conducting polymers for biosensing applications. Biomaterials 28, 791–805.

Ahvenainen, R., 2003. Novel Food Packaging Techniques. CRC Press, Boca Raton.

Arora, K., Chand, S., Malhotra, B.D., 2006. Recent developments in bio-molecular electronics techniques for food pathogens. Anal. Chim. Acta 568, 259–274.

Arshak, K., Adley, C., Moore, E., Cunniffe, C., Campion, M., Harris, J., 2007. Characterisation of polymer nanocomposite sensors for quantification of bacterial cultures. Sens. Actuat. B 126, 226–231.

Barreca, D., Comini, E., Ferrucci, A.P., Gasparotto, A., Maccato, C., Maragno, C., Sberveglieri, G., Tondello, E., 2007. First example of $ZnO-TiO_2$ nanocomposites by chemical vapor deposition: structure, morphology, composition, and gas sensing performances. Chem. Mater. 19, 5642–5649.

Butler, P., 2001. Smart packaging: intelligent packaging for food, beverages, pharmaceuticals and household products. Mater. World 9, 11–13.

Cao, Q., Rogers, J.A., 2009. Ultrathin films of single-walled carbon nanotubes for electronics and sensors: a review of fundamental and applied aspects. Adv. Mater. 21, 29–53.

Cao, C., Gontard, L.C., Thuy Tram, L.L., Wolff, A., Bang, D.D., 2011. Dual enlargement of gold nanoparticles: from mechanism to scanometric detection of pathogenic bacteria. Small 7, 1701–1708.

CDC, 2011. CDC 2011 estimates: findings. CDC Centers for Disease Control and Prevention, USA. Available from: http://www.cdc.gov/foodborneburden/2011-foodborne-estimates.html

Cederquist, K.B., Kelley, S.O., 2012. Nanostructured biomolecular detectors: pushing performance at the nanoscale. Curr. Opin. Chem. Biol. 16, 415–421.

Charych, D., Cheng, Q., Reichert, A., Kuziemko, G., Stroh, M., Nagy, J.O., Spevak, W., Stevens, R.C., 1996. A 'litmus test' for molecular recognition using artificial membranes. Chem. Biol. 3, 113–120.

Che, Y., Yang, X., Loser, S., Zang, L., 2008. Expedient vapor probing of organic amines using fluorescent nanofibers fabricated from an n-type organic semiconductor. Nano Lett. 8, 2219–2223.

Chen, S.-H., Wu, V.C.H., Chuang, Y.-C., Lin, C.-S., 2008. Using oligonucleotide-functionalized Au nanoparticles to rapidly detect foodborne pathogens on a piezoelectric biosensor. J. Microbiol. Methods 73, 7–17.

Comini, E., Faglia, G., Sberveglieri, G., Calestani, D., Zanotti, L., Zha, M., 2005. Tin oxide nanobelts electrical and sensing properties. Sens. Actuat. B, 2–6, 111–112.

de Azeredo, H.M.C., 2013. Antimicrobial nanostructures in food packaging. Trends Food Sci. Technol. 30, 56–69.

Driskell, J.D., Tripp, R.A., 2009. Emerging technologies in nanotechnology-based pathogen detection. Clin. Microbiol. Newslett. 31, 137–144.

Dudak, F.C., Boyaci, .H., 2009. Multiplex detection of *Escherichia Coli* and *Salmonella Enteritidis* by using quantum dot-labeled antibodies. J. Rap. Methods Automat. Microbiol. 17, 315–327.

Duncan, T., 2011. Applications of nanotechnology in food packaging and food safety: barrier materials, antimicrobials and sensors. J. Coll. Interf. Sci. 363, 1–24.

Eggins, B.R., 2002. Chemical Sensors and Biosensors. Wiley, Chichester, UK.

Farahi, R.H., Passian, A., Tetard, L., Thundat, T., 2012. Critical issues in sensor science to aid food and water safety. ACS Nano 6, 4548–4556.

Gerst, M., Bunduchi, R., Graham, I., 2005. Current Issues in RFID Standardisation. University of Edinburgh, Edinburgh.

Gilmartin, N., O'Kennedy, R., 2012. Nanobiotechnologies for the detection and reduction of pathogens. Enzy. Micro. Technol. 50, 87–95.

Guven, B., Basaran-Akgul, N., Temur, E., Tamer, U., Boyac, I.H., 2011. SERS-based sandwich immunoassay using antibody coated magnetic nanoparticles for *Escherichia coli* enumeration. Analyst 136, 740–748.

Han, J.H., 2014. A review of food packaging technologies and innovations. In: Han, J.H. (Ed.), Innovations in Food Packaging. second ed. Academic Press, San Diego, pp. 3–12.

Harrop, P., 2012. Radio-frequency identifiction (RFID) for food and beverage packaging applications. In: Yam, K.L., Lee, D.S. (Eds.), Emerging Food Packaging Technologies. Woodhead Publishing, Cambridge, UK, pp. 153–174.

Hong, H., Kim, J., Jung, S., Park, C., 2005. Thick film planar CO_2 sensors based on Na β-alumina solid electrolyte. J. Electroceram. 15, 151–157.
Huang, J.-Y., Li, X., Zhou, W., 2015. Safety assessment of nanocomposite for food packaging application. Trends Food Sci. Technol. 45, 187–199.
Hutton, T., 2003. Food Packaging: An Introduction: Key Topics in Food Science and Technology. Campden and Chorleywood Food Research Association Group, Gloucestershire.
Ileš, D., Martinović, G., Kozak, D., 2011. Review of potential use, benefits and risks of nanosensors and nanotechnologies in food. J. Theory Appl. Mech. Eng. 53, 127–136.
Jelinek, R., Kolusheva, S., Mann, E., Aviram, M., 2015. Chromatic polymer assays for the analysis of lipid and lipoprotein peroxidation. Lip. Technol. 27, 86–89.
Kerry, J.P., O'Grady, M.N., Hogan, S.A., 2006. Past, current and potential utilisation of active and intelligent packaging systems for meat and muscle-based products: a review. Meat Sci. 74, 113–130.
Khanna, V.K., 2008. New-generation nano-engineered biosensors, enabling nanotechnologies and nanomaterials. Sens. Rev. 28, 39–45.
Koo, Y.-E.L., Cao, Y., Kopelman, R., Koo, S.M., Brasuel, M., Philbert, M.A., 2004. Real-time measurements of dissolved oxygen inside live cells by organically modified silicate fluorescent nanosensors. Anal. Chem. 76, 2498–2505.
Koskela, J., Sarfraz, J., Ihalainen, P., Määttänen, A., Pulkkinen, P., Tenhu, H., Nieminen, T., Kilpelä, A., Peltonen, J., 2015. Monitoring the quality of raw poultry by detecting hydrogen sulfide with printed sensors. Sens. Actuat. B 218, 89–96.
Kowalczyk, J.W., 2013. Tracing Emerging Irreversibilities and Development Pathways in the Field of Nanosensor Technologies (Dissertation). Utrecht University, Utrecht, NL.
Kress-Rogers, E., 2001. Chemosensors, biosensors, immunosensors and DNA probes: the base devices. In: Kress-Rogers, E., Brimelow, C.J.B. (Eds.), Instrumentation and Sensors for the Food Industry. second ed. Woodhead Publishing, Cambridge, UK, pp. 623–713.
Kumar, P., Reinitz, H.W., Simunovic, J., Sandeep, K.P., Franzon, P.D., 2009. Overview of RFID technology and its applications in the food industry. J. Food Sci. 74, R101–R106.
Kuorwel, K.K., Cran, M.J., Orbell, J.D., Buddhadasa, S., Bigger, S.W., 2015. Review of mechanical properties, migration, and potential applications in active food packaging systems containing nanoclays and nanosilver. Comp. Rev. Food Sci. Food Saf. 14, 411–430.
Kuswandi, B., Wicaksono, Y., Jayus, Abdullah, A., Heng, L., Ahmad, M., 2011. Smart packaging: sensors for monitoring of food quality and safety. Sens. Instrum. Food Qual. Saf. 5, 137–146.
Lalpuria, M., Anantheswaran, R., Floros, J., 2012. Packaging technologies and their role in food safety. In: Demirci, A., Ngadi, M.O. (Eds.), Microbial Decontamination in the Food Industry. Woodhead Publishing, Cambridge, UK, pp. 701–745.
Lam, H., Kostov, Y., 2010. Optical instrumentation for bioprocess monitoring. In: Rao, G. (Ed.), Optical Sensor Systems in Biotechnology. Springer Berlin, Heidelberg, pp. 125–142.
Law, J.W.-F., Ab Mutalib, N.-S., Chan, K.-G., Lee, L.-H., 2014. Rapid methods for the detection of foodborne bacterial pathogens: principles, applications, advantages and limitations. Front. Microbiol. 5, 770.
Lee, S.J., Rahman, A.T.M.M., 2014. Intelligent packaging for food products. In: Han, J.H. (Ed.), Innovations in Food Packaging. second ed. Academic Press, San Diego, pp. 171–209.

Lee, J.B., Campolongo, M.J., Kahn, J.S., Roh, Y.H., Hartman, M.R., Luo, D., 2010. DNA-based nanostructures for molecular sensing. Nanoscale 2, 188–197.
Li, Y., Cu, Y.T.H., Luo, D., 2005. Multiplexed detection of pathogen DNA with DNA-based fluorescence nanobarcodes. Nat. Biotechnol. 23, 885–889.
López-Campos, G., Martínez-Suárez, J., Aguado-Urda, M., López-Alonso, V., 2012. Detection, Identification, and Analysis of Foodborne Pathogens, Microarray Detection and Characterization of Bacterial Foodborne Pathogens. Springer, USA, pp. 13–32.
McEvoy, A.K., Von Bueltzingsloewen, C., McDonagh, C.M., MacCraith, B.D., Klimant, I., Wolfbeis, O.S., 2003. Optical Sensors for Application in Intelligent Food-Packaging Technology. SPIE Proceed, Ireland, pp. 806–815.
Metak, A.M., Nabhani, F., Connolly, S.N., 2015. Migration of engineered nanoparticles from packaging into food products. LWT Food Sci. Technol. 64, 781–787.
Mills, A., 2005. Oxygen indicators and intelligent inks for packaging food. Chem. Soc. Rev. 34, 1003–1011.
Morrison, D.W.G., Dokmeci, M.R., Demirci, U., Khademhosseini, A., 2007. Clinical Applications of Micro- and Nanoscale Biosensors, Biomedical Nanostructures. John Wiley & Sons, Hoboken, NJ, USA, 439–460.
Nickolaj, F.S., 2013. Biosensors for the Control of Biochemical Parameters in the Diagnostics of Diseases, Portable Biosensing of Food Toxicants and Environmental Pollutants. CRC Press, Boca Raton, pp. 743–774.
Papkovsky, D.B., Ponomarev, G.V., Trettnak, W., O'Leary, P., 1995. Phosphorescent complexes of porphyrin ketones: optical properties and application to oxygen sensing. Anal. Chem. 67, 4112–4117.
Park, C.H., Kim, J.P., Lee, S.W., Jeon, N.L., Yoo, P.J., Sim, S.J., 2009. A direct, multiplex biosensor platform for pathogen detection based on cross-linked polydiacetylene (PDA) supramolecules. Adv. Funct. Mater. 19, 3703–3710.
Pires, A.S., Soares, N., da Silva, L., de Andrade, N., Silveira, M., de Carvalho, A., 2010. Polydiacetylene as a biosensor: fundamentals and applications in the food industry. Food Bioproc. Technol. 3, 172–181.
Potyrailo, R.A., Surman, C., Go, S., Lee, Y., Sivavec, T., Morris, W.G., 2009. Development of radio-frequency identification sensors based on organic electronic sensing materials for selective detection of toxic vapors. J. Appl. Phys. 106, 124902.
Puligundla, P., Jung, J., Ko, S., 2012. Carbon dioxide sensors for intelligent food packaging applications. Food Cont. 25, 328–333.
Rajesh, Takashima, W., Kaneto, K., 2004. Amperometric phenol biosensor based on covalent immobilization of tyrosinase onto an electrochemically prepared novel copolymer poly (*N*-3-aminopropyl pyrrole-*co*-pyrrole) film. Sens. Actuat. B 102, 271–277.
Ranjan, S., Dasgupta, N., Chakraborty, A., Melvin Samuel, S., Ramalingam, C., Shanker, R., Kumar, A., 2014. Nanoscience and nanotechnologies in food industries: opportunities and research trends. J. Nanopart. Res. 16, 1–23.
Retama, J.R., Mecerreyes, D., Lopez-Ruiz, B., Lopez-Cabarcos, E., 2005. Synthesis and characterization of semiconducting polypyrrole/polyacrylamide microparticles with GOx for biosensor applications. Colloid. Surf. A 270–271, 239–244.
Rhim, J.-W., Park, H.-M., Ha, C.-S., 2013. Bio-nanocomposites for food packaging applications. Prog. Polym. Sci. 38, 1629–1652.
Rooney, M.L., 1995. Active Food Packaging. Springer, USA.
Ruiz-Garcia, L., Lunadei, L., Barreiro, P., Robla, J.I., 2009. A review of wireless sensor technologies and applications in agriculture and food industry: state of the art and current trends. Sensors 9, 4728–4750.

Sandhya, 2010. Modified atmosphere packaging of fresh produce: current status and future needs. LWT Food Sci. Technol. 43, 381–392.
Sanjib, K.P., Jatindra, K.S., 2014. Nanotechnology in Food Processing, Introduction to Advanced Food Process Engineering. CRC Press, Boca Raton, pp. 315–342.
Sharrock, K.R., 2012. Advances in freshness and safety indicators in food and beverage packaging. In: Yam, K.L., Lee, D.S. (Eds.), Emerging Food Packaging Technologies. Woodhead Publishing, Cambridge, UK, pp. 175–197.
Siegrist, M., Cousin, M.-E., Kastenholz, H., Wiek, A., 2007. Public acceptance of nanotechnology foods and food packaging: the influence of affect and trust. Appetite 49, 459–466.
Silbert, L., Ben Shlush, I., Israel, E., Porgador, A., Kolusheva, S., Jelinek, R., 2006. Rapid chromatic detection of bacteria by use of a new biomimetic polymer sensor. Appl. Environ. Microbiol. 72, 7339–7344.
Silvestre, C., Duraccio, D., Cimmino, S., 2011. Food packaging based on polymer nanomaterials. Prog. Polym. Sci. 36, 1766–1782.
Siró, I., 2014. Intelligent Packaging and Food Safety, Practical Food Safety. John Wiley & Sons, Chichester, UK, pp. 375–394.
Smits, E., Schram, J., Nagelkerke, M., Kusters, R., Heck, G.V., Acht, V.V., Koetse, M., Brand, J.V.D., Gelinck, G., 2012. Development of printed RFID sensor tags for smart food packaging. IMCS 2012–The 14th International Meeting on Chemical Sensors, Germany, pp. 403–406.
Smolander, M., Hurme, E., Ahvenainen, R., 1997. Leak indicators for modified-atmosphere packages. Trends Food Sci. Technol. 8, 101–106.
Snow, E.S., Perkins, F.K., 2005. Capacitance and conductance of single-walled carbon nanotubes in the presence of chemical vapors. Nano Lett. 5, 2414–2417.
Snow, E.S., Perkins, F.K., Houser, E.J., Badescu, S.C., Reinecke, T.L., 2005. Chemical detection with a single-walled carbon nanotube capacitor. Science 307, 1942–1945.
Sozer, N., Kokini, J.L., 2009. Nanotechnology and its applications in the food sector. Trends Biotechnol. 27, 82–89.
Su, Y.-l., Li, J.-R., Jiang, L., Cao, J., 2005. Biosensor signal amplification of vesicles functionalized with glycolipid for colorimetric detection of *Escherichia coli*. J. Coll. Interf. Sci. 284, 114–119.
Summers, L., 1992. Intelligent Packaging. Centre for Exploitation of Science and Technology, London.
Thompson, R.B., Lakowicz, J.R., 1993. Fiber optic pH sensor based on phase fluorescence lifetimes. Anal. Chem. 65, 853–856.
Tieke, B., Graf, H.J., Wegner, G., Naegele, B., Ringsdorf, H., Banerjie, A., Day, D., Lando, J.B., 1977. Polymerization of mono- and multilayer forming diacetylenes. Coll. Polym. Sci. 255, 521–531.
Tombelli, S., 2012. Piezoelectric biosensors for medical applications. In: Higson, S. (Ed.), Biosensors for Medical Applications. Woodhead Publishing, Cambridge, UK, pp. 41–64.
Valdés, M., Valdés González, A., García Calzón, J., Díaz-García, M., 2009. Analytical nanotechnology for food analysis. Microchim. Acta 166, 1–19.
Vamvakaki, V., Chaniotakis, N.A., 2007. Pesticide detection with a liposome-based nano-biosensor. Biosens. Bioelectron. 22, 2848–2853.
Vanderroost, M., Ragaert, P., Devlieghere, F., De Meulenaer, B., 2014. Intelligent food packaging: the next generation. Trends Food Sci. Technol. 39, 47–62.
Vergara, A., Llobet, E., Ramírez, J.L., Ivanov, P., Fonseca, L., Zampolli, S., Scorzoni, A., Becker, T., Marco, S., Wöllenstein, J., 2007. An RFID reader with onboard sensing capability for monitoring fruit quality. Sens. Actuat. B 127, 143–149.

Wang, L., Song, S., Pan, D., Li, D., Fan, C., 2010a. Gold Nanoparticle-Based Sensing Strategies for Biomolecular Detection. Pure Appl. Chem., p. 81.

Wang, S., Singh, A.K., Senapati, D., Neely, A., Yu, H., Ray, P.C., 2010b. Rapid colorimetric identification and targeted photothermal lysis of Salmonella bacteria by using bioconjugated oval-shaped gold nanoparticles. Chem.—Eur. J. 16, 5600–5606.

Wang, Z., Duan, N., Li, J., Ye, J., Ma, S., Le, G., 2011. Ultrasensitive chemiluminescent immunoassay of *Salmonella* with silver enhancement of nanogold labels. Luminescence 26, 136–141.

Weinstein, R., 2005. RFID: a technical overview and its application to the enterprise. IT Prof. 7, 27–33.

Wiziack, N.K.L., Paterno, L.G., Fonseca, F.J., Mattoso, L.H.C., 2007. Effect of film thickness and different electrode geometries on the performance of chemical sensors made of nanostructured conducting polymer films. Sens. Actuat. B 122, 484–492.

Wolfbeis, O.S., 2008. Fiber-optic chemical sensors and biosensors. Anal. Chem. 80, 4269–4283.

Xu, K., Huang, J., Ye, Z., Ying, Y., Li, Y., 2009. Recent development of nano-materials used in DNA biosensors. Sensors 9, 5534.

Yam, K.L., Takhistov, P.T., Miltz, J., 2005. Intelligent packaging: concepts and applications. J. Food Sci. 70, R1–R10.

Yang, B., Aksak, B., Lin, Q., Sitti, M., 2006. Compliant and low-cost humidity nanosensors using nanoporous polymer membranes. Sens. Actuat. B 114, 254–262.

Zhang, W.-H., Zhang, W.-D., 2008. Fabrication of SnO_2–ZnO nanocomposite sensor for selective sensing of trimethylamine and the freshness of fishes. Sens. Actuat. B 134, 403–408.

Zhao, Y., Ye, M., Chao, Q., Jia, N., Ge, Y., Shen, H., 2009. Simultaneous detection of multifood-borne pathogenic bacteria based on functionalized quantum dots coupled with immunomagnetic separation in food samples. J. Agr. Food Chem. 57, 517–524.

Zhao, Y., Hao, C., Ma, W., Yong, Q., Yan, W., Kuang, H., Wang, L., Xu, C., 2011. Magnetic bead-based multiplex DNA sequence detection of genetically modified organisms using quantum dot-encoded silicon dioxide nanoparticles. J. Phys. Chem. C 115, 20134–20140.

Zhao, X., Lin, C.W., Wang, J., Oh, D.H., 2014. Advances in rapid detection methods for foodborne pathogens. J. Microbiol. Biotechnol. 24, 297–312.

Zourob, M., Elwary, S., Turner, A.P.F., 2008. Principles of Bacterial Detection: Biosensors, Recognition Receptors and Microsystems. Springer, New York.

19

NANOMATERIAL-BASED BIOSENSORS FOR FOOD CONTAMINANT ASSESSMENT

Rocio B. Dominguez*, Akhtar Hayat, Gustavo A. Alonso†, Juan M. Gutiérrez‡, Roberto Muñoz‡, Jean-Louis Marty§**

**Advanced Materials Research Center, S.C., Chihuahua, Chihuahua, Mexico; **COMSATS Institute of Information Technology (CIIT), Interdisciplinary Research Centre in Biomedical Materials (IRCBM), Lahore, Pakistan; †Autonomous University of Guerrero, Chilpancingo, Guerrero, Mexico; ‡CINVESTAV-IPN, Department of Electrical Engineering, Bioelectronics Section, Mexico DF, Mexico; §Université de Perpignan Via Domitia, IMAGES, Perpignan Cedex, France*

1 Introduction

Major food safety issues, such as the milk contamination scandal in China and the Irish pork crisis, have highlighted the need for reliable analytical detection methodologies to ensure food safety and protect consumers' health. The problem of food contamination is a challenging issue, and is considered to be as old as the beginnings of food processing and production technologies (Ellis et al., 2012). The first person to take initiative regarding food adulteration was the German analytical chemist Frederick Accum, who studied food alteration and culinary poison (Accum, 1820). This study enabled researchers to identify the risks associated with some of the food contaminants and their serious toxic effects on humans. This pioneer work subsequently led to further studies of food contaminants in thousands of beverages and foodstuff samples.

Further research was followed by rigorous regulation in some countries such as the United Kingdom, establishing the foundation for many modern food regulation authorities all over the world. By the end of this century, it was evident that food safety did not concern only a specific country but was of interest all around the world with no international boundaries since all life on earth planet relies on pure and healthful foodstuffs (Ellis et al., 2012).

Nanobiosensors. http://dx.doi.org/10.1016/B978-0-12-804301-1.00019-9

Food for human necessities is provided through a worldwide complex network and, in addition to being vulnerable to biological contaminants, can be attacked by a large variety of organic compounds produced by anthropogenic phenomena or natural processes. These can be classified as residues that are present in food or the environment as a result of their application, and contaminants that are not related to human activity but are produced as a result of natural or anthropogenic activities (Food and Agriculture Organization of the United Nations, 2015). These compounds can arise during production (ie, pesticides or veterinary drugs), during storage, and/or during food processing stages (ie, mycotoxins). Modern societies have great concerns regarding the presence of harmful compounds in foodstuffs, particularly after witnessing a number of contamination incidents all around the world (Malik et al., 2010). Apart from its impact on public health, the occurrence of such contaminants has a negative impact on the economy of affected countries, limiting or reducing the marketing of food products in national and international markets. In the aforementioned consequence, the monitoring of residues and contaminants is an issue of vital importance in food safety, and public and private organizations have put in much effort to overcome this risk.

The total elimination of food contaminants is impossible, though good manufacturing practice and Hazard Analysis and Critical Control Point systems can be a way to prevent or reduce food contamination impact (Marvin et al., 2009). Therefore, the best possible approach toward this problem can be to keep harmful chemical residue below risk level. As described earlier, to protect consumer health against the risk associated with harmful residues, national and international organizations have set the regulations to control the presence of contaminants to ensure food safety. The European Union (EU) is leading the world to provide up-to-date regulations based on scientific assessment of harmful impact and then subsequently assuring a safe food supply. In this context, the EU has formed the European Food Safety Authority (EFSA) to provide scientific advice on the nature of existing and emerging food contaminants. Similarly, the rapid alert system for food and feed (RASFF) has also been introduced, along with establishment of maximum residue levels for many food born contaminants. The regulation for some food contaminants still needs to be formed and a situation of zero tolerance policy is considered for such analytes. In the United States, two agencies including Food and Drug Administration (FDA) and the Environmental Protection Agency (EPA), regulate food safety issues. Such strategies are also implemented in other countries. Similarly, the Codex Alimentarius (an international body) is working to ensure the implementation of food safety–related regulations to protect consumer health all over the world (World Health Organization, 2015).

1.1 Current Method for Food Analysis

The high number of food contaminants and residues has generated an increasing demand for the analysis of possible contamination sources such as potential ingredients, raw materials (animal or plant origin), and food commodities. In this context, the mixture of contaminants with their continuous changing nature poses serious risk, and it is very tricky challenge to monitor them (Farré and Barceló, 2013). Therefore, it is of vital importance to introduce novel and reliable methodologies in order to monitor a wide range of analytes simultaneously (Díaz et al., 2012). Generally, food contaminant analysis methodologies rely on three steps:

- Sample treatment, which involves the extraction of desired analyte from the matrix, integrating a cleanup and preconcentration step.
- Separation of the analyte to ensure its highly selective and specific detection.
- Qualitative/quantitative analysis to have insight on the level of the analyte in the given sample.

In order to fulfill these requirements, and to ensure food safety requirements, different analytical methodologies have been designed and adopted to perform routine laboratory analysis.

1.1.1 Chromatographic Methods

Chromatographic methods are the most commonly used techniques in routine laboratories for simultaneous detection of many classes of contaminants and residues (Haider et al., 2014a,b). Gas chromatography has been employed for the detection of nonpolar compounds, but recently, liquid chromatography has been extensively used for the detection of more polar compounds along with nonpolar compounds. The best-known example in this regard is the use of ultra high performance liquid chromatography (UHPLC), which is based on stationary phases with particle size lower than 2 μm. This technique offers many advantages such as improved sensitivity, speed, and peak resolution and is the most adequate choice in many applications related to food safety. The other advantages include its high throughput due to considerable reduction in the analysis time, allowing the elution of more than 100 compounds in less than 11 min. These characteristics make it a widely used technique in routine laboratories for generic chromatographic conditions (Núñez et al., 2012). The combination of UHPLC with appropriate extraction procedure has been proven to reduce the overall analysis time, elevate the sample throughput, permitting the simultaneous determination of a wide variety of compounds in a large number of samples per a day, and thus reducing the cost per analysis. Despite many advantages, there are still some problems associated with UHPLC, such as the need of high back-pressure due to the low particle size of the applied stationary phase. Capillary electrophoresis has been proposed as an alternative to the UHPLC due to its different selectivity, reduced volume of sample and reagents, and its high separation output. Several attempts to apply CE for food safety have been carried out, but its integration in routine laboratories is still rare and not well established (Kumar et al., 2010).

1.1.2 Mass Spectrometry

Mass spectrometry is considered one of the most suitable techniques employed in food analysis to ensure consumer health, and is recommended as the most appropriate analysis method for food residue and organic adulteration. This technique is suitable for both targeted and nontargeted analysis (Wang et al., 2013). High-resolution mass spectrometry is an advanced type of mass spectrometry, but was associated with high instrumental complexity a few years back, limiting its application to only the most demanding analytes. However, the integration of modern analyzers such as time of flight has significantly reduced its complexity, making it suitable for routine analysis. Some of advantages of this technique include but are not limited to: (1) suitability for simultaneous analysis of a large number of compounds; (2) both targeted and nontargeted analysis can be performed with one injection analysis; (3) isotopic pattern and fragments are helpful for the confirmation process and reliable for retrospective analysis (Kaufmann, 2012). On the other hand, disadvantages include: (1) requirement of high inversion in standards for quantitative analysis; (2) validation process is time consuming; (3) difficult to maintain quality system due to high number of compounds; (4) integration of complex software to obtain the final information; (5) heavy raw data and skilled personnel needed to operate analyzers (Romero-González, 2015).

2 Biosensors as Emerging Tools for Analysis

Conventional methods are expensive and require long procedures and skilled staff, which limit their use for routine analysis. As an alternative to classical methods, there is increasing interest in the use of biosensors as analytical tools in medicine, agriculture, food safety, bioprocessing, and industrial and environmental monitoring. Because of their exceptional performance capabilities, which include high specificity and sensitivity, rapid response, low cost, relatively compact size, and user-friendly operation, biosensors have become an important tool for detection of chemical and biological components. Human beings have performed bioanalysis since a long time ago, with the use of nerve cells of the nose to detect scents or the enzymatic reaction in the tongue to taste food. With progress in understanding about the function of living organisms, scientists have sought to apply them in man-made reactions to detect trace amounts of biochemicals in complex systems. Using bioreceptors from biological organisms, researchers have developed biosensors as a new mean of analytical and chemical analysis. The field of biosensors originated in the works by Clark and Lyons (1962), Guilbault et al. (1962), Updike and Hicks (1967), and Guilbault and Montalvo (1969). Later on, a mediated electrochemical biosensor using ferrocene to alleviate electroactive species was described by Di Gleria et al. (1986). This work led to the successful commercialization of a glucose pen by Medisens. Since these first reports, there was an extensive growth in the field of biosensor with a wide variety of applications in food monitoring and environmental sensing.

According to the National Research Council, a biosensor is a detection device with a living organism or a product derived from living system as recognition element (ie, bioreceptor) and a transducer to convert biological reaction into a measurable signal or indication. The most important properties, such as specificity and sensitivity of a biosensor, are dependent on the bioreceptor itself, the immobilization method, and the transduction method. The high specificity of the system can be successfully achieved only if there is a strong efficient coupling reaction between the biological and transducer components. A general representation of biosensor is depicted in Fig. 19.1.

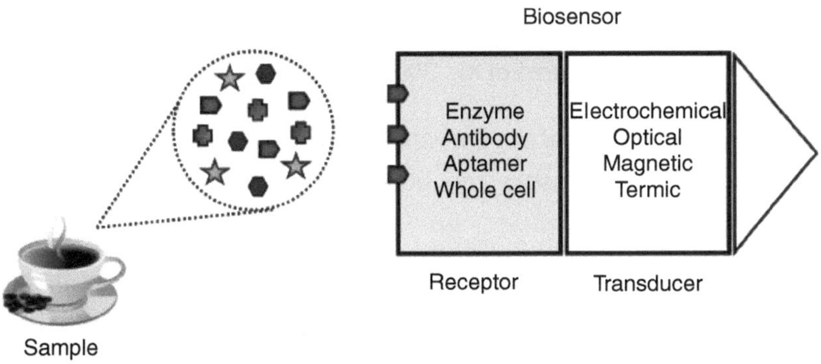

Figure 19.1. General representation of a biosensor and their interaction with sample analytes.

Bioreceptors are the key factors in the fabrication of biosensors and can take many forms. Numerous biorecognition element have been used in biosensing, however, they can be classified into following major categories.

1. Enzymes that catalyze specific biochemical reactions.
2. Antibodies (also known as immunoglobulins), which bind a particular substance with high affinity. They form an important part of a biological group termed binding proteins.
3. Nucleic acid/DNA/aptamers are ligands selected to have high binding affinity and specificity to the target molecules.
4. Whole cells contain a wide range of biomolecules that respond to many types of chemicals.
5. Receptors that interact with specific chemicals such as hormones.

The biosensors based on bioreceptors including enzymes, antibodies, and aptamers have been discussed in detail later, due to their wider applications in the field of food biosensors.

2.1 Enzyme-Based Biosensor

A major portion of the research and development has been devoted to enzyme-based sensors in the biosensors technology. The response of the enzymatic biosensor is determined by the concentration of the enzymatic product, and this reaction is controlled by the rates of two simultaneous processe. These are the conversion of the substrate and the diffusion of the product from the enzyme layer. As enzymes are not consumed in the reaction, they are very suitable for use in biosensors for the continuous monitoring of the analyte concentration. Clark and Lyons (1962) developed the first enzymatic biosensor with glucose oxidase (GOD) for quantification of glucose within a sample. Since this achievement, a broad variety of enzymatic biosensors have been developed for determination of several substances (D'Orazio, 2003; Wang, 2002b; Wilson and Gifford, 2005; Arya et al., 2008; Andreescu et al., 2002; Zhou and Zhi, 2006; Yonemori et al., 2009).

2.1.1 Classification of the Enzymatic Biosensor

Enzymatic biosensors can be categorized into the following four types:

1. The basis for the majority of the biosensor is the conversion of the desired analyte into a detectable product. This product is then quantified by its properties such as electro-active or optical; in such a type of biosensor, an analyte works as a substrate of a specific enzyme (Yonemori et al., 2009; Zhou and Zhi, 2006).

2. The second approach is enzyme thermistor, where heat evolved by the enzyme catalyzed reaction of the analyte is used to quantify analyte concentration (Cooney et al., 1974; Mosbach and Danielsson, 1974).
3. The next type is where the analyte inhibits the enzymatic activity. In this biosensor, certain analytes block a specific enzyme reaction, decreasing the original signal (Istamboulie et al., 2007).
4. A novel enzymatic biosensor strategy is enzyme optodes, in which the optical signal originates from the enzyme itself or from associated biomolecules such as cofactors instead of from enzymatic product (Lubbers and Opitz, 1975; Völkl et al., 1980).

The main benefit of an enzymatic biosensor is the high specificity and sensitivity of the enzyme used as biological recognition element. The first developed enzymatic biosensors were mediator-free enzyme electrodes, which were composed of a suitable enzyme immobilized close to surface of an ion-selective electrode. The enzymatic reaction leading to consumption or production of the specific ion was monitored amperometrically or potentiometrically as a function of analyte concentration (Turner, 1987). However, the first commercially available enzymatic biosensor was based on the mediated electron transfer mechanism (Cass et al., 1984).

2.2 Antibody-Based Biosensor

Immunoassay with immunological proteins was first described by Yalow and Berson (1959) and applied for plasma insulin in human subjects. Antibodies are very well known as biorecognition elements due to their exquisite specificities and strong affinities for the target antigen. Antibodies have been successfully applied in the area of clinical diagnostic, in addition to the exploitation of the monoclonal and polyclonal antibodies in the biosensor technology. Polyclonal antibodies are the product of immunization of a living host organism, usually a mammal. Polyclonal antiserum (the basic body product obtained) contains many different antibodies, but only a small fraction is likely to recognize the immunogen. Polyclonal antibodies are synthesized by multiple plasma cells. The major drawback with the use of polyclonal antibodies is the highly heterogeneous nature of the polyclonal antiserum. A method for the preparation of highly homogeneous antibodies (termed *monoclonal antibodies* or Mabs), was developed in 1975 and has made a major impact in the field of immunology. Mabs often possess very high specificity, and are obtained by growing hybridoma cell lines in culture with relatively high purity (Byfield

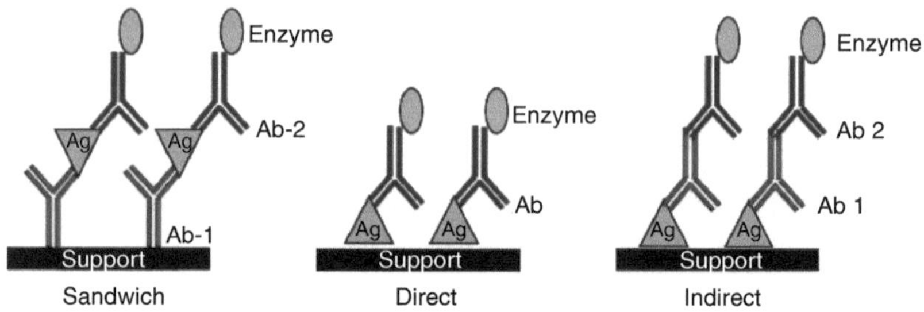

Figure 19.2. Formats for ELISA assay.

and Abuknesha, 1994). The key parameters for the use of antibodies in the biosensor applications are sensitivity, selectivity, stability, immobilization, labelling, and antibody size.

Immunoassays are the bioanalytical methods in which the quantization of the analyte depends on the reaction of an antigen (analyte) and an antibody. Principally, these methods are based on a competitive and noncompetitive binding reaction between sample analyte and limited binding sites on a highly specific anti-analyte antibody. When these immunoanalytical reagents are mixed and incubated, the formation of an immune complex takes place. Analysis is achieved by measuring the label activity (eg, radiation, fluorescence, or enzyme) or by the change in other chemical or physical properties on the transducer surface. A standard calibration curve as a function of the concentration of the analyte in the sample is constructed. Various formats of antibody-based assays have been employed in the biosensor field. In Fig. 19.2 three of the most common formats found in literature for enzyme-linked immunosorbent assay (ELISA) are presented. In the sandwich format, first antibody is immobilized over a support surface (eg, microtiter plate, electrode) and antigen is trapped between the first and the labeled second antibody. For direct format, antigen is immobilized on the support and a labeled antibody reveals the formed immune complex. In the indirect format, antigen is also immobilized but the labeled second antibody reveals the immune complex.

2.3 Aptamer-Based Biosensor

Aptamers are functional molecules engineered in vitro through a procedure known as systematic evolution of ligands by exponential enrichment (SELEX) first reported in 1990 (Tuerk and Gold, 1990; Ellington and Szostak, 1990). They are based on short sequences of either DNA or RNA that fold into 3D structures and bind by complementary shape interaction.

The SELEX process is a technique for screening very large combinatorial libraries of oligonucleotides by an iterative process of in vitro selection and amplification. In the screening process, a random sequence oligonucleotide library, usually consisting of approximately 10^{14}–10^{16} oligonucleotide strands with a random sequence of defined size, is incubated with the target of interest. Sequences that bind to the target are separated from the unbound species using a suitable partitioning method, and then, the sequence of these candidates is amplified using polymerase chain reaction (PCR). This population of selected sequences represents a mixture of oligomers with variable affinity toward the target analyte. The single-strand population obtained after the purification step is incubated with a fresh sample of the target for the next round of selection. Iteration of the aforementioned protocol results in the isolation of a pool of nucleotide sequences displaying sequential motifs, which after 8–15 iterative SELEX runs converge to one or a few binding sequences. Once the sequence is identified, an aptamer is produced by chemical synthesis (Stoltenburg et al., 2007).

One main advantage of aptamers over antibodies is their chemical stability. Since their fabrication does not involve the use of animals, aptamers are less expensive and tedious compared with antibodies. A wide range of molecules such as fluorescent dyes, enzymes, biotin, and labels such as NPs can be attached to aptamers for different detection methodologies (Paniel et al., 2013). In addition, unlike antibodies, aptamers can be regenerated and reused. Different configurations, such as single site binding and dual site binding with a variety of formats can be implemented for aptamer-based assays detection. Plenty of format assays based on aptamer as bioreceptor have been well reported for diverse analytical applications such as health, environmental, food, and clinical diagnosis (Tombelli et al., 2007).

2.4 Role of Nanomaterials for Biosensors

Nanomaterials exhibit highly tunable size and shape as well as controllable surface charge and physiochemical characteristics (Asefa et al., 2009). Nanomaterials also show unique surface chemistry, thermal and electrical properties, and high surface area per unit mass, which is desirable to enhance biosensor sensitivity (Hayat et al., 2014). There is extensive literature material concerning the development and application of nanomaterials to biosensor design and its results on environmental, clinical, and food analysis have been widely reported (Pérez-López and Merkoçi, 2011a; Wang, 2007; Xu and Wang, 2012; Periasamy et al., 2009). Due to their importance, we present some of the main roles of nanomaterials applied to biosensors.

2.4.1 Immobilization Platform for Biomolecules

One of the critical steps on biosensor design is the immobilization strategy for the receptor biomolecule (Sassolas et al., 2012). Usually, direct surface adsorption of biomolecules onto bulk materials results in denaturation and loss of bioactivity (Li et al., 2010). Thus, in recent years traditional approaches for biosensor fabrication have been greatly improved by the inclusion of nanomaterials resulting in the reduction of diffusion limits as well as the increasing of biomolecule loading. The most commonly found nanomaterials involved in biomolecule immobilization are metal nanoparticles (MNPs) and magnetic beads (MBs) (Xu and Wang, 2012; Arduini et al., 2013; Putzbach and Ronkainen, 2013).

Metal and metal oxide nanoparticles, such as silver and gold nanoparticles (AuNPs) as well as zinc oxide, aluminum oxide, zirconium oxide, manganese oxide, and nickel oxide have been explored for the immobilization of biomolecules, specifically enzymes. However, among MNPs, AuNPs are the most extensively reported, given their outstanding characteristics such as high surface-to-volume ratio and high surface energy, which allows a stable immobilization (Li et al., 2010). AuNPs allow fast and direct electron transfer between a wide range of electroactive species and electrode materials. In addition, AuNPs exhibit light-scattering properties and enhancement of the local electromagnetic field. Thus, AuNPs have been reported for the development of a variety of biosensors with electrochemical, optical and piezoelectric techniques (Li et al., 2010). Immobilization with AuNPs has been prerformed through covalent bonds, amine groups and thiol chemistry. However, other nanopaticles (NPs) such as nonmetallic polystyrene nanoparticles have also been used in the literature to improve the biomolecule immobilization efficiency (Hayat et al., 2014).

MBs are the most extensive nanomaterials found in literature for biomolecule immobilization (Palecek and Fojta, 2007; Stanciu et al., 2009; Xu and Wang, 2012). A wide variety of magnetic nanoparticles applied as magnetic supports for biosensor fabrication have been reported. These include diverse encapsulated biomaterials and copolymers along with magnetic nanoparticles. As MNPs, magnetic nanoparticles exhibit high surface-to-volume ratio, which allows high binding capacity and enhances catalytic specificity of the conjugated biomolecules. In addition, immobilization with MBs can be operated with an external magnetic field to capture NPs over the transducer surface. For affinity immobilization, magnetic nanoparticles with streptavidin, amine and carboxylic functional group and protein G are commercially available (Hayat et al., 2014). Additionally to the available functional groups for immobilization, bioreceptors can be artificially modified to achieve the desired orientation using genetically modified enzymes with added artificial tags (Putzbach and Ronkainen, 2013; Campàs et al., 2009).

2.4.2 Signal Amplification

Actually the detection of lower analyte concentration is mandatory in a variety of fields, such as clinical diagnosis, environmental monitoring and food safety assessment. A strength output signal induced by the analyte concentration, even at low levels, can be helpful to improve the sensitivity of the protocol. Thus there are many efforts to develop new strategies with highly sensitive detection with, for example, the inclusion of new electroactive labels (Lei and Ju, 2012). The introduction of nanotechnology allows the possibility of nanomaterial-based signal amplification for high sensitivity and selectivity of analytes. MNPs, semiconductor nanoparticles, and carbon nanosized structures are among the nanomaterial for electrochemical signal amplification.

2.4.2.1 Signal Amplification Through Catalytic Reaction

Enhancing catalytic activity with bio-functionalized NPs can produce synergistic effects to achieve lower limits of detection. This effect can also extend the linear correlation to a wider range of operation. For example, nanoceria, platinum, copper, and nickel NPs have been applied to modify the performance of electrochemical biosensors given their redox capabilities. These modified electrodes have been applied for the detection of target analytes, such as amikacin, for enhancing the sensitivity of detection of well-known biomolecules such as alkaline phosphatase and as electrocatalytic labels to enhance electrochemical detection (Xu et al., 2003; Hayat and Andreescu, 2013; Polsky et al., 2006). Other NPs reported in literature for enhancing the catalytic properties of electrochemical biosensors are silver and gold nanoparticles. These biosensors have been applied for the detection of glucose (Lin et al., 2009). Besides NPs, electrochemical biosensors have been modified with carbon-based materials such as single-walled and multiwalled carbon nanotubes (SWCNT) to achieve signal amplification (Putzbach and Ronkainen, 2013).

2.4.2.2 Nanomaterials as Mediator

For electrochemical biosensors construction, the electrical contact between electrode surface and redox biomolecules is of great importance. However, this direct communication between biomolecule (eg, enzyme active site) and the electrode surface can be obstructed by the insulating protein shell resulting in problems for final electrochemical measurements (Li et al., 2010). Thus, nanomaterials and especially the conductive properties of MNPs have been employed to enhance the electron transfer rate between the biomolecules and the transducer surface. In this sense, MNPs act as mediators to facilitate the electron transfer and can replace other commonly found mediators in electrochemical biosensors (Putzbach and Ronkainen, 2013). Besides MNPs, some nonmetallic NPs such as oxide nanoparticles and semiconductor nanoparticles have been also included to improve the electron transfer rate between biomolecules and electrode surface. The improvement in electron flow can be related with the conductivity of NPs as well as the arrangement between nanoparticles and biomolecules (Hayat et al., 2014).

In a different approach, nanomaterials can be applied as mediators for deposition of electrochemically active species. The nanomaterial-modified surfaces of electrodes can be used as seeds to for the deposition of metals through a growth mechanism (Wang et al., 2001; Guo and Wang, 2007). NPs over the electrode surface create nucleation sites to improve the metal deposition, which result in higher loading of electroactive species and transducer surface communication. The increased communication can be explained by the fact that surface-deposited nanoparticles are closer to the electrode surface as compared to the nanoparticle labels (Hayat et al., 2014). For example, this feature is of great importance for competitive labeled immunosensors and aptasensors, since the dimension of analyte and biomolecules cause a larger distance between label and electrode surface (Kienberger et al., 2005). In addition, seed-mediated nanoparticles exhibit high electroactivity since they are free-form stabilizers, which makes them suitable for the design of amperometric biosensors (Hayat et al., 2014). Seed method can also be used to deposit electrocatalyst material. In this approach, the transducer surface is incubated in a high concentrated substrate solution, which can be electrochemically oxidized or reduced by the deposit electrocatalyst. Concentration of electrocatalyst can be measured indirectly by the resulting catalytic current from electrocatalysis of the substrate (Pei et al., 2013). Since a clean surface can provide efficient binding of substrates, the catalytic properties of these particle-modified electrode surfaces are expected to be very high (Hayat et al., 2014).

2.4.3 Nanocomposites

Nanocomposites manufactured by combining different nanostructures exhibit properties of both materials (ie, the electronic structure and characteristics of the individual nanomaterials). These nanocomposites have attracted great attention in the field of sensors, especially electrochemical sensors. Nanocomposites have been employed in electrochemical biosensors as immobilization supports, signal amplifiers and electrochemical signal generating probe (Wu et al., 2009; Xian et al., 2006; Wang et al., 2011).

2.5 Nanomaterial-Based Biosensors in Flow Injection Analysis

Automation and continuous monitoring are desirable characteristics for foodstuffs analysis. The main techniques for liquid handling in chemical analysis are flow injection analysis (FIA) and sequential analysis (SIA) (Ruzicka and Hansen, 2008; Ruzicka and Marshall, 1990). Briefly, for FIA operation a defined amount of sample is injected in a continuous, nonsegmented carrier stream. The flow moves the sample through the manifold and a combination of diffusion and convection process allows the mixing of samples with the reagents on the carrier stream. Then the resulting reactions can be monitored with a suitable flow-through cell in a variety of detectors (ie, photometers, pH meters, ion-selective electrode). On the other hand, SIA was developed to avoid large amount of reagents used during FIA continuous operation and to provide more facilities for multireagent assays. Unlike FIA, fluids on SIA are computer controlled and flow direction can be altered according to assay necessities. Also, the inclusion of elements such as the microsyringe pump allows the injection of even smaller volumes at the microliter level.

The most common nanomaterial found on flow systems are MBs. In this sense, new flow configurations such as bead injection analysis (BIA) and lab on a valve (LOV) have been presented (Hansen and Miro, 2007; Chen and Wang, 2007). However, to take advantage of continuous injection, flow-through reaction chambers have been integrated to flow systems. Nanomaterials can be used as solid support for immobilization in such chambers. For example, proteins can be immobilized by magnetic nanoparticles using a constant magnetic field into the chamber or by using CNTs as support. With these systems, only a limited and accurate amount of enzymes can be used. Therefore, the reaction chambers are designed, depending on which position will be placed in the system, as shown in Fig. 19.3.

Enzyme can be immobilized in suitable chamber or nanoparticles (NPs) can be held within flow-through reaction channels forming bead-packed reactors. The amount of the enzyme can be varied by the size of the chosen reactor type. As the accessibility

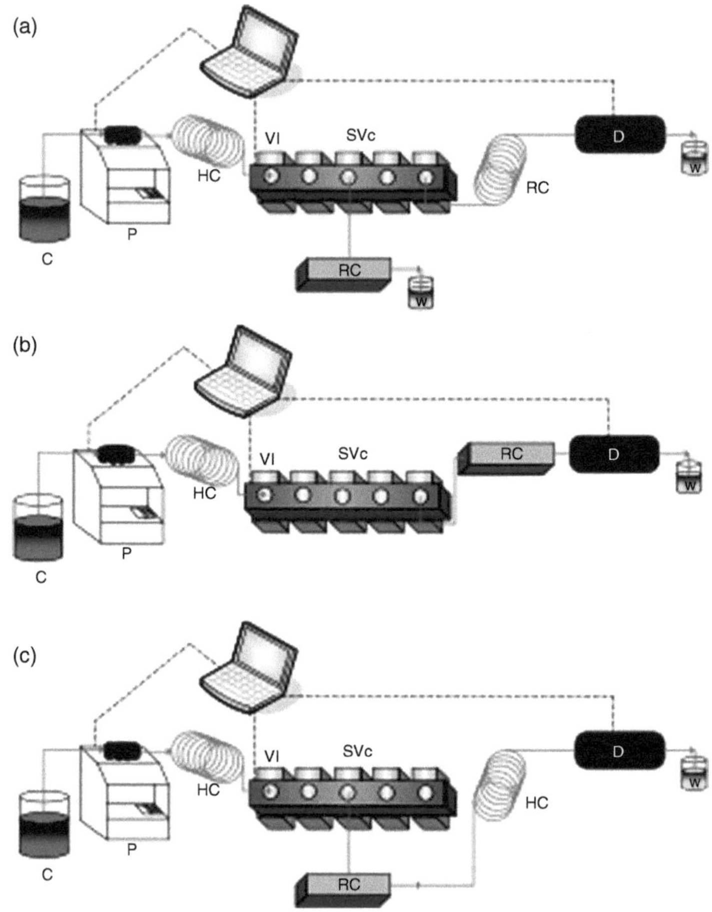

Figure 19.3. Placement of immobilized enzyme reactors in SIA manifolds. (a) Reactor is placed in one of the lateral ports of the selection valves. (b) Chamber is placed between the selection valves and the detector. (c) Chamber is located between the selection valves and reaction coil. *C*, Carrier; *P*, pump; *HC*, holding coil; *Vi*, input valve; *SVs*, selection valves; *RC*, reaction chamber; *RC*, reaction coil; *D*, detector; *W*, waste.

of the enzyme for the analyte is most efficient in the bead-packed reactor, this is the preferred design. The NPs have to fulfill several requirements, some of which are previously mentioned:

- NPs with functional groups for binding should be used. The immobilization must be sufficiently stable, even when the composition of the solution changes due to the presence of the sample.
- Suitable mechanical properties to be integrated in a continuous flowing stream.
- Analyte and products from the reaction should not be absorbed by the NPs.

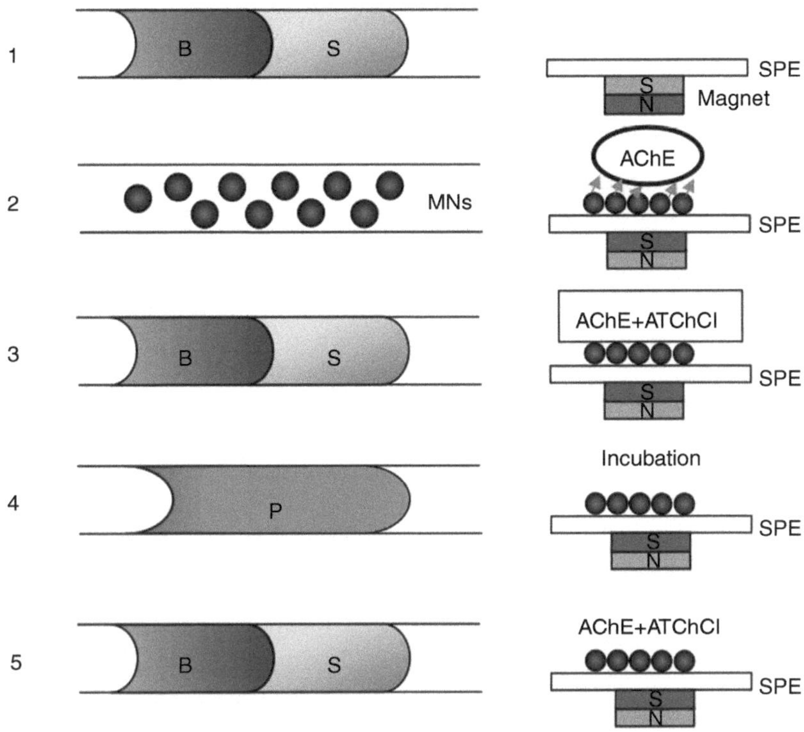

Figure 19.4. Experimental protocol for pesticide detection using enzymatic biosensors based on magnetic nanoparticles in flow system.

Nowadays these reactors can be fabricated using microstructuring technologies, with the bioreceptor being immobilized on different NPs. This allows the fabrication of highly integrated online detection systems very low limits of detection and high reproducibility (Wang, 2002a).

By using nanomaterials such as NPs, the immobilization process can be controlled with the direction and magnitude of a magnetic field (MF) (Pérez-López and Merkoçi, 2011b). Due to small size of magnetic nanoparticles, just a small quantity of bioreceptors can be attached to the nanoparticle and deposition of the magnetic nanomaterial onto the transducer can be controlled by magnitude of a magnetic field. A high magnitude of MF will attract more magnetic nanoparticles while with a low magnitude of MF fewer nanoparticles will be attracted. By optimizing MF, the energy from catalyst of the reaction will be transmitted in an optimized way due to all the bioreceptor can be spread over all the transducer homogeneously as seen Fig. 19.4.

The five-step protocol in Fig. 19.4 can be explained as follows: (1) buffer (B) and substrate (S) are pumped toward the surface of screen-printed electrode (SPE) in order to wash the transducer, (2) magnetic nanoparticles (MNPs) with immobilized enzyme are injected to surface of the SPE. (3) Buffer and substrate are pumped to the newly created biosensor and initial response is generated. (4) Analyte (A) is sent to the biosensor for a given period of time (ie, incubation time). (5) Buffer and substrate are pumped to the biosensor in order to find the remaining activity. But following this sequence a new target concentration can be registered since in every sequence a new biosensor is created. When the magnet is inverted any nanoparticle can be attached to the support (step 1). In order to perform a complete calibration curve for a given analyte (eg, pesticide) all previous processes should be repeated using different concentrations of the target analyte.

The previous described process can be used for different target analytes (eg, pesticides, mycotoxins) toward different immobilized bioreceptors (eg, enzymes, antibody, aptamer). Some conditions must be taken into account: the bioreceptor should have a chemical treatment in order to be attached to the magnetic nanoparticles. Any toxins can be tested in an aqueous form. The magnet should move around its own axis with the aim to spread uniformly the bioreceptor over the transducer (Dominguez et al., 2015).

3 Applications of Nanomaterial-Based Biosensors for Food Analysis

3.1 Nanomaterials-Based Biosensors for the Detection of Pesticides

Biosensors based on the immobilization of acetylcholinesterse (AChE) or organophosphorus hydrolase (OP) enzymes have been successfully applied in the detection of pesticides. However, size, stability, reproducibility, and high sensitivity of biosensors are always challenges to be achieved. With the inclusion of nanomaterials such as CNTs, AuNPs, quantum dots (QDs), and magnetic nanoparticles in biosensor fabrication, pesticide detection has been significantly enhanced on features such as size, stability, and detection limits (Periasamy et al., 2009; Zhang et al., 2014).

For these biosensors, the most common procedure is the immobilization of bioreceptor with the nanomaterial (ie, nanoparticles surface) via chemical bonding (Zhang et al., 2014). Thus, the bioreceptor is not trapped in a membrane and the pesticide can arrive directly into the active site of the AChE as depicted on Fig. 19.5.

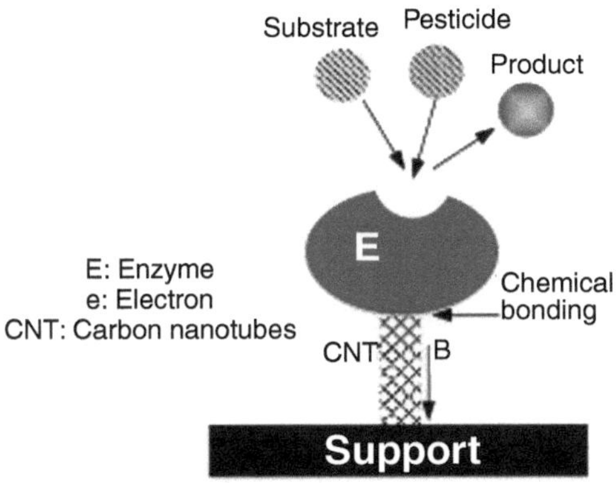

Figure 19.5. Carbon Nanotubes linked to enzyme.

There are many advantages that nanomaterial-based biosensors (NBBs) have over traditional biosensors. For example, modified electrodes with CNTs enhanced anodic detection of *p*-nitro phenol (a product of the OPH reaction). In this study, NBBs exhibited higher sensitivity and stability compared with traditional electrodes. However, the inclusion of nanomaterial should be suited to the targeted pesticide (OP, carbamate, pyrethroid). Parathion is one of the most studied pesticides with moderate toxicity and a regulated permissible concentration set by the European Union in foods of 0.05 mg/kg. Recently, a composite of ZrO_2/Au film was used in order to perform voltammetric detection of parathion. The nanofilm showed strong affinity to parathion and a limit of detection (LOD) lower than 4 ng/mL was achieved (Wang and Li, 2008). An outstanding advantage of NNBs is the size reduction, which is associated with low reagent consumption and the possibility of full integration in analytical systems such as lab on chips (LOC). In this sense, Llopis et al. (2009) presented a LOC for carbofuran detection with immobilized AChE over magnetic beads. The device featured low volumes of reagents, shorter incubation times and detection at ppb level (Llopis et al., 2009).

As an illustration of NBB for pesticide detection, in this section we present an enzymatic NBB incorporated into an automatic system for the on line detection of chlorpyriphos-oxon (CPO). Immobilization of enzyme is based on the affinity between functionalized magnetic beads and a genetically engineered AChE. The development, characterization, and performance of the system in the detection of CPO are presented.

3.1.1 Affinity Immobilization of AChE on Magnetic Beads

Immobilization technique based on metal-chelate affinity was used to bind genetically modified B131 AChE with histidine tag to preactivated Ni-IDA magnetic beads (Istamboulie et al., 2007). Here, immobilization protocol is briefly described: 30 μL of colloidal suspension of Ni–IDA beads were placed in a 1.5 mL tube and washed twice with 300 μL of binding buffer. The tube was placed on the Adem-Mag SV magnetic support to hold magnetic beads while liquid solution was discharged. Then, beads were taken along with 1 mL of enzymatic solution and gently mixed for 15 min. The tube was placed on the magnetic support to eliminate the supernatant. Finally magnetic beads were washed twice with 200 μL of binding buffer to eliminate unbound enzyme.

3.1.2 Flow Operation of Enzymatic Nanomaterial-Based Biosensor

As stated before, the AChE-coated magnetic beads (AChE-MBs) were incorporated into continuous flow to perform online detection of CPO. The custom flow system was assembled with an automatic microsyringe coupled with a multiport valve controlled by a graphical custom interface developed in LabVIEW 8.5. SPE was integrated to the system with a magnetic custom flow cell. Operational parameters such as flow rate, incubation time, washing step, injection volumes, and signal measurements were also verified by the same interface. All reagents during flow assay were injected or aspirated by controlling a bidirectional syringe pump and valves. Amperometric measurements were registered using the PGSTAT Tacussel potentiostat connected to the PMD1208FS data acquisition card (Mishra et al., 2012).

The flow system was used to perform automated amperometric measurements of NBB against CPO. To find the best NBB response, 1 μL of AChE-MBs was diluted in 300 μL of PBS buffer and gently mixed during 1 min. Injections of 50, 100, 150, 200, 250, and 300 μL of AChE-MBs suspension were dispensed with the syringe to the flow cell and placed over the surface of SPE. A volume of 200 μL was chosen for biosensor fabrication as it exhibited good amplitude (650 nA) with a low RSD. of 2.97% after three repetitions. Flow rates in the range of 100 μL/min to 600 μL/min were studied. For the final working conditions a flow rate of 440 μL/min was chosen, according with the RSD of 4.78% for $N = 10$ assays. The enzymatic NBB amplitude was registered by adding 500 μL of 1 mM of ATChCl substrate and 500 μL of PBS buffer for washing. Incubation time of CPO was set in 10 min and 900 μL of pesticides solution in the range of 10^{-6}–10^{-11} was injected to the flow cell. Inhibited AChE-MBs were discharged after each measurement and new AChE-MBs were injected for a new

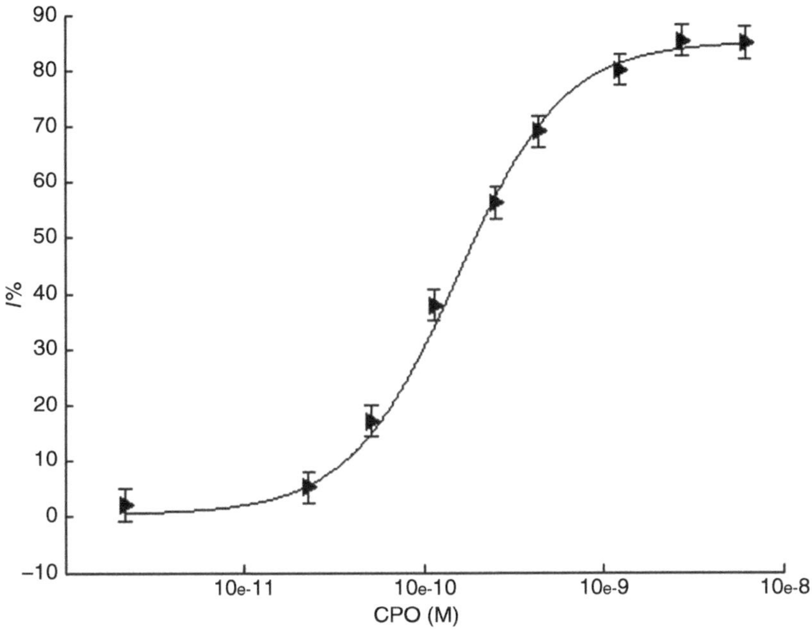

Figure 19.6. Calibration curve of enzymatic NBB for CPO detection.

measurement. Real samples in the range of range of 10^{-7}–10^{-10} M of CPO were also tested with the system (Dominguez et al., 2015).

3.1.3 Results for the Quantification of CPO

Once the stability of biosensor was verified without major variations, the system was applied to analyze the inhibitory effect of CPO over B131 enzyme. The high stability achieved on biosensor response ensured that signal reduction was induced by pesticide inhibition and not by enzyme leaking. Fig. 19.6 shows the calibration curve obtained by plotting the calculated inhibition (I%) against the corresponding CPO concentration. Each experimental point in the calibration curve was measured by triplicate. The range evaluated for CPO was 1×10^{-6} to 7.8×10^{-11} M. The achieved LOD was 1×10^{-11} M of CPO and the average RSD for the whole range was 1.92%. The applicability of the automatic system on pesticide detection in real samples was evaluated by analyzing lake water. Unspiked water samples were incubated for 10 min without any pretreatment in the range of 1.1×10^{-8} to 1×10^{-10} M. The LOD for this assay was 1.56×10^{-10} M with RSD of 3.59% and $R^2 = 0.9926$ (Dominguez et al., 2015).

Previous study showed the higher performance of NBB over biosensor fabricated with traditional immobilization techniques

(Istamboulie et al., 2007). The inclusion of AChE-MBs allowed a renewable biosensing layer, avoiding additional steps such as reactivation. Lower LOD was reached and by including AChE-MBs into a flow system automated detection of target analyte CPO was achieved on buffer samples and real samples.

3.2 Nanomaterial-Based Immunosensor for Okadaic Acid Detection

Okadaic acid (OA) is a lipophilic marine biotoxin produced by *Dinophysis* and *Prorocentrum* dinoflagellates. For humans there is a risk of poisoning by consuming contaminated shellfish such as bivalve mussels. The most well-known effect of OA is diarrheic shellfish poisoning (DSP), but its potential cancerogenic, mutagenic, and immunotoxic effects have been studied for a while. Thus, the European Commission has established a maximum OA-permitted level of 160 mg/kg of mussels, and even a level of 45 mg/kg of mussels has been considered (Hayat et al., 2011). Mouse bioassay was previously recommended as reference method for OA detection, but was replaced with liquid chromatography coupled with fluorescence because of its inherent ethical problems, poor selectivity, and poor accuracy. Regardless of the current techniques, there is still a need for fast, simple, autonomous, and potentially continuous OA detection method.

Biosensors have been proposed as alternative tools for OA detection. For example, an enzymatic biosensor based on protein phosphatase 2A (PP2A) was developed by Campas et al. (2008). The biosensor showed a very low limit of detection (LOD) but the enzyme was very expensive and exhibited poor stability and matrix effect in addition to false positives. A different approach is the development of immunosensors since numerous immuno-based test kits have validated the specificity of antibody for OA detection. Recently magnetic beads (MBs) have been proposed as adaptable supports for immunosensing platforms. MBs-based immunoassays have many advantages over conventional ones: (1) increase of surface-to-volume ratio between biological component and the coated nanobeads because; (2) easy manipulation toward an external magnetic field; (3) shorter periods of coating, competition, and blocking time; and (4) adaptability of MBs allowing different strategies of immobilization. In addition, MBs can be handled along with flow injection analysis (FIA) representing an alternative for rapid, sensitive, and online detection of target molecules such as OA. This section presents part of our group experience in the development of a MBs-based immunosensor for OA detection as well as its implementation on continuous flow format.

3.2.1 Immobilization of Okadaic Acid on Magnetic Beads

An immunosensor based on competitive indirect enzyme-linked assay was developed with MBs as support for OA. The strategy for immobilization of OA was the strong affinity between biotin and streptavidin. In this case, biotinylated OA was synthetized according to reported protocol (Hayat et al., 2011). Briefly, 100 μL of OA at 0.5 g/L and 60 μL of a 26 mM solution of NHS and EDC were added together in a tube with 0.1 M MES buffer, pH 5.5. This mixture was stirred for 1 h at room temperature. After, 40 μL of 40 mM biotin-LC-PEO_3-amine in MES buffer were added and stirred for 24 h at room temperature. 600 μL of PBS (1×) were added to the reaction mixture. The reaction mixture was transferred into a dialysis membrane and dialyzed against 0.05 M carbonate buffer for 4 h and finally over night against milliQ water at 4°C. The resulting biotinylated OA was stored at 4°C.

For immobilization of biotinylated OA streptavidin coated MBs were used. 50 μL of beads (10 mg/mL) were taken in 1.5 mL microtube and washed six times with 50 μL of PBS (1×). The MBs were then incubated with 50 μL of biotinylated OA for 1 h at room temperature; then MBs were washed three times with 50 μL of PBS (1×). The biotinylated magnetic beads (OA-MBs) were resuspended in 50 μL of PBS (1×) and stored at 4°C. During all procedures MBs were attached with the magnetic support Adem-Mag SV and for biosensor fabrication OA-MBs were hold on the surface of a SPE by the application of an external magnetic field.

3.2.2 Continuous-Flow Immunosensor Based on Magnetic Beads

The prepared OA-MBs were incorporated into a custom flow system to perform a flow immunoassay. The custom flow system was similar to that presented in Section 3.1.2. Prior to flow operation, 100 μL of 1% BSA were incubated over the surface of SPE for 1 h at room temperature to prevent nonspecific binding. Once blocked, SPE was disposed into the flow cell and connected to the potentiostat. The prepared OA-MBs were diluted with working buffer (PBS 1× pH 7.4) and injected to the flow cell at 0.20 μL/s, where the magnetic field trapped the OA-MBs over the SPE surface.

Aspiration of OA standard solutions at different concentrations or spiked samples and anti-OA monoclonal antibody (anti OA-MAb) solution was used for competition. Then ALP labeled goat anti-mouse IgG secondary antibody solution was also aspirated. Both the competition mixture and the secondary antibody solution were incubated for 30 min at room temperature. Finally, a mixture containing 10% DEA buffer (pH 9.5) and 5 mg/mL 1-NP

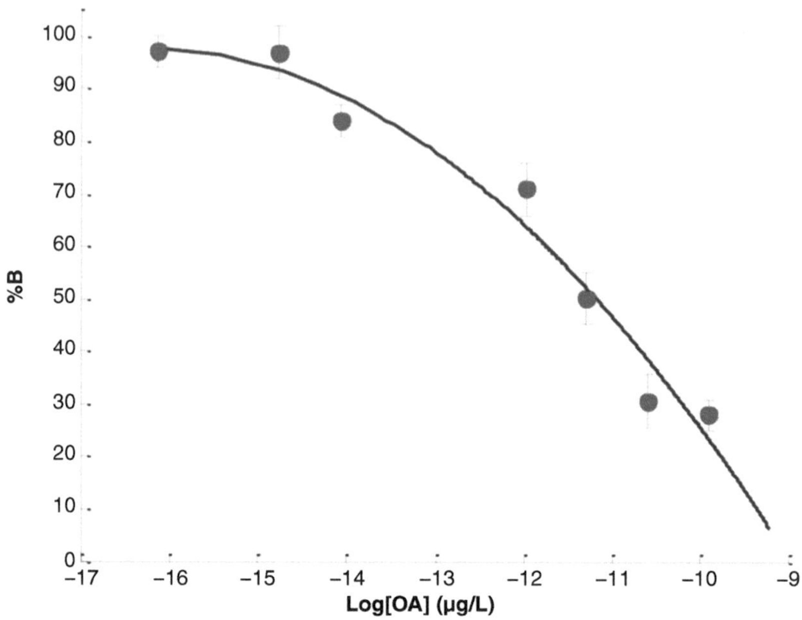

Figure 19.7. Calibration curve of flow nanomaterial based immunosensor for OA detection.

was continuously passed over the immunosensor surface for 5 min at 1.6 μL/s. Between incubation of both antibodies washing steps were performed by injecting washing buffer (PBS 1×, 0.5% tween) at 0.83 μL/s. Finally, a potential of 200 mV was applied to the immunosensor for amperometric detection. The resulting height of the electrochemical signal was recorded and plotted against OA concentration to give a calibration curve. During all the experiments, modified MBs were attached to the SPE by applying an external magnetic field.

3.2.3 Results of Flow Nanomaterial-Based Immunosensor for OA Detection

The described method with optimized conditions was applied to determine OA in spiked and mussel real samples (Dominguez et al., 2012). Mussel samples were prepared according to our previous reported protocol and they were spiked after filtration to study the OA recovery rate and the matrix effect over the flow immunosensor performance. Fig. 19.7 shows the calibration curve obtained for flow nanomaterial-based immunosensor.

Then mussel extracts were spiked with two different known concentrations of the OA. The calibration curve shown in Fig. 19.7 was used to determine the recovery values. No significant difference in

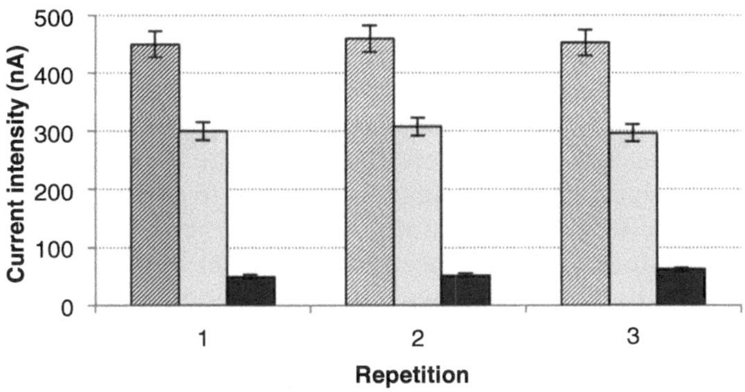

Figure 19.8. Triplicate measurement of flow nanomaterial based immunosensor. Diagonal-lined bar represents test value, light gray represents competition value, and darker bar represents control.

spiked and obtained values was observed. As already demonstrated in our previous study, the low matrix effect could be attributed to the specificity of the antibody–antigen reaction and the high number of washing steps. This great advantage demonstrated that our fully automated system could be used to quantify OA in real mussel samples, compared to the enzymatic biosensor that suffered a high matrix effect. The recovery results for the developed flow system validated the suitability of the method even at extremely lower concentration of OA than the highest level allowed by the European Commission (160 μg/kg). The precision of the immunoassays was evaluated by the relative standard deviation ((%RSD for the triplicate measurements.

In Fig. 19.8 the values of control, competition and test for a triplicate measurement are shown. RSD with this method was 5% indicating acceptable and precision and fabrication reproducibility. The spiked OA concentrations were 40 and 10 μg/kg, while the values obtained with this method were 19.25 and 9.6 μg/kg respectively (Dominguez et al., 2012). These results indicated that the measured values are in good agreement with the spiked values. Thus, the present method could satisfy the need of automated immunoassay for OA determination in real mussel samples.

4 Applications of Sensor Arrays for Food Analysis

The importance of human senses for food analysis in quality control and safety assessment has also been emphasized (Ghasemi-Varnamkhasti et al., 2010). In past sections detection was performed only with the interaction of a specific analyte with a single biosensor, but approaches concerning a higher number of sensing elements have also been extensively studied. The so-called biomimetic systems are formed by a sensor array (analogs to the thousands of receptors found in taste and smell senses) coupled with a data processing tool (analogs to brain) for interpretation. Some well-known examples of biomimetic systems are the artificial vision, the electronic noses (e-nose) and the electronic tongues (e-tongue). These two last, have been proved in pharmaceutical, environmental and biomedical fields but there is an increasing interest for their potential applications in food industry for monitoring tasks, quantification purposes and quality control (Escuder-Gilabert and Peris, 2010; Mimendia et al., 2010).

4.1 Principles of Sensor Arrays

Chemical sensors transform chemical information within a sample into a useful analytical signal by the interaction of the specific analyte with the receptor. Receptor can be either synthetic (chemosensor) or biological (biosensor) but is always intended to bring the maximum allowed selectivity and sensitivity to the target analyte. However, in practice, most sensors show some cross-reactivity to other compounds in the sample (Tisch and Haick, 2010). By mimicking mammal senses, e-noses and e-tongues incorporate chemical sensors with partial or low selectivity (cross-selectivity) instead of sensors with high selectivity. The whole array then respond to several components at the same time providing a unique complex pattern or "digital fingerprint" for each sample. This concept of chemical sensing, known as global selectivity, allows the description of a given sample by the complete response of all sensor array rather than by the value of a single element (Vlasov et al., 2005).

According with literature most e-tongues are formed by electrochemical sensors and, from this group, a substantial number are formed by ion selective electrodes (ISE), followed by voltammetric sensor arrays and impedimetric e-tongues (del Valle, 2010). There are also optical sensor arrays that, based on their operation, can be considered e-tongues. The purposes of analysis for e-tongues can be classified as:

1. E-tongues for multianalyte detection. This has been shown on arrays based on ISEs to quantify ions on binary, ternary, and cuaternary mixtures, sensor arrays to solve heavy metal mixtures, and enzymatic biosensor arrays to determine several species of pesticides within a single sample (Calvo et al., 2007; Valdes-Ramirez et al., 2009; Mimendia et al., 2009).

2. Qualitative assessment. This has been specially applied to differentiate specific characteristics of foodstuffs such as geographical origin, fabrication process or ageing (Domínguez et al., 2014; Cetó et al., 2011; Gutiérrez et al., 2010).
3. Taste sensor. Along with evaluation panels, e-tongue has been applied to evaluate organoleptic characteristics of foodstuff products (Song et al., 2014).

Given the importance of information given by sensor array, the modification of sensors properties has been attracting attention to improve the analysis of samples. In this sense, nanomaterials are playing an important role by changing the catalytic properties of electrochemical sensors by including different nanomaterials such as MNPs, nanowires (NWs), and nanofibers (NFs) on their fabrication. Other nanomaterials such as metal oxide nanoparticles, can be suited for sensor arrays by changing their base material, size, and doping level. In case of CNTs, their selectivity can be modified by immobilization of different bioreceptors. As a result, sensor arrays could bring more information and distinctive patterns can be associated with complex samples, such as those present in foodstuffs. For example, Niu et al. modified a ceramic chip with nine catalytic nanomaterials (MgO, CaO, BaO, MgO–Al_2O_3, CaO–Al_2O_3, BaO–Al_2O_3, MgO–Al_2O_3–Eu, CaO–Al_2O_3–Eu, and BaO–Al_2O_3–Eu) to form a chemiluminescence sensor array (Niu and Liu, 2014). The e-tongue was applied to the identification of eight vineyards from different Chinese regions, achieving a good discrimination between groups. In a different application, Martín et al. (2012) prepared bisphthalocyaninates NWs and grouped together on a sensor array to detect phenolic antioxidants such as caffeic, gallic, vanillic, and ferulic acids. A sensor array of modified graphite-epoxy voltammetric sensors with copper and platinum NPs along with polyaniline- and polypyrrole-based sensors was able to discriminate between varieties of Cava wines (Cetó et al., 2011). Even the inclusion of nanomaterials in sensor arrays is still increasing and there is evidence that improvements on sensor array can lead to advantages such as size and reagent consumption reduction, better electrochemical characteristics, as well as an improvement on the general performance (eg, better discrimination and/or quantification).

4.2 Data Analysis and Pattern Recognition Methods for Sensor Arrays

In the past decades, chemical field experienced a revolution with the development of sophisticated complex equipment along with high-speed computer facilities, resulting in large amount of data (chemical data) to be analyzed (Rodionova and Pomerantsev, 2006). There was an urgent need to analyze such raw data with proper models to extract meaningful information. This promoted the development of chemometrics, which is dedicated to find relevant information from a large database of measured chemical data. Sensor arrays, from e-nose to e-tongues, have always been related with pattern recognition and chemometrics since all their elements contribute to create a multidimensional pattern for each sample.

Usually when a sensor or biosensor is calibrated, a linear model for description is preferred; however, for the complete sensor array linear and nonlinear models can be applied. Among the most commonly linear methods applied for data analysis on e-tongue are principal component analysis (PCA), linear discriminant analysis (LDA), and partial least square (PLS).

PCA is a method applied for dimensionality reduction and to visualize the grouping tendency of similar variables for classification purposes. PCA projects the data in a reduce hyperspace, defined by the principal components. These are linear combinations of the original variables (eg, sensor array measurements) with the first principal component having the largest variance, the second principal component having the second largest variance, and so on. Thus, by selecting the first significant component, the data dimension is reduced and information preserved (Hanson and Thaler, 2005).

LDA has been applied for classification of samples but, unlike PCA, is a supervised learning method and requires a proper training and data set. PLS is preferred when sensor array is intended for quantification purposes.

When the interactions between the elements of the sensor arrays are described by nonlinear relationships methods such as artificial neural networks (ANN) and support vector machines (SVM) are preferred over the typical linear models. From these models, ANNs are of great interest as modeling and calibration tools because of their ability to mimic biological neural networks functioning in a simplified way. Even historically ANNs are related with artificial intelligence, they have been successfully applied in several branches of analytical chemistry and even there is an increasing interest in the foodstuff industry (Marini, 2009).

For ANN, the most commonly architectures found in chemical literature are multilayer perceptron (MLP) and radial basis function (RBF). MLP is usually found in a three-layer architecture but configurations with more layers can be found as well. For modeling a sensor array, the number of neurons in the input layer is usually the number of elements in the array while the elements in the output layer are the expected values for target analytes in the case of quantification or the expected classes (ie, specific characteristics) for classification (Marini, 2009).

Finally, SVM modeling represent a new approach to pattern classification that has attracted a lot of attention in many real-world applications ranging from data mining, chemistry, and biotechnology. Its principle comes from the framework of the statistical learning theory, which is appropriate for approaching classification and regression problems. SVM are based in the premise of finding the optimal decision boundaries to separate data from each input class. Generally these data can not be separated by a straight line, but by applying a number of mathematical nonlinear functions known as kernel functions the input data is transferred into a linearly separable feature space. SVM models are less susceptible to the well-known problem of overfitting, since they operate on the induction principle of structural risk minimization, which minimizes an upper bound on the generalization error. During training the values of the capacity constant (C) and variable parameter (ζ) must be determined by the user. These parameters regulate the relevance of classification error during training. The major advantage of SVM is related to their global and unique solution avoiding the multiple local minima problem of models such as ANN. In addition, the final complexity of SVM models does not depend on the dimensionality of the input space (ie, sensor array) (Sadik et al., 2004; Smola and Schölkopf, 2004).

4.3 Current Applications of Sensor Arrays for Foodstuff Analysis

When it concerns to foodstuffs, there are characteristics that can be better studied by the inclusion of higher number of descriptors. For that purpose, the imitation of human senses into their electronic counterparts has been largely studied with the development of artificial e-noses and e- tongues have in the last years. Since the first e-tongue described Toko et al. as a taste sensor to the later device presented by Vlasov et al., there has always been a key relationship between e-tongues and food analysis. In the literature, there are extensive reviews about the results of foodstuffs processed by e-tongue. However, this section emphasizes the current applications of nanotechnology on sensor array to improve the performance of e-tongues on food analysis. In the next table, these results are summarized, as well as the role of nanomaterial applied and the math tool employed for data analysis (Table 19.1).

Table 19.1 Applications of Sensor Arrays Modified with Different Nanomaterials for the Analysis of Foodstuff Samples

Foodstuff Sample	Nanomaterial	Data Analysis	Purpose of e-Tongue	References
Aspartame, sucrose and cellobiose	SWCNT with FET technology	Data fitting	E-tongue was applied as taste sensor to identify the sweetness of each edulcorant	Song et al. (2014)
Brandy	Modified voltammetric electrodes with silver copper and platinum nanoparticles	FFT with LDA	Discrimination of brandies samples and identification of potential undesirable compounds for brandies storage	Cetó et al. (2013)
Honey	CPE modified with NiO nanoparticles	PCA	Classification of honey according its botanical origin	Tiwari et al. (2014)
Vinegar	Chemiluninescence sensor array with catalytic nanomaterials MgO, CaO, BaO, MgO–Al_2O_3, CaO–Al_2O_3, BaO–Al_2O_3, MgO–Al_2O_3–Eu, CaO–Al_2O_3–Eu and BaO–Al_2O_3–Eu	LDA	Discrimination of vinegars from different Chinese regions	Niu and Liu (2014)
Cava wine	Modified sensor array with copper and platinum nanoparticles	PCA, ANN, and calibration curve for sugar	Classification of cava according with its variety and quantification of contained sugar	Cetó et al. (2011)
—	Bisphthalocyanine nanowires	PCA and PLS-DA	Detection of phenolic antioxidants of interest in the food industry such as caffeic, gallic, vanillic and ferulic acids	Martín et al. (2012)
Wine	Sensor array with copper and platinum nanoparticles	PCA, ANN	Classification of wines its variety and monitoring of undesirable compounds produced by storage or careless elaboration	Gutiérrez et al. (2010)

Table 19.1 Applications of Sensor Arrays Modified with Different Nanomaterials for the Analysis of Foodstuff Samples (*cont.*)

Foodstuff Sample	Nanomaterial	Data Analysis	Purpose of e-Tongue	References
Chinese tea	Electrodes modified with SnO_2, ZnO, TiO_2, Bi_2O_3 nanoparticles	PCA and SVM	Classification of tea samples either on green tea or black tea	Liu et al. (2014)
Vegetable samples	Carboxylated poly-pyrrole nanotube (CPNT) with FET	—	Taste sensor to identify bitterness on food	Song et al. (2012)

Cetó, X., Gutiérrez, J. M., Moreno-Barón, L., Alegret, S., Del Valle, M., 2011. Voltammetric electronic tongue in the analysis of cava wines. Electroanalysis 23 (1), 72–78.
Cetó, X., Llobet, M., Marco, J., del Valle, M., 2013. Application of an electronic tongue towards the analysis of brandies. Anal. Methods 5 (5), 1120–1129.
Gutiérrez, J. M., Moreno-Barón, L., Pividori, M. I., Alegret, S. del Valle, M., 2010. A voltammetric electronic tongue made of modified epoxy-graphite electrodes for the qualitative analysis of wine. Microchimica Acta 169 (3–4), 261–268.
Liu, N., Liang, Y., Bin, J., Zhang, Z., Huang, J., Shu, R., Yang, K., 2014. Classification of green and black teas by PCA and SVM analysis of cyclic voltammetric signals from metallic oxide-modified electrode. Food Anal. Methods. 7 (2),. 472–480.
Martín, M.G., de Saja, J. A., Muñoz, R. Rodríguez-Méndez, M. L., 2012. Multisensor system based on bisphthalocyanine nanowires for the detection of antioxidants. Electrochimica Acta 68, 88–94.
Niu, W. and Liu, Y., 2014. Discrimination of vinegars using a catalytic nanomaterials-based chemiluminescence sensor array. Luminescence 29 (2), 138–142.
Song, H. S., Jin, H. J., Ahn, S. R., Kim, D., Lee, S. H., Kim, U.-K., Simons, C. T., Hong, S. Park, T. H., 2014. Bioelectronic tongue using heterodimeric human taste receptor for the discrimination of sweeteners with human-like performance. ACS Nano 8 (10), 9781–9789.
Song, H. S., Kwon, O. S., Lee, S. H., Park, S. J., Kim, U.-K., Jang, J. Park, T. H., 2012. Human taste receptor-functionalized field effect transistor as a human-like nanobioelectronic tongue. Nano Lett. 13 (1), 172–178.
Tiwari, K., Biswas, S., Tudu, B., Bandhopadhyay, R., Chatterjee, A. Pramanik, P., 2014. Voltammetric technique for honey analysis using NiO/Nps modified carbon paste electrode. Control, Instrumentation, Energy and Communication (CIEC), International Conference on: IEEE, 152–155.

5 Conclusions

Current methods for food analysis present challenging issues when fast, easy, and online monitoring of food samples is pursued. Biosensors represent an increasing trend for food analysis and the integration of nanomaterial technology allows the possibility of new features such as new platforms for biomolecule immobilization, facilities for continuous monitoring, and higher sensitivity. Some applications of nanomaterials to biosensor field for food detection were presented, for both single elements and sensor arrays (eg, e-tongue). However, there are still unexplored possibilities and several potential improvements when it comes to novel nanomaterials applied to biosensor field.

References

Accum, F.C., 1820. A Treatise on Adulterations of Food, and Culinary Poisons, Exhibiting the Fraudulent Sophistications of Bread, Beer, Wine, and Other Articles Employed in Domestic Economy and Methods of Detecting Them. Longman & Company, Philadelphia.

Andreescu, S., Barthelmebs, L., Marty, J.L., 2002. Immobilization of acetylcholinesterase on screen-printed electrodes: comparative study between three immobilization methods and applications to the detection of organophosphorus insecticides. Anal. Chim. Acta 464 (2), 171–180.

Arduini, F., Guidone, S., Amine, A., Palleschi, G., Moscone, D., 2013. Acetylcholinesterase biosensor based on self-assembled monolayer-modified gold-screen printed electrodes for organophosphorus insecticide detection. Sens. Actuat. B 179, 201–208.

Arya, S.K., Datta, M., Malhotra, B.D., 2008. Recent advances in cholesterol biosensor. Biosens. Bioelectron. 23 (7), 1083–1100.

Asefa, T., Duncan, C.T., Sharma, K.K., 2009. Recent advances in nanostructured chemosensors and biosensors. Analyst 134 (10), 1980–1990.

Byfield, M., Abuknesha, R., 1994. Biochemical aspects of biosensors. Biosens. Bioelectron. 9 (4), 373–399.

Calvo, D., Grossl, M., Cortina, M., del Valle, M., 2007. Automated SIA system using an array of potentiometric sensors for determining alkaline-earth ions in water. Electroanalysis 19 (6), 644–651.

Campas, M., de la Iglesia, P., Le Berre, M., Kane, M., Diogene, J., Marty, J.L., 2008. Enzymatic recycling-based amperometric immunosensor for the ultrasensitive detection of okadaic acid in shellfish. Biosens. Bioelectron. 24 (4), 716–722.

Campàs, M., Prieto-Simón, B., Marty, J.-L., 2009. A review of the use of genetically engineered enzymes in electrochemical biosensors. Semin. Cell Dev. Biol. 20 (1), 3–9.

Cass, A.E., Davis, G., Francis, G.D., Hill, H.A.O., Aston, W.J., Higgins, I.J., Plotkin, E.V., Scott, L.D., Turner, A.P., 1984. Ferrocene-mediated enzyme electrode for amperometric determination of glucose. Anal. Chem. 56 (4), 667–671.

Cetó, X., Gutiérrez, J.M., Moreno-Barón, L., Alegret, S., Del Valle, M., 2011. Voltammetric electronic tongue in the analysis of Cava wines. Electroanalysis 23 (1), 72–78.

Cetó, X., Llobet, M., Marco, J., del Valle, M., 2013. Application of an electronic tongue towards the analysis of brandies. Anal. Methods 5 (5), 1120–1129.

Chen, X.W., Wang, J.H., 2007. The miniaturization of bioanalytical assays and sample pretreatments by exploiting meso-fluidic lab-on-valve configurations: a review. Anal. Chim. Acta 602, 173–180.

Clark, L.C., Lyons, C., 1962. Electrode systems for continuous monitoring in cardiovascular surgery. Ann. NY Acad. Sci. 102 (1), 29–45.

Cooney, C., Weaver, J., Tannehbaum, S., Faller, D., Shields, A., Jahnke, M., 1974. The thermal enzyme probe—a novel approach to chemical analysis. Enzyme Engineeringvol. 2Springer, Plenum, New York, pp. 411–417.

del Valle, M., 2010. Electronic tongues employing electrochemical sensors. Electroanalysis 22 (14), 1539–1555.

Di Gleria, K., Green, M.J., Hill, H.A.O., McNeil, C.J., 1986. Homogeneous ferrocene-mediated amperometric immunoassay. Anal. Chem. 58 (6), 1203–1205.

Díaz, R., Ibáñez, M., Sancho, J.V., Hernández, F., 2012. Target and nontarget screening strategies for organic contaminants, residues and illicit substances in food, environmental and human biological samples by UHPLC-QTOF-MS. Anal. Methods 4 (1), 196–209.

Dominguez, R.B., Hayat, A., Sassolas, A., Alonso, G.A., Munoz, R., Marty, J.-L., 2012. Automated flow-through amperometric immunosensor for highly sensitive and online detection of okadaic acid in mussel sample. Talanta 99, 232–237.

Dominguez, R.B., Alonso, G.A., Muñoz, R., Hayat, A., Marty, J.-L., 2015. Design of a novel magnetic particles based electrochemical biosensor for organophosphate insecticide detection in flow injection analysis. Sens. Actuat.B 208, 491–496.

Domínguez, R.B., Moreno-Barón, L., Muñoz, R., Gutiérrez, J.M., 2014. Voltammetric electronic tongue and support vector machines for identification of selected features in Mexican coffee. Sensors 14 (9), 17770–17785.

D'Orazio, P., 2003. Biosensors in clinical chemistry. Clin. Chim. Acta 334 (1), 41–69.

Ellington, A.D., Szostak, J.W., 1990. In vitro selection of RNA molecules that bind specific ligands. Nature 346 (6287), 818–822.

Ellis, D.I., Brewster, V.L., Dunn, W.B., Allwood, J.W., Golovanov, A.P., Goodacre, R., 2012. rsquo;Fingerprinting food: current technologies for the detection of food adulteration and contamination. Chem. Soc. Rev. 41 (17), 5706–5727.

Escuder-Gilabert, L., Peris, M., 2010. Review: highlights in recent applications of electronic tongues in food analysis. Anal. Chim. Acta 665 (1), 15–25.

Farré, M., Barceló, D., 2013. Analysis of emerging contaminants in food. Trends Anal. Chem. 43, 240–253.

Food and Agriculture Organization of the United Nations, 2015. Definitions for the Purposes of the Codex Alimentarius. Available from: http://www.fao.org/docrep/005/y2200e/y2200e07.htm

Ghasemi-Varnamkhasti, M., Mohtasebi, S.S., Siadat, M., 2010. Biomimetic-based odor and taste sensing systems to food quality and safety characterization: an overview on basic principles and recent achievements. J. Food Eng. 100 (3), 377–387.

Guilbault, G.G., Montalvo, Jr., J.G., 1969. Urea-specific enzyme electrode. J. Am. Chem. Soc. 91 (8), 2164–2165.

Guilbault, G., Kramer, D., Cannon, Jr., P., 1962. Electrical determination of organophosphorous compounds. Anal. Chem. 34 (11), 1437–1439.

Guo, S., Wang, E., 2007. Synthesis and electrochemical applications of gold nanoparticles. Anal. Chim. Acta 598 (2), 181–192.

Gutiérrez, J.M., Moreno-Barón, L., Pividori, M.I., Alegret, S., del Valle, M., 2010. A voltammetric electronic tongue made of modified epoxy-graphite electrodes for the qualitative analysis of wine. Microchim. Acta 169 (3–4), 261–268.

Haider, W., Barillier, D., Hayat, A., Gaillard, J.-L., Ledauphin, J., 2014a. Rapid quantification and comparison of major volatile compounds of ciders from France (Normandy and Brittany) using microextraction by packed sorbent (MEPS). Anal. Methods 6 (5), 1364–1376.

Haider, W., Delahaie, S., Saint-Clair, J.-F., Hayat, A., Barillier, D., Ledauphin, J., 2014b. Physicochemical characterization of Pommeau de Normandie: establishing a relationship between metal contents and turbidity of the Pommeau samples. Eur. Food Res. Technol. 239 (4), 567–575.

Hansen, E.H., Miro, M., 2007. How flow-injection analysis (FIA) over the past 25 years has changed our way of performing chemical analyses. Trends Anal. Chem. 26 (1), 18–26.

Hanson, C.W., Thaler, E.R., 2005. Electronic nose prediction of a clinical pneumonia score: biosensors and microbes. J. Am. Soc. Anesthesiol. 102 (1), 63–68, 63-68.

Hayat, A., Andreescu, S., 2013. Nanoceria particles as catalytic amplifiers for alkaline phosphatase assays. Anal. Chem. 85 (21), 10028–10032.

Hayat, A., Barthelmebs, L., Marty, J.-L., 2011. Enzyme-linked immunosensor based on super paramagnetic nanobeads for easy and rapid detection of okadaic acid. Anal. Chim. Acta 690 (2), 248–252.

Hayat, A., Catanante, G., Marty, J.L., 2014. Current trends in nanomaterial-based amperometric biosensors. Sensors 14 (12), 23439–23461.

Istamboulie, G., Andreescu, S., Marty, J.-L., Noguer, T., 2007. Highly sensitive detection of organophosphorus insecticides using magnetic microbeads and genetically engineered acetylcholinesterase. Biosens. Bioelectron. 23 (4), 506–512.

Kaufmann, A., 2012. The current role of high-resolution mass spectrometry in food analysis. Anal. Bioanal. Chem. 403 (5), 1233–1249.

Kienberger, F., Kada, G., Mueller, H., Hinterdorfer, P., 2005. Single molecule studies of antibody–antigen interaction strength versus intra-molecular antigen stability. J. Mol. Biol. 347 (3), 597–606.

Kumar, A., Malik, A.K., Picó, Y., 2010. Sample preparation methods for the determination of pesticides in foods using CE-UV/MS. Electrophoresis 31 (13), 2115–2125.

Lei, J., Ju, H., 2012. Signal amplification using functional nanomaterials for biosensing. Chem. Soc. Rev. 41 (6), 2122–2134.

Li, Y., Schluesener, H.J., Xu, S., 2010. Gold nanoparticle-based biosensors. Gold Bull. 43 (1), 29–41.

Lin, J., He, C., Zhao, Y., Zhang, S., 2009. One-step synthesis of silver nanoparticles/carbon nanotubes/chitosan film and its application in glucose biosensor. Sens. Actuat. B 137 (2), 768–773.

Liu, N., Liang, Y., Bin, J., Zhang, Z., Huang, J., Shu, R., Yang, K., 2014. Classification of green and black teas by PCA and SVM analysis of cyclic voltammetric signals from metallic oxide-modified electrode. Food Anal. Methods. 7 (2), 472–480.

Llopis, X., Pumera, M., Alegret, S., Merkoci, A., 2009. Lab-on-a-chip for ultrasensitive detection of carbofuran by enzymatic inhibition with replacement of enzyme using magnetic beads. Lab Chip 9 (2), 213–218.

Lubbers, D., Opitz, N., 1975. The pCO_2/pO_2 optrode: a new pCO_2, pO_2 device for the measurement of pCO_2 or pO_2 in gases or fluids. Z. Naturforsch. C 30, 532–533.

Malik, A.K., Blasco, C., Picó, Y., 2010. Liquid chromatography–mass spectrometry in food safety. J. Chromatogr. A 1217 (25), 4018–4040.

Marini, F., 2009. Artificial neural networks in foodstuff analyses: trends and perspectives: a review. Anal. Chim. Acta 635 (2), 121–131.

Martín, M.G., de Saja, J.A., Muñoz, R., Rodríguez-Méndez, M.L., 2012. Multisensor system based on bisphthalocyanine nanowires for the detection of antioxidants. Electrochim. Acta 68, 88–94.

Marvin, H., Kleter, G., Frewer, L., Cope, S., Wentholt, M., Rowe, G., 2009. A working procedure for identifying emerging food safety issues at an early stage: implications for European and international risk management practices. Food Control 20 (4), 345–356.

Mimendia, A., Legin, A., Merkoci, A., del Valle, M., 2009. Use of sequential injection analysis to construct a potentiometric electronic tongue: application to the multidetermination of heavy metals. Olfact. Electron. Nose Proceed. 1137, 239–242.

Mimendia, A., Gutierrez, J.M., Leija, L., Hernandez, P.R., Favari, L., Munoz, R., del Valle, M., 2010. A review of the use of the potentiometric electronic tongue in the monitoring of environmental systems. Environ. Model. Softw. 25 (9), 1023–1030.

Mishra, R.K., Dominguez, R.B., Bhand, S., Munoz, R., Marty, J.-L., 2012. A novel automated flow-based biosensor for the determination of organophosphate pesticides in milk. Biosens. Bioelectron. 32 (1), 56–61.
Mosbach, K., Danielsson, B., 1974. An enzyme thermistor. Biochim. Biophys. Acta (BBA) Enzymol. 364 (1), 140–145.
Niu, W., Liu, Y., 2014. Discrimination of vinegars using a catalytic nanomaterials-based chemiluminescence sensor array. Luminescence 29 (2), 138–142.
Núñez, O., Gallart-Ayala, H., Martins, C.P., Lucci, P., 2012. New trends in fast liquid chromatography for food and environmental analysis. J. Chromatog. A 1228, 298–323.
Palecek, E., Fojta, M., 2007. Magnetic beads as versatile tools for electrochemical DNA and protein biosensing. Talanta 74 (3), 276–290.
Paniel, N., Baudart, J., Hayat, A., Barthelmebs, L., 2013. Aptasensor and genosensor methods for detection of microbes in real world samples. Methods 64 (3), 229–240.
Pei, X., Zhang, B., Tang, J., Liu, B., Lai, W., Tang, D., 2013. Sandwich-type immunosensors and immunoassays exploiting nanostructure labels: a review. Anal. Chim. Acta 758, 1–18.
Pérez-López, B., Merkoçi, A., 2011a. Nanomaterials based biosensors for food analysis applications. Trends Food Sci. Technol. 22 (11), 625–639.
Pérez-López, B., Merkoçi, A., 2011b. Magnetic nanoparticles modified with carbon nanotubes for electrocatalytic magnetoswitchable biosensing applications. Adv. Funct. Mater. 21 (2), 255–260.
Periasamy, A.P., Umasankar, Y., Chen, S.M., 2009. Nanomaterials: acetylcholinesterase enzyme matrices for organophosphorus pesticides electrochemical sensors: a review'. Sensors 9 (6), 4034–4055.
Polsky, R., Gill, R., Kaganovsky, L., Willner, I., 2006. Nucleic acid-functionalized Pt nanoparticles: catalytic labels for the amplified electrochemical detection of biomolecules. Anal. Chem. 78 (7), 2268–2271.
Putzbach, W., Ronkainen, N.J., 2013. Immobilization techniques in the fabrication of nanomaterial-based electrochemical biosensors: a review. Sensors 13 (4), 4811–4840.
Rodionova, O.Y., Pomerantsev, A.L., 2006. Chemometrics: achievements and prospects. Russ. Chem. Rev. 75 (4), 271.
Romero-González, R., 2015. Food safety: how analytical chemists ensure it. Anal. Methods.
Ruzicka, J., Hansen, E.H., 2008. Retro-review of flow-injection analysis. Trends Anal. Chem. 27 (5), 390–393.
Ruzicka, J., Marshall, G.D., 1990. Sequential injection: a new concept for chemical sensors, process analysis, and laboratory assays. Anal. Chim. Acta 237, 329–343.
Sadik, O., Land, W.H., Wanekaya, A.K., Uematsu, M., Embrechts, M.J., Wong, L., Leibensperger, D., Volykin, A., 2004. Detection and classification of organophosphate nerve agent simulants using support vector machines with multiarray sensors. J. Chem. Info. Comp. Sci. 44 (2), 499–507.
Sassolas, A., Blum, L.J., Leca-Bouvier, B.D., 2012. Immobilization strategies to develop enzymatic biosensors. Biotechnol. Adv. 30 (3), 489–511.
Smola, A.J., Schölkopf, B., 2004. A tutorial on support vector regression. Statist. Comput. 14 (3), 199–222.
Song, H.S., Jin, H.J., Ahn, S.R., Kim, D., Lee, S.H., Kim, U.-K., Simons, C.T., Hong, S., Park, T.H., 2014. Bioelectronic tongue using heterodimeric human taste receptor for the discrimination of sweeteners with human-like performance. ACS Nano 8 (10), 9781–9789.

Song, H.S., Kwon, O.S., Lee, S.H., Park, S.J., Kim, U.-K., Jang, J., Park, T.H., 2012. Human taste receptor-functionalized field effect transistor as a human-like nanobioelectronic tongue. Nano Lett. 13 (1), 172–178.

Stanciu, L., Won, Y.-H., Ganesana, M., Andreescu, S., 2009. Magnetic particle-based hybrid platforms for bioanalytical sensors. Sensors 9 (4), 2976–2999.

Stoltenburg, R., Reinemann, C., Strehlitz, B., 2007. SELEX—a (r) evolutionary method to generate high-affinity nucleic acid ligands. Biomol. Eng. 24 (4), 381–403.

Tisch, U., Haick, H., 2010. Nanomaterials for cross-reactive sensor arrays. MRS Bull. 35 (10), 797–803.

Tiwari, K., Biswas, S., Tudu, B., Bandhopadhyay, R., Chatterjee, A., Pramanik, P., 2014. Voltammetric technique for honey analysis using NiO/Nps modified carbon paste electrode. Control, Instrumentation, Energy and Communication (CIEC), International Conference on: IEEE, 152–155.

Tombelli, S., Minunni, M., Mascini, M., 2007. Aptamers-based assays for diagnostics, environmental, and food analysis. Biomol. Eng. 24 (2), 191–200.

Tuerk, C., Gold, L., 1990. Systematic evolution of ligands by exponential enrichment: RNA ligands to bacteriophage T4 DNA polymerase. Science 249 (4968), 505–510.

Turner, A., Karube, I., Wilson, G.S., 1987. Biosensors: Fundamentals and Applications. Oxford University Press, New York.

Updike, S., Hicks, G., 1967. The enzyme electrode. Nature 214, 986–988.

Valdes-Ramirez, G., Gutierrez, M., del Valle, M., Ramirez-Silva, M.T., Fournier, D., Marty, J.L., 2009. Automated resolution of dichlorvos and methylparaoxon pesticide mixtures employing a flow injection system with an inhibition electronic tongue. Biosens. Bioelectron. 24 (5), 1103–1108.

Vlasov, Y., Legin, A., Rudnitskaya, A., Di Natale, C., D'amico, A., 2005. Nonspecific sensor arrays ("electronic tongue") for chemical analysis of liquids (IUPAC Technical Report). Pure Appl. Chem. 77 (11), 1965–1983.

Völkl, K.-P., Opitz, N., Lübbers, D., 1980. Continuous measurement of concentrations of alcohol using a fluorescence-photometric enzymatic method. Fr. J. Anal. Chem. 301 (2), 162–163.

Wang, J., 2002a. Electrochemical detection for microscale analytical systems: a review. Talanta 56 (2), 223–231.

Wang, J., 2002b. Glucose biosensors: 40 years of advances and challenges. Sens. Update 10 (1), 107–119.

Wang, J., 2007. Nanoparticle-based electrochemical bioassays of proteins. Electroanalysis 19 (7–8), 769–776.

Wang, M., Li, Z., 2008. Nano-composite ZrO_2/Au film electrode for voltammetric detection of parathion. Sens. Actuat. B 133 (2), 607–612.

Wang, J., Xu, D., Kawde, A.N., Polsky, R., 2001. Metal nanoparticle-based electrochemical stripping potentiometric detection of DNA hybridization. Anal. Chem. 73 (22), 5576–5581.

Wang, Q., Zhang, B., Lin, X., Weng, W., 2011. Hybridization biosensor based on the covalent immobilization of probe DNA on chitosan–mutiwalled carbon nanotubes nanocomposite by using glutaraldehyde as an arm linker. Sens. Actuat. B 156 (2), 599–605.

Wang, X., Wang, S., Cai, Z., 2013. The latest developments and applications of mass spectrometry in food-safety and quality analysis. Trends Anal. Chem. 52, 170–185.

Wilson, G.S., Gifford, R., 2005. Biosensors for real-time in vivo measurements. Biosens. Bioelectron. 20 (12), 2388–2403.

World Health Organization (FAO/WHO), 2015. Codex International Food Standards. Available from: http://www.codexalimentarius.org/standards/en/

Wu, H., Wang, J., Kang, X., Wang, C., Wang, D., Liu, J., Aksay, I.A., Lin, Y., 2009. Glucose biosensor based on immobilization of glucose oxidase in platinum nanoparticles/graphene/chitosan nanocomposite film. Talanta 80 (1), 403–406.

Xian, Y., Hu, Y., Liu, F., Xian, Y., Wang, H., Jin, L., 2006. Glucose biosensor based on Au nanoparticles–conductive polyaniline nanocomposite. Biosens. Bioelectron. 21 (10), 1996–2000.

Xu, Y., Wang, E., 2012. Electrochemical biosensors based on magnetic micro/nano particles. Electrochim. Acta 84, 62–73.

Xu, J.-Z., Zhu, J.-J., Wang, H., Chen, H.-Y., 2003. Nano-sized copper oxide modified carbon paste electrodes as an amperometric sensor for amikacin. Anal. Lett. 36 (13), 2723–2733.

Yalow, R.S., Berson, S.A., 1959. Assay of plasma insulin in human subjects by immunological methods. Nature 184, 1648–1649.

Yonemori, Y., Takahashi, E., Ren, H., Hayashi, T., Endo, H., 2009. Biosensor system for continuous glucose monitoring in fish. Anal. Chim. Acta 633 (1), 90–96.

Zhang, W., Asiri, A.M., Liu, D., Du, D., Lin, Y., 2014. Nanomaterial-based biosensors for environmental and biological monitoring of organophosphorus pesticides and nerve agents. Trends Anal. Chem. 54, 1–10.

Zhou, Y., Zhi, J., 2006. Development of an amperometric biosensor based on covalent immobilization of tyrosinase on a boron-doped diamond electrode. Electrochem. Commun. 8 (12), 1811–1816.

20

FLUORESCENT NANOSENSORS: RAPID TOOL FOR DETECTION OF FOOD CONTAMINANTS

Bhaswati Bhattacharya*, Siddhartha Singha*, Santanu Basu**

**National Institute of Food Technology Entrepreneurship and Management, Kundli, Haryana, India; **Panjab University, Dr. S. S. Bhatnagar University Institute of Chemical Engineering & Technology, Chandigarh, India*

1 Introduction

Fluorescence is emission of characteristic UV-visible electromagnetic radiation by molecules excited by electromagnetic radiation of a particular wavelength. This phenomenon has been the basis of various analytical techniques across the field. For example, fluorescent probes like tailor-made dyes are quite popular for in vivo marking of cells for disease detection, efficacy study of drug molecules, structural evaluation of biological materials, and so forth. However, the traditional fluorescent molecules are not always useful in complex biological system due to their susceptibility toward molecular environment and their tendency of photo-bleaching. As an alternative, nanoparticles with capability of fluorescence have been found useful in bioassays.

Currently the food industry is feeling pressing needs of employing sensitive, selective, and robust nanosensors owing to conscious customers, demanding regulatory bodies, issues like bioterrorism. Therefore, food processing and food science is driving the study of a potential fluorescent nanoparticles, a.k.a fluorescent nanosensors. Miniaturized biosensors based on fluorescence principles are slowly gaining commercial importance for rapid detection of food contaminants, such as, pesticide residues, foodborne pathogens, and their toxins.

In past two decades the spectroscopic method or imaging techniques have seen a very rapid pace of development. Evolution of laser-based imaging systems such as stimulated emission

Nanobiosensors. http://dx.doi.org/10.1016/B978-0-12-804301-1.00020-5

depletion, photo-activated localization microscopy, and stochastic optical reconstruction microscopy have enabled the measurement of various parameters of fluorescence such as intensity, decay time, and polarization, effects of resonance energy transfer, quenching, or photo-induced electron transfer. In sync with the development of the measurement devices, the capacity of synthesizing novel nanomaterials with customized fluorescence properties has expanded to a great extent. A large number of nanoparticles with silica or modified silica, hydrophobic and hydrophilic organic polymers, semiconducting organic polymers, quantum dots, carbon (quantum) dots, nanoclusters and nanotubes, nanodiamonds, metal and metal oxides particles, and so forth, have been synthesized and characterized by different workers. They differ in the method of synthesis, surface plus fluorescence properties, and application. Hence, an elaborate discussion of all of these nanosensors is beyond the scope of a single chapter.

This chapter on fluorescent nanosensors thematically discusses the fluorescent-based quantum dots for food contaminant detection and is broadly divided into two thematic areas. The first part of the chapter introduces the concept of fluorescence for smooth sailing into the main concept of fluorescent nanosensors. The second part elaborates synthesis, characterization, and application of variety of quantum dots with fluorescence.

2 What Is Fluorescence?

2.1 Phenomena

The emission of light from the electronically excited states of molecules is termed as luminescence. Depending upon the nature of the excited states from where the emission is taking place, luminescence can be either fluorescence and/or phosphorescence. In case of excited singlet states, the electron is oppositely spin-paired with the electron in the ground-state orbital and henceforth, transition takes place from excited singlet state to ground state, which is spin allowed resulting in photon emission.

In most cases, fluorescence is observed due to aromatic compounds. A few common examples of fluorescent compounds generally termed as *fluorophores* are depicted in Fig. 20.1. One of the most explored fluorophores is quinine, which is available in tonic water. If the tonic water is exposed to sunlight then a faint blue glow is observed on the surface. This light is more prominent if the observation is taken at right angle with respect to the direction of sunlight and in a solvent with low dielectric constant. Sir John Frederick William Herschel was the first person to report in 1845

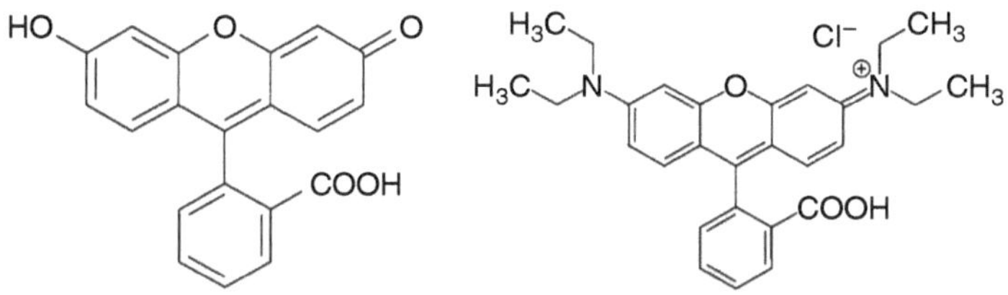

Figure 20.1. Typical fluorophores used for sensing: fluorescein and rhodamine.

(Herschel, 1845) that quinine solution gives fluorescence. There are several other fluorophores that are encountered in daily life. Antifreeze substances show a green or red/orange glow, which is due to the presence of trace amount of fluorescein or rhodamine (Fig. 20.1). Compounds, such as, anthracene and perylene also shows structured fluorescence and these compounds can be used to monitor oil pollution.

Fluorescence is known for high sensitivity detection. In 1877, fluorescence was exploited to establish that the Danube and Rhine rivers are interconnected by underground streams (Berlman, 1971) Fluorescein was added in Danube and after several hours green fluorescence of fluorescein appeared in a small river that flows to the Rhine. The history of fluorescence is provided in detail by Berlman (1971).

The plot of fluorescence intensity against wavelength (nanometers) or wavenumber (cm^{-1}) would give a fluorescence emission spectrum. Fig. 20.2 depicts two types of fluorescence emission spectra. The chemical structure of the fluorophore as well as the solvent in which the fluorophore is dissolved dictates the nature of the emission spectra. In the case of perylene,

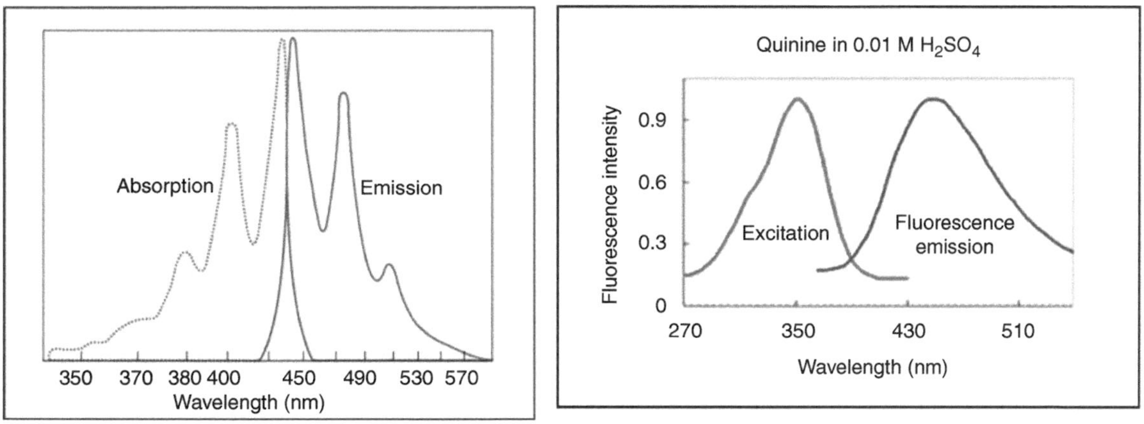

Figure 20.2. Emission spectra of perylene and quinine.

the emission spectra is structured due to the distinct vibrational energy levels of the ground and excited states, whereas other compounds like quinine show spectra devoid of any kind of vibrational structure.

2.2 Principle and Characteristics

2.2.1 Jablonski Diagram

The different molecular processes that can occur between the different electronic states of a molecule can be represented by the Jablonski diagram (Jablonski, 1935). The diagram can be of different forms to illustrate the different photophysical processes that occur. Professor Alexander Jablonski has been regarded as the father of fluorescence spectroscopy after this. Fig. 20.3 shows a representative Jablonski diagram. The ground, first, and second singlet electronic states are depicted as S_0, S_1, and S_2 respectively. Again, the fluorophores can exist in the different vibrational energy levels of a single electronic energy level. The transitions between the different energy states are shown as vertical lines to depict that absorption

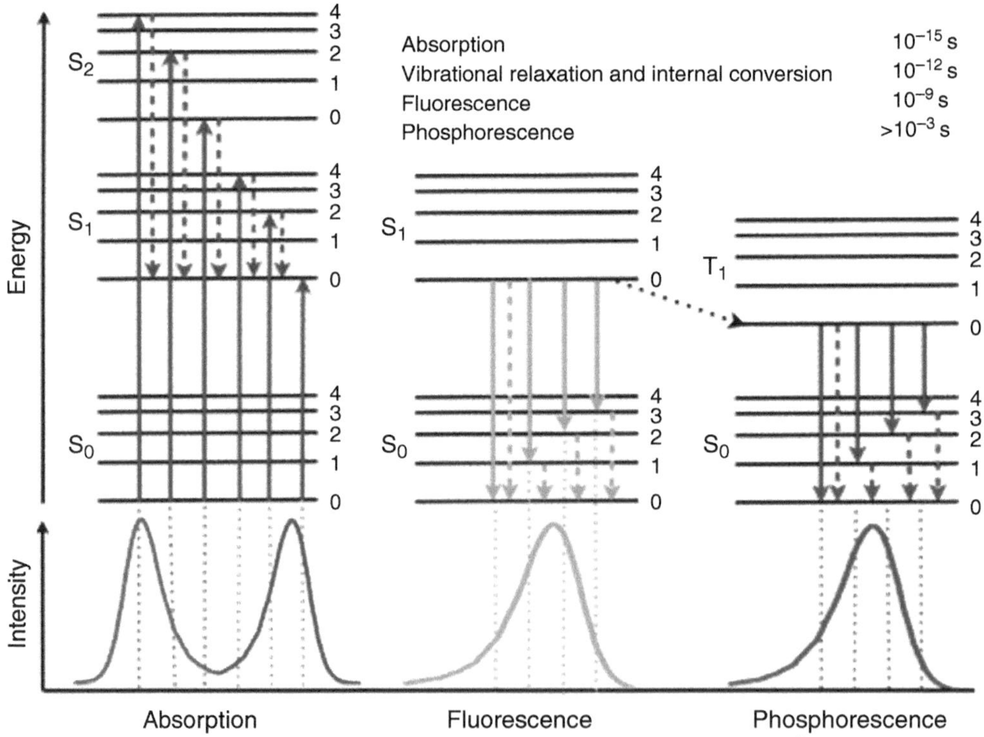

Figure 20.3. Jablonski diagram depicting different electronic states and photophysical phenomena.

of light is instantaneous in nature. The transitions are in the order of femtoseconds, which is a measurement too small for significant displacement of nuclei as per Franck–Condon principle.

Both of the processes—that is, absorption and emission—occur from the lowest vibrational energy state. Thermal energy is not sufficient to populate the excited vibrational states at room temperature, so light energy is required to induce fluorescence from S_1. There are several processes that can occur after absorption of light by the molecule. In most cases, a fluorophore is excited to higher vibrational levels of either S_1 or S_2. The molecules then immediately relax to the lowest vibrational level of S_1 in a time period of 10^{-12} s or even less and this process is termed as internal conversion. Internal conversion takes place prior to emission as the fluorescence lifetime is in the order of 10^{-8} s. Thus, emission occurs from the lowest vibrational energy level of S_1, which is a thermally equilibrated excited state.

There are possibilities of spin inversion where the molecules from S_1 state can go to the first triplet state T_1. The emission from T_1 is termed as phosphorescence and appears at longer wavelengths with respect to fluorescence. The process of movement of the molecules from singlet state to triplet state is called as intersystem crossing. Rate constants of triplet emission are much smaller than fluorescence as transition from T_1 to S_0 is spin forbidden. Phosphorescence quantum yields can be increased by the presence of heavy atoms, such as, bromine and iodine.

2.2.2 Stokes Shift

Fluorescence is observed at lower energies or longer wavelengths as compared to absorption and is shown in Jablonski diagram (Stokes, 1852). In case of all fluorescent molecules in solution there is considerable energy loss between excitation and emission. The molecules once excited rapidly decay to the lowest vibrational level of S_1 that results in Stokes shift. In addition to the above, the molecules further loose excess vibrational energy as they decay to lowest vibrational levels of S_0 from higher vibrational levels of singlet ground state. Further Stokes shift can be observed due to the effect of solvents, reactions in the excited states, complex formation, energy transfer, and so forth.

Michael Kasha (Kasha, 1950) stated that the fluorescence spectrum remains same irrespective of the excitation wavelength. After the fluorophores are excited to the higher electronic and vibrational energy levels, they immediately return to the lowest vibrational level of S_1, dissipating the excess energy. Hence, due to this fast relaxation process, the fluorescence emission spectra are normally independent of the excitation wavelength.

2.2.3 Fluorescence Quenching

The decrease in the intensity of fluorescence is called quenching. In the excited state when a fluorophore is deactivated by some other molecule (commonly known as a quencher) due to contact or collision, then it is referred to as collisional quenching. The fluorophore returns to the ground state on interaction with quencher molecule and the process is diffusion-controlled. There is no chemical alternation of the molecules during such quenching phenomena. The Stern–Volmer Eq. 20.1 describes collisional quenching:

$$\frac{F_0}{F} = 1 + K[Q] = 1 + k_q \tau_0 [Q] \tag{20.1}$$

Here, K = Stern–Volmer quenching constant, k_q = the bimolecular quenching constant, τ_0 = initial lifetime, $[Q]$ = the concentration of the quencher. The Stern–Volmer quenching constant K highlights the level of interaction of the fluorophore with the quencher. When a fluorophore is embedded deep inside a macromolecule then it is highly inaccessible to the water-soluble quenchers and then the K value is lower, whereas if the fluorophore is free or on the surface of macromolecules, then K is quite high.

The quenching mechanism depends on the fluorophore and the quencher. When acrylamide quenches the emission of indole group in tryptophan then the quenching is due to the electron transfer from indole to acrylamide in the excited state. Halogen and heavy atoms cause quenching through spin-orbit coupling and intersystem. Apart from collisional quenching there exist other processes through which fluorescence quenching can take place. Sometimes, fluorophore molecules form complexes with the quenchers in the ground state that are nonfluorescent. This type of quenching is called static quenching as it occurs in the ground state and is not affected by diffusion or molecular collisions.

2.2.4 Resonance Energy Transfer (RET)

This process occurs when the emission spectrum of a fluorophore called as donor overlaps with the absorption spectrum of another molecule known as acceptor (Förster, 1948) illustrated in Fig. 20.4.

The aforementioned absorption process is dependent on the concentration of the acceptor. In the RET process, there is no involvement of intermediate photons. The donor and acceptor molecules are coupled by a dipole–dipole interaction and the extent of energy transfer is given by the spectral overlap and

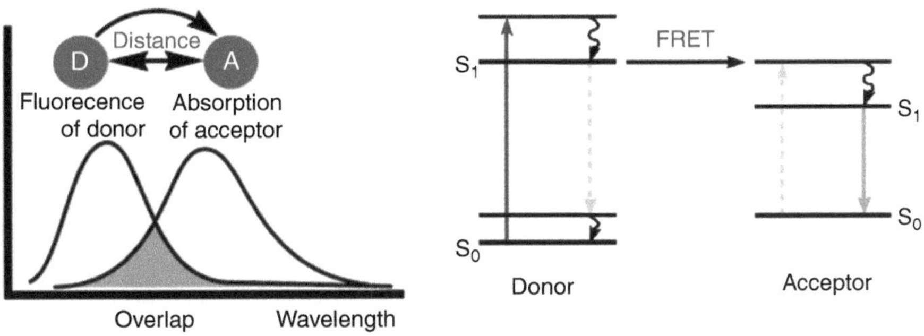

Figure 20.4. Illustration of FRET spectrally and through Jablonski diagram.

the distance between the donor and acceptor. The rate of energy transfer is expressed as Eq. 20.2:

$$k_T(r) = \frac{1}{\tau_D}\left(\frac{R_0}{r}\right)^6 \quad (20.2)$$

Here r is the distance between the donor (D) and the acceptor (A) while τ_D is the lifetime of the donor without energy transfer. Spectral overlap is described in terms of the Förster distance (R_0). The efficiency of energy transfer for a single donor–acceptor pair at a certain fixed distance is given as Eq. 20.3:

$$E = \frac{R_0^6}{R_0^6 + r^6} \quad (20.3)$$

The aforementioned expression suggests that the extent of energy transfer depends on distance (r). The Förster distances are generally comparable in size to the biological macromolecules and thereby this RET has been used as a spectroscopic ruler for measurements of distances between the different active sites on proteins (Fig. 20.5).

Depending upon the nature and linkage of donors and acceptors, the theory of RET can vary. Donors and acceptors can be covalently linked, can be free moving in solutions, or may be

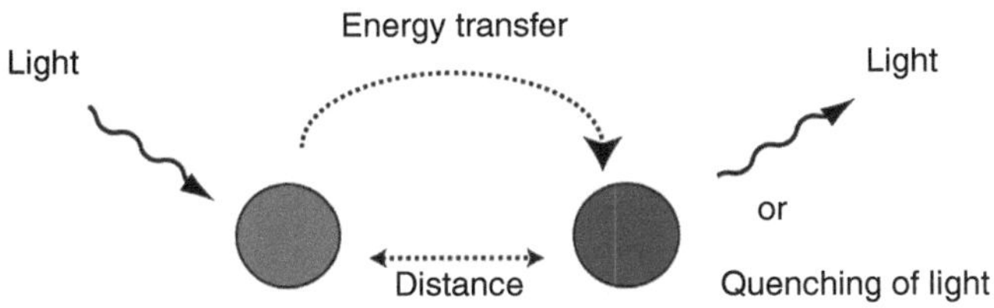

Figure 20.5. Illustration of FRET as a "spectroscopic ruler."

confined in macromolecules. The extent of energy transfer may increase depending on the lifetime of the donor molecule.

2.2.5 Fluorescence Lifetime and Quantum Yield

The two most significant features of a fluorophore molecule is the quantum yield and fluorescence lifetime. The lifetime determines the time available for the fluorophore to interact with or diffuse in its environment. The rate of fluorescence emission of the fluorophore (Γ) and the rate of its nonradiative decay to S_0 (k_{nr}) gives a measure of the quantum yield of the fluorophore. Both the rate constants Γ and k_{nr} depopulate the excited state. The fluorescence quantum yield is given by the ratio of number of photons that are emitted to the number of photons absorbed. A fraction of the fluorophores relaxes through emission and the quantum yield is expressed as Eq. 20.4:

$$Q = \frac{\Gamma}{\Gamma + k_{nr}} \tag{20.4}$$

The quantum yield can be close to unity if the nonradiative decay rate is substantially smaller than the radiative decay rate, that is, $k_{nr} < \Gamma$. There can be different nonradiative processes, such as, internal conversion, intersystem crossing, and all these nonradiative constants, which are grouped as k_{nr}. Stokes losses decrease the energy yield of fluorescence less than unity.

The fluorescence lifetime can be defined as the average time that the molecule resides in the excited state prior to return to the ground state. It can be represented as Eq. 20.5:

$$\tau = \frac{1}{\Gamma + k_{nr}} \tag{20.5}$$

In absence of any nonradiative processes, the fluorescence lifetime measured is termed *intrinsic lifetime* and is given as Eq. 20.6:

$$\tau_n = \frac{1}{\Gamma} \tag{20.6}$$

Theoretically, the natural lifetime τ_n of a fluorophore can be calculated from its absorption spectra, extinction coefficient and the emission spectra. In reality, natural lifetime can be determined from the measured lifetime (τ) and quantum yield as given in Eq. 20.7:

$$\tau_n = \frac{\tau}{Q} \tag{20.7}$$

In case if the rate constants (Γ or k_{nr}) are affected by certain factors, then both fluorescence lifetime and quantum yield get modified. For example, a large rate of internal conversion or slow rate of emission can make a molecule nonfluorescent. Biochemical fluorophores behave in an unpredictable manner unlike their unsubstituted aromatic molecules, if a fraction of fluorophores is located closer to quencher groups, as happens in the case of tryptophan residues in proteins, then there will be inevitable discrepancies. This leads to poor agreement between calculated τ from Eq. 20.7 and that calculated from its absorption and emission spectra.

2.2.6 Steady State and Time-Resolved Fluorescence

In steady-state fluorescence measurements, the sample is illuminated with a continuous beam of light, and the intensity or emission spectrum is recorded. Most of the measurements are steady state due to the nanosecond time scale of fluorescence. A molecule quickly reaches to the steady state once it is exposed to light. The second kind of measurement is time-resolved for measuring intensity decays and/or anisotropy decays. In the case of time-resolved studies, the sample is irradiated with a light pulse, where the pulse width is shorter than the decay time of the molecule (Fig. 20.6) and this time-resolved intensity is then recorded with a high-speed detection system, where intensity is measured in the nanosecond timescale.

The intensity decay of a fluorophore that have a single decay time (τ) is given as Eq. 20.8:

$$I(t) = I_0 e^{-t/\tau} \tag{20.8}$$

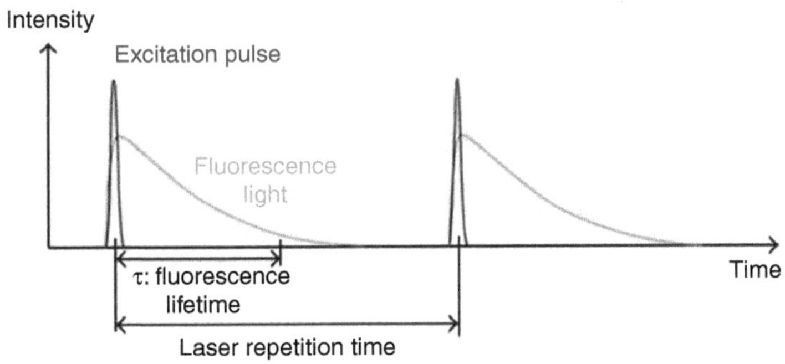

Figure 20.6. Illustration of fluorescence life-time principle. The short-pulsed excitation light [*dark gray* (red in the web version)] and the longer time duration fluorescence emission light [*light gray* (green in the web version)] is shown as a function of time.

Where I_0 is the intensity at $t = 0$ immediately following the excitation pulse. Again, steady-state intensity is the average taken over the time-resolved intensity decays of the sample and can be expressed as Eq. 20.9:

$$I_{ss} = \int_0^\infty I_0 e^{-t/\tau}\, dt = I_0 \tau \tag{20.9}$$

The value of I_0 depends on the concentration of fluorophore and other instrumental parameters. Eq. (20.9) reflects that the steady-state intensity is proportional to the lifetime of a fluorophore.

2.3 Fluorescence Imaging and Sensing

Fluorophores act as an indicator (or probe) to determine intrinsically nonfluorescent species and other parameters, such as, pH, oxygen, or metal ions. The photoluminescent properties (fluorescence or phosphorescence intensity, anisotropy, lifetime, or emission wavelength) respond to the chemical composition of the environment. Luminescence probes can be configured on different basic principles and can be categorized as per sensing mechanism: (1) quenchable probes, (2) fluorogenic probes, (3) dual wavelength probes, (4) FRET-based probes, and (5) photo-induced electron transfer (PET) systems. The functionality of the fluorophore determines the kind of imaging method. Fluorescence quenching is one approach in use for chemical sensing and imaging. In due course of a chemical reaction, luminescence amplification is observed for fluorogenic probes. The "turn on" luminescent probes have proven to be advantageous over quenchable probes. There are also dual-wavelength probes, which are categorized under sensitive dyes that can be applied to intrinsically referenced assays. There are PET systems where the probes exhibit shifts in their absorption and/or emission maxima on protonation or formation of a complex with metal ions. Typically, chemical sensors are known to display two important features: (1) these sensors should contain a chemical or molecular recognition system (receptor), which is connected in series with a physico-chemical transducer (Thevenot et al., 1999); and (2) they should be miniaturized devices that are able to disclose real-time and on-line information of the presence of certain specific compounds or ions even in complex biological samples (Cammann et al., 1996).

Luminescent probes can often be encapsulated in nanoparticles to enhance the delivery module and thus also could be detected by a remote monitoring methodology. Thus, optical chemical sensors should consist of the recognition unit (in a few cases, the chromophore itself can act both as a receptor and the transducer), a light source, and a photodetector. Fluorescent sensors are made of polymer materials where the indicator is incorporated and permeable to the target analyte.

2.4 Instrumental Methods for Sensory Imaging

The instrumentation for sensory imaging consists of: (1) an indicator or a sensor layer, (2) an electronic system based on optics to record the photoluminescence of the indicator or the sensor layer, and (3) a control unit connected to the electronic system with image processing software. Such kind of imaging systems has wide application range from macro- to microscopic level. These systems have three components, a light source, such as a halogen lamp, light-emitting diodes, or laser diodes and optical filters in order to separate the short wavelength excitation light from the long wavelength luminescence light. In some compact instruments a beam splitter (dichroic mirror) is used, which reflects the short wavelength excitation light in sample direction while transmitting the long wavelength luminescence to the detector. Laser scanning and imaging, these are the two basic approaches that are in use for detection of analyte. The scanning principle is exploited in the confocal microscopes and in microarray readers. In general, scanners are equipped with various lasers for different excitation wavelength, a movable x/y stage and a photomultiplier tube (PMT) as a detector. On the other hand, a typical imaging system would consist of a CCD camera as the detector along with a LED array or white light source in combination with a set of appropriate filters (Schäferling, 2012).

The use of fluorescent sensors in imaging is sometimes associated with certain disadvantages, such as, photobleaching of the dyes, background fluorescence of the sample, inhomogeneous illumination and instability of the light source, scattering of light, nonuniform distribution of the fluorescent probes in the sample analyte and also a varying thickness of the sample or sensor layer. Hence, the utmost requirement is that the optical sensor system should be thoroughly calibrated and referenced. Internally referenced techniques like fluorescence lifetime imaging (FLIM) are frequently employed, as it records the fluorescence lifetime, which is probably the most attractive intrinsically referenced parameter. The fluorescence lifetime as per Eq. 20.8 is not affected by the concentration of the fluorophores, static quenching effects, or the brightness of the light source used. In contrast, dynamic quenching, resonance energy transfer, and temperature have a strong influence on the fluorescence decay. Thereby, the fluorescence lifetime is a preferred parameter in fluorescence sensing and imaging.

In recent times, nanosensors based on fluorescence came into the field with wide applicability, which extended even to the detection of contaminants in the food material. In the forthcoming sections, sensors based on nanomaterials will be discussed with special focus on semiconductor-based nanomaterials, which are termed as quantum dots.

3 Nanotechnology-Enabled Sensors: Quantum Dots

To meet the increasing demands of industry, new approaches based on nanotechnology have completely changed the area of sensor technology. This technology is enabling the development of small, inexpensive, and highly efficient sensors with wide applications. It is envisaged that by enhancing the interactions that occur at the nanoscale, nanosensors offer significant advantages over conventional sensors. This may be in terms of greater sensitivity and selectivity, lower production costs, and reduced power consumption, as well as improved stability. The unique properties of nanoscale materials make them ideal for sensing and such materials can be integrated into existing sensing technologies or can be made into new devices. This technology not only enhances the properties of existing materials but also enables the fabrication of novel materials and whose properties can be tailored specifically for sensing applications.

There are possibilities for developing nano-bioorganic elements that are suitable for intracellular measurements. For sensing applications, nanotechnology allows development of nanostructures and the possibility of forming features, and which cannot be imagined with conventional microtechnologies. Such sensors are more favorable for sensing than the classically fabricated systems. Sensitivity may increase in such cases due to tailored conduction properties, the limits of detection may be lowered, infinitely small quantities of samples can be analyzed, direct analyte detection becomes possible sometimes without using labels and specificity may get improved. Improved sensitivity is a major attraction for developing nanotechnology coupled sensors. At the extreme nanoscale limit, there exists the potential to detect single molecule or atom. The small size, light weight, and high *surface-to-volume ratio* of nanostructures are the best candidates for the improvement in the detection of chemical and biological species with high sensitivity. Additionally in nanostructures the entire structure can be affected by the analyte and not only the surface as conventional sensors. As far as selectivity is concerned, nanoscale sensors and materials may not implicitly result in greater selectivity; however, nanostructuring materials and applying surface modifications and functionalization may greatly assist. The speed with which species can be detected is affected by the sensor's dimensions. Nanostructures minimize the time taken for a measure and to diffuse into and out of that volume. Hence, nanoscale modifications present the opportunity for also improving the sensor's dynamic performance. Nanosized spherical luminescent materials are applied as biomolecular labels in FRET assays. The incorporation of such photoluminescent materials into rigid polymeric matrices laid the platform for new biomolecular labels with higher brightness, good photostability, and substantial chemical stability. Moreover, with these kinds of modifications, the fluorophores get protective shielding from oxygen and other metal ions present in the environment that can cause quenching of the fluorophore. Such probes are also less affected by changes in pH and other chemical interferences.

3.1 Optical Nanosensors

These are modern analytical tools that bring together the advantages of customary sensor technology with the flexibility of using dissolved indicators. They are defined as devices with dimensions smaller than 1000 nm that have the capability to incessantly monitor chemical or biological parameters by optically converting the available information into signal useful for analytical purposes. The dimensions of optical NSs vary from a few nanometers, for instance, for macromolecules with sensitivity up to 1 μm. Miniaturization of many fiber-optic sensors to submicron size has been achieved. Moreover, optical fibers impregnated with nanomaterials like carbon nanotubes have been successfully applied. Wide-ranging interest in surface-plasmon resonance (SPR) and localized surface plasmon laser (LSPR) spectroscopy has been reawakened by the availability of new fabrication methods for plasmonic materials. Optical nanosensors can be based on macromolecules, polymer materials, and sol–gels, multifunctional core–shell systems, multifunctional magnetic beads, metal beads, and quantum dots (Borisov and Klimant, 2008).

3.2 Nanosensor Materials

Nanoparticles (NPs) can be classified on the basis of their composition as shown in Table 20.1 (Khanna, 2012).

Table 20.1 Classification of Nanomaterials According to Composition

Sl No	Class	Examples
1	Metal-based	Metallic NPs, such as, Ag, Au, metal oxides such as zinc, or titanium oxides.
2	Carbon-based	Carbon nanotubes, fullerenes
3	Semiconductor-based	Quantum dots
4	Polymer-based	Dendrimers
5	Composite	Nanoclays, biomolecules combined with nanoparticles to form nanosized biocomposite

3.2.1 Core/Shell Structured Nanoparticles

These are nanostructures that have the core made of a material coated with another material (Fig. 20.7). They have a size range of 20–200 nm. The necessity to shift to core/shell NPs arises from the quest for improvement in the properties of base NPs. Taking into consideration of the size of the NPs, the shell material is chosen is such that the agglomeration of particles is prevented that results in improving the monodispersity of the particles. The core/shell structure also enhances the thermal and chemical stability of the NPs, improves their solubility, makes them less cytotoxic (poisonous to living cells), and allows conjugation

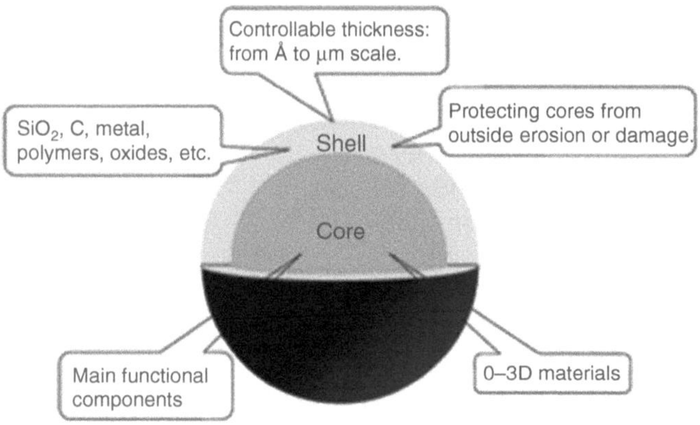

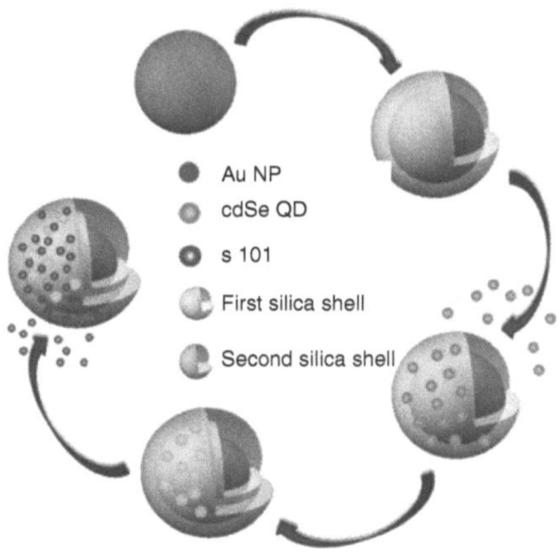

Figure 20.7. Single multilayer core–shell nanostructures. From Liwei Su, Yu Ling, Zhen Zhou, 2011. Nanoscale 3, 3967–3983; Kochuveedu, S.T., Son, T., Lee, Y., Lee, M., Kim, D., Kim, D.H., 2014. Sci. Rep. 4, 4735.

of other molecules to these particles. In some cases, the shell also prevents the oxidation of the core material. When a core NP is coated with a polymeric layer or an inorganic layer like silica (SiO_2), synergistically emerged functions can be envisioned because the polymeric or the inorganic layer would endow the hybrid structure with an additional function/property on top of the function/property of the core.

The most widely used core/shell nanocomposites are gold or silver core with silica shell. The gold/silica NPs are used in optical sensing, and the thickness of the silica coat alters the property of gold nanoparticles. In semiconductor NPs, the core is made of semiconductor material, semiconductor alloy, or metal oxide with shell made of semiconductor material, metal oxide, or an inorganic material like silica. These structures can be binary (with a core and shell) or ternary (with a core and two shells). Quantum dots (QDs) are one of the most common binary structures. These are alloys of group III or group V metals or group IV and group VI metals: CdSe/Cds, CdSe/ZnS, ZnSe/ZnS, CdTe/CdS, and so forth; the shell thickness determines the emission range of these particles.

3.3 Quantum Dots

These represent promising alternatives to organic fluorophores (chemical groups responsible for fluorescence). These are colloidal semiconductor nanocrystals, 1–10 nm in diameter, containing typically 10^3–10^4 atoms, because quantum confinement takes place in all three spatial dimensions (Wang and Herrron, 1991; Bruchez et al., 1998; Murphy and Coffer, 2002). The reduction of the size and dimensionality of metals results in a drastic change in the electronic properties as the spatial length scale of the electronic motion is reduced with decreasing size. When one dimension of a metallic material becomes comparable to the de Broglie wavelength (the wavelength of a particle given by $\lambda = h/p$, where h is Planck's constant and p is momentum), the quantum confinements appear.

In a semiconductor particle of this size, when an electron is promoted to the conduction band from the valence band, a hole is generated in the valence band with its own charge (+1) and effective mass. The electron and hole are *bound* to each other via coulombic forces of attraction, and this quasiparticle is known as exciton. The exciton is considered a hydrogen-like system and the spatial separation of the electron–hole pair can be calculated using the Bohr approximation of hydrogen atom.

$$r = \frac{\varepsilon h^2}{\pi m_r e^2} \qquad (20.10)$$

Where r is the radius of the sphere, ε is the dielectric constant of the semiconductor, m_r is the reduced mass of the electron–hole pair, h is Planck's constant, and e is the charge on electron. For many semiconductors, the masses of the electron and hole have been determined by ion cyclotron resonance and are generally in the range 0.1–3 m_e (m_e is the mass of the electron). For typical semiconductor dielectric constants, the calculation suggests that the electron–hole pair spatial separation is ~1–10 for most semiconductors (Gaponenko, 1998). The electronic structure of these quantum dots becomes intermediate between localized and delocalized bands, and there is a change of electronic and optical properties of the material due to such quantum confinement The quantum dot is a good example of "particle-in-a-box" and the energies of the particle in the box depend on the size of the box. In the quantum dot, the band-gap energy becomes size dependent and is evident in simple absorption spectra of quantum dots in solutions (Fig. 20.8). The band-gap energy of the semiconductor spectra from such spectra is generally taken as the absorption energy onset. As the particle size decreases the absorption onset, blue-shifts indicate an increase in band-gap energy.

QDs possess good brightness, high photostability (unchanged by the influence of light), broad absorption (radiation enters in a wide range of wavelengths) that allows for their simultaneous excitation, and relatively narrow emission. Application of QDs in

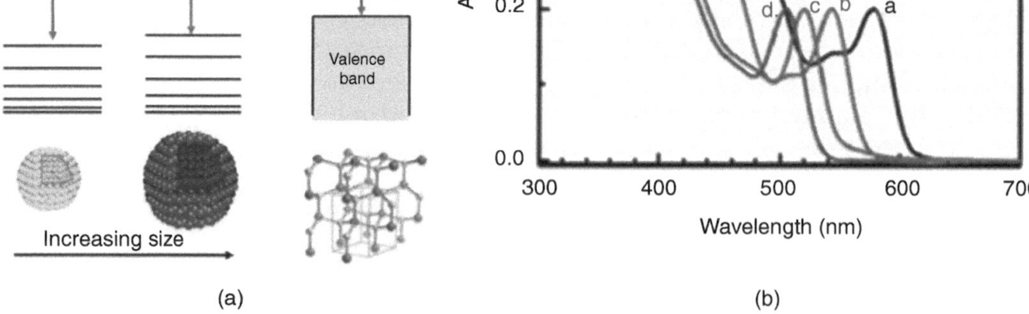

Figure 20.8. (a) The energy band structure of Cdse crystals as size increases from small quantum dot and then to bulk semiconductor. (b) Absorption spectra of different-sized CdSe quantum dots. (Part a) From Kilmov, V.I., 2003. Los Alamos Sci. 28, 214–220. (Part b) From Tian J., Cao, G., 2013. Nano Rev. 4, 22578.

optical sensors can, however, be compromised by a high degree of nonspecific binding and by the fact that many substances act as quenchers (decreasing the fluorescence intensity of a given substance). Minimizing the interferences QDs can be used to successfully design sensors.

3.3.1 *Classification of Quantum Dots*

Composition: On the basis of composition, quantum dots are classified as primary (first stage), binary (composed of two parts or two pieces), and ternary (composed of three items). Primary QDs are those made of single elements such as silicon QDs. Binary QDs are composed of two elements, such as, CdS, CdSe, and so forth. Ternary QDs contain three elements, such as, CdSeTe, ZnCdSe, and so forth.

The use of semiconductor nanocrystals is potentially problematic because the high surface area of the nanocrystal might lead to reduced luminescence efficiency (ratio of energy of emitted photon to that of exciting photon, assuming that every exciting photon yields an emitted photon), and photochemical degradation. Controlling or altering the band-gap of a nanomaterial by changing the composition has led to the development of core–shell nanocrystals with high quantum yields at room temperature and with much improved photochemical stability. By enclosing a core nanocrystal of one material with a shell of another having a larger band-gap, the excitation can be confined to the core, eliminating nonradiative relaxation pathways and preventing photochemical degradation. Provision of chemical protection to the core, particularly against oxidation, is an added advantage of the shell.

Surfaces of the quantum dots are sometimes capped with molecules or ligands to serve a few purposes: (1) partial passivation of dangling bonds on the surfaces of QDs; (2) preventing agglomeration of QDs through steric hindrance (the prevention or retardation of inter- or intramolecular interactions by the blockage of access to a reactive site by nearby groups; it arises from the crowding resulting from spatial structure of a molecule); (3) imparting solubility to the QDs. The ligands generally have two functional groups, one group binding with the QD and the other group interacting with the environment. A QD coated with surfactant molecules presents hydrocarbon chains to the environment while the polar head groups of the surfactant associate with the QD surface. Consequently, a surfactant coated QD is water insoluble but is easily dispersed in nonpolar organic solvents like toluene or hexane.

3.3.2 *Optical Properties: Quantum Dots as Fluorescent Labels*

If the surface of the quantum dot is sufficiently passivated, it can have quantum yields of light emission that approaches to those of organic dyes (~0.6). Labels or tags constitute the foundations of luminescence and fluorescence imaging and sensing. Typical QDs are core–shell or core only structures functionalized with different coatings. Their properties depend to a considerable extent on particle synthesis and surface modification. Addition of a passivation shell often has the effect of producing a small red shift in absorption and emission relative to the core QD due to tunneling of charge carriers into the shell. QDs have the striking property of an absorption that gradually increases toward shorter wavelengths and a narrow emission band of mostly symmetric shape. The spectral position of absorption and emission are tunable by particle size (*quantum size effect*).

The general principle for QD biofunctionalization is to make QDs water dispersible and then to be bound with biomolecules. Currently, only a few standard protocols are available for labeling biomolecules with QDs and the choice of suitable coupling chemistry depends on surface functionalization. QDs are promising in microarrays (multiplex lab-on-a-chip device), immunoassays and fluorescence in situ hybridization (the use of labeled nucleic acid sequence probes for the visualization of specific DNA or RNA sequences). Owing to the optical properties of organic dyes, their suitability for multicolor signaling at single-wavelength excitation is limited. QDs are the perfect contenders for spectral multiplexing at a single excitation wavelength because of their inimitable flexibility in excitation and their very narrow and symmetric emission bands, which simplify color discrimination.

Organic dyes and the majority of fluorophores have poor photostability; for QDs, the microenvironment effect on spectroscopic features is mainly governed by the accessibility of the core surface. This again depends on the ligand and the strength of its binding to the QD surface and the shell quality. If no ligand desorption occurs, properly shelled QDs are minimally sensitive to microenvironment polarity. QD emission is scarcely responsive to viscosity, contrary to that of many organic dyes and QDs are also less susceptible to aggregation-induced fluorescence quenching.

3.4 Quantum Dots as FRET-Based Probes

FRET involves the nonradiative transfer of fluorescence energy from an excited donor particle to an acceptor via dipole–dipole interaction through space (Fig. 20.9). FRET is commonly suited to measuring changes in distance, instead of absolute distances, appropriate for measuring protein conformational changes, monitoring protein interactions, and assaying of enzyme activity. QD-FRET sensors are based on the interactions between QDs serving

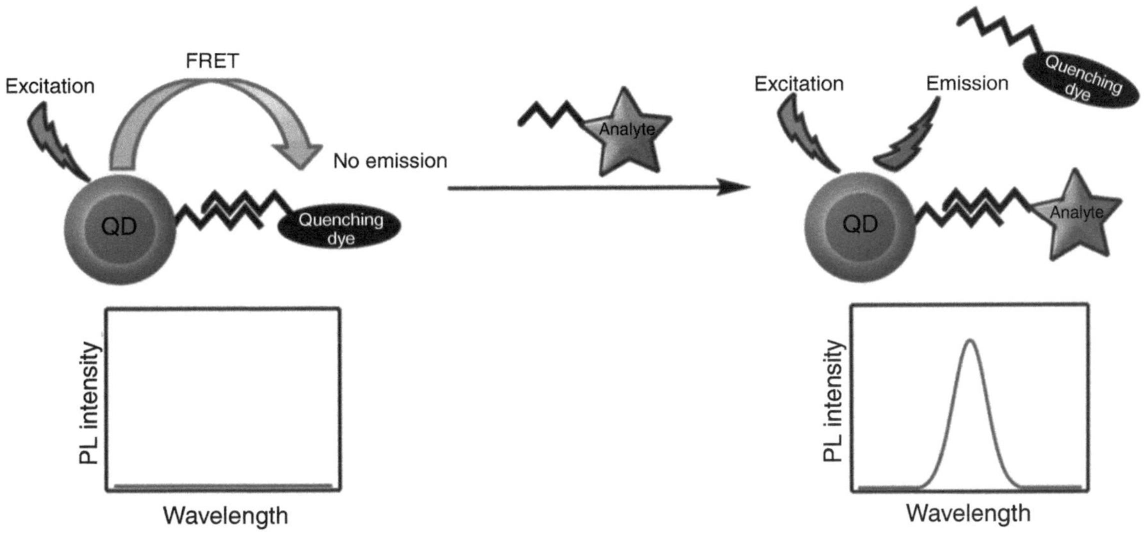

Figure 20.9. QD-based FRET probes displaying the displacement of a quenching dye by an analyte thus resulting in recovery of QD emission.* From Shamirian A., Ghai A., Snee P.T., 2015. Sensors 15(6), 13028–13051.

as donors and molecular fluorophores attached to the QD surface, acting as fluorescent acceptors. QDs have been exploited as FRET donors with organic dyes as acceptors, with the QD emission size-tuned to match the absorption band of the acceptor dye. Through FRET mechanism, QDs respond indirectly to environmental changes without any chemical interaction that could affect their photophysical properties and degrade their brightness.

Several QD-based FRET strategies have been developed for detecting nucleic acid, proteins, and other small molecules. Owing to the free choice of the QD excitation wavelength, cross-talk is circumvented in such FRET pairs. However, the distance dependence of FRET means that both the size of the QD itself and that of the surface coating change the FRET efficiency and that typically renders QD-FRET less efficient as compared to FRET with organic dyes. Owing to the substantial size of QDs, this limitation can be partly overcome by increasing the number of adjoining small organic acceptor dyes.

3.4.1 Synthesis of Quantum Dots

The most common and general method is to make colloidal solutions or to grow on solid substrates. In the colloidal approach, precursors of the material are reacted in the presence of a stabilizing agent that will restrict the growth of the particle and keep it within quantum confinement limits. For example, aqueous Cd(II) salts can be mixed with anionic or Lewis basic polymers such as sodium polyphosphate or polyamines and the subsequent addition of a sulfide source produces CdS nanoparticles that are in the 1–10 nm size range. QDs may be manufactured with diameters ranging between broad limits starting from a few nanometers to as large as few micrometers. Size tuning is possible by controlling relative concentrations and rates of addition. QD synthesis can be tailored to meet specific requirements, with core, shell, and coating characteristics all affecting photochemical properties. Size distributions are controllable within 2% using accurate growth techniques, involving high annealing temperatures.

3.4.2 CdSe/ZnS Core/Shell QDs

CdSe-ZnS core/shell QDs are prepared using synthetic techniques involving growth and annealing of organometallic precursors at high temperature. This method (Wang et al., 2009) is as follows:

1. A mixture of tri-*n*-octylphosphine oxide and hexadecyl amine is heated. Cadmium acetate is added to the solution.
2. The stock solution of trioctylphosphine-selenid is prepared by dissolving 0.2 g selenium in 5 g of tri-*n*-octylphosphine (TOP).
3. This stock solution is injected quickly into the reaction solution along with vigorous stirring of the same. Then nucleation of CdSe nanocrystals takes place.
4. Zinc acetate and bis(trimethylsilyl) sulfide are added for the inorganic epitaxial growth of the shell on the surface of the core. This epitaxial layer growth is done for about 2 h. CdSe/ZnS QDs with diameter of ~3 nm, thus synthesized, dissolved in chloroform, and preserved in a sealed condition.

Riegler et al. (2008) described another method: Cadmium stearate and TOP-Se are reacted at temperatures above 200°C by fast injection of TOP-Se into a mixture of tri-*n*-octylphosphine oxide and cadmium stearate. The CdSe cores are passivated and annealed by growing a shell of two additional monolayers of ZnS on their surface.

3.4.3 Quantum Dots and Organic Fluorophores

The potential of detection to a great extent is determined by the physicochemical properties of the chromophore in use (Mason, 1999; Waggoner, 2006; Lakowicz, 2006) and that includes its chemical nature and size, its biocompatibility, and the effective interplay between the chromophore and the biological unit. Fluorophore properties affect the detection limit and the dynamic range of the method, and the aptness for multiplexing, that is, parallel detection of different target analytes. Vital properties of a good label should be as follows (Resch-Genger et al., 2008):

1. It should be conveniently excitable, without excitation of the biological matrix at that instant, and the label should easily be detectable.
2. It should have a high molar absorption coefficient at the excitation wavelength with a high fluorescence quantum yield.
3. It should be soluble in relevant buffer solutions, cell culture media or body fluids.
4. It must be stable under conditions of usage.
5. The label should have necessary functional groups for site-specific labelling.
6. Its availability in reproducible quality must be ensured.
7. Whenever required, its deliverability into cells should be possible without any toxic effects.

Organic fluorophore-based diagnostics is a versatile and widely practiced technology but there are certain difficulties faced with conventional fluorophores in multiplexed environment. As mentioned earlier, in comparison to organic dyes, QDs have an absorption that increases toward shorter wavelengths (below the first excitonic absorption band) along with a narrow symmetric emission band. The spectral positions of absorption and emission are tunable by particle size and width of the emission peak is determined by QD size distribution. The broad absorption enables free selection of the excitation wavelength and clear separation of excitation and emission. The size-dependent molar extinction coefficients at the first absorption band of QDs (100,000–1,000,000 $M^{-1}cm^{-1}$) (Yu et al., 2003) are generally large as compared to organic fluorophores (25,000–250,000 $M^{-1}cm^{-1}$). Fluorescence quantum yields of adequately surface-passivated QDs are normally high (0.65–0.85 for CdSe (Wang et al., 2003; Talapin et al., 2004), $\leq$0.6 for CdS (Spanhel et al., 1987), and 0.1–0.4 for InP (Xu et al., 2006). There are also examples of emitters, such as, CdTe and CdHgTe that have high fluorescence quantum yield in the visible-NIR wavelength (Shavel et al., 2006) and PbS and PbSe for the NIR wavelength region (Hinds et al., 2007; Du et al., 2002). On the other hand organic dyes have high fluorescence quantum yield in the visible range but they fall short in the NIR wavelength range with poor photo stability. Moreover in case of QDs, long lifetimes as compared to organic dyes enable temporal discrimination of the signal from autofluorescence and scattered excitation light by time-gated measurements, thus enhancing sensitivity.

One of the important aspect of labels that it should not precipitate or aggregate under experimental conditions. In case of organic dyes solubility can be tuned via substituents without hampering the optical properties and other relevant features of the dye. On the other hand, QD dispersibility is controlled by the chemical nature of the surface coating. Although, for example, CdTe is inherently dispersible in water but high-quality CdSe needs to be made water-dispersible using small charged ligands (Chan and Nie, 1998), such as, mercaptopropionic acid or cystamine (Fig. 20.10). Charged surfactants can also be used that can intercalate with the hydrophobic ligands. QDs can also be dispersed in aqueous solution by attaching bulky polymeric surface ligands such as polyethylene glycol (Fig. 20.10). Electrostatically stabilized QDs are much smaller than sterically stabilized QDs but the former is more effective for biological applications as they can easily enter into the matrix but they aggregate in buffers of high ionic strength, whereas sterically stabilized QDs are less sensitive to ionic strength. Thus, the best possible way is to use smaller but still bulky charged ligands (Nann, 2005), such as polyethyleneimine or polyelectrolytes (Mattheakis et al., 2004), or an additional amphiphilic inorganic shell such as silica, which can further be functionalized (Figure 20.10).

Attachment of organic dyes to biomolecules, such as, peptides, proteins, or oligonucleotides requires suitable functional groups for covalent binding. The advantage of organic dyes in this regard is that there is availability of functionalized dyes with established labelling protocols, purification and characterization techniques for dye bioconjugates (Waggoner, 2006). Further, small size of the organic dye minimizes steric hindrance and removes the possibility of interference with functions of biomolecules. Several fluorophores can be attached to a single biomolecule to maximize the fluorescence signal, but site-specificity is a problem with these dyes, also. Again, high-label densities can result in fluorescence quenching, depending on the structure of the dye, charge, and hydrophilicity (Mujumdar et al., 1993; Gruber et al., 2000). Excepting fluorescence quenching, QDs face similar challenges like the organic dyes. Till date there are no optimized methods for labeling biomolecules with QDs (Xing et al., 2007). The general principle is to make QDs water dispersible and then attaching them to biomolecules via electrostatically or covalent binding (Fig. 20.10). Moreover QDs have a tendency to aggregate if surface passivation is improper. When several biomolecules are tagged to a single QD (Goldman et al., 2004), then orientation of biomolecules become uncontrollable sometimes. This in turn may affect spectroscopic properties and colloidal stability of QDs and also functionality of biomolecules. Targeting cells may become difficult due to larger size of QDs. Hydrothermal synthesis of CdTe and their bioconjugation with Bovine Serum Albumin (BSA) (Li et al., 2007) is reported

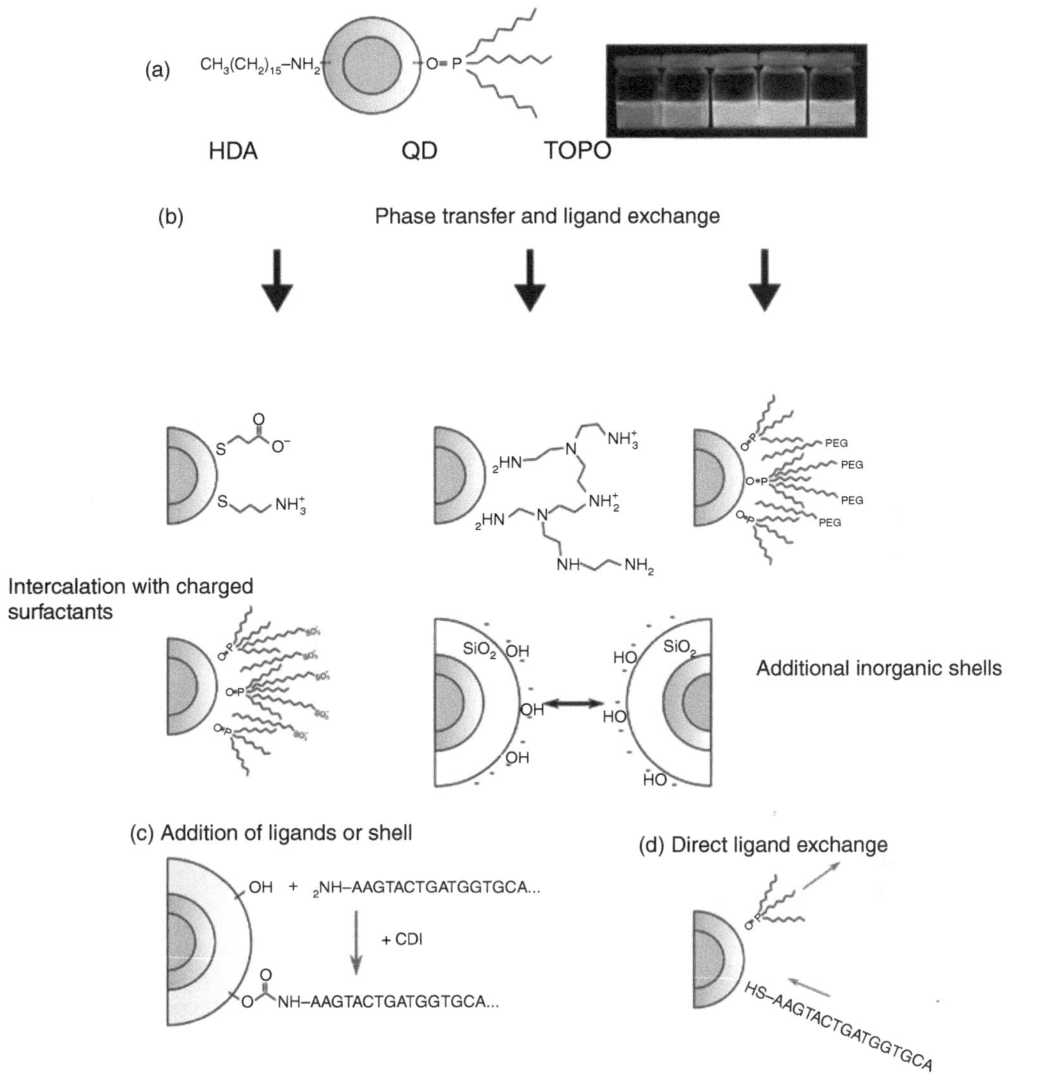

Figure 20.10. Overview of strategies to prepare water-dispersible QDs and QD bioconjugates. (a) QDs bearing hydrophobic ligands after preparation in organic solvent. HDA, hexadecylamine; TOPO, and trioctylphosphineoxide. (b) Ligand-exchange strategies to generate water-dispersible QDs. Illustrated are electrostatic colloidal stabilization (left), electrostatic and steric stabilization (middle), and steric stabilization of colloid (right). (c) Coupling of water-dispersible QDs to biomolecules, oligonucleotides. (d) Alternatively, the QDs bearing hydrophobic ligands can be subjected to direct ligand and exchange.

in a recent study where the crucial role of temperature has been stated in the colloidal synthesis of the QDs of various sizes keeping the pH and concentration ratio of ligand to monomer constant. The absorption spectrum of the bioconjugate differs from the standard absorption spectra of QDs and proteins, which confirms the successful conjugation of QDs with protein molecules.

The microenvironment affects spectral position, fluorescence quantum yields, fluorescence lifetime of an organic fluorophore. The different parameters of microenvironment include polarity, proticity, viscosity, pH, ionic strength. The presence of surfactants, fluorescence quenchers like oxygen also affects the spectral and temporal properties of an organic dye acting as a label. In case of QDs, the microenvironment effect on the spectroscopic features is majorly governed by the accessibility of the core surface (Medintz et al., 2005). The QDs that are properly shelled with ligands are less sensitive to the microenvironment polarity provided that there is no ligand desorption (Ji et al., 2008). Emissions from QDs are not affected by the viscosity but with change in ionic strength electrostatically stabilized QDs tend to aggregate. Not many studies are available on the effect of the microenvironment on the spectroscopic properties of QDs as there is a large variety of the coatings used in QDs.

4 Quantum Dots: Sensing Foodborne Pathogens and Toxicants

4.1 Contaminants: Farm to Fork

Public concern over food safety and food quality has gained immense attention over the past decade. Food can be contaminated in different stages and operations involved in production, processing, distribution, storage, and handling of a food and food ingredients from primary production to consumption. Monitoring contaminants (chemical or biological) in food is important as such contaminant affects human life and animal health, as well as international trade of agricultural products and food commodities. Industrial agricultural practices have evolved as a result of globalization; this, on one hand, has increased the food production but on the other hand also increased the potential likelihood of food contamination. There is a recent report that 31 major foodborne pathogens are the cause of sickness of 9.4 million people, 55,961 cases of hospitalizations, and 1,351 deaths every year in the United States (Schallan et al., 2011). The major cause of food contamination across globe is use of pesticides and herbicides.

Other major chemical contaminants like mycotoxins are worldwide contaminants of food and feed that cause health problems and economic losses. Biotoxins are toxic substances produced by and derived from plants and animals. Pesticides are substances or mixture of substances intended for preventing, destroying, repelling, or mitigating any pests or weeds. The enormous use of pesticides resulted in significant groundwater contamination due to their accumulation in the food chain, which gets filtered through various food formulations and drinking water. On the other hand the most common foodborne pathogens are *Escherichia coli* O157:H7, some strains of *Staphylococcus aureus*, *Bacillus anthracis*, *Shigella* spp,. *Salmonella* spp., *Campylobacter jejuni*, *Clostridium botulinum*, *Clostridium perfringens*, *Listeria monocytogenes*, *Yersinis enterocolitica*, *Vibrio cholera*, and *Coxiella burnetti* (Moss and Adams, 2008; Feng, 2001). In recent times, there are reports of several mycotoxin outbreaks that pose a serious threat to human and animal health. Lately pharmaceutical residues have become an increasing problem due to the use of antibiotics. They are easily transferred from the animals to humans and the effectiveness of antibiotics for common treatments gets highly diminished.

4.2 Sensing Methods

Although there are several standard methods available for the detection and quantification of contaminants in food samples, but all those are laborious, time-consuming with limited specificity and sensitivity. Sensing utilizing QDs is one of the nondestructive methods that is quite effective for the detection of foodborne pathogens and toxins. For QDs to act as biosensors they need to be water dispersible, and so aqueous methods are preferred over other chemical synthesis methods. The concentrations of the precursors, the capping material, and the solvents are the primary factors for the colloidal synthesis of QDs. At considerably high temperatures, the precursors convert them to monomers and undergo supersaturation favoring nanocrystal growth through the process of nucleation. Temperature being one of the crucial factors in the growth of nanocrystals should be high enough such that it enables the rearrangement and annealing of atoms during the synthesis process. The smaller particles grow faster at high monomer concentrations, resulting in monodisperse particles, and the process is termed *focusing*. Colloidal synthesis results in highly monodisperse-size tunable nanomaterials with suitable optical properties for biosensor applications. Biocompatibility needs to be addressed in the first place, before conjugating QDs with biomolecules, as most of the biological reactions (except those that involve lipids) are hydrophilic in nature. In most of the cases, capping materials with a thiol group, such as, mercapto-acetic acid, thioglycerol, or mercaptoethanesulfonate, along with peptides that contain cysteine as the terminal amino acid group or having multiple cysteine groups are found suitable to be used as surfactants for the dispersal of QDs in aqueous solutions. A broad, well-defined absorption spectrum and a narrow emission band with full-width half-maxima (FWHM) of 50–60 nm for QDs are preferred for bioconjugation. Three basic methods are prevalent for conjugation of QDs to biomolecules. There can be possible cross-linking of the amine, carboxyl, or thiol group, which are available on the surface of the QDs, secondly directly attaching the biomolecules to the surface atoms of the QD through bonding of thiol groups or metals. The third method is via electrostatic interactions between the oppositely charged biomolecules and the QD surfaces. The method adopted for conjugation will depend upon the target biomolecule to be detected. Again each and every biomolecule have certain advantages and disadvantages for its use in specific targeting for detection of foodborne pathogens and toxins. The performance of each biomolecule for detecting a specific bacterial target is highly dependent on the complex food matrix and the growth of the bacteria itself. First, the food matrix is quite complex in nature; in addition to that, the different components present may cause possible degradation or nonspecific binding to the biomolecule, which will result in erroneous results. There are different biomolecules that can be attached to water-soluble QDs for specific biological targeting (Table 20.2) (Burris and Stewart, 2012), such as, nucleic acid, antibodies, carbohydrates, and so forth.

The nucleic acid consisting of 15–25 nucleotides with a fluorescent molecule at one end and a quencher molecule at the other end can serve as a probe for sensing. In an unbound condition the fluorescence remains quenched due to the structural orientation as the fluorescent species reside in very close proximity to the quencher molecule. In presence of the target molecule the nucleic acid finds its complementary nucleic acid sequence, binds to it and fluorescence is switched on for possible detection. This methodology was first developed by Tyagi and Kramer (1996) and further modifications of this technology have led to new, improved methods for detection of foodborne pathogens in vitro using q-PCR. Liming and Bhagwat (2004) detected single Salmonella species in fruits and vegetables using the aforementioned technique. Antibodies are also in frequent use as biomolecules coupled with QDs to act as detectors for foodborne pathogens and toxins. Monoclonal and Polyclonal are two types of antibodies used in

Table 20.2 QD-Based Sensors Coupling Biomolecules for Foodborne Pathogen Detection

Sensor Type	Pathogen(s)	Target Biomolecule	Detection Level	Food Matrix	Detection Time	References
Nucleic acid	*S. typhimurium*	Nucleic acid	10 pM DNA	N/A	N/D	Vasudev et al. (2011)
	E. coli 0157:H7	Nucleic acid	9.72 fM DNA	N/A	N/D	Kim and Son (2010)
Immunoassay	*E. coli* 0157:H7	Antibody	7.3 log CFU/mL	N/A	N/D	Hahn et al. (2005)
	E. coli 0157:H7	Antibody	<10 CFU/mL	N/A	2 h	Sanvicens et al. (2011)
	E. coli 0157:H7	Antibody	3 log CFU/mL	N/A	<2 h	Su and Li (2004)
	E. coli 0157:H7 *S. typhimurium*	Antibody	4 log CFU/mL	N/A	<2 h	Yang and Li (2006)
	S. typhimurium,	Antibody	5.5 log CFU/mL	Milk (diluted)	<2 h	Zhao et al. (2009)
	S. flexneri, and *E. coli* 0157:H7		3 log CFU/mL	Apple juice	<2 h	
Phage-based	*E. coli*	Protein	20 cells/mL	Water	<1 h	Edgar et al. (2006)
PCR	*E. coli*	Lectin	4 log CFU/mL	N/A	N/D	Mukhopadhyay et al. (2009)
Flow cytometry	*E. coli* 0157:H7	Antibody	6 log CFU/mL	Water	N/D	Hahn et al. (2008)
Impedimetric	*S. enteriditis*	Antibody	4 log CFU/mL	PBS	3 min	Kim et al. (2007)
			5 log CFU/mL	Milk	3 min	

N/A, not applied to a food system; N/D, Not determined.

immunoassays. And monoclonal antibodies are found to be more specific than polyclonal antibodies on the basis of their action mechanism. Although there are certain disadvantages associated with the use of antibodies, their ability for simultaneous detection and separation of bacteria and toxins has kept them in active use as one of the tools for foodborne pathogen detection. Carbohydrates possess the capacity to bind with proteins and/or lipids, which enables them to detect pathogens and their toxins. The pathogens and toxins have the ability to recognize certain specific carbohydrate sequences and attach to their epithelial cells, this property of bacterial adhesion is utilized for the detection of pathogens and their toxins (Sharon, 2006). Unlike nucleic acids and antibodies, carbohydrates have an affinity for widely different organisms, which gives these carbohydrates broad specificity. This is an added advantage for the carbohydrates as it enables them to target organisms that have gone though mutation and are different from the original organism. Furthermore, carbohydrates do not denature or lose their activity under extreme conditions like high temperature and/or pH changes, and this makes them appropriate to detect pathogens and toxins in food matrices.

These QDs can be used in bioassays for detection of pesticides as these are highly photostable and the spectral properties of the QDs make them suitable for multianalyte detection. Nichkova et al. (2007) developed a multiplexed immunoassay using QDs for detection of atrazine mercapturate (AM) and 3-phenoxybenzoic acid (PBA). In case of pyrethroid insecticide and herbicide atrazine, PBA and AM serve as biomarkers. The working principle of these immonoassays is based on the immunointeractions between anti-PBA antibody and biotinylated anti-AM antibody and their corresponding antigens. Both the antigens, BSA-PBA and BSA-AM, are immobilized and are allowed to react with anti-PBA antibody and biotinylated anti-AM antibody, respectively. The above complexes are further incubated with antirabbit secondary antibodies that are conjugated to QD_{580} and streptavidin-QD_{620} and then successively were utilized to obtain fluorescence signals. In another study a rapid detection method for the herbicide 2,4-D was developed using CdTe QDs (Vinayaka et al., 2009). The quantitative estimation of the pesticides depends on the competitive binding of the free pesticide and a known concentration of QD-bioconjugated pesticide toward the immobilized antibody in the immunoreactor column.

There are considerable number of reports and studies where QDs have been potentially used for the detection of foodborne pathogens and their toxins in processed and unprocessed food and have gained importance in recent times. Bacterial toxins are of two types, endotoxins that cause lysis of the cell wall and exotoxins that cause disruption of normal cellular metabolism. Foodborne botulism occurs due to ingestion of botulinum toxin produced by the bacterium *Clostridium botulinum*. Warner et al. (2009) have designed and developed a fluorescence sandwich immunoassay for the detection of botulinum neurotoxin serotype A (BoNT/A). In this particular assay QDs and high affinity antibodies are utilized as reporter molecules. The antibodies bind to nonoverlapping epitopes (part of antigen) present in toxin and the recombinant fragment. The toxin gets quantified through the fluorescence obtained from QD. *E. coli* produces two types of toxins, heat-stable and heat-labile. The cell surface of *E. coli* contains mannose-specific lectin. A carbohydrate functionalized CdS quantum dots was synthesized for the detection of *E. coli* (Mukhopadhyay et al., 2009). CdS QDs are coated with thiolated mannose that specifically can recognize the mannose-specific lectin in *E. coli.* Goldman et al. (2002) have proposed a multiplexed fluoroimmunoassay that is based on CdSe-ZnS core–shell QDs for the detection of cholera toxin (CT), *Shiga*-like toxin (SLT), ricin, and Staphylococcal enterotoxin (SEB). The detection of SEB in food products is crucial as it has the ability to cause toxic-shock syndrome. In order to attain specificity in the assay, toxin-specific antibodies were used as recognition molecules. Nonspecific interactions and background fluorescence from the QDs along with antigen–antibody cross-reactivity interfere with the assay. Interactions between individual QDs may cause spectral overlap and may interfere with the multiplexed analysis. A multicolor quantum dot nanoprobe was designed by Ho et al. (2005) for the detection of a pathogenic strain of *Bacillus anthracis.* The detection principle is suitably based on the identification of rpoB, which is a chromosomal gene marker and can differentiate *Bacillus anthracis* from other *Bacillus* species and also plasmid markers like pagA and capC known to determine anthrax toxin. Oligonucleotide probes known to be complementary to each gene marker are being designed specifically and conjugated to QDs with different emission wavelengths. The probes are also designed to form a sandwich nanoassembly that binds simultaneously to the target DNA molecule. The fluorescence signal obtained from each conjugated QD determines the presence of rpoB, pagA, or capC.

4.3 Toxicity and Future Potential of QDs

The cytotoxicity of QDs in vivo studies is an important issue. Toxicity due to heavy metals due to the possible interactions of the QDs with cells should be taken care of before proceeding for vitro and vivo studies. The cytotoxicity that is due to cadmium present in many of these nanocrystals is well known. It is critical to recognize that whether these cytotoxic substances can leak out of the QD particles over time, upon illumination or oxidation (Zhang et al., 2006; Hoshino et al., 2004; Ma et al., 2006), in addition whether ligands or coatings are cytotoxic (Lewinski et al., 2008). QDs are chemically synthesized and the physicochemical properties and toxicity of these chemicals should be tested in organisms. There can be problem due to the posttreatment of the sample and QDs disposal when used as biosensors and that may subsequently lead to toxicological effects on the whole ecosystem. For in vivo applications, a biocompatibility factor needs to be addressed when QDs are bioconjugated with adequate biomolecular capping materials like immortalized cell lines, liposomes and phospholipid vesicles. Biosensors based on nanofabrication are proving to be emerging tools for detection of food contaminants, such as, pesticide residues, foodborne pathogens, and their toxins. These sensors have additional advantages in terms of cost per assay, minimum sample volume, time taken for analysis, and potential for multiple-analyte analysis. Bioconjugated QDs are applied for multianalyte detection of DNA, proteins, pesticides, and so forth. Biosensors utilizing optical properties of QDs in nucleic acid assays can detect target DNA sequences, polymorphic regions, and gene expression. On the other hand, there is a considerable progress in the field of chemically modified QDs and their applications for optical sensing analytes like ions (cations or anions), monitoring pH changes. QD conjugates are implemented for different optical-sensing systems employing fluorescence, resonance energy transfer (FRET), and luminescence electron transfer (ET). QDs have a bright future in NIR fluorescence in vivo imaging, which requires labels that exhibit high fluorescence quantum yield in the 650–900 nm window, have adequate stability, good aqueous solubility, and low cytotoxicity in conjunction with large two-photon action cross-sections as desired for deep tissue imaging. Moreover QDs are attractive candidates for the development of multifunctional composite materials for the combination of two or more biomedical imaging modalities, like NIR fluorescence-magnetic resonance imaging. Although there is substantial achievement in the area of molecular sensing with QDs, still there are certain challenges. Till now, in most of the studies, the photophysical properties of QDs and electrical, optical, binding properties of the chemically modified layers are separately studied. In the QDs with core–shell nanostructures, the shell layer can actually participate in the binding or the activation of the analyte toward its interaction with QDs. Again, catalytic materials can be incorporated into the capping layer in order to activate the substrate analyte for its interaction with QD.

QD-based detection system can bring revolution in the area of food safety and quality as they ensure rapid, reliable and sensitive detection of contaminants in food samples.

5 Conclusions

In order to ensure safer production, distribution, and storage of food materials require cheaper, accurate, and robust chemical sensors. Fluorescent nanosensors are promising candidates for analyzing food contaminants, such as, pesticide residues, foodborne pathogens, and their toxins. However, most of such sensors do not qualify the requirement of mass production. There is an array of inorganic and organic materials to synthesize quantum dots that possess posses interesting electrical and optical properties and chemo/stereo specificity. A continuous study of these sensors is going on to make them suitable for mass production using different synthetic routes and reaction engineering approach. Applications of nanosensors are also being focused. In addition to that a hurdle for usage of these sensors in commercial space is their toxicity issues. The synthetic particles due to their nanosize react very differently from their bulk counterpart both in physiological and in environmental condition. Hence, a considerable number of research projects are dedicated to toxicological evaluation for these nanosensors.

References

Berlman, I.B., 1971. Handbook of Fluorescence Spectra of Aromatic Molecules, second ed. Academic Press, New York.

Borisov, S.M., Klimant, I., 2008. Optical nanosensors: smart tools in bioanalytics. Analyst 133, 1302–1307.

Bruchez, M., Moronne, Jr.M., Gin, P., Weiss, S., Alivisatos, A.P., 1998. Semiconductor nanocrystals as fluorescent biological labels. Science 281, 2013–2016.

Burris, K.P., Stewart, Jr., C.N., 2012. Fluorescent nanoparticles: sensing pathogens and toxins in foods and crops. Trends Food Sci. Technol. 28, 143–152.

Cammann, K., Guibault, E.A., Hall, H., Kellner, R., Wolfbeis, O.S., 1996. The Cambridge definition of chemical sensors. Proceedings of the Cambridge Workshop on Chemical Sensors and Biosensors. Cambridge University Press, New York.

Chan, W.C.W., Nie, S., 1998. Quantum dot bioconjugates for ultrasensitive nonisotopic detection. Science 281, 2016–2018.

Du, H., et al., 2002. Optical properties of colloidal PbSe nanocrystals. Nano Lett. 2, 1321–1324.

Edgar, R., et al., 2006. High-sensitivity bacterial detection using biotin-tagged phage and quantum-dot nanocomplexes. Proc. Natl. Acad. Sci. USA 103, 4841–4845.

Feng, P., 2001. In: Labbe, R.G., Garcia, S. (Eds.), Guide to Foodborne Pathogens. John Wiley & Sons, New York, pp. 143–162.

Förster, Th., 1948. Intermolecular energy migration and fluorescence (Transl. R.S. Knox). Ann. Phys. 2, 55–75.

Gaponenko, S.V., 1998. Optical Properties of Semiconductor Nanocrystals. Cambridge University Press, New York.

Goldman, E.R., Anderson, G.P., Tran, P.T., Mattoussi, H., Charles, P.T., Mauro, J.M., 2002. Conjugation of luminescent quantum dots with antibodies using an engineered adaptor protein to provide new reagents for fluoroimmunoassays. Anal. Chem. 74, 841–847.

Goldman, E.R., et al., 2004. Multiplexed toxin analysis using four colors of quantum dot fluororeagents. Anal. Chem. 76, 684–688.

Gruber, H.J., et al., 2000. Anomalous fluorescence enhancement of Cy3 and Cy3.5 versus anomalous fluorescence loss of Cy5 and Cy7 upon covalently linking to IgC and noncovalent binding to avidin. Bioconj. Chem. 11, 696–704.

Hahn, M.A., et al., 2005. Detection of single bacterial pathogens with semiconductor quantum dots. Anal. Chem. 77, 4861–4869.

Hahn, M.A., et al., 2008. Flow cytometric analysis to detect pathogens in bacterial cell mixtures using semiconductor quantum dots. Anal. Chem. 80, 864–872.

Herschel, Sir J.F.W., 1845. On a case of superficial color presented by a homogeneous liquid internally colourless. Phil. Trans. Roy. Soc. 135, 143–145.

Hinds, S., et al., 2007. NIR-emitting colloidal quantum dots having 26% luminescence quantum yield in buffer solution. J. Am. Chem. Soc. 129, 7218–7219.

Ho, Y.P., Kung, M.C., Yang, S., Wang, T.H., 2005. Multiplexed hybridization detection with multicolor colocalization of quantum dot nanoprobes. Nano Lett. 5, 1693–1697.

Hoshino, A., et al., 2004. Physicochemical properties and cellular toxicity of nanocrystal quantum dots depend on their surface modification. Nano Lett. 4, 2163–2169.

Jablonski, A., 1935. Uber den Mechanisms des Photolumineszenz von Farbstoffphosphoren. Z Phys. 94, 38–46.

Ji, X., Copenhaver, D., Sichmeller, C., Peng, X., 2008. Ligand bonding and dynamics on colloidal nanocrystals at room temperature: the case of alkylamines on CdSe nanocrystals. J. Am. Chem. Soc. 130, 5726–5735.

Kasha, M., 1950. Characterization of electronic transitions in complex molecules. Disc. Faraday Soc. 9, 14–19.

Khanna, V., 2012. Nanosensors: Physical, Chemical and Biological. CRC Press, Taylor & Francis Group, Washington.

Kim, G.-H., Son, A., 2010. Quantitative detection of *E. Coli* O157:H7 *eaeA* gene using quantum dots and magnetic particles. Biotechnol. Bioprocess Eng. 15, 1084–1093.

Kim, G., et al., 2007. Nano-particle enhanced impedimetric biosensor for detection of foodborne pathogens. J. Phys. 61, 555.

Lakowicz, J.R., 2006. Principles of Fluorescence Spectroscopy, third ed. Springer-Science + Business Media, New York.

Lewinski, N., Colvin, V., Drezek, R., 2008. Cytotoxicity of nanoparticles. Small 4, 26–49.

Li, M., Ge, Y., Chen, Q., Xu, S., Wang, N., Zhang, X., 2007. Hydrothermal synthesis of highly luminescent CdTe quantum dots by adjusting precursors' concentration and their conjunction with BSA as biological fluorescent probes. Talanta 72, 89–94.

Liming, S.H., Bhagwat, A.A., 2004. Application of a molecular beacon-real-time PCR technology to detect *Salmonella* species contaminating fruits and vegetables. Int. J. Food Microbiol. 95, 177–187.

Ma, J., et al., 2006. Photostability of thiol-capped CdTe quantum dots in living cells: the effect of photooxidation. Nanotechnology 17, 2083–2089.

Mason, W.T., 1999. Fluorescent and Luminescent Probes for Biological Activity, second ed. Academic Press, London.

Mattheakis, L.C., et al., 2004. Optical coding of mammalian cells using semiconductor quantum dots. Anal. Biochem. 327, 200–208.

Medintz, I.L., Uyeda, H.T., Goldman, E.R., Mattoussi, H., 2005. QD bioconjugates for imaging, labelling, and sensing. Nat. Mater. 4, 435–446.

Moss, M.M., Adams, M.R., 2008. Food Microbiology. Royal Society of Chemistry, Cambridge.

Mujumdar, R.B., Ernst, L.A., Mujumdar, S.R., Lewis, C.J., Waggoner, A.S., 1993. Cyanine dye labeling agents: sulfoindocyanine succidimidyl esters. Bioconj. Chem. 4, 105–111.

Mukhopadhyay, B., Martins, M.B., Karamanska, R., Russell, D.A., Field, R.A., 2009. Bacterial detection using carbohydrate-functionalised CdS quantum dots: a model study exploiting *E. coli* recognition of mannosides. Tetrahedron Lett. 50, 886–889.
Murphy, C.J., Coffer, J.L., 2002. Quantum dots: a primer. Appl. Spectro. 56, 16A–27A.
Nann, T., 2005. Phase-transfer of CdSe@ZnS quantum dots using amphiphilic hyperbranched polyethylenimine. Chem. Commun. 13, 1735–1736.
Nichkova, M., Dosev, D., Davis, A.E., Gee, S.J., Kennedy, I.M., Hammock, B.D., 2007. Quantum dots as reporters in multiplexed immunoassays for biomarkers of exposure to agrochemicals. Anal. Lett. 40, 1423–1433.
Resch-Genger, U., Grabolle, M., Cavaliere-Jaricot, S., Nitschke, R., Nann, T., 2008. Quantum dots versus organic dyes as fluorescent labels. Nature Method. 5, 763–773.
Riegler, J., Ditengou, F., Palme, K., Nann, T., 2008. Blue shift of CdSe/ZnS nanocrystal-labels upon DNA-hybridization. J. Nanobiotechnol. 6, 7.
Sanvicens, N., et al., 2011. Quantum dot-based array for sensitive detection of *Escherichia coli*. Anal. Bioanal. Chem. 399, 2755–2762.
Schäferling, M., 2012. The art of fluorescence imaging with chemical sensors. Angew Chem. Int. Ed. 51, 3532–3554.
Schallan, E., Hoekstra, R.M., Angulo, F.J., Tauxe, R.V., Widdowson, M.A., Roy, S.L., et al., 2011. Fodborne illness acquired in the United States: major pathogens. Emerg. Infect. Dis. 17, 7–15.
Sharon, N., 2006. Carbohydrates as future anti-adhesion drugs for infectious diseases. Biochim. Biophys. Acta. 1760, 527–537.
Shavel, A., Gaponik, N., Eychmuller, A., 2006. Factors governing the quality of aqueous CdTe nanocrystals. J. Phys. Chem. B 110, 19280–19284.
Spanhel, L., Haase, M., Weller, H., Henglein, A., 1987. Photochemistry of colloidal semiconductors. 20. Surface modification and stability of strong luminescing CdS particles. J. Am. Chem. Soc. 109, 5649–5655.
Stokes, G.G., 1852. On the change of refrangibility of light. Philos. Trans. Roy. Soc. 142, 463–562.
Su, X.L., Li, Y., 2004. Quantum dot biolabeling coupled with immunomagnetic separation for detection of *Escherichia coli* O157:H7. Anal. Chem. 76, 4806–4810.
Talapin, D.V., et al., 2004. CdSe/CdS/ZnS and CdSe/ZnSE/ZnS core-shell-shell nanocrystals. J. Phys. Chem. B 108, 18826–18831.
Thevenot, D.R., Tóth, K., Durst, R.A., Wilson, G.S., 1999. Electrochemical biosensors: recommended definitions and classifications. Pure Appl. Chem. 71, 2333–2348.
Tyagi, S., Kramer, F.R., 1996. Molecular beacons: probes that fluoresce upon hybridization. Nature Biotechnol. 14, 303–308.
Vasudev, M., et al., 2011. Optoelectronic signatures of DNA based hybrid nanostructures. IEEE Trans. Nanotechnol. 10, 35–43.
Vinayaka, A.C., Basheer, S., Thakur, M.S., 2009. Bioconjugation of CdTe quantum dot for the detection of 2,4-dichlorophenoxyacetic acid by competitive fluoroimmunoassay based biosensor. Biosens. Bioelectron. 24, 1615–1620.
Waggoner, A., 2006. Fluorescent labels for proteomics and genomics. Curr. Opin. Chem. Biol. 10, 62–66.
Wang, Y., Herrron, N., 1991. Nanometer-sized semiconductor clusters: materials synthesis, quantum size effects, and photophysical properties. J. Phys. Chem. 95, 525–532.
Wang, X., Qu, L., Zhang, J., Peng, X., Xiao, M., 2003. Surface-related emission in highly luminescent CdSe QDs. Nano Lett. 3, 1103–1106.

Wang, Z., Xu, Q., Wang, H.-Q., Yang, Q., Yu, J.-H., Zhao, Y.-D., 2009. Hydrogen peroxide biosensor based on direct electron transfer of horseradish peroxidise with vapor-deposited quantum dots. Sens. Actuat. B-Chem. 138, 278–282.

Warner, M.G., Grate, J.W., Tyler, A., Ozanich, R.M., Miller, K.D., Lou, J., Marks, J.D., Bruckner-lea, C.J., 2009. Quantum dot immunoassays in renewable surface column and 96-well plate formats for the fluorescence detection of botulinum neurotoxin using high-affinity antibodies. Biosens. Bioelectron. 25, 179–184.

Xing, Y., et al., 2007. Bioconjugated quantum dots for multiplexed and quantitative immunohistochemistry. Nat. Protoc. 2, 1152–1165.

Xu, S., Kumar, S., Nann, T., 2006. Rapid synthesis of high-quality InP nanocrystals. J. Am. Chem. Soc. 128, 1054–1055.

Yang, L., Li, Y., 2006. Simultaneous detection of *Escherichia coli* O157:H7 and Salmonella Typhimurium using quantum dots as fluorescence labels. Analyst 131, 394–401.

Yu, W.W., Qu, L., Guo, W., Peng, X., 2003. Experimental determination of the extinction coefficient of CdTe, CdSe and CdS nanocrystals. Chem. Mater. 15, 2854–2860.

Zhang, Y., et al., 2006. Time-dependent photoluminescence blue shift of the quantum dots in living cells: effect of oxidation by singlet oxygen. J. Am. Chem. Soc. 128, 13396–13401.

Zhao, Y., et al., 2009. Simultaneous detection of multifood-borne pathogenic bacteria based on functionalized quantum dots coupled with immunomagnetic separation in food samples. J. Agric. Food Chem. 57, 517–524.

SUBJECT INDEX

FSC
www.fsc.org
MIX
Papier | Fördert
gute Waldnutzung
FSC® C083411

Kolibri 360 GmbH
Ferdinand-Jühlke-Straße 7
99095 Erfurt, Deutschland
produktsicherheit@kolibri360.de